Elementary and Intermediate Algebra: A Combined Course

FOURTH EDITION

- **Ron Larson**
  The Pennsylvania State University
  The Behrend College

- **Robert P. Hostetler**
  The Pennsylvania State University
  The Behrend College

With the assistance of
Patrick M. Kelly
Mercyhurst College

Houghton Mifflin Company   Boston   New York

Vice President and Publisher: Jack Shira
Associate Sponsoring Editor: Cathy Cantin
Development Manager: Maureen Ross
Associate Editor: Marika Hoe
Assistant Editor: James Cohen
Supervising Editor: Karen Carter
Senior Project Editor: Patty Bergin
Editorial Assistant: Allison Seymour
Production Technology Supervisor: Gary Crespo
Executive Marketing Manager: Michael Busnach
Senior Marketing Manager: Ben Rivera
Marketing Assistant: Lisa Lawler
Senior Manufacturing Coordinator: Priscilla Bailey
Composition and Art: Meridian Creative Group
Cover Design Manager: Diana Coe

Cover art © by Dale Chihuly
20,000 Pounds of Ice and Neon, detail
1992
Contemporary Arts Center, Cincinnati, OH
Photo: Russell Johnson

We have included examples and exercises that use real-life data as well as technology output
from a variety of software. This would not have been possible without the help of many people
and organizations. Our wholehearted thanks go to all their time and effort.

Trademark acknowledgement: TI is a registered trademark of Texas Instruments, Inc.

Printed in the U.S.A.

Library of Congress Catalog Card Number: 2003110125

ISBN: 0-618-38836-2

123456789–DOW–08 07 06 05 04

Common Formulas

Distance

$d = rt$

d = distance traveled
t = time
r = rate

Temperature

$F = \dfrac{9}{5}C + 32$

F = degrees Fahrenheit
C = degrees Celsius

Simple Interest

$I = Prt$

I = interest
P = principal
r = annual interest rate
t = time in years

Compound Interest

$A = P\left(1 + \dfrac{r}{n}\right)^{nt}$

A = balance
P = principal
r = annual interest rate
n = compoundings per year
t = time in years

Coordinate Plane: Midpoint Formula

Midpoint of line segment joining (x_1, y_1) and (x_2, y_2)

$\left(\dfrac{x_1 + x_2}{2}, \dfrac{y_1 + y_2}{2}\right)$

Coordinate Plane: Distance Formula

d = distance between points (x_1, y_1) and (x_2, y_2)

$d = \sqrt{(x_2 - x_1)^2 + (y_2 - y_1)^2}$

Quadratic Formula

Solutions of $ax^2 + bx + c = 0$

$$x = \dfrac{-b \pm \sqrt{b^2 - 4ac}}{2a}$$

Rules of Exponents

(Assume $a \neq 0$ and $b \neq 0$.)

$a^0 = 1$

$a^m \cdot a^n = a^{m+n}$

$(ab)^m = a^m \cdot b^m$

$(a^m)^n = a^{mn}$

$\dfrac{a^m}{a^n} = a^{m-n}$

$\left(\dfrac{a}{b}\right)^m = \dfrac{a^m}{b^m}$

$a^{-n} = \dfrac{1}{a^n}$

$\left(\dfrac{a}{b}\right)^{-n} = \dfrac{b^n}{a^n}$

Basic Rules of Algebra

Commutative Property of Addition

$a + b = b + a$

Commutative Property of Multiplication

$ab = ba$

Associative Property of Addition

$(a + b) + c = a + (b + c)$

Associative Property of Multiplication

$(ab)c = a(bc)$

Left Distributive Property

$a(b + c) = ab + ac$

Right Distributive Property

$(a + b)c = ac + bc$

Additive Identity Property

$a + 0 = 0 + a = a$

Multiplicative Identity Property

$a \cdot 1 = 1 \cdot a = a$

Additive Inverse Property

$a + (-a) = 0$

Multiplicative Inverse Property

$a \cdot \dfrac{1}{a} = 1, \quad a \neq 0$

Properties of Equality

Addition Property of Equality

If $a = b$, then $a + c = b + c$.

Multiplication Property of Equality

If $a = b$, then $ac = bc$.

Cancellation Property of Addition

If $a + c = b + c$, then $a = b$.

Cancellation Property of Multiplication

If $ac = bc$, and $c \neq 0$, then $a = b$.

Zero Factor Property

If $ab = 0$, then $a = 0$ or $b = 0$.

Contents

CONTENTS

Appendices

*Appendices C, D, E, F, and G are available on the textbook website.
Go to* math.college.hmco.com/students *and link to* **Elementary and Intermediate Algebra: A Combined Course,** *Fourth Edition.*

A Word from the Authors

Welcome to *Elementary and Intermediate Algebra: A Combined Course*, Fourth Edition. In this revision, we have continued to focus on developing students' proficiency and conceptual understanding of algebra. We hope you enjoy the Fourth Edition.

In response to suggestions from elementary and intermediate algebra instructors, we have revised and reorganized the coverage of topics for the Fourth Edition. "Operations with Integers" (formerly Section 1.2) has been split into two sections: "Adding and Subtracting Integers" (Section 1.2) and "Multiplying and Dividing Integers" (Section 1.3). Section 5.3 "Negative Exponents and Scientific Notation" from the Third Edition has been retitled "Integer Exponents and Scientific Notation" and is now Section 5.1 in the Fourth Edition. Old Section 5.4 "Dividing Polynomials" is now "Dividing Polynomials and Synthetic Division." The chapter on "Systems of Equations" (retitled as "Systems of Equations and Inequalities") has been moved to follow the chapter on "Rational Expressions, Equations, and Functions" (Chapter 7). In addition to these changes, several new sections and a chapter have been added to the Fourth Edition. "Complex Fractions" (Section 7.4) and "Applications and Variation" (Section 7.6) have been added to Chapter 7 "Rational Expressions, Equations, and Functions." "Systems of Linear Inequalities" (Section 8.6) has been added to Chapter 8 "Systems of Equations and Inequalities." "Adding and Subtracting Radical Expressions" (Section 9.3) has been added to Chapter 9 "Radicals and Complex Numbers." A chapter on "Conics" has been added (Chapter 12), as has a new appendix, "Review of Elementary Algebra Topics" (Appendix A). This appendix reviews the first six chapters of the text and is intended to facilitate the transition from Elementary Algebra to Intermediate Algebra. Finally, two appendices that were sections in the Third Edition are now available on the student resource website.

In order to address the diverse needs and abilities of students, we offer a straightforward approach to the presentation of difficult concepts. In the Fourth Edition, the emphasis is on helping students learn a variety of techniques—symbolic, numeric, and visual—for solving problems. We are committed to providing students with a successful and meaningful course of study.

Our approach begins with *Motivating the Chapter*, a feature that introduces each chapter. These multipart problems are designed to show students the relevance of algebra to the world around them. Each *Motivating the Chapter* feature is a real-life application that requires students to apply the concepts of the chapter in order to solve each part of the problem. Problem-solving and critical thinking skills are emphasized here and throughout the text in applications that appear in the examples and exercise sets.

To improve the usefulness of the text as a study tool, we have added two new, paired features to the beginning of each section: *What You Should Learn* lists the main objectives that students will encounter throughout the section, and *Why You Should Learn It* provides a motivational explanation for learning the given objectives. To help keep students focused as they read the section, each objective presented in *What You Should Learn* is restated in the margin at the point where the concept is introduced.

In this edition, the *Study Tip, Technology: Tip,* and *Technology: Discovery* features have been revised. *Study Tip* features provide hints, cautionary notes, and words of advice for students as they learn the material. *Technology: Tip* features provide point-of-use instruction for using a graphing calculator, whereas *Technology: Discovery* features encourage students to explore mathematical concepts using their graphing or scientific calculators. All technology features are highlighted and can easily be omitted without loss of continuity in coverage of material.

The new chapter summary feature *What Did You Learn?* highlights important mathematical vocabulary (*Key Terms*) and primary concepts (*Key Concepts*) from the chapter. For easy reference, the *Key Terms* are correlated to the chapter by page number and the *Key Concepts* by section number.

As students proceed through each chapter, they have many opportunities to assess their understanding and practice skills. A set of *Exercises*, located at the end of each section, correlates to the *Examples* found within the section. *Mid-Chapter Quizzes* and *Chapter Tests* offer students self-assessment tools halfway through and at the conclusion of each chapter. *Review Exercises*, organized by section, restate the *What You Should Learn* objectives so that students may refer back to the appropriate topic discussion when working through the exercises. In addition, the *Review: Concepts, Skills, and Problem Solving* exercises that precede each exercise set, and the *Cumulative Tests* that follow Chapters 3, 6, 9, and 12, give students more opportunities to revisit and review previously learned concepts.

To show students the practical uses of algebra, we highlight the connections between the mathematical concepts and the real world in the multitude of applications found throughout the text. We believe that students can overcome their difficulties in mathematics if they are encouraged and supported throughout the learning process. Too often, students become frustrated and lose interest in the material when they cannot follow the text. With this in mind, every effort has been made to write a readable text that can be understood by every student. We hope that your students find our approach engaging and effective.

Ron Larson

Ron Larson

Robert P. Hostetler

Robert P. Hostetler

Features

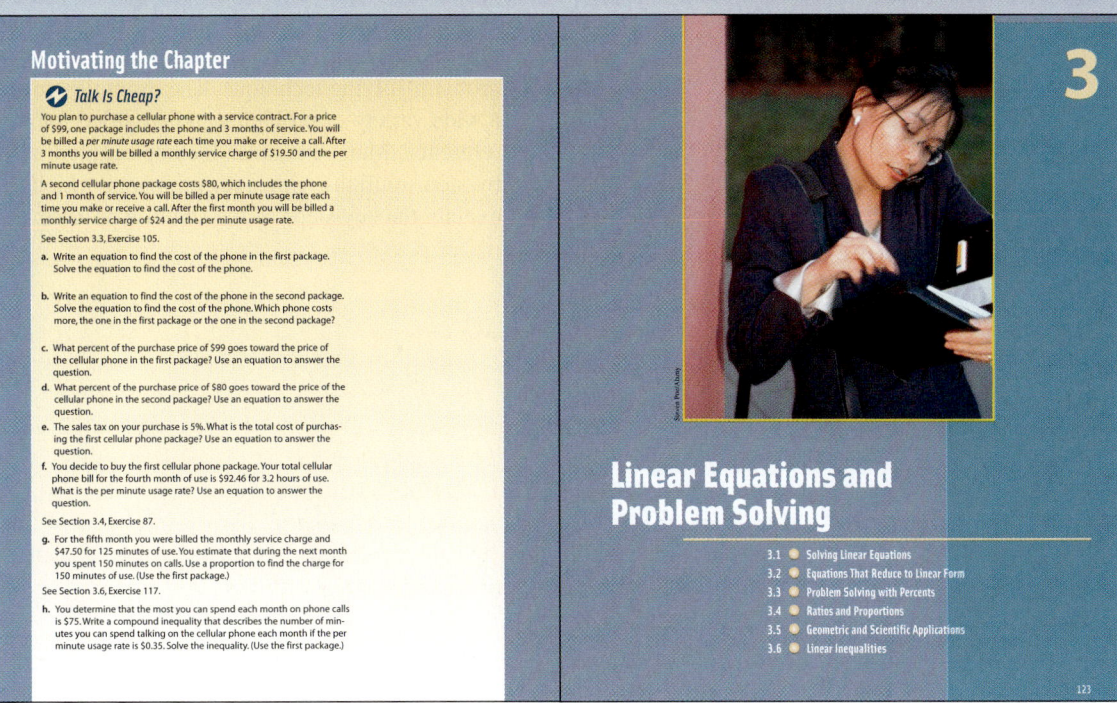

Motivating the Chapter

Talk Is Cheap?

You plan to purchase a cellular phone with a service contract. For a price of $99, one package includes the phone and 3 months of service. You will be billed a *per minute usage rate* each time you make or receive a call. After 3 months you will be billed a monthly service charge of $19.50 and the per minute usage rate.

A second cellular phone package costs $80, which includes the phone and 1 month of service. You will be billed a per minute usage rate each time you make or receive a call. After the first month you will be billed a monthly service charge of $24 and the per minute usage rate.

See Section 3.3, Exercise 105.

a. Write an equation to find the cost of the phone in the first package. Solve the equation to find the cost of the phone.

b. Write an equation to find the cost of the phone in the second package. Solve the equation to find the cost of the phone. Which phone costs more, the one in the first package or the one in the second package?

c. What percent of the purchase price of $99 goes toward the price of the cellular phone in the first package? Use an equation to answer the question.

d. What percent of the purchase price of $80 goes toward the price of the cellular phone in the second package? Use an equation to answer the question.

e. The sales tax on your purchase is 5%. What is the total cost of purchasing the first cellular phone package? Use an equation to answer the question.

f. You decide to buy the first cellular phone package. Your total cellular phone bill for the fourth month of use is $92.46 for 3.2 hours of use. What is the per minute usage rate? Use an equation to answer the question.

See Section 3.4, Exercise 87.

g. For the fifth month you were billed the monthly service charge and $47.50 for 125 minutes of use. You estimate that during the next month you spent 150 minutes on calls. Use a proportion to find the charge for 150 minutes of use. (Use the first package.)

See Section 3.6, Exercise 117.

h. You determine that the most you can spend each month on phone calls is $75. Write a compound inequality that describes the number of minutes you can spend talking on the cellular phone each month if the per minute usage rate is $0.35. Solve the inequality. (Use the first package.)

3

Linear Equations and Problem Solving

3.1 ● Solving Linear Equations
3.2 ● Equations That Reduce to Linear Form
3.3 ● Problem Solving with Percents
3.4 ● Ratios and Proportions
3.5 ● Geometric and Scientific Applications
3.6 ● Linear Inequalities

123

Chapter Opener

Every chapter opens with *Motivating the Chapter*. These multipart problems use concepts discussed in the chapter and present them in the context of a single real-world application. *Motivating the Chapter* problems are correlated to specific sections and can be assigned as part of an exercise set or as an individual or group project. The icon ⟲ identifies an exercise that relates back to *Motivating the Chapter*.

Section Opener   *New*

Every section begins with a list of learning objectives called *What You Should Learn*. Each objective is restated in the margin at the point where it is covered. *Why You Should Learn It* provides a motivational explanation for learning the given objectives.

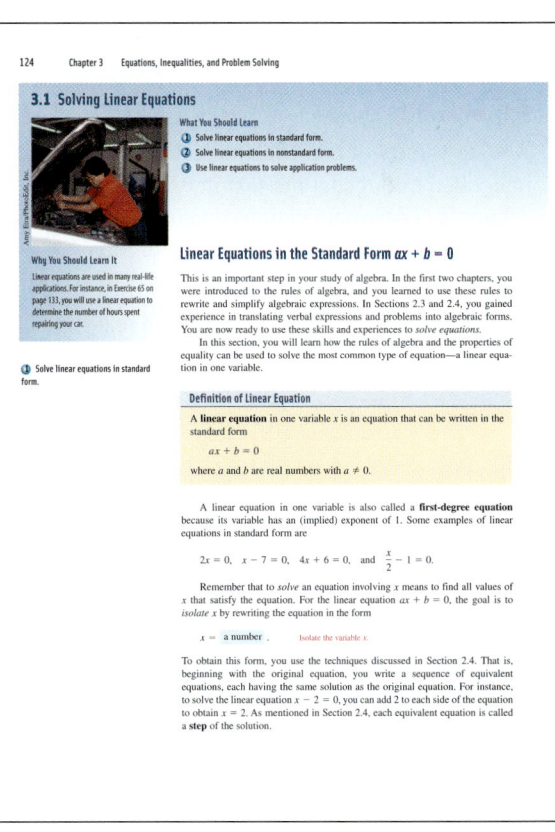

124   Chapter 3   Equations, Inequalities, and Problem Solving

3.1 Solving Linear Equations

What You Should Learn

1 Solve linear equations in standard form.
2 Solve linear equations in nonstandard form.
3 Use linear equations to solve application problems.

Why You Should Learn It

Linear equations are used in many real-life applications. For instance, in Exercise 65 on page 133, you will use a linear equation to determine the number of hours spent repairing your car.

1 Solve linear equations in standard form.

Linear Equations in the Standard Form $ax + b = 0$

This is an important step in your study of algebra. In the first two chapters, you were introduced to the rules of algebra, and you learned to use these rules to rewrite and simplify algebraic expressions. In Sections 2.3 and 2.4, you gained experience in translating verbal expressions and problems into algebraic forms. You are now ready to use these skills and experiences to *solve equations*.

In this section, you will learn how the rules of algebra and the properties of equality can be used to solve the most common type of equation—a linear equation in one variable.

Definition of Linear Equation

A **linear equation** in one variable x is an equation that can be written in the standard form

$$ax + b = 0$$

where a and b are real numbers with $a \neq 0$.

A linear equation in one variable is also called a **first-degree equation** because its variable has an (implied) exponent of 1. Some examples of linear equations in standard form are

$$2x = 0, \quad x - 7 = 0, \quad 4x + 6 = 0, \quad \text{and} \quad \frac{x}{2} - 1 = 0.$$

Remember that to *solve* an equation involving x means to find all values of x that satisfy the equation. For the linear equation $ax + b = 0$, the goal is to *isolate* x by rewriting the equation in the form

$$x = \text{a number}. \quad \text{Isolate the variable } x.$$

To obtain this form, you use the techniques discussed in Section 2.4. That is, beginning with the original equation, you write a sequence of equivalent equations, each having the same solution as the original equation. For instance, to solve the linear equation $x - 2 = 0$, you can add 2 to each side of the equation to obtain $x = 2$. As mentioned in Section 2.4, each equivalent equation is called a **step** of the solution.

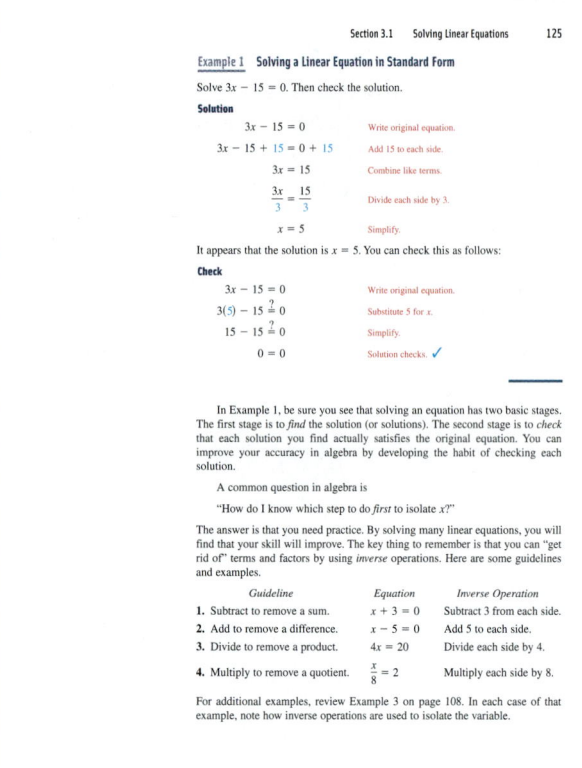

Examples

Each example has been carefully chosen to illustrate a particular mathematical concept or problem-solving technique. The examples cover a wide variety of problems and are titled for easy reference. Many examples include detailed, step-by-step solutions with side comments, which explain the key steps of the solution process.

Applications

A wide variety of real-life applications are integrated throughout the text in examples and exercises. These applications demonstrate the relevance of algebra in the real world. Many of the applications use current, real data. The icon indicates an example involving a real-life application.

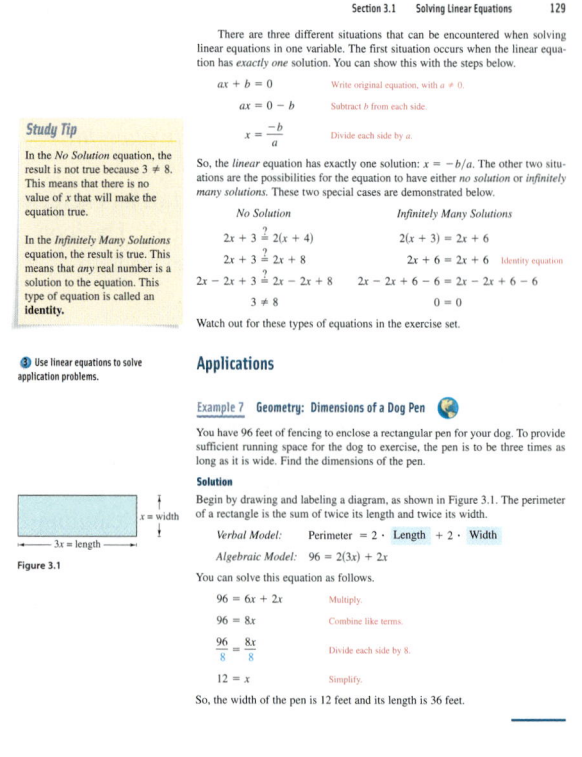

Panel 1 (top left):

136   Chapter 3   Equations, Inequalities, and Problem Solving

Example 2   Solving a Linear Equation Involving Parentheses

Solve $3(2x - 1) + x = 11$. Then check your solution.

Solution

$$3(2x - 1) + x = 11 \qquad \text{Write original equation.}$$
$$3 \cdot 2x - 3 \cdot 1 + x = 11 \qquad \text{Distributive Property}$$
$$6x - 3 + x = 11 \qquad \text{Simplify.}$$
$$6x + x - 3 = 11 \qquad \text{Group like terms.}$$
$$7x - 3 = 11 \qquad \text{Combine like terms.}$$
$$7x - 3 + 3 = 11 + 3 \qquad \text{Add 3 to each side.}$$
$$7x = 14 \qquad \text{Combine like terms.}$$
$$\frac{7x}{7} = \frac{14}{7} \qquad \text{Divide each side by 7.}$$
$$x = 2 \qquad \text{Simplify.}$$

Check

$$3(2x - 1) + x = 11 \qquad \text{Write original equation.}$$
$$3[2(2) - 1] + 2 \stackrel{?}{=} 11 \qquad \text{Substitute 2 for } x.$$
$$3(4 - 1) + 2 \stackrel{?}{=} 11 \qquad \text{Simplify.}$$
$$3(3) + 2 \stackrel{?}{=} 11 \qquad \text{Simplify.}$$
$$9 + 2 \stackrel{?}{=} 11 \qquad \text{Simplify.}$$
$$11 = 11 \qquad \text{Solution checks.} \checkmark$$

The solution is $x = 2$.

Example 3   Solving a Linear Equation Involving Parentheses

Solve $5(x + 2) = 2(x - 1)$.

Solution

$$5(x + 2) = 2(x - 1) \qquad \text{Write original equation.}$$
$$5x + 10 = 2x - 2 \qquad \text{Distributive Property}$$
$$5x - 2x + 10 = 2x - 2x - 2 \qquad \text{Subtract } 2x \text{ from each side.}$$
$$3x + 10 = -2 \qquad \text{Combine like terms.}$$
$$3x + 10 - 10 = -2 - 10 \qquad \text{Subtract 10 from each side.}$$
$$3x = -12 \qquad \text{Combine like terms.}$$
$$x = -4 \qquad \text{Divide each side by 3.}$$

The solution is $x = -4$. Check this in the original equation.

Panel 2 (top right):

Problem Solving

This text provides many opportunities for students to sharpen their problem-solving skills. In both the examples and the exercises, students are asked to apply verbal, numerical, analytical, and graphical approaches to problem solving. In the spirit of the AMATYC and NCTM standards, students are taught a five-step strategy for solving applied problems, which begins with constructing a verbal model and ends with checking the answer.

Geometry

The Fourth Edition continues to provide coverage and integration of geometry in examples and exercises. The icon ▲ indicates an exercise involving geometry.

Panel 3 (bottom left):

Section 3.4   Ratios and Proportions   161

Example 6   Geometry: Similar Triangles

A triangular lot has perpendicular sides of lengths 100 feet and 210 feet. You are to make a proportional sketch of this lot using 8 inches as the length of the shorter side. How long should you make the other side?

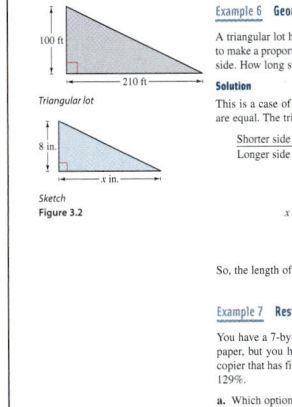

Triangular lot

Sketch
Figure 3.2

Solution

This is a case of similar triangles in which the ratios of the corresponding sides are equal. The triangles are shown in Figure 3.2.

$$\frac{\text{Shorter side of lot}}{\text{Longer side of lot}} = \frac{\text{Shorter side of sketch}}{\text{Longer side of sketch}} \qquad \text{Proportion for similar triangles}$$

$$\frac{100}{210} = \frac{8}{x} \qquad \text{Substitute.}$$

$$x \cdot 100 = 210 \cdot 8 \qquad \text{Cross-multiply.}$$

$$x = \frac{1680}{100} = 16.8 \qquad \text{Divide each side by 100.}$$

So, the length of the longer side of the sketch should be 16.8 inches.

Example 7   Resizing a Picture

You have a 7-by-8-inch picture of a graph that you want to paste into a research paper, but you have only a 6-by-6-inch space in which to put it. You go to the copier that has five options for resizing your graph: 64%, 78%, 100%, 121%, and 129%.

a. Which option should you choose?

b. What are the measurements of the resized picture?

Solution

a. Because the longest side must be reduced from 8 inches to no more than 6 inches, consider the proportion

$$\frac{\text{New length}}{\text{Old length}} = \frac{\text{New percent}}{\text{Old percent}} \qquad \text{Original proportion}$$

$$\frac{6}{8} = \frac{x}{100} \qquad \text{Substitute.}$$

$$\frac{6}{8} \cdot 100 = x \qquad \text{Multiply each side by 100.}$$

$$75 = x. \qquad \text{Simplify.}$$

To guarantee a fit, you should choose the 64% option, because 78% is greater than the required 75%.

b. To find the measurements of the resized picture, multiply by 64% or 0.64.

Length $= 0.64(8) = 5.12$ inches   Width $= 0.64(7) = 4.48$ inches

The size of the reduced picture is 5.12 inches by 4.48 inches.

Panel 4 (bottom right):

Section 3.1   Solving Linear Equations   133

19. $9x = -21$ **20.** $-14x = 42$

21. $8x - 4 = 20$ **22.** $-7x + 24 = 3$

23. $25x - 4 = 46$ **24.** $15x - 18 = 12$

25. $10 - 4x = -6$ **26.** $15 - 3x = -15$

27. $6x - 4 = 0$ **28.** $8z - 2 = 0$

29. $3y - 2 = 2y$ **30.** $2s - 13 = 28s$

31. $4 - 7x = 5x$ **32.** $24 - 5x = x$

33. $4 - 5t = 16 + t$ **34.** $3x + 4 = x + 10$

35. $-3t + 5 = -3t$ **36.** $4z + 2 = 4z$

37. $15x - 3 = 15 - 3x$ **38.** $2x - 5 = 7x + 10$

39. $7a - 18 = 3a - 2$ **40.** $4x - 2 = 3x + 1$

41. $7x + 9 = 3x + 1$ **42.** $6t - 3 = 8t + 1$

43. $4x - 6 = 4x - 6$ **44.** $5 - 3x = 5 - 3x$

45. $2x + 4 = -3(x - 2)$ **46.** $4(y + 1) = -y + 5$

47. $2x = -3x$ **48.** $6t = 9t$

49. $2x - 5 + 10x = 3$ **50.** $-4x + 10 + 10x = 4$

51. $\frac{x}{3} = 10$ **52.** $-\frac{x}{2} = 3$

53. $x - \frac{1}{3} = \frac{4}{3}$ **54.** $x + \frac{4}{5} = \frac{9}{5}$

55. $t - \frac{1}{3} = \frac{1}{2}$ **56.** $z + \frac{4}{5} = -\frac{3}{10}$

57. $5t - 4 + 3t = 4(2t - 1)$

58. $7z - 5z - 8 = 2(z - 4)$

59. $2(y - 9) = -5y - 4$

60. $6 - 21x = 3(4 - 7x)$

Solving Problems

61. ▲ *Geometry* The perimeter of a rectangle is 240 inches. The length is twice its width. Find the dimensions of the rectangle.

62. ▲ *Geometry* The length of a tennis court is 6 feet more than twice the width (see figure). Find the width of the court if the length is 78 feet.

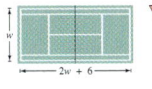

Figure for 62

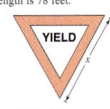

YIELD

Figure for 63

63. ▲ *Geometry* The sign in the figure has the shape of an equilateral triangle (sides have the same length). The perimeter of the sign is 225 centimeters. Find the length of its sides.

64. ▲ *Geometry* You are asked to cut a 12-foot board into three pieces. Two pieces are to have the same length and the third is to be twice as long as the others. How long are the pieces?

65. *Car Repair* The bill (including parts and labor) for the repair of your car is shown. Some of the bill is unreadable. From what is given, can you determine how many hours were spent on labor? Explain.

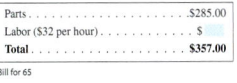

Parts	.$285.00
Labor ($32 per hour)	$
Total	**$357.00**

Bill for 65

66. *Car Repair* The bill for the repair of your car was $439. The cost for parts was $265. The cost for labor was $29 per hour. How many hours did the repair work take?

67. *Ticket Sales* Tickets for a community theater are $10 for main floor seats and $8 for balcony seats. There are 400 seats on the main floor, and these were sold out for the evening performance. The total revenue from ticket sales was $5200. How many balcony seats were sold?

68. *Ticket Sales* Tickets for a marching band competition are $5 for 50-yard-line seats and $3 for bleacher seats. Eight hundred 50-yard-line seats were sold. The total revenue from ticket sales was $5500. How many bleacher seats were sold?

69. *Summer Jobs* You have two summer jobs. In the first job, you work 40 hours a week and earn $9.25 an hour at a coffee shop. In the second job, you tutor for $7.50 an hour and can work as many hours as you want. You want to earn a combined total of $425 a week. How many hours must you tutor?

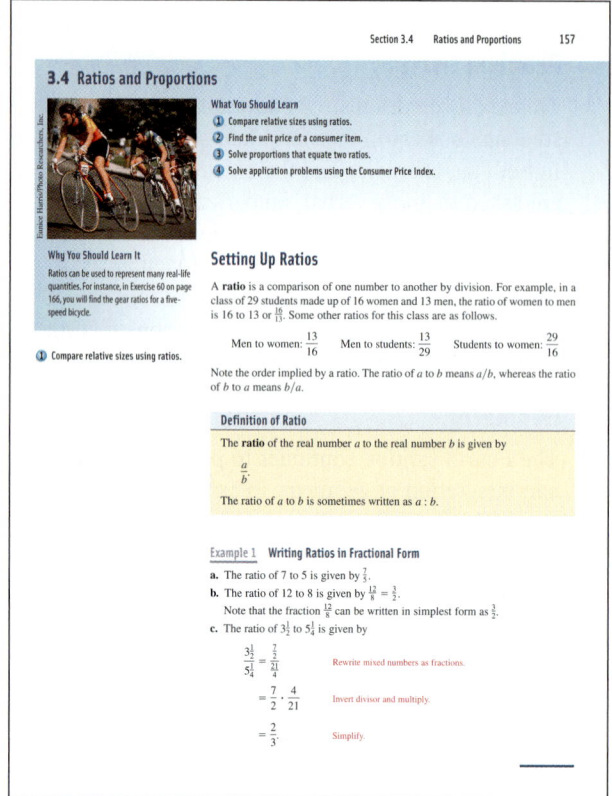

Definitions and Rules

All important definitions, rules, formulas, properties, and summaries of solution methods are highlighted for emphasis. Each of these features is also titled for easy reference.

Study Tips

Study Tips offer students specific point-of-use suggestions for studying algebra, as well as pointing out common errors and discussing alternative solution methods. They appear in the margins.

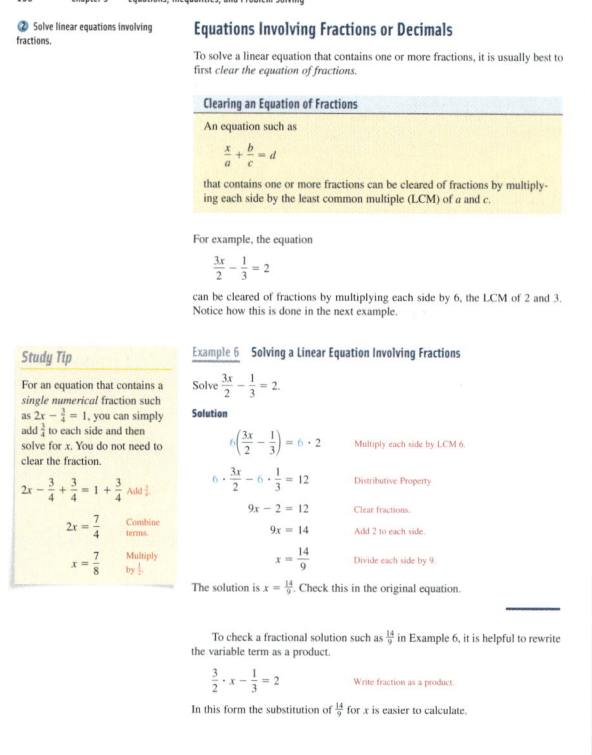

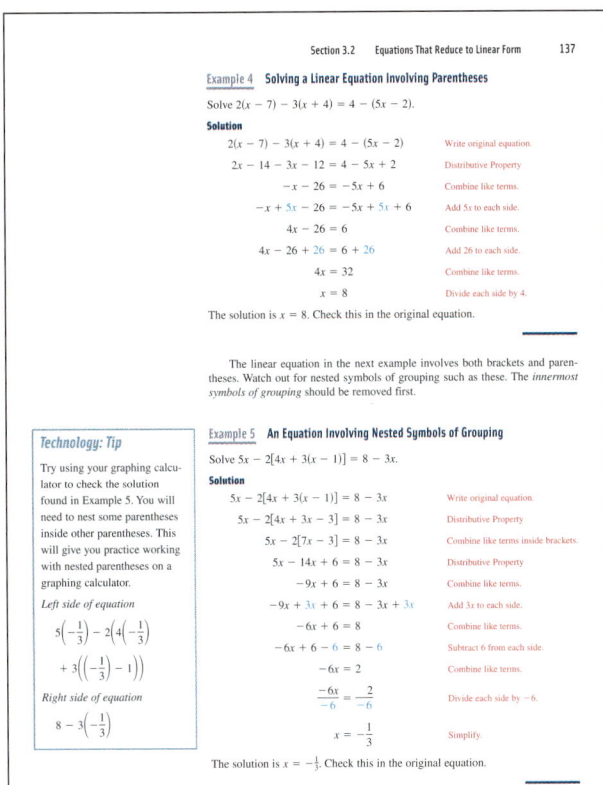

Graphics

Visualization is a critical problem-solving skill. To encourage the development of this skill, students are shown how to use graphs to reinforce algebraic and numeric solutions and to interpret data. The numerous figures in examples and exercises throughout the text were computer-generated for accuracy.

Technology: Tips

Point-of-use instructions for using graphing calculators appear in the margins. These features encourage the use of graphing technology as a tool for visualization of mathematical concepts, for verification of other solution methods, and for facilitation of computations. The *Technology: Tips* can easily be omitted without loss of continuity in coverage.

Technology: Discovery

Technology: Discovery features invite students to engage in active exploration of mathematical concepts and discovery of mathematical relationships through the use of scientific or graphing calculators. These activities encourage students to utilize their critical thinking skills and help them develop an intuitive understanding of theoretical concepts. *Technology: Discovery* features can easily be omitted without loss of continuity in coverage.

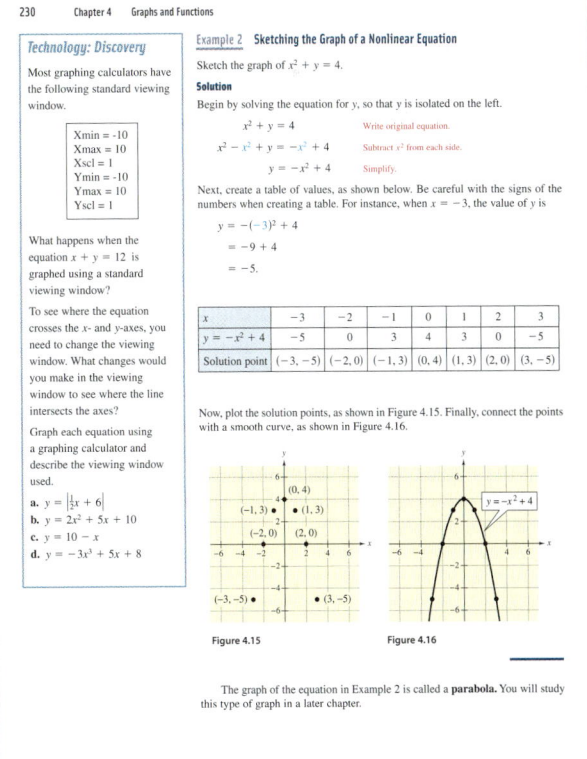

FEATURES

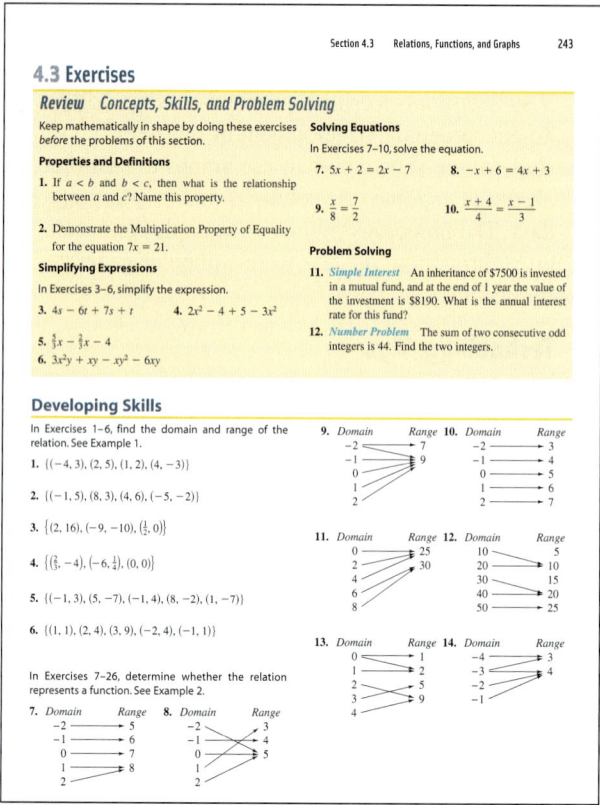

Review: Concepts, Skills, and Problem Solving

Each exercise set (except in Chapter 1) is preceded by these review exercises that are designed to help students keep up with concepts and skills learned in previous chapters. Answers to all *Review: Concepts, Skills, and Problem Solving* exercises are given in the back of the student text.

Exercises

The exercise sets are grouped into three categories: *Developing Skills, Solving Problems,* and *Explaining Concepts.* The exercise sets offer a diverse variety of computational, conceptual, and applied problems to accommodate many teaching and learning styles. Designed to build competence, skill, and understanding, each exercise set is graded in difficulty to allow students to gain confidence as they progress. Detailed solutions to all odd-numbered exercises are given in the *Student Solutions Guide,* and answers to all odd-numbered exercises are given in the back of the student text.

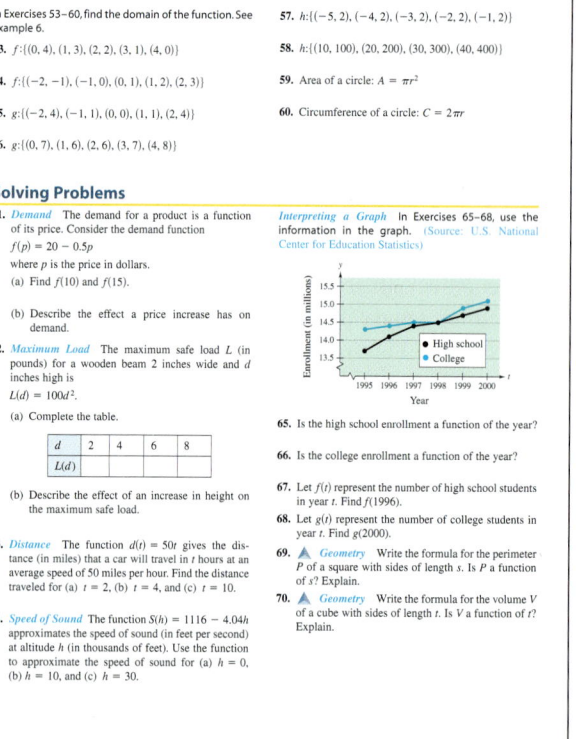

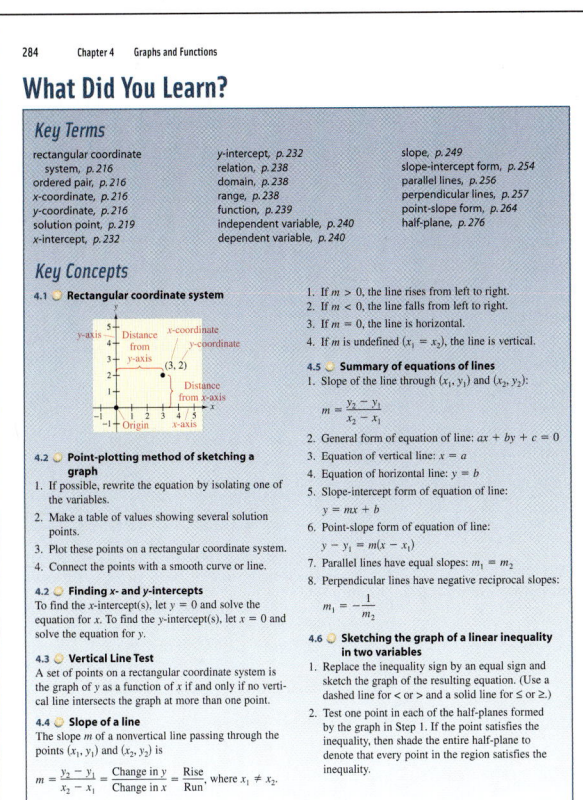

What Did You Learn? (Chapter Summary)

Located at the end of every chapter, *What Did You Learn?* summarizes the *Key Terms* (referenced by page) and the *Key Concepts* (referenced by section) presented in the chapter. This effective study tool aids students as they review concepts and prepare for exams.

Review Exercises

The *Review Exercises* at the end of each chapter have been reorganized in the Fourth Edition. All skill-building and application exercises are first ordered by section, then grouped according to the objectives stated within *What You Should Learn.* This organization allows students to easily identify the appropriate sections and concepts for study and review.

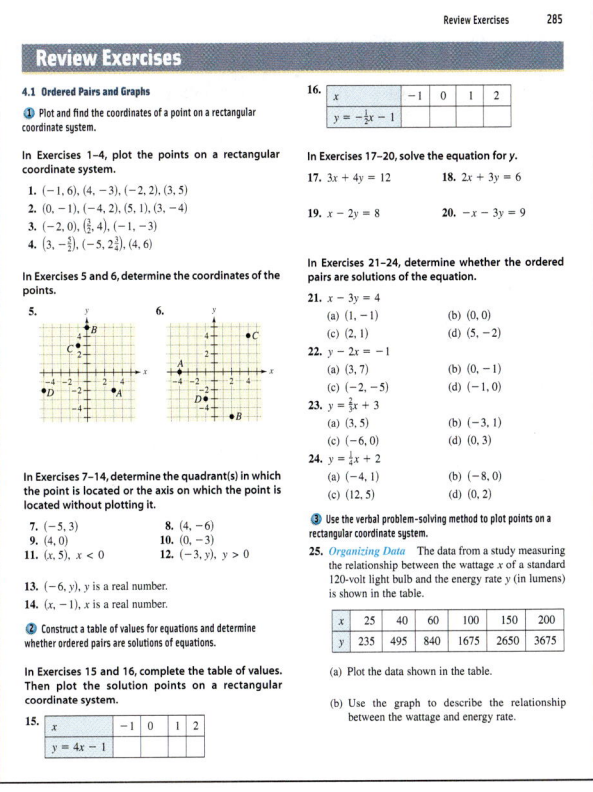

---

168   Chapter 3   Equations, Inequalities, and Problem Solving

Mid-Chapter Quiz

Take this quiz as you would take a quiz in class. After you are done, check
your work against the answers in the back of the book.

In Exercises 1–10, solve the equation.

1. $74 - 12x = 2$
2. $10(y - 8) = 0$
3. $3x + 1 = x + 20$
4. $6x + 8 = 8 - 2x$
5. $-10x + \frac{2}{3} = \frac{7}{3} - 5x$
6. $\frac{x}{5} + \frac{x}{8} = 1$
7. $\frac{9 + x}{3} = 15$
8. $7 - 2(5 - x) = -7$
9. $\frac{x + 3}{6} = \frac{4}{3}$
10. $\frac{x + 7}{5} = \frac{x + 9}{7}$

**In Exercises 11 and 12, solve the equation. Round your answer to two
decimal places. In your own words, explain how to check the solution.**

11. $32.86 - 10.5x = 11.25$
12. $\frac{x}{5.45} + 3.2 = 12.6$

13. What number is 62% of 25?
14. What number is $\frac{1}{2}$% of 8400?
15. 300 is what percent of 150?
16. 145.6 is 32% of what number?

17. You have two jobs. In the first job, you work 40 hours a week at a candy store
and earn $7.50 per hour. In the second job, you earn $6.00 per hour
babysitting and can work as many hours as you want. You want to earn $360
a week. How many hours must you work at the second job?

18. A region has an area of 42 square meters. It must be divided into three
subregions so that the second has twice the area of the first, and the third has
twice the area of the second. Find the area of each subregion.

19. To get an A in a psychology course, you must have an average of at least 90
points for three tests of 100 points each. For the first two tests, your scores
are 84 and 93. What must you score on the third test to earn a 90% average
for the course?

20. The circle graph at the left shows the number of endangered wildlife and
plant species for the year 2001. What percent of the total endangered wildlife
and plant species were birds? (Source: U.S. Fish and Wildlife Service)

21. Two people can paint a room in t hours, where t must satisfy the equation
$t/4 + t/12 = 1$. How long will it take for the two people to paint the room?

22. A large round pizza has a radius of $r = 15$ inches, and a small round pizza
has a radius of $r = 8$ inches. Find the ratio of the area of the large pizza to
the area of the small pizza. (*Hint:* The area of a circle is $A = \pi r^2$.)

23. A car uses 30 gallons of gasoline for a trip of 800 miles. How many gallons
would be used on a trip of 700 miles?

**Endangered Wildlife and
Plant Species**

Plants 593
Mammals 314
Other 169
Birds 253
Reptiles 78
Fishes 81

Figure for 20

Chapter Test

Take this test as you would take a test in class. After you are done, check your
work against the answers in the back of the book.

1. Plot the points $(-1, 2)$, $(1, 4)$, and $(2, -1)$ on a rectangular coordinate
system. Connect the points with line segments to form a right triangle.

2. Determine whether the ordered pairs are solutions of $y = |x| + |x - 2|$.
   (a) $(0, -2)$ (b) $(0, 2)$ (c) $(-4, 10)$ (d) $(-2, -2)$

3. What is the y-coordinate of any point on the x-axis?
4. Find the x- and y-intercepts of the graph of $3x - 4y + 12 = 0$.

5. Complete the table at the left and use the results to sketch the graph of the
equation $x - 2y = 6$.

x	-2	-1	0	1	2
y					

Table for 5

In Exercises 6–9, sketch the graph of the equation.

6. $x + 2y = 6$
7. $y = \frac{1}{4}x - 1$
8. $y = |x + 2|$
9. $y = (x - 3)^2$

10. Does the table at the left represent y as a function of x? Explain.

Input, x	0	1	2	1	0
Output, y	4	5	8	-3	-1

Table for 10

11. Does the graph at the left represent y as a function of x? Explain.
12. Evaluate $f(x) = x^3 - 2x^2$ as indicated, and simplify.
   (a) $f(0)$ (b) $f(2)$ (c) $f(-2)$ (d) $f(\frac{1}{2})$
13. Find the slope of the line passing through the points $(-5, 0)$ and $(2, \frac{3}{2})$.
14. A line with slope $m = -2$ passes through the point $(-3, 4)$. Plot the point
and use the slope to find two additional points on the line. (There are many
correct answers.)
15. Find the slope of a line *perpendicular* to the line $3x - 5y + 2 = 0$.
16. Find an equation of the line that passes through the point $(0, 6)$ with slope
$m = -\frac{3}{8}$.
17. Write an equation of the vertical line that passes through the point $(3, -7)$.
18. Determine whether the points are solutions of $3x + 5y \leq 16$.
   (a) $(2, 2)$ (b) $(6, -1)$ (c) $(-2, 4)$ (d) $(7, -1)$

Figure for 11

In Exercises 19–22, sketch the graph of the linear inequality.

19. $y \geq -2$
20. $y < 5 - 2x$
21. $x \geq 2$
22. $y \leq 5$

23. The sales y of a product are modeled by $y = 230x + 5000$, where x is time
in years. Interpret the meaning of the slope in this model.

291

Mid-Chapter Quiz

Each chapter contains a *Mid-Chapter Quiz*.
Answers to all questions in the *Mid-Chapter Quiz*
are given in the back of the student text.

Chapter Test

Each chapter ends with a *Chapter Test*. Answers
to all questions in the *Chapter Test* are given in
the back of the student text.

Cumulative Test

The *Cumulative Tests* that follow Chapters 3, 6, 9,
and 12 provide a comprehensive self-assessment
tool that helps students check their mastery of
previously covered material. Answers to all
questions in the *Cumulative Tests* are given in
the back of the student text.

Cumulative Test: Chapters 1–3

Take this test as you would take a test in class. After you are done, check your
work against the answers in the back of the book.

1. Place the correct symbol (< or >) between the numbers: $-\frac{3}{4}$ $\left|-\frac{7}{8}\right|$.

In Exercises 2–7, evaluate the expression.

2. $(-200)(2)(-3)$
3. $\frac{3}{8} - \frac{5}{6}$
4. $-\frac{2}{9} \div \frac{8}{75}$
5. $-(-2)^3$
6. $3 + 2(6) - 1$
7. $24 + 12 \div 3$

In Exercises 8 and 9, evaluate the expression when $x = -2$ and $y = 3$.

8. $-3x - (2y)^2$
9. $4y - x^3$

10. Use exponential form to write the product $3 \cdot (x + y) \cdot (x + y) \cdot 3 \cdot 3$.

11. Use the Distributive Property to expand $-2x(x - 3)$.
12. Identify the property of real numbers illustrated by
$$2 + (3 + x) = (2 + 3) + x.$$

In Exercises 13–16, simplify the expression.

13. $(3x^3)(5x^4)$
14. $(a^3b^2)(ab)$
15. $2x^2 - 3x + 5x^2 - (2 + 3x)$
16. $3(x^2 + x) - 2(2x - x^2)$

17. Determine whether the value of x is a solution of $x + 1 = 4(x - 2)$.
   (a) $x = -8$ (b) $x = 3$

In Exercises 18–21, solve the equation and check your solution.

18. $12x - 3 = 7x + 27$
19. $2x - \frac{5x}{4} = 13$
20. $2(x - 3) + 3 = 12 - x$
21. $|3x + 1| = 5$

In Exercises 22–25, solve and graph the inequality.

22. $12 - 3x \leq -15$
23. $-1 \leq \frac{x + 3}{2} < 2$
24. $-4x + 1 \leq 5$ or $-5x + 1 \geq 7$
25. $|8x - 3| \geq 13$

212

Elementary and Intermediate Algebra: A Combined Course, Fourth Edition, by Larson and Hostetler is accompanied by a comprehensive supplements package, which includes resources for both students and instructors. All items are keyed to the text.

Printed Resources

For Students

Student Solutions Guide by Carolyn F. Neptune, Johnson County Community College, and Gerry C. Fitch, Louisiana State University   (0-618-38838-9)

- Detailed, step-by-step solutions to all Review: Concepts, Skills, and Problem Solving exercises and to all odd-numbered exercises in the section exercise sets and in the review exercises
- Detailed, step-by-step solutions to all Mid-Chapter Quiz, Chapter Test, and Cumulative Test questions

For Instructors

Instructor's Annotated Edition   (0-618-38837-0)

Instructor's Resource Guide by Carolyn F. Neptune, Gerry C. Fitch, and Ann Rutledge Kraus, The Pennsylvania State University, The Behrend College (0-618-38839-7)

Technology Resources

For Students

HM mathSpace® Student CD-ROM   (0-618-38844-3)

Website (http://math.college.hmco.com/students)

Houghton Mifflin Instructional Videos and DVDs by Dana Mosely (Video ISBN: 0-618-38840-0; DVD ISBN: 0-618-38841-9)

SMARTHINKING™ Live, Online Tutoring   Houghton Mifflin has partnered with SMARTHINKING to provide an easy-to-use and effective online tutorial service. *Whiteboard Simulations* and *Practice Area* promote real-time visual interaction. Three levels of service are offered.

- **Text-Specific Tutoring** provides real-time, one-on-one instruction with a specially qualified "e-structor."
- *Questions Any Time* allows students to submit questions to the tutor outside the scheduled hours and receive a reply within 24 hours.
- *Independent Study Resources* connect students with around-the-clock access to additional educational services, including interactive websites, diagnostic tests, and Frequently Asked Questions posed to SMARTHINKING e-structors.

For Instructors

HMClassPrep™ with HM Testing   (0-618-38842-7)

Website (http://math.college.hmco.com/instructors)

Acknowledgments

We would like to thank the many people who have helped us revise the various editions of this text. Their encouragement, criticisms, and suggestions have been invaluable to us.

Reviewers

Mary Kay Best, Coastal Bend College; Patricia K. Bezona, Valdosta State University; Connie L. Buller, Metropolitan Community College; Mistye R. Canoy, Holmes Community College; Maggie W. Flint, Northeast State Technical Community College; William Hoard, Front Range Community College; Andrew J. Kaim, DePaul University; Jennifer L. Laveglia, Bellevue Community College; Aaron Montgomery, Purdue University North Central; William Naegele, South Suburban College; Jeanette O'Rourke, Middlesex County College; Judith Pranger, Binghamton University; Kent Sandefer, Mohave Community College; Robert L. Sartain, Howard Payne University; Jon W. Scott, Montgomery College; John Seims, Mesa Community College; Ralph Selensky, Eastern Arizona College; Charles I. Sherrill, Community College of Aurora; Kay Stroope, Phillips Community College of the University of Arkansas; Bettie Truitt, Black Hawk College; Betsey S. Whitman, Framingham State College; George J. Witt, Glendale Community College.

We would also like to thank the staff of Larson Texts, Inc. and the staff of Meridian Creative Group, who assisted in preparing the manuscript, rendering the art package, and typesetting and proofreading the pages and the supplements.

On a personal level, we are grateful to our wives, Deanna Gilbert Larson and Eloise Hostetler, for their love, patience, and support. Also, a special thanks goes to R. Scott O'Neil.

If you have suggestions for improving this text, please feel free to write to us. Over the past two decades we have received many useful comments from both instructors and students, and we value these comments very much.

Ron Larson
Robert P. Hostetler

ACKNOWLEDGMENTS

How to Study Algebra

Your success in algebra depends on your active participation both in class and outside of class. Because the material you learn each day builds on the material you learned previously, it is important that you keep up with the course work every day and develop a clear plan of study. To help you learn how to study algebra, we have prepared a set of guidelines that highlight key study strategies.

Preparing for Class

The syllabus your instructor provides is an invaluable resource that outlines the major topics to be covered in the course. Use it to help you prepare. As a general rule, you should set aside two to four hours of study time for each hour spent in class. Being prepared is the first step toward success in algebra. Before class,

- Review your notes from the previous class.

- Read the portion of the text that will be covered in class.

- Use the *What You Should Learn* objectives listed at the beginning of each section to keep you focused on the main ideas of the section.

- Pay special attention to the definitions, rules, and concepts highlighted in boxes. Also, be sure you understand the meanings of mathematical symbols and of terms written in boldface type. Keep a vocabulary journal for easy reference.

- Read through the solved examples. Use the side comments that accompany the solution steps to help you follow the solution process. Also, read the *Study Tips* given in the margins.

- Make notes of anything you do not understand as you read through the text. If you still do not understand after your instructor covers the topic in question, ask questions before your instructor moves on to a new topic.

- If you are using technology in this course, read the *Technology*: *Tips* and try the *Technology*: *Discovery* exercises.

Keeping Up

Another important step toward success in algebra involves your ability to keep up with the work. It is very easy to fall behind, especially if you miss a class. To keep up with the course work, be sure to

- Attend every class. Bring your text, a notebook, and a pen or pencil. If you miss a class, get the notes from a classmate as soon as possible and review them carefully.

- Take notes in class. After class, read through your notes and add explanations so that your notes make sense to *you*.

- Reread the portion of the text that was covered in class. This time, work each example *before* reading through the solution.

- Do your homework as soon as possible, while concepts are still fresh in your mind.

Use your notes from class, the text discussion, the examples, and the *Study Tips* as you do your homework. Many exercises are keyed to specific examples in the text for easy reference.

Getting Extra Help

It can be very frustrating when you do not understand concepts and are unable to complete homework assignments. However, there are many resources available to help you with your study of algebra.

Your instructor may have office hours. If you are feeling overwhelmed and need help, make an appointment to discuss your difficulties with your instructor.

Find a study partner or a study group. Sometimes it helps to work through problems with another person.

Arrange to get regular assistance from a tutor. Many colleges have a math resource center available on campus as well.

Consult one of the many ancillaries available with this text: the *Student Solutions Guide,* HM mathSpace™ Student CD-ROM, videotapes, DVDs, and additional study resources available at our website at *http://math.college.hmco.com/students.*

Preparing for an Exam

The last step toward success in algebra lies in how you prepare for and complete exams. If you have followed the suggestions given above, then you are almost ready for exams. Do not assume that you can cram for the exam the night before—this seldom works. As a final preparation for the exam,

Read the *What Did You Learn?* chapter summary, which is keyed to each section, and review the concepts and terms.

Work through the *Review Exercises* if you need extra practice on material from a particular section.

Take the *Mid-Chapter Quiz* and the *Chapter Test* as if you were in class. You should set aside at least one hour per test. Check your answers against the answers given in the back of the book.

Review your notes and the portion of the text that will be covered on the exam.

Avoid studying up until the last minute. This will only make you anxious.

Once the exam begins, read through the directions and the entire exam before beginning. Work the problems that you know how to do first to avoid spending too much time on any one problem. Time management is extremely important when taking an exam.

If you finish early, use the remaining exam time to go over your work.

When you get an exam back, review it carefully and go over your errors. Rework the problems you answered incorrectly. Discovering the mistakes you made will help you improve your test-taking ability.

STUDY PLAN

Motivating the Chapter

 December in Lexington, Virginia

In December 2001, the city of Lexington, Virginia, had an average daily high temperature of 5.8°C. The daily average temperatures and the daily high temperatures for the last 14 days of December 2001 are shown in the table.   (Source: WREL Weather Station, Lexington, Virginia)

Day	18	19	20	21	22	23	24
Average temperature (°C)	7.9°	2.1°	3.6°	1.4°	−1.3°	$\frac{5}{9}°$	2.2°
High temperature (°C)	11.5°	12°	7.6°	6.8°	7.4°	7.3°	6.2°

Day	25	26	27	28	29	30	31
Average temperature (°C)	−2.6°	−2.7°	$-2\frac{1}{2}°$	1.8°	2.6°	−3.9°	−4.3°
High temperature (°C)	2.6°	2.4°	2.2°	7°	6.8°	$-\frac{2}{9}°$	$1\frac{7}{10}°$

Here are some of the types of questions you will be able to answer as you study this chapter. You will be asked to answer parts (a)–(f) in Section 1.1, Exercise 79.

a. Write the set *A* of *integer* average and high temperatures.

b. Write the set *B* of *rational* high temperatures.

c. Write the set *C* of *nonnegative* average temperatures.

d. Write the high temperatures in *increasing* order.

e. Write the average temperatures in *decreasing* order.

f. What day(s) had average and high temperatures that were opposite numbers?

You will be asked to answer parts (g)–(k) in Section 1.4, Exercise 165.

g. What day had a high temperature of greatest departure from the monthly average high temperature of 5.8?

h. What successive days had the greatest change in average temperature?

i. What successive days had the greatest change in high temperature?

j. Find the average of the average temperatures for the 14 days.

k. In which of the preceding problems is the concept of absolute value used?

Owaki-Kulla/Corbis

The Real Number System

1.1 Real Numbers: Order and Absolute Value

© George B. Diebold/Corbis

What You Should Learn

① Define sets and use them to classify numbers as natural, integer, rational, or irrational.

② Plot numbers on the real number line.

③ Use the real number line and inequality symbols to order real numbers.

④ Find the absolute value of a number.

Why You Should Learn It

Understanding sets and subsets of real numbers will help you to analyze real-life situations accurately.

① Define sets and use them to classify numbers as natural, integer, rational, or irrational.

Sets and Real Numbers

The ability to communicate precisely is an essential part of a modern society, and it is the primary goal of this text. Specifically, this section introduces the language used to communicate numerical concepts.

The formal term that is used in mathematics to talk about a collection of objects is the word **set.** For instance, the set $\{1, 2, 3\}$ contains the three numbers 1, 2, and 3. Note that a pair of braces $\{\ \ \}$ is used to list the members of the set. Parentheses $(\ \ )$ and brackets $[\ \ ]$ are used to represent other ideas.

The set of numbers that is used in arithmetic is called the set of **real numbers.** The term *real* distinguishes real numbers from *imaginary* numbers—a type of number that is used in some mathematics courses. You will not study imaginary numbers in Elementary Algebra.

If each member of a set A is also a member of a set B, then A is called a **subset** of B. The set of real numbers has many important subsets, each with a special name. For instance, the set

$$\{1, 2, 3, 4, \ldots\}$$ A subset of the set of real numbers

is the set of **natural numbers** or **positive integers.** Note that the three dots indicate that the pattern continues. For instance, the set also contains the numbers 5, 6, 7, and so on. Every positive integer is a real number, but there are many real numbers that are not positive integers. For example, the numbers -2, 0, and $\frac{1}{2}$ are real numbers, but they are not positive integers.

Positive integers can be used to describe many things that you encounter in everyday life. For instance, you might be taking four classes this term, or you might be paying $180 a month for rent. But even in everyday life, positive integers cannot describe some concepts accurately. For instance, you could have a zero balance in your checking account, or the temperature could be $-5°\text{F}$. To describe such quantities you need to expand the set of positive integers to include **zero** and the **negative integers.** The expanded set is called the set of **integers.**

$$\underbrace{\{\ldots, -3, -2, -1,}_{\text{Negative integers}} \overset{\text{Zero}}{0,} \underbrace{1, 2, 3, \ldots\}}_{\text{Positive integers}}$$ Set of integers

The set of integers is also a subset of the set of real numbers.

Even with the set of integers, there are still many quantities in everyday life that you cannot describe accurately. The costs of many items are not in whole-dollar amounts, but in parts of dollars, such as $1.19 or $39.98. You might work $8\frac{1}{2}$ hours, or you might miss the first half of a movie. To describe such quantities, you can expand the set of integers to include **fractions.** The expanded set is called the set of **rational numbers.** In the formal language of mathematics, a real number is **rational** if it can be written as a ratio of two integers. So, $\frac{3}{4}$ is a rational number; so is 0.5 $\left(\text{it can be written as } \frac{1}{2}\right)$; and so is every integer. A real number that is not rational is called **irrational** and cannot be written as the ratio of two integers. One example of an irrational number is $\sqrt{2}$, which is read as the positive square root of 2. Another example is π (the Greek letter pi), which represents the ratio of the circumference of a circle to its diameter. Each of the sets of numbers mentioned—natural numbers, integers, rational numbers, and irrational numbers—is a subset of the set of real numbers, as shown in Figure 1.1.

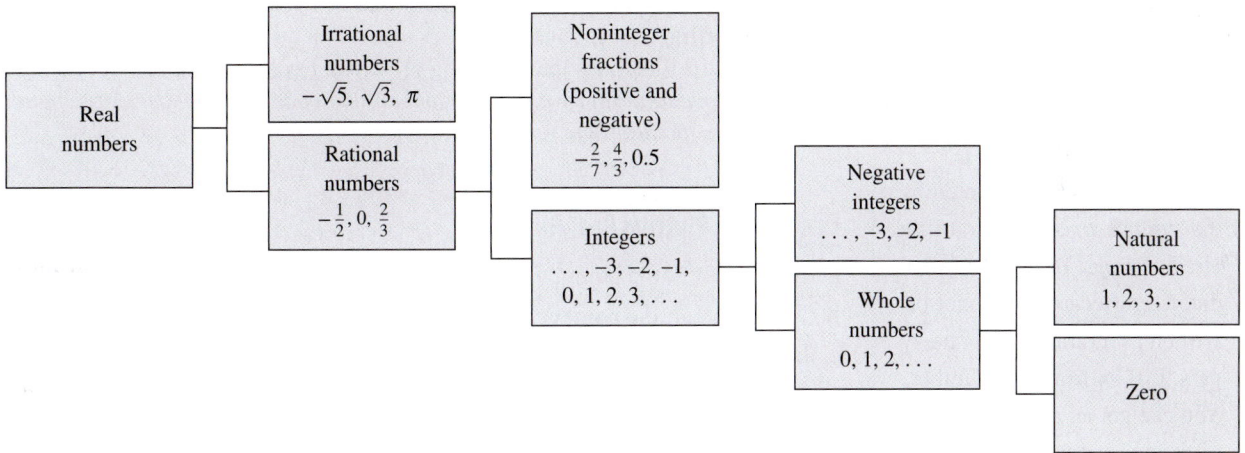

Figure 1.1   Subsets of Real Numbers

Example 1   Classifying Real Numbers

Which of the numbers in the following set are (a) natural numbers, (b) integers, (c) rational numbers, and (d) irrational numbers?

$$\left\{ \frac{1}{2}, -1, 0, 4, -\frac{5}{8}, \frac{4}{2}, -\frac{3}{1}, 0.86, \sqrt{2}, \sqrt{9} \right\}$$

Solution

a. Natural numbers: $\left\{ 4, \frac{4}{2} = 2, \sqrt{9} = 3 \right\}$

b. Integers: $\left\{ -1, 0, 4, \frac{4}{2} = 2, -\frac{3}{1} = -3, \sqrt{9} = 3 \right\}$

c. Rational numbers: $\left\{ \frac{1}{2}, -1, 0, 4, -\frac{5}{8}, \frac{4}{2} = 2, -\frac{3}{1} = -3, 0.86, \sqrt{9} = 3 \right\}$

d. Irrational number: $\left\{ \sqrt{2} \right\}$

The Real Number Line

The diagram used to represent the real numbers is called the **real number line.** It consists of a horizontal line with a point (the **origin**) labeled 0. Numbers to the left of 0 are **negative** and numbers to the right of 0 are **positive**, as shown in Figure 1.2. The real number zero is neither positive nor negative. So, the term **nonnegative** implies that a number may be positive or zero.

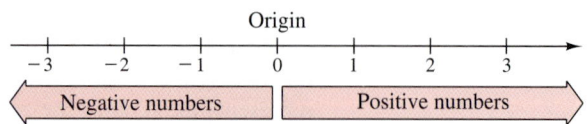

Figure 1.2   The Real Number Line

Drawing the point on the real number line that corresponds to a real number is called **plotting** the real number.

Example 2 illustrates the following principle. *Each point on the real number line corresponds to exactly one real number, and each real number corresponds to exactly one point on the real number line.*

Technology: Tip

The Greek letter pi, denoted by the symbol π, is the ratio of the circumference of a circle to its diameter. Because π cannot be written as a ratio of two integers, it is an irrational number. You can get an approximation of π on a scientific or graphing calculator by using the following keystroke.

Keystroke   *Display*

  $\boxed{\pi}$ 3.141592654

Between which two integers would you plot π on the real number line?

Example 2   Plotting Real Numbers

a. In Figure 1.3, the point corresponds to the real number $-\frac{1}{2}$.

b. In Figure 1.4, the point corresponds to the real number 2.

c. In Figure 1.5, the point corresponds to the real number $-\frac{3}{2}$.

d. In Figure 1.6, the point corresponds to the real number 1.

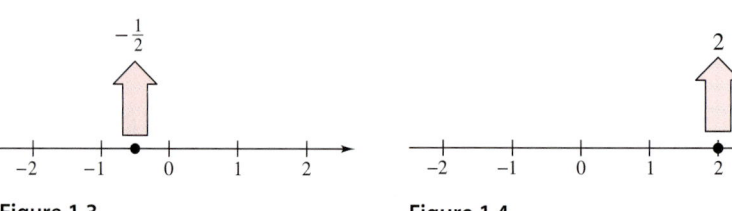

Figure 1.3   **Figure 1.4**

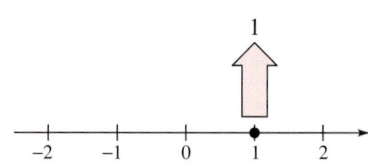

Figure 1.5   **Figure 1.6**

③ Use the real number line and inequality symbols to order real numbers.

Ordering Real Numbers

The real number line provides you with a way of comparing any two real numbers. For instance, if you choose any two (different) numbers on the real number line, one of the numbers must be to the left of the other number. The number to the left is **less than** the number to the right. Similarly, the number to the right is **greater than** the number to the left. For example, from Figure 1.7 you can see that -3 is less than 2 because -3 lies to the left of 2 on the number line. A "less than" comparison is denoted by the **inequality symbol** $<$. For instance, "-3 is less than 2" is denoted by $-3 < 2$.

Similarly, the inequality symbol $>$ is used to denote a "greater than" comparison. For instance, "2 is greater than -3" is denoted by $2 > -3$. The inequality symbol $\leq$ means **less than or equal to,** and the inequality symbol $\geq$ means **greater than or equal to.**

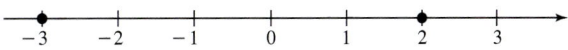

Figure 1.7 -3 lies to the left of 2.

When you are asked to **order** two numbers, you are simply being asked to say which of the two numbers is greater.

Example 3   Ordering Integers

Place the correct inequality symbol ($<$ or $>$) between each pair of numbers.

a. 3   5   **b.** -3 -5 **c.** 4   0

d. -2 2   **e.** 1 -4

Solution

a. $3 < 5$, because 3 lies to the *left* of 5.   See Figure 1.8.

b. $-3 > -5$, because -3 lies to the *right* of -5.   See Figure 1.9.

c. $4 > 0$, because 4 lies to the *right* of 0.   See Figure 1.10.

d. $-2 < 2$, because -2 lies to the *left* of 2.   See Figure 1.11.

e. $1 > -4$, because 1 lies to the *right* of -4.   See Figure 1.12.

Figure 1.8

Figure 1.9

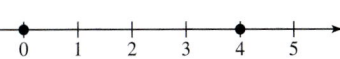

Figure 1.10

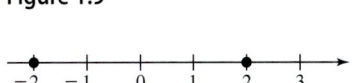

Figure 1.11

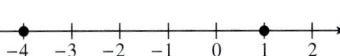

Figure 1.12

There are two ways to order fractions: you can write both fractions with the same denominator, or you can rewrite both fractions in decimal form. Here are two examples.

$$\frac{1}{3} = \frac{4}{12} \quad \text{and} \quad \frac{1}{4} = \frac{3}{12} \quad \Longrightarrow \quad \frac{1}{3} > \frac{1}{4}$$

$$\frac{11}{131} \approx 0.084 \quad \text{and} \quad \frac{19}{209} \approx 0.091 \quad \Longrightarrow \quad \frac{11}{131} < \frac{19}{209}$$

The symbol $\approx$ means "is approximately equal to."

Example 4   Ordering Fractions

Place the correct inequality symbol ($<$ or $>$) between each pair of numbers.

a. $\frac{1}{3}$ ⬚ $\frac{1}{5}$ **b.** $-\frac{3}{2}$ ⬚ $\frac{1}{2}$

Solution

a. $\frac{1}{3} > \frac{1}{5}$, because $\frac{1}{3} = \frac{5}{15}$ lies to the *right* of $\frac{1}{5} = \frac{3}{15}$ (see Figure 1.13).

b. $-\frac{3}{2} < \frac{1}{2}$, because $-\frac{3}{2}$ lies to the *left* of $\frac{1}{2}$ (see Figure 1.14).

Figure 1.13

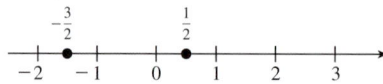

Figure 1.14

Example 5   Ordering Decimals

Place the correct inequality symbol ($<$ or $>$) between each pair of numbers.

a. -3.1 ⬚ 2.8 **b.** -1.09 ⬚ -1.90

Solution

a. $-3.1 < 2.8$, because -3.1 lies to the *left* of 2.8 (see Figure 1.15).

b. $-1.09 > -1.90$, because -1.09 lies to the *right* of -1.90 (see Figure 1.16).

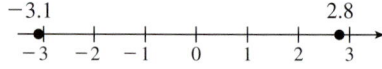

Figure 1.15

Figure 1.16

④ Find the absolute value of a number.

Absolute Value

Two real numbers are **opposites** of each other if they lie the same distance from, but on opposite sides of, zero. For example, -2 is the opposite of 2, and 4 is the opposite of -4, as shown in Figure 1.17.

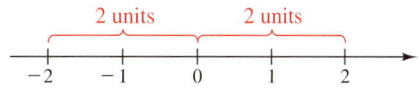

-2 **is the opposite of 2.**

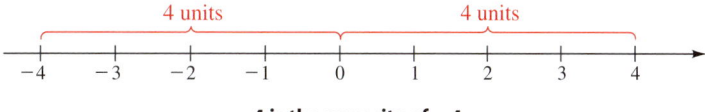

4 is the opposite of -4.

Figure 1.17

Parentheses are useful for denoting the opposite of a negative number. For example, $-(-3)$ means the opposite of -3, which you know to be 3. That is,

$$-(-3) = 3.$$
The opposite of -3 is 3.

For any real number, its distance from zero on the real number line is its **absolute value.** A pair of vertical bars, $|\ \ |$, is used to denote absolute value. Here are two examples.

$$|5| = \text{``distance between 5 and 0''} = 5$$

$$|-8| = \text{``distance between } -8 \text{ and 0''} = 8 \qquad \text{See Figure 1.18.}$$

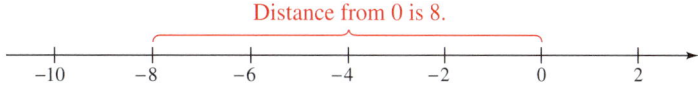

Figure 1.18

Because opposite numbers lie the same distance from zero on the real number line, they have the same absolute value. So, $|5| = 5$ and $|-5| = 5$ (see Figure 1.19).

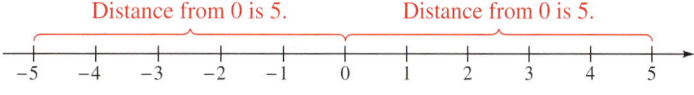

Figure 1.19

You can write this more simply as $|5| = |-5| = 5$.

Definition of Absolute Value

If a is a real number, then the **absolute value** of a is

$$|a| = \begin{cases} a, & \text{if } a \geq 0 \\ -a, & \text{if } a < 0 \end{cases}.$$

The absolute value of a real number is either positive or zero (never negative). For instance, by definition, $|-3| = -(-3) = 3$. Moreover, zero is the only real number whose absolute value is 0. That is, $|0| = 0$.

The word **expression** means a collection of numbers and symbols such as $3 + 5$ or $|-4|$. When asked to **evaluate** an expression, you are to find the *number* that is equal to the expression.

Example 6   Evaluating Absolute Values

Evaluate each expression.

a. $|-10|$

b. $\left|\dfrac{3}{4}\right|$

c. $|-3.2|$

d. $-|-6|$

Solution

a. $|-10| = 10$, because the distance between -10 and 0 is 10.

b. $\left|\dfrac{3}{4}\right| = \dfrac{3}{4}$, because the distance between $\dfrac{3}{4}$ and 0 is $\dfrac{3}{4}$.

c. $|-3.2| = 3.2$, because the distance between -3.2 and 0 is 3.2.

d. $-|-6| = -(6) = -6$

Note in Example 6(d) that $-|-6| = -6$ does not contradict the fact that the absolute value of a real number cannot be negative. The expression $-|-6|$ calls for the *opposite* of an absolute value and so it must be negative.

Example 7   Comparing Absolute Values

Place the correct symbol ($<$, $>$, or $=$) between each pair of numbers.

a. $|-9|$ ⬚ $|9|$

b. $|-3|$ ⬚ 5

c. 0 ⬚ $|-7|$

d. -4 ⬚ $-|-4|$

e. $|12|$ ⬚ $|-15|$

f. 2 ⬚ $-|-2|$

Solution

a. $|-9| = |9|$, because $|-9| = 9$ and $|9| = 9$.

b. $|-3| < 5$, because $|-3| = 3$ and 3 is less than 5.

c. $0 < |-7|$, because $|-7| = 7$ and 0 is less than 7.

d. $-4 = -|-4|$, because $-|-4| = -4$ and -4 is equal to -4.

e. $|12| < |-15|$, because $|12| = 12$ and $|-15| = 15$, and 12 is less than 15.

f. $2 > -|-2|$, because $-|-2| = -2$ and 2 is greater than -2.

1.1 Exercises

Developing Skills

In Exercises 1–4, determine which of the numbers in the set are (a) natural numbers, (b) integers, and (c) rational numbers. See Example 1.

1. $\left\{-3, 20, -\frac{3}{2}, \frac{9}{3}, 4.5\right\}$

2. $\left\{10, -82, -\frac{24}{3}, -8.2, \frac{1}{5}\right\}$

3. $\left\{-\frac{5}{2}, 6.5, -4.5, \frac{8}{4}, \frac{3}{4}\right\}$

4. $\left\{8, -1, \frac{4}{3}, -3.25, -\frac{10}{2}\right\}$

In Exercises 5–8, plot the numbers on the real number line. See Example 2.

5. $-7, 1.5$

6. $4, -3.2$

7. $\frac{1}{4}, 0, -2$

8. $-\frac{3}{2}, 5, 1$

In Exercises 9–18, plot each real number as a point on the real number line and place the correct inequality symbol ($<$ or $>$) between the pair of real numbers. See Examples 3 and 4.

9. $3 \quad\rule{1cm}{0.4pt}\quad -4$
10. $6 \quad\rule{1cm}{0.4pt}\quad -2$
11. $4 \quad\rule{1cm}{0.4pt}\quad -\frac{7}{2}$
12. $2 \quad\rule{1cm}{0.4pt}\quad \frac{3}{2}$
13. $0 \quad\rule{1cm}{0.4pt}\quad -\frac{7}{16}$
14. $-\frac{7}{3} \quad\rule{1cm}{0.4pt}\quad -\frac{7}{2}$
15. $-4.6 \quad\rule{1cm}{0.4pt}\quad 1.5$
16. $28.60 \quad\rule{1cm}{0.4pt}\quad -3.75$
17. $\frac{7}{16} \quad\rule{1cm}{0.4pt}\quad \frac{5}{8}$
18. $-\frac{3}{8} \quad\rule{1cm}{0.4pt}\quad -\frac{5}{8}$

In Exercises 19–22, find the distance between a and zero on the real number line.

19. $a = 2$

20. $a = 5$

21. $a = -4$

22. $a = -10$

In Exercises 23–28, find the opposite of the number. Plot the number and its opposite on the real number line. What is the distance of each from 0?

23. 5
24. 2
25. -3.8
26. -7.5
27. $-\frac{5}{2}$
28. $-\frac{3}{4}$

In Exercises 29–32, find the absolute value of the real number and its distance from 0.

29.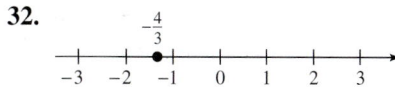

30.

31.

32.

In Exercises 33–46, evaluate the expression. See Example 6.

33. $|7|$
34. $|6|$
35. $|-11|$
36. $|-15|$
37. $|-3.4|$
38. $|-16.2|$
39. $\left|-\frac{7}{2}\right|$
40. $\left|-\frac{9}{16}\right|$
41. $-|4.09|$
42. $-|91.3|$
43. $-|-23.6|$
44. $-|-43.8|$
45. $|0|$
46. $|\pi|$

In Exercises 47–58, place the correct symbol ($<$, $>$, or $=$) between the pair of real numbers. See Example 7.

47. $|-15| \quad\rule{1cm}{0.4pt}\quad |15|$
48. $|525| \quad\rule{1cm}{0.4pt}\quad |-525|$
49. $|-4| \quad\rule{1cm}{0.4pt}\quad |3|$
50. $|16| \quad\rule{1cm}{0.4pt}\quad |-25|$

51. $|32|$ ___ $|-50|$

52. $|-1026|$ ___ $|800|$

53. $\left|\frac{3}{16}\right|$ ___ $\left|\frac{3}{2}\right|$

54. $\left|\frac{7}{8}\right|$ ___ $\left|\frac{4}{3}\right|$

55. $-|-48.5|$ ___ $|-48.5|$

56. $-|-64|$ ___ $|-64|$

57. $|-\pi|$ ___ $-|-2\pi|$

58. $-|-4.9|$ ___ $|-10.2|$

In Exercises 59–62, plot the numbers on the real number line.

59. $\frac{5}{2}, \pi, -2, -|-3|$

60. $3.7, \frac{16}{3}, -|-1.9|, -\frac{1}{2}$

61. $-4, \frac{7}{3}, |-3|, 0, -|4.5|$

62. $|-2.3|, 3.2, -2.3, -|3.2|$

In Exercises 63–68, find all real numbers whose distance from a is given by d.

63. $a = 8, d = 12$ **64.** $a = 6, d = 7$

65. $a = 21.3, d = 6$ **66.** $a = 42.5, d = 7$

67. $a = -2, d = 3.5$ **68.** $a = -7, d = 7.2$

Solving Problems

In Exercises 69–78, give three examples of numbers that satisfy the given conditions.

69. A real number that is a negative integer

70. A real number that is a whole number

71. A real number that is not a rational number

72. A real number that is not an irrational number

73. An integer that is a rational number

74. A rational number that is not an integer

75. A rational number that is not a negative number

76. A real number that is not a positive rational number

77. A real number that is not an integer

78. An integer that is not a whole number

Explaining Concepts

79. ⚡ Answer parts (a)–(f) of Motivating the Chapter.

80. *Writing* ✏ Explain why $\frac{8}{4}$ is a natural number, but $\frac{7}{4}$ is not.

81. *Writing* ✏ How many numbers are three units from 0 on the real number line? Explain your answer.

82. *Writing* ✏ Explain why the absolute value of every real number is positive.

83. *Writing* ✏ Which real number lies farther from 0 on the real number line?

(a) -25 (b) 10

Explain your answer.

84. *Writing* ✏ Which real number lies farther from -7 on the real number line?

(a) 3   (b) -10

Explain your answer.

The symbol ⚡ indicates an exercise in which you are asked to answer parts of the Motivating the Chapter problem found on the Chapter Opener pages.

85. *Writing* ✏ Explain how to determine the smaller of two different real numbers.

86. *Writing* ✏ Select the smaller real number and explain your answer.

(a) $\frac{3}{8}$ (b) 0.35

True or False? In Exercises 87–96, decide whether the statement is true or false. Justify your answer.

87. $-5 > -13$

88. $-10 > -2$

89. $6 < -17$

90. $4 < -9$

91. The absolute value of any real number is always positive.

92. The absolute value of a number is equal to the absolute value of its opposite.

93. The absolute value of a rational number is a rational number.

94. A given real number corresponds to exactly one point on the real number line.

95. The opposite of a positive number is a negative number.

96. Every rational number is an integer.

1.2 Adding and Subtracting Integers

© Chuck Savage/Corbis

What You Should Learn

1. Add integers using a number line.
2. Add integers with like signs and with unlike signs.
3. Subtract integers with like signs and with unlike signs.

Why You Should Learn It

Real numbers are used to represent many real-life quantities. For instance, in Exercise 101 on page 19, you will use real numbers to find the increase in enrollment at private and public schools in the United States.

1. Add integers using a number line.

Adding Integers Using a Number Line

In this and the next section, you will study the four operations of arithmetic (addition, subtraction, multiplication, and division) on the set of integers. There are many examples of these operations in real life. For example, your business had a gain of $550 during one week and a loss of $600 the next week. Over the two-week period, your business had a combined profit of

$$550 + (-600) = -50$$

which means you had an overall loss of $50.

The number line is a good visual model for demonstrating addition of integers. To add two integers, $a + b$, using a number line, start at 0. Then move left or right a units depending on whether a is positive or negative. From that position, move left or right b units depending on whether b is positive or negative. The final position is called the **sum.**

Example 1   Adding Integers with Like Signs Using a Number Line

Find each sum.

a. $5 + 2$ **b.** $-3 + (-5)$

Solution

a. Start at zero and move five units to the right. Then move two more units to the right, as shown in Figure 1.20. So, $5 + 2 = 7$.

b. Start at zero and move three units to the left. Then move five more units to the left, as shown in Figure 1.21. So, $-3 + (-5) = -8$.

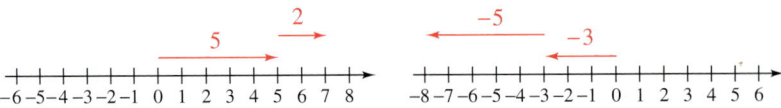

Figure 1.20   **Figure 1.21**

Example 2   Adding Integers with Unlike Signs Using a Number Line

Find each sum.

a. $-5 + 2$ **b.** $7 + (-3)$ **c.** $-4 + 4$

Solution

a. Start at zero and move five units to the left. Then move two units to the right, as shown in Figure 1.22.

Figure 1.22

So, $-5 + 2 = -3$.

b. Start at zero and move seven units to the right. Then move three units to the left, as shown in Figure 1.23.

Figure 1.23

So, $7 + (-3) = 4$.

c. Start at zero and move four units to the left. Then move four units to the right, as shown in Figure 1.24.

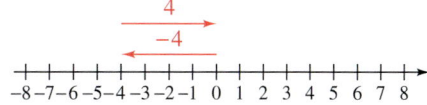

Figure 1.24

So, $-4 + 4 = 0$.

In Example 2(c), notice that the sum of -4 and 4 is 0. Two numbers whose sum is zero are called **opposites** (or **additive inverses**) of each other, So, -4 is the opposite of 4 and 4 is the opposite of -4.

2 Add integers with like signs and with unlike signs.

Adding Integers

Examples 1 and 2 illustrated a *graphical* approach to adding integers. It is more common to use an *algebraic* approach to adding integers, as summarized in the following rules.

Addition of Integers

1. To **add** two integers with *like* signs, add their absolute values and attach the common sign to the result.

2. To **add** two integers with *unlike* signs, subtract the smaller absolute value from the larger absolute value and attach the sign of the integer with the larger absolute value.

Example 3   Adding Integers

a. Unlike signs: $22 + (-17) = |22| - |-17| = 22 - 17 = 5$

b. Unlike signs: $-84 + 14 = -(|-84| - |-14|) = -(84 - 14) = -70$

c. Like signs: $-18 + (-62) = -(|-18| + |-62|) = -(18 + 62) = -80$

$$
\begin{array}{r}
1\ 1 \\
1\ 4\ 8 \\
6\ 2 \\
+\ 5\ 3\ 6 \\
\hline
7\ 4\ 6
\end{array}
$$

Figure 1.25   Carrying Algorithm

There are different ways to add three or more integers. You can use the **carrying algorithm** with a vertical format with nonnegative integers, as shown in Figure 1.25, or you can add them two at a time, as illustrated in Example 4.

Example 4   Account Balance

At the beginning of a month, your account balance was \$28. During the month you deposited \$60 and withdrew \$40. What was your balance at the end of the month?

Solution

$$
\begin{aligned}
\$28 + \$60 + (-\$40) &= (\$28 + \$60) + (-\$40) \\
&= \$88 + (-\$40) \\
&= \$48 \qquad\qquad \text{Balance}
\end{aligned}
$$

3 Subtract integers with like signs and with unlike signs.

Subtracting Integers

Subtraction can be thought of as "taking away." For instance, $8 - 5$ can be thought of as "8 take away 5," which leaves 3. Moreover, note that $8 + (-5) = 3$, which means that

$$8 - 5 = 8 + (-5).$$

In other words, $8 - 5$ can also be accomplished by "adding the opposite of 5 to 8."

Subtraction of Integers

To **subtract** one integer from another, add the opposite of the integer being subtracted to the other integer. The result is called the **difference** of the two integers.

Example 5 Subtracting Integers

a. $3 - 8 = 3 + (-8) = -5$ Add opposite of 8.

b. $10 - (-13) = 10 + 13 = 23$ Add opposite of -13.

c. $-5 - 12 = -5 + (-12) = -17$ Add opposite of 12.

d. $-4 - (-17) - 23 = -4 + 17 + (-23) = -10$ Add opposite of -17 and opposite of 23.

Be sure you understand that the terminology involving subtraction is not the same as that used for negative numbers. For instance, -5 is read as "negative 5," but $8 - 5$ is read as "8 subtract 5." It is important to distinguish between the operation and the signs of the numbers involved. For instance, in $-3 - 5$ the operation is subtraction and the numbers are -3 and 5.

For subtraction problems involving only two nonnegative integers, you can use the **borrowing algorithm** shown in Figure 1.26.

$$
\begin{array}{r}
3\ 10\ 15 \\
\cancel{4}\ \cancel{1}\ 5 \\
-2\ \ 7\ \ 6 \\
\hline
1\ \ 3\ \ 9
\end{array}
$$

Figure 1.26   Borrowing Algorithm

Example 6 Subtracting Integers

a. Subtract 10 from -4 means: $-4 - 10 = -4 + (-10) = -14$.

b. -3 subtract -8 means: $-3 - (-8) = -3 + 8 = 5$.

To evaluate expressions that contain a series of additions and subtractions, write the subtractions as equivalent additions and simplify from left to right, as shown in Example 7.

Example 7 Evaluating Expressions

Evaluate each expression.

a. $-13 - 7 + 11 - (-4)$ **b.** $5 - (-9) - 12 + 2$

c. $-1 - 3 - 4 + 6$ **d.** $5 - 1 - 8 + 3 + 4 - (-10)$

Solution

a. $-13 - 7 + 11 - (-4) = -13 + (-7) + 11 + 4$ Add opposites.

$\qquad\qquad\qquad\qquad\quad = -20 + 15$ Add two numbers at a time.

$\qquad\qquad\qquad\qquad\quad = -5$ Add.

b. $5 - (-9) - 12 + 2 = 5 + 9 + (-12) + 2$ Add opposites.

$\qquad\qquad\qquad\qquad = 14 + (-10)$ Add two numbers at a time.

$\qquad\qquad\qquad\qquad = 4$ Add.

c. $-1 - 3 - 4 + 6 = -1 + (-3) + (-4) + 6$ Add opposites.

$\qquad\qquad\qquad\quad = -4 + 2$ Add two numbers at a time.

$\qquad\qquad\qquad\quad = -2$ Add.

d. $5 - 1 - 8 + 3 + 4 - (-10) = 5 + (-1) + (-8) + 3 + 4 + 10$

$\qquad\qquad\qquad\qquad\qquad\quad = 4 + (-5) + 14 = 13$

Example 8   Temperature Change

The temperature in Minneapolis, Minnesota at 4 P.M. was 15°F. By midnight, the temperature had decreased by 18°. What was the temperature in Minneapolis at midnight?

Solution

To find the temperature at midnight, subtract 18 from 15.

$$15 - 18 = 15 + (-18)$$
$$= -3$$

The temperature in Minneapolis at midnight was -3°F.

This text includes several examples and exercises that use a calculator. As each new calculator application is encountered, you will be given general instructions for using a calculator. These instructions, however, may not agree precisely with the steps required by *your* calculator, so be sure you are familiar with the use of the keys on your own calculator.

For each of the calculator examples in the text, two possible keystroke sequences are given: one for a standard *scientific* calculator and one for a *graphing* calculator.

Example 9   Evaluating Expressions with a Calculator

Evaluate each expression with a calculator.

a. $-4 - 5$ **b.** $2 - (3 - 9)$

Keystrokes	*Display*	
a. 4 +/− − 5 =	−9	Scientific
(−) 4 − 5 ENTER	−9	Graphing

Keystrokes	*Display*	
b. 2 − (3 − 9) =	8	Scientific
2 − (3 − 9) ENTER	8	Graphing

Technology: Tip

The keys +/− and (−) change a number to its opposite and − is the subtraction key. For instance, the keystrokes − 4 − 5 ENTER will not produce the result shown in Example 9(a).

1.2 Exercises

Developing Skills

In Exercises 1–8, find the sum and demonstrate the addition on the real number line. See Examples 1 and 2.

1. $2 + 7$

2. $3 + 9$

3. $10 + (-3)$

4. $14 + (-8)$

5. $-6 + 4$

6. $-12 + 5$

7. $(-8) + (-3)$

8. $(-4) + (-7)$

In Exercises 9–42, find the sum. See Example 3.

9. $6 + 10$

10. $8 + 3$

11. $14 + (-14)$

12. $10 + (-10)$

13. $-45 + 45$

14. $-23 + 23$

15. $14 + 13$

16. $20 + 19$

17. $-23 + (-4)$

18. $-32 + (-16)$

19. $18 + (-12)$

20. $34 + (-16)$

21. $75 + 100$

22. $54 + 68$

23. $9 + (-14)$

24. $18 + (-26)$

25. $10 - 6 + 34$

26. $7 - 4 + 1$

27. $-15 + (-3) + 8$

28. $-82 + (-36) + 82$

29. $8 + 16 + (-3)$

30. $2 + (-51) + 13$

31. $17 + (-2) + 5$

32. $24 + 1 + (-19)$

33. $-13 + 12 + 4$

34. $-31 + 20 + 15$

35. $15 + (-75) + (-75)$

36. $32 + (-32) + (-16)$

37. $104 + 203 + (-613) + (-214)$

38. $4365 + (-2145) + (-1873) + 40,084$

39. $312 + (-564) + (-100)$

40. $1200 + (-1300) + (-275)$

41. $-890 + 90 + (-82)$

42. $-770 + (-383) + 492$

In Exercises 43–76, find the difference. See Example 5.

43. $12 - 9$

44. $55 - 20$

45. $39 - 13$

46. $45 - 35$

47. $4 - (-1)$

48. $9 - (-6)$

49. $18 - (-7)$

50. $27 - (-12)$

51. $32 - (-4)$

52. $47 - (-43)$

53. $19 - (-31)$

54. $12 - (-5)$

55. $27 - 57$

56. $18 - 32$

57. $61 - 85$

58. $53 - 74$

59. $22 - 131$

60. $48 - 222$

61. $2 - 11$

62. $3 - 15$

63. $13 - 24$

64. $26 - 34$

65. $-135 - (-114)$

66. $-63 - (-8)$

67. $-4 - (-4)$

68. $-942 - (-942)$

69. $-10 - (-4)$

70. $-12 - (-7)$

71. $-71 - 32$

72. $-84 - 106$

73. $-210 - 400$

74. $-120 - 142$

75. $-110 - (-30)$

76. $-2500 - (-600)$

77. Subtract 15 from -6.

78. Subtract 24 from -17.

79. Subtract -120 from 380.

80. Subtract -80 from 140.

81. *Think About It*   What number must be added to 10 to obtain -5?

82. *Think About It*   What number must be added to 36 to obtain -12?

83. *Think About It*   What number must be subtracted from -12 to obtain 24?

84. *Think About It*   What number must be subtracted from -20 to obtain 15?

In Exercises 85–90, evaluate the expression. See Example 7.

85. $-1 + 3 - (-4) + 10$

86. $12 - 6 + 3 - (-8)$

87. $6 + 7 - 12 - 5$

88. $-3 + 2 - 20 + 9$

89. $-(-5) + 7 - 18 + 4$

90. $-15 - (-2) + 4 - 6$

Solving Problems

91. *Temperature Change*   The temperature at 6 A.M. was $-10°$F. By noon, the temperature had increased by $22°$F. What was the temperature at noon?

92. *Account Balance*   A credit card owner charged $142 worth of goods on her account. Find the balance after a payment of $87 was made.

93. *Sports*   A hiker hiked 847 meters down the Grand Canyon. He climbed back up 385 meters and then rested. Find his distance down the canyon where he rested.

94. *Sports*   A fisherman dropped his line 27 meters below the surface of the water. Because the fish were not biting there, he decided to raise his line by 8 meters. How far below the surface of the water was his line?

95. *Profit*   A telephone company lost $650,000 during the first 6 months of the year. By the end of the year, the company had an overall profit of $362,000. What was the company's profit during the second 6 months of the year?

96. *Altitude*   An airplane flying at an altitude of 31,000 feet is instructed to descend to an altitude of 24,000 feet. How many feet must the airplane descend?

97. *Account Balance*   At the beginning of a month, your account balance was $2750. During the month you withdrew $350 and $500, deposited $450, and earned interest of $6.42. What was your balance at the end of the month?

98. *Account Balance*   At the beginning of a month, your account balance was $1204. During the month, you withdrew $725 and $821, deposited $150 and $80, and earned interest of $2.02. What was your balance at the end of the month?

99. *Temperature Change*   When you left for class in the morning, the temperature was $25°$C. By the time class ended, the temperature had increased by $4°$. While you studied, the temperature increased by $3°$. During your soccer practice, the temperature decreased by $9°$. What was the temperature after your soccer practice?

100. *Temperature Change*   When you left for class in the morning, the temperature was $40°$F. By the time class ended, the temperature had increased by $13°$. While you studied, the temperature decreased by $5°$. During your club meeting, the temperature decreased by $6°$. What was the temperature after your club meeting?

101. *Education*   The bar graph shows the total enrollment (in millions) at public and private schools in the United States for the years 1995 to 2001. (Source: U.S. National Center for Education Statistics)

(a) Find the increase in enrollment from 1996 to 2001.

(b) Find the increase in enrollment from 1999 to 2001.

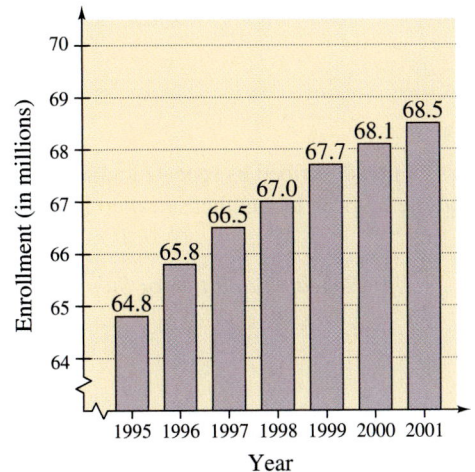

102. *Retail Price*   The bar graph shows the average retail price of a half-gallon of ice cream in the United States for the years 1996 to 2000. (Source: U.S. Bureau of Labor Statistics)

(a) Find the increase in the average retail price of ice cream from 1997 to 1998.

(b) Find the increase in the average retail price of ice cream from 1998 to 1999.

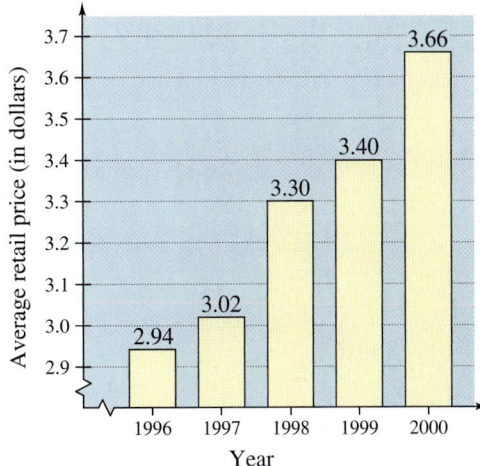

Figure for 102

In Exercises 103 and 104, an addition problem is shown visually on the real number line. (a) Write the addition problem and find the sum. (b) State the rule for the addition of integers demonstrated.

103.

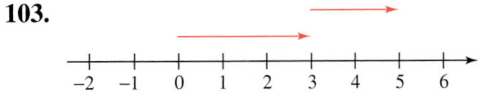

104.

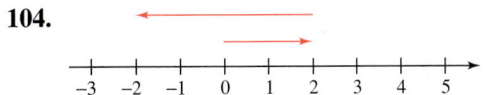

Explaining Concepts

105. *Writing*   Explain why the sum of two negative integers is a negative integer.

106. *Writing*   In your own words, write the rule for adding two integers of opposite signs. How do you determine the sign of the sum?

107. Write an expression that illustrates 8 subtracted from 5.

108. Write an expression that illustrates -6 subtracted from -4.

109. Write an expression using addition that can be used to subtract 12 from 9.

110. Write a simplified expression that can be used to evaluate $-9 - (-15)$.

1.3 Multiplying and Dividing Integers

Joe Sohm/The Image Works

What You Should Learn

1 Multiply integers with like signs and with unlike signs.

2 Divide integers with like signs and with unlike signs.

3 Find factors and prime factors of an integer.

4 Represent the definitions and rules of arithmetic symbolically.

Why You Should Learn It

You can multiply integers to solve real-life problems. For instance, in Exercise 107 on page 31, you will multiply integers to find the area of a football field.

1 Multiply integers with like signs and with unlike signs.

Multiplying Integers

Multiplication of two integers can be described as repeated addition or subtraction. The result of multiplying one number by another is called a **product.** Here are three examples.

Multiplication	*Repeated Addition or Subtraction*
$3 \times 5 = 15$	$5 + 5 + 5 = 15$

Add 5 three times.

$4 \times (-2) = -8$ $\qquad (-2) + (-2) + (-2) + (-2) = -8$

Add -2 four times.

$(-3) \times (-4) = 12$ $\qquad -(-4) - (-4) - (-4) = 12$

Subtract -4 three times.

Multiplication is denoted in a variety of ways. For instance,

$$7 \times 3, \quad 7 \cdot 3, \quad 7(3), \quad (7)3, \quad \text{and} \quad (7)(3)$$

all denote the product of "7 times 3," which is 21.

Rules for Multiplying Integers

1. The product of an integer and zero is 0.

2. The product of two integers with *like* signs is *positive.*

3. The product of two integers with *different* signs is *negative.*

To find the product of more than two numbers, first find the product of their absolute values. If there is an *even* number of negative factors, then the product is positive. If there is an *odd* number of negative factors, then the product is negative. For instance,

$$5(-3)(-4)(7) = 420. \qquad \text{Even number of negative factors}$$

Example 1   Multiplying Integers

a. $4(10) = 40$ (Positive) · (positive) = positive

b. $-6 \cdot 9 = -54$ (Negative) · (positive) = negative

c. $(-5)(-7) = 35$ (Negative) · (negative) = positive

d. $3(-12) = -36$ (Positive) · (negative) = negative

e. $-12 \cdot 0 = 0$ (Negative) · (zero) = zero

f. $(-2)(8)(-3)(-1) = -(2 \cdot 8 \cdot 3 \cdot 1)$ Odd number of negative factors

$\qquad\qquad\qquad = -48$ Answer is negative.

Be careful to distinguish properly between expressions such as $3(-5)$ and $3 - 5$ or $-3(-5)$ and $-3 - 5$. The first of each pair is a *multiplication* problem, whereas the second is a *subtraction* problem.

Multiplication	*Subtraction*
$3(-5) = -15$	$3 - 5 = -2$
$-3(-5) = 15$	$-3 - 5 = -8$

To multiply two integers having two or more digits, we suggest the **vertical multiplication algorithm** demonstrated in Figure 1.27. The sign of the product is determined by the usual multiplication rule.

$$
\begin{array}{r}
47 \\
\times \quad 23 \\
\hline
141 \\
94 \\
\hline
1081
\end{array}
$$

$141 \;\Leftarrow\;$ Multiply 3 times 47.

$94 \;\Leftarrow\;$ Multiply 2 times 47.

$1081 \;\Leftarrow\;$ Add columns.

Figure 1.27   Vertical Multiplication Algorithm

Example 2   Geometry: Volume of a Box

Find the volume of the rectangular box shown in Figure 1.28.

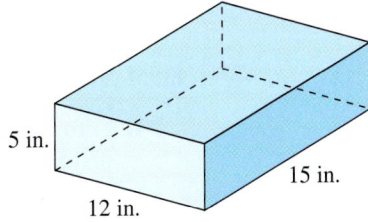

5 in.
12 in.
15 in.

Figure 1.28

Study Tip

Formulas from geometry can be found on the inside front cover of this text.

Solution

To find the volume, multiply the length, width, and height of the box.

Volume $=$ (Length) · (Width) · (Height)

$\qquad\quad = (15 \text{ inches}) \cdot (12 \text{ inches}) \cdot (5 \text{ inches})$

$\qquad\quad = 900 \text{ cubic inches}$

So, the box has a volume of 900 cubic inches.

2 Divide integers with like signs and with unlike signs.

Dividing Integers

Just as subtraction can be expressed in terms of addition, you can express division in terms of multiplication. Here are some examples.

Division		Related Multiplication
$15 \div 3 = 5$	because	$15 = 5 \cdot 3$
$-15 \div 3 = -5$	because	$-15 = -5 \cdot 3$
$15 \div (-3) = -5$	because	$15 = (-5) \cdot (-3)$
$-15 \div (-3) = 5$	because	$-15 = 5 \cdot (-3)$

The result of dividing one integer by another is called the **quotient** of the integers. **Division** is denoted by the symbol $\div$, or by $/$, or by a horizontal line. For example,

$$30 \div 6, \quad 30/6, \quad \text{and} \quad \frac{30}{6}$$

all denote the quotient of 30 and 6, which is 5. Using the form $30 \div 6$, 30 is called the **dividend** and 6 is the **divisor**. In the forms $30/6$ and $\frac{30}{6}$, 30 is the **numerator** and 6 is the **denominator.**

It is important to know how to use 0 in a division problem. Zero divided by a nonzero integer is always 0. For instance,

$$\frac{0}{13} = 0 \quad \text{because} \quad 0 = 0 \cdot 13.$$

On the other hand, division by zero, $13 \div 0$, is *undefined*.

Because division can be described in terms of multiplication, the rules for dividing two integers with like or unlike signs are the same as those for multiplying such integers.

Technology: Discovery

Does $\frac{1}{0} = 0$? Does $\frac{2}{0} = 0$? Write each division above in terms of multiplication. What does this tell you about division by zero? What does your calculator display when you perform the division?

Rules for Dividing Integers

1. Zero divided by a nonzero integer is 0, whereas a nonzero integer divided by zero is *undefined*.

2. The quotient of two nonzero integers with *like* signs is *positive*.

3. The quotient of two nonzero integers with *different* signs is *negative*.

Example 3   Dividing Integers

a. $\dfrac{-42}{-6} = 7$ because $-42 = 7(-6)$.

b. $36 \div (-9) = -4$ because $(-4)(-9) = 36$.

c. $0 \div (-13) = 0$ because $(0)(-13) = 0$.

d. $-105 \div 7 = -15$ because $(-15)(7) = -105$.

e. $-97 \div 0$ is undefined.

$$\begin{array}{r} 27 \\ 13\overline{)351} \\ \underline{26} \\ 91 \\ \underline{91} \end{array}$$

Figure 1.29 Long Division Algorithm

When dividing large numbers, the **long division algorithm** can be used. For instance, the long division algorithm shown in Figure 1.29 shows that

$$351 \div 13 = 27.$$

Remember that division can be checked by multiplying the answer by the divisor. So it is true that

$$351 \div 13 = 27 \quad \text{because} \quad 27(13) = 351.$$

All four operations on integers (addition, subtraction, multiplication, and division) are used in the following real-life example.

Example 4 **Stock Purchase**

On Monday you bought \$500 worth of stock in a company. During the rest of the week, you recorded the gains and losses in your stock's value as shown in the table.

Tuesday	Wednesday	Thursday	Friday
Gained \$15	Lost \$18	Lost \$23	Gained \$10

a. What was the value of the stock at the close of Wednesday?

b. What was the value of the stock at the end of the week?

c. What would the total loss have been if Thursday's loss had occurred each of the four days?

d. What was the average daily gain (or loss) for the four days recorded?

Solution

a. The value at the close of Wednesday was

$$500 + 15 - 18 = \$497.$$

b. The value of the stock at the end of the week was

$$500 + 15 - 18 - 23 + 10 = \$484.$$

c. The loss on Thursday was \$23. If this loss had occurred each day, the total loss would have been

$$4(23) = \$92.$$

d. To find the average daily gain (or loss), add the gains and losses of the four days and divide by 4. So, the average is

$$\text{Average} = \frac{15 + (-18) + (-23) + 10}{4} = \frac{-16}{4} = -4.$$

This means that during the four days, the stock had an average loss of \$4 per day.

Factors and Prime Numbers

The set of positive integers

$$\{1, 2, 3, \ldots\}$$

is one subset of the real numbers that has intrigued mathematicians for many centuries.

Historically, an important number concept has been *factors* of positive integers. From experience, you know that in a multiplication problem such as $3 \cdot 7 = 21$, the numbers 3 and 7 are called *factors* of 21.

$$3 \cdot 7 = 21$$

Factors   Product

It is also correct to call the numbers 3 and 7 *divisors* of 21, because 3 and 7 each divide evenly into 21.

Definition of Factor (or Divisor)

If a and b are positive integers, then a is a **factor** (or **divisor**) of b if and only if there is a positive integer c such that $a \cdot c = b$.

The concept of factors allows you to classify positive integers into three groups: *prime* numbers, *composite* numbers, and the number 1.

Definitions of Prime and Composite Numbers

1. A positive integer greater than 1 with no factors other than itself and 1 is called a **prime number,** or simply a **prime.**

2. A positive integer greater than 1 with more than two factors is called a **composite number,** or simply a **composite.**

The numbers 2, 3, 5, 7, and 11 are primes because they have only themselves and 1 as factors. The numbers 4, 6, 8, 9, and 10 are composites because each has more than two factors. The number 1 is neither prime nor composite because 1 is its only factor.

Every composite number can be expressed as a *unique* product of prime factors. Here are some examples.

$$6 = 2 \cdot 3, \ 15 = 3 \cdot 5, \ 18 = 2 \cdot 3 \cdot 3, \ 42 = 2 \cdot 3 \cdot 7, \ 124 = 2 \cdot 2 \cdot 31$$

According to the definition of a prime number, is it possible for any negative number to be prime? Consider the number -2. Is it prime? Are its only factors one and itself? No, because

$$-2 = 1(-2),$$

$$-2 = (-1)(2),$$

or $-2 = (-1)(1)(2).$

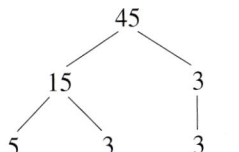

Figure 1.30   Tree Diagram

One strategy for factoring a composite number into prime numbers is to begin by finding the smallest prime number that is a factor of the composite number. Dividing this factor into the number yields a *companion* factor. For instance, 3 is the smallest prime number that is a factor of 45 and its companion factor is $15 = 45 \div 3$. Because 15 is also a composite number, continue hunting for factors and companion factors until each factor is prime. As shown in Figure 1.30, a *tree diagram* is a nice way to record your work. From the tree diagram, you can see that the prime factorization of 45 is

$$45 = 3 \cdot 3 \cdot 5.$$

Example 5   Prime Factorization

Write the prime factorization for each number.

a. 84   **b.** 78   **c.** 133   **d.** 43

Solution

a. 2 is a recognized divisor of 84. So, $84 = 2 \cdot 42 = 2 \cdot 2 \cdot 21 = 2 \cdot 2 \cdot 3 \cdot 7$.

b. 2 is a recognized divisor of 78. So, $78 = 2 \cdot 39 = 2 \cdot 3 \cdot 13$.

c. If you do not recognize a divisor of 133, you can start by dividing any of the prime numbers 2, 3, 5, 7, 11, 13, etc., into 133. You will find 7 to be the first prime to divide 133. So, $133 = 7 \cdot 19$ (19 is prime).

d. In this case, none of the primes less than 43 divides 43. So, 43 is prime.

Other aids to finding prime factors of a number n include the following divisibility tests.

Divisibility Tests

		Example
1.	A number is divisible by 2 if it is *even*.	364 is divisible by 2 because it is even.
2.	A number is divisible by 3 if the sum of its digits is divisible by 3.	261 is divisible by 3 because $2 + 6 + 1 = 9$.
3.	A number is divisible by 9 if the sum of its digits is divisible by 9.	738 is divisible by 9 because $7 + 3 + 8 = 18$.
4.	A number is divisible by 5 if its units digit is 0 or 5.	325 is divisible by 5 because its units digit is 5.
5.	A number is divisible by 10 if its units digit is 0.	120 is divisible by 10 because its units digit is 0.

When a number is **divisible** by 2, it means that 2 divides into the number without leaving a remainder.

④ Represent the definitions and rules of arithmetic symbolically.

Summary of Definitions and Rules

So far in this chapter, rules and procedures have been described more with words than with symbols. For instance, subtraction is verbally defined as "adding the opposite of the number being subtracted." As you move to higher and higher levels of mathematics, it becomes more and more convenient to use symbols to describe rules and procedures. For instance, subtraction is symbolically defined as

$$a - b = a + (-b).$$

At its simplest level, algebra is a symbolic form of arithmetic. This arithmetic–algebra connection can be illustrated in the following way.

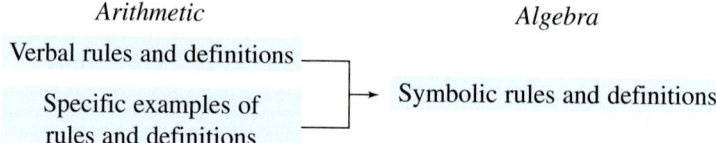

An illustration of this connection is shown in Example 6.

Example 6   Writing a Rule of Arithmetic in Symbolic Form

Write an example and an algebraic description of the arithmetic rule: *The product of two integers with unlike signs is negative.*

Solution

Example

For the integers -3 and 7,

$$(-3) \cdot 7 = 3 \cdot (-7)$$
$$= -(3 \cdot 7)$$
$$= -21.$$

Algebraic Description

If a and b are positive integers, then

$$(-a) \cdot b = a \cdot (-b) = -(a \cdot b).$$

Unlike signs   Unlike signs   Negative product

The list on the following page summarizes the algebraic versions of important definitions and rules of arithmetic. In each case a specific example is included for clarification.

Arithmetic Summary

Definitions: Let a, b, and c be integers.

Definition	*Example*
1. Subtraction: $$a - b = a + (-b)$$	$5 - 7 = 5 + (-7)$
2. Multiplication: (a is a positive integer) $$a \cdot b = \underbrace{b + b + \cdots + b}_{a \text{ terms}}$$	$3 \cdot 5 = 5 + 5 + 5$
3. Division: ($b \neq 0$) $$a \div b = c, \text{ if and only if } a = c \cdot b.$$	$12 \div 4 = 3$ because $12 = 3 \cdot 4.$
4. Less than: $a < b$ if there is a positive real number c such that $a + c = b.$	$-2 < 1$ because $-2 + 3 = 1.$
5. Absolute value: $|a| = \begin{cases} a, & \text{if } a \geq 0 \\ -a, & \text{if } a < 0 \end{cases}$	$|-3| = -(-3) = 3$
6. Divisor: a is a divisor of b if and only if there is an integer c such that $a \cdot c = b.$	7 is a divisor of 21 because $7 \cdot 3 = 21.$

Rules: Let a and b be integers.

Rule	*Example*
1. Addition: 　(a) To add two integers with *like* signs, add their absolute values and attach the common sign to the result.	$3 + 7 = |3| + |7| = 10$
(b) To add two integers with *unlike* signs, subtract the smaller absolute value from the larger absolute value and attach the sign of the integer with the larger absolute value.	$\begin{aligned} -5 + 8 &= |8| - |-5| \\ &= 8 - 5 \\ &= 3 \end{aligned}$
2. Multiplication: 　(a) $a \cdot 0 = 0 = 0 \cdot a$	$3 \cdot 0 = 0 = 0 \cdot 3$
(b) Like signs: $a \cdot b > 0$	$(-2)(-5) = 10$
(c) Different signs: $a \cdot b < 0$	$(2)(-5) = -10$
3. Division: 　(a) $\dfrac{0}{a} = 0$	$\dfrac{0}{4} = 0$
(b) $\dfrac{a}{0}$ is undefined.	$\dfrac{6}{0}$ is undefined.
(c) Like signs: $\dfrac{a}{b} > 0$	$\dfrac{-2}{-3} = \dfrac{2}{3}$
(d) Different signs: $\dfrac{a}{b} < 0$	$\dfrac{-5}{7} = -\dfrac{5}{7}$

Example 7 Using Definitions and Rules

a. Use the definition of subtraction to complete the statement.

$$4 - 9 = $$

b. Use the definition of multiplication to complete the statement.

$$6 + 6 + 6 + 6 = $$

c. Use the definition of absolute value to complete the statement.

$$|-9| = $$

d. Use the rule for adding integers with unlike signs to complete the statement.

$$-7 + 3 = $$

e. Use the rule for multiplying integers with unlike signs to complete the statement.

$$-9 \times 2 = $$

Solution

a. $4 - 9 = 4 + (-9) = -5$

b. $6 + 6 + 6 + 6 = 4 \cdot 6 = 24$

c. $|-9| = -(-9) = 9$

d. $-7 + 3 = -(|-7| - |3|) = -4$

e. $-9 \times 2 = -18$

Example 8 Finding a Pattern

Complete each pattern. Decide which rules the patterns demonstrate.

a. $3 \cdot (3) \ = 9$ **b.** $-3 \cdot (3) \ = \ -9$

$3 \cdot (2) \ = 6$ $-3 \cdot (2) \ = \ -6$

$3 \cdot (1) \ = 3$ $-3 \cdot (1) \ = \ -3$

$3 \cdot (0) \ = 0$ $-3 \cdot (0) \ = \ \ \ 0$

$3 \cdot (-1) = $ $-3 \cdot (-1) = $

$3 \cdot (-2) = $ $-3 \cdot (-2) = $

$3 \cdot (-3) = $ $-3 \cdot (-3) = $

Solution

a. $3 \cdot (-1) = -3$ **b.** $-3 \cdot (-1) = 3$

$3 \cdot (-2) = -6$ $-3 \cdot (-2) = 6$

$3 \cdot (-3) = -9$ $-3 \cdot (-3) = 9$

The product of integers with unlike signs is negative and the product of integers with like signs is positive.

1.3 Exercises

Developing Skills

In Exercises 1–4, write each multiplication as repeated addition or subtraction and find the product.

1. $3 \cdot 2$

2. 4×5

3. $5 \times (-3)$

4. $(-6)(-2)$

In Exercises 5–30, find the product. See Example 1.

5. 7×3

6. 6×4

7. $0 \cdot 2$

8. $13 \cdot 0$

9. $4(-8)$

10. $10(-5)$

11. $(310)(-3)$

12. $(125)(-4)$

13. $-7(5)$

14. $-9(3)$

15. $(-6)(-12)$

16. $(-20)(-8)$

17. $(-500)(-6)$

18. $(-350)(-4)$

19. $5(-3)(-6)$

20. $6(-2)(-4)$

21. $-7(3)(-1)$

22. $-2(5)(-3)$

23. $(-2)(-3)(-5)$

24. $(-10)(-4)(-2)$

25. $|(-3)4|$

26. $|8(-9)|$

27. $|3(-5)(6)|$

28. $|8(-3)(5)|$

29. $|6(20)(4)|$

30. $|9(12)(2)|$

In Exercises 31–40, use the vertical multiplication algorithm to find the product.

31. 26×13

32. 14×9

33. $(-14) \times 24$

34. $(-8) \times 30$

35. $75(-63)$

36. $(-72)(866)$

37. $(-13)(-20)$

38. $(-11)(-24)$

39. $(-21)(-429)$

40. $(-14)(-585)$

In Exercises 41–60, perform the division, if possible. If not possible, state the reason. See Example 3.

41. $27 \div 9$

42. $35 \div 7$

43. $72 \div (-12)$

44. $54 \div (-9)$

45. $(-28) \div 4$

46. $(-108) \div 9$

47. $-35 \div (-5)$

48. $(-24) \div (-4)$

49. $\frac{8}{0}$

50. $\frac{17}{0}$

51. $\frac{0}{8}$

52. $\frac{0}{17}$

53. $\frac{-81}{-3}$

54. $\frac{-125}{-25}$

55. $\frac{6}{-1}$

56. $\frac{-33}{1}$

57. $\frac{-28}{4}$

58. $\dfrac{72}{-12}$

59. $(-27) \div (-27)$

60. $(-83) \div (-83)$

In Exercises 61–70, use the long division algorithm to find the quotient.

61. $1440 \div 45$

62. $936 \div 52$

63. $1440 \div (-45)$

64. $936 \div (-52)$

65. $-1312 \div 16$

66. $-5152 \div 23$

67. $2750 \div 25$

68. $22,010 \div 71$

69. $-9268 \div (-28)$

70. $-6804 \div (-36)$

In Exercises 71–74, use a calculator to perform the specified operation(s).

71. $\dfrac{44,290}{515}$

72. $\dfrac{33,511}{47}$

73. $\dfrac{169,290}{162}$

74. $\dfrac{1,027,500}{250}$

Mental Math   In Exercises 75–78, find the product mentally. Explain your strategy.

75. $72(8)(25)$

76. $64(5)(20)$

77. $(-2)(532)(500)$

78. $(-4)(262)(50)$

In Exercises 79–88, decide whether the number is prime or composite.

79. 240

80. 533

81. 643

82. 257

83. 3911

84. 1321

85. 1281

86. 1323

87. 3555

88. 8324

In Exercises 89–98, write the prime factorization of the number. See Example 5.

89. 12

90. 52

91. 561

92. 245

93. 210

94. 525

95. 2535

96. 1521

97. 192

98. 264

In Exercises 99–102, complete the statement using the indicated definition or rule. See Example 7.

99. Definition of division: $12 \div 4 = \boxed{}$

100. Definition of absolute value: $|8| = \boxed{}$

101. Rule for multiplying integers by 0:

$6 \cdot 0 = \boxed{} = 0 \cdot 6$

102. Rule for dividing integers with unlike signs:

$\dfrac{30}{-10} = \boxed{}$

Solving Problems

103. *Temperature Change*   The temperature measured by a weather balloon is decreasing approximately 3° for each 1000-foot increase in altitude. The balloon rises 8000 feet. What is the total temperature change?

104. *Stock Price*   The Dow Jones average loses 11 points on each of four consecutive days. What is the cumulative loss during the four days?

105. *Savings Plan*   After you save $50 per month for 10 years, what is the total amount you have saved?

106. *Loss Leaders*   To attract customers, a grocery store runs a sale on bananas. The bananas are *loss leaders,* which means the store loses money on the bananas but hopes to make it up on other items. The store sells 800 pounds at a loss of 26 cents per pound. What is the total loss?

107. ▲ *Geometry*   Find the area of the football field.

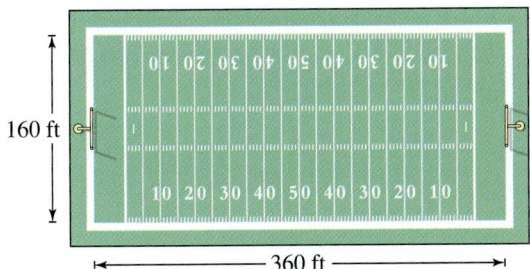

108. ▲ *Geometry*   Find the area of the garden.

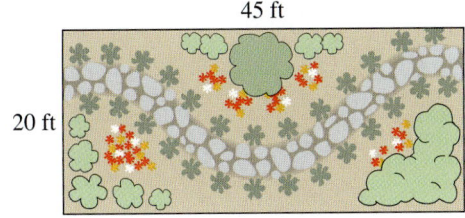

109. *Average Speed*   A commuter train travels a distance of 195 miles between two cities in 3 hours. What is the average speed of the train in miles per hour?

110. *Average Speed*   A jogger runs a race that is 6 miles long in 54 minutes. What is the average speed of the jogger in minutes per mile?

111. *Exam Scores*   A student has a total of 328 points after four 100-point exams.

(a) What is the average number of points scored per exam?

(b) The scores on the four exams are 87, 73, 77, and 91. Plot each of the scores and the average score on the real number line.

(c) Find the difference between each score and the average score. Find the sum of these distances and give a possible explanation of the result.

112. *Sports*   A football team gains a total of 20 yards after four downs.

(a) What is the average number of yards gained per down?

(b) The gains on the four downs are 8 yards, 4 yards, 2 yards, and 6 yards. Plot each of the gains and the average gain on the real number line.

(c) Find the difference between each gain and the average gain. Find the sum of these distances and give a possible explanation of the result.

▲ *Geometry*   In Exercises 113 and 114, find the volume of the rectangular solid. The volume is found by multiplying the length, width, and height of the solid. See Example 2.

113.

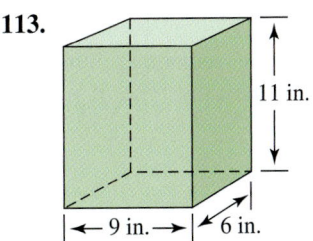

114.

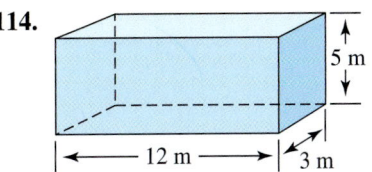

Explaining Concepts

115. *Writing* ✏ What is the only even prime number? Explain why there are no other even prime numbers.

116. *Investigation* Twin primes are prime numbers that differ by 2. For instance, 3 and 5 are twin primes. How many other twin primes are less than 100?

117. *Writing* ✏ The number 1997 is not divisible by a prime number that is less than 45. Explain why this implies that 1997 is a prime number.

118. *Think About It* If a negative number is used as a factor 25 times, what is the sign of the product?

119. *Think About It* If a negative number is used as a factor 16 times, what is the sign of the product?

120. *Writing* ✏ Write a verbal description of what is meant by $3(-5)$.

121. *Writing* ✏ In your own words, write the rules for determining the sign of the product or quotient of real numbers.

122. *The Sieve of Eratosthenes* Write the integers from 1 through 100 in 10 lines of 10 numbers each.

(a) Cross out the number 1. Cross out all multiples of 2 other than 2 itself. Do the same for 3, 5, and 7.

(b) Of what type are the remaining numbers? Explain why this is the only type of number left.

123. *Writing* ✏ Explain why the product of an even integer and any other integer is even. What can you conclude about the product of two odd integers?

124. *Writing* ✏ Explain how to check the result of a division problem.

125. *Think About It* An integer n is divided by 2 and the quotient is an even integer. What does this tell you about n? Give an example.

126. Which of the following is (are) undefined: $\frac{1}{1}, \frac{0}{1}, \frac{1}{0}$?

127. *Investigation* The **proper factors** of a number are all its factors less than the number itself. A number is **perfect** if the sum of its proper factors is equal to the number. A number is **abundant** if the sum of its proper factors is greater than the number. Which numbers less than 25 are perfect? Which are abundant? Try to find the first perfect number greater than 25.

Mid-Chapter Quiz

Take this quiz as you would take a quiz in class. After you are done, check your work against the answers in the back of the book.

In Exercises 1–4, plot each real number as a point on the real line and place the correct inequality symbol ($<$ or $>$) between the real numbers.

1. $\frac{3}{16}$ ▭ $\frac{3}{8}$

2. -2.5 ▭ -4

3. -7 ▭ 3

4. 2π ▭ 6

In Exercises 5 and 6, evaluate the expression.

5. $-|-0.75|$

6. $\left|-\frac{17}{19}\right|$

In Exercises 7 and 8, place the correct symbol ($<$, $>$, or $=$) between the real numbers.

7. $\left|\frac{7}{2}\right|$ ▭ $|-3.5|$

8. $\left|\frac{3}{4}\right|$ ▭ $-|0.75|$

9. Subtract -13 from -22.

10. Find the absolute value of the sum of -54 and 26.

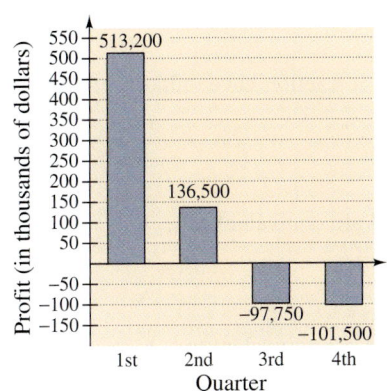

Figure for 24

In Exercises 11–22, evaluate the expression.

11. $34 + 65$

12. $-24 + (-51)$

13. $-15 - 12$

14. $-35 - (-10)$

15. $25 + (-75)$

16. $72 - 134$

17. $12 + (-6) - 8 + 10$

18. $-9 - 17 + 36 + (-15)$

19. $-6(10)$

20. $(-7)(-13)$

21. $\dfrac{-45}{-3}$

22. $\dfrac{-24}{6}$

23. Write the prime factorization of 144.

24. An electronics manufacturer's quarterly profits are shown in the bar graph at the left. What is the manufacturer's total profit for the year?

25. A cord of wood is a pile 8 feet long, 4 feet wide, and 4 feet high. The volume of a rectangular solid is its length times its width times its height. Find the number of cubic feet in a cord of wood.

26. It is necessary to cut a 90-foot rope into six pieces of equal length. What is the length of each piece?

27. At the beginning of a month your account balance was $738. During the month, you withdrew $550, deposited $189, and payed a fee of $10. What was your balance at the end of the month?

28. When you left for class in the morning, the temperature was 60°F. By the time class ended, the temperature had increased by 15°. While you studied, the temperature increased by 2°. During your work study, the temperature decreased by 12°. What was the temperature after your work study?

1.4 Operations with Rational Numbers

Lon C. Diehl/PhotoEdit

What You Should Learn

1️⃣ Rewrite fractions as equivalent fractions.

2️⃣ Add and subtract fractions.

3️⃣ Multiply and divide fractions.

4️⃣ Add, subtract, multiply, and divide decimals.

Why You Should Learn It

Rational numbers are used to represent many real-life quantities. For instance, in Exercise 149 on page 46, you will use rational numbers to find the increase in the Dow Jones Industrial Average.

1️⃣ Rewrite fractions as equivalent fractions.

Rewriting Fractions

A **fraction** is a number that is written as a quotient, with a *numerator* and a *denominator*. The terms *fraction* and *rational number* are related, but are not exactly the same. The term *fraction* refers to a number's form, whereas the term *rational number* refers to its classification. For instance, the number 2 is a fraction when it is written as $\frac{2}{1}$, but it is a rational number regardless of how it is written.

Rules of Signs for Fractions

1. If the numerator and denominator of a fraction have like signs, the value of the fraction is positive.

2. If the numerator and denominator of a fraction have unlike signs, the value of the fraction is negative.

All of the following fractions are positive and are equivalent to $\frac{2}{3}$.

$$\frac{2}{3}, \frac{-2}{-3}, -\frac{-2}{3}, -\frac{2}{-3}$$

All of the following fractions are negative and are equivalent to $-\frac{2}{3}$.

$$-\frac{2}{3}, \frac{-2}{3}, \frac{2}{-3}, -\frac{-2}{-3}$$

In both arithmetic and algebra, it is often beneficial to write a fraction in **simplest form** or reduced form, which means that the numerator and denominator have no common factors (other than 1). By finding the prime factors of the numerator and the denominator, you can determine what common factor(s) to divide out.

Writing a Fraction in Simplest Form

To write a fraction in simplest form, divide both the numerator and denominator by their greatest common factor (GCF).

(See Figure 1.31.)

Study Tip

To find the **greatest common factor** (or **GCF**) of two natural numbers, write the prime factorization of each number. The greatest common factor is the product of the common factors. For instance, from the prime factorizations

$$18 = 2 \cdot 3 \cdot 3$$

and

$$42 = 2 \cdot 3 \cdot 7$$

you can see that the common factors of 18 and 42 are 2 and 3. So, it follows that the greatest common factor is $2 \cdot 3$ or 6.

Example 1   Writing Fractions in Simplest Form

Write each fraction in simplest form.

a. $\dfrac{18}{24}$ **b.** $\dfrac{35}{21}$ **c.** $\dfrac{24}{72}$

Solution

a. $\dfrac{18}{24} = \dfrac{2 \cdot \overset{1}{3} \cdot \overset{1}{3}}{2 \cdot 2 \cdot 2 \cdot \underset{1}{3}} = \dfrac{3}{4}$ Divide out GCF of 6.

b. $\dfrac{35}{21} = \dfrac{5 \cdot \overset{1}{7}}{3 \cdot \underset{1}{7}} = \dfrac{5}{3}$ Divide out GCF of 7.

c. $\dfrac{24}{72} = \dfrac{\overset{1}{2} \cdot \overset{1}{2} \cdot \overset{1}{2} \cdot \overset{1}{3}}{\underset{1}{2} \cdot \underset{1}{2} \cdot \underset{1}{2} \cdot \underset{1}{3} \cdot 3} = \dfrac{1}{3}$ Divide out GCF of 24.

You can obtain an **equivalent fraction** by multiplying the numerator and denominator by the same nonzero number or by dividing the numerator and denominator by the same nonzero number. Here are some examples.

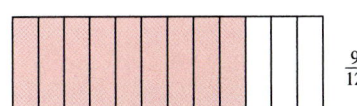

$\dfrac{9}{12}$

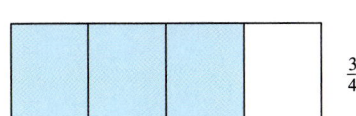

$\dfrac{3}{4}$

Figure 1.31   Equivalent Fractions

Fraction		Equivalent Fraction	Operation
$\dfrac{9}{12}$	$= \dfrac{\overset{1}{3} \cdot 3}{\underset{1}{3} \cdot 4}$	$\dfrac{3}{4}$	Divide numerator and denominator by 3. (See Figure 1.31.)
$\dfrac{6}{5}$	$= \dfrac{6 \cdot 2}{5 \cdot 2}$	$\dfrac{12}{10}$	Multiply numerator and denominator by 2.
$-\dfrac{8}{12}$	$= -\dfrac{\overset{1}{2} \cdot \overset{1}{2} \cdot 2}{\underset{1}{2} \cdot \underset{1}{2} \cdot 3}$	$-\dfrac{2}{3}$	Divide numerator and denominator by GCF of 4.

Example 2   Writing Equivalent Fractions

Write an equivalent fraction with the indicated denominator.

a. $\dfrac{2}{3} = \dfrac{\quad}{15}$ **b.** $\dfrac{4}{7} = \dfrac{\quad}{42}$ **c.** $\dfrac{9}{15} = \dfrac{\quad}{35}$

Solution

a. $\dfrac{2}{3} = \dfrac{2 \cdot 5}{3 \cdot 5} = \dfrac{10}{15}$ Multiply numerator and denominator by 5.

b. $\dfrac{4}{7} = \dfrac{4 \cdot 6}{7 \cdot 6} = \dfrac{24}{42}$ Multiply numerator and denominator by 6.

c. $\dfrac{9}{15} = \dfrac{\overset{}{3} \cdot 3}{\underset{}{3} \cdot 5} = \dfrac{3 \cdot 7}{5 \cdot 7} = \dfrac{21}{35}$ Reduce first, then multiply numerator and denominator by 7.

2 Add and subtract fractions.

Adding and Subtracting Fractions

To add fractions with *like* denominators such as $\frac{3}{12}$ and $\frac{4}{12}$, add the numerators and write the sum over the like denominator.

$$\frac{3}{12} + \frac{4}{12} = \frac{3+4}{12}$$

$$= \frac{7}{12} \qquad \text{Add the numbers in the numerator.}$$

To add fractions with *unlike* denominators such as $\frac{1}{4}$ and $\frac{1}{3}$, rewrite the fractions as equivalent fractions with a common denominator.

$$\frac{1}{4} + \frac{1}{3} = \frac{1 \cdot 3}{4 \cdot 3} + \frac{1 \cdot 4}{3 \cdot 4} \qquad \text{Rewrite fractions in equivalent form.}$$

$$= \frac{3}{12} + \frac{4}{12} \qquad \text{Rewrite with like denominators.}$$

$$= \frac{7}{12} \qquad \text{Add numerators.}$$

To find a common denominator for two or more fractions, find the **least common multiple** (LCM) of their denominators. For instance, for the fractions $\frac{3}{8}$ and $-\frac{5}{12}$, the least common multiple of their denominators, 8 and 12, is 24. To see this, consider all multiples of 8 (8, 16, 24, 32, 40, 48, . . .) and all multiples of 12 (12, 24, 36, 48, . . .). The numbers 24 and 48 are common multiples, and the number 24 is the smallest of the common multiples.

$$\frac{3}{8} + \frac{-5}{12} = \frac{3(3)}{8(3)} + \frac{(-5)(2)}{12(2)} \qquad \text{LCM of 8 and 12 is 24.}$$

$$= \frac{9}{24} + \frac{-10}{24} \qquad \text{Rewrite with like denominators.}$$

$$= \frac{9-10}{24} \qquad \text{Add numerators.}$$

$$= \frac{-1}{24} \qquad \text{Simplify.}$$

$$= -\frac{1}{24}$$

Study Tip

Adding fractions with unlike denominators is an example of a basic problem-solving strategy that is used in mathematics—rewriting a given problem in a simpler or more familiar form.

Addition and Subtraction of Fractions

Let a, b, and c be integers with $c \neq 0$.

1. *With like denominators:*

$$\frac{a}{c} + \frac{b}{c} = \frac{a+b}{c} \quad \text{or} \quad \frac{a}{c} - \frac{b}{c} = \frac{a-b}{c}$$

2. *With unlike denominators:* rewrite both fractions so that they have like denominators. Then use the rule for adding and subtracting fractions with like denominators.

Example 3   Adding Fractions

Add: $1\frac{4}{5} + \frac{11}{15}$.

Solution

To begin, rewrite the **mixed number** $1\frac{4}{5}$ as a fraction.

$$1\frac{4}{5} = 1 + \frac{4}{5} = \frac{5}{5} + \frac{4}{5} = \frac{9}{5}$$

Then add the two fractions as follows.

$$1\frac{4}{5} + \frac{11}{15} = \frac{9}{5} + \frac{11}{15} \qquad \text{Rewrite } 1\frac{4}{5} \text{ as } \frac{9}{5}.$$

$$= \frac{9(3)}{5(3)} + \frac{11}{15} \qquad \text{LCM of 5 and 15 is 15.}$$

$$= \frac{27}{15} + \frac{11}{15} \qquad \text{Rewrite with like denominators.}$$

$$= \frac{38}{15} \qquad \text{Add numerators.}$$

Study Tip

In Example 3, a common short-cut for writing $1\frac{4}{5}$ as $\frac{9}{5}$ is to multiply 1 by 5, add the result to 4, and then divide by 5, as follows.

$$1\frac{4}{5} = \frac{1(5) + 4}{5} = \frac{9}{5}$$

Example 4   Subtracting Fractions

Subtract: $\frac{7}{9} - \frac{11}{12}$.

Solution

$$\frac{7}{9} - \frac{11}{12} = \frac{7(4)}{9(4)} + \frac{-11(3)}{12(3)} \qquad \text{LCM of 9 and 12 is 36.}$$

$$= \frac{28}{36} + \frac{-33}{36} \qquad \text{Rewrite with like denominators.}$$

$$= \frac{-5}{36} \qquad \text{Add numerators.}$$

$$= -\frac{5}{36}$$

You can add or subtract *two* fractions, without first finding a common denominator, by using the following rule.

Alternative Rule for Adding and Subtracting Two Fractions

If a, b, c, and d are integers with $b \neq 0$ and $d \neq 0$, then

$$\frac{a}{b} + \frac{c}{d} = \frac{ad + bc}{bd} \quad \text{or} \quad \frac{a}{b} - \frac{c}{d} = \frac{ad - bc}{bd}.$$

On page 36, the sum of $\frac{3}{8}$ and $-\frac{5}{12}$ was found using the least common multiple of 8 and 12. Compare those solution steps with the following steps, which use the alternative rule for adding or subtracting two fractions.

$$\frac{3}{8} + \frac{-5}{12} = \frac{3(12) + 8(-5)}{8(12)}$$ Apply alternative rule.

$$= \frac{36 - 40}{96}$$ Simplify.

$$= \frac{-4}{96}$$ Simplify.

$$= -\frac{1}{24}$$ Write in simplest form.

Technology: Tip

When you use a scientific or graphing calculator to add or subtract fractions, your answer may appear in decimal form. An answer such as 0.583333333 is not as exact as $\frac{7}{12}$ and may introduce roundoff error. Refer to the user's manual for your calculator for instructions on adding and subtracting fractions and displaying answers in fraction form.

Example 5   Subtracting Fractions

$$\frac{5}{16} - \left(-\frac{7}{30}\right) = \frac{5}{16} + \frac{7}{30}$$ Add the opposite.

$$= \frac{5(30) + 16(7)}{16(30)}$$ Apply alternative rule.

$$= \frac{150 + 112}{480}$$ Simplfy.

$$= \frac{262}{480}$$ Simplify.

$$= \frac{131}{240}$$ Write in simplest form.

Example 6   Combining Three or More Fractions

Evaluate $\dfrac{5}{6} - \dfrac{7}{15} + \dfrac{3}{10} - 1$.

Solution

The least common denominator of 6, 15, and 10 is 30. So, you can rewrite the original expression as follows.

$$\frac{5}{6} - \frac{7}{15} + \frac{3}{10} - 1 = \frac{5(5)}{6(5)} + \frac{(-7)(2)}{15(2)} + \frac{3(3)}{10(3)} + \frac{(-1)(30)}{30}$$

$$= \frac{25}{30} + \frac{-14}{30} + \frac{9}{30} + \frac{-30}{30}$$ Rewrite with like denominators.

$$= \frac{25 - 14 + 9 - 30}{30}$$ Add numerators.

$$= \frac{-10}{30} = -\frac{1}{3}$$ Simplify.

③ Multiply and divide fractions.

Multiplying and Dividing Fractions

The procedure for multiplying fractions is simpler than those for adding and subtracting fractions. Regardless of whether the fractions have like or unlike denominators, you can find the product of two fractions by multiplying the numerators and multiplying the denominators.

Multiplication of Fractions

Let a, b, c, and d be integers with $b \neq 0$ and $d \neq 0$. Then the product of $\dfrac{a}{b}$ and $\dfrac{c}{d}$ is

$$\frac{a}{b} \cdot \frac{c}{d} = \frac{a \cdot c}{b \cdot d}.$$ Multiply numerators and denominators.

Example 7   Multiplying Fractions

a. $\dfrac{5}{8} \cdot \dfrac{3}{2} = \dfrac{5(3)}{8(2)}$ Multiply numerators and denominators.

$\qquad\quad = \dfrac{15}{16}$ Simplify.

b. $\left(-\dfrac{7}{9}\right)\left(-\dfrac{5}{21}\right) = \dfrac{7}{9} \cdot \dfrac{5}{21}$ Product of two negatives is positive.

$\qquad\qquad\qquad = \dfrac{7(5)}{9(21)}$ Multiply numerators and denominators.

$\qquad\qquad\qquad = \dfrac{7(5)}{9(3)(7)}$ Divide out common factors.

$\qquad\qquad\qquad = \dfrac{5}{27}$ Write in simplest form.

Example 8   Multiplying Three Fractions

$\left(3\dfrac{1}{5}\right)\left(-\dfrac{7}{6}\right)\left(\dfrac{5}{3}\right) = \left(\dfrac{16}{5}\right)\left(-\dfrac{7}{6}\right)\left(\dfrac{5}{3}\right)$ Rewrite mixed number as a fraction.

$\qquad\qquad\qquad\qquad = \dfrac{16(-7)(5)}{5(6)(3)}$ Multiply numerators and denominators.

$\qquad\qquad\qquad\qquad = -\dfrac{(8)(2)(7)(5)}{(5)(3)(2)(3)}$ Divide out common factors.

$\qquad\qquad\qquad\qquad = -\dfrac{56}{9}$ Write in simplest form.

The **reciprocal** or **multiplicative inverse** of a number is the number by which it must be multiplied to obtain 1. For instance, the reciprocal of 3 is $\frac{1}{3}$ because $3\left(\frac{1}{3}\right) = 1$. Similarly, the reciprocal of $-\frac{2}{3}$ is $-\frac{3}{2}$ because

$$\left(-\frac{2}{3}\right)\left(-\frac{3}{2}\right) = 1.$$

To divide two fractions, multiply the first fraction by the *reciprocal* of the second fraction. Another way of saying this is "invert the divisor and multiply."

Division of Fractions

Let a, b, c, and d be integers with $b \neq 0$, $c \neq 0$, and $d \neq 0$. Then the quotient of $\dfrac{a}{b}$ and $\dfrac{c}{d}$ is

$$\frac{a}{b} \div \frac{c}{d} = \frac{a}{b} \cdot \frac{d}{c}. \qquad \text{Invert divisor and multiply.}$$

Example 9   Dividing Fractions

a. $\dfrac{5}{8} \div \dfrac{20}{12}$ **b.** $\dfrac{6}{13} \div \left(-\dfrac{9}{26}\right)$ **c.** $-\dfrac{1}{4} \div (-3)$

Solution

a.
$$\frac{5}{8} \div \frac{20}{12} = \frac{5}{8} \cdot \frac{12}{20} \qquad \text{Invert divisor and multiply.}$$

$$= \frac{(5)(12)}{(8)(20)} \qquad \text{Multiply numerators and denominators.}$$

$$= \frac{(5)(3)(4)}{(8)(4)(5)} \qquad \text{Divide out common factors.}$$

$$= \frac{3}{8} \qquad \text{Write in simplest form.}$$

b.
$$\frac{6}{13} \div \left(-\frac{9}{26}\right) = \frac{6}{13} \cdot \left(-\frac{26}{9}\right) \qquad \text{Invert divisor and multiply.}$$

$$= -\frac{(6)(26)}{(13)(9)} \qquad \text{Multiply numerators and denominators.}$$

$$= -\frac{(2)(3)(2)(13)}{(13)(3)(3)} \qquad \text{Divide out common factors.}$$

$$= -\frac{4}{3} \qquad \text{Write in simplest form.}$$

c.
$$-\frac{1}{4} \div (-3) = -\frac{1}{4} \cdot \left(-\frac{1}{3}\right) \qquad \text{Invert divisor and multiply.}$$

$$= \frac{(-1)(-1)}{(4)(3)} \qquad \text{Multiply numerators and denominators.}$$

$$= \frac{1}{12} \qquad \text{Write in simplest form.}$$

④ Add, subtract, multiply, and divide decimals.

Operations with Decimals

Rational numbers can be represented as **terminating** or **repeating decimals.** Here are some examples.

Terminating Decimals	*Repeating Decimals*
$\dfrac{1}{4} = 0.25$	$\dfrac{1}{6} = 0.1666\ldots$ or $0.1\overline{6}$
$\dfrac{3}{8} = 0.375$	$\dfrac{1}{3} = 0.3333\ldots$ or $0.\overline{3}$
$\dfrac{2}{10} = 0.2$	$\dfrac{1}{12} = 0.0833\ldots$ or $0.08\overline{3}$
$\dfrac{5}{16} = 0.3125$	$\dfrac{8}{33} = 0.2424\ldots$ or $0.\overline{24}$

Note that the bar notation is used to indicate the *repeated* digit (or digits) in the decimal notation. You can obtain the decimal representation of any fraction by long division. For instance, the decimal representation of $\frac{5}{12}$ is $0.41\overline{6}$, as can be seen from the following long division algorithm.

$$
\begin{array}{r}
0.4166\ \ldots = 0.41\overline{6} \\
12\overline{)\,5.0000} \\
\underline{4\,8} \\
20 \\
\underline{12} \\
80 \\
\underline{72} \\
80
\end{array}
$$

For calculations involving decimals such as $0.41666\ldots$, you must **round the decimal.** For instance, rounded to two decimal places, the number $0.41666\ldots$ is 0.42. Similarly, rounded to three decimal places, the number $0.41666\ldots$ is 0.417.

Technology: Tip

You can use a calculator to round decimals. For instance, to round 0.9375 to two decimal places on a scientific calculator, enter

$\boxed{\text{FIX}}\ \boxed{2}\ .9375\ \boxed{=}$

On a graphing calculator, enter

round (.9375, 2) $\boxed{\text{ENTER}}$

Without using a calculator, round -0.88247 to three decimal places. Verify your answer with a calculator. Name the rounding and decision digits.

Rounding a Decimal

1. Determine the number of digits of accuracy you wish to keep. The digit in the last position you keep is called the **rounding digit,** and the digit in the first position you discard is called the **decision digit.**

2. If the decision digit is 5 or greater, round up by adding 1 to the rounding digit.

3. If the decision digit is 4 or less, round down by leaving the rounding digit unchanged.

Given Decimal	*Rounded to Three Places*
0.9763	0.976
0.9768	0.977
0.9765	0.977

Example 10   Operations with Decimals

a. Add 0.583, 1.06, and 2.9104.

b. Multiply -3.57 and 0.032.

Solution

a. To add decimals, align the decimal points and proceed as in integer addition.

$$
\begin{array}{r}
\overset{1\ 1}{\ 0.583} \\
1.06 \\
+\ \ 2.9104 \\
\hline
4.5534
\end{array}
$$

b. To multiply decimals, use integer multiplication and then place the decimal point (in the product) so that the number of decimal places equals the sum of the decimal places in the two factors.

$$
\begin{array}{r}
-3.57 \\
\times\quad 0.032 \\
\hline
714 \\
1071 \\
\hline
-0.11424
\end{array}
$$

Two decimal places
Three decimal places

Five decimal places

Example 11   Dividing Decimal Fractions

Divide 1.483 by 0.56. Round the answer to two decimal places.

Solution

To divide 1.483 by 0.56, convert the divisor to an integer by moving its decimal point to the right. Move the decimal point in the dividend an equal number of places to the right. Place the decimal point in the quotient directly above the new decimal point in the dividend and then divide as with integers.

$$
\begin{array}{r}
2.648 \\
56\,\overline{)\,148.300} \\
\underline{112} \\
36\ 3 \\
\underline{33\ 6} \\
2\ 70 \\
\underline{2\ 24} \\
460 \\
\underline{448}
\end{array}
$$

Rounded to two decimal places, the answer is 2.65. This answer can be written as

$$
\frac{1.483}{0.56} \approx 2.65
$$

where the symbol $\approx$ means **is approximately equal to.**

Example 12   Physical Fitness

To satisfy your health and fitness requirement, you decide to take a tennis class. You learn that you burn about 400 calories per hour playing tennis. In one week, you played tennis for $\frac{3}{4}$ hour on Tuesday, 2 hours on Wednesday, and $1\frac{1}{2}$ hours on Thursday. How many total calories did you burn playing tennis in one week? What was the average number of calories you burned playing tennis for the three days?

Solution

The total number of calories you burned playing tennis in one week was

$$400\left(\frac{3}{4}\right) + 400(2) + 400\left(1\frac{1}{2}\right) = 300 + 800 + 600 = 1700 \text{ calories.}$$

The average number of calories you burned playing tennis for the three days was

$$\frac{1700}{3} \approx 566.67 \text{ calories.}$$

Summary of Rules for Fractions

Let a, b, c, and d be real numbers.

Rule	Example

1. Rules of signs for fractions:

$$\frac{-a}{-b} = \frac{a}{b}$$

$$\frac{-a}{b} = \frac{a}{-b} = -\frac{a}{b}$$

$$\frac{-12}{-4} = \frac{12}{4}$$

$$\frac{-12}{4} = \frac{12}{-4} = -\frac{12}{4}$$

2. Equivalent fractions:

$$\frac{a}{b} = \frac{a \cdot c}{b \cdot c}, b \neq 0, c \neq 0$$

$$\frac{1}{4} = \frac{3}{12} \text{ because } \frac{1}{4} = \frac{1 \cdot 3}{4 \cdot 3} = \frac{3}{12}$$

3. Addition of fractions:

$$\frac{a}{b} + \frac{c}{d} = \frac{ad + bc}{bd}, \quad b \neq 0, \quad d \neq 0$$

$$\frac{1}{3} + \frac{2}{7} = \frac{1 \cdot 7 + 3 \cdot 2}{3 \cdot 7} = \frac{13}{21}$$

4. Subtraction of fractions:

$$\frac{a}{b} - \frac{c}{d} = \frac{ad - bc}{bd}, \quad b \neq 0, \quad d \neq 0$$

$$\frac{1}{3} - \frac{2}{7} = \frac{1 \cdot 7 - 3 \cdot 2}{3 \cdot 7} = \frac{1}{21}$$

5. Multiplication of fractions:

$$\frac{a}{b} \cdot \frac{c}{d} = \frac{a \cdot c}{b \cdot d}, \quad b \neq 0, \quad d \neq 0$$

$$\frac{1}{3} \cdot \frac{2}{7} = \frac{1(2)}{3(7)} = \frac{2}{21}$$

6. Division of fractions:

$$\frac{a}{b} \div \frac{c}{d} = \frac{a}{b} \cdot \frac{d}{c}, b \neq 0, c \neq 0, d \neq 0$$

$$\frac{1}{3} \div \frac{2}{7} = \frac{1}{3} \cdot \frac{7}{2} = \frac{7}{6}$$

1.4 Exercises

Developing Skills

In Exercises 1–12, find the greatest common factor.

1. 6, 10

2. 6, 9

3. 20, 45

4. 48, 64

5. 45, 90

6. 27, 54

7. 18, 84, 90

8. 84, 98, 192

9. 240, 300, 360

10. 117, 195, 507

11. 134, 225, 315, 945

12. 80, 144, 214, 504

In Exercises 13–20, write the fraction in simplest form. See Example 1.

13. $\frac{2}{8}$

14. $\frac{3}{18}$

15. $\frac{12}{18}$

16. $\frac{16}{56}$

17. $\frac{60}{192}$

18. $\frac{45}{225}$

19. $\frac{28}{350}$

20. $\frac{88}{154}$

In Exercises 21–24, each figure is divided into regions of equal area. Write a fraction that represents the shaded portion of the figure. Then write the fraction in simplest form.

21.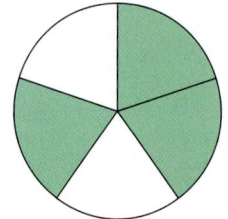

22.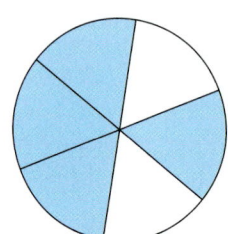

23.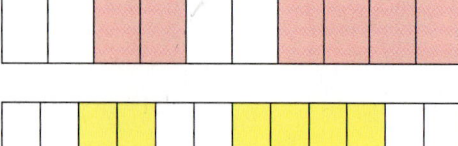

24.

In Exercises 25–28, write an equivalent fraction with the indicated denominator. See Example 2.

25. $\frac{3}{8} = \frac{}{16}$

26. $\frac{4}{5} = \frac{}{15}$

27. $\frac{6}{15} = \frac{}{25}$

28. $\frac{21}{49} = \frac{}{28}$

In Exercises 29–42, find the sum or difference. Write the result in simplest form.

29. $\frac{7}{15} + \frac{2}{15}$

30. $\frac{13}{35} + \frac{5}{35}$

31. $\frac{9}{11} + \frac{5}{11}$

32. $\frac{5}{6} + \frac{13}{6}$

33. $\frac{9}{16} - \frac{3}{16}$

34. $\frac{15}{32} - \frac{7}{32}$

35. $-\frac{23}{11} + \frac{12}{11}$

36. $-\frac{39}{23} - \frac{11}{23}$

37. $\frac{3}{4} - \frac{5}{4}$

38. $\frac{3}{8} - \frac{5}{8}$

39. $\frac{7}{10} + \left(-\frac{3}{10}\right)$

40. $\frac{11}{15} + \left(-\frac{2}{15}\right)$

41. $\frac{2}{5} + \frac{4}{5} + \frac{1}{5}$

42. $\frac{2}{9} + \frac{4}{9} + \frac{1}{9}$

In Exercises 43–66, evaluate the expression. Write the result in simplest form. See Examples 3, 4, and 5.

43. $\frac{1}{2} + \frac{1}{3}$

44. $\frac{3}{5} + \frac{1}{2}$

45. $\frac{1}{4} - \frac{1}{3}$

46. $\frac{2}{3} - \frac{1}{6}$

47. $\frac{3}{16} + \frac{3}{8}$

48. $\frac{2}{3} + \frac{4}{9}$

49. $-\frac{1}{8} - \frac{1}{6}$

50. $-\frac{13}{8} - \frac{3}{4}$

51. $4 - \frac{8}{3}$

52. $2 - \frac{17}{25}$

53. $-\frac{7}{8} - \frac{5}{6}$

54. $-\frac{5}{12} - \frac{1}{9}$

55. $\frac{3}{4} - \frac{2}{5}$

56. $\frac{5}{8} - \frac{1}{6}$

57. $-\frac{5}{6} - \left(-\frac{3}{4}\right)$

58. $-\frac{1}{9} - \left(-\frac{3}{5}\right)$

59. $3\frac{1}{2} + 5\frac{2}{3}$

60. $5\frac{3}{4} + 8\frac{1}{10}$

61. $1\frac{3}{16} - 2\frac{1}{4}$

62. $5\frac{7}{8} - 2\frac{1}{2}$

63. $15\frac{5}{6} - 20\frac{1}{4}$

64. $6 - 3\frac{5}{8}$

65. $-5\frac{2}{3} - 4\frac{5}{12}$

66. $-2\frac{3}{4} - 3\frac{1}{5}$

In Exercises 67–72, evaluate the expression. Write the result in simplest form. See Example 6.

67. $\frac{5}{12} - \frac{3}{8} + \frac{5}{16}$

68. $-\frac{3}{7} + \frac{5}{14} + \frac{3}{4}$

69. $3 + \frac{12}{3} + \frac{1}{9}$

70. $1 + \frac{2}{3} - \frac{5}{6}$

71. $2 - \frac{25}{6} - \frac{3}{4}$

72. $2 - \frac{15}{16} - \frac{7}{8}$

In Exercises 73–76, determine the unknown fractional part of the circle graph.

73.
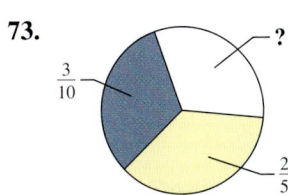

$\frac{3}{10}$? $\frac{2}{5}$

74.
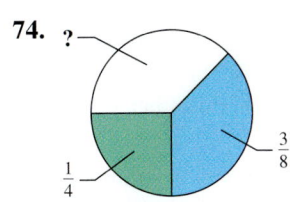

? $\frac{3}{8}$ $\frac{1}{4}$

75.
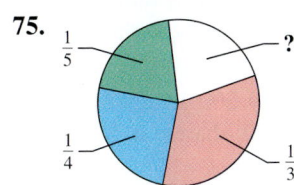

$\frac{1}{5}$? $\frac{1}{4}$ $\frac{1}{3}$

76.
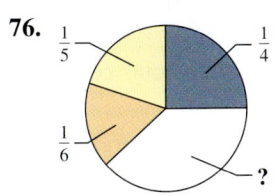

$\frac{1}{5}$ $\frac{1}{4}$ $\frac{1}{6}$?

In Exercises 77–100, evaluate the expression. Write the result in simplest form. See Examples 7 and 8.

77. $\frac{1}{2} \cdot \frac{3}{4}$

78. $\frac{3}{5} \cdot \frac{1}{2}$

79. $-\frac{2}{3} \cdot \frac{5}{7}$

80. $-\frac{5}{6} \cdot \frac{1}{2}$

81. $\frac{2}{3}\left(-\frac{9}{16}\right)$

82. $\left(\frac{5}{3}\right)\left(-\frac{3}{5}\right)$

83. $\left(-\frac{3}{4}\right)\left(-\frac{4}{9}\right)$

84. $\left(-\frac{7}{16}\right)\left(-\frac{12}{5}\right)$

85. $\left(\frac{5}{18}\right)\left(\frac{3}{4}\right)$

86. $\left(\frac{3}{28}\right)\left(\frac{7}{8}\right)$

87. $\left(\frac{11}{12}\right)\left(-\frac{9}{44}\right)$

88. $\left(\frac{5}{12}\right)\left(-\frac{6}{25}\right)$

89. $\left(-\frac{3}{11}\right)\left(-\frac{11}{3}\right)$

90. $\left(-\frac{7}{15}\right)\left(-\frac{15}{7}\right)$

91. $9\left(\frac{4}{15}\right)$

92. $24\left(\frac{7}{18}\right)$

93. $\left(-\frac{3}{2}\right)\left(-\frac{15}{16}\right)\left(\frac{12}{25}\right)$

94. $\left(\frac{1}{2}\right)\left(-\frac{4}{15}\right)\left(-\frac{5}{24}\right)$

95. $6\left(\frac{3}{4}\right)\left(\frac{2}{9}\right)$

96. $8\left(\frac{5}{12}\right)\left(\frac{3}{10}\right)$

97. $2\frac{3}{4} \cdot 3\frac{2}{3}$

98. $2\frac{4}{5} \cdot 6\frac{2}{3}$

99. $-5\frac{2}{3} \cdot 4\frac{1}{2}$

100. $-8\frac{1}{2} \cdot 3\frac{2}{5}$

In Exercises 101–104, find the reciprocal of the number. Show that the product of the number and its reciprocal is 1.

101. 7

102. 14

103. $\frac{4}{7}$

104. $-\frac{5}{9}$

In Exercises 105–122, evaluate the expression and write the result in simplest form. If it is not possible, explain why. See Example 9.

105. $\frac{3}{8} \div \frac{3}{4}$

106. $\frac{5}{16} \div \frac{25}{8}$

107. $-\frac{5}{12} \div \frac{45}{32}$

108. $-\frac{16}{21} \div \frac{12}{27}$

109. $\frac{3}{5} \div \frac{7}{5}$

110. $\frac{7}{8} \div \frac{3}{8}$

111. $\left(-\frac{5}{6}\right) \div \left(-\frac{8}{10}\right)$

112. $\left(-\frac{14}{15}\right) \div \left(-\frac{24}{25}\right)$

113. $-10 \div \frac{1}{9}$

114. $-6 \div \frac{1}{3}$

115. $0 \div (-21)$

116. $0 \div (-33)$

117. $\frac{3}{5} \div 0$

118. $\frac{11}{13} \div 0$

119. $3\frac{3}{4} \div 1\frac{1}{2}$

120. $2\frac{4}{9} \div 5\frac{1}{3}$

121. $3\frac{3}{4} \div 2\frac{5}{8}$

122. $1\frac{5}{6} \div 2\frac{1}{3}$

In Exercises 123–132, write the fraction in decimal form. (Use the bar notation for repeating digits.)

123. $\frac{3}{4}$

124. $\frac{5}{8}$

125. $\frac{9}{16}$

126. $\frac{7}{20}$

127. $\frac{2}{3}$

128. $\frac{5}{6}$

129. $\frac{7}{12}$

130. $\frac{8}{15}$

131. $\frac{5}{11}$

132. $\frac{5}{21}$

In Exercises 133–146, evaluate the expression. Round your answer to two decimal places. See Examples 10 and 11.

133. $132.1 + (-25.45)$

134. $408.9 + (-13.12)$

135. $1.21 + 4.06 - 3.00$

136. $3.4 + 1.062 - 5.13$

137. $-0.0005 - 2.01 + 0.111$

138. $-1.0012 - 3.25 + 0.2$

139. $(-6.3)(9.05)$

140. $3.7(-14.8)$

141. $(-0.05)(-85.95)$

142. $(-0.09)(-0.45)$

143. $4.69 \div 0.12$

144. $7.14 \div 0.94$

145. $1.062 \div (-2.1)$

146. $2.011 \div (-3.3)$

Estimation In Exercises 147 and 148, estimate the sum to the nearest integer.

147. $\frac{3}{11} + \frac{7}{10}$

148. $\frac{5}{8} + \frac{9}{7}$

Solving Problems

149. *Stock Price* On August 7, 2002, the Dow Jones Industrial Average closed at 8456.20 points. On August 8, 2002, it closed at 8712.00 points. Determine the increase in the Dow Jones Industrial Average.

150. *Sewing* A pattern requires $3\frac{1}{6}$ yards of material to make a skirt and an additional $2\frac{3}{4}$ yards to make a matching jacket. Find the total amount of material required.

151. *Agriculture* During the months of January, February, and March, a farmer bought $8\frac{3}{4}$ tons, $7\frac{1}{5}$ tons, and $9\frac{3}{8}$ tons of feed, respectively. Find the total amount of feed purchased during the first quarter of the year.

152. *Cooking* You are making a batch of cookies. You have placed 2 cups of flour, $\frac{1}{3}$ cup butter, $\frac{1}{2}$ cup brown sugar, and $\frac{1}{3}$ cup granulated sugar in a mixing bowl. How many cups of ingredients are in the mixing bowl?

153. *Construction Project* The highway workers have a sign beside a construction project indicating what fraction of the work has been completed. At the beginnings of May and June the fractions of work completed were $\frac{5}{16}$ and $\frac{2}{3}$, respectively. What fraction of the work was completed during the month of May?

154. *Fund Drive* A charity is raising funds and has a display showing how close they are to reaching their goal. At the end of the first week of the fund drive, the display shows $\frac{1}{9}$ of the goal. At the end of the second week, the display shows $\frac{3}{5}$ of the goal. What fraction of the goal was gained during the second week?

155. *Consumer Awareness* At a convenience store you buy two gallons of milk at $2.59 per gallon and three loaves of bread at $1.68 per loaf. You give the clerk a 20-dollar bill. How much change will you receive? (Assume there is no sales tax.)

156. *Consumer Awareness* A cellular phone company charges $1.16 for the first minute and $0.85 for each additional minute. Find the cost of a seven-minute cellular phone call.

157. *Cooking* You make 60 ounces of dough for breadsticks. Each breadstick requires $\frac{5}{4}$ ounces of dough. How many breadsticks can you make?

158. *Unit Price* A $2\frac{1}{2}$-pound can of food costs $4.95. What is the cost per pound?

159. *Consumer Awareness* The sticker on a new car gives the fuel efficiency as 22.3 miles per gallon. The average cost of fuel is $1.259 per gallon. Estimate the annual fuel cost for a car that will be driven approximately 12,000 miles per year.

160. *Walking Time* Your apartment is $\frac{3}{4}$ mile from the subway. You walk at the rate of $3\frac{1}{4}$ miles per hour. How long does it take you to walk to the subway?

161. *Stock Purchase* You buy 200 shares of stock at $23.63 per share and 300 shares at $86.25 per share.

(a) Estimate the total cost of the stock.

(b) Use a calculator to find the total cost of the stock.

162. *Music* Each day for a week, you practiced the saxophone for $\frac{2}{3}$ hour.

(a) Explain how to use mental math to estimate the number of hours of practice in a week.

(b) Determine the actual number of hours you practiced during the week. Write the result in decimal form, rounding to one decimal place.

163. *Consumer Awareness* The prices per gallon of regular unleaded gasoline at three service stations are $1.259, $1.369, and $1.279, respectively. Find the average price per gallon.

164. *Consumer Awareness* The prices of a 16-ounce bottle of soda at three different convenience stores are $1.09, $1.25, and $1.10, respectively. Find the average price for the bottle of soda.

Explaining Concepts

165. ⚡ Answer parts (g)–(l) of Motivating the Chapter.

166. *Writing* ✎ Is it true that the sum of two fractions of like signs is positive? If not, give an example that shows the statement is false.

167. *Writing* ✎ Does $\frac{2}{3} + \frac{3}{2} = (2+3)/(3+2) = 1$? Explain your answer.

168. *Writing* ✎ In your own words, describe the rule for determining the sign of the product of two fractions.

169. *Writing* ✎ Is it true that $\frac{2}{3} = 0.67$? Explain your answer.

170. *Writing* ✎ Use the figure to determine how many one-fourths are in 3. Explain how to obtain the same result by division.

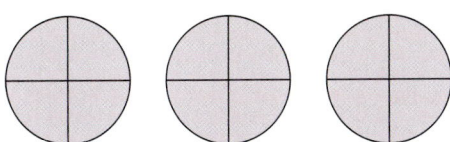

171. *Writing* ✎ Use the figure to determine how many one-sixths are in $\frac{2}{3}$. Explain how to obtain the same result by division.

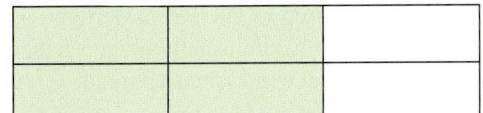

172. *Investigation* When using a calculator to perform operations with decimals, you should try to get in the habit of rounding your answers *only* after all the calculations are done. If you round the answer at a preliminary stage, you can introduce unnecessary roundoff error. The dimensions of a box are $l = 5.24$, $w = 3.03$, and $h = 2.749$. Find the volume, $l \cdot w \cdot h$, by multiplying the numbers and then rounding the answer to one decimal place. Now use a second method, first rounding each dimension to one decimal place and then multiplying the numbers. Compare your answers, and explain which of these techniques produces the more accurate answer.

True or False? In Exercises 173–178, decide whether the statement is true or false. Justify your answer.

173. The reciprocal of every nonzero integer is an integer.

174. The reciprocal of every nonzero rational number is a rational number.

175. The product of two nonzero rational numbers is a rational number.

176. The product of two positive rational numbers is greater than either factor.

177. If $u > v$, then $u - v > 0$.

178. If $u > 0$ and $v > 0$, then $u - v > 0$.

179. *Estimation* Use mental math to determine whether $\left(5\frac{3}{4}\right) \times \left(4\frac{1}{8}\right)$ is less than 20. Explain your reasoning.

180. Determine the placement of the digits 3, 4, 5, and 6 in the following addition problem so that you obtain the specified sum. Use each number only once.

$$\frac{\quad\rule{1cm}{0.4pt}\quad}{\quad\rule{1cm}{0.4pt}\quad} + \frac{\quad\rule{1cm}{0.4pt}\quad}{\quad\rule{1cm}{0.4pt}\quad} = \frac{13}{10}$$

181. If the fractions represented by the points P and R are multiplied, what point on the number line best represents their product: M, S, N, P, or T? (Source: National Council of Teachers of Mathematics)

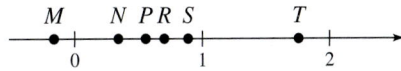

1.5 Exponents, Order of Operations, and Properties of Real Numbers

Michael Newman/PhotoEdit

What You Should Learn

1 Rewrite repeated mutiplication in exponential form and evaluate exponential expressions.

2 Evaluate expressions using order of operations.

3 Identify and use the properties of real numbers.

Why You Should Learn It

Properties of real numbers can be used to solve real-life problems. For instance, in Exercise 124 on page 57, you will use the Distributive Property to find the amount paid for a new truck.

1 Rewrite repeated multiplication in exponential form and evaluate exponential expressions.

Exponents

In Section 1.3, you learned that multiplication by a positive integer can be described as repeated addition.

Repeated Addition   *Multiplication*

$$\underbrace{7 + 7 + 7 + 7}_{\text{4 terms of 7}}$$ $$4 \times 7$$

In a similar way, repeated multiplication can be described in **exponential form.**

Repeated Multiplication   *Exponential Form*

$$\underbrace{7 \cdot 7 \cdot 7 \cdot 7}_{\text{4 factors of 7}}$$ $$7^4$$

In the exponential form 7^4, 7 is the **base** and it specifies the repeated factor. The number 4 is the **exponent** and it indicates how many times the base occurs as a factor.

When you write the exponential form 7^4, you can say that you are raising 7 to the fourth **power.** When a number is raised to the *first* power, you usually do not write the exponent 1. For instance, you would usually write 5 rather than 5^1. Here are some examples of how exponential expressions are read.

Exponential Expression   *Verbal Statement*

7^2 "seven to the second power" or "seven squared"

4^3 "four to the third power" or "four cubed"

$(-2)^4$ "negative two to the fourth power"

-2^4 "the opposite of two to the fourth power"

It is important to recognize how exponential forms such as $(-2)^4$ and -2^4 differ.

$(-2)^4 = (-2)(-2)(-2)(-2)$ The negative sign is part of the base.

$\qquad = 16$ The value of the expression is positive.

$-2^4 = -(2 \cdot 2 \cdot 2 \cdot 2)$ The negative sign is not part of the base.

$\qquad = -16$ The value of the expression is negative.

Technology: Discovery

When a negative number is raised to a power, the use of parentheses is very important. To discover why, use a calculator to evaluate $(-5)^4$ and -5^4. Write a statement explaining the results. Then use a calculator to evaluate $(-5)^3$ and -5^3. If necessary, write a new statement explaining your discoveries.

Keep in mind that an exponent applies only to the factor (number) directly preceding it. Parentheses are needed to include a negative sign or other factors as part of the base.

Example 1   Evaluating Exponential Expressions

a. $2^5 = 2 \cdot 2 \cdot 2 \cdot 2 \cdot 2$ Rewrite expression as a product.

$\quad = 32$ Simplify.

b. $\left(\dfrac{2}{3}\right)^4 = \dfrac{2}{3} \cdot \dfrac{2}{3} \cdot \dfrac{2}{3} \cdot \dfrac{2}{3}$ Rewrite expression as a product.

$\quad = \dfrac{2 \cdot 2 \cdot 2 \cdot 2}{3 \cdot 3 \cdot 3 \cdot 3}$ Multiply fractions.

$\quad = \dfrac{16}{81}$ Simplify.

Example 2   Evaluating Exponential Expressions

a. $(-4)^3 = (-4)(-4)(-4)$ Rewrite expression as a product.

$\quad = -64$ Simplify.

b. $(-3)^4 = (-3)(-3)(-3)(-3)$ Rewrite expression as a product.

$\quad = 81$ Simplify.

c. $-3^4 = -(3 \cdot 3 \cdot 3 \cdot 3)$ Rewrite expression as a product.

$\quad = -81$ Simplify.

In parts (a) and (b) of Example 2, note that when a negative number is raised to an odd power, the result is *negative,* and when a negative number is raised to an even power, the result is *positive.*

Example 3   Transporting Capacity

A truck can transport a load of motor oil that is 6 cases high, 6 cases wide, and 6 cases long. Each case contains 6 quarts of motor oil. How many quarts can the truck transport?

Solution

A sketch can help you solve this problem. From Figure 1.32, there are $6 \cdot 6 \cdot 6$ cases of motor oil and each case contains 6 quarts. You can see that 6 occurs as a factor four times, which implies that the total number of quarts is

$$(6 \cdot 6 \cdot 6) \cdot 6 = 6^4 = 1296.$$

So, the truck can transport 1296 quarts of oil.

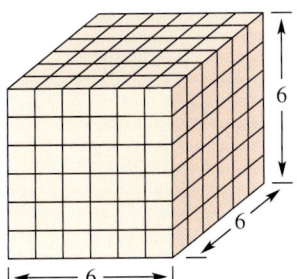

Figure 1.32

2 Evaluate expressions using order of operations.

Order of Operations

Up to this point in the text, you have studied five operations of arithmetic—addition, subtraction, multiplication, division, and exponentiation (repeated multiplication). When you use more than one operation in a given problem, you face the question of which operation to do first. For example, without further guidelines, you could evaluate $4 + 3 \cdot 5$ in two ways.

Add First	Multiply First
$4 + 3 \cdot 5 \stackrel{?}{=} (4 + 3) \cdot 5$	$4 + 3 \cdot 5 \stackrel{?}{=} 4 + (3 \cdot 5)$
$= 7 \cdot 5$	$= 4 + 15$
$= 35$	$= 19$

According to the established **order of operations,** the second evaluation is correct. The reason for this is that multiplication has a higher priority than addition. The accepted priorities for order of operations are summarized below.

Order of Operations

1. Perform operations inside *symbols of grouping*—( ) or [ ]— or *absolute value symbols,* starting with the innermost symbols.

2. Evaluate all *exponential* expressions.

3. Perform all *multiplications* and *divisions* from left to right.

4. Perform all *additions* and *subtractions* from left to right.

In the priorities for order of operations, note that the highest priority is given to **symbols of grouping** such as parentheses or brackets. This means that when you want to be sure that you are communicating an expression correctly, you can insert symbols of grouping to specify which operations you intend to be performed first. For instance, if you want to make sure that $4 + 3 \cdot 5$ will be evaluated correctly, you can write it as $4 + (3 \cdot 5)$.

Technology: Discovery

To discover if your calculator performs the established order of operations, evaluate $7 + 5 \cdot 3 - 2^4 \div 4$ exactly as it appears. Does your calculator display 5 or 18? If your calculator performs the established order of operations, it will display 18.

Study Tip

When you use symbols of grouping in an expression, you should alternate between parentheses and brackets. For instance, the expression

$$10 - (3 - [4 - (5 + 7)])$$

is easier to understand than $10 - (3 - (4 - (5 + 7)))$.

Example 4 Order of Operations

a. $7 - [(5 \cdot 3) + 2^3] = 7 - [15 + 2^3]$ Multiply inside the parentheses.

$= 7 - [15 + 8]$ Evaluate exponential expression.

$= 7 - 23$ Add inside the brackets.

$= -16$ Subtract.

b. $36 \div (3^2 \cdot 2) - 6 = 36 \div (9 \cdot 2) - 6$ Evaluate exponential expression.

$= 36 \div 18 - 6$ Multiply inside the parentheses.

$= 2 - 6$ Divide.

$= -4$ Subtract.

Example 5   Order of Operations

a. $\dfrac{3}{7} \div \dfrac{8}{7} + \left(-\dfrac{3}{5}\right)\left(\dfrac{1}{3}\right) = \dfrac{3}{7} \cdot \dfrac{7}{8} + \left(-\dfrac{3}{5}\right)\left(\dfrac{1}{3}\right)$ Invert divisor and multiply.

$$= \dfrac{3}{8} + \left(-\dfrac{1}{5}\right)$$ Multiply fractions.

$$= \dfrac{15}{40} + \dfrac{-8}{40}$$ Find common denominator.

$$= \dfrac{7}{40}$$ Add fractions.

b. $\dfrac{8}{3}\left(\dfrac{1}{6} + \dfrac{1}{4}\right) = \dfrac{8}{3}\left(\dfrac{2}{12} + \dfrac{3}{12}\right)$ Find common denominator.

$$= \dfrac{8}{3}\left(\dfrac{5}{12}\right)$$ Add inside the parentheses.

$$= \dfrac{40}{36}$$ Multiply fractions.

$$= \dfrac{10}{9}$$ Simplify.

Example 6   Order of Operations

Evaluate the expression $6 + \dfrac{8 + 7}{3^2 - 4} - (-5)$.

Solution

Using the established order of operations, you can evaluate the expression as follows.

$$6 + \dfrac{8 + 7}{3^2 - 4} - (-5) = 6 + \dfrac{8 + 7}{9 - 4} - (-5)$$ Evaluate exponential expression.

$$= 6 + \dfrac{15}{9 - 4} - (-5)$$ Add in numerator.

$$= 6 + \dfrac{15}{5} - (-5)$$ Subtract in denominator.

$$= 6 + 3 - (-5)$$ Divide.

$$= 9 + 5$$ Add.

$$= 14$$ Add.

In Example 6, note that a fraction bar acts as a symbol of grouping. For instance,

$$\dfrac{8 + 7}{3^2 - 4} \quad \text{means} \quad (8 + 7) \div (3^2 - 4), \quad \text{not} \quad 8 + 7 \div 3^2 - 4.$$

3 Identify and use the properties of real numbers.

Properties of Real Numbers

You are now ready for the symbolic versions of the properties that are true about operations with real numbers. These properties are referred to as **properties of real numbers.** The table shows a verbal description and an illustrative example for each property. Keep in mind that the letters a, b, c, etc., represent real numbers, even though only rational numbers have been used to this point.

Properties of Real Numbers: Let a, b, and c be real numbers.

Property	*Example*
1. *Commutative Property of Addition:* Two real numbers can be added in either order. $a + b = b + a$	$3 + 5 = 5 + 3$
2. *Commutative Property of Multiplication:* Two real numbers can be multiplied in either order. $ab = ba$	$4 \cdot (-7) = -7 \cdot 4$
3. *Associative Property of Addition:* When three real numbers are added, it makes no difference which two are added first. $(a + b) + c = a + (b + c)$	$(2 + 6) + 5 = 2 + (6 + 5)$
4. *Associative Property of Multiplication:* When three real numbers are multiplied, it makes no difference which two are multiplied first. $(ab)c = a(bc)$	$(3 \cdot 5) \cdot 2 = 3 \cdot (5 \cdot 2)$
5. *Distributive Property:* Multiplication distributes over addition. $a(b + c) = ab + ac$ $(a + b)c = ac + bc$	$3(8 + 5) = 3 \cdot 8 + 3 \cdot 5$ $(3 + 8)5 = 3 \cdot 5 + 8 \cdot 5$
6. *Additive Identity Property:* The sum of zero and a real number equals the number itself. $a + 0 = 0 + a = a$	$3 + 0 = 0 + 3 = 3$
7. *Multiplicative Identity Property:* The product of 1 and a real number equals the number itself. $a \cdot 1 = 1 \cdot a = a$	$4 \cdot 1 = 1 \cdot 4 = 4$
8. *Additive Inverse Property:* The sum of a real number and its opposite is zero. $a + (-a) = 0$	$3 + (-3) = 0$
9. *Multiplicative Inverse Property:* The product of a nonzero real number and its reciprocal is 1. $a \cdot \dfrac{1}{a} = 1, \ a \neq 0$	$8 \cdot \dfrac{1}{8} = 1$

Example 7   Identifying Properties of Real Numbers

Identify the property of real numbers illustrated by each statement.

a. $3(a + 2) = 3 \cdot a + 3 \cdot 2$

b. $5 \cdot \dfrac{1}{5} = 1$

c. $7 + (5 + b) = (7 + 5) + b$

d. $(b + 3) + 0 = b + 3$

e. $4a = a(4)$

Solution

a. This statement illustrates the Distributive Property.

b. This statement illustrates the Multiplicative Inverse Property.

c. This statement illustrates the Associative Property of Addition.

d. This statement illustrates the Additive Identity Property.

e. This statement illustrates the Commutative Property of Multiplication.

Example 8   Using the Properties of Real Numbers

Complete each statement using the specified property of real numbers.

a. Commutative Property of Addition:

$$5 + a = $$

b. Associative Property of Multiplication:

$$2(7c) = $$

c. Distributive Property

$$3 \cdot a + 3 \cdot 4 = $$

Solution

a. By the Commutative Property of Addition, you can write

$$5 + a = a + 5.$$

b. By the Associative Property of Multiplication, you can write

$$2(7c) = (2 \cdot 7)c.$$

c. By the Distributive Property, you can write

$$3 \cdot a + 3 \cdot 4 = 3(a + 4).$$

One of the distinctive things about algebra is that its rules make sense. You don't have to accept them on "blind faith"—instead, you can learn the reasons that the rules work. For instance, the next example looks at some basic differences among the operations of addition, multiplication, subtraction, and division.

Example 9   Properties of Real Numbers

In the summary of properties of real numbers on page 52, why are all the properties listed in terms of addition and multiplication and not subtraction and division?

Solution

The reason for this is that subtraction and division lack many of the properties listed in the summary. For instance, subtraction and division are not commutative. To see this, consider the following.

$$7 - 5 \neq 5 - 7 \quad \text{and} \quad 12 \div 4 \neq 4 \div 12$$

Similarly, subtraction and division are not associative.

$$9 - (5 - 3) \neq (9 - 5) - 3 \quad \text{and} \quad 12 \div (4 \div 2) \neq (12 \div 4) \div 2$$

Example 10   Geometry: Area

You measure the width of a billboard and find that it is 60 feet. You are told that its height is 22 feet less than its width.

a. Write an expression for the area of the billboard.

b. Use the Distributive Property to rewrite the expression.

c. Find the area of the billboard.

Solution

a. Begin by drawing and labeling a diagram, as shown in Figure 1.33. To find an expression for the area of the billboard, multiply the width by the height.

$$\text{Area} = \text{Width} \times \text{Height}$$
$$= 60(60 - 22)$$

b. To rewrite the expression $60(60 - 22)$ using the Distributive Property, distribute 60 over the subtraction.

$$60(60 - 22) = 60(60) - 60(22)$$

c. To find the area of the billboard, evaluate the expression from part (b) as follows.

$$60(60) - 60(22) = 3600 - 1320 \qquad \text{Multiply.}$$
$$= 2280 \qquad \text{Subtract.}$$

So, the area of the billboard is 2280 square feet.

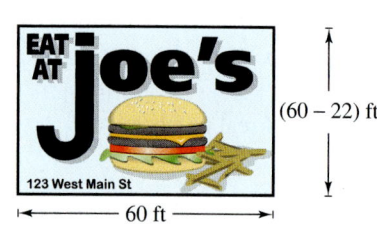

$(60 - 22)$ ft

60 ft

Figure 1.33

From Example 10(b) you can see that the Distributive Property is also true for subtraction. For instance, the "subtraction form" of $a(b + c) = ab + ac$ is

$$a(b - c) = a[b + (-c)]$$
$$= ab + a(-c)$$
$$= ab - ac.$$

1.5 Exercises

Developing Skills

In Exercises 1–8, rewrite in exponential form.

1. $2 \cdot 2 \cdot 2 \cdot 2 \cdot 2$

2. $4 \cdot 4 \cdot 4 \cdot 4 \cdot 4 \cdot 4$

3. $(-5) \cdot (-5) \cdot (-5) \cdot (-5)$

4. $(-3) \cdot (-3) \cdot (-3) \cdot (-3)$

5. $\left(-\frac{1}{4}\right) \cdot \left(-\frac{1}{4}\right) \cdot \left(-\frac{1}{4}\right)$

6. $\left(-\frac{3}{5}\right) \cdot \left(-\frac{3}{5}\right) \cdot \left(-\frac{3}{5}\right) \cdot \left(-\frac{3}{5}\right)$

7. $(1.6) \cdot (1.6) \cdot (1.6) \cdot (1.6) \cdot (1.6)$

8. $(8.7) \cdot (8.7) \cdot (8.7)$

In Exercises 9–16, rewrite as a product.

9. $(-3)^6$

10. $(-8)^4$

11. $\left(\frac{3}{8}\right)^5$

12. $\left(\frac{3}{11}\right)^4$

13. $\left(-\frac{1}{2}\right)^5$

14. $\left(-\frac{4}{5}\right)^6$

15. $(9.8)^3$

16. $(0.01)^8$

In Exercises 17–28, evaluate the expression. See Examples 1 and 2.

17. 3^2

18. 4^3

19. 2^6

20. 5^3

21. $(-5)^3$

22. $(-4)^2$

23. -4^2

24. $-(-6)^3$

25. $\left(\frac{1}{4}\right)^3$

26. $\left(\frac{4}{5}\right)^3$

27. $(-1.2)^3$

28. $(-1.5)^4$

In Exercises 29–70, evaluate the expression. If it is not possible, state the reason. Write fractional answers in simplest form. See Examples 4, 5, and 6.

29. $4 - 6 + 10$

30. $8 + 9 - 12$

31. $5 - (8 - 15)$

32. $13 - (12 - 3)$

33. $-|2 - (6 + 5)|$

34. $125 - |10 - (25 - 3)|$

35. $15 + 3 \cdot 4$

36. $9 - 5 \cdot 2$

37. $25 - 32 \div 4$

38. $16 + 24 \div 8$

39. $(16 - 5) \div (3 - 5)$

40. $(19 - 4) \div (7 - 2)$

41. $(10 - 16) \cdot (20 - 26)$

42. $(14 - 17) \cdot (13 - 19)$

43. $(45 \div 10) \cdot 2$

44. $(38 \div 5) \cdot 3$

45. $[360 - (8 + 12)] \div 10$

46. $[127 - (13 + 4)] \div 11$

47. $5 + (2^2 \cdot 3)$

48. $181 - (13 \cdot 3^2)$

49. $(-6)^2 - (5^2 \cdot 4)$

50. $(-3)^3 + (12 \div 2^2)$

51. $\left(3 \cdot \frac{5}{9}\right) + 1 - \frac{1}{3}$

52. $\frac{2}{3}\left(\frac{3}{4}\right) + 2 - \frac{1}{2}$

53. $18\left(\frac{1}{2} + \frac{2}{3}\right)$

54. $4\left(-\frac{2}{3} + \frac{4}{3}\right)$

55. $\frac{7}{25}\left(\frac{7}{16} - \frac{1}{8}\right)$

56. $\frac{3}{2}\left(\frac{2}{3} + \frac{1}{6}\right)$

57. $\dfrac{3 + [15 \div (-3)]}{16}$

58. $\dfrac{5 + [(-12) \div 4]}{24}$

59. $\dfrac{3 \cdot 6 - 4 \cdot 6}{5 + 1}$

60. $\dfrac{5 \cdot 3 + 5 \cdot 6}{7 - 2}$

61. $\frac{7}{3}\left(\frac{2}{3}\right) \div \frac{28}{15}$

62. $\frac{3}{8}\left(\frac{1}{5}\right) \div \frac{25}{32}$

63. $\dfrac{1 - 3^2}{-2}$

64. $\dfrac{3^2 + 4^2}{5}$

65. $\dfrac{3^2 - 4^2}{0}$

66. $\dfrac{0}{3^2 - 4^2}$

67. $\dfrac{5^2 + 12^2}{13}$

68. $\dfrac{4^2 - 2^3}{4}$

69. $\dfrac{0}{5^2 + 1}$

70. $\dfrac{3^2 + 1}{0}$

In Exercises 71–74, use a calculator to evaluate the expression. Round your answer to two decimal places.

71. $300\left(1 + \dfrac{0.1}{12}\right)^{24}$

72. $1000 \div \left(1 + \dfrac{0.09}{4}\right)^8$

73. $\dfrac{1.32 + 4(3.68)}{1.5}$

74. $\dfrac{4.19 - 7(2.27)}{14.8}$

In Exercises 75–92, identify the property of real numbers illustrated by the statement. See Example 7.

75. $6(-3) = -3(6)$

76. $16 + 10 = 10 + 16$

77. $x + 10 = 10 + x$

78. $8x = x(8)$

79. $0 + 15 = 15$

80. $1 \cdot 4 = 4$

81. $-16 + 16 = 0$

82. $(2 \cdot 3)4 = 2(3 \cdot 4)$

83. $(10 + 3) + 2 = 10 + (3 + 2)$

84. $25 + (-25) = 0$

85. $4(3 \cdot 10) = (4 \cdot 3)10$

86. $(32 + 8) + 5 = 32 + (8 + 5)$

87. $7\left(\frac{1}{7}\right) = 1$

88. $14 + (-14) = 0$

89. $6(3 + x) = 6 \cdot 3 + 6x$

90. $(14 + 2)3 = 14 \cdot 3 + 2 \cdot 3$

91. $\frac{1}{a}(3 + y) = \frac{1}{a}(3) + \frac{1}{a}(y)$

92. $[(x + y)u]v = (x + y)(uv)$

In Exercises 93–104, complete the statement using the specified property of real numbers. See Example 8.

93. Commutative Property of Addition:
   $y + 5 = $

94. Commutative Property of Addition:
   $3 + x = $

95. Commutative Property of Multiplication:
   $10(-3) = $

96. Commutative Property of Multiplication:
   $5(u + v) = $

97. Distributive Property:
   $6(x + 2) = $

98. Distributive Property:
   $5(u + v) = $

99. Distributive Property:
   $(4 + y)25 = $

100. Distributive Property:
   $(4 - y)12 = $

101. Associative Property of Addition:
   $3x + (2y + 5) = $

102. Associative Property of Addition:
   $10 + (x + 2y) = $

103. Associative Property of Multiplication:
   $12(3 \cdot 4) = $

104. Associative Property of Multiplication:
   $(6x)y = $

In Exercises 105–112, find (a) the additive inverse and (b) the multiplicative inverse of the quantity.

105. 50

106. 12

107. -1

108. $-\frac{1}{2}$

109. $2x$

110. $5y$

111. ab

112. uv

In Exercises 113–116, simplify the expression using (a) the Distributive Property and (b) order of operations.

113. $3(6 + 10)$

114. $4(8 - 3)$

115. $\frac{2}{3}(9 + 24)$

116. $\frac{1}{2}(4 - 2)$

In Exercises 117–120, identify the property of real numbers used to justify each step.

117. $7x + 9 + 2x$

$= 7x + 2x + 9$

$= (7x + 2x) + 9$

$= (7 + 2)x + 9$

$= 9x + 9$

$= 9(x + 1)$

118. $19 + 5x + 24$

$= 19 + 24 + 5x$

$= (19 + 24) + 5x$

$= 43 + 5x$

119. $3 + 10(x + 1)$

$= 3 + 10x + 10$

$= 3 + 10 + 10x$

$= (3 + 10) + 10x$

$= 13 + 10x$

120. $2(x + 3) + x$

$= 2x + 2 \cdot 3 + x$

$= 2x + x + 6$

$= (2 + 1)x + 6$

$= 3x + 6$

$= 3(x + 2)$

Solving Problems

▲ *Geometry*   In Exercises 121 and 122, find the area of the region.

121.

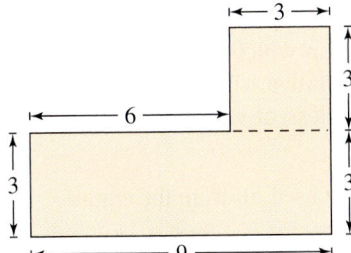

122.

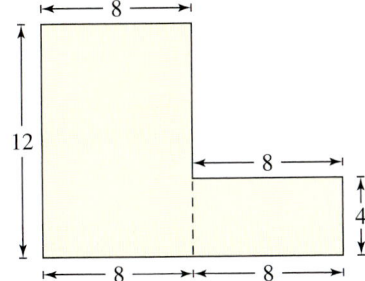

123. *Sales Tax*   You purchase a sweater for x dollars. There is a 6% sales tax, which implies that the total amount you must pay is $x + 0.06x$.

(a) Use the Distributive Property to rewrite the expression.

(b) The sweater costs $25.95. How much must you pay for the sweater including sales tax?

124. *Cost of a Truck*   A new truck can be paid for by 48 monthly payments of x dollars each plus a down payment of 2.5 times the amount of the monthly payment. This implies that the total amount paid for the truck is $2.5x + 48x$.

(a) Use the Distributive Property to rewrite the expression.

(b) What is the total amount paid for a truck that has a monthly payment of $435?

125. ▲ *Geometry*   The width of a movie screen is 30 feet and its height is 8 feet less than the width. Write an expression for the area of the movie screen. Use the Distributive Property to rewrite the expression.

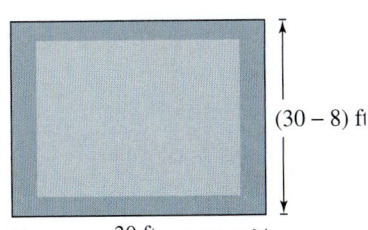

126. ▲ *Geometry*   A picture frame is 36 inches wide and its height is 9 inches less than its width. Write an expression for the area of the picture frame. Use the Distributive Property to rewrite the expression.

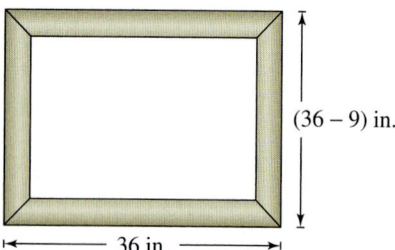

(36 − 9) in.

36 in.

▲ *Geometry*   In Exercises 127 and 128, write an expression for the perimeter of the triangle shown in the figure. Use the properties of real numbers to simplify the expression.

127.

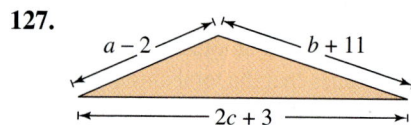

$a − 2$ $b + 11$

$2c + 3$

128.

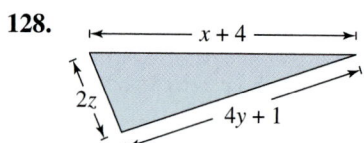

$x + 4$

$2z$

$4y + 1$

▲ *Geometry*   In Exercises 129 and 130, find the area of the shaded rectangle in two ways. Explain how the results are related to the Distributive Property.

129.

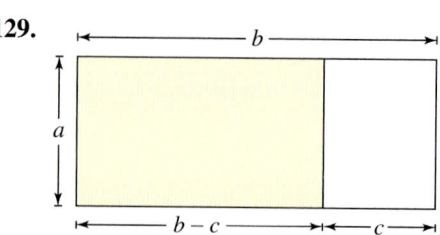

b

a

$b − c$ c

130.

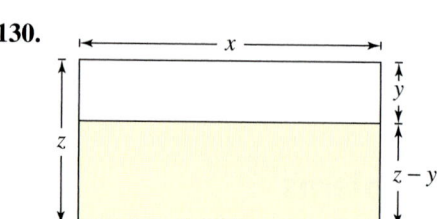

x

y

z

$z − y$

Think About It   In Exercises 131 and 132, determine whether the order in which the two activities are performed is "commutative." That is, do you obtain the same result regardless of which activity is performed first?

131. (a) "Drain the used oil from the engine."

(b) "Fill the crankcase with 5 quarts of new oil."

132. (a) "Weed the flower beds."

(b) "Mow the lawn."

Explaining Concepts

133. Consider the expression 3^5.

(a) What is the number 3 called?

(b) What is the number 5 called?

134. *Writing*   Are $−6^2$ and $(−6)^2$ equal? Explain.

135. *Writing*   Are $2 \cdot 5^2$ and 10^2 equal? Explain.

136. *Writing*   In your own words, describe the priorities for the established order of operations.

137. *Writing*   In your own words, state the Associative Properties of Addition and Multiplication. Give an example of each.

138. *Writing* ✏ In your own words, state the Commutative Properties of Addition and Multiplication. Give an example of each.

Writing ✏ In Exercises 139–142, explain why the statement is true. (The symbol ≠ means "is not equal to.")

139. $4 \cdot 6^2 \neq 24^2$

140. $4 - (6 - 2) \neq 4 - 6 - 2$

141. $-3^2 \neq (-3)(-3)$ **142.** $\dfrac{8 - 6}{2} \neq 4 - 6$

143. *Error Analysis*   Describe the error.

$$-9 + \frac{9 + 20}{3(5)} - (-3) = -9 + \frac{9}{3} + \frac{20}{5} - (-3)$$
$$= -9 + 3 + 4 - (-3)$$
$$= 1$$

144. *Error Analysis*   Describe the error.

$$7 - 3(8 + 1) - 15 = 4(8 + 1) - 15$$
$$= 4(9) - 15$$
$$= 36 - 15$$
$$= 21$$

Writing ✏ In Exercises 145–148, explain why the statement is true.

145. $5(x + 3) \neq 5x + 3$ **146.** $7(x - 2) \neq 7x - 2$

147. $\dfrac{8}{0} \neq 0$ **148.** $5\left(\dfrac{1}{5}\right) \neq 0$

149. Match each expression in the first column with its value in the second column.

Expression	Value
$(6 + 2) \cdot (5 + 3)$	19
$(6 + 2) \cdot 5 + 3$	22
$6 + 2 \cdot 5 + 3$	64
$6 + 2 \cdot (5 + 3)$	43

150. Using the established order of operations, which of the following expressions has a value of 72? For those that don't, decide whether you can insert parentheses into the expression so that its value is 72.

(a) $4 + 2^3 - 7$ (b) $4 + 8 \cdot 6$

(c) $93 - 25 - 4$ (d) $70 + 10 \div 5$

(e) $60 + 20 \div 2 + 32$

(f) $35 \cdot 2 + 2$

151. Consider the rectangle shown in the figure.

(a) Find the area of the rectangle by adding the areas of regions I and II.

(b) Find the area of the rectangle by multiplying its length by its width.

(c) Explain how the results of parts (a) and (b) relate to the Distributive Property.

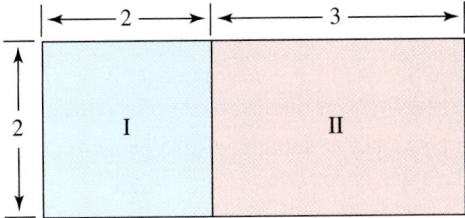

What Did You Learn?

Key Terms

real numbers, *p. 2*
natural numbers, *p. 2*
integers, *p. 2*
rational numbers, *p. 3*
irrational numbers, *p. 3*
real number line, *p. 4*

inequality symbol, *p. 5*
opposites, *p. 7*
absolute value, *p. 7*
expression, *p. 8*
evaluate, *p. 8*
additive inverse, *p. 13*

factor, *p. 24*
prime number, *p. 24*
greatest common factor,
   p. 35
reciprocal, *p. 40*
exponent, *p. 48*

Key Concepts

1.1 Ordering of real numbers
Use the real number line and an inequality symbol (<, >, ≤, or ≥) to order real numbers.

1.1 Absolute value
The absolute value of a number is its distance from zero on the real number line. The absolute value is either positive or zero.

1.2 Addition and subtraction of integers
To add two integers with like signs, add their absolute values and attach the common sign to the result.

To add two integers with different signs, subtract the smaller absolute value from the larger absolute value and attach the sign of the integer with the larger absolute value.

To subtract one integer from another, add the opposite of the integer being subtracted to the other integer.

1.3 Rules for multiplying and dividing integers
The product of an integer and zero is 0.

Zero divided by a nonzero integer is 0, whereas a nonzero integer divided by zero is undefined.

The product or quotient of two nonzero integers with like signs is positive.

The product or quotient of two nonzero integers with different signs is negative.

1.4 Addition and subtraction of fractions
1. Add or subtract two fractions with like denominators:

$$\frac{a}{c} + \frac{b}{c} = \frac{a+b}{c} \text{ or } \frac{a}{c} - \frac{b}{c} = \frac{a-b}{c}, c \neq 0$$

2. To add two fractions with unlike denominators, rewrite both fractions so that they have like denominators. Then use the rule for adding and subtracting fractions with like denominators.

1.4 Multiplication of fractions
$$\frac{a}{b} \cdot \frac{c}{d} = \frac{a \cdot c}{b \cdot d}, \ b \neq 0, d \neq 0$$

1.4 Division of fractions
$$\frac{a}{b} \div \frac{c}{d} = \frac{a}{b} \cdot \frac{d}{c}, \ b \neq 0, c \neq 0, d \neq 0$$

1.5 Order of operations
1. Perform operations inside symbols of grouping— ( ) or [ ]—or absolute value symbols, starting with the innermost symbols.

2. Evaluate all exponential expressions.

3. Perform all multiplications and divisions from left to right.

4. Perform all additions and subtractions from left to right.

1.5 Properties of real numbers
Commutative Property of Addition $a + b = b + a$
Commutative Property of Multiplication $ab = ba$
Associative Property of Addition
   $(a + b) + c = a + (b + c)$
Associative Property of Multiplication $(ab)c = a(bc)$
Distributive Property
   $a(b + c) = ab + ac$ $a(b - c) = ab - ac$
   $(a + b)c = ac + bc$ $(a - b)c = ac - bc$
Additive Identity Property $a + 0 = a$
Multiplicative Identity Property $a \cdot 1 = a$
Additive Inverse Property $a + (-a) = 0$

Multiplicative Inverse Property $a \cdot \dfrac{1}{a} = 1, \quad a \neq 0$

Review Exercises

1.1 Real Numbers: Order and Absolute Value

1 Define sets and use them to classify numbers as natural, integer, rational, or irrational.

In Exercises 1 and 2, determine which of the numbers in the set are (a) natural numbers, (b) integers, (c) rational numbers, and (d) irrational numbers.

1. $\left\{-1, 4.5, \frac{2}{5}, -\frac{1}{7}, \sqrt{4}, \sqrt{5}\right\}$

2. $\left\{10, -3, \frac{4}{5}, \pi, -3.\overline{16}, -\frac{19}{11}\right\}$

2 Plot numbers on the real number line.

In Exercises 3–8, plot the numbers on the real number line.

3. $-3, 5$

4. $-8, 11$

5. $-6, \frac{5}{4}$

6. $-\frac{7}{2}, 9$

7. $-1, 0, \frac{1}{2}$

8. $-2, -\frac{1}{3}, 5$

3 Use the real number line and inequality symbols to order real numbers.

In Exercises 9–12, plot each real number as a point on the real number line and place the correct inequality symbol (< or >) between the pair of real numbers.

9. $-\frac{1}{10}$ 4

10. $\frac{25}{3}$ $\frac{5}{3}$

11. -3 -7

12. 10.6 -3.5

13. Which is smaller: $\frac{2}{3}$ or 0.6?

14. Which is smaller: $-\frac{1}{3}$ or -0.3?

4 Find the absolute value of a number.

In Exercises 15–18, find the opposite of the number, and determine the distance of the number and its opposite from 0.

15. 152

16. -10.4

17. $-\frac{7}{3}$

18. $\frac{2}{3}$

In Exercises 19–22, evaluate the expression.

19. $|-8.5|$

20. $|3.4|$

21. $-|-8.5|$

22. $|-9.6|$

In Exercises 23–26, place the correct symbol (<, >, or =) between the pair of real numbers.

23. $|-84|$ $|84|$

24. $|-10|$ $|4|$

25. $\left|\frac{3}{10}\right|$ $-\left|\frac{4}{5}\right|$

26. $|2.3|$ $-|2.3|$

1.2 Adding and Subtracting Integers

1 Add integers using a number line.

In Exercises 27–30, find the sum and demonstrate the addition on the real number line.

27. $4 + 3$

28. $15 + (-6)$

29. $-1 + (-4)$

30. $-6 + (-2)$

2 Add integers with like signs and with unlike signs.

In Exercises 31–40, find the sum.

31. $16 + (-5)$

32. $25 + (-10)$

33. $-125 + 30$

34. $-54 + 12$

35. $-13 + (-76)$

36. $-24 + (-25)$

37. $-10 + 21 + (-6)$

38. $-23 + 4 + (-11)$

39. $-17 + (-3) + (-9)$

40. $-16 + (-2) + (-8)$

41. *Profit*   A small software company had a profit of $95,000 in January, a loss of $64,400 in February, and a profit of $51,800 in March. What was the company's overall profit (or loss) for the three months?

42. *Account Balance*   At the beginning of a month, your account balance was $3090. During the month, you withdrew $870 and $465, deposited $109, and earned interest of $10.05. What was your balance at the end of the month?

43. *Writing* Is the sum of two integers, one negative and one positive, negative? Explain.

44. *Writing* Is the sum of two negative integers negative? Explain.

③ Subtract integers with like signs and with unlike signs.

In Exercises 45–54, find the difference.

45. $28 - 7$

46. $43 - 12$

47. $8 - 15$

48. $17 - 26$

49. $14 - (-19)$

50. $28 - (-4)$

51. $-18 - 4$

52. $-37 - 14$

53. $-12 - (-7)$

54. $-26 - (-8)$

55. Subtract -549 from 613.

56. What number must be subtracted from -83 to obtain 43?

1.3 Multiplying and Dividing Integers

① Multiply integers with like signs and with unlike signs.

In Exercises 57–68, find the product.

57. $15 \cdot 3$

58. $21 \cdot 4$

59. $-3 \cdot 24$

60. $-2 \cdot 44$

61. $6(-8)$

62. $12(-5)$

63. $(-5)(-9)$

64. $(-10)(-81)$

65. $3(-6)(3)$

66. $15(-2)(7)$

67. $(-4)(-5)(-2)$

68. $(-12)(-2)(-6)$

② Divide integers with like signs and with unlike signs.

In Exercises 69–78, perform the division, if possible. If not possible, state the reason.

69. $72 \div 8$

70. $63 \div 9$

71. $\dfrac{-72}{6}$

72. $\dfrac{-162}{9}$

73. $75 \div (-5)$

74. $48 \div (-4)$

75. $\dfrac{-52}{-4}$

76. $\dfrac{-64}{-4}$

77. $815 \div 0$

78. $135 \div 0$

79. *Automobile Maintenance* You rotate the tires on your truck, including the spare, so that all five tires are used equally. After 40,000 miles, how many miles has each tire been driven?

80. *Unit Price* At a garage sale, you buy a box of six glass canisters for a total of $78. All the canisters are of equal value. How much is each one worth?

③ Find factors and prime factors of an integer.

In Exercises 81–84, decide whether the number is prime or composite.

81. 839

82. 909

83. 1764

84. 1847

In Exercises 85–88, write the prime factorization of the number.

85. 378

86. 858

87. 1612

88. 1787

④ Represent the definitions and rules of arithmetic symbolically.

In Exercises 89–92, complete the statement using the indicated definition or rule.

89. Rule for multiplying integers with unlike signs:
$12 \times (-3) =$ ▢

90. Definition of multiplication:
$(-4) + (-4) + (-4) =$ ▢

91. Definition of absolute value: $|-7| =$ ▢

92. Rule for adding integers with unlike signs:
$-9 + 5 =$ ▢

1.4 Operations with Rational Numbers

① Rewrite fractions as equivalent fractions.

In Exercises 93–96, find the greatest common factor.

93. 54, 90

94. 154, 220

95. 63, 84, 441

96. 99, 132, 253

In Exercises 97–100, write an equivalent fraction with the indicated denominator.

97. $\dfrac{2}{3} = \dfrac{}{15}$

98. $\dfrac{3}{7} = \dfrac{}{28}$

99. $\dfrac{6}{10} = \dfrac{}{25}$

100. $\dfrac{9}{12} = \dfrac{}{16}$

2 Add and subtract fractions.

In Exercises 101–112, evaluate the expression. Write the result in simplest form.

101. $\dfrac{3}{25} + \dfrac{7}{25}$

102. $\dfrac{9}{64} + \dfrac{7}{64}$

103. $\dfrac{27}{16} - \dfrac{15}{16}$

104. $-\dfrac{5}{12} + \dfrac{1}{12}$

105. $-\dfrac{5}{9} + \dfrac{2}{3}$

106. $\dfrac{7}{15} - \dfrac{2}{25}$

107. $-\dfrac{25}{32} + \left(-\dfrac{7}{24}\right)$

108. $-\dfrac{7}{8} - \dfrac{11}{12}$

109. $5 - \dfrac{15}{4}$

110. $\dfrac{12}{5} - 3$

111. $5\dfrac{3}{4} - 3\dfrac{5}{8}$

112. $-3\dfrac{7}{10} + 1\dfrac{1}{20}$

113. *Meteorology* The table shows the amount of rainfall (in inches) during a five-day period. What was the total amount of rainfall for the five days?

Day	Mon	Tue	Wed	Thu	Fri
Rainfall (in inches)	$\frac{3}{8}$	$\frac{1}{2}$	$\frac{1}{8}$	$1\frac{1}{4}$	$\frac{1}{2}$

114. *Fuel Consumption* The morning and evening readings of the fuel gauge on a car were $\frac{7}{8}$ and $\frac{1}{3}$, respectively. What fraction of the tank of fuel was used that day?

3 Multiply and divide fractions.

In Exercises 115–126, evaluate the expression and write the result in simplest form. If it is not possible, explain why.

115. $\dfrac{5}{8} \cdot \dfrac{-2}{15}$

116. $\dfrac{3}{32} \cdot \dfrac{32}{3}$

117. $35\left(\frac{1}{35}\right)$

118. $(-6)\left(\frac{5}{36}\right)$

119. $\frac{3}{8}\left(-\frac{2}{27}\right)$

120. $-\frac{5}{12}\left(-\frac{4}{25}\right)$

121. $\frac{5}{14} \div \frac{15}{28}$

122. $-\frac{7}{10} \div \frac{4}{15}$

123. $\left(-\frac{3}{4}\right) \div \left(-\frac{7}{8}\right)$

124. $\frac{15}{32} \div \left(-\frac{5}{4}\right)$

125. $-\frac{5}{9} \div 0$

126. $0 \div \frac{1}{12}$

127. *Meteorology* During an eight-hour period, $6\frac{3}{4}$ inches of snow fell. What was the average rate of snowfall per hour?

128. *Sports* In three strokes on the golf course, you hit your ball a total distance of $64\frac{7}{8}$ meters. What is your average distance per stroke?

4 Add, subtract, multiply, and divide decimals.

In Exercises 129–136, evaluate the expression. Round your answer to two decimal places.

129. $4.89 + 0.76$

130. $1.29 + 0.44$

131. $3.815 - 5.19$

132. $7.234 - 8.16$

133. $(1.49)(-0.5)$

134. $(2.34)(-1.2)$

135. $5.25 \div 0.25$

136. $10.18 \div 1.6$

137. *Consumer Awareness* A telephone company charges \$0.64 for the first minute and \$0.72 for each additional minute. Find the cost of a five-minute call.

138. *Consumer Awareness* A television costs \$120.75 plus \$27.56 each month for 18 months. Find the total cost of the television.

1.5 Exponents, Order of Operations, and Properties of Real Numbers

1 Rewrite repeated multiplication in exponential form and evaluate exponential expressions.

In Exercises 139 and 140, rewrite in exponential form.

139. $6 \cdot 6 \cdot 6 \cdot 6 \cdot 6$

140. $(-3) \cdot (-3) \cdot (-3)$

In Exercises 141 and 142, rewrite as a product.

141. $(-7)^4$

142. $\left(\frac{1}{2}\right)^5$

In Exercises 143–146, evaluate the expression.

143. 2^4

144. $(-6)^2$

145. $\left(-\frac{3}{4}\right)^3$

146. $\left(\frac{2}{3}\right)^2$

2 Evaluate expressions using order of operations.

In Exercises 147–166, evaluate the expression. Write fractional answers in simplest form.

147. $12 - 2 \cdot 3$

148. $1 + 7 \cdot 3 - 10$

149. $18 \div 6 \cdot 7$

150. $3^2 \cdot 4 \div 2$

151. $20 + (8^2 \div 2)$

152. $(8 - 3) \div 15$

153. $240 - (4^2 \cdot 5)$ **154.** $5^2 - (625 \cdot 5^2)$

155. $3^2(5 - 2)^2$ **156.** $-5(10 - 7)^3$

157. $\left(\frac{3}{4}\right)\left(\frac{5}{6}\right) + 4$ **158.** $75 - 24 \div 2^3$

159. $122 - [45 - (32 + 8) - 23]$

160. $-58 - (48 - 12) - (-30 - 4)$

161. $\dfrac{6 \cdot 4 - 36}{4}$ **162.** $\dfrac{144}{2 \cdot 3 \cdot 3}$

163. $\dfrac{54 - 4 \cdot 3}{6}$ **164.** $\dfrac{3 \cdot 5 + 125}{10}$

165. $\dfrac{78 - |-78|}{5}$ **166.** $\dfrac{300}{15 - |-15|}$

In Exercises 167–170, use a calculator to evaluate the expression. Round your answer to two decimal places.

167. $(5.8)^4 - (3.2)^5$ **168.** $\dfrac{(15.8)^3}{(2.3)^8}$

169. $\dfrac{3000}{(1.05)^{10}}$ **170.** $500\left(1 + \dfrac{0.07}{4}\right)^{40}$

171. *Depreciation*   After 3 years, the value of a $16,000 car is given by $16{,}000\left(\frac{3}{4}\right)^3$.

(a) What is the value of the car after 3 years?

(b) How much has the car depreciated during the 3 years?

172. ▲ *Geometry*   The volume of water in a hot tub is given by $V = 6^2 \cdot 3$ (see figure). How many cubic feet of water will the hot tub hold? Find the total weight of the water in the tub. (Use the fact that 1 cubic foot of water weighs 62.4 pounds.)

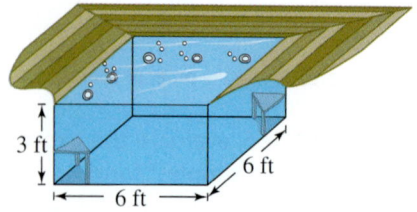

3 ft   6 ft   6 ft

3 Identify and use the properties of real numbers.

In Exercises 173–180, identify the property of real numbers illustrated by the statement.

173. $123 - 123 = 0$

174. $9 \cdot \frac{1}{9} = 1$

175. $14(3) = 3(14)$

176. $5(3x) = (5 \cdot 3)x$

177. $17 \cdot 1 = 17$

178. $10 + 6 = 6 + 10$

179. $-2(7 + x) = -2 \cdot 7 + (-2)x$

180. $2 + (3 + x) = (2 + 3) + x$

In Exercises 181–184, complete the statement using the specified property of real numbers.

181. Additive Identity Property:

$(z + 1) + 0 =$

182. Distributive Property:

$8(x + 2) =$

183. Commutative Property of Addition:

$2y + 1 =$

184. Associative Property of Multiplication:

$9(4x) =$

185. ▲ *Geometry*   Find the area of the shaded rectangle in two ways. Explain how the results are related to the Distributive Property.

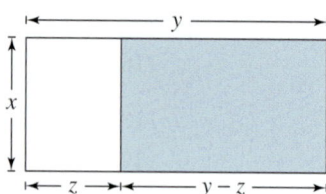

y x z $y - z$

Chapter Test

Take this test as you would take a test in class. After you are done, check your work against the answers in the back of the book.

1. Which of the following are (a) natural numbers, (b) integers, and (c) rational numbers?
$$\left\{4, -6, \tfrac{1}{2}, 0, \pi, \tfrac{7}{9}\right\}$$

2. Place the correct inequality symbol ($<$ or $>$) between the real numbers.

$$-\frac{3}{5} \qquad -|-2|$$

In Exercises 3–18, evaluate the expression. Write fractional answers in simplest form.

3. $16 + (-20)$

4. $-50 - (-60)$

5. $7 + |-3|$

6. $64 - (25 - 8)$

7. $-5(32)$

8. $\dfrac{-72}{-9}$

9. $\dfrac{-15 + 6}{3}$

10. $-\dfrac{(-2)(5)}{10}$

11. $\frac{5}{6} - \frac{1}{8}$

12. $\left(-\frac{9}{50}\right)\left(-\frac{20}{27}\right)$

13. $\dfrac{7}{16} \div \dfrac{21}{28}$

14. $\dfrac{-8.1}{0.3}$

15. $-(0.8)^2$

16. $35 - (50 \div 5^2)$

17. $5(3 + 4)^2 - 10$

18. $18 - 7 \cdot 4 + 2^3$

In Exercises 19–22, identify the property of real numbers illustrated by the statement.

19. $3(4 + 6) = 3 \cdot 4 + 3 \cdot 6$

20. $5 \cdot \frac{1}{5} = 1$

21. $3 + (4 + 8) = (3 + 4) + 8$

22. $3(x + 2) = (x + 2)3$

23. Write the fraction $\frac{36}{162}$ in simplest form.

24. Write the prime factorization of 324.

25. A jogger runs a race that is 8 miles long in 58 minutes. What is the average speed of the jogger in minutes per mile?

26. At the grocery store, you buy two cartons of eggs at $1.59 a carton and three bottles of soda at $1.50 a bottle. You give the clerk a 20-dollar bill. How much change will you receive? (Assume there is no sales tax.)

Motivating the Chapter

Beachwood Rental

Beachwood Rental is a rental company specializing in equipment for parties and special events. A wedding ceremony is to be held under a canopy that contains 15 rows of 12 chairs.

See Section 2.1, Exercise 91.

a. Let c represent the rental cost of a chair. Write an expression that represents the cost of renting all of the chairs under the canopy. The table at the right lists the rental prices for two types of chairs. Use the expression you wrote to find the cost of renting the plastic chairs and the cost of renting the wood chairs.

Chair rental	
Plastic	$1.95
Wood	$2.95

b. The table at the right lists the available canopy sizes. The rental rate for a canopy is $115 + 0.25t$ dollars, where t represents the size of the canopy in square feet. Find the cost of each canopy. (*Hint:* The total area under a 20 by 20 foot canopy is $20 \cdot 20 = 400$ square feet.)

Canopy sizes	
Canopy 1	20 by 20 feet
Canopy 2	20 by 30 feet
Canopy 3	30 by 40 feet
Canopy 4	30 by 60 feet
Canopy 5	40 by 60 feet

The figure at the right shows the arrangement of the chairs under the canopy. Beachwood Rental recommends the following.

Width of center aisle—Three times the space between rows
Width of side aisle—Two times the space between rows
Depth of rear aisle—Two times the space between rows
Depth of front region—Seven feet more than three times the space between rows

See Section 2.3, Exercise 85.

c. Let x represent the space (in feet) between rows of chairs. Write an expression for the width of the center aisle. Write an expression for the width of a side aisle.

d. Each chair is 14 inches wide. Convert the width of a chair to feet. Write an expression for the width of the canopy.

e. Write an expression for the depth of the rear aisle. Write an expression for the depth of the front region.

f. Each chair is 12 inches deep. Convert the depth of a chair to feet. Write an expression for the depth of the canopy.

g. When $x = 2$ feet, what is the width of the center aisle? What are the width and depth of the canopy? What size canopy do you need? What is the total rental cost of the canopy and chairs if the wood chairs are used?

h. What could be done to save on the rental cost?

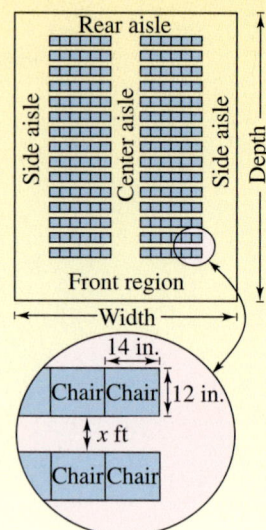

Mark Gibson/Unicorn Stock Photos

Fundamentals of Algebra

2.1 Writing and Evaluating Algebraic Expressions

Rubberball Production/Getty Images

What You Should Learn

1 Define and identify terms, variables, and coefficients of algebraic expressions.

2 Define exponential form and interpret exponential expressions.

3 Evaluate algebraic expressions using real numbers.

Why You Should Learn It

Algebraic expressions can be used to represent real-life quantities, such as weekly income from a part-time job. See Example 1.

Variables and Algebraic Expressions

One of the distinguishing characteristics of algebra is its use of symbols to represent quantities whose numerical values are unknown. Here is a simple example.

Example 1   Writing an Algebraic Expression

You accept a part-time job for $7 per hour. The job offer states that you will be expected to work between 15 and 30 hours a week. Because you don't know how many hours you will work during a week, your total income for a week is unknown. Moreover, your income will probably *vary* from week to week. By representing the variable quantity (the number of hours worked) by the letter x, you can represent the weekly income by the following algebraic expression.

$7 per   Number of
hour   hours worked

$$7x$$

In the product $7x$, the number 7 is a *constant* and the letter x is a *variable*.

1 Define and identify terms, variables, and coefficients of algebraic expressions.

Definition of Algebraic Expression

A collection of letters (**variables**) and real numbers (**constants**) combined by using addition, subtraction, multiplication, or division is an **algebraic expression.**

Some examples of algebraic expressions are

$$3x + y, \quad -5a^3, \quad 2W - 7, \quad \frac{x}{y + 3}, \quad \text{and} \quad x^2 - 4x + 5.$$

The **terms** of an algebraic expression are those parts that are separated by *addition*. For example, the expression $x^2 - 4x + 5$ has three terms: x^2, $-4x$, and 5. Note that $-4x$, rather than $4x$, is a term of $x^2 - 4x + 5$ because

$$x^2 - 4x + 5 = x^2 + (-4x) + 5. \qquad \text{To subtract, add the opposite.}$$

For variable terms such as x^2 and $-4x$, the numerical factor is the **coefficient** of the term. Here, the coefficient of x^2 is 1 and the coefficient of $-4x$ is -4.

Example 2   Identifying the Terms of an Algebraic Expression

Identify the terms of each algebraic expression.

a. $x + 2$

b. $3x + \dfrac{1}{2}$

c. $2y - 5x - 7$

d. $5(x - 3) + 3x - 4$

e. $4 - 6x + \dfrac{x + 9}{3}$

Solution

Algebraic Expression	Terms
a. $x + 2$	$x, 2$
b. $3x + \dfrac{1}{2}$	$3x, \dfrac{1}{2}$
c. $2y - 5x - 7$	$2y, -5x, -7$
d. $5(x - 3) + 3x - 4$	$5(x - 3), 3x, -4$
e. $4 - 6x + \dfrac{x + 9}{3}$	$4, -6x, \dfrac{x + 9}{3}$

The terms of an algebraic expression depend on the way the expression is written. Rewriting the expression can (and, in fact, usually does) change its terms. For instance, the expression $2 + 4 - x$ has three terms, but the equivalent expression $6 - x$ has only two terms.

Example 3   Identifying Coefficients

Identify the coefficient of each term.

a. $-5x^2$

b. x^3

c. $\dfrac{2x}{3}$

d. $-\dfrac{x}{4}$

e. $-x^3$

Solution

Term	Coefficient	Comment
a. $-5x^2$	-5	Note that $-5x^2 = (-5)x^2$.
b. x^3	1	Note that $x^3 = 1 \cdot x^3$.
c. $\dfrac{2x}{3}$	$\dfrac{2}{3}$	Note that $\dfrac{2x}{3} = \dfrac{2}{3}(x)$.
d. $-\dfrac{x}{4}$	$-\dfrac{1}{4}$	Note that $-\dfrac{x}{4} = -\dfrac{1}{4}(x)$.
e. $-x^3$	-1	Note that $-x^3 = (-1)x^3$.

2 Define exponential form and interpret exponential expressions.

Exponential Form

You know from Section 1.5 that a number raised to a power can be evaluated by repeated multiplication. For example, 7^4 represents the product obtained by multiplying 7 by itself four times.

$$\overset{\text{Exponent}}{7^{\underset{\underset{\text{Base}}{|}}{4}} = \underbrace{7 \cdot 7 \cdot 7 \cdot 7}_{\text{4 factors}}}$$

In general, for any positive integer n and any real number a, you have

$$a^n = \underbrace{a \cdot a \cdot a \cdots a}_{n \text{ factors}}.$$

This rule applies to factors that are variables as well as to factors that are *algebraic expressions.*

Study Tip

Be sure you understand the difference between repeated addition

$$\underbrace{x + x + x + x}_{\text{4 terms}} = 4x$$

and repeated multiplication

$$\underbrace{x \cdot x \cdot x \cdot x}_{\text{4 factors}} = x^4.$$

Definition of Exponential Form

Let n be a positive integer and let a be a real number, a variable, or an algebraic expression.

$$a^n = \underbrace{a \cdot a \cdot a \cdots a}_{n \text{ factors}}$$

In this definition, remember that the letter a can be a number, a variable, or an algebraic expression. It may be helpful to think of a as a box into which you can place any algebraic expression.

$$\boxed{}^{\,n} = \boxed{} \cdot \boxed{} \cdots \boxed{}$$

The box may contain a number, a variable, or an algebraic expression.

Example 4   Interpreting Exponential Expressions

a. $3^4 = 3 \cdot 3 \cdot 3 \cdot 3$ **b.** $3x^4 = 3 \cdot x \cdot x \cdot x \cdot x$

c. $(-3x)^4 = (-3x)(-3x)(-3x)(-3x) = (-3)(-3)(-3)(-3) \cdot x \cdot x \cdot x \cdot x$

d. $(y + 2)^3 = (y + 2)(y + 2)(y + 2)$

e. $(5x)^2 y^3 = (5x)(5x)y \cdot y \cdot y = 5 \cdot 5 \cdot x \cdot x \cdot y \cdot y \cdot y$

Be sure you understand the priorities for order of operations involving exponents. Here are some examples that tend to cause problems.

Expression	Correct Evaluation	Incorrect Evaluation
-3^2	$-(3 \cdot 3) = -9$	$~~(-3)(-3) = 9~~$
$(-3)^2$	$(-3)(-3) = 9$	$~~-(3 \cdot 3) = -9~~$
$3x^2$	$3 \cdot x \cdot x$	$~~(3x)(3x)~~$
$-3x^2$	$-3 \cdot x \cdot x$	$~~-(3x)(3x)~~$
$(-3x)^2$	$(-3x)(-3x)$	$~~-(3x)(3x)~~$

3 Evaluate algebraic expressions using real numbers.

Evaluating Algebraic Expressions

In applications of algebra, you are often required to **evaluate** an algebraic expression. This means you are to find the value of an expression when its variables are substituted by real numbers. For instance, when $x = 2$, the value of the expression $2x + 3$ is as follows.

Expression	Substitute 2 for x.	Value of Expression
$2x + 3$	$2(2) + 3$	7

When finding the value of an algebraic expression, be sure to replace every occurrence of the specified variable with the appropriate real number. For instance, when $x = -2$, the value of $x^2 - x + 3$ is

$$(-2)^2 - (-2) + 3 = 4 + 2 + 3 = 9.$$

Example 5   Evaluating Algebraic Expressions

Evaluate each expression when $x = -3$ and $y = 5$.

a. $-x$

b. $x - y$

c. $3x + 2y$

d. $y - 2(x + y)$

e. $y^2 - 3y$

Solution

a. When $x = -3$, the value of $-x$ is

$$-x = -(-3) \qquad \text{Substitute } -3 \text{ for } x.$$
$$= 3. \qquad \text{Simplify.}$$

b. When $x = -3$ and $y = 5$, the value of $x - y$ is

$$x - y = -3 - 5 \qquad \text{Substitute } -3 \text{ for } x \text{ and } 5 \text{ for } y.$$
$$= -8. \qquad \text{Simplify.}$$

c. When $x = -3$ and $y = 5$, the value of $3x + 2y$ is

$$3x + 2y = 3(-3) + 2(5) \qquad \text{Substitute } -3 \text{ for } x \text{ and } 5 \text{ for } y.$$
$$= -9 + 10 \qquad \text{Simplify.}$$
$$= 1. \qquad \text{Simplify.}$$

d. When $x = -3$ and $y = 5$, the value of $y - 2(x + y)$ is

$$y - 2(x + y) = 5 - 2[(-3) + 5] \qquad \text{Substitute } -3 \text{ for } x \text{ and } 5 \text{ for } y.$$
$$= 5 - 2(2) \qquad \text{Simplify.}$$
$$= 1. \qquad \text{Simplify.}$$

e. When $y = 5$, the value of $y^2 - 3y$ is

$$y^2 - 3y = (5)^2 - 3(5) \qquad \text{Substitute } 5 \text{ for } y.$$
$$= 25 - 15 \qquad \text{Simplify.}$$
$$= 10. \qquad \text{Simplify.}$$

Study Tip

As shown in parts (a), (c), and (d) of Example 5, it is a good idea to use parentheses when substituting a negative number for a variable.

Example 6   Evaluating Algebraic Expressions

Evaluate each expression when $x = 4$ and $y = -6$.

a. y^2 **b.** $-y^2$ **c.** $y - x$ **d.** $|y - x|$ **e.** $|x - y|$

Solution

a. When $y = -6$, the value of the expression y^2 is
$$y^2 = (-6)^2 = 36.$$

b. When $y = -6$, the value of the expression $-y^2$ is
$$-y^2 = -(y^2) = -(-6)^2 = -36.$$

c. When $x = 4$ and $y = -6$, the value of the expression $y - x$ is
$$y - x = (-6) - 4 = -6 - 4 = -10.$$

d. When $x = 4$ and $y = -6$, the value of the expression $|y - x|$ is
$$|y - x| = |-6 - 4| = |-10| = 10.$$

e. When $x = 4$ and $y = -6$, the value of the expression $|x - y|$ is
$$|x - y| = |4 - (-6)| = |4 + 6| = |10| = 10.$$

Example 7   Evaluating Algebraic Expressions

Evaluate each expression when $x = -5$, $y = -2$, and $z = 3$.

a. $\dfrac{y + 2z}{5y - xz}$

b. $(y + 2z)(z - 3y)$

Solution

a. When $x = -5$, $y = -2$, and $z = 3$, the value of the expression is

$$\frac{y + 2z}{5y - xz} = \frac{-2 + 2(3)}{5(-2) - (-5)(3)} \qquad \text{Substitute for } x, y, \text{ and } z.$$

$$= \frac{-2 + 6}{-10 + 15} \qquad \text{Simplify.}$$

$$= \frac{4}{5}. \qquad \text{Simplify.}$$

b. When $y = -2$ and $z = 3$, the value of the expression is

$$(y + 2z)(z - 3y) = [(-2) + 2(3)][3 - 3(-2)] \qquad \text{Substitute for } y \text{ and } z.$$

$$= (-2 + 6)(3 + 6) \qquad \text{Simplify.}$$

$$= (4)(9) \qquad \text{Simplify.}$$

$$= 36. \qquad \text{Simplify.}$$

Technology: Tip

If you have a graphing calculator, try using it to store and evaluate the expression from Example 8. You can use the following steps to evaluate $-9x + 6$ when $x = 2$.

• Store the expression as Y_1.
• Store 2 in X.

  2 [STO▸] [X,T,Θ,n] [ENTER]

• Display Y_1.

  [VARS] [Y-VARS] [ENTER]
  [ENTER]

  and then press [ENTER] again.

On occasion you may need to evaluate an algebraic expression for *several* values of x. In such cases, a table format is a useful way to organize the values of the expression.

Example 8   Repeated Evaluation of an Expression

Complete the table by evaluating the expression $5x + 2$ for each value of x shown in the table.

x	-1	0	1	2
$5x + 2$				

Solution

Begin by substituting each value of x into the expression.

When $x = -1$: $5x + 2 = 5(-1) + 2 = -5 + 2 = -3$
When $x = 0$: $5x + 2 = 5(0) + 2 = 0 + 2 = 2$
When $x = 1$: $5x + 2 = 5(1) + 2 = 5 + 2 = 7$
When $x = 2$: $5x + 2 = 5(2) + 2 = 10 + 2 = 12$

Once you have evaluated the expression for each value of x, fill in the table with the values.

x	-1	0	1	2
$5x + 2$	-3	2	7	12

Example 9   Geometry: Area

Write an expression for the area of the rectangle shown in Figure 2.1. Then evaluate the expression to find the area of the rectangle when $x = 7$.

Solution

Area of a rectangle = Length · Width

$$= (x + 5) \cdot x \qquad \text{Substitute.}$$

To evaluate the expression when $x = 7$, substitute 7 for x in the expression for the area of the rectangle.

$$(x + 5) \cdot x = (7 + 5) \cdot 7 \qquad \text{Substitute 7 for } x.$$

$$= 12 \cdot 7 \qquad \text{Simplify.}$$

$$= 84 \qquad \text{Simplify.}$$

So, the area of the rectangle is 84 square units.

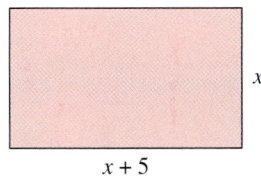

$x + 5$

Figure 2.1

2.1 Exercises

Review   Concepts, Skills, and Problem Solving

Keep mathematically in shape by doing these exercises *before* the problems of this section.

Properties and Definitions

In Exercises 1–4, identify the property of real numbers illustrated by the statement.

1. $x(5) = 5x$

2. $10 - 10 = 0$

3. $3(t + 2) = 3 \cdot t + 3 \cdot 2$

4. $7 + (8 + z) = (7 + 8) + z$

Simplifying Expressions

In Exercises 5–10, evaluate the expression.

5. $10 - |-7|$

6. $6 - (10 - 12)$

7. $\dfrac{3 - (5 - 20)}{4}$

8. $\dfrac{6}{7} - \dfrac{4}{7}$

9. $-\dfrac{3}{4}\left(\dfrac{28}{33}\right)$

10. $\dfrac{5}{8} \div \dfrac{3}{16}$

Problem Solving

11. *Savings* You plan to save $50 per month for 10 years. How much money will you set aside during the 10 years?

12. ▲ *Geometry* It is necessary to cut a 120-foot rope into eight pieces of equal length. What is the length of each piece?

Developing Skills

In Exercises 1–4, write an algebraic expression for the statement. See Example 1.

1. The distance traveled in t hours if the average speed is 60 miles per hour

2. The cost of an amusement park ride for a family of n people if the cost per person is $1.25

3. The cost of m pounds of meat if the cost per pound is $2.19

4. The total weight of x bags of fertilizer if each bag weighs 50 pounds

In Exercises 5–10, identify the variables and constants in the expression.

5. $x + 3$

6. $y + 1$

7. $x + z$

8. $a + b$

9. $2^3 + x$

10. $3^2 + z$

In Exercises 11–24, identify the terms of the expression. See Example 2.

11. $4x + 3$

12. $3x^2 + 5$

13. $6x - 1$

14. $5 - 3t^2$

15. $\frac{5}{3} - 3y^3$

16. $6x - \frac{2}{3}$

17. $a^2 + 4ab + b^2$

18. $x^2 + 18xy + y^2$

19. $3(x + 5) + 10$

20. $16 - (x + 1)$

21. $15 + \dfrac{5}{x}$

22. $\dfrac{6}{t} + 22$

23. $\dfrac{3}{x + 2} - 3x + 4$

24. $\dfrac{5}{x - 5} - 7x^2 + 18$

In Exercises 25–34, identify the coefficient of the term. See Example 3.

25. $14x$

26. $25y$

27. $-\frac{1}{3}y$

28. $-\frac{2}{3}n$

29. $\dfrac{2x}{5}$

30. $\dfrac{3x}{4}$

31. $2\pi x^2$

32. πt^4

33. $-3.06u$

34. $-5.32b$

In Exercises 35–52, expand the expression as a product of factors. See Example 4.

35. y^5

36. x^6

37. $2^2 x^4$

38. $5^3 x^2$

39. $4y^2 z^3$

40. $3uv^4$

41. $(a^2)^3$

42. $(z^3)^3$

43. $4x^3 \cdot x^4$

44. $a^2 y^2 \cdot y^3$

45. $9(ab)^3$

46. $2(xz)^4$

47. $(x + y)^2$

48. $(s - t)^5$

49. $\left(\dfrac{a}{3s}\right)^4$

50. $\left(\dfrac{2}{5x}\right)^3$

51. $[2(a - b)^3][2(a - b)^2]$

52. $[3(r + s)^2][3(r + s)]^2$

In Exercises 53–62, rewrite the product in exponential form.

53. $2 \cdot u \cdot u \cdot u \cdot u$

54. $\frac{1}{3} \cdot x \cdot x \cdot x \cdot x \cdot x$

55. $(2u) \cdot (2u) \cdot (2u) \cdot (2u)$

56. $\frac{1}{3}x \cdot \frac{1}{3}x \cdot \frac{1}{3}x \cdot \frac{1}{3}x \cdot \frac{1}{3}x$

57. $a \cdot a \cdot a \cdot b \cdot b$

58. $y \cdot y \cdot z \cdot z \cdot z \cdot z$

59. $3 \cdot (x - y) \cdot (x - y) \cdot 3 \cdot 3$

60. $(u - v) \cdot (u - v) \cdot 8 \cdot 8 \cdot 8 \cdot (u - v)$

61. $\dfrac{x + y}{4} \cdot \dfrac{x + y}{4} \cdot \dfrac{x + y}{4}$

62. $\dfrac{r - s}{5} \cdot \dfrac{r - s}{5} \cdot \dfrac{r - s}{5} \cdot \dfrac{r - s}{5}$

In Exercises 63–80, evaluate the algebraic expression for the given values of the variable(s). If it is not possible, state the reason. See Examples 5, 6, and 7.

Expression	Values
63. $2x - 1$	(a) $x = \frac{1}{2}$ (b) $x = -4$
64. $3x - 2$	(a) $x = \frac{4}{3}$ (b) $x = -1$
65. $2x^2 - 5$	(a) $x = -2$
	(b) $x = 3$
66. $64 - 16t^2$	(a) $t = 2$ (b) $t = -3$
67. $3x - 2y$	(a) $x = 4, y = 3$
	(b) $x = \frac{2}{3}, y = -1$
68. $10u - 3v$	(a) $u = 3, v = 10$
	(b) $u = -2, v = \frac{4}{7}$
69. $x - 3(x - y)$	(a) $x = 3, y = 3$
	(b) $x = 4, y = -4$
70. $-3x + 2(x + y)$	(a) $x = -2, y = 2$
	(b) $x = 0, y = 5$
71. $b^2 - 4ab$	(a) $a = 2, \ b = -3$
	(b) $a = 6, b = -4$
72. $a^2 + 2ab$	(a) $a = -2, b = 3$
	(b) $a = 4, b = -2$
73. $\dfrac{x - 2y}{x + 2y}$	(a) $x = 4, y = 2$
	(b) $x = 4, y = -2$
74. $\dfrac{5x}{y - 3}$	(a) $x = 2, y = 4$
	(b) $x = 2, y = 3$
75. $\dfrac{-y}{x^2 + y^2}$	(a) $x = 0, y = 5$
	(b) $x = 1, y = -3$
76. $\dfrac{2x - y}{y^2 + 1}$	(a) $x = 1, y = 2$
	(b) $x = 1, y = 3$

77. *Area of a Triangle*

$\frac{1}{2}bh$ (a) $b = 3, h = 5$

(b) $b = 2, h = 10$

78. *Distance Traveled*

rt (a) $r = 50, t = 3.5$

(b) $r = 35, t = 4$

79. *Volume of a Rectangular Prism*

lwh (a) $l = 4, w = 2, h = 9$

(b) $l = 100, w = 0.8, h = 4$

	Expression		*Values*

80. *Simple Interest*

Prt (a) $P = 1000, r = 0.08, t = 3$

(b) $P = 500, r = 0.07, t = 5$

81. *Finding a Pattern*

(a) Complete the table by evaluating the expression $3x - 2$. See Example 8.

x	-1	0	1	2	3	4
$3x - 2$						

(b) Use the table to find the increase in the value of the expression for each one-unit increase in x.

(c) From the pattern of parts (a) and (b), predict the increase in the algebraic expression $\frac{2}{3}x + 4$ for each one-unit increase in x. Then verify your prediction.

82. *Finding a Pattern*

(a) Complete the table by evaluating the expression $3 - 2x$. See Example 8.

x	-1	0	1	2	3	4
$3 - 2x$						

(b) Use the table to find the change in the value of the expression for each one-unit increase in x.

(c) From the pattern of parts (a) and (b), predict the change in the algebraic expression $4 - \frac{3}{2}x$ for each one-unit increase in x. Then verify your prediction.

Solving Problems

▲ *Geometry* In Exercises 83–86, find an expression for the area of the figure. Then evaluate the expression for the given value(s) of the variable(s).

83. $n = 8$

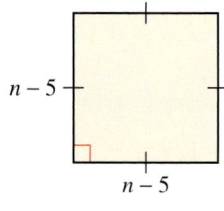

84. $x = 10, y = 3$

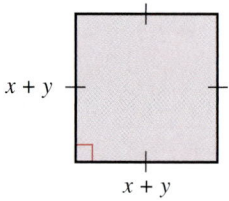

85. $a = 5, b = 4$

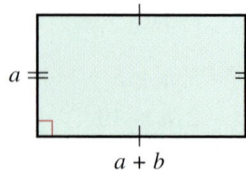

86. $x = 9$

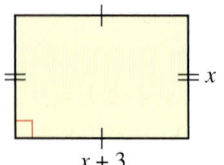

87. *Exploration* For any natural number n, the sum of the numbers $1, 2, 3, \ldots, n$ is equal to

$$\frac{n(n + 1)}{2}, \quad n \geq 1$$

Verify the formula for (a) $n = 3$, (b) $n = 6$, and (c) $n = 10$.

88. *Exploration* A convex polygon with n sides has

$$\frac{n(n - 3)}{2}, \quad n \geq 4$$

diagonals. Verify the formula for (a) a square (two diagonals), (b) a pentagon (five diagonals), and (c) a hexagon (nine diagonals).

89. ⊞ *Iteration and Exploration*   Once an expression has been evaluated for a specified value, the expression can be repeatedly evaluated by using the result of the preceding evaluation as the input for the next evaluation.

(a) The procedure for repeated evaluation of the algebraic expression $\frac{1}{2}x + 3$ can be accomplished on a graphing calculator, as follows.

- Clear the display.
- Enter 2 in the display and press ENTER.
- Enter $\frac{1}{2} *$ ANS $+ 3$ and press ENTER.
- Each time ENTER is pressed, the calculator will evaluate the expression at the value of x obtained in the preceding computation. Continue the process six more times. What value does the expression appear to be approaching? If necessary, round your answers to three decimal places.

(b) Repeat part (a) starting with $x = 12$.

90. ⊞ *Exploration*   Repeat Exercise 89 using the expression $\frac{3}{4}x + 2$. If necessary, round your answers to three decimal places.

Explaining Concepts

91. ⚙ Answer parts (a) and (b) of Motivating the Chapter on page 66.

92. *Writing* ✎   Discuss the difference between terms and factors.

93. *Writing* ✎   Is $3x$ a term of $4 - 3x$? Explain.

94. In the expression $(10x)^3$, what is $10x$ called? What is 3 called?

95. *Writing* ✎   Is it possible to evaluate the expression

$$\frac{x + 2}{y - 3}$$

when $x = 5$ and $y = 3$? Explain.

96. *Writing* ✎   Explain why the formulas in Exercises 87 and 88 will always yield natural numbers.

97. *Writing* ✎   You are teaching an algebra class and one of your students hands in the following problem. Evaluate $y - 2(x - y)$ when $x = 2$ and $y = -4$.

$$y - 2(x - y) = -4 - 2(2 - 4)$$
$$= -4 - 2(-2)$$
$$= -4 + 4$$
$$= 0$$

What is the error in this work? What are some possible related errors? Discuss ways of helping students avoid these types of errors.

The symbol ⊞ indicates an exercise in which you are instructed to use a graphing calculator.

2.2 Simplifying Algebraic Expressions

Bill Pogue/Getty Images

What You Should Learn

① Use the properties of algebra.

② Combine like terms of an algebraic expression.

③ Simplify an algebraic expression by rewriting the terms.

④ Use the Distributive Property to remove symbols of grouping.

Why You Should Learn It

You can use an algebraic expression to find the area of a house lot, as shown in Exercise 157 on page 89.

① Use the properties of algebra.

Study Tip

You'll discover as you review the table of properties at the right that they are the same as the properties of real numbers on page 52. The only difference is that the input for algebra rules can be real numbers, variables, or algebraic expressions.

Properties of Algebra

You are now ready to combine algebraic expressions using the properties below.

Properties of Algebra

Let a, b, and c represent real numbers, variables, or algebraic expressions.

Property	*Example*
Commutative Property of Addition:	
$a + b = b + a$	$3x + x^2 = x^2 + 3x$
Commutative Property of Multiplication:	
$ab = ba$	$(5 + x)x = x(5 + x)$
Associative Property of Addition:	
$(a + b) + c = a + (b + c)$	$(2x + 7) + x^2 = 2x + (7 + x^2)$
Associative Property of Multiplication:	
$(ab)c = a(bc)$	$(2x \cdot 5y) \cdot 7 = 2x \cdot (5y \cdot 7)$
Distributive Property:	
$a(b + c) = ab + ac$	$4x(7 + 3x) = 4x \cdot 7 + 4x \cdot 3x$
$(a + b)c = ac + bc$	$(2y + 5)y = 2y \cdot y + 5 \cdot y$
Additive Identity Property:	
$a + 0 = 0 + a = a$	$3y^2 + 0 = 0 + 3y^2 = 3y^2$
Multiplicative Identity Property:	
$a \cdot 1 = 1 \cdot a = a$	$(-2x^3) \cdot 1 = 1 \cdot (-2x^3) = -2x^3$
Additive Inverse Property:	
$a + (-a) = 0$	$3y^2 + (-3y^2) = 0$
Multiplicative Inverse Property:	
$a \cdot \dfrac{1}{a} = 1, \quad a \neq 0$	$(x^2 + 2) \cdot \dfrac{1}{x^2 + 2} = 1$

Example 1   Applying the Basic Rules of Algebra

Use the indicated rule to complete each statement.

a. Additive Identity Property: $(x - 2) +$ ⬚ $= x - 2$

b. Commutative Property of Multiplication: $5(y + 6) =$ ⬚

c. Commutative Property of Addition: $5(y + 6) =$ ⬚

d. Distributive Property: $5(y + 6) =$ ⬚

e. Associative Property of Addition: $(x^2 + 3) + 7 =$ ⬚

f. Additive Inverse Property: ⬚ $+ 4x = 0$

Solution

a. $(x - 2) + 0 = x - 2$

b. $5(y + 6) = (y + 6)5$

c. $5(y + 6) = 5(6 + y)$

d. $5(y + 6) = 5y + 5(6)$

e. $(x^2 + 3) + 7 = x^2 + (3 + 7)$

f. $-4x + 4x = 0$

Example 2 illustrates some common uses of the Distributive Property. Study this example carefully. Such uses of the Distributive Property are very important in algebra. Applying the Distributive Property as illustrated in Example 2 is called **expanding** an algebraic expression.

Example 2   Using the Distributive Property

Use the Distributive Property to expand each expression.

a. $2(7 - x)$ **b.** $(10 - 2y)3$ **c.** $2x(x + 4y)$ **d.** $-(1 - 2y + x)$

Solution

a. $2(7 - x) = 2 \cdot 7 - 2 \cdot x$

$= 14 - 2x$

b. $(10 - 2y)3 = 10(3) - 2y(3)$

$= 30 - 6y$

c. $2x(x + 4y) = 2x(x) + 2x(4y)$

$= 2x^2 + 8xy$

d. $-(1 - 2y + x) = (-1)(1 - 2y + x)$

$= (-1)(1) - (-1)(2y) + (-1)(x)$

$= -1 + 2y - x$

Study Tip

In Example 2(d), the negative sign is distributed over each term in the parentheses by multiplying each term by -1.

In the next example, note how area can be used to demonstrate the Distributive Property.

Example 3　The Distributive Property and Area

Write the area of each component of the figure. Then demonstrate the Distributive Property by writing the total area of each figure in two ways.

a.

	2	4
3		

$\vdash$ 2 + 4 $\dashv$

b.

	a	b
a		

$\vdash$ $a + b$ $\dashv$

c.

	d	$3a$	c
$2b$			

$\vdash$ $d + 3a + c$ $\dashv$

Solution

a.

	2	4
3	6	12

The total area is $3(2 + 4) = 3 \cdot 2 + 3 \cdot 4 = 6 + 12 = 18$.

b.

	a	b
a	a^2	ab

The total area is $a(a + b) = a \cdot a + a \cdot b = a^2 + ab$.

c.

	d	$3a$	c
$2b$	$2bd$	$6ab$	$2bc$

The total area is $2b(d + 3a + c) = 2bd + 6ab + 2bc$.

② Combine like terms of an algebraic expression.

Combining Like Terms

Two or more terms of an algebraic expression can be combined only if they are **like terms.**

> ### Definition of Like Terms
>
> In an algebraic expression, two terms are said to be **like terms** if they are both constant terms or if they have the same variable factor(s). Factors such as x in $5x$ and ab in $6ab$ are called **variable factors.**

　　The terms $5x$ and $-3x$ are like terms because they have the same variable factor, x. Similarly, $3x^2y$, $-x^2y$, and $\frac{1}{3}(x^2y)$ are like terms because they have the same variable factors, x^2 and y.

Study Tip

Notice in Example 4(b) that x^2 and $3x$ are *not* like terms because the variable x is not raised to the same power in both terms.

Example 4　Identifying Like Terms in Expressions

Expression	*Like Terms*
a. $5xy + 1 - xy$	$5xy$ and $-xy$
b. $12 - x^2 + 3x - 5$	12 and -5
c. $7x - 3 - 2x + 5$	$7x$ and $-2x$, -3 and 5

To combine like terms in an algebraic expression, you can simply add their respective coefficients and attach the common variable factor. This is actually an application of the Distributive Property, as shown in Example 5.

Example 5   Combining Like Terms

Simplify each expression by combining like terms.

a. $5x + 2x - 4$ **b.** $-5 + 8 + 7y - 5y$ **c.** $2y - 3x - 4x$

Solution

a. $5x + 2x - 4 = (5 + 2)x - 4$ Distributive Property

$\qquad\qquad\qquad = 7x - 4$ Simplest form

b. $-5 + 8 + 7y - 5y = (-5 + 8) + (7 - 5)y$ Distributive Property

$\qquad\qquad\qquad\qquad\quad = 3 + 2y$ Simplest form

c. $2y - 3x - 4x = 2y - x(3 + 4)$ Distributive Property

$\qquad\qquad\qquad\quad = 2y - x(7)$ Simplify.

$\qquad\qquad\qquad\quad = 2y - 7x$ Simplest form

Often, you need to use other rules of algebra before you can apply the Distributive Property to combine like terms. This is illustrated in the next example.

Example 6   Using Rules of Algebra to Combine Like Terms

Simplify each expression by combining like terms.

a. $7x + 3y - 4x$ **b.** $12a - 5 - 3a + 7$ **c.** $y - 4x - 7y + 9y$

Solution

a. $7x + 3y - 4x = 3y + 7x - 4x$ Commutative Property

$\qquad\qquad\qquad = 3y + (7x - 4x)$ Associative Property

$\qquad\qquad\qquad = 3y + (7 - 4)x$ Distributive Property

$\qquad\qquad\qquad = 3y + 3x$ Simplest form

b. $12a - 5 - 3a + 7 = 12a - 3a - 5 + 7$ Commutative Property

$\qquad\qquad\qquad\qquad = (12a - 3a) + (-5 + 7)$ Associative Property

$\qquad\qquad\qquad\qquad = (12 - 3)a + (-5 + 7)$ Distributive Property

$\qquad\qquad\qquad\qquad = 9a + 2$ Simplest form

c. $y - 4x - 7y + 9y = -4x + (y - 7y + 9y)$ Group like terms.

$\qquad\qquad\qquad\qquad = -4x + (1 - 7 + 9)y$ Distributive Property

$\qquad\qquad\qquad\qquad = -4x + 3y$ Simplest form

Study Tip

As you gain experience with the rules of algebra, you may want to combine some of the steps in your work. For instance, you might feel comfortable listing only the following steps to solve part (b) of Example 6.

$12a - 5 - 3a + 7$

$= (12a - 3a) + (-5 + 7)$

$= 9a + 2$

3 Simplify an algebraic expression by rewriting the terms.

Simplifying Algebraic Expressions

Simplifying an algebraic expression by rewriting it in a more usable form is one of the three most frequently used skills in algebra. You will study the other two—solving an equation and sketching the graph of an equation—later in this text.

To **simplify an algebraic expression** generally means to remove symbols of grouping and combine like terms. For instance, the expression $x + (3 + x)$ can be simplified as $2x + 3$.

Example 7   Simplifying Algebraic Expressions

Simplify each expression.

a. $-3(-5x)$ **b.** $7(-x)$

Solution

a. $-3(-5x) = (-3)(-5)x$ Associative Property

$\qquad\qquad = 15x$ Simplest form

b. $7(-x) = 7(-1)(x)$ Coefficient of $-x$ is -1.

$\qquad\qquad = -7x$ Simplest form

Example 8   Simplifying Algebraic Expressions

Simplify each expression.

a. $\dfrac{5x}{3} \cdot \dfrac{3}{5}$ **b.** $x^2(-2x^3)$ **c.** $(-2x)(4x)$ **d.** $(2rs)(r^2s)$

Solution

a. $\dfrac{5x}{3} \cdot \dfrac{3}{5} = \left(\dfrac{5}{3} \cdot x\right) \cdot \dfrac{3}{5}$ Coefficient of $\dfrac{5x}{3}$ is $\dfrac{5}{3}$.

$\qquad\qquad = \left(\dfrac{5}{3} \cdot \dfrac{3}{5}\right) \cdot x$ Commutative and Associative Properties

$\qquad\qquad = 1 \cdot x$ Multiplicative Inverse

$\qquad\qquad = x$ Multiplicative Identity

b. $x^2(-2x^3) = (-2)(x^2 \cdot x^3)$ Commutative and Associative Properties

$\qquad\qquad = -2 \cdot x \cdot x \cdot x \cdot x \cdot x$ Repeated multiplication

$\qquad\qquad = -2x^5$ Exponential form

c. $(-2x)(4x) = (-2 \cdot 4)(x \cdot x)$ Commutative and Associative Properties

$\qquad\qquad = -8x^2$ Exponential form

d. $(2rs)(r^2s) = 2(r \cdot r^2)(s \cdot s)$ Commutative and Associative Properties

$\qquad\qquad = 2 \cdot r \cdot r \cdot r \cdot s \cdot s$ Repeated multiplication

$\qquad\qquad = 2r^3s^2$ Exponential form

④ Use the Distributive Property to remove symbols of grouping.

Symbols of Grouping

The main tool for removing symbols of grouping is the Distributive Property, as illustrated in Example 9. You may want to review order of operations in Section 1.5.

Example 9   Removing Symbols of Grouping

Simplify each expression.

a. $-(2y - 7)$ **b.** $5x + (x - 7)2$

c. $-2(4x - 1) + 3x$ **d.** $3(y - 5) - (2y - 7)$

Solution

a. $-(2y - 7) = -2y + 7$ Distributive Property

b. $5x + (x - 7)2 = 5x + 2x - 14$ Distributive Property

$\qquad\qquad\qquad = 7x - 14$ Combine like terms.

c. $-2(4x - 1) + 3x = -8x + 2 + 3x$ Distributive Property

$\qquad\qquad\qquad = -8x + 3x + 2$ Commutative Property

$\qquad\qquad\qquad = -5x + 2$ Combine like terms.

d. $3(y - 5) - (2y - 7) = 3y - 15 - 2y + 7$ Distributive Property

$\qquad\qquad\qquad = (3y - 2y) + (-15 + 7)$ Group like terms.

$\qquad\qquad\qquad = y - 8$ Combine like terms.

Example 10   Removing Nested Symbols of Grouping

Simplify each expression.

a. $5x - 2[4x + 3(x - 1)]$

b. $-7y + 3[2y - (3 - 2y)] - 5y + 4$

Solution

a. $5x - 2[4x + 3(x - 1)]$

$\quad = 5x - 2[4x + 3x - 3]$ Distributive Property

$\quad = 5x - 2[7x - 3]$ Combine like terms.

$\quad = 5x - 14x + 6$ Distributive Property

$\quad = -9x + 6$ Combine like terms.

b. $-7y + 3[2y - (3 - 2y)] - 5y + 4$

$\quad = -7y + 3[2y - 3 + 2y] - 5y + 4$ Distributive Property

$\quad = -7y + 3[4y - 3] - 5y + 4$ Combine like terms.

$\quad = -7y + 12y - 9 - 5y + 4$ Distributive Property

$\quad = (-7y + 12y - 5y) + (-9 + 4)$ Group like terms.

$\quad = -5$ Combine like terms.

Study Tip

When a parenthetical expression is preceded by a *plus* sign, you can remove the parentheses without changing the signs of the terms inside.

$3y + (-2y + 7)$

$= 3y - 2y + 7$

When a parenthetical expression is preceded by a *minus* sign, however, you must change the sign of each term to remove the parentheses.

$3y - (2y - 7)$

$= 3y - 2y + 7$

Remember that $-(2y - 7)$ is equal to $(-1)(2y - 7)$, and the Distributive Property can be used to "distribute the minus sign" to obtain $-2y + 7$.

Example 11   Simplifying an Algebraic Expression

Simplify $2x(x + 3y) + 4(5 - xy)$.

Solution

$$2x(x + 3y) + 4(5 - xy) = 2x \cdot x + 6xy + 20 - 4xy \qquad \text{Distributive Property}$$

$$= 2x^2 + 6xy - 4xy + 20 \qquad \text{Commutative Property}$$

$$= 2x^2 + 2xy + 20 \qquad \text{Combine like terms.}$$

The next example illustrates the use of the Distributive Property with a fractional expression.

Example 12   Simplifying a Fractional Expression

Simplify $\dfrac{x}{4} + \dfrac{2x}{7}$.

Solution

$$\frac{x}{4} + \frac{2x}{7} = \frac{1}{4}x + \frac{2}{7}x \qquad \text{Write with fractional coefficients.}$$

$$= \left(\frac{1}{4} + \frac{2}{7}\right)x \qquad \text{Distributive Property}$$

$$= \left[\frac{1(7)}{4(7)} + \frac{2(4)}{7(4)}\right]x \qquad \text{Common denominator}$$

$$= \frac{15}{28}x \qquad \text{Simplest form}$$

Example 13   Geometry: Perimeter and Area

Using Figure 2.2, write and simplify an expression for (a) the perimeter and (b) the area of the triangle.

Solution

a. Perimeter of a Triangle = Sum of the Three Sides

$$= 2x + (2x + 4) + (x + 5) \qquad \text{Substitute.}$$

$$= (2x + 2x + x) + (4 + 5) \qquad \text{Group like terms.}$$

$$= 5x + 9 \qquad \text{Combine like terms.}$$

b. Area of a Triangle $= \frac{1}{2} \cdot$ Base $\cdot$ Height

$$= \frac{1}{2}(x + 5)(2x) \qquad \text{Substitute.}$$

$$= \frac{1}{2}(2x)(x + 5) \qquad \text{Commutative Property}$$

$$= x(x + 5) \qquad \text{Multiply.}$$

$$= x^2 + 5x \qquad \text{Distributive Property}$$

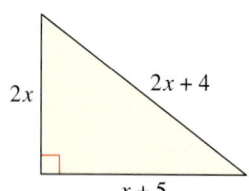

$2x$ $2x + 4$

$x + 5$

Figure 2.2

2.2 Exercises

Review   *Concepts, Skills, and Problem Solving*

Keep in mathematical shape by doing these exercises *before* the problems of this section.

Properties and Definitions

1. *Writing* ✏ Explain what it means to find the prime factorization of a number.

2. Identify the property of real numbers illustrated by the statement: $\frac{1}{2}(4x + 10) = 2x + 5$.

Simplifying Expressions

In Exercises 3–10, perform the operation.

3. $0 - (-12)$ **4.** $60 - (-60)$

5. $-12 - 2 + |-3|$

6. $-730 + 1820 + 3150 + (-10,000)$

7. Find the sum of 72 and -37.

8. Subtract 600 from 250.

9. $\frac{5}{16} - \frac{3}{10}$ **10.** $\frac{9}{16} + 2\frac{3}{12}$

Problem Solving

11. *Profit*   An athletic shoe company showed a loss of $1,530,000 during the first 6 months of 2003. The company ended the year with an overall profit of $832,000. What was the profit during the last two quarters of the year?

12. *Average Speed*   A family on vacation traveled 676 miles in 13 hours. Determine their average speed in miles per hour.

Developing Skills

In Exercises 1–22, identify the property (or properties) of algebra illustrated by the statement. See Example 1.

1. $3a + 5b = 5b + 3a$

2. $x + 2y = 2y + x$

3. $-10(xy^2) = (-10x)y^2$

4. $(9x)y = 9(xy)$

5. $rt + 0 = rt$

6. $-8x + 0 = -8x$

7. $(x^2 + y^2) \cdot 1 = x^2 + y^2$

8. $1 \cdot (5z + 12) = 5z + 12$

9. $(3x + 2y) + z = 3x + (2y + z)$

10. $-4a + (b^2 + 2c) = (-4a + b^2) + 2c$

11. $2zy = 2yz$

12. $-7a^2c = -7ca^2$

13. $-5x(y + z) = -5xy - 5xz$

14. $x(y + z) = xy + xz$

15. $(5m + 3) - (5m + 3) = 0$

16. $(2x - 10) - (2x - 10) = 0$

17. $16xy \cdot \dfrac{1}{16xy} = 1, \quad xy \neq 0$

18. $(x + y) \cdot \dfrac{1}{(x + y)} = 1, \quad x + y \neq 0$

19. $(x + 2)(x + y) = x(x + y) + 2(x + y)$

20. $(a + 6)(b + 2c) = (a + 6)b + (a + 6)2c$

21. $x^2 + (y^2 - y^2) = x^2$

22. $3y + (z^3 - z^3) = 3y$

In Exercises 23–34, complete the statement. Then state the property of algebra that you used. See Example 1.

23. $(-5r)s = -5()$

24. $(7x)y^2 = 7()$

25. $v(2) = $

26. $(2x - y)(-3) = -3$

27. $5(t - 2) = 5($ $) + 5($ $)$

28. $x(y + 4) = x($ $) + x($ $)$

29. $(2z - 3) +$ $= 0$

30. $(x + 10) +$ $= 0$

31. $5x($ $) = 1, \quad x \neq 0$

32. $4z^2($ $) = 1, z \neq 0$

33. $12 + (8 - x) =$ $- x$

34. $-11 + (5 + 2y) =$ $+ 2y$

In Exercises 35–62, use the Distributive Property to expand the expression. See Example 2.

35. $2(16 + 8z)$

36. $5(7 + 3x)$

37. $8(-3 + 5m)$

38. $12(-2 + y)$

39. $10(9 - 6x)$

40. $3(7 - 4a)$

41. $-8(2 + 5t)$

42. $-9(4 + 2b)$

43. $-5(2x - y)$

44. $-3(11y - 4)$

45. $(x + 2)(3)$

46. $(r + 12)(2)$

47. $(4 - t)(-6)$

48. $(3 - x)(-5)$

49. $4(x + xy + y^2)$

50. $6(r - t + s)$

51. $3(x^2 + x)$

52. $9(a^2 + a)$

53. $4(2y^2 - y)$

54. $5(3x^2 - x)$

55. $-z(5 - 2z)$

56. $-t(12 - 4t)$

57. $-4y(3y - 4)$

58. $-6s(6s - 1)$

59. $-(u - v)$

60. $-(x + y)$

61. $x(3x - 4y)$

62. $r(2r^2 - t)$

In Exercises 63–66, write the area of each component of the figure. Then demonstrate the Distributive Property by writing the total area of each figure in two ways. See Example 3.

63.

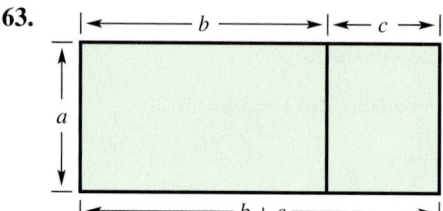

64.

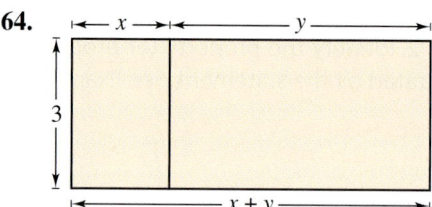

65.

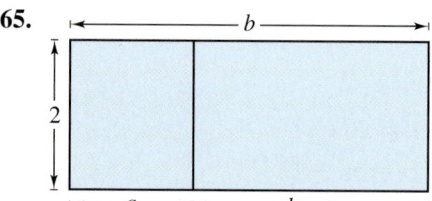

66.

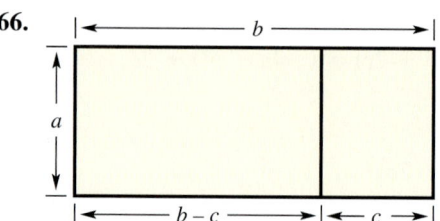

In Exercises 67–70, identify the terms of the expression and the coefficient of each term.

67. $6x^2 - 3xy + y^2$

68. $4a^2 - 9ab + b^2$

69. $-ab + 5ac - 7bc$

70. $-4xy + 2xz - yz$

In Exercises 71–76, identify the like terms. See Example 4.

71. $16t^3 + 4t - 5t + 3t^3$

72. $-\frac{1}{4}x^2 - 3x + \frac{3}{4}x^2 + x$

73. $4rs^2 - 2r^2s + 12rs^2$

74. $6x^2y + 2xy - 4x^2y$

75. $x^3 + 4x^2y - 2y^2 + 5xy^2 + 10x^2y + 3x^3$

76. $a^2 + 5ab^2 - 3b^2 + 7a^2b - ab^2 + a^2$

In Exercises 77–96, simplify the expression by combining like terms. See Examples 5 and 6.

77. $3y - 5y$

78. $-16x + 25x$

79. $x + 5 - 3x$

80. $7s + 3 - 3s$

81. $2x + 9x + 4$

82. $10x - 4 - 5x$

83. $5r + 6 - 2r + 1$

84. $2t - 4 + 8t + 9$

85. $x^2 - 2xy + 4 + xy$

86. $r^2 + 3rs - 6 - rs$

87. $5z - 5 + 10z + 2z + 16$

88. $7x - 4x + 8 + 3x - 6$

89. $z^3 + 2z^2 + z + z^2 + 2z + 1$

90. $3x^2 - x^2 + 4x + 3x^2 - x + x^2$

91. $2x^2y + 5xy^2 - 3x^2y + 4xy + 7xy^2$

92. $6rt - 3r^2t + 2rt^2 - 4rt - 2r^2t$

93. $3\left(\dfrac{1}{x}\right) - \dfrac{1}{x} + 8$

94. $1.2\left(\dfrac{1}{x}\right) + 3.8\left(\dfrac{1}{x}\right) - 4x$

95. $5\left(\dfrac{1}{t}\right) + 6\left(\dfrac{1}{t}\right) - 2t$

96. $16\left(\dfrac{a}{b}\right) - 6\left(\dfrac{a}{b}\right) + \dfrac{3}{2} - \dfrac{1}{2}$

True or False? In Exercises 97–100, determine whether the statement is true or false. Justify your answer.

97. $3(x - 4) \overset{?}{=} 3x - 4$

98. $-3(x - 4) \overset{?}{=} -3x - 12$

99. $6x - 4x \overset{?}{=} 2x$

100. $12y^2 + 3y^2 \overset{?}{=} 36y^2$

Mental Math In Exercises 101–108, use the Distributive Property to perform the required arithmetic *mentally*. For example, you work as a mechanic where the wage is $14 per hour and time-and-one-half for overtime. So, your hourly wage for overtime is

$$14(1.5) = 14\left(1 + \tfrac{1}{2}\right) = 14 + 7 = \$21.$$

101. $8(52) = 8(50 + 2)$

102. $7(33) = 7(30 + 3)$

103. $9(48) = 9(50 - 2)$

104. $6(29) = 6(30 - 1)$

105. $-4(59) = -4(60 - 1)$

106. $-6(28) = -6(30 - 2)$

107. $5(7.98) = 5(8 - 0.02)$

108. $12(11.95) = 12(12 - 0.05)$

In Exercises 109–122, simplify the expression. See Examples 7 and 8.

109. $2(6x)$

110. $7(5a)$

111. $-(4x)$

112. $-(5t)$

113. $(-2x)(-3x)$

114. $-4(-3y)$

115. $(-5z)(2z^2)$

116. $(10t)(-4t^2)$

117. $\dfrac{18a}{5} \cdot \dfrac{15}{6}$

118. $\dfrac{5x}{8} \cdot \dfrac{16}{5}$

119. $\left(-\dfrac{3x^2}{2}\right)\left(\dfrac{4x}{2}\right)$

120. $\left(\dfrac{4x}{3}\right)\left(\dfrac{3x}{2}\right)$

121. $(12xy^2)(-2x^3y^2)$

122. $(7r^2s^3)(3rs)$

In Exercises 123–142, simplify the expression by removing symbols of grouping and combining like terms. See Examples 9, 10, and 11.

123. $2(x - 2) + 4$

124. $3(x - 5) - 2$

125. $6(2s - 1) + s + 4$

126. $(2x - 1)(2) + x$

127. $m - 3(m - 5)$

128. $5l - 6(3l - 5)$

129. $-6(1 - 2x) + 10(5 - x)$

130. $3(r - 2s) - 5(3r - 5s)$

131. $\frac{2}{3}(12x + 15) + 16$

132. $\frac{3}{8}(4 - y) - \frac{5}{2} + 10$

133. $3 - 2[6 + (4 - x)]$

134. $10x + 5[6 - (2x + 3)]$

135. $7x(2 - x) - 4x$

136. $-6x(x - 1) + x^2$

137. $4x^2 + x(5 - x)$

138. $-z(z - 2) + 3z^2 + 5$

139. $-3t(4 - t) + t(t + 1)$

140. $-2x(x - 1) + x(3x - 2)$

141. $3t[4 - (t - 3)] + t(t + 5)$

142. $4y[5 - (y + 1)] + 3y(y + 1)$

In Exercises 143–150, use the Distributive Property to simplify the expression. See Example 12.

143. $\dfrac{2x}{3} - \dfrac{x}{3}$

144. $\dfrac{7y}{8} - \dfrac{3y}{8}$

145. $\dfrac{4z}{5} + \dfrac{3z}{5}$

146. $\dfrac{5t}{12} + \dfrac{7t}{12}$

147. $\dfrac{x}{3} - \dfrac{5x}{4}$

148. $\dfrac{5x}{7} + \dfrac{2x}{3}$

149. $\dfrac{3x}{10} - \dfrac{x}{10} + \dfrac{4x}{5}$

150. $\dfrac{3z}{4} - \dfrac{z}{2} - \dfrac{z}{3}$

Solving Problems

 Geometry In Exercises 151 and 152, write an expression for the perimeter of the triangle shown in the figure. Use the properties of algebra to simplify the expression.

151.

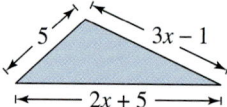

152.

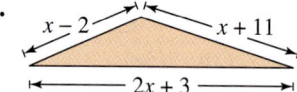

 Geometry In Exercises 153 and 154, write and simplify an expression for (a) the perimeter and (b) the area of the rectangle.

153.

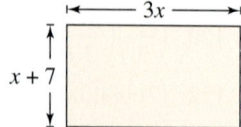

154.

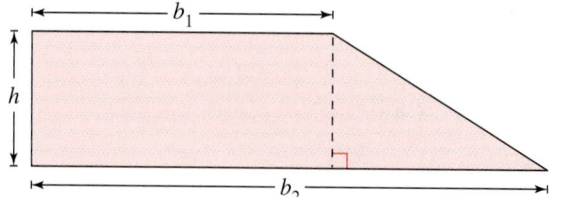

155. *Geometry* The area of a trapezoid with parallel bases of lengths b_1 and b_2 and height h is $\frac{1}{2}h(b_1 + b_2)$ (see figure).

(a) Show that the area can also be expressed as $b_1h + \frac{1}{2}(b_2 - b_1)h$, and give a geometric explanation for the area represented by each term in this expression.

(b) Find the area of a trapezoid with $b_1 = 7$, $b_2 = 12$, and $h = 3$.

156. ▲ *Geometry* The remaining area of a square with side length x after a smaller square with side length y has been removed (see figure) is $(x + y)(x - y)$.

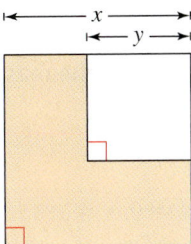

(a) Show that the remaining area can also be expressed as $x(x - y) + y(x - y)$, and give a geometric explanation for the area represented by each term in this expression.

(b) Find the remaining area of a square with side length 9 after a square with side length 5 has been removed.

▲ *Geometry* In Exercises 157 and 158, use the formula for the area of a trapezoid, $\frac{1}{2}h(b_1 + b_2)$, to find the area of the trapezoidal house lot and tile.

157.

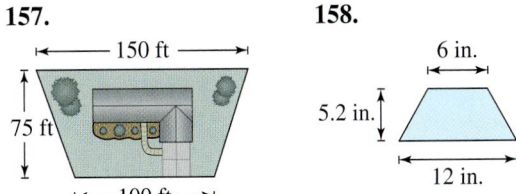

158.

6 in.

5.2 in.

12 in.

Explaining Concepts

159. *Writing* Discuss the difference between $(6x)^4$ and $6x^4$.

160. The expressions $4x$ and x^4 each represent repeated operations. What are the operations? Write the expressions showing the repeated operations.

161. *Writing* In your own words, state the definition of like terms. Give an example of like terms and an example of unlike terms.

162. *Writing* Describe how to combine like terms. What operations are used? Give an example of an expression that can be simplified by combining like terms.

Writing In Exercises 163 and 164, explain why the two expressions are not like terms.

163. $\frac{1}{2}x^2y$, $\frac{5}{2}xy^2$

164. $-16x^2y^3$, $7x^2y$

165. *Error Analysis* Describe the error.

$$\frac{x}{3} + \frac{4x}{3} = \frac{5x}{6}$$

166. *Writing* In your own words, describe the procedure for removing nested symbols of grouping.

167. *Writing* Does the expression $[x - (3 \cdot 4)] \div 5$ change if the parentheses are removed? Does it change if the brackets are removed? Explain.

168. *Writing* In your own words, describe the priorities for order of operations.

Mid-Chapter Quiz

Take this quiz as you would take a quiz in class. After you are done, check your work against the answers in the back of the book.

In Exercises 1 and 2, evaluate the algebraic expression for the specified values of the variable(s). If it is not possible, state the reason.

1. $x^2 - 3x$ (a) $x = 3$ (b) $x = -2$
              (c) $x = 0$

2. $\dfrac{x}{y - 3}$ (a) $x = 2, y = 4$ (b) $x = 0, y = -1$
             (c) $x = 5, y = 3$

In Exercises 3 and 4, identify the terms and coefficients of the expression.

3. $4x^2 - 2x$ **4.** $5x + 3y - 12z$

5. Rewrite each expression in exponential form.
    (a) $3y \cdot 3y \cdot 3y \cdot 3y$ (b) $2 \cdot (x - 3) \cdot (x - 3) \cdot 2 \cdot 2$

In Exercises 6–9, simplify the expression.

6. $-4(-5y^2)$ **7.** $\dfrac{6}{7} \cdot \dfrac{7x}{6}$ **8.** $(-3y)^2 y^3$ **9.** $\dfrac{2z^2}{3y} \cdot \dfrac{5z}{7}$

In Exercises 10–13, identify the property of algebra illustrated by the statement.

10. $-3(2y) = (-3 \cdot 2)y$ **11.** $(x + 2)y = xy + 2y$

12. $3y \cdot \dfrac{1}{3y} = 1, \quad y \neq 0$ **13.** $x - x^2 + 2 = -x^2 + x + 2$

In Exercises 14 and 15, use the Distributive Property to expand the expression.

14. $2x(3x - 1)$ **15.** $-4(2y - 3)$

In Exercises 16 and 17, simplify the expression by combining like terms.

16. $y^2 - 3xy + y + 7xy$ **17.** $10\left(\dfrac{1}{u}\right) - 7\left(\dfrac{1}{u}\right) + 3u$

In Exercises 18 and 19, simplify the expression by removing symbols of grouping and combining like terms.

18. $5(a - 2b) + 3(a + b)$ **19.** $4x + 3[2 - 4(x + 6)]$

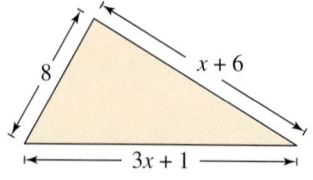

Figure for 20

20. Write and simplify an expression for the perimeter of the triangle (see figure).

21. Evaluate the expression $4 \cdot 10^4 + 5 \cdot 10^3 + 7 \cdot 10^2$.

2.3 Algebra and Problem Solving

James Marshall/The Image Works

Why You Should Learn It

Translating verbal sentences and phrases into algebraic expressions enables you to solve real-life problems. For instance, in Exercise 58 on page 102, you will find an expression for the total distance traveled by an airplane.

① Define algebra as a problem-solving language.

Study Tip

As you study this text, it is helpful to view algebra from the "big picture" as shown in Figure 2.3. The ability to write algebraic expressions and equations is needed in the major components of algebra — *simplifying* expressions, *solving* equations, and *graphing* functions.

What You Should Learn

① Define algebra as a problem-solving language.

② Construct verbal mathematical models from written statements.

③ Translate verbal phrases into algebraic expressions.

④ Identify hidden operations when constructing algebraic expressions.

⑤ Use problem-solving strategies to solve application problems.

What Is Algebra?

Algebra is a problem-solving language that is used to solve real-life problems. It has four basic components, which tend to nest within each other, as indicated in Figure 2.3.

1. Symbolic representations and applications of the rules of arithmetic

2. Rewriting (reducing, simplifying, factoring) algebraic expressions into equivalent forms

3. Creating and solving equations

4. Studying relationships among variables by the use of functions and graphs

---

1. Rules of arithmetic

2. Algebraic expressions: rewriting into equivalent forms

3. Algebraic equations: creating and solving

4. Functions and graphs: relationships among variables

Figure 2.3

Notice that one of the components deals with expressions and another deals with equations. As you study algebra, it is important to understand the difference between simplifying or rewriting an algebraic *expression,* and solving an algebraic *equation.* In general, remember that a mathematical expression *has no equal sign,* whereas a mathematical equation *must have an equal sign.*

When you use an equal sign to *rewrite* an expression, you are merely indicating the *equivalence* of the new expression and the previous one.

Original Expression	*equals*	*Equivalent Expression*
$(a + b)c$	$=$	$ac + bc$

② Construct verbal mathematical models from written statements.

Constructing Verbal Models

In the first two sections of this chapter, you studied techniques for rewriting and simplifying algebraic expressions. In this section you will study ways to construct algebraic expressions from written statements by first constructing a **verbal mathematical model.**

Take another look at Example 1 in Section 2.1 (page 68). In that example, you are paid $7 per hour and your weekly pay can be represented by the verbal model

$$\boxed{\begin{array}{c}\text{Pay per}\\\text{hour}\end{array}} \cdot \boxed{\begin{array}{c}\text{Number}\\\text{of hours}\end{array}} = \boxed{7 \text{ dollars}} \cdot \boxed{x \text{ hours}} = 7x.$$

Note the hidden operation of multiplication in this expression. Nowhere in the verbal problem does it say you are to multiply 7 times x. It is *implied* in the problem. This is often the case when algebra is used to solve real-life problems.

Example 1   Constructing an Algebraic Expression

You are paid 5¢ for each aluminum soda can and 3¢ for each glass soda bottle you collect. Write an algebraic expression that represents the total weekly income for this recycling activity.

Solution

Before writing an algebraic expression for the weekly income, it is helpful to construct an informal verbal model. For instance, the following verbal model could be used.

$$\boxed{\begin{array}{c}\text{Pay per}\\\text{can}\end{array}} \cdot \boxed{\begin{array}{c}\text{Number}\\\text{of cans}\end{array}} + \boxed{\begin{array}{c}\text{Pay per}\\\text{bottle}\end{array}} \cdot \boxed{\begin{array}{c}\text{Number of}\\\text{bottles}\end{array}}$$

Note that the word *and* in the problem indicates addition. Because both the number of cans and the number of bottles can vary from week to week, you can use the two variables c and b, respectively, to write the following algebraic expression.

$$\boxed{5 \text{ cents}} \cdot \boxed{c \text{ cans}} + \boxed{3 \text{ cents}} \cdot \boxed{b \text{ bottles}} = 5c + 3b$$

In 2000, about 1 million tons of aluminum containers were recycled. This accounted for about 45% of all aluminum containers produced. (Source: Franklin Associates, Ltd.)

In Example 1, notice that c is used to represent the number of *cans* and b is used to represent the number of *bottles*. When writing algebraic expressions, choose variables that can be identified with the unknown quantities.

The number of one kind of item can be expressed in terms of the number of another kind of item. Suppose the number of cans in Example 1 was said to be "three times the number of bottles." In this case, only one variable would be needed and the model could be written as

$$\boxed{5 \text{ cents}} \cdot \boxed{3 \cdot b \text{ cans}} + \boxed{3 \text{ cents}} \cdot \boxed{b \text{ bottles}} = 5(3b) + 3b$$

$$= 15b + 3b$$

$$= 18b.$$

③ Translate verbal phrases into algebraic expressions.

Translating Phrases

When translating verbal sentences and phrases into algebraic expressions, it is helpful to watch for key words and phrases that indicate the four different operations of arithmetic. The following list shows several examples.

Translating Phrases into Algebra Expressions

Key Words and Phrases	Verbal Description	Expression
Addition:		
Sum, plus, greater than, increased by, more than,	The sum of 6 and x	$6 + x$
exceeds, total of	Eight more than y	$y + 8$
Subtraction:		
Difference, minus, less than, decreased by,	Five decreased by a	$5 - a$
subtracted from, reduced by, the remainder	Four less than z	$z - 4$
Multiplication:		
Product, multiplied by, twice, times, percent of	Five times x	$5x$
Division:		
Quotient, divided by, ratio, per	The ratio of x and 3	$\dfrac{x}{3}$

Example 2   Translating Phrases Having Specified Variables

Translate each phrase into an algebraic expression.

a. Three less than m **b.** y decreased by 10

c. The product of 5 and x **d.** The quotient of n and 7

Solution

a. Three less than m

   $m - 3$ Think: 3 subtracted from what?

b. y decreased by 10

   $y - 10$ Think: What is subtracted from y?

c. The product of 5 and x

   $5x$ Think: 5 times what?

d. The quotient of n and 7

   $\dfrac{n}{7}$ Think: n is divided by what?

Example 3   Translating Phrases Having Specified Variables

Translate each phrase into an algebraic expression.

a. Six times the sum of x and 7

b. The product of 4 and x, divided by 3

c. k decreased by the product of 8 and m

Solution

a. Six times the sum of x and 7

$$6(x + 7)$$ Think: 6 multiplied by what?

b. The product of 4 and x, divided by 3

$$\frac{4x}{3}$$ Think: What is divided by 3?

c. k decreased by the product of 8 and m

$$k - 8m$$ Think: What is subtracted from k?

In most applications of algebra, the variables are not specified and it is your task to assign variables to the *appropriate* quantities. Although similar to the translations in Examples 2 and 3, the translations in the next example may seem more difficult because variables have not been assigned to the unknown quantities.

Example 4   Translating Phrases Having No Specified Variables

Translate each phrase into a variable expression.

a. The sum of 3 and a number

b. Five decreased by the product of 3 and a number

c. The difference of a number and 3, all divided by 12

Solution

In each case, let x be the unspecified number.

a. The sum of 3 and a number

$$3 + x$$ Think: 3 added to what?

b. Five decreased by the product of 3 and a number

$$5 - 3x$$ Think: What is subtracted from 5?

c. The difference of a number and 3, all divided by 12

$$\frac{x - 3}{12}$$ Think: What is divided by 12?

> **Study Tip**
>
> Any variable, such as b, k, n, r, or x, can be chosen to represent an unspecified number. The choice is a matter of preference. In Example 4, x was chosen as the variable.

A good way to learn algebra is to do it *forward* and *backward*. In the next example, algebraic expressions are translated into verbal form. Keep in mind that other key words could be used to describe the operations in each expression. Your goal is to use key words or phrases that keep the verbal expressions clear and concise.

Example 5   Translating Algebraic Expressions into Verbal Form

Without using a variable, write a verbal description for each expression.

a. $7x - 12$ **b.** $7(x - 12)$ **c.** $5 + \dfrac{x}{2}$ **d.** $\dfrac{5 + x}{2}$ **e.** $(3x)^2$

Solution

a. *Algebraic expression:* $7x - 12$
  Primary operation: Subtraction
  Terms: $7x$ and 12
  Verbal description: Twelve less than the product of 7 and a number

b. *Algebraic expression:* $7(x - 12)$
  Primary operation: Multiplication
  Factors: 7 and $(x - 12)$
  Verbal description: Seven times the difference of a number and 12

c. *Algebraic expression:* $5 + \dfrac{x}{2}$
  Primary operation: Addition
  Terms: 5 and $\dfrac{x}{2}$
  Verbal description: Five added to the quotient of a number and 2

d. *Algebraic expression:* $\dfrac{5 + x}{2}$
  Primary operation: Division
  Numerator, denominator: Numerator is $5 + x$; denominator is 2
  Verbal description: The sum of 5 and a number, all divided by 2

e. *Algebraic expression:* $(3x)^2$
  Primary operation: Raise to a power
  Base, power: $3x$ is the base, 2 is the power
  Verbal description: The square of the product of 3 and x

Translating algebraic expressions into verbal phrases is more difficult than it may appear. It is easy to write a phrase that is ambiguous. For instance, what does the phrase "the sum of 5 and a number times 2" mean? Without further information, this phrase could mean

$$5 + 2x \quad \text{or} \quad 2(5 + x).$$

④ Identify hidden operations when constructing algebraic expressions.

Verbal Models with Hidden Operations

Most real-life problems do not contain verbal expressions that clearly identify all the arithmetic operations involved. You need to rely on past experience and the physical nature of the problem in order to identify the operations hidden in the problem statement. Multiplication is the operation most commonly hidden in real life applications. Watch for *hidden operations* in the next two examples.

Example 6   Discovering Hidden Operations

a. A cash register contains n nickels and d dimes. Write an expression for this amount of money in cents.

b. A person riding a bicycle travels at a constant rate of 12 miles per hour. Write an expression showing how far the person can ride in t hours.

c. A person paid x dollars plus 6% sales tax for an automobile. Write an expression for the total cost of the automobile.

Solution

a. The amount of money is a sum of products.

Verbal Model:	Value of nickel	·	Number of nickels	+	Value of dime	·	Number of dimes

Labels:	Value of nickel = 5	(cents)
	Number of nickels = n	(nickels)
	Value of dime = 10	(cents)
	Number of dimes = d	(dimes)

Expression:	$5n + 10d$	(cents)

b. The distance traveled is a product.

Verbal Model:	Rate of travel · Time traveled

Labels:	Rate of travel = 12	(miles per hour)
	Time traveled = t	(hours)

Expression:	$12t$	(miles)

c. The total cost is a sum.

Verbal Model:	Cost of automobile	+	Percent of sales tax	·	Cost of automobile

Labels:	Percent of sales tax = 0.06	(decimal form)
	Cost of automobile = x	(dollars)

Expression:	$x + 0.06x = (1 + 0.06)x$
	$= 1.06x$

Study Tip

In Example 6(b), the final answer is listed in terms of miles. This makes sense as described below.

$$12\,\frac{\text{miles}}{\text{hours}} \cdot t\,\text{hours}$$

Note that the hours "divide out," leaving miles as the unit of measure. This technique is called *unit analysis* and can be very helpful in determining the final unit of measure.

Notice in part (c) of Example 6 that the equal sign is used to denote the equivalence of the three expressions. It is not an equation to be solved.

⑤ Use problem-solving strategies to solve application problems.

Additional Problem-Solving Strategies

In addition to constructing verbal models, there are other problem-solving strategies that can help you succeed in this course.

Summary of Additional Problem-Solving Strategies

1. **Guess, Check, and Revise** Guess a reasonable solution based on the given data. Check the guess, and revise it, if necessary. Continue guessing, checking, and revising until a correct solution is found.

2. **Make a Table/Look for a Pattern** Make a table using the data in the problem. Look for a number pattern. Then use the pattern to complete the table or find a solution.

3. **Draw a Diagram** Draw a diagram that shows the facts from the problem. Use the diagram to visualize the action of the problem. Use algebra to find a solution. Then check the solution against the facts.

4. **Solve a Simpler Problem** Construct a simpler problem that is similar to the original problem. Solve the simpler problem. Then use the same procedure to solve the original problem.

Study Tip

The most common errors made when solving algebraic problems are arithmetic errors. Be sure to check your arithmetic when solving algebraic problems.

Example 7   Guess, Check, and Revise

You deposit $500 in an account that earns 6% interest compounded annually. The balance in the account after t years is $A = 500(1 + 0.06)^t$. How long will it take for your investment to double?

Solution

You can solve this problem using a guess, check, and revise strategy. For instance, you might guess that it takes 10 years for your investment to double. The balance in 10 years is

$$A = 500(1 + 0.06)^{10} \approx \$895.42.$$

Because the amount has not yet doubled, you increase your guess to 15 years.

$$A = 500(1 + 0.06)^{15} \approx \$1198.28$$

Because this amount is more than double the investment, your next guess should be a number between 10 and 15. After trying several more numbers, you can determine that your balance doubles in about 11.9 years.

Another strategy that works well for a problem such as Example 7 is to make a table of data values. You can use a calculator to create the following table.

t	2	4	6	8	10	12
A	561.80	631.24	709.26	796.92	895.42	1006.10

Example 8   Make a Table/Look for a Pattern

Find each product. Then describe the pattern and use your description to find the product of 14 and 16.

$$1 \cdot 3, \ 2 \cdot 4, \ 3 \cdot 5, \ 4 \cdot 6, \ 5 \cdot 7, \ 6 \cdot 8, \ 7 \cdot 9$$

Solution

One way to help find a pattern is to organize the results in a table.

Numbers	$1 \cdot 3$	$2 \cdot 4$	$3 \cdot 5$	$4 \cdot 6$	$5 \cdot 7$	$6 \cdot 8$	$7 \cdot 9$
Product	3	8	15	24	35	48	63

From the table, you can see that each of the products is 1 less than a perfect square. For instance, 3 is 1 less than 2^2 or 4, 8 is 1 less than 3^2 or 9, 15 is 1 less than 4^2 or 16, and so on.

If this pattern continues for other numbers, you can hypothesize that the product of 14 and 16 is 1 less than 15^2 or 225. That is,

$$14 \cdot 16 = 15^2 - 1$$

$$= 224.$$

You can confirm this result by actually multiplying 14 and 16.

Example 9   Draw a Diagram

The outer dimensions of a rectangular apartment are 25 feet by 40 feet. The combination living room, dining room, and kitchen areas occupy two-fifths of the apartment's area. Find the area of the remaining rooms.

Solution

For this problem, it helps to draw a diagram, as shown in Figure 2.4. From the figure, you can see that the total area of the apartment is

$$\text{Area} = (\text{Length})(\text{Width})$$

$$= (40)(25)$$

$$= 1000 \text{ square feet.}$$

The area occupied by the living room, dining room, and kitchen is

$$\frac{2}{5}(1000) = 400 \text{ square feet.}$$

This implies that the remaining rooms must have a total area of

$$1000 - 400 = 600 \text{ square feet.}$$

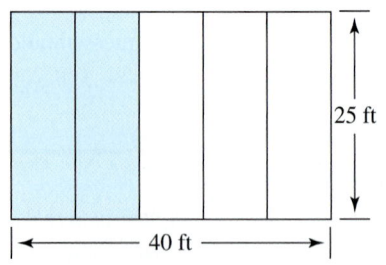

25 ft

40 ft

Figure 2.4

Example 10   Solve a Simpler Problem

You are driving on an interstate highway and are traveling at an average speed of 60 miles per hour. How far will you travel in $12\frac{1}{2}$ hours?

Solution

One way to solve this problem is to use the formula that relates distance, rate, and time. Suppose, however, that you have forgotten the formula. To help you remember, you could solve some simpler problems.

- If you travel 60 miles per hour for 1 hour, you will travel 60 miles.
- If you travel 60 miles per hour for 2 hours, you will travel 120 miles.
- If you travel 60 miles per hour for 3 hours, you will travel 180 miles.

From these examples, it appears that you can find the total miles traveled by multiplying the rate times the time. So, if you travel 60 miles per hour for $12\frac{1}{2}$ hours, you will travel a distance of

$$(60)(12.5) = 750 \text{ miles.}$$

Hidden operations are often involved when variable names (labels) are assigned to unknown quantities. A good strategy is to use a *specific* case to help you write a model for the *general* case. For instance, a specific case of finding three consecutive integers

$$3, 3 + 1, \text{ and } 3 + 2$$

may help you write a general case for finding three consecutive integers n, $n + 1$, and $n + 2$. This strategy is illustrated in Examples 11 and 12.

Example 11   Using a Specific Case to Find a General Case

In each of the following, use the variable to label the unknown quantity.

a. A person's weekly salary is d dollars. What is the annual salary?

b. A person's annual salary is y dollars. What is the monthly salary?

Solution

a. There are 52 weeks in a year.

   Specific case: If the weekly salary is $200, then the annual salary (in dollars) is $52 \cdot 200$.

   General case: If the weekly salary is d dollars, then the annual salary (in dollars) is $52 \cdot d$ or $52d$.

b. There are 12 months in a year.

   Specific case: If the annual salary is $24,000, then the monthly salary (in dollars) is $24{,}000 \div 12$.

   General case: If the annual salary is y dollars, then the monthly salary (in dollars) is $y \div 12$ or $y/12$.

Example 12   Using a Specific Case to Find a General Case

In each of the following, use the variable to label the unknown quantity.

a. You are k inches shorter than a friend. You are 60 inches tall. How tall is your friend?

b. A consumer buys g gallons of gasoline for a total of d dollars. What is the price per gallon?

c. A person drives on the highway at an average speed of 60 miles per hour for t hours. How far has the person traveled?

Solution

a. You are k inches shorter than a friend.

Specific case: If you are 10 inches shorter than your friend, then your friend is $60 + 10$ inches tall.

General case: If you are k inches shorter than your friend, then your friend is $60 + k$ inches tall.

b. To obtain the price per gallon, divide the price by the number of gallons.

Specific case: If the total price is $11.50 and the total number of gallons is 10, then the price per gallon is $11.50 \div 10$ dollars per gallon.

General case: If the total price is d dollars and the total number of gallons is g, then the price per gallon is $d \div g$ or d/g dollars per gallon.

c. To obtain the distance driven, multiply the speed by the number of hours.

Specific case: If the person has driven for 2 hours at a speed of 60 miles per hour, then the person has traveled $60 \cdot 2$ miles.

General case: If the person has driven for t hours at a speed of 60 miles per hour, then the person has traveled $60t$ miles.

Most of the verbal problems you encounter in a mathematics text have precisely the right amount of information necessary to solve the problem. In real life, however, you may need to collect additional information, as shown in Example 13.

Example 13   Enough Information?

Decide what additional information is needed to solve the following problem.

During a given week, a person worked 48 hours for the same employer. The hourly rate for overtime is $14. Write an expression for the person's gross pay for the week, including any pay received for overtime.

Solution

To solve this problem, you would need to know how much the person is normally paid per hour. You would also need to be sure that the person normally works 40 hours per week and that overtime is paid on time worked beyond 40 hours.

2.3 Exercises

Review   Concepts, Skills, and Problem Solving

Keep mathematically in shape by doing these exercises *before* the problems of this section.

Properties and Definitions

1. The product of two real numbers is -35 and one of the factors is 5. What is the sign of the other factor?

2. Determine the sum of the digits of 744. Since this sum is divisible by 3, the number 744 is divisible by what numbers?

3. *True or False?* -4^2 is positive.

4. *True or False?* $(-4)^2$ is positive.

Simplifying Expressions

In Exercises 5–10, evaluate the expression.

5. $(-6)(-13)$

6. $|4(-6)(5)|$

7. $\left(-\frac{4}{3}\right)\left(-\frac{9}{16}\right)$

8. $\frac{7}{8} \div \frac{3}{16}$

9. $\left|-\frac{5}{9}\right| + 2$

10. $-7\frac{3}{5} - 3\frac{1}{2}$

Problem Solving

11. *Consumerism*   A coat costs $133.50, including tax. You save $30 a week. How many weeks must you save in order to buy the coat? How much money will you have left?

12. ▲ *Geometry*   The length of a rectangle is $1\frac{1}{2}$ times its width. Its width is 8 meters. Find its perimeter.

Developing Skills

In Exercises 1–6, match the verbal phrase with the correct algebraic expression.

(a) $11 + \frac{1}{3}x$ (b) $3x - 12$

(c) $3(x - 12)$ (d) $12 - 3x$

(e) $11x + \frac{1}{3}$ (f) $12x + 3$

1. Twelve decreased by 3 times a number
2. Eleven more than $\frac{1}{3}$ of a number
3. Eleven times a number plus $\frac{1}{3}$
4. Three increased by 12 times a number
5. The difference between 3 times a number and 12
6. Three times the difference of a number and 12

In Exercises 7–30, translate the phrase into an algebraic expression. Let x represent the real number. See Examples 1, 2, 3, and 4.

7. A number increased by 5
8. 17 more than a number
9. A number decreased by 25
10. A number decreased by 7
11. Six less than a number
12. Ten more than a number
13. Twice a number
14. The product of 30 and a number
15. A number divided by 3
16. A number divided by 100
17. The ratio of a number to 50
18. One-half of a number
19. Three-tenths of a number
20. Twenty-five hundredths of a number
21. A number is tripled and the product is increased by 5.
22. A number is increased by 5 and the sum is tripled.
23. Eight more than 5 times a number
24. The quotient of a number and 5 is decreased by 15.
25. Ten times the sum of a number and 4
26. Seventeen less than 4 times a number
27. The absolute value of the sum of a number and 4
28. The absolute value of 4 less than twice a number

29. The square of a number, increased by 1

30. Twice the square of a number, increased by 4

In Exercises 31–44, write a verbal description of the algebraic expression, without using a variable. (There is more than one correct answer.) See Example 5.

31. $x - 10$

32. $x + 9$

33. $3x + 2$

34. $4 - 7x$

35. $\frac{1}{2}x - 6$

36. $9 - \frac{1}{4}x$

37. $3(2 - x)$

38. $-10(t - 6)$

39. $\dfrac{t + 1}{2}$

40. $\dfrac{y - 3}{4}$

41. $\dfrac{1}{2} - \dfrac{t}{5}$

42. $\dfrac{1}{4} + \dfrac{x}{8}$

43. $x^2 + 5$

44. $x^3 - 1$

In Exercises 45–52, translate the phrase into a mathematical expression. Simplify the expression.

45. The sum of x and 3 is multiplied by x.

46. The sum of 6 and n is multiplied by 5.

47. The sum of 25 and x is added to x.

48. The sum of 4 and x is added to the sum of x and -8.

49. Nine is subtracted from x and the result is multiplied by 3.

50. The square of x is added to the product of x and $x + 1$.

51. The product of 8 times the sum of x and 24 is divided by 2.

52. Fifteen is subtracted from x and the difference is multiplied by 4.

Solving Problems

53. *Money* A cash register contains d dimes. Write an algebraic expression that represents the total amount of money (in dollars). See Example 6.

54. *Money* A cash register contains d dimes and q quarters. Write an algebraic expression that represents the total amount of money (in dollars).

55. *Sales Tax* The sales tax on a purchase of L dollars is 6%. Write an algebraic expression that represents the total amount of sales tax. (*Hint:* Use the decimal form of 6%.)

56. *Income Tax* The state income tax on a gross income of I dollars in Pennsylvania is 2.8%. Write an algebraic expression that represents the total amount of income tax. (*Hint:* Use the decimal form of 2.8%.)

57. *Travel Time* A truck travels 100 miles at an average speed of r miles per hour. Write an algebraic expression that represents the total travel time.

58. *Distance* An airplane travels at the rate of r miles per hour for 3 hours. Write an algebraic expression that represents the total distance traveled by the airplane.

59. *Consumerism* A campground charges $15 for adults and $2 for children. Write an algebraic expression that represents the total camping fee for m adults and n children.

60. *Hourly Wage* The hourly wage for an employee is $12.50 per hour plus 75 cents for each of the q units produced during the hour. Write an algebraic expression that represents the total hourly earnings for the employee.

Guess, Check, and Revise   In Exercises 61–64, an expression for the balance in an account is given. Guess, check, and revise to determine the time (in years) necessary for the investment of $1000 to double. See Example 7.

61. Interest rate: 7%

$1000(1 + 0.07)^t$

62. Interest rate: 5%

$1000(1 + 0.05)^t$

63. Interest rate: 6%

$1000(1 + 0.06)^t$

64. Interest rate: 8%

$1000(1 + 0.08)^t$

Finding a Pattern   In Exercises 65 and 66, complete the table. The third row in the table is the difference between consecutive entries of the second row. Describe the pattern of the third row. See Example 8.

65.

n	0	1	2	3	4	5
$2n - 1$						
Differences						

66.

n	0	1	2	3	4	5
$7n + 5$						
Differences						

Exploration   In Exercises 67 and 68, find values for a and b such that the expression $an + b$ yields the table values.

67.

n	0	1	2	3	4	5
$an + b$	4	9	14	19	24	29

68.

n	0	1	2	3	4	5
$an + b$	1	5	9	13	17	21

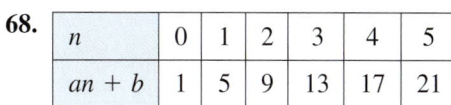 *Geometry*   In Exercises 69–74, write an algebraic expression that represents the area of the region. Use the rules of algebra to simplify the expression.

69.

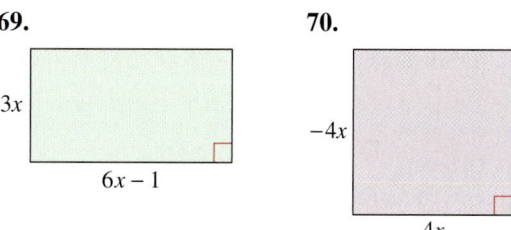

70.

71.

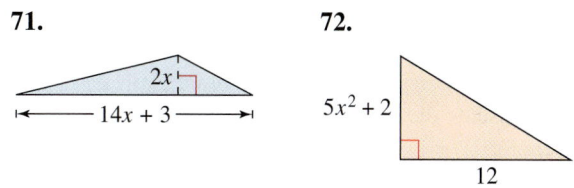

72.

73.

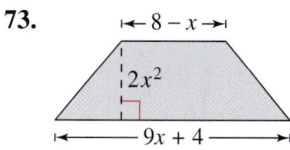

74.

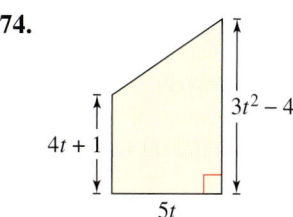

Drawing a Diagram   In Exercises 75 and 76, draw figures satisfying the specified conditions. See Example 9.

75. The sides of a square have length a centimeters. Draw the square. Draw the rectangle obtained by extending two parallel sides of the square 6 centimeters. Find expressions for the perimeter and area of each figure.

76. The dimensions of a rectangular lawn are 150 feet by 250 feet. The property owner has the option of buying a rectangular strip x feet wide along one 250-foot side of the lawn. Draw diagrams representing the lawn before and after the purchase. Write an expression for the area of each.

77. ▲ *Geometry*   A rectangle has sides of length $3w$ and w. Write an algebraic expression that represents the area of the rectangle.

78. ▲ *Geometry*   A square has sides of length s. Write an algebraic expression that represents the perimeter of the square.

79. ▲ *Geometry*   Write an algebraic expression that represents the perimeter of the picture frame in the figure.

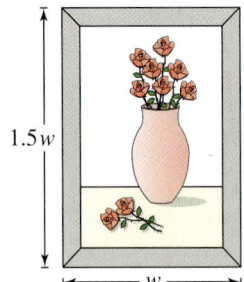

1.5w

w

80. ▲ *Geometry*   A computer screen has sides of length s inches (see figure). Write an algebraic expression that represents the area of the screen. Write the area using the correct unit of measure.

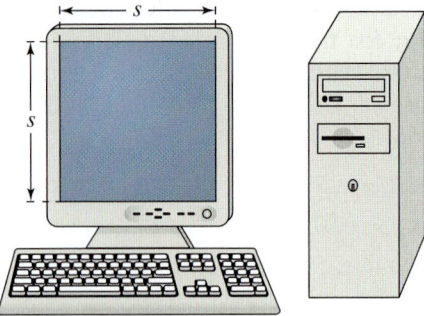

Figure for 80

In Exercises 81–84, decide what additional information is needed to solve the problem. (Do not solve the problem.) See Example 13.

81. *Distance*   A family taking a Sunday drive through the country travels at an average speed of 45 miles per hour. How far have they traveled by 3:00 P.M.?

82. *Consumer Awareness*   You purchase an MP3 player during a sale at an electronics store. The MP3 player is discounted by 15%. What is the sale price of the player?

83. *Consumerism*   You decide to budget your money so that you can afford a new computer. The cost of the computer is $975. You put half of your weekly paycheck into your savings account to pay for the computer. How many hours will you have to work at your job in order to be able to afford the computer?

84. *Painting*   A painter is going to paint a rectangular room that is twice as long as it is wide. One gallon of paint covers 100 square feet. How much money will he have to spend on paint?

Explaining Concepts

85. 🕖   Answer parts (c)–(h) of Motivating the Chapter on page 66.

86. The word *difference* indicates what operation?

87. The word *quotient* indicates what operation?

88. Determine which phrase(s) is (are) equivalent to the expression $n + 4$.

  (a) 4 more than n (b) the sum of n and 4

  (c) n less than 4   (d) the ratio of n to 4

  (e) the total of 4 and n

89. *Writing*   Determine whether order is important when translating each phrase into an algebraic expression. Explain.

  (a) x increased by 10
  (b) 10 decreased by x

  (c) the product of x and 10

  (d) the quotient of x and 10

90. Give two interpretations of "the quotient of 5 and a number times 3."

2.4 Introduction to Equations

Byron Aughenbaugh/Getty Images

What You Should Learn

1 Distinguish between an algebraic expression and an algebraic equation.

2 Check whether a given value is a solution of an equation.

3 Use properties of equality to solve equations.

4 Use a verbal model to construct an algebraic equation.

Why You Should Learn It

You can use verbal models to write algebraic equations that model real-life situations. For instance, in Exercise 64 on page 114, you will write an equation to determine how far away a lightning strike is after hearing the thunder.

1 Distinguish between an algebraic expression and an algebraic equation.

Equations

An **equation** is a statement that two algebraic expressions are equal. For example,

$$x = 3, \quad 5x - 2 = 8, \quad \frac{x}{4} = 7, \quad \text{and} \quad x^2 - 9 = 0$$

are equations. To **solve** an equation involving the variable x means to find all values of x that make the equation true. Such values are called **solutions.** For instance, $x = 2$ is a solution of the equation

$$5x - 2 = 8$$

because

$$5(2) - 2 = 8$$

is a true statement. The solutions of an equation are said to **satisfy** the equation.

Be sure that you understand the distinction between an algebraic expression and an algebraic equation. The differences are summarized in the following table.

Algebraic Expression	Algebraic Equation
• Example: $4(x - 1)$ • Contains *no* equal sign • Can sometimes be *simplified* to an equivalent form: $4(x - 1)$ simplifies to $4x - 4$ • Can be evaluated for any real number for which the expression is defined	• Example: $4(x - 1) = 12$ • Contains an equal sign and is true for only certain values of the variable • Solution is found by forming equivalent equations using the properties of equality: $4(x - 1) = 12$ $4x - 4 = 12$ $4x = 16$ $x = 4$

② Check whether a given value is a
solution of an equation.

Checking Solutions of Equations

To **check** whether a given solution is a solution to an equation, substitute the given value into the original equation. If the substitution results in a true statement, then the value is a solution of the equation. If the substitution results in a false statement, then the value is not a solution of the equation. This process is illustrated in Examples 1 and 2.

Example 1   Checking a Solution of an Equation

Determine whether $x = -2$ is a solution of $x^2 - 5 = 4x + 7$.

Solution

$$x^2 - 5 = 4x + 7 \qquad \text{Write original equation.}$$
$$(-2)^2 - 5 \overset{?}{=} 4(-2) + 7 \qquad \text{Substitute } -2 \text{ for } x.$$
$$4 - 5 \overset{?}{=} -8 + 7 \qquad \text{Simplify.}$$
$$-1 = -1 \qquad \text{Solution checks. } \checkmark$$

Because the substitution results in a true statement, you can conclude that $x = -2$ is a solution of the original equation.

Study Tip

When checking a solution, you should write a question mark over the equal sign to indicate that you are not sure of the validity of the equation.

Just because you have found one solution of an equation, you should not conclude that you have found all of the solutions. For instance, you can check that $x = 6$ is also a solution of the equation in Example 1 as follows.

$$x^2 - 5 = 4x + 7 \qquad \text{Write original equation.}$$
$$(6)^2 - 5 \overset{?}{=} 4(6) + 7 \qquad \text{Substitute 6 for } x.$$
$$36 - 5 \overset{?}{=} 24 + 7 \qquad \text{Simplify.}$$
$$31 = 31 \qquad \text{Solution checks. } \checkmark$$

Example 2   A Trial Solution That Does Not Check

Determine whether $x = 2$ is a solution of $x^2 - 5 = 4x + 7$.

Solution

$$x^2 - 5 = 4x + 7 \qquad \text{Write original equation.}$$
$$(2)^2 - 5 \overset{?}{=} 4(2) + 7 \qquad \text{Substitute 2 for } x.$$
$$4 - 5 \overset{?}{=} 8 + 7 \qquad \text{Simplify.}$$
$$-1 \neq 15 \qquad \text{Solution does not check. } \times$$

Because the substitution results in a false statement, you can conclude that $x = 2$ is not a solution of the original equation.

③ Use properties of equality to solve equations.

Forming Equivalent Equations

It is helpful to think of an equation as having two sides that are in balance. Consequently, when you try to solve an equation, you must be careful to maintain that balance by performing the same operation on each side.

Two equations that have the same set of solutions are called **equivalent.** For instance, the equations

$$x = 3 \quad \text{and} \quad x - 3 = 0$$

are equivalent because both have only one solution—the number 3. When any one of the operations in the following list is applied to an equation, the resulting equation is equivalent to the original equation.

Forming Equivalent Equations: Properties of Equality

An equation can be transformed into an *equivalent equation* using one or more of the following procedures.

	Original Equation	*Equivalent Equation(s)*
1. *Simplify either side:* Remove symbols of grouping, combine like terms, or simplify fractions on one or both sides of the equation.	$3x - x = 8$	$2x = 8$
2. *Apply the Addition Property of Equality:* Add (or subtract) the same quantity to (from) *each* side of the equation.	$x - 2 = 5$	$x - 2 + 2 = 5 + 2$ $x = 7$
3. *Apply the Multiplication Property of Equality:* Multiply (or divide) each side of the equation by the same *nonzero* quantity.	$3x = 9$	$\dfrac{3x}{3} = \dfrac{9}{3}$ $x = 3$
4. *Interchange the two sides of the equation.*	$7 = x$	$x = 7$

The second and third operations in this list can be used to eliminate terms or factors in an equation. For example, to solve the equation $x - 5 = 1$, you need to eliminate the term -5 on the left side. This is accomplished by adding its opposite, 5, to each side.

$$x - 5 = 1 \qquad \text{Write original equation.}$$

$$x - 5 + 5 = 1 + 5 \qquad \text{Add 5 to each side.}$$

$$x + 0 = 6 \qquad \text{Combine like terms.}$$

$$x = 6 \qquad \text{Solution}$$

These four equations are equivalent, and they are called the **steps** of the solution.

The next example shows how the properties of equality can be used to solve equations. You will get many more opportunities to practice these skills in the next chapter. For now, your goal should be to understand why each step in the solution is valid. For instance, the second step in part (a) of Example 3 is valid because the Addition Property of Equality states that you can add the same quantity to each side of an equation.

Example 3   Operations Used to Solve Equations

Identify the property of equality used to solve each equation.

a. $x - 5 = 0$ Original equation

$x - 5 + 5 = 0 + 5$ Add 5 to each side.

$x = 5$ Solution

b. $\dfrac{x}{5} = -2$ Original equation

$\dfrac{x}{5}(5) = -2(5)$ Multiply each side by 5.

$x = -10$ Solution

c. $4x = 9$ Original equation

$\dfrac{4x}{4} = \dfrac{9}{4}$ Divide each side by 4.

$x = \dfrac{9}{4}$ Solution

d. $\dfrac{5}{3}x = 7$ Original equation

$\dfrac{3}{5} \cdot \dfrac{5}{3}x = \dfrac{3}{5} \cdot 7$ Multiply each side by $\frac{3}{5}$.

$x = \dfrac{21}{5}$ Solution

> **Study Tip**
>
> In Example 3(c), each side of the equation is divided by 4 to eliminate the coefficient 4 on the left side. You could just as easily *multiply* each side by $\frac{1}{4}$. Both techniques are legitimate—which one you decide to use is a matter of personal preference.

Solution

a. The Addition Property of Equality is used to add 5 to each side of the equation in the second step. Adding 5 eliminates the term -5 from the left side of the equation.

b. The Multiplication Property of Equality is used to multiply each side of the equation by 5 in the second step. Multiplying by 5 eliminates the denominator from the left side of the equation.

c. The Multiplication Property of Equality is used to divide each side of the equation by 4 $\left(\text{or multiply each side by } \frac{1}{4}\right)$ in the second step. Dividing by 4 eliminates the coefficient from the left side of the equation.

d. The Multiplication Property of Equality is used to multiply each side of the equation by $\frac{3}{5}$ in the second step. Multiplying by the *reciprocal* of the fraction $\frac{5}{3}$ eliminates the fraction from the left side of the equation.

④ Use a verbal model to construct an algebraic equation.

Constructing Equations

It is helpful to use two phases in constructing equations that model real-life situations, as shown below.

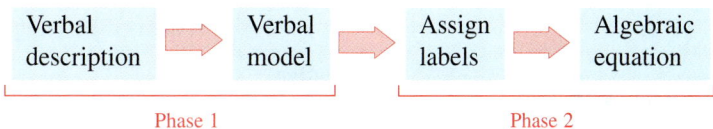

Phase 1   Phase 2

In the first phase, you translate the verbal description into a *verbal model.* In the second phase, you assign labels and translate the verbal model into a *mathematical model* or *algebraic equation.* Here are two examples of verbal models.

1. The sale price of a basketball is $28. The sale price is $7 less than the original price. What is the original price?

Verbal Model: Sale price = Original price − Discount

$28 = Original price − $7

2. The original price of a basketball is $35. The original price is discounted by $7. What is the sale price?

Verbal Model: Sale price = Original price − Discount

Sale price = $35 − $7

Example 4   Using a Verbal Model to Construct an Equation

Write an algebraic equation for the following problem.

The total income that an employee received in 2003 was $31,550. How much was the employee paid each week? Assume that each weekly paycheck contained the same amount, and that the year consisted of 52 weeks.

Solution

Verbal Model: Income for year = 52 · Weekly pay

Labels: Income for year = 31,550   (dollars)
Weekly pay = x (dollars)

Algebraic Model: $31{,}550 = 52x$

When you construct an equation, be sure to check that both sides of the equation represent the *same* unit of measure. For instance, in Example 4, both sides of the equation $31{,}550 = 52x$ represent dollar amounts.

Example 5   Using a Verbal Model to Construct an Equation

Write an algebraic equation for the following problem.

Returning to college after spring break, you travel 3 hours and stop for lunch. You know that it takes 45 minutes to complete the last 36 miles of the 180-mile trip. What is the average speed during the first 3 hours of the trip?

Solution

Verbal Model:

$$\text{Distance} = \text{Rate} \cdot \text{Time}$$

Labels:

Distance = $180 - 36 = 144$	(miles)
Rate = r	(miles per hour)
Time = 3	(hours)

Algebraic Model:

$$144 = 3r$$

> **Study Tip**
>
> In Example 5, the information that it takes 45 minutes to complete the last part of the trip is unnecessary information. This type of unnecessary information in an applied problem is sometimes called a *red herring*.

Example 6   Using a Verbal Model to Construct an Equation

Write an algebraic equation for the following problem.

Tickets for a concert cost $45 for each floor seat and $30 for each stadium seat. There were 800 seats on the main floor, and these were sold out. The total revenue from ticket sales was $54,000. How many stadium seats were sold?

Solution

Verbal Model:

$$\boxed{\text{Total revenue}} = \boxed{\text{Revenue from floor seats}} + \boxed{\text{Revenue from stadium seats}}$$

Labels:

Total revenue = $54,000$	(dollars)
Price per floor seat = 45	(dollars per seat)
Number of floor seats = 800	(seats)
Price per stadium seat = 30	(dollars per seat)
Number of stadium seats = x	(seats)

Algebraic Model:

$$54{,}000 = 45(800) + 30x$$

In Example 6, you can use the following *unit analysis* to check that both sides of the equation are measured in dollars.

$$54{,}000 \text{ dollars} = \left(\frac{45 \text{ dollars}}{\text{seat}} \right)(800 \text{ seats}) + \left(\frac{30 \text{ dollars}}{\text{seat}} \right)(x \text{ seats})$$

In Section 3.1, you will study techniques for solving the equations constructed in Examples 4, 5, and 6.

2.4 Exercises

Review   Concepts, Skills, and Problem Solving

Keep mathematically in shape by doing these exercises *before* the problems of this section.

Properties and Definitions

1. If the numerator and denominator of a fraction have unlike signs, the sign of the fraction is _____.

2. *Writing* ✏️   If a negative number is used as a factor eight times, what is the sign of the product? Explain.

3. Complete the Commutative Property:

   $6 + 10 = $ _____ .

4. Identify the property of real numbers illustrated by $6\left(\frac{1}{6}\right) = 1$.

Simplifying Expressions

In Exercises 5–10, simplify the expression.

5. $t^2 \cdot t^5$

6. $(-3y^3)y^2$

7. $6x + 9x$

8. $4 - 3t + t$

9. $-(-8b)$

10. $7(-10x)$

Graphs and Models

▲ *Geometry*   In Exercises 11 and 12, write and simplify expressions for the perimeter and area of the figure.

11.

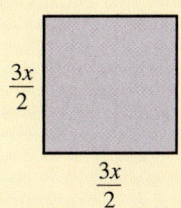

$\frac{3x}{2}$

$\frac{3x}{2}$

12.

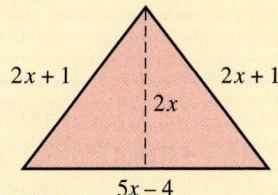

$2x+1$　$2x$　$2x+1$

$5x-4$

Developing Skills

In Exercises 1–16, determine whether each value of x is a solution of the equation. See Examples 1 and 2.

	Equation		*Values*	
1.	$4x - 12 = 0$		(a) $x = 3$	(b) $x = 5$
2.	$3x - 3 = 0$		(a) $x = 4$	(b) $x = 1$
3.	$6x + 1 = -11$		(a) $x = 2$	(b) $x = -2$
4.	$2x + 5 = -15$		(a) $x = -10$	(b) $x = 5$
5.	$x + 5 = 2x$		(a) $x = -1$	(b) $x = 5$
6.	$2x - 3 = 5x$		(a) $x = 0$	(b) $x = -1$
7.	$x + 3 = 2(x - 4)$		(a) $x = 11$	(b) $x = -5$
8.	$5x - 1 = 3(x + 5)$		(a) $x = 8$	(b) $x = -2$

	Equation		*Values*	
9.	$2x + 10 = 7(x + 1)$		(a) $x = \frac{3}{5}$	(b) $x = -\frac{2}{3}$
10.	$3(3x + 2) = 9 - x$		(a) $x = -\frac{3}{4}$	(b) $x = \frac{3}{10}$
11.	$x^2 - 4 = x + 2$		(a) $x = 3$	(b) $x = -2$
12.	$x^2 = 8 - 2x$		(a) $x = 2$	(b) $x = -4$
13.	$\dfrac{2}{x} - \dfrac{1}{x} = 1$		(a) $x = 0$	(b) $x = \dfrac{1}{3}$
14.	$\dfrac{4}{x} + \dfrac{2}{x} = 1$		(a) $x = 0$	(b) $x = 6$
15.	$\dfrac{5}{x - 1} + \dfrac{1}{x} = 5$		(a) $x = 3$	(b) $x = \dfrac{1}{6}$
16.	$\dfrac{3}{x - 2} = x$		(a) $x = -1$	(b) $x = 3$

In Exercises 17–26, use a calculator to determine whether the value of x is a solution of the equation.

Equation	*Values*
17. $x + 1.7 = 6.5$	(a) $x = -3.1$
	(b) $x = 4.8$
18. $7.9 - x = 14.6$	(a) $x = -6.7$
	(b) $x = 5.4$
19. $40x - 490 = 0$	(a) $x = 12.25$
	(b) $x = -12.25$
20. $20x - 560 = 0$	(a) $x = 27.5$
	(b) $x = -27.5$
21. $2x^2 - x - 10 = 0$	(a) $x = \frac{5}{2}$
	(b) $x = -1.09$
22. $22x - 5x^2 = 17$	(a) $x = 1$
	(b) $x = 3.4$
23. $\dfrac{1}{x} - \dfrac{9}{x - 4} = 1$	(a) $x = 0$
	(b) $x = -2$
24. $x = \dfrac{3}{4x + 1}$	(a) $x = -0.25$
	(b) $x = 0.75$
25. $x^3 - 1.728 = 0$	(a) $x = \frac{6}{5}$
	(b) $x = -\frac{6}{5}$
26. $4x^2 - 10.24 = 0$	(a) $x = \frac{8}{5}$
	(b) $x = -\frac{8}{5}$

In Exercises 27–34, justify each step of the solution. See Example 3.

27.
$$5x + 12 = 22$$
$$5x + 12 - 12 = 22 - 12$$
$$5x = 10$$
$$\frac{5x}{5} = \frac{10}{5}$$
$$x = 2$$

28.
$$14 - 3x = 5$$
$$14 - 3x - 14 = 5 - 14$$
$$14 - 14 - 3x = -9$$
$$-3x = -9$$
$$\frac{-3x}{-3} = \frac{-9}{-3}$$
$$x = 3$$

29. $\dfrac{2}{3}x = 12$
$$\frac{3}{2}\left(\frac{2}{3}x\right) = \frac{3}{2}(12)$$
$$x = 18$$

30. $\dfrac{4}{5}x = -28$
$$\frac{5}{4}\left(\frac{4}{5}x\right) = \frac{5}{4}(-28)$$
$$x = -35$$

31. $\quad 2(x - 1) = x + 3$
$$2x - 2 = x + 3$$
$$2x - x - 2 = x - x + 3$$
$$x - 2 = 3$$
$$x - 2 + 2 = 3 + 2$$
$$x = 5$$

32. $\quad x + 6 = -6(4 - x)$
$$x + 6 = -24 + 6x$$
$$x - x + 6 = -24 + 6x - x$$
$$6 = 5x - 24$$
$$6 + 24 = 5x - 24 + 24$$
$$30 = 5x$$
$$\frac{30}{5} = \frac{5x}{5}$$
$$6 = x$$

33. $\quad x = -2(x + 3)$
$$x = -2x - 6$$
$$x + 2x = -2x + 2x - 6$$
$$3x = 0 - 6$$
$$3x = -6$$
$$\frac{3x}{3} = \frac{-6}{3}$$
$$x = -2$$

34. $\dfrac{x}{3} = x + 1$

$$3\left(\dfrac{x}{3}\right) = 3(x + 1)$$

$$x = 3x + 3$$

$$x - 3x = 3x - 3x + 3$$

$$-2x = 0 + 3$$

$$-2x = 3$$

$$\dfrac{-2x}{-2} = \dfrac{3}{-2}$$

$$x = -\dfrac{3}{2}$$

In Exercises 35–38, use a property of equality to solve the equation. Check your solution. See Examples 1, 2, and 3.

35. $x - 8 = 5$

36. $x + 3 = 19$

37. $3x = 30$

38. $\dfrac{x}{4} = 12$

Solving Problems

In Exercises 39–44, write a verbal description of the algebraic equation without using a variable. (There is more than one correct answer.)

39. $2x + 5 = 21$

40. $3x - 2 = 7$

41. $10(x - 3) = 8x$

42. $2(x - 5) = 12$

43. $\dfrac{x + 1}{3} = 8$

44. $\dfrac{x - 2}{10} = 6$

In Exercises 45–68, construct an equation for the word problem. Do *not* solve the equation. See Examples 4, 5, and 6.

45. The sum of a number and 12 is 45. What is the number?

46. The sum of 3 times a number and 4 is 16. What is the number?

47. Four times the sum of a number and 6 is 100. What is the number?

48. Find a number such that 6 times the number subtracted from 120 is 96.

49. Find a number such that 2 times the number decreased by 14 equals the number divided by 3.

50. The sum of a number and 8, divided by 4, is 32. What is the number?

51. *Test Score*   After your instructor added 6 points to each student's test score, your score is 94. What was your original score?

52. *Meteorology*   With the 1.2-inch rainfall today, the total for the month is 4.5 inches. How much had been recorded for the month before today's rainfall?

53. *Consumerism*   You have $1044 saved for the purchase of a new computer that will cost $1926. How much more must you save?

54. *List Price*   The sale price of a coat is $225.98. The discount is $64. What is the list (original) price?

55. *Travel Costs*   A company pays its sales representatives 35 cents per mile if they use their personal cars. A sales representative submitted a bill to be reimbursed for $148.05 for driving. How many miles did the sales representative drive?

56. *Money*   A student has n quarters and seven $1 bills totaling $8.75. How many quarters does the student have?

57. ▲ *Geometry*   The base of a rectangular box is 4 feet by 6 feet and its volume is 72 cubic feet (see figure). What is the height of the box?

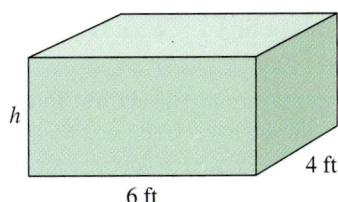

h

4 ft

6 ft

58. ▲ *Geometry* The width of a rectangular mirror is one-third its length, as shown in the figure. The perimeter of the mirror is 96 inches. What are the dimensions of the mirror?

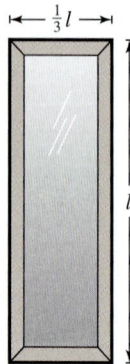

59. *Average Speed* After traveling for 3 hours, your family is still 25 miles from completing a 160-mile trip (see figure). What was the average speed during the first 3 hours of the trip?

25 miles

160 miles

60. *Average Speed* After traveling for 4 hours, you are still 24 miles from completing a 200-mile trip. It requires one-half hour to travel the last 24 miles. What was the average speed during the first 4 hours of the trip?

61. *Average Speed* A group of students plans to take two cars to a soccer tournament. The first car leaves on time, travels at an average speed of 45 miles per hour, and arrives at the destination in 3 hours. The second car leaves one-half hour after the first car and arrives at the tournament at the same time as the students in the first car. What is the average speed of the second car?

62. *Dow Jones Average* The Dow Jones average fell 58 points during a week and was 8695 at the close of the market on Friday. What was the average at the close of the market on the previous Friday?

63. *Consumer Awareness* The price of a gold ring has increased by $45 over the past year. It is now selling for $375. What was the price one year ago?

64. *Meteorology* You hear thunder 3 seconds after seeing the lightning. The speed of sound is 1100 feet per second. How far away is the lightning?

65. *Depreciation* A textile corporation buys equipment with an initial purchase price of $750,000. It is estimated that its useful life will be 3 years and at that time its value will be $75,000. The total depreciation is divided equally among the three years. (Depreciation is the difference between the initial price of an item and its current value.) What is the total amount of depreciation declared each year?

66. *Car Payments* You make 48 monthly payments of $158 each to buy a used car. The total amount financed is $6000. What is the total amount of interest that you paid?

67. *Fund Raising* A student group is selling boxes of greeting cards at a profit of $1.75 each. The group needs $2000 more to have enough money for a trip to Washington, D.C. How many boxes does the group need to sell to earn $2000?

68. *Consumer Awareness* The price of a compact car increased $1432 over the past year. The price of the car was $9850 two years ago and $10,120 one year ago. What is its current price?

Unit Analysis In Exercises 69–76, simplify the expression. State the units of the simplified value.

69. $\dfrac{3 \text{ dollars}}{\text{unit}} \cdot (5 \text{ units})$

70. $\dfrac{25 \text{ miles}}{\text{gallon}} \cdot (15 \text{ gallons})$

71. $\dfrac{50 \text{ pounds}}{\text{foot}} \cdot (3 \text{ feet})$

72. $\dfrac{3 \text{ dollars}}{\text{pound}} \cdot (5 \text{ pounds})$

73. $\dfrac{5 \text{ feet}}{\text{second}} \cdot \dfrac{60 \text{ seconds}}{\text{minute}} \cdot (20 \text{ minutes})$

74. $\dfrac{12 \text{ dollars}}{\text{hour}} \cdot \dfrac{1 \text{ hour}}{60 \text{ minutes}} \cdot (45 \text{ minutes})$

75. $\dfrac{100 \text{ centimeters}}{\text{meter}} \cdot (2.4 \text{ meters})$

76. $\dfrac{1000 \text{ milliliters}}{\text{liter}} \cdot (5.6 \text{ liters})$

Explaining Concepts

77. *Writing* Explain how to decide whether a real number is a solution of an equation. Give an example of an equation with a solution that checks and one that does not check.

78. *Writing* In your own words, explain what is meant by the term *equivalent equations*.

79. *Writing* Explain the difference between simplifying an expression and solving an equation. Give an example of each.

80. *Writing* Describe a real-life problem that uses the following verbal model.

$$\boxed{\begin{array}{c}\text{Revenue}\\\text{of }\$840\end{array}} = \boxed{\begin{array}{c}\$35 \text{ per}\\\text{case}\end{array}} \cdot \boxed{\begin{array}{c}\text{Number}\\\text{of cases}\end{array}}$$

81. *Writing* Describe, from memory, the steps that can be used to transform an equation into an equivalent equation.

What Did You Learn?

Key Terms

variables, *p. 68*
constants, *p. 68*
algebraic expression, *p. 68*
terms, *p. 68*
coefficient, *p. 68*

evaluate an algebraic
    expression, *p. 71*
expanding an algebraic
    expression, *p. 79*
like terms, *p. 80*
simplify an algebraic
    expression, *p. 82*

verbal mathematical
    model, *p. 92*
equation, *p. 105*
solutions, *p. 105*
satisfy, *p. 105*
equivalent equations,
    p. 107

Key Concepts

2.1 ◯ Exponential form

Repeated multiplication can be expressed in exponential form using a base a and an exponent n, where a is a real number, variable, or algebraic expression and n is a positive integer.

$$a^n = a \cdot a \cdots a$$

2.1 ◯ Evaluating algebraic expressions

To evaluate an algebraic expression, substitute every occurrence of the variable in the expression with the appropriate real number and perform the operation(s).

2.2 ◯ Properties of algebra

Commutative Property:
Addition $a + b = b + a$
Multiplication $ab = ba$

Associative Property:
Addition $(a + b) + c = a + (b + c)$
Multiplication $(ab)c = a(bc)$

Distributive Property:
$a(b + c) = ab + ac$ $a(b - c) = ab - ac$
$(a + b)c = ac + bc$ $(a - b)c = ac - bc$

Identities:
Additive $a + 0 = 0 + a = a$
Multiplicative $a \cdot 1 = 1 \cdot a = a$

Inverses:
Additive $a + (-a) = 0$
Multiplicative $a \cdot \dfrac{1}{a} = 1, \; a \neq 0$

2.2 ◯ Combining like terms

To combine like terms in an algebraic expression, add their respective coefficients and attach the common variable factor.

2.2 ◯ Simplifying an algebraic expression

To simplify an algebraic expression, remove symbols of grouping and combine like terms.

2.3 ◯ Additional problem-solving strategies

Additional problem-solving strategies are listed below.

1. Guess, check, and revise
2. Make a table/look for a pattern
3. Draw a diagram
4. Solve a simpler problem

2.4 ◯ Checking solutions of equations

To check a solution, substitute the given solution for each occurrence of the variable in the original equation. Evaluate each side of the equation. If both sides are equivalent, the solution checks.

2.4 ◯ Properties of equality

Addition: Add (or subtract) the same quantity to (from) each side of the equation.

Multiplication: Multiply (or divide) each side of the equation by the same nonzero quantity.

2.4 ◯ Constructing equations

From the verbal description, write a verbal mathematical model. Assign labels to the known and unknown quantities, and write an algebraic model.

Review Exercises

2.1 Writing and Evaluating Algebraic Expressions

① Define and identify terms, variables, and coefficients of algebraic expressions.

In Exercises 1 and 2, identify the variable and the constant in the expression.

1. $15 - x$

2. $t - 5^2$

In Exercises 3–8, identify the terms and the coefficients of the expression.

3. $12y + y^2$

4. $4x - \frac{1}{2}x^3$

5. $5x^2 - 3xy + 10y^2$

6. $y^2 - 10yz + \frac{2}{3}z^2$

7. $\frac{2y}{3} - \frac{4x}{y}$

8. $-\frac{4b}{9} + \frac{11a}{b}$

② Define exponential form and interpret exponential expressions.

In Exercises 9–12, rewrite the product in exponential form.

9. $5z \cdot 5z \cdot 5z$

10. $\frac{3}{8}y \cdot \frac{3}{8}y \cdot \frac{3}{8}y \cdot \frac{3}{8}y$

11. $(b - c) \cdot (b - c) \cdot 6 \cdot 6$

12. $2 \cdot (a + b) \cdot 2 \cdot (a + b) \cdot 2$

③ Evaluate algebraic expressions using real numbers.

In Exercises 13–18, evaluate the algebraic expression for the given values of the variable(s).

Expression	*Values*
13. $x^2 - 2x + 5$	(a) $x = 0$ (b) $x = 2$
14. $x^3 - 8$	(a) $x = 2$ (b) $x = 4$
15. $x^2 - x(y + 1)$	(a) $x = 2, y = -1$ (b) $x = 1, y = 2$
16. $2r + r(t^2 - 3)$	(a) $r = 3, t = -2$ (b) $r = -2, t = 3$
17. $\dfrac{x + 5}{y}$	(a) $x = -5, y = 3$ (b) $x = 2, y = -1$

Expression	*Values*
18. $\dfrac{a - 9}{2b}$	(a) $a = 7, b = -3$ (b) $a = -4, b = 5$

2.2 Simplifying Algebraic Expressions

① Use the properties of algebra.

In Exercises 19–24, identify the property of algebra illustrated by the statement.

19. $xy \cdot \dfrac{1}{xy} = 1$

20. $u(vw) = (uv)w$

21. $(x - y)(2) = 2(x - y)$

22. $(a + b) + 0 = a + b$

23. $2x + (3y - z) = (2x + 3y) - z$

24. $x(y + z) = xy + xz$

In Exercises 25–32, use the Distributive Property to expand the expression.

25. $4(x + 3y)$

26. $3(8s - 12t)$

27. $-5(2u - 3v)$

28. $-3(-2x - 8y)$

29. $x(8x + 5y)$

30. $-u(3u - 10v)$

31. $-(-a + 3b)$

32. $(7 - 2j)(-6)$

② Combine like terms of an algebraic expression.

In Exercises 33–44, simplify the expression by combining like terms.

33. $3a - 5a$

34. $6c - 2c$

35. $3p - 4q + q + 8p$

36. $10x - 4y - 25x + 6y$

37. $\frac{1}{4}s - 6t + \frac{7}{2}s + t$

38. $\frac{2}{3}a + \frac{3}{5}a - \frac{1}{2}b + \frac{2}{3}b$

39. $x^2 + 3xy - xy + 4$

40. $uv^2 + 10 - 2uv^2 + 2$

41. $5x - 5y + 3xy - 2x + 2y$

42. $y^3 + 2y^2 + 2y^3 - 3y^2 + 1$

43. $5\left(1 + \dfrac{r}{n}\right)^2 - 2\left(1 + \dfrac{r}{n}\right)^2$

44. $-7\left(\dfrac{1}{u}\right) + 4\left(\dfrac{1}{u^2}\right) + 3\left(\dfrac{1}{u}\right)$

(3) Simplify an algebraic expression by rewriting the terms.

In Exercises 45–52, simplify the expression.

45. $12(4t)$

46. $8(7x)$

47. $-5(-9x^2)$

48. $-10(-3b^3)$

49. $(-6x)(2x^2)$

50. $(-3y^2)(15y)$

51. $\dfrac{12x}{5} \cdot \dfrac{10}{3}$

52. $\dfrac{4z}{15} \cdot \dfrac{9}{2}$

(4) Use the Distributive Property to remove symbols of grouping.

In Exercises 53–64, simplify the expression by removing symbols of grouping and combining like terms.

53. $5(u - 4) + 10$

54. $16 - 3(v + 2)$

55. $3s - (r - 2s)$

56. $50x - (30x + 100)$

57. $-3(1 - 10z) + 2(1 - 10z)$

58. $8(15 - 3y) - 5(15 - 3y)$

59. $\frac{1}{3}(42 - 18z) - 2(8 - 4z)$

60. $\frac{1}{4}(100 + 36s) - (15 - 4s)$

61. $10 - [8(5 - x) + 2]$

62. $3[2(4x - 5) + 4] - 3$

63. $2[x + 2(y - x)]$

64. $2t[4 - (3 - t)] + 5t$

65. *Depreciation* You pay P dollars for new equipment. Its value after 5 years is given by

$$P\left(\dfrac{9}{10}\right)\left(\dfrac{9}{10}\right)\left(\dfrac{9}{10}\right)\left(\dfrac{9}{10}\right)\left(\dfrac{9}{10}\right).$$

Simplify the expression.

66. ▲ *Geometry* The height of a triangle is $1\frac{1}{2}$ times its base. Its area is given by $\frac{1}{2}b\left(\frac{3}{2}b\right)$. Simplify the expression.

67. Simplify the algebraic expression that represents the sum of three consecutive odd integers, $2n - 1$, $2n + 1$, and $2n + 3$.

68. Simplify the algebraic expression that represents the sum of three consecutive even integers, $2n$, $2n + 2$, $2n + 4$.

69. ▲ *Geometry* The face of a DVD player has the dimensions shown in the figure. Write an algebraic expression that represents the area of the face of the DVD player excluding the compartment holding the disc.

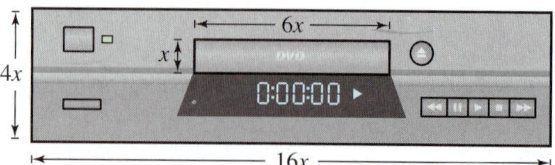

70. ▲ *Geometry* Write an expression for the perimeter of the figure. Use the rules of algebra to simplify the expression.

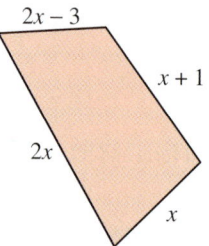

2.3 Algebra and Problem Solving

(2) Construct verbal mathematical models from written statements.

In Exercises 71 and 72, construct a verbal model and then write an algebraic expression that represents the specified quantity.

71. The total hourly wage for an employee when the base pay is $8.25 per hour and an additional $0.60 is paid for each unit produced per hour

72. The total cost for a family to stay one night at a campground if the charge is $18 for the parents plus $3 for each of the children

③ Translate verbal phrases into algebraic expressions.

In Exercises 73–82, translate the phrase into an algebraic expression. Let x represent the real number.

73. Two-thirds of a number, plus 5

74. One hundred, decreased by 5 times a number

75. Ten less than twice a number

76. The ratio of a number to 10

77. Fifty increased by the product of 7 and a number

78. Ten decreased by the quotient of a number and 2

79. The sum of a number and 10 all divided by 8

80. The product of 15 and a number, decreased by 2

81. The sum of the square of a real number and 64

82. The absolute value of the sum of a number and -10

In Exercises 83–86, write a verbal description of the expression without using a variable. (There is more than one correct answer.)

83. $x + 3$

84. $3x - 2$

85. $\dfrac{y - 2}{3}$

86. $4(x + 5)$

④ Identify hidden operations when constructing algebraic expressions.

87. *Commission*   A salesperson earns 5% commission on his total weekly sales, x. Write an algebraic expression that represents the amount in commissions that the salesperson earns in a week.

88. *Sale Price*   A cordless phone is advertised for 20% off the list price of L dollars. Write an algebraic expression that represents the sale price of the phone.

89. *Rent*   The monthly rent for your apartment is $625 for n months. Write an algebraic expression that represents the total rent.

90. *Distance*   A car travels for 10 hours at an average speed of s miles per hour. Write an algebraic expression that represents the total distance traveled by the car.

⑥ Use problem-solving strategies to solve application problems.

91. *Finding a Pattern*

(a) Complete the table. The third row in the table is the difference between consecutive entries of the second row. The fourth row is the difference between consecutive entries of the third row.

n	0	1	2	3	4	5
$n^2 + 3n + 2$						
Differences						
Differences						

(b) Describe the patterns of the third and fourth rows.

92. *Finding a Pattern*   Find values for a and b such that the expression $an + b$ yields the table values.

n	0	1	2	3	4	5
$an + b$	4	9	14	19	24	29

2.4 Introduction to Equations

② Check whether a given value is a solution of an equation.

In Exercises 93–102, determine whether each value of x is a solution of the equation.

	Equation	*Values*
93.	$5x + 6 = 36$	(a) $x = 3$ (b) $x = 6$
94.	$17 - 3x = 8$	(a) $x = 3$ (b) $x = -3$
95.	$3x - 12 = x$	(a) $x = -1$ (b) $x = 6$
96.	$8x + 24 = 2x$	(a) $x = 0$ (b) $x = -4$

Equation　　　　　　　　　*Values*

97. $4(2 - x) = 3(2 + x)$　(a) $x = \dfrac{2}{7}$　(b) $x = -\dfrac{2}{3}$

98. $5x + 2 = 3(x + 10)$　(a) $x = 14$　(b) $x = -10$

99. $\dfrac{4}{x} - \dfrac{2}{x} = 5$　　　　　(a) $x = -1$　(b) $x = \dfrac{2}{5}$

100. $\dfrac{x}{3} + \dfrac{x}{6} = 1$　　　　　(a) $x = \dfrac{2}{9}$　(b) $x = -\dfrac{2}{9}$

101. $x(x - 7) = -12$　　　(a) $x = 3$　(b) $x = 4$

102. $x(x + 1) = 2$　　　　(a) $x = 1$　(b) $x = -2$

3 Use properties of equality to solve equations.

In Exercises 103 and 104, justify each step of the solution.

103.
$$3(x - 2) = x + 2$$
$$3x - 6 = x + 2$$
$$3x - x - 6 = x - x + 2$$
$$2x - 6 = 2$$
$$2x - 6 + 6 = 2 + 6$$
$$2x = 8$$
$$\frac{2x}{2} = \frac{8}{2}$$
$$x = 4$$

104.
$$x = -(x - 14)$$
$$x = -x + 14$$
$$x + x = -x + x + 14$$
$$2x = 14$$
$$\frac{2x}{2} = \frac{14}{2}$$
$$x = 7$$

4 Use a verbal model to construct an algebraic equation.

In Exercises 105–108, construct an equation for the word problem. Do *not* solve the equation.

105. The sum of a number and its reciprocal is $\dfrac{37}{6}$. What is the number?

106. *Distance*　A car travels 135 miles in t hours with an average speed of 45 miles per hour (see figure). How many hours did the car travel?

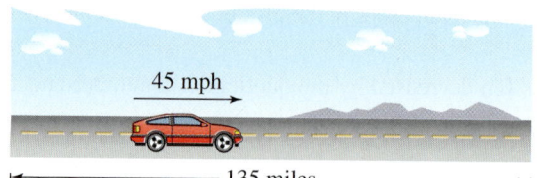

107. ▲ *Geometry*　The area of the shaded region in the figure is 24 square inches. What is the length of the rectangle?

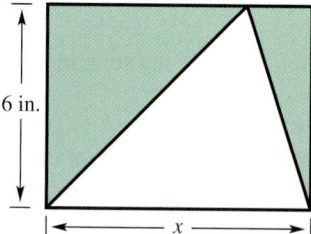

108. ▲ *Geometry*　The perimeter of the face of a rectangular traffic light is 72 inches (see figure). What are the dimensions of the traffic light?

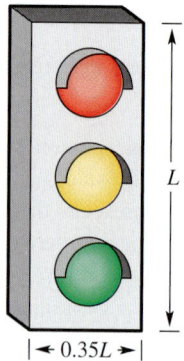

Chapter Test

Take this test as you would take a test in class. After you are done, check your work against the answers in the back of the book.

1. Identify the terms and coefficients of the expression.

$$2x^2 - 7xy + 3y^3$$

2. Rewrite the product in exponential form.

$$x \cdot (x + y) \cdot x \cdot (x + y) \cdot x$$

In Exercises 3–6, identify the property of algebra illustrated by the statement.

3. $(5x)y = 5(xy)$

4. $2 + (x - y) = (x - y) + 2$

5. $7xy + 0 = 7xy$

6. $(x + 5) \cdot \dfrac{1}{(x + 5)} = 1$

In Exercises 7–10, use the Distributive Property to expand the expression.

7. $3(x + 8)$

8. $5(4r - s)$

9. $-y(3 - 2y)$

10. $-9(4 + 2x + x^2)$

In Exercises 11–14, simplify the expression.

11. $3b - 2a + a - 10b$

12. $15(u - v) - 7(u - v)$

13. $3z - (4 + z)$

14. $2[10 - (t + 1)]$

15. Evaluate the expression when $x = 3$ and $y = -12$.

    (a) $x^3 - 2$ (b) $x^2 + 4(y + 2)$

16. Explain why it is not possible to evaluate $\dfrac{a + 2b}{3a - b}$ when $a = 2$ and $b = 6$.

17. Translate the phrase, "one-fifth of a number, increased by two," into an algebraic expression. Let n represent the number.

18. (a) Write expressions for the perimeter and area of the rectangle at the left.

    (b) Simplify the expressions.

    (c) Identify the unit of measure for each expression.

    (d) Evaluate each expression when $w = 12$ feet.

w

$2w - 4$

Figure for 18

19. The prices of concert tickets for adults and children are \$15 and \$10, respectively. Write an algebraic expression that represents the total income from the concert for m adults and n children.

20. Determine whether the values of x are solutions of $6(3 - x) - 5(2x - 1) = 7$.

    (a) $x = -2$ (b) $x = 1$

Motivating the Chapter

⚡ Talk Is Cheap?

You plan to purchase a cellular phone with a service contract. For a price of $99, one package includes the phone and 3 months of service. You will be billed a *per minute usage rate* each time you make or receive a call. After 3 months you will be billed a monthly service charge of $19.50 and the per minute usage rate.

A second cellular phone package costs $80, which includes the phone and 1 month of service. You will be billed a per minute usage rate each time you make or receive a call. After the first month you will be billed a monthly service charge of $24 and the per minute usage rate.

See Section 3.3, Exercise 105.

a. Write an equation to find the cost of the phone in the first package. Solve the equation to find the cost of the phone.

b. Write an equation to find the cost of the phone in the second package. Solve the equation to find the cost of the phone. Which phone costs more, the one in the first package or the one in the second package?

c. What percent of the purchase price of $99 goes toward the price of the cellular phone in the first package? Use an equation to answer the question.

d. What percent of the purchase price of $80 goes toward the price of the cellular phone in the second package? Use an equation to answer the question.

e. The sales tax on your purchase is 5%. What is the total cost of purchasing the first cellular phone package? Use an equation to answer the question.

f. You decide to buy the first cellular phone package. Your total cellular phone bill for the fourth month of use is $92.46 for 3.2 hours of use. What is the per minute usage rate? Use an equation to answer the question.

See Section 3.4, Exercise 87.

g. For the fifth month you were billed the monthly service charge and $47.50 for 125 minutes of use. You estimate that during the next month you spent 150 minutes on calls. Use a proportion to find the charge for 150 minutes of use. (Use the first package.)

See Section 3.6, Exercise 117.

h. You determine that the most you can spend each month on phone calls is $75. Write a compound inequality that describes the number of minutes you can spend talking on the cellular phone each month if the per minute usage rate is $0.35. Solve the inequality. (Use the first package.)

Stephen Poe/Alamy

3

Equations, Inequalities, and Problem Solving

3.1 Solving Linear Equations

Amy Etra/PhotoEdit, Inc.

What You Should Learn

1 Solve linear equations in standard form.

2 Solve linear equations in nonstandard form.

3 Use linear equations to solve application problems.

Why You Should Learn It

Linear equations are used in many real-life applications. For instance, in Exercise 65 on page 133, you will use a linear equation to determine the number of hours spent repairing your car.

1 Solve linear equations in standard form.

Linear Equations in the Standard Form $ax + b = 0$

This is an important step in your study of algebra. In the first two chapters, you were introduced to the rules of algebra, and you learned to use these rules to rewrite and simplify algebraic expressions. In Sections 2.3 and 2.4, you gained experience in translating verbal expressions and problems into algebraic forms. You are now ready to use these skills and experiences to *solve equations*.

In this section, you will learn how the rules of algebra and the properties of equality can be used to solve the most common type of equation—a linear equation in one variable.

Definition of Linear Equation

A **linear equation** in one variable x is an equation that can be written in the standard form

$$ax + b = 0$$

where a and b are real numbers with $a \neq 0$.

A linear equation in one variable is also called a **first-degree equation** because its variable has an (implied) exponent of 1. Some examples of linear equations in standard form are

$$2x = 0, \quad x - 7 = 0, \quad 4x + 6 = 0, \quad \text{and} \quad \frac{x}{2} - 1 = 0.$$

Remember that to *solve* an equation involving x means to find all values of x that satisfy the equation. For the linear equation $ax + b = 0$, the goal is to *isolate* x by rewriting the equation in the form

$$x = \boxed{\text{a number}} \, . \qquad \text{\textcolor{red}{Isolate the variable } } x.$$

To obtain this form, you use the techniques discussed in Section 2.4. That is, beginning with the original equation, you write a sequence of equivalent equations, each having the same solution as the original equation. For instance, to solve the linear equation $x - 2 = 0$, you can add 2 to each side of the equation to obtain $x = 2$. As mentioned in Section 2.4, each equivalent equation is called a **step** of the solution.

Example 1   Solving a Linear Equation in Standard Form

Solve $3x - 15 = 0$. Then check the solution.

Solution

$$3x - 15 = 0 \qquad \text{Write original equation.}$$

$$3x - 15 + 15 = 0 + 15 \qquad \text{Add 15 to each side.}$$

$$3x = 15 \qquad \text{Combine like terms.}$$

$$\frac{3x}{3} = \frac{15}{3} \qquad \text{Divide each side by 3.}$$

$$x = 5 \qquad \text{Simplify.}$$

It appears that the solution is $x = 5$. You can check this as follows:

Check

$$3x - 15 = 0 \qquad \text{Write original equation.}$$

$$3(5) - 15 \overset{?}{=} 0 \qquad \text{Substitute 5 for } x.$$

$$15 - 15 \overset{?}{=} 0 \qquad \text{Simplify.}$$

$$0 = 0 \qquad \text{Solution checks.} \checkmark$$

In Example 1, be sure you see that solving an equation has two basic stages. The first stage is to *find* the solution (or solutions). The second stage is to *check* that each solution you find actually satisfies the original equation. You can improve your accuracy in algebra by developing the habit of checking each solution.

A common question in algebra is

"How do I know which step to do *first* to isolate x?"

The answer is that you need practice. By solving many linear equations, you will find that your skill will improve. The key thing to remember is that you can "get rid of" terms and factors by using *inverse* operations. Here are some guidelines and examples.

Guideline	*Equation*	*Inverse Operation*
1. Subtract to remove a sum.	$x + 3 = 0$	Subtract 3 from each side.
2. Add to remove a difference.	$x - 5 = 0$	Add 5 to each side.
3. Divide to remove a product.	$4x = 20$	Divide each side by 4.
4. Multiply to remove a quotient.	$\frac{x}{8} = 2$	Multiply each side by 8.

For additional examples, review Example 3 on page 108. In each case of that example, note how inverse operations are used to isolate the variable.

Example 2   Solving a Linear Equation in Standard Form

Solve $2x + 3 = 0$. Then check the solution.

Solution

$2x + 3 = 0$	Write original equation.
$2x + 3 - 3 = 0 - 3$	Subtract 3 from each side.
$2x = -3$	Combine like terms.
$\dfrac{2x}{2} = -\dfrac{3}{2}$	Divide each side by 2.
$x = -\dfrac{3}{2}$	Simplify.

Check

$2x + 3 = 0$	Write original equation.
$2\left(-\dfrac{3}{2}\right) + 3 \stackrel{?}{=} 0$	Substitute $-\frac{3}{2}$ for x.
$-3 + 3 \stackrel{?}{=} 0$	Simplify.
$0 = 0$	Solution checks. ✓

So, the solution is $x = -\frac{3}{2}$.

Example 3   Solving a Linear Equation in Standard Form

Solve $5x - 12 = 0$. Then check the solution.

Solution

$5x - 12 = 0$	Write original equation.
$5x - 12 + 12 = 0 + 12$	Add 12 to each side.
$5x = 12$	Combine like terms.
$\dfrac{5x}{5} = \dfrac{12}{5}$	Divide each side by 5.
$x = \dfrac{12}{5}$	Simplify.

Check

$5x - 12 = 0$	Write original equation.
$5\left(\dfrac{12}{5}\right) - 12 \stackrel{?}{=} 0$	Substitute $\frac{12}{5}$ for x.
$12 - 12 \stackrel{?}{=} 0$	Simplify.
$0 = 0$	Solution checks. ✓

So, the solution is $x = \frac{12}{5}$.

② Solve linear equations in nonstandard form.

Study Tip

To eliminate a fractional coefficient, it may be easier to multiply each side by the *reciprocal* of the fraction than to divide by the fraction itself. Here is an example.

$$-\frac{2}{3}x = 4$$

$$\left(-\frac{3}{2}\right)\left(-\frac{2}{3}\right)x = \left(-\frac{3}{2}\right)4$$

$$x = -\frac{12}{2}$$

$$x = -6$$

Example 4   Solving a Linear Equation in Standard Form

Solve $\frac{x}{3} + 3 = 0$. Then check the solution.

Solution

$\frac{x}{3} + 3 = 0$	Write original equation.
$\frac{x}{3} + 3 - 3 = 0 - 3$	Subtract 3 from each side.
$\frac{x}{3} = -3$	Combine like terms.
$3\left(\frac{x}{3}\right) = 3(-3)$	Multiply each side by 3.
$x = -9$	Simplify.

Check

$\frac{x}{3} + 3 = 0$	Write original equation.
$\frac{-9}{3} + 3 \overset{?}{=} 0$	Substitute -9 for x.
$-3 + 3 \overset{?}{=} 0$	Simplify.
$0 = 0$	Solution checks. ✓

So, the solution is $x = -9$.

Technology: Tip

Remember to check your solution in the original equation. This can be done efficiently with a graphing calculator.

As you gain experience in solving linear equations, you will probably find that you can perform some of the solution steps in your head. For instance, you might solve the equation given in Example 4 by writing only the following steps.

$\frac{x}{3} + 3 = 0$	Write original equation.
$\frac{x}{3} = -3$	Subtract 3 from each side.
$x = -9$	Multiply each side by 3.

Solving a Linear Equation in Nonstandard Form

The definition of linear equation contains the phrase "that can be written in the standard form $ax + b = 0$." This suggests that some linear equations may come in nonstandard or disguised form.

A common form of linear equations is one in which the variable terms are not combined into one term. In such cases, you can begin the solution by *combining like terms*. Note how this is done in the next two examples.

Example 5   Solving a Linear Equation in Nonstandard Form

Solve $3y + 8 - 5y = 4$. Then check your solution.

Solution

$3y + 8 - 5y = 4$	Write original equation.
$3y - 5y + 8 = 4$	Group like terms.
$-2y + 8 = 4$	Combine like terms.
$-2y + 8 - 8 = 4 - 8$	Subtract 8 from each side.
$-2y = -4$	Combine like terms.
$\dfrac{-2y}{-2} = \dfrac{-4}{-2}$	Divide each side by -2.
$y = 2$	Simplify.

Study Tip

In Example 5, note that the variable in the equation doesn't always have to be x. Any letter can be used.

Check

$3y + 8 - 5y = 4$	Write original equation.
$3(2) + 8 - 5(2) \overset{?}{=} 4$	Substitute 2 for y.
$6 + 8 - 10 \overset{?}{=} 4$	Simplify.
$4 = 4$	Solution checks. ✓

So, the solution is $y = 2$.

The solution for Example 5 began by collecting like terms. You can use any of the properties of algebra to attain your goal of "isolating the variable." The next example shows how to solve a linear equation using the Distributive Property.

Example 6   Using the Distributive Property

Solve $x + 6 = 2(x - 3)$.

Solution

$x + 6 = 2(x - 3)$	Write original equation.
$x + 6 = 2x - 6$	Distributive Property
$x - 2x + 6 = 2x - 2x - 6$	Subtract $2x$ from each side.
$-x + 6 = -6$	Combine like terms.
$-x + 6 - 6 = -6 - 6$	Subtract 6 from each side.
$-x = -12$	Combine like terms.
$(-1)(-x) = (-1)(-12)$	Multiply each side by -1.
$x = 12$	Simplify.

Study Tip

You can isolate the variable term on either side of the equal sign. For instance, Example 6 could have been solved in the following way.

$$x + 6 = 2(x - 3)$$
$$x + 6 = 2x - 6$$
$$x - x + 6 = 2x - x - 6$$
$$6 = x - 6$$
$$6 + 6 = x - 6 + 6$$
$$12 = x$$

The solution is $x = 12$. Check this in the original equation.

There are three different situations that can be encountered when solving linear equations in one variable. The first situation occurs when the linear equation has *exactly one* solution. You can show this with the steps below.

$$ax + b = 0 \qquad \text{Write original equation, with } a \neq 0.$$

$$ax = 0 - b \qquad \text{Subtract } b \text{ from each side.}$$

$$x = \frac{-b}{a} \qquad \text{Divide each side by } a.$$

So, the *linear* equation has exactly one solution: $x = -b/a$. The other two situations are the possibilities for the equation to have either *no solution* or *infinitely many solutions*. These two special cases are demonstrated below.

No Solution

$$2x + 3 \overset{?}{=} 2(x + 4)$$

$$2x + 3 \overset{?}{=} 2x + 8$$

$$2x - 2x + 3 \overset{?}{=} 2x - 2x + 8$$

$$3 \neq 8$$

Infinitely Many Solutions

$$2(x + 3) = 2x + 6$$

$$2x + 6 = 2x + 6 \qquad \text{Identity equation}$$

$$2x - 2x + 6 - 6 = 2x - 2x + 6 - 6$$

$$0 = 0$$

Watch out for these types of equations in the exercise set.

Study Tip

In the *No Solution* equation, the result is not true because $3 \neq 8$. This means that there is no value of x that will make the equation true.

In the *Infinitely Many Solutions* equation, the result is true. This means that *any* real number is a solution to the equation. This type of equation is called an **identity.**

③ Use linear equations to solve application problems.

Applications

Example 7   Geometry: Dimensions of a Dog Pen

You have 96 feet of fencing to enclose a rectangular pen for your dog. To provide sufficient running space for the dog to exercise, the pen is to be three times as long as it is wide. Find the dimensions of the pen.

Solution

Begin by drawing and labeling a diagram, as shown in Figure 3.1. The perimeter of a rectangle is the sum of twice its length and twice its width.

Verbal Model: Perimeter $= 2 \cdot$ Length $+ 2 \cdot$ Width

Algebraic Model: $96 = 2(3x) + 2x$

You can solve this equation as follows.

$$96 = 6x + 2x \qquad \text{Multiply.}$$

$$96 = 8x \qquad \text{Combine like terms.}$$

$$\frac{96}{8} = \frac{8x}{8} \qquad \text{Divide each side by 8.}$$

$$12 = x \qquad \text{Simplify.}$$

So, the width of the pen is 12 feet and its length is 36 feet.

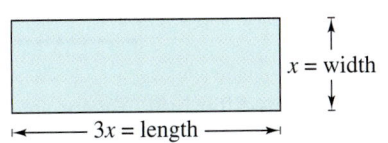

$x = $ width

$3x = $ length

Figure 3.1

Example 8   Ticket Sales

Tickets for a concert are $40 for each floor seat and $20 for each stadium seat. There are 800 seats on the main floor, and these are sold out. The total revenue from ticket sales is $92,000. How many stadium seats were sold?

Solution

Verbal Model: | Total revenue | = | Revenue from floor seats | + | Revenue from stadium seats |

Labels:
Total revenue = 92,000   (dollars)
Price per floor seat = 40   (dollars per seat)
Number of floor seats = 800   (seats)
Price per stadium seat = 20   (dollars per seat)
Number of stadium seats = x (seats)

Algebraic Model: $92{,}000 = 40(800) + 20x$

Now that you have written an algebraic equation to represent the problem, you can solve the equation as follows.

$$92{,}000 = 40(800) + 20x$$ Write original equation.

$$92{,}000 = 32{,}000 + 20x$$ Simplify.

$$92{,}000 - 32{,}000 = 32{,}000 - 32{,}000 + 20x$$ Subtract 32,000 from each side.

$$60{,}000 = 20x$$ Combine like terms.

$$\frac{60{,}000}{20} = \frac{20x}{20}$$ Divide each side by 20.

$$3000 = x$$ Simplify.

There were 3000 stadium seats sold. To check this solution, you should go back to the original statement of the problem and substitute 3000 stadium seats into the equation. You will find that the total revenue is $92,000.

Two integers are called **consecutive integers** if they differ by 1. So, for any integer n, its next two larger consecutive integers are $n + 1$ and $(n + 1) + 1$ or $n + 2$. You can denote three consecutive integers by $n, n + 1,$ and $n + 2$.

Expressions for Special Types of Integers

Let n be an integer. Then the following expressions can be used to denote even integers, odd integers, and consecutive integers, respectively.

1. $2n$ denotes an *even* integer.

2. $2n - 1$ and $2n + 1$ denote *odd* integers.

3. The set $\{n, n + 1, n + 2\}$ denotes three *consecutive* integers.

Example 9   Consecutive Integers

Find three consecutive integers whose sum is 48.

Solution

Verbal Model: First integer + Second integer + Third integer = 48

Labels: First integer = n
Second integer = $n + 1$
Third integer = $n + 2$

Equation: $n + (n + 1) + (n + 2) = 48$ Original equation

$3n + 3 = 48$ Combine like terms.

$3n + 3 - 3 = 48 - 3$ Subtract 3 from each side.

$3n = 45$ Combine like terms.

$\dfrac{3n}{3} = \dfrac{45}{3}$ Divide each side by 3.

$n = 15$ Simplify.

So, the first integer is 15, the second integer is $15 + 1 = 16$, and the third integer is $15 + 2 = 17$. Check this in the original statement of the problem.

Example 10   Consecutive Even Integers

Find two consecutive even integers such that the sum of the first even integer and three times the second is 78.

Solution

Verbal Model: First even integer + 3 · Second even integer = 78

Labels: First even integer = $2n$
Second even integer = $2n + 2$

Equation: $2n + 3(2n + 2) = 78$ Original equation

$2n + 6n + 6 = 78$ Distributive Property

$8n + 6 = 78$ Combine like terms.

$8n + 6 - 6 = 78 - 6$ Subtract 6 from each side.

$8n = 72$ Combine like terms.

$\dfrac{8n}{8} = \dfrac{72}{8}$ Divide each side by 8.

$n = 9$ Simplify.

So, the first even integer is $2 \cdot 9 = 18$, and the second even integer is $2 \cdot 9 + 2 = 20$. Check this in the original statement of the problem.

3.1 Exercises

Review   Concepts, Skills, and Problem Solving

Keep mathematically in shape by doing these exercises *before* the problems of this section.

Properties and Definitions

1. Identify the property of real numbers illustrated by $(3y - 9) \cdot 1 = 3y - 9$.

2. Identify the property of real numbers illustrated by $(2x + 5) + 8 = 2x + (5 + 8)$.

Simplifying Expressions

In Exercises 3–10, simplify the expression.

3. $3 - 2x + 14 + 7x$

4. $4a + 2ab - b^2 + 5ab - b^2$

5. $-3(x - 5)^2(x - 5)^3$

6. $(4rs)(-5r^2)(2s^3)$

7. $\dfrac{2m^2}{3n} \cdot \dfrac{3m}{5n^3}$

8. $\dfrac{5(x + 3)^2}{10(x + 8)}$

9. $-3(3x - 2y) + 5y$

10. $3v - (4 - 5v)$

Problem Solving

11. *Distance*   The length of a relay race is $\frac{3}{4}$ mile. The last change of runners occurs at the $\frac{2}{3}$ mile marker. How far does the last person run?

12. *Agriculture*   During the months of January, February, and March, a farmer bought $10\frac{1}{3}$ tons, $7\frac{3}{5}$ tons, and $12\frac{5}{6}$ tons of soybeans, respectively. Find the total amount of soybeans purchased during the first quarter of the year.

Developing Skills

Mental Math   In Exercises 1–8, solve the equation mentally.

1. $x + 6 = 0$

2. $a + 5 = 0$

3. $x - 9 = 4$

4. $u - 3 = 8$

5. $7y = 28$

6. $4s = 12$

7. $4z = -36$

8. $9z = -63$

In Exercises 9–12, justify each step of the solution. See Examples 1–6.

9.
$$5x + 15 = 0$$
$$5x + 15 - 15 = 0 - 15$$
$$5x = -15$$
$$\frac{5x}{5} = \frac{-15}{5}$$
$$x = -3$$

10.
$$7x - 14 = 0$$
$$7x - 14 + 14 = 0 + 14$$
$$7x = 14$$
$$\frac{7x}{7} = \frac{14}{7}$$
$$x = 2$$

11.
$$-2x + 5 = 13$$
$$-2x + 5 - 5 = 13 - 5$$
$$-2x = 8$$
$$\frac{-2x}{-2} = \frac{8}{-2}$$
$$x = -4$$

12.
$$22 - 3x = 10$$
$$22 - 3x + 3x = 10 + 3x$$
$$22 = 10 + 3x$$
$$22 - 10 = 10 + 3x - 10$$
$$12 = 3x$$
$$\frac{12}{3} = \frac{3x}{3}$$
$$4 = x$$

In Exercises 13–60, solve the equation and check your solution. (Some equations have no solution.) See Examples 1–6.

13. $8x - 16 = 0$

14. $4x - 24 = 0$

15. $3x + 21 = 0$

16. $2x + 52 = 0$

17. $5x = 30$

18. $12x = 18$

19. $9x = -21$ **20.** $-14x = 42$

21. $8x - 4 = 20$ **22.** $-7x + 24 = 3$

23. $25x - 4 = 46$ **24.** $15x - 18 = 12$

25. $10 - 4x = -6$ **26.** $15 - 3x = -15$

27. $6x - 4 = 0$ **28.** $8z - 2 = 0$

29. $3y - 2 = 2y$ **30.** $2s - 13 = 28s$

31. $4 - 7x = 5x$ **32.** $24 - 5x = x$

33. $4 - 5t = 16 + t$ **34.** $3x + 4 = x + 10$

35. $-3t + 5 = -3t$ **36.** $4z + 2 = 4z$

37. $15x - 3 = 15 - 3x$ **38.** $2x - 5 = 7x + 10$

39. $7a - 18 = 3a - 2$ **40.** $4x - 2 = 3x + 1$

41. $7x + 9 = 3x + 1$ **42.** $6t - 3 = 8t + 1$

43. $4x - 6 = 4x - 6$ **44.** $5 - 3x = 5 - 3x$

45. $2x + 4 = -3(x - 2)$ **46.** $4(y + 1) = -y + 5$

47. $2x = -3x$ **48.** $6t = 9t$

49. $2x - 5 + 10x = 3$ **50.** $-4x + 10 + 10x = 4$

51. $\dfrac{x}{3} = 10$ **52.** $-\dfrac{x}{2} = 3$

53. $x - \dfrac{1}{3} = \dfrac{4}{3}$ **54.** $x + \dfrac{5}{2} = \dfrac{9}{2}$

55. $t - \dfrac{1}{3} = \dfrac{1}{2}$ **56.** $z + \dfrac{2}{5} = -\dfrac{3}{10}$

57. $5t - 4 + 3t = 4(2t - 1)$

58. $7z - 5z - 8 = 2(z - 4)$

59. $2(y - 9) = -5y - 4$

60. $6 - 21x = 3(4 - 7x)$

Solving Problems

61. ▲ *Geometry*   The perimeter of a rectangle is 240 inches. The length is twice its width. Find the dimensions of the rectangle.

62. ▲ *Geometry*   The length of a tennis court is 6 feet more than twice the width (see figure). Find the width of the court if the length is 78 feet.

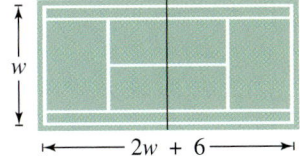

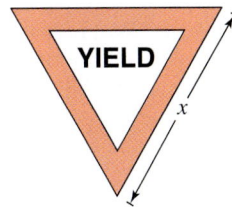

YIELD

x

Figure for 62   Figure for 63

63. ▲ *Geometry*   The sign in the figure has the shape of an equilateral triangle (sides have the same length). The perimeter of the sign is 225 centimeters. Find the length of its sides.

64. ▲ *Geometry*   You are asked to cut a 12-foot board into three pieces. Two pieces are to have the same length and the third is to be twice as long as the others. How long are the pieces?

65. *Car Repair*   The bill (including parts and labor) for the repair of your car is shown. Some of the bill is unreadable. From what is given, can you determine how many hours were spent on labor? Explain.

Parts	.$285.00
Labor ($32 per hour)	$
Total	**$357.00**

Bill for 65

66. *Car Repair*   The bill for the repair of your car was $439. The cost for parts was $265. The cost for labor was $29 per hour. How many hours did the repair work take?

67. *Ticket Sales*   Tickets for a community theater are $10 for main floor seats and $8 for balcony seats. There are 400 seats on the main floor, and these were sold out for the evening performance. The total revenue from ticket sales was $5200. How many balcony seats were sold?

68. *Ticket Sales*   Tickets for a marching band competition are $5 for 50-yard-line seats and $3 for bleacher seats. Eight hundred 50-yard-line seats were sold. The total revenue from ticket sales was $5500. How many bleacher seats were sold?

69. *Summer Jobs*   You have two summer jobs. In the first job, you work 40 hours a week and earn $9.25 an hour at a coffee shop. In the second job, you tutor for $7.50 an hour and can work as many hours as you want. You want to earn a combined total of $425 a week. How many hours must you tutor?

70. *Summer Jobs*   You have two summer jobs. In the first job, you work 30 hours a week and earn $8.75 an hour at a gas station. In the second job, you work as a landscaper for $11.00 an hour and can work as many hours as you want. You want to earn a combined total of $400 a week. How many hours must you work at the second job?

71. *Number Problem*   Five times the sum of a number and 16 is 100. Find the number.

72. *Number Problem*   Find a number such that the sum of twice that number and 31 is 69.

73. *Number Problem*   The sum of two consecutive odd integers is 72. Find the two integers.

74. *Number Problem*   The sum of two consecutive even integers is 154. Find the two integers.

75. *Number Problem*   The sum of three consecutive odd integers is 159. Find the three integers.

76. *Number Problem*   The sum of three consecutive even integers is 192. Find the three integers.

Explaining Concepts

77. The scale below is balanced. Each blue box weighs 1 ounce. How much does the red box weigh? If you removed three blue boxes from each side, would the scale still balance? What property of equality does this illustrate?

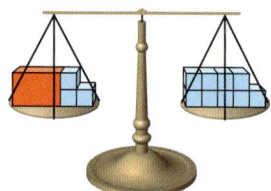

78. *Writing*   In your own words, describe the steps that can be used to transform an equation into an equivalent equation.

79. *Writing*   Explain how to solve the equation $x + 5 = 32$. What property of equality are you using?

80. *Writing*   Explain how to solve the equation $3x = 5$. What property of equality are you using?

81. *True or False?*   Subtracting 0 from each side of an equation yields an equivalent equation. Justify your answer.

82. *True or False?*   Multiplying each side of an equation by 0 yields an equivalent equation. Justify your answer.

83. *Finding a Pattern*   The length of a rectangle is t times its width (see figure). The rectangle has a perimeter of 1200 meters, which implies that $2w + 2(tw) = 1200$, where w is the width of the rectangle.

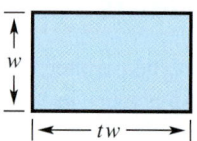

(a) Complete the table.

t	1	1.5	2
Width			
Length			
Area			

t	3	4	5
Width			
Length			
Area			

(b) Use the completed table to draw a conclusion concerning the area of a rectangle of given perimeter as the length increases relative to its width.

3.2 Equations That Reduce to Linear Form

Merrit Vincent/PhotoEdit, Inc.

What You Should Learn

1. Solve linear equations containing symbols of grouping.
2. Solve linear equations involving fractions.
3. Solve linear equations involving decimals.

Why You Should Learn It

Many real-life applications can be modeled with linear equations involving decimals. For instance, Exercise 81 on page 144 shows how a linear equation can model the projected number of persons 65 years and older in the United States.

1. Solve linear equations containing symbols of grouping.

Equations Containing Symbols of Grouping

In this section you will continue your study of linear equations by looking at more complicated types of linear equations. To solve a linear equation that contains symbols of grouping, use the following guidelines.

1. Remove symbols of grouping from each side by using the Distributive Property.
2. Combine like terms.
3. Isolate the variable in the usual way using properties of equality.
4. Check your solution in the original equation.

Example 1   Solving a Linear Equation Involving Parentheses

Solve $4(x - 3) = 8$. Then check your solution.

Solution

$4(x - 3) = 8$	Write original equation.
$4 \cdot x - 4 \cdot 3 = 8$	Distributive Property
$4x - 12 = 8$	Simplify.
$4x - 12 + 12 = 8 + 12$	Add 12 to each side.
$4x = 20$	Combine like terms.
$\dfrac{4x}{4} = \dfrac{20}{4}$	Divide each side by 4.
$x = 5$	Simplify.

Check

$4(5 - 3) \stackrel{?}{=} 8$	Substitute 5 for x in original equation.
$4(2) \stackrel{?}{=} 8$	Simplify.
$8 = 8$	Solution checks. ✔

The solution is $x = 5$.

Study Tip

Notice in the check of Example 1 that you do not need to use the Distributive Property to remove the parentheses. Simply evaluate the expression within the parentheses and then multiply.

Example 2 Solving a Linear Equation Involving Parentheses

Solve $3(2x - 1) + x = 11$. Then check your solution.

Solution

$$3(2x - 1) + x = 11 \qquad \text{Write original equation.}$$
$$3 \cdot 2x - 3 \cdot 1 + x = 11 \qquad \text{Distributive Property}$$
$$6x - 3 + x = 11 \qquad \text{Simplify.}$$
$$6x + x - 3 = 11 \qquad \text{Group like terms.}$$
$$7x - 3 = 11 \qquad \text{Combine like terms.}$$
$$7x - 3 + 3 = 11 + 3 \qquad \text{Add 3 to each side.}$$
$$7x = 14 \qquad \text{Combine like terms.}$$
$$\frac{7x}{7} = \frac{14}{7} \qquad \text{Divide each side by 7.}$$
$$x = 2 \qquad \text{Simplify.}$$

Check

$$3(2x - 1) + x = 11 \qquad \text{Write original equation.}$$
$$3[2(2) - 1] + 2 \stackrel{?}{=} 11 \qquad \text{Substitute 2 for } x.$$
$$3(4 - 1) + 2 \stackrel{?}{=} 11 \qquad \text{Simplify.}$$
$$3(3) + 2 \stackrel{?}{=} 11 \qquad \text{Simplify.}$$
$$9 + 2 \stackrel{?}{=} 11 \qquad \text{Simplify.}$$
$$11 = 11 \qquad \text{Solution checks. } \checkmark$$

The solution is $x = 2$.

Example 3 Solving a Linear Equation Involving Parentheses

Solve $5(x + 2) = 2(x - 1)$.

Solution

$$5(x + 2) = 2(x - 1) \qquad \text{Write original equation.}$$
$$5x + 10 = 2x - 2 \qquad \text{Distributive Property}$$
$$5x - 2x + 10 = 2x - 2x - 2 \qquad \text{Subtract } 2x \text{ from each side.}$$
$$3x + 10 = -2 \qquad \text{Combine like terms.}$$
$$3x + 10 - 10 = -2 - 10 \qquad \text{Subtract 10 from each side.}$$
$$3x = -12 \qquad \text{Combine like terms.}$$
$$x = -4 \qquad \text{Divide each side by 3.}$$

The solution is $x = -4$. Check this in the original equation.

Example 4   Solving a Linear Equation Involving Parentheses

Solve $2(x - 7) - 3(x + 4) = 4 - (5x - 2)$.

Solution

$$2(x - 7) - 3(x + 4) = 4 - (5x - 2)$$ Write original equation.

$$2x - 14 - 3x - 12 = 4 - 5x + 2$$ Distributive Property

$$-x - 26 = -5x + 6$$ Combine like terms.

$$-x + 5x - 26 = -5x + 5x + 6$$ Add $5x$ to each side.

$$4x - 26 = 6$$ Combine like terms.

$$4x - 26 + 26 = 6 + 26$$ Add 26 to each side.

$$4x = 32$$ Combine like terms.

$$x = 8$$ Divide each side by 4.

The solution is $x = 8$. Check this in the original equation.

The linear equation in the next example involves both brackets and parentheses. Watch out for nested symbols of grouping such as these. The *innermost symbols of grouping* should be removed first.

Example 5   An Equation Involving Nested Symbols of Grouping

Solve $5x - 2[4x + 3(x - 1)] = 8 - 3x$.

Solution

$$5x - 2[4x + 3(x - 1)] = 8 - 3x$$ Write original equation.

$$5x - 2[4x + 3x - 3] = 8 - 3x$$ Distributive Property

$$5x - 2[7x - 3] = 8 - 3x$$ Combine like terms inside brackets.

$$5x - 14x + 6 = 8 - 3x$$ Distributive Property

$$-9x + 6 = 8 - 3x$$ Combine like terms.

$$-9x + 3x + 6 = 8 - 3x + 3x$$ Add $3x$ to each side.

$$-6x + 6 = 8$$ Combine like terms.

$$-6x + 6 - 6 = 8 - 6$$ Subtract 6 from each side.

$$-6x = 2$$ Combine like terms.

$$\frac{-6x}{-6} = \frac{2}{-6}$$ Divide each side by -6.

$$x = -\frac{1}{3}$$ Simplify.

The solution is $x = -\frac{1}{3}$. Check this in the original equation.

② Solve linear equations involving fractions.

Equations Involving Fractions or Decimals

To solve a linear equation that contains one or more fractions, it is usually best to first *clear the equation of fractions.*

Clearing an Equation of Fractions

An equation such as

$$\frac{x}{a} + \frac{b}{c} = d$$

that contains one or more fractions can be cleared of fractions by multiplying each side by the least common multiple (LCM) of a and c.

For example, the equation

$$\frac{3x}{2} - \frac{1}{3} = 2$$

can be cleared of fractions by multiplying each side by 6, the LCM of 2 and 3. Notice how this is done in the next example.

Study Tip

For an equation that contains a *single numerical* fraction such as $2x - \frac{3}{4} = 1$, you can simply add $\frac{3}{4}$ to each side and then solve for x. You do not need to clear the fraction.

$$2x - \frac{3}{4} + \frac{3}{4} = 1 + \frac{3}{4} \quad \text{Add } \frac{3}{4}.$$

$$2x = \frac{7}{4} \quad \text{Combine terms.}$$

$$x = \frac{7}{8} \quad \text{Multiply by } \frac{1}{2}.$$

Example 6 Solving a Linear Equation Involving Fractions

Solve $\dfrac{3x}{2} - \dfrac{1}{3} = 2$.

Solution

$$6\left(\frac{3x}{2} - \frac{1}{3}\right) = 6 \cdot 2 \quad \text{Multiply each side by LCM 6.}$$

$$6 \cdot \frac{3x}{2} - 6 \cdot \frac{1}{3} = 12 \quad \text{Distributive Property}$$

$$9x - 2 = 12 \quad \text{Clear fractions.}$$

$$9x = 14 \quad \text{Add 2 to each side.}$$

$$x = \frac{14}{9} \quad \text{Divide each side by 9.}$$

The solution is $x = \frac{14}{9}$. Check this in the original equation.

To check a fractional solution such as $\frac{14}{9}$ in Example 6, it is helpful to rewrite the variable term as a product.

$$\frac{3}{2} \cdot x - \frac{1}{3} = 2 \quad \text{Write fraction as a product.}$$

In this form the substitution of $\frac{14}{9}$ for x is easier to calculate.

Example 7   Solving a Linear Equation Involving Fractions

Solve $\dfrac{x}{5} + \dfrac{3x}{4} = 19$. Then check your solution.

Solution

$$\dfrac{x}{5} + \dfrac{3x}{4} = 19 \qquad \text{Write original equation.}$$

$$20\left(\dfrac{x}{5}\right) + 20\left(\dfrac{3x}{4}\right) = 20(19) \qquad \text{Multiply each side by LCM 20.}$$

$$4x + 15x = 380 \qquad \text{Simplify.}$$

$$19x = 380 \qquad \text{Combine like terms.}$$

$$x = 20 \qquad \text{Divide each side by 19.}$$

Check

$$\dfrac{20}{5} + \dfrac{3(20)}{4} \overset{?}{=} 19 \qquad \text{Substitute 20 for } x \text{ in original equation.}$$

$$4 + 15 \overset{?}{=} 19 \qquad \text{Simplify.}$$

$$19 = 19 \qquad \text{Solution checks. } \checkmark$$

The solution is $x = 20$.

Example 8   Solving a Linear Equation Involving Fractions

Solve $\dfrac{2}{3}\left(x + \dfrac{1}{4}\right) = \dfrac{1}{2}$.

> **Study Tip**
>
> Notice in Example 8 that to clear all fractions in the equation, you multiply by 12, which is the LCM of 3, 4, and 2.

Solution

$$\dfrac{2}{3}\left(x + \dfrac{1}{4}\right) = \dfrac{1}{2} \qquad \text{Write original equation.}$$

$$\dfrac{2}{3}x + \dfrac{2}{12} = \dfrac{1}{2} \qquad \text{Distributive Property}$$

$$12 \cdot \dfrac{2}{3}x + 12 \cdot \dfrac{2}{12} = 12 \cdot \dfrac{1}{2} \qquad \text{Multiply each side by LCM 12.}$$

$$8x + 2 = 6 \qquad \text{Simplify.}$$

$$8x = 4 \qquad \text{Subtract 2 from each side.}$$

$$x = \dfrac{4}{8} \qquad \text{Divide each side by 8.}$$

$$x = \dfrac{1}{2} \qquad \text{Simplify.}$$

The solution is $x = \frac{1}{2}$. Check this in the original equation.

A common type of linear equation is one that equates two fractions. To solve such an equation, consider the fractions to be **equivalent** and use **cross-multiplication.** That is, if

$$\frac{a}{b} = \frac{c}{d}, \quad \text{then} \quad a \cdot d = b \cdot c.$$

Note how cross-multiplication is used in the next example.

Example 9   Using Cross-Multiplication

Use cross-multiplication to solve $\dfrac{x + 2}{3} = \dfrac{8}{5}$. Then check your solution.

Solution

$$\frac{x + 2}{3} = \frac{8}{5} \qquad \text{Write original equation.}$$

$$5(x + 2) = 3(8) \qquad \text{Cross multiply.}$$

$$5x + 10 = 24 \qquad \text{Distributive Property}$$

$$5x = 14 \qquad \text{Subtract 10 from each side.}$$

$$x = \frac{14}{5} \qquad \text{Divide each side by 5.}$$

Check

$$\frac{x + 2}{3} = \frac{8}{5} \qquad \text{Write original equation.}$$

$$\frac{\left(\frac{14}{5} + 2\right)}{3} \overset{?}{=} \frac{8}{5} \qquad \text{Substitute } \tfrac{14}{5} \text{ for } x.$$

$$\frac{\left(\frac{14}{5} + \frac{10}{5}\right)}{3} \overset{?}{=} \frac{8}{5} \qquad \text{Write 2 as } \tfrac{10}{5}.$$

$$\frac{\frac{24}{5}}{3} \overset{?}{=} \frac{8}{5} \qquad \text{Simplify.}$$

$$\frac{24}{5}\left(\frac{1}{3}\right) \overset{?}{=} \frac{8}{5} \qquad \text{Invert and multiply.}$$

$$\frac{8}{5} = \frac{8}{5} \qquad \text{Solution checks. } \checkmark$$

The solution is $x = \tfrac{14}{5}$.

Bear in mind that cross-multiplication can be used only with equations written in a form that equates two fractions. Try rewriting the equation in Example 6 in this form and then use cross-multiplication to solve for x.

More extensive applications of cross-multiplication will be discussed when you study ratios and proportions later in this chapter.

③ Solve linear equations involving decimals.

Many real-life applications of linear equations involve decimal coefficients. To solve such an equation, you can clear it of decimals in much the same way you clear an equation of fractions. Multiply each side by a power of 10 that converts all decimal coefficients to integers, as shown in the next example.

Example 10   Solving a Linear Equation Involving Decimals

Solve $0.3x + 0.2(10 - x) = 0.15(30)$. Then check your solution.

Solution

$0.3x + 0.2(10 - x) = 0.15(30)$	Write original equation.
$0.3x + 2 - 0.2x = 4.5$	Distributive Property
$0.1x + 2 = 4.5$	Combine like terms.
$10(0.1x + 2) = 10(4.5)$	Multiply each side by 10.
$x + 20 = 45$	Clear decimals.
$x = 25$	Subtract 20 from each side.

Check

$0.3(25) + 0.2(10 - 25) \stackrel{?}{=} 0.15(30)$	Substitute 25 for x in original equation.
$0.3(25) + 0.2(-15) \stackrel{?}{=} 0.15(30)$	Perform subtraction within parentheses.
$7.5 - 3.0 \stackrel{?}{=} 4.5$	Multiply.
$4.5 = 4.5$	Solution checks. ✓

The solution is $x = 25$.

Study Tip

There are other ways to solve the decimal equation in Example 10. You could first clear the equation of decimals by multiplying each side by 100. Or, you could keep the decimals and use a graphing calculator to do the arithmetic operations. The method you choose is a matter of personal preference.

Example 11   ACT Participants

The number y (in thousands) of students who took the ACT from 1996 to 2002 can be approximated by the linear model $y = 30.5t + 746$, where t represents the year, with $t = 6$ corresponding to 1996. Assuming that this linear pattern continues, find the year in which there will be 1234 thousand students taking the ACT. (Source: The ACT, Inc.)

Solution

To find the year in which there will be 1234 thousand students taking the ACT, substitute 1234 for y in the original equation and solve the equation for t.

$1234 = 30.5t + 746$	Substitute 1234 for y in original equation.
$488 = 30.5t$	Subtract 746 from each side.
$16 = t$	Divide each side by 30.5.

Because $t = 6$ corresponds to 1996, $t = 16$ must represent 2006. So, from this model, there will be 1234 thousand students taking the ACT in 2006. Check this in the original statement of the problem.

3.2 Exercises

Review   Concepts, Skills, and Problem Solving

Keep mathematically in shape by doing these exercises *before* the problems of this section.

Properties and Definitions

1. *Writing* ✏️   In your own words, describe how you add the following fractions.

   (a) $\frac{1}{5} + \frac{7}{5}$

   (b) $\frac{1}{5} + \frac{7}{3}$

2. Create two examples of algebraic expressions.

Simplifying Expressions

In Exercises 3–10, simplify the expression.

3. $(-2x)^2 x^4$

4. $-y^2(-2y)^3$

5. $5z^3(z^2)$

6. $a^2 + 3a + 4 - 2a - 6$

7. $\frac{5x}{3} - \frac{2x}{3} - 4$

8. $2x^2 - 4 + 5 - 3x^2$

9. $-y^2(y^2 + 4) + 6y^2$

10. $5t(2 - t) + t^2$

Problem Solving

11. *Fuel Usage*   At the beginning of the day, a gasoline tank was full. The tank holds 20 gallons. At the end of the day, the fuel gauge indicates that the tank is $\frac{5}{8}$ full. How many gallons of gasoline were used?

12. *Consumerism*   You buy a pickup truck for $1800 down and 36 monthly payments of $625 each.

   (a) What is the total amount you will pay?

   (b) The final cost of the pickup is $19,999. How much extra did you pay in finance charges and other fees?

Developing Skills

In Exercises 1–52, solve the equation and check your solution. (Some of the equations have no solution.) See Examples 1–8.

1. $2(y - 4) = 0$

2. $9(y - 7) = 0$

3. $-5(t + 3) = 10$

4. $-3(x + 1) = 18$

5. $25(z - 2) = 60$

6. $2(x - 3) = 4$

7. $7(x + 5) = 49$

8. $4(x + 1) = 24$

9. $4 - (z + 6) = 8$

10. $25 - (y + 3) = 15$

11. $3 - (2x - 4) = 3$

12. $16 - (3x - 10) = 5$

13. $12(x - 3) = 0$

14. $4(z - 2) = 0$

15. $3(2x - 1) = 3(2x + 5)$

16. $4(z - 2) = 2(2z - 4)$

17. $-3(x + 4) = 4(x + 4)$

18. $-8(x - 6) = 3(x - 6)$

19. $7 = 3(x + 2) - 3(x - 5)$

20. $24 = 12(z + 1) - 3(4z - 2)$

21. $7x - 2(x - 2) = 12$

22. $15(x + 1) - 8x = 29$

23. $6 = 3(y + 1) - 4(1 - y)$

24. $100 = 4(y - 6) - (y - 1)$

25. $-6(3 + x) + 2(3x + 5) = 0$

26. $-3(5x + 2) + 5(1 + 3x) = 0$

27. $2[(3x + 5) - 7] = 3(5x - 2)$

28. $3[(5x + 1) - 4] = 4(2x - 3)$

29. $4x + 3[x - 2(2x - 1)] = 4 - 3x$

30. $16 + 4[5x - 4(x + 2)] = 7 - 2x$

31. $\frac{y}{5} = \frac{3}{5}$

32. $\frac{z}{3} = \frac{10}{3}$

33. $\frac{y}{5} = -\frac{3}{10}$

34. $\frac{v}{4} = -\frac{7}{8}$

35. $\frac{6x}{25} = \frac{3}{5}$

36. $\frac{8x}{9} = \frac{2}{3}$

37. $\frac{5x}{4} + \frac{1}{2} = 0$

38. $\frac{3z}{7} + \frac{6}{11} = 0$

39. $\dfrac{x}{5} - \dfrac{1}{2} = 3$

40. $\dfrac{y}{4} - \dfrac{5}{8} = 2$

41. $\dfrac{x}{5} - \dfrac{x}{2} = 1$

42. $\dfrac{x}{3} + \dfrac{x}{4} = 1$

43. $2s + \dfrac{3}{2} = 2s + 2$

44. $\dfrac{3}{4} + 5s = -2 + 5s$

45. $3x + \dfrac{1}{4} = \dfrac{3}{4}$

46. $2x - \dfrac{3}{8} = \dfrac{5}{8}$

47. $\dfrac{1}{5}x + 1 = \dfrac{3}{10}x - 4$

48. $\dfrac{1}{8}x + 3 = \dfrac{1}{4}x + 5$

49. $\dfrac{2}{3}(z + 5) - \dfrac{1}{4}(z + 24) = 0$

50. $\dfrac{3x}{2} + \dfrac{1}{4}(x - 2) = 10$

51. $\dfrac{100 - 4u}{3} = \dfrac{5u + 6}{4} + 6$

52. $\dfrac{8 - 3x}{2} - 4 = \dfrac{x}{6}$

In Exercises 53–62, use cross-multiplication to solve the equation. See Example 9.

53. $\dfrac{t + 4}{6} = \dfrac{2}{3}$

54. $\dfrac{x - 6}{10} = \dfrac{3}{5}$

55. $\dfrac{x - 2}{5} = \dfrac{2}{3}$

56. $\dfrac{2x + 1}{3} = \dfrac{5}{2}$

57. $\dfrac{5x - 4}{4} = \dfrac{2}{3}$

58. $\dfrac{10x + 3}{6} = \dfrac{1}{2}$

59. $\dfrac{x}{4} = \dfrac{1 - 2x}{3}$

60. $\dfrac{x + 1}{6} = \dfrac{3x}{10}$

61. $\dfrac{10 - x}{2} = \dfrac{x + 4}{5}$

62. $\dfrac{2x + 3}{5} = \dfrac{3 - 4x}{8}$

In Exercises 63–72, solve the equation. Round your answer to two decimal places. See Example 10.

63. $0.2x + 5 = 6$

64. $4 - 0.3x = 1$

65. $0.234x + 1 = 2.805$

66. $275x - 3130 = 512$

67. $0.02x - 0.96 = 1.50$

68. $1.35x + 14.50 = 6.34$

69. $\dfrac{x}{3.25} + 1 = 2.08$

70. $\dfrac{x}{4.08} + 7.2 = 5.14$

71. $\dfrac{x}{3.155} = 2.850$

72. $\dfrac{3x}{4.5} = \dfrac{1}{8}$

Solving Problems

73. *Time to Complete a Task*   Two people can complete 80% of a task in t hours, where t must satisfy the equation $t/10 + t/15 = 0.8$. How long will it take for the two people to complete 80% of the task?

74. *Time to Complete a Task*   Two machines can complete a task in t hours, where t must satisfy the equation $t/10 + t/15 = 1$. How long will it take for the two machines to complete the task?

75. *Course Grade*   To get an A in a course, you must have an average of at least 90 points for four tests of 100 points each. For the first three tests, your scores are 87, 92, and 84. What must you score on the fourth exam to earn a 90% average for the course?

76. *Course Grade*   Repeat Exercise 75 if the fourth test is weighted so that it counts for twice as much as each of the first three tests.

In Exercises 77–80, use the equation and solve for x.

$$p_1 x + p_2(a - x) = p_3 a$$

77. *Mixture Problem*   Determine the number of quarts of a 10% solution that must be mixed with a 30% solution to obtain 100 quarts of a 25% solution. ($p_1 = 0.1$, $p_2 = 0.3$, $p_3 = 0.25$, and $a = 100$.)

78. *Mixture Problem*   Determine the number of gallons of a 25% solution that must be mixed with a 50% solution to obtain 5 gallons of a 30% solution. ($p_1 = 0.25$, $p_2 = 0.5$, $p_3 = 0.3$, and $a = 5$.)

79. *Mixture Problem*   An eight-quart automobile cooling system is filled with coolant that is 40% antifreeze. Determine the amount that must be withdrawn and replaced with pure antifreeze so that the 8 quarts of coolant will be 50% antifreeze. ($p_1 = 1$, $p_2 = 0.4$, $p_3 = 0.5$, and $a = 8$.)

80. *Mixture Problem*   A grocer mixes two kinds of nuts costing $2.49 per pound and $3.89 per pound to make 100 pounds of a mixture costing $3.19 per pound. How many pounds of the nuts costing $2.49 per pound must be put into the mixture? ($p_1 = 2.49$, $p_2 = 3.89$, $p_3 = 3.19$, and $a = 100$.)

81. *Data Analysis*   The table shows the projected numbers N (in millions) of persons 65 years of age or older in the United States. (Source: U.S. Census Bureau)

Year	2005	2015	2025	2035
N	36.4	46.0	62.6	74.8

A model for the data is

$$N = 1.32t + 28.6$$

where t represents time in years, with $t = 5$ corresponding to the year 2005. According to the model, in what year will the population of those 65 or older exceed 80 million?

82. *Fireplace Construction*   A fireplace is 93 inches wide. Each brick in the fireplace has a length of 8 inches and there is $\frac{1}{2}$ inch of mortar between adjoining bricks (see figure). Let n be the number of bricks per row.

(a) Explain why the number of bricks per row is the solution of the equation $8n + \frac{1}{2}(n - 1) = 93$.

(b) Find the number of bricks per row in the fireplace.

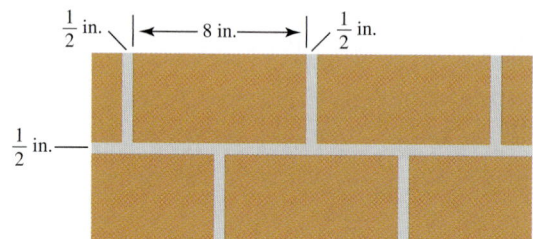

Explaining Concepts

83. *Writing*   In your own words, describe the procedure for removing symbols of grouping. Give some examples.

84. You could solve $3(x - 7) = 15$ by applying the Distributive Property as the first step. However, there is another way to begin. What is it?

85. *Error Analysis*   Describe the error.

$$-2(x - 5) = 8$$
$$-2x - 5 = 8$$

86. *Writing*   Explain what happens when you divide each side of an equation by a variable factor.

87. *Writing*   What is meant by the least common multiple of the denominators of two or more fractions? Discuss the method for finding the least common multiple of the denominators of fractions.

88. *Writing*   When solving an equation that contains fractions, explain what is accomplished by multiplying each side of the equation by the least common multiple of the denominators of the fractions.

89. *Writing*   When simplifying an algebraic *expression* involving fractions, why can't you simplify the expression by multiplying by the least common multiple of the denominators?

3.3 Problem Solving with Percents

Bob Mahoney/The Image Works

What You Should Learn

1. Convert percents to decimals and fractions and convert decimals and fractions to percents.
2. Solve linear equations involving percents.
3. Solve application problems involving markups and discounts.

Why You Should Learn It

Real-life data can be organized using circle graphs and percents. For instance, in Exercise 101 on page 156, a circlegraph is used to show the percents of Americans in different age groups visiting office-based physicians.

1 Convert percents to decimals and fractions and convert decimals and fractions to percents.

Percents

In applications involving percents, you usually must convert the percents to decimal (or fractional) form before performing any arithmetic operations. Consequently, you need to be able to convert from percents to decimals (or fractions), and vice versa. The following verbal model can be used to perform the conversions.

$$\boxed{\text{Decimal or fraction}} \cdot \boxed{100\%} = \boxed{\text{Percent}}$$

For example, the decimal 0.38 corresponds to 38 percent. That is,

$$0.38(100\%) = 38\%.$$

Example 1   Converting Decimals and Fractions to Percents

Convert each number to a percent.

a. $\dfrac{3}{5}$ **b.** 1.20

Solution

a. *Verbal Model:* $\boxed{\text{Fraction}} \cdot \boxed{100\%} = \boxed{\text{Percent}}$

Equation: $\dfrac{3}{5}(100\%) = \dfrac{300}{5}\%$

$$= 60\%$$

So, the fraction $\frac{3}{5}$ corresponds to 60%.

b. *Verbal Model:* $\boxed{\text{Decimal}} \cdot \boxed{100\%} = \boxed{\text{Percent}}$

Equation: $(1.20)(100\%) = 120\%$

So, the decimal 1.20 corresponds to 120%.

Study Tip

Note in Example 1(b) that it is possible to have percents that are larger than 100%. It is also possible to have percents that are less than 1%, such as $\frac{1}{2}\%$ or 0.78%.

Example 2   Converting Percents to Decimals and Fractions

a. Convert 3.5% to a decimal.

b. Convert 55% to a fraction.

Solution

a. *Verbal Model:* $\boxed{\text{Decimal}} \cdot \boxed{100\%} = \boxed{\text{Percent}}$

Label: $x = \text{decimal}$

Equation: $x(100\%) = 3.5\%$ Original equation

$$x = \frac{3.5\%}{100\%}$$ Divide each side by 100%.

$$x = 0.035$$ Simplify.

So, 3.5% corresponds to the decimal 0.035.

b. *Verbal Model:* $\boxed{\text{Fraction}} \cdot \boxed{100\%} = \boxed{\text{Percent}}$

Label: $x = \text{fraction}$

Equation: $x(100\%) = 55\%$ Original equation

$$x = \frac{55\%}{100\%}$$ Divide each side by 100%.

$$x = \frac{11}{20}$$ Simplify.

So, 55% corresponds to the fraction $\frac{11}{20}$.

Some percents occur so commonly that it is helpful to memorize their conversions. For instance, 100% corresponds to 1 and 200% corresponds to 2. The table below shows the decimal and fraction conversions for several percents.

Percent	10%	$12\frac{1}{2}\%$	20%	25%	$33\frac{1}{3}\%$	50%	$66\frac{2}{3}\%$	75%
Decimal	0.1	0.125	0.2	0.25	$0.\overline{3}$	0.5	$0.\overline{6}$	0.75
Fraction	$\frac{1}{10}$	$\frac{1}{8}$	$\frac{1}{5}$	$\frac{1}{4}$	$\frac{1}{3}$	$\frac{1}{2}$	$\frac{2}{3}$	$\frac{3}{4}$

Percent means *per hundred* or *parts of 100*. (The Latin word for 100 is *centum*.) For example, 20% means 20 parts of 100, which is equivalent to the fraction 20/100 or $\frac{1}{5}$. In applications involving percent, many people like to state percent in terms of a portion. For instance, the statement "20% of the population lives in apartments" is often stated as "1 out of every 5 people lives in an apartment."

2 Solve linear equations involving percents.

The Percent Equation

The primary use of percents is to compare two numbers. For example, 2 is 50% of 4, and 5 is 25% of 20. The following model is helpful.

Verbal Model: $a = p$ percent of b

Labels: b = base number
p = percent (in decimal form)
a = number being compared to b

Equation: $a = p \cdot b$

Example 3   Solving Percent Equations

a. What number is 30% of 70?

b. Fourteen is 25% of what number?

c. One hundred thirty-five is what percent of 27?

Solution

a. *Verbal Model:*   What number $=$ 30% of 70

Label: a = unknown number

Equation: $a = (0.3)(70) = 21$

So, 21 is 30% of 70.

b. *Verbal Model:*   14 $=$ 25% of what number

Label: b = unknown number

Equation: $14 = 0.25b$

$$\frac{14}{0.25} = b$$

$$56 = b$$

So, 14 is 25% of 56.

c. *Verbal Model:*   135 $=$ What percent of 27

Label: p = unknown percent (in decimal form)

Equation: $135 = p(27)$

$$\frac{135}{27} = p$$

$$5 = p$$

So, 135 is 500% of 27.

From Example 3, you can see that there are three basic types of percent problems. Each can be solved by substituting the two given quantities into the percent equation and solving for the third quantity.

Question	*Given*	*Percent Equation*
a is what percent of b?	a and b	Solve for p.
What number is p percent of b?	p and b	Solve for a.
a is p percent of what number?	a and p	Solve for b.

For instance, part (b) of Example 3 fits the form "a is p percent of what number?"

In most real-life applications, the base number b and the number a are much more disguised than they are in Example 3. It sometimes helps to think of a as a "new" amount and b as the "original" amount.

Example 4   Real Estate Commission

A real estate agency receives a commission of $8092.50 for the sale of a $124,500 house. What percent commission is this?

Solution

Verbal Model:

$$\text{Commission} = \frac{\text{Percent}}{\text{(in decimal form)}} \cdot \text{Sale price}$$

Labels:

Commission = 8092.50	(dollars)
Percent = p	(in decimal form)
Sale price = 124,500	(dollars)

Equation:

$8092.50 = p \cdot (124,500)$ Original equation

$\dfrac{8092.50}{124,500} = p$ Divide each side by 124,500.

$0.065 = p$ Simplify.

So, the real estate agency receives a commission of 6.5%.

Example 5   Cost-of-Living Raise

A union negotiates for a cost-of-living raise of 7%. What is the raise for a union member whose salary is $23,240? What is this person's new salary?

Solution

Verbal Model:

$$\text{Raise} = \frac{\text{Percent}}{\text{(in decimal form)}} \cdot \text{Salary}$$

Labels:

Raise = a	(dollars)
Percent = 7% = 0.07	(in decimal form)
Salary = 23,240	(dollars)

Equation: $a = 0.07(23,240) = 1626.80$

So, the raise is $1626.80 and the new salary is

$23,240.00 + 1626.80 = \$24,866.80.$

Example 6   Course Grade

You missed an A in your chemistry course by only three points. Your point total for the course is 402. How many points were possible in the course? (Assume that you needed 90% of the course total for an A.)

Solution

Verbal Model:

$$\boxed{\text{Your points}} + \boxed{\text{3 points}} = \boxed{\text{Percent (in decimal form)}} \cdot \boxed{\text{Total points}}$$

Labels: Your points $= 402$ (points)
            Percent $= 90\% = 0.9$ (in decimal form)
            Total points for course $= b$ (points)

Equation: $402 + 3 = 0.9b$ Original equation

            $405 = 0.9b$ Add.

            $\dfrac{405}{0.9} = b$ Divide each side by 0.9.

            $450 = b$ Simplify.

So, there were 450 total points for the course. You can check your solution as follows.

$402 + 3 = 0.9b$ Write original equation.

$402 + 3 \overset{?}{=} 0.9(450)$ Substitute 450 for b.

$405 = 405$ Solution checks. ✓

Example 7   Percent Increase

The monthly basic cable TV rate was $7.69 in 1980 and $30.08 in 2000. Find the percent increase in the monthly basic cable TV rate from 1980 to 2000. (Source: Paul Kagan Associates, Inc.)

Solution

Verbal Model:

$$\boxed{\text{2000 price}} = \boxed{\text{1980 price}} \cdot \boxed{\text{Percent increase (in decimal form)}} + \boxed{\text{1980 price}}$$

Labels: 2000 price $= 30.08$ (dollars)
            Percent increase $= p$ (in decimal form)
            1980 price $= 7.69$ (dollars)

Equation: $30.08 = 7.69p + 7.69$ Original equation

            $22.39 = 7.69p$ Subtract 7.69 from each side.

            $2.91 \approx p$ Divide each side by 7.69.

So, the percent increase in the monthly basic cable TV rate from 1980 to 2000 is approximately 291%. Check this in the original statement of the problem.

③ Solve application problems involving markups and discounts.

Markups and Discounts

You may have had the experience of buying an item at one store and later finding that you could have paid less for the same item at another store. The basic reason for this price difference is **markup,** which is the difference between the **cost** (the amount a retailer pays for the item) and the **price** (the amount at which the retailer sells the item to the consumer). A verbal model for this problem is as follows.

$$\boxed{\text{Selling price}} = \boxed{\text{Cost}} + \boxed{\text{Markup}}$$

In such a problem, the markup may be known or it may be expressed as a percent of the cost. This percent is called the **markup rate.**

$$\boxed{\text{Markup}} = \boxed{\text{Markup rate}} \cdot \boxed{\text{Cost}}$$

Markup is one of those "hidden operations" referred to in Section 2.3.

In business and economics, the terms *cost* and *price* do not mean the same thing. The cost of an item is the amount a business pays for the item. The price of an item is the amount for which the business sells the item.

Example 8   Finding the Selling Price

A sporting goods store uses a markup rate of 55% on all items. The cost of a golf bag is $45. What is the selling price of the bag?

Solution

Verbal Model: $\boxed{\text{Selling price}} = \boxed{\text{Cost}} + \boxed{\text{Markup}}$

Labels:
Selling price = x (dollars)
Cost = 45   (dollars)
Markup rate = 0.55   (rate in decimal form)
Markup = (0.55)(45)   (dollars)

Equation: $x = 45 + (0.55)(45)$ Original equation.

$= 45 + 24.75$ Multiply.

$= \$69.75$ Simplify.

The selling price is $69.75. You can check your solution as follows:

$x = 45 + (0.55)(45)$ Write original equation.

$69.75 \overset{?}{=} 45 + (0.55)(45)$ Substitute 69.75 for x.

$69.75 = 69.75$ Solution checks. ✔

In Example 8, you are given the cost and are asked to find the selling price. Example 9 illustrates the reverse problem. That is, in Example 9 you are given the selling price and are asked to find the cost.

Example 9   Finding the Cost of an Item

The selling price of a pair of ski boots is $98. The markup rate is 60%. What is the cost of the boots?

Solution

Verbal Model:

Selling price $=$ Cost $+$ Markup

Labels:

Selling price $= 98$ (dollars)
Cost $= x$ (dollars)
Markup rate $= 0.60$ (rate in decimal form)
Markup $= 0.60x$ (dollars)

Equation:

$98 = x + 0.60x$ Original equation

$98 = 1.60x$ Combine like terms.

$61.25 = x$ Divide each side by 1.60.

The cost is $61.25. Check this in the original statement of the problem.

Example 10   Finding the Markup Rate

A pair of walking shoes sells for $60. The cost of the walking shoes is $24. What is the markup rate?

Solution

Verbal Model:

Selling price $=$ Cost $+$ Markup

Labels:

Selling price $= 60$ (dollars)
Cost $= 24$ (dollars)
Markup rate $= p$ (rate in decimal form)
Markup $= p(24)$ (dollars)

Equation:

$60 = 24 + p(24)$ Original equation

$36 = 24p$ Subtract 24 from each side.

$1.5 = p$ Divide each side by 24.

Because $p = 1.5$, it follows that the markup rate is 150%.

The mathematics of a discount is similar to that of a markup. The model for this situation is

Selling price $=$ List price $-$ Discount

where the **discount** is given in dollars, and the **discount rate** is given as a percent of the list price. Notice the "hidden operation" in the discount.

Discount $=$ Discount rate $\cdot$ List price

Example 11   Finding the Discount Rate

During a midsummer sale, a lawn mower listed at $199.95 is on sale for $139.95. What is the discount rate?

Solution

Verbal Model:

$$\boxed{\text{Discount}} = \boxed{\begin{array}{c}\text{Discount}\\\text{rate}\end{array}} \cdot \boxed{\begin{array}{c}\text{List}\\\text{price}\end{array}}$$

Labels:

Discount = $199.95 - 139.95 = 60$ (dollars)
List price = 199.95 (dollars)
Discount rate = p (rate in decimal form)

Equation:

$60 = p(199.95)$ Original equation

$0.30 \approx p$ Divide each side by 199.95.

Because $p \approx 0.30$, it follows that the discount rate is approximately 30%.

Example 12   Finding the Sale Price

A drug store advertises 40% off the prices of all summer tanning products. A bottle of suntan oil lists for $3.49. What is the sale price?

Solution

Verbal Model:

$$\boxed{\begin{array}{c}\text{Sale}\\\text{price}\end{array}} = \boxed{\begin{array}{c}\text{List}\\\text{price}\end{array}} - \boxed{\text{Discount}}$$

Labels:

List price = 3.49 (dollars)
Discount rate = 0.4 (rate in decimal form)
Discount = $0.4(3.49)$ (dollars)
Sale price = x (dollars)

Equation: $x = 3.49 - (0.4)(3.49) \approx \$2.09$

The sale price is $2.09. Check this in the original statement of the problem.

The following guidelines summarize the problem-solving strategy that you should use when solving word problems.

Guidelines for Solving Word Problems

1. Write a *verbal model* that describes the problem.

2. Assign *labels* to fixed quantities and variable quantities.

3. Rewrite the verbal model as an *algebraic equation* using the assigned labels.

4. *Solve* the resulting algebraic equation.

5. *Check* to see that your solution satisfies the original problem as stated.

3.3 Exercises

Review   Concepts, Skills, and Problem Solving

Keep mathematically in shape by doing these exercises *before* the problems of this section.

Properties and Definitions

1. *Writing* Explain how to put the two numbers 63 and -28 in order.

2. For any real number, its distance from ▒ on the real number line is its absolute value.

Simplifying Expressions

In Exercises 3–6, evaluate the expression.

3. $8 - |-7 + 11| + (-4)$

4. $34 - [54 - (-16 + 4) + 6]$

5. Subtract 230 from -300.

6. Find the absolute value of the difference of 17 and -12.

In Exercises 7 and 8, use the Distributive Property to expand the expression.

7. $4(2x - 5)$ 8. $-z(xz - 2y^2)$

In Exercises 9 and 10, evaluate the algebraic expression for the specified values of the variables. (If not possible, state the reason.)

9. $x^2 - y^2$
   (a) $x = 4, y = 3$
   (b) $x = -5, y = 3$

10. $\dfrac{z^2 + 2}{x^2 - 1}$
    (a) $x = 1, z = 1$
    (b) $x = 2, z = 2$

Problem Solving

11. *Consumer Awareness* A telephone company charges \$1.37 for the first minute of a long-distance telephone call and \$0.95 for each additional minute. Find the cost of a 15-minute telephone call.

12. *Distance* A train travels at the rate of r miles per hour for 5 hours. Write an algebraic expression that represents the total distance traveled by the train.

Developing Skills

In Exercises 1–12, complete the table showing the equivalent forms of a percent. See Examples 1 and 2.

	Percent	Parts out of 100	Decimal	Fraction
1.	40%			
2.	16%			
3.	7.5%			
4.	75%			
5.		63		
6.		10.5		
7.			0.155	
8.			0.80	
9.				$\frac{3}{5}$
10.				$\frac{3}{20}$
11.	150%			
12.			1.25	

In Exercises 13–20, convert the decimal to a percent. See Example 1.

13. 0.62

14. 0.57

15. 0.20

16. 0.38

17. 0.075

18. 0.005

19. 2.38

20. 1.75

In Exercises 21–28, convert the percent to a decimal. See Example 2.

21. 12.5%   22. 95%

23. 125%   24. 250%

25. 8.5%   26. 0.3%

27. $\frac{3}{4}\%$ 28. $4\frac{4}{5}\%$

In Exercises 29–36, convert the fraction to a percent. See Example 1.

29. $\frac{4}{5}$ **30.** $\frac{1}{4}$

31. $\frac{5}{4}$ **32.** $\frac{6}{5}$

33. $\frac{5}{6}$ **34.** $\frac{2}{3}$

35. $\frac{21}{20}$ **36.** $\frac{5}{2}$

In Exercises 37–40, what percent of the figure is shaded? (There are a total of 360° in a circle.)

37.   **38.**

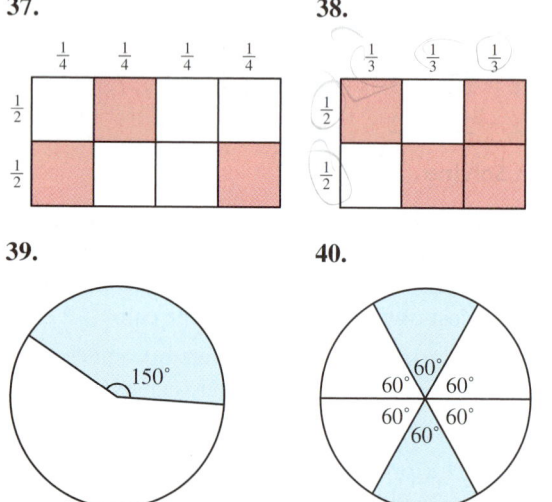

39.   **40.**

In Exercises 41–64, solve the percent equation. See Example 3.

41. What number is 30% of 150?

42. What number is 62% of 1200?

43. What number is $66\frac{2}{3}\%$ of 816?

44. What number is $33\frac{1}{3}\%$ of 516?

45. What number is 0.75% of 56?

46. What number is 0.2% of 100,000?

47. What number is 200% of 88?

48. What number is 325% of 450?

49. 903 is 43% of what number?

50. 425 is 85% of what number?

51. 275 is $12\frac{1}{2}\%$ of what number?

52. 814 is $66\frac{2}{3}\%$ of what number?

53. 594 is 450% of what number?

54. 210 is 250% of what number?

55. 2.16 is 0.6% of what number?

56. 51.2 is 0.08% of what number?

57. 576 is what percent of 800?

58. 1950 is what percent of 5000?

59. 45 is what percent of 360?

60. 38 is what percent of 5700?

61. 22 is what percent of 800?

62. 110 is what percent of 110?

63. 1000 is what percent of 200?

64. 148.8 is what percent of 960?

In Exercises 65–74, find the missing quantities. See Examples 8, 9, and 10.

	Cost	Selling Price	Markup	Markup Rate
65.	$26.97	$49.95		
66.	$71.97	$119.95		
67.		$74.38		81.5%
68.		$69.99		55.5%
69.		$125.98	$56.69	
70.		$350.00	$80.77	
71.		$15,900.00	$2650.00	
72.		$224.87	$75.08	
73.	$107.97			85.2%
74.	$680.00			$33\frac{1}{3}\%$

In Exercises 75–84, find the missing quantities. See Examples 11 and 12.

	List Price	Sale Price	Discount	Discount Rate
75.	$39.95	$29.95		
76.	$50.99	$45.99		
77.		$18.95		20%
78.		$189.00		40%
79.	$189.99	159.99	$30.00	
80.	$18.95		$8.00	
81.	$119.96			50%
82.	$84.95			65%
83.		$695.00	$300.00	
84.		$259.97	$135.00	

Solving Problems

85. *Rent*   You spend 17% of your monthly income of $3200 for rent. What is your monthly payment?

86. *Cost of Housing*   You budget 30% of your annual after-tax income for housing. Your after-tax income is $38,500. What amount can you spend on housing?

87. *Retirement Plan*   You budget $7\frac{1}{2}$% of your gross income for an individual retirement plan. Your annual gross income is $45,800. How much will you put in your retirement plan each year?

88. *Enrollment*   In the fall of 2001, 41% of the students enrolled at Alabama State University were freshmen. The enrollment of the college was 5590. Find the number of freshmen enrolled in the fall of 2001. (Source: Alabama State University)

89. *Meteorology*   During the winter of 2000–2001, 33.6 inches of snow fell in Detroit, Michigan. Of that amount, 25.1 inches fell in December. What percent of the total snowfall amount fell in December? (Source: National Weather Service)

90. *Inflation Rate*   You purchase a lawn tractor for $3750 and 1 year later you note that the cost has increased to $3900. Determine the inflation rate (as a percent) for the tractor.

91. *Unemployment Rate*   During a recession, 72 out of 1000 workers in the population were unemployed. Find the unemployment rate (as a percent).

92. *Layoff*   Because of slumping sales, a small company laid off 30 of its 153 employees.

(a) What percent of the work force was laid off?

(b) Complete the statement: "About 1 out of every ▨ workers was laid off."

93. *Original Price*   A coat sells for $250 during a 20% off storewide clearance sale. What was the original price of the coat?

94. *Course Grade*   You were six points shy of a B in your mathematics course. Your point total for the course was 394. How many points were possible in the course? (Assume that you needed 80% of the course total for a B.)

95. *Consumer Awareness*   The price of a new van is approximately 110% of what it was 3 years ago. The current price is $26,850. What was the approximate price 3 years ago?

96. *Membership Drive*   Because of a membership drive for a public television station, the current membership is 125% of what it was a year ago. The current number of members is 7815. How many members did the station have last year?

97. *Eligible Voters*   The news media reported that 6432 votes were cast in the last election and that this represented 63% of the eligible voters of a district. How many eligible voters are in the district?

98. *Quality Control*   A quality control engineer tested several parts and found two to be defective. The engineer reported that 2.5% were defective. How many were tested?

99. ▲ *Geometry*   A rectangular plot of land measures 650 feet by 825 feet (see figure). A square garage with sides of length 24 feet is built on the plot of land. What percentage of the plot of land is occupied by the garage?

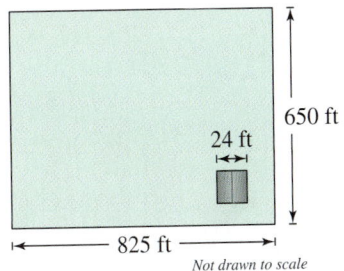

Not drawn to scale

100. ▲ *Geometry*   A circular target is attached to a rectangular board, as shown in the figure. The radius of the circle is $4\frac{1}{2}$ inches, and the measurements of the board are 12 inches by 15 inches. What percentage of the board is covered by the target? (The area of a circle is $A = \pi r^2$, where r is the radius of the circle.)

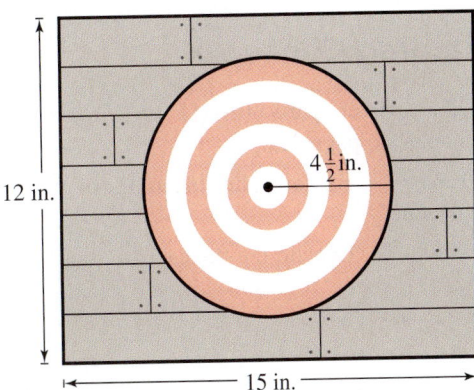

101. *Data Analysis*   In 1999 there were 841.3 million visits to office-based physicians. The circle graph classifies the age groups of those making the visits. Approximate the number of Americans in each of the classifications.   (Source: U.S. National Center for Health Statistics)

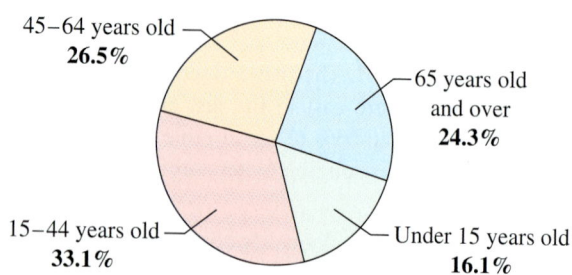

45–64 years old
26.5%

65 years old and over
24.3%

15–44 years old
33.1%

Under 15 years old
16.1%

102. *Graphical Estimation*   The bar graph shows the numbers (in thousands) of criminal cases commenced in the United States District Courts from 1997 through 2001.   (Source: Administrative Office of the U.S. Courts)

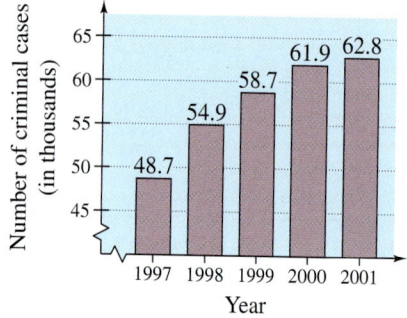

(a) Determine the percent increase in cases from 1997 to 1998.

(b) Determine the percent increase in cases from 1998 to 2001.

103. *Interpreting a Table*   The table shows the numbers of women scientists and the percents of women scientists in the United States in three fields for the years 1983 and 2000.   (Source: U.S. Bureau of Labor Statistics)

Field	1983		2000	
	Number	%	Number	%
Math/Computer	137,048	29.6%	651,236	31.4%
Chemistry	22,834	23.3%	46,359	30.3%
Biology	22,440	40.8%	51,756	45.4%

(a) Find the total number of mathematicians and computer scientists (men and women) in 2000.

(b) Find the total number of chemists (men and women) in 1983.

(c) Find the total number of biologists (men and women) in 2000.

104. *Data Analysis*   The table shows the approximate population (in millions) of Bangladesh for each decade from 1960 through 2000. Approximate the percent growth rate for each decade. If the growth rate of the 1990s continued until the year 2020, approximate the population in 2020.   (Source: U.S. Bureau of the Census, International Data Base)

Year	1960	1970	1980	1990	2000
Population	54.6	67.4	88.1	109.9	129.2

Explaining Concepts

105.   Answer parts (a)–(f) of Motivating the Chapter on page 122.

106. *Writing*   Explain the meaning of the word "percent."

107. *Writing*   Explain the concept of "rate."

108. *Writing*   Can any positive decimal be written as a percent? Explain.

109. *Writing*   Is it true that $\frac{1}{2}\% = 50\%$? Explain.

3.4 Ratios and Proportions

Eunice Harris/Photo Researchers, Inc.

Why You Should Learn It

Ratios can be used to represent many real-life quantities. For instance, in Exercise 60 on page 166, you will find the gear ratios for a five-speed bicycle.

What You Should Learn

1. Compare relative sizes using ratios.
2. Find the unit price of a consumer item.
3. Solve proportions that equate two ratios.
4. Solve application problems using the Consumer Price Index.

1 Compare relative sizes using ratios.

Setting Up Ratios

A **ratio** is a comparison of one number to another by division. For example, in a class of 29 students made up of 16 women and 13 men, the ratio of women to men is 16 to 13 or $\frac{16}{13}$. Some other ratios for this class are as follows.

$$\text{Men to women: } \frac{13}{16} \qquad \text{Men to students: } \frac{13}{29} \qquad \text{Students to women: } \frac{29}{16}$$

Note the order implied by a ratio. The ratio of a to b means a/b, whereas the ratio of b to a means b/a.

Definition of Ratio

The **ratio** of the real number a to the real number b is given by

$$\frac{a}{b}.$$

The ratio of a to b is sometimes written as $a : b$.

Example 1   Writing Ratios in Fractional Form

a. The ratio of 7 to 5 is given by $\frac{7}{5}$.

b. The ratio of 12 to 8 is given by $\frac{12}{8} = \frac{3}{2}$.

Note that the fraction $\frac{12}{8}$ can be written in simplest form as $\frac{3}{2}$.

c. The ratio of $3\frac{1}{2}$ to $5\frac{1}{4}$ is given by

$$\frac{3\frac{1}{2}}{5\frac{1}{4}} = \frac{\frac{7}{2}}{\frac{21}{4}} \qquad \text{Rewrite mixed numbers as fractions.}$$

$$= \frac{7}{2} \cdot \frac{4}{21} \qquad \text{Invert divisor and multiply.}$$

$$= \frac{2}{3}. \qquad \text{Simplify.}$$

There are many real-life applications of ratios. For instance, ratios are used to describe opinion surveys (for/against), populations (male/female, unemployed/employed), and mixtures (oil/gasoline, water/alcohol).

When comparing two *measurements* by a ratio, you should use the same unit of measurement in both the numerator and the denominator. For example, to find the ratio of 4 feet to 8 inches, you could convert 4 feet to 48 inches (by multiplying by 12) to obtain

$$\frac{4 \text{ feet}}{8 \text{ inches}} = \frac{48 \text{ inches}}{8 \text{ inches}} = \frac{48}{8} = \frac{6}{1}$$

or you could convert 8 inches to $\frac{8}{12}$ feet (by dividing by 12) to obtain

$$\frac{4 \text{ feet}}{8 \text{ inches}} = \frac{4 \text{ feet}}{\frac{8}{12} \text{ feet}} = 4 \cdot \frac{12}{8} = \frac{6}{1}.$$

If you use different units of measurement in the numerator and denominator, then you *must* include the units. If you use the same units of measurement in the numerator and denominator, then it is not necessary to write the units. A list of common conversion factors is found on the inside back cover.

Example 2   Comparing Measurements

Find ratios to compare the relative sizes of the following.

a. 5 gallons to 7 gallons   **b.** 3 meters to 40 centimeters

c. 200 cents to 3 dollars   **d.** 30 months to $1\frac{1}{2}$ years

Solution

a. Because the units of measurement are the same, the ratio is $\frac{5}{7}$.

b. Because the units of measurement are different, begin by converting meters to centimeters *or* centimeters to meters. Here, it is easier to convert meters to centimeters by multiplying by 100.

$$\frac{3 \text{ meters}}{40 \text{ centimeters}} = \frac{3(100) \text{ centimeters}}{40 \text{ centimeters}} \qquad \text{Convert meters to centimeters.}$$

$$= \frac{300}{40} \qquad \text{Multiply numerator.}$$

$$= \frac{15}{2} \qquad \text{Simplify.}$$

c. Because 200 cents is the same as 2 dollars, the ratio is

$$\frac{200 \text{ cents}}{3 \text{ dollars}} = \frac{2 \text{ dollars}}{3 \text{ dollars}} = \frac{2}{3}.$$

d. Because $1\frac{1}{2}$ years $= 18$ months, the ratio is

$$\frac{30 \text{ months}}{1\frac{1}{2} \text{ years}} = \frac{30 \text{ months}}{18 \text{ months}} = \frac{30}{18} = \frac{5}{3}.$$

② Find the unit price of a consumer item.

Unit Prices

As a consumer, you must be able to determine the unit prices of items you buy in order to make the best use of your money. The **unit price** of an item is given by the ratio of the total price to the total units.

$$\frac{\text{Unit}}{\text{price}} = \frac{\text{Total price}}{\text{Total units}}$$

The word *per* is used to state unit prices. For instance, the unit price for a particular brand of coffee might be 4.69 dollars *per* pound, or $4.69 per pound.

Example 3   Finding a Unit Price

Find the unit price (in dollars per ounce) for a five-pound, four-ounce box of detergent that sells for $4.62.

Solution

Begin by writing the weight in ounces. That is,

$$5 \text{ pounds} + 4 \text{ ounces} = 5 \text{ pounds}\left(\frac{16 \text{ ounces}}{1 \text{ pound}}\right) + 4 \text{ ounces}$$

$$= 80 \text{ ounces} + 4 \text{ ounces}$$

$$= 84 \text{ ounces}.$$

Next, determine the unit price as follows.

Verbal
Model: $\dfrac{\text{Unit}}{\text{price}} = \dfrac{\text{Total price}}{\text{Total units}}$

Unit Price: $\dfrac{\$4.62}{84 \text{ ounces}} = \$0.055 \text{ per ounce}$

Example 4   Comparing Unit Prices

Which has the lower unit price: a 12-ounce box of breakfast cereal for $2.69 or a 16-ounce box of the same cereal for $3.49?

Solution

The unit price for the smaller box is

$$\text{Unit price} = \frac{\text{Total price}}{\text{Total units}} = \frac{\$2.69}{12 \text{ ounces}} \approx \$0.224 \text{ per ounce}.$$

The unit price for the larger box is

$$\text{Unit price} = \frac{\text{Total price}}{\text{Total units}} = \frac{\$3.49}{16 \text{ ounces}} \approx \$0.218 \text{ per ounce}.$$

So, the larger box has a slightly lower unit price.

③ Solve proportions that equate two ratios.

Solving Proportions

A **proportion** is a statement that equates two ratios. For example, if the ratio of a to b is the same as the ratio of c to d, you can write the proportion as

$$\frac{a}{b} = \frac{c}{d}.$$

In typical applications, you know the values of three of the letters (quantities) and are required to find the value of the fourth. To solve such a fractional equation, you can use the *cross-multiplication* procedure introduced in Section 3.2.

Solving a Proportion

If

$$\frac{a}{b} = \frac{c}{d}$$

then $ad = bc$. The quantities a and d are called the **extremes** of the proportion, whereas b and c are called the **means** of the proportion.

Example 5   Solving Proportions

Solve each proportion.

a. $\dfrac{50}{x} = \dfrac{2}{28}$ **b.** $\dfrac{x}{3} = \dfrac{10}{6}$

Solution

a.
$$\frac{50}{x} = \frac{2}{28} \qquad \text{Write original proportion.}$$

$$50(28) = 2x \qquad \text{Cross-multiply.}$$

$$\frac{1400}{2} = x \qquad \text{Divide each side by 2.}$$

$$700 = x \qquad \text{Simplify.}$$

So, the ratio of 50 to 700 is the same as the ratio of 2 to 28.

b.
$$\frac{x}{3} = \frac{10}{6} \qquad \text{Write original proportion.}$$

$$x = \frac{30}{6} \qquad \text{Multiply each side by 3.}$$

$$x = 5 \qquad \text{Simplify.}$$

So, the ratio of 5 to 3 is the same as the ratio of 10 to 6.

To solve an equation, you want to isolate the variable. In Example 5(b), this was done by multiplying each side by 3 instead of cross-multiplying. In this case, multiplying each side by 3 was the only step needed to isolate the x-variable. However, either method is valid for solving the equation.

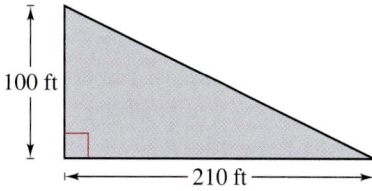

100 ft

210 ft

Triangular lot

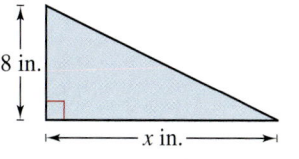

8 in.

x in.

Sketch

Figure 3.2

Example 6   Geometry: Similar Triangles

A triangular lot has perpendicular sides of lengths 100 feet and 210 feet. You are to make a proportional sketch of this lot using 8 inches as the length of the shorter side. How long should you make the other side?

Solution

This is a case of similar triangles in which the ratios of the corresponding sides are equal. The triangles are shown in Figure 3.2.

$$\frac{\text{Shorter side of lot}}{\text{Longer side of lot}} = \frac{\text{Shorter side of sketch}}{\text{Longer side of sketch}}$$
Proportion for similar triangles

$$\frac{100}{210} = \frac{8}{x}$$
Substitute.

$$x \cdot 100 = 210 \cdot 8$$
Cross-multiply.

$$x = \frac{1680}{100} = 16.8$$
Divide each side by 100.

So, the length of the longer side of the sketch should be 16.8 inches.

Example 7   Resizing a Picture

You have a 7-by-8-inch picture of a graph that you want to paste into a research paper, but you have only a 6-by-6-inch space in which to put it. You go to the copier that has five options for resizing your graph: 64%, 78%, 100%, 121%, and 129%.

a. Which option should you choose?

b. What are the measurements of the resized picture?

Solution

a. Because the longest side must be reduced from 8 inches to no more than 6 inches, consider the proportion

$$\frac{\text{New length}}{\text{Old length}} = \frac{\text{New percent}}{\text{Old percent}}$$
Original proportion

$$\frac{6}{8} = \frac{x}{100}$$
Substitute.

$$\frac{6}{8} \cdot 100 = x$$
Multiply each side by 100.

$$75 = x.$$
Simplify.

To guarantee a fit, you should choose the 64% option, because 78% is greater than the required 75%.

b. To find the measurements of the resized picture, multiply by 64% or 0.64.

$$\text{Length} = 0.64(8) = 5.12 \text{ inches} \qquad \text{Width} = 0.64(7) = 4.48 \text{ inches}$$

The size of the reduced picture is 5.12 inches by 4.48 inches.

Example 8   Gasoline Cost

You are driving from New York to Phoenix, a trip of 2450 miles. You begin the trip with a full tank of gas and after traveling 424 miles, you refill the tank for $24.00. How much should you plan to spend on gasoline for the entire trip?

Solution

Verbal Model:

$$\frac{\text{Cost for trip}}{\text{Cost for tank}} = \frac{\text{Miles for trip}}{\text{Miles for tank}}$$

Labels:

Cost of gas for entire trip $= x$	(dollars)
Cost of gas for tank $= 24$	(dollars)
Miles for entire trip $= 2450$	(miles)
Miles for tank $= 424$	(miles)

Proportion: $\dfrac{x}{24} = \dfrac{2450}{424}$ Original proportion

$x = 24\left(\dfrac{2450}{424}\right)$ Multiply each side by 24.

$x \approx 138.68$ Simplify.

You should plan to spend approximately $138.68 for gasoline on the trip. Check this in the original statement of the problem.

---

④ Solve application problems using the Consumer Price Index.

The Consumer Price Index

The rate of inflation is important to all of us. Simply stated, *inflation* is an economic condition in which the price of a fixed amount of goods or services increases. So, a fixed amount of money buys less in a given year than in previous years.

The most widely used measurement of inflation in the United States is the *Consumer Price Index* (CPI), often called the *Cost-of-Living Index*. The table below shows the "All Items" or general index for the years 1970 to 2001. (Source: Bureau of Labor Statistics)

Year	CPI	Year	CPI	Year	CPI	Year	CPI
1970	38.8	1978	65.2	1986	109.6	1994	148.2
1971	40.5	1979	72.6	1987	113.6	1995	152.4
1972	41.8	1980	82.4	1988	118.6	1996	156.9
1973	44.4	1981	90.9	1989	124.0	1997	160.5
1974	49.3	1982	96.5	1990	130.7	1998	163.0
1975	53.8	1983	99.6	1991	136.2	1999	166.6
1976	56.9	1984	103.9	1992	140.3	2000	172.2
1977	60.6	1985	107.6	1993	144.5	2001	177.1

To determine (from the CPI) the change in the buying power of a dollar from one year to another, use the following proportion.

$$\frac{\text{Price in year } n}{\text{Price in year } m} = \frac{\text{Index in year } n}{\text{Index in year } m}$$

Example 9   Using the Consumer Price Index

You paid $35,000 for a house in 1971. What is the amount you would pay for the same house in 2000?

Solution

Verbal Model: $\dfrac{\text{Price in 2000}}{\text{Price in 1971}} = \dfrac{\text{Index in 2000}}{\text{Index in 1971}}$

Labels: Price in 2000 $= x$ (dollars)
Price in 1971 $= 35,000$ (dollars)
Index in 2000 $= 172.2$
Index in 1971 $= 40.5$

Proportion: $\dfrac{x}{35,000} = \dfrac{172.2}{40.5}$ Original proportion

$x = \dfrac{172.2}{40.5} \cdot 35,000$ Multiply each side by 35,000.

$x \approx \$148,815$ Simplify.

So, you would pay approximately $148,815 for the house in 2000. Check this in the original statement of the problem.

Example 10   Using the Consumer Price Index

You inherited a diamond pendant from your grandmother in 1999. The pendant was appraised at $1300. What was the value of the pendant when your grandmother bought it in 1973?

Solution

Verbal Model: $\dfrac{\text{Price in 1999}}{\text{Price in 1973}} = \dfrac{\text{Index in 1999}}{\text{Index in 1973}}$

Labels: Price in 1999 $= 1300$ (dollars)
Price in 1973 $= x$ (dollars)
Index in 1999 $= 166.6$
Index in 1973 $= 44.4$

Proportion: $\dfrac{1300}{x} = \dfrac{166.6}{44.4}$ Original proportion

$57,720 = 166.6x$ Cross-multiply.

$346 \approx x$ Divide each side by 166.6.

So, the value of the pendant in 1973 was approximately $346. Check this in the original statement of the problem.

3.4 Exercises

Review *Concepts, Skills, and Problem Solving*

Keep mathematically in shape by doing these exercises *before* the problems of this section.

Properties and Definitions

1. *Writing* Explain how to write $\frac{15}{12}$ in simplest form.

2. *Writing* Explain how to divide $\frac{3}{5}$ by $\frac{x}{2}$.

3. Complete the Associative Property: $(3x)y =$ ☐ .

4. Identify the property of real numbers illustrated by $x^2 + 0 = x^2$.

Simplifying Expressions

In Exercises 5–10, evaluate the expression.

5. $3^2 - (-4)$ 6. $(-5)^3 + 3$

7. 9.3×10^6

8. $\dfrac{-|7 + 3^2|}{4}$

9. $(-4)^2 - (30 \div 50)$

10. $(8 \cdot 9) + (-4)^3$

Writing Models

In Exercises 11 and 12, translate the phrase into an algebraic expression.

11. Twice the difference of a number and 10

12. The area of a triangle with base b and height $\frac{1}{2}(b + 6)$

Developing Skills

In Exercises 1–8, write the ratio as a fraction in simplest form. See Example 1.

1. 36 to 9
2. 45 to 15
3. 27 to 54
4. 27 to 63
5. 14 : 21
6. 12 : 30
7. 144 : 16
8. 60 : 45

In Exercises 9–26, find a ratio that compares the relative sizes of the quantities. (Use the same units of measurement for both quantities.) See Example 2.

9. Forty-two inches to 21 inches
10. Eighty-one feet to 27 feet
11. Forty dollars to $60
12. Twenty-four pounds to 30 pounds
13. One quart to 1 gallon
14. Three inches to 2 feet
15. Seven nickels to 3 quarters
16. Twenty-four ounces to 3 pounds
17. Three hours to 90 minutes
18. Twenty-one feet to 35 yards
19. Seventy-five centimeters to 2 meters
20. Three meters to 128 centimeters
21. Sixty milliliters to 1 liter
22. Fifty cubic centimeters to 1 liter
23. Ninety minutes to 2 hours
24. Five and one-half pints to 2 quarts
25. Three thousand pounds to 5 tons
26. Twelve thousand pounds to 2 tons

In Exercises 27–30, find the unit price (in dollars per ounce). See Example 3.

27. A 20-ounce can of pineapple for 98¢
28. An 18-ounce box of cereal for $4.29
29. A one-pound, four-ounce loaf of bread for $1.46
30. A one-pound package of cheese for $3.08

In Exercises 31–36, which product has the lower unit price? See Example 4.

31. (a) A $27\frac{3}{4}$-ounce can of spaghetti sauce for $1.68
    (b) A 32-ounce jar of spaghetti sauce for $1.87

32. (a) A 16-ounce package of margarine quarters for $1.54

    (b) A three-pound tub of margarine for $3.62

33. (a) A 10-ounce package of frozen green beans for 72¢

    (b) A 16-ounce package of frozen green beans for 93¢

34. (a) An 18-ounce jar of peanut butter for $1.92

    (b) A 28-ounce jar of peanut butter for $3.18

35. (a) A two-liter bottle (67.6 ounces) of soft drink for $1.09

    (b) Six 12-ounce cans of soft drink for $1.69

36. (a) A one-quart container of oil for $2.12

    (b) A 2.5-gallon container of oil for $19.99

In Exercises 37–52, solve the proportion. See Example 5.

37. $\dfrac{5}{3} = \dfrac{20}{y}$ 38. $\dfrac{9}{x} = \dfrac{18}{5}$

39. $\dfrac{4}{t} = \dfrac{2}{25}$ 40. $\dfrac{5}{x} = \dfrac{3}{2}$

41. $\dfrac{y}{25} = \dfrac{12}{10}$ 42. $\dfrac{z}{35} = \dfrac{5}{14}$

43. $\dfrac{8}{3} = \dfrac{t}{6}$ 44. $\dfrac{x}{6} = \dfrac{7}{12}$

45. $\dfrac{0.5}{0.8} = \dfrac{n}{0.3}$ 46. $\dfrac{2}{4.5} = \dfrac{t}{0.5}$

47. $\dfrac{x+1}{5} = \dfrac{3}{10}$ 48. $\dfrac{z-3}{8} = \dfrac{3}{16}$

49. $\dfrac{x+6}{3} = \dfrac{x-5}{2}$ 50. $\dfrac{x-2}{4} = \dfrac{x+10}{10}$

51. $\dfrac{x+2}{8} = \dfrac{x-1}{3}$ 52. $\dfrac{x-4}{5} = \dfrac{x}{6}$

Solving Problems

In Exercises 53–62, express the statement as a ratio in simplest form. (Use the same units of measurement for both quantities.)

53. *Study Hours* You study 4 hours per day and are in class 6 hours per day. Find the ratio of the number of study hours to class hours.

54. *Income Tax* You have $16.50 of state tax withheld from your paycheck per week when your gross pay is $750. Find the ratio of tax to gross pay.

55. *Consumer Awareness* Last month, you used your cellular phone for 36 long-distance minutes and 184 local minutes. Find the ratio of local minutes to long-distance minutes.

56. *Education* There are 2921 students and 127 faculty members at your school. Find the ratio of the number of students to the number of faculty members.

57. *Compression Ratio* The *compression ratio* of an engine is the ratio of the expanded volume of gas in one of its cylinders to the compressed volume of gas in the cylinder (see figure). A cylinder in a diesel engine has an expanded volume of 345 cubic centimeters and a compressed volume of 17.25 cubic centimeters. What is the compression ratio of this engine?

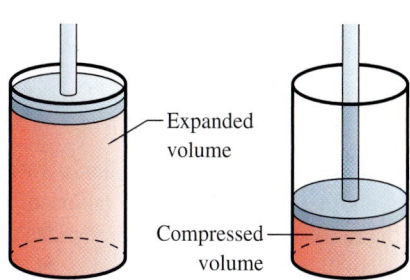

Figure for 57

58. *Turn Ratio* The *turn ratio* of a transformer is the ratio of the number of turns on the secondary winding to the number of turns on the primary winding (see figure). A transformer has a primary winding with 250 turns and a secondary winding with 750 turns. What is its turn ratio?

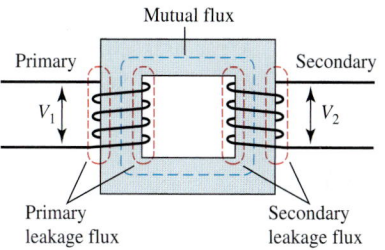

59. *Gear Ratio*   The gear ratio of two gears is the ratio of the number of teeth on one gear to the number of teeth on the other gear. Find the gear ratio of the larger gear to the smaller gear for the gears shown in the figure.

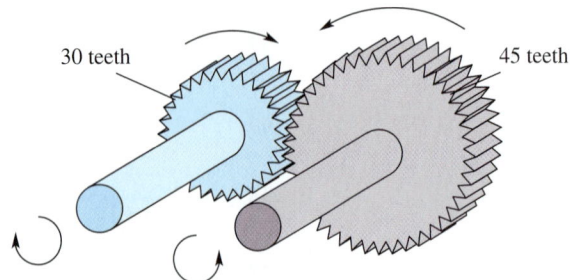

30 teeth   45 teeth

60. *Gear Ratio*   On a five-speed bicycle, the ratio of the pedal gear to the axle gear depends on which axle gear is engaged. Use the table to find the gear ratios for the five different gears. For which gear is it easiest to pedal? Why?

Gear	1st	2nd	3rd	4th	5th
Teeth on pedal gear	52	52	52	52	52
Teeth on axle gear	28	24	20	17	14

61. ▲ *Geometry*   A large pizza has a radius of 10 inches and a small pizza has a radius of 7 inches. Find the ratio of the area of the large pizza to the area of the small pizza. (*Note:* The area of a circle is $A = \pi r^2$.)

62. *Specific Gravity*   The *specific gravity* of a substance is the ratio of its weight to the weight of an equal volume of water. Kerosene weighs 0.82 gram per cubic centimeter and water weighs 1 gram per cubic centimeter. What is the specific gravity of kerosene?

63. *Gasoline Cost*   A car uses 20 gallons of gasoline for a trip of 500 miles. How many gallons would be used on a trip of 400 miles?

64. *Amount of Fuel*   A tractor requires 4 gallons of diesel fuel to plow for 90 minutes. How many gallons of fuel would be required to plow for 8 hours?

65. *Building Material*   One hundred cement blocks are required to build a 16-foot wall. How many blocks are needed to build a 40-foot wall?

66. *Force on a Spring*   A force of 50 pounds stretches a spring 4 inches. How much force is required to stretch the spring 6 inches?

67. *Real Estate Taxes*   The tax on a property with an assessed value of $65,000 is $825. Find the tax on a property with an assessed value of $90,000.

68. *Real Estate Taxes*   The tax on a property with an assessed value of $65,000 is $1100. Find the tax on a property with an assessed value of $90,000.

69. *Polling Results*   In a poll, 624 people from a sample of 1100 indicated they would vote for the republican candidate. How many votes can the candidate expect to receive from 40,000 votes cast?

70. *Quality Control*   A quality control engineer found two defective units in a sample of 50. At this rate, what is the expected number of defective units in a shipment of 10,000 units?

71. *Pumping Time*   A pump can fill a 750-gallon tank in 35 minutes. How long will it take to fill a 1000-gallon tank with this pump?

72. *Recipe*   Two cups of flour are required to make one batch of cookies. How many cups are required for $2\frac{1}{2}$ batches?

73. *Amount of Gasoline*   The gasoline-to-oil ratio for a two-cycle engine is 40 to 1. How much gasoline is required to produce a mixture that contains one-half pint of oil?

74. *Building Material*   The ratio of cement to sand in an 80-pound bag of dry mix is 1 to 4. Find the number of pounds of sand in the bag. (*Note:* Dry mix is composed of only cement and sand.)

75. *Map Scale*   On a map, $1\frac{1}{4}$ inch represents 80 miles. Estimate the distance between two cities that are 6 inches apart on the map.

76. *Map Scale*   On a map, $1\frac{1}{2}$ inches represents 40 miles. Estimate the distance between two cities that are 4 inches apart on the map.

▲ *Geometry*   In Exercises 77 and 78, find the length *x* of the side of the larger triangle. (Assume that the two triangles are similar, and use the fact that corresponding sides of similar triangles are proportional.)

77.

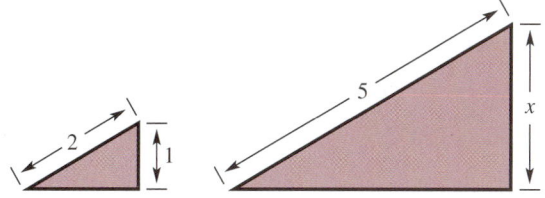

78.

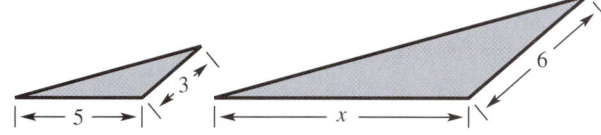

79. ▲ *Geometry*   Find the length of the shadow of the man shown in the figure. (*Hint:* Use similar triangles to create a proportion.)

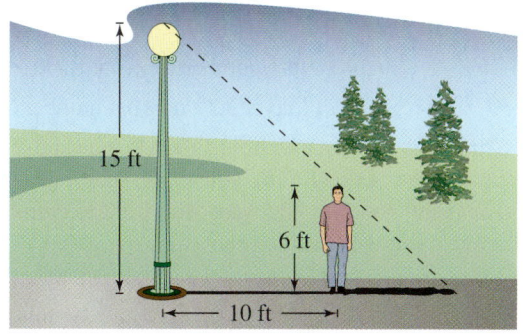

80. ▲ *Geometry*   Find the height of the tree shown in the figure. (*Hint:* Use similar triangles to create a proportion.)

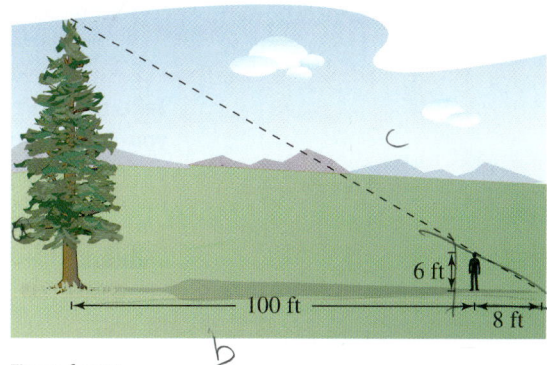

Figure for 80

81. *Resizing a Picture*   You have an 8-by-10-inch photo of a soccer player that must be reduced to a size of 1.6 by 2 inches for the school yearbook. What percent does the photo need to be reduced by in order to fit the allotted space?

82. *Resizing a Picture*   You have a 7-by-5-inch photo of the math club that must be reduced to a size of 5.6 by 4 inches for the school yearbook. What percent does the photo need to be reduced by in order to fit the allotted space?

In Exercises 83–86, use the Consumer Price Index table on page 162 to estimate the price of the item in the indicated year.

83. The 1999 price of a lawn tractor that cost $2875 in 1978

84. The 2000 price of a watch that cost $158 in 1988

85. The 1970 price of a gallon of milk that cost $2.75 in 1996

86. The 1980 price of a coat that cost $225 in 2001

Explaining Concepts

87. ❷   Answer part (g) of Motivating the Chapter on page 122.

88. *Writing*   In your own words, describe the term *ratio*.

89. *Writing*   You are told that the ratio of men to women in a class is 2 to 1. Does this information tell you the total number of people in the class? Explain.

90. *Writing*   Explain the following statement. "When setting up a ratio, be sure you are comparing apples to apples and not apples to oranges."

91. *Writing*   In your own words, describe the term *proportion*.

92. Create a proportion problem. Exchange problems with another student and solve the problem you receive.

Mid-Chapter Quiz

Take this quiz as you would take a quiz in class. After you are done, check your work against the answers in the back of the book.

In Exercises 1–10, solve the equation.

1. $74 - 12x = 2$

2. $10(y - 8) = 0$

3. $3x + 1 = x + 20$

4. $6x + 8 = 8 - 2x$

5. $-10x + \dfrac{2}{3} = \dfrac{7}{3} - 5x$

6. $\dfrac{x}{5} + \dfrac{x}{8} = 1$

7. $\dfrac{9 + x}{3} = 15$

8. $7 - 2(5 - x) = -7$

9. $\dfrac{x + 3}{6} = \dfrac{4}{3}$

10. $\dfrac{x + 7}{5} = \dfrac{x + 9}{7}$

In Exercises 11 and 12, solve the equation. Round your answer to two decimal places. In your own words, explain how to check the solution.

11. $32.86 - 10.5x = 11.25$

12. $\dfrac{x}{5.45} + 3.2 = 12.6$

13. What number is 62% of 25?

14. What number is $\frac{1}{2}$% of 8400?

15. 300 is what percent of 150?

16. 145.6 is 32% of what number?

17. You have two jobs. In the first job, you work 40 hours a week at a candy store and earn $7.50 per hour. In the second job, you earn $6.00 per hour babysitting and can work as many hours as you want. You want to earn $360 a week. How many hours must you work at the second job?

18. A region has an area of 42 square meters. It must be divided into three subregions so that the second has twice the area of the first, and the third has twice the area of the second. Find the area of each subregion.

19. To get an A in a psychology course, you must have an average of at least 90 points for three tests of 100 points each. For the first two tests, your scores are 84 and 93. What must you score on the third test to earn a 90% average for the course?

20. The circle graph at the left shows the number of endangered wildlife and plant species for the year 2001. What percent of the total endangered wildlife and plant species were birds? (Source: U.S. Fish and Wildlife Service)

21. Two people can paint a room in t hours, where t must satisfy the equation $t/4 + t/12 = 1$. How long will it take for the two people to paint the room?

22. A large round pizza has a radius of $r = 15$ inches, and a small round pizza has a radius of $r = 8$ inches. Find the ratio of the area of the large pizza to the area of the small pizza. (*Hint:* The area of a circle is $A = \pi r^2$.)

23. A car uses 30 gallons of gasoline for a trip of 800 miles. How many gallons would be used on a trip of 700 miles?

Endangered Wildlife and Plant Species

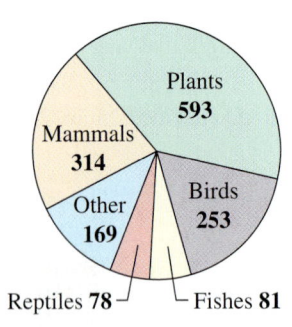

Figure for 20

3.5 Geometric and Scientific Applications

What You Should Learn

1. Use common formulas to solve application problems.
2. Solve mixture problems involving hidden products.
3. Solve work-rate problems.

Esbin-Anderson/The Image Works

Why You Should Learn It

The formula for distance can be used whenever you decide to take a road trip. For instance, in Exercise 52 on page 179, you will use the formula for distance to find the travel time for an automobile trip.

Using Formulas

Some formulas occur so frequently in problem solving that it is to your benefit to memorize them. For instance, the following formulas for area, perimeter, and volume are often used to create verbal models for word problems. In the geometric formulas below, A represents area, P represents perimeter, C represents circumference, and V represents volume.

1 Use common formulas to solve application problems.

Study Tip

When solving problems involving perimeter, area, or volume, be sure you list the units of measurement for your answers.

Common Formulas for Area, Perimeter, and Volume

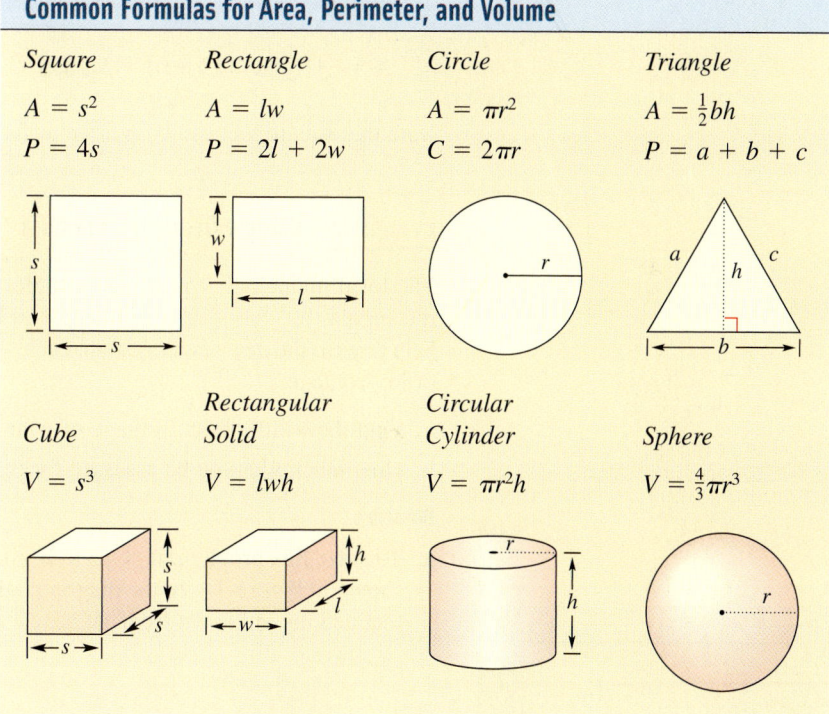

Square	Rectangle	Circle	Triangle
$A = s^2$	$A = lw$	$A = \pi r^2$	$A = \frac{1}{2}bh$
$P = 4s$	$P = 2l + 2w$	$C = 2\pi r$	$P = a + b + c$

Cube	Rectangular Solid	Circular Cylinder	Sphere
$V = s^3$	$V = lwh$	$V = \pi r^2 h$	$V = \frac{4}{3}\pi r^3$

- *Perimeter* is always measured in linear units, such as inches, feet, miles, centimeters, meters, and kilometers.
- *Area* is always measured in square units, such as square inches, square feet, square centimeters, and square meters.
- *Volume* is always measured in cubic units, such as cubic inches, cubic feet, cubic centimeters, and cubic meters.

Figure 3.3

Example 1　Using a Geometric Formula

A sailboat has a triangular sail with an area of 96 square feet and a base that is 16 feet long, as shown in Figure 3.3. What is the height of the sail?

Solution

Because the sail is triangular, and you are given its area, you should begin with the formula for the area of a triangle.

$$A = \frac{1}{2}bh \qquad\qquad \text{Area of a triangle}$$

$$96 = \frac{1}{2}(16)h \qquad\qquad \text{Substitute 96 for } A \text{ and 16 for } b.$$

$$96 = 8h \qquad\qquad \text{Simplify.}$$

$$12 = h \qquad\qquad \text{Divide each side by 8.}$$

The height of the sail is 12 feet.

In Example 1, notice that b and h are measured in feet. When they are multiplied in the formula $\frac{1}{2}bh$, the resulting area is measured in *square* feet.

$$A = \frac{1}{2}(16 \text{ feet})(12 \text{ feet}) = 96 \text{ feet}^2$$

Note that square feet can be written as feet^2.

Example 2　Using a Geometric Formula

The local municipality is planning to develop the street along which you own a rectangular lot that is 500 feet deep and has an area of 100,000 square feet. To help pay for the new sewer system, each lot owner will be assessed $5.50 per foot of lot frontage.

a. Find the width of the frontage of your lot.

b. How much will you be assessed for the new sewer system?

Solution

a. To solve this problem, it helps to begin by drawing a diagram such as the one shown in Figure 3.4. In the diagram, label the depth of the property as $l = 500$ feet and the unknown frontage as w.

$$A = lw \qquad\qquad \text{Area of a rectangle}$$

$$100{,}000 = 500(w) \qquad\qquad \text{Substitute 100,000 for } A \text{ and 500 for } l.$$

$$200 = w \qquad\qquad \text{Divide each side by 500 and simplify.}$$

The frontage of the rectangular lot is 200 feet.

b. If each foot of frontage costs $5.50, then your total assessment will be 200(5.50) = $1100.

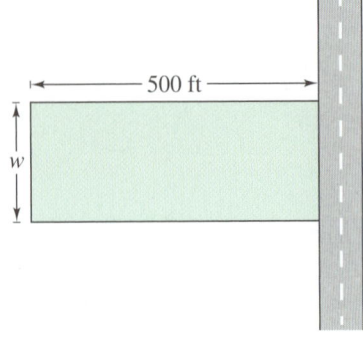

Figure 3.4

Miscellaneous Common Formulas

Temperature: F = degrees Fahrenheit, C = degrees Celsius

$$F = \frac{9}{5}C + 32$$

Simple Interest: I = interest, P = principal, r = interest rate, t = time

$$I = Prt$$

Distance: d = distance traveled, r = rate, t = time

$$d = rt$$

In some applications, it helps to rewrite a common formula by solving for a different variable. For instance, using the common formula for temperature you can obtain a formula for C (degrees Celsius) in terms of F (degrees Fahrenheit) as follows.

$$F = \frac{9}{5}C + 32 \qquad \text{Temperature formula}$$

$$F - 32 = \frac{9}{5}C \qquad \text{Subtract 32 from each side.}$$

$$\frac{5}{9}(F - 32) = C \qquad \text{Multiply each side by } \tfrac{5}{9}.$$

$$C = \frac{5}{9}(F - 32) \qquad \text{Formula}$$

Example 3   Simple Interest

An amount of $5000 is deposited in an account paying simple interest. After 6 months, the account has earned $162.50 in interest. What is the annual interest rate for this account?

Solution

$$I = Prt \qquad \text{Simple interest formula}$$

$$162.50 = 5000(r)\left(\frac{1}{2}\right) \qquad \text{Substitute for } I, P, \text{ and } t.$$

$$162.50 = 2500r \qquad \text{Simplify.}$$

$$\frac{162.50}{2500} = r \qquad \text{Divide each side by 2500.}$$

$$0.065 = r \qquad \text{Simplify.}$$

The annual interest rate is $r = 0.065$ (or 6.5%). Check this solution in the original statement of the problem.

One of the most familiar rate problems and most often used formulas in real life is the one that relates distance, rate (or speed), and time: $d = rt$. For instance, if you travel at a constant (or average) rate of 50 miles per hour for 45 minutes, the total distance traveled is given by

$$\left(50 \, \frac{\text{miles}}{\text{hour}}\right)\left(\frac{45}{60} \, \text{hour}\right) = 37.5 \text{ miles.}$$

As with all problems involving applications, be sure to check that the units in the model make sense. For instance, in this problem the rate is given in *miles per hour*. So, in order for the solution to be given in *miles*, you must convert the time (from minutes) to *hours*. In the model, you can think of dividing out the 2 "hours," as follows.

$$\left(50 \, \frac{\text{miles}}{\text{hour}}\right)\left(\frac{45}{60} \, \text{hour}\right) = 37.5 \text{ miles}$$

Karl Weatherly/Getty Images

In 2001, about 24.5 million Americans ran or jogged on a regular basis. Almost three times that many walked regularly for exercise.

Example 4   A Distance-Rate-Time Problem

You jog at an average rate of 8 kilometers per hour. How long will it take you to jog 14 kilometers?

Solution

Verbal Model: Distance = Rate · Time

Labels: Distance = 14   (kilometers)
               Rate = 8   (kilometers per hour)
               Time = t (hours)

Equation: $14 = 8(t)$

                  $\dfrac{14}{8} = t$

                  $1.75 = t$

It will take you 1.75 hours (or 1 hour and 45 minutes). Check this in the original statement of the problem.

If you are having trouble solving a distance-rate-time problem, consider making a table such as that shown below for Example 4.

Distance = Rate · Time

Rate (km/hr)	8	8	8	8	8	8	8	8
Time (hours)	0.25	0.50	0.75	1.00	1.25	1.50	1.75	2.00
Distance (kilometers)	2	4	6	8	10	12	14	16

② Solve mixture problems involving hidden products.

Solving Mixture Problems

Many real-world problems involve combinations of two or more quantities that make up a new or different quantity. Such problems are called **mixture problems.** They are usually composed of the sum of two or more "hidden products" that involve *rate factors.* Here is the generic form of the verbal model for mixture problems.

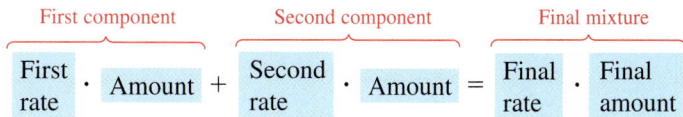

The rate factors are usually expressed as *percents* or *percents of measure* such as dollars per pound, jobs per hour, or gallons per minute.

Example 5   A Nut Mixture Problem

A grocer wants to mix cashew nuts worth $7 per pound with 15 pounds of peanuts worth $2.50 per pound. To obtain a nut mixture worth $4 per pound, how many pounds of cashews are needed? How many pounds of mixed nuts will be produced for the grocer to sell?

Solution

In this problem, the rates are the *unit prices* of the nuts.

Verbal Model:	$\boxed{\text{Total cost of cashews}} + \boxed{\text{Total cost of peanuts}} = \boxed{\text{Total cost of mixed nuts}}$	

Labels:
Unit price of cashews = 7   (dollars per pound)
Unit price of peanuts = 2.5   (dollars per pound)
Unit price of mixed nuts = 4   (dollars per pound)
Amount of cashews = x (pounds)
Amount of peanuts = 15   (pounds)
Amount of mixed nuts = $x + 15$ (pounds)

Equation:
$$7(x) + 2.5(15) = 4(x + 15)$$
$$7x + 37.5 = 4x + 60$$
$$3x = 22.5$$
$$x = \frac{22.5}{3} = 7.5$$

The grocer needs 7.5 pounds of cashews. This will result in $x + 15 = 7.5 + 15 = 22.5$ pounds of mixed nuts. You can check these results as follows.

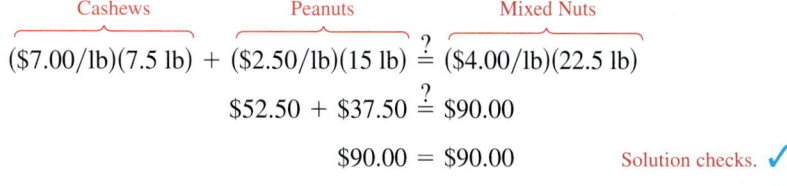

$$(\$7.00/\text{lb})(7.5 \text{ lb}) + (\$2.50/\text{lb})(15 \text{ lb}) \overset{?}{=} (\$4.00/\text{lb})(22.5 \text{ lb})$$
$$\$52.50 + \$37.50 \overset{?}{=} \$90.00$$
$$\$90.00 = \$90.00 \qquad \text{Solution checks.} ✔$$

Example 6   A Solution Mixture Problem

A pharmacist needs to strengthen a 15% alcohol solution with a pure alcohol solution to obtain a 32% solution. How much pure alcohol should be added to 100 milliliters of the 15% solution (see Figure 3.5)?

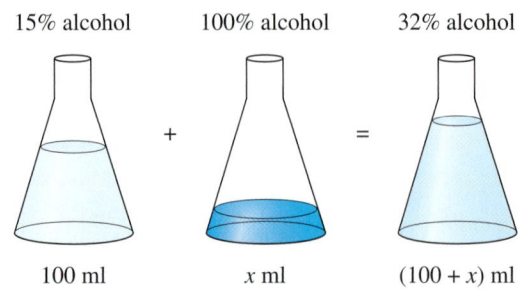

15% alcohol   100% alcohol   32% alcohol

100 ml x ml $(100 + x)$ ml

Figure 3.5

Solution

In this problem, the rates are the alcohol *percents* of the solutions.

Verbal Model:	Amount of 15% alcohol solution	+	Amount of 100% alcohol solution	=	Amount of final alcohol solution

Labels: 15% solution: Percent alcohol = 0.15   (decimal form)
                      Amount of alcohol solution = 100   (milliliters)
         100% solution: Percent alcohol = 1.00   (decimal form)
                      Amount of alcohol solution = x (milliliters)
          32% solution: Percent alcohol = 0.32   (decimal form)
                      Amount of alcohol solution = $100 + x$ (milliliters)

Equation: $0.15(100) + 1.00(x) = 0.32(100 + x)$

$$15 + x = 32 + 0.32x$$

$$0.68x = 17$$

$$x = \frac{17}{0.68}$$

$$= 25 \text{ ml}$$

So, the pharmacist should add 25 milliliters of pure alcohol to the 15% solution. You can check this in the original statement of the problem as follows.

15% solution   100% solution Final solution

$$0.15(100) + 1.00(25) \overset{?}{=} 0.32(125)$$

$$15 + 25 \overset{?}{=} 40$$

$$40 = 40 \qquad \text{Solution checks. } ✔$$

Remember that mixture problems are sums of two or more hidden products that involve different rates. Watch for such problems in the exercises.

Mixture problems can also involve a "mix" of investments, as shown in the next example.

Example 7   Investment Mixture

You invested a total of $10,000 at $4\frac{1}{2}\%$ and $5\frac{1}{2}\%$ simple interest. During 1 year the two accounts earned $508.75. How much did you invest in each account?

Solution

Verbal Model:

$$\boxed{\begin{array}{c}\text{Interest earned}\\\text{from } 4\frac{1}{2}\%\end{array}} + \boxed{\begin{array}{c}\text{Interest earned}\\\text{from } 5\frac{1}{2}\%\end{array}} = \boxed{\begin{array}{c}\text{Total interest}\\\text{earned}\end{array}}$$

Labels:

Amount invested at $4\frac{1}{2}\% = x$	(dollars)
Amount invested at $5\frac{1}{2}\% = 10{,}000 - x$	(dollars)
Interest earned from $4\frac{1}{2}\% = (x)(0.045)(1)$	(dollars)
Interest earned from $5\frac{1}{2}\% = (10{,}000 - x)(0.055)(1)$	(dollars)
Total interest earned $= 508.75$	(dollars)

Equation:

$$0.045x + 0.055(10{,}000 - x) = 508.75$$
$$0.045x + 550 - 0.055x = 508.75$$
$$550 - 0.01x = 508.75$$
$$-0.01x = -41.25$$
$$x = 4125$$

So, you invested $4125 at $4\frac{1}{2}\%$ and $10{,}000 - x = 10{,}000 - 4125 = \$5875$ at $5\frac{1}{2}\%$. Check this in the original statement of the problem.

---

3 Solve work-rate problems.

Solving Work-Rate Problems

Although not generally referred to as such, most **work-rate problems** are actually *mixture* problems because they involve two or more rates. Here is the generic form of the verbal model for work-rate problems.

$$\boxed{\begin{array}{c}\text{First}\\\text{rate}\end{array}} \cdot \boxed{\text{Time}} + \boxed{\begin{array}{c}\text{Second}\\\text{rate}\end{array}} \cdot \boxed{\text{Time}} = \boxed{\begin{array}{c}1\\\text{(one whole job}\\\text{completed)}\end{array}}$$

In work-rate problems, the work rate is the *reciprocal* of the time needed to do the entire job. For instance, if it takes 7 hours to complete a job, the per-hour work rate is

$$\frac{1}{7} \text{ job per hour.}$$

Similarly, if it takes $4\frac{1}{2}$ minutes to complete a job, the per-minute rate is

$$\frac{1}{4\frac{1}{2}} = \frac{1}{\frac{9}{2}} = \frac{2}{9} \text{ job per minute.}$$

Example 8   A Work-Rate Problem

Consider two machines in a paper manufacturing plant. Machine 1 can complete one job (2000 pounds of paper) in 3 hours. Machine 2 is newer and can complete one job in $2\frac{1}{2}$ hours. How long will it take the two machines working together to complete one job?

Solution

Verbal Model: Portion done by machine 1 $+$ Portion done by machine 2 $=$ 1 (one whole job completed)

Labels:
One whole job completed $= 1$ (job)
Rate (machine 1) $= \frac{1}{3}$ (job per hour)
Time (machine 1) $= t$ (hours)
Rate (machine 2) $= \frac{2}{5}$ (job per hour)
Time (machine 2) $= t$ (hours)

Equation:
$$\left(\tfrac{1}{3}\right)(t) + \left(\tfrac{2}{5}\right)(t) = 1$$
$$\left(\tfrac{1}{3} + \tfrac{2}{5}\right)(t) = 1$$
$$\left(\tfrac{11}{15}\right)(t) = 1$$
$$t = \tfrac{15}{11}$$

It will take $\frac{15}{11}$ hours (or about 1.36 hours) for the machines to complete the job working together. Check this solution in the original statement of the problem.

Study Tip

Note in Example 8 that the "2000 pounds" of paper is unnecessary information. The 2000 pounds is represented as "one complete job." This unnecessary information is called a *red herring*.

Example 9   A Fluid-Rate Problem

An above-ground swimming pool has a capacity of 15,600 gallons, as shown in Figure 3.6. A drain pipe can empty the pool in $6\frac{1}{2}$ hours. At what rate (in gallons per minute) does the water flow through the drain pipe?

Solution

To begin, change the time from hours to minutes by multiplying by 60. That is, $6\frac{1}{2}$ hours is equal to $(6.5)(60)$ or 390 minutes.

Verbal Model: Volume of pool $=$ Rate $\cdot$ Time

Labels:
Volume $= 15{,}600$ (gallons)
Rate $= r$ (gallons per minute)
Time $= 390$ (minutes)

Equation:
$$15{,}600 = r(390)$$
$$\frac{15{,}600}{390} = r$$
$$40 = r$$

The water is flowing through the drain pipe at a rate of 40 gallons per minute. Check this solution in the original statement of the problem.

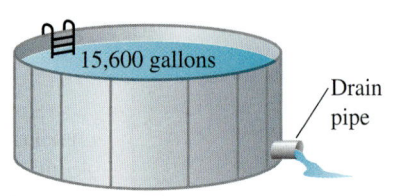

15,600 gallons

Drain pipe

Figure 3.6

3.5 Exercises

Review   Concepts, Skills, and Problem Solving

Keep mathematically in shape by doing these exercises *before* the problems of this section.

Properties and Definitions

1. If n is an integer, distinguish between $2n$ and $2n + 1$.

2. Demonstrate the Addition Property of Equality for the equation $2x - 3 = 10$.

Simplifying Expressions

In Exercises 3–10, simplify the expression.

3. $(-3.5y^2)(8y)$

4. $(-3x^2)(x^4)$

5. $\left(\dfrac{24u}{15}\right)\left(\dfrac{25u^2}{6}\right)$

6. $12\left(\dfrac{3y}{18}\right)$

7. $5x(2 - x) + 3x$

8. $3t - 4(2t - 8)$

9. $3(v - 4) + 7(v - 4)$

10. $5[6 - 2(x - 3)]$

Problem Solving

11. *Sales Tax*   You buy a computer for $1150 and your total bill is $1219. Find the sales tax rate.

12. *Consumer Awareness*   A mail-order catalog lists an area rug for $109.95, plus a shipping charge of $14.25. A local store has a sale on the same rug with 20% off a list price of $139.99. Which is the better bargain?

Developing Skills

In Exercises 1–14, solve for the specified variable.

1. Solve for h: $A = \frac{1}{2}bh$

2. Solve for R: $E = IR$

3. Solve for r: $A = P + Prt$

4. Solve for L: $P = 2L + 2W$

5. Solve for l: $V = lwh$

6. Solve for r: $C = 2\pi r$

7. Solve for C: $S = C + RC$

8. Solve for L: $S = L - RL$

9. Solve for m_2: $F = \alpha\dfrac{m_1 m_2}{r^2}$

10. Solve for b: $V = \frac{4}{3}\pi a^2 b$

11. Solve for b: $A = \frac{1}{2}(a + b)h$

12. Solve for r: $V = \frac{1}{3}\pi h^2(3r - h)$

13. Solve for a: $h = v_0 t + \frac{1}{2}at^2$

14. Solve for a: $S = \dfrac{n}{2}[2a + (n - 1)d]$

In Exercises 15–18, evaluate the formula for the specified values of the variables. (List the *units* of the answer.)

15. *Volume of a Right Circular Cylinder:* $V = \pi r^2 h$
$r = 5$ meters, $h = 4$ meters

16. *Body Mass Index:* $B = \dfrac{703w}{h^2}$
$w = 127$ pounds, $h = 61$ inches

17. *Electric Power:* $I = \dfrac{P}{V}$
$P = 1500$ watts, $V = 110$ volts

18. *Statistical z-score:* $z = \dfrac{x - m}{s}$
$x = 100$ points, $m = 80$ points, $s = 10$ points

In Exercises 19–24, find the missing distance, rate, or time. See Example 4.

	Distance, d	Rate, r	Time, t
19.		4 m/min	12 min
20.		62 mi/hr	$2\frac{1}{2}$ hr
21.	128 km	8 km/hr	
22.	210 mi	50 mi/hr	
23.	2054 m		18 sec
24.	482 ft		40 min

Solving Problems

In Exercises 25–32, use a common geometric formula to solve the problem. See Examples 1 and 2.

25. ▲ *Geometry*   Each room in the floor plan of a house is square (see figure). The perimeter of the bathroom is 32 feet. The perimeter of the kitchen is 80 feet. Find the area of the living room.

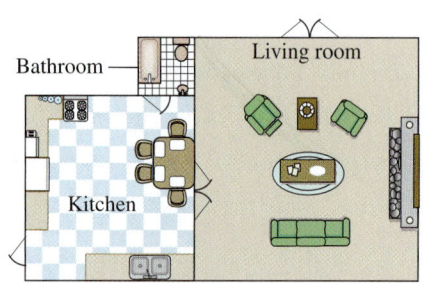

26. ▲ *Geometry*   A rectangle has a perimeter of 10 feet and a width of 2 feet. Find the length of the rectangle.

27. ▲ *Geometry*   A triangle has an area of 48 square meters and a height of 12 meters. Find the length of the base.

28. ▲ *Geometry*   The perimeter of a square is 48 feet. Find its area.

29. ▲ *Geometry*   The circumference of a wheel is 30π inches. Find the diameter of the wheel.

30. ▲ *Geometry*   A circle has a circumference of 15 meters. What is the radius of the circle? Round your answer to two decimal places.

31. ▲ *Geometry*   A circle has a circumference of 25 meters. Find the radius and area of the circle. Round your answers to two decimal places.

32. ▲ *Geometry*   The volume of a right circular cylinder is $V = \pi r^2 h$. Find the volume of a right circular cylinder that has a radius of 2 meters and a height of 3 meters. List the units of measurement for your result.

▲ *Geometry*   In Exercises 33–36, use the closed rectangular box shown in the figure to solve the problem.

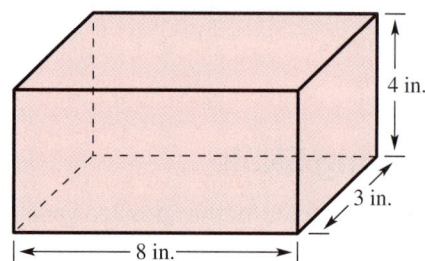

33. Find the area of the base.

34. Find the perimeter of the base.

35. Find the volume of the box.

36. Find the surface area of the box. (*Note:* This is the combined area of the six surfaces.)

Simple Interest   In Exercises 37–44, use the formula for simple interest. See Example 3.

37. Find the interest on a $1000 bond paying an annual rate of 9% for 6 years.

38. A $1000 corporate bond pays an annual rate of $7\frac{1}{2}\%$. The bond matures in $3\frac{1}{2}$ years. Find the interest on the bond.

39. You borrow $15,000 for $\frac{1}{2}$ year. You promise to pay back the principal and the interest in one lump sum. The annual interest rate is 13%. What is your payment?

40. You have a balance of $650 on your credit card that you cannot pay this month. The annual interest rate on an unpaid balance is 19%. Find the lump sum of principal and interest due in 1 month.

41. Find the annual rate on a savings account that earns $110 interest in 1 year on a principal of $1000.

42. Find the annual interest rate on a certificate of deposit that earned $128.98 interest in 1 year on a principal of $1500.

43. How long must $700 be invested at an annual interest rate of 6.25% to earn $460 interest?

44. How long must $1000 be invested at an annual interest rate of $7\frac{1}{2}\%$ to earn $225 interest?

In Exercises 45–54, use the formula for distance to solve the problem. See Example 4.

45. *Space Shuttle*   The speed of the space shuttle (see figure) is 17,500 miles per hour. How long will it take the shuttle to travel a distance of 3000 miles?

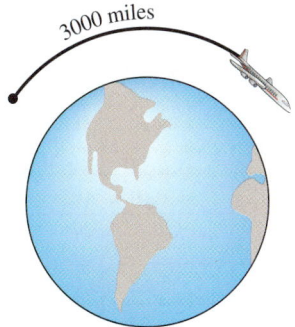

3000 miles

46. *Speed of Light*   The speed of light is 670,616,629.4 miles per hour, and the distance between Earth and the sun is 93,000,000 miles. How long does it take light from the sun to reach Earth?

47. *Average Speed*   Determine the average speed of an experimental plane that can travel 3000 miles in 2.6 hours.

48. *Average Speed*   Determine the average speed of an Olympic runner who completes the 10,000-meter race in 27 minutes and 45 seconds.

49. *Distance*   Two cars start at a given point and travel in the same direction at average speeds of 45 miles per hour and 52 miles per hour (see figure). How far apart will they be in 4 hours?

45 mph   52 mph

d

Figure for 49

50. *Distance*   Two planes leave Orlando International Airport at approximately the same time and fly in opposite directions (see figure). Their speeds are 510 miles per hour and 600 miles per hour. How far apart will the planes be after $1\frac{1}{2}$ hours?

510 mph   600 mph

d

51. *Travel Time*   Two cars start at the same point and travel in the same direction at average speeds of 40 miles per hour and 55 miles per hour. How much time must elapse before the two cars are 5 miles apart?

52. *Travel Time*   On the first part of a 225-mile automobile trip you averaged 55 miles per hour. On the last part of the trip you averaged 48 miles per hour because of increased traffic congestion. The total trip took 4 hours and 15 minutes. Find the travel time for each part of the trip.

53. *Think About It*   A truck traveled at an average speed of 60 miles per hour on a 200-mile trip to pick up a load of freight. On the return trip, with the truck fully loaded, the average speed was 40 miles per hour.

(a) Guess the average speed for the round trip.

(b) Calculate the average speed for the round trip. Is the result the same as in part (a)? Explain.

54. *Time*   A jogger leaves a point on a fitness trail running at a rate of 4 miles per hour. Ten minutes later a second jogger leaves from the same location running at 5 miles per hour. How long will it take the second jogger to overtake the first? How far will each have run at that point?

Mixture Problem   In Exercises 55–58, determine the numbers of units of solutions 1 and 2 required to obtain the desired amount and percent alcohol concentration of the final solution. See Example 6.

	Concentration Solution 1	Concentration Solution 2	Concentration Final Solution	Amount of Final Solution
55.	10%	30%	25%	100 gal
56.	25%	50%	30%	5 L
57.	15%	45%	30%	10 qt
58.	70%	90%	75%	25 gal

59. *Number of Stamps*   You have 100 stamps that have a total value of $31.02. Some of the stamps are worth 24¢ each and the others are worth 37¢ each. How many stamps of each type do you have?

60. *Number of Stamps*   You have 20 stamps that have a total value of $6.62. Some of the stamps are worth 24¢ each and others are worth 37¢ each. How many stamps of each type do you have?

61. *Number of Coins*   A person has 20 coins in nickels and dimes with a combined value of $1.60. Determine the number of coins of each type.

62. *Number of Coins*   A person has 50 coins in dimes and quarters with a combined value of $7.70. Determine the number of coins of each type.

63. *Nut Mixture*   A grocer mixes two kinds of nuts that cost $2.49 and $3.89 per pound to make 100 pounds of a mixture that costs $3.47 per pound. How many pounds of each kind of nut are put into the mixture? See Example 5.

64. *Flower Order*   A floral shop receives a $384 order for roses and carnations. The prices per dozen for the roses and carnations are $18 and $12, respectively. The order contains twice as many roses as carnations. How many of each type of flower are in the order?

65. *Antifreeze*   The cooling system in a truck contains 4 gallons of coolant that is 30% antifreeze. How much must be withdrawn and replaced with 100% antifreeze to bring the coolant in the system to 50% antifreeze?

66. *Ticket Sales*   Ticket sales for a play total $1700. The number of tickets sold to adults is three times the number sold to children. The prices of the tickets for adults and children are $5 and $2, respectively. How many of each type were sold?

67. *Investment Mixture*   You divided $6000 between two investments earning 7% and 9% simple interest. During 1 year the two accounts earned $500. How much did you invest in each account? See Example 7.

68. *Investment Mixture*   You divided an inheritance of $30,000 into two investments earning 8.5% and 10% simple interest. During 1 year, the two accounts earned $2700. How much did you invest in each account?

69. *Interpreting a Table*   An agricultural corporation must purchase 100 tons of cattle feed. The feed is to be a mixture of soybeans, which cost $200 per ton, and corn, which costs $125 per ton.

(a) Complete the table, where x is the number of tons of corn in the mixture.

Corn, x	Soybeans, $100 - x$	Price per ton of the mixture
0		
20		
40		
60		
80		
100		

(b) How does an increase in the number of tons of corn affect the number of tons of soybeans in the mixture?

(c) How does an increase in the number of tons of corn affect the price per ton of the mixture?

(d) If there were equal weights of corn and soybeans in the mixture, how would the price of the mixture relate to the price of each component?

70. *Interpreting a Table* A metallurgist is making 5 ounces of an alloy of metal A, which costs $52 per ounce, and metal B, which costs $16 per ounce.

(a) Complete the table, where x is the number of ounces of metal A in the alloy.

Metal A, x	Metal B, $5 - x$	Price per ounce of the alloy
0		
1		
2		
3		
4		
5		

(b) How does an increase in the number of ounces of metal A in the alloy affect the number of ounces of metal B in the alloy?

(c) How does an increase in the number of ounces of metal A in the alloy affect the price of the alloy?

(d) If there were equal amounts of metal A and metal B in the alloy, how would the price of the alloy relate to the price of each of the components?

71. *Work Rate* You can mow a lawn in 2 hours using a riding mower, and in 3 hours using a push mower. Using both machines together, how long will it take you and a friend to mow the lawn? See Example 8.

72. *Work Rate* One person can complete a typing project in 6 hours, and another can complete the same project in 8 hours. If they both work on the project, in how many hours can it be completed?

73. *Work Rate* One worker can complete a task in m minutes while a second can complete the task in $9m$ minutes. Show that by working together they can complete the task in $t = \frac{9}{10}m$ minutes.

74. *Work Rate* One worker can complete a task in h hours while a second can complete the task in $3h$ hours. Show that by working together they can complete the task in $t = \frac{3}{4}h$ hours.

75. *Age Problem* A mother was 30 years old when her son was born. How old will the son be when his age is $\frac{1}{3}$ his mother's age?

76. *Age Problem* The difference in age between a father and daughter is 32 years. Determine the age of the father when his age is twice that of his daughter.

77. *Poll Results* One thousand people were surveyed in an opinion poll. Candidates A and B received approximately the same number of votes. Candidate C received twice as many votes as either of the other two candidates. How many votes did each candidate receive?

78. *Poll Results* One thousand people were surveyed in an opinion poll. The numbers of votes for candidates A, B, and C had ratios 5 to 3 to 2, respectively. How many people voted for each candidate?

Explaining Concepts

79. *Writing* In your own words, describe the units of measure used for perimeter, area, and volume. Give examples of each.

80. *Writing* If the height of a triangle is doubled, does the area of the triangle double? Explain.

81. *Writing* If the radius of a circle is doubled, does its circumference double? Does its area double? Explain.

82. *Writing* It takes you 4 hours to drive 180 miles. Explain how to use mental math to find your average speed. Then explain how your method is related to the formula $d = rt$.

83. It takes you 5 hours to complete a job. What portion do you complete each hour?

3.6 Linear Inequalities

Royalty-Free/Corbis

What You Should Learn

1. Sketch the graphs of inequalities.
2. Identify the properties of inequalities that can be used to create equivalent inequalities.
3. Solve linear inequalities.
4. Solve compound inequalities.
5. Solve application problems involving inequalities.

Why You Should Learn It

Linear inequalities can be used to model and solve real-life problems. For instance, Exercises 115 and 116 on page 195 show how to use linear inequalities to analyze air pollutant emissions.

1. Sketch the graphs of inequalities.

Intervals on the Real Number Line

In this section you will study **algebraic inequalities,** which are inequalities that contain one or more variable terms. Some examples are

$$x \le 4, \quad x \ge -3, \quad x + 2 < 7, \quad \text{and} \quad 4x - 6 < 3x + 8.$$

As with an equation, you **solve** an inequality in the variable x by finding all values of x for which the inequality is true. Such values are called **solutions** and are said to **satisfy** the inequality. The set of all solutions of an inequality is the **solution set** of the inequality. The **graph** of an inequality is obtained by plotting its solution set on the real number line. Often, these graphs are intervals—either bounded or unbounded.

Bounded Intervals on the Real Number Line

Let a and b be real numbers such that $a < b$. The following intervals on the real number line are called **bounded intervals.** The numbers a and b are the **endpoints** of each interval. A bracket indicates that the endpoint is included in the interval, and a parenthesis indicates that the endpoint is excluded.

Notation	Interval Type	Inequality	Graph
$[a, b]$	Closed	$a \le x \le b$	
(a, b)	Open	$a < x < b$	
$[a, b)$		$a \le x < b$	
$(a, b]$		$a < x \le b$	

The **length** of the interval $[a, b]$ is the distance between its endpoints: $b - a$. The lengths of $[a, b]$, (a, b), $[a, b)$, and $(a, b]$ are the same. The reason that these four types of intervals are called "bounded" is that each has a finite length. An interval that *does not* have a finite length is **unbounded** (or **infinite**).

Unbounded Intervals on the Real Number Line

Let a and b be real numbers. The following intervals on the real number line are called **unbounded intervals**.

Notation	Interval Type	Inequality	Graph
$[a, \infty)$		$x \geq a$	
(a, ∞)	Open	$x > a$	
$(-\infty, b]$		$x \leq b$	
$(-\infty, b)$	Open	$x < b$	
$(-\infty, \infty)$	Entire real line		

The symbols ∞ (**positive infinity**) and $-\infty$ (**negative infinity**) do not represent real numbers. They are simply convenient symbols used to describe the unboundedness of an interval such as $(-5, \infty)$. This is read as the interval from -5 to infinity.

Example 1   Graphs of Inequalities

Sketch the graph of each inequality.

a. $-3 < x \leq 1$ **b.** $0 < x < 2$

c. $-3 < x$ **d.** $x \leq 2$

Study Tip

In Example 1(c), the inequality $-3 < x$ can also be written as $x > -3$. In other words, saying "-3 is less than x" is the same as saying "x is greater than -3."

Solution

a. The graph of $-3 < x \leq 1$ is a bounded interval.

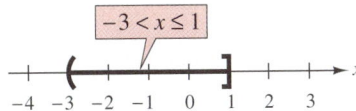

b. The graph of $0 < x < 2$ is a bounded interval.

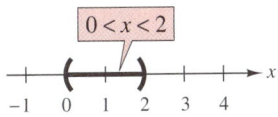

c. The graph of $-3 < x$ is an unbounded interval.

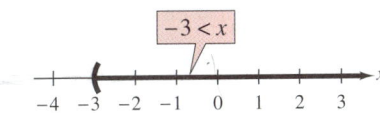

d. The graph of $x \leq 2$ is an unbounded interval.

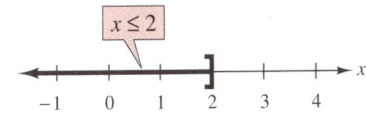

② Identify the properties of inequalities that can be used to create equivalent inequalities.

Properties of Inequalities

Solving a linear inequality is much like solving a linear equation. You isolate the variable, using the **properties of inequalities.** These properties are similar to the properties of equality, but there are two important exceptions. *When each side of an inequality is multiplied or divided by a negative number, the direction of the inequality symbol must be reversed.* Here is an example.

$$-2 < 5 \qquad \text{Original inequality}$$

$$(-3)(-2) > (-3)(5) \qquad \text{Multiply each side by } -3 \text{ and reverse the inequality.}$$

$$6 > -15 \qquad \text{Simplify.}$$

Two inequalities that have the same solution set are **equivalent inequalities.** The following list of operations can be used to create equivalent inequalities.

Properties of Inequalities

1. *Addition and Subtraction Properties*

Adding the same quantity to, or subtracting the same quantity from, each side of an inequality produces an equivalent inequality.

If $a < b$, then $a + c < b + c$.

If $a < b$, then $a - c < b - c$.

2. *Multiplication and Division Properties: Positive Quantities*

Multiplying or dividing each side of an inequality by a positive quantity produces an equivalent inequality.

If $a < b$ and c is positive, then $ac < bc$.

If $a < b$ and c is positive, then $\dfrac{a}{c} < \dfrac{b}{c}$.

3. *Multiplication and Division Properties: Negative Quantities*

Multiplying or dividing each side of an inequality by a negative quantity produces an equivalent inequality in which the inequality symbol is reversed.

If $a < b$ and c is negative, then $ac > bc$.   Reverse inequality

If $a < b$ and c is negative, then $\dfrac{a}{c} > \dfrac{b}{c}$.   Reverse inequality

4. *Transitive Property*

Consider three quantities for which the first quantity is less than the second, and the second is less than the third. It follows that the first quantity must be less than the third quantity.

If $a < b$ and $b < c$, then $a < c$.

These properties remain true if the symbols $<$ and $>$ are replaced by $\leq$ and $\geq$. Moreover, a, b, and c can represent real numbers, variables, or expressions. Note that you cannot multiply or divide each side of an inequality by zero.

3 Solve linear inequalities.

Solving a Linear Inequality

An inequality in one variable is a **linear inequality** if it can be written in one of the following forms.

$$ax + b \leq 0, \quad ax + b < 0, \quad ax + b \geq 0, \quad ax + b > 0$$

The solution set of a linear inequality can be written in set notation. For the solution $x > 1$, the set notation is $\{x \mid x > 1\}$ and is read "the set of all x such that x is greater than 1."

As you study the following examples, pay special attention to the steps in which the inequality symbol is reversed. *Remember that when you multiply or divide an inequality by a negative number, you must reverse the inequality symbol.*

Study Tip

Checking the solution set of an inequality is not as simple as checking the solution set of an equation. (There are usually too many x-values to substitute back into the original inequality.) You can, however, get an indication of the validity of a solution set by substituting a few convenient values of x. For instance, in Example 2, try checking that $x = 0$ satisfies the original inequality, whereas $x = 4$ does not.

Example 2   Solving a Linear Inequality

$x + 6 < 9$	Original inequality
$x + 6 - 6 < 9 - 6$	Subtract 6 from each side.
$x < 3$	Combine like terms.

The solution set consists of all real numbers that are less than 3. The solution set in interval notation is $(-\infty, 3)$ and in set notation is $\{x \mid x < 3\}$. The graph is shown in Figure 3.7.

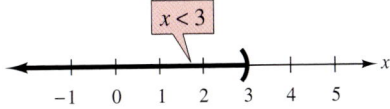

Figure 3.7

Example 3   Solving a Linear Inequality

$8 - 3x \leq 20$	Original inequality
$8 - 8 - 3x \leq 20 - 8$	Subtract 8 from each side.
$-3x \leq 12$	Combine like terms.
$\dfrac{-3x}{-3} \geq \dfrac{12}{-3}$	Divide each side by -3 and reverse the inequality symbol.
$x \geq -4$	Simplify.

The solution set in interval notation is $[-4, \infty)$ and in set notation is $\{x \mid x \geq -4\}$. The graph is shown in Figure 3.8.

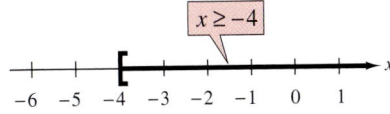

Figure 3.8

Technology: Tip

Most graphing calculators can graph a linear inequality. Consult your user's guide for specific instructions. The graph below shows the solution of the inequality in Example 4. Notice that the graph occurs as an interval above the x-axis.

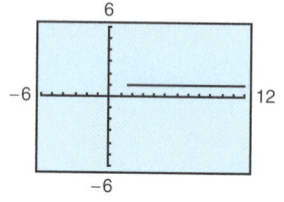

Example 4   Solving a Linear Inequality

$7x - 3 > 3(x + 1)$	Original inequality
$7x - 3 > 3x + 3$	Distributive Property
$7x - 3x - 3 > 3x - 3x + 3$	Subtract $3x$ from each side.
$4x - 3 > 3$	Combine like terms.
$4x - 3 + 3 > 3 + 3$	Add 3 to each side.
$4x > 6$	Combine like terms.
$\dfrac{4x}{4} > \dfrac{6}{4}$	Divide each side by 4.
$x > \dfrac{3}{2}$	Simplify.

The solution set consists of all real numbers that are greater than $\frac{3}{2}$. The solution set in interval notation is $\left(\frac{3}{2}, \infty\right)$ and in set notation is $\left\{x \mid x > \frac{3}{2}\right\}$. The graph is shown in Figure 3.9.

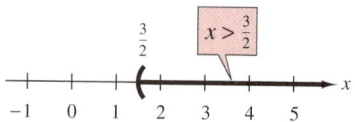

Figure 3.9

Example 5   Solving a Linear Inequality

$\dfrac{2x}{3} + 12 < \dfrac{x}{6} + 18$	Original inequality
$6 \cdot \left(\dfrac{2x}{3} + 12\right) < 6 \cdot \left(\dfrac{x}{6} + 18\right)$	Multiply each side by LCD of 6.
$4x + 72 < x + 108$	Distributive Property
$4x - x < 108 - 72$	Subtract x and 72 from each side.
$3x < 36$	Combine like terms.
$x < 12$	Divide each side by 3.

Study Tip

An inequality can be cleared of fractions in the same way an equation can be cleared of fractions—by multiplying each side by the least common denominator. This is shown in Example 5.

The solution set consists of all real numbers that are less than 12. The solution set in interval notation is $(-\infty, 12)$ and in set notation is $\{x \mid x < 12\}$. The graph is shown in Figure 3.10.

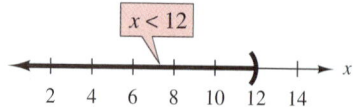

Figure 3.10

④ Solve compound inequalities.

Solving a Compound Inequality

Two inequalities joined by the word *and* or the word *or* constitute a **compound inequality.** When two inequalities are joined by the word *and*, the solution set consists of all real numbers that satisfy *both* inequalities. The solution set for the compound inequality $-4 \leq 5x - 2$ *and* $5x - 2 < 7$ can be written more simply as the **double inequality**

$$-4 \leq 5x - 2 < 7.$$

A compound inequality formed by the word *and* is called **conjunctive** and is the only kind that has the potential to form a double inequality. A compound inequality joined by the word *or* is called **disjunctive** and cannot be re-formed into a double inequality.

Example 6   Solving a Double Inequality

Solve the double inequality $-7 \leq 5x - 2 < 8$.

Solution

$-7 \leq 5x - 2 < 8$	Write original inequality.
$-7 + 2 \leq 5x - 2 + 2 < 8 + 2$	Add 2 to all three parts.
$-5 \leq 5x < 10$	Combine like terms.
$\dfrac{-5}{5} \leq \dfrac{5x}{5} < \dfrac{10}{5}$	Divide each part by 5.
$-1 \leq x < 2$	Simplify.

The solution set consists of all real numbers that are greater than or equal to -1 and less than 2. The solution set in interval notation is $[-1, 2)$ and in set notation is $\{x \mid -1 \leq x < 2\}$. The graph is shown in Figure 3.11.

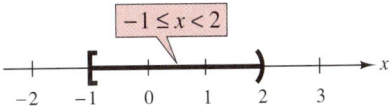

Figure 3.11

The double inequality in Example 6 could have been solved in two parts, as follows.

$$-7 \leq 5x - 2 \qquad \text{and} \qquad 5x - 2 < 8$$
$$-5 \leq 5x \qquad\qquad\qquad\quad 5x < 10$$
$$-1 \leq x \qquad\qquad\qquad\quad\ x < 2$$

The solution set consists of all real numbers that satisfy both inequalities. In other words, the solution set is the set of all values of x for which $-1 \leq x < 2$.

Example 7   Solving a Conjunctive Inequality

Solve the compound inequality $-1 \leq 2x - 3$ and $2x - 3 < 5$.

Solution

Begin by writing the conjunctive inequality as a double inequality.

$$-1 \leq 2x - 3 < 5 \qquad \text{Write as double inequality.}$$

$$-1 + 3 \leq 2x - 3 + 3 < 5 + 3 \qquad \text{Add 3 to all three parts.}$$

$$2 \leq 2x < 8 \qquad \text{Combine like terms.}$$

$$\frac{2}{2} \leq \frac{2x}{2} < \frac{8}{2} \qquad \text{Divide each part by 2.}$$

$$1 \leq x < 4 \qquad \text{Solution set}$$

The solution set is $1 \leq x < 4$ or, in set notation, $\{x \mid 1 \leq x < 4\}$. The graph of the solution set is shown in Figure 3.12.

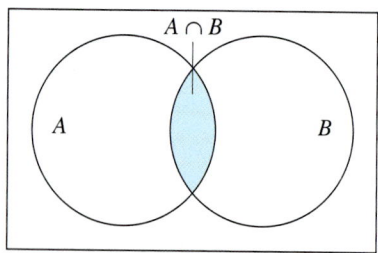

Figure 3.12

Example 8   Solving a Disjunctive Inequality

Solve the compound inequality $-3x + 6 \leq 2$ or $-3x + 6 \geq 7$.

Solution

$-3x + 6 \leq 2$	or $\qquad -3x + 6 \geq 7$	Write original inequality.
$-3x + 6 - 6 \leq 2 - 6$	$-3x + 6 - 6 \geq 7 - 6$	Subtract 6 from all parts.
$-3x \leq -4$	$-3x \geq 1$	Combine like terms.
$\dfrac{-3x}{-3} \geq \dfrac{-4}{-3}$	$\dfrac{-3x}{-3} \leq \dfrac{1}{-3}$	Divide all parts by -3 and reverse both inequality symbols.
$x \geq \dfrac{4}{3}$	$x \leq -\dfrac{1}{3}$	Simplify.

The solution set is $x \leq -\frac{1}{3}$ or $x \geq \frac{4}{3}$ or, in set notation, $\left\{x \mid x \leq -\frac{1}{3} \text{ or } x \geq \frac{4}{3}\right\}$. The graph of the solution set is shown in Figure 3.13.

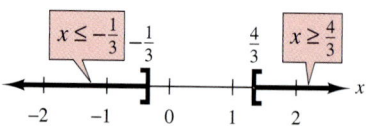

Figure 3.13

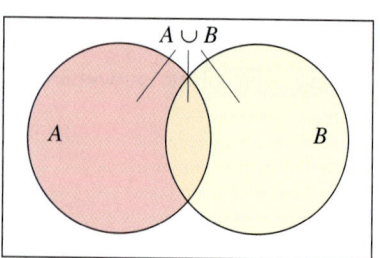

Intersection of two sets

Union of two sets

Figure 3.14

Compound inequalities can be written using *symbols*. For compound inequalities, the word *and* is represented by the symbol ∩, which is read as **intersection.** The word *or* is represented by the symbol ∪, which is read as **union.** Graphical representations are shown in Figure 3.14. If A and B are sets, then x is in $A \cap B$ if it is in both A and B. Similarly, x is in $A \cup B$ if it is in A or B, or possibly both.

Example 9   Writing a Solution Set Using Union

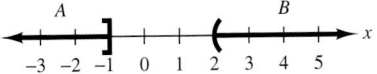

Figure 3.15

A solution set is shown on the number line in Figure 3.15.

a. Write the solution set as a compound inequality.

b. Write the solution set using the union symbol.

Solution

a. As a compound inequality, you can write the solution set as $x \leq -1$ or $x > 2$.

b. Using set notation, you can write the left interval as $A = \{x | x \leq -1\}$ and the right interval as $B = \{x | x > 2\}$. So, using the union symbol, the entire solution set can be written as $A \cup B$.

Example 10   Writing a Solution Set Using Intersection

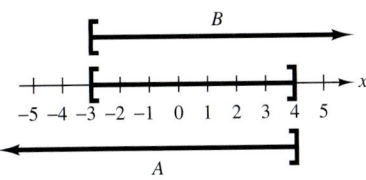

Figure 3.16

Write the compound inequality using the intersection symbol.

$$-3 \leq x \leq 4$$

Solution

Consider the two sets $A = \{x | x \leq 4\}$ and $B = \{x | x \geq -3\}$. These two sets overlap, as shown on the number line in Figure 3.16. The compound inequality $-3 \leq x \leq 4$ consists of all numbers that are in $x \leq 4$ and $x \geq -3$, which means that it can be written as $A \cap B$.

⑤ Solve application problems involving inequalities.

Applications

Linear inequalities in real-life problems arise from statements that involve phrases such as "at least," "no more than," "minimum value," and so on. Study the meanings of the key phrases in the next example.

Example 11   Translating Verbal Statements

Verbal Statement	*Inequality*	
a. x is at most 3.	$x \leq 3$	"at most" means "less than or equal to."
b. x is no more than 3.	$x \leq 3$	
c. x is at least 3.	$x \geq 3$	"at least" means "greater than or equal to."
d. x is no less than 3.	$x \geq 3$	
e. x is more than 3.	$x > 3$	
f. x is less than 3.	$x < 3$	
g. x is a minimum of 3.	$x \geq 3$	
h. x is at least 2, but less than 7.	$2 \leq x < 7$	
i. x is greater than 2, but no more than 7.	$2 < x \leq 7$	

To solve real-life problems involving inequalities, you can use the same "verbal-model approach" you use with equations.

Example 12   Finding the Maximum Width of a Package

An overnight delivery service will not accept any package whose combined length and minimum girth (perimeter of a cross section) exceeds 132 inches. You are sending a rectangular package that has square cross sections. The length of the package is 68 inches. What is the maximum width of the sides of its square cross sections?

Solution

First make a sketch. In Figure 3.17, the length of the package is 68 inches, and each side is x inches wide because the package has a square cross section.

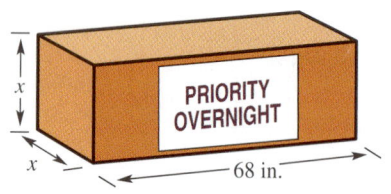

Figure 3.17

Verbal Model: Length + Girth ≤ 132 inches

Labels: Width of a side = x (inches)
Length = 68   (inches)
Girth = $4x$ (inches)

Inequality: $68 + 4x \le 132$

$$4x \le 64$$

$$x \le 16$$

The width of each side of the package must be less than or equal to 16 inches.

Example 13   Comparing Costs

A subcompact car can be rented from Company A for $240 per week with no extra charge for mileage. A similar car can be rented from Company B for $100 per week plus an additional 25 cents for each mile driven. How many miles must you drive in a week so that the rental fee for Company B is more than that for Company A?

Solution

Miles driven	Company A	Company B
520	$240.00	$230.00
530	$240.00	$232.50
540	$240.00	$235.00
550	$240.00	$237.50
560	$240.00	$240.00
570	$240.00	$242.50

Verbal Model: Weekly cost for Company B > Weekly cost for Company A

Labels: Number of miles driven in one week = m (miles)
Weekly cost for Company A = 240   (dollars)
Weekly cost for Company B = $100 + 0.25m$ (dollars)

Inequality: $100 + 0.25m > 240$

$$0.25m > 140$$

$$m > 560$$

So, the car from Company B is more expensive if you drive more than 560 miles in a week. The table shown at the left helps confirm this conclusion.

3.6 Exercises

Review   Concepts, Skills, and Problem Solving

Keep mathematically in shape by doing these exercises *before* the problems of this section.

Properties and Definitions

In Exercises 1–4, identify the property of real numbers illustrated by the statement.

1. $3yx = 3xy$

2. $3xy - 3xy = 0$

3. $6(x - 2) = 6x - 6 \cdot 2$

4. $3x + 0 = 3x$

Evaluating Expressions

In Exercises 5–10, evaluate the algebraic expression for the specified values of the variables. If not possible, state the reason.

5. $x^2 - y^2$

   $x = 4,\quad y = 3$

6. $4s + st$

   $s = 3,\quad t = -4$

7. $\dfrac{x}{x^2 + y^2}$

   $x = 0,\quad y = 3$

8. $\dfrac{z^2 + 2}{x^2 - 1}$

   $x = 2,\quad z = -1$

9. $\dfrac{a}{1 - r}$

   $a = 2,\quad r = \frac{1}{2}$

10. $2l + 2w$

   $l = 3,\quad w = 1.5$

Problem Solving

▲ *Geometry*   In Exercises 11 and 12, find the area of the trapezoid. The area of a trapezoid with parallel bases b_1 and b_2 and height h is $A = \frac{1}{2}(b_1 + b_2)h$.

11.

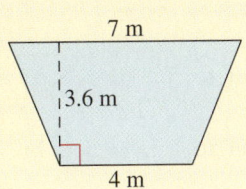

7 m
3.6 m
4 m

12.

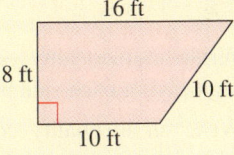

16 ft
8 ft
10 ft
10 ft

Developing Skills

In Exercises 1–4, determine whether each value of x satisfies the inequality.

Inequality		Values	

1. $7x - 10 > 0$

   (a) $x = 3$ (b) $x = -2$

   (c) $x = \frac{5}{2}$ (d) $x = \frac{1}{2}$

2. $3x + 2 < \dfrac{7x}{5}$

   (a) $x = 0$ (b) $x = 4$

   (c) $x = -4$ (d) $x = -1$

3. $0 < \dfrac{x + 5}{6} < 2$

   (a) $x = 10$ (b) $x = 4$

   (c) $x = 0$ (d) $x = -6$

4. $-2 < \dfrac{3 - x}{2} \le 2$

   (a) $x = 0$ (b) $x = 3$

   (c) $x = 9$ (d) $x = -12$

In Exercises 5–10, match the inequality with its graph. [The graphs are labeled (a), (b), (c), (d), (e), and (f).]

(a)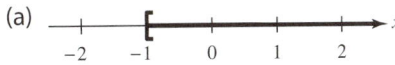

(b)

(c)

(d)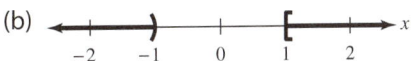

(e)

(f)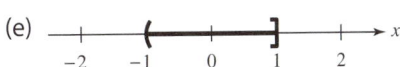

5. $x \ge -1$

6. $-1 < x \le 1$

7. $x \le -1$ or $x \ge 2$

8. $x < -1$ or $x \ge 1$

9. $-2 < x < 1$

10. $x < 2$

In Exercises 11–24, sketch the graph of the inequality. See Example 1.

11. $x \le 2$

12. $x > -6$

13. $x > 3.5$

14. $x \le -2.5$

15. $-5 < x \le 3$

16. $-1 < x \le 5$

17. $4 > x \ge 1$

18. $9 \ge x \ge 3$

19. $\frac{3}{2} \ge x > 0$

20. $-\frac{15}{4} < x < -\frac{5}{2}$

21. $x < -5$ or $x \ge -1$

22. $x \le -4$ or $x > 0$

23. $x \le 3$ or $x > 7$

24. $x \le -1$ or $x \ge 1$

25. Write an inequality equivalent to $5 - \frac{1}{3}x > 8$ by multiplying each side by -3.

26. Write an inequality equivalent to $5 - \frac{1}{3}x > 8$ by adding $\frac{1}{3}x$ to each side.

In Exercises 27–74, solve the inequality and sketch the solution on the real number line. See Examples 2–8.

27. $x - 4 \ge 0$

28. $x + 1 < 0$

29. $x + 7 \le 9$

30. $z - 4 > 0$

31. $2x < 8$

32. $3x \ge 12$

33. $-9x \ge 36$

34. $-6x \le 24$

35. $-\frac{3}{4}x < -6$

36. $-\frac{1}{5}x > -2$

37. $5 - x \le -2$

38. $1 - y \ge -5$

39. $2x - 5.3 > 9.8$

40. $1.6x + 4 \le 12.4$

41. $5 - 3x < 7$

42. $12 - 5x > 5$

43. $3x - 11 > -x + 7$

44. $21x - 11 \le 6x + 19$

45. $-3x + 7 < 8x - 13$

46. $6x - 1 > 3x - 11$

47. $\frac{x}{4} > 2 - \frac{x}{2}$

48. $\frac{x}{6} - 1 \le \frac{x}{4}$

49. $\frac{x - 4}{3} + 3 \le \frac{x}{8}$

50. $\frac{x + 3}{6} + \frac{x}{8} \ge 1$

51. $\frac{3x}{5} - 4 < \frac{2x}{3} - 3$

52. $\frac{4x}{7} + 1 > \frac{x}{2} + \frac{5}{7}$

53. $0 < 2x - 5 < 9$

54. $-6 \le 3x - 9 < 0$

55. $8 < 6 - 2x \le 12$

56. $-10 \le 4 - 7x < 10$

57. $-1 < -0.2x < 1$

58. $-2 < -0.5s \le 0$

59. $-3 < \frac{2x - 3}{2} < 3$

60. $0 \le \frac{x - 5}{2} < 4$

61. $1 > \frac{x - 4}{-3} > -2$

62. $-\frac{2}{3} < \frac{x - 4}{-6} \le \frac{1}{3}$

63. $2x - 4 \le 4$ and $2x + 8 > 6$

64. $7 + 4x < -5 + x$ and $2x + 10 \le -2$

65. $8 - 3x > 5$ and $x - 5 \ge -10$

66. $9 - x \le 3 + 2x$ and $3x - 7 \le -22$

67. $6.2 - 1.1x > 1$ or $1.2x - 4 > 2.7$

68. $0.4x - 3 \le 8.1$ or $4.2 - 1.6x \le 3$

69. $7x + 11 < 3 + 4x$ or $\frac{5}{2}x - 1 \ge 9 - \frac{3}{2}x$

70. $3x + 10 \le -x - 6$ or $\frac{x}{2} + 5 < \frac{5}{2}x - 4$

71. $-3(y + 10) \geq 4(y + 10)$

72. $2(4 - z) \geq 8(1 + z)$

73. $-4 \leq 2 - 3(x + 2) < 11$

74. $16 < 4(y + 2) - 5(2 - y) \leq 24$

In Exercises 75–80, write the solution set as a compound inequality. Then write the solution using set notation and the union or intersection symbol. See Example 9.

75.

76.

77.

78.

79.

80.

In Exercises 81–86, write the compound inequality using set notation and the union or intersection symbol. See Example 10.

81. $-7 \leq x < 0$

82. $2 < x < 8$

83. $x < -5$ or $x > 3$

84. $x \geq -1$ or $x < -6$

85. $-\frac{9}{2} < x \leq -\frac{3}{2}$

86. $x < 0$ or $x \geq \frac{2}{3}$

In Exercises 87–92, rewrite the statement using inequality notation. See Example 11.

87. x is nonnegative.   **88.** y is more than -2.

89. z is at least 8.   **90.** m is at least 4.

91. n is at least 10, but no more than 16.

92. x is at least 450, but no more than 500.

In Exercises 93–98, write a verbal description of the inequality.

93. $x \geq \frac{5}{2}$

94. $t < 4$

95. $3 \leq y < 5$

96. $-4 \leq t \leq 4$

97. $0 < z \leq \pi$

98. $-2 < x \leq 5$

Solving Problems

99. *Budget*   A student group has $4500 budgeted for a field trip. The cost of transportation for the trip is $1900. To stay within the budget, all other costs C must be no more than what amount?

100. *Budget*   You have budgeted $1800 per month for your total expenses. The cost of rent per month is $600 and the cost of food is $350. To stay within your budget, all other costs C must be no more than what amount?

101. *Meteorology*   Miami's average temperature is greater than the average temperature in Washington, DC, and the average temperature in Washington, DC is greater than the average temperature in New York City. How does the average temperature in Miami compare with the average temperature in New York City?

102. *Elevation*   The elevation (above sea level) of San Francisco is less than the elevation of Dallas, and the elevation of Dallas is less than the elevation of Denver. How does the elevation of San Francisco compare with the elevation of Denver?

103. *Operating Costs*   A utility company has a fleet of vans. The annual operating cost per van is

$$C = 0.35m + 2900$$

where m is the number of miles traveled by a van in a year. What is the maximum number of miles that will yield an annual operating cost that is less than $12,000?

104. *Operating Costs*   A fuel company has a fleet of trucks. The annual operating cost per truck is

$$C = 0.58m + 7800$$

where m is the number of miles traveled by a truck in a year. What is the maximum number of miles that will yield an annual operating cost that is less than $25,000?

Cost, Revenue, and Profit   In Exercises 105 and 106, the revenue R from selling x units and the cost C of producing x units of a product are given. In order to obtain a profit, the revenue must be greater than the cost. For what values of x will this product produce a profit?

105. $R = 89.95x$

$C = 61x + 875$

106. $R = 105.45x$

$C = 78x + 25{,}850$

107. *Long-Distance Charges*   The cost of an international long-distance telephone call is $0.96 for the first minute and $0.75 for each additional minute. The total cost of the call cannot exceed $5. Find the interval of time that is available for the call.

108. *Long-Distance Charges*   The cost of an international long-distance telephone call is $1.45 for the first minute and $0.95 for each additional minute. The total cost of the call cannot exceed $15. Find the interval of time that is available for the call.

109. ▲ *Geometry*   The length of a rectangle is 16 centimeters. The perimeter of the rectangle must be at least 36 centimeters and not more than 64 centimeters. Find the interval for the width x.

110. ▲ *Geometry*   The width of a rectangle is 14 meters. The perimeter of the rectangle must be at least 100 meters and not more than 120 meters. Find the interval for the length x.

111. *Number Problem*   Four times a number n must be at least 12 and no more than 30. What interval contains this number?

112. *Number Problem*   Determine all real numbers n such that $\frac{1}{3}n$ must be more than 7.

113. *Hourly Wage*   Your company requires you to select one of two payment plans. One plan pays a straight $12.50 per hour. The second plan pays $8.00 per hour plus $0.75 per unit produced per hour. Write an inequality for the number of units that must be produced per hour so that the second option yields the greater hourly wage. Solve the inequality.

114. *Monthly Wage*   Your company requires you to select one of two payment plans. One plan pays a straight $3000 per month. The second plan pays $1000 per month plus a commission of 4% of your gross sales. Write an inequality for the gross sales per month for which the second option yields the greater monthly wage. Solve the inequality.

Environment In Exercises 115 and 116, use the following equation, which models the air pollutant emissions *y* (in millions of metric tons) of methane caused by landfills in the continental United States from 1994 to 2000 (see figure).

$$y = -0.434t + 12.23, \text{ for } 4 \le t \le 10$$

In this model, *t* represents the year, with *t* = 4 corresponding to 1994. (Source: U.S. Energy Information Administration)

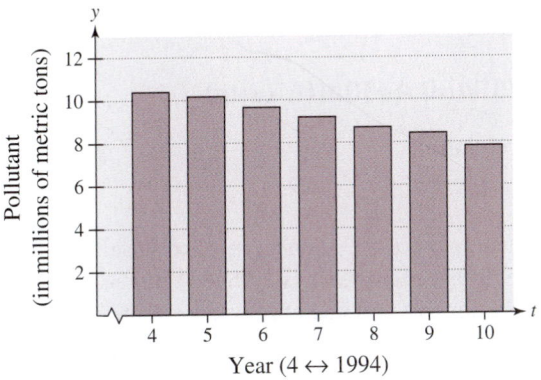

Pollutant (in millions of metric tons)

Year (4 ↔ 1994)

115. During which years was the air pollutant emission of methane caused by landfills greater than 10 million metric tons?

116. During which years was the air pollutant emission of methane caused by landfills less than 8.5 million metric tons?

Explaining Concepts

117. ⚡ Answer part (h) of Motivating the Chapter on page 122.

118. *Writing* ✏️ Is adding −5 to each side of an inequality the same as subtracting 5 from each side? Explain.

119. *Writing* ✏️ Is dividing each side of an inequality by 5 the same as multiplying each side by $\frac{1}{5}$? Explain.

120. *Writing* ✏️ Describe any differences between properties of equalities and properties of inequalities.

121. Give an example of "reversing an inequality symbol."

122. If $-3 \le x \le 10$, then $-x$ must be in what interval?

3.7 Absolute Value Equations and Inequalities

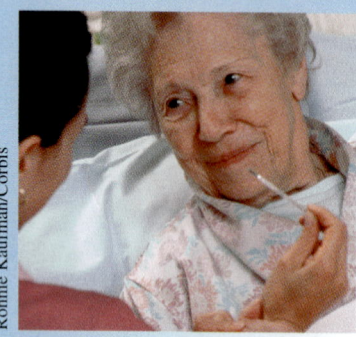

Ronnie Kaufman/Corbis

What You Should Learn

1 Solve absolute value equations.

2 Solve inequalities involving absolute value.

Why You Should Learn It

Absolute value equations and inequalities can be used to model and solve real-life problems. For instance, in Exercise 125 on page 205, you will use an absolute value inequality to describe the normal body temperature range.

Solving Equations Involving Absolute Value

Consider the **absolute value equation**

$$|x| = 3.$$

The only solutions of this equation are $x = -3$ and $x = 3$, because these are the only two real numbers whose distance from zero is 3. (See Figure 3.18.) In other words, the absolute value equation $|x| = 3$ has exactly two solutions: $x = -3$ and $x = 3$.

1 Solve absolute value equations.

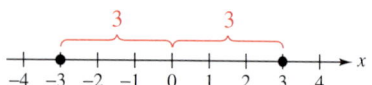

Figure 3.18

Solving an Absolute Value Equation

Let x be a variable or an algebraic expression and let a be a real number such that $a \geq 0$. The solutions of the equation $|x| = a$ are given by $x = -a$ and $x = a$. That is,

$$|x| = a \implies x = -a \quad \text{or} \quad x = a.$$

Study Tip

The strategy for solving absolute value equations is to *rewrite* the equation in *equivalent forms* that can be solved by previously learned methods. This is a common strategy in mathematics. That is, when you encounter a new type of problem, you try to rewrite the problem so that it can be solved by techniques you already know.

Example 1   Solving Absolute Value Equations

Solve each absolute value equation.

a. $|x| = 10$ **b.** $|x| = 0$ **c.** $|y| = -1$

Solution

a. This equation is equivalent to the two linear equations

$$x = -10 \quad \text{and} \quad x = 10. \quad \text{Equivalent linear equations}$$

So, the absolute value equation has two solutions: $x = -10$ and $x = 10$.

b. This equation is equivalent to the two linear equations

$$x = 0 \quad \text{and} \quad x = 0. \quad \text{Equivalent linear equations}$$

Because both equations are the same, you can conclude that the absolute value equation has only one solution: $x = 0$.

c. This absolute value equation has *no solution* because it is not possible for the absolute value of a real number to be negative.

Example 2   Solving Absolute Value Equations

Solve $|3x + 4| = 10$.

Solution

$$|3x + 4| = 10 \qquad \text{Write original equation.}$$

$$3x + 4 = -10 \quad \text{or} \quad 3x + 4 = 10 \qquad \text{Equivalent equations}$$

$$3x + 4 - 4 = -10 - 4 \quad 3x + 4 - 4 = 10 - 4 \qquad \text{Subtract 4 from each side.}$$

$$3x = -14 \qquad\qquad 3x = 6 \qquad \text{Combine like terms.}$$

$$x = -\frac{14}{3} \qquad\qquad x = 2 \qquad \text{Divide each side by 3.}$$

Check

$$|3x + 4| = 10 \qquad\qquad |3x + 4| = 10$$

$$\left|3\left(-\tfrac{14}{3}\right) + 4\right| \overset{?}{=} 10 \qquad\qquad |3(2) + 4| \overset{?}{=} 10$$

$$|-14 + 4| \overset{?}{=} 10 \qquad\qquad |6 + 4| \overset{?}{=} 10$$

$$|-10| = 10 \ \checkmark \qquad\qquad |10| = 10 \ \checkmark$$

When solving absolute value equations, remember that it is possible that they have no solution. For instance, the equation $|3x + 4| = -10$ has no solution because the absolute value of a real number cannot be negative. Do not make the mistake of trying to solve such an equation by writing the "equivalent" linear equations as $3x + 4 = -10$ and $3x + 4 = 10$. These equations have solutions, but they are both extraneous.

The equation in the next example is not given in the **standard form**

$$|ax + b| = c, \quad c \geq 0.$$

Notice that the first step in solving such an equation is to write it in standard form.

Example 3   An Absolute Value Equation in Nonstandard Form

Solve $|2x - 1| + 3 = 8$.

Solution

$$|2x - 1| + 3 = 8 \qquad \text{Write original equation.}$$

$$|2x - 1| = 5 \qquad \text{Write in standard form.}$$

$$2x - 1 = -5 \quad \text{or} \quad 2x - 1 = 5 \qquad \text{Equivalent equations}$$

$$2x = -4 \qquad\qquad 2x = 6 \qquad \text{Add 1 to each side.}$$

$$x = -2 \qquad\qquad x = 3 \qquad \text{Divide each side by 2.}$$

The solutions are $x = -2$ and $x = 3$. Check these in the original equation.

If two algebraic expressions are equal in absolute value, they must either be equal to each other or be the *opposites* of each other. So, you can solve equations of the form

$$|ax + b| = |cx + d|$$

by forming the two linear equations

Expressions equal   Expressions opposite

$$ax + b = cx + d \quad \text{and} \quad ax + b = -(cx + d).$$

Example 4   Solving an Equation Involving Two Absolute Values

Solve $|3x - 4| = |7x - 16|$.

Solution

$$|3x - 4| = |7x - 16| \qquad \text{Write original equation.}$$

$3x - 4 = 7x - 16 \quad$ or $\quad 3x - 4 = -(7x - 16) \qquad$ Equivalent equations

$$-4x - 4 = -16 \qquad\qquad 3x - 4 = -7x + 16$$

$$-4x = -12 \qquad\qquad\qquad 10x = 20$$

$$x = 3 \qquad\qquad\qquad\qquad x = 2 \qquad \text{Solutions}$$

The solutions are $x = 3$ and $x = 2$. Check these in the original equation.

Example 5   Solving an Equation Involving Two Absolute Values

Solve $|x + 5| = |x + 11|$.

Solution

By equating the expression $(x + 5)$ to the opposite of $(x + 11)$, you obtain

$$x + 5 = -(x + 11) \qquad \text{Equivalent equation}$$

$$x + 5 = -x - 11 \qquad \text{Distributive Property}$$

$$2x + 5 = -11 \qquad \text{Add } x \text{ to each side.}$$

$$2x = -16 \qquad \text{Subtract 5 from each side.}$$

$$x = -8. \qquad \text{Divide each side by 2.}$$

However, by setting the two expressions equal to each other, you obtain

$$x + 5 = x + 11 \qquad \text{Equivalent equation}$$

$$x = x + 6 \qquad \text{Subtract 5 from each side.}$$

$$0 = 6 \qquad \text{Subtract } x \text{ from each side.}$$

which is a false statement. So, the original equation has only one solution: $x = -8$. Check this solution in the original equation.

Study Tip

When solving equations of the form

$$|ax + b| = |cx + d|$$

it is possible that one of the resulting equations will not have a solution. Note this occurrence in Example 5.

② Solve inequalities involving absolute value.

Solving Inequalities Involving Absolute Value

To see how to solve inequalities involving absolute value, consider the following comparisons.

$$|x| = 2$$
$$x = -2 \text{ and } x = 2$$

$$|x| < 2$$
$$-2 < x < 2$$

$$|x| > 2$$
$$x < -2 \text{ or } x > 2$$

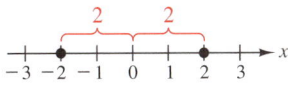   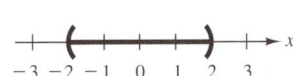   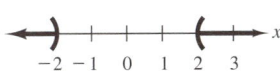

These comparisons suggest the following rules for solving inequalities involving absolute value.

Solving an Absolute Value Inequality

Let x be a variable or an algebraic expression and let a be a real number such that $a > 0$.

1. The solutions of $|x| < a$ are all values of x that lie between $-a$ and a. That is,

   $$|x| < a \quad \text{if and only if} \quad -a < x < a.$$

2. The solutions of $|x| > a$ are all values of x that are less than $-a$ or greater than a. That is,

   $$|x| > a \quad \text{if and only if} \quad x < -a \text{ or } x > a.$$

These rules are also valid if $<$ is replaced by $\leq$ and $>$ is replaced by $\geq$.

Example 6   Solving an Absolute Value Inequality

Solve $|x - 5| < 2$.

Solution

$|x - 5| < 2$	Write original inequality.
$-2 < x - 5 < 2$	Equivalent double inequality
$-2 + 5 < x - 5 + 5 < 2 + 5$	Add 5 to all three parts.
$3 < x < 7$	Combine like terms.

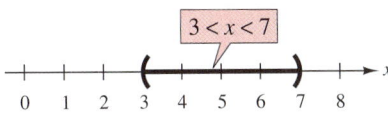

Figure 3.19

The solution set consists of all real numbers that are greater than 3 and less than 7. The solution set in interval notation is $(3, 7)$ and in set notation is $\{x \mid 3 < x < 7\}$. The graph of this solution set is shown in Figure 3.19.

To verify the solution of an absolute value inequality, you need to check values in the solution set and outside of the solution set. For instance, in Example 6 you can check that $x = 4$ is in the solution set and that $x = 2$ and $x = 8$ are not in the solution set.

> **Study Tip**
>
> In Example 6, note that an absolute value inequality of the form $|x| < a$ (or $|x| \le a$) can be solved with a double inequality. An inequality of the form $|x| > a$ (or $|x| \ge a$) cannot be solved with a double inequality. Instead, you must solve two separate inequalities, as demonstrated in Example 7.

Example 7   Solving an Absolute Value Inequality

Solve $|3x - 4| \ge 5$.

Solution

$$|3x - 4| \ge 5 \qquad\qquad\qquad\qquad \text{Write original inequality.}$$

$$3x - 4 \le -5 \quad \text{or} \quad 3x - 4 \ge 5 \qquad \text{Equivalent inequalities}$$

$$3x - 4 + 4 \le -5 + 4 \quad 3x - 4 + 4 \ge 5 + 4 \qquad \text{Add 4 to all parts.}$$

$$3x \le -1 \qquad\qquad\qquad 3x \ge 9 \qquad \text{Combine like terms.}$$

$$\frac{3x}{3} \le \frac{-1}{3} \qquad\qquad\qquad \frac{3x}{3} \ge \frac{9}{3} \qquad \text{Divide each side by 3.}$$

$$x \le -\frac{1}{3} \qquad\qquad\qquad x \ge 3 \qquad \text{Simplify.}$$

The solution set consists of all real numbers that are less than or equal to $-\frac{1}{3}$ or greater than or equal to 3. The solution set in interval notation is $\left(-\infty, -\frac{1}{3}\right] \cup [3, \infty)$ and in set notation is $\left\{x \mid x \le -\frac{1}{3} \text{ or } x \ge 3\right\}$. The graph is shown in Figure 3.20.

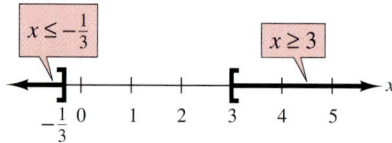

Figure 3.20

Example 8   Solving an Absolute Value Inequality

$$\left|2 - \frac{x}{3}\right| \le 0.01 \qquad\qquad \text{Original inequality}$$

$$-0.01 \le 2 - \frac{x}{3} \le 0.01 \qquad\qquad \text{Equivalent double inequality}$$

$$-2.01 \le -\frac{x}{3} \le -1.99 \qquad\qquad \text{Subtract 2 from all three parts.}$$

$$-2.01(-3) \ge -\frac{x}{3}(-3) \ge -1.99(-3) \qquad \begin{array}{l}\text{Multiply all three parts by } -3 \text{ and}\\ \text{reverse both inequality symbols.}\end{array}$$

$$6.03 \ge x \ge 5.97 \qquad\qquad \text{Simplify.}$$

$$5.97 \le x \le 6.03 \qquad\qquad \text{Solution set in standard form}$$

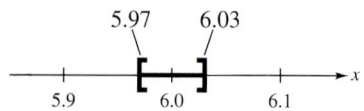

Figure 3.21

The solution set consists of all real numbers that are greater than or equal to 5.97 *and* less than or equal to 6.03. The solution set in interval notation is $[5.97, 6.03]$ and in set notation is $\{x \mid 5.97 \le x \le 6.03\}$. The graph is shown in Figure 3.21.

Technology: Tip

Most graphing calculators can graph absolute value inequalities. Consult your user's guide for specific instructions. The graph below shows the solution of the inequality $|x - 5| < 2$. Notice that the graph occurs as an interval above the *x*-axis.

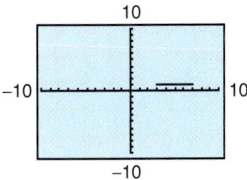

Example 9   Creating a Model

To test the accuracy of a rattlesnake's "pit-organ sensory system," a biologist blindfolded a rattlesnake and presented the snake with a warm "target." Of 36 strikes, the snake was on target 17 times. In fact, the snake was within 5 degrees of the target for 30 of the strikes. Let *A* represent the number of degrees by which the snake is off target. Then $A = 0$ represents a strike that is aimed directly at the target. Positive values of *A* represent strikes to the right of the target and negative values of *A* represent strikes to the left of the target. Use the diagram shown in Figure 3.22 to write an absolute value inequality that describes the interval in which the 36 strikes occurred.

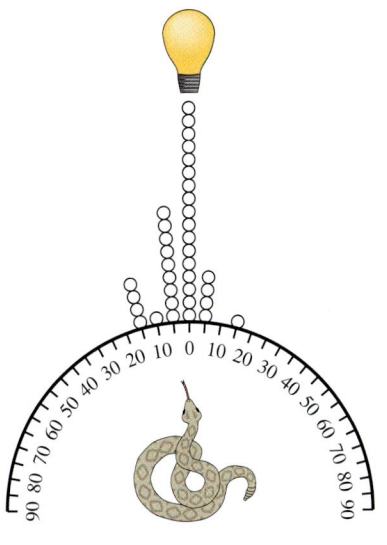

Figure 3.22

Solution

From the diagram, you can see that the snake was never off by more than 15 degrees in either direction. As a compound inequality, this can be represented by $-15 \le A \le 15$. As an absolute value inequality, the interval in which the strikes occurred can be represented by $|A| \le 15$.

Example 10   Production

The estimated daily production at an oil refinery is given by the absolute value inequality $|x - 200{,}000| \le 25{,}000$, where *x* is measured in barrels of oil. Solve the inequality to determine the maximum and minimum production levels.

Solution

$$|x - 200{,}000| \le 25{,}000 \qquad \text{Write original inequality.}$$

$$-25{,}000 \le x - 200{,}000 \le 25{,}000 \qquad \text{Equivalent double inequality}$$

$$175{,}000 \le x \le 225{,}000 \qquad \text{Add 200,000 to all three parts.}$$

So, the oil refinery can produce a maximum of 225,000 barrels of oil and a minimum of 175,000 barrels of oil per day.

3.7 Exercises

Review *Concepts, Skills, and Problem Solving*

Keep mathematically in shape by doing these exercises *before* the problems of this section.

Properties and Definitions

1. *Writing* If n is an integer, how do the numbers $2n$ and $2n - 1$ differ? Explain.

2. *Writing* Are $-2x^4$ and $(-2x)^4$ equal? Explain.

3. *Writing* Show how to write $\frac{35}{14}$ in simplest form.

4. *Writing* Show how to divide $\frac{4}{5}$ by $\frac{z}{3}$.

Order of Real Numbers

In Exercises 5–10, place the correct inequality symbol ($<$ or $>$) between the two real numbers.

5. -5.6 ____ -4.8

6. $-|7.2|$ ____ $-|-13.1|$

7. $-\frac{3}{4}$ ____ -5

8. $-\frac{1}{5}$ ____ $-\frac{1}{3}$

9. π ____ -3

10. 6 ____ $\frac{13}{2}$

Problem Solving

Budget In Exercises 11 and 12, determine whether there is more or less than a $500 variance between the budgeted amount and the actual expense.

11. Wages
    Budgeted: $162,700
    Actual: $163,356

12. Taxes
    Budgeted: $42,640
    Actual: $42,335

Developing Skills

In Exercises 1–4, determine whether the value is a solution of the equation.

Equation	Value
1. $|4x + 5| = 10$	$x = -3$
2. $|2x - 16| = 10$	$x = 3$
3. $|6 - 2w| = 2$	$w = 4$
4. $\left|\frac{1}{2}t + 4\right| = 8$	$t = 6$

In Exercises 5–8, transform the absolute value equation into two linear equations.

5. $|x - 10| = 17$

6. $|7 - 2t| = 5$

7. $|4x + 1| = \frac{1}{2}$

8. $|22k + 6| = 9$

In Exercises 9–52, solve the equation. (Some equations have no solution.) See Examples 1–5.

9. $|x| = 4$

10. $|x| = 3$

11. $|t| = -45$

12. $|s| = 16$

13. $|h| = 0$

14. $|x| = -82$

15. $|5x| = 15$

16. $\left|\frac{1}{3}x\right| = 2$

17. $|x + 1| = 5$

18. $|x + 5| = 7$

19. $\left|\frac{2s + 3}{5}\right| = 5$

20. $\left|\frac{7a + 6}{4}\right| = 2$

21. $|32 - 3y| = 16$

22. $|3 - 5x| = 13$

23. $|3x + 4| = -16$

24. $|20 - 5t| = 50$

25. $|4 - 3x| = 0$

26. $|3x - 2| = -5$

27. $\left|\frac{2}{3}x + 4\right| = 9$

28. $\left|3 - \frac{4}{5}x\right| = 1$

29. $|0.32x - 2| = 4$

30. $|2 - 1.5x| = 2$

31. $|5x - 3| + 8 = 22$

32. $|6x - 4| - 7 = 3$

33. $|3x + 9| - 12 = -8$

34. $|5 - 2x| + 10 = 6$

35. $\left|\dfrac{x - 3}{4}\right| + 4 = 4$

36. $\left|\dfrac{x - 3}{4}\right| - 3 = 2$

37. $\left|\dfrac{5x - 3}{2}\right| + 2 = 6$

38. $\left|\dfrac{3z + 5}{6}\right| - 3 = 6$

39. $-2|7 - 4x| = -16$

40. $4|5x + 1| = 24$

41. $3|2x - 5| + 4 = 7$

42. $2|4 - 3x| - 6 = -2$

43. $|x + 8| = |2x + 1|$

44. $|10 - 3x| = |x + 7|$

45. $|x + 2| = |3x - 1|$

46. $|x - 2| = |2x - 15|$

47. $|45 - 4x| = |32 - 3x|$

48. $|5x + 4| = |3x + 25|$

49. $\left|\frac{3}{4}x - 2\right| = \left|\frac{1}{2}x + 5\right|$

50. $\left|\frac{3}{2}r + 2\right| = \left|\frac{1}{2}r - 3\right|$

51. $|4x - 10| = 2|2x + 3|$

52. $3|2 - 3x| = |9x + 21|$

Think About It In Exercises 53 and 54, write an absolute value equation that represents the verbal statement.

53. The distance between x and 5 is 3.

54. The distance between t and -2 is 6.

In Exercises 55–58, determine whether each *x*-value is a solution of the inequality.

Inequality		*Values*	
55. $|x| < 3$	(a) $x = 2$	(b) $x = -4$	
	(c) $x = 4$	(d) $x = -1$	
56. $|x| \le 5$	(a) $x = -7$	(b) $x = -4$	
	(c) $x = 4$	(d) $x = 9$	
57. $|x - 7| \ge 3$	(a) $x = 9$	(b) $x = -4$	
	(c) $x = 11$	(d) $x = 6$	
58. $|x - 3| > 5$	(a) $x = 16$	(b) $x = 3$	
	(c) $x = -2$	(d) $x = -3$	

In Exercises 59–62, transform the absolute value inequality into a double inequality or two separate inequalities.

59. $|y + 5| < 3$

60. $|6x + 7| \le 5$

61. $|7 - 2h| \ge 9$

62. $|8 - x| > 25$

In Exercises 63–66, sketch a graph that shows the real numbers that satisfy the statement.

63. All real numbers greater than -2 *and* less than 5

64. All real numbers greater than or equal to 3 *and* less than 10

65. All real numbers less than or equal to 4 *or* greater than 7

66. All real numbers less than -6 *or* greater than or equal to 6

In Exercises 67–104, solve the inequality. See Examples 6–8.

67. $|y| < 4$

68. $|x| < 6$

69. $|x| \ge 6$

70. $|y| \geq 4$

71. $|2x| < 14$

72. $|4z| \leq 9$

73. $\left|\dfrac{y}{3}\right| \leq 3$

74. $\left|\dfrac{t}{2}\right| < 4$

75. $|y - 2| \leq 4$ **76.** $|x - 3| \leq 6$

77. $|x + 6| > 10$ **78.** $|x - 4| \geq 3$

79. $|2x - 1| \leq 7$ **80.** $|3x + 4| < 2$

81. $|6t + 15| \geq 30$ **82.** $|3t + 1| > 5$

83. $|2 - 5x| > -8$ **84.** $|8 - 7x| < -6$

85. $|3x + 10| < -1$ **86.** $|4x - 5| > -3$

87. $\dfrac{|x + 2|}{10} \leq 8$ **88.** $\dfrac{|s - 3|}{5} > 4$

89. $\dfrac{|y - 16|}{4} < 30$ **90.** $\dfrac{|a + 6|}{2} \geq 16$

91. $\left|\dfrac{z}{10} - 3\right| > 8$ **92.** $\left|\dfrac{x}{8} + 1\right| < 0$

93. $\left|\dfrac{3x + 4}{5}\right| > \dfrac{7}{5}$ **94.** $\left|\dfrac{3 - 2x}{4}\right| \geq 5$

95. $|0.2x - 3| < 4$ **96.** $|1.5t - 8| \leq 16$

97. $\left|6 - \dfrac{3}{5}x\right| \leq 0.4$ **98.** $\left|3 - \dfrac{x}{4}\right| > 0.15$

99. $-2|3x + 6| < 4$ **100.** $-4|2x - 7| > -12$

101. $\left|9 - \dfrac{x}{2}\right| - 7 \leq 4$ **102.** $\left|8 - \dfrac{2}{3}x\right| + 6 \geq 10$

103. $\left|\dfrac{3x - 2}{4}\right| + 5 \geq 5$ **104.** $\left|\dfrac{2x - 4}{5}\right| - 9 \leq 3$

In Exercises 105–110, use a graphing calculator to solve the inequality.

105. $|3x + 2| < 4$ **106.** $|2x - 1| \leq 3$

107. $|2x + 3| > 9$ **108.** $|7r - 3| > 11$

109. $|x - 5| + 3 \leq 5$ **110.** $|a + 1| - 4 < 0$

In Exercises 111–114, match the inequality with its graph. [The graphs are labeled (a), (b), (c), and (d).]

(a)

(b)

(c)

(d)

111. $|x - 4| \leq 4$ **112.** $|x - 4| < 1$

113. $\frac{1}{2}|x - 4| > 4$ **114.** $|2(x - 4)| \geq 4$

In Exercises 115–118, write an absolute value inequality that represents the interval.

115.

116.

117.

118.

The symbol   indicates an exercise in which you are instructed to use a graphing calculator.

In Exercises 119–122, write an absolute value inequality that represents the verbal statement.

119. The set of all real numbers x whose distance from 0 is less than 3.

120. The set of all real numbers x whose distance from 0 is more than 2.

121. The set of all real numbers x whose distance from 5 is more than 6.

122. The set of all real numbers x whose distance from 16 is less than 5.

Solving Problems

123. *Temperature*   The operating temperature of an electronic device must satisfy the inequality $|t - 72| \leq 10$, where t is given in degrees Fahrenheit. Sketch the graph of the solution of the inequality. What are the maximum and minimum temperatures?

124. *Time Study*   A time study was conducted to determine the length of time required to perform a task in a manufacturing process. The times required by approximately two-thirds of the workers in the study satisfied the inequality

$$\left| \frac{t - 15.6}{1.9} \right| \leq 1$$

where t is time in minutes. Sketch the graph of the solution of the inequality. What are the maximum and minimum times?

125. *Body Temperature*   Physicians consider an adult's body temperature x to be normal if it is between 97.6°F and 99.6°F. Write an absolute value inequality that describes this normal temperature range.

126. *Accuracy of Measurements*   In woodshop class, you must cut several pieces of wood to within $\frac{3}{16}$ inch of the teacher's specifications. Let $(s - x)$ represent the difference between the specification s and the measured length x of a cut piece.

(a) Write an absolute value inequality that describes the values of x that are within specifications.

(b) The length of one piece of wood is specified to be $s = 5\frac{1}{8}$ inches. Describe the acceptable lengths for this piece.

Explaining Concepts

127. Give a graphical description of the absolute value of a real number.

128. Give an example of an absolute value equation that has only one solution.

129. *Writing*   In your own words, explain how to solve an absolute value equation. Illustrate your explanation with an example.

130. The graph of the inequality $|x - 3| < 2$ can be described as *all real numbers that are within two units of 3*. Give a similar description of $|x - 4| < 1$.

131. Complete $|2x - 6| \leq$ ▢   so that the solution is $0 \leq x \leq 6$.

132. When you buy a 16-ounce bag of chips, you probably expect to get *precisely* 16 ounces. Suppose the actual weight w (in ounces) of a "16-ounce" bag of chips is given by $|w - 16| \leq \frac{1}{2}$. You buy four 16-ounce bags. What is the greatest amount you can expect to get? What is the least? Explain.

133. *Writing*   You are teaching a class in algebra and one of your students hands in the following solution. What is wrong with this solution? What could you say to help your students avoid this type of error?

$$|3x - 4| \geq -5$$

$$3x - 4 \leq -5 \quad \text{or} \quad 3x - 4 \geq 5$$

$$3x \leq -1 \qquad\qquad 3x \geq 9$$

$$x \leq -\tfrac{1}{3} \qquad\qquad x \geq 3$$

What Did You Learn?

Key Terms

linear equation, *p. 124*

consecutive integers, *p. 130*

cross-multiplication, *p. 140*

markup, *p. 150*

discount, *p. 151*

ratio, *p. 157*

unit price, *p. 159*

proportion, *p. 160*

mixture problems, *p. 173*

work-rate problems, *p. 175*

linear inequality, *p. 185*

compound inequality, *p. 187*

intersection, *p. 188*

union, *p. 188*

absolute value equation, *p. 196*

Key Concepts

3.1 ○ Solving a linear equation

Solve a linear equation using inverse operations to isolate the variable.

3.1 ○ Expressions for special types of integers

Let n be an integer.

1. $2n$ denotes an *even* integer.

2. $2n - 1$ and $2n + 1$ denote *odd* integers.

3. The set $\{n, n + 1, n + 2\}$ denotes three *consecutive* integers.

3.2 ○ Solving equations containing symbols of grouping

1. Remove symbols of grouping from each side by using the Distributive Property.

2. Combine like terms.

3. Isolate the variable in the usual way using properties of equality.

4. Check your solution in the original equation.

3.2 ○ Equations involving fractions or decimals

1. Clear an equation of fractions by multiplying each side by the least common multiple (LCM) of the denominators.

2. Use cross-multiplication to solve a linear equation that equates two fractions. That is, if

$$\frac{a}{b} = \frac{c}{d}, \text{ then } a \cdot d = b \cdot c.$$

3. To solve a linear equation with decimal coefficients, multiply each side by a power of 10 that converts all decimal coefficients to integers.

3.3 ○ The percent equation

The percent equation $a = p \cdot b$ compares two numbers, where b is the base number, p is the percent in decimal form, and a is the number being compared to b.

3.3 ○ Guidelines for solving word problems

1. Write a *verbal model* that describes the problem.

2. Assign *labels* to fixed quantities and variable quantities.

3. Rewrite the verbal model as an *algebraic equation* using the assigned labels.

4. *Solve* the resulting algebraic equation.

5. *Check* to see that your solution satisfies the original problem as stated.

3.4 ○ Solving a proportion

If $\dfrac{a}{b} = \dfrac{c}{d}$, then $ad = bc$.

3.6 ○ Properties of inequalities

Let a, b, and c be real numbers, variables, or algebraic expressions.

Addition: If $a < b$, then $a + c < b + c$.

Subtraction: If $a < b$, then $a - c < b - c$.

Multiplication: If $a < b$ and $c > 0$, then $ac < bc$.

If $a < b$ and $c < 0$, then $ac > bc$.

Division: If $a < b$ and $c > 0$, then $\dfrac{a}{c} < \dfrac{b}{c}$.

If $a < b$ and $c < 0$, then $\dfrac{a}{c} > \dfrac{b}{c}$.

Transitive: If $a < b$ and $b < c$, then $a < c$.

3.6 ○ Solving a linear inequality or a compound inequality

Solve a linear inequality by performing inverse operations on all parts of the inequality.

3.7 ○ Solving an absolute value equation or inequality

Solve an absolute value equation by rewriting as two linear equations. Solve an absolute inequality by rewriting as a compound inequality.

Review Exercises

3.1 Solving Linear Equations

1 Solve linear equations in standard form.

In Exercises 1–4, solve the equation and check your solution.

1. $2x - 10 = 0$

2. $12y + 72 = 0$

3. $-3y - 12 = 0$

4. $-7x + 21 = 0$

2 Solve linear equations in nonstandard form.

In Exercises 5–18, solve the equation and check your solution.

5. $x + 10 = 13$

6. $x - 3 = 8$

7. $5 - x = 2$

8. $3 = 8 - x$

9. $10x = 50$

10. $-3x = 21$

11. $8x + 7 = 39$

12. $12x - 5 = 43$

13. $24 - 7x = 3$

14. $13 + 6x = 61$

15. $15x - 4 = 16$

16. $3x - 8 = 2$

17. $\dfrac{x}{5} = 4$

18. $-\dfrac{x}{14} = \dfrac{1}{2}$

3 Use linear equations to solve application problems.

19. *Hourly Wage*   Your hourly wage is $8.30 per hour plus 60 cents for each unit you produce. How many units must you produce in an hour so that your hourly wage is $15.50?

20. *Consumer Awareness*   A long-distance carrier's connection fee for a phone call is $1.25. There is also a charge of $0.10 per minute. How long was a phone call that cost $3.05?

21. ▲ *Geometry*   The perimeter of a rectangle is 260 meters. The length is 30 meters greater than its width. Find the dimensions of the rectangle.

22. ▲ *Geometry*   A 10-foot board is cut so that one piece is 4 times as long as the other. Find the length of each piece.

3.2 Equations That Reduce to Linear Form

1 Solve linear equations containing symbols of grouping.

In Exercises 23–28, solve the equation and check your solution.

23. $3x - 2(x + 5) = 10$

24. $4x + 2(7 - x) = 5$

25. $2(x + 3) = 6(x - 3)$

26. $8(x - 2) = 3(x + 2)$

27. $7 - [2(3x + 4) - 5] = x - 3$

28. $14 + [3(6x - 15) + 4] = 5x - 1$

2 Solve linear equations involving fractions.

In Exercises 29–36, solve the equation and check your solution.

29. $\dfrac{2}{3}x - \dfrac{1}{6} = \dfrac{9}{2}$

30. $\dfrac{1}{8}x + \dfrac{3}{4} = \dfrac{5}{2}$

31. $\dfrac{x}{3} - \dfrac{1}{9} = 2$

32. $\dfrac{1}{2} - \dfrac{x}{8} = 7$

33. $\dfrac{u}{10} + \dfrac{u}{5} = 6$

34. $\dfrac{x}{3} + \dfrac{x}{5} = 1$

35. $\dfrac{2x}{9} = \dfrac{2}{3}$

36. $\dfrac{5y}{13} = \dfrac{2}{5}$

3 Solve linear equations involving decimals.

In Exercises 37–40, solve the equation. Round your answer to two decimal places.

37. $5.16x - 87.5 = 32.5$

38. $2.825x + 3.125 = 12.5$

39. $\dfrac{x}{4.625} = 48.5$

40. $5x + \dfrac{1}{4.5} = 18.125$

3.3 Problem Solving with Percents

① Convert percents to decimals and fractions and convert decimals and fractions to percents.

In Exercises 41 and 42, complete the table showing the equivalent forms of a percent.

Percent	Parts out of 100	Decimal	Fraction
41. 35%			
42.			$\frac{4}{5}$

② Solve linear equations involving percents.

In Exercises 43–48, solve the percent equation.

43. What number is 125% of 16?

44. What number is 0.8% of 3250?

45. 150 is $37\frac{1}{2}$% of what number?

46. 323 is 95% of what number?

47. 150 is what percent of 250?

48. 130.6 is what percent of 3265?

③ Solve application problems involving markups and discounts.

49. *Selling Price*   An electronics store uses a markup rate of 62% on all items. The cost of a CD player is $48. What is the selling price of the CD player?

50. *Sale Price*   A clothing store advertises 30% off the list price of all sweaters. A turtleneck sweater has a list price of $120. What is the sale price?

51. *Sales*   The sales (in millions) for the Yankee Candle Company in the years 2000 and 2001 were $338.8 and $379.8, respectively. Determine the percent increase in sales from 2000 to 2001.   (Source: The Yankee Candle Company)

52. *Price Increase*   The manufacturer's suggested retail price for a car is $18,459. Estimate the price of a comparably equipped car for the next model year if the price will increase by $4\frac{1}{2}$%.

3.4 Ratios and Proportions

① Compare relative sizes using ratios.

In Exercises 53–56, find a ratio that compares the relative sizes of the quantities. (Use the same units of measurement for both quantities.)

53. Eighteen inches to 4 yards

54. One pint to 2 gallons

55. Two hours to 90 minutes

56. Four meters to 150 centimeters

② Find the unit price of a consumer item.

In Exercises 57 and 58, which product has the lower unit price?

57. (a) An 18-ounce container of cooking oil for $0.89

(b) A 24-ounce container of cooking oil for $1.12

58. (a) A 17.4-ounce box of pasta noodles for $1.32

(b) A 32-ounce box of pasta noodles for $2.62

③ Solve proportions that equate two ratios.

In Exercises 59–64, solve the proportion.

59. $\dfrac{7}{16} = \dfrac{z}{8}$

60. $\dfrac{x}{12} = \dfrac{5}{4}$

61. $\dfrac{x+2}{4} = -\dfrac{1}{3}$

62. $\dfrac{x-4}{1} = \dfrac{9}{4}$

63. $\dfrac{x-3}{2} = \dfrac{x+6}{5}$

64. $\dfrac{x+1}{3} = \dfrac{x+2}{4}$

④ Solve application problems using the Consumer Price Index.

In Exercises 65 and 66, use the Consumer Price Index table on page 162 to estimate the price of the item in the indicated year.

65. The 2001 price of a recliner chair that cost $78 in 1984

66. The 1986 price of a microwave oven that cost $120 in 1999

3.5 Geometric and Scientific Applications

① Use common formulas to solve application problems.

In Exercises 67 and 68, solve for the specified variable.

67. Solve for w: $P = 2l + 2w$

68. Solve for t: $I = Prt$

In Exercises 69–72, find the missing distance, rate, or time.

Distance, d	Rate, r	Time, t
69.	65 mi/hr	8 hr

70. 855 m   5 m/min

71. 3000 mi   50 hr

72. 1000 km   25 hr

73. *Distance* An airplane has an average speed of 475 miles per hour. How far will it travel in $2\frac{1}{3}$ hours?

74. *Average Speed* You can walk 20 kilometers in 3 hours and 47 minutes. What is your average speed?

75. ▲ *Geometry* The width of a rectangular swimming pool is 4 feet less than its length. The perimeter of the pool is 112 feet. Find the dimensions of the pool.

76. ▲ *Geometry* The perimeter of an isosceles triangle is 65 centimeters. Find the length of the two equal sides if each is 10 centimeters longer than the third side. (An isosceles triangle has two sides of equal length.)

Simple Interest **In Exercises 77 and 78, use the simple interest formula.**

77. Find the total interest you will earn on a $1000 corporate bond that matures in 5 years and has an annual interest rate of 9.5%.

78. Find the annual interest rate on a certificate of deposit that pays $60 per year in interest on a principal of $750.

② Solve mixture problems involving hidden products.

79. *Number of Coins* You have 30 coins in dimes and quarters with a combined value of $5.55. Determine the number of coins of each type.

80. *Birdseed Mixture* A pet store owner mixes two types of birdseed that cost $1.25 per pound and $2.20 per pound to make 20 pounds of a mixture that costs $1.65 per pound. How many pounds of each kind of birdseed are in the mixture?

③ Solve work-rate problems.

81. *Work Rate* One person can complete a task in 5 hours, and another can complete the same task in 6 hours. How long will it take both people working together to complete the task?

82. *Work Rate* The person in Exercise 81 who can complete the task in 5 hours has already worked 1 hour when the second person starts. How long will they work together to complete the task?

3.6 Linear Inequalities

① Sketch the graphs of inequalities.

In Exercises 83–86, sketch the graph of the inequality.

83. $-3 \le x < 1$ **84.** $4 < x < 5.5$

85. $-7 < x$ **86.** $x \ge -2$

③ Solve linear inequalities.

In Exercises 87–98, solve the inequality and sketch the solution on the real number line.

87. $x - 5 \le -1$ **88.** $x + 8 > 5$

89. $-5x < 30$

90. $-11x \ge 44$

91. $5x + 3 > 18$

92. $3x - 11 \le 7$

93. $8x + 1 \ge 10x - 11$

94. $12 - 3x < 4x - 2$

95. $\frac{1}{3} - \frac{1}{2}y < 12$

96. $\frac{x}{4} - 2 < \frac{3x}{8} + 5$

97. $-4(3 - 2x) \le 3(2x - 6)$

98. $3(2 - y) \ge 2(1 + y)$

④ Solve compound inequalities.

In Exercises 99–104, solve the compound inequality and sketch the solution on the real number line.

99. $-6 \le 2x + 8 < 4$

100. $-13 \le 3 - 4x < 13$

101. $5 > \dfrac{x + 1}{-3} > 0$

102. $12 \ge \dfrac{x - 3}{2} > 1$

103. $5x - 4 < 6$ and $3x + 1 > -8$

104. $6 - 2x \le 1$ or $10 - 4x > -6$

⑤ Solve application problems involving inequalities.

105. *Sales Goal*   The weekly salary of an employee is $150 plus a 6% commission on total sales. The employee needs a minimum salary of $650 per week. How much must be sold to produce this salary?

106. *Long-Distance Charges*   The cost of an international long-distance telephone call is $0.99 for the first minute and $0.49 for each additional minute. The total cost of the call cannot exceed $7.50. Find the interval of time that is available for the call.

3.7 Absolute Value Equations and Inequalities

① Solve absolute value equations.

In Exercises 107–118, solve the equation.

107. $|x| = 6$

108. $|x| = -4$

109. $|4 - 3x| = 8$

110. $|2x + 3| = 7$

111. $|5x + 4| - 10 = -6$

112. $|x - 2| - 2 = 4$

113. $|2x + 10| = |x|$

114. $|5x - 8| = |x|$

115. $|3x - 4| = |x + 2|$

116. $|5x + 6| = |2x - 1|$

117. $|12 - x| = |4x + 7|$

118. $|1 - 2x| = |16 - 3x|$

② Solve inequalities involving absolute value.

In Exercises 119–130, solve the inequality.

119. $|x - 4| > 3$

120. $|t + 3| > 2$

121. $|x - 9| \le 15$

122. $|n + 1| \ge 4$

123. $|3x| > 9$

124. $\left|\dfrac{t}{3}\right| < 1$

125. $|2x - 7| < 15$

126. $|5x - 1| < 9$

127. $|4m - 2| \ge 2$

128. $|3a + 8| \le 22$

129. $|b + 2| - 6 > 1$

130. $|2y - 1| + 4 < -1$

⊞ **In Exercises 131 and 132, use a graphing calculator to solve the inequality.**

131. $|2x - 5| \ge 1$

132. $|5(1 - x)| \le 25$

In Exercises 133–136, write an absolute value inequality that represents the interval.

133.

134.

135.

136.

137. *Temperature*   The storage temperature of a computer must satisfy the inequality

$$|t - 78.3| \le 38.3$$

where t is given in degrees Fahrenheit. Sketch the graph of the solution of the inequality. What are the maximum and minimum temperatures?

138. *Temperature*   The operating temperature of a computer must satisfy the inequality

$$|t - 77| \le 27$$

where t is given in degrees Fahrenheit. Sketch the graph of the solution of the inequality. What are the maximum and minimum temperatures?

Chapter Test

Take this test as you would take a test in class. After you are done, check your
work against the answers in the back of the book.

In Exercises 1–8, solve the equation and check your solution.

1. $8x + 104 = 0$

2. $4x - 3 = 18$

3. $5 - 3x = -2x - 2$

4. $10 - (2 - x) = 2x + 1$

5. $\dfrac{3x}{4} = \dfrac{5}{2} + x$

6. $\dfrac{t + 2}{3} = \dfrac{2t}{5}$

7. $|2x + 6| = 16$

8. $|3x - 5| = |6x - 1|$

**In Exercises 9–14, solve each inequality and sketch the solution on the real
number line.**

9. $3x + 12 \geq -6$

10. $1 + 2x > 7 - x$

11. $0 \leq \dfrac{1 - x}{4} < 2$

12. $-7 < 4(2 - 3x) \leq 20$

13. $|x - 3| \leq 2$

14. $|5x - 3| > 12$

15. Solve $4.08(x + 10) = 9.50(x - 2)$. Round your answer to two decimal places.

16. The bill (including parts and labor) for the repair of a home appliance is
    $142. The cost of parts is $62 and the cost of labor is $32 per hour. How
    many hours were spent repairing the appliance?

17. Write the fraction $\frac{3}{8}$ as a percent and as a decimal.

18. 324 is 27% of what number?   19. 90 is what percent of 250?

20. Write the ratio of 40 inches to 2 yards as a fraction in simplest form. Use the
    same units for both quantities, and explain how you made this conversion.

21. Solve the proportion $\dfrac{2x}{3} = \dfrac{x + 4}{5}$.

22. Find the length x of the side of the larger triangle shown in the figure at the
    left. (Assume that the two triangles are similar, and use the fact that corre-
    sponding sides of similar triangles are proportional.)

23. You traveled 264 miles in $5\frac{1}{2}$ hours. What was your average speed?

24. You can paint a building in 9 hours. Your friend can paint the same building
    in 12 hours. Working together, how long will it take the two of you to paint
    the building?

25. Solve for R in the formula $S = C + RC$.

26. How much must you deposit in an account to earn $500 per year at 8% simple
    interest?

27. Translate the statement "t is at least 8" into a linear inequality.

28. A utility company has a fleet of vans. The annual operating cost per van
    is $C = 0.37m + 2700$, where m is the number of miles traveled by a van
    in a year. What is the maximum number of miles that will yield an annual
    operating cost that is less than or equal to $11,950?

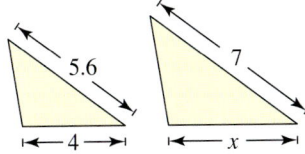

Figure for 24

211

Cumulative Test: Chapters 1–3

Take this test as you would take a test in class. After you are done, check your work against the answers in the back of the book.

1. Place the correct symbol ($<$ or $>$) between the numbers: $-\frac{3}{4}$ ▢ $\left|-\frac{7}{8}\right|$.

In Exercises 2–7, evaluate the expression.

2. $(-200)(2)(-3)$

3. $\frac{3}{8} - \frac{5}{6}$

4. $-\frac{2}{9} \div \frac{8}{75}$

5. $-(-2)^3$

6. $3 + 2(6) - 1$

7. $24 + 12 \div 3$

In Exercises 8 and 9, evaluate the expression when $x = -2$ and $y = 3$.

8. $-3x - (2y)^2$

9. $4y - x^3$

10. Use exponential form to write the product $3 \cdot (x + y) \cdot (x + y) \cdot 3 \cdot 3$.

11. Use the Distributive Property to expand $-2x(x - 3)$.

12. Identify the property of real numbers illustrated by
$$2 + (3 + x) = (2 + 3) + x.$$

In Exercises 13–16, simplify the expression.

13. $(3x^3)(5x^4)$

14. $(a^3b^2)(ab)$

15. $2x^2 - 3x + 5x^2 - (2 + 3x)$

16. $3(x^2 + x) - 2(2x - x^2)$

17. Determine whether the value of x is a solution of $x + 1 = 4(x - 2)$.

(a) $x = -8$

(b) $x = 3$

In Exercises 18–21, solve the equation and check your solution.

18. $12x - 3 = 7x + 27$

19. $2x - \dfrac{5x}{4} = 13$

20. $2(x - 3) + 3 = 12 - x$

21. $|3x + 1| = 5$

In Exercises 22–25, solve and graph the inequality.

22. $12 - 3x \le -15$

23. $-1 \le \dfrac{x + 3}{2} < 2$

24. $-4x + 1 \le 5$ or $-5x + 1 \ge 7$

25. $|8x - 3| \ge 13$

26. The sticker on a new car gives the fuel efficiency as 28.3 miles per gallon. In your own words, explain how to estimate the annual fuel cost for the buyer if the car will be driven approximately 15,000 miles per year and the fuel cost is $1.179 per gallon.

27. The perimeter of a rectangle is 60 meters. The length is $1\frac{1}{2}$ times its width. Find the dimensions of the rectangle.

28. The price of a television set is approximately 108% of what it was 2 years ago. The current price is $535. What was the approximate price 2 years ago?

29. Write the ratio "24 ounces to 2 pounds" as a fraction in simplest form.

30. The sum of two consecutive even integers is 494. Find the two numbers.

31. The suggested retail price of a digital camcorder is $1150. The camcorder is on sale for "20% off" the list price. Find the sale price.

32. The selling price of a box of cereal is $4.68. The markup rate for the grocery store is 40%. What is the cost of the cereal?

33. The figure below shows two pieces of property. The assessed values of the properties are proportional to their areas. The value of the larger piece is $95,000. What is the value of the smaller piece?

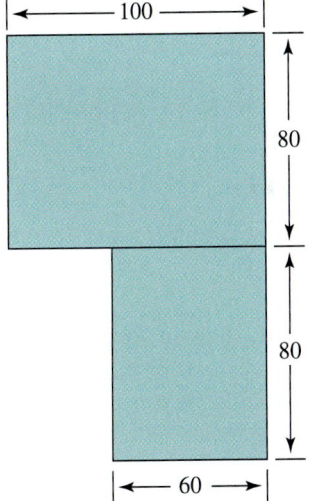

34. A train's average speed is 60 miles per hour. How long will it take the train to travel 562 miles?

35. For the first hour of a 350-mile trip, your average speed is 40 miles per hour. You want the average speed for the entire trip to be 50 miles per hour. Determine the average speed that must be maintained for the remainder of the trip.

Motivating the Chapter

⟲ Salary Plus Commission

You work as a sales representative for an advertising agency. You are paid a weekly salary, plus a commission on all ads placed by your accounts. The table shows your sales and your total weekly earnings.

	Week 1	Week 2	Week 3	Week 4
Weekly sales	$24,000	$7000	$0	$36,000
Weekly earnings	$980	$640	$500	$1220

See Section 4.1, Exercise 77.

a. Rewrite the data as a set of ordered pairs.

b. Plot the ordered pairs on a rectangular coordinate system.

See Section 4.3, Exercise 73.

c. Does the table represent a function? If so, identify the dependent and independent variables.

d. Describe what you consider to be appropriate domain and range values.

See Section 4.5, Exercise 109.

e. Explain how to determine whether the data in the table follows a linear pattern.

f. Determine the slope of the line passing through the ordered pairs for week 1 and week 2. (Let x represent the weekly sales and let y represent the weekly earnings.) What is the *rate* at which the weekly pay increases for each unit increase in ad sales? What is the rate called in the context of the problem?

g. Write an equation that describes the linear relationship between weekly sales and weekly earnings.

h. Sketch a graph of the equation. Identify the y-intercept and explain its meaning in the context of the problem. Identify the x-intercept. Does the x-intercept have any meaning in the context of the problem? If so, what is it?

See Section 4.6, Exercise 71.

i. What amount of ad sales is needed to guarantee a weekly pay of at least $840?

Michael Newman/PhotoEdit, Inc.

Graphs and Functions

4.1 Ordered Pairs and Graphs

Bill E. Barnes/PhotoEdit, Inc.

What You Should Learn

1 Plot and find the coordinates of a point on a rectangular coordinate system.

2 Construct a table of values for equations and determine whether ordered pairs are solutions of equations.

3 Use the verbal problem-solving method to plot points on a rectangular coordinate system.

Why You Should Learn It

The Cartesian plane can be used to represent relationships between two variables. For instance, Exercises 67–70 on page 226 show how to represent graphically the number of new privately owned housing starts in the United States.

1 Plot and find the coordinates of a point on a rectangular coordinate system.

The Rectangular Coordinate System

Just as you can represent real numbers by points on the real number line, you can represent **ordered pairs** of real numbers by points in a plane. This plane is called a **rectangular coordinate system** or the **Cartesian plane,** after the French mathematician René Descartes (1596–1650).

A rectangular coordinate system is formed by two real lines intersecting at right angles, as shown in Figure 4.1. The horizontal number line is usually called the **x-axis** and the vertical number line is usually called the **y-axis.** (The plural of axis is *axes.*) The point of intersection of the two axes is called the **origin,** and the axes separate the plane into four regions called **quadrants.**

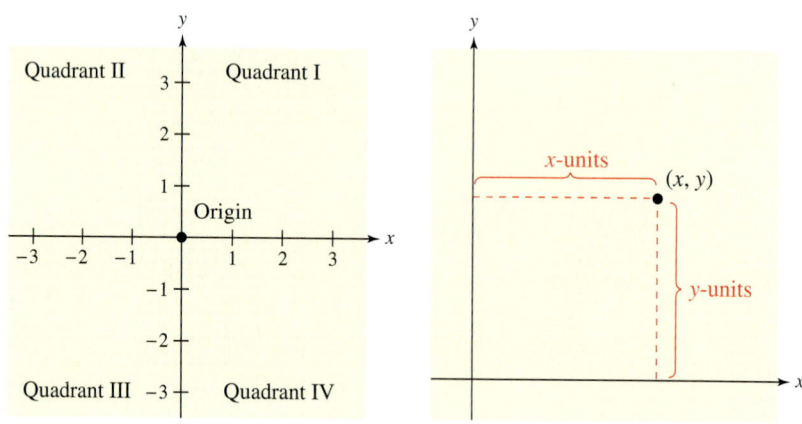

Figure 4.1   Figure 4.2

Each point in the plane corresponds to an **ordered pair** (x, y) of real numbers x and y, called the **coordinates** of the point. The first number (or **x-coordinate**) tells how far to the left or right the point is from the vertical axis, and the second number (or **y-coordinate**) tells how far up or down the point is from the horizontal axis, as shown in Figure 4.2.

A positive x-coordinate implies that the point lies to the *right* of the vertical axis; a negative x-coordinate implies that the point lies to the *left* of the vertical axis; and an x-coordinate of zero implies that the point lies *on* the vertical axis. Similarly, a positive y-coordinate implies that the point lies *above* the horizontal axis, and a negative y-coordinate implies that the point lies *below* the horizontal axis.

Locating a point in a plane is called **plotting** the point. This procedure is demonstrated in Example 1.

Example 1   Plotting Points on a Rectangular Coordinate System

Plot the points $(-1, 2)$, $(3, 0)$, $(2, -1)$, $(3, 4)$, $(0, 0)$, and $(-2, -3)$ on a rectangular coordinate system.

Solution

The point $(-1, 2)$ is one unit to the *left* of the vertical axis and two units *above* the horizontal axis.

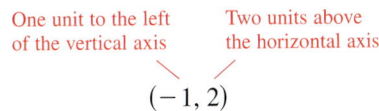

Similarly, the point $(3, 0)$ is three units to the *right* of the vertical axis and *on* the horizontal axis. (It is on the horizontal axis because the y-coordinate is zero.) The other four points can be plotted in a similar way, as shown in Figure 4.3.

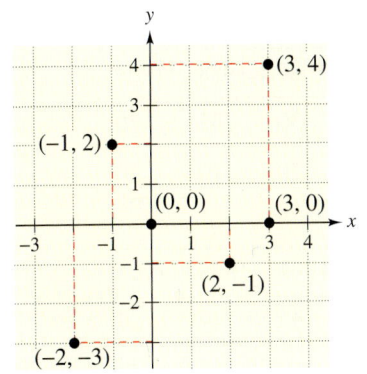

Figure 4.3

In Example 1 you were given the coordinates of several points and were asked to plot the points on a rectangular coordinate system. Example 2 looks at the reverse problem—that is, you are given points on a rectangular coordinate system and are asked to determine their coordinates.

Example 2   Finding Coordinates of Points

Determine the coordinates of each of the points shown in Figure 4.4.

Solution

Point A lies three units to the *left* of the vertical axis and two units *above* the horizontal axis. So, point A must be given by the ordered pair $(-3, 2)$. The coordinates of the other four points can be determined in a similar way, and the results are summarized as follows.

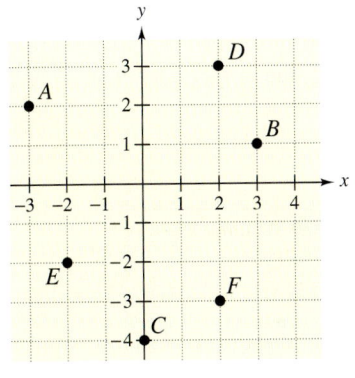

Figure 4.4

Point	Position	Coordinates
A	Three units *left,* two units *up*	$(-3, 2)$
B	Three units *right,* one unit *up*	$(3, 1)$
C	Zero units *left* (or right), four units *down*	$(0, -4)$
D	Two units *right,* three units *up*	$(2, 3)$
E	Two units *left,* two units *down*	$(-2, -2)$
F	Two units *right,* three units *down*	$(2, -3)$

In Example 2, note that point A $(-3, 2)$ and point F $(2, -3)$ are different points. The order in which the numbers appear in an ordered pair is important. Notice that because point C lies on the y-axis, it has an x-coordinate of 0.

Each year since 1967, the winners of the American Football Conference and the National Football Conference have played in the Super Bowl. The first Super Bowl was played between the Green Bay Packers and the Kansas City Chiefs.

Example 3   Super Bowl Scores

The scores of the winning and losing football teams in the Super Bowl games from 1983 through 2003 are shown in the table. Plot these points on a rectangular coordinate system.   (Source: National Football League)

Year	1983	1984	1985	1986	1987	1988	1989
Winning score	27	38	38	46	39	42	20
Losing score	17	9	16	10	20	10	16

Year	1990	1991	1992	1993	1994	1995	1996
Winning score	55	20	37	52	30	49	27
Losing score	10	19	24	17	13	26	17

Year	1997	1998	1999	2000	2001	2002	2003
Winning score	35	31	34	23	34	20	48
Losing score	21	24	19	16	7	17	21

Solution

The *x*-coordinates of the points represent the year of the game, and the *y*-coordinates represent either the winning score or the losing score. In Figure 4.5, the winning scores are shown as black dots, and the losing scores are shown as blue dots. Note that the break in the *x*-axis indicates that the numbers between 0 and 1983 have been omitted.

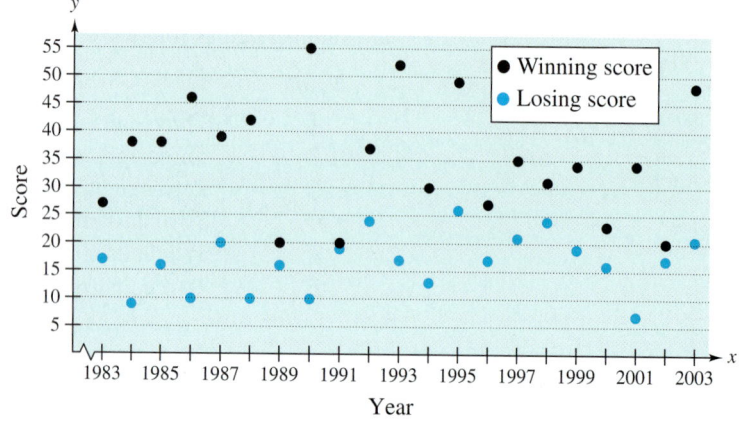

Figure 4.5

② Construct a table of values for equations and determine whether ordered pairs are solutions of equations.

Ordered Pairs as Solutions of Equations

In Example 3, the relationship between the year and the Super Bowl scores was given by a **table of values.** In mathematics, the relationship between the variables x and y is often given by an equation. From the equation, you can construct your own table of values. For instance, consider the equation

$$y = 2x + 1.$$

To construct a table of values for this equation, choose several x-values and then calculate the corresponding y-values. For example, if you choose $x = 1$, the corresponding y-value is

$$y = 2(1) + 1 \qquad \text{Substitute 1 for } x.$$

$$y = 3. \qquad \text{Simplify.}$$

The corresponding ordered pair $(x, y) = (1, 3)$ is a **solution point** (or **solution**) of the equation. The table below is a table of values (and the corresponding solution points) using x-values of $-3, -2, -1, 0, 1, 2,$ and 3. These x-values are arbitrary. You should try to use x-values that are convenient and simple to use.

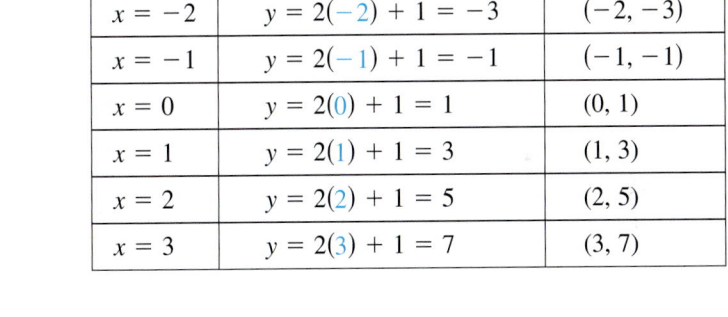

Choose x	Calculate y from $y = 2x + 1$	Solution point
$x = -3$	$y = 2(-3) + 1 = -5$	$(-3, -5)$
$x = -2$	$y = 2(-2) + 1 = -3$	$(-2, -3)$
$x = -1$	$y = 2(-1) + 1 = -1$	$(-1, -1)$
$x = 0$	$y = 2(0) + 1 = 1$	$(0, 1)$
$x = 1$	$y = 2(1) + 1 = 3$	$(1, 3)$
$x = 2$	$y = 2(2) + 1 = 5$	$(2, 5)$
$x = 3$	$y = 2(3) + 1 = 7$	$(3, 7)$

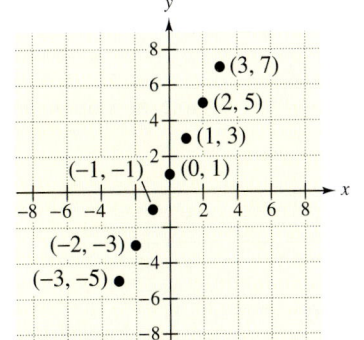

Figure 4.6

Once you have constructed a table of values, you can get a visual idea of the relationship between the variables x and y by plotting the solution points on a rectangular coordinate system. For instance, the solution points shown in the table are plotted in Figure 4.6.

In many places throughout this course, you will see that approaching a problem in different ways can help you understand the problem better. For instance, the discussion above looks at solutions of an equation in three ways.

Three Approaches to Problem Solving

1. **Algebraic Approach** Use algebra to find several solutions.

2. **Numerical Approach** Construct a table that shows several solutions.

3. **Graphical Approach** Draw a graph that shows several solutions.

Technology: Tip

Consult the user's guide for your graphing calculator to see if your graphing calculator has a *table* feature. By using the *table* feature in the *ask* mode, you can create a table of values for an equation.

When constructing a table of values for an equation, it is helpful first to solve the equation for y. For instance, the equation $4x + 2y = -8$ can be solved for y as follows.

$4x + 2y = -8$	Write original equation.
$4x - 4x + 2y = -8 - 4x$	Subtract $4x$ from each side.
$2y = -8 - 4x$	Combine like terms.
$\dfrac{2y}{2} = \dfrac{-8 - 4x}{2}$	Divide each side by 2.
$y = -4 - 2x$	Simplify.

This procedure is further demonstrated in Example 4.

Example 4 Constructing a Table of Values

Construct a table of values showing five solution points for the equation

$$6x - 2y = 4.$$

Then plot the solution points on a rectangular coordinate system. Choose x-values of $-2, -1, 0, 1,$ and 2.

Solution

$6x - 2y = 4$	Write original equation.
$6x - 6x - 2y = 4 - 6x$	Subtract $6x$ from each side.
$-2y = -6x + 4$	Combine like terms.
$\dfrac{-2y}{-2} = \dfrac{-6x + 4}{-2}$	Divide each side by -2.
$y = 3x - 2$	Simplify.

Now, using the equation $y = 3x - 2$, you can construct a table of values, as shown below.

x	-2	-1	0	1	2
$y = 3x - 2$	-8	-5	-2	1	4
Solution point	$(-2, -8)$	$(-1, -5)$	$(0, -2)$	$(1, 1)$	$(2, 4)$

Finally, from the table you can plot the five solution points on a rectangular coordinate system, as shown in Figure 4.7.

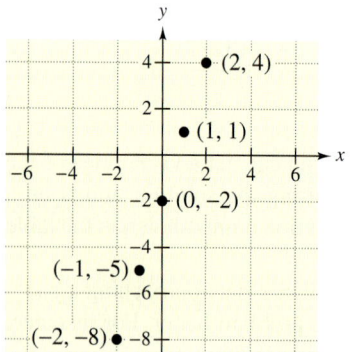

Figure 4.7

In the next example, you are given several ordered pairs and are asked to determine whether they are solutions of the original equation. To do this, you need to substitute the values of x and y into the equation. If the substitution produces a true equation, the ordered pair (x, y) is a solution and is said to **satisfy** the equation.

Guidelines for Verifying Solutions

To verify that an ordered pair (x, y) is a solution to an equation with variables x and y, use the following steps.

1. Substitute the values of x and y into the equation.

2. Simplify each side of the equation.

3. If each side simplifies to the same number, the ordered pair is a solution. If the two sides yield different numbers, the ordered pair is not a solution.

Example 5   Verifying Solutions of an Equation

Determine whether each of the ordered pairs is a solution of $x + 3y = 6$.

a. $(1, 2)$ **b.** $\left(-2, \frac{8}{3}\right)$ **c.** $(0, 2)$

Solution

a. For the ordered pair $(1, 2)$, substitute $x = 1$ and $y = 2$ into the original equation.

$$x + 3y = 6 \qquad \text{Write original equation.}$$
$$1 + 3(2) \stackrel{?}{=} 6 \qquad \text{Substitute 1 for } x \text{ and 2 for } y.$$
$$7 \neq 6 \qquad \text{Is not a solution.} \ ✗$$

Because the substitution does not satisfy the original equation, you can conclude that the ordered pair $(1, 2)$ is *not* a solution of the original equation.

b. For the ordered pair $\left(-2, \frac{8}{3}\right)$, substitute $x = -2$ and $y = \frac{8}{3}$ into the original equation.

$$x + 3y = 6 \qquad \text{Write original equation.}$$
$$(-2) + 3\left(\tfrac{8}{3}\right) \stackrel{?}{=} 6 \qquad \text{Substitute } -2 \text{ for } x \text{ and } \tfrac{8}{3} \text{ for } y.$$
$$-2 + 8 \stackrel{?}{=} 6 \qquad \text{Simplify.}$$
$$6 = 6 \qquad \text{Is a solution.} \ ✓$$

Because the substitution satisfies the original equation, you can conclude that the ordered pair $\left(-2, \frac{8}{3}\right)$ *is* a solution of the original equation.

c. For the ordered pair $(0, 2)$, substitute $x = 0$ and $y = 2$ into the original equation.

$$x + 3y = 6 \qquad \text{Write original equation.}$$
$$0 + 3(2) \stackrel{?}{=} 6 \qquad \text{Substitute 0 for } x \text{ and 2 for } y.$$
$$6 = 6 \qquad \text{Is a solution.} \ ✓$$

Because the substitution satisfies the original equation, you can conclude that the ordered pair $(0, 2)$ *is* a solution of the original equation.

③ Use the verbal problem-solving method to plot points on a rectangular coordinate system.

Application

Example 6   Total Cost

You set up a small business to assemble computer keyboards. Your initial cost is $120,000, and your unit cost to assemble each keyboard is $40. Write an equation that relates your total cost to the number of keyboards produced. Then plot the total costs of producing 1000, 2000, 3000, 4000, and 5000 keyboards.

Solution

The total cost equation must represent both the unit cost and the initial cost. A verbal model for this problem is as follows.

Verbal Model: $\boxed{\text{Total cost}} = \boxed{\text{Unit cost}} \cdot \boxed{\text{Number of keyboards}} + \boxed{\text{Initial cost}}$

Labels: Total cost = C (dollars)
            Unit cost = 40   (dollars per keyboard)
            Number of keyboards = x (keyboards)
            Initial cost = 120,000   (dollars)

Algebraic Model: $C = 40x + 120,000$

Using this equation, you can construct the following table of values.

x	1,000	2,000	3,000	4,000	5,000
$C = 40x + 120,000$	160,000	200,000	240,000	280,000	320,000

From the table, you can plot the ordered pairs, as shown in Figure 4.8.

Figure 4.8

Although graphs can help you visualize relationships between two variables, they can also be misleading. The graphs shown in Figure 4.9 and Figure 4.10 represent the yearly profits for a truck rental company. The graph in Figure 4.9 is misleading. The scale on the vertical axis makes it appear that the change in profits from 1998 to 2002 is dramatic, but the total change is only $3000, which is small in comparison with $3,000,000.

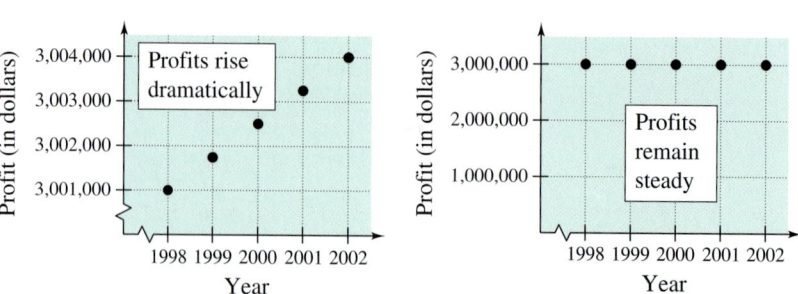

Figure 4.9   **Figure 4.10**

4.1 Exercises

Review   Concepts, Skills, and Problem Solving

Keep mathematically in shape by doing these exercises *before* the problems of this section.

Properties and Definitions

1. *Writing* ✎   Is $3x = 7$ a linear equation? Explain. Is $x^2 + 3x = 2$ a linear equation? Explain.

2. *Writing* ✎   Explain how to check whether $x = 3$ is a solution to the equation $5x - 4 = 11$.

Solving Equations

In Exercises 3–10, solve the equation.

3. $-y = 10$

4. $10 - t = 6$

5. $3x - 42 = 0$

6. $64 - 16x = 0$

7. $125(r - 1) = 625$

8. $2(3 - y) = 7y + 5$

9. $20 - \frac{1}{9}x = 4$

10. $0.35x = 70$

Problem Solving

11. *Cost*   The total cost of a lot and house is $154,000. The cost of constructing the house is 7 times the cost of the lot. What is the cost of the lot?

12. *Summer Jobs*   You have two summer jobs. In the first job, you work 40 hours a week and earn $9.50 an hour. In the second job, you work as many hours as you want and earn $8 an hour. You want to earn $450 a week. How many hours a week should you work at the second job?

Developing Skills

In Exercises 1–10, plot the points on a rectangular coordinate system. See Example 1.

1. $(3, 2), (-4, 2), (2, -4)$

2. $(-1, 6), (-1, -6), (4, 6)$

3. $(-10, -4), (4, -4), (0, 0)$

4. $(-6, 4), (0, 0), (3, -2)$

5. $(-3, 4), (0, -1), (2, -2), (5, 0)$

6. $(-1, 3), (0, 2), (-4, -4), (-1, 0)$

7. $\left(\frac{3}{2}, -1\right), \left(-3, \frac{3}{4}\right), \left(\frac{1}{2}, -\frac{1}{2}\right)$

8. $\left(-\frac{2}{3}, 4\right), \left(\frac{1}{2}, -\frac{5}{2}\right), \left(-4, -\frac{5}{4}\right)$

9. $(3, -4), \left(\frac{5}{2}, 0\right), (0, 3)$ 10. $\left(\frac{5}{2}, 2\right), \left(-3, \frac{4}{3}\right), \left(\frac{3}{4}, \frac{9}{4}\right)$

In Exercises 11–14, determine the coordinates of the points. See Example 2.

11.

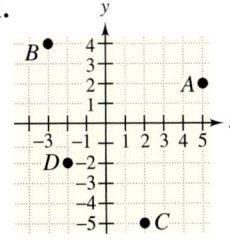

12.
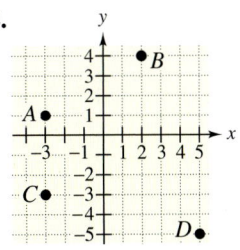

13.
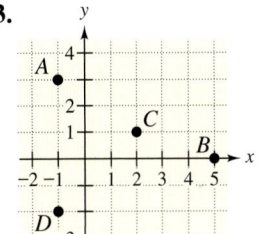

14.
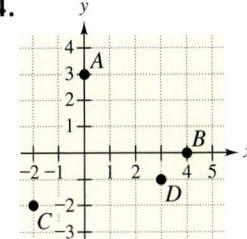

In Exercises 15–20, determine the quadrant in which the point is located without plotting it.

15. $(-3, 1)$

16. $(4, -3)$

17. $\left(-\frac{1}{8}, -\frac{2}{7}\right)$

18. $\left(\frac{3}{11}, \frac{7}{8}\right)$

19. $(-100, -365.6)$

20. $(-157.4, 305.6)$

In Exercises 21–26, determine the quadrant(s) in which the point is located without plotting it. Assume $x \neq 0$ and $y \neq 0$.

21. $(-5, y)$, y is a real number.

22. $(6, y)$, y is a real number.

23. $(x, -2)$, x is a real number.

24. $(x, 3)$, x is a real number.

25. (x, y), $xy < 0$

26. (x, y), $xy > 0$

In Exercises 27–34, plot the points and connect them with line segments to form the figure.

27. Triangle: $(-1, 1)$, $(2, -1)$, $(3, 4)$

28. Triangle: $(0, 3)$, $(-1, -2)$, $(4, 8)$

29. Square: $(2, 4)$, $(5, 1)$, $(2, -2)$, $(-1, 1)$

30. Rectangle: $(2, 1)$, $(4, 2)$, $(-1, 7)$, $(1, 8)$

31. Parallelogram: $(5, 2)$, $(7, 0)$, $(1, -2)$, $(-1, 0)$

32. Parallelogram: $(-1, 1)$, $(0, 4)$, $(4, -2)$, $(5, 1)$

33. Rhombus: $(0, 0)$, $(3, 2)$, $(2, 3)$, $(5, 5)$

34. Rhombus: $(0, 0)$, $(1, 2)$, $(2, 1)$, $(3, 3)$

In Exercises 35–40, complete the table of values. Then plot the solution points on a rectangular coordinate system. See Example 4.

35.

x	-2	0	2	4	6
$y = 3x - 4$					

36.

x	-2	0	2	4	6
$y = 2x + 1$					

37.

x	-4	-2	4	6	8
$y = -\frac{3}{2}x + 5$					

38.

x	-4	-2	0	2	4
$y = -\frac{1}{2}x + 3$					

39.

x	-2	-1	0	1	2
$y = 2x - 1$					

40.

x	-2	0	$\frac{1}{2}$	2	4
$y = -\frac{7}{2}x + 3$					

In Exercises 41–50, solve the equation for y.

41. $7x + y = 8$

42. $2x + y = 1$

43. $10x - y = 2$

44. $12x - y = 7$

45. $6x - 3y = 3$

46. $15x - 5y = 25$

47. $x + 4y = 8$

48. $x - 2y = -6$

49. $4x - 5y = 3$

50. $4y - 3x = 7$

In Exercises 51–58, determine whether the ordered pairs are solutions of the equation. See Example 5.

51. $y = 2x + 4$ (a) $(3, 10)$ (b) $(-1, 3)$
                        (c) $(0, 0)$ (d) $(-2, 0)$

52. $y = 5x - 2$ (a) $(2, 0)$ (b) $(-2, -12)$
                        (c) $(6, 28)$ (d) $(1, 1)$

53. $2y - 3x + 1 = 0$ (a) $(1, 1)$ (b) $(5, 7)$
                            (c) $(-3, -1)$ (d) $(-3, -5)$

54. $x - 8y + 10 = 0$ (a) $(-2, 1)$ (b) $(6, 2)$
                            (c) $(0, -1)$ (d) $(2, -4)$

55. $y = \frac{2}{3}x$ (a) $(6, 6)$ (b) $(-9, -6)$
                            (c) $(0, 0)$ (d) $\left(-1, \frac{2}{3}\right)$

56. $y = -\frac{7}{8}x$ (a) $(-5, -2)$ (b) $(0, 0)$
                             (c) $(8, 8)$ (d) $\left(\frac{3}{5}, 1\right)$

57. $y = 3 - 4x$ (a) $\left(-\frac{1}{2}, 5\right)$ (b) $(1, 7)$
                       (c) $(0, 0)$ (d) $\left(-\frac{3}{4}, 0\right)$

58. $y = \frac{3}{2}x + 1$ (a) $\left(0, \frac{3}{2}\right)$ (b) $(4, 7)$
                                (c) $\left(\frac{2}{3}, 2\right)$ (d) $(-2, -2)$

Solving Problems

59. *Organizing Data*   The distance y (in centimeters) a spring is compressed by a force x (in kilograms) is given by $y = 0.066x$. Complete a table of values for $x = 20, 40, 60, 80,$ and 100 to determine the distance the spring is compressed for each of the specified forces. Plot the results on a rectangular coordinate system.

60. *Organizing Data*   A company buys a new copier for $9500. Its value y after x years is given by $y = -800x + 9500$. Complete a table of values for $x = 0, 2, 4, 6,$ and 8 to determine the value of the copier at the specified times. Plot the results on a rectangular coordinate system.

61. *Organizing Data*   With an initial cost of $5000, a company will produce x units of a video game at $25 per unit. Write an equation that relates the total cost of producing x units to the number of units produced. Plot the cost for producing 100, 150, 200, 250, and 300 units.

62. *Organizing Data*   An employee earns $10 plus $0.50 for every x units produced per hour. Write an equation that relates the employee's total hourly wage to the number of units produced. Plot the hourly wage for producing 2, 5, 8, 10, and 20 units per hour.

63. *Organizing Data*   The table shows the normal temperatures y (in degrees Fahrenheit) for Anchorage, Alaska for each month x of the year, with $x = 1$ corresponding to January. (Source: National Climatic Data Center)

x	1	2	3	4	5	6
y	15	19	26	36	47	54

x	7	8	9	10	11	12
y	58	56	48	35	21	16

(a) Plot the data shown in the table. Did you use the same scale on both axes? Explain.

(b) Using the graph, find the three consecutive months when the normal temperature changes the least.

64. *Organizing Data*   The table shows the speed of a car x (in miles per hour) and the approximate stopping distance y (in feet).

x	20	30	40	50	60
y	63	109	164	229	303

(a) Plot the data in the table.

(b) The x-coordinates increase at equal increments of 10 miles per hour. Describe the pattern for the y-coordinates. What are the implications for the driver?

65. *Graphical Interpretation*   The table shows the numbers of hours x that a student studied for five different algebra exams and the resulting scores y.

x	3.5	1	8	4.5	0.5
y	72	67	95	81	53

(a) Plot the data in the table.

(b) Use the graph to describe the relationship between the number of hours studied and the resulting exam score.

66. *Graphical Interpretation*   The table shows the net income per share of common stock y (in dollars) for the Dow Chemical Company for the years 1991 through 2000, where x represents the year. (Source: Dow Chemical Company 2000 Annual Report)

x	1991	1992	1993	1994	1995
y	1.15	0.33	0.78	1.12	2.57

x	1996	1997	1998	1999	2000
y	2.57	2.60	1.94	2.01	2.24

(a) Plot the data in the table.

(b) Use the graph to determine the year that had the greatest increase and the year that had the greatest decrease in the net income per share of common stock.

Graphical Estimation   In Exercises 67–70, use the scatter plot, which shows new privately owned housing unit starts (in thousands) in the United States from 1988 through 2000.   (Source: U.S. Census Bureau)

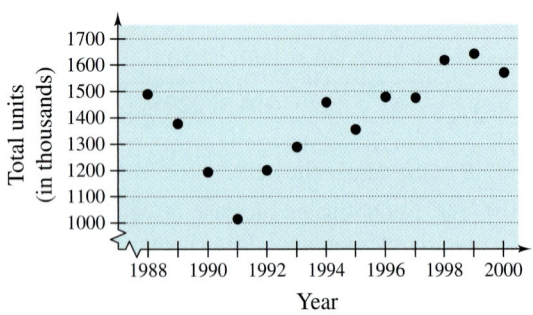

67. Estimate the number of new housing unit starts in 1989.

68. Estimate the number of new housing unit starts in 1994.

69. Estimate the increase and the percent increase in housing unit starts from 1997 to 1998.

70. Estimate the decrease and the percent decrease in housing unit starts from 1999 to 2000.

Graphical Estimation   In Exercises 71–74, use the scatter plot, which shows the per capita personal income in the United States from 1993 through 2000. (Source: U.S. Bureau of Economic Analysis)

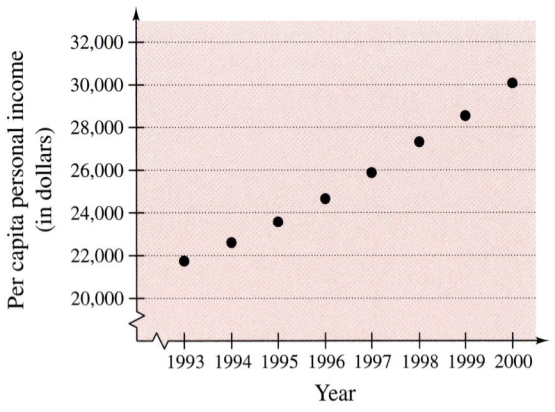

71. Estimate the per capita personal income in 1994.

72. Estimate the per capita personal income in 1995.

73. Estimate the percent increase in per capita personal income from 1999 to 2000.

74. The per capita personal income in 1980 was $10,205. Estimate the percent increase in per capita personal income from 1980 to 1993.

Graphical Estimation   In Exercises 75 and 76, use the bar graph, which shows the percents of gross domestic product spent on health care in several countries in 2000.   (Source: Organization for Economic Cooperation and Development)

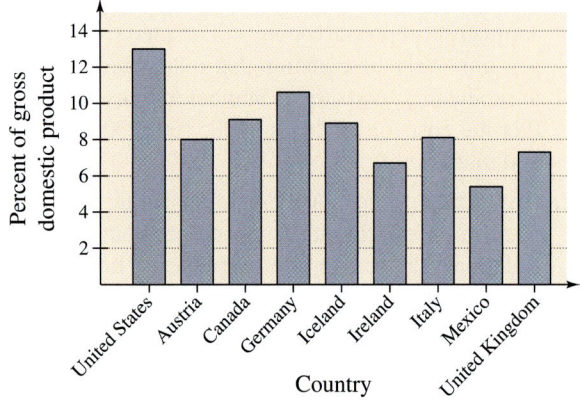

75. Estimate the percent of gross domestic product spent on health care in Mexico.

76. Estimate the percent of gross domestic product spent on health care in the United States.

Explaining Concepts

77. ✪ Answer parts (a) and (b) of Motivating the Chapter on page 214.

78. What is the x-coordinate of any point on the y-axis? What is the y-coordinate of any point on the x-axis?

79. *Writing*✐ Describe the signs of the x- and y-coordinates of points that lie in the first and second quadrants.

80. *Writing* ✐   Describe the signs of the *x*- and *y*-coordinates of points that lie in the third and fourth quadrants.

81. (a) Plot the points $(3, 2)$, $(-5, 4)$, and $(6, -4)$ on a rectangular coordinate system.

   (b) Change the sign of the *y*-coordinate of each point plotted in part (a). Plot the three new points on the same rectangular coordinate system used in part (a).

   (c) What can you infer about the location of a point when the sign of its *y*-coordinate is changed?

82. (a) Plot the points $(3, 2)$, $(-5, 4)$, and $(6, -4)$ on a rectangular coordinate system.

   (b) Change the sign of the *x*-coordinate of each point plotted in part (a). Plot the three new points on the same rectangular coordinate system used in part (a).

   (c) What can you infer about the location of a point when the sign of its *x*-coordinate is changed?

83. *Writing* ✐   Discuss the significance of the word "ordered" when referring to an ordered pair (x, y).

84. *Writing* ✐   When the point (x, y) is plotted, what does the *x*-coordinate measure? What does the *y*-coordinate measure?

85. In a rectangular coordinate system, must the scales on the *x*-axis and *y*-axis be the same? If not, give an example in which the scales differ.

86. *Writing* ✐   Review the tables in Exercises 35–40 and observe that in some cases the *y*-coordinates of the solution points increase and in others the *y*-coordinates decrease. What factor in the equation causes this? Explain.

4.2 Graphs of Equations in Two Variables

What You Should Learn

① Sketch graphs of equations using the point-plotting method.

② Find and use x- and y-intercepts as aids to sketching graphs.

③ Use the verbal problem-solving method to write an equation and sketch its graph.

Why You Should Learn It

The graph of an equation can help you see relationships between real-life quantities. For instance, in Exercise 83 on page 237, a graph can be used to illustrate the change over time in the life expectancy for a child at birth.

① Sketch graphs of equations using the point-plotting method.

The Graph of an Equation in Two Variables

You have already seen that the solutions of an equation involving two variables can be represented by points on a rectangular coordinate system. The set of *all* such points is called the **graph** of the equation.

To see how to sketch a graph, consider the equation

$$y = 2x - 1.$$

To begin sketching the graph of this equation, construct a table of values, as shown at the left. Next, plot the solution points on a rectangular coordinate system, as shown in Figure 4.11. Finally, find a pattern for the plotted points and use the pattern to connect the points with a smooth curve or line, as shown in Figure 4.12.

x	$y = 2x - 1$	Solution point
-3	-7	$(-3, -7)$
-2	-5	$(-2, -5)$
-1	-3	$(-1, -3)$
0	-1	$(0, -1)$
1	1	$(1, 1)$
2	3	$(2, 3)$
3	5	$(3, 5)$

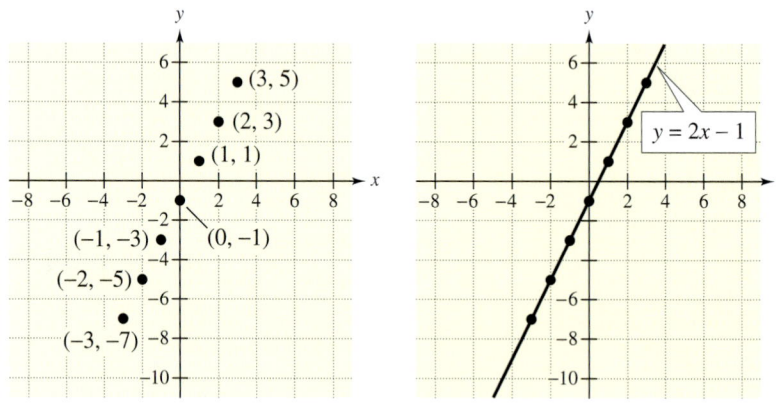

Figure 4.11   Figure 4.12

The Point-Plotting Method of Sketching a Graph

1. If possible, rewrite the equation by isolating one of the variables.

2. Make a table of values showing several solution points.

3. Plot these points on a rectangular coordinate system.

4. Connect the points with a smooth curve or line.

Example 1   Sketching the Graph of an Equation

Sketch the graph of $3x + y = 5$.

Solution

Begin by solving the equation for y, so that y is isolated on the left.

$$3x + y = 5 \qquad \text{Write original equation.}$$

$$3x - 3x + y = -3x + 5 \qquad \text{Subtract } 3x \text{ from each side.}$$

$$y = -3x + 5 \qquad \text{Simplify.}$$

Next, create a table of values, as shown below.

x	-2	-1	0	1	2	3
$y = -3x + 5$	11	8	5	2	-1	-4
Solution point	$(-2, 11)$	$(-1, 8)$	$(0, 5)$	$(1, 2)$	$(2, -1)$	$(3, -4)$

Now, plot the solution points, as shown in Figure 4.13. It appears that all six points lie on a line, so complete the sketch by drawing a line through the points, as shown in Figure 4.14.

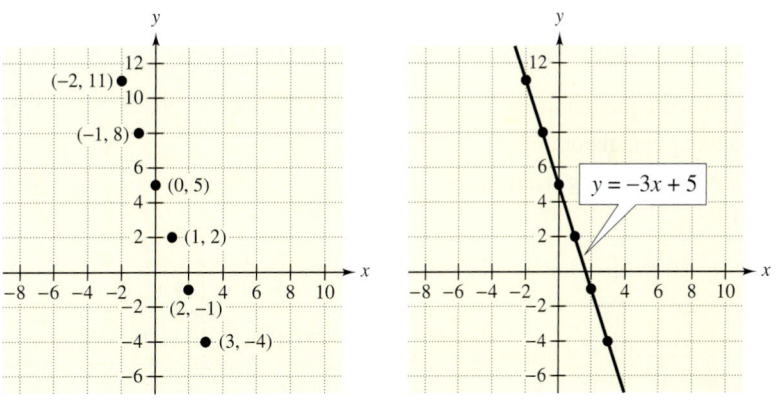

Figure 4.13   **Figure 4.14**

When creating a table of values, you are generally free to choose any x-values. When doing this, however, remember that the more x-values you choose, the easier it will be to recognize a pattern.

The equation in Example 1 is an example of a **linear equation in two variables**—the variables are raised to the first power and the graph of the equation is a line. As shown in the next two examples, graphs of nonlinear equations are not lines.

Example 2   Sketching the Graph of a Nonlinear Equation

Sketch the graph of $x^2 + y = 4$.

Solution

Begin by solving the equation for y, so that y is isolated on the left.

$$x^2 + y = 4 \qquad \text{Write original equation.}$$
$$x^2 - x^2 + y = -x^2 + 4 \qquad \text{Subtract } x^2 \text{ from each side.}$$
$$y = -x^2 + 4 \qquad \text{Simplify.}$$

Next, create a table of values, as shown below. Be careful with the signs of the numbers when creating a table. For instance, when $x = -3$, the value of y is

$$y = -(-3)^2 + 4$$
$$= -9 + 4$$
$$= -5.$$

x	-3	-2	-1	0	1	2	3
$y = -x^2 + 4$	-5	0	3	4	3	0	-5
Solution point	$(-3, -5)$	$(-2, 0)$	$(-1, 3)$	$(0, 4)$	$(1, 3)$	$(2, 0)$	$(3, -5)$

Now, plot the solution points, as shown in Figure 4.15. Finally, connect the points with a smooth curve, as shown in Figure 4.16.

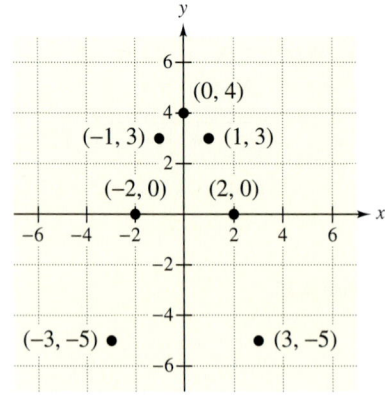

Figure 4.15

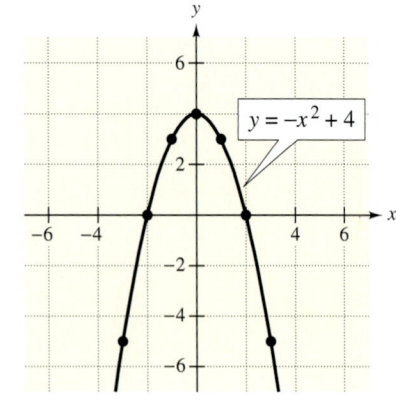

Figure 4.16

The graph of the equation in Example 2 is called a **parabola.** You will study this type of graph in a later chapter.

Example 3 examines the graph of an equation that involves an absolute value. Remember that the absolute value of a number is its distance from zero on the real number line. For instance, $|-5| = 5$, $|2| = 2$, and $|0| = 0$.

Example 3   The Graph of an Absolute Value Equation

Sketch the graph of $y = |x - 1|$.

Solution

This equation is already written in a form with y isolated on the left. You can begin by creating a table of values, as shown below. Be sure to check the values in this table to make sure that you understand how the absolute value is working. For instance, when $x = -2$, the value of y is

$$y = |-2 - 1|$$
$$= |-3|$$
$$= 3.$$

Similarly, when $x = 2$, the value of y is $|2 - 1| = 1$.

x	-2	-1	0	1	2	3	4
$y = |x - 1|$	3	2	1	0	1	2	3
Solution point	$(-2, 3)$	$(-1, 2)$	$(0, 1)$	$(1, 0)$	$(2, 1)$	$(3, 2)$	$(4, 3)$

Plot the solution points, as shown in Figure 4.17. It appears that the points lie in a "V-shaped" pattern, with the point $(1, 0)$ lying at the bottom of the "V." Following this pattern, connect the points to form the graph shown in Figure 4.18.

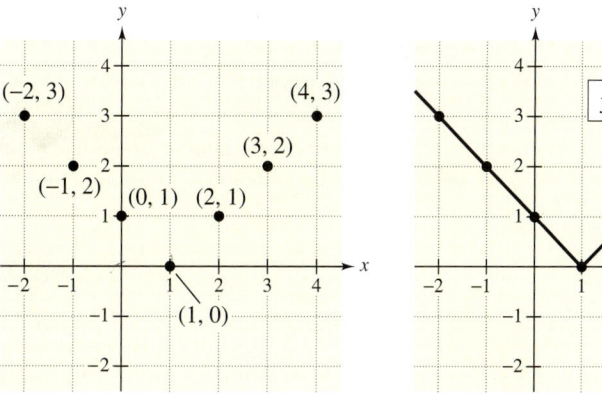

Figure 4.17   Figure 4.18

2 Find and use *x*- and *y*-intercepts as aids to sketching graphs.

Intercepts: Aids to Sketching Graphs

Two types of solution points that are especially useful are those having zero as either the *x*- or *y*-coordinate. Such points are called **intercepts** because they are the points at which the graph intersects the *x*- or *y*-axis.

Definition of Intercepts

The point $(a, 0)$ is called an **x-intercept** of the graph of an equation if it is a solution point of the equation. To find the *x*-intercept(s), let $y = 0$ and solve the equation for *x*.

The point $(0, b)$ is called a **y-intercept** of the graph of an equation if it is a solution point of the equation. To find the *y*-intercept(s), let $x = 0$ and solve the equation for *y*.

Example 4   Finding the Intercepts of a Graph

Find the intercepts and sketch the graph of $y = 2x - 5$.

Solution

To find any *x*-intercepts, let $y = 0$ and solve the resulting equation for *x*.

$$y = 2x - 5 \qquad \text{Write original equation.}$$

$$0 = 2x - 5 \qquad \text{Let } y = 0.$$

$$\frac{5}{2} = x \qquad \text{Solve equation for } x.$$

To find any *y*-intercepts, let $x = 0$ and solve the resulting equation for *y*.

$$y = 2x - 5 \qquad \text{Write original equation.}$$

$$y = 2(0) - 5 \qquad \text{Let } x = 0.$$

$$y = -5 \qquad \text{Solve equation for } y.$$

So, the graph has one *x*-intercept, which occurs at the point $\left(\frac{5}{2}, 0\right)$, and one *y*-intercept, which occurs at the point $(0, -5)$. To sketch the graph of the equation, create a table of values. (Include the intercepts in the table.) Then plot the points and connect the points with a line, as shown in Figure 4.19.

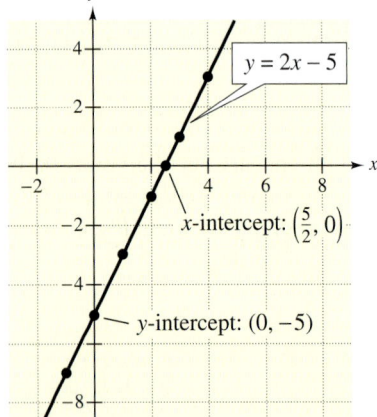

Figure 4.19

x	-1	0	1	2	$\frac{5}{2}$	3	4
$y = 2x - 5$	-7	-5	-3	-1	0	1	3
Solution point	$(-1, -7)$	$(0, -5)$	$(1, -3)$	$(2, -1)$	$\left(\frac{5}{2}, 0\right)$	$(3, 1)$	$(4, 3)$

When you create a table of values, include any intercepts you have found. You should also include points to the left and to the right of the intercepts. This helps to give a more complete view of the graph.

③ Use the verbal problem-solving method to write an equation and sketch its graph.

Application

Example 5   Depreciation

The value of a $35,500 sport utility vehicle (SUV) depreciates over 10 years (the depreciation is the same each year). At the end of the 10 years, the salvage value is expected to be $5500.

a. Find an equation that relates the value of the SUV to the number of years.

b. Sketch the graph of the equation.

c. What is the *y*-intercept of the graph and what does it represent in the context of the problem?

Solution

a. The total depreciation over the 10 years is $35,500 - 5500 = \$30,000$. Because the same amount is depreciated each year, it follows that the annual depreciation is $30,000/10 = \$3000$.

Verbal Model:	Value after *t* years	=	Original value	−	Annual depreciation	·	Number of years

Labels: Value after *t* years = *y*   (dollars)
            Original value = 35,500   (dollars)
            Annual depreciation = 3000   (dollars per year)
            Number of years = *t*   (years)

Algebraic Model: $y = 35,500 - 3000t$

b. A sketch of the graph of the depreciation equation is shown in Figure 4.20.

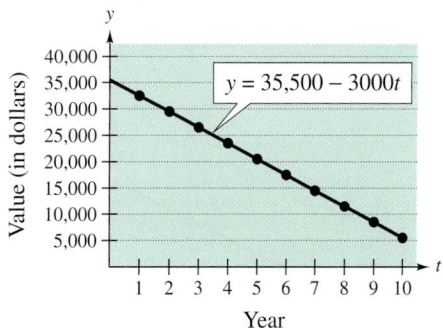

Figure 4.20

c. To find the *y*-intercept of the graph, let $t = 0$ and solve the equation for *y*.

$$y = 35,500 - 3000t \qquad \text{Write original equation.}$$

$$y = 35,000 - 3000(0) \qquad \text{Substitute 0 for } t.$$

$$y = 35,500 \qquad \text{Simplify.}$$

So, the *y*-intercept is (0, 35,500), and it corresponds to the original value of the SUV.

4.2 Exercises

Properties and Definitions

1. *Writing* ✏ If $x - 2 > 5$ and c is an algebraic expression, then what is the relationship between $x - 2 + c$ and $5 + c$?

2. *Writing* ✏ If $x - 2 < 5$ and $c < 0$, then what is the relationship between $(x - 2)c$ and $5c$?

3. Complete the Multiplicative Inverse Property: $x(1/x) = $ ☐.

4. Identify the property of real numbers illustrated by $x + y = y + x$.

Simplifying Expressions

In Exercises 5–10, simplify the expression.

5. $-3(3x - 2y) + 5y$ 6. $3z - (4 - 5z)$

7. $-y^2(y^2 + 4) + 6y^2$ 8. $5t(2 - t) + t^2$

9. $3[6x - 5(x - 2)]$ 10. $5(t - 2) - 5(t - 2)$

Problem Solving

11. *Company Reimbursement*   A company reimburses its sales representatives $30 per day plus 35 cents per mile for the use of their personal cars. A sales representative submits a bill for $52.75 for driving her own car.

    (a) How many miles did she drive?

    (b) How many days did she drive? Explain.

12. ▲ *Geometry*   The width of a rectangular mirror is $\frac{3}{5}$ its length. The perimeter of the mirror is 80 inches. What are the measurements of the mirror?

Developing Skills

In Exercises 1–8, match the equation with its graph. [The graphs are labeled (a), (b), (c), (d), (e), (f), (g), and (h).]

(a)

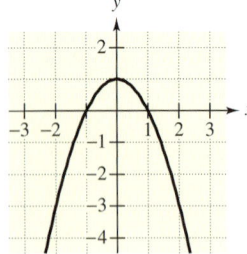

(b)

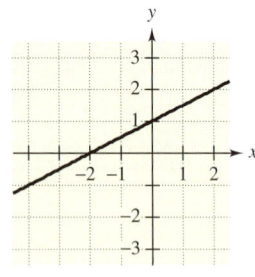

(c)

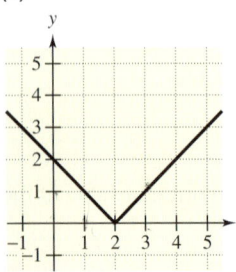

(d)

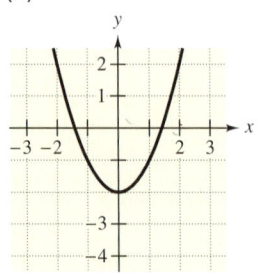

(e)

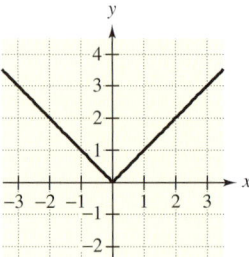

(f)

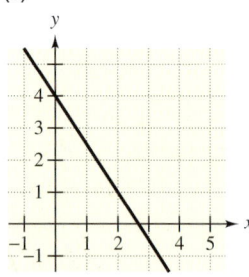

(g)

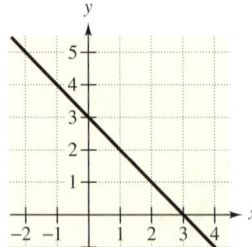

(h)

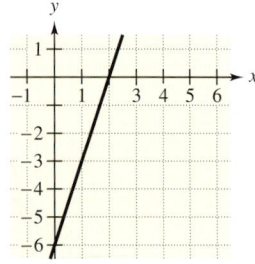

1. $y = 3 - x$ 2. $y = \frac{1}{2}x + 1$

3. $y = -x^2 + 1$ 4. $y = |x|$

5. $y = 3x - 6$

6. $y = |x - 2|$

7. $y = x^2 - 2$

8. $y = 4 - \frac{3}{2}x$

In Exercises 9–16, complete the table and use the results to sketch the graph of the equation. See Examples 1–3.

9. $y = 9 - x$

x	-2	-1	0	1	2
y					

10. $y = x - 1$

x	-2	-1	0	1	2
y					

11. $x + 2y = 4$

x	-2	0	2	4	6
y					

12. $3x - 2y = 6$

x	-2	0	2	4	6
y					

13. $y = (x - 1)^2$

x	-1	0	1	2	3
y					

14. $y = x^2 + 3$

x	-2	-1	0	1	2
y					

15. $y = |x + 1|$

x	-3	-2	-1	0	1
y					

16. $y = |x| - 2$

x	-2	-1	0	1	2
y					

In Exercises 17–24, graphically estimate the x- and y-intercepts of the graph. Then check your results algebraically.

17. $4x - 2y = -8$

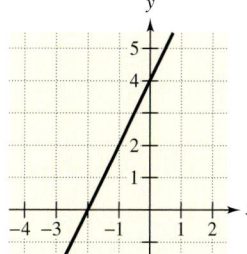

18. $5y - 2x = 10$

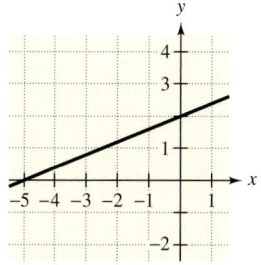

19. $x + 3y = 6$

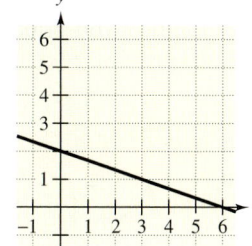

20. $4x + 3y = 12$

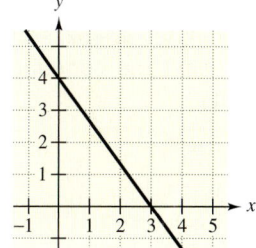

21. $y = |x| - 3$

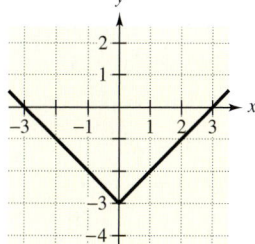

22. $y = 4 - |x|$

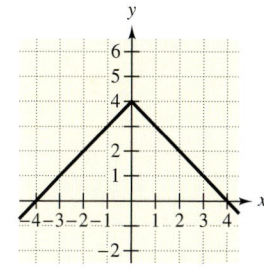

23. $y = 16 - x^2$

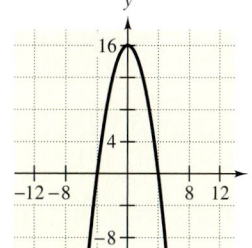

24. $y = x^2 - 4$

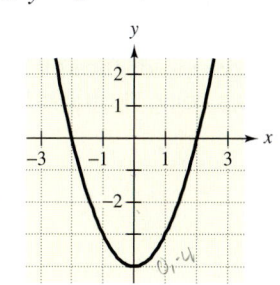

In Exercises 25–36, find the x- and y-intercepts (if any) of the graph of the equation. See Example 4.

25. $y = -2x + 7$

26. $y = 5x - 3$

27. $y = \frac{1}{2}x - 1$

28. $y = -\frac{1}{2}x + 3$

29. $x - y = 1$

30. $x + y = 10$

31. $2x + y = 4$

32. $3x - 2y = 1$

33. $2x + 6y - 9 = 0$

34. $2x - 5y + 50 = 0$

35. $\frac{3}{4}x - \frac{1}{2}y = 3$

36. $\frac{1}{2}x + \frac{2}{3}y = 1$

In Exercises 37–62, sketch the graph of the equation and label the coordinates of at least three solution points.

37. $y = 2 - x$ **38.** $y = 12 - x$

39. $y = x - 1$ **40.** $y = x + 8$

41. $y = 3x$ **42.** $y = -2x$

43. $4x + y = 6$ **44.** $2x + y = -2$

45. $10x + 5y = 20$ **46.** $7x + 7y = 14$

47. $4x - y = 2$ **48.** $2x - y = 5$

49. $y = \frac{3}{8}x + 15$ **50.** $y = 14 - \frac{2}{3}x$

51. $y = \frac{2}{3}x - 5$ **52.** $y = \frac{3}{2}x + 3$

53. $y = x^2$ **54.** $y = -x^2$

55. $y = -x^2 + 9$ **56.** $y = x^2 - 1$

57. $y = (x - 3)^2$ **58.** $y = -(x + 2)^2$

59. $y = |x - 5|$ **60.** $y = |x + 3|$

61. $y = 5 - |x|$ **62.** $y = |x| - 3$

In Exercises 63–66, use a graphing calculator to graph both equations in the same viewing window. Are the graphs identical? If so, what property of real numbers is being illustrated?

63. $y_1 = \frac{1}{3}x - 1$
$y_2 = -1 + \frac{1}{3}x$

64. $y_1 = 3\left(\frac{1}{4}x\right)$
$y_2 = \left(3 \cdot \frac{1}{4}\right)x$

65. $y_1 = 2(x - 2)$
$y_2 = 2x - 4$

66. $y_1 = 2 + (x + 4)$
$y_2 = (2 + x) + 4$

In Exercises 67–74, use a graphing calculator to graph the equation. (Use a standard viewing window.)

67. $y = 4x$ **68.** $y = -2x$

69. $y = -\frac{1}{3}x$ **70.** $y = \frac{1}{2}x$

71. $y = -2x^2 + 5$ **72.** $y = x^2 - 7$

73. $y = |x + 1| - 2$ **74.** $y = 4 - |x - 2|$

In Exercises 75 and 76, use a graphing calculator and the given viewing window to graph the equation.

75. $y = 25 - 5x$

```
Xmin = -5
Xmax = 7
Xscl = 1
Ymin = -5
Ymax = 30
Yscl = 5
```

76. $y = 1.7 - 0.1x$

```
Xmin = -10
Xmax = 25
Xscl = 5
Ymin = -5
Ymax = 5
Yscl = .5
```

In Exercises 77–80, use a graphing calculator to graph the equation and find a viewing window that yields a graph that matches the one shown.

77. $y = \frac{1}{2}x + 2$

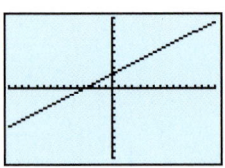

78. $y = 2x - 1$

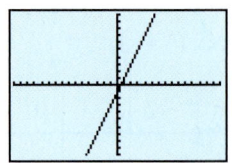

79. $y = \frac{1}{4}x^2 - 4x + 12$

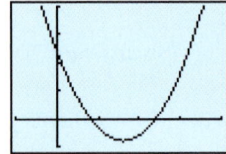

80. $y = 16 - 4x - x^2$

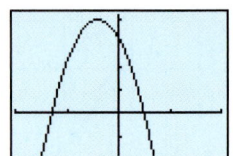

Solving Problems

81. *Creating a Model*   Let *y* represent the distance traveled by a car that is moving at a constant speed of 35 miles per hour. Let *t* represent the number of hours the car has traveled. Write an equation that relates *y* to *t* and sketch its graph.

82. *Creating a Model*   The cost of printing a book is $500, plus $5 per book. Let *C* represent the total cost and let *x* represent the number of books. Write an equation that relates *C* and *x* and sketch its graph.

83. *Life Expectancy*   The table shows the life expectancy *y* (in years) in the United States for a child at birth for the years 1994 through 1999.

Year	1995	1996	1997	1998	1999	2000
y	75.8	76.1	76.5	76.7	76.7	76.9

A model for this data is $y = 0.21t + 74.8$, where *t* is the time in years, with $t = 5$ corresponding to 1995. (Source: U.S. National Center for Health Statistics and U.S. Census Bureau)

(a) Plot the data and graph the model on the same set of coordinate axes.

(b) Use the model to predict the life expectancy for a child born in 2010.

84. *Graphical Comparisons*   The graphs of two types of depreciation are shown. In one type, called *straight-line depreciation,* the value depreciates by the same amount each year. In the other type, called *declining balances,* the value depreciates by the same percent each year. Which is which?

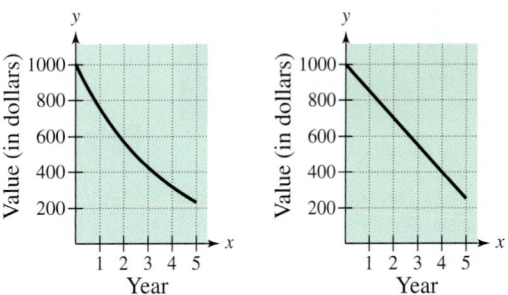

85. *Graphical Interpretation*   In Exercise 84, what is the original cost of the equipment that is being depreciated?

86. *Writing*   Compare the benefits and disadvantages of the two types of depreciation shown in Exercise 84.

Explaining Concepts

87. *Writing*   In your own words, define what is meant by the *graph* of an equation.

88. *Writing*   In your own words, describe the point-plotting method of sketching the graph of an equation.

89. *Writing*   In your own words, describe how you can check that an ordered pair (*x*, *y*) is a solution of an equation.

90. *Writing*   Explain how to find the *x*- and *y*-intercepts of a graph.

91. You are walking toward a tree. Let *x* represent the time (in seconds) and let *y* represent the distance (in feet) between you and the tree. Sketch a possible graph that shows how *x* and *y* are related.

92. How many solution points can an equation in two variables have? How many points do you need to determine the general shape of the graph?

4.3 Relations, Functions, and Graphs

Robert Grubbs/Photo Network

What You Should Learn

① Identify the domain and range of a relation.

② Determine if relations are functions by inspection or by using the Vertical Line Test.

③ Use function notation and evaluate functions.

④ Identify the domain of a function.

Why You Should Learn It

Relations and functions can be used to describe real-life situations. For instance, in Exercise 71 on page 247, a relation is used to model the length of time between sunrise and sunset for Erie, Pennsylvania.

Relations

Many everyday occurrences involve pairs of quantities that are matched with each other by some rule of correspondence. For instance, each person is matched with a birth month (person, month); the number of hours worked is matched with a paycheck (hours, pay); an instructor is matched with a course (instructor, course); and the time of day is matched with the outside temperature (time, temperature). In each instance, sets of ordered pairs can be formed. Such sets of ordered pairs are called **relations.**

① Identify the domain and range of a relation.

Definition of Relation

A **relation** is any set of ordered pairs. The set of first components in the ordered pairs is the **domain** of the relation. The set of second components is the **range** of the relation.

In mathematics, relations are commonly described by ordered pairs of *numbers*. The set of *x*-coordinates is the domain, and the set of *y*-coordinates is the range. In the relation $\{(3, 5), (1, 2), (4, 4), (0, 3)\}$, the domain D and range R are the sets $D = \{3, 1, 4, 0\}$ and $R = \{5, 2, 4, 3\}$.

Example 1   Analyzing a Relation

Find the domain and range of the relation $\{(0, 1), (1, 3), (2, 5), (3, 5), (0, 3)\}$. Then sketch a graphical representation of the relation.

Solution

The domain is the set of all first components of the relation, and the range is the set of all second components.

$$D = \{0, 1, 2, 3\} \quad \text{and} \quad R = \{1, 3, 5\}$$

A graphical representation is shown in Figure 4.21.

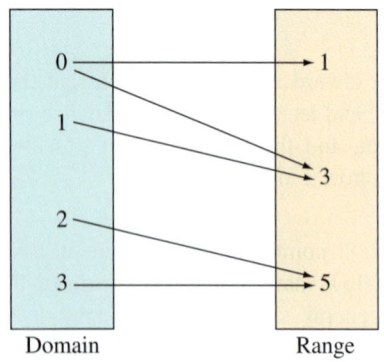

Domain   Range

Figure 4.21

You should note that it is not necessary to list repeated components of the domain and range of a relation.

② Determine if relations are functions by inspection or by using the Vertical Line Test.

Functions

In the study of mathematics and its applications, the focus is mainly on a special type of relation, called a **function.**

Definition of Function

A **function** is a relation in which no two ordered pairs have the same first component and different second components.

This definition means that a given first component cannot be paired with two different second components. For instance, the pairs $(1, 3)$ and $(1, -1)$ could not be ordered pairs of a function.

Consider the relations described at the beginning of this section.

Relation	Ordered Pairs	Sample Relation
1	(person, month)	$\{(A, May), (B, Dec), (C, Oct), . . .\}$
2	(hours, pay)	$\{(12, 84), (4, 28), (6, 42), (15, 105), . . .\}$
3	(instructor, course)	$\{(A, MATH001), (A, MATH002), . . .\}$
4	(time, temperature)	$\{(8, 70°), (10, 78°), (12, 78°), . . .\}$

The first relation *is* a function because each person has only one birth month. The second relation *is* a function because the number of hours worked at a particular job can yield only *one* paycheck amount. The third relation *is not* a function because an instructor can teach more than one course. The fourth relation *is* a function. Note that the ordered pairs $(10, 78°)$ and $(12, 78°)$ do not violate the definition of a function.

Study Tip

The ordered pairs of a relation can be thought of in the form (input, output). For a *function*, a given input cannot yield two different outputs. For instance, if the input is a person's name and the output is that person's month of birth, then your name as the input can yield only your month of birth as the output.

Example 2   Testing Whether a Relation Is a Function

Decide whether the relation represents a function.

a. Input: a, b, c

Output: 2, 3, 4

$\{(a, 2), (b, 3), (c, 4)\}$

b.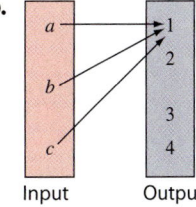

c.

same ___

Input x	Output y	(x, y)
3	1	(3, 1)
4	3	(4, 3)
5	4	(5, 4)
3	2	(3, 2)

Solution

a. This set of ordered pairs *does* represent a function. No first component has two different second components.

b. This diagram *does* represent a function. No first component has two different second components.

c. This table *does not* represent a function. The first component 3 is paired with two different second components, 1 and 2.

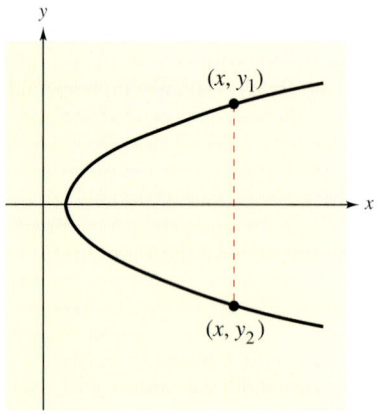

Figure 4.22

In algebra, it is common to represent functions by equations in two variables rather than by ordered pairs. For instance, the equation $y = x^2$ represents the variable y as a function of x. The variable x is the **independent variable** (the input) and y is the **dependent variable** (the output). In this context, the domain of the function is the set of all *allowable* values for x, and the range is the *resulting* set of all values taken on by the dependent variable y.

From the graph of an equation, it is easy to determine whether the equation represents y as a function of x. The graph in Figure 4.22 *does not* represent a function of x because the indicated value of x is paired with two y-values. Graphically, this means that a vertical line intersects the graph more than once.

Vertical Line Test

A set of points on a rectangular coordinate system is the graph of y as a function of x if and only if no vertical line intersects the graph at more than one point.

Example 3　Using the Vertical Line Test for Functions

Use the Vertical Line Test to determine whether y is a function of x.

a.

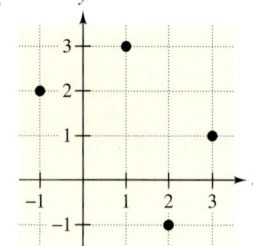

b.

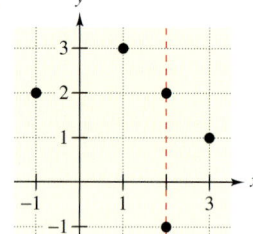

c.

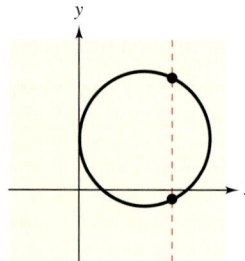

d.

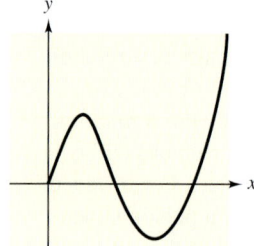

Solution

a. From the graph, you can see that no vertical line intersects more than one point on the graph. So, the relation *does* represent y as a function of x.

b. From the graph, you can see that a vertical line intersects more than one point on the graph. So, the relation *does not* represent y as a function of x.

c. From the graph, you can see that a vertical line intersects more than one point on the graph. So, the relation *does not* represent y as a function of x.

d. From the graph, you can see that no vertical line intersects more than one point on the graph. So, the relation *does* represent y as a function of x.

3 Use function notation and evaluate functions.

Function Notation

To discuss functions represented by equations, it is common practice to give them names using **function notation.** For instance, the function

$$y = 2x - 6$$

can be given the name "f" and written in function notation as

$$f(x) = 2x - 6.$$

Function Notation

In the notation $f(x)$:

> f is the **name** of the function.
> x is a **domain** (or input) value.
> $f(x)$ is a **range** (or output) value y for a given x.

The symbol $f(x)$ is read as **the value of f at x or simply f of x.**

The process of finding the value of $f(x)$ for a given value of x is called **evaluating a function.** This is accomplished by substituting a given x-value (input) into the equation to obtain the value of $f(x)$ (output). Here is an example.

Function	*x-Values (input)*	*Function Values (output)*
$f(x) = 4 - 3x$	$x = -2$	$f(-2) = 4 - 3(-2) = 4 + 6 = 10$
	$x = -1$	$f(-1) = 4 - 3(-1) = 4 + 3 = 7$
	$x = 0$	$f(0) = 4 - 3(0) = 4 - 0 = 4$
	$x = 2$	$f(2) = 4 - 3(2) = 4 - 6 = -2$
	$x = 3$	$f(3) = 4 - 3(3) = 4 - 9 = -5$

Although f and x are often used as a convenient function name and independent (input) variable, you can use other letters. For instance, the equations

$$f(x) = x^2 - 3x + 5, \quad f(t) = t^2 - 3t + 5, \quad \text{and} \quad g(s) = s^2 - 3s + 5$$

all define the same function. In fact, the letters used are just "placeholders" and this same function is well described by the form

$$f(\quad) = (\quad)^2 - 3(\quad) + 5$$

where the parentheses are used in place of a letter. To evaluate $f(-2)$, simply place -2 in each set of parentheses, as follows.

$$f(-2) = (-2)^2 - 3(-2) + 5$$
$$= 4 + 6 + 5$$
$$= 15$$

It is important to put parentheses around the x-value (input) and then simplify the result.

Example 4   Evaluating a Function

Let $f(x) = x^2 + 1$. Find each value of the function.

a. $f(-2)$ **b.** $f(0)$

Solution

a. $f(x) = x^2 + 1$ Write original function.

    $f(-2) = (-2)^2 + 1$ Substitute -2 for x.

    $= 4 + 1 = 5$ Simplify.

b. $f(x) = x^2 + 1$ Write original function.

    $f(0) = (0)^2 + 1$ Substitute 0 for x.

    $= 0 + 1 = 1$ Simplify.

Example 5   Evaluating a Function

Let $g(x) = 3x - x^2$. Find each value of the function.

a. $g(2)$ **b.** $g(0)$

Solution

a. Substituting 2 for x produces $g(2) = 3(2) - (2)^2 = 6 - 4 = 2$.
b. Substituting 0 for x produces $g(0) = 3(0) - (0)^2 = 0 - 0 = 0$.

④ Identify the domain of a function.

Finding the Domain of a Function

The domain of a function may be explicitly described along with the function, or it may be *implied* by the context in which the function is used. For instance, if weekly pay is a function of hours worked (for a 40-hour work week), the implied domain is typically the interval $0 \le x \le 40$. Certainly x cannot be negative in this context.

Example 6   Finding the Domain of a Function

Find the domain of each function.

a. $f:\{(-3, 0), (-1, 2), (0, 4), (2, 4), (4, -1)\}$
b. Area of a square: $A = s^2$

Solution

a. The domain of f consists of all first components in the set of ordered pairs. So, the domain is $\{-3, -1, 0, 2, 4\}$.
b. For the area of a square, you must choose positive values for the side s. So, the domain is the set of all real numbers s such that $s > 0$.

4.3 Exercises

Review   Concepts, Skills, and Problem Solving

Keep mathematically in shape by doing these exercises *before* the problems of this section.

Properties and Definitions

1. If $a < b$ and $b < c$, then what is the relationship between a and c? Name this property.

2. Demonstrate the Multiplication Property of Equality for the equation $7x = 21$.

Simplifying Expressions

In Exercises 3–6, simplify the expression.

3. $4s - 6t + 7s + t$ **4.** $2x^2 - 4 + 5 - 3x^2$

5. $\frac{5}{3}x - \frac{2}{3}x - 4$

6. $3x^2y + xy - xy^2 - 6xy$

Solving Equations

In Exercises 7–10, solve the equation.

7. $5x + 2 = 2x - 7$ **8.** $-x + 6 = 4x + 3$

9. $\frac{x}{8} = \frac{7}{2}$ **10.** $\frac{x+4}{4} = \frac{x-1}{3}$

Problem Solving

11. *Simple Interest*   An inheritance of $7500 is invested in a mutual fund, and at the end of 1 year the value of the investment is $8190. What is the annual interest rate for this fund?

12. *Number Problem*   The sum of two consecutive odd integers is 44. Find the two integers.

Developing Skills

In Exercises 1–6, find the domain and range of the relation. See Example 1.

1. $\{(-4, 3), (2, 5), (1, 2), (4, -3)\}$

2. $\{(-1, 5), (8, 3), (4, 6), (-5, -2)\}$

3. $\{(2, 16), (-9, -10), (\frac{1}{2}, 0)\}$

4. $\{(\frac{2}{3}, -4), (-6, \frac{1}{4}), (0, 0)\}$

5. $\{(-1, 3), (5, -7), (-1, 4), (8, -2), (1, -7)\}$

6. $\{(1, 1), (2, 4), (3, 9), (-2, 4), (-1, 1)\}$

In Exercises 7–26, determine whether the relation represents a function. See Example 2.

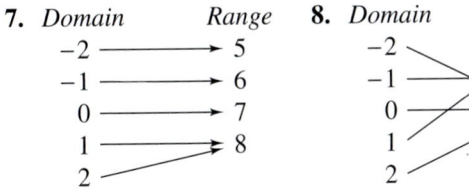

9.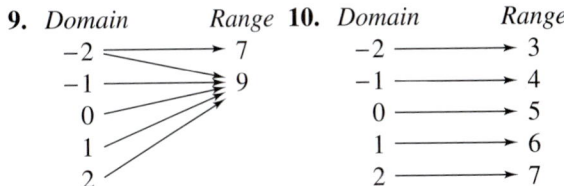

11.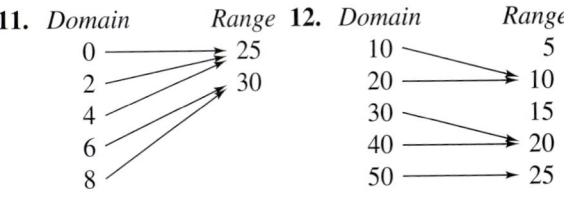

13.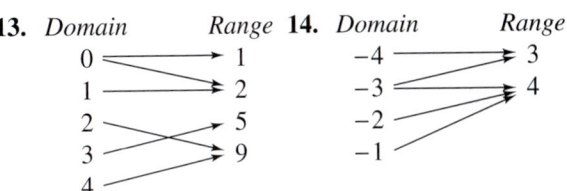

15. *Domain*   *Range*

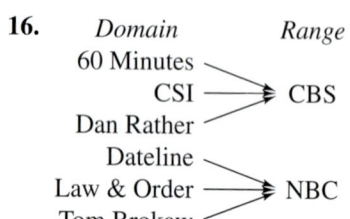

CBS ——→ 60 Minutes
CBS ——→ CSI
——→ Dan Rather
NBC ——→ Dateline
NBC ——→ Law & Order
——→ Tom Brokaw

16.   *Domain*   *Range*

60 Minutes ——↘
CSI ——→ CBS
Dan Rather ——↗
Dateline ——↘
Law & Order ——→ NBC
Tom Brokaw ——↗

17. *Domain*   *Range*

	Single women in the labor force
Year	(in percent)
1997 ——→	67.9
1998 ——→	68.5
1999 ——→	68.7
2000 ——→	69.0

(Source: U.S. Bureau of Labor Statistics)

18.   *Domain*   *Range*

Percent daily value
of vitamin C   Cereal
per serving

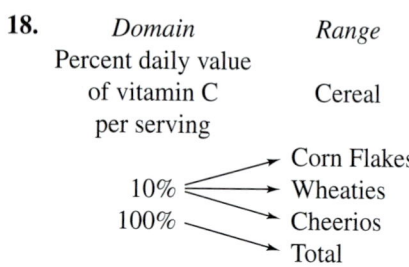

10% ——→ Corn Flakes
10% ——→ Wheaties
100% ——→ Cheerios
——→ Total

19.

Input x	Output y	(x, y)
0	2	(0, 2)
1	4	(1, 4)
2	6	(2, 6)
3	8	(3, 8)
4	10	(4, 10)

20.

Input x	Output y	(x, y)
0	2	(0, 2)
1	4	(1, 4)
2	6	(2, 6)
1	8	(1, 8)
0	10	(0, 10)

21.

Input x	Output y	(x, y)
1	1	(1, 1)
3	2	(3, 2)
5	3	(5, 3)
3	4	(3, 4)
1	5	(1, 5)

22.

Input x	Output y	(x, y)
2	1	(2, 1)
4	1	(4, 1)
6	1	(6, 1)
8	1	(8, 1)
10	1	(10, 1)

23. {(0, 25), (2, 25), (4, 30), (6, 30), (8, 30)}

24. {(10, 5), (20, 10), (30, 15), (40, 20), (50, 25)}

25. Input: a, b, c
Output: 0, 1, 2
{$(a, 0), (b, 1), (c, 2)$}

26. Input: 3, 5, 7
Output: d, e, f
{$(3, d) (5, e), (7, f), (7, d)$}

In Exercises 27–36, use the Vertical Line Test to determine whether *y* is a function of *x*. See Example 3.

27.

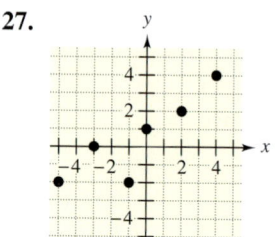

28.

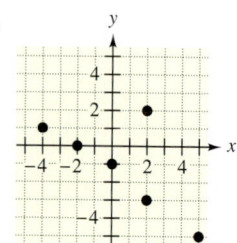

29.

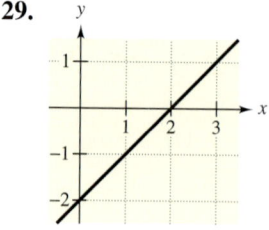

30.

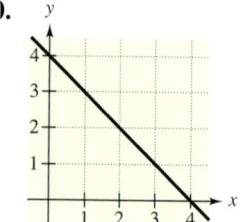

31.

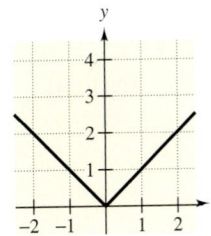

32.

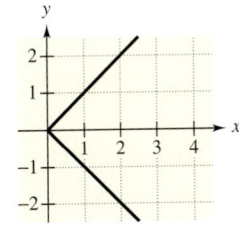

33.

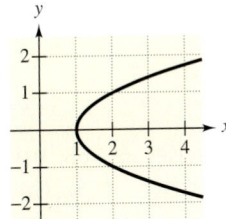

34.

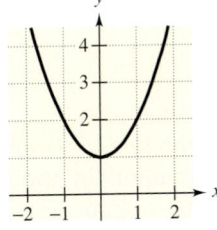

35.

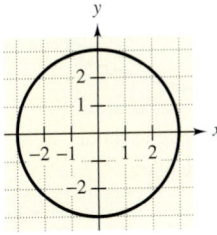

36.

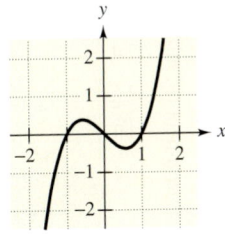

In Exercises 37–52, evaluate the function as indicated, and simplify. See Examples 4 and 5.

37. $f(x) = \frac{1}{2}x$ (a) $f(2)$ (b) $f(5)$
    (c) $f(-4)$ (d) $f\left(-\frac{2}{3}\right)$

38. $g(x) = -\frac{4}{5}x$ (a) $g(5)$ (b) $g(0)$
    (c) $g(-3)$ (d) $g\left(-\frac{5}{4}\right)$

39. $f(x) = 2x - 1$ (a) $f(0)$ (b) $f(3)$
    (c) $f(-3)$ (d) $f\left(-\frac{1}{2}\right)$

40. $f(t) = 3 - 4t$ (a) $f(0)$ (b) $f(1)$
    (c) $f(-2)$ (d) $f\left(\frac{3}{4}\right)$

41. $f(x) = 4x + 1$ (a) $f(1)$ (b) $f(-1)$
    (c) $f(-4)$ (d) $f\left(-\frac{4}{3}\right)$

42. $g(t) = 5 - 2t$ (a) $g\left(\frac{5}{2}\right)$ (b) $g(-10)$
    (c) $g(0)$ (d) $g\left(\frac{3}{4}\right)$

43. $h(t) = \frac{1}{4}t - 1$ (a) $h(200)$ (b) $h(-12)$
    (c) $h(8)$ (d) $h\left(-\frac{5}{2}\right)$

44. $f(s) = 4 - \frac{2}{3}s$ (a) $f(60)$ (b) $f(-15)$
    (c) $f(-18)$ (d) $f\left(\frac{1}{2}\right)$

45. $f(v) = \frac{1}{2}v^2$ (a) $f(-4)$ (b) $f(4)$
    (c) $f(0)$ (d) $f(2)$

46. $g(u) = -2u^2$ (a) $g(0)$ (b) $g(2)$
    (c) $g(3)$ (d) $g(-4)$

47. $g(x) = 2x^2 - 3x + 1$ (a) $g(0)$ (b) $g(-2)$
    (c) $g(1)$ (d) $g\left(\frac{1}{2}\right)$

48. $h(x) = x^2 + 4x - 1$ (a) $h(0)$ (b) $h(-4)$
    (c) $h(10)$ (d) $h\left(\frac{3}{2}\right)$

49. $g(u) = |u + 2|$ (a) $g(2)$ (b) $g(-2)$
    (c) $g(10)$ (d) $g\left(-\frac{5}{2}\right)$

50. $h(s) = |s| + 2$ (a) $h(4)$ (b) $h(-10)$
    (c) $h(-2)$ (d) $h\left(\frac{3}{2}\right)$

51. $h(x) = x^3 - 1$ (a) $h(0)$ (b) $h(1)$
    (c) $h(3)$ (d) $h\left(\frac{1}{2}\right)$

52. $f(x) = 16 - x^4$ (a) $f(-2)$ (b) $f(2)$
    (c) $f(1)$ (d) $f(3)$

In Exercises 53–60, find the domain of the function. See Example 6.

53. $f:\{(0, 4), (1, 3), (2, 2), (3, 1), (4, 0)\}$

54. $f:\{(-2, -1), (-1, 0), (0, 1), (1, 2), (2, 3)\}$

55. $g:\{(-2, 4), (-1, 1), (0, 0), (1, 1), (2, 4)\}$

56. $g:\{(0, 7), (1, 6), (2, 6), (3, 7), (4, 8)\}$

57. $h:\{(-5, 2), (-4, 2), (-3, 2), (-2, 2), (-1, 2)\}$

58. $h:\{(10, 100), (20, 200), (30, 300), (40, 400)\}$

59. Area of a circle: $A = \pi r^2$

60. Circumference of a circle: $C = 2\pi r$

Solving Problems

61. *Demand*   The demand for a product is a function of its price. Consider the demand function

$$f(p) = 20 - 0.5p$$

where p is the price in dollars.

(a) Find $f(10)$ and $f(15)$.

(b) Describe the effect a price increase has on demand.

62. *Maximum Load*   The maximum safe load L (in pounds) for a wooden beam 2 inches wide and d inches high is

$$L(d) = 100d^2.$$

(a) Complete the table.

d	2	4	6	8
$L(d)$				

(b) Describe the effect of an increase in height on the maximum safe load.

63. *Distance*   The function $d(t) = 50t$ gives the distance (in miles) that a car will travel in t hours at an average speed of 50 miles per hour. Find the distance traveled for (a) $t = 2$, (b) $t = 4$, and (c) $t = 10$.

64. *Speed of Sound*   The function $S(h) = 1116 - 4.04h$ approximates the speed of sound (in feet per second) at altitude h (in thousands of feet). Use the function to approximate the speed of sound for (a) $h = 0$, (b) $h = 10$, and (c) $h = 30$.

Interpreting a Graph   In Exercises 65–68, use the information in the graph. (Source: U.S. National Center for Education Statistics)

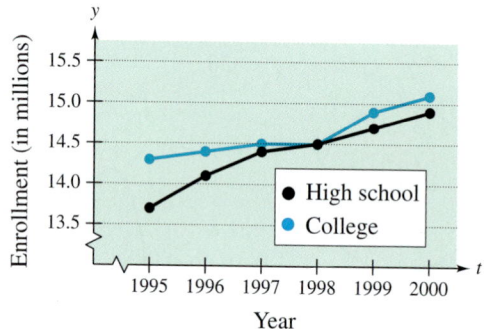

65. Is the high school enrollment a function of the year?

66. Is the college enrollment a function of the year?

67. Let $f(t)$ represent the number of high school students in year t. Find $f(1996)$.

68. Let $g(t)$ represent the number of college students in year t. Find $g(2000)$.

69. ▲ *Geometry*   Write the formula for the perimeter P of a square with sides of length s. Is P a function of s? Explain.

70. ▲ *Geometry*   Write the formula for the volume V of a cube with sides of length t. Is V a function of t? Explain.

71. *Sunrise and Sunset*   The graph approximates the length of time L (in hours) between sunrise and sunset in Erie, Pennsylvania, for the year 2002. The variable t represents the day of the year.   (Source: Fly-By-Day Consulting, Inc.)

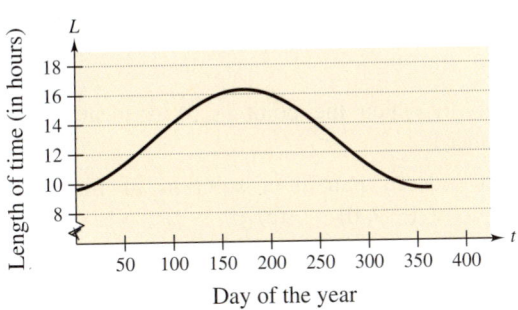

Day of the year

(a) Is the length of time L a function of the day of the year t?

(b) Estimate the range for this relation.

72. *SAT Scores and Grade-Point Average*   The graph shows the SAT score x and the grade-point average y for 12 students.

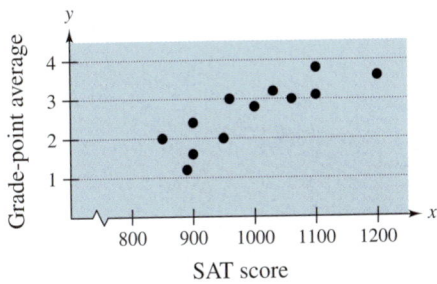

SAT score

(a) Is the grade-point average y a function of the SAT score x?

(b) Estimate the range for this relation.

Explaining Concepts

73. ⊘ Answer parts (c) and (d) of Motivating the Chapter on page 214.

74. *Writing* Explain the difference between a relation and a function. Give an example of a relation that is not a function.

75. Is it possible to find a function that is not a relation? If it is, find one.

76. *Writing* Explain the meaning of the terms *domain* and *range* in the context of a function.

77. *Writing* In your own words, explain how to use the Vertical Line Test.

78. *Writing* Describe some advantages of using function notation.

79. *Writing* Is it possible for the number of elements in the domain of a relation to be greater than the number of elements in the range of the relation? Explain.

80. *Writing* Determine whether the statement uses the word *function* in a way that is mathematically correct. Explain your reasoning.

(a) The amount of money in your savings account is a function of your salary.

(b) The speed at which a free-falling baseball strikes the ground is a function of the height from which it is dropped.

Mid-Chapter Quiz

Take this quiz as you would take a quiz in class. After you are done, check your work against the answers in the back of the book.

1. Plot the points $(4, -2)$ and $\left(-1, -\frac{5}{2}\right)$ on a rectangular coordinate system.

2. Determine the quadrant(s) in which the point $(x, 5)$ is located without plotting it. (x is a real number.)

3. Determine whether each ordered pair is a solution of the equation $y = 9 - |x|$: (a) $(2, 7)$ (b) $(-9, 0)$ (c) $(0, -9)$.

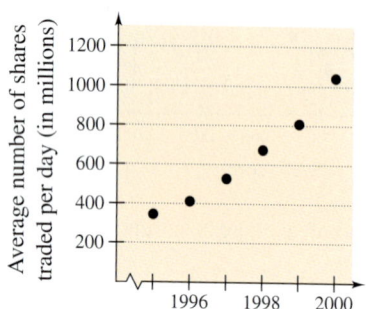

Figure for 4

4. The scatter plot at the left shows the average number of shares traded per day (in millions) on the New York Stock Exchange for the years 1995 through 2000. Estimate the average number of shares traded per day for each year from 1995 to 2000.   (Source: The New York Stock Exchange, Inc.)

In Exercises 5 and 6, find the x- and y-intercepts of the graph of the equation.

5. $x - 3y = 12$

6. $y = -7x + 2$

In Exercises 7–9, sketch the graph of the equation.

7. $y = 5 - 2x$ **8.** $y = (x + 2)^2$ **9.** $y = |x + 3|$

In Exercises 10 and 11, find the domain and range of the relation.

10. $\{(1, 4), (2, 6), (3, 10), (2, 14), (1, 0)\}$

11. $\{(-3, 6), (-2, 6), (-1, 6), (0, 6)\}$

12. Determine whether the relation in the figure is a function of x using the Vertical Line Test.

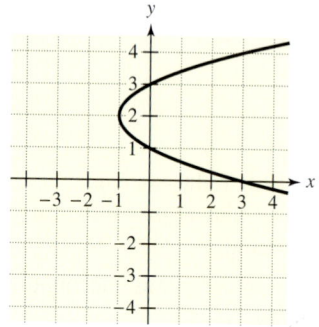

Figure for 12

In Exercises 13 and 14, evaluate the function as indicated, and simplify.

13. $f(x) = 3(x + 2) - 4$

(a) $f(0)$ (b) $f(-3)$

14. $g(x) = 4 - x^2$

(a) $g(-1)$ (b) $g(8)$

15. Find the domain of the function f: $\{(10, 1), (15, 3), (20, 9), (25, 27)\}$.

16. ▦ Use a graphing calculator to graph $y = 3.6x - 2.4$. Graphically estimate the intercepts of the graph. Explain how to verify your estimates algebraically.

17. A new computer system sells for approximately $3000 and depreciates at the rate of $500 per year for 4 years.

(a) Find an equation that relates the value of the computer system to the number of years t.

(b) Sketch the graph of the equation.

(c) What is the y-intercept of the graph, and what does it represent in the context of the problem?

4.4 Slope and Graphs of Linear Equations

Kent Meireis/The Image Works

What You Should Learn

1 Determine the slope of a line through two points.

2 Write linear equations in slope-intercept form and graph the equations.

3 Use slopes to determine whether lines are parallel, perpendicular, or neither.

Why You Should Learn It

Slopes of lines can be used in many business applications. For instance, in Exercise 92 on page 261, you will interpret the meaning of the slopes of linear equations that model the predicted profit for an outerwear manufacturer.

1 Determine the slope of a line through two points.

The Slope of a Line

The **slope** of a nonvertical line is the number of units the line rises or falls vertically for each unit of horizontal change from left to right. For example, the line in Figure 4.23 rises two units for each unit of horizontal change from left to right, and so this line has a slope of $m = 2$.

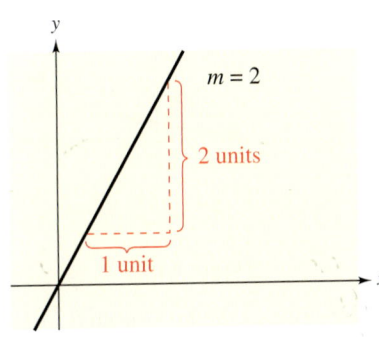

Figure 4.23   Figure 4.24

Study Tip

In the definition at the right, the *rise* is the vertical change between the points and the *run* is the horizontal change between the points.

Definition of the Slope of a Line

The **slope** m of a nonvertical line passing through the points (x_1, y_1) and (x_2, y_2) is

$$m = \frac{y_2 - y_1}{x_2 - x_1} = \frac{\text{Change in } y}{\text{Change in } x} = \frac{\text{Rise}}{\text{Run}}$$

where $x_1 \neq x_2$ (see Figure 4.24).

When the formula for slope is used, the *order of subtraction* is important. Given two points on a line, you are free to label either of them (x_1, y_1) and the other (x_2, y_2). However, once this has been done, you must form the numerator and denominator using the same order of subtraction.

$$m = \frac{y_2 - y_1}{x_2 - x_1} \qquad m = \frac{y_1 - y_2}{x_1 - x_2} \qquad m = \frac{y_2 - y_1}{x_1 - x_2}$$

Correct   Correct   Incorrect

Example 1 Finding the Slope of a Line Through Two Points

Find the slope of the line passing through each pair of points.

a. $(-2, 0)$ and $(3, 1)$ **b.** $(0, 0)$ and $(1, -1)$

Solution

a. Let $(x_1, y_1) = (-2, 0)$ and $(x_2, y_2) = (3, 1)$. The slope of the line through these points is

$$m = \frac{y_2 - y_1}{x_2 - x_1}$$

$$= \frac{1 - 0}{3 - (-2)} \qquad \text{Difference in } y\text{-values}$$
$$\qquad\qquad\qquad\quad \text{Difference in } x\text{-values}$$

$$= \frac{1}{5}. \qquad\qquad \text{Simplify.}$$

The graph of the line is shown in Figure 4.25.

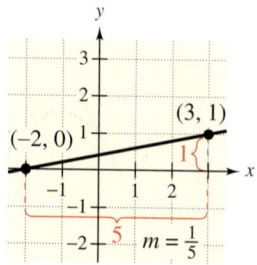

Figure 4.25

b. The slope of the line through $(0, 0)$ and $(1, -1)$ is

$$m = \frac{-1 - 0}{1 - 0} \qquad \text{Difference in } y\text{-values}$$
$$\qquad\qquad\qquad\quad \text{Difference in } x\text{-values}$$

$$= \frac{-1}{1} \qquad\qquad \text{Simplify.}$$

$$= -1. \qquad\qquad \text{Simplify.}$$

The graph of the line is shown in Figure 4.26.

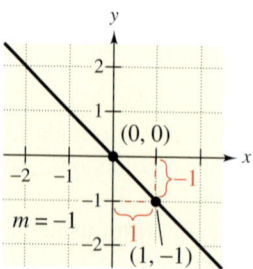

Figure 4.26

Example 2   Horizontal and Vertical Lines and Slope

Find the slope of the line passing through each pair of points.

a. $(-1, 2)$ and $(2, 2)$ **b.** $(2, 4)$ and $(2, 1)$

Solution

a. The line through $(-1, 2)$ and $(2, 2)$ is horizontal because its y-coordinates are the same. The slope of this horizontal line is

$$m = \frac{2 - 2}{2 - (-1)} \qquad \text{Difference in } y\text{-values}$$
$$ \qquad\qquad\quad \text{Difference in } x\text{-values}$$

$$= \frac{0}{3} \qquad\qquad\quad \text{Simplify.}$$

$$= 0. \qquad\qquad\quad \text{Simplify.}$$

The graph of the line is shown in Figure 4.27.

b. The line through $(2, 4)$ and $(2, 1)$ is vertical because its x-coordinates are the same. Applying the formula for slope, you have

$$\frac{4 - 1}{2 - 2} = \frac{3}{0}. \qquad \text{Division by 0 is undefined.}$$

Because division by zero is not defined, the slope of a vertical line is not defined. The graph of the line is shown in Figure 4.28.

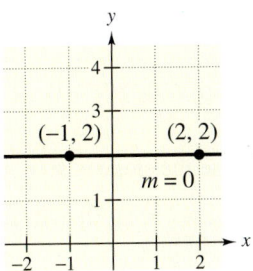

Figure 4.27

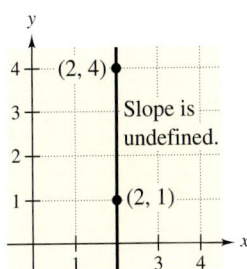

Figure 4.28

From the slopes of the lines shown in Figures 4.25–4.28, you can make several generalizations about the slope of a line.

Slope of a Line

1. A line with positive slope $(m > 0)$ *rises* from left to right.

2. A line with negative slope $(m < 0)$ *falls* from left to right.

3. A line with zero slope $(m = 0)$ is *horizontal*.

4. A line with undefined slope is *vertical*.

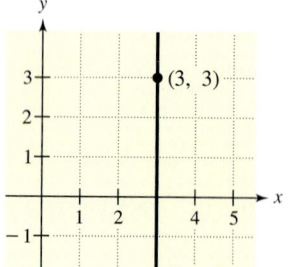

Vertical line: undefined slope
Figure 4.29

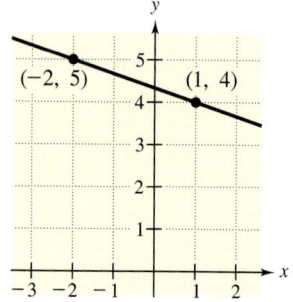

Line falls: negative slope
Figure 4.30

Example 3   Using Slope to Describe Lines

Describe the line through each pair of points.

a. $(3, -2)$ and $(3, 3)$ **b.** $(-2, 5)$ and $(1, 4)$

Solution

a. Let $(x_1, y_1) = (3, -2)$ and $(x_2, y_2) = (3, 3)$.

$$m = \frac{3 - (-2)}{3 - 3} = \frac{5}{0} \qquad \text{Undefined slope (See Figure 4.29.)}$$

Because the slope is undefined, the line is vertical.

b. Let $(x_1, y_1) = (-2, 5)$ and $(x_2, y_2) = (1, 4)$.

$$m = \frac{4 - 5}{1 - (-2)} = -\frac{1}{3} < 0 \qquad \text{Negative slope (See Figure 4.30.)}$$

Because the slope is negative, the line falls from left to right.

Example 4   Using Slope to Describe Lines

Describe the line through each pair of points.

a. $(-4, -3)$ and $(0, -3)$ **b.** $(1, 0)$ and $(4, 6)$

Solution

a. Let $(x_1, y_1) = (-4, -3)$ and $(x_2, y_2) = (0, -3)$.

$$m = \frac{-3 - (-3)}{0 - (-4)} = \frac{0}{4} = 0 \qquad \text{Zero slope (See Figure 4.31.)}$$

Because the slope is zero, the line is horizontal.

b. Let $(x_1, y_1) = (1, 0)$ and $(x_2, y_2) = (4, 6)$.

$$m = \frac{6 - 0}{4 - 1} = \frac{6}{3} = 2 > 0 \qquad \text{Positive slope (See Figure 4.32.)}$$

Because the slope is positive, the line rises from left to right.

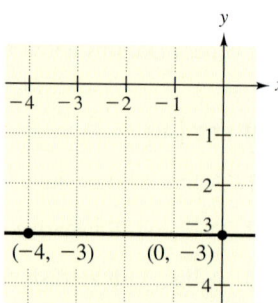

Horizontal line: zero slope
Figure 4.31

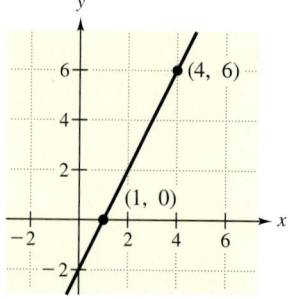

Line rises: positive slope
Figure 4.32

Any two points on a nonvertical line can be used to calculate its slope. This is demonstrated in the next two examples.

Example 5 Finding the Slope of a Ladder

Find the slope of the ladder leading up to the tree house in Figure 4.33.

Solution

Consider the tree trunk as the y-axis and the level ground as the x-axis. The endpoints of the ladder are $(0, 12)$ and $(5, 0)$. So, the slope of the ladder is

$$m = \frac{y_2 - y_1}{x_2 - x_1} = \frac{0 - 12}{5 - 0} = -\frac{12}{5}.$$

Figure 4.33

Example 6 Finding the Slope of a Line

Sketch the graph of the line $3x - 2y = 4$. Then find the slope of the line. (Choose two different pairs of points on the line and show that the same slope is obtained from either pair.)

Solution

Begin by solving the equation for y.

$$y = \frac{3}{2}x - 2$$

Then, construct a table of values, as shown below.

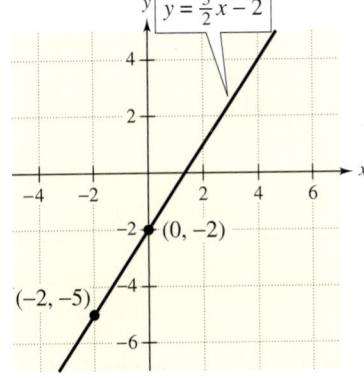

(a)

x	-2	0	2	4
$y = \frac{3}{2}x - 2$	-5	-2	1	4
Solution point	$(-2, -5)$	$(0, -2)$	$(2, 1)$	$(4, 4)$

From the solution points shown in the table, sketch the line, as shown in Figure 4.34. To calculate the slope of the line using two different sets of points, first use the points $(-2, -5)$ and $(0, -2)$, as shown in Figure 4.34(a), and obtain a slope of

$$m = \frac{-2 - (-5)}{0 - (-2)} = \frac{3}{2}.$$

Next, use the points $(2, 1)$ and $(4, 4)$, as shown in Figure 4.34(b), and obtain a slope of

$$m = \frac{4 - 1}{4 - 2} = \frac{3}{2}.$$

Try some other pairs of points on the line to see that you obtain a slope of $m = \frac{3}{2}$ regardless of which two points you use.

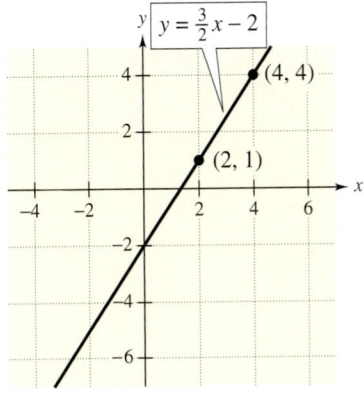

(b)

Figure 4.34

② Write linear equations in slope-intercept form and graph the equations.

Slope as a Graphing Aid

You saw in Section 4.1 that before creating a table of values for an equation, it is helpful first to solve the equation for y. When you do this for a linear equation, you obtain some very useful information. Consider the results of Example 6.

$$3x - 2y = 4 \qquad \text{Write original equation.}$$

$$3x - 3x - 2y = -3x + 4 \qquad \text{Subtract } 3x \text{ from each side.}$$

$$-2y = -3x + 4 \qquad \text{Simplify.}$$

$$\frac{-2y}{-2} = \frac{-3x + 4}{-2} \qquad \text{Divide each side by } -2.$$

$$y = \frac{3}{2}x - 2 \qquad \text{Simplify.}$$

Observe that the coefficient of x is the slope of the graph of this equation (see Example 6). Moreover, the constant term, -2, gives the y-intercept of the graph.

$$y = \underset{\text{slope}}{\boxed{\tfrac{3}{2}}} \, x + \underset{\text{y-intercept } (0, -2)}{\boxed{-2}}$$

This form is called the **slope-intercept form** of the equation of the line.

Slope-Intercept Form of the Equation of a Line

The graph of the equation

$$y = mx + b$$

is a line whose slope is m and whose y-intercept is $(0, b)$. (See Figure 4.35.)

Study Tip

Remember that slope is a *rate of change*. In the slope-intercept equation

$$y = mx + b$$

the slope m is the rate of change of y with respect to x.

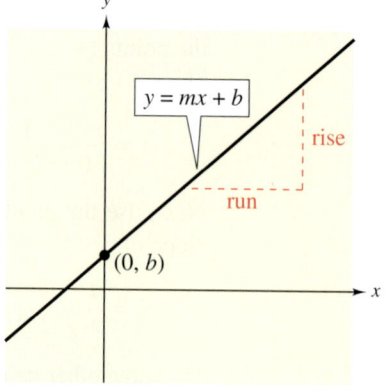

Figure 4.35

So far, you have been plotting several points to sketch the equation of a line. However, now that you can recognize equations of lines, you don't have to plot as many points—two points are enough. (You might remember from geometry that *two points are all that are necessary to determine a line.*) The next example shows how to use the slope to help sketch a line.

Example 7   Using the Slope and *y*-Intercept to Sketch a Line

Use the slope and *y*-intercept to sketch the graph of

$$x - 3y = -6.$$

Solution

First, write the equation in slope-intercept form.

$x - 3y = -6$	Write original equation.
$-3y = -x - 6$	Subtract x from each side.
$y = \dfrac{-x - 6}{-3}$	Divide each side by -3.
$y = \dfrac{1}{3}x + 2$	Simplify to slope-intercept form.

So, the slope of the line is $m = \frac{1}{3}$ and the *y*-intercept is $(0, b) = (0, 2)$. Now you can sketch the graph of the equation. First, plot the *y*-intercept, as shown in Figure 4.36. Then, using a slope of $\frac{1}{3}$,

$$m = \frac{1}{3} = \frac{\text{Change in } y}{\text{Change in } x}$$

locate a second point on the line by moving three units to the right and one unit up (or one unit up and three units to the right), also shown in Figure 4.36. Finally, obtain the graph by drawing a line through the two points (see Figure 4.37).

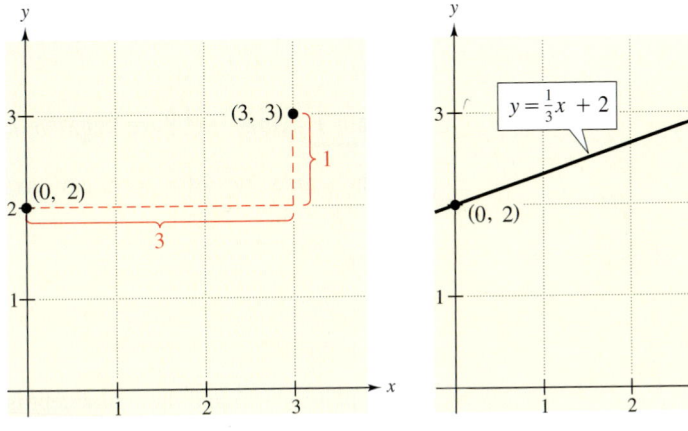

Figure 4.36   **Figure 4.37**

3 Use slopes to determine whether lines are parallel, perpendicular, or neither.

Parallel and Perpendicular Lines

You know from geometry that two lines in a plane are **parallel** if they do not intersect. What this means in terms of their slopes is shown in Example 8.

Example 8   Lines That Have the Same Slope

On the same set of coordinate axes, sketch the lines $y = 3x$ and $y = 3x - 4$.

Solution

For the line

$$y = 3x$$

the slope is $m = 3$ and the y-intercept is $(0, 0)$. For the line

$$y = 3x - 4$$

the slope is also $m = 3$ and the y-intercept is $(0, -4)$. The graphs of these two lines are shown in Figure 4.38.

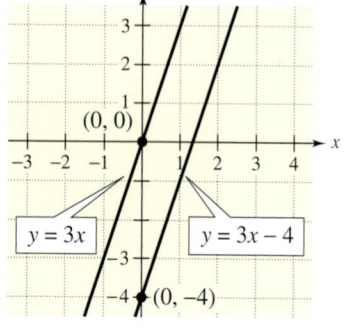

Figure 4.38

In Example 8, notice that the two lines have the same slope and that the two lines appear to be parallel. The following rule states that this is always the case.

Parallel Lines

Two distinct nonvertical lines are parallel if and only if they have the same slope.

The phrase "if and only if" in this rule is used in mathematics as a way to write two statements in one. The first statement says that *if two distinct nonvertical lines have the same slope, they must be parallel.* The second (or reverse) statement says that *if two distinct nonvertical lines are parallel, they must have the same slope.*

Example 9   Lines That Have Negative Reciprocal Slopes

On the same set of coordinate axes, sketch the lines $y = 5x + 2$ and $y = -\frac{1}{5}x - 4$.

Solution

For the line

$$y = 5x + 2$$

the slope is $m = 5$ and the y-intercept is $(0, 2)$. For the line

$$y = -\frac{1}{5}x - 4$$

the slope is $m = -\frac{1}{5}$ and the y-intercept is $(0, -4)$. The graphs of these two lines are shown in Figure 4.39.

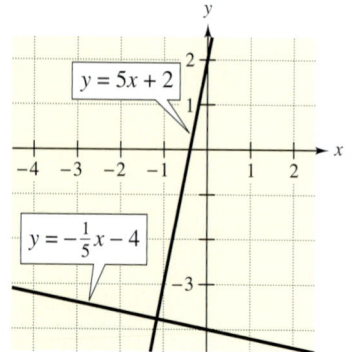

Figure 4.39

In Example 9, notice that the two lines have slopes that are negative reciprocals of each other and that the two lines appear to be perpendicular. Another rule from geometry is that two lines in a plane are **perpendicular** if they intersect at right angles. In terms of their slopes, this means that two nonvertical lines are perpendicular if their slopes are negative reciprocals of each other.

Perpendicular Lines

Consider two nonvertical lines whose slopes are m_1 and m_2. The two lines are perpendicular if and only if their slopes are *negative reciprocals* of each other. That is,

$$m_1 = -\frac{1}{m_2}$$

or, equivalently,

$$m_1 \cdot m_2 = -1.$$

Example 10   Parallel or Perpendicular?

Determine whether the pairs of lines are parallel, perpendicular, or neither.

a. $y = -3x - 2$, $y = \frac{1}{3}x + 1$

b. $y = \frac{1}{2}x + 1$, $y = \frac{1}{2}x - 1$

Solution

a. The first line has a slope of $m_1 = -3$ and the second line has a slope of $m_2 = \frac{1}{3}$. Because these slopes are negative reciprocals of each other, the two lines must be perpendicular, as shown in Figure 4.40.

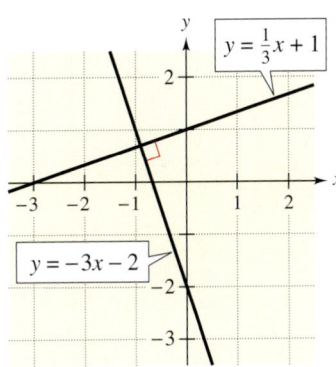

Figure 4.40

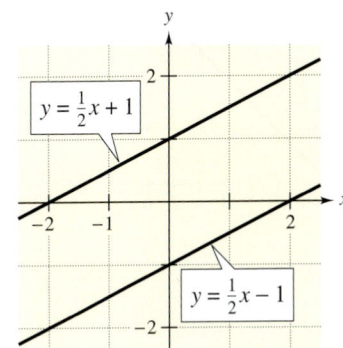

Figure 4.41

b. Both lines have a slope of $m = \frac{1}{2}$. So, the two lines must be parallel, as shown in Figure 4.41.

4.4 Exercises

Keep mathematically in shape by doing these exercises *before* the problems of this section.

Properties and Definitions

1. Two equations that have the same set of solutions are called _____.

2. Use the Addition Property of Equality to fill in the blank.

   $5x - 2 = 6$

   $5x = 6 + \boxed{}$

Simplifying Expressions

In Exercises 3–10, simplify the expression.

3. $x^2 \cdot x^3$

4. $z^2 \cdot z^2$

5. $(-y^2)y$

6. $5x^2(x^5)$

7. $(25x^3)(2x^2)$

8. $(3yz)(6yz^3)$

9. $x^2 - 2x - x^2 + 3x + 2$

10. $x^2 - 5x - 2 + x$

Problem Solving

11. *Carpentry*   A carpenter must cut a 10-foot board into three pieces. Two are to have the same length and the third is to be three times as long as the two of equal length. Find the lengths of the three pieces.

12. *Repair Bill*   The bill for the repair of your dishwasher was $113. The cost for parts was $65. The cost for labor was $32 per hour. How many hours did the repair work take?

Developing Skills

In Exercises 1–10, estimate the slope (if it exists) of the line from its graph.

1.
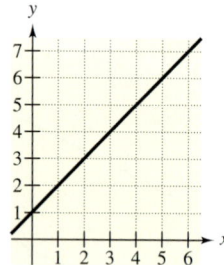

2.
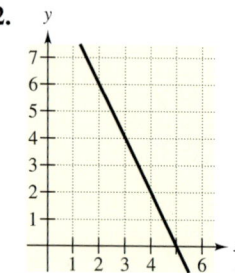

3.
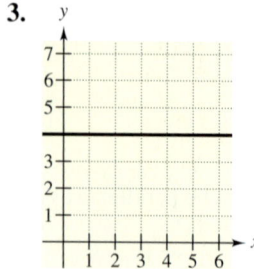

4.
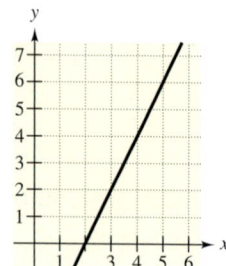

5.
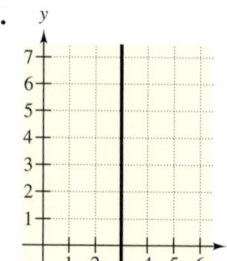

6.
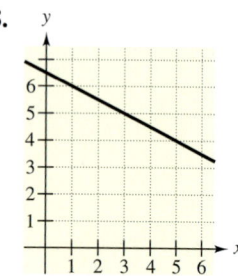

7.

8.

9.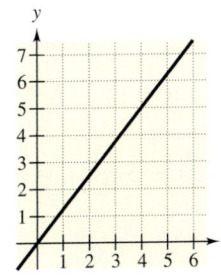

10.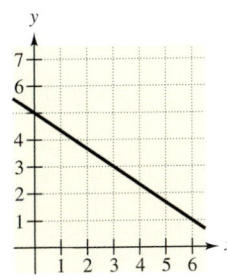

23. $(3, -4), (8, -4)$ **24.** $(1, -2), (-2, -2)$

25. $\left(\frac{1}{4}, \frac{3}{2}\right), \left(\frac{9}{2}, -3\right)$ **26.** $\left(\frac{5}{4}, \frac{1}{4}\right), \left(\frac{7}{8}, 2\right)$

27. $(3.2, -1), (-3.2, 4)$ **28.** $(1.4, 3), (-1.4, 5)$

29. $(3.5, -1), (5.75, 4.25)$ **30.** $(6, 6.4), (-3.1, 5.2)$

31. $(a, 3), (4, 3), \ a \neq 4$ **32.** $(4, a), (4, 2), \ a \neq 2$

In Exercises 11 and 12, identify the line in the figure that has each slope.

11. (a) $m = \frac{3}{2}$

    (b) $m = 0$

    (c) $m = -\frac{2}{3}$

    (d) $m = -2$

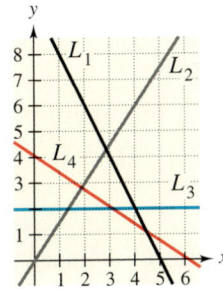

In Exercises 33 and 34, complete the table. Use two different pairs of solution points to show that the same slope is obtained using either pair. See Example 6.

x		-2	0	2	4
y					
Solution point					

33. $y = -2x - 2$ **34.** $y = 3x + 4$

12. (a) $m = -\frac{3}{4}$

    (b) $m = \frac{1}{2}$

    (c) m is undefined.

    (d) $m = 3$

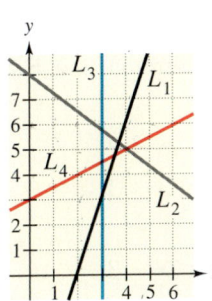

In Exercises 35–38, use the formula for slope to find the value of y such that the line through the points has the given slope.

35. Points: $(3, -2), (0, y)$ **36.** Points: $(-3, y), (8, 2)$
    Slope: $m = -8$ Slope: $m = 2$

37. Points: $(-4, y), (7, 6)$ **38.** Points: $(0, 10), (6, y)$
    Slope: $m = \frac{5}{2}$ Slope: $m = -\frac{1}{3}$

In Exercises 13–32, plot the points and find the slope (if possible) of the line passing through the points. State whether the line rises, falls, is horizontal, or is vertical. See Examples 1–4.

13. $(0, 0), (4, 5)$ **14.** $(0, 0), (3, 6)$

15. $(0, 0), (8, -4)$ **16.** $(0, 0), (-1, 3)$

17. $(0, 6), (8, 0)$ **18.** $(5, 0), (0, 7)$

19. $(-3, -2), (1, 6)$ **20.** $(2, 4), (-4, -4)$

21. $(-6, -1), (-6, 4)$ **22.** $(-4, -10), (-4, 0)$

In Exercises 39–50, a point on a line and the slope of the line are given. Plot the point and use the slope to find two additional points on the line. (There are many correct answers.)

39. $(2, 1), \ m = 0$ **40.** $(5, 10), \ m = 0$

41. $(1, -6), \ m = 2$ **42.** $(-2, 4), \ m = 1$

43. $(0, 1), \ m = -2$ **44.** $(5, 6), \ m = -3$

45. $(-4, 0),\ m = \frac{2}{3}$ **46.** $(-1, -1),\ m = \frac{1}{4}$

47. $(3, 5),\ m = -\frac{1}{2}$ **48.** $(1, 3),\ m = -\frac{4}{3}$

49. $(-8, 1)$
m is undefined.

50. $(6, -4)$
m is undefined.

In Exercises 51–56, sketch the graph of a line through the point $(0, 2)$ having the given slope.

51. $m = 0$ **52.** m is undefined.
53. $m = 3$ **54.** $m = -1$
55. $m = -\frac{2}{3}$ **56.** $m = \frac{3}{4}$

In Exercises 57–62, plot the x- and y-intercepts and sketch the graph of the line.

57. $2x + 3y + 6 = 0$ **58.** $3x + 4y + 12 = 0$
59. $5x - 2y - 10 = 0$ **60.** $3x - 7y - 21 = 0$
61. $6x - 4y + 12 = 0$ **62.** $2x - 5y - 20 = 0$

In Exercises 63–76, write the equation in slope-intercept form. Use the slope and y-intercept to sketch the line. See Example 7.

63. $x - y = 0$ **64.** $x + y = 0$

65. $\frac{1}{2}x + y = 0$ **66.** $\frac{3}{4}x - y = 0$

67. $2x - y - 3 = 0$ **68.** $x - y + 2 = 0$

69. $x - 3y + 6 = 0$ **70.** $3x - 2y - 2 = 0$

71. $x + 2y - 2 = 0$ **72.** $10x + 6y - 3 = 0$

73. $3x - 4y + 2 = 0$ **74.** $2x - 3y + 1 = 0$

75. $y + 5 = 0$ **76.** $y - 3 = 0$

In Exercises 77–80, determine whether the lines L_1 and L_2 passing through the pairs of points are parallel, perpendicular, or neither.

77. L_1: $(0, -1),\ (5, 9)$ **78.** L_1: $(-2, -1),\ (1, 5)$
  L_2: $(0, 3),\ (4, 1)$ L_2: $(1, 3),\ (5, 5)$

79. L_1: $(3, 6),\ (-6, 0)$ **80.** L_1: $(4, 8),\ (-4, 2)$
  L_2: $(0, -1),\ \left(5, \frac{7}{3}\right)$ L_2: $(3, -5),\ \left(-1, \frac{1}{3}\right)$

In Exercises 81–84, sketch the graphs of the two lines on the same rectangular coordinate system. Determine whether the lines are parallel, perpendicular, or neither. Use a graphing calculator to verify your result. (Use a square setting.) See Examples 8–10.

81. $y_1 = 2x - 3$ **82.** $y_1 = -\frac{1}{3}x - 3$
  $y_2 = 2x + 1$ $y_2 = \frac{1}{3}x + 1$

83. $y_1 = 2x - 3$ **84.** $y_1 = -\frac{1}{3}x - 3$
  $y_2 = -\frac{1}{2}x + 1$ $y_2 = 3x + 1$

Solving Problems

85. *Roof Pitch* Determine the slope, or pitch, of the roof of the house shown in the figure.

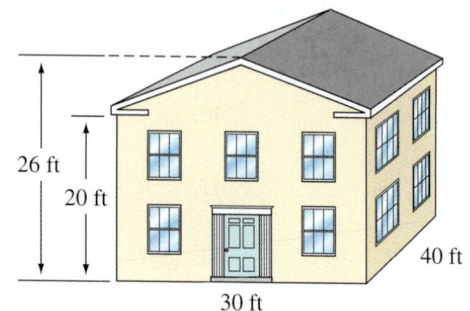

26 ft
20 ft
40 ft
30 ft

86. *Ladder* Find the slope of the ladder shown in the figure.

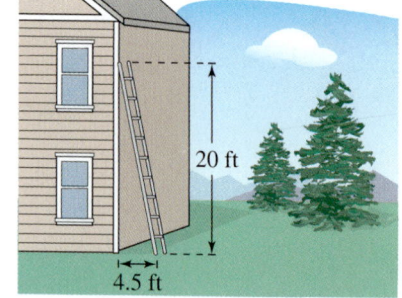

20 ft
4.5 ft

87. *Subway Track*   A subway track rises 3 feet over a 200-foot horizontal distance.

   (a) Draw a diagram of the track and label the rise and run.

   (b) Find the slope of the track.

   (c) Would the slope be steeper if the track rose 3 feet over a distance of 100 feet? Explain.

88. *Water-Ski Ramp*   In tournament water-ski jumping, the ramp rises to a height of 6 feet on a raft that is 21 feet long.

   (a) Draw a diagram of the ramp and label the rise and run.

   (b) Find the slope of the ramp.

   (c) Would the slope be steeper if the ramp rose 6 feet over a distance of 24 feet? Explain.

89. *Flight Path*   An airplane leaves an airport. As it flies over a town, its altitude is 4 miles. The town is about 20 miles from the airport. Approximate the slope of the linear path followed by the airplane during takeoff.

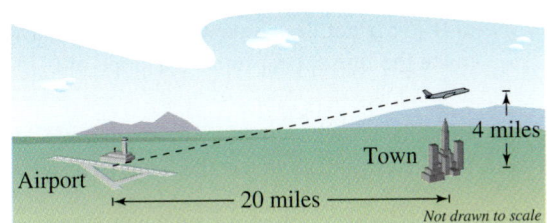

90. *Slide*   The ladder of a straight slide in a playground is 8 feet high. The distance along the ground from the ladder to the foot of the slide is 12 feet. Approximate the slope of the slide.

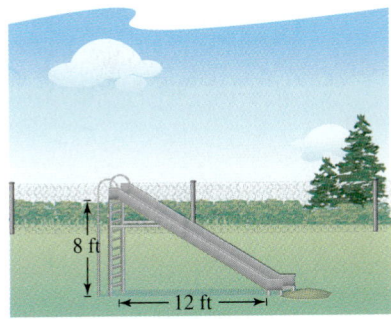

91. *Net Sales*   The graph shows the net sales (in billions of dollars) for Wal-Mart for the years 1996 through 2000. (Source: 2000 Wal-Mart Annual Report)

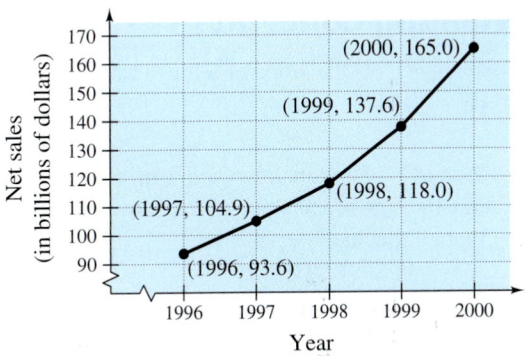

Figure for 91

   (a) Find the slopes of the four line segments.

   (b) Find the slope of the line segment connecting the years 1996 and 2000. Interpret the meaning of this slope in the context of the problem.

92. *Profit*   Based on different assumptions, the marketing department of an outerwear manufacturer develops two linear models to predict the annual profit of the company over the next 10 years. The models are $P_1 = 0.2t + 2.4$ and $P_2 = 0.3t + 2.4$, where P_1 and P_2 represent profit in millions of dollars and t is time in years $(0 \le t \le 10)$.

   (a) Interpret the slopes of the two linear models in the context of the problem.

   (b) Which model predicts a faster increase in profits?

   (c) Use each model to predict profits when $t = 10$.

   (d) ▦ Use a graphing calculator to graph the models in the same viewing window.

Rate of Change   In Exercises 93–98, the slopes of lines representing annual sales y in terms of time t in years are given. Use the slopes to determine any change in annual sales for a 1-year increase in time t.

93. $m = 76$

94. $m = 0$

95. $m = 18$

96. $m = 0.5$

97. $m = -14$

98. $m = -4$

Explaining Concepts

99. *Writing*   Is the slope of a line a ratio? Explain.

100. *Writing*   Explain how you can visually determine the sign of the slope of a line by observing the graph of the line.

101. *True or False?*   If both the *x*- and *y*-intercepts of a line are positive, then the slope of the line is positive. Justify your answer.

102. *Writing*   Which slope is steeper: -5 or 2? Explain.

103. *Writing*   Is it possible to have two perpendicular lines with positive slopes? Explain.

104. *Writing*   The slope of a line is $\frac{3}{2}$. *x* is increased by eight units. How much will *y* change? Explain.

105. When a quantity *y* is increasing or decreasing at a constant rate over time *t*, the graph of *y* versus *t* is a line. What is another name for the rate of change?

106. *Writing*   Explain how to use slopes to determine if the points $(-2, -3)$, $(1, 1)$, and $(3, 4)$ lie on the same line.

107. *Writing*   When determining the slope of the line through two points, does the order of subtracting coordinates of the points matter? Explain.

108. *Misleading Graphs*

(a) Use a graphing calculator to graph the line $y = 0.75x - 2$ for each viewing window.

Xmin = -10	Xmin = 0
Xmax = 10	Xmax = 1
Xscl = 2	Xscl = 0.5
Ymin = -100	Ymin = -2
Ymax = 100	Ymax = -1.5
Yscl = 10	Yscl = 0.1

(b) Do the lines appear to have the same slope?

(c) Does either of the lines appear to have a slope of 0.75? If not, find a viewing window that will make the line appear to have a slope of 0.75.

(d) Describe real-life situations in which it would be to your advantage to use the two given settings.

4.5 Equations of Lines

Davis Barber/PhotoEdit, Inc.

What You Should Learn

① Write equations of lines using the point-slope form.

② Write the equations of horizontal and vertical lines.

③ Use linear models to solve application problems.

Why You Should Learn It

Real-life problems can be modeled and solved using linear equations. For instance, in Example 8 on page 269, a linear equation is used to model the relationship between the time and the height of a mountain climber.

① Write equations of lines using the point-slope form.

The Point-Slope Form of the Equation of a Line

In Sections 4.1 through 4.4, you studied analytic (or coordinate) geometry. Analytic geometry uses a coordinate plane to give visual representations of algebraic concepts, such as equations or functions.

There are two basic types of problems in analytic geometry.

1. Given an equation, sketch its graph.

Algebra Geometry

2. Given a graph, write its equation.

Geometry Algebra

In Section 4.4, you worked primarily with the first type of problem. In this section, you will study the second type. Specifically, you will learn how to write the equation of a line when you are given its slope and a point on the line. Before a general formula for doing this is given, consider the following example.

Example 1   Writing an Equation of a Line

A line has a slope of $\frac{5}{3}$ and passes through the point $(2, 1)$. Find its equation.

Solution

Begin by sketching the line, as shown in Figure 4.42. The slope of a line is the same through any two points on the line. So, using *any* representative point (x, y) and the given point $(2, 1)$, it follows that the slope of the line is

$$m = \frac{y - 1}{x - 2}.$$

Difference in *y*-coordinates

Difference in *x*-coordinates

By substituting $\frac{5}{3}$ for m, you obtain the equation of the line.

$$\frac{5}{3} = \frac{y - 1}{x - 2} \qquad \text{Slope formula}$$

$$5(x - 2) = 3(y - 1) \qquad \text{Cross-multiply.}$$

$$5x - 10 = 3y - 3 \qquad \text{Distributive Property}$$

$$5x - 3y = 7 \qquad \text{Equation of line}$$

Figure 4.42

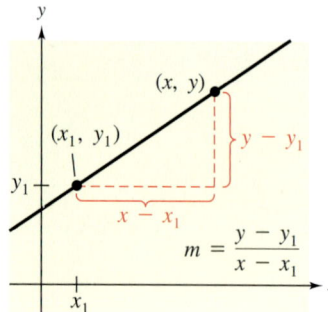

Figure 4.43

The procedure in Example 1 can be used to derive a *formula* for the equation of a line given its slope and a point on the line. In Figure 4.43, let (x_1, y_1) be a given point on a line whose slope is m. If (x, y) is any *other* point on the line, it follows that

$$\frac{y - y_1}{x - x_1} = m.$$

This equation in variables x and y can be rewritten in the form

$$y - y_1 = m(x - x_1)$$

which is called the **point-slope form** of the equation of a line.

Point-Slope Form of the Equation of a Line

The **point-slope form** of the equation of a line with slope m and passing through the point (x_1, y_1) is

$$y - y_1 = m(x - x_1).$$

Example 2   The Point-Slope Form of the Equation of a Line

Write an equation of the line that passes through the point $(1, -2)$ and has slope $m = 3$.

Solution

Use the point-slope form with $(x_1, y_1) = (1, -2)$ and $m = 3$.

$$y - y_1 = m(x - x_1) \qquad \text{Point-slope form}$$
$$y - (-2) = 3(x - 1) \qquad \text{Substitute } -2 \text{ for } y_1, 1 \text{ for } x_1, \text{ and 3 for } m.$$
$$y + 2 = 3x - 3 \qquad \text{Simplify.}$$
$$y = 3x - 5 \qquad \text{Equation of line}$$

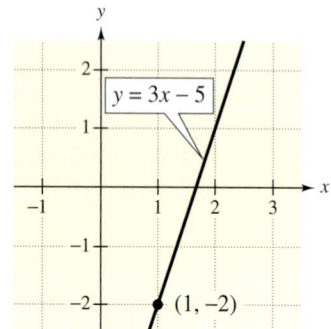

Figure 4.44

So, an equation of the line is $y = 3x - 5$. Note that this is the slope-intercept form of the equation. The graph of this line is shown in Figure 4.44.

In Example 2, note that it was concluded that $y = 3x - 5$ is "an" equation of the line rather than "the" equation of the line. The reason for this is that every equation can be written in many equivalent forms. For instance,

$$y = 3x - 5, \quad 3x - y = 5, \quad \text{and} \quad 3x - y - 5 = 0$$

are all equations of the line in Example 2. The first of these equations $(y = 3x - 5)$ is in the slope-intercept form

$$y = mx + b \qquad \text{Slope-intercept form}$$

and it provides the most information about the line. The last of these equations $(3x - y - 5 = 0)$ is in the general form of the equation of a line.

$$ax + by + c = 0 \qquad \text{General form}$$

The point-slope form can be used to find an equation of a line passing through any two points (x_1, y_1) and (x_2, y_2). First, use the formula for the slope of a line passing through these two points.

$$m = \frac{y_2 - y_1}{x_2 - x_1}$$

Then, knowing the slope, use the point-slope form to obtain the equation

$$y - y_1 = \frac{y_2 - y_1}{x_2 - x_1}(x - x_1). \qquad \text{Two-point form}$$

This is sometimes called the **two-point form** of the equation of a line.

Example 3 An Equation of a Line Passing Through Two Points

Write an equation of the line that passes through the points $(3, 1)$ and $(-3, 4)$.

Solution

Let $(x_1, y_1) = (3, 1)$ and $(x_2, y_2) = (-3, 4)$. The slope of a line passing through these points is

$$m = \frac{y_2 - y_1}{x_2 - x_1} \qquad \text{Formula for slope}$$

$$= \frac{4 - 1}{-3 - 3} \qquad \text{Substitute for } x_1, y_1, x_2, \text{ and } y_2.$$

$$= \frac{3}{-6} \qquad \text{Simplify.}$$

$$= -\frac{1}{2}. \qquad \text{Simplify.}$$

Now, use the point-slope form to find an equation of the line.

$$y - y_1 = m(x - x_1) \qquad \text{Point-slope form}$$

$$y - 1 = -\frac{1}{2}(x - 3) \qquad \text{Substitute 1 for } y_1, 3 \text{ for } x_1, \text{ and } -\frac{1}{2} \text{ for } m.$$

$$y - 1 = -\frac{1}{2}x + \frac{3}{2} \qquad \text{Simplify.}$$

$$y = -\frac{1}{2}x + \frac{5}{2} \qquad \text{Equation of line}$$

The graph of this line is shown in Figure 4.45.

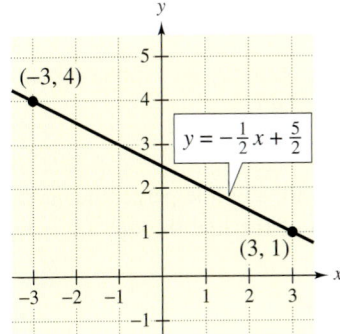

Figure 4.45

In Example 3, it does not matter which of the two points is labeled (x_1, y_1) and which is labeled (x_2, y_2). Try switching these labels to $(x_1, y_1) = (-3, 4)$ and $(x_2, y_2) = (3, 1)$ and reworking the problem to see that you obtain the same equation.

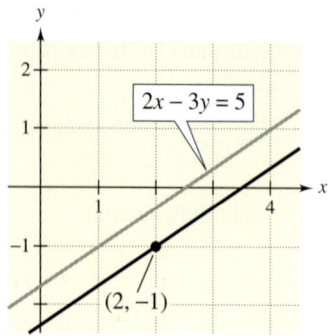

Figure 4.46

Example 4   Equations of Parallel Lines

Write an equation of the line that passes through the point $(2, -1)$ and is parallel to the line

$$2x - 3y = 5$$

as shown in Figure 4.46.

Solution

To begin, write the original equation in slope-intercept form.

$2x - 3y = 5$	Write original equation.
$-3y = -2x + 5$	Subtract $2x$ from each side.
$y = \dfrac{2}{3}x - \dfrac{5}{3}$	Divide each side by -3.

Because the line has a slope of $m = \frac{2}{3}$, it follows that any parallel line must have the same slope. So, an equation of the line through $(2, -1)$, parallel to the original line, is

$y - y_1 = m(x - x_1)$	Point-slope form
$y - (-1) = \dfrac{2}{3}(x - 2)$	Substitute -1 for y_1, 2 for x_1, and $\frac{2}{3}$ for m.
$y + 1 = \dfrac{2}{3}x - \dfrac{4}{3}$	Distributive Property
$y = \dfrac{2}{3}x - \dfrac{7}{3}.$	Equation of parallel line

Example 5   Equations of Perpendicular Lines

Write an equation of the line that passes through the point $(2, -1)$ and is perpendicular to the line $2x - 3y = 5$, as shown in Figure 4.47.

Solution

From Example 4, the original line has a slope of $\frac{2}{3}$. So, any line perpendicular to this line must have a slope of $-\frac{3}{2}$. So, an equation of the line through $(2, -1)$, perpendicular to the original line, is

$y - y_1 = m(x - x_1)$	Point-slope form
$y - (-1) = -\dfrac{3}{2}(x - 2)$	Substitute -1 for y_1, 2 for x_1, and $-\frac{3}{2}$ for m.
$y + 1 = -\dfrac{3}{2}x + 3$	Distributive Property
$y = -\dfrac{3}{2}x + 2.$	Equation of perpendicular line

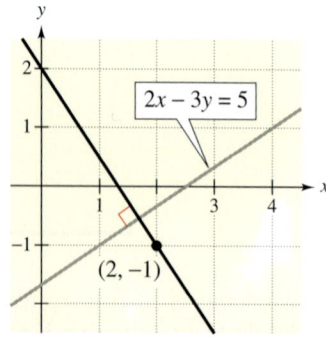

Figure 4.47

② Write the equations of horizontal and vertical lines.

Equations of Horizontal and Vertical Lines

Recall from Section 4.4 that a horizontal line has a slope of zero. From the slope-intercept form of the equation of a line, you can see that a horizontal line has an equation of the form

$$y = (0)x + b \quad \text{or} \quad y = b. \qquad \text{Horizontal line}$$

This is consistent with the fact that each point on a horizontal line through $(0, b)$ has a y-coordinate of b. Similarly, each point on a vertical line through $(a, 0)$ has an x-coordinate of a. Because you know that a vertical line has an undefined slope, you know that it has an equation of the form

$$x = a. \qquad \text{Vertical line}$$

Every line has an equation that can be written in the **general form**

$$ax + by + c = 0 \qquad \text{General form}$$

where a and b are not *both* zero.

Example 6 Writing Equations of Horizontal and Vertical Lines

Write an equation for each line.

a. Vertical line through $(-3, 2)$

b. Line passing through $(-1, 2)$ and $(4, 2)$

c. Line passing through $(0, 2)$ and $(0, -2)$

d. Horizontal line through $(0, -4)$

Solution

a. Because the line is vertical and passes through the point $(-3, 2)$, every point on the line has an x-coordinate of -3. So, the equation of the line is

$$x = -3. \qquad \text{Vertical line}$$

b. Because both points have the same y-coordinate, the line through $(-1, 2)$ and $(4, 2)$ is horizontal. So, its equation is

$$y = 2. \qquad \text{Horizontal line}$$

c. Because both points have the same x-coordinate, the line through $(0, 2)$ and $(0, -2)$ is vertical. So, its equation is

$$x = 0. \qquad \text{Vertical line (y-axis)}$$

d. Because the line is horizontal and passes through the point $(0, -4)$, every point on the line has a y-coordinate of -4. So, the equation of the line is

$$y = -4. \qquad \text{Horizontal line}$$

The graphs of the lines are shown in Figure 4.48.

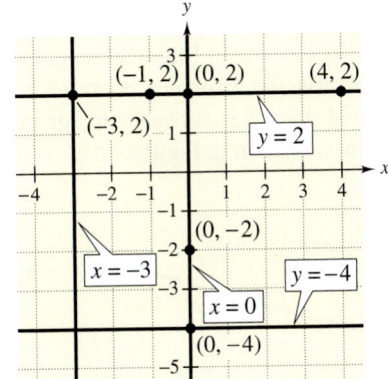

Figure 4.48

In Example 6(c), note that the equation $x = 0$ represents the y-axis. In a similar way, you can show that the equation $y = 0$ represents the x-axis.

③ Use linear models to solve application problems.

Applications

Example 7 Predicting Sales

Harley-Davidson, Inc. had total sales of $2452.9 million in 1999 and $2906.4 million in 2000. Using only this information, write a linear equation that models the sales in terms of the year. Then predict the sales for 2001.   (Source: Harley-Davidson, Inc.)

Solution

Let $t = 9$ represent 1999. Then the two given values are represented by the data points $(9, 2452.9)$ and $(10, 2906.4)$. The slope of the line through these points is

$$m = \frac{2906.4 - 2452.9}{10 - 9}$$

$$= 453.5.$$

Using the point-slope form, you can find the equation that relates the sales y and the year t to be

$y - y_1 = m(t - t_1)$	Point-slope form
$y - 2452.9 = 453.5(t - 9)$	Substitute for y_1, m, and t_1.
$y - 2452.9 = 453.5t - 4081.5$	Distributive Property
$y = 453.5t - 1628.6.$	Write in slope-intercept form.

Using this equation, a prediction of the sales in 2001 $(t = 11)$ is

$$y = 453.5(11) - 1628.6 = \$3359.9 \text{ million.}$$

In this case, the prediction is quite good—the actual sales in 2001 were $3363.4 million. The graph of this equation is shown in Figure 4.49.

The graph shows: $y = 453.5t - 1628.6$, with points $(11, 3359.9)$, $(10, 2906.4)$, and $(9, 2452.9)$. Sales (in millions of dollars) on the y-axis from 2400 to 3600; Year (9 ↔ 1999) on the t-axis at 9, 10, 11.

Figure 4.49

The prediction method illustrated in Example 7 is called **linear extrapolation.** Note in Figure 4.50 that for linear extrapolation, the estimated point lies *to the right* of the given points. When the estimated point lies *between* two given points, the method is called **linear interpolation,** as shown in Figure 4.51.

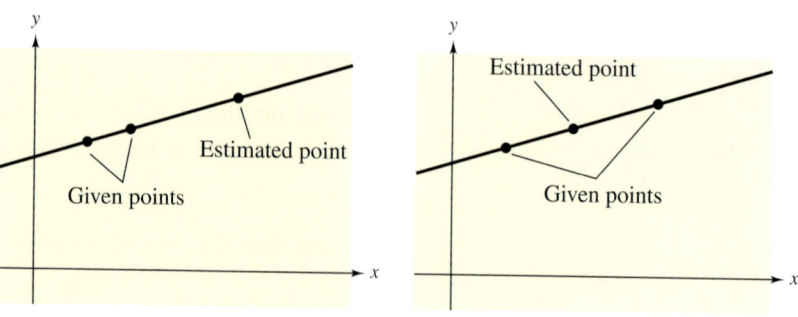

Linear Extrapolation
Figure 4.50

Linear Interpolation
Figure 4.51

In the linear equation $y = mx + b$, you know that m represents the slope of the line. In applications, the slope of a line can often be interpreted as the *rate of change of y with respect to x*. Rates of change should always be described with appropriate units of measure.

Example 8   Using Slope as a Rate of Change

A mountain climber is climbing up a 500-foot cliff. By 1 P.M., the mountain climber has climbed 115 feet up the cliff. By 4 P.M., the climber has reached a height of 280 feet, as shown in Figure 4.52. Find the average rate of change of the climber and use this rate of change to find a linear model that relates the height of the climber to the time.

Solution

Let y represent the height of the climber and let t represent the time. Then the two points that represent the climber's two positions are $(t_1, y_1) = (1, 115)$ and $(t_2, y_2) = (4, 280)$. So, the average rate of change of the climber is

$$\text{Average rate of change} = \frac{y_2 - y_1}{t_2 - t_1} = \frac{280 - 115}{4 - 1} = 55 \text{ feet per hour.}$$

So, an equation that relates the height of the climber to the time is

$$y - y_1 = m(t - t_1) \qquad \text{\color{red}{Point-slope form}}$$

$$y - 115 = 55(t - 1) \qquad \text{\color{red}{Substitute } y_1 = 115, t_1 = 1, \text{ and } m = 55.}$$

$$y = 55t + 60. \qquad \text{\color{red}{Linear model}}$$

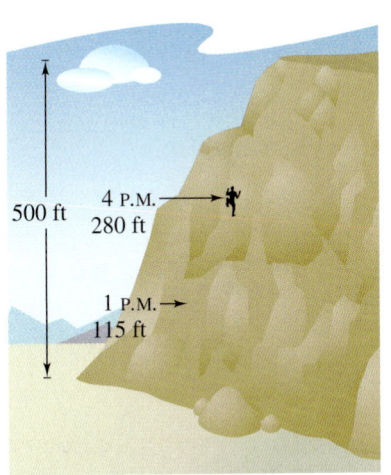

Figure 4.52

You have now studied several formulas that relate to equations of lines. In the summary below, remember that the formulas that deal with slope cannot be applied to vertical lines. For instance, the lines $x = 2$ and $y = 3$ are perpendicular, but they do not follow the "negative reciprocal property" of perpendicular lines because the line $x = 2$ is vertical (and has no slope).

Summary of Equations of Lines

1. Slope of the line through (x_1, y_1) and (x_2, y_2): $m = \dfrac{y_2 - y_1}{x_2 - x_1}$

2. General form of equation of line: $ax + by + c = 0$

3. Equation of vertical line: $x = a$

4. Equation of horizontal line: $y = b$

5. Slope-intercept form of equation of line: $y = mx + b$

6. Point-slope form of equation of line: $y - y_1 = m(x - x_1)$

7. Parallel lines have *equal* slopes: $m_1 = m_2$

8. Perpendicular lines have *negative reciprocal* slopes: $m_1 = -\dfrac{1}{m_2}$

Study Tip

The slope-intercept form of the equation of a line is better suited for *sketching a line.* On the other hand, the point-slope form of the equation of a line is better suited for *creating the equation of a line,* given its slope and a point on the line.

4.5 Exercises

Review *Concepts, Skills, and Problem Solving*

Keep mathematically in shape by doing these exercises *before* the problems of this section.

Properties and Definitions

1. *Writing* ✐   Find the greatest common factor of 180 and 300 and explain how you arrived at your answer.

2. *Writing* ✐   Find the least common multiple of 180 and 300 and explain how you arrived at your answer.

Simplifying Expressions

In Exercises 3–6, simplify the expression.

3. $4(3 - 2x)$

4. $x^2(xy^3)$

5. $3x - 2(x - 5)$

6. $u - [3 + (u - 4)]$

Solving Equations

In Exercises 7–10, solve for y in terms of x.

7. $3x + y = 4$

8. $4 - y + x = 0$

9. $4x - 5y = -2$

10. $3x + 4y - 5 = 0$

Developing Skills

In Exercises 1–14, write an equation of the line that passes through the point and has the specified slope. Sketch the line. See Example 1.

1. $(0, 0), m = -2$

2. $(0, -2), m = 3$

3. $(6, 0), m = \frac{1}{2}$

4. $(0, 10), m = -\frac{1}{4}$

5. $(-2, 1), m = 2$

6. $(3, -5), m = -1$

7. $(-8, -1), m = -\frac{1}{4}$

8. $(12, 4), m = -\frac{2}{3}$

9. $\left(\frac{1}{2}, -3\right), m = 0$

10. $\left(-\frac{5}{4}, 6\right), m = 0$

11. $\left(0, \frac{3}{2}\right), m = \frac{2}{3}$

12. $\left(0, -\frac{5}{2}\right), m = \frac{3}{4}$

13. $(2, 4), m = -0.8$

14. $(6, -3), m = 0.67$

In Exercises 15–26, use the point-slope form to write an equation of the line that passes through the point and has the specified slope. Write the equation in slope-intercept form. See Example 2.

15. $(0, -4), m = 3$

16. $(-7, 0), m = 2$

17. $(-3, 6), m = -2$

18. $(-4, 1), m = -4$

19. $(9, 0), m = -\frac{1}{3}$

20. $(0, 2), m = \frac{3}{5}$

21. $(-10, 4), m = 0$

22. $(-2, -5), m = 0$

23. $(8, 1), m = -\frac{3}{4}$

24. $(1, 10), m = -\frac{1}{3}$

25. $(-2, 1), m = \frac{2}{3}$

26. $(1, -3), m = \frac{1}{2}$

In Exercises 27–38, determine the slope of the line. If it is not possible, explain why.

27. $y = \frac{3}{8}x - 4$

28. $y = -\frac{3}{5}x - 2$

29. $y - 2 = 5(x + 3)$

30. $y + 3 = -2(x - 6)$

31. $y + \frac{5}{6} = \frac{2}{3}(x + 4)$

32. $y - \frac{1}{4} = \frac{5}{8}\left(x - \frac{13}{5}\right)$

33. $y + 9 = 0$

34. $y - 6 = 0$

35. $x - 12 = 0$

36. $x + 5 = 0$

37. $3x - 2y + 10 = 0$

38. $5x + 4y - 8 = 0$

In Exercises 39–42, write the slope-intercept form of the line that has the specified y-intercept and slope.

39.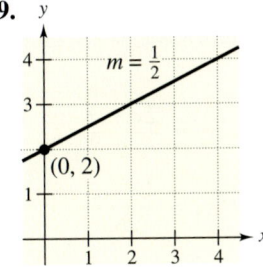

40.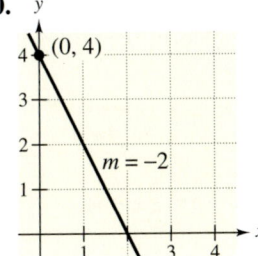

41.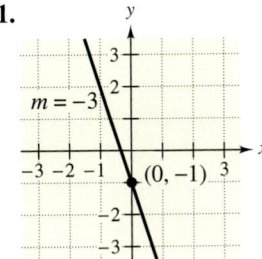

42.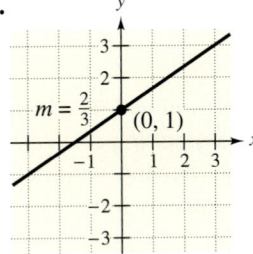

In Exercises 43–46, write the point-slope form of the equation of the line.

43.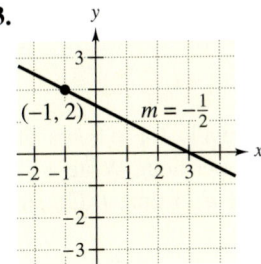

44.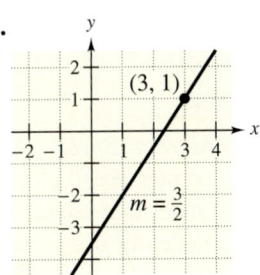

45.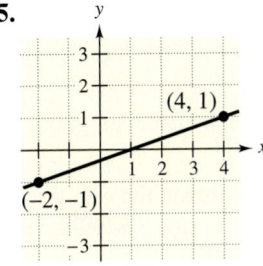

46.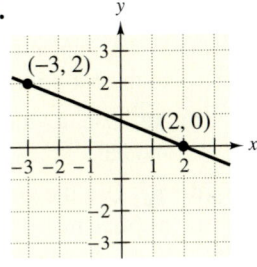

In Exercises 47–58, write an equation of the line that passes through the points. When possible, write the equation in slope-intercept form. Sketch the line. See Example 3.

47. $(0, 0)$, $(4, -4)$

48. $(0, 0)$, $(-2, 4)$

49. $(0, 0)$, $(2, -4)$

50. $(6, -1)$, $(3, 3)$

51. $(-2, 3)$, $(6, -5)$

52. $(-4, 6)$, $(-2, 3)$

53. $(-6, 2)$, $(3, 5)$

54. $(-9, 7)$, $(-4, 4)$

55. $(5, -1)$, $(3, 2)$

56. $(0, 3)$, $(5, 3)$

57. $\left(\frac{5}{2}, -1\right)$, $\left(\frac{9}{2}, 7\right)$

58. $\left(4, \frac{5}{3}\right)$, $\left(-1, \frac{2}{3}\right)$

In Exercises 59–72, write an equation of the line passing through the points. Write the equation in general form.

59. $(0, 3)$, $(3, 0)$

60. $(0, -2)$, $(-2, 0)$

61. $(5, -1)$, $(-5, 5)$

62. $(4, 3)$, $(-4, 5)$

63. $(5, 4)$, $(1, -4)$

64. $(-5, 7)$, $(-2, 1)$

65. $(5, -1)$, $(7, -4)$

66. $(3, 5)$, $(1, 6)$

67. $(-3, 8)$, $(2, 5)$

68. $(9, -9)$, $(7, -5)$

69. $\left(2, \frac{1}{2}\right)$, $\left(\frac{1}{2}, \frac{5}{2}\right)$

70. $\left(\frac{1}{4}, 1\right)$, $\left(-\frac{3}{4}, -\frac{2}{3}\right)$

71. $(1, 0.6)$, $(2, -0.6)$

72. $(-8, 0.6)$, $(2, -2.4)$

In Exercises 73–82, write equations of the lines through the point (a) parallel and (b) perpendicular to the given line. See Examples 4 and 5.

73. $(2, 1)$

$x - y = 3$

74. $(-3, 2)$

$x + y = 7$

75. $(-12, 4)$

$3x + 4y = 7$

76. $(15, -2)$

$5x + 3y = 0$

77. $(1, 3)$
$2x + y = 0$

78. $(5, -2)$
$x + 5y = 3$

79. $(-1, 0)$
$y + 3 = 0$

80. $(2, 5)$
$x - 4 = 0$

81. $(4, -1)$
$3y - 2x = 7$

82. $(-6, 5)$
$4x - 5y = 2$

In Exercises 83–90, write an equation of the line. See Example 6.

83. Vertical line through $(-2, 4)$

84. Horizontal line through $(7, 3)$

85. Horizontal line through $\left(\frac{1}{2}, \frac{2}{3}\right)$

86. Vertical line through $\left(\frac{1}{4}, 0\right)$

87. Line passing through $(4, 1)$ and $(4, 8)$

88. Line passing through $(-1, 5)$ and $(6, 5)$

89. Line passing through $(1, -8)$ and $(7, -8)$

90. Line passing through $(3, 0)$ and $(3, 5)$

Graphical Exploration   In Exercises 91–94, use a graphing calculator to graph the lines in the same viewing window. Use the square setting. Are the lines parallel, perpendicular, or neither?

91. $y_1 = -0.4x + 3$

$y_2 = \frac{5}{2}x - 1$

92. $y_1 = \frac{2x - 3}{3}$

$y_2 = \frac{4x + 3}{6}$

93. $y_1 = 0.4x + 1$

$y_2 = x + 2.5$

94. $y_1 = \frac{3}{4}x - 5$

$y_2 = -\frac{3}{4}x + 2$

Graphical Exploration   In Exercises 95 and 96, use a graphing calculator to graph the equations in the same viewing window. Use the square setting. What can you conclude?

95. $y_1 = \frac{1}{3}x + 2$

$y_2 = -3x + 2$

96. $y_1 = 4x + 2$

$y_2 = -\frac{1}{4}x + 2$

Solving Problems

97. *Wages*   A sales representative receives a monthly salary of $2000 plus a commission of 2% of the total monthly sales. Write a linear model that relates total monthly wages W to sales S.

98. *Wages*   A sales representative receives a salary of $2300 per month plus a commission of 3% of the total monthly sales. Write a linear model that relates wages W to sales S.

99. *Reimbursed Expenses*   A sales representative is reimbursed $225 per day for lodging and meals plus $0.35 per mile driven. Write a linear model that relates the daily cost C to the number of miles driven x.

100. *Reimbursed Expenses*   A sales representative is reimbursed $250 per day for lodging and meals plus $0.30 per mile driven. Write a linear model that relates the daily cost C to the number of miles driven x.

101. *Average Speed*   A car travels for t hours at an average speed of 50 miles per hour. Write a linear model that relates distance d to time t. Graph the model for $0 \le t \le 5$.

102. *Discount*   A department store is offering a 20% discount on all items in its inventory.

(a) Write a linear model that relates the sale price S to the list price L.

(b) Use a graphing calculator to graph the model.

(c) Use the graph to estimate the sale price of a coffee maker whose list price is $49.98. Verify your estimate algebraically.

103. *Depreciation*   A school district purchases a high-volume printer, copier, and scanner for $25,000. After 1 year, its depreciated value is $22,700. The depreciation is linear. See Example 7.

(a) Write a linear model that relates the value V of the equipment to the time t in years.

(b) Use the model to estimate the value of the equipment after 3 years.

104. *Depreciation*   A sub shop purchases a used pizza oven for $875. After 1 year, its depreciated value is $790. The depreciation is linear.

(a) Write a linear model that relates the value V of the oven to the time t in years.

(b) Use the model to estimate the value of the oven after 5 years.

105. *Rental Demand*   A real estate office handles an apartment complex with 50 units. When the rent per unit is $580 per month, all 50 units are occupied. However, when the rent is $625 per month, the average number of occupied units drops to 47. Assume that the relationship between the monthly rent p and the demand x is linear.

(a) Represent the given information as two ordered pairs of the form (x, p). Plot these ordered pairs.

(b) Write a linear model that relates the monthly rent p to the demand x. Graph the model and describe the relationship between the rent and the demand.

(c) *Linear Extrapolation*   Use the model in part (b) to predict the number of units occupied if the rent is raised to $655.

(d) *Linear Interpolation*   Use the model in part (b) to estimate the number of units occupied if the rent is $595.

106. *Soft Drink Demand*   When soft drinks sold for $0.80 per can at football games, approximately 6000 cans were sold. When the price was raised to $1.00 per can, the demand dropped to 4000. Assume that the relationship between the price p and the demand x is linear.

(a) Represent the given information as two ordered pairs of the form (x, p). Plot these ordered pairs.

(b) Write a linear model that relates the price p to the demand x. Graph the model and describe the relationship between the price and the demand.

(c) *Linear Extrapolation*   Use the model in part (b) to predict the number of soft drinks sold if the price is raised to $1.10.

(d) *Linear Interpolation*   Use the model in part (b) to estimate the number of soft drinks sold if the price is $0.90.

107. *Graphical Interpretation*   Match each situation labeled (a), (b), (c), and (d) with one of the graphs labeled (e), (f), (g), and (h). Then determine the slope of each line and interpret the slope in the context of the real-life situation.

(a) A friend is paying you $10 per week to repay a $100 loan.

(b) An employee is paid $12.50 per hour plus $1.50 for each unit produced per hour.

(c) A sales representative receives $40 per day for food plus $0.32 for each mile traveled.

(d) A television purchased for $600 depreciates $100 per year.

(e) 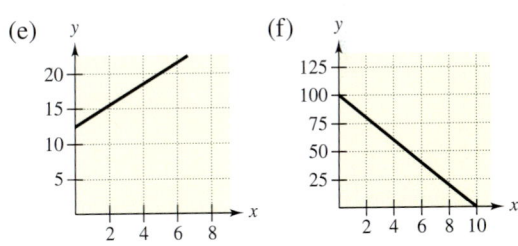   (f)

(g) 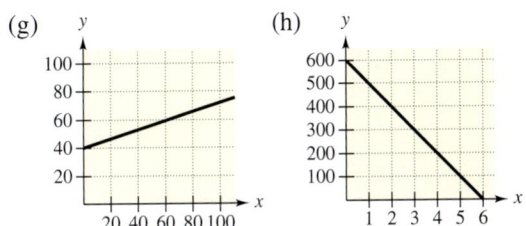   (h)

108. *Rate of Change* You are given the dollar value of a product in 2005 and the rate at which the value is expected to change during the next 5 years. Use this information to write a linear equation that gives the dollar value V of the product in terms of the year t. (Let $t = 5$ represent 2005.)

2005 Value	Rate
(a) $2540	$125 increase per year
(b) $156	$4.50 increase per year

2005 Value	Rate
(c) $20,400	$2000 decrease per year
(d) $45,000	$2300 decrease per year
(e) $31	$0.75 increase per year
(f) $4500	$800 decrease per year

Explaining Concepts

109. ⊘ Answer parts (e)–(h) of Motivating the Chapter on page 214.

110. *Writing*✍ Can any pair of points on a line be used to calculate the slope of the line? Explain.

111. *Writing*✍ Can the equation of a vertical line be written in slope-intercept form? Explain.

112. In the equation $y = mx + b$, what do m and b represent?

113. In the equation $y - y_1 = m(x - x_1)$, what do x_1 and y_1 represent?

114. *Writing*✍ Explain how to find analytically the x-intercept of the line given by $y = mx + b$.

115. *Think About It* Find the slope of the line for the equation $5x + 7y - 21 = 0$. Use the same process to find a formula for the slope of the line $ax + by + c = 0$ where $b \neq 0$.

116. What is implied about the graphs of the lines $a_1x + b_1y + c_1 = 0$ and $a_2x + b_2y + c_2 = 0$ if $\dfrac{a_1}{b_1} = \dfrac{a_2}{b_2}$?

117. *Research Project* Use a newspaper or weekly news magazine to find an example of data that is *increasing* linearly with time. Find a linear model that relates the data to time. Repeat the project for data that is decreasing.

4.6 Graphs of Linear Inequalities

Rachel Epstein/PhotoEdit, Inc.

What You Should Learn

1. Determine whether an ordered pair is a solution of a linear inequality in two variables.
2. Sketch graphs of linear inequalities in two variables.
3. Use linear inequalities to model and solve real-life problems.

Why You Should Learn It

Linear inequalities can be used to model and solve real-life problems. For instance, in Exercise 70 on page 283, you will use a linear inequality to analyze the components of dietary supplements.

1. **Determine whether an ordered pair is a solution of a linear inequality in two variables.**

Linear Inequalities in Two Variables

A **linear inequality in two variables,** x and y, is an inequality that can be written in one of the forms below (where a and b are not both zero).

$$ax + by < c, \quad ax + by > c, \quad ax + by \le c, \quad ax + by \ge c$$

Some examples include: $x - y > 2$, $3x - 2y \le 6$, $x \ge 5$, and $y < -1$. An ordered pair (x_1, y_1) is a **solution** of a linear inequality in x and y if the inequality is true when x_1 and y_1 are substituted for x and y, respectively. For instance, the ordered pair $(3, 2)$ is a solution of the inequality $x - y > 0$ because $3 - 2 > 0$ is a true statement.

Example 1   Verifying Solutions of Linear Inequalities

Determine whether each point is a solution of $3x - y \ge -1$.

a. $(0, 0)$ **b.** $(1, 4)$ **c.** $(-1, 2)$

Solution

a. $3x - y \ge -1$ Write original inequality.

   $3(0) - 0 \overset{?}{\ge} -1$ Substitute 0 for x and 0 for y.

   $0 \ge -1$ Inequality is satisfied. ✓

Because the inequality is satisfied, the point $(0, 0)$ *is* a solution.

b. $3x - y \ge -1$ Write original inequality.

   $3(1) - 4 \overset{?}{\ge} -1$ Substitute 1 for x and 4 for y.

   $-1 \ge -1$ Inequality is satisfied. ✓

Because the inequality is satisfied, the point $(1, 4)$ *is* a solution.

c. $3x - y \ge -1$ Write original inequality.

   $3(-1) - 2 \overset{?}{\ge} -1$ Substitute -1 for x and 2 for y.

   $-5 \not\ge -1$ Inequality is not satisfied. ✗

Because the inequality is not satisfied, the point $(-1, 2)$ *is not* a solution.

2 Sketch graphs of linear inequalities in two variables.

The Graph of a Linear Inequality in Two Variables

The **graph** of an inequality is the collection of all solution points of the inequality. To sketch the graph of a linear inequality such as

$$3x - 2y < 6 \qquad \text{Original linear inequality}$$

begin by sketching the graph of the *corresponding linear equation*

$$3x - 2y = 6. \qquad \text{Corresponding linear equation}$$

Use *dashed* lines for the inequalities < and > and *solid* lines for the inequalities ≤ and ≥. The graph of the equation separates the plane into two regions, called **half-planes.** In each half-plane, one of the following *must* be true.

1. All points in the half-plane are solutions of the inequality.

2. No point in the half-plane is a solution of the inequality.

So, you can determine whether the points in an entire half-plane satisfy the inequality by simply testing *one* point in the region. This graphing procedure is summarized as follows.

Sketching the Graph of a Linear Inequality in Two Variables

1. Replace the inequality sign by an equal sign and sketch the graph of the resulting equation. (Use a dashed line for < or > and a solid line for ≤ or ≥.)

2. Test one point in each of the half-planes formed by the graph in Step 1. If the point satisfies the inequality, then shade the entire half-plane to denote that every point in the region satisfies the inequality.

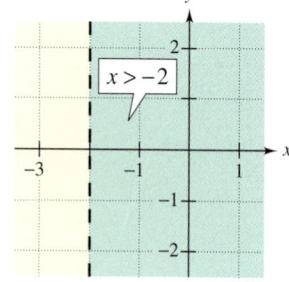

Figure 4.53

Example 2   Sketching the Graph of a Linear Inequality

Sketch the graph of each linear inequality.

a. $x > -2$

b. $y \leq 3$

Solution

a. The graph of the corresponding equation $x = -2$ is a vertical line. The points (x, y) that satisfy the inequality $x > -2$ are those lying to the right of this line, as shown in Figure 4.53.

b. The graph of the corresponding equation $y = 3$ is a horizontal line. The points (x, y) that satisfy the inequality $y \leq 3$ are those lying below (or on) this line, as shown in Figure 4.54.

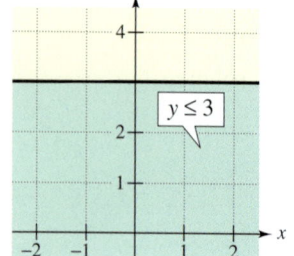

Figure 4.54

Notice that a dashed line is used for the graph of $x > -2$ and a solid line is used for the graph of $y \leq 3$.

Example 3   Sketching the Graph of a Linear Inequality

Sketch the graph of the linear inequality

$$x - y < 2.$$

Solution

The graph of the corresponding equation

$$x - y = 2 \qquad \text{Write corresponding linear equation.}$$

is a line, as shown in Figure 4.55. Because the origin $(0, 0)$ does not lie on the line, use it as the test point.

$$x - y < 2 \qquad \text{Write original inequality.}$$

$$0 - 0 \overset{?}{<} 2 \qquad \text{Substitute 0 for } x \text{ and 0 for } y.$$

$$0 < 2 \qquad \text{Inequality is satisfied. } \checkmark$$

Because $(0, 0)$ satisfies the inequality, the graph consists of the half-plane lying above the line. Try checking a point below the line. Regardless of the point you choose, you will see that it does not satisfy the inequality.

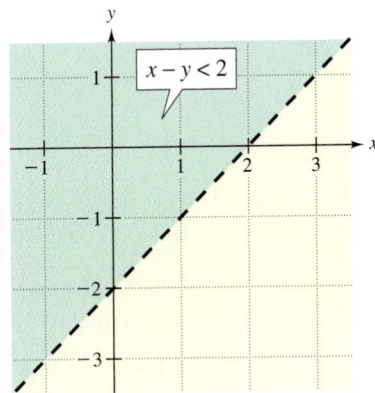

Figure 4.55

Technology: Tip

Many graphing calculators are capable of graphing linear inequalities. Consult the user's guide of your graphing calculator for specific instructions.

   The graph of $y \leq -x + 2$ is shown at the right.

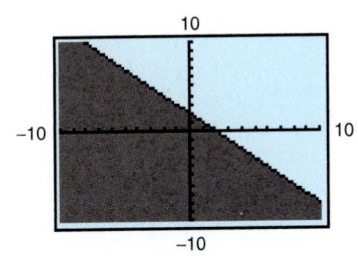

For a linear inequality in two variables, you can sometimes simplify the graphing procedure by writing the inequality in *slope-intercept* form. For instance, by writing $x - y < 2$ in the form $y > x - 2$, you can see that the solution points lie *above* the line $y = x - 2$, as shown in Figure 4.55. Similarly, by writing the inequality $3x - 2y > 5$ in the form

$$y < \frac{3}{2}x - \frac{5}{2}$$

you can see that the solutions lie *below* the line $y = \frac{3}{2}x - \frac{5}{2}$, as shown in Figure 4.56.

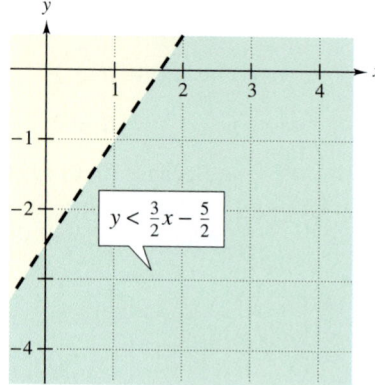

Figure 4.56

Example 4   Sketching the Graph of a Linear Inequality

Use the slope-intercept form of a linear equation as an aid in sketching the graph of the inequality $5x + 4y \le 12$.

Solution

To begin, rewrite the inequality in slope-intercept form.

$5x + 4y \le 12$ Write original inequality.

$4y \le -5x + 12$ Subtract $5x$ from each side.

$y \le -\dfrac{5}{4}x + 3$ Write in slope-intercept form.

From this form, you can conclude that the solution is the half-plane lying *on* or *below* the line $y = -\frac{5}{4}x + 3$. The graph is shown in Figure 4.57. You can verify this by testing the solution point $(0, 0)$.

$5x + 4y \le 12$ Write original inequality.

$5(0) + 4(0) \overset{?}{\le} 12$ Substitute 0 for x and 0 for y.

$0 \le 12$ Inequality is satisfied. ✔

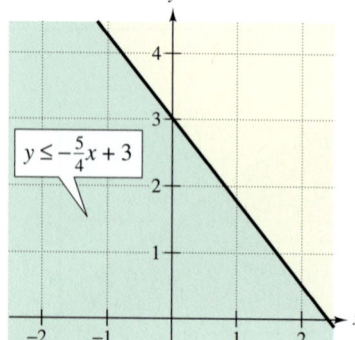

Figure 4.57

③ Use linear inequalities to model and solve real-life problems.

Application

Example 5   Working to Meet a Budget

Your budget requires you to earn *at least* $160 per week. You work two part-time jobs. One is at a fast-food restaurant, which pays $6 per hour, and the other is tutoring for $8 per hour. Let x represent the number of hours worked at the restaurant and let y represent the number of hours tutoring.

a. Write an inequality that represents the number of hours worked at each job in order to meet your budget requirements.

b. Graph the inequality and identify at least two ordered pairs (x, y) that identify the number of hours you must work at each job in order to meet your budget requirements.

Solution

a. To write the inequality, use the problem-solving method.

Verbal Model:	Hourly pay at fast-food restaurant	·	Number of hours at fast-food restaurant	+	Hourly pay tutoring	·	Number of hours tutoring	≥	Earnings in a week

Labels: Hourly pay at fast-food restaurant = 6   (dollars per hour)
Number of hours at fast-food restaurant = x (hours)
Hourly pay tutoring = 8   (dollars per hour)
Number of hours tutoring = y (hours)
Earnings in a week = 160   (dollars)

Algebraic Inequality: $6x + 8y \geq 160$

b. To sketch the graph, rewrite the inequality in slope-intercept form.

$$6x + 8y \geq 160 \qquad \text{Write original inequality.}$$

$$8y \geq -6x + 160 \qquad \text{Subtract } 6x \text{ from each side.}$$

$$y \geq -\frac{3}{4}x + 20 \qquad \text{Divide each side by 8.}$$

Graph the corresponding equation

$$y = -\frac{3}{4}x + 20$$

and shade the half-plane lying above the line, as shown in Figure 4.58. From the graph, you can see that two solutions that will yield the desired weekly earnings of at least $160 are $(8, 14)$ and $(15, 10)$. In other words, you could work 8 hours at the restaurant and 14 hours as a tutor, or 15 hours at the restaurant and 10 hours as a tutor, to meet your budget requirements. There are many other solutions.

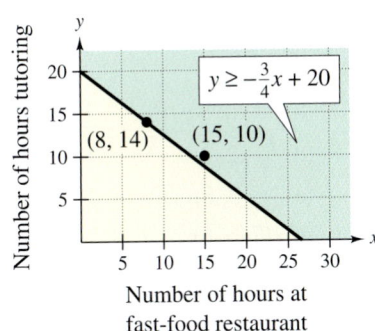

Number of hours at fast-food restaurant

Figure 4.58

4.6 Exercises

Review   *Concepts, Skills, and Problem Solving*

Keep mathematically in shape by doing these exercises *before* the problems of this section.

Properties and Definitions

In Exercises 1–4, complete the property of inequalities by inserting the correct inequality symbol. (Let a, b, and c be real numbers, variables, or algebraic expressions.)

1. If $a < b$, then $a + 5$ ___ $b + 5$.
2. If $a < b$, then $2a$ ___ $2b$.
3. If $a < b$, then $-3a$ ___ $-3b$.
4. If $a < b$ and $b < c$, then a ___ c.

Solving Inequalities

In Exercises 5–10, solve the inequality and sketch the solution on the real number line.

5. $x + 3 > 0$ 6. $2 - x \geq 0$

7. $2t - 11 \leq 5$
8. $\frac{3}{2}y + 8 < 20$
9. $2(x - 5) > 13$
10. $-4(x + 7) \geq -36$

Problem Solving

11. *Sales Commission*   A sales representative receives a commission of 4.5% of the total monthly sales. Determine the sales of a representative who earned \$544.50 as a sales commission.

12. *Work Rate*   One person can complete a typing project in 3 hours, and another can complete the same project in 4 hours. If they both work on the project, in how many hours can it be completed?

Developing Skills

In Exercises 1–8, determine whether the points are solutions of the inequality. See Example 1.

Inequality	*Points*
1. $x + 4y > 10$	(a) $(0, 0)$
	(b) $(3, 2)$
	(c) $(1, 2)$
	(d) $(-2, 4)$
2. $2x + 3y > 9$	(a) $(0, 0)$
	(b) $(1, 1)$
	(c) $(2, 2)$
	(d) $(-2, 5)$
3. $-3x + 5y \leq 12$	(a) $(1, 2)$
	(b) $(2, -3)$
	(c) $(1, 3)$
	(d) $(2, 8)$
4. $5x + 3y < 100$	(a) $(25, 10)$
	(b) $(6, 10)$
	(c) $(0, -12)$
	(d) $(4, 5)$

Inequality	*Points*
5. $3x - 2y < 2$	(a) $(1, 3)$
	(b) $(2, 0)$
	(c) $(0, 0)$
	(d) $(3, -5)$
6. $y - 2x > 5$	(a) $(4, 13)$
	(b) $(8, 1)$
	(c) $(0, 7)$
	(d) $(1, -3)$
7. $5x + 4y \geq 6$	(a) $(-2, 4)$
	(b) $(5, 5)$
	(c) $(7, 0)$
	(d) $(-2, 5)$
8. $5y + 8x \leq 14$	(a) $(-3, 8)$
	(b) $(7, -6)$
	(c) $(1, 1)$
	(d) $(3, 0)$

In Exercises 9–12, state whether the boundary of the graph of the inequality should be dashed or solid.

9. $2x + 3y < 6$

10. $2x + 3y \leq 6$

11. $2x + 3y \geq 6$

12. $2x + 3y > 6$

In Exercises 13–16, match the inequality with its graph. [The graphs are labeled (a), (b), (c), and (d).]

(a)

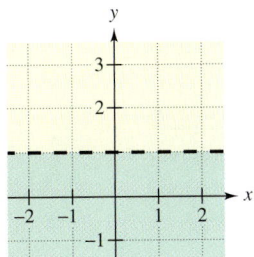

(b)

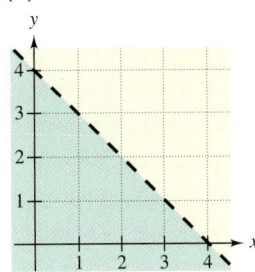

(c)

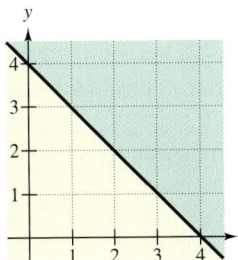

(d)

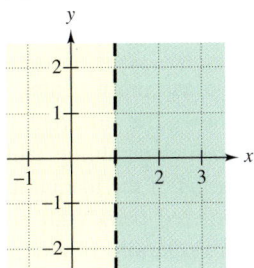

13. $x + y < 4$

14. $x + y \geq 4$

15. $x > 1$

16. $y < 1$

In Exercises 17–20, match the inequality with its graph. [The graphs are labeled (a), (b), (c), and (d).]

(a)

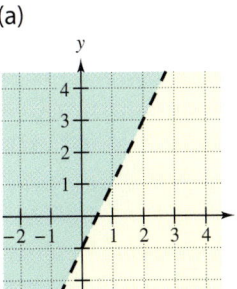

(b)

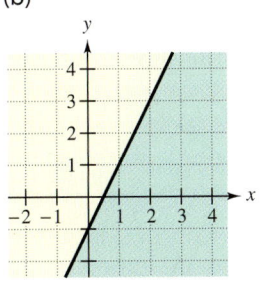

(c)

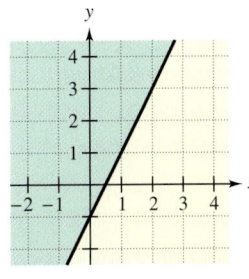

(d)

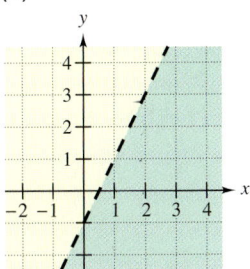

17. $2x - y \leq 1$

18. $2x - y < 1$

19. $2x - y \geq 1$

20. $2x - y > 1$

In Exercises 21–50, sketch the graph of the linear inequality. See Examples 2–4.

21. $y \geq 3$ **22.** $x \leq 0$

23. $x > -4$ **24.** $y < -2$

25. $y < 3x$ **26.** $y > 5x$

27. $x - y < 0$ **28.** $x + y > 0$

29. $y \leq 2x - 1$ **30.** $y \geq -x + 3$

31. $y \leq 2 - x$ **32.** $y \geq 2x + 1$

33. $y > 2 - x$ **34.** $y < -x + 3$

35. $y > -2x + 10$ **36.** $y < 3x + 1$

37. $y \geq \frac{2}{3}x + \frac{1}{3}$

38. $y \leq -\frac{3}{4}x + 2$

39. $-3x + 2y - 6 < 0$

40. $x - 2y + 6 \leq 0$

41. $2x + y - 3 \geq 3$

42. $x + 4y + 2 \geq 2$

43. $5x + 2y < 5$

44. $5x + 2y > 5$

45. $x \geq 3y - 5$

46. $x > -2y + 10$

47. $y - 3 < \frac{1}{2}(x - 4)$

48. $y + 1 < -2(x - 3)$

49. $\dfrac{x}{3} + \dfrac{y}{4} < 1$

50. $\dfrac{x}{-2} + \dfrac{y}{2} > 1$

In Exercises 51–58, use a graphing calculator to graph the linear inequality.

51. $y \geq 2x - 1$

52. $y \leq 4 - 0.5x$

53. $y \leq -2x + 4$

54. $y \geq x - 3$

55. $y \geq \frac{1}{2}x + 2$

56. $y \leq -\frac{2}{3}x + 6$

57. $6x + 10y - 15 \leq 0$

58. $3x - 2y + 4 \geq 0$

In Exercises 59–64, write an inequality for the shaded region shown in the figure.

59.

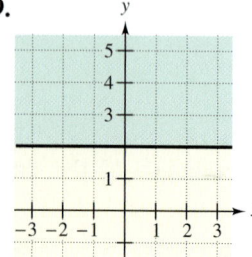

60.

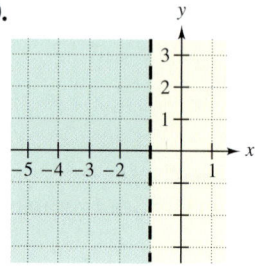

61.

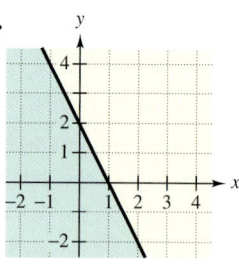

62.

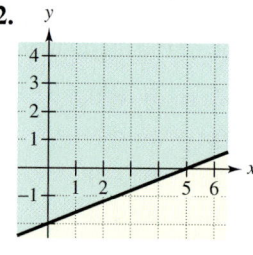

63.

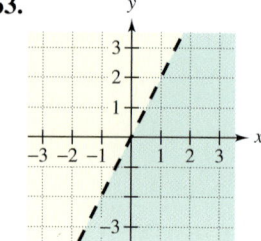

64.

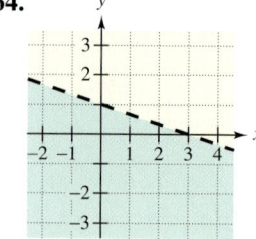

Solving Problems

65. *Part-Time Jobs*   You work two part-time jobs. One is at a grocery store, which pays $9 per hour, and the other is mowing lawns, which pays $6 per hour. Between the two jobs, you want to earn at least $150 a week. Write a linear inequality that shows the different numbers of hours you can work at each job, and sketch the graph of the inequality. From the graph, find several ordered pairs with positive integer coordinates that are solutions of the inequality.

66. *Money*   A cash register must have at least $25 in change consisting of d dimes and q quarters. Write a linear inequality that shows the different numbers of coins that can be in the cash register, and sketch the graph of the inequality. From the graph, find several ordered pairs with positive integer coordinates that are solutions of the inequality.

67. *Manufacturing*   Each table produced by a furniture company requires 1 hour in the assembly center. The matching chair requires $1\frac{1}{2}$ hours in the assembly center. A total of 12 hours per day is available in

the assembly center. Write a linear inequality that shows the different numbers of hours that can be spent assembling tables and chairs, and sketch the graph of the inequality. From the graph, find several ordered pairs with positive integer coordinates that are solutions of the inequality.

68. *Inventory*   A store sells two models of computers. The costs to the store of the two models are $2000 and $3000, and the owner of the store does not want more than $30,000 invested in the inventory for these two models. Write a linear inequality that represents the different numbers of each model that can be held in inventory, and sketch the graph of the inequality. From the graph, find several ordered pairs with positive integer coordinates that are solutions of the inequality.

69. *Sports*   Your hockey team needs at least 60 points for the season in order to advance to the playoffs. Your team finishes with w wins, each worth 2 points, and t ties, each worth 1 point. Write a linear inequality that shows the different numbers of points your team can score to advance to the playoffs, and sketch the graph of the inequality. From the graph, find several ordered pairs with positive integer coordinates that are solutions of the inequality.

70. *Nutrition*   A dietitian is asked to design a special dietary supplement using two different foods. Each ounce of food X contains 20 units of calcium and each ounce of food Y contains 10 units of calcium. The minimum daily requirement in the diet is 300 units of calcium. Write a linear inequality that shows the different numbers of units of food X and food Y required, and sketch the graph of the inequality. From the graph, find several ordered pairs with positive integer coordinates that are solutions of the inequality.

Explaining Concepts

71. ⊘   Answer part (i) of Motivating the Chapter on page 214.

72. List the four forms of a linear inequality in variables x and y.

73. What is meant by saying that (x_1, y_1) is a solution of a linear inequality in x and y?

74. *Writing*   Explain the difference between graphs that have dashed lines and those that have solid lines.

75. *Writing*   After graphing the boundary, explain how you determine which half-plane is the graph of a linear inequality.

76. *Writing*   Explain the difference between graphing the solution to the inequality $x \geq 1$ (a) on the real number line and (b) on a rectangular coordinate system.

77. Write the inequality whose graph consists of all points above the x-axis.

78. *Writing*   Does $2x < 2y$ have the same graph as $y > x$? Explain.

79. Write an inequality whose graph has no points in the first quadrant.

What Did You Learn?

Key Terms

rectangular coordinate
  system, *p. 216*
ordered pair, *p. 216*
x-coordinate, *p. 216*
y-coordinate, *p. 216*
solution point, *p. 219*
x-intercept, *p. 232*

y-intercept, *p. 232*
relation, *p. 238*
domain, *p. 238*
range, *p. 238*
function, *p. 239*
independent variable, *p. 240*
dependent variable, *p. 240*

slope, *p. 249*
slope-intercept form, *p. 254*
parallel lines, *p. 256*
perpendicular lines, *p. 257*
point-slope form, *p. 264*
half-plane, *p. 276*

Key Concepts

4.1 ◐ Rectangular coordinate system

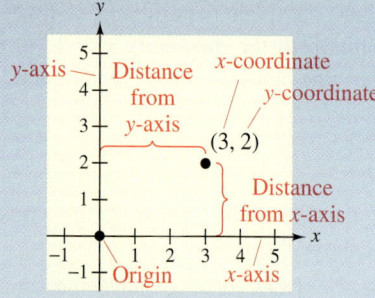

4.2 ◐ Point-plotting method of sketching a graph

1. If possible, rewrite the equation by isolating one of the variables.

2. Make a table of values showing several solution points.

3. Plot these points on a rectangular coordinate system.

4. Connect the points with a smooth curve or line.

4.2 ◐ Finding *x*- and *y*-intercepts

To find the *x*-intercept(s), let $y = 0$ and solve the equation for *x*. To find the *y*-intercept(s), let $x = 0$ and solve the equation for *y*.

4.3 ◐ Vertical Line Test

A set of points on a rectangular coordinate system is the graph of *y* as a function of *x* if and only if no vertical line intersects the graph at more than one point.

4.4 ◐ Slope of a line

The slope *m* of a nonvertical line passing through the points (x_1, y_1) and (x_2, y_2) is

$$m = \frac{y_2 - y_1}{x_2 - x_1} = \frac{\text{Change in } y}{\text{Change in } x} = \frac{\text{Rise}}{\text{Run}}, \text{ where } x_1 \neq x_2.$$

1. If $m > 0$, the line rises from left to right.
2. If $m < 0$, the line falls from left to right.
3. If $m = 0$, the line is horizontal.
4. If *m* is undefined $(x_1 = x_2)$, the line is vertical.

4.5 ◐ Summary of equations of lines

1. Slope of the line through (x_1, y_1) and (x_2, y_2):

$$m = \frac{y_2 - y_1}{x_2 - x_1}$$

2. General form of equation of line: $ax + by + c = 0$

3. Equation of vertical line: $x = a$

4. Equation of horizontal line: $y = b$

5. Slope-intercept form of equation of line:

$$y = mx + b$$

6. Point-slope form of equation of line:

$$y - y_1 = m(x - x_1)$$

7. Parallel lines have equal slopes: $m_1 = m_2$

8. Perpendicular lines have negative reciprocal slopes:

$$m_1 = -\frac{1}{m_2}$$

4.6 ◐ Sketching the graph of a linear inequality in two variables

1. Replace the inequality sign by an equal sign and sketch the graph of the resulting equation. (Use a dashed line for $<$ or $>$ and a solid line for $\leq$ or $\geq$.)

2. Test one point in each of the half-planes formed by the graph in Step 1. If the point satisfies the inequality, then shade the entire half-plane to denote that every point in the region satisfies the inequality.

Review Exercises

4.1 Ordered Pairs and Graphs

1 Plot and find the coordinates of a point on a rectangular coordinate system.

In Exercises 1–4, plot the points on a rectangular coordinate system.

1. $(-1, 6), (4, -3), (-2, 2), (3, 5)$

2. $(0, -1), (-4, 2), (5, 1), (3, -4)$

3. $(-2, 0), \left(\frac{3}{2}, 4\right), (-1, -3)$

4. $\left(3, -\frac{5}{2}\right), \left(-5, 2\frac{3}{4}\right), (4, 6)$

In Exercises 5 and 6, determine the coordinates of the points.

5.

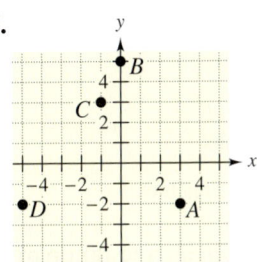

6.

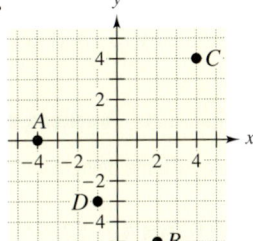

In Exercises 7–14, determine the quadrant(s) in which the point is located or the axis on which the point is located without plotting it.

7. $(-5, 3)$ **8.** $(4, -6)$

9. $(4, 0)$ **10.** $(0, -3)$

11. $(x, 5), \ x < 0$ **12.** $(-3, y), \ y > 0$

13. $(-6, y)$, y is a real number.

14. $(x, -1)$, x is a real number.

2 Construct a table of values for equations and determine whether ordered pairs are solutions of equations.

In Exercises 15 and 16, complete the table of values. Then plot the solution points on a rectangular coordinate system.

15.

x	-1	0	1	2
$y = 4x - 1$				

16.

x	-1	0	1	2
$y = -\frac{1}{2}x - 1$				

In Exercises 17–20, solve the equation for y.

17. $3x + 4y = 12$ **18.** $2x + 3y = 6$

19. $x - 2y = 8$ **20.** $-x - 3y = 9$

In Exercises 21–24, determine whether the ordered pairs are solutions of the equation.

21. $x - 3y = 4$
  (a) $(1, -1)$ (b) $(0, 0)$
  (c) $(2, 1)$ (d) $(5, -2)$

22. $y - 2x = -1$
  (a) $(3, 7)$ (b) $(0, -1)$
  (c) $(-2, -5)$ (d) $(-1, 0)$

23. $y = \frac{2}{3}x + 3$
  (a) $(3, 5)$ (b) $(-3, 1)$
  (c) $(-6, 0)$ (d) $(0, 3)$

24. $y = \frac{1}{4}x + 2$
  (a) $(-4, 1)$ (b) $(-8, 0)$
  (c) $(12, 5)$ (d) $(0, 2)$

3 Use the verbal problem-solving method to plot points on a rectangular coordinate system.

25. *Organizing Data*   The data from a study measuring the relationship between the wattage x of a standard 120-volt light bulb and the energy rate y (in lumens) is shown in the table.

x	25	40	60	100	150	200
y	235	495	840	1675	2650	3675

(a) Plot the data shown in the table.

(b) Use the graph to describe the relationship between the wattage and energy rate.

26. *Organizing Data* The table shows the average salaries (in thousands of dollars) for professional baseball players in the United States for the years 1997 through 2002, where x represents the year. (Source: Major League Baseball and the Associated Press)

x	1997	1998	1999	2000	2001	2002
y	1314	1385	1572	1834	2089	2341

(a) Plot the data shown in the table.

(b) Use the graph to describe the relationship between the year and the average salary.

(c) Find the percent increase in average salaries for baseball players from 1997 to 2002.

4.2 Graphs of Equations in Two Variables

1 Sketch graphs of equations using the point-plotting method.

In Exercises 27–38, sketch the graph of the equation using the point-plotting method.

27. $y = 7$

28. $x = -2$

29. $y = 3x$

30. $y = -2x$

31. $y = 4 - \frac{1}{2}x$

32. $y = \frac{3}{2}x - 3$

33. $y - 2x - 4 = 0$

34. $3x + 2y + 6 = 0$

35. $y = 2x - 1$

36. $y = 5 - 4x$

37. $y = \frac{1}{4}x + 2$

38. $y = -\frac{2}{3}x - 2$

2 Find and use x- and y-intercepts as aids to sketching graphs.

In Exercises 39–46, find the x- and y-intercepts (if any) of the graph of the equation. Then sketch the graph of the equation and label the x- and y-intercepts.

39. $y = 6x + 2$

40. $y = -3x + 5$

41. $y = \frac{2}{5}x - 2$

42. $y = \frac{1}{3}x + 1$

43. $2x - y = 4$

44. $3x - y = 10$

45. $4x + 2y = 8$

46. $9x + 3y = 6$

3 Use the verbal problem-solving method to write an equation and sketch its graph.

47. *Creating a Model* The cost of producing a DVD is $125, plus $3 per DVD. Let C represent the total cost and let x represent the number of DVDs. Write an equation that relates C and x and sketch its graph.

48. *Creating a Model* Let y represent the distance traveled by a train that is moving at a constant speed of 80 miles per hour. Let t represent the number of hours the train has traveled. Write an equation that relates y to t and sketch its graph.

4.3 Relations, Functions, and Graphs

1 Identify the domain and range of a relation.

In Exercises 49–52, find the domain and range of the relation.

49. $\{(8, 3), (-2, 7), (5, 1), (3, 8)\}$

50. $\{(0, 1), (-1, 3), (4, 6), (-7, 5)\}$

51. $\{(2, -3), (-2, 3), (7, 0), (-4, -2)\}$

52. $\{(1, 7), (-3, 4), (6, 5), (-2, -9)\}$

2 Determine if relations are functions by inspection or by using the Vertical Line Test.

In Exercises 53–56, determine whether the relation represents a function.

53.

Domain	Range
1	2
2	5
3	7
4	9
5	

54.

Domain	Range
5	5
7	9
9	13
11	17
13	19

55.

Input x	Output y	(x, y)
0	0	$(0, 0)$
2	1	$(2, 1)$
4	1	$(4, 1)$
6	2	$(6, 2)$
2	3	$(2, 3)$

56.

Input x	Output y	(x, y)
-6	1	$(-6, 1)$
-3	0	$(-3, 0)$
0	1	$(0, 1)$
3	4	$(3, 4)$
6	2	$(6, 2)$

In Exercises 57–62, use the Vertical Line Test to determine whether y is a function of x.

57.

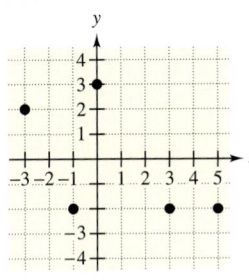

58.

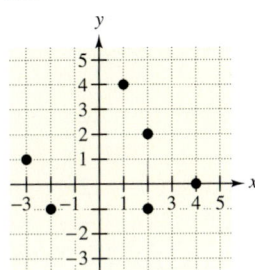

59.

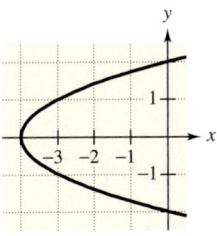

60.

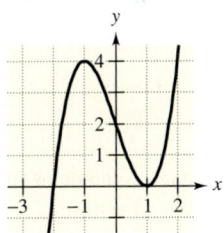

61.

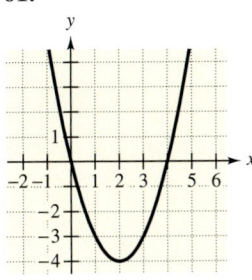

62.

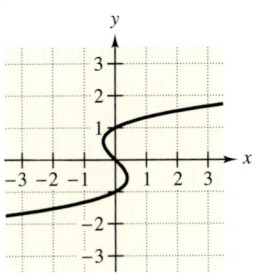

③ Use function notation and evaluate functions.

In Exercises 63–68, evaluate the function as indicated, and simplify.

63. $f(x) = 25x$ (a) $f(-1)$ (b) $f(7)$
                        (c) $f(10)$ (d) $f\left(-\frac{4}{3}\right)$

64. $f(x) = 2x - 7$ (a) $f(-1)$ (b) $f(3)$
                          (c) $f\left(\frac{1}{2}\right)$ (d) $f(-4)$

65. $g(t) = -16t^2 + 64$ (a) $g(0)$ (b) $g\left(\frac{1}{4}\right)$
                                (c) $g(1)$ (d) $g(2)$

66. $h(u) = u(u - 3)^2$ (a) $h(0)$ (b) $h(3)$
                               (c) $h(-1)$ (d) $h\left(\frac{3}{2}\right)$

67. $f(x) = |2x + 3|$ (a) $f(0)$ (b) $f(5)$
                            (c) $f(-4)$ (d) $f\left(-\frac{3}{2}\right)$

68. $f(x) = |x| - 4$ (a) $f(-1)$ (b) $f(1)$
                           (c) $f(-4)$ (d) $f(2)$

69. *Demand* The demand for a product is a function of its price. Consider the demand function

$$f(p) = 40 - 0.2p$$

where p is the price in dollars. Find the demand for (a) $p = 10$, (b) $p = 50$, and (c) $p = 100$.

70. *Profit* The profit for a product is a function of the amount spent on advertising for the product. Consider the profit function

$$f(x) = 8000 + 2000x - 50x^2$$

where x is the amount (in hundreds of dollars) spent on advertising. Find the profit for (a) $x = 5$, (b) $x = 10$, and (c) $x = 20$.

④ Identify the domain of a function.

In Exercises 71–74, find the domain of the function.

71. $f:\{(1, 5), (2, 10), (3, 15), (4, -10), (5, -15)\}$

72. $g:\{(-3, 6), (-2, 4), (-1, 2), (0, 0), (1, -2)\}$

73. $h:\{(-2, 12), (-1, 10), (0, 8), (1, 10), (2, 12)\}$

74. $f:\{(0, 7), (1, 7), (2, 5), (3, 7), (4, 7)\}$

4.4 Slope and Graphs of Linear Equations

1 Determine the slope of a line through two points.

In Exercises 75 and 76, estimate the slope of the line from its graph.

75.

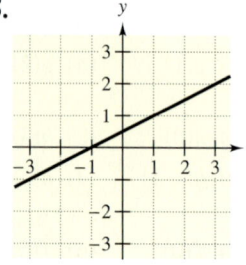

76.

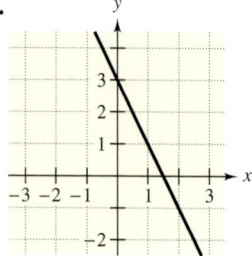

In Exercises 77–88, plot the points and find the slope (if possible) of the line passing through the points. State whether the line rises, falls, is horizontal, or is vertical.

77. $(2, 1), (14, 6)$

78. $(-2, 2), (3, -10)$

79. $(-1, 0), (6, 2)$

80. $(1, 6), (4, 2)$

81. $(4, 0), (4, 6)$

82. $(1, 3), (4, 3)$

83. $(-2, 5), (1, 1)$

84. $(-6, 1), (10, 5)$

85. $(1, -4), (5, 10)$

86. $(-3, 3), (8, 6)$

87. $\left(0, \frac{5}{2}\right), \left(\frac{5}{6}, 0\right)$

88. $(0, 0), \left(3, \frac{4}{5}\right)$

89. *Truck*   The floor of a truck is 4 feet above ground level. The end of the ramp used in loading the truck rests on the ground 6 feet behind the truck. Determine the slope of the ramp.

90. *Flight Path*   An aircraft is on its approach to an airport. Radar shows its altitude to be 15,000 feet when it is 10 miles from touchdown. Approximate the slope of the linear path followed by the aircraft during landing.

2 Write linear equations in slope-intercept form and graph the equations.

In Exercises 91–98, write the equation in slope-intercept form. Use the slope and *y*-intercept to sketch the line.

91. $2x - y = -1$

92. $-4x + y = -2$

93. $12x + 4y = 8$

94. $2x - 2y = 12$

95. $3x + 6y = 12$

96. $7x + 21y = -14$

97. $5y - 2x = 5$

98. $3y - x = 6$

3 Use slopes to determine whether lines are parallel, perpendicular, or neither.

In Exercises 99–102, determine whether lines L_1 and L_2 passing through the pairs of points are parallel, perpendicular, or neither.

99. $L_1: (0, 3), (-2, 1)$
   $L_2: (-8, -3), (4, 9)$

100. $L_1: (-3, -1), (2, 5)$
   $L_2: (2, 11), (8, 6)$

101. $L_1: (3, 6), (-1, -5)$
   $L_2: (-2, 3), (4, 7)$

102. $L_1: (-1, 2), (-1, 4)$
   $L_2: (7, 3), (4, 7)$

4.5 Equations of Lines

① Write equations of lines using the point-slope form.

In Exercises 103–112, use the point-slope form to write an equation of the line that passes through the point and has the specified slope. Write the equation in slope-intercept form.

103. $(4, -1)$, $m = 2$

104. $(-5, 2)$, $m = 3$

105. $(1, 2)$, $m = -4$

106. $(7, -3)$, $m = -1$

107. $(-5, -2)$, $m = \frac{4}{5}$

108. $(12, -4)$, $m = -\frac{1}{6}$

109. $(-1, 3)$, $m = -\frac{8}{3}$

110. $(4, -2)$, $m = \frac{8}{5}$

111. $(3, 8)$, m is undefined.

112. $(-4, 6)$, $m = 0$

In Exercises 113–120, write an equation of the line passing through the points. Write the equation in general form.

113. $(-4, 0)$, $(0, -2)$

114. $(-4, -2)$, $(4, 6)$

115. $(0, 8)$, $(6, 8)$

116. $(2, -6)$, $(2, 5)$

117. $(-1, 2)$, $(4, 7)$

118. $\left(0, \frac{4}{3}\right)$, $(3, 0)$

119. $(2.4, 3.3)$, $(6, 7.8)$

120. $(-1.4, 0)$, $(3.2, 9.2)$

In Exercises 121–124, write equations of the lines through the point (a) parallel and (b) perpendicular to the given line.

121. $(-6, 3)$
     $2x + 3y = 1$

122. $\left(\frac{1}{5}, -\frac{4}{5}\right)$
     $5x + y = 2$

123. $\left(\frac{3}{8}, 4\right)$
     $4x + 3y = 16$

124. $(-2, 1)$
     $5x = 2$

② Write the equations of horizontal and vertical lines.

In Exercises 125–128, write an equation of the line.

125. Horizontal line through $(-4, 5)$

126. Horizontal line through $(3, -7)$

127. Vertical line through $(5, -1)$

128. Vertical line through $(-10, 4)$

③ Use linear models to solve application problems.

129. *Wages*   A pharmaceutical salesperson receives a monthly salary of $2500 plus a commission of 7% of the total monthly sales. Write a linear model that relates total monthly wages W to sales S.

130. *Rental Demand*   A real estate office handles an apartment complex with 50 units. When the rent per unit is $380 per month, all 50 units are occupied. However, when the rent is $425 per month, the average number of occupied units drops to 47. Assume that the relationship between the monthly rent p and the demand x is linear.

(a) Represent the given information as two ordered pairs of the form (x, p). Plot these ordered pairs.

(b) Write a linear model that relates the monthly rent p to the demand x. Graph the model and describe the relationship between the rent and the demand.

(c) *Linear Extrapolation* Use the model in part (b) to predict the number of units occupied if the rent is raised to $485.

(d) *Linear Interpolation* Use the model in part (b) to estimate the number of units occupied if the rent is $410.

4.6 Graphs of Linear Inequalities

① Determine whether an ordered pair is a solution of a linear inequality in two variables.

In Exercises 131 and 132, determine whether the points are solutions of the inequality.

131. $x - y > 4$

    (a) $(-1, -5)$

    (b) $(0, 0)$

    (c) $(3, -2)$

    (d) $(8, 1)$

132. $y - 2x \leq -1$

    (a) $(0, 0)$

    (b) $(-2, 1)$

    (c) $(-3, 4)$

    (d) $(-1, -6)$

② Sketch graphs of linear inequalities in two variables.

In Exercises 133–138, sketch the graph of the linear inequality.

133. $x - 2 \geq 0$ **134.** $y + 3 < 0$

135. $2x + y < 1$ **136.** $3x - 4y > 2$

137. $x \leq 4y - 2$ **138.** $x \geq 3 - 2y$

In Exercises 139–142, write an inequality for the shaded region shown in the figure.

139.

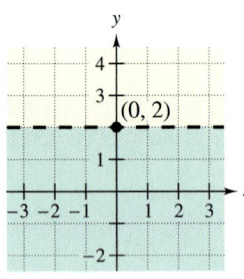

140.

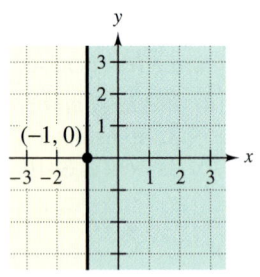

141.

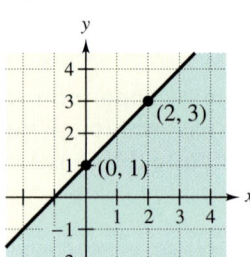

142.

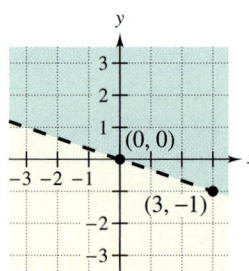

③ Use linear inequalities to model and solve real-life problems.

143. *Manufacturing* Each VCR produced by an electronics manufacturer requires 2 hours in the assembly center. Each camcorder produced by the same manufacturer requires 3 hours in the assembly center. A total of 120 hours per week is available in the assembly center. Write a linear inequality that shows the different numbers of hours that can be spent assembling VCRs and camcorders, and sketch the graph of the inequality. From the graph, find several ordered pairs with positive integer coordinates that are solutions of the inequality.

144. *Manufacturing* A company produces two types of wood chippers, Economy and Deluxe. The Deluxe model requires 3 hours in the assembly center and the Economy model requires $1\frac{1}{2}$ hours in the assembly center. A total of 24 hours per day is available in the assembly center. Write a linear inequality that shows the different numbers of hours that can be spent assembling the two models, and sketch the graph of the inequality. From the graph, find several ordered pairs with positive integer coordinates that are solutions of the inequality.

Chapter Test

Take this test as you would take a test in class. After you are done, check your work against the answers in the back of the book.

1. Plot the points $(-1, 2)$, $(1, 4)$, and $(2, -1)$ on a rectangular coordinate system. Connect the points with line segments to form a right triangle.

2. Determine whether the ordered pairs are solutions of $y = |x| + |x - 2|$.
   (a) $(0, -2)$ (b) $(0, 2)$ (c) $(-4, 10)$ (d) $(-2, -2)$

3. What is the y-coordinate of any point on the x-axis?

4. Find the x- and y-intercepts of the graph of $3x - 4y + 12 = 0$.

5. Complete the table at the left and use the results to sketch the graph of the equation $x - 2y = 6$.

x	-2	-1	0	1	2
y					

Table for 5

In Exercises 6–9, sketch the graph of the equation.

6. $x + 2y = 6$

7. $y = \frac{1}{4}x - 1$

8. $y = |x + 2|$

9. $y = (x - 3)^2$

Input, x	0	1	2	1	0
Output, y	4	5	8	-3	-1

Table for 10

10. Does the table at the left represent y as a function of x? Explain.

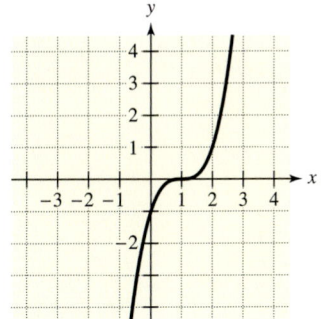

Figure for 11

11. Does the graph at the left represent y as a function of x? Explain.

12. Evaluate $f(x) = x^3 - 2x^2$ as indicated, and simplify.
    (a) $f(0)$ (b) $f(2)$ (c) $f(-2)$ (d) $f\left(\frac{1}{2}\right)$

13. Find the slope of the line passing through the points $(-5, 0)$ and $\left(2, \frac{3}{2}\right)$.

14. A line with slope $m = -2$ passes through the point $(-3, 4)$. Plot the point and use the slope to find two additional points on the line. (There are many correct answers.)

15. Find the slope of a line *perpendicular* to the line $3x - 5y + 2 = 0$.

16. Find an equation of the line that passes through the point $(0, 6)$ with slope $m = -\frac{3}{8}$.

17. Write an equation of the vertical line that passes through the point $(3, -7)$.

18. Determine whether the points are solutions of $3x + 5y \leq 16$.
    (a) $(2, 2)$ (b) $(6, -1)$ (c) $(-2, 4)$ (d) $(7, -1)$

In Exercises 19–22, sketch the graph of the linear inequality.

19. $y \geq -2$

20. $y < 5 - 2x$

21. $x \geq 2$

22. $y \leq 5$

23. The sales y of a product are modeled by $y = 230x + 5000$, where x is time in years. Interpret the meaning of the slope in this model.

291

Motivating the Chapter

⚡ Packaging Restrictions

A shipping company has the following restrictions on the dimensions and weight of packages.

1. The maximum weight is 150 pounds.
2. The maximum length is 108 inches.
3. The sum of the length and girth can be at most 130 inches.

The girth of a package is the minimum distance around the package, as shown in the figure.

$$\text{Girth} = 2(\text{Height} + \text{Width})$$

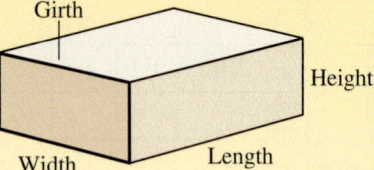

You are shipping a package that has a height of x inches. The length of the package is twice the square of the height, and the width is 5 inches more than 3 times the height.

See Section 5.2, Exercise 103.

a. Write an expression for the length of the package in terms of the height x. Write an expression for the width of the package in terms of the height x.

b. Write an expression for the *perimeter* of the base of the package. Simplify the expression.

c. Write an expression for the *girth* of the package. Simplify the expression. Write an expression for the sum of the length and the girth. If the height of the package is 5 inches, does the package meet the second and third restrictions? Explain.

See Section 5.3, Exercise 133.

d. Write an expression for the *surface area* of the package. Simplify the expression. (The surface area is the sum of the areas of the six sides of the package.)

e. The length of the package is changed to match its width (5 inches more than 3 times its height). Write an expression for the area of the base. Simplify the expression.

f. Write an expression for the *volume* of the package in part (e). Simplify the expression.

Najlah Feanny/Corbis SABA

5

Exponents and Polynomials

5.1 Integer Exponents and Scientific Notation

Thad Samuels II Abell/Getty Images

What You Should Learn

1. Use the rules of exponents to simplify expressions.

2. Rewrite exponential expressions involving negative and zero exponents.

3. Write very large and very small numbers in scientific notation.

Why You Should Learn It

Scientific notation can be used to represent very large real-life quantities. For instance, in Exercise 140 on page 303, you will use scientific notation to represent the average amount of poultry produced per person.

1. **Use the rules of exponents to simplify expressions.**

Rules of Exponents

Recall from Section 1.5 that *repeated multiplication* can be written in what is called **exponential form.** Let n be a positive integer and let a be a real number. Then the product of n factors of a is given by

$$a^n = \underbrace{a \cdot a \cdot a \cdots a}_{n \text{ factors}}.$$ *a* is the base and *n* is the exponent.

When multiplying two exponential expressions that have the *same base*, you add exponents. To see why this is true, consider the product $a^3 \cdot a^2$. Because the first expression represents three factors of a and the second represents two factors of a, the product of the two expressions represents five factors of a, as follows.

$$a^3 \cdot a^2 = \underbrace{(a \cdot a \cdot a)}_{3 \text{ factors}} \cdot \underbrace{(a \cdot a)}_{2 \text{ factors}} = \underbrace{(a \cdot a \cdot a \cdot a \cdot a)}_{5 \text{ factors}} = a^{3+2} = a^5$$

Rules of Exponents

Let m and n be positive integers, and let a and b represent real numbers, variables, or algebraic expressions.

	Rule	*Example*
1. Product:	$a^m \cdot a^n = a^{m+n}$	$x^5(x^4) = x^{5+4} = x^9$
2. Product-to-Power:	$(ab)^m = a^m \cdot b^m$	$(2x)^3 = 2^3(x^3) = 8x^3$
3. Power-to-Power:	$(a^m)^n = a^{mn}$	$(x^2)^3 = x^{2 \cdot 3} = x^6$
4. Quotient:	$\dfrac{a^m}{a^n} = a^{m-n}, m > n, a \neq 0$	$\dfrac{x^5}{x^3} = x^{5-3} = x^2, x \neq 0$
5. Quotient-to-Power:	$\left(\dfrac{a}{b}\right)^m = \dfrac{a^m}{b^m}, b \neq 0$	$\left(\dfrac{x}{4}\right)^2 = \dfrac{x^2}{4^2} = \dfrac{x^2}{16}$

The product rule and the product-to-power rule can be extended to three or more factors. For example,

$$a^m \cdot a^n \cdot a^k = a^{m+n+k} \quad \text{and} \quad (abc)^m = a^m b^m c^m.$$

Study Tip

In the expression $x + 5$, the coefficient of x is understood to be 1. Similarly, the power (or exponent) of x is also understood to be 1. So

$$x^4 \cdot x \cdot x^2 = x^{4+1+2} = x^7.$$

Note such occurrences in Examples 1(a) and 2(b).

Example 1 Using Rules of Exponents

Simplify: **a.** $(x^2y^4)(3x)$ **b.** $-2(y^2)^3$ **c.** $(-2y^2)^3$ **d.** $(3x^2)(-5x)^3$

Solution

a. $(x^2y^4)(3x) = 3(x^2 \cdot x)(y^4) = 3(x^{2+1})(y^4) = 3x^3y^4$

b. $-2(y^2)^3 = (-2)(y^{2\cdot3}) = -2y^6$

c. $(-2y^2)^3 = (-2)^3(y^2)^3 = -8(y^{2\cdot3}) = -8y^6$

d. $(3x^2)(-5x)^3 = 3(-5)^3(x^2 \cdot x^3) = 3(-125)(x^{2+3}) = -375x^5$

Example 2 Using Rules of Exponents

Simplify: **a.** $\dfrac{14a^5b^3}{7a^2b^2}$ **b.** $\left(\dfrac{x^2}{2y}\right)^3$ **c.** $\dfrac{x^ny^{3n}}{x^2y^4}$ **d.** $\dfrac{(2a^2b^3)^2}{a^3b^2}$

Solution

a. $\dfrac{14a^5b^3}{7a^2b^2} = 2(a^{5-2})(b^{3-2}) = 2a^3b$

b. $\left(\dfrac{x^2}{2y}\right)^3 = \dfrac{(x^2)^3}{(2y)^3} = \dfrac{x^{2\cdot3}}{2^3y^3} = \dfrac{x^6}{8y^3}$

c. $\dfrac{x^ny^{3n}}{x^2y^4} = x^{n-2}y^{3n-4}$

d. $\dfrac{(2a^2b^3)^2}{a^3b^2} = \dfrac{2^2(a^{2\cdot2})(b^{3\cdot2})}{a^3b^2} = \dfrac{4a^4b^6}{a^3b^2} = 4(a^{4-3})(b^{6-2}) = 4ab^4$

Integer Exponents

2 Rewrite exponential expressions involving negative and zero exponents.

The definition of an exponent can be extended to include zero and negative integers. If a is a real number such that $a \neq 0$, then a^0 is defined as 1. Moreover, if m is an integer, then a^{-m} is defined as the reciprocal of a^m.

Definitions of Zero Exponents and Negative Exponents

Let a and b be real numbers such that $a \neq 0$ and $b \neq 0$, and let m be an integer.

1. $a^0 = 1$ **2.** $a^{-m} = \dfrac{1}{a^m}$ **3.** $\left(\dfrac{a}{b}\right)^{-m} = \left(\dfrac{b}{a}\right)^m$

These definitions are consistent with the rules of exponents given on page 294. For instance, consider the following.

$$x^0 \cdot x^m = x^{0+m} = x^m = 1 \cdot x^m$$

(x^0 is the same as 1)

Example 3   Zero Exponents and Negative Exponents

Rewrite each expression without using zero exponents or negative exponents.

a. 3^0 **b.** 3^{-2} **c.** $\left(\frac{3}{4}\right)^{-1}$

Solution

a. $3^0 = 1$ — Definition of zero exponents

b. $3^{-2} = \frac{1}{3^2} = \frac{1}{9}$ — Definition of negative exponents

c. $\left(\frac{3}{4}\right)^{-1} = \left(\frac{4}{3}\right)^{1} = \frac{4}{3}$ — Definition of negative exponents

> **Study Tip**
>
> Because the expression a^0 is equal to 1 for any real number a such that $a \neq 0$, zero cannot have a zero exponent. So, 0^0 is undefined.

The following rules are valid for all integer exponents, including integer exponents that are zero or negative. (The first five rules were listed on page 294.)

Summary of Rules of Exponents

Let m and n be integers, and let a and b represent real numbers, variables, or algebraic expressions. (All denominators and bases are nonzero.)

	Product and Quotient Rules	*Example*
1.	$a^m \cdot a^n = a^{m+n}$	$x^4(x^3) = x^{4+3} = x^7$
2.	$\dfrac{a^m}{a^n} = a^{m-n}$	$\dfrac{x^3}{x} = x^{3-1} = x^2$
	Power Rules	
3.	$(ab)^m = a^m \cdot b^m$	$(3x)^2 = 3^2(x^2) = 9x^2$
4.	$(a^m)^n = a^{mn}$	$(x^3)^3 = x^{3 \cdot 3} = x^9$
5.	$\left(\dfrac{a}{b}\right)^m = \dfrac{a^m}{b^m}$	$\left(\dfrac{x}{3}\right)^2 = \dfrac{x^2}{3^2} = \dfrac{x^2}{9}$
	Zero and Negative Exponent Rules	
6.	$a^0 = 1$	$(x^2 + 1)^0 = 1$
7.	$a^{-m} = \dfrac{1}{a^m}$	$x^{-2} = \dfrac{1}{x^2}$
8.	$\left(\dfrac{a}{b}\right)^{-m} = \left(\dfrac{b}{a}\right)^m$	$\left(\dfrac{x}{3}\right)^{-2} = \left(\dfrac{3}{x}\right)^2 = \dfrac{3^2}{x^2} = \dfrac{9}{x^2}$

Example 4   Using Rules of Exponents

a. $2x^{-1} = 2(x^{-1}) = 2\left(\frac{1}{x}\right) = \frac{2}{x}$ — Use negative exponent rule and simplify.

b. $(2x)^{-1} = \frac{1}{(2x)^1} = \frac{1}{2x}$ — Use negative exponent rule and simplify.

Study Tip

As you become accustomed to working with negative exponents, you will probably not write as many steps as shown in Example 5. For instance, to rewrite a fraction involving exponents, you might use the following simplified rule. *To move a factor from the numerator to the denominator or vice versa, change the sign of its exponent.* You can apply this rule to the expression in Example 5(a) by "moving" the factor x^{-2} to the numerator and changing the exponent to 2. That is,

$$\frac{3}{x^{-2}} = 3x^2.$$

Remember, you can move only *factors* in this manner, not terms.

Example 5   Using Rules of Exponents

Rewrite each expression using only positive exponents. (For each expression, assume that $x \neq 0$).

a. $\dfrac{3}{x^{-2}} = \dfrac{3}{\left(\dfrac{1}{x^2}\right)}$ Negative exponent rule

$\qquad\quad = 3\left(\dfrac{x^2}{1}\right)$ Invert divisor and multiply.

$\qquad\quad = 3x^2$ Simplify.

b. $\dfrac{1}{(3x)^{-2}} = \dfrac{1}{\left[\dfrac{1}{(3x)^2}\right]}$ Use negative exponent rule.

$\qquad\quad = \dfrac{1}{\left(\dfrac{1}{(9x^2)}\right)}$ Use product-to-power rule and simplify.

$\qquad\quad = (1)\left(\dfrac{9x^2}{1}\right)$ Invert divisor and multiply.

$\qquad\quad = 9x^2$ Simplify.

Example 6   Using Rules of Exponents

Rewrite each expression using only positive exponents. (For each expression, assume that $x \neq 0$ and $y \neq 0$.)

a. $(-5x^{-3})^2 = (-5)^2(x^{-3})^2$ Product-to-power rule

$\qquad\qquad\quad = 25x^{-6}$ Power-to-product rule

$\qquad\qquad\quad = \dfrac{25}{x^6}$ Negative exponent rule

b. $-\left(\dfrac{7x}{y^2}\right)^{-2} = -\left(\dfrac{y^2}{7x}\right)^2$ Negative exponent rule

$\qquad\qquad\quad = -\dfrac{(y^2)^2}{(7x)^2}$ Quotient-to-power rule

$\qquad\qquad\quad = -\dfrac{y^4}{49x^2}$ Power-to-power and product-to-power rules

c. $\dfrac{12x^2y^{-4}}{6x^{-1}y^2} = 2(x^{2-(-1)})(y^{-4-2})$ Quotient rule

$\qquad\qquad\quad = 2x^3y^{-6}$ Simplify.

$\qquad\qquad\quad = \dfrac{2x^3}{y^6}$ Negative exponent rule

Example 7   Using Rules of Exponents

Rewrite each expression using only positive exponents. (For each expression, assume that $x \neq 0$ and $y \neq 0$.)

a. $\left(\dfrac{8x^{-1}y^4}{4x^3y^2}\right)^{-3} = \left(\dfrac{2y^2}{x^4}\right)^{-3}$ Simplify.

$= \left(\dfrac{x^4}{2y^2}\right)^{3}$ Negative exponent rule

$= \dfrac{x^{12}}{2^3 y^6}$ Quotient-to-power rule

$= \dfrac{x^{12}}{8y^6}$ Simplify.

b. $\dfrac{3xy^0}{x^2(5y)^0} = \dfrac{3x(1)}{x^2(1)} = \dfrac{3}{x}$ Zero exponent rule

---

③ Write very large and very small numbers in scientific notation.

Scientific Notation

Exponents provide an efficient way of writing and computing with very large and very small numbers. For instance, a drop of water contains more than 33 billion billion molecules—that is, 33 followed by 18 zeros. It is convenient to write such numbers in **scientific notation.** This notation has the form $c \times 10^n$, where $1 \leq c < 10$ and n is an integer. So, the number of molecules in a drop of water can be written in scientific notation as follows.

$$33,000,000,000,000,000,000 = 3.3 \times 10^{19}$$

19 places

The *positive* exponent 19 indicates that the number being written in scientific notation is *large* (10 or more) and that the decimal point has been moved 19 places. A *negative* exponent in scientific notation indicates that the number is *small* (less than 1).

Example 8   Writing Scientific Notation

Write each number in scientific notation.

a. 0.0000684   b. 937,200,000

Solution

a. $0.0000684 = 6.84 \times 10^{-5}$ Small number ➡ negative exponent

Five places

b. $937,200,000.0 = 9.372 \times 10^8$ Large number ➡ positive exponent

Eight places

Example 9   Writing Decimal Notation

Write each number in decimal notation.

a. 2.486×10^2 **b.** 1.81×10^{-6}

Solution

a. $2.486 \times 10^2 = 248.6$

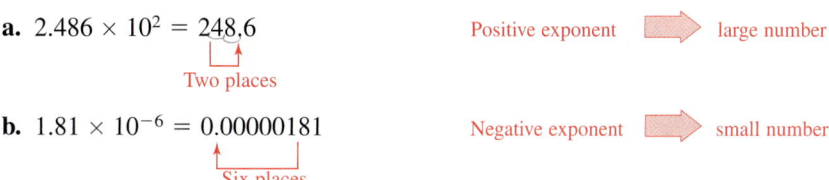

Two places

Positive exponent → large number

b. $1.81 \times 10^{-6} = 0.00000181$

Six places

Negative exponent → small number

Example 10   Using Scientific Notation

Rewrite the factors in scientific notation and then evaluate

$$\frac{(2{,}400{,}000{,}000)(0.0000045)}{(0.00003)(1500)}.$$

Solution

$$\frac{(2{,}400{,}000{,}000)(0.0000045)}{(0.00003)(1500)} = \frac{(2.4 \times 10^9)(4.5 \times 10^{-6})}{(3.0 \times 10^{-5})(1.5 \times 10^3)}$$

$$= \frac{(2.4)(4.5)(10^3)}{(4.5)(10^{-2})}$$

$$= (2.4)(10^5)$$

$$= 240{,}000$$

Technology: Tip

Most scientific and graphing calculators automatically switch to scientific notation when they are showing large or small numbers that exceed the display range.

To *enter* numbers in scientific notation, your calculator should have an exponential entry key labeled EE or EXP. Consult the user's guide of your calculator for instructions on keystrokes and how numbers in scientific notation are displayed.

Example 11   Using Scientific Notation with a Calculator

Use a calculator to evaluate each expression.

a. $65{,}000 \times 3{,}400{,}000{,}000$ **b.** $0.000000348 \div 870$

Solution

a. 6.5 EXP 4 × 3.4 EXP 9 =   Scientific

6.5 EE 4 × 3.4 EE 9 ENTER   Graphing

The calculator display should read 2.21E 14, which implies that

$$(6.5 \times 10^4)(3.4 \times 10^9) = 2.21 \times 10^{14} = 221{,}000{,}000{,}000{,}000.$$

b. 3.48 EXP 7 +/− ÷ 8.7 EXP 2 =   Scientific

3.48 EE (−) 7 ÷ 8.7 EE 2 ENTER   Graphing

The calculator display should read 4E −10, which implies that

$$\frac{3.48 \times 10^{-7}}{8.7 \times 10^2} = 4.0 \times 10^{-10} = 0.0000000004.$$

5.1 Exercises

Review   *Concepts, Skills, and Problem Solving*

Keep mathematically in shape by doing these exercises *before* the problems of this section.

Properties and Definitions

1. *Writing* ✐   In your own words, describe the graph of an equation.

2. *Writing* ✐   Describe the point-plotting method of graphing an equation.

3. Find the coordinates of two points on the graph of $g(x) = \sqrt{x - 2}$.

4. *Writing* ✐   Describe the procedure for finding the x- and y-intercepts of the graph of an equation.

Evaluating Functions

In Exercises 5–8, evaluate the function as indicated, and simplify.

5. $f(x) = 3x - 9$
   (a) $f(-2)$
   (b) $f\left(\frac{1}{2}\right)$

6. $f(x) = x^2 + x$
   (a) $f(4)$
   (b) $f(-2)$

7. $f(x) = 6x - x^2$
   (a) $f(0)$ (b) $f(t + 1)$

8. $f(x) = \dfrac{x - 2}{x + 2}$
   (a) $f(10)$ (b) $f(4 - z)$

Graphing Equations

⊞ In Exercises 9–12, use a graphing calculator to graph the function. Identify any intercepts.

9. $f(x) = 5 - 2x$

10. $h(x) = \frac{1}{2}x + |x|$

11. $g(x) = x^2 - 4x$

12. $f(x) = 2\sqrt{x + 1}$

Developing Skills

In Exercises 1–20, use the rules of exponents to simplify the expression (if possible). See Examples 1 and 2.

1. (a) $-3x^3 \cdot x^5$
   (b) $(-3x)^2 \cdot x^5$

2. (a) $5^2y^4 \cdot y^2$
   (b) $(5y)^2 \cdot y^4$

3. (a) $(-5z^2)^3$
   (b) $(-5z^4)^2$

4. (a) $(-5z^3)^2$
   (b) $(-5z)^4$

5. (a) $(u^3v)(2v^2)$
   (b) $(-4u^4)(u^5v)$

6. (a) $(6xy^7)(-x)$
   (b) $(x^5y^3)(2y^3)$

7. (a) $5u^2 \cdot (-3u^6)$
   (b) $(2u)^4(4u)$

8. (a) $(3y)^3(2y^2)$
   (b) $3y^3 \cdot 2y^2$

9. (a) $-(m^5n)^3(-m^2n^2)^2$
   (b) $(-m^5n)(m^2n^2)$

10. (a) $-(m^3n^2)(mn^3)$
    (b) $-(m^3n^2)^2(-mn^3)$

11. (a) $\dfrac{27m^5n^6}{9mn^3}$
    (b) $\dfrac{-18m^3n^6}{-6mn^3}$

12. (a) $\dfrac{28x^2y^3}{2xy^2}$
    (b) $\dfrac{24xy^2}{8y}$

13. (a) $\left(\dfrac{3x}{4y}\right)^2$
    (b) $\left(\dfrac{5u}{3v}\right)^3$

14. (a) $\left(\dfrac{2a}{3y}\right)^5$
    (b) $-\left(\dfrac{2a}{3y}\right)^2$

15. (a) $-\dfrac{(-2x^2y)^3}{9x^2y^2}$
    (b) $-\dfrac{(-2xy^3)^2}{6y^2}$

16. (a) $\dfrac{(-4xy)^3}{8xy^2}$
    (b) $\dfrac{(-xy)^4}{-3(xy)^2}$

17. (a) $\left[\dfrac{(-5u^3v)^2}{10u^2v}\right]^2$
    (b) $\left[\dfrac{-5(u^3v)^2}{10u^2v}\right]^2$

18. (a) $\left[\dfrac{(3x^2)(2x)^2}{(-2x)(6x)}\right]^2$
    (b) $\left[\dfrac{(3x^2)(2x)^4}{(-2x)^2(6x)}\right]^2$

19. (a) $\dfrac{x^{2n+4} y^{4n}}{x^5 y^{2n+1}}$ (b) $\dfrac{x^{6n} y^{n-7}}{x^{4n+2} y^5}$

20. (a) $\dfrac{x^{3n} y^{2n-1}}{x^n y^{n+3}}$ (b) $\dfrac{x^{4n-6} y^{n+10}}{x^{2n-5} y^{n-2}}$

In Exercises 21–50, evaluate the expression. See Example 3.

21. 5^{-2}

22. 2^{-4}

23. -10^{-3}

24. -20^{-2}

25. $(-3)^0$

26. 25^0

27. $\dfrac{1}{4^{-3}}$

28. $\dfrac{1}{-8^{-2}}$

29. $\dfrac{1}{(-2)^{-5}}$

30. $-\dfrac{1}{6^{-2}}$

31. $\left(\tfrac{2}{3}\right)^{-1}$

32. $\left(\tfrac{4}{5}\right)^{-3}$

33. $\left(\tfrac{3}{16}\right)^{0}$

34. $\left(-\tfrac{5}{8}\right)^{-2}$

35. $27 \cdot 3^{-3}$

36. $4^2 \cdot 4^{-3}$

37. $\dfrac{3^4}{3^{-2}}$

38. $\dfrac{5^{-1}}{5^2}$

39. $\dfrac{10^3}{10^{-2}}$

40. $\dfrac{10^{-5}}{10^{-6}}$

41. $(4^2 \cdot 4^{-1})^{-2}$

42. $(5^3 \cdot 5^{-4})^{-3}$

43. $(2^{-3})^2$

44. $(-4^{-1})^{-2}$

45. $2^{-3} + 2^{-4}$

46. $4 - 3^{-2}$

47. $\left(\tfrac{3}{4} + \tfrac{5}{8}\right)^{-2}$

48. $\left(\tfrac{1}{2} - \tfrac{2}{3}\right)^{-1}$

49. $(5^0 - 4^{-2})^{-1}$

50. $(32 + 4^{-3})^0$

In Exercises 51–90, rewrite the expression using only positive exponents, and simplify. (Assume that any variables in the expression are nonzero.) See Examples 4–7.

51. $y^4 \cdot y^{-2}$

52. $x^{-2} \cdot x^{-5}$

53. $z^5 \cdot z^{-3}$

54. $t^{-1} \cdot t^{-6}$

55. $7x^{-4}$

56. $3y^{-3}$

57. $(4x)^{-3}$

58. $(5u)^{-2}$

59. $\dfrac{1}{x^{-6}}$

60. $\dfrac{4}{y^{-1}}$

61. $\dfrac{8a^{-6}}{6a^{-7}}$

62. $\dfrac{6u^{-2}}{15u^{-1}}$

63. $\dfrac{(4t)^0}{t^{-2}}$

64. $\dfrac{(5u)^{-4}}{(5u)^0}$

65. $(2x^2)^{-2}$

66. $(4a^{-2})^{-3}$

67. $(-3x^{-3}y^2)(4x^2y^{-5})$

68. $(5s^5t^{-5})(-6s^{-2}t^4)$

69. $(3x^2y^{-2})^{-2}$

70. $(-4y^{-3}z)^{-3}$

71. $\left(\dfrac{x}{10}\right)^{-1}$

72. $\left(\dfrac{4}{z}\right)^{-2}$

73. $\dfrac{6x^3y^{-3}}{12x^{-2}y}$

74. $\dfrac{2y^{-1}z^{-3}}{4yz^{-3}}$

75. $\left(\dfrac{3u^2v^{-1}}{3^3u^{-1}v^3}\right)^{-2}$

76. $\left(\dfrac{5^2x^3y^{-3}}{125xy}\right)^{-1}$

77. $\left(\dfrac{a^{-2}}{b^{-2}}\right)\left(\dfrac{b}{a}\right)^3$

78. $\left(\dfrac{a^{-3}}{b^{-3}}\right)\left(\dfrac{b}{a}\right)^3$

79. $(2x^3y^{-1})^{-3}(4xy^{-6})$

80. $(ab)^{-2}(a^2b^2)^{-1}$

81. $u^4(6u^{-3}v^0)(7v)^0$

82. $x^5(3x^0y^4)(7y)^0$

83. $[(x^{-4}y^{-6})^{-1}]^2$

84. $[(2x^{-3}y^{-2})^2]^{-2}$

85. $\dfrac{(2a^{-2}b^4)^3b}{(10a^3b)^2}$

86. $\dfrac{(5x^2y^{-5})^{-1}}{2x^{-5}y^4}$

87. $(u + v^{-2})^{-1}$

88. $x^{-2}(x^2 + y^2)$

89. $\dfrac{a + b}{ba^{-1} - ab^{-1}}$

90. $\dfrac{u^{-1} - v^{-1}}{u^{-1} + v^{-1}}$

In Exercises 91–104, write the number in scientific notation. See Example 8.

91. 3,600,000

92. 98,100,000

93. 47,620,000

94. 956,300,000

95. 0.00031

96. 0.00625

97. 0.0000000381

98. 0.0007384

99. *Land Area of Earth:* 57,300,000 square miles

100. *Water Area of Earth:* 139,500,000 square miles

101. *Light Year:* 9,460,800,000,000 kilometers

102. *Thickness of Soap Bubble:* 0.0000001 meter

103. *Relative Density of Hydrogen:* 0.0899 grams per milliliter.

104. *One Micron (Millionth of Meter):* 0.00003937 inch

In Exercises 105–114, write the number in decimal notation. See Example 9.

105. 6×10^7

106. 5.05×10^{12}

107. 1.359×10^{-7}

108. 8.6×10^{-9}

109. *2001 Merrill Lynch Revenues:* $\$3.8757 \times 10^{10}$ (Source: 2001 Merrill Lynch Annual Report)

110. *Number of Air Sacs in Lungs:* 3.5×10^8

111. *Interior Temperature of Sun:* 1.5×10^7 degrees Celsius

112. *Width of Air Molecule:* 9.0×10^{-9} meter

113. *Charge of Electron:* 4.8×10^{-10} electrostatic unit

114. *Width of Human Hair:* 9.0×10^{-4} meter

In Exercises 115–124, evaluate the expression without a calculator. See Example 10.

115. $(2 \times 10^9)(3.4 \times 10^{-4})$

116. $(6.5 \times 10^6)(2 \times 10^4)$

117. $(5 \times 10^4)^2$

118. $(4 \times 10^6)^3$

119. $\dfrac{3.6 \times 10^{12}}{6 \times 10^5}$

120. $\dfrac{2.5 \times 10^{-3}}{5 \times 10^2}$

121. $(4,500,000)(2,000,000,000)$

122. $(62,000,000)(0.0002)$

123. $\dfrac{64,000,000}{0.00004}$

124. $\dfrac{72,000,000,000}{0.00012}$

In Exercises 125–132, evaluate with a calculator. Write the answer in scientific notation, $c \times 10^n$, with c rounded to two decimal places. See Example 11.

125. $\dfrac{(0.0000565)(2,850,000,000,000)}{0.00465}$

126. $\dfrac{(3,450,000,000)(0.000125)}{(52,000,000)(0.000003)}$

127. $\dfrac{1.357 \times 10^{12}}{(4.2 \times 10^2)(6.87 \times 10^{-3})}$

128. $\dfrac{(3.82 \times 10^5)^2}{(8.5 \times 10^4)(5.2 \times 10^{-3})}$

129. $72,400 \times 2,300,000,000$

130. $(8.67 \times 10^4)^7$

131. $\dfrac{(5,000,000)^3(0.000037)^2}{(0.005)^4}$

132. $\dfrac{(6,200,000)(0.005)^3}{(0.00035)^5}$

Solving Problems

133. *Distance*　The distance from Earth to the sun is approximately 93 million miles. Write this distance in scientific notation.

134. *Electrons*　A cube of copper with an edge of 1 centimeter has approximately 8.483×10^{22} free electrons. Write this real number in decimal notation.

135. *Light Year*　One light year (the distance light can travel in 1 year) is approximately 9.46×10^{15} meters. Approximate the time (in minutes) for light to travel from the sun to Earth if that distance is approximately 1.50×10^{11} meters.

136. *Distance*　The star Alpha Andromeda is approximately 95 light years from Earth. Determine this distance in meters. (See Exercise 135 for the definition of a light year.)

137. *Masses of Earth and Sun*　The masses of Earth and the sun are approximately 5.98×10^{24} kilograms and 1.99×10^{30} kilograms, respectively. The mass of the sun is approximately how many times that of Earth?

138. *Metal Expansion*　When the temperature of an iron steam pipe 200 feet long is increased by 75°C, the length of the pipe will increase by an amount $75(200)(1.1 \times 10^{-5})$. Find this amount of increase in length.

139. *Federal Debt*　In July 2000, the estimated population of the United States was 275 million people, and the estimated federal debt was 5629 billion dollars. Use these two numbers to determine the amount each person would have to pay to eliminate the debt. (Source: U.S. Census Bureau and U.S. Office of Management and Budget)

140. *Poultry Production*　In 2000, the estimated population of the world was 6 billion people, and the world-wide production of poultry meat was 58 million metric tons. Use these two numbers to determine the average amount of poultry produced per person in 2000. (Source: U.S. Census Bureau and U.S. Department of Agriculture)

Explaining Concepts

141. In $(3x)^4$, what is $3x$ called? What is 4 called?

142. *Writing*　Discuss any differences between $(-2x)^{-4}$ and $-2x^{-4}$.

143. *Writing*　In your own words, describe how you can "move" a factor from the numerator to the denominator or vice versa.

144. *Writing*　Is the number 32.5×10^5 written in scientific notation? Explain.

145. *Writing*　When is scientific notation an efficient way of writing and computing real numbers?

5.2 Adding and Subtracting Polynomials

David Lassman/The Image Works

What You Should Learn

1 Identify the degrees and leading coefficients of polynomials.

2 Add polynomials using a horizontal or vertical format.

3 Subtract polynomials using a horizontal or vertical format.

Why You Should Learn It

Polynomials can be used to model and solve real-life problems. For instance, in Exercise 101 on page 312, polynomials are used to model the numbers of daily morning and evening newspapers in the United States.

1 Identify the degrees and leading coefficients of polynomials.

Basic Definitions

Recall from Section 2.1 that the *terms* of an algebraic expression are those parts separated by addition. An algebraic expression whose terms are all of the form ax^k, where a is any real number and k is a nonnegative integer, is called a **polynomial in one variable,** or simply a **polynomial.** Here are some examples of polynomials in one variable.

$$2x + 5, \quad x^2 - 3x + 7, \quad 9x^5, \quad \text{and} \quad x^3 + 8$$

In the term ax^k, a is the **coefficient** of the term and k is the **degree** of the term. Note that the degree of the term ax is 1, and the degree of a constant term is 0. Because a polynomial is an algebraic *sum*, the coefficients take on the signs between the terms. For instance,

$$x^4 + 2x^3 - 5x^2 + 7 = (1)x^4 + 2x^3 + (-5)x^2 + (0)x + 7$$

has coefficients 1, 2, -5, 0, and 7. For this polynomial, the last term, 7, is the **constant term.** Polynomials are usually written in the order of descending powers of the variable. This is called **standard form.** Here are three examples.

Nonstandard Form	Standard Form
$4 + x$	$x + 4$
$3x^2 - 5 - x^3 + 2x$	$-x^3 + 3x^2 + 2x - 5$
$18 - x^2 + 3$	$-x^2 + 21$

The **degree of a polynomial** is the degree of the term with the highest power, and the coefficient of this term is the **leading coefficient** of the polynomial. For instance, the polynomial

Degree

$$-3x^4 + 4x^2 + x + 7$$

Leading coefficient

is of fourth degree, and its leading coefficient is -3. The reasons why the degree of a polynomial is important will become clear as you study factoring and problem solving in Chapter 6.

Definition of a Polynomial in x

Let $a_n, a_{n-1}, \ldots, a_2, a_1, a_0$ be real numbers and let n be a nonnegative integer. A **polynomial in x** is an expression of the form

$$a_n x^n + a_{n-1} x^{n-1} + \cdots + a_2 x^2 + a_1 x + a_0$$

where $a_n \neq 0$. The polynomial is of **degree n,** and the number a_n is called the **leading coefficient.** The number a_0 is called the **constant term.**

Example 1   Identifying Polynomials

Identify which of the following are polynomials, and for any that are not polynomials, state why.

a. $3x^4 - 8x + x^{-1}$ **b.** $x^2 - 3x + 1$

c. $x^3 + 3x^{1/2}$ **d.** $-\dfrac{1}{3}x + \dfrac{x^3}{4}$

Solution

a. $3x^4 - 8x + x^{-1}$ is *not* a polynomial because the third term, x^{-1}, has a negative exponent.

b. $x^2 - 3x + 1$ is a polynomial of degree 2 with integer coefficients.

c. $x^3 + 3x^{1/2}$ is *not* a polynomial because the exponent in the second term, $3x^{1/2}$, is not an integer.

d. $-\dfrac{1}{3}x + \dfrac{x^3}{4}$ is a polynomial of degree 3 with rational coefficients.

Example 2   Determining Degrees and Leading Coefficients

Write each polynomial in standard form and identify the degree and leading coefficient.

	Polynomial	Standard Form	Degree	Leading Coefficient
a.	$4x^2 - 5x^7 - 2 + 3x$	$-5x^7 + 4x^2 + 3x - 2$	7	-5
b.	$4 - 9x^2$	$-9x^2 + 4$	2	-9
c.	8	8	0	8
d.	$2 + x^3 - 5x^2$	$x^3 - 5x^2 + 2$	3	1

In part (c), note that a polynomial with only a constant term has a degree of zero.

A polynomial with only one term is called a **monomial.** Polynomials with two *unlike* terms are called **binomials,** and those with three *unlike* terms are called **trinomials.** For example, $3x^2$ is a *monomial,* $-3x + 1$ is a *binomial,* and $4x^3 - 5x + 6$ is a *trinomial.*

② Add polynomials using a horizontal or vertical format.

Adding Polynomials

As with algebraic expressions, the key to adding two polynomials is to recognize *like* terms—those having the *same degree*. By the Distributive Property, you can then combine the like terms using either a horizontal or a vertical format of terms. For instance, the polynomials $2x^2 + 3x + 1$ and $x^2 - 2x + 2$ can be added horizontally to obtain

$$(2x^2 + 3x + 1) + (x^2 - 2x + 2) = (2x^2 + x^2) + (3x - 2x) + (1 + 2)$$
$$= 3x^2 + x + 3$$

or they can be added vertically to obtain the same result.

$$2x^2 + 3x + 1 \qquad \text{Vertical format}$$
$$\underline{x^2 - 2x + 2}$$
$$3x^2 + x + 3$$

Example 3 Adding Polynomials Horizontally

Use a horizontal format to find each sum.

a. $(2x^2 + 4x - 1) + (x^2 - 3)$ — Original polynomials

$\quad = (2x^2 + x^2) + (4x) + (-1 - 3)$ — Group like terms.

$\quad = 3x^2 + 4x - 4$ — Combine like terms.

b. $(x^3 + 2x^2 + 4) + (3x^2 - x + 5)$ — Original polynomials

$\quad = (x^3) + (2x^2 + 3x^2) + (-x) + (4 + 5)$ — Group like terms.

$\quad = x^3 + 5x^2 - x + 9$ — Combine like terms.

c. $(2x^2 - x + 3) + (4x^2 - 7x + 2) + (-x^2 + x - 2)$ — Original polynomials

$\quad = (2x^2 + 4x^2 - x^2) + (-x - 7x + x) + (3 + 2 - 2)$ — Group like terms.

$\quad = 5x^2 - 7x + 3$ — Combine like terms.

Example 4 Adding Polynomials Vertically

Use a vertical format to find each sum.

a. $(-4x^3 - 2x^2 + x - 5) + (2x^3 + 3x + 4)$

b. $(5x^3 + 2x^2 - x + 7) + (3x^2 - 4x + 7) + (-x^3 + 4x^2 - 2x - 8)$

Solution

a. $-4x^3 - 2x^2 + x - 5$

$\underline{2x^3 + 3x + 4}$

$-2x^3 - 2x^2 + 4x - 1$

b. $5x^3 + 2x^2 - x + 7$

$3x^2 - 4x + 7$

$\underline{-x^3 + 4x^2 - 2x - 8}$

$4x^3 + 9x^2 - 7x + 6$

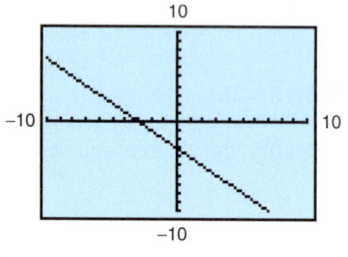
Study Tip

When you use a vertical format to add polynomials, be sure that you line up the *like terms*.

Subtracting Polynomials

To subtract one polynomial from another, you *add the opposite* by changing the
sign of each term of the polynomial that is being subtracted and then adding the
resulting like terms. Note how $(x^2 - 1)$ is subtracted from $(2x^2 - 4)$.

$$(2x^2 - 4) - (x^2 - 1) = 2x^2 - 4 - x^2 + 1 \qquad \text{Distributive Property}$$

$$= (2x^2 - x^2) + (-4 + 1) \qquad \text{Group like terms.}$$

$$= x^2 - 3 \qquad \text{Combine like terms.}$$

Recall from the Distributive Property that

$$-(x^2 - 1) = (-1)(x^2 - 1) = -x^2 + 1.$$

Example 5 Subtracting Polynomials Horizontally

Use a horizontal format to find each difference.

a. $(2x^2 + 3) - (3x^2 - 4)$

b. $(3x^3 - 4x^2 + 3) - (x^3 + 3x^2 - x - 4)$

Solution

a. $(2x^2 + 3) - (3x^2 - 4) = 2x^2 + 3 - 3x^2 + 4 \qquad$ Distributive Property

$$= (2x^2 - 3x^2) + (3 + 4) \qquad \text{Group like terms.}$$

$$= -x^2 + 7 \qquad \text{Combine like terms.}$$

b. $(3x^3 - 4x^2 + 3) - (x^3 + 3x^2 - x - 4) \qquad$ Original polynomials

$$= 3x^3 - 4x^2 + 3 - x^3 - 3x^2 + x + 4 \qquad \text{Distributive Property}$$

$$= (3x^3 - x^3) + (-4x^2 - 3x^2) + (x) + (3 + 4) \qquad \text{Group like terms.}$$

$$= 2x^3 - 7x^2 + x + 7 \qquad \text{Combine like terms.}$$

Example 6 Combining Polynomials Horizontally

Use a horizontal format to perform the indicated operations.

$$(x^2 - 2x + 1) - [(x^2 + x - 3) + (-2x^2 - 4x)]$$

Solution

$$(x^2 - 2x + 1) - [(x^2 + x - 3) + (-2x^2 - 4x)] \qquad \text{Original polynomials}$$

$$= (x^2 - 2x + 1) - [(x^2 - 2x^2) + (x - 4x) + (-3)] \qquad \text{Group like terms.}$$

$$= (x^2 - 2x + 1) - [-x^2 - 3x - 3] \qquad \text{Combine like terms.}$$

$$= x^2 - 2x + 1 + x^2 + 3x + 3 \qquad \text{Distributive Property}$$

$$= (x^2 + x^2) + (-2x + 3x) + (1 + 3) \qquad \text{Group like terms.}$$

$$= 2x^2 + x + 4 \qquad \text{Combine like terms.}$$

Be especially careful to use the correct signs when subtracting one polynomial from another. One of the most common mistakes in algebra is to forget to change signs correctly when subtracting one expression from another. Here is an example.

Wrong sign

$$(x^2 + 3) - (x^2 + 2x - 2) \neq x^2 + 3 - x^2 + 2x - 2 \qquad \text{Common error}$$

Wrong sign

Note that the error is forgetting to change *all* of the signs in the polynomial that is being subtracted. Here is the correct way to perform the subtraction.

Correct sign

$$(x^2 + 3) - (x^2 + 2x - 2) = x^2 + 3 - x^2 - 2x + 2 \qquad \text{Correct}$$

Correct sign

Just as you did for addition, you can use a vertical format to subtract one polynomial from another. (The vertical format does not work well with subtractions involving three or more polynomials.) When using a vertical format, write the polynomial being subtracted underneath the one from which it is being subtracted. Be sure to line up like terms in vertical columns.

Example 7   Subtracting Polynomials Vertically

Use a vertical format to find each difference.

a. $(3x^2 + 7x - 6) - (3x^2 + 7x)$

b. $(5x^3 - 2x^2 + x) - (4x^2 - 3x + 2)$

c. $(4x^4 - 2x^3 + 5x^2 - x + 8) - (3x^4 - 2x^3 + 3x - 4)$

Solution

a.
$$\begin{array}{r} (3x^2 + 7x - 6) \\ -(3x^2 + 7x \qquad) \\ \hline \end{array} \qquad \Longrightarrow \qquad \begin{array}{r} 3x^2 + 7x - 6 \\ -3x^2 - 7x \qquad \\ \hline -6 \end{array} \qquad \text{Change signs and add.}$$

b.
$$\begin{array}{r} (5x^3 - 2x^2 + \ x \qquad) \\ -(\qquad 4x^2 - 3x + 2) \\ \hline \end{array} \qquad \Longrightarrow \qquad \begin{array}{r} 5x^3 - 2x^2 + \ x \qquad \\ - 4x^2 + 3x - 2 \\ \hline 5x^3 - 6x^2 + 4x - 2 \end{array} \qquad \text{Change signs and add.}$$

c.
$$\begin{array}{r} (4x^4 - 2x^3 + 5x^2 - \ x + 8) \\ -(3x^4 - 2x^3 \qquad + 3x - 4) \\ \hline \end{array} \qquad \Longrightarrow \qquad \begin{array}{r} 4x^4 - 2x^3 + 5x^2 - \ x + \ 8 \\ -3x^4 + 2x^3 \qquad - 3x + \ 4 \\ \hline x^4 \qquad + 5x^2 - 4x + 12 \end{array}$$

In Example 7, try using a horizontal arrangement to perform the subtractions.

Example 8　Combining Polynomials

Perform the indicated operations and simplify.

$$(3x^2 - 7x + 2) - (4x^2 + 6x - 1) + (-x^2 + 4x + 5)$$

Solution

$$(3x^2 - 7x + 2) - (4x^2 + 6x - 1) + (-x^2 + 4x + 5)$$
$$= 3x^2 - 7x + 2 - 4x^2 - 6x + 1 - x^2 + 4x + 5$$
$$= (3x^2 - 4x^2 - x^2) + (-7x - 6x + 4x) + (2 + 1 + 5)$$
$$= -2x^2 - 9x + 8$$

Example 9　Combining Polynomials

Perform the indicated operations and simplify.

$$(-2x^2 + 4x - 3) - [(4x^2 - 5x + 8) - 2(-x^2 + x + 3)]$$

Solution

$$(-2x^2 + 4x - 3) - [(4x^2 - 5x + 8) - 2(-x^2 + x + 3)]$$
$$= (-2x^2 + 4x - 3) - [4x^2 - 5x + 8 + 2x^2 - 2x - 6]$$
$$= (-2x^2 + 4x - 3) - [(4x^2 + 2x^2) + (-5x - 2x) + (8 - 6)]$$
$$= (-2x^2 + 4x - 3) - [6x^2 - 7x + 2]$$
$$= -2x^2 + 4x - 3 - 6x^2 + 7x - 2$$
$$= (-2x^2 - 6x^2) + (4x + 7x) + (-3 - 2)$$
$$= -8x^2 + 11x - 5$$

Example 10　Geometry: Area of a Region

Find an expression for the area of the shaded region shown in Figure 5.1.

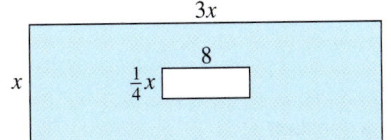

Figure 5.1

Solution

To find a polynomial that represents the area of the shaded region, subtract the area of the inner rectangle from the area of the outer rectangle, as follows.

$$\begin{array}{ccc} \text{Area of} & = & \text{Area of} & - & \text{Area of} \\ \text{shaded region} & & \text{outer rectangle} & & \text{inner rectangle} \end{array}$$

$$= 3x(x) - 8\left(\frac{1}{4}x\right)$$
$$= 3x^2 - 2x$$

5.2 Exercises

Review   Concepts, Skills, and Problem Solving

Keep mathematically in shape by doing these exercises *before* the problems of this section.

Properties and Definitions

1. *Writing*   In your own words, state the definition of an algebraic expression.

2. *Writing*   State the definition of the terms of an algebraic expression.

Simplifying Expressions

In Exercises 3–6, use the Distributive Property to expand the expression.

3. $10(x - 1)$

4. $4(3 - 2z)$

5. $-\frac{1}{2}(4 - 6x)$

6. $-25(2x - 3)$

In Exercises 7–10, simplify the expression.

7. $8y - 2x + 7x - 10y$

8. $\frac{5}{6}x - \frac{2}{3}x + 8$

9. $10(x - 1) - 3(x + 2)$

10. $-3[x + (2 + 3x)]$

Graphing Equations

In Exercises 11 and 12, graph the equation. Use a graphing calculator to verify your graph.

11. $y = 2 + \frac{3}{2}x$

12. $y = |x - 1| + x$

Developing Skills

In Exercises 1–8, determine whether the expression is a polynomial. If it is not, explain why. See Example 1.

1. $9 - z$

2. $t^2 - 4$

3. $x^{2/3} + 8$

4. $9 - z^{1/2}$

5. $6x^{-1}$

6. $1 - 4x^{-2}$

7. $z^2 - 3z + \frac{1}{4}$

8. $t^3 - 3t + 4$

In Exercises 9–18, write the polynomial in standard form. Then identify its degree and leading coefficient. See Example 2.

9. $12x + 9$

10. $4 - 7y$

11. $7x - 5x^2 + 10$

12. $5 - x + 15x^2$

13. $8x + 2x^5 - x^2 - 1$

14. $5x^3 - 3x^2 + 10$

15. 10

16. -32

17. $v_0 t - 16t^2$ (v_0 is a constant.)

18. $64 - \frac{1}{2}at^2$ (a is a constant.)

In Exercises 19–24, determine whether the polynomial is a monomial, binomial, or trinomial.

19. $14y - 2$

20. -16

21. $93z^2$

22. $a^2 + 2a - 9$

23. $4x + 18x^2 - 5$

24. $6x^2 + x$

In Exercises 25–30, give an example of a polynomial in one variable satisfying the condition. (*Note:* There are many correct answers.)

25. A binomial of degree 3

26. A trinomial of degree 4

27. A monomial of degree 2

28. A binomial of degree 5

29. A trinomial of degree 6

30. A monomial of degree 0

In Exercises 31–42, use a horizontal format to find the sum. See Example 3.

31. $(11x - 2) + (3x + 8)$

32. $(-2x + 4) + (x - 6)$

33. $(3z^2 - z + 2) + (z^2 - 4)$

34. $(6x^4 + 8x) + (4x - 6)$

35. $b^2 + (b^3 - 2b^2 + 3) + (b^3 - 3)$

36. $(3x^2 - x) + 5x^3 + (-4x^3 + x^2 - 8)$

37. $(2ab - 3) + (a^2 - 2ab) + (4b^2 - a^2)$

38. $(uv - 3) + (4uv + 1)$

39. $\left(\frac{2}{3}y^2 - \frac{3}{4}\right) + \left(\frac{5}{6}y^2 + 2\right)$

40. $\left(\frac{3}{4}x^3 - \frac{1}{2}\right) + \left(\frac{1}{8}x^3 + 3\right)$

41. $(0.1t^3 - 3.4t^2) + (1.5t^3 - 7.3)$

42. $(0.7x^2 - 0.2x + 2.5) + (7.4x - 3.9)$

In Exercises 43–56, use a vertical format to find the sum. See Example 4.

43. $2x + 5$
  $3x + 8$

44. $10x - 7$
  $6x + 4$

45. $-2x + 10$
  $x - 38$

46. $4x^2 + 13$
  $3x^2 - 11$

47. $(-x^3 + 3) + (3x^3 + 2x^2 + 5)$

48. $(2z^3 + 3z - 2) + (z^2 - 2z)$

49. $(3x^4 - 2x^3 - 4x^2 + 2x - 5) + (x^2 - 7x + 5)$

50. $(x^5 - 4x^3 + x + 9) + (2x^4 + 3x^3 - 3)$

51. $(x^2 - 4) + (2x^2 + 6)$

52. $(x^3 + 2x - 3) + (4x + 5)$

53. $(x^2 - 2x + 2) + (x^2 + 4x) + 2x^2$

54. $(5y + 10) + (y^2 - 3y - 2) + (2y^2 + 4y - 3)$

55. Add $8y^3 + 7$ to $5 - 3y^3$.

56. Add $2z - 8z^2 - 3$ to $z^2 + 5z$.

In Exercises 57–66, use a horizontal format to find the difference. See Examples 5 and 6.

57. $(11x - 8) - (2x + 3)$

58. $(9x + 2) - (15x - 4)$

59. $(x^2 - x) - (x - 2)$

60. $(x^2 - 4) - (x^2 - 4)$

61. $(4 - 2x - x^3) - (3 - 2x + 2x^3)$

62. $(t^4 - 2t^2) - (3t^2 - t^4 - 5)$

63. $10 - (u^2 + 5)$

64. $(z^3 + z^2 + 1) - z^2$

65. $(x^5 - 3x^4 + x^3 - 5x + 1) - (4x^5 - x^3 + x - 5)$

66. $(t^4 + 5t^3 - t^2 + 8t - 10) -$
  $(t^4 + t^3 + 2t^2 + 4t - 7)$

In Exercises 67–80, use a vertical format to find the difference. See Example 7.

67. $2x - 2$
  $- (x - 1)$

68. $9x + 7$
  $- (3x + 9)$

69. $2x^2 - x + 2$
  $- (3x^2 + x - 1)$

70. $y^4 - 2$
  $- (y^4 + 2)$

71. $(-3x^3 - 4x^2 + 2x - 5) - (2x^4 + 2x^3 - 4x + 5)$

72. $(12x^3 + 25x^2 - 15) - (-2x^3 + 18x^2 - 3x)$

73. $(2 - x^3) - (2 + x^3)$

74. $(4z^3 - 6) - (-z^3 + z - 2)$

75. $(4t^3 - 3t + 5) - (3t^2 - 3t - 10)$

76. $(-s^2 - 3) - (2s^2 + 10s)$

77. $(6x^3 - 3x^2 + x) - [(x^3 + 3x^2 + 3) + (x - 3)]$

78. $(y^2 - y) - [(2y^2 + y) - (4y^2 - y + 2)]$

79. Subtract $7x^3 - 4x + 5$ from $10x^3 + 15$.

80. Subtract $y^5 - y^4$ from $y^2 + 3y^4$.

In Exercises 81–94, perform the indicated operations and simplify. See Examples 8 and 9.

81. $(6x - 5) - (8x + 15)$

82. $(2x^2 + 1) + (x^2 - 2x + 1)$

83. $-(x^3 - 2) + (4x^3 - 2x)$

84. $-(5x^2 - 1) - (-3x^2 + 5)$

85. $2(x^4 + 2x) + (5x + 2)$

86. $(z^4 - 2z^2) + 3(z^4 + 4)$

87. $(15x^2 - 6) - (-8x^3 - 14x^2 - 17)$

88. $(15x^4 - 18x - 19) - (-13x^4 - 5x + 15)$

89. $5z - [3z - (10z + 8)]$

90. $(y^3 + 1) - [(y^2 + 1) + (3y - 7)]$

91. $2(t^2 + 5) - 3(t^2 + 5) + 5(t^2 + 5)$

92. $-10(u + 1) + 8(u - 1) - 3(u + 6)$

93. $8v - 6(3v - v^2) + 10(10v + 3)$

94. $3(x^2 - 2x + 3) - 4(4x + 1) - (3x^2 - 2x)$

Solving Problems

▲ *Geometry*   In Exercises 95 and 96, find an expression for the perimeter of the figure.

95.

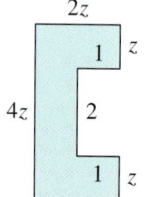

96.

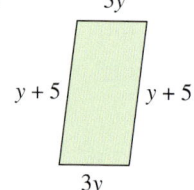

▲ *Geometry*   In Exercises 97–100, find an expression for the area of the shaded region of the figure. See Example 10.

97.

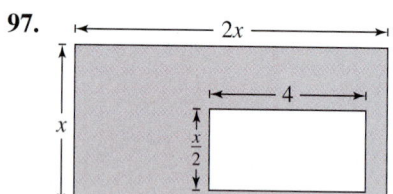

98.

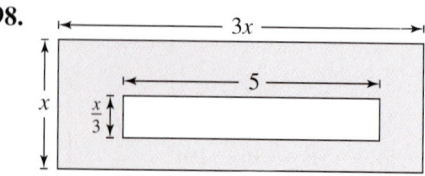

99.

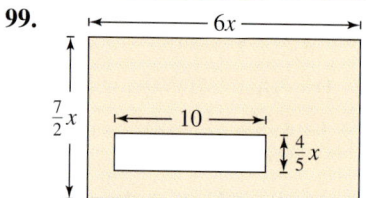

100.

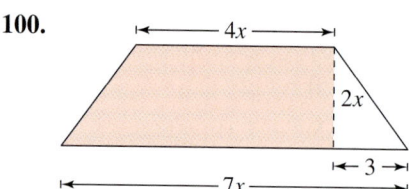

101. *Comparing Models*   The numbers of daily morning M and evening E newspapers for the years 1995 through 2000 can be modeled by

$$M = -0.29t^2 + 24.7t + 543, \quad 5 \le t \le 10$$

and

$$E = -31.8t + 1042, \quad 5 \le t \le 10$$

where t represents the year, with $t = 5$ corresponding to 1995.   (Source: Editor & Publisher Co.)

(a) Add the polynomials to find a model for the total number T of daily newspapers.

(b) ▦ Use a graphing calculator to graph all three models.

(c) ▦ Use the graphs from part (b) to determine whether the numbers of morning, evening, and total newspapers are increasing or decreasing.

102. *Cost, Revenue, and Profit* The cost C of producing x units of a product is $C = 100 + 30x$. The revenue R for selling x units is $R = 90x - x^2$, where $0 \le x \le 40$. The profit P is the difference between revenue and cost.

(a) Perform the subtraction required to find the polynomial representing profit P.

(b) ⊞ Use a graphing calculator to graph the polynomial representing profit.

(c) ⊞ Determine the profit when 30 units are produced and sold. Use the graph in part (b) to predict the change in profit if x is some value other than 30.

Explaining Concepts

103. ◔ Answer parts (a)–(c) of Motivating the Chapter on page 292.

104. *Writing*✐ Explain the difference between the degree of a term of a polynomial and the degree of a polynomial.

105. *Writing*✐ Determine which of the two statements is always true. Is the statement not selected always false? Explain.

(a) "A polynomial is a trinomial."

(b) "A trinomial is a polynomial."

106. *Writing*✐ In your own words, define "like terms." What is the only factor of like terms that can differ?

107. *Writing*✐ Describe how to combine like terms. What operations are used?

108. *Writing*✐ Is a polynomial an algebraic expression? Explain.

109. *Writing*✐ Is the sum of two binomials always a binomial? Explain.

110. *Writing*✐ Write a paragraph that explains how the adage "You can't add apples and oranges" might relate to adding two polynomials. Include several examples to illustrate the applicability of this statement.

111. *Writing*✐ In your own words, explain how to subtract polynomials. Give an example.

Mid-Chapter Quiz

Take this quiz as you would take a quiz in class. After you are done, check your work against the answers in the back of the book.

In Exercises 1–4, simplify the expression. (Assume that no denominator is zero.)

1. $(3a^2b)^2$

2. $(-3xy)^2(2x^2y)^3$

3. $\dfrac{-12x^3y}{9x^5y^2}$

4. $\dfrac{3t^3}{(-6t)^2}$

In Exercises 5 and 6, rewrite the expression using only positive exponents.

5. $5x^{-2}y^{-3}$

6. $\dfrac{3x^{-2}y}{5z^{-1}}$

In Exercises 7 and 8, use rules of exponents to simplify the expression using only positive exponents. (Assume that no variable is zero.)

7. $(3a^{-3}b^2)^{-2}$

8. $(4t^{-3})^0$

9. Write the number 9,460,000,000 in scientific notation.

10. Write the number 5.021×10^{-8} in decimal notation.

11. Explain why $x^2 + 2x - 3x^{-1}$ is not a polynomial.

12. Determine the degree and the leading coefficient of the polynomial $-3x^4 + 2x^2 - x$.

13. Give an example of a trinomial in one variable of degree 5.

In Exercises 14–17, perform the indicated operations and simplify.

14. $(y^2 + 3y - 1) + (4 + 3y)$

15. $(3v^2 - 5) - (v^3 + 2v^2 - 6v)$

16. $9s - [6 - (s - 5) + 7s]$

17. $-3(4 - x) + 4(x^2 + 2) - (x^2 - 2x)$

In Exercises 18 and 19, use a vertical format to find the sum.

18. $5x^4 \quad + 2x^2 + \quad x - 3$
$\quad \quad \underline{3x^3 - 2x^2 - 3x + 5}$

19. $2x^3 + x^2 \quad \quad - 8$
$\quad \quad \underline{\quad \quad 5x^2 - 3x - 9}$

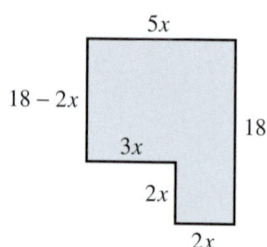

Figure for 20

20. Find an expression for the perimeter of the figure.

5.3 Multiplying Polynomials: Special Products

Holly Harris/Getty Images

What You Should Learn

1️⃣ Find products with monomial multipliers.

2️⃣ Multiply binomials using the Distributive Property and the FOIL Method.

3️⃣ Multiply polynomials using a horizontal or vertical format.

4️⃣ Identify and use special binomial products.

Why You Should Learn It

Multiplying polynomials enables you to model and solve real-life problems. For instance, in Exercise 129 on page 327, you will multiply polynomials to find the total consumption of milk in the United States.

1️⃣ Find products with monomial multipliers.

Monomial Multipliers

To multiply polynomials, you use many of the rules for simplifying algebraic expressions. You may want to review these rules from Section 2.2 and Section 5.1.

1. The Distributive Property

2. Combining like terms

3. Removing symbols of grouping

4. Rules of exponents

The simplest type of polynomial multiplication involves a monomial multiplier. The product is obtained by direct application of the Distributive Property. For instance, to multiply the monomial x by the polynomial $(2x + 5)$, multiply *each* term of the polynomial by x.

$$(x)(2x + 5) = (x)(2x) + (x)(5) = 2x^2 + 5x$$

Example 1   Finding Products with Monomial Multipliers

Find each product.

a. $(3x - 7)(-2x)$ **b.** $3x^2(5x - x^3 + 2)$ **c.** $(-x)(2x^2 - 3x)$

Solution

a. $(3x - 7)(-2x) = 3x(-2x) - 7(-2x)$ Distributive Property

$$= -6x^2 + 14x$$ Write in standard form.

b. $3x^2(5x - x^3 + 2)$

$$= (3x^2)(5x) - (3x^2)(x^3) + (3x^2)(2)$$ Distributive Property

$$= 15x^3 - 3x^5 + 6x^2$$ Rules of exponents

$$= -3x^5 + 15x^3 + 6x^2$$ Write in standard form.

c. $(-x)(2x^2 - 3x) = (-x)(2x^2) - (-x)(3x)$ Distributive Property

$$= -2x^3 + 3x^2$$ Write in standard form.

2 Multiply binomials using the Distributive Property and the FOIL Method.

Multiplying Binomials

To multiply two binomials, you can use both (left and right) forms of the Distributive Property. For example, if you treat the binomial $(5x + 7)$ as a single quantity, you can multiply $(3x - 2)$ by $(5x + 7)$ as follows.

$$(3x - 2)(5x + 7) = 3x(5x + 7) - 2(5x + 7)$$
$$= (3x)(5x) + (3x)(7) - (2)(5x) - 2(7)$$
$$= 15x^2 + 21x - 10x - 14$$

Product of First terms	Product of Outer terms	Product of Inner terms	Product of Last terms

$$= 15x^2 + 11x - 14$$

With practice, you should be able to multiply two binomials without writing out all of the steps above. In fact, the four products in the boxes above suggest that you can write the product of two binomials in just one step. This is called the **FOIL Method.** Note that the words *first, outer, inner,* and *last* refer to the positions of the terms in the original product.

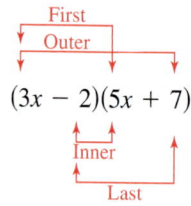

$$(3x - 2)(5x + 7)$$

Technology: Tip

Remember that you can use a graphing calculator to check whether you have performed a polynomial operation correctly. For instance, to check if $(x - 1)(x + 5) = x^2 + 4x - 5$ you can "graph the left side and graph the right side" in the same viewing window, as shown below. Because both graphs are the same, you can conclude that the multiplication was performed correctly.

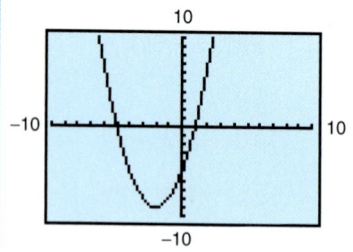

Example 2   Multiplying with the Distributive Property

Use the Distributive Property to find each product.

a. $(x - 1)(x + 5)$

b. $(2x + 3)(x - 2)$

Solution

a. $(x - 1)(x + 5) = x(x + 5) - 1(x + 5)$ Right Distributive Property
$$= x^2 + 5x - x - 5$$ Left Distributive Property
$$= x^2 + (5x - x) - 5$$ Group like terms.
$$= x^2 + 4x - 5$$ Combine like terms.

b. $(2x + 3)(x - 2) = 2x(x - 2) + 3(x - 2)$ Right Distributive Property
$$= 2x^2 - 4x + 3x - 6$$ Left Distributive Property
$$= 2x^2 + (-4x + 3x) - 6$$ Group like terms.
$$= 2x^2 - x - 6$$ Combine like terms.

Example 3   Multiplying Binomials Using the FOIL Method

Use the FOIL Method to find each product.

a. $(x + 4)(x - 4)$

b. $(3x + 5)(2x + 1)$

Solution

$$\qquad\qquad\qquad\quad \text{F}\quad \text{O}\quad \text{I}\quad \text{L}$$

a. $(x + 4)(x - 4) = x^2 - 4x + 4x - 16$

$$= x^2 - 16 \qquad\qquad\qquad \text{Combine like terms.}$$

$$\qquad\qquad\qquad\quad \text{F}\quad \text{O}\quad \text{I}\quad \text{L}$$

b. $(3x + 5)(2x + 1) = 6x^2 + 3x + 10x + 5$

$$= 6x^2 + 13x + 5 \qquad\qquad \text{Combine like terms.}$$

In Example 3(a), note that the outer and inner products add up to zero.

Example 4   A Geometric Model of a Polynomial Product

Use the geometric model shown in Figure 5.2 to show that

$$x^2 + 3x + 2 = (x + 1)(x + 2).$$

Solution

The top of the figure shows that the sum of the areas of the six rectangles is

$$x^2 + (x + x + x) + (1 + 1) = x^2 + 3x + 2.$$

The bottom of the figure shows that the area of the rectangle is

$$(x + 1)(x + 2) = x^2 + 2x + x + 2$$

$$= x^2 + 3x + 2.$$

So, $x^2 + 3x + 2 = (x + 1)(x + 2)$.

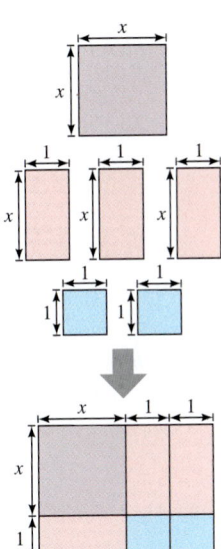

Figure 5.2

Example 5   Simplifying a Polynomial Expression

Simplify the expression and write the result in standard form.

$$(4x + 5)^2$$

Solution

$$(4x + 5)^2 = (4x + 5)(4x + 5) \qquad\qquad \text{Repeated multiplication}$$

$$= 16x^2 + 20x + 20x + 25 \qquad\qquad \text{Use FOIL Method.}$$

$$= 16x^2 + 40x + 25 \qquad\qquad \text{Combine like terms.}$$

Example 6   Simplifying a Polynomial Expression

Simplify the expression and write the result in standard form.

$$(3x^2 - 2)(4x + 7) - (4x)^2$$

Solution

$$(3x^2 - 2)(4x + 7) - (4x)^2$$

$$= 12x^3 + 21x^2 - 8x - 14 - (4x)^2 \qquad \text{Use FOIL Method.}$$

$$= 12x^3 + 21x^2 - 8x - 14 - 16x^2 \qquad \text{Square monomial.}$$

$$= 12x^3 + 5x^2 - 8x - 14 \qquad \text{Combine like terms.}$$

3 Multiply polynomials using a horizontal or vertical format.

Multiplying Polynomials

The FOIL Method for multiplying two binomials is simply a device for guaranteeing that *each term of one binomial is multiplied by each term of the other binomial.*

$$(ax + b)(cx + d) = ax(cx) + ax(d) + b(cx) + b(d)$$

$$\text{F} \qquad \text{O} \qquad \text{I} \qquad \text{L}$$

This same rule applies to the product of any two polynomials: *each term of one polynomial must be multiplied by each term of the other polynomial.* This can be accomplished using either a horizontal or a vertical format.

Example 7   Multiplying Polynomials (Horizontal Format)

Use a horizontal format to find each product.

a. $(x - 4)(x^2 - 4x + 2)$ **b.** $(2x^2 - 7x + 1)(4x + 3)$

Solution

a. $(x - 4)(x^2 - 4x + 2)$

$$= x(x^2 - 4x + 2) - 4(x^2 - 4x + 2) \qquad \text{Distributive Property}$$

$$= x^3 - 4x^2 + 2x - 4x^2 + 16x - 8 \qquad \text{Distributive Property}$$

$$= x^3 - 8x^2 + 18x - 8 \qquad \text{Combine like terms.}$$

b. $(2x^2 - 7x + 1)(4x + 3)$

$$= (2x^2 - 7x + 1)(4x) + (2x^2 - 7x + 1)(3) \qquad \text{Distributive Property}$$

$$= 8x^3 - 28x^2 + 4x + 6x^2 - 21x + 3 \qquad \text{Distributive Property}$$

$$= 8x^3 - 22x^2 - 17x + 3 \qquad \text{Combine like terms.}$$

Example 8   Multiplying Polynomials (Vertical Format)

Use a vertical format to find the product of $(3x^2 + x - 5)$ and $(2x - 1)$.

Solution

With a vertical format, line up like terms in the same vertical columns, just as you align digits in whole number multiplication.

$$
\begin{array}{r}
3x^2 + x - 5 \\
\times 2x - 1 \\
\hline
-3x^2 - x + 5 \\
6x^3 + 2x^2 - 10x \\
\hline
6x^3 - x^2 - 11x + 5
\end{array}
$$

Place polynomial with most terms on top.

Line up like terms.

$-1(3x^2 + x - 5)$

$2x(3x^2 + x - 5)$

Combine like terms in columns.

Example 9   Multiplying Polynomials (Vertical Format)

Use a vertical format to find the product of $(4x^3 + 8x - 1)$ and $(2x^2 + 3)$.

Solution

$$
\begin{array}{r}
4x^3 + 8x - 1 \\
\times 2x^2 + 3 \\
\hline
12x^3 + 24x - 3 \\
8x^5 + 16x^3 - 2x^2 \\
\hline
8x^5 + 28x^3 - 2x^2 + 24x - 3
\end{array}
$$

Place polynomial with most terms on top.

Line up like terms.

$3(4x^3 + 8x - 1)$

$2x^2(4x^3 + 8x - 1)$

Combine like terms in columns.

When multiplying two polynomials, it is best to write each in standard form before using either the horizontal or vertical format. This is illustrated in the next example.

Example 10   Multiplying Polynomials (Vertical Format)

Write the polynomials in standard form and use a vertical format to find the product of $(x + 3x^2 - 4)$ and $(5 + 3x - x^2)$.

Solution

$$
\begin{array}{r}
3x^2 + x - 4 \\
\times -x^2 + 3x + 5 \\
\hline
15x^2 + 5x - 20 \\
9x^3 + 3x^2 - 12x \\
-3x^4 - x^3 + 4x^2 \\
\hline
-3x^4 + 8x^3 + 22x^2 - 7x - 20
\end{array}
$$

Write in standard form.

Write in standard form.

$5(3x^2 + x - 4)$

$3x(3x^2 + x - 4)$

$-x^2(3x^2 + x - 4)$

Combine like terms.

Example 11   Multiplying Polynomials

Multiply $(x - 3)^3$.

Solution

To raise $(x - 3)$ to the third power, you can use two steps. First, because $(x - 3)^3 = (x - 3)^2(x - 3)$, find the product $(x - 3)^2$.

$$(x - 3)^2 = (x - 3)(x - 3) \qquad \text{Repeated multiplication}$$
$$= x^2 - 3x - 3x + 9 \qquad \text{Use FOIL Method.}$$
$$= x^2 - 6x + 9 \qquad \text{Combine like terms.}$$

Now multiply $x^2 - 6x + 9$ by $x - 3$, as follows.

$$(x^2 - 6x + 9)(x - 3) = (x^2 - 6x + 9)(x) - (x^2 - 6x + 9)(3)$$
$$= x^3 - 6x^2 + 9x - 3x^2 + 18x - 27$$
$$= x^3 - 9x^2 + 27x - 27.$$

So, $(x - 3)^3 = x^3 - 9x^2 + 27x - 27$.

---

4 Identify and use special binomial products.

Special Products

Some binomial products, such as those in Examples 3(a) and 5, have special forms that occur frequently in algebra. The product

$$(x + 4)(x - 4)$$

is called a **product of the sum and difference of two terms.** With such products, the two middle terms cancel, as follows.

$$(x + 4)(x - 4) = x^2 - 4x + 4x - 16 \qquad \text{Sum and difference of two terms}$$
$$= x^2 - 16 \qquad \text{Product has no middle term.}$$

Another common type of product is the **square of a binomial.**

$$(4x + 5)^2 = (4x + 5)(4x + 5) \qquad \text{Square of a binomial}$$
$$= 16x^2 + 20x + 20x + 25 \qquad \text{Use FOIL Method.}$$
$$= 16x^2 + 40x + 25 \qquad \begin{array}{l}\text{Middle term is twice the product}\\ \text{of the terms of the binomial.}\end{array}$$

Study Tip

You should learn to recognize the patterns of the two special products at the right. The FOIL Method can be used to verify each rule.

In general, when a binomial is squared, the resulting middle term is always twice the product of the two terms.

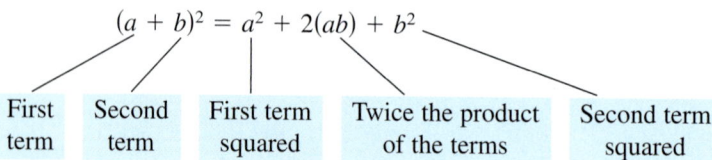

$$(a + b)^2 = a^2 + 2(ab) + b^2$$

| First term | Second term | First term squared | Twice the product of the terms | Second term squared |

Be sure to include the middle term. For instance, $(a + b)^2$ is not equal to $a^2 + b^2$.

Special Products

Let a and b be real numbers, variables, or algebraic expressions.

| *Special Product* | *Example* |

Sum and Difference of Two Terms:

$$(a + b)(a - b) = a^2 - b^2 \qquad\qquad (2x - 5)(2x + 5) = 4x^2 - 25$$

Square of a Binomial:

$$(a + b)^2 = a^2 + 2ab + b^2 \qquad (3x + 4)^2 = 9x^2 + 2(3x)(4) + 16$$
$$= 9x^2 + 24x + 16$$

$$(a - b)^2 = a^2 - 2ab + b^2 \qquad (x - 7)^2 = x^2 - 2(x)(7) + 49$$
$$= x^2 - 14x + 49$$

Example 12   Finding the Product of the Sum and Difference of Two Terms

Multiply $(x + 2)(x - 2)$.

Solution

$$(x + 2)(x - 2) = (x)^2 - (2)^2$$
$$= x^2 - 4$$

Example 13   Finding the Product of the Sum and Difference of Two Terms

Multiply $(5x - 6)(5x + 6)$.

Solution

$$(5x - 6)(5x + 6) = (5x)^2 - (6)^2$$
$$= 25x^2 - 36$$

Example 14   Squaring a Binomial

Multiply $(4x - 9)^2$.

Solution

$$(4x - 9)^2 = (4x)^2 - 2(4x)(9) + (9)^2$$
$$= 16x^2 - 72x + 81$$

Example 15 Squaring a Binomial

Multiply $(3x + 7)^2$.

Solution

2nd term   Twice the product of the terms
1st term   (1st term)^2 (2nd term)^2

$$(3x + 7)^2 = (3x)^2 + 2(3x)(7) + (7)^2$$
$$= 9x^2 + 42x + 49$$

Example 16 Squaring a Binomial

Multiply $(6 - 5x^2)^2$.

Solution

2nd term   Twice the product of the terms
1st term   (1st term)^2 (2nd term)^2

$$(6 - 5x^2)^2 = (6)^2 - 2(6)(5x^2) + (5x^2)^2$$
$$= 36 - 60x^2 + (5)^2(x^2)^2$$
$$= 36 - 60x^2 + 25x^4$$

Example 17 Finding the Dimensions of a Golf Tee

A landscaper wants to reshape a square tee area for the ninth hole of a golf course. The new tee area is to have one side 2 feet longer and the adjacent side 6 feet longer than the original tee. (See Figure 5.3.) The new tee has 204 square feet more area than the original tee. What are the dimensions of the original tee?

Solution

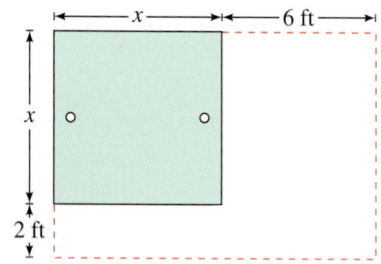

Figure 5.3

Verbal Model: New area = Old area + 204

Labels: Original length = original width = x (feet)
New length = $x + 6$ (feet)
New width = $x + 2$ (feet)

Equation: $(x + 6)(x + 2) = x^2 + 204$ x^2 is original area.

$x^2 + 8x + 12 = x^2 + 204$ Multiply factors.

$8x + 12 = 204$ Subtract x^2 from each side.

$8x = 192$ Subtract 12 from each side.

$x = 24$ Simplify.

The original tee measured 24 feet by 24 feet.

5.3 Exercises

Review *Concepts, Skills, and Problem Solving*

Keep mathematically in shape by doing these exercises *before* the problems of this section.

Properties and Definitions

1. *Writing* ✏ Relative to the *x*- and *y*-axes, explain the meaning of each coordinate of the point $(3, -2)$.

2. A point lies four units from the *x*-axis and three units from the *y*-axis. Give the ordered pair for such a point in each quadrant.

Simplifying Expressions

In Exercises 3–8, simplify the expression.

3. $\frac{3}{4}x - \frac{5}{2} + \frac{3}{2}x$

4. $4 - 2(3 - x)$

5. $2(x - 4) + 5x$

6. $4(3 - y) + 2(y + 1)$

7. $-3(z - 2) - (z - 6)$

8. $(u - 2) - 3(2u + 1)$

Problem Solving

9. *Sales Commission* Your sales commission rate is 5.5%. Your commission is $1600. How much did you sell?

10. *Distance* A jogger leaves a location on a fitness trail running at a rate of 4 miles per hour. Fifteen minutes later, a second jogger leaves from the same location running at 5 miles per hour. How long will it take the second jogger to overtake the first, and how far will each have run at that point? Use a diagram to help solve the problem.

Graphing Equations

🖩 In Exercises 11 and 12, use a graphing calculator to graph the equation. Identify any intercepts.

11. $y = 4 - \frac{1}{2}x$

12. $y = x(x - 4)$

Developing Skills

In Exercises 1–50, perform the multiplication and simplify. See Examples 1–3, 5, and 6.

1. $x(-2x)$

2. $y(-3y)$

3. $t^2(4t)$

4. $3u(u^4)$

5. $\left(\frac{x}{4}\right)(10x)$

6. $9x\left(\frac{x}{12}\right)$

7. $(-2b^2)(-3b)$

8. $(-4x)(-5x)$

9. $y(3 - y)$

10. $z(z - 3)$

11. $-x(x^2 - 4)$

12. $-t(10 - 9t^2)$

13. $-3x(2x^2 + 5)$

14. $-5u(u^2 + 4)$

15. $-4x(3 + 3x^2 - 6x^3)$

16. $5v(5 - 4v + 5v^2)$

17. $3x(x^2 - 2x + 1)$

18. $y(4y^2 + 2y - 3)$

19. $2x(x^2 - 2x + 8)$

20. $3y(y^2 - y + 5)$

21. $4t^3(t - 3)$

22. $-2t^4(t + 6)$

23. $x^2(4x^2 - 3x + 1)$

24. $y^2(2y^2 + y - 5)$

25. $-3x^3(4x^2 - 6x + 2)$

26. $5u^4(2u^3 - 3u + 3)$

27. $-2x(-3x)(5x + 2)$

28. $4x(-2x)(x^2 - 1)$

29. $-2x(-6x^4) - 3x^2(2x^2)$

30. $-8y(-5y^4) - 2y^2(5y^3)$

31. $(x + 3)(x + 4)$

32. $(x - 5)(x + 10)$

33. $(3x - 5)(2x + 1)$

34. $(7x - 2)(4x - 3)$

35. $(2x - y)(x - 2y)$

36. $(x + y)(x + 2y)$

37. $(2x + 4)(x + 1)$

38. $(4x + 3)(2x - 1)$

39. $(6 - 2x)(4x + 3)$

40. $(8x - 6)(5 - 4x)$

41. $(3x - 2y)(x - y)$

42. $(7x + 5y)(x + y)$

43. $(3x^2 - 4)(x + 2)$

44. $(5x^2 - 2)(x - 1)$

45. $(2x^2 + 4)(x^2 + 6)$

46. $(7x^2 - 3)(2x^2 - 4)$

47. $(3s + 1)(3s + 4) - (3s)^2$

48. $(2t + 5)(4t - 2) - (2t)^2$

49. $(4x^2 - 1)(2x + 8) + (-x^2)^3$

50. $(3 - 3x^2)(4 - 5x^2) - (-x^3)^2$

In Exercises 51–64, use a horizontal format to find the product. See Example 7.

51. $(x + 10)(x + 2)$

52. $(x - 1)(x + 3)$

53. $(2x - 5)(x + 2)$

54. $(3x - 2)(2x - 3)$

55. $(x + 1)(x^2 + 2x - 1)$

56. $(x - 3)(x^2 - 3x + 4)$

57. $(x^3 - 2x + 1)(x - 5)$

58. $(x + 1)(x^2 - x + 1)$

59. $(x - 2)(x^2 + 2x + 4)$

60. $(x + 9)(x^2 - x - 4)$

61. $(x^2 + 3)(x^2 - 6x + 2)$

62. $(x^2 + 3)(x^2 - 2x + 3)$

63. $(3x^2 + 1)(x^2 - 4x - 2)$

64. $(x^2 + 2x + 5)(4x^3 - 2)$

In Exercises 65–80, use a vertical format to find the product. See Examples 8–10.

65. $x + 3$
       $\underline{\times\ x - 2}$

66. $2x - 1$
       $\underline{\times\ 5x + 1}$

67. $4x^4 - 6x^2 + 9$
       $\underline{\times\qquad 2x\ + 3}$

68. $x^2 - 3x + 9$
       $\underline{\times\qquad\quad x + 3}$

69. $(x^2 - x + 2)(x^2 + x - 2)$

70. $(x^2 + 2x + 5)(2x^2 - x - 1)$

71. $(x^3 + x + 3)(x^2 + 5x - 4)$

72. $(x^3 - x - 1)(x^2 + x + 1)$

73. $(x - 2)^3$ **74.** $(x + 3)^3$

75. $(x - 1)^2(x - 1)^2$

76. $(x + 4)^2(x + 4)^2$

77. $(x + 2)^2(x - 4)$ **78.** $(x - 4)^2(x - 1)$

79. $(u - 1)(2u + 3)(2u + 1)$

80. $(2x + 5)(x - 2)(5x - 3)$

In Exercises 81–110, use a special product pattern to find the product. See Examples 12–16.

81. $(x + 3)(x - 3)$ **82.** $(x - 5)(x + 5)$

83. $(x + 20)(x - 20)$ **84.** $(y + 9)(y - 9)$

85. $(2u + 3)(2u - 3)$ **86.** $(3z + 4)(3z - 4)$

87. $(4t - 6)(4t + 6)$ **88.** $(3u + 7)(3u - 7)$

89. $(2x + 3y)(2x - 3y)$

90. $(5u + 12v)(5u - 12v)$

91. $(4u - 3v)(4u + 3v)$

92. $(8a - 5b)(8a + 5b)$

93. $(2x^2 + 5)(2x^2 - 5)$

94. $(4t^2 + 6)(4t^2 - 6)$

95. $(x + 6)^2$

96. $(a - 2)^2$

97. $(t - 3)^2$

98. $(x + 10)^2$

99. $(3x + 2)^2$

100. $(2x - 8)^2$

101. $(8 - 3z)^2$

102. $(1 - 5t)^2$

103. $(2x - 5y)^2$

104. $(4s + 3t)^2$

105. $(6t + 5s)^2$

106. $(3u - 8v)^2$

107. $[(x + 1) + y]^2$

108. $[(x - 3) - y]^2$

109. $[u - (v - 3)]^2$

110. $[2u + (v + 1)]^2$

In Exercises 111 and 112, perform the multiplication and simplify.

111. $(x + 2)^2 - (x - 2)^2$ 112. $(u + 5)^2 + (u - 5)^2$

Think About It   In Exercises 113 and 114, is the equation an identity? Explain.

113. $(x + y)^3 = x^3 + 3x^2y + 3xy^2 + y^3$

114. $(x - y)^3 = x^3 - 3x^2y + 3xy^2 - y^3$

In Exercises 115 and 116, use the results of Exercise 113 to find the product.

115. $(x + 2)^3$ 116. $(x + 1)^3$

Solving Problems

117. ▲ *Geometry*   The base of a triangular sail is $2x$ feet and its height is $(x + 10)$ feet (see figure). Find an expression for the area of the sail.

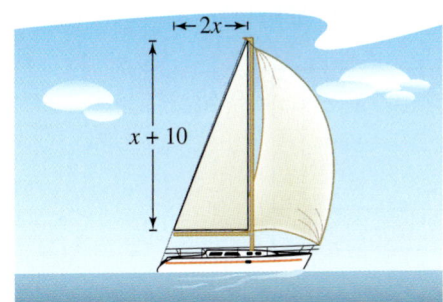

118. ▲ *Geometry*   The height of a rectangular sign is twice its width w (see figure). Find an expression for (a) the perimeter and (b) the area of the sign.

▲ *Geometry*   In Exercises 119–122, what polynomial product is represented by the geometric model? Explain. See Example 4.

119.

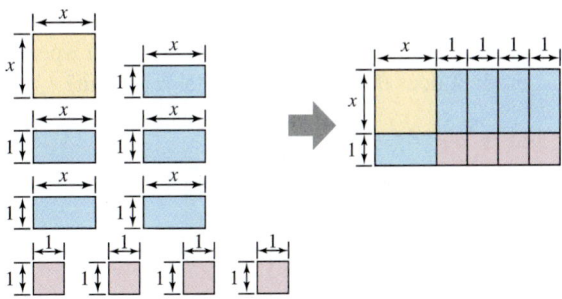

120.

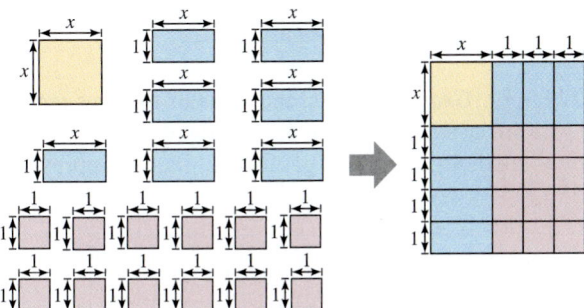

121.

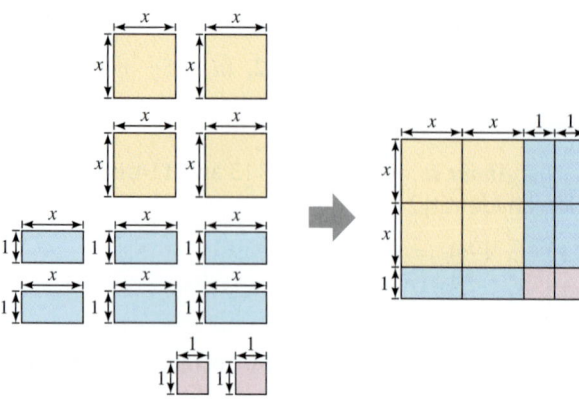

122.

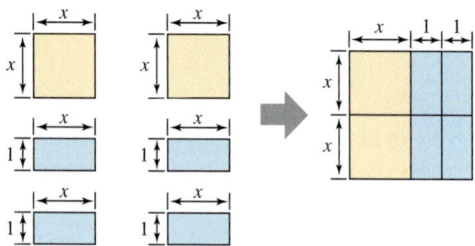

123. ▲ *Geometry*   Add the areas of the four rectangular regions shown in the figure. What special product does the geometric model represent?

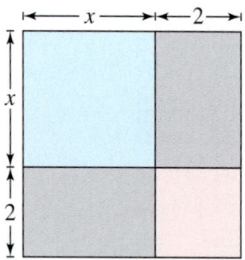

124. ▲ *Geometry*   Add the areas of the four rectangular regions shown in the figure. What special method does the geometric model represent?

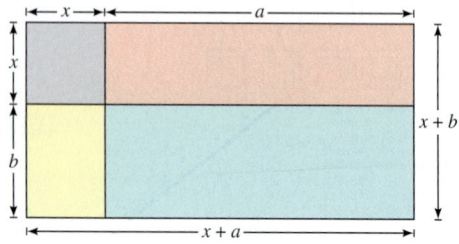

▲ *Geometry*   In Exercises 125 and 126, find a polynomial product that represents the area of the region. Then simplify the product.

125.

x 4

x

5

126.

$x + 1$ $x + 1$

x

x

▲ *Geometry*   In Exercises 127 and 128, find two different expressions that represent the area of the shaded portion of the figure.

127.

x 3

x

4

128.

z 4

z

5

129. *Milk Consumption*   The per capita consumption (average consumption per person) of milk M (in gallons) in the United States for the years 1990 through 2000 is given by

$$M = -0.32t + 24.3, \quad 0 \le t \le 10.$$

The population P (in millions) of the United States during the same time period is given by

$$P = -0.016t^2 + 2.69t + 250.1, \quad 0 \le t \le 10.$$

In both models, t represents the year, with $t = 0$ corresponding to 1990.   (Source: USDA/Economic Research Service and U.S. Census Bureau)

(a) Multiply the polynomials to find a model for the total consumption of milk T in the United States.

(b) 🖩 Use a graphing calculator to graph the model from part (a).

(c) 🖩 Use the graph from part (b) to estimate the total consumption of milk in 1998.

130. 🖩 *Interpreting Graphs*   When x units of a home theater system are sold, the revenue R is given by

$$R = x(900 - 0.5x).$$

(a) Use a graphing calculator to graph the equation.

(b) Multiply the factors in the expression for revenue and use a graphing calculator to graph the product in the same viewing window you used in part (a). Verify that the graph is the same as in part (a).

(c) Find the revenue when 500 units are sold. Use the graph to determine if revenue would increase or decrease if more units were sold.

131. *Compound Interest*   After 2 years, an investment of $500 compounded annually at interest rate r will yield an amount $500(1 + r)^2$. Find this product.

132. *Compound Interest*   After 2 years, an investment of $1200 compounded annually at interest rate r will yield an amount $1200(1 + r)^2$. Find this product.

Explaining Concepts

133. ⚡ Answer parts (d)–(f) of Motivating the Chapter on page 292.

134. *Writing* ✎   Explain why an understanding of the Distributive Property is essential in multiplying polynomials. Illustrate your explanation with an example.

135. *Writing* ✎   Describe the rules of exponents that are used to multiply polynomials. Give examples.

136. *Writing* ✎   Discuss any differences between the expressions $(3x)^2$ and $3x^2$.

137. *Writing* ✎   Explain the meaning of each letter of "FOIL" as it relates to multiplying two binomials.

138. *Writing* ✎   What is the degree of the product of two polynomials of degrees m and n? Explain.

139. *Writing* ✎   A polynomial with m terms is multiplied by a polynomial with n terms. How many monomial-by-monomial products must be found? Explain.

140. *True or False?*   Because the product of two monomials is a monomial, it follows that the product of two binomials is a binomial. Justify your answer.

141. *Finding a Pattern*   Perform each multiplication.

(a) $(x - 1)(x + 1)$

(b) $(x - 1)(x^2 + x + 1)$

(c) $(x - 1)(x^3 + x^2 + x + 1)$

(d) From the pattern formed in the first three products, can you predict the product of

$$(x - 1)(x^4 + x^3 + x^2 + x + 1)?$$

Verify your prediction by multiplying.

5.4 Dividing Polynomials and Synthetic Division

What You Should Learn

1. Divide polynomials by monomials and write in simplest form.
2. Use long division to divide polynomials by polynomials.
3. Use synthetic division to divide polynomials by polynomials of the form $x - k$.
4. Use synthetic division to factor polynomials.

Why You Should Learn It

Division of polynomials is useful in higher-level mathematics when factoring and finding zeros of polynomials.

Dividing a Polynomial by a Monomial

To divide a polynomial by a monomial, *reverse* the procedure used to add or subtract two rational expressions. Here is an example.

$$2 + \frac{1}{x} = \frac{2x}{x} + \frac{1}{x} = \frac{2x + 1}{x}$$ Add fractions.

$$\frac{2x + 1}{x} = \frac{2x}{x} + \frac{1}{x} = 2 + \frac{1}{x}$$ Divide by monomial.

1. Divide polynomials by monomials and write in simplest form.

Dividing a Polynomial by a Monomial

Let u, v, and w represent real numbers, variables, or algebraic expressions such that $w \neq 0$.

1. $\dfrac{u + v}{w} = \dfrac{u}{w} + \dfrac{v}{w}$ **2.** $\dfrac{u - v}{w} = \dfrac{u}{w} - \dfrac{v}{w}$

When dividing a polynomial by a monomial, remember to write the resulting expressions in simplest form, as illustrated in Example 1.

Example 1   Dividing a Polynomial by a Monomial

Perform the division and simplify.

$$\frac{12x^2 - 20x + 8}{4x}$$

Solution

$$\frac{12x^2 - 20x + 8}{4x} = \frac{12x^2}{4x} - \frac{20x}{4x} + \frac{8}{4x}$$ Divide each term in the numerator by $4x$.

$$= \frac{3(4x)(x)}{4x} - \frac{5(4x)}{4x} + \frac{2(4)}{4x}$$ Divide out common factors.

$$= 3x - 5 + \frac{2}{x}$$ Simplified form

2 Use long division to divide polynomials by polynomials.

Long Division

In the previous example, you learned how to divide one polynomial by another by factoring and dividing out common factors. For instance, you can divide $x^2 - 2x - 3$ by $x - 3$ as follows.

$$(x^2 - 2x - 3) \div (x - 3) = \frac{x^2 - 2x - 3}{x - 3} \qquad \text{Write as fraction.}$$

$$= \frac{(x + 1)(x - 3)}{x - 3} \qquad \text{Factor numerator.}$$

$$= \frac{(x + 1)(x - 3)}{x - 3} \qquad \text{Divide out common factor.}$$

$$= x + 1, \quad x \neq 3 \qquad \text{Simplified form}$$

This procedure works well for polynomials that factor easily. For those that do not, you can use a more general procedure that follows a "long division algorithm" similar to the algorithm used for dividing positive integers, which is reviewed in Example 2.

Example 2   Long Division Algorithm for Positive Integers

Use the long division algorithm to divide 6584 by 28.

Solution

$$
\begin{array}{r}
235 \\
28 \overline{)6584} \\
\underline{56} \\
98 \\
\underline{84} \\
144 \\
\underline{140} \\
4
\end{array}
$$

Think $\frac{65}{28} \approx 2$.

Think $\frac{98}{28} \approx 3$.

Think $\frac{144}{28} \approx 5$.

Multiply 2 by 28.

Subtract and bring down 8.

Multiply 3 by 28.

Subtract and bring down 4.

Multiply 5 by 28.

Remainder

So, you have

$$6584 \div 28 = 235 + \frac{4}{28}$$

$$= 235 + \frac{1}{7}.$$

In Example 2, the numerator 6584 is the **dividend,** 28 is the **divisor,** 235 is the **quotient,** and 4 is the **remainder.**

In the next several examples, you will see how the long division algorithm can be extended to cover the division of one polynomial by another.

Along with the long division algorithm, follow the steps below when performing long division of polynomials.

Long Division of Polynomials

1. Write the dividend and divisor in descending powers of the variable.

2. Insert placeholders with zero coefficients for missing powers of the variable. (See Example 5.)

3. Perform the long division of the polynomials as you would with integers.

4. Continue the process until the degree of the remainder is less than that of the divisor.

Example 3 Long Division Algorithm for Polynomials

Think $x^2/x = x.$

Think $3x/x = 3.$

$$
\begin{array}{r}
x + 3 \\
x - 1 \overline{)\, x^2 + 2x + 4\ } \\
\underline{x^2 - \ \ x} \\
3x + 4 \\
\underline{3x - 3} \\
7
\end{array}
$$

Multiply x by $(x - 1)$.

Subtract and bring down 4.

Multiply 3 by $(x - 1)$.

Subtract.

The remainder is a fractional part of the divisor, so you can write

Dividend   Quotient   Remainder

$$
\frac{x^2 + 2x + 4}{x - 1} = x + 3 + \frac{7}{x - 1}.
$$

Divisor   Divisor

You can check a long division problem by multiplying by the divisor. For instance, you can check the result of Example 3 as follows.

$$
\frac{x^2 + 2x + 4}{x - 1} \stackrel{?}{=} x + 3 + \frac{7}{x - 1}
$$

$$
(x - 1)\left(\frac{x^2 + 2x + 4}{x - 1}\right) \stackrel{?}{=} (x - 1)\left(x + 3 + \frac{7}{x - 1}\right)
$$

$$
x^2 + 2x + 4 \stackrel{?}{=} (x + 3)(x - 1) + 7
$$

$$
x^2 + 2x + 4 \stackrel{?}{=} (x^2 + 2x - 3) + 7
$$

$$
x^2 + 2x + 4 = x^2 + 2x + 4 \quad \checkmark
$$

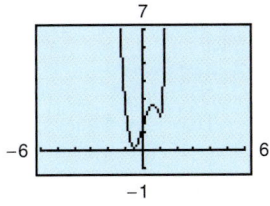
Example 4   Writing in Standard Form Before Dividing

Divide $-13x^3 + 10x^4 + 8x - 7x^2 + 4$ by $3 - 2x$.

Solution

First write the divisor and dividend in standard polynomial form.

$$
\begin{array}{r}
-5x^3 - x^2 + 2x - 1 \\
-2x + 3\overline{)\,10x^4 - 13x^3 - 7x^2 + 8x + 4} \\
\underline{10x^4 - 15x^3} \\
2x^3 - 7x^2 \\
\underline{2x^3 - 3x^2} \\
-4x^2 + 8x \\
\underline{-4x^2 + 6x} \\
2x + 4 \\
\underline{2x - 3} \\
7
\end{array}
$$

Multiply $-5x^3$ by $(-2x + 3)$.
Subtract and bring down $-7x^2$.
Multiply $-x^2$ by $(-2x + 3)$.
Subtract and bring down $8x$.
Multiply $2x$ by $(-2x + 3)$.
Subtract and bring down 4.
Multiply -1 by $(-2x + 3)$.

This shows that

$$
\overbrace{\underbrace{\frac{10x^4 - 13x^3 - 7x^2 + 8x + 4}{-2x + 3}}_{\text{Divisor}}}^{\text{Dividend}} = \overbrace{-5x^3 - x^2 + 2x - 1}^{\text{Quotient}} + \overbrace{\frac{7}{\underbrace{-2x + 3}_{\text{Divisor}}}}^{\text{Remainder}}.
$$

When the dividend is missing one or more powers of x, the long division algorithm requires that you account for the missing powers, as shown in Example 5.

Example 5   Accounting for Missing Powers of x

Divide $x^3 - 2$ by $x - 1$.

Solution

To account for the missing x^2- and x-terms, insert $0x^2$ and $0x$.

$$
\begin{array}{r}
x^2 + x + 1 \\
x - 1\overline{)\,x^3 + 0x^2 + 0x - 2} \\
\underline{x^3 - x^2} \\
x^2 + 0x \\
\underline{x^2 - x} \\
x - 2 \\
\underline{x - 1} \\
-1
\end{array}
$$

Insert $0x^2$ and $0x$.
Multiply x^2 by $(x - 1)$.
Subtract and bring down $0x$.
Multiply x by $(x - 1)$.
Subtract and bring down -2.
Multiply 1 by $(x - 1)$.
Subtract.

So, you have

$$
\frac{x^3 - 2}{x - 1} = x^2 + x + 1 - \frac{1}{x - 1}.
$$

In each of the long division examples presented so far, the divisor has been a first-degree polynomial. The long division algorithm works just as well with polynomial divisors of degree two or more, as shown in Example 6.

Example 6 A Second-Degree Divisor

Divide $x^4 + 6x^3 + 6x^2 - 10x - 3$ by $x^2 + 2x - 3$.

Solution

$$
\begin{array}{r}
x^2 + 4x + 1 \\
x^2 + 2x - 3 \overline{\smash{\big)}\, x^4 + 6x^3 + 6x^2 - 10x - 3} \\
\underline{x^4 + 2x^3 - 3x^2} \\
4x^3 + 9x^2 - 10x \\
\underline{4x^3 + 8x^2 - 12x} \\
x^2 + 2x - 3 \\
\underline{x^2 + 2x - 3} \\
0
\end{array}
$$

Multiply x^2 by $(x^2 + 2x - 3)$.
Subtract and bring down $-10x$.
Multiply $4x$ by $(x^2 + 2x - 3)$.
Subtract and bring down -3.
Multiply 1 by $(x^2 + 2x - 3)$.
Subtract.

So, $x^2 + 2x - 3$ divides evenly into $x^4 + 6x^3 + 6x^2 - 10x - 3$. That is,

$$
\frac{x^4 + 6x^3 + 6x^2 - 10x - 3}{x^2 + 2x - 3} = x^2 + 4x + 1, \ x \neq -3, \ x \neq 1.
$$

> **Study Tip**
>
> If the remainder of a division problem is zero, the divisor is said to **divide evenly** into the dividend.

3 Use synthetic division to divide polynomials by polynomials of the form $x - k$.

Synthetic Division

There is a nice shortcut for division by polynomials of the form $x - k$. It is called **synthetic division** and is outlined for a third-degree polynomial as follows.

Synthetic Division of a Third-Degree Polynomial

Use synthetic division to divide $ax^3 + bx^2 + cx + d$ by $x - k$, as follows.

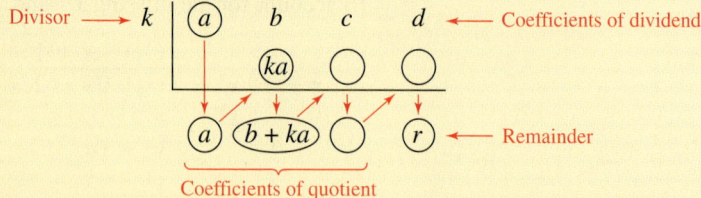

Vertical Pattern: Add terms.
Diagonal Pattern: Multiply by k.

Keep in mind that synthetic division works *only* for divisors of the form $x - k$. Remember that $x + k = x - (-k)$. Moreover, the degree of the quotient is always one less than the degree of the dividend.

Example 7   Using Synthetic Division

Use synthetic division to divide $x^3 + 3x^2 - 4x - 10$ by $x - 2$.

Solution

The coefficients of the dividend form the top row of the synthetic division array. Because you are dividing by $x - 2$, write 2 at the top left of the array. To begin the algorithm, bring down the first coefficient. Then multiply this coefficient by 2, write the result in the second row, and add the two numbers in the second column. By continuing this pattern, you obtain the following.

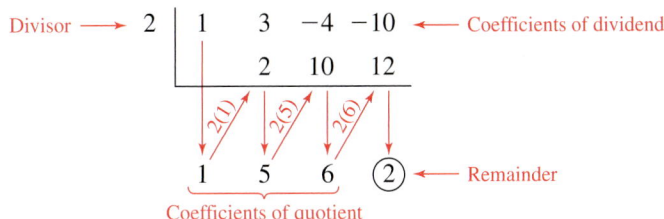

The bottom row shows the coefficients of the quotient. So, the quotient is

$$1x^2 + 5x + 6$$

and the remainder is 2. So, the result of the division problem is

$$\frac{x^3 + 3x^2 - 4x - 10}{x - 2} = x^2 + 5x + 6 + \frac{2}{x - 2}.$$

Factoring and Division

④ **Use synthetic division to factor polynomials.**

Synthetic division (or long division) can be used to factor polynomials. If the remainder in a synthetic division problem is zero, you know that the divisor divides *evenly* into the dividend. So, the original polynomial can be factored as the product of two polynomials of lesser degrees, as in Example 8. You will learn more about factoring in the next chapter.

Example 8   Factoring a Polynomial

The polynomial $x^3 - 7x + 6$ can be factored using synthetic division. Because $x - 1$ is a factor of the polynomial, you can divide as follows.

$$
\begin{array}{r|rrrr}
1 & 1 & 0 & -7 & 6 \\
  &   & 1 & 1 & -6 \\
\hline
  & 1 & 1 & -6 & \boxed{0}
\end{array}
\quad \leftarrow \text{Remainder}
$$

Because the remainder is zero, the divisor divides evenly into the dividend:

$$\frac{x^3 - 7x + 6}{x - 1} = x^2 + x - 6.$$

From this result, you can factor the original polynomial as follows.

$$x^3 - 7x + 6 = (x - 1)(x^2 + x - 6)$$

5.4 Exercises

Review   Concepts, Skills, and Problem Solving

Keep mathematically in shape by doing these exercises *before* the problems of this section.

Properties and Definitions

1. *Writing*   Show how to write the fraction $120y/90$ in simplified form.

2. Write an algebraic expression that represents the product of two consecutive odd integers, the first of which is $2n + 1$.

3. Write an algebraic expression that represents the sum of two consecutive odd integers, the first of which is $2n + 1$.

4. Write an algebraic expression that represents the product of two consecutive even integers, the first of which is $2n$.

Solving Equations

In Exercises 5–10, solve the equation.

5. $3(2 - x) = 5x$

6. $125 - 50x = 0$

7. $8y^2 - 50 = 0$

8. $t^2 - 8t = 0$

9. $x^2 + x - 42 = 0$

10. $x(10 - x) = 25$

Models and Graphs

11. *Monthly Wages*   You receive a monthly salary of $1500 plus a commission of 12% of sales. Find a model for the monthly wages y as a function of sales x. Graph the model.

12. *Education*   In the year 2003, a college had an enrollment of 3680 students. Enrollment was projected to increase by 60 students per year. Find a model for the enrollment N as a function of time t in years. (Let $t = 3$ represent the year 2003.) Graph the function for the years 2003 through 2013.

Developing Skills

In Exercises 1–14, perform the division. See Example 1.

1. $(7x^3 - 2x^2) \div x$

2. $(6a^2 + 7a) \div a$

3. $(4x^2 - 2x) \div (-x)$

4. $(5y^3 + 6y^2 - 3y) \div (-y)$

5. $(m^4 + 2m^2 - 7) \div m$

6. $(x^3 + x - 2) \div x$

7. $\dfrac{50z^3 + 30z}{-5z}$

8. $\dfrac{18c^4 - 24c^2}{-6c}$

9. $\dfrac{8z^3 + 3z^2 - 2z}{2z}$

10. $\dfrac{6x^4 + 8x^3 - 18x^2}{3x^2}$

11. $\dfrac{4x^5 - 6x^4 + 12x^3 - 8x^2}{4x^2}$

12. $\dfrac{15x^{12} - 5x^9 + 30x^6}{5x^6}$

13. $(5x^2y - 8xy + 7xy^2) \div 2xy$

14. $(-14s^4t^2 + 7s^2t^2 - 18t) \div 2s^2t$

In Exercises 15–52, perform the division. See Examples 2–6.

15. $\dfrac{x^2 - 8x + 15}{x - 3}$

16. $\dfrac{t^2 - 18t + 72}{t - 6}$

17. $(x^2 + 15x + 50) \div (x + 5)$

18. $(y^2 - 6y - 16) \div (y + 2)$

19. Divide $x^2 - 5x + 8$ by $x - 2$.

20. Divide $x^2 + 10x - 9$ by $x - 3$.

21. Divide $21 - 4x - x^2$ by $3 - x$.

22. Divide $5 + 4x - x^2$ by $1 + x$.

23. $\dfrac{5x^2 + 2x + 3}{x + 2}$

24. $\dfrac{2x^2 + 13x + 15}{x + 5}$

25. $\dfrac{12x^2 + 17x - 5}{3x + 2}$

26. $\dfrac{8x^2 + 2x + 3}{4x - 1}$

27. $(12t^2 - 40t + 25) \div (2t - 5)$

28. $(15 - 14u - 8u^2) \div (5 + 2u)$

29. Divide $2y^2 + 7y + 3$ by $2y + 1$.

30. Divide $10t^2 - 7t - 12$ by $2t - 3$.

31. $\dfrac{x^3 - 2x^2 + 4x - 8}{x - 2}$

32. $\dfrac{x^3 + 4x^2 + 7x + 28}{x + 4}$

33. $\dfrac{9x^3 - 3x^2 - 3x + 4}{3x + 2}$

34. $\dfrac{4y^3 + 12y^2 + 7y - 3}{2y + 3}$

35. $(2x + 9) \div (x + 2)$

36. $(12x - 5) \div (2x + 3)$

37. $\dfrac{x^2 + 16}{x + 4}$

38. $\dfrac{y^2 + 8}{y + 2}$

39. $\dfrac{6z^2 + 7z}{5z - 1}$

40. $\dfrac{8y^2 - 2y}{3y + 5}$

41. $\dfrac{16x^2 - 1}{4x + 1}$

42. $\dfrac{81y^2 - 25}{9y - 5}$

43. $\dfrac{x^3 + 125}{x + 5}$

44. $\dfrac{x^3 - 27}{x - 3}$

45. $(x^3 + 4x^2 + 7x + 6) \div (x^2 + 2x + 3)$

46. $(2x^3 + 2x^2 - 2x - 15) \div (2x^2 + 4x + 5)$

47. $(4x^4 - 3x^2 + x - 5) \div (x^2 - 3x + 2)$

48. $(8x^5 + 6x^4 - x^3 + 1) \div (2x^3 - x^2 - 3)$

49. Divide $x^6 - 1$ by $x - 1$.

50. Divide x^3 by $x - 1$.

51. $x^5 \div (x^2 + 1)$

52. $x^4 \div (x - 2)$

In Exercises 53–56, simplify the expression.

53. $\dfrac{4x^4}{x^3} - 2x$

54. $\dfrac{15x^3y}{10x^2} + \dfrac{3xy^2}{2y}$

55. $\dfrac{8u^2v}{2u} + \dfrac{3(uv)^2}{uv}$

56. $\dfrac{x^2 + 2x - 3}{x - 1} - (3x - 4)$

In Exercises 57–68, use synthetic division to divide. See Example 7.

57. $(x^2 + x - 6) \div (x - 2)$

58. $(x^2 + 5x - 6) \div (x + 6)$

59. $\dfrac{x^3 + 3x^2 - 1}{x + 4}$

60. $\dfrac{x^4 - 4x^2 + 6}{x - 4}$

61. $\dfrac{x^4 - 4x^3 + x + 10}{x - 2}$

62. $\dfrac{2x^5 - 3x^3 + x}{x - 3}$

63. $\dfrac{5x^3 - 6x^2 + 8}{x - 4}$

64. $\dfrac{5x^3 + 6x + 8}{x + 2}$

65. $\dfrac{10x^4 - 50x^3 - 800}{x - 6}$

66. $\dfrac{x^5 - 13x^4 - 120x + 80}{x + 3}$

67. $\dfrac{0.1x^2 + 0.8x + 1}{x - 0.2}$

68. $\dfrac{x^3 - 0.8x + 2.4}{x + 0.1}$

In Exercises 69–76, factor the polynomial into two polynomials of lesser degrees given one of its factors. See Example 8.

	Polynomial	Factor
69.	$x^3 - 13x + 12$	$x - 3$
70.	$x^3 + x^2 - 32x - 60$	$x + 5$
71.	$6x^3 - 13x^2 + 9x - 2$	$x - 1$
72.	$9x^3 - 3x^2 - 56x - 48$	$x - 3$
73.	$x^4 + 7x^3 + 3x^2 - 63x - 108$	$x + 3$
74.	$x^4 - 6x^3 - 8x^2 + 96x - 128$	$x - 4$
75.	$15x^2 - 2x - 8$	$x - \frac{4}{5}$
76.	$18x^2 - 9x - 20$	$x + \frac{5}{6}$

In Exercises 77 and 78, find the constant c such that the denominator divides evenly into the numerator.

77. $\dfrac{x^3 + 2x^2 - 4x + c}{x - 2}$

78. $\dfrac{x^4 - 3x^2 + c}{x + 6}$

In Exercises 79 and 80, use a graphing calculator to graph the two equations in the same viewing window. Use the graphs to verify that the expressions are equivalent. Verify the results algebraically.

79. $y_1 = \dfrac{x + 4}{2x}$

$y_2 = \dfrac{1}{2} + \dfrac{2}{x}$

80. $y_1 = \dfrac{x^2 + 2}{x + 1}$

$y_2 = x - 1 + \dfrac{3}{x + 1}$

In Exercises 81 and 82, perform the division assuming that n is a positive integer.

81. $\dfrac{x^{3n} + 3x^{2n} + 6x^n + 8}{x^n + 2}$

82. $\dfrac{x^{3n} - x^{2n} + 5x^n - 5}{x^n - 1}$

Think About It   In Exercises 83 and 84, the divisor, quotient, and remainder are given. Find the dividend.

	Divisor	Quotient	Remainder
83.	$x - 6$	$x^2 + x + 1$	-4
84.	$x + 3$	$x^3 + x^2 - 4$	8

Finding a Pattern   In Exercises 85 and 86, complete the table for the function. The first row is completed for Exercise 85. What conclusion can you draw as you compare the values of $f(k)$ with the remainders? (Use synthetic division to find the remainders.)

85. $f(x) = x^3 - x^2 - 2x$

86. $f(x) = 2x^3 - x^2 - 2x + 1$

k	$f(k)$	Divisor $(x - k)$	Remainder
-2	-8	$x + 2$	-8
-1			
0			
$\frac{1}{2}$			
1			
2			

Solving Problems

87. ▲ *Geometry* The area of a rectangle is $2x^3 + 3x^2 - 6x - 9$ and its length is $2x + 3$. Find the width of the rectangle.

88. ▲ *Geometry* A rectangular house has a volume of $x^3 + 55x^2 + 650x + 2000$ cubic feet (the space in the attic is not included). The height of the house is $x + 5$ feet (see figure). Find the number of square feet of floor space *on the first floor* of the house.

89. $V = x^3 + 18x^2 + 80x + 96$

▲ *Geometry* In Exercises 89 and 90, you are given the expression for the volume of the solid shown. Find the expression for the missing dimension.

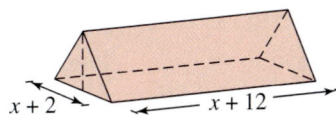

90. $V = h^4 + 3h^3 + 2h^2$

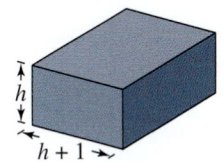

Explaining Concepts

91. *Error Analysis* Describe the error.

$$\frac{6x + 5y}{x} = \frac{6x + 5y}{x} = 6 + 5y$$

92. Create a polynomial division problem and identify the dividend, divisor, quotient, and remainder.

93. *Writing* Explain what it means for a divisor to divide *evenly* into a dividend.

94. *Writing* Explain how you can check polynomial division.

95. *True or False?* If the divisor divides evenly into the dividend, the divisor and quotient are factors of the dividend. Justify your answer.

96. *Writing* For synthetic division, what form must the divisor have?

97. ▦ *Writing* Use a graphing calculator to graph each polynomial in the same viewing window using the standard setting. Use the *zero* or *root* feature to find the x-intercepts. What can you conclude about the polynomials? Verify your conclusion algebraically.
  (a) $y = (x - 4)(x - 2)(x + 1)$
  (b) $y = (x^2 - 6x + 8)(x + 1)$
  (c) $y = x^3 - 5x^2 + 2x + 8$

98. ▦ *Writing* Use a graphing calculator to graph the function

$$f(x) = \frac{x^3 - 5x^2 + 2x + 8}{x - 2}.$$

Use the *zero* or *root* feature to find the x-intercepts. Why does this function have only two x-intercepts? To what other function does the graph of $f(x)$ appear to be equivalent? What is the difference between the two graphs?

What Did You Learn?

Key Terms

exponential form, *p. 294*
scientific notation, *p. 298*
polynomial, *p. 304*
constant term, *p. 304*
standard form of a
   polynomial, *p. 304*

degree of a polynomial, *p. 304*
leading coefficient, *p. 304*
monomial, *p. 305*
binomial, *p. 305*
trinomial, *p. 305*
FOIL Method, *p. 316*

dividend, *p. 329*
divisor, *p. 329*
quotient, *p. 329*
remainder, *p. 329*
synthetic division, *p. 332*

Key Concepts

5.1 ⬤ Summary of rules of exponents

Let m and n be integers, and let a and b represent real numbers, variables, or algebraic expressions. (All denominators and bases are nonzero.)

1. Product Rule: $a^m a^n = a^{m+n}$

2. Quotient Rule: $\dfrac{a^m}{a^n} = a^{m-n}$

3. Product-to-Power Rule; $(ab)^m = a^m b^m$

4. Power-to-Power Rule: $(a^m)^n = a^{mn}$

5. Quotient-to-Power Rule: $\left(\dfrac{a}{b}\right)^m = \dfrac{a^m}{b^m}$

6. Zero Exponent Rule: $a^0 = 1$

7. Negative Exponent Rule: $a^{-m} = \dfrac{1}{a^m}$

8. Negative Exponent Rule: $\left(\dfrac{a}{b}\right)^{-m} = \left(\dfrac{b}{a}\right)^m$

5.2 ⬤ Polynomial in *x*

Let $a_n, a_{n-1}, \ldots, a_2, a_1, a_0$ be real numbers and let n be a nonnegative integer. A polynomial in x is an expression of the form

$$a_n x^n + a_{n-1} x^{n-1} + \cdots + a_2 x^2 + a_1 x + a_0$$

where $a_n \neq 0$. The polynomial is of degree n, and the number a_n is called the leading coefficient. The number a_0 is called the constant term.

5.2 ⬤ Adding polynomials

To add polynomials, you combine like terms (those having the same degree) by using the Distributive Property.

5.2 ⬤ Subtracting polynomials

To subtract one polynomial from another, you add the opposite by changing the sign of each term of the polynomial that is being subtracted and then adding the resulting like terms.

5.3 ⬤ Multiplying polynomials

1. To multiply a polynomial by a monomial, apply the Distributive Property.

2. To multiply two binomials, use the FOIL Method. Combine the product of the **F**irst terms, the product of the **O**uter terms, the product of the **I**nner terms, and the product of the **L**ast terms.

3. To multiply two polynomials, use the Distributive Property to multiply each term of one polynomial by each term of the other polynomial.

5.3 ⬤ Special products

Let a and b be real numbers, variables, or algebraic expressions.

Sum and Difference of Two Terms:

$$(a + b)(a - b) = a^2 - b^2$$

Square of a Binomial:

$$(a + b)^2 = a^2 + 2ab + b^2$$

$$(a - b)^2 = a^2 - 2ab + b^2$$

5.4 ⬤ Dividing polynomials

1. To divide a polynomial by a monomial, divide each term of the polynomial by the monomial.

2. To divide a polynomial by a binomial, follow the long division algorithm used for dividing whole numbers.

3. Use synthetic division to divide a polynomial by a binomial of the form $x - k$. [Remember that $x + k = x - (-k)$.]

Review Exercises

5.1 Integer Exponents and Scientific Notation

① Use the rules of exponents to simplify expressions.

In Exercises 1–14, use the rules of exponents to simplify the expression (if possible).

1. $x^2 \cdot x^3$

2. $-3y^2 \cdot y^4$

3. $(u^2)^3$

4. $(v^4)^2$

5. $(-2z)^3$

6. $(-3y)^2(2)$

7. $-(u^2v)^2(-4u^3v)$

8. $(12x^2y)(3x^2y^4)^2$

9. $\dfrac{12z^5}{6z^2}$

10. $\dfrac{15m^3}{25m}$

11. $\dfrac{120u^5v^3}{15u^3v}$

12. $-\dfrac{(-2x^2y^3)^2}{-3xy^2}$

13. $\left(\dfrac{72x^4}{6x^2}\right)^2$

14. $\left(-\dfrac{y^2}{2}\right)^3$

② Rewrite exponential expressions involving negative and zero exponents.

In Exercises 15–18, evaluate the expression.

15. $(2^3 \cdot 3^2)^{-1}$

16. $(2^{-2} \cdot 5^2)^{-2}$

17. $\left(\dfrac{2}{5}\right)^{-3}$

18. $\left(\dfrac{1}{3^{-2}}\right)^2$

In Exercises 19–30, rewrite the expression using only positive exponents, and simplify. (Assume that any variables in the expression are nonzero.)

19. $(6y^4)(2y^{-3})$

20. $4(-3x)^{-3}$

21. $\dfrac{4x^{-2}}{2x}$

22. $\dfrac{15t^5}{24t^{-3}}$

23. $(x^3y^{-4})^0$

24. $(5x^{-2}y^4)^{-2}$

25. $\dfrac{2a^{-3}b^4}{4a^5b^{-5}}$

26. $\dfrac{2u^0v^{-2}}{10u^{-1}v^{-3}}$

27. $\left(\dfrac{3x^{-1}y^2}{12x^5y^{-3}}\right)^{-1}$

28. $\left(\dfrac{4x^{-3}z^{-1}}{8x^4z}\right)^{-2}$

29. $u^3(5u^0v^{-1})(9u)^2$

30. $a^4(2a^{-1}b^2)(ab)^0$

③ Write very large and very small numbers in scientific notation.

In Exercises 31 and 32, write the number in scientific notation.

31. 0.0000538

32. 30,296,000,000

In Exercises 33 and 34, write the number in decimal form.

33. 4.833×10^8

34. 2.74×10^{-4}

In Exercises 35–38, evaluate the expression without a calculator.

35. $(6 \times 10^3)^2$

36. $(3 \times 10^{-3})(8 \times 10^7)$

37. $\dfrac{3.5 \times 10^7}{7 \times 10^4}$

38. $\dfrac{1}{(6 \times 10^{-3})^2}$

5.2 Adding and Subtracting Polynomials

① Identify the degrees and leading coefficients of polynomials.

In Exercises 39–46, write the polynomial in standard form. Then identify its degree and leading coefficient.

39. $10x - 4 - 5x^3$

40. $2x^2 + 9$

41. $4x^3 - 2x + 5x^4 - 7x^2$

42. $6 - 3x + 6x^2 - x^3$

43. $7x^4 - 1$

44. $12x^2 + 2x - 8x^5 + 1$

45. -2

46. $\frac{1}{4}t^2$

In Exercises 47–50, give an example of a polynomial in one variable that satisfies the condition. (There are many correct answers.)

47. A trinomial of degree 4

48. A monomial of degree 2

49. A binomial of degree 1

50. A trinomial of degree 5

2 Add polynomials using a horizontal or vertical format.

In Exercises 51–62, perform the addition.

51. $(2x + 3) + (x - 4)$ **52.** $(5x + 7) + (x - 2)$

53. $\left(\frac{1}{2}x + \frac{2}{3}\right) + \left(4x + \frac{1}{3}\right)$ **54.** $\left(\frac{3}{4}y + 2\right) + \left(\frac{1}{2}y - \frac{2}{5}\right)$

55. $(2x^3 - 4x^2 + 3) + (x^3 + 4x^2 - 2x)$

56. $(3y^3 + 5y^2 - 9y) + (2y^3 - 3y + 10)$

57. $-4(6 - x + x^2) + (3x^2 + x)$

58. $(4 - x^2) + 2(x - 2)$

59. $(3u + 4u^2) + 5(u + 1) + 3u^2$

60. $6(u^2 + 2) + 12u + (u^2 - 5u + 2)$

61. $\begin{aligned}-x^4 - 2x^2 + 3 \\ 3x^4 - 5x^2 \\ \hline\end{aligned}$

62. $\begin{aligned}5z^3 - 4z - 7 \\ z^2 - 2z \\ \hline\end{aligned}$

63. ▲ *Geometry* The length of a rectangular wall is x units, and its height is $(x - 3)$ units (see figure). Find an expression for the perimeter of the wall.

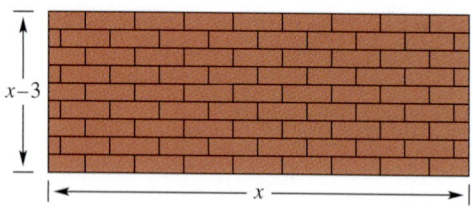

64. ▲ *Geometry* A rectangular garden has length $(t + 5)$ feet and width $2t$ feet (see figure). Find an expression for the perimeter of the garden.

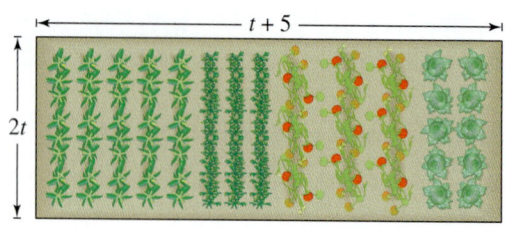

Figure for 64

3 Subtract polynomials using a horizontal or vertical format.

In Exercises 65–78, find the difference.

65. $(t - 5) - (3t - 1)$ **66.** $(y + 3) - (y - 9)$

67. $\left(\frac{1}{2}x + 5\right) - \left(\frac{3}{4}x - \frac{1}{3}\right)$ **68.** $\left(2x - \frac{1}{5}\right) - \left(\frac{1}{4}x + \frac{1}{4}\right)$

69. $(6x^2 - 9x - 5) - (4x^2 - 6x + 1)$

70. $(3y^2 + 2y - 9) - (5y^2 - y + 7)$

71. $3(2x^2 - 4) - (2x^2 - 5)$

72. $(5t^2 + 2) - 2(4t^2 + 1)$

73. $(z^2 + 6z) - 3(z^2 + 2z)$

74. $(-x^3 - 3x) - 2(2x^3 + x + 1)$

75. $4y^2 - [y - 3(y^2 + 2)]$

76. $(6a^3 + 3a) - 2[a - (a^3 + 2)]$

77. $\begin{aligned}5x^2 + 2x - 27 \\ -(2x^2 - 2x - 13) \\ \hline\end{aligned}$

78. $\begin{aligned}12y^4 - 15y^2 + 7 \\ -(18y^4 + 4y^2 - 9) \\ \hline\end{aligned}$

79. *Cost, Revenue, and Profit* The cost C of producing x units of a product is $C = 15 + 26x$. The revenue R for selling x units is $R = 40x - \frac{1}{2}x^2$, $0 \le x \le 20$. The profit P is the difference between revenue and cost.

(a) Perform the subtraction required to find the polynomial representing profit P.

(b) ▦ Use a graphing calculator to graph the polynomial representing profit.

(c) ▦ Determine the profit when 14 units are produced and sold. Use the graph in part (b) to describe the profit when x is less than or greater than 14.

80. ▦ *Comparing Models*   The table shows population projections (in millions) for the United States for selected years from 2005 to 2030. There are three series of projections: lowest P_L, middle P_M, and highest P_H.   (Source: U.S. Census Bureau)

Year	2005	2010	2015	2020	2025	2030
P_L	284.0	291.4	298.0	303.7	308.2	311.7
P_M	287.7	299.9	312.3	324.9	337.8	351.1
P_H	292.3	310.9	331.6	354.6	380.4	409.6

In the following models for the data, $t = 5$ corresponds to the year 2005.

$$P_L = -0.020t^2 + 1.81t + 275.4$$

$$P_M = 2.53t + 274.6$$

$$P_H = 0.052t^2 + 2.84t + 277.0$$

(a) Use a graphing calculator to plot the data and graph the models in the same viewing window.

(b) Find $(P_L + P_H)/2$. Use a graphing calculator to graph this polynomial and state which graph from part (a) it most resembles. Does this seem reasonable? Explain.

(c) Find $P_H - P_L$. Use a graphing calculator to graph this polynomial. Explain why it is increasing.

5.3 Multiplying Polynomials: Special Products

① Find products with monomial multipliers.

In Exercises 81–84, perform the multiplication and simplify.

81. $2x(x + 4)$

82. $3y(y + 1)$

83. $(4x - 2)(-3x^2)$

84. $(5 - 7y)(-6y^2)$

② Multiply binomials using the Distributive Property and the FOIL Method.

In Exercises 85–90, perform the multiplication and simplify.

85. $(x - 4)(x + 6)$ **86.** $(u + 5)(u - 2)$

87. $(x + 3)(2x - 4)$ **88.** $(y + 2)(4y - 3)$

89. $(4x - 3)(3x + 4)$ **90.** $(6 - 2x)(7x + 10)$

③ Multiply polynomials using a horizontal or vertical format.

In Exercises 91–100, perform the multiplication and simplify.

91. $(x^2 + 5x + 2)(2x + 3)$

92. $(s^2 + 4s - 3)(s - 3)$

93. $(2t - 1)(t^2 - 3t + 3)$

94. $(4x + 2)(x^2 + 6x - 5)$

95.
$$\begin{array}{r} 3x^2 + x - 2 \\ \times\quad 2x - 1 \\ \hline \end{array}$$

96.
$$\begin{array}{r} 5y^2 - 2y + 9 \\ \times\quad 3y + 4 \\ \hline \end{array}$$

97.
$$\begin{array}{r} y^2 - 4y + 5 \\ \times\quad y^2 + 2y - 3 \\ \hline \end{array}$$

98.
$$\begin{array}{r} x^2 + 8x - 12 \\ \times\quad x^2 - 9x + 2 \\ \hline \end{array}$$

99. $(2x + 1)^3$

100. $(3y - 2)^3$

101. ▲ *Geometry*   The width of a rectangular window is $(2x + 6)$ inches and its height is $(3x + 10)$ inches (see figure). Find an expression for the area of the window.

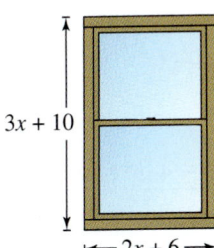

$3x + 10$

$\longleftarrow 2x + 6 \longrightarrow$

102. ▲ *Geometry*   The width of a rectangular parking lot is $(x + 25)$ meters and its length is $(x + 30)$ meters (see figure). Find an expression for the area of the parking lot.

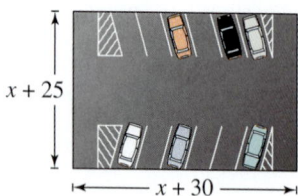

$x + 25$

$x + 30$

4 Identify and use special binomial products.

In Exercises 103–114, use a special product pattern to find the product.

103. $(x + 3)^2$

104. $(x - 5)^2$

105. $(4x - 7)^2$

106. $(9 - 2x)^2$

107. $\left(\frac{1}{2}x - 4\right)^2$

108. $(4 + 3b)^2$

109. $(u - 6)(u + 6)$

110. $(r + 3)(r - 3)$

111. $(2x - y)^2$

112. $(3a + b)^2$

113. $(2x - 4y)(2x + 4y)$

114. $(4u + 5v)(4u - 5v)$

5.4 Dividing Polynomials and Synthetic Division

1 Divide polynomials by monomials and write in simplest form.

In Exercises 115–118, perform the division.

115. $(4x^3 - x) \div (2x)$

116. $(10x + 15) \div (5x)$

117. $\dfrac{3x^3y^2 - x^2y^2 + x^2y}{x^2y}$

118. $\dfrac{6a^3b^3 + 2a^2b - 4ab^2}{2ab}$

2 Use long division to divide polynomials by polynomials.

In Exercises 119–124, perform the division.

119. $\dfrac{6x^3 + 2x^2 - 4x + 2}{3x - 1}$

120. $\dfrac{4x^4 - x^3 - 7x^2 + 18x}{x - 2}$

121. $\dfrac{x^4 - 3x^2 + 2}{x^2 - 1}$

122. $\dfrac{x^4 - 4x^3 + 3x}{x^2 - 1}$

123. $\dfrac{x^5 - 3x^4 + x^2 + 6}{x^3 - 2x^2 + x - 1}$

124. $\dfrac{x^6 + 4x^5 - 3x^2 + 5x}{x^3 + x^2 - 4x + 3}$

3 Use synthetic division to divide polynomials by polynomials of the form $x - k$.

In Exercises 125–128, use synthetic division to divide.

125. $\dfrac{x^3 + 7x^2 + 3x - 14}{x + 2}$

126. $\dfrac{x^4 - 2x^3 - 15x^2 - 2x + 10}{x - 5}$

127. $(x^4 - 3x^2 - 25) \div (x - 3)$

128. $(2x^3 + 5x - 2) \div \left(x + \frac{1}{2}\right)$

4 Use synthetic division to factor polynomials.

In Exercises 129 and 130, factor the polynomial into two polynomials of lesser degrees given one of its factors.

Polynomial	*Factor*
129. $x^3 + 2x^2 - 5x - 6$	$x - 2$
130. $2x^3 + x^2 - 2x - 1$	$x + 1$

Chapter Test

Take this test as you would take a test in class. After you are done, check your work against the answers in the back of the book.

1. Determine the degree and leading coefficient of $-5.2x^3 + 3x^2 - 8$.

2. Explain why the expression is not a polynomial: $\dfrac{4}{x^2 + 2}$.

3. (a) Write 0.000032 in scientific notation.

   (b) Write 6.04×10^7 in decimal notation.

In Exercises 4 and 5, rewrite each expression using only positive exponents, and simplify. (Assume that any variables in the expression are nonzero.)

4. (a) $\dfrac{4x^{-2}y^3}{5^{-1}x^3y^{-2}}$

   (b) $\left(\dfrac{-2x^2y}{z^{-3}}\right)^{-2}$

5. (a) $\left(-\dfrac{2u^2}{v^{-1}}\right)^3\left(\dfrac{3v^2}{u^{-3}}\right)$

   (b) $\dfrac{(-3x^2y^{-1})^4}{6x^2y^0}$

6. (a) $\dfrac{6x^{-7}}{(-2x^2)^{-3}}$

   (b) $\left(\dfrac{4y^2}{5x}\right)^{-2}$

In Exercises 7–12, perform the indicated operations and simplify.

7. (a) $(5a^2 - 3a + 4) + (a^2 - 4)$

   (b) $(16 - y^2) - (16 + 2y + y^2)$

8. (a) $-2(2x^4 - 5) + 4x(x^3 + 2x - 1)$

   (b) $4t - [3t - (10t + 7)]$

9. (a) $-3x(x - 4)$

   (b) $(2x - 3y)(x + 5y)$

10. (a) $(x - 1)[2x + (x - 3)]$

    (b) $(2s - 3)(3s^2 - 4s + 7)$

11. (a) $\dfrac{3x^2 + 5x - 9}{x}$

    (b) $\dfrac{6y^3 - 2y^2 + 8}{-2y}$

12. (a) $\dfrac{t^4 + t^2 - 6t}{t^2 - 2}$

    (b) $\dfrac{2x^4 - 15x^2 - 7}{x - 3}$

13. Write an expression for the area of the shaded region in the figure. Then simplify the expression.

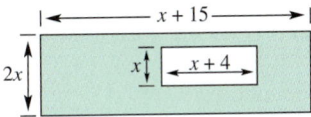

14. The area of a rectangle is $x^2 - 2x - 3$ and its length is $x + 1$. Find the width of the rectangle.

15. The revenue R from the sale of x computer desks is given by $R = x^2 - 35x$. The cost C of producing x computer desks is given by $C = 150 + 12x$. Perform the subtraction required to find the polynomial representing the profit P.

343

Motivating the Chapter

 Dimensions of a Potato Storage Bin

A bin used to store potatoes has the form of a rectangular solid with a volume (in cubic feet) given by the polynomial

$$12x^3 + 64x^2 - 48x.$$

See Section 6.3, Exercise 117.

a. The height of the bin is $4x$ feet. Write an expression for the area of the base of the bin.

b. Factor the expression for the area of the base of the bin. Use the result to write expressions for the length and width of the bin.

See Section 6.5, Exercise 113.

c. The area of the base of the bin is 32 square feet. What are the dimensions of the bin?

d. You are told that the bin has a volume of 256 cubic feet. Can you find the dimensions of the bin? Explain your reasoning.

e. A polynomial that represents the volume of the truck bin in cubic feet is

$$6x^3 + 32x^2 - 24x.$$

How many truckloads does it take to fill the bin? Explain your reasoning.

Nik Wheeler/Corbis

Factoring and Solving Equations

345

6.1 Factoring Polynomials with Common Factors

Dave G. Houser/Corbis

What You Should Learn

1 Find the greatest common factor of two or more expressions.

2 Factor out the greatest common monomial factor from polynomials.

3 Factor polynomials by grouping.

Why You Should Learn It

In some cases, factoring a polynomial enables you to determine unknown quantities. For example, in Exercise 118 on page 353, you will factor the expression for the revenue from selling pool tables to determine an expression for the price of the pool tables.

1 **Find the greatest common factor of two or more expressions.**

Greatest Common Factor

In Chapter 5, you used the Distributive Property to multiply polynomials. In this chapter, you will study the *reverse* process, which is **factoring.**

Multiplying Polynomials

$$2x(7 - 3x) \implies 14x - 6x^2$$

Factor Factor   Product

Factoring Polynomials

$$14x - 6x^2 \implies 2x(7 - 3x)$$

Product   Factor Factor

To factor an expression efficiently, you need to understand the concept of the *greatest common factor* of two (or more) integers or terms. In Section 1.4, you learned that the **greatest common factor** of two or more integers is the greatest integer that is a factor of each integer. For example, the greatest common factor of $12 = 2 \cdot 2 \cdot 3$ and $30 = 2 \cdot 3 \cdot 5$ is $2 \cdot 3 = 6$.

Example 1   Finding the Greatest Common Factor

To find the greatest common factor of $5x^2y^2$ and $30x^3y$, first factor each term.

$$5x^2y^2 = 5 \cdot x \cdot x \cdot y \cdot y = (5x^2y)(y)$$

$$30x^3y = 2 \cdot 3 \cdot 5 \cdot x \cdot x \cdot x \cdot y = (5x^2y)(6x)$$

So, you can conclude that the greatest common factor is $5x^2y$.

Example 2   Finding the Greatest Common Factor

To find the greatest common factor of $8x^5$, $20x^3$, and $16x^4$, first factor each term.

$$8x^5 = 2 \cdot 2 \cdot 2 \cdot x \cdot x \cdot x \cdot x \cdot x = (4x^3)(2x^2)$$

$$20x^3 = 2 \cdot 2 \cdot 5 \cdot x \cdot x \cdot x = (4x^3)(5)$$

$$16x^4 = 2 \cdot 2 \cdot 2 \cdot 2 \cdot x \cdot x \cdot x \cdot x = (4x^3)(4x)$$

So, you can conclude that the greatest common factor is $4x^3$.

② Factor out the greatest common monomial factor from polynomials.

Common Monomial Factors

Consider the three terms listed in Example 2 as terms of the polynomial

$$8x^5 + 16x^4 + 20x^3.$$

The greatest common factor, $4x^3$, of these terms is the **greatest common monomial factor** of the polynomial. When you use the Distributive Property to remove this factor from each term of the polynomial, you are **factoring out** the greatest common monomial factor.

$$8x^5 + 16x^4 + 20x^3 = 4x^3(2x^2) + 4x^3(4x) + 4x^3(5)$$ Factor each term.

$$= 4x^3(2x^2 + 4x + 5)$$ Factor out common monomial factor.

Study Tip

To find the greatest common monomial factor of a polynomial, answer these two questions.

1. What is the greatest integer factor common to each coefficient of the polynomial?

2. What is the highest-power variable factor common to each term of the polynomial?

Example 3   Greatest Common Monomial Factor

Factor out the greatest common monomial factor from $6x - 18$.

Solution

The greatest common integer factor of $6x$ and 18 is 6. There is no common variable factor.

$$6x - 18 = 6(x) - 6(3)$$ Greatest common monomial factor is 6.

$$= 6(x - 3)$$ Factor 6 out of each term.

Example 4   Greatest Common Monomial Factor

Factor out the greatest common monomial factor from $10y^3 - 25y^2$.

Solution

For the terms $10y^3$ and $25y^2$, 5 is the greatest common integer factor and y^2 is the highest-power common variable factor.

$$10y^3 - 25y^2 = (5y^2)(2y) - (5y^2)(5)$$ Greatest common factor is $5y^2$.

$$= 5y^2(2y - 5)$$ Factor $5y^2$ out of each term.

Example 5   Greatest Common Monomial Factor

Factor out the greatest common monomial factor from $45x^3 - 15x^2 - 15$.

Solution

The greatest common integer factor of $45x^3$, $15x^2$, and 15 is 15. There is no common variable factor.

$$45x^3 - 15x^2 - 15 = 15(3x^3) - 15(x^2) - 15(1)$$

$$= 15(3x^3 - x^2 - 1)$$

Example 6   Greatest Common Monomial Factor

Factor out the greatest common monomial factor from $35y^3 - 7y^2 - 14y$.

Solution

$$35y^3 - 7y^2 - 14y = 7y(5y^2) - 7y(y) - 7y(2) \qquad \text{Greatest common factor is } 7y.$$
$$= 7y(5y^2 - y - 2) \qquad \text{Factor } 7y \text{ out of each term.}$$

Example 7   Greatest Common Monomial Factor

Factor out the greatest common monomial factor from $3xy^2 - 15x^2y + 12xy$.

Solution

$$3xy^2 - 15x^2y + 12xy = 3xy(y) - 3xy(5x) + 3xy(4) \qquad \text{Greatest common factor is } 3xy.$$
$$= 3xy(y - 5x + 4) \qquad \text{Factor } 3xy \text{ out of each term.}$$

The greatest common monomial factor of the terms of a polynomial is usually considered to have a positive coefficient. However, sometimes it is convenient to factor a negative number out of a polynomial.

Example 8   A Negative Common Monomial Factor

Factor the polynomial $-2x^2 + 8x - 12$ in two ways.

a. Factor out a common monomial factor of 2.

b. Factor out a common monomial factor of -2.

Solution

a. To factor out the common monomial factor of 2, write the following.

$$-2x^2 + 8x - 12 = 2(-x^2) + 2(4x) + 2(-6) \qquad \text{Factor each term.}$$
$$= 2(-x^2 + 4x - 6) \qquad \text{Factored form}$$

b. To factor -2 out of the polynomial, write the following.

$$-2x^2 + 8x - 12 = -2(x^2) + (-2)(-4x) + (-2)(6) \qquad \text{Factor each term.}$$
$$= -2(x^2 - 4x + 6) \qquad \text{Factored form}$$

Check this result by multiplying $(x^2 - 4x + 6)$ by -2. When you do, you will obtain the original polynomial.

With experience, you should be able to omit writing the first step shown in Examples 6, 7, and 8. For instance, to factor -2 out of $-2x^2 + 8x - 12$, you could simply write

$$-2x^2 + 8x - 12 = -2(x^2 - 4x + 6).$$

③ Factor polynomials by grouping.

Factoring by Grouping

There are occasions when the common factor of an expression is not simply a monomial. For instance, the expression.

$$x^2(x - 2) + 3(x - 2)$$

has the common *binomial* factor $(x - 2)$. Factoring out this common factor produces

$$x^2(x - 2) + 3(x - 2) = (x - 2)(x^2 + 3).$$

This type of factoring is part of a more general procedure called **factoring by grouping.**

Example 9   Common Binomial Factors

Factor each expression.

a. $5x^2(7x - 1) - 3(7x - 1)$ **b.** $2x(3x - 4) + (3x - 4)$
c. $3y^2(y - 3) + 4(3 - y)$

Solution

a. Each of the terms of this expression has a binomial factor of $(7x - 1)$.

$$5x^2(7x - 1) - 3(7x - 1) = (7x - 1)(5x^2 - 3)$$

b. Each of the terms of this expression has a binomial factor of $(3x - 4)$.

$$2x(3x - 4) + (3x - 4) = (3x - 4)(2x + 1)$$

Be sure you see that when $(3x - 4)$ is factored out of itself, you are left with the factor 1. This follows from the fact that $(3x - 4)(1) = (3x - 4)$.

c. $3y^2(y - 3) + 4(3 - y) = 3y^2(y - 3) - 4(y - 3)$ Write $4(3 - y)$ as $-4(y - 3)$.

$$\qquad\qquad\qquad\qquad = (y - 3)(3y^2 - 4)\qquad\text{Common factor is } (y - 3).$$

In Example 9, the polynomials were already grouped so that it was easy to determine the common binomial factors. In practice, you will have to do the grouping as well as the factoring. To see how this works, consider the expression

$$x^3 + 2x^2 + 3x + 6$$

and try to factor it. Note first that there is no common monomial factor to take out of all four terms. But suppose you *group* the first two terms together and the last two terms together.

$$x^3 + 2x^2 + 3x + 6 = (x^3 + 2x^2) + (3x + 6)\qquad\text{Group terms.}$$

$$\qquad\qquad\qquad = x^2(x + 2) + 3(x + 2)\qquad\text{Factor out common monomial factor in each group.}$$

$$\qquad\qquad\qquad = (x + 2)(x^2 + 3)\qquad\text{Factored form}$$

When factoring by grouping, be sure to group terms that have a common monomial factor. For example, in the polynomial above, you should not group the first term x^3 with the fourth term 6.

Example 10   Factoring by Grouping

Factor $x^3 + 2x^2 + x + 2$.

Solution

$$x^3 + 2x^2 + x + 2 = (x^3 + 2x^2) + (x + 2)$$ Group terms.

$$= x^2(x + 2) + (x + 2)$$ Factor out common monomial factor in each group.

$$= (x + 2)(x^2 + 1)$$ Factored form

Note that in Example 10 the polynomial is factored by grouping the first and second terms and the third and fourth terms. You could just as easily have grouped the first and third terms and the second and fourth terms, as follows.

$$x^3 + 2x^2 + x + 2 = (x^3 + x) + (2x^2 + 2)$$

$$= x(x^2 + 1) + 2(x^2 + 1) = (x^2 + 1)(x + 2)$$

Example 11   Factoring by Grouping

Factor $3x^2 - 12x - 5x + 20$.

Solution

$$3x^2 - 12x - 5x + 20 = (3x^2 - 12x) + (-5x + 20)$$ Group terms.

$$= 3x(x - 4) - 5(x - 4)$$ Factor out common monomial factor in each group.

$$= (x - 4)(3x - 5)$$ Factored form

Note how a -5 is factored out so that the common binomial factor $x - 4$ appears.

You can always check to see that you have factored an expression correctly by multiplying the factors and comparing the result with the original expression. Try using multiplication to check the results of Examples 10 and 11.

Study Tip

Notice in Example 12 that the polynomial is not written in standard form. You could have rewritten the polynomial before factoring and still obtained the same result.

$2x^3 + 4x - x^2 - 2$

$= 2x^3 - x^2 + 4x - 2$

$= (2x^3 - x^2) + (4x - 2)$

$= x^2(2x - 1) + 2(2x - 1)$

$= (2x - 1)(x^2 + 2)$

Example 12   Geometry: Area of a Rectangle

The area of a rectangle of width $2x - 1$ is given by the polynomial $2x^3 + 4x - x^2 - 2$, as shown in Figure 6.1. Factor this expression to determine the length of the rectangle.

Solution

$$2x^3 + 4x - x^2 - 2 = (2x^3 + 4x) + (-x^2 - 2)$$ Group terms.

$$= 2x(x^2 + 2) - (x^2 + 2)$$ Factor out common monomial factor in each group.

$$= (x^2 + 2)(2x - 1)$$ Factored form

You can see that the length of the rectangle is $x^2 + 2$.

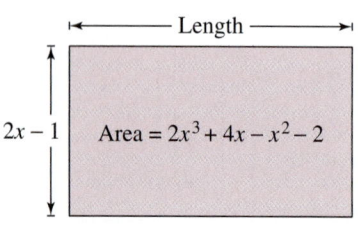

Length

$2x - 1$ Area $= 2x^3 + 4x - x^2 - 2$

Figure 6.1

6.1 Exercises

Review   Concepts, Skills, and Problem Solving

Keep mathematically in shape by doing these exercises *before* the problems of this section.

Properties and Definitions

1. *Writing* ✏ Find the greatest common factor of 18 and 42. Explain how you arrived at your answer.

2. *Writing* ✏ Find the greatest common factor of 30, 45, and 135. Explain how you arrived at your answer.

Simplifying Expressions

In Exercises 3–6, simplify the expression.

3. $2x - (x - 5) + 4(3 - x)$

4. $3(x - 2) - 2(4 - x) - 7x$

5. $\left(\dfrac{3x^2y^3}{2x^5y^2}\right)^2$

6. $\left(\dfrac{-a^3b^{-1}}{4a^{-2}b^3}\right)^2$

Graphing Equations

In Exercises 7–10, graph the equation and show the coordinates of at least three solution points, including any intercepts.

7. $y = 8 - 4x$

8. $3x - y = 6$

9. $y = -\frac{1}{2}x^2$

10. $y = |x + 2|$

Problem Solving

11. *Commission Rate* Determine the commission rate for an employee who earned $1620 in commissions on sales of $54,000.

12. *Work Rate* One person can complete a typing project in 10 hours, and another can complete the same project in 6 hours. Working together, how long will they take to complete the project?

Developing Skills

In Exercises 1–16, find the greatest common factor of the expressions. See Examples 1 and 2.

1. $z^2, -z^6$

2. t^4, t^7

3. $2x^2, 12x$

4. $36x^4, 18x^3$

5. u^2v, u^3v^2

6. $r^6s^4, -rs$

7. $9yz^2, -12y^2z^3$

8. $-15x^6y^3, 45xy^3$

9. $14x^2, 1, 7x^4$

10. $5y^4, 10x^2y^2, 1$

11. $28a^4b^2, 14a^3b^3, 42a^2b^5$

12. $16x^2y, 12xy^2, 36x^2y^2$

13. $2(x + 3), 3(x + 3)$

14. $4(x - 5), 3x(x - 5)$

15. $x(7x + 5), 7x + 5$

16. $x - 4, y(x - 4)$

In Exercises 17–60, factor the polynomial. (*Note:* Some of the polynomials have no common monomial factor.) See Examples 3–7.

17. $3x + 3$

18. $5y + 5$

19. $6z - 6$

20. $3x - 3$

21. $8t - 16$

22. $3u + 12$

23. $-25x - 10$

24. $-14y - 7$

25. $24y^2 - 18$

26. $7z^3 + 21$

27. $x^2 + x$

28. $-s^3 - s$

29. $25u^2 - 14u$

30. $36t^4 + 24t^2$

31. $2x^4 + 6x^3$

32. $9z^6 + 27z^4$

33. $7s^2 + 9t^2$

34. $12x^2 - 5y^3$

35. $12x^2 - 2x$

36. $12u + 9u^2$

37. $-10r^3 - 35r$

38. $-144a^2 + 24a$

39. $16a^3b^3 + 24a^4b^3$

40. $6x^4y + 12x^2y$

41. $10ab + 10a^2b$

42. $21x^2z - 35xz$

43. $12x^2 + 16x - 8$

44. $9 - 3y - 15y^2$

45. $100 + 75z - 50z^2$ **46.** $42t^3 - 21t^2 + 7$

47. $9x^4 + 6x^3 + 18x^2$ **48.** $32a^5 - 2a^3 + 6a$

49. $5u^2 + 5u^2 + 5u$ **50.** $11y^3 - 22y^2 + 11y^2$

51. $x(x - 3) + 5(x - 3)$ **52.** $x(x + 6) + 3(x + 6)$

53. $t(s + 10) - 8(s + 10)$ **54.** $y(q - 5) - 10(q - 5)$

55. $a^2(b + 2) - b(b + 2)$ **56.** $x^3(y + 4) + y(y + 4)$

57. $z^3(z + 5) + z^2(z + 5)$
58. $x^3(x - 2) + x(x - 2)$
59. $(a + b)(a - b) + a(a + b)$
60. $(x + y)(x - y) - x(x - y)$

In Exercises 61–68, factor a negative real number from the polynomial and then write the polynomial factor in standard form. See Example 8.

61. $5 - 10x$ **62.** $3 - 6x$
63. $3000 - 3x$ **64.** $9 - 2x^2$

65. $4 + 2x - x^2$ **66.** $18 - 12x - 6x^2$

67. $4 + 12x - 2x^2$ **68.** $8 - 4x - 12x^2$

In Exercises 69–100, factor the polynomial by grouping. See Examples 9–11.

69. $x^2 + 10x + x + 10$
70. $x^2 - 5x + x - 5$
71. $a^2 - 4a + a - 4$
72. $x^2 + 25x + x + 25$
73. $x^2 + 3x + 4x + 12$
74. $x^2 - x + 3x - 3$
75. $x^2 + 2x + 5x + 10$
76. $x^2 - 6x + 5x - 30$
77. $x^2 + 3x - 5x - 15$
78. $x^2 + 4x + x + 4$
79. $4x^2 - 14x + 14x - 49$
80. $4x^2 - 6x + 6x - 9$

81. $6x^2 + 3x - 2x - 1$
82. $5x^2 + 20x - x - 4$
83. $8x^2 + 32x + x + 4$
84. $8x^2 - 4x - 2x + 1$
85. $3x^2 - 2x + 3x - 2$
86. $12x^2 + 42x - 10x - 35$
87. $2x^2 - 4x - 3x + 6$
88. $35x^2 - 40x + 21x - 24$
89. $ky^2 - 4ky + 2y - 8$
90. $ay^2 + 3ay + 3y + 9$
91. $t^3 - 3t^2 + 2t - 6$
92. $3s^3 + 6s^2 + 2s + 4$
93. $x^3 + 2x^2 + x + 2$
94. $x^3 - 5x^2 + x - 5$
95. $6z^3 + 3z^2 - 2z - 1$
96. $4u^3 - 2u^2 - 6u + 3$
97. $x^3 - 3x - x^2 + 3$
98. $x^3 + 7x - 3x^2 - 21$
99. $4x^2 - x^3 - 8 + 2x$
100. $5x^2 + 10x^3 + 4 + 8x$

In Exercises 101–106, fill in the missing factor.

101. $\frac{1}{4}x + \frac{3}{4} = \frac{1}{4}($ ⬚ $)$
102. $\frac{5}{6}x - \frac{1}{6} = \frac{1}{6}($ ⬚ $)$
103. $2y - \frac{1}{5} = \frac{1}{5}($ ⬚ $)$
104. $3z + \frac{3}{4} = \frac{1}{4}($ ⬚ $)$
105. $\frac{7}{8}x + \frac{5}{16}y = \frac{1}{16}($ ⬚ $)$
106. $\frac{5}{12}u - \frac{5}{8}v = \frac{1}{24}($ ⬚ $)$

⊞ In Exercises 107–110, use a graphing calculator to graph both equations in the same viewing window. What can you conclude?

107. $y_1 = 9 - 3x$ **108.** $y_1 = x^2 - 4x$
$\quad\ \ y_2 = -3(x - 3)$ $\qquad\ \ y_2 = x(x - 4)$

109. $y_1 = 6x - x^2$
$\qquad\ \ y_2 = x(6 - x)$
110. $y_1 = x(x + 2) - 3(x + 2)$
$\qquad\ \ y_2 = (x + 2)(x - 3)$

Solving Problems

▲ *Geometry*   In Exercises 111 and 112, factor the polynomial to find an expression for the length of the rectangle. See Example 12.

111. Area $= 2x^2 + 2x$

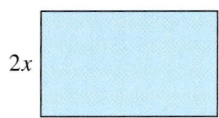

112. Area $= x^2 + 2x + 10x + 20$

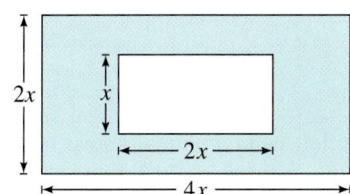

▲ *Geometry*   In Exercises 113 and 114, write an expression for the area of the shaded region and factor the expression if possible.

113.

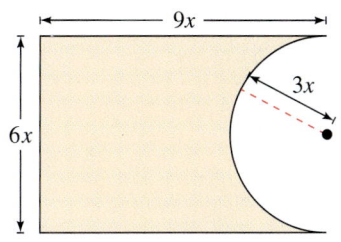

114.

115. ▲ *Geometry*   The surface area of a right circular cylinder is given by

$$2\pi r^2 + 2\pi rh$$

where r is the radius of the base of the cylinder and h is the height of the cylinder. Factor the expression for the surface area.

116. *Simple Interest*   The amount after t years when a principal of P dollars is invested at $r\%$ simple interest is given by

$$P + Prt.$$

Factor the expression for simple interest.

117. *Chemical Reaction*   The rate of change in a chemical reaction is

$$kQx - kx^2$$

where Q is the original amount, x is the new amount, and k is a constant of proportionality. Factor the expression.

118. *Unit Price*   The revenue R from selling x units of a product at a price of p dollars per unit is given by $R = xp$. For a pool table the revenue is

$$R = 900x - 0.1x^2.$$

Factor the revenue model and determine an expression that represents the price p in terms of x.

Explaining Concepts

119. Give an example of a polynomial that is written in factored form.

120. Give an example of a trinomial whose greatest common monomial factor is $3x$.

121. *Writing*✐   How do you check your result when factoring a polynomial?

122. *Writing*✐   In your own words, describe a method for finding the greatest common factor of a polynomial.

123. *Writing*✐   Explain how the word *factor* can be used as a noun and as a verb.

124. Give several examples of the use of the Distributive Property in factoring.

125. Give an example of a polynomial with four terms that can be factored by grouping.

6.2 Factoring Trinomials

What You Should Learn

1 Factor trinomials of the form $x^2 + bx + c$.

2 Factor trinomials in two variables.

3 Factor trinomials completely.

Why You Should Learn It

The techniques for factoring trinomials will help you in solving quadratic equations in Section 6.5.

1 Factor trinomials of the form $x^2 + bx + c$.

Factoring Trinomials of the Form $x^2 + bx + c$

From Section 5.3, you know that the product of two binomials is often a trinomial. Here are some examples.

$$\textit{Factored Form} \quad \text{F} \quad \text{O} \quad \text{I} \quad \text{L} \quad \textit{Trinomial Form}$$
$$(x - 1)(x + 5) = x^2 + 5x - x - 5 = x^2 + 4x - 5$$
$$(x - 3)(x - 3) = x^2 - 3x - 3x + 9 = x^2 - 6x + 9$$
$$(x + 5)(x + 1) = x^2 + x + 5x + 5 = x^2 + 6x + 5$$
$$(x - 2)(x - 4) = x^2 - 4x - 2x + 8 = x^2 - 6x + 8$$

Try covering the factored forms in the left-hand column above. Can you determine the factored forms from the trinomial forms? In this section, you will learn how to factor trinomials of the form $x^2 + bx + c$. To begin, consider the following factorization.

$$(x + m)(x + n) = x^2 + nx + mx + mn$$
$$= x^2 + \underbrace{(n + m)}x + \underbrace{mn}$$

Sum of terms   Product of terms

$$= x^2 + \boxed{b}\, x + \boxed{c}$$

So, to *factor* a trinomial $x^2 + bx + c$ into a product of two binomials, you must find two numbers m and n whose product is c and whose sum is b.

There are many different techniques that can be used to factor trinomials. The most common technique is to use *guess, check, and revise* with mental math. For example, try factoring the trinomial

$$x^2 + 5x + 6.$$

You need to find two numbers whose product is 6 and whose sum is 5. Using mental math, you can determine that the numbers are 2 and 3.

The product of 2 and 3 is 6.

$$x^2 + 5x + 6 = (x + 2)(x + 3)$$

The sum of 2 and 3 is 5.

Example 1   Factoring a Trinomial

Factor the trinomial $x^2 + 5x - 6$.

Solution

You need to find two numbers whose product is -6 and whose sum is 5.

The product of -1 and 6 is -6.

$$x^2 + 5x - 6 = (x - 1)(x + 6)$$

The sum of -1 and 6 is 5.

Example 2   Factoring a Trinomial

Factor the trinomial $x^2 - x - 6$.

Solution

You need to find two numbers whose product is -6 and whose sum is -1.

The product of -3 and 2 is -6.

$$x^2 - x - 6 = (x - 3)(x + 2)$$

The sum of -3 and 2 is -1.

Example 3   Factoring a Trinomial

Factor the trinomial $x^2 - 5x + 6$.

Solution

You need to find two numbers whose product is 6 and whose sum is -5.

The product of -2 and -3 is 6.

$$x^2 - 5x + 6 = (x - 2)(x - 3)$$

The sum of -2 and -3 is -5.

Example 4   Factoring a Trinomial

Factor the trinomial $14 + 5x - x^2$.

Solution

It is helpful first to factor out -1 and write the polynomial factor in standard form.

$$14 + 5x - x^2 = -1(x^2 - 5x - 14)$$

Now you need two numbers -7 and 2 whose product is -14 and whose sum is -5. So,

$$14 + 5x - x^2 = -(x^2 - 5x - 14) = -(x - 7)(x + 2).$$

If you have trouble factoring a trinomial, it helps to make a list of all the distinct pairs of factors and then check each sum. For instance, consider the trinomial

$$x^2 - 5x - 24 = (x +)(x -).$$ Opposite signs

In this trinomial the constant term is negative, so you need to find two numbers with opposite signs whose product is -24 and whose sum is -5.

Factors of -24	*Sum*	
$1, -24$	-23	
$-1, 24$	23	
$2, -12$	-10	
$-2, 12$	10	
$3, -8$	-5	Correct choice
$-3, 8$	5	
$4, -6$	-2	
$-4, 6$	2	

So, $x^2 - 5x - 24 = (x + 3)(x - 8)$.

With experience, you will be able to narrow the list of possible factors *mentally* to only two or three possibilities whose sums can then be tested to determine the correct factorization. Here are some suggestions for narrowing the list.

Guidelines for Factoring $x^2 + bx + c$

To factor $x^2 + bx + c$, you need to find two numbers m and n whose product is c and whose sum is b.

$$x^2 + bx + c = (x + m)(x + n)$$

1. If c is positive, then m and n have like signs that match the sign of b.

2. If c is negative, then m and n have unlike signs.

3. If $|b|$ is small relative to $|c|$, first try those factors of c that are closest to each other in absolute value.

Study Tip

Notice that factors may be written in any order. For example,

$(x - 5)(x + 3) = (x + 3)(x - 5)$

and

$(x + 2)(x + 18) = (x + 18)(x + 2)$

because of the Commutative Property of Multiplication.

Example 5   Factoring a Trinomial

Factor the trinomial $x^2 - 2x - 15$.

Solution

You need to find two numbers whose product is -15 and whose sum is -2.

The product of -5 and 3 is -15.

$$x^2 - 2x - 15 = (x - 5)(x + 3)$$

The sum of -5 and 3 is -2.

Study Tip

Not all trinomials are factorable using integer factors. For instance, $x^2 - 2x - 6$ is not factorable using integer factors because there is no pair of factors of -6 whose sum is -2. Such nonfactorable trinomials are called **prime polynomials**.

Example 6　Factoring a Trinomial

Factor the trinomial $x^2 + 7x - 30$.

Solution

You need to find two numbers whose product is -30 and whose sum is 7.

The product of -3 and 10 is -30.

$$x^2 + 7x - 30 = (x - 3)(x + 10)$$

The sum of -3 and 10 is 7.

---

2 Factor trinomials in two variables.

Factoring Trinomials in Two Variables

So far, the examples in this section have all involved trinomials of the form

$x^2 + bx + c.$　　　　Trinomial in one variable

The next three examples show how to factor trinomials of the form

$x^2 + bxy + cy^2.$　　　　Trinomial in two variables

Note that this trinomial has two variables, x and y. However, from the factorization

$$x^2 + bxy + cy^2 = (x + my)(x + ny) = x^2 + (m + n)xy + mny^2$$

you can see that you still need to find two factors of c whose sum is b.

Study Tip

With *any* factoring problem, remember that you can check your result by multiplying. For instance, in Example 7, you can check the result by multiplying $(x - 4y)$ by $(x + 3y)$ to see that you obtain $x^2 - xy - 12y^2$.

Example 7　Factoring a Trinomial in Two Variables

Factor the trinomial $x^2 - xy - 12y^2$.

Solution

You need to find two numbers whose product is -12 and whose sum is -1.

The product of -4 and 3 is -12.

$$x^2 - xy - 12y^2 = (x - 4y)(x + 3y)$$

The sum of -4 and 3 is -1.

Example 8　Factoring a Trinomial in Two Variables

Factor the trinomial $y^2 - 6xy + 8x^2$.

Solution

You need to find two numbers whose product is 8 and whose sum is -6.

The product of -2 and -4 is 8.

$$y^2 - 6xy + 8x^2 = (y - 2x)(y - 4x)$$

The sum of -2 and -4 is -6.

Example 9 Factoring a Trinomial in Two Variables

Factor the trinomial $x^2 + 11xy + 10y^2$.

Solution

You need to find two numbers whose product is 10 and whose sum is 11.

The product of 1 and 10 is 10.

$$x^2 + 11xy + 10y^2 = (x + y)(x + 10y)$$

The sum of 1 and 10 is 11.

---

3 Factor trinomials completely.

Factoring Completely

Some trinomials have a common monomial factor. In such cases you should first factor out the common monomial factor. Then you can try to factor the resulting trinomial by the methods of this section. This "multiple-stage factoring process" is called **factoring completely.** The trinomial below is completely factored.

$$2x^2 - 4x - 6 = 2(x^2 - 2x - 3)$$

Factor out common monomial factor 2.

$$= 2(x - 3)(x + 1)$$

Factor trinomial.

Example 10 Factoring Completely

Factor the trinomial $2x^2 - 12x + 10$ completely.

Solution

$$2x^2 - 12x + 10 = 2(x^2 - 6x + 5)$$

Factor out common monomial factor 2.

$$= 2(x - 5)(x - 1)$$

Factor trinomial.

Example 11 Factoring Completely

Factor the trinomial $3x^3 - 27x^2 + 54x$ completely.

Solution

$$3x^3 - 27x^2 + 54x = 3x(x^2 - 9x + 18)$$

Factor out common monomial factor $3x$.

$$= 3x(x - 3)(x - 6)$$

Factor trinomial.

Example 12 Factoring Completely

Factor the trinomial $4y^4 + 32y^3 + 28y^2$ completely.

Solution

$$4y^4 + 32y^3 + 28y^2 = 4y^2(y^2 + 8y + 7)$$

Factor out common monomial factor $4y^2$.

$$= 4y^2(y + 1)(y + 7)$$

Factor trinomial.

6.2 Exercises

Review   Concepts, Skills, and Problem Solving

Keep mathematically in shape by doing these exercises *before* the problems of this section.

Properties and Definitions

1. *Writing* ✎   Explain what is meant by the intercepts of a graph and explain how to find the intercepts of a graph.

2. What is the leading coefficient of the polynomial $3x - 7x^2 + 4x^3 - 4$?

Rewriting Algebraic Expressions

In Exercises 3–8, find the product.

3. $y(y + 2)$

4. $-a^2(a - 1)$

5. $(x - 2)(x - 5)$

6. $(v - 4)(v + 7)$

7. $(2x + 5)(2x - 5)$

8. $x^2(x + 1) - 5(x^2 - 2)$

Problem Solving

9. *Profit*   A consulting company showed a loss of $2,500,000 during the first 6 months of 2002. The company ended the year with an overall profit of $1,475,000. What was the profit during the second 6 months of the year?

10. *Cost*   Computer printer ink cartridges cost $11.95 per cartridge. There are 12 cartridges per box, and five boxes were ordered. Determine the total cost of the order.

11. *Cost, Revenue, and Profit*   The revenue R from selling x units of a product is $R = 75x$. The cost C of producing x units is $C = 62.5x + 570$. In order to obtain a profit P, the revenue must be greater than the cost. For what values of x will this product produce a profit?

12. *Distance Traveled*   The minimum and maximum speeds on an interstate highway are 40 miles per hour and 65 miles per hour. You travel nonstop for $3\frac{1}{2}$ hours on this highway. Assuming that you stay within the speed limits, write an inequality for the distance you travel.

Explaining Concepts

In Exercises 1–8, fill in the missing factor. Then check your answer by multiplying the factors.

1. $x^2 + 4x + 3 = (x + 3)()$

2. $x^2 + 5x + 6 = (x + 3)()$

3. $a^2 + a - 6 = (a + 3)()$

4. $c^2 + 2c - 3 = (c + 3)()$

5. $y^2 - 2y - 15 = (y + 3)()$

6. $y^2 - 4y - 21 = (y + 3)()$

7. $z^2 - 5z + 6 = (z - 3)()$

8. $z^2 - 4z + 3 = (z - 3)()$

In Exercises 9–14, find all possible products of the form $(x + m)(x + n)$ where $m \cdot n$ is the specified product. (Assume that m and n are integers.)

9. $m \cdot n = 11$

10. $m \cdot n = 5$

11. $m \cdot n = 14$

12. $m \cdot n = 10$

13. $m \cdot n = 12$

14. $m \cdot n = 18$

In Exercises 15–44, factor the trinomial. (*Note:* Some of the trinomials may be prime.) See Examples 1–9.

15. $x^2 + 6x + 8$

16. $x^2 + 13x + 12$

17. $x^2 - 13x + 40$

18. $x^2 - 9x + 14$

19. $z^2 - 7z + 12$

20. $x^2 + 10x + 24$

21. $y^2 + 5y + 11$

22. $s^2 - 7s - 25$

23. $x^2 - x - 6$

24. $x^2 + x - 6$

25. $x^2 + 2x - 15$

26. $b^2 - 2b - 15$

27. $y^2 - 6y + 10$

28. $c^2 - 6c + 10$

29. $u^2 - 22u - 48$

30. $x^2 - x - 36$

31. $x^2 + 19x + 60$

32. $x^2 + 3x - 70$

33. $x^2 - 17x + 72$

34. $x^2 + 21x + 108$

35. $x^2 - 8x - 240$

36. $r^2 - 30r + 216$

37. $x^2 + xy - 2y^2$

38. $x^2 - 5xy + 6y^2$

39. $x^2 + 8xy + 15y^2$

40. $u^2 - 4uv - 5v^2$

41. $x^2 - 7xz - 18z^2$

42. $x^2 + 15xy + 50y^2$

43. $a^2 + 2ab - 15b^2$

44. $y^2 + 4yz - 60z^2$

In Exercises 45–64, factor the trinomial completely. (*Note*: Some of the trinomials may be prime.) See Examples 10–12.

45. $3x^2 + 21x + 30$

46. $4x^2 - 32x + 60$

47. $4y^2 - 8y - 12$

48. $5x^2 - 20x - 25$

49. $3z^2 + 5z + 6$

50. $7x^2 + 5x + 10$

51. $9x^2 + 18x - 18$

52. $6x^2 - 24x - 6$

53. $x^3 - 13x^2 + 30x$

54. $x^3 + x^2 - 2x$

55. $x^4 - 5x^3 + 6x^2$

56. $x^4 + 3x^3 - 10x^2$

57. $-3y^2x - 9yx + 54x$

58. $-5x^2z + 15xz + 50z$

59. $x^3 + 5x^2y + 6xy^2$

60. $x^2y - 6xy^2 + y^3$

61. $2x^3y + 4x^2y^2 - 6xy^3$

62. $2x^3y - 10x^2y^2 + 6xy^3$

63. $x^4y^2 + 3x^3y^3 + 2x^2y^4$

64. $x^4y^2 + x^3y^3 - 2x^2y^4$

In Exercises 65–70, find all integers b such that the trinomial can be factored.

65. $x^2 + bx + 18$

66. $x^2 + bx + 10$

67. $x^2 + bx - 21$

68. $x^2 + bx - 7$

69. $x^2 + bx + 36$

70. $x^2 + bx - 48$

In Exercises 71–76, find two integers c such that the trinomial can be factored. (There are many correct answers.)

71. $x^2 + 3x + c$

72. $x^2 + 5x + c$

73. $x^2 - 4x + c$

74. $x^2 - 15x + c$

75. $x^2 - 9x + c$

76. $x^2 + 12x + c$

Graphical Reasoning   In Exercises 77–80, use a graphing calculator to graph the two equations in the same viewing window. What can you conclude?

77. $y_1 = x^2 - x - 6$
$y_2 = (x + 2)(x - 3)$

78. $y_1 = x^2 - 10x + 16$
$y_2 = (x - 2)(x - 8)$

79. $y_1 = x^3 + x^2 - 20x$
$y_2 = x(x - 4)(x + 5)$

80. $y_1 = 2x - x^2 - x^3$
$y_2 = x(1 - x)(2 + x)$

Solving Problems

81. *Exploration*   An open box is to be made from a four-foot-by-six-foot sheet of metal by cutting equal squares from the corners and turning up the sides (see figure). The volume of the box can be modeled by $V = 4x^3 - 20x^2 + 24x,\quad 0 < x < 2.$

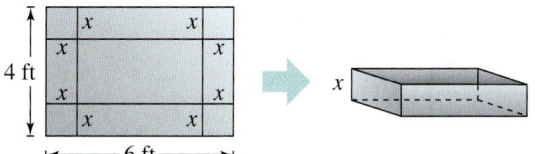

(a) Factor the trinomial that models the volume of the box. Use the factored form to explain how the model was found.

(b) 🖩 Use a graphing calculator to graph the trinomial over the specified interval. Use the graph to approximate the size of the squares to be cut from the corners so that the volume of the box is greatest.

82. *Exploration*   If the box in Exercise 81 is to be made from a six-foot-by-eight-foot sheet of metal, the volume of the box would be modeled by

$V = 4x^3 - 28x^2 + 48x,\quad 0 < x < 3.$

(a) Factor the trinomial that models the volume of the box. Use the factored form to explain how the model was found.

(b) 🖩 Use a graphing calculator to graph the trinomial over the specified interval. Use the graph to approximate the size of the squares to be cut from the corners so that the volume of the box is greatest.

83. 🔺 *Geometry*   The area of the rectangle shown in the figure is $x^2 + 30x + 200$. What is the area of the shaded region?

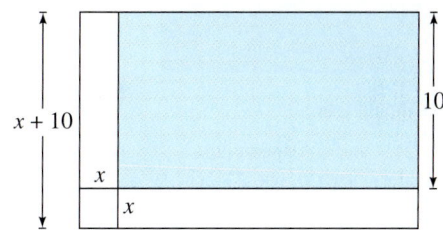

84. 🔺 *Geometry*   The area of the rectangle shown in the figure is $x^2 + 17x + 70$. What is the area of the shaded region?

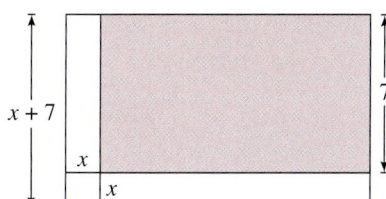

Explaining Concepts

85. State which of the following are factorizations of $2x^2 + 6x - 20$. For each correct factorization, state whether or not it is complete.

(a) $(2x - 4)(x + 5)$ (b) $(2x - 4)(2x + 10)$

(c) $(x - 2)(x + 5)$ (d) $2(x - 2)(x + 5)$

86. *Writing*   In factoring $x^2 - 4x + 3$, why is it unnecessary to test $(x - 1)(x + 3)$ and $(x + 1)(x - 3)$?

87. *Writing*   In your own words, explain how to factor a trinomial of the form $x^2 + bx + c$. Give examples with your explanation.

88. *Writing*   What is meant by a prime trinomial?

89. *Writing*   Can you completely factor a trinomial into two different sets of prime factors? Explain.

90. *Writing*   In factoring the trinomial $x^2 + bx + c$, is the process easier if c is a prime number such as 5 or a composite number such as 120? Explain.

6.3 More About Factoring Trinomials

What You Should Learn

1. Factor trinomials of the form $ax^2 + bx + c$.
2. Factor trinomials completely.
3. Factor trinomials by grouping.

Why You Should Learn It

Trinomials can be used in many geometric applications. For example, in Exercise 112 on page 369, you will factor the expression for the volume of a swimming pool to determine an expression for the width of the swimming pool.

1 Factor trinomials of the form $ax^2 + bx + c$.

Factoring Trinomials of the Form $ax^2 + bx + c$

In this section, you will learn how to factor a trinomial whose leading coefficient is *not* 1. To see how this works, consider the following.

$$\text{Factors of } a$$

$$ax^2 + bx + c = (x +)(x +)$$

$$\text{Factors of } c$$

The goal is to find a combination of factors of a and c such that the outer and inner products add up to the middle term bx.

Example 1 Factoring a Trinomial of the Form $ax^2 + bx + c$

Factor the trinomial $4x^2 - 4x - 3$.

Solution

First, observe that $4x^2 - 4x - 3$ has no common monomial factor. For this trinomial, $a = 4$ and $c = -3$. You need to find a combination of the factors of 4 and -3 such that the outer and inner products add up to $-4x$. The possible combinations are as follows.

Factors	$O + I$	
Inner product $= 4x$		
$(x + 1)(4x - 3)$	$-3x + 4x = x$	x does not equal $-4x$.
Outer product $= -3x$		
$(x - 1)(4x + 3)$	$3x - 4x = -x$	$-x$ does not equal $-4x$.
$(x + 3)(4x - 1)$	$-x + 12x = 11x$	$11x$ does not equal $-4x$.
$(x - 3)(4x + 1)$	$x - 12x = -11x$	$-11x$ does not equal $-4x$.
$(2x + 1)(2x - 3)$	$-6x + 2x = -4x$	$-4x$ equals $-4x$. ✔
$(2x - 1)(2x + 3)$	$6x - 2x = 4x$	$4x$ does not equal $-4x$.

So, the correct factorization is $4x^2 - 4x - 3 = (2x + 1)(2x - 3)$.

Example 2 **Factoring a Trinomial of the Form $ax^2 + bx + c$**

Factor the trinomial $6x^2 + 5x - 4$.

Solution

First, observe that $6x^2 + 5x - 4$ has no common monomial factor. For this trinomial, $a = 6$ and $c = -4$. You need to find a combination of the factors of 6 and -4 such that the outer and inner products add up to $5x$.

Factors	$O + I$	
$(x + 1)(6x - 4)$	$-4x + 6x = 2x$	2x does not equal 5x.
$(x - 1)(6x + 4)$	$4x - 6x = -2x$	$-2x$ does not equal 5x.
$(x + 4)(6x - 1)$	$-x + 24x = 23x$	23x does not equal 5x.
$(x - 4)(6x + 1)$	$x - 24x = -23x$	$-23x$ does not equal 5x.
$(x + 2)(6x - 2)$	$-2x + 12x = 10x$	10x does not equal 5x.
$(x - 2)(6x + 2)$	$2x - 12x = -10x$	$-10x$ does not equal 5x.
$(2x + 1)(3x - 4)$	$-8x + 3x = -5x$	$-5x$ does not equal 5x.
$(2x - 1)(3x + 4)$	$8x - 3x = 5x$	5x equals 5x. ✓
$(2x + 4)(3x - 1)$	$-2x + 12x = 10x$	10x does not equal 5x.
$(2x - 4)(3x + 1)$	$2x - 12x = -10x$	$-10x$ does not equal 5x.
$(2x + 2)(3x - 2)$	$-4x + 6x = 2x$	2x does not equal 5x.
$(2x - 2)(3x + 2)$	$4x - 6x = -2x$	$-2x$ does not equal 5x.

So, the correct factorization is $6x^2 + 5x - 4 = (2x - 1)(3x + 4)$.

Study Tip

If the original trinomial has no common monomial factors, then its binomial factors cannot have common monomial factors. So, in Example 2, you don't have to test factors such as $(6x - 4)$ that have a common monomial factor of 2.

The following guidelines can help shorten the list of possible factorizations.

Guidelines for Factoring $ax^2 + bx + c\,(a > 0)$

1. If the trinomial has a common monomial factor, you should factor out the common factor before trying to find the binomial factors.

2. Because the resulting trinomial has no common monomial factors, you do not have to test any binomial factors that have a common monomial factor.

3. Switch the signs of the factors of c when the middle term $(O + I)$ is correct except in sign.

Using these guidelines, you can shorten the list in Example 2 to the following.

$(x + 4)(6x - 1) = 6x^2 + 23x - 4$		23x does not equal 5x.
$(2x + 1)(3x - 4) = 6x^2 - 5x - 4$		Opposite sign
$(2x - 1)(3x + 4) = 6x^2 + 5x - 4$	⬅	Correct factorization

Technology: Tip

As with other types of factoring, you can use a graphing calculator to check your results. For instance, graph

$y_1 = 2x^2 + x - 15$ and

$y_2 = (2x - 5)(x + 3)$

in the same viewing window, as shown below. Because both graphs are the same, you can conclude that

$2x^2 + x - 15$

$= (2x - 5)(x + 3).$

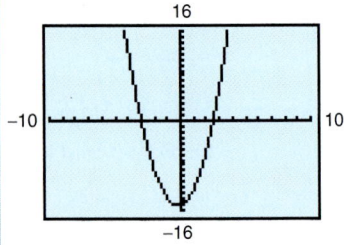

2 Factor trinomials completely.

Example 3   Factoring a Trinomial of the Form $ax^2 + bx + c$

Factor the trinomial $2x^2 + x - 15$.

Solution

First, observe that $2x^2 + x - 15$ has no common monomial factor. For this trinomial, $a = 2$, which factors as $(1)(2)$, and $c = -15$, which factors as $(1)(-15)$, $(-1)(15)$, $(3)(-5)$, and $(-3)(5)$.

$$(2x + 1)(x - 15) = 2x^2 - 29x - 15$$

$$(2x + 15)(x - 1) = 2x^2 + 13x - 15$$

$$(2x + 3)(x - 5) = 2x^2 - 7x - 15$$

$$(2x + 5)(x - 3) = 2x^2 - x - 15 \qquad \text{Middle term has opposite sign.}$$

$$(2x - 5)(x + 3) = 2x^2 + x - 15 \quad \text{} \quad \text{Correct factorization}$$

So, the correct factorization is

$$2x^2 + x - 15 = (2x - 5)(x + 3).$$

Notice in Example 3 that when the middle term has the incorrect sign, you need only to change the signs of the second terms of the two factors.

Factoring Completely

Remember that if a trinomial has a common monomial factor, the common monomial factor should be factored out first. The complete factorization will then show all monomial and binomial factors.

Example 4   Factoring Completely

Factor $4x^3 - 30x^2 + 14x$ completely.

Solution

Begin by factoring out the common monomial factor.

$$4x^3 - 30x^2 + 14x = 2x(2x^2 - 15x + 7)$$

Now, for the new trinomial $2x^2 - 15x + 7$, $a = 2$ and $c = 7$. The possible factorizations of this trinomial are as follows.

$$(2x - 7)(x - 1) = 2x^2 - 9x + 7$$

$$(2x - 1)(x - 7) = 2x^2 - 15x + 7 \quad \text{} \quad \text{Correct factorization}$$

So, the complete factorization of the original trinomial is

$$4x^3 - 30x^2 + 14x = 2x(2x^2 - 15x + 7)$$

$$= 2x(2x - 1)(x - 7).$$

In factoring a trinomial with a negative leading coefficient, first factor -1 out of the trinomial, as demonstrated in Example 5.

Example 5   A Negative Leading Coefficient

Factor the trinomial $-5x^2 + 7x + 6$.

Solution

This trinomial has a negative leading coefficient, so you should begin by factoring -1 out of the trinomial.

$$-5x^2 + 7x + 6 = (-1)(5x^2 - 7x - 6)$$

Now, for the new trinomial $5x^2 - 7x - 6$, you have $a = 5$ and $c = -6$. After testing the possible factorizations, you can conclude that

$$(x - 2)(5x + 3) = 5x^2 - 7x - 6.$$ Correct factorization

So, a correct factorization is

$$-5x^2 + 7x + 6 = (-1)(x - 2)(5x + 3)$$

$$= (-x + 2)(5x + 3).$$ Distributive Property

Another correct factorization is $(x - 2)(-5x - 3)$.

---

③ Factor trinomials by grouping.

Factoring by Grouping

The examples in this and the preceding section have shown how to use the *guess, check, and revise* strategy to factor trinomials. An alternative technique to use is factoring by grouping. Recall from Section 6.1 that the polynomial

$$x^3 + 2x^2 + 3x + 6$$

was factored by first grouping terms and then applying the Distributive Property.

$$x^3 + 2x^2 + 3x + 6 = (x^3 + 2x^2) + (3x + 6)$$ Group terms.

$$= x^2(x + 2) + 3(x + 2)$$ Factor out common monomial factor in each group.

$$= (x + 2)(x^2 + 3)$$ Distributive Property

By rewriting the middle term of the trinomial $2x^2 + x - 15$ as

$$2x^2 + x - 15 = 2x^2 + 6x - 5x - 15$$

you can group the first two terms and the last two terms and factor the trinomial as follows.

$$2x^2 + x - 15 = 2x^2 + 6x - 5x - 15$$ Rewrite middle term.

$$= (2x^2 + 6x) + (-5x - 15)$$ Group terms.

$$= 2x(x + 3) - 5(x + 3)$$ Factor out common monomial factor in each group.

$$= (x + 3)(2x - 5)$$ Distributive Property

Guidelines for Factoring $ax^2 + bx + c$ by Grouping

1. If necessary, write the trinomial in standard form.

2. Choose factors of the product ac that add up to b.

3. Use these factors to rewrite the middle term as a sum or difference.

4. Group and remove any common monomial factors from the first two terms and the last two terms.

5. If possible, factor out the common binomial factor.

Example 6 Factoring a Trinomial by Grouping

Use factoring by grouping to factor the trinomial $2x^2 + 5x - 3$.

Solution

In the trinomial $2x^2 + 5x - 3$, $a = 2$ and $c = -3$, which implies that the product ac is -6. Now, because -6 factors as $(6)(-1)$, and $6 - 1 = 5 = b$, you can rewrite the middle term as $5x = 6x - x$. This produces the following.

$$
\begin{aligned}
2x^2 + 5x - 3 &= 2x^2 + 6x - x - 3 && \text{Rewrite middle term.} \\
&= (2x^2 + 6x) + (-x - 3) && \text{Group terms.} \\
&= 2x(x + 3) - (x + 3) && \text{Factor out common monomial factor in each group.} \\
&= (x + 3)(2x - 1) && \text{Factor out common binomial factor.}
\end{aligned}
$$

So, the trinomial factors as

$$2x^2 + 5x - 3 = (x + 3)(2x - 1).$$

Example 7 Factoring a Trinomial by Grouping

Use factoring by grouping to factor the trinomial $6x^2 - 11x - 10$.

Solution

In the trinomial $6x^2 - 11x - 10$, $a = 6$ and $c = -10$, which implies that the product ac is -60. Now, because -60 factors as $(-15)(4)$ and $-15 + 4 = -11 = b$, you can rewrite the middle term as $-11x = -15x + 4x$. This produces the following.

$$
\begin{aligned}
6x^2 - 11x - 10 &= 6x^2 - 15x + 4x - 10 && \text{Rewrite middle term.} \\
&= (6x^2 - 15x) + (4x - 10) && \text{Group terms.} \\
&= 3x(2x - 5) + 2(2x - 5) && \text{Factor out common monomial factor in each group.} \\
&= (2x - 5)(3x + 2) && \text{Factor out common binomial factor.}
\end{aligned}
$$

So, the trinomial factors as

$$6x^2 - 11x - 10 = (2x - 5)(3x + 2).$$

6.3 Exercises

Review   Concepts, Skills, and Problem Solving

Keep mathematically in shape by doing these exercises *before* the problems of this section.

Properties and Definitions

1. Is 29 prime or composite?

2. Without dividing 255 by 3, how can you tell whether it is divisible by 3?

Simplifying Expressions

In Exercises 3–6, write the prime factorization.

3. 500

4. 315

5. 792

6. 2275

In Exercises 7 and 8, multiply and simplify.

7. $(2x - 5)(x + 7)$

8. $(3x - 2)^2$

Graphing Equations

In Exercises 9 and 10, graph the equation and identify any intercepts.

9. $y = (3 + x)(3 - x)$

10. $3x + 6y - 12 = 0$

11. *Stretching a Spring*   An equation for the distance y (in inches) a spring is stretched from its equilibrium point when a force of x pounds is applied is $y = 0.066x$.

(a) Graph the model.

(b) Estimate y when a force of 100 pounds is applied.

Developing Skills

In Exercises 1–18, fill in the missing factor.

1. $2x^2 + 7x - 4 = (2x - 1)()$

2. $3x^2 + x - 4 = (3x + 4)()$

3. $3t^2 + 4t - 15 = (3t - 5)()$

4. $5t^2 + t - 18 = (5t - 9)()$

5. $7x^2 + 15x + 2 = (7x + 1)()$

6. $3x^2 + 4x + 1 = (3x + 1)()$

7. $5x^2 + 18x + 9 = (x + 3)()$

8. $5x^2 + 19x + 12 = (x + 3)()$

9. $5a^2 + 12a - 9 = (a + 3)()$

10. $5c^2 + 11c - 12 = (c + 3)()$

11. $4z^2 - 13z + 3 = (z - 3)()$

12. $6z^2 - 23z + 15 = (z - 3)()$

13. $6x^2 - 23x + 7 = (3x - 1)()$

14. $6x^2 - 13x + 6 = (2x - 3)()$

15. $9a^2 - 6a - 8 = (3a + 2)()$

16. $4a^2 - 4a - 15 = (2a + 3)()$

17. $18t^2 + 3t - 10 = (6t + 5)()$

18. $12x^2 - 31x + 20 = (3x - 4)()$

In Exercises 19–22, find all possible products of the form $(5x + m)(x + n)$, where $m \cdot n$ is the specified product. (Assume that m and n are integers.)

19. $m \cdot n = 3$

20. $m \cdot n = 21$

21. $m \cdot n = 12$

22. $m \cdot n = 36$

In Exercises 23–50, factor the trinomial. (*Note:* Some of the trinomials may be prime.) See Examples 1–3.

23. $2x^2 + 5x + 3$

24. $3x^2 + 7x + 2$

25. $4y^2 + 5y + 1$

26. $3x^2 + 5x - 2$

27. $2y^2 - 3y + 1$

28. $3a^2 - 5a + 2$

29. $2x^2 - x - 3$

30. $3z^2 - z - 2$

31. $5x^2 - 2x + 1$

32. $4z^2 - 8z + 1$

33. $2x^2 + x + 3$

34. $6x^2 - 10x + 5$

35. $5s^2 - 10s + 6$

36. $6v^2 + v - 2$

37. $4x^2 + 13x - 12$

38. $6y^2 - 7y - 20$

39. $9x^2 - 18x + 8$

40. $4a^2 - 16a + 15$

41. $18u^2 - 9u - 2$

42. $24s^2 + 37s - 5$

43. $15a^2 + 14a - 8$

44. $12x^2 - 8x - 15$

45. $10t^2 - 3t - 18$

46. $10t^2 + 43t - 9$

47. $15m^2 + 16m - 15$

48. $21b^2 - 40b - 21$

49. $16z^2 - 34z + 15$

50. $12x^2 - 41x + 24$

In Exercises 51–60, factor the trinomial. (*Note:* The leading coefficient is negative.) See Example 5.

51. $-2x^2 + x + 3$

52. $-5x^2 + x + 4$

53. $4 - 4x - 3x^2$

54. $-4x^2 + 17x + 15$

55. $-6x^2 + 7x + 10$

56. $2 + x - 6x^2$

57. $1 - 4x - 60x^2$

58. $2 + 5x - 12x^2$

59. $16 - 8x - 15x^2$

60. $20 + 17x - 10x^2$

In Exercises 61–82, factor the polynomial completely. (*Note:* Some of the polynomials may be prime.) See Examples 4 and 5.

61. $6x^2 - 3x$

62. $3a^4 - 9a^3$

63. $15y^2 + 18y$

64. $24y^3 - 16y$

65. $u(u - 3) + 9(u - 3)$

66. $x(x - 8) - 2(x - 8)$

67. $2v^2 + 8v - 42$

68. $4z^2 - 12z - 40$

69. $-3x^2 - 3x - 60$

70. $5y^2 + 40y + 35$

71. $9z^2 - 24z + 15$

72. $6x^2 + 8x - 8$

73. $4x^2 + 4x + 2$

74. $6x^2 - 6x - 36$

75. $-15x^4 - 2x^3 + 8x^2$

76. $15y^2 - 7y^3 - 2y^4$

77. $3x^3 + 4x^2 + 2x$

78. $5x^3 - 3x^2 - 4x$

79. $6x^3 + 24x^2 - 192x$

80. $35x + 28x^2 - 7x^3$

81. $18u^4 + 18u^3 - 27u^2$

82. $12x^5 - 16x^4 + 8x^3$

In Exercises 83–88, find all integers b such that the trinomial can be factored.

83. $3x^2 + bx + 10$

84. $4x^2 + bx + 3$

85. $2x^2 + bx - 6$

86. $5x^2 + bx - 6$

87. $6x^2 + bx + 20$

88. $8x^2 + bx - 18$

In Exercises 89–94, find two integers c such that the trinomial can be factored. (There are many correct answers.)

89. $4x^2 + 3x + c$

90. $2x^2 + 5x + c$

91. $3x^2 - 10x + c$

92. $8x^2 - 3x + c$

93. $6x^2 - 5x + c$

94. $4x^2 - 9x + c$

In Exercises 95–110, factor the trinomial by grouping. See Examples 6 and 7.

95. $3x^2 + 7x + 2$

96. $2x^2 + 5x + 2$

97. $2x^2 + x - 3$

98. $5x^2 - 14x - 3$

99. $6x^2 + 5x - 4$

100. $12y^2 + 11y + 2$

101. $15x^2 - 11x + 2$

102. $12x^2 - 13x + 1$

103. $3a^2 + 11a + 10$

104. $3z^2 - 4z - 15$

105. $16x^2 + 2x - 3$

106. $20c^2 + 19c - 1$

107. $12x^2 - 17x + 6$

108. $10y^2 - 13y - 30$

109. $6u^2 - 5u - 14$

110. $12x^2 + 28x + 15$

Solving Problems

111. ▲ *Geometry* The sandbox shown in the figure has a height of x and a width of $x + 2$. The volume of the sandbox is $2x^3 + 7x^2 + 6x$. Find the length l of the sandbox.

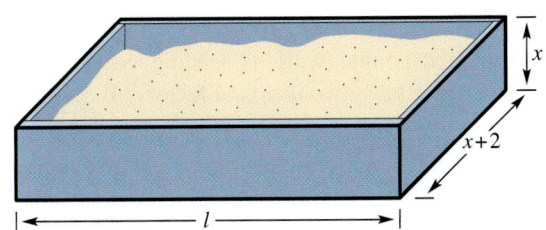

112. ▲ *Geometry* The swimming pool shown in the figure has a depth of d and a length of $5d + 2$. The volume of the swimming pool is $15d^3 - 14d^2 - 8d$. Find the width w of the swimming pool.

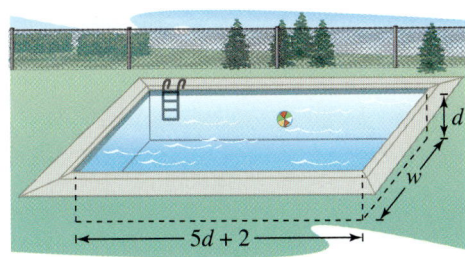

113. ▲ *Geometry* The area of the rectangle shown in the figure is $2x^2 + 9x + 10$. What is the area of the shaded region?

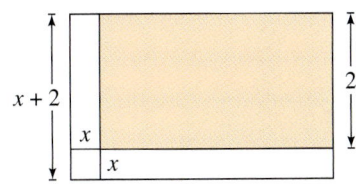

114. ▲ *Geometry* The area of the rectangle shown in the figure is $3x^2 + 10x + 3$. What is the area of the shaded region?

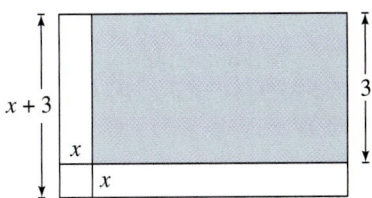

115. *Graphical Exploration* Consider the equations

$$y_1 = 2x^3 + 3x^2 - 5x$$

and

$$y_2 = x(2x + 5)(x - 1).$$

(a) Factor the trinomial represented by y_1. What is the relationship between y_1 and y_2?

(b) ▦ Demonstrate your answer to part (a) graphically by using a graphing calculator to graph y_1 and y_2 in the same viewing window.

(c) ▦ Identify the x- and y-intercepts of the graphs of y_1 and y_2.

116. *Beam Deflection*   A cantilever beam of length l is fixed at the origin. A load weighing W pounds is attached to the end of the beam (see figure). The deflection y of the beam x units from the origin is given by

$$y = -\frac{1}{10}x^2 - \frac{1}{120}x^3, \ 0 \le x \le 3.$$

Factor the expression for the deflection. (Write the binomial factor with positive integer coefficients.)

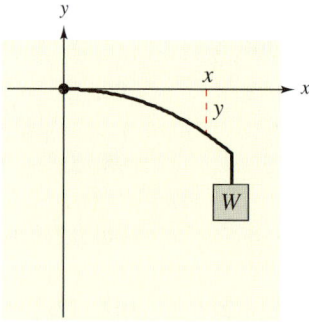

Figure for 116

Explaining Concepts

117. ⚡ Answer parts (a) and (b) of Motivating the Chapter on page 344.

118. *Writing* ✏ Explain the meaning of each letter of FOIL.

119. *Writing* ✏ Without multiplying the factors, explain why $(2x + 3)(x + 5)$ is not a factorization of $2x^2 + 7x - 15$?

120. *Error Analysis*   Describe the error.

$$9x^2 - 9x - 54 = (3x + 6)(3x - 9)$$
$$= 3(x + 2)(x - 3)$$

121. *Writing* ✏ In factoring $ax^2 + bx + c$, how many possible factorizations must be tested if a and c are prime? Explain your reasoning.

122. Give an example of a prime trinomial that is of the form $ax^2 + bx + c$.

123. Give an example of a trinomial of the form $ax^3 + bx^2 + cx$ that has a common monomial factor of $2x$.

124. Can a trinomial with its leading coefficient not equal to 1 have two identical factors? If so, give an example.

125. *Writing* ✏ Many people think the technique of factoring a trinomial by grouping is more efficient than the *guess, check, and revise* strategy, especially when the coefficients a and c have many factors. Try factoring $6x^2 - 13x + 6$, $2x^2 + 5x - 12$, and $3x^2 + 11x - 4$ using both methods. Which method do you prefer? Explain the advantages and disadvantages of each method.

Mid-Chapter Quiz

Take this quiz as you would take a quiz in class. After you are done, check your work against the answers in the back of the book.

In Exercises 1–4, fill in the missing factor.

1. $\frac{2}{3}x - 1 = \frac{1}{3}($ ⬚ $)$

2. $x^2y - xy^2 = xy($ ⬚ $)$

3. $y^2 + y - 42 = (y + 7)($ ⬚ $)$

4. $3y^2 - y - 30 = (3y - 10)($ ⬚ $)$

In Exercises 5–16, factor the polynomial completely.

5. $10x^2 + 70$

6. $2a^3b - 4a^2b^2$

7. $x(x + 2) - 3(x + 2)$

8. $t^3 - 3t^2 + t - 3$

9. $y^2 + 11y + 30$

10. $u^2 + u - 30$

11. $x^3 - x^2 - 30x$

12. $2x^2y + 8xy - 64y$

13. $2y^2 - 3y - 27$

14. $6 - 13z - 5z^2$

15. $6x^2 - x - 2$

16. $10s^4 - 14s^3 + 2s^2$

17. Find all integers b such that the trinomial

$$x^2 + bx + 12$$

can be factored. Describe the method you used.

18. Find two integers c such that the trinomial

$$x^2 - 10x + c$$

can be factored. Describe the method you used. (There are many correct answers.)

19. Find all possible products of the form

$$(3x + m)(x + n)$$

such that $m \cdot n = 6$. Describe the method you used.

20. The area of the rectangle shown in the figure is $3x^2 + 38x + 80$. What is the area of the shaded region?

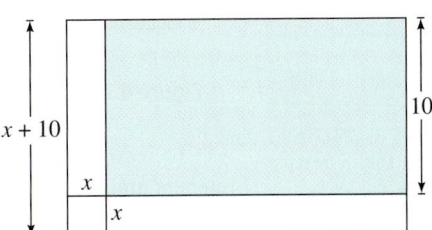

21. Use a graphing calculator to graph $y_1 = -2x^2 + 11x - 12$ and $y_2 = (3 - 2x)(x - 4)$ in the same viewing window. What can you conclude?

6.4 Factoring Polynomials with Special Forms

Carol Simowitz Photography

What You Should Learn

1 Factor the difference of two squares.

2 Recognize repeated factorization.

3 Identify and factor perfect square trinomials.

4 Factor the sum or difference of two cubes.

Why You Should Learn It

You can factor polynomials with special forms that model real-life situations. For instance, in Example 12 on page 377, an expression that models the safe working load for a piano lifted by a rope is factored.

1 Factor the difference of two squares.

Difference of Two Squares

One of the easiest special polynomial forms to recognize and to factor is the form $a^2 - b^2$. It is called a **difference of two squares,** and it factors according to the following pattern.

> ### Difference of Two Squares
>
> Let a and b be real numbers, variables, or algebraic expressions.
>
> $$a^2 - b^2 = (a + b)(a - b)$$
>
> Difference   Opposite signs

Technology: Discovery

Use your calculator to verify the special polynomial form called the "difference of two squares." To do so, evaluate the equation when $a = 16$ and $b = 9$. Try more values, including negative values. What can you conclude?

This pattern can be illustrated geometrically, as shown in Figure 6.2. The area of the shaded region on the left is represented by $a^2 - b^2$ (the area of the larger square minus the area of the smaller square). On the right, the *same* area is represented by a rectangle whose width is $a + b$ and whose length is $a - b$.

Figure 6.2

Study Tip

Note in the following that x-terms of higher power can be perfect squares.

$$25 - 64x^4 = (5)^2 - (8x^2)^2$$

$$= (5 + 8x^2)(5 - 8x^2)$$

To recognize perfect square terms, look for coefficients that are squares of integers and for variables raised to *even* powers. Here are some examples.

Original Polynomial		Difference of Squares		Factored Form
$x^2 - 1$		$(x)^2 - (1)^2$		$(x + 1)(x - 1)$
$4x^2 - 9$		$(2x)^2 - (3)^2$		$(2x + 3)(2x - 3)$

Study Tip

When factoring a polynomial, remember that you can check your result by multiplying the factors. For instance, you can check the factorization in Example 1(a) as follows.

$$(x + 6)(x - 6)$$
$$= x^2 - 6x + 6x - 36$$
$$= x^2 - 36$$

Example 1   Factoring the Difference of Two Squares

Factor each polynomial.

a. $x^2 - 36$ **b.** $x^2 - \frac{4}{25}$ **c.** $81x^2 - 49$

Solution

a. $x^2 - 36 = x^2 - 6^2$ Write as difference of two squares.

             $= (x + 6)(x - 6)$ Factored form

b. $x^2 - \frac{4}{25} = x^2 - \left(\frac{2}{5}\right)^2$ Write as difference of two squares.

             $= \left(x + \frac{2}{5}\right)\left(x - \frac{2}{5}\right)$ Factored form

c. $81x^2 - 49 = (9x)^2 - 7^2$ Write as difference of two squares.

             $= (9x + 7)(9x - 7)$ Factored form

Check your results by using the FOIL Method.

The rule $a^2 - b^2 = (a + b)(a - b)$ applies to polynomials or expressions in which a and b are themselves expressions.

Example 2   Factoring the Difference of Two Squares

Factor the polynomial $(x + 1)^2 - 4$.

Solution

$$(x + 1)^2 - 4 = (x + 1)^2 - 2^2 \quad \text{Write as difference of two squares.}$$
$$= [(x + 1) + 2][(x + 1) - 2] \quad \text{Factored form}$$
$$= (x + 3)(x - 1) \quad \text{Simplify.}$$

Check your result by using the FOIL Method.

Sometimes the difference of two squares can be hidden by the presence of a common monomial factor. Remember that with all factoring techniques, you should first remove any common monomial factors.

Example 3   Removing a Common Monomial Factor First

Factor the polynomial $20x^3 - 5x$.

Solution

$$20x^3 - 5x = 5x(4x^2 - 1) \quad \text{Factor out common monomial factor } 5x.$$
$$= 5x[(2x)^2 - 1^2] \quad \text{Write as difference of two squares.}$$
$$= 5x(2x + 1)(2x - 1) \quad \text{Factored form}$$

② Recognize repeated factorization.

Repeated Factorization

To factor a polynomial completely, you should always check to see whether the factors obtained might themselves be factorable. That is, can any of the factors be factored? For instance, after factoring the polynomial $(x^4 - 1)$ once as the difference of two squares

$$x^4 - 1 = (x^2)^2 - 1^2 \qquad \text{Write as difference of two squares.}$$

$$= (x^2 + 1)(x^2 - 1) \qquad \text{Factored form}$$

you can see that the second factor is itself the difference of two squares. So, to factor the polynomial *completely*, you must continue the factoring process.

$$x^4 - 1 = (x^2 + 1)(x^2 - 1) \qquad \text{Factor as difference of two squares.}$$

$$= (x^2 + 1)(x + 1)(x - 1) \qquad \text{Factor completely.}$$

Another example of repeated factoring is shown in the next example.

Example 4   Factoring Completely

Factor the polynomial $x^4 - 16$ completely.

Solution

Recognizing $x^4 - 16$ as a difference of two squares, you can write

$$x^4 - 16 = (x^2)^2 - 4^2 \qquad \text{Write as difference of two squares.}$$

$$= (x^2 + 4)(x^2 - 4). \qquad \text{Factored form}$$

Note that the second factor $(x^2 - 4)$ is itself a difference of two squares and so

$$x^4 - 16 = (x^2 + 4)(x^2 - 4) \qquad \text{Factor as difference of two squares.}$$

$$= (x^2 + 4)(x + 2)(x - 2). \qquad \text{Factor completely.}$$

Example 5   Factoring Completely

Factor $48x^4 - 3$ completely.

Solution

Start by removing the common monomial factor.

$$48x^4 - 3 = 3(16x^4 - 1) \qquad \text{Remove common monomial factor 3.}$$

Recognizing $16x^4 - 1$ as the difference of two squares, you can write

$$48x^4 - 3 = 3(16x^4 - 1) \qquad \text{Factor out common monomial.}$$

$$= 3[(4x^2)^2 - 1^2] \qquad \text{Write as difference of two squares.}$$

$$= 3(4x^2 + 1)(4x^2 - 1) \qquad \text{Recognize } 4x^2 - 1 \text{ as a difference of two squares.}$$

$$= 3(4x^2 + 1)[(2x)^2 - 1^2] \qquad \text{Write as difference of two squares.}$$

$$= 3(4x^2 + 1)(2x + 1)(2x - 1). \qquad \text{Factor completely.}$$

Study Tip

Note in Example 4 that no attempt was made to factor the *sum of two squares*. A second-degree polynomial that is the sum of two squares cannot be factored as the product of binomials (using integer coefficients). For instance, the second-degree polynomials

$$x^2 + 4$$

and

$$4x^2 + 9$$

cannot be factored using integer coefficients. In general, *the sum of two squares is not factorable*.

③ Identify and factor perfect square trinomials.

Perfect Square Trinomials

A **perfect square trinomial** is the square of a binomial. For instance,

$$x^2 + 4x + 4 = (x + 2)(x + 2)$$
$$= (x + 2)^2$$

is the square of the binomial $(x + 2)$. Perfect square trinomials come in two forms: one in which the middle term is positive and the other in which the middle term is negative. In both cases, the first and last terms are positive perfect squares.

Perfect Square Trinomials

Let a and b be real numbers, variables, or algebraic expressions.

1. $a^2 + 2ab + b^2 = (a + b)^2$ **2.** $a^2 - 2ab + b^2 = (a - b)^2$

Same sign   Same sign

Example 6   Identifying Perfect Square Trinomials

Which of the following are perfect square trinomials?

a. $m^2 - 4m + 4$

b. $4x^2 - 2x + 1$

c. $y^2 + 6y - 9$

d. $x^2 + x + \frac{1}{4}$

Solution

a. This polynomial *is* a perfect square trinomial. It factors as $(m - 2)^2$.

b. This polynomial *is not* a perfect square trinomial because the middle term is not twice the product of $2x$ and 1.

c. This polynomial *is not* a perfect square trinomial because the last term, -9, is not positive.

d. This polynomial *is* a perfect square trinomial. The first and last terms are perfect squares, x^2 and $\left(\frac{1}{2}\right)^2$, and it factors as $\left(x + \frac{1}{2}\right)^2$.

Study Tip

To recognize a perfect square trinomial, remember that the first and last terms must be perfect squares and positive, and the middle term must be twice the product of a and b. (The middle term can be positive or negative.) Watch for squares of fractions.

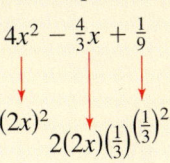

$$4x^2 - \frac{4}{3}x + \frac{1}{9}$$
$$(2x)^2 \quad 2(2x)\left(\frac{1}{3}\right) \quad \left(\frac{1}{3}\right)^2$$

Example 7   Factoring a Perfect Square Trinomial

Factor the trinomial $y^2 - 6y + 9$.

Solution

$$y^2 - 6y + 9 = y^2 - 2(3y) + 3^2 \qquad \text{Recognize the pattern.}$$
$$= (y - 3)^2 \qquad \text{Write in factored form.}$$

Example 8 Factoring a Perfect Square Trinomial

Factor the trinomial $16x^2 + 40x + 25$.

Solution

$$16x^2 + 40x + 25 = (4x)^2 + 2(4x)(5) + 5^2 \qquad \text{Recognize the pattern.}$$
$$= (4x + 5)^2 \qquad \text{Write in factored form.}$$

Example 9 Factoring a Perfect Square Trinomial

Factor the trinomial $9x^2 - 24xy + 16y^2$.

Solution

$$9x^2 - 24xy + 16y^2 = (3x)^2 - 2(3x)(4y) + (4y)^2 \qquad \text{Recognize the pattern.}$$
$$= (3x - 4y)^2 \qquad \text{Write in factored form.}$$

④ **Factor the sum or difference of two cubes.**

Sum or Difference of Two Cubes

The last type of special factoring presented in this section is the sum or difference of two *cubes*. The patterns for these two special forms are summarized below.

Study Tip

When using either of the factoring patterns at the right, pay special attention to the signs. Remembering the "like" and "unlike" patterns for the signs is helpful.

Sum or Difference of Two Cubes

Let a and b be real numbers, variables, or algebraic expressions.

Like signs

1. $a^3 + b^3 = (a + b)(a^2 - ab + b^2)$

Unlike signs

Like signs

2. $a^3 - b^3 = (a - b)(a^2 + ab + b^2)$

Unlike signs

Example 10 Factoring a Sum of Two Cubes

Factor the polynomial $y^3 + 27$.

Solution

$$y^3 + 27 = y^3 + 3^3 \qquad \text{Write as sum of two cubes.}$$
$$= (y + 3)[y^2 - (y)(3) + 3^2] \qquad \text{Factored form}$$
$$= (y + 3)(y^2 - 3y + 9) \qquad \text{Simplify.}$$

Study Tip

It is easy to make arithmetic errors when applying the patterns for factoring the sum or difference of two cubes. When you use these patterns, be sure to check your work by multiplying the factors.

Example 11   Factoring Differences of Two Cubes

Factor each polynomial.

a. $64 - x^3$ **b.** $2x^3 - 16$

Solution

a. $64 - x^3 = 4^3 - x^3$ Write as difference of two cubes.

$\qquad\qquad = (4 - x)[4^2 + (4)(x) + x^2]$ Factored form

$\qquad\qquad = (4 - x)(16 + 4x + x^2)$ Simplify.

b. $2x^3 - 16 = 2(x^3 - 8)$ Factor out common monomial factor 2.

$\qquad\qquad = 2(x^3 - 2^3)$ Write as difference of two cubes.

$\qquad\qquad = 2(x - 2)[x^2 + (x)(2) + 2^2]$ Factored form

$\qquad\qquad = 2(x - 2)(x^2 + 2x + 4)$ Simplify.

Example 12   Safe Working Load

An object lifted with a rope should not weigh more than the safe working load for the rope. To lift a 600-pound piano, the safe working load for a natural fiber rope is given by $150c^2 - 600$, where c is the circumference of the rope (in inches). Factor this expression.

Solution

$150c^2 - 600 = 150(c^2 - 4)$ Factor out common monomial factor.

$\qquad\qquad = 150(c^2 - 2^2)$ Write as difference of two squares.

$\qquad\qquad = 150(c + 2)(c - 2)$ Factored form

The following guidelines are steps for applying the various procedures involved in factoring polynomials.

Guidelines for Factoring Polynomials

1. Factor out any common factors.

2. Factor according to one of the special polynomial forms: difference of two squares, sum or difference of two cubes, or perfect square trinomials.

3. Factor trinomials, $ax^2 + bx + c$, with $a = 1$ or $a \neq 1$.

4. Factor by grouping—for polynomials with four terms.

5. Check to see whether the factors themselves can be factored.

6. Check the results by multiplying the factors.

6.4 Exercises

Review   Concepts, Skills, and Problem Solving

Keep mathematically in shape by doing these exercises *before* the problems of this section.

Properties and Definitions

In Exercises 1 and 2, determine the quadrant or quadrants in which the point must be located.

1. $(-5, 2)$

2. $(x, 3)$, x is a real number.

3. Find the coordinates of the point on the x-axis and four units to the left of the y-axis.

4. Find the coordinates of the point nine units to the right of the y-axis and six units below the x-axis.

Solving Equations

In Exercises 5–10, solve the equation and check your solution.

5. $7 + 5x = 7x - 1$

6. $2 - 5(x - 1) = 2[x + 10(x - 1)]$

7. $2(x + 1) = 0$

8. $\frac{3}{4}(12x - 8) = 10$

9. $\frac{x}{5} + \frac{1}{5} = \frac{7}{10}$

10. $\frac{3x}{4} + \frac{1}{2} = 8$

Problem Solving

11. *Membership Drive* Because of a membership drive for a public television station, the current membership is 120% of what it was a year ago. The current membership is 8345. How many members did the station have last year?

12. *Budget* You budget 26% of your annual after-tax income for housing. Your after-tax income is $46,750. What amount can you spend on housing?

Developing Skills

In Exercises 1–22, factor the difference of two squares. See Examples 1 and 2.

1. $x^2 - 36$

2. $y^2 - 49$

3. $u^2 - 64$

4. $x^2 - 4$

5. $49 - x^2$

6. $81 - x^2$

7. $u^2 - \frac{1}{4}$

8. $t^2 - \frac{1}{16}$

9. $v^2 - \frac{4}{9}$

10. $u^2 - \frac{25}{81}$

11. $16y^2 - 9$

12. $9z^2 - 25$

13. $100 - 49x^2$

14. $16 - 81x^2$

15. $(x - 1)^2 - 4$

16. $(t + 2)^2 - 9$

17. $25 - (z + 5)^2$

18. $16 - (a + 2)^2$

19. $x^2 - y^2$

20. $x^2 - a^2$

21. $9y^2 - 25z^2$

22. $100x^2 - 81y^2$

In Exercises 23–36, factor the polynomial completely. See Examples 3–5.

23. $2x^2 - 72$

24. $3x^2 - 27$

25. $4x - 25x^3$

26. $a^3 - 16a$

27. $8y^2 - 50z^2$

28. $20x^2 - 180y^2$

29. $y^4 - 81$

30. $z^4 - 16$

31. $1 - x^4$

32. $256 - u^4$

33. $3x^4 - 48$

34. $18 - 2x^4$

35. $81x^4 - 16y^4$

36. $81x^4 - z^4$

In Exercises 37–54, factor the perfect square trinomial. See Examples 6–9.

37. $x^2 - 4x + 4$

38. $x^2 + 10x + 25$

39. $z^2 + 6z + 9$

40. $a^2 - 12a + 36$

41. $4t^2 + 4t + 1$

42. $9x^2 - 12x + 4$

43. $25y^2 - 10y + 1$

44. $16z^2 + 24z + 9$

45. $b^2 + b + \frac{1}{4}$

46. $x^2 + \frac{2}{5}x + \frac{1}{25}$

47. $4x^2 - x + \frac{1}{16}$

48. $4t^2 - \frac{4}{3}t + \frac{1}{9}$

49. $x^2 - 6xy + 9y^2$

50. $16x^2 - 8xy + y^2$

51. $4y^2 + 20yz + 25z^2$

52. $u^2 + 8uv + 16v^2$

53. $9a^2 - 12ab + 4b^2$

54. $49m^2 - 28mn + 4n^2$

Think About It In Exercises 55–60, find two real numbers b such that the expression is a perfect square trinomial.

55. $x^2 + bx + 1$

56. $x^2 + bx + 100$

57. $x^2 + bx + \frac{16}{25}$

58. $y^2 + by + \frac{1}{9}$

59. $4x^2 + bx + 81$

60. $4x^2 + bx + 9$

In Exercises 61–64, find a real number c such that the expression is a perfect square trinomial.

61. $x^2 + 6x + c$

62. $x^2 + 10x + c$

63. $y^2 - 4y + c$

64. $z^2 - 14z + c$

In Exercises 65–76, factor the sum or difference of two cubes. See Examples 10 and 11.

65. $x^3 - 8$

66. $x^3 - 27$

67. $y^3 + 64$

68. $z^3 + 125$

69. $1 + 8t^3$

70. $27s^3 + 1$

71. $27u^3 - 8$

72. $64v^3 - 125$

73. $x^3 - y^3$

74. $a^3 - b^3$

75. $27x^3 + 64y^3$

76. $27y^3 + 125z^3$

In Exercises 77–118, factor the polynomial completely. (*Note:* Some of the polynomials may be prime.)

77. $6x - 36$

78. $8t + 48$

79. $u^2 + 3u$

80. $x^3 - 4x^2$

81. $5y^2 - 25y$

82. $12a^2 - 24a$

83. $5y^2 - 125$

84. $6x^2 - 54y^2$

85. $y^4 - 25y^2$

86. $y^4 - 49y^2$

87. $x^2 - 4xy + 4y^2$

88. $9y^2 - 6yz + z^2$

89. $x^2 - 2x + 1$

90. $16 + 6x - x^2$

91. $9x^2 + 10x + 1$

92. $4x^3 + 3x^2 + x$

93. $2x^3 - 2x^2y - 4xy^2$

94. $2y^3 - 7y^2z - 15yz^2$

95. $9t^2 - 16$

96. $16t^2 - 144$

97. $36 - (z + 6)^2$

98. $(t - 4)^2 - 9$

99. $(t - 1)^2 - 121$

100. $(x - 3)^2 - 100$

101. $u^3 + 2u^2 + 3u$

102. $u^3 + 2u^2 - 3u$

103. $x^2 + 81$

104. $x^2 + 16$

105. $2t^3 - 16$

106. $24x^3 - 3$

107. $2a^3 - 16b^3$

108. $54x^3 - 2y^3$

109. $x^4 - 81$

110. $2x^4 - 32$

111. $x^4 - y^4$

112. $81y^4 - z^4$

113. $x^3 - 4x^2 - x + 4$

114. $y^3 + 3y^2 - 4y - 12$

115. $x^4 + 3x^3 - 16x^2 - 48x$

116. $36x + 18x^2 - 4x^3 - 2x^4$

117. $64 - y^6$

118. $1 - y^8$

🖩 *Graphical Reasoning*   In Exercises 119–122, use a graphing calculator to graph the two equations in the same viewing window. What can you conclude?

119. $y_1 = x^2 - 36$

$y_2 = (x + 6)(x - 6)$

120. $y_1 = x^2 - 8x + 16$

$y_2 = (x - 4)^2$

121. $y_1 = x^3 - 6x^2 + 9x$

$y_2 = x(x - 3)^2$

122. $y_1 = x^3 + 27$

$y_2 = (x + 3)(x^2 - 3x + 9)$

Mental Math   In Exercises 123–126, evaluate the quantity mentally using the two samples as models.

$29^2 = (30 - 1)^2$

$= 30^2 - 2 \cdot 30 \cdot 1 + 1^2$

$= 900 - 60 + 1 = 841$

$48 \cdot 52 = (50 - 2)(50 + 2)$

$= 50^2 - 2^2 = 2496$

123. 21^2

124. 49^2

125. $59 \cdot 61$

126. $28 \cdot 32$

Solving Problems

127. 🔺 *Geometry*   An annulus is the region between two concentric circles. The area of the annulus shown in the figure is $\pi R^2 - \pi r^2$. Give the complete factorization of the expression for the area.

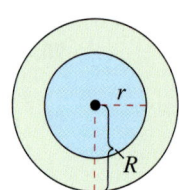

128. *Free-Falling Object*   The height of an object that is dropped from the top of the USX Tower in Pittsburgh is given by the expression $-16t^2 + 841$, where t is the time in seconds. Factor this expression.

In Exercises 129 and 130, write the polynomial as the difference of two squares. Use the result to factor the polynomial completely.

129. $x^2 + 6x + 8 = (x^2 + 6x + 9) - 1$

$= \boxed{}^2 - \boxed{}^2$

130. $x^2 + 8x + 12 = (x^2 + 8x + 16) - 4$

$= \boxed{}^2 - \boxed{}^2$

131. *Writing* ✏️   The figure below shows two cubes: a large cube whose volume is a^3 and a smaller cube whose volume is b^3. If the smaller cube is removed from the larger, the remaining solid has a volume of $a^3 - b^3$ and is composed of three rectangular boxes, labeled Box 1, Box 2, and Box 3. Find the volume of each box and describe how these results are related to the following special product pattern.

$a^3 - b^3 = (a - b)(a^2 + ab + b^2)$

$= (a - b)a^2 + (a - b)ab + (a - b)b^2$

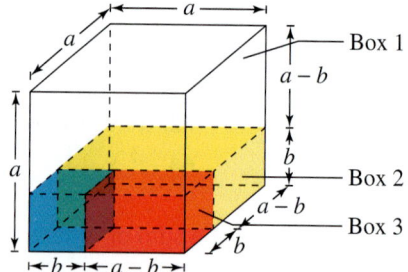

132. ▲ *Geometry*   From the eight vertices of a cube of dimension x, cubes of dimension y are removed (see figure).

(a) Write an expression for the volume of the solid that remains after the eight cubes at the vertices are removed.

(b) Factor the expression for the volume in part (a).

(c) In the context of this problem, y must be less than what multiple of x? Explain your answer geometrically and from the result of part (b).

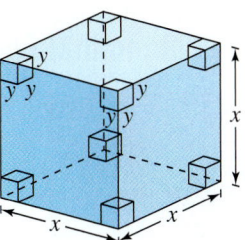

Figure for 132

Explaining Concepts

133. *Writing* ✐   Explain how to identify and factor the difference of two squares.

134. *Writing* ✐   Explain how to identify and factor a perfect square trinomial.

135. Is the expression $x(x + 2) - 2(x + 2)$ completely factored? If not, rewrite it in factored form.

136. *Writing* ✐   Is $x^2 + 4$ equal to $(x + 2)^2$? Explain.

137. *True or False?*   Because the sum of two squares cannot be factored, it follows that the sum of two cubes cannot be factored. Justify your answer.

138. *Writing* ✐   In your own words, state the guidelines for factoring polynomials.

6.5 Solving Polynomial Equations by Factoring

Carol Havens/Corbis

What You Should Learn

1 Use the Zero-Factor Property to solve equations.

2 Solve quadratic equations by factoring.

3 Solve higher-degree polynomial equations by factoring.

4 Solve application problems by factoring.

Why You Should Learn It

Quadratic equations can be used to model and solve real-life problems. For instance, Exercise 103 on page 390 shows how a quadratic equation can be used to model the time it takes an object thrown from the Royal Gorge Bridge to reach the ground.

1 Use the Zero-Factor Property to solve equations.

The Zero-Factor Property

You have spent the first four sections of this chapter developing skills for *rewriting* (simplifying and factoring) polynomials. In this section you will use these skills, together with the **Zero-Factor Property,** to solve polynomial equations.

> ### Zero-Factor Property
>
> Let a and b be real numbers, variables, or algebraic expressions. If a and b are factors such that
>
> $$ab = 0$$
>
> then $a = 0$ or $b = 0$. This property also applies to three or more factors.

Study Tip

The Zero-Factor Property is just another way of saying that the only way the product of two or more factors can be zero is if one (or more) of the factors is zero.

The Zero-Factor Property is the primary property for solving equations in algebra. For instance, to solve the equation

$$(x - 1)(x + 2) = 0 \qquad \text{Original equation}$$

you can use the Zero-Factor Property to conclude that either $(x - 1)$ or $(x + 2)$ must be zero. Setting the first factor equal to zero implies that $x = 1$ is a solution.

$$x - 1 = 0 \quad \Longrightarrow \quad x = 1 \qquad \text{First solution}$$

Similarly, setting the second factor equal to zero implies that $x = -2$ is a solution.

$$x + 2 = 0 \quad \Longrightarrow \quad x = -2 \qquad \text{Second solution}$$

So, the equation $(x - 1)(x + 2) = 0$ has exactly two solutions: $x = 1$ and $x = -2$. Check these solutions by substituting them in the original equation.

$$
\begin{aligned}
(x - 1)(x + 2) &= 0 && \text{Write original equation.} \\
(1 - 1)(1 + 2) &\stackrel{?}{=} 0 && \text{Substitute 1 for } x. \\
(0)(3) &= 0 && \text{First solution checks. } \checkmark \\
(-2 - 1)(-2 + 2) &\stackrel{?}{=} 0 && \text{Substitute } -2 \text{ for } x. \\
(-3)(0) &= 0 && \text{Second solution checks. } \checkmark
\end{aligned}
$$

② Solve quadratic equations by factoring.

Solving Quadratic Equations by Factoring

Definition of Quadratic Equation

A **quadratic equation** is an equation that can be written in the general form

$$ax^2 + bx + c = 0 \qquad \text{Quadratic equation}$$

where a, b, and c are real numbers with $a \neq 0$.

Here are some examples of quadratic equations.

$$x^2 - 2x - 3 = 0, \quad 2x^2 + x - 1 = 0, \quad x^2 - 5x = 0$$

In the next four examples, note how you can combine your factoring skills with the Zero-Factor Property to solve quadratic equations.

Example 1   Using Factoring to Solve a Quadratic Equation

Solve $x^2 - x - 6 = 0$.

Solution

First, make sure that the right side of the equation is zero. Next, factor the left side of the equation. Finally, apply the Zero-Factor Property to find the solutions.

$x^2 - x - 6 = 0$	Write original equation.
$(x + 2)(x - 3) = 0$	Factor left side of equation.
$x + 2 = 0 \quad \Longrightarrow \quad x = -2$	Set 1st factor equal to 0 and solve for x.
$x - 3 = 0 \quad \Longrightarrow \quad x = 3$	Set 2nd factor equal to 0 and solve for x.

The equation has two solutions: $x = -2$ and $x = 3$.

Check

$(-2)^2 - (-2) - 6 \overset{?}{=} 0$	Substitute -2 for x in original equation.
$4 + 2 - 6 \overset{?}{=} 0$	Simplify.
$0 = 0$	Solution checks. ✔
$(3)^2 - 3 - 6 \overset{?}{=} 0$	Substitute 3 for x in original equation.
$9 - 3 - 6 \overset{?}{=} 0$	Simplify.
$0 = 0$	Solution checks. ✔

Study Tip

In Section 3.1, you learned that the general strategy for solving a linear equation is to *isolate the variable.* Notice in Example 1 that the general strategy for solving a quadratic equation is to factor the equation into linear factors.

Factoring and the Zero-Factor Property allow you to solve a quadratic equation by converting it into two *linear* equations, which you already know how to solve. This is a common strategy of algebra—to break down a given problem into simpler parts, each of which can be solved by previously learned methods.

In order for the Zero-Factor Property to be used, a polynomial equation *must* be written in **general form.** That is, the polynomial must be on one side of the equation and zero must be the only term on the other side of the equation. To write $x^2 - 3x = 10$ in general form, subtract 10 from each side of the equation.

$$x^2 - 3x = 10 \qquad \text{Write original equation.}$$

$$x^2 - 3x - 10 = 10 - 10 \qquad \text{Subtract 10 from each side.}$$

$$x^2 - 3x - 10 = 0 \qquad \text{General form}$$

To solve this equation, factor the left side as $(x - 5)(x + 2)$, then form the linear equations $x - 5 = 0$ and $x + 2 = 0$. The solutions of these two linear equations are $x = 5$ and $x = -2$, respectively. Be sure you see that the Zero-Factor Property can be applied only to a product that is equal to *zero*. For instance, you cannot factor the left side as $x(x - 3) = 10$ and assume that $x = 10$ and $x - 3 = 10$ yield solutions. For instance, if you substitute $x = 10$ into the original equation you obtain the false statement $70 = 10$. Similarly, when $x = 13$ is substituted into the original equation you obtain another false statement, $130 = 10$. The general strategy for solving a quadratic equation by factoring is summarized in the following guidelines.

Guidelines for Solving Quadratic Equations

1. Write the quadratic equation in general form.

2. Factor the left side of the equation.

3. Set each factor with a variable equal to zero.

4. Solve each linear equation.

5. Check each solution in the original equation.

Example 2 Solving a Quadratic Equation by Factoring

Solve $2x^2 + 5x = 12$.

Solution

$$2x^2 + 5x = 12 \qquad \text{Write original equation.}$$

$$2x^2 + 5x - 12 = 0 \qquad \text{Write in general form.}$$

$$(2x - 3)(x + 4) = 0 \qquad \text{Factor left side of equation.}$$

$$2x - 3 = 0 \qquad \text{Set 1st factor equal to 0.}$$

$$x = \tfrac{3}{2} \qquad \text{Solve for } x.$$

$$x + 4 = 0 \qquad \text{Set 2nd factor equal to 0.}$$

$$x = -4 \qquad \text{Solve for } x.$$

The solutions are $x = \tfrac{3}{2}$ and $x = -4$. Check these solutions in the original equation.

In Examples 1 and 2, the original equations each involved a second-degree (quadratic) polynomial and each had *two different* solutions. You will sometimes encounter second-degree polynomial equations that have only one (repeated) solution. This occurs when the left side of the general form of the equation is a perfect square trinomial, as shown in Example 3.

Example 3   A Quadratic Equation with a Repeated Solution

Solve $x^2 - 2x + 16 = 6x$.

Solution

$x^2 - 2x + 16 = 6x$	Write original equation.
$x^2 - 8x + 16 = 0$	Write in general form.
$(x - 4)^2 = 0$	Factor.
$x - 4 = 0$ or $x - 4 = 0$	Set factors equal to 0.
$x = 4$	Solve for x.

Note that even though the left side of this equation has two factors, the factors are the same. So, the only solution of the equation is $x = 4$. This solution is called a **repeated solution.**

Check

$x^2 - 2x + 16 = 6x$	Write original equation.
$(4)^2 - 2(4) + 16 \stackrel{?}{=} 6(4)$	Substitute 4 for x.
$16 - 8 + 16 \stackrel{?}{=} 24$	Simplify.
$24 = 24$	Solution checks. ✔

Example 4   Solving a Quadratic Equation by Factoring

Solve $(x + 3)(x + 6) = 4$.

Solution

Begin by multiplying the factors on the left side.

$(x + 3)(x + 6) = 4$	Write original equation.
$x^2 + 9x + 18 = 4$	Multiply factors.
$x^2 + 9x + 14 = 0$	Write in general form.
$(x + 2)(x + 7) = 0$	Factor.
$x + 2 = 0$ ⟹ $x = -2$	Set 1st factor equal to 0 and solve for x.
$x + 7 = 0$ ⟹ $x = -7$	Set 2nd factor equal to 0 and solve for x.

The equation has two solutions: $x = -2$ and $x = -7$. Check these in the original equation.

Technology: Discovery

Write the equation in Example 3 in general form. Graph this equation on your graphing calculator.

$$y = x^2 - 8x + 16$$

What are the x-intercepts of the graph of the equation?

Write the equation in Example 4 in general form. Graph this equation on your graphing calculator.

$$y = x^2 + 9x + 14$$

What are the x-intercepts of the graph of the equation?

How do the x-intercepts relate to the solutions of the equations? What can you conclude about the solutions to the equations and the x-intercepts of the graphs of the equations?

③ Solve higher-degree polynomial equations by factoring.

Solving Higher-Degree Equations by Factoring

Example 5 Solving a Polynomial Equation with Three Factors

Solve $3x^3 = 15x^2 + 18x$.

Solution

$$3x^3 = 15x^2 + 18x \qquad \text{Write original equation.}$$
$$3x^3 - 15x^2 - 18x = 0 \qquad \text{Write in general form.}$$
$$3x(x^2 - 5x - 6) = 0 \qquad \text{Factor out common factor.}$$
$$3x(x - 6)(x + 1) = 0 \qquad \text{Factor.}$$
$$3x = 0 \quad \Longrightarrow \quad x = 0 \qquad \text{Set 1st factor equal to 0.}$$
$$x - 6 = 0 \quad \Longrightarrow \quad x = 6 \qquad \text{Set 2nd factor equal to 0.}$$
$$x + 1 = 0 \quad \Longrightarrow \quad x = -1 \qquad \text{Set 3rd factor equal to 0.}$$

So, $x = 0$, $x = 6$, and $x = -1$. Check these three solutions.

Technology: Discovery

Use a graphing calculator to graph

$$y = x^2 + 3x - 40.$$

From the graph, determine the number of solutions of the equation. Explain how to use a graphing calculator to solve

$$2x^3 - 3x^2 - 5x + 1 = 0.$$

How many solutions does the equation have? How does the number of solutions relate to the degree of the equation?

Notice that the equation in Example 5 is a third-degree equation and has three solutions. This is not a coincidence. In general, a polynomial equation can have *at most* as many solutions as its degree. For instance, a second-degree equation can have zero, one, or two solutions. Notice that the equation in Example 6 is a fourth-degree equation and has four solutions.

Example 6 Solving a Polynomial Equation with Four Factors

Solve $x^4 + x^3 - 4x^2 - 4x = 0$.

Solution

$$x^4 + x^3 - 4x^2 - 4x = 0 \qquad \text{Write original equation.}$$
$$x(x^3 + x^2 - 4x - 4) = 0 \qquad \text{Factor out common factor.}$$
$$x[(x^3 + x^2) + (-4x - 4)] = 0 \qquad \text{Group terms.}$$
$$x[x^2(x + 1) - 4(x + 1)] = 0 \qquad \text{Factor grouped terms.}$$
$$x[(x + 1)(x^2 - 4)] = 0 \qquad \text{Distributive Property}$$
$$x(x + 1)(x + 2)(x - 2) = 0 \qquad \text{Difference of two squares}$$
$$x = 0 \quad \Longrightarrow \quad x = 0$$
$$x + 1 = 0 \quad \Longrightarrow \quad x = -1$$
$$x + 2 = 0 \quad \Longrightarrow \quad x = -2$$
$$x - 2 = 0 \quad \Longrightarrow \quad x = 2$$

So, $x = 0$, $x = -1$, $x = -2$, and $x = 2$. Check these four solutions.

4 Solve application problems by factoring.

Applications

Example 7   Geometry: Dimensions of a Room

A rectangular room has an area of 192 square feet. The length of the room is 4 feet more than its width, as shown in Figure 6.3. Find the dimensions of the room.

Solution

Verbal Model:

$$\boxed{\text{Length}} \cdot \boxed{\text{Width}} = \boxed{\text{Area}}$$

Labels: Length $= x + 4$ (feet)
Width $= x$ (feet)
Area $= 192$ (square feet)

Equation: $(x + 4)x = 192$

$$x^2 + 4x - 192 = 0$$

$$(x + 16)(x - 12) = 0$$

$$x = -16 \quad \text{or} \quad x = 12$$

Because the negative solution does not make sense, choose the positive solution $x = 12$. When the width of the room is 12 feet, the length of the room is

Length $= x + 4 = 12 + 4 = 16$ feet.

So, the dimensions of the room are 12 feet by 16 feet. Check this solution in the original statement of the problem.

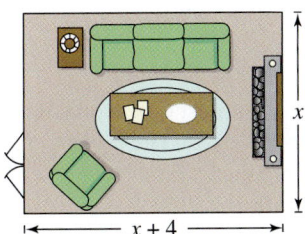

Figure 6.3

Example 8   Free-Falling Object

The height of a rock dropped into a well that is 64 feet deep above the water level is given by the position function $h(t) = -16t^2 + 64$, where the height is measured in feet and the time t is measured in seconds. (See Figure 6.4.) How long will it take the rock to hit the water at the bottom of the well?

Solution

In Figure 6.4, note that the water level of the well corresponds to a height of 0 feet. So, substitute a height of 0 for $h(t)$ in the equation and solve for t.

$0 = -16t^2 + 64$	Substitute 0 for $h(t)$.
$16t^2 - 64 = 0$	Write in general form.
$16(t^2 - 4) = 0$	Factor out common factor.
$16(t + 2)(t - 2) = 0$	Difference of two squares
$t = -2 \quad \text{or} \quad t = 2$	Solutions using Zero-Factor Property

Because a time of -2 seconds does not make sense in this problem, choose the positive solution $t = 2$, and conclude that the rock hits the water 2 seconds after it is dropped. Check this solution in the original statement of the problem.

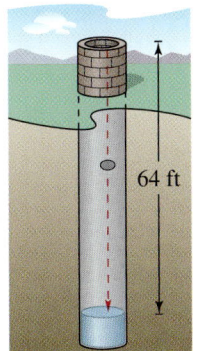

Figure 6.4

6.5 Exercises

Review   Concepts, Skills, and Problem Solving

Keep mathematically in shape by doing these exercises *before* the problems of this section.

Properties and Definitions

In Exercises 1–4, identify the property of real numbers illustrated by the statement.

1. $3uv - 3uv = 0$

2. $5z \cdot 1 = 5z$

3. $2s(1 - s) = 2s - 2s^2$

4. $(3x)y = 3(xy)$

Solving Equations

In Exercises 5–10, solve the equation.

5. $4 - \frac{1}{2}x = 6$

6. $500 - 0.75x = 235$

7. $4(x - 3) - (4x + 5) = 0$

8. $12(3 - x) = 5 - 7(2x + 1)$

9. $\dfrac{12 + x}{4} = 13$

10. $8(t - 24) = 0$

Problem Solving

11. *Cost, Revenue, and Profit*   The cost C of producing x units of a product is $C = 12 + 8x$. The revenue R from selling x units of the product is $R = 16x - \frac{1}{4}x^2$, where $0 \le x \le 20$. The profit P is $P = R - C$.

(a) Perform the subtraction required to find the polynomial representing profit.

(b) 📟 Use a graphing calculator to graph the polynomial representing profit.

(c) Determine the profit when $x = 16$ units are produced and sold.

12. *Real Estate Taxes*   The tax on a property with an assessed value of $125,000 is $1300. Find the tax on a property with an assessed value of $80,000.

Developing Skills

In Exercises 1–12, use the Zero-Factor Property to solve the equation.

1. $2x(x - 8) = 0$ **2.** $z(z + 6) = 0$

3. $(y - 3)(y + 10) = 0$ **4.** $(s - 16)(s + 15) = 0$

5. $25(a + 4)(a - 2) = 0$ **6.** $17(t - 3)(t + 8) = 0$

7. $(2t + 5)(3t + 1) = 0$ **8.** $(5x - 3)(x - 8) = 0$

9. $4x(2x - 3)(2x + 25) = 0$

10. $\frac{1}{5}x(x - 2)(3x + 4) = 0$

11. $(x - 3)(2x + 1)(x + 4) = 0$

12. $(y - 39)(2y + 7)(y + 12) = 0$

In Exercises 13–78, solve the equation by factoring. See Examples 1–6.

13. $5y - y^2 = 0$ **14.** $3x^2 + 9x = 0$

15. $9x^2 + 15x = 0$ **16.** $4x^2 - 6x = 0$

17. $2x^2 = 32x$ **18.** $8x^2 = 5x$

19. $5y^2 = 15y$ **20.** $3x^2 = 7x$

21. $x^2 - 25 = 0$ **22.** $x^2 - 121 = 0$

23. $3y^2 - 48 = 0$ **24.** $25z^2 - 100 = 0$

25. $x^2 - 3x - 10 = 0$ **26.** $x^2 - x - 12 = 0$

27. $x^2 - 10x + 24 = 0$ **28.** $20 - 9x + x^2 = 0$

29. $4x^2 + 15x = 25$ **30.** $14x^2 + 9x = -1$

31. $7 + 13x - 2x^2 = 0$ **32.** $11 + 32y - 3y^2 = 0$

33. $3y^2 - 2 = -y$ **34.** $-2x - 15 = -x^2$

35. $-13x + 36 = -x^2$ **36.** $x^2 - 15 = -2x$

37. $m^2 - 8m + 18 = 2$ **38.** $a^2 + 4a + 10 = 6$

39. $x^2 + 16x + 57 = -7$

40. $x^2 - 12x + 21 = -15$

41. $4z^2 - 12z + 15 = 6$

42. $16t^2 + 48t + 40 = 4$

43. $x(x + 2) - 10(x + 2) = 0$

44. $x(x - 15) + 3(x - 15) = 0$

45. $u(u - 3) + 3(u - 3) = 0$

46. $x(x + 10) - 2(x + 10) = 0$

47. $x(x - 5) = 36$ **48.** $s(s + 4) = 96$

49. $y(y + 6) = 72$ **50.** $x(x - 4) = 12$

51. $t(2t - 3) = 35$ **52.** $3u(3u + 1) = 20$

53. $(a + 2)(a + 5) = 10$ **54.** $(x - 8)(x - 7) = 20$

55. $(x - 4)(x + 5) = 10$

56. $(u - 6)(u + 4) = -21$

57. $(t - 2)^2 = 16$ **58.** $(s + 4)^2 = 49$

59. $9 = (x + 2)^2$ **60.** $1 = (y + 3)^2$

61. $(x - 3)^2 - 25 = 0$ **62.** $1 - (x + 1)^2 = 0$

63. $81 - (x + 4)^2 = 0$ **64.** $(s + 5)^2 - 49 = 0$

65. $x^3 - 19x^2 + 84x = 0$ **66.** $x^3 + 18x^2 + 45x = 0$

67. $6t^3 = t^2 + t$ **68.** $3u^3 = 5u^2 + 2u$

69. $z^2(z + 2) - 4(z + 2) = 0$

70. $16(3 - u) - u^2(3 - u) = 0$

71. $a^3 + 2a^2 - 9a - 18 = 0$

72. $x^3 - 2x^2 - 4x + 8 = 0$

73. $c^3 - 3c^2 - 9c + 27 = 0$

74. $v^3 + 4v^2 - 4v - 16 = 0$

75. $x^4 - 3x^3 - x^2 + 3x = 0$

76. $x^4 + 2x^3 - 9x^2 - 18x = 0$

77. $8x^4 + 12x^3 - 32x^2 - 48x = 0$

78. $9x^4 - 15x^3 - 9x^2 + 15x = 0$

Graphical Reasoning In Exercises 79–82, determine the x-intercepts of the graph and explain how the x-intercepts correspond to the solutions of the polynomial equation when $y = 0$.

79. $y = x^2 - 9$ **80.** $y = x^2 - 4x + 4$

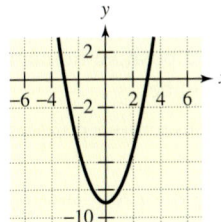 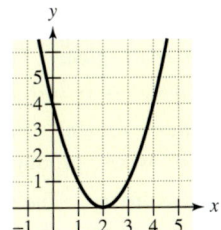

81. $y = x^3 - 6x^2 + 9x$ **82.** $y = x^3 - 3x^2 - x + 3$

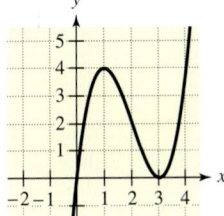 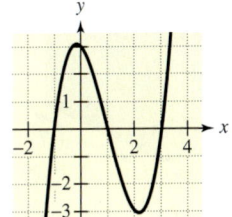

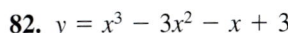 In Exercises 83–90, use a graphing calculator to graph the equation and find any x-intercepts of the graph. Verify algebraically that any x-intercepts are solutions of the polynomial equation when $y = 0$.

83. $y = x^2 - 6x$ **84.** $y = x^2 - 11x + 28$

85. $y = x^2 - 8x + 12$ **86.** $y = (x - 2)^2 - 9$

87. $y = 2x^2 + 5x - 12$ **88.** $y = x^3 - 4x$

89. $y = 2x^3 - 5x^2 - 12x$ **90.** $y = 2 + x - 2x^2 - x^3$

91. Let a and b be real numbers such that $a \neq 0$. Find the solutions of $ax^2 + bx = 0$.

92. Let a be a nonzero real number. Find the solutions of $ax^2 - ax = 0$.

Solving Problems

Think About It   In Exercises 93 and 94, find a quadratic equation with the given solutions.

93. $x = -3,$ $x = 5$

94. $x = 1,$ $x = 6$

95. *Number Problem*   The sum of a positive number and its square is 240. Find the number.

96. *Number Problem*   Find two consecutive positive integers whose product is 132.

97. ▲ *Geometry*   The rectangular floor of a storage shed has an area of 330 square feet. The length of the floor is 7 feet more than its width (see figure). Find the dimensions of the floor.

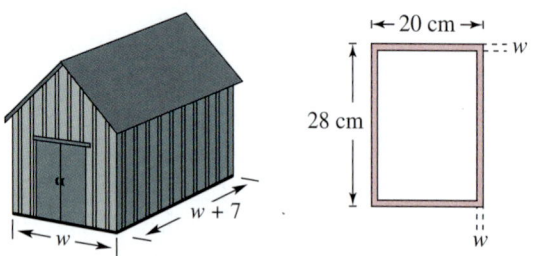

Figure for 97   Figure for 98

98. ▲ *Geometry*   The outside dimensions of a picture frame are 28 centimeters and 20 centimeters (see figure). The area of the exposed part of the picture is 468 square centimeters. Find the width w of the frame.

99. ▲ *Geometry*   A triangle has an area of 48 square inches. The height of the triangle is $1\frac{1}{2}$ times its base. Find the base and height of the triangle.

100. ▲ *Geometry*   The height of a triangle is 4 inches less than its base. The area of the triangle is 70 square inches. Find the base and height of the triangle.

101. ▲ *Geometry*   An open box is to be made from a rectangular piece of material that is 5 meters long and 4 meters wide. The box is made by cutting squares of dimension x from the corners and turning up the sides, as shown in the figure. The volume V of a rectangular solid is the product of its length, width, and height.

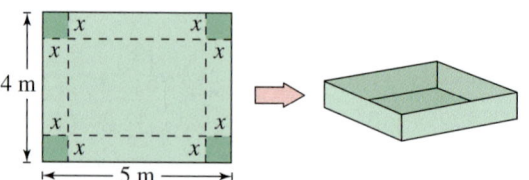

Figure for 101

(a) Show algebraically that the volume of the box is given by $V = (5 - 2x)(4 - 2x)x$.

(b) Determine the values of x for which $V = 0$. Determine an appropriate domain for the function V in the context of this problem.

(c) Complete the table.

x	0.25	0.50	0.75	1.00	1.25	1.50	1.75
V							

(d) Use the table to determine x when $V = 3$. Verify the result algebraically.

(e) 🖩 Use a graphing calculator to graph the volume function. Use the graph to approximate the value of x that yields the box of greatest volume.

102. ▲ *Geometry*   An open box with a square base is to be constructed from 880 square inches of material. The height of the box is 6 inches. What are the dimensions of the base?   (*Hint:* The surface area is given by $S = x^2 + 4xh$.)

103. *Free-Falling Object*   An object is thrown upward from the Royal Gorge Bridge in Colorado, 1053 feet above the Arkansas River, with an initial velocity of 48 feet per second. The height h (in feet) of the object is modeled by the position equation $h = -16t^2 + 48t + 1053$ where t is the time measured in seconds. How long will it take for the object to reach the ground?

104. *Free-Falling Object*   A hammer is dropped from a construction project 576 feet above the ground. The height h (in feet) of the hammer is modeled by the position equation $h = -16t^2 + 576$ where t is the time in seconds. How long will it take for the hammer to reach the ground?

105. *Free-Falling Object*   A penny is dropped from the roof of a building 256 feet above the ground. The height h (in feet) of the penny after t seconds is modeled by the equation $h = -16t^2 + 256$. How long will it take for the penny to reach the ground?

106. *Free-Falling Object*   An object is thrown upward from a height of 32 feet with an initial velocity of 16 feet per second. The height h (in feet) of the object after t seconds is modeled by the equation $h = -16t^2 + 32$. How long will it take for the object to reach the ground?

107. *Free-Falling Object*   An object falls from the roof of a building 194 feet above the ground toward a balcony 50 feet above the ground. The height h (in feet) of the object after t seconds is modeled by the equation $h = -16t^2 + 194$. How long will it take for the object to reach the balcony?

108. *Free-Falling Object*   Your friend stands 96 feet above you on a cliff. You throw an object upward with an initial velocity of 80 feet per second. The height h (in feet) of the object after t seconds is modeled by the equation $h = -16t^2 + 80t$. How long will it take for the object to reach your friend on the way up? On the way down?

109. *Break-Even Analysis*   The revenue R from the sale of x VCRs is given by $R = 90x - x^2$. The cost of producing x VCRs is given by $C = 200 + 60x$. How many VCRs must be produced and sold in order to break even?

110. *Break-Even Analysis*   The revenue R from the sale of x cameras is given by $R = 60x - x^2$. The cost of producing x cameras is given by $C = 75 + 40x$. How many cameras must be produced and sold in order to break even?

111. *Investigation*   Solve the equation

$$2(x + 3)^2 + (x + 3) - 15 = 0$$

in the following two ways.

(a) Let $u = x + 3$, and solve the resulting equation for u. Then find the corresponding values of x that are solutions of the original equation.

(b) Expand and collect like terms in the original equation, and solve the resulting equation for x.

(c) Which method is easier? Explain.

112. *Investigation*   Solve each equation using both methods described in Exercise 107.

(a) $3(x + 6)^2 - 10(x + 6) - 8 = 0$

(b) $8(x + 2)^2 - 18(x + 2) + 9 = 0$

Explaining Concepts

113.   Answer parts (c)–(e) of Motivating the Chapter on page 344.

114. Give an example of how the Zero-Factor Property can be used to solve a quadratic equation.

115. *True or False?*   If $(2x - 5)(x + 4) = 1$, then $2x - 5 = 1$ or $x + 4 = 1$. Justify your answer.

116. *Writing*   Is it possible for a quadratic equation to have only one solution? Explain.

117. What is the maximum number of solutions of an nth-degree polynomial equation? Give an example of a third-degree equation that has only one real number solution.

118.   The polynomial equation $x^3 - x - 3 = 0$ *cannot* be solved algebraically using any of the techniques described in this book. It does, however, have one solution that is a real number.

(a) *Graphical Solution:* Use a graphing calculator to graph the equation and estimate the solution.

(b) *Numerical Solution:* Use the *table* feature of a graphing calculator to create a table and estimate the solution.

What Did You Learn?

Key Terms

factoring, *p. 346*

greatest common factor, *346*

greatest common monomial factor, *p. 347*

factoring out, *p. 347*

prime polynomials, *p. 357*

factoring completely, *p. 358*

quadratic equation, *p. 383*

general form, *p. 384*

repeated solution, *p. 385*

Key Concepts

6.1 ◯ Factoring out common monomial factors

Use the Distributive Property to remove the greatest common monomial factor from each term of a polynomial.

6.1 ◯ Factoring polynomials by grouping

For polynomials with four terms, group the first two terms together and the last two terms together. Factor these two groupings and then look for a common binomial factor.

6.2 ◯ Guidelines for factoring $x^2 + bx + c$

To factor $x^2 + bx + c$, you need to find two numbers m and n whose product is c and whose sum is b.

$$x^2 + bx + c = (x + m)(x + n)$$

1. If c is positive, then m and n have like signs that match the sign of b.

2. If c is negative, then m and n have unlike signs.

3. If $|b|$ is small relative to $|c|$, first try those factors of c that are closest to each other in absolute value.

6.3 ◯ Guidelines for factoring $ax^2 + bx + c\,(a > 0)$

1. If the trinomial has a common monomial factor, you should factor out the common factor before trying to find the binomial factors.

2. You do not have to test any binomial factors that have a common monomial factor.

3. Switch the signs of the factors of c when the middle term (O + I) is correct except in sign.

6.3 ◯ Guidelines for factoring $ax^2 + bx + c$ by grouping

1. If necessary, write the trinomial in standard form.

2. Choose factors of the product ac that add up to b.

3. Use these factors to rewrite the middle term as a sum or difference.

4. Group and remove any common monomial factors from the first two terms and the last two terms.

5. If possible, factor out the common binomial factor.

6.4 ◯ Difference of two squares

Let a and b be real numbers, variables, or algebraic expressions. Then the expression $a^2 - b^2$ can be factored as follows: $a^2 - b^2 = (a + b)(a - b)$.

6.4 ◯ Perfect square trinomials

Let a and b be real numbers, variables, or algebraic expressions. Then the expressions $a^2 \pm 2ab + b^2$ can be factored as follows: $a^2 \pm 2ab + b^2 = (a \pm b)^2$.

6.4 ◯ Sum or difference of two cubes

Let a and b be real numbers, variables, or algebraic expressions. Then the expressions $a^3 \pm b^3$ can be factored as follows: $a^3 \pm b^3 = (a \pm b)(a^2 \mp ab + b^2)$.

6.4 ◯ Guidelines for factoring polynomials

1. Factor out any common factors.

2. Factor according to one of the special polynomial forms: difference of two squares, sum or difference of two cubes, or perfect square trinomials.

3. Factor trinomials, $ax^2 + bx + c$, with $a = 1$ or $a \neq 1$.

4. Factor by grouping—for polynomials with four terms.

5. Check to see whether the factors themselves can be factored.

6. Check the results by multiplying the factors.

6.5 ◯ Zero-Factor Property

Let a and b be real numbers, variables, or algebraic expressions. If a and b are factors such that $ab = 0$, then $a = 0$ or $b = 0$. This property also applies to three or more factors.

6.5 ◯ Solving a quadratic equation

To solve a quadratic equation, write the equation in general form. Factor the quadratic into linear factors and apply the Zero-Factor Property.

Review Exercises

6.1 Factoring Polynomials with Common Factors

① Find the greatest common factor of two or more expressions.

In Exercises 1–8, find the greatest common factor of the expressions.

1. t^2, t^5

2. $-y^3, y^8$

3. $3x^4, 21x^2$

4. $14z^2, 21z$

5. $14x^2y^3, -21x^3y^5$

6. $-15y^2z^2, 5y^2z$

7. $8x^2y, 24xy^2, 4xy$

8. $27ab^5, 9ab^6, 18a^2b^3$

② Factor out the greatest common monomial factor from polynomials.

In Exercises 9–18, factor the polynomial.

9. $3x - 6$

10. $7 + 21x$

11. $3t - t^2$

12. $u^2 - 6u$

13. $5x^2 + 10x^3$

14. $7y - 21y^4$

15. $8a - 12a^3$

16. $14x - 26x^4$

17. $5x^3 + 5x^2 - 5x$

18. $6u - 9u^2 + 15u^3$

▲ *Geometry* In Exercises 19 and 20, write an expression for the area of the shaded region and factor the expression.

19.

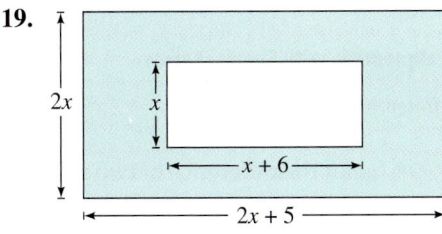

20.

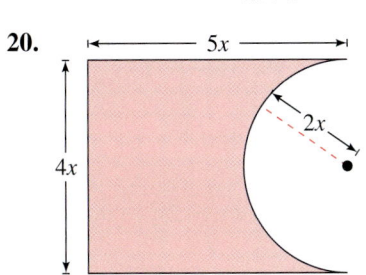

③ Factor polynomials by grouping.

In Exercises 21–30, factor the polynomial by grouping.

21. $x(x + 1) - 3(x + 1)$

22. $5(y - 3) - y(y - 3)$

23. $2u(u - 2) + 5(u - 2)$

24. $7(x + 8) + 3x(x + 8)$

25. $y^3 + 3y^2 + 2y + 6$

26. $z^3 - 5z^2 + z - 5$

27. $x^3 + 2x^2 + x + 2$

28. $x^3 - 5x^2 + 5x - 25$

29. $x^2 - 4x + 3x - 12$

30. $2x^2 + 6x - 5x - 15$

6.2 Factoring Trinomials

① Factor trinomials of the form $x^2 + bx + c$.

In Exercises 31–38, factor the trinomial.

31. $x^2 - 3x - 28$

32. $x^2 - 3x - 40$

33. $u^2 + 5u - 36$

34. $y^2 + 15y + 56$

35. $x^2 - 2x - 24$

36. $x^2 + 8x + 15$

37. $y^2 + 10y + 21$

38. $a^2 - 7a + 12$

In Exercises 39–42, find all integers b such that the trinomial can be factored.

39. $x^2 + bx + 9$

40. $y^2 + by + 25$

41. $z^2 + bz + 11$

42. $x^2 + bx + 14$

② Factor trinomials in two variables.

In Exercises 43–48, factor the trinomial.

43. $x^2 + 9xy - 10y^2$

44. $u^2 + uv - 5v^2$

45. $y^2 - 6xy - 27x^2$

46. $v^2 + 18uv + 32u^2$

47. $x^2 - 2xy - 8y^2$

48. $a^2 - ab - 30b^2$

③ Factor trinomials completely.

In Exercises 49–54, factor the trinomial completely.

49. $4x^2 - 24x + 32$ **50.** $3u^2 - 6u - 72$

51. $x^3 + 9x^2 + 18x$ **52.** $y^3 - 8y^2 + 15y$

53. $4x^3 + 36x^2 + 56x$ **54.** $2y^3 - 4y^2 - 30y$

6.3 More About Factoring Trinomials

① Factor trinomials of the form $ax^2 + bx + c$.

In Exercises 55–68, factor the trinomial.

55. $5 - 2x - 3x^2$ **56.** $8x^2 - 18x + 9$

57. $50 - 5x - x^2$ **58.** $7 + 5x - 2x^2$

59. $6x^2 + 7x + 2$ **60.** $16x^2 + 13x - 3$

61. $4y^2 - 3y - 1$ **62.** $5x^2 - 12x + 7$

63. $3x^2 + 7x - 6$ **64.** $45y^2 - 8y - 4$

65. $3x^2 + 5x - 2$ **66.** $7x^2 - 4x - 3$

67. $2x^2 - 3x + 1$ **68.** $3x^2 + 8x + 4$

In Exercises 69 and 70, find all integers b such that the trinomial can be factored.

69. $x^2 + bx - 24$ **70.** $2x^2 + bx - 16$

In Exercises 71 and 72, find two integers c such that the trinomial can be factored. (There are many correct answers.)

71. $2x^2 - 4x + c$ **72.** $5x^2 + 6x + c$

② Factor trinomials completely.

In Exercises 73–78, factor the polynomial completely.

73. $6u^3 + 3u^2 - 30u$ **74.** $8x^3 - 8x^2 - 30x$

75. $8y^3 - 20y^2 + 12y$ **76.** $14x^3 + 26x^2 - 4x$

77. $6x^3 + 14x^2 - 12x$ **78.** $12y^3 + 36y^2 + 15y$

79. ▲ *Geometry*   The cake box shown in the figure has a height of x and a width of $x + 1$. The volume of the box is $3x^3 + 4x^2 + x$. Find the length l of the box.

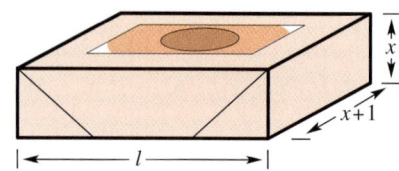

80. ▲ *Geometry*   The area of the rectangle shown in the figure is $2x^2 + 5x + 3$. What is the area of the shaded region?

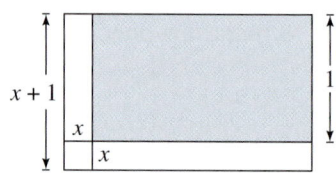

③ Factor trinomials by grouping.

In Exercises 81–86, factor the trinomial by grouping.

81. $2x^2 - 13x + 21$ **82.** $3a^2 - 13a - 10$

83. $4y^2 + y - 3$ **84.** $6z^2 - 43z + 7$

85. $6x^2 + 11x - 10$ **86.** $21x^2 - 25x - 4$

6.4 Factoring Polynomials with Special Forms

① Factor the difference of two squares.

In Exercises 87–94, factor the difference of two squares.

87. $a^2 - 100$ **88.** $36 - b^2$

89. $25 - 4y^2$ **90.** $16b^2 - 1$

91. $12x^2 - 27$ **92.** $100x^2 - 64$

93. $(u + 1)^2 - 4$ **94.** $(y - 2)^2 - 9$

2 Recognize repeated factorization.

In Exercises 95 and 96, fill in the missing factors.

95. $x^3 - x = x()()$

96. $u^4 - v^4 = (u^2 + v^2)()()$

In Exercises 97–100, factor the polynomial completely.

97. $s^3t - st^3$

98. $5x^3 - 20xy^2$

99. $x^4 - y^4$

100. $2a^4 - 32$

3 Identify and factor perfect square trinomials.

In Exercises 101–106, factor the perfect square trinomial.

101. $x^2 - 8x + 16$

102. $y^2 + 24y + 144$

103. $9s^2 + 12s + 4$

104. $16x^2 - 40x + 25$

105. $y^2 + 4yz + 4z^2$

106. $u^2 - 2uv + v^2$

4 Factor the sum or difference of two cubes.

In Exercises 107–112, factor the sum or difference of two cubes.

107. $a^3 + 1$

108. $z^3 + 8$

109. $27 - 8t^3$

110. $z^3 - 125$

111. $8x^3 + y^3$

112. $125a^3 - 27b^3$

6.5 Solving Polynomial Equations by Factoring

1 Use the Zero-Factor Property to solve equations.

In Exercises 113–118, use the Zero-Factor Property to solve the equation.

113. $4x(x - 2) = 0$

114. $-7x(2x + 5) = 0$

115. $(2x + 1)(x - 3) = 0$

116. $(x - 7)(3x - 8) = 0$

117. $(x + 10)(4x - 1)(5x + 9) = 0$

118. $3x(x + 8)(2x - 7) = 0$

2 Solve quadratic equations by factoring.

In Exercises 119–126, solve the quadratic equation by factoring.

119. $3s^2 - 2s - 8 = 0$

120. $x^2 - 25x = -150$

121. $10x(x - 3) = 0$

122. $3x(4x + 7) = 0$

123. $z(5 - z) + 36 = 0$

124. $(x + 3)^2 - 25 = 0$

125. $v^2 - 100 = 0$

126. $x^2 - 121 = 0$

3 Solve higher-degree polynomial equations by factoring.

In Exercises 127–134, solve the polynomial equation by factoring.

127. $2y^4 + 2y^3 - 24y^2 = 0$

128. $9x^4 - 15x^3 - 6x^2 = 0$

129. $x^3 - 11x^2 + 18x = 0$

130. $x^3 + 20x^2 + 36x = 0$

131. $b^3 - 6b^2 - b + 6 = 0$

132. $x^3 + 3x^2 - 5x - 15 = 0$

133. $x^4 - 5x^3 - 9x^2 + 45x = 0$

134. $2x^4 + 6x^3 - 50x^2 - 150x = 0$

4 Solve application problems by factoring.

135. *Number Problem*   Find two consecutive positive odd integers whose product is 195.

136. *Number Problem*   Find two consecutive positive even integers whose product is 224.

137. ▲ *Geometry*   A rectangle has an area of 900 square inches. The length of the rectangle is $2\frac{1}{4}$ times its width. Find the dimensions of the rectangle.

138. *Free-Falling Object*   An object is thrown upward from the Trump Tower in New York City, which is 664 feet tall, with an initial velocity of 45 feet per second. The height h (in feet) of the object is modeled by the position equation $h = -16t^2 + 45t + 664$, where t is the time (in seconds). How long will it take the object to reach the ground?

Chapter Test

Take this test as you would take a test in class. After you are done, check your work against the answers in the back of the book.

In Exercises 1–10, factor the polynomial completely .

1. $7x^2 - 14x^3$

2. $z(z + 7) - 3(z + 7)$

3. $t^2 - 4t - 5$

4. $6x^2 - 11x + 4$

5. $3y^3 + 72y^2 - 75y$

6. $4 - 25v^2$

7. $4x^2 - 20x + 25$

8. $16 - (z + 9)^2$

9. $x^3 + 2x^2 - 9x - 18$

10. $16 - z^4$

11. Fill in the missing factor: $\dfrac{2}{5}x - \dfrac{3}{5} = \dfrac{1}{5}($ 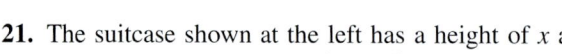 $).$

12. Find all integers b such that $x^2 + bx + 5$ can be factored.

13. Find a real number c such that $x^2 + 12x + c$ is a perfect square trinomial.

14. Explain why $(x + 1)(3x - 6)$ is not a complete factorization of $3x^2 - 3x - 6$.

In Exercises 15–20, solve the equation.

15. $(x + 4)(2x - 3) = 0$

16. $3x^2 + 7x - 6 = 0$

17. $y(2y - 1) = 6$

18. $2x^2 - 3x = 8 + 3x$

19. $2x^3 - 8x^2 - 24x = 0$

20. $y^4 + 7y^3 - 3y^2 - 21y = 0$

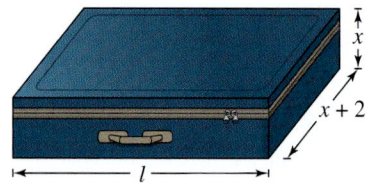

Figure for 21

21. The suitcase shown at the left has a height of x and a width of $x + 2$. The volume of the suitcase is $x^3 + 6x^2 + 8x$. Find the length l of the suitcase.

22. The width of a rectangle is 5 inches less than its length. The area of the rectangle is 84 square inches. Find the dimensions of the rectangle.

23. An object is thrown upward from the top of the AON Center in Chicago, with an initial velocity of 14 feet per second at a height of 1136 feet. The height h (in feet) of the object is modeled by the position equation

$$h = -16t^2 + 14t + 1136$$

where t is the time measured in seconds. How long will it take for the object to reach the ground? How long will it take the object to fall to a height of 806 feet?

24. Find two consecutive positive even integers whose product is 624.

25. The perimeter of a rectangular storage lot at a car dealership is 800 feet. The lot is surrounded by fencing that costs \$15 per foot for the front side and \$10 per foot for the remaining three sides. The total cost of the fencing is \$9500. Find the dimensions of the storage lot.

Take this test as you would take a test in class. After you are done, check your work against the answers in the back of the book.

1. Describe how to identify the quadrants in which the points $(-2, y)$ must be located. (y is a real number.)

2. Determine whether the ordered pairs are solution points of the equation $9x - 4y + 36 = 0$.
   - (a) $(-1, -1)$
   - (b) $(8, 27)$
   - (c) $(-4, 0)$
   - (d) $(3, -2)$

In Exercises 3 and 4, sketch the graph of the equation and determine any intercepts of the graph.

3. $y = 2 - |x|$

4. $x + 2y = 8$

5. Determine whether the relation at the left represents a function.

6. The slope of a line is $-\frac{1}{4}$ and a point on the line is $(2, 1)$. Find the coordinates of a second point on the line. Explain why there are many correct answers.

7. Find an equation of the line through $\left(0, -\frac{3}{2}\right)$ with slope $m = \frac{5}{6}$.

Domain   *Range*

Figure for 5

In Exercises 8 and 9, sketch the lines and determine whether they are parallel, perpendicular, or neither.

8. $y_1 = \frac{2}{3}x - 3,\ y_2 = -\frac{3}{2}x + 1$

9. $y_1 = 2 - 0.4x,\ y_2 = -\frac{2}{5}x$

10. Subtract: $(x^3 - 3x^2) - (x^3 + 2x^2 - 5)$.

11. Multiply: $(6z)(-7z)(z^2)$.

12. Multiply: $(3x + 5)(x - 4)$.

13. Multiply: $(5x - 3)(5x + 3)$.

14. Expand: $(5x + 6)^2$.

15. Divide: $(6x^2 + 72x) \div 6x$.

16. Divide: $\dfrac{x^2 - 3x - 2}{x - 4}$.

17. Simplify: $\dfrac{(3xy^2)^{-2}}{6x^{-3}}$.

18. Factor: $2u^2 - 6u$.

19. Factor and simplify: $(x - 2)^2 - 16$.

20. Factor completely: $x^3 + 8x^2 + 16x$.

21. Factor completely: $x^3 + 2x^2 - 4x - 8$.

22. Solve: $u(u - 12) = 0$.

23. Solve: $x(x + 3)(x - 7) = 0$.

24. A sales representative is reimbursed $125 per day for lodging and meals, plus $0.35 per mile driven. Write a linear equation giving the daily cost C to the company in terms of x, the number of miles driven. Explain the reasoning you used to write the model. Find the cost for a day when the representative drives 70 miles.

25. The cost of operating a pizza delivery car is $0.70 per mile after an initial investment of $9000. What mileage on the car will keep the cost at or below $36,400?

Motivating the Chapter

⚡ A Canoe Trip

You and a friend are planning a canoe trip on a river. You want to travel 10 miles upstream and 10 miles back downstream during daylight hours. You know that in still water you are able to paddle the canoe at an average speed of 5 miles per hour. While traveling upstream your average speed will be 5 miles per hour minus the speed of the current, and while traveling downstream your average speed will be 5 miles per hour plus the speed of the current.

See Section 7.3, Exercise 85.

a. Write an expression that represents the time it will take to travel upstream in terms of the speed x (in miles per hour) of the current. Write an expression that represents the time it will take to travel downstream in terms of the speed of the current.

b. Write a function f for the entire time (in hours) of the trip in terms of x.

c. Write the rational function f as a single fraction.

See Section 7.5, Exercise 87.

d. The time for the entire trip is $6\frac{1}{4}$ hours. What is the speed of the current? Explain.

e. The speed of the current is 4 miles per hour. Can you and your friend make the trip during 12 hours of daylight? Explain.

Larry Prosor/Superstock, Inc.

Rational Expressions, Equations, and Functions

7.1 Rational Expressions and Functions

Paul Barton/Corbis

What You Should Learn

1 Find the domain of a rational function.

2 Simplify rational expressions.

Why You Should Learn It

Rational expression can be used to solve real-life problems. For instance, in Exercise 91 on page 411, you will find a rational expression that models the average cable television revenue per subscriber.

1 Find the domain of a rational function.

The Domain of a Rational Function

A fraction whose numerator and denominator are polynomials is called a **rational expression.** Some examples are

$$\frac{3}{x+4}, \quad \frac{2x}{x^2 - 4x + 4}, \quad \text{and} \quad \frac{x^2 - 5x}{x^2 + 2x - 3}.$$

Because division by zero is undefined, the denominator of a rational expression cannot be zero. So, in your work with rational expressions, you must assume that all real number values of the variable that make the denominator zero are excluded. For the three fractions above, $x = -4$ is excluded from the first fraction, $x = 2$ from the second, and both $x = 1$ and $x = -3$ from the third. The set of *usable* values of the variable is called the **domain** of the rational expression.

Definition of a Rational Expression

Let u and v be polynomials. The algebraic expression

$$\frac{u}{v}$$

is a **rational expression.** The **domain** of this rational expression is the set of all real numbers for which $v \neq 0$.

Like polynomials, rational expressions can be used to describe functions. Such functions are called **rational functions.**

Study Tip

Every polynomial is also a rational expression because you can consider the denominator to be 1. The domain of every polynomial is the set of all real numbers.

Definition of a Rational Function

Let $u(x)$ and $v(x)$ be polynomial functions. The function

$$f(x) = \frac{u(x)}{v(x)}$$

is a **rational function.** The **domain** of f is the set of all real numbers for which $v(x) \neq 0$.

Example 1   Finding the Domains of Rational Functions

Find the domain of each rational function.

a. $f(x) = \dfrac{4}{x - 2}$ **b.** $g(x) = \dfrac{2x + 5}{8}$

Solution

a. The denominator is zero when $x - 2 = 0$ or $x = 2$. So, the domain is all real values of x such that $x \neq 2$. In interval notation, you can write the domain as

Domain $= (-\infty, 2) \cup (2, \infty)$.

b. The denominator, 8, is never zero, so the domain is the set of *all* real numbers. In interval notation, you can write the domain as

Domain $= (-\infty, \infty)$.

Technology: Discovery

Use a graphing calculator to graph the equation that corresponds to part (a) of Example 1. Then use the *trace* or *table* feature of the calculator to determine the behavior of the graph near $x = 2$. Graph the equation that corresponds to part (b) of Example 1. How does this graph differ from the graph in part (a)?

Example 2   Finding the Domains of Rational Functions

Find the domain of each rational function.

a. $f(x) = \dfrac{5x}{x^2 - 16}$ **b.** $h(x) = \dfrac{3x - 1}{x^2 - 2x - 3}$

Solution

a. The denominator is zero when $x^2 - 16 = 0$. Solving this equation by factoring, you find that the denominator is zero when $x = -4$ or $x = 4$. So, the domain is all real values of x such that $x \neq -4$ and $x \neq 4$. In interval notation, you can write the domain as

Domain $= (-\infty, -4) \cup (-4, 4) \cup (4, \infty)$.

b. The denominator is zero when $x^2 - 2x - 3 = 0$. Solving this equation by factoring, you find that the denominator is zero when $x = 3$ or when $x = -1$. So, the domain is all real values of x such that $x \neq 3$ and $x \neq -1$. In interval notation, you can write the domain as

Domain $= (-\infty, -1) \cup (-1, 3) \cup (3, \infty)$.

Study Tip

Remember that when interval notation is used, the symbol $\cup$ means *union* and the symbol $\cap$ means *intersection*.

In applications involving rational functions, it is often necessary to restrict the domain further. To indicate such a restriction, you should write the domain to the right of the fraction. For instance, the domain of the rational function

$$f(x) = \frac{x^2 + 20}{x + 4}, \qquad x > 0$$

is the set of *positive* real numbers, as indicated by the inequality $x > 0$. Note that the normal domain of this function would be all real values of x such that $x \neq -4$. However, because "$x > 0$" is listed to the right of the function, the domain is further restricted by this inequality.

Example 3   An Application Involving a Restricted Domain

You have started a small business that manufactures lamps. The initial investment for the business is $120,000. The cost of each lamp that you manufacture is $15. So, your total cost of producing x lamps is

$$C = 15x + 120{,}000. \qquad \text{Cost function}$$

Your average cost per lamp depends on the number of lamps produced. For instance, the average cost per lamp $\overline{C}$ for producing 100 lamps is

$$\overline{C} = \frac{15(100) + 120{,}000}{100} \qquad \text{Substitute 100 for } x.$$

$$= \$1215. \qquad \text{Average cost per lamp for 100 lamps}$$

The average cost per lamp decreases as the number of lamps increases. For instance, the average cost per lamp $\overline{C}$ for producing 1000 lamps is

$$\overline{C} = \frac{15(1000) + 120{,}000}{1000} \qquad \text{Substitute 1000 for } x.$$

$$= \$135. \qquad \text{Average cost per lamp for 1000 lamps}$$

In general, the average cost of producing x lamps is

$$\overline{C} = \frac{15x + 120{,}000}{x}. \qquad \text{Average cost per lamp for } x \text{ lamps}$$

What is the domain of this rational function?

Solution

If you were considering this function from only a mathematical point of view, you would say that the domain is all real values of x such that $x \neq 0$. However, because this function is a mathematical model representing a real-life situation, you must decide which values of x make sense in real life. For this model, the variable x represents the number of lamps that you produce. Assuming that you cannot produce a fractional number of lamps, you can conclude that the domain is the set of positive integers—that is,

$$\text{Domain} = \{1, 2, 3, 4, \dots\}.$$

② Simplify rational expressions.

Simplifying Rational Expressions

As with numerical fractions, a rational expression is said to be in **simplified** (or **reduced**) **form** if its numerator and denominator have no common factors (other than ± 1). To simplify rational expressions, you can apply the rule below.

Simplifying Rational Expressions

Let u, v, and w represent real numbers, variables, or algebraic expressions such that $v \neq 0$ and $w \neq 0$. Then the following is valid.

$$\frac{uw}{vw} = \frac{u\cancel{w}}{v\cancel{w}} = \frac{u}{v}$$

Be sure you divide out only *factors*, not *terms*. For instance, consider the expressions below.

$$\frac{2 \cdot 2}{2(x+5)}$$ You *can* divide out the common factor 2.

$$\frac{3+x}{3+2x}$$ You *cannot* divide out the common term 3.

Simplifying a rational expression requires two steps: (1) completely factor the numerator and denominator and (2) divide out any *factors* that are common to both the numerator and denominator. So, your success in simplifying rational expressions actually lies in your ability to *factor completely* the polynomials in both the numerator and denominator.

Example 4   Simplifying a Rational Expression

Simplify the rational expression $\dfrac{2x^3 - 6x}{6x^2}$.

Solution

First note that the domain of the rational expression is all real values of x such that $x \neq 0$. Then, completely factor both the numerator and denominator.

$$\frac{2x^3 - 6x}{6x^2} = \frac{2x(x^2 - 3)}{2x(3x)}$$ Factor numerator and denominator.

$$= \frac{2\cancel{x}(x^2 - 3)}{2\cancel{x}(3x)}$$ Divide out common factor $2x$.

$$= \frac{x^2 - 3}{3x}$$ Simplified form

In simplified form, the domain of the rational expression is the same as that of the original expression—all real values of x such that $x \neq 0$.

Technology: Tip

Use the *table* feature of a graphing calculator to compare the two functions in Example 5.

$$y_1 = \frac{x^2 + 2x - 15}{3x - 9}$$

$$y_2 = \frac{x + 5}{3}$$

Set the increment value of the table to 1 and compare the values at $x = 0, 1, 2, 3, 4,$ and 5. Next set the increment value to 0.1 and compare the values at $x = 2.8, 2.9, 3.0, 3.1,$ and 3.2. From the table you can see that the functions differ only at $x = 3$. This shows why $x \neq 3$ must be written as part of the simplified form of the original expression.

Example 5 Simplifying a Rational Expression

Simplify the rational expression $\dfrac{x^2 + 2x - 15}{3x - 9}$.

Solution

The domain of the rational expression is all real values of x such that $x \neq 3$.

$$\frac{x^2 + 2x - 15}{3x - 9} = \frac{(x + 5)(x - 3)}{3(x - 3)} \qquad \text{Factor numerator and denominator.}$$

$$= \frac{(x + 5)(x - 3)}{3(x - 3)} \qquad \text{Divide out common factor } (x - 3).$$

$$= \frac{x + 5}{3}, \ x \neq 3 \qquad \text{Simplified form}$$

Dividing out common factors from the numerator and denominator of a rational expression can change its domain. For instance, in Example 5 the domain of the original expression is all real values of x such that $x \neq 3$. So, the original expression is equal to the simplified expression for all real numbers *except* 3.

Example 6 Simplifying a Rational Expression

Simplify the rational expression $\dfrac{x^3 - 16x}{x^2 - 2x - 8}$.

Solution

The domain of the rational expression is all real values of x such that $x \neq -2$ and $x \neq 4$.

$$\frac{x^3 - 16x}{x^2 - 2x - 8} = \frac{x(x^2 - 16)}{(x + 2)(x - 4)} \qquad \text{Partially factor.}$$

$$= \frac{x(x + 4)(x - 4)}{(x + 2)(x - 4)} \qquad \text{Factor completely.}$$

$$= \frac{x(x + 4)(x - 4)}{(x + 2)(x - 4)} \qquad \text{Divide out common factor } (x - 4).$$

$$= \frac{x(x + 4)}{x + 2}, \ x \neq 4 \qquad \text{Simplified form}$$

When simplifying a rational expression, be aware of the domain. If the domain in the original expression is no longer the same as the domain in the simplified expression, it is important to list the domain next to the simplified expression so that both the original and simplified expressions are equal. For instance, in Example 6 the restriction $x \neq 4$ is listed so that the domains agree for the original and simplified expressions. The example does not list $x \neq -2$ because it is apparent by looking at either expression.

Example 7   Simplification Involving a Change in Sign

Simplify the rational expression $\dfrac{2x^2 - 9x + 4}{12 + x - x^2}$.

Solution

The domain of the rational expression is all real values of x such that $x \neq -3$ and $x \neq 4$.

$$\frac{2x^2 - 9x + 4}{12 + x - x^2} = \frac{(2x - 1)(x - 4)}{(4 - x)(3 + x)}$$ Factor numerator and denominator.

$$= \frac{(2x - 1)(x - 4)}{-(x - 4)(3 + x)}$$ $(4 - x) = -(x - 4)$

$$= \frac{(2x - 1)(x - 4)}{-(x - 4)(3 + x)}$$ Divide out common factor $(x - 4)$.

$$= -\frac{2x - 1}{3 + x}, \quad x \neq 4$$ Simplified form

The simplified form is equivalent to the original expression for all values of x such that $x \neq 4$. Note that $x = -3$ is excluded from the domains of both the original and simplified expressions.

In Example 7, be sure you see that when dividing the numerator and denominator by the common factor of $(x - 4)$, you keep the minus sign. In the simplified form of the fraction, this text uses the convention of moving the minus sign out in front of the fraction. However, this is a personal preference. All of the following forms are equivalent.

$$-\frac{2x - 1}{3 + x} = \frac{-(2x - 1)}{3 + x} = \frac{-2x + 1}{3 + x} = \frac{2x - 1}{-3 - x} = \frac{2x - 1}{-(3 + x)}$$

In the next three examples, rational expressions that involve more than one variable are simplified.

Example 8   A Rational Expression Involving Two Variables

Simplify the rational expression $\dfrac{3xy + y^2}{2y}$.

Solution

The domain of the rational expression is all real values of y such that $y \neq 0$.

$$\frac{3xy + y^2}{2y} = \frac{y(3x + y)}{2y}$$ Factor numerator and denominator.

$$= \frac{y(3x + y)}{2y}$$ Divide out common factor y.

$$= \frac{3x + y}{2}, \quad y \neq 0$$ Simplified form

Study Tip

Be sure to factor *completely* the numerator and denominator of a rational expression before concluding that there is no common factor. This may involve a change in signs. Remember that the Distributive Property allows you to write $(b - a)$ as $-(a - b)$. Watch for this in Example 7.

Example 9   A Rational Expression Involving Two Variables

$$\frac{2x^2 + 2xy - 4y^2}{5x^3 - 5xy^2} = \frac{2(x - y)(x + 2y)}{5x(x - y)(x + y)} \qquad \text{Factor numerator and denominator.}$$

$$= \frac{2(x - y)(x + 2y)}{5x(x - y)(x + y)} \qquad \text{Divide out common factor } (x - y).$$

$$= \frac{2(x + 2y)}{5x(x + y)}, \; x \neq y \qquad \text{Simplified form}$$

The domain of the original rational expression is all real numbers such that $x \neq 0$ and $x \neq \pm y$.

Example 10   A Rational Expression Involving Two Variables

$$\frac{4x^2y - y^3}{2x^2y - xy^2} = \frac{(2x - y)(2x + y)y}{(2x - y)xy} \qquad \text{Factor numerator and denominator.}$$

$$= \frac{(2x - y)(2x + y)y}{(2x - y)xy} \qquad \text{Divide out common factors } (2x - y) \text{ and } y.$$

$$= \frac{2x + y}{x}, \; y \neq 0, \; y \neq 2x \qquad \text{Simplified form}$$

The domain of the original rational expression is all real numbers such that $x \neq 0$, $y \neq 0$, and $y \neq 2x$.

Example 11   Geometry: Area

Find the ratio of the area of the shaded portion of the triangle to the total area of the triangle. (See Figure 7.1.)

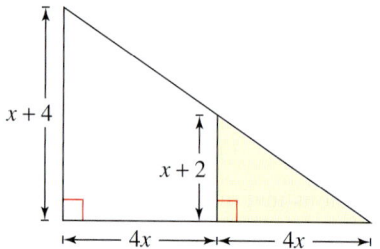

Figure 7.1

Solution

The area of the shaded portion of the triangle is given by

$$\text{Area} = \tfrac{1}{2}(4x)(x + 2) = \tfrac{1}{2}(4x^2 + 8x) = 2x^2 + 4x.$$

The total area of the triangle is given by

$$\text{Area} = \tfrac{1}{2}(4x + 4x)(x + 4) = \tfrac{1}{2}(8x)(x + 4) = \tfrac{1}{2}(8x^2 + 32x) = 4x^2 + 16x.$$

So, the ratio of the area of the shaded portion of the triangle to the total area of the triangle is

$$\frac{2x^2 + 4x}{4x^2 + 16x} = \frac{2x(x + 2)}{4x(x + 4)} = \frac{x + 2}{2(x + 4)}, \quad x \neq 0.$$

As you study the examples and work the exercises in this section and the next four sections, keep in mind that you are *rewriting expressions in simpler forms*. You are not solving equations. Equal signs are used in the steps of the simplification process only to indicate that the new form of the expression is *equivalent* to the original form.

7.1 Exercises

Review   Concepts, Skills, and Problem Solving

Keep mathematically in shape by doing these exercises *before* the problems of this section.

Properties and Definitions

1. *Writing* ✎   Define the slope of the line through the points (x_1, y_1) and (x_2, y_2).

2. *Writing* ✎   Make a statement about the slope m of the line for each condition.

   (a) The line rises from left to right.

   (b) The line falls from left to right.

   (c) The line is horizontal.

   (d) The line is vertical.

Simplifying Expressions

In Exercises 3–8, simplify the expression.

3. $2(x + 5) - 3 - (2x - 3)$

4. $3(y + 4) + 5 - (3y + 5)$

5. $4 - 2[3 + 4(x + 1)]$ 6. $5x + x[3 - 2(x - 3)]$

7. $\left(\dfrac{5}{x^2}\right)^2$ 8. $-\dfrac{(2u^2v)^2}{-3uv^2}$

Problem Solving

9. *Mixture Problem*   Determine the numbers of gallons of a 30% solution and a 60% solution that must be mixed to obtain 20 gallons of a 40% solution.

10. *Original Price*   A suit sells for $375 during a 25% off storewide clearance sale. What was the original price of the suit?

Developing Skills

In Exercises 1–20, find the domain of the rational function. See Examples 1 and 2.

1. $f(x) = \dfrac{x^2 + 9}{4}$

2. $f(y) = \dfrac{y^2 - 3}{7}$

3. $f(x) = \dfrac{7}{x - 5}$

4. $g(x) = \dfrac{-3}{x - 9}$

5. $f(x) = \dfrac{12x}{4 - x}$

6. $h(y) = \dfrac{2y}{6 - y}$

7. $g(x) = \dfrac{2x}{x + 10}$

8. $f(x) = \dfrac{4x}{x + 1}$

9. $h(x) = \dfrac{x}{x^2 + 4}$

10. $h(x) = \dfrac{4x}{x^2 + 16}$

11. $f(y) = \dfrac{y - 4}{y(y + 3)}$

12. $f(z) = \dfrac{z + 2}{z(z - 4)}$

13. $f(t) = \dfrac{5t}{t^2 - 16}$

14. $f(x) = \dfrac{x}{x^2 - 4}$

15. $g(y) = \dfrac{y + 5}{y^2 - 3y}$

16. $g(t) = \dfrac{t - 6}{t^2 + 5t}$

17. $g(x) = \dfrac{x + 1}{x^2 - 5x + 6}$

18. $h(t) = \dfrac{3t^2}{t^2 - 2t - 3}$

19. $f(u) = \dfrac{u^2}{3u^2 - 2u - 5}$

20. $g(y) = \dfrac{y + 5}{4y^2 - 5y - 6}$

In Exercises 21–26, evaluate the rational function as indicated and simplify. If not possible, state the reason.

21. $f(x) = \dfrac{4x}{x + 3}$

    (a) $f(1)$ (b) $f(-2)$

    (c) $f(-3)$ (d) $f(0)$

22. $f(x) = \dfrac{x - 10}{4x}$

    (a) $f(10)$ (b) $f(0)$

    (c) $f(-2)$ (d) $f(12)$

23. $g(x) = \dfrac{x^2 - 4x}{x^2 - 9}$

    (a) $g(0)$ (b) $g(4)$

    (c) $g(3)$ (d) $g(-3)$

24. $g(t) = \dfrac{t - 2}{2t - 5}$

    (a) $g(2)$ (b) $g\left(\frac{5}{2}\right)$

    (c) $g(-2)$ (d) $g(0)$

25. $h(s) = \dfrac{s^2}{s^2 - s - 2}$

    (a) $h(10)$ (b) $h(0)$

    (c) $h(-1)$ (d) $h(2)$

26. $f(x) = \dfrac{x^3 + 1}{x^2 - 6x + 9}$

    (a) $f(-1)$ (b) $f(3)$

    (c) $f(-2)$ (d) $f(2)$

In Exercises 27–32, describe the domain. See Example 3.

27. ▲ *Geometry* A rectangle of length x inches has an area of 500 square inches. The perimeter P of the rectangle is given by

$$P = 2\left(x + \frac{500}{x}\right).$$

28. *Cost* The cost C in millions of dollars for the government to seize $p\%$ of an illegal drug as it enters the country is given by

$$C = \frac{528p}{100 - p}.$$

29. *Inventory Cost* The inventory cost I when x units of a product are ordered from a supplier is given by

$$I = \frac{0.25x + 2000}{x}.$$

30. *Average Cost* The average cost $\overline{C}$ for a manufacturer to produce x units of a product is given by

$$\overline{C} = \frac{1.35x + 4570}{x}.$$

31. *Pollution Removal* The cost C in dollars of removing $p\%$ of the air pollutants in the stack emission of a utility company is given by the rational function

$$C = \frac{80{,}000p}{100 - p}.$$

32. *Consumer Awareness* The average cost of a movie video rental $\overline{M}$ when you consider the cost of purchasing a video cassette recorder and renting x movie videos at $3.49 per movie is

$$\overline{M} = \frac{75 + 3.49x}{x}.$$

In Exercises 33–40, fill in the missing factor.

33. $\dfrac{5(\quad\quad\quad)}{6(x + 3)} = \dfrac{5}{6}, \quad x \neq -3$

34. $\dfrac{7(\quad\quad\quad)}{15(x - 10)} = \dfrac{7}{15}, \quad x \neq 10$

35. $\dfrac{3x(x + 16)^2}{2(\quad\quad\quad)} = \dfrac{x}{2}, \quad x \neq -16$

36. $\dfrac{25x^2(x - 10)}{12(\quad\quad\quad)} = \dfrac{5x}{12}, \quad x \neq 10, \quad x \neq 0$

37. $\dfrac{(x + 5)(\quad\quad\quad)}{3x^2(x - 2)} = \dfrac{x + 5}{3x}, \quad x \neq 2$

38. $\dfrac{(3y - 7)(\quad\quad\quad)}{y^2 - 4} = \dfrac{3y - 7}{y + 2}, \quad y \neq 2$

39. $\dfrac{8x(\quad\quad\quad)}{x^2 - 3x - 10} = \dfrac{8x}{x - 5}, \quad x \neq -2$

40. $\dfrac{(3 - z)(\quad\quad\quad)}{z^3 + 2z^2} = \dfrac{3 - z}{z^2}, \quad z \neq -2$

In Exercises 41–78, simplify the rational expression. See Examples 4–10.

41. $\dfrac{5x}{25}$

42. $\dfrac{32y}{24}$

43. $\dfrac{12y^2}{2y}$

44. $\dfrac{15z^3}{15z^3}$

45. $\dfrac{18x^2y}{15xy^4}$

46. $\dfrac{16y^2z^2}{60y^5z}$

47. $\dfrac{3x^2 - 9x}{12x^2}$

48. $\dfrac{8x^3 + 4x^2}{20x}$

49. $\dfrac{x^2(x - 8)}{x(x - 8)}$

50. $\dfrac{a^2b(b - 3)}{b^3(b - 3)^2}$

51. $\dfrac{2x - 3}{4x - 6}$

52. $\dfrac{y^2 - 81}{2y - 18}$

53. $\dfrac{5 - x}{3x - 15}$

54. $\dfrac{x^2 - 36}{6 - x}$

55. $\dfrac{a + 3}{a^2 + 6a + 9}$

56. $\dfrac{u^2 - 12u + 36}{u - 6}$

57. $\dfrac{x^2 - 7x}{x^2 - 14x + 49}$

58. $\dfrac{z^2 + 22z + 121}{3z + 33}$

59. $\dfrac{y^3 - 4y}{y^2 + 4y - 12}$

60. $\dfrac{x^2 - 7x}{x^2 - 4x - 21}$

61. $\dfrac{x^3 - 4x}{x^2 - 5x + 6}$

62. $\dfrac{x^4 - 25x^2}{x^2 + 2x - 15}$

63. $\dfrac{3x^2 - 7x - 20}{12 + x - x^2}$

64. $\dfrac{2x^2 + 3x - 5}{7 - 6x - x^2}$

65. $\dfrac{2x^2 + 19x + 24}{2x^2 - 3x - 9}$

66. $\dfrac{2y^2 + 13y + 20}{2y^2 + 17y + 30}$

67. $\dfrac{15x^2 + 7x - 4}{25x^2 - 16}$

68. $\dfrac{56z^2 - 3z - 20}{49z^2 - 16}$

69. $\dfrac{3xy^2}{xy^2 + x}$

70. $\dfrac{x + 3x^2y}{3xy + 1}$

71. $\dfrac{y^2 - 64x^2}{5(3y + 24x)}$

72. $\dfrac{x^2 - 25z^2}{x + 5z}$

73. $\dfrac{5xy + 3x^2y^2}{xy^3}$

74. $\dfrac{4u^2v - 12uv^2}{18uv}$

75. $\dfrac{u^2 - 4v^2}{u^2 + uv - 2v^2}$

76. $\dfrac{x^2 + 4xy}{x^2 - 16y^2}$

77. $\dfrac{3m^2 - 12n^2}{m^2 + 4mn + 4n^2}$

78. $\dfrac{x^2 + xy - 2y^2}{x^2 + 3xy + 2y^2}$

In Exercises 79 and 80, complete the table. What can you conclude?

79.

x	-2	-1	0	1	2	3	4
$\dfrac{x^2 - x - 2}{x - 2}$							
$x + 1$							

80.

x	-2	-1	0	1	2	3	4
$\dfrac{x^2 + 5x}{x}$							
$x + 5$							

Solving Problems

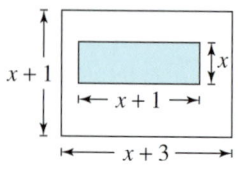

▲ *Geometry*   In Exercises 81–84, find the ratio of the area of the shaded portion to the total area of the figure. See Example 11.

81.

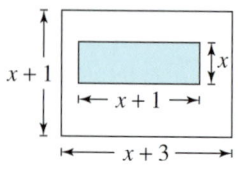

82.

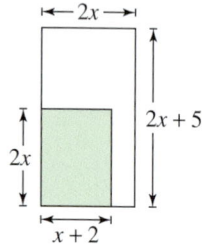

83.

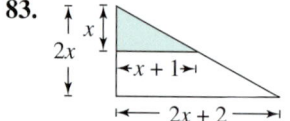

84.

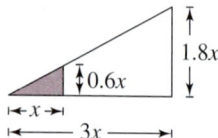

85. *Average Cost*   A machine shop has a setup cost of $2500 for the production of a new product. The cost of labor and material for producing each unit is $9.25.

(a) Write the total cost C as a function of x, the number of units produced.

(b) Write the average cost per unit $\overline{C} = C/x$ as a function of x, the number of units produced.

(c) Determine the domain of the function in part (b).

(d) Find the value of $\overline{C}(100)$.

86. *Average Cost*   A greeting card company has an initial investment of $60,000. The cost of producing one dozen cards is $6.50.

(a) Write the total cost C as a function of x, the number of cards in dozens produced.

(b) Write the average cost per dozen $\overline{C} = C/x$ as a function of x, the number of cards in dozens produced.

(c) Determine the domain of the function in part (b).

(d) Find the value of $\overline{C}(11,000)$.

87. *Distance Traveled*   A van starts on a trip and travels at an average speed of 45 miles per hour. Three hours later, a car starts on the same trip and travels at an average speed of 60 miles per hour.

(a) Find the distance each vehicle has traveled when the car has been on the road for t hours.

(b) Use the result of part (a) to write the distance between the van and the car as a function of t.

(c) Write the ratio of the distance the car has traveled to the distance the van has traveled as a function of t.

88. *Distance Traveled*   A car starts on a trip and travels at an average speed of 55 miles per hour. Two hours later, a second car starts on the same trip and travels at an average speed of 65 miles per hour.

(a) Find the distance each vehicle has traveled when the second car has been on the road for t hours.

(b) Use the result of part (a) to write the distance between the first car and the second car as a function of t.

(c) Write the ratio of the distance the second car has traveled to the distance the first car has traveled as a function of t.

89. ▲ *Geometry*   One swimming pool is circular and another is rectangular. The rectangular pool's width is three times its depth. Its length is 6 feet more than its width. The circular pool has a diameter that is twice the width of the rectangular pool, and it is 2 feet deeper. Find the ratio of the circular pool's volume to the rectangular pool's volume.

90. ▲ *Geometry*   A circular pool has a radius five times its depth. A rectangular pool has the same depth as the circular pool. Its width is 4 feet more than three times its depth and its length is 2 feet less than six times its depth. Find the ratio of the rectangular pool's volume to the circular pool's volume.

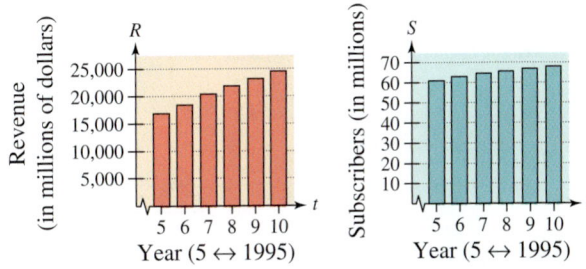

Figures for 91 and 92

Cable TV Revenue   In Exercises 91 and 92, use the following polynomial models, which give the total basic cable television revenue R (in millions of dollars) and the number of basic cable subscribers S (in millions) from 1995 through 2000 (see figures).

$R = 1531.1t + 9358, \quad 5 \le t \le 10$

$S = 1.33t + 54.6, \quad 5 \le t \le 10$

In these models, t represents the year, with $t = 5$ corresponding to 1995. (Source: Paul Kagen Associates, Inc.)

91. Find a rational model that represents the average basic cable television revenue per subscriber during the years 1995 to 2000.

92. Use the model found in Exercise 91 to complete the table, which shows the average basic cable television revenue per subscriber.

Year	1995	1996	1997	1998	1999	2000
Average revenue						

Explaining Concepts

93. *Writing*   Define the term *rational expression*.

94. Give an example of a rational function whose domain is the set of all real numbers.

95. *Writing*   How do you determine whether a rational expression is in simplified form?

96. *Writing*   Can you divide out common terms from the numerator and denominator of a rational expression? Explain.

97. *Error Analysis*   Describe the error.

$$\frac{2x^2}{x^2+4} = \frac{2x^2}{x^2+4} = \frac{2}{1+4} = \frac{2}{5}$$

98. *Writing*   Is the following statement true? Explain.

$$\frac{6x-5}{5-6x} = -1$$

99. You are the instructor of an algebra course. One of your students turns in the following incorrect solutions. Find the errors, discuss the student's misconceptions, and construct correct solutions.

a. $\dfrac{3x^2 + 5x - 4}{x} = 3x + 5 - 4 = 3x + 1$

b. $\dfrac{x^2 + 7x}{x+7} = \dfrac{x^2}{x} + \dfrac{7x}{7} = x + x = 2x$

7.2 Multiplying and Dividing Rational Expressions

What You Should Learn

1 Multiply rational expressions and simplify.

2 Divide rational expressions and simplify.

Why You Should Learn It

Multiplication and division of rational expressions can be used to solve real-life applications. For instance, Example 9 on page 416 shows how a rational expression is used to model the amount Americans spent per person on books and maps from 1995 to 2000.

1 Multiply rational expressions and simplify.

Multiplying Rational Expressions

The rule for multiplying rational expressions is the same as the rule for multiplying numerical fractions. That is, you *multiply numerators, multiply denominators, and write the new fraction in simplified form.*

$$\frac{3}{4} \cdot \frac{7}{6} = \frac{21}{24} = \frac{\cancel{3} \cdot 7}{\cancel{3} \cdot 8} = \frac{7}{8}$$

Multiplying Rational Expressions

Let u, v, w, and z represent real numbers, variables, or algebraic expressions such that $v \neq 0$ and $z \neq 0$. Then the product of u/v and w/z is

$$\frac{u}{v} \cdot \frac{w}{z} = \frac{uw}{vz}.$$

In order to recognize common factors in the product, write the numerators and denominators in completely factored form, as demonstrated in Example 1.

Example 1   Multiplying Rational Expressions

Multiply the rational expressions $\dfrac{4x^3y}{3xy^4} \cdot \dfrac{-6x^2y^2}{10x^4}$.

Solution

$$\frac{4x^3y}{3xy^4} \cdot \frac{-6x^2y^2}{10x^4} = \frac{(4x^3y) \cdot (-6x^2y^2)}{(3xy^4) \cdot (10x^4)} \qquad \text{Multiply numerators and denominators.}$$

$$= \frac{-24x^5y^3}{30x^5y^4} \qquad \text{Simplify.}$$

$$= \frac{-4(6)(x^5)(y^3)}{5(6)(x^5)(y^3)(y)} \qquad \text{Factor and divide out common factors.}$$

$$= -\frac{4}{5y}, \quad x \neq 0 \qquad \text{Simplified form}$$

Example 2   Multiplying Rational Expressions

Multiply the rational expressions.

$$\frac{x}{5x^2 - 20x} \cdot \frac{x - 4}{2x^2 + x - 3}$$

Solution

$$\frac{x}{5x^2 - 20x} \cdot \frac{x - 4}{2x^2 + x - 3}$$

$$= \frac{x \cdot (x - 4)}{(5x^2 - 20x) \cdot (2x^2 + x - 3)} \qquad \text{Multiply numerators and denominators.}$$

$$= \frac{x(x - 4)}{5x(x - 4)(x - 1)(2x + 3)} \qquad \text{Factor.}$$

$$= \frac{x(x - 4)}{5x(x - 4)(x - 1)(2x + 3)} \qquad \text{Divide out common factors.}$$

$$= \frac{1}{5(x - 1)(2x + 3)}, \ x \neq 0, \ x \neq 4 \qquad \text{Simplified form}$$

Example 3   Multiplying Rational Expressions

Multiply the rational expressions.

$$\frac{4x^2 - 4x}{x^2 + 2x - 3} \cdot \frac{x^2 + x - 6}{4x}$$

Solution

$$\frac{4x^2 - 4x}{x^2 + 2x - 3} \cdot \frac{x^2 + x - 6}{4x}$$

$$= \frac{4x(x - 1)(x + 3)(x - 2)}{(x - 1)(x + 3)(4x)} \qquad \text{Multiply and factor.}$$

$$= \frac{4x(x - 1)(x + 3)(x - 2)}{(x - 1)(x + 3)(4x)} \qquad \text{Divide out common factors.}$$

$$= x - 2, \ x \neq 0, \ x \neq 1, \ x \neq -3 \qquad \text{Simplified form}$$

The rule for multiplying rational expressions can be extended to cover products involving expressions that are not in fractional form. To do this, rewrite the (nonfractional) expression as a fraction whose denominator is 1. Here is a simple example.

$$\frac{x + 3}{x - 2} \cdot (5x) = \frac{x + 3}{x - 2} \cdot \frac{5x}{1}$$

$$= \frac{(x + 3)(5x)}{x - 2}$$

$$= \frac{5x(x + 3)}{x - 2}$$

Technology: Tip

You can use a graphing calculator to check your results when multiplying rational expressions. For instance, in Example 3, try graphing the equations

$$y_1 = \frac{4x^2 - 4x}{x^2 + 2x - 3} \cdot \frac{x^2 + x - 6}{4x}$$

and

$$y_2 = x - 2$$

in the same viewing window and use the *table* feature to create a table of values for the two equations. If the two graphs coincide, and the values of y_1 and y_2 are the same in the table except where a common factor has been divided out, as shown below, you can conclude that the solution checks.

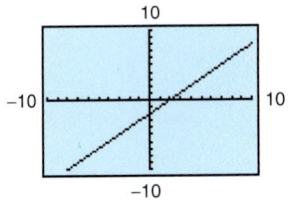

In the next example, note how to divide out a factor that differs only in sign. The Distributive Property is used in the step in which $(y - x)$ is rewritten as $(-1)(x - y)$.

Example 4 Multiplying Rational Expressions

Multiply the rational expressions.

$$\frac{x - y}{y^2 - x^2} \cdot \frac{x^2 - xy - 2y^2}{3x - 6y}$$

Solution

$$\frac{x - y}{y^2 - x^2} \cdot \frac{x^2 - xy - 2y^2}{3x - 6y}$$

$$= \frac{(x - y)(x - 2y)(x + y)}{(y + x)(y - x)(3)(x - 2y)} \qquad \text{Multiply and factor.}$$

$$= \frac{(x - y)(x - 2y)(x + y)}{(y + x)(-1)(x - y)(3)(x - 2y)} \qquad (y - x) = -1(x - y)$$

$$= \frac{(x - y)(x - 2y)(x + y)}{(x + y)(-1)(x - y)(3)(x - 2y)} \qquad \text{Divide out common factors.}$$

$$= -\frac{1}{3}, \ x \neq y, \ x \neq -y, \ x \neq 2y \qquad \text{Simplified form}$$

The rule for multiplying rational expressions can be extended to cover products of three or more expressions, as shown in Example 5.

Example 5 Multiplying Three Rational Expressions

Multiply the rational expressions.

$$\frac{x^2 - 3x + 2}{x + 2} \cdot \frac{3x}{x - 2} \cdot \frac{2x + 4}{x^2 - 5x}$$

Solution

$$\frac{x^2 - 3x + 2}{x + 2} \cdot \frac{3x}{x - 2} \cdot \frac{2x + 4}{x^2 - 5x}$$

$$= \frac{(x - 1)(x - 2)(3)(x)(2)(x + 2)}{(x + 2)(x - 2)(x)(x - 5)} \qquad \text{Multiply and factor.}$$

$$= \frac{(x - 1)(x - 2)(3)(x)(2)(x + 2)}{(x + 2)(x - 2)(x)(x - 5)} \qquad \text{Divide out common factors.}$$

$$= \frac{6(x - 1)}{x - 5}, \ x \neq 0, \ x \neq 2, \ x \neq -2 \qquad \text{Simplified form}$$

2 Divide rational expressions and simplify.

Dividing Rational Expressions

To divide two rational expressions, multiply the first expression by the *reciprocal* of the second. That is, *invert the divisor and multiply*.

Dividing Rational Expressions

Let u, v, w, and z represent real numbers, variables, or algebraic expressions such that $v \neq 0$, $w \neq 0$, and $z \neq 0$. Then the quotient of u/v and w/z is

$$\frac{u}{v} \div \frac{w}{z} = \frac{u}{v} \cdot \frac{z}{w} = \frac{uz}{vw}.$$

Example 6   Dividing Rational Expressions

Divide the rational expressions.

$$\frac{x}{x+3} \div \frac{4}{x-1}$$

Solution

$$\frac{x}{x+3} \div \frac{4}{x-1} = \frac{x}{x+3} \cdot \frac{x-1}{4} \qquad \text{Invert divisor and multiply.}$$

$$= \frac{x(x-1)}{(x+3)(4)} \qquad \text{Multiply numerators and denominators.}$$

$$= \frac{x(x-1)}{4(x+3)}, \ x \neq 1 \qquad \text{Simplify.}$$

Example 7   Dividing Rational Expressions

$$\frac{2x}{3x-12} \div \frac{x^2-2x}{x^2-6x+8} \qquad \text{Original expressions}$$

$$= \frac{2x}{3x-12} \cdot \frac{x^2-6x+8}{x^2-2x} \qquad \text{Invert divisor and multiply.}$$

$$= \frac{(2)(x)(x-2)(x-4)}{(3)(x-4)(x)(x-2)} \qquad \text{Factor.}$$

$$= \frac{(2)(x)(x-2)(x-4)}{(3)(x-4)(x)(x-2)} \qquad \text{Divide out common factors.}$$

$$= \frac{2}{3}, \ x \neq 0, \ x \neq 2, \ x \neq 4 \qquad \text{Simplified form}$$

Remember that the original expression is equivalent to $\frac{2}{3}$ except for $x = 0$, $x = 2$, and $x = 4$.

Example 8 Dividing Rational Expressions

Divide the rational expressions.

$$\frac{x^2 - y^2}{2x + 2y} \div \frac{2x^2 - 3xy + y^2}{6x + 2y}$$

Solution

$$\frac{x^2 - y^2}{2x + 2y} \div \frac{2x^2 - 3xy + y^2}{6x + 2y}$$

$$= \frac{x^2 - y^2}{2x + 2y} \cdot \frac{6x + 2y}{2x^2 - 3xy + y^2} \qquad \text{Invert divisor and multiply.}$$

$$= \frac{(x + y)(x - y)(2)(3x + y)}{(2)(x + y)(2x - y)(x - y)} \qquad \text{Factor.}$$

$$= \frac{(x + y)(x - y)(2)(3x + y)}{(2)(x + y)(2x - y)(x - y)} \qquad \text{Divide out common factors.}$$

$$= \frac{3x + y}{2x - y}, x \neq y, x \neq -y \qquad \text{Simplified form}$$

Example 9 Amount Spent on Books and Maps

The amount A (in millions of dollars) Americans spent on books and maps and the population P (in millions) of the United States for the period 1995 through 2000 can be modeled by

$$A = \frac{-24.86t + 17{,}862.7}{-0.05t + 1.0}, \quad 5 \leq t \leq 10$$

and

$$P = 2.46t + 250.8, \quad 5 \leq t \leq 10$$

where t represents the year, with $t = 5$ corresponding to 1995. Find a model T for the amount Americans spent *per person* on books and maps. (Source: U.S. Bureau of Economic Analysis and U.S. Census Bureau)

Solution

To find a model T for the amount Americans spent per person on books and maps, divide the total amount by the population.

$$T = \frac{-24.86t + 17{,}862.7}{-0.05t + 1.0} \div 2.46t + 250.8 \qquad \text{Divide amount spent by population.}$$

$$= \frac{-24.86t + 17{,}862.7}{-0.05t + 1.0} \cdot \frac{1}{2.46t + 250.8} \qquad \text{Invert divisor and multiply.}$$

$$= \frac{-24.86t + 17{,}862.7}{(-0.05t + 1.0)(2.46t + 250.8)}, \quad 5 \leq t \leq 10 \qquad \text{Model}$$

7.2 Exercises

Review — Concepts, Skills, and Problem Solving

Keep mathematically in shape by doing these exercises *before* the problems of this section.

Properties and Definitions

1. *Writing* ✎   Explain how to factor the difference of two squares $9t^2 - 4$.

2. *Writing* ✎   Explain how to factor the perfect square trinomial $4x^2 - 12x + 9$.

3. *Writing* ✎   Explain how to factor the sum of two cubes $8x^3 + 64$.

4. *Writing* ✎   Factor $3x^2 + 13x - 10$, and explain how you can check your answer.

Algebraic Operations

In Exercises 5–10, factor the expression completely.

5. $5x - 20x^2$

6. $64 - (x - 6)^2$

7. $15x^2 - 16x - 15$

8. $16t^2 + 8t + 1$

9. $y^3 - 64$

10. $8x^3 + 1$

Graphs

In Exercises 11 and 12, sketch the line through the point with each indicated slope on the same set of coordinate axes.

	Point	Slopes	
11.	$(2, -3)$	(a) 0	(b) Undefined
		(c) 2	(d) $-\frac{1}{3}$
12.	$(-1, 4)$	(a) 2	(b) -1
		(c) $\frac{1}{2}$	(d) Undefined

Developing Skills

In Exercises 1–8, fill in the missing factor.

1. $\dfrac{7x^2}{3y()} = \dfrac{7}{3y}, \quad x \neq 0$

2. $\dfrac{14x(x - 3)^2}{(x - 3)()} = \dfrac{2x}{x - 3}, \quad x \neq 3$

3. $\dfrac{3x(x + 2)^2}{(x - 4)()} = \dfrac{3x}{x - 4}, \quad x \neq -2$

4. $\dfrac{(x + 1)^3}{x()} = \dfrac{x + 1}{x}, \quad x \neq -1$

5. $\dfrac{3u()}{7v(u + 1)} = \dfrac{3u}{7v}, \quad u \neq -1$

6. $\dfrac{(3t + 5)()}{5t^2(3t - 5)} = \dfrac{3t + 5}{t}, \quad t \neq \dfrac{5}{3}$

7. $\dfrac{13x()}{4 - x^2} = \dfrac{13x}{x - 2}, \quad x \neq -2$

8. $\dfrac{x^2()}{x^2 - 10x} = \dfrac{x^2}{10 - x}, \quad x \neq 0$

In Exercises 9–36, multiply and simplify. See Examples 1–5.

9. $7x \cdot \dfrac{9}{14x}$

10. $\dfrac{6}{5a} \cdot (25a)$

11. $\dfrac{8s^3}{9s} \cdot \dfrac{6s^2}{32s}$

12. $\dfrac{3x^4}{7x} \cdot \dfrac{8x^2}{9}$

13. $16u^4 \cdot \dfrac{12}{8u^2}$

14. $25x^3 \cdot \dfrac{8}{35x}$

15. $\dfrac{8}{3 + 4x} \cdot (9 + 12x)$

16. $(6 - 4x) \cdot \dfrac{10}{3 - 2x}$

17. $\dfrac{8u^2v}{3u + v} \cdot \dfrac{u + v}{12u}$

18. $\dfrac{1 - 3xy}{4x^2y} \cdot \dfrac{46x^4y^2}{15 - 45xy}$

19. $\dfrac{12 - r}{3} \cdot \dfrac{3}{r - 12}$ **20.** $\dfrac{8 - z}{8 + z} \cdot \dfrac{z + 8}{z - 8}$

21. $\dfrac{(2x - 3)(x + 8)}{x^3} \cdot \dfrac{x}{3 - 2x}$

22. $\dfrac{x + 14}{x^3(10 - x)} \cdot \dfrac{x(x - 10)}{5}$

23. $\dfrac{4r - 12}{r - 2} \cdot \dfrac{r^2 - 4}{r - 3}$

24. $\dfrac{5y - 20}{5y + 15} \cdot \dfrac{2y + 6}{y - 4}$

25. $\dfrac{2t^2 - t - 15}{t + 2} \cdot \dfrac{t^2 - t - 6}{t^2 - 6t + 9}$

26. $\dfrac{y^2 - 16}{y^2 + 8y + 16} \cdot \dfrac{3y^2 - 5y - 2}{y^2 - 6y + 8}$

27. $(x^2 - 4y^2) \cdot \dfrac{xy}{(x - 2y)^2}$

28. $(u - 2v)^2 \cdot \dfrac{u + 2v}{u - 2v}$

29. $\dfrac{x^2 + 2xy - 3y^2}{(x + y)^2} \cdot \dfrac{x^2 - y^2}{x + 3y}$

30. $\dfrac{(x - 2y)^2}{x + 2y} \cdot \dfrac{x^2 + 7xy + 10y^2}{x^2 - 4y^2}$

31. $\dfrac{x + 5}{x - 5} \cdot \dfrac{2x^2 - 9x - 5}{3x^2 + x - 2} \cdot \dfrac{x^2 - 1}{x^2 + 7x + 10}$

32. $\dfrac{t^2 + 4t + 3}{2t^2 - t - 10} \cdot \dfrac{t}{t^2 + 3t + 2} \cdot \dfrac{2t^2 + 4t^3}{t^2 + 3t}$

33. $\dfrac{9 - x^2}{2x + 3} \cdot \dfrac{4x^2 + 8x - 5}{4x^2 - 8x + 3} \cdot \dfrac{6x^4 - 2x^3}{8x^2 + 4x}$

34. $\dfrac{16x^2 - 1}{4x^2 + 9x + 5} \cdot \dfrac{5x^2 - 9x - 18}{x^2 - 12x + 36} \cdot \dfrac{12 + 4x - x^2}{4x^2 - 13x + 3}$

35. $\dfrac{x^3 + 3x^2 - 4x - 12}{x^3 - 3x^2 - 4x + 12} \cdot \dfrac{x^2 - 9}{x}$

36. $\dfrac{xu - yu + xv - yv}{xu + yu - xv - yv} \cdot \dfrac{xu + yu + xv + yv}{xu - yu - xv + yv}$

In Exercises 37–50, divide and simplify. See Examples 6–8.

37. $x^2 \div \dfrac{3x}{4}$ **38.** $\dfrac{u}{10} \div u^2$

39. $\dfrac{2x}{5} \div \dfrac{x^2}{15}$ **40.** $\dfrac{3y^2}{20} \div \dfrac{y}{15}$

41. $\dfrac{7xy^2}{10u^2v} \div \dfrac{21x^3}{45uv}$ **42.** $\dfrac{25x^2y}{60x^3y^2} \div \dfrac{5x^4y^3}{16x^2y}$

43. $\dfrac{3(a + b)}{4} \div \dfrac{(a + b)^2}{2}$

44. $\dfrac{x^2 + 9}{5(x + 2)} \div \dfrac{x + 3}{5(x^2 - 4)}$

45. $\dfrac{(x^3y)^2}{(x + 2y)^2} \div \dfrac{x^2y}{(x + 2y)^3}$

46. $\dfrac{x^2 - y^2}{2x^2 - 8x} \div \dfrac{(x - y)^2}{2xy}$

47. $\dfrac{y^2 - 2y - 15}{y^2 - 9} \div \dfrac{12 - 4y}{y^2 - 6y + 9}$

48. $\dfrac{x + 3}{x^2 + 7x + 10} \div \dfrac{x^2 + 6x + 9}{x^2 + 5x + 6}$

49. $\dfrac{x^2 + 2x - 15}{x^2 + 11x + 30} \div \dfrac{x^2 - 8x + 15}{x^2 + 2x - 24}$

50. $\dfrac{y^2 + 5y - 14}{y^2 + 10y + 21} \div \dfrac{y^2 + 5y + 6}{y^2 + 7y + 12}$

In Exercises 51–58, perform the operations and simplify. (In Exercises 57 and 58, n is a positive integer.)

51. $\left[\dfrac{x^2}{9} \cdot \dfrac{3(x+4)}{x^2+2x}\right] \div \dfrac{x}{x+2}$

52. $\left(\dfrac{x^2+6x+9}{x^2} \cdot \dfrac{2x+1}{x^2-9}\right) \div \dfrac{4x^2+4x+1}{x^2-3x}$

53. $\left[\dfrac{xy+y}{4x} \div (3x+3)\right] \div \dfrac{y}{3x}$

54. $\dfrac{3u^2-u-4}{u^2} \div \dfrac{3u^2+12u+4}{u^4-3u^3}$

55. $\dfrac{2x^2+5x-25}{3x^2+5x+2} \cdot \dfrac{3x^2+2x}{x+5} \div \left(\dfrac{x}{x+1}\right)^2$

56. $\dfrac{t^2-100}{4t^2} \cdot \dfrac{t^3-5t^2-50t}{t^4+10t^3} \div \dfrac{(t-10)^2}{5t}$

57. $x^3 \cdot \dfrac{x^{2n}-9}{x^{2n}+4x^n+3} \div \dfrac{x^{2n}-2x^n-3}{x}$

58. $\dfrac{x^{n+1}-8x}{x^{2n}+2x^n+1} \cdot \dfrac{x^{2n}-4x^n-5}{x} \div x^n$

In Exercises 59 and 60, use a graphing calculator to graph the two equations in the same viewing window. Use the graphs and a table of values to verify that the expressions are equivalent. Verify the results algebraically.

59. $y_1 = \dfrac{x^2-10x+25}{x^2-25} \cdot \dfrac{x+5}{2}$

$y_2 = \dfrac{x-5}{2}, \quad x \ne \pm 5$

60. $y_1 = \dfrac{3x+15}{x^4} \div \dfrac{x+5}{x^2}$

$y_2 = \dfrac{3}{x^2}, \quad x \ne -5$

Solving Problems

▲ *Geometry*   In Exercises 61 and 62, write and simplify an expression for the area of the shaded region.

61.

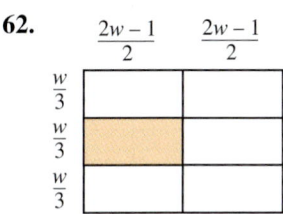

62.

63.

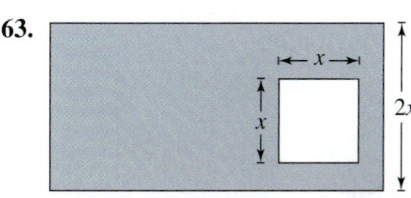

Probability   In Exercises 63–66, consider an experiment in which a marble is tossed into a rectangular box with dimensions $2x$ centimeters by $4x+2$ centimeters.

The probability that the marble will come to rest in the unshaded portion of the box is equal to the ratio of the unshaded area to the total area of the figure. Find the probability in simplified form.

64.

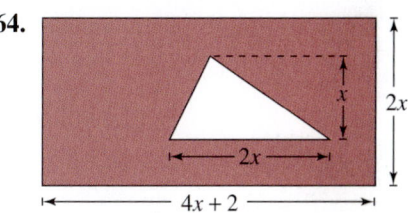

65.

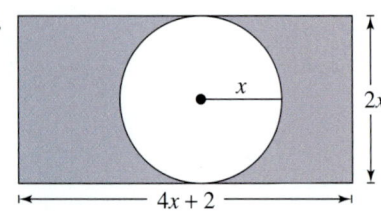

66.

67. *Photocopy Rate*   A photocopier produces copies at a rate of 20 pages per minute.

(a) Determine the time required to copy 1 page.

(b) Determine the time required to copy x pages.

(c) Determine the time required to copy 35 pages.

68. *Pumping Rate*   The rate for a pump is 15 gallons per minute.

(a) Determine the time required to pump 1 gallon.

(b) Determine the time required to pump x gallons.

(c) Determine the time required to pump 130 gallons.

Explaining Concepts

69. *Writing*   In your own words, explain how to divide rational expressions.

70. *Writing*   Explain how to divide a rational expression by a polynomial.

71. *Error Analysis*   Describe the error.

$$\frac{x^2 - 4}{5x} \div \frac{x + 2}{x - 2} = \frac{5x}{x^2 - 4} \cdot \frac{x + 2}{x - 2}$$

$$= \frac{5x}{(x + 2)(x - 2)} \cdot \frac{x + 2}{x - 2}$$

$$= \frac{5x}{(x - 2)^2}$$

72. *Writing*   Complete the table for the given values of x.

x	60	100	1000
$\dfrac{x - 10}{x + 10}$			
$\dfrac{x + 50}{x - 50}$			
$\dfrac{x - 10}{x + 10} \cdot \dfrac{x + 50}{x - 50}$			

x	10,000	100,000	1,000,000
$\dfrac{x - 10}{x + 10}$			
$\dfrac{x + 50}{x - 50}$			
$\dfrac{x - 10}{x + 10} \cdot \dfrac{x + 50}{x - 50}$			

What kind of pattern do you see? Try to explain what is going on. Can you see why?

7.3 Adding and Subtracting Rational Expressions

Tom Carter/PhotoEdit

What You Should Learn

1 Add or subtract rational expressions with like denominators and simplify.

2 Add or subtract rational expressions with unlike denominators and simplify.

Why You Should Learn It

Addition and subtraction of rational expressions can be used to solve real-life applications. For instance, in Exercise 83 on page 429, you will find a rational expression that models the number of participants in high school athletic programs.

1 Add or subtract rational expressions with like denominators and simplify.

Adding or Subtracting with Like Denominators

As with numerical fractions, the procedure used to add or subtract two rational expressions depends on whether the expressions have *like* or *unlike* denominators. To add or subtract two rational expressions with *like* denominators, simply combine their numerators and place the result over the common denominator.

Adding or Subtracting with Like Denominators

If u, v, and w are real numbers, variables, or algebraic expressions, and $w \neq 0$, the following rules are valid.

1. $\dfrac{u}{w} + \dfrac{v}{w} = \dfrac{u + v}{w}$ Add fractions with like denominators.

2. $\dfrac{u}{w} - \dfrac{v}{w} = \dfrac{u - v}{w}$ Subtract fractions with like denominators.

Example 1   Adding and Subtracting with Like Denominators

a. $\dfrac{x}{4} + \dfrac{5 - x}{4} = \dfrac{x + (5 - x)}{4} = \dfrac{5}{4}$ Add numerators.

b. $\dfrac{7}{2x - 3} - \dfrac{3x}{2x - 3} = \dfrac{7 - 3x}{2x - 3}$ Subtract numerators.

Example 2   Subtracting Rational Expressions and Simplifying

$$\dfrac{x}{x^2 - 2x - 3} - \dfrac{3}{x^2 - 2x - 3} = \dfrac{x - 3}{x^2 - 2x - 3}$$ Subtract numerators.

$$= \dfrac{(1)(x - 3)}{(x - 3)(x + 1)}$$ Factor.

$$= \dfrac{1}{x + 1}, \quad x \neq 3$$ Simplified form

Study Tip

After adding or subtracting two (or more) rational expressions, check the resulting fraction to see if it can be simplified, as illustrated in Example 2.

The rules for adding and subtracting rational expressions with like denominators can be extended to cover sums and differences involving three or more rational expressions, as illustrated in Example 3.

Example 3   Combining Three Rational Expressions

$$\frac{x^2 - 26}{x - 5} - \frac{2x + 4}{x - 5} + \frac{10 + x}{x - 5}$$ Original expressions

$$= \frac{(x^2 - 26) - (2x + 4) + (10 + x)}{x - 5}$$ Write numerator over common denominator.

$$= \frac{x^2 - 26 - 2x - 4 + 10 + x}{x - 5}$$ Distributive Property

$$= \frac{x^2 - x - 20}{x - 5}$$ Combine like terms.

$$= \frac{(x - 5)(x + 4)}{x - 5}$$ Factor and divide out common factor.

$$= x + 4, \quad x \neq 5$$ Simplified form

2 Add or subtract rational expressions with unlike denominators and simplify.

Adding or Subtracting with Unlike Denominators

To add or subtract rational expressions with *unlike* denominators, you must first rewrite each expression using the **least common multiple (LCM)** of the denominators of the individual expressions. The least common multiple of two (or more) polynomials is the simplest polynomial that is a multiple of each of the original polynomials. This means that the LCM must contain all the *different* factors in the polynomials and each of these factors must be repeated the maximum number of times it occurs in any one of the polynomials.

Example 4   Finding Least Common Multiples

a. The least common multiple of

$$6x = 2 \cdot 3 \cdot x, \quad 2x^2 = 2 \cdot x \cdot x, \quad \text{and} \quad 9x^3 = 3 \cdot 3 \cdot x \cdot x \cdot x$$

is $2 \cdot 3 \cdot 3 \cdot x \cdot x \cdot x = 18x^3$.

b. The least common multiple of

$$x^2 - x = x(x - 1) \quad \text{and} \quad 2x - 2 = 2(x - 1)$$

is $2x(x - 1)$.

c. The least common multiple of

$$3x^2 + 6x = 3x(x + 2) \quad \text{and} \quad x^2 + 4x + 4 = (x + 2)^2$$

is $3x(x + 2)^2$.

To add or subtract rational expressions with *unlike* denominators, you must first rewrite the rational expressions so that they have *like* denominators. The like denominator that you use is the least common multiple of the original denominators and is called the **least common denominator (LCD)** of the original rational expressions. Once the rational expressions have been written with like denominators, you can simply add or subtract these rational expressions using the rules given at the beginning of this section.

Example 5   Adding with Unlike Denominators

Add the rational expressions: $\dfrac{7}{6x} + \dfrac{5}{8x}$.

Solution

By factoring the denominators, $6x = 2 \cdot 3 \cdot x$ and $8x = 2^3 \cdot x$, you can conclude that the least common denominator is $2^3 \cdot 3 \cdot x = 24x$.

$$\dfrac{7}{6x} + \dfrac{5}{8x} = \dfrac{7(4)}{6x(4)} + \dfrac{5(3)}{8x(3)}$$ Rewrite expressions using LCD of $24x$.

$$= \dfrac{28}{24x} + \dfrac{15}{24x}$$ Like denominators

$$= \dfrac{28 + 15}{24x} = \dfrac{43}{24x}$$ Add fractions and simplify.

Example 6   Subtracting with Unlike Denominators

Subtract the rational expressions: $\dfrac{3}{x - 3} - \dfrac{5}{x + 2}$.

Solution

The only factors of the denominators are $x - 3$ and $x + 2$. So, the least common denominator is $(x - 3)(x + 2)$.

$$\dfrac{3}{x - 3} - \dfrac{5}{x + 2}$$ Write original expressions.

$$= \dfrac{3(x + 2)}{(x - 3)(x + 2)} - \dfrac{5(x - 3)}{(x - 3)(x + 2)}$$ Rewrite expressions using LCD of $(x - 3)(x + 2)$.

$$= \dfrac{3x + 6}{(x - 3)(x + 2)} - \dfrac{5x - 15}{(x - 3)(x + 2)}$$ Distributive Property

$$= \dfrac{(3x + 6) - (5x - 15)}{(x - 3)(x + 2)}$$ Subtract fractions.

$$= \dfrac{3x + 6 - 5x + 15}{(x - 3)(x + 2)}$$ Distributive Property

$$= \dfrac{-2x + 21}{(x - 3)(x + 2)}$$ Simplified form

Technology: Tip

You can use a graphing calculator to check your results when adding or subtracting rational expressions. For instance, in Example 5, try graphing the equations

$$y_1 = \dfrac{7}{6x} + \dfrac{5}{8x}$$

and

$$y_2 = \dfrac{43}{24x}$$

in the same viewing window. If the two graphs coincide, as shown below, you can conclude that the solution checks.

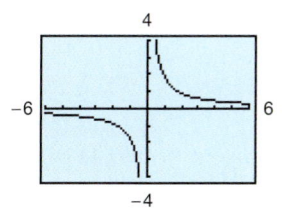

Study Tip

In Example 7, notice that the denominator $2 - x$ is rewritten as $(-1)(x - 2)$ and then the problem is changed from addition to subtraction.

Example 7   Adding with Unlike Denominators

$$\frac{6x}{x^2 - 4} + \frac{3}{2 - x}$$ Original expressions

$$= \frac{6x}{(x + 2)(x - 2)} + \frac{3}{(-1)(x - 2)}$$ Factor denominators.

$$= \frac{6x}{(x + 2)(x - 2)} - \frac{3(x + 2)}{(x + 2)(x - 2)}$$ Rewrite expressions using LCD of $(x + 2)(x - 2)$.

$$= \frac{6x}{(x + 2)(x - 2)} - \frac{3x + 6}{(x + 2)(x - 2)}$$ Distributive Property

$$= \frac{6x - (3x + 6)}{(x + 2)(x - 2)}$$ Subtract.

$$= \frac{6x - 3x - 6}{(x + 2)(x - 2)}$$ Distributive Property

$$= \frac{3x - 6}{(x + 2)(x - 2)}$$ Simplify.

$$= \frac{3(x - 2)}{(x + 2)(x - 2)}$$ Factor and divide out common factor.

$$= \frac{3}{x + 2}, \quad x \neq 2$$ Simplified form

Example 8   Subtracting with Unlike Denominators

$$\frac{x}{x^2 - 5x + 6} - \frac{1}{x^2 - x - 2}$$ Original expressions

$$= \frac{x}{(x - 3)(x - 2)} - \frac{1}{(x - 2)(x + 1)}$$ Factor denominators.

$$= \frac{x(x + 1)}{(x - 3)(x - 2)(x + 1)} - \frac{1(x - 3)}{(x - 3)(x - 2)(x + 1)}$$ Rewrite expressions using LCD of $(x - 3)(x - 2)(x + 1)$.

$$= \frac{x^2 + x}{(x - 3)(x - 2)(x + 1)} - \frac{x - 3}{(x - 3)(x - 2)(x + 1)}$$ Distributive Property

$$= \frac{(x^2 + x) - (x - 3)}{(x - 3)(x - 2)(x + 1)}$$ Subtract fractions.

$$= \frac{x^2 + x - x + 3}{(x - 3)(x - 2)(x + 1)}$$ Distributive Property

$$= \frac{x^2 + 3}{(x - 3)(x - 2)(x + 1)}$$ Simplified form

Example 9   Combining Three Rational Expressions

$$\frac{4x}{x^2 - 16} + \frac{x}{x + 4} - \frac{2}{x} = \frac{4x}{(x + 4)(x - 4)} + \frac{x}{x + 4} - \frac{2}{x}$$

$$= \frac{4x(x)}{x(x + 4)(x - 4)} + \frac{x(x)(x - 4)}{x(x + 4)(x - 4)} - \frac{2(x + 4)(x - 4)}{x(x + 4)(x - 4)}$$

$$= \frac{4x^2 + x^2(x - 4) - 2(x^2 - 16)}{x(x + 4)(x - 4)}$$

$$= \frac{4x^2 + x^3 - 4x^2 - 2x^2 + 32}{x(x + 4)(x - 4)}$$

$$= \frac{x^3 - 2x^2 + 32}{x(x + 4)(x - 4)}$$

To add or subtract two rational expressions, you can use the LCD method or the basic definition

$$\frac{a}{b} \pm \frac{c}{d} = \frac{ad \pm bc}{bd}, \quad b \neq 0, d \neq 0.$$ Basic definition

This definition provides an efficient way of adding or subtracting two rational expressions that have no common factors in their denominators.

Example 10   Dog Registrations

For the years 1997 through 2001, the number of rottweilers R (in thousands) and the number of collies C (in thousands) registered with the American Kennel Club can be modeled by

$$R = \frac{-1.9t^2 + 3726}{t^2} \quad \text{and} \quad C = \frac{-1.04t + 5}{-0.18t + 1}, \quad 7 \leq t \leq 11$$

where t represents the year, with $t = 7$ corresponding to 1997. Find a rational model T for the number of rottweilers and collies registered with the American Kennel Club.   (Source: American Kennel Club)

Solution

To find a model for T, find the sum of R and C.

$$T = \frac{-1.9t^2 + 3726}{t^2} + \frac{(-1.04t + 5)}{(-0.18t + 1)}$$ Sum of R and C.

$$= \frac{(-1.9t^2 + 3726)(-0.18t + 1) + t^2(-1.04t + 5)}{t^2(-0.18t + 1)}$$ Basic definition

$$= \frac{0.342t^3 - 1.9t^2 - 670.68t + 3726 - 1.04t^3 + 5t^2}{t^2(-0.18t + 1)}$$ FOIL Method and Distributive Property

$$= \frac{-0.698t^3 + 3.1t^2 - 670.68t + 3726}{t^2(-0.18t + 1)}$$ Combine like terms.

7.3 Exercises

Review   Concepts, Skills, and Problem Solving

Keep mathematically in shape by doing these exercises *before* the problems of this section.

Properties and Definitions

1. Write the equation $5y - 3x - 4 = 0$ in the following forms.

    (a) Slope-intercept form

    (b) Point-slope form (many correct answers)

2. *Writing* Explain how you can visually determine the sign of the slope of a line by observing its graph.

Simplifying Expressions

In Exercises 3–10, perform the multiplication and simplify.

3. $-6x(10 - 7x)$ **4.** $(2 - y)(3 + 2y)$

5. $(11 - x)(11 + x)$ **6.** $(4 - 5z)(4 + 5z)$

7. $(x + 1)^2$ **8.** $t(t^2 + 1) - t(t^2 - 1)$

9. $(x - 2)(x^2 + 2x + 4)$

10. $t(t - 4)(2t + 3)$

Creating Expressions

▲ *Geometry*   In Exercises 11 and 12, write and simplify expressions for the perimeter and area of the figure.

11.

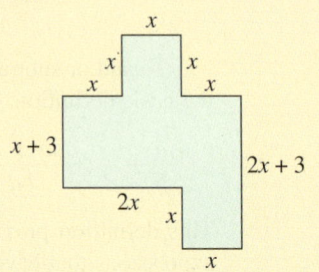

12.

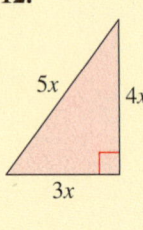

Developing Skills

In Exercises 1–16, combine and simplify. See Examples 1–3.

1. $\dfrac{5x}{8} - \dfrac{7x}{8}$ **2.** $\dfrac{7y}{12} + \dfrac{9y}{12}$

3. $\dfrac{2}{3a} - \dfrac{11}{3a}$ **4.** $\dfrac{6}{19x} - \dfrac{7}{19x}$

5. $\dfrac{x}{9} - \dfrac{x + 2}{9}$ **6.** $\dfrac{4 - y}{4} + \dfrac{3y}{4}$

7. $\dfrac{z^2}{3} + \dfrac{z^2 - 2}{3}$ **8.** $\dfrac{10x^2 + 1}{3} - \dfrac{10x^2}{3}$

9. $\dfrac{2x + 5}{3x} + \dfrac{1 - x}{3x}$ **10.** $\dfrac{16 + z}{5z} - \dfrac{11 - z}{5z}$

11. $\dfrac{3y}{3} - \dfrac{3y - 3}{3} - \dfrac{7}{3}$ **12.** $\dfrac{-16u}{9} - \dfrac{27 - 16u}{9} + \dfrac{2}{9}$

13. $\dfrac{3y - 22}{y - 6} - \dfrac{2y - 16}{y - 6}$

14. $\dfrac{5x - 1}{x + 4} + \dfrac{5 - 4x}{x + 4}$

15. $\dfrac{2x - 1}{x(x - 3)} + \dfrac{1 - x}{x(x - 3)}$

16. $\dfrac{7s - 5}{2s + 5} + \dfrac{3(s + 10)}{2s + 5}$

In Exercises 17–28, find the least common multiple of the expressions. See Example 4.

17. $5x^2, 20x^3$ **18.** $14t^2, 42t^5$

19. $9y^3, 12y$ **20.** $44m^2, 10m$

21. $15x^2, 3(x + 5)$ **22.** $6x^2, 15x(x - 1)$

23. $63z^2(z + 1), 14(z + 1)^4$

24. $18y^3, 27y(y - 3)^2$

25. $8t(t + 2), 14(t^2 - 4)$

26. $2y^2 + y - 1, 4y^2 - 2y$

27. $6(x^2 - 4), 2x(x + 2)$

28. $t^3 + 3t^2 + 9t, 2t^2(t^2 - 9)$

41. $\dfrac{x - 8}{x^2 - 25}, \dfrac{9x}{x^2 - 10x + 25}$

42. $\dfrac{3y}{y^2 - y - 12}, \dfrac{y - 4}{y^2 + 3y}$

In Exercises 29–34, fill in the missing factor.

29. $\dfrac{7x^2}{4a()} = \dfrac{7}{4a}, \quad x \neq 0$

30. $\dfrac{3y(x - 3)^2}{(x - 3)()} = \dfrac{21y}{x - 3}$

31. $\dfrac{5r()}{3v(u + 1)} = \dfrac{5r}{3v}, \quad u \neq -1$

32. $\dfrac{(3t + 5)()}{10t^2(3t - 5)} = \dfrac{3t + 5}{2t}, \quad t \neq \dfrac{5}{3}$

33. $\dfrac{7y()}{4 - x^2} = \dfrac{7y}{x - 2}, \quad x \neq -2$

34. $\dfrac{4x^2()}{x^2 - 10x} = \dfrac{4x^2}{10 - x}, \quad x \neq 0$

In Exercises 35–42, find the least common denominator of the two fractions and rewrite each fraction using the least common denominator.

35. $\dfrac{n + 8}{3n - 12}, \dfrac{10}{6n^2}$

36. $\dfrac{8s}{(s + 2)^2}, \dfrac{3}{s^3 + s^2 - 2s}$

37. $\dfrac{2}{x^2(x - 3)}, \dfrac{5}{x(x + 3)}$

38. $\dfrac{5t}{2t(t - 3)^2}, \dfrac{4}{t(t - 3)}$

39. $\dfrac{v}{2v^2 + 2v}, \dfrac{4}{3v^2}$

40. $\dfrac{4x}{(x + 5)^2}, \dfrac{x - 2}{x^2 - 25}$

In Exercises 43–76, combine and simplify. See Examples 5–9.

43. $\dfrac{5}{4x} - \dfrac{3}{5}$

44. $\dfrac{10}{b} + \dfrac{1}{10b}$

45. $\dfrac{7}{a} + \dfrac{14}{a^2}$

46. $\dfrac{1}{6u^2} - \dfrac{2}{9u}$

47. $\dfrac{20}{x - 4} + \dfrac{20}{4 - x}$

48. $\dfrac{15}{2 - t} - \dfrac{7}{t - 2}$

49. $\dfrac{3x}{x - 8} - \dfrac{6}{8 - x}$

50. $\dfrac{1}{y - 6} + \dfrac{y}{6 - y}$

51. $25 + \dfrac{10}{x + 4}$

52. $\dfrac{100}{x - 10} - 8$

53. $\dfrac{3x}{3x - 2} + \dfrac{2}{2 - 3x}$

54. $\dfrac{y}{5y - 3} - \dfrac{3}{3 - 5y}$

55. $\dfrac{7}{2x} + \dfrac{1}{x - 2}$

56. $\dfrac{3}{y - 1} + \dfrac{5}{4y}$

57. $\dfrac{x}{x + 3} - \dfrac{5}{x - 2}$

58. $\dfrac{1}{x + 4} - \dfrac{1}{x + 2}$

59. $\dfrac{12}{x^2 - 9} - \dfrac{2}{x - 3}$

60. $\dfrac{12}{x^2 - 4} - \dfrac{3}{x + 2}$

61. $\dfrac{3}{x - 5} + \dfrac{2}{x + 5}$

62. $\dfrac{7}{2x - 3} + \dfrac{3}{2x + 3}$

63. $\dfrac{4}{x^2} - \dfrac{4}{x^2 + 1}$

64. $\dfrac{2}{y^2 + 2} + \dfrac{1}{2y^2}$

65. $\dfrac{x}{x^2 - 9} + \dfrac{3}{x^2 - 5x + 6}$

66. $\dfrac{x}{x^2 - x - 30} - \dfrac{1}{x + 5}$

67. $\dfrac{4}{x - 4} + \dfrac{16}{(x - 4)^2}$

68. $\dfrac{3}{x - 2} - \dfrac{1}{(x - 2)^2}$

69. $\dfrac{y}{x^2 + xy} - \dfrac{x}{xy + y^2}$

70. $\dfrac{5}{x + y} + \dfrac{5}{x^2 - y^2}$

71. $\dfrac{4}{x} - \dfrac{2}{x^2} + \dfrac{4}{x + 3}$

72. $\dfrac{5}{2} - \dfrac{1}{2x} - \dfrac{3}{x + 1}$

73. $\dfrac{3u}{u^2 - 2uv + v^2} + \dfrac{2}{u - v} - \dfrac{u}{u - v}$

74. $\dfrac{1}{x - y} - \dfrac{3}{x + y} + \dfrac{3x - y}{x^2 - y^2}$

75. $\dfrac{x + 2}{x - 1} - \dfrac{2}{x + 6} - \dfrac{14}{x^2 + 5x - 6}$

76. $\dfrac{x}{x^2 + 15x + 50} + \dfrac{7}{x + 10} - \dfrac{x - 1}{x + 5}$

In Exercises 77 and 78, use a graphing calculator to graph the two equations in the same viewing window. Use the graphs to verify that the expressions are equivalent. Verify the results algebraically.

77. $y_1 = \dfrac{2}{x} + \dfrac{4}{x - 2}, \ y_2 = \dfrac{6x - 4}{x(x - 2)}$

78. $y_1 = 3 - \dfrac{1}{x - 1}, \ y_2 = \dfrac{3x - 4}{x - 1}$

Solving Problems

79. *Work Rate*   After working together for t hours on a common task, two workers have completed fractional parts of the job equal to $t/4$ and $t/6$. What fractional part of the task has been completed?

80. *Work Rate*   After working together for t hours on a common task, two workers have completed fractional parts of the job equal to $t/3$ and $t/5$. What fractional part of the task has been completed?

81. *Rewriting a Fraction*   The fraction $4/(x^3 - x)$ can be rewritten as a sum of three fractions, as follows.

$$\dfrac{4}{x^3 - x} = \dfrac{A}{x} + \dfrac{B}{x + 1} + \dfrac{C}{x - 1}$$

The numbers A, B, and C are the solutions of the system

$$\begin{cases} A + B + C = 0 \\ \ \ \ \ \ \ - B + C = 0 \\ -A \ \ \ \ \ \ \ \ \ \ = 4. \end{cases}$$

Solve the system and verify that the sum of the three resulting fractions is the original fraction.

82. *Rewriting a Fraction*   The fraction

$$\dfrac{x + 1}{x^3 - x^2}$$

can be rewritten as a sum of three fractions, as follows.

$$\dfrac{x + 1}{x^3 - x^2} = \dfrac{A}{x} + \dfrac{B}{x^2} + \dfrac{C}{x - 1}$$

The numbers A, B, and C are the solutions of the system

$$\begin{cases} A \ \ \ \ \ \ \ + C = 0 \\ -A + B \ \ \ \ \ \ = 1 \\ \ \ \ \ \ - B \ \ \ \ \ \ = 1. \end{cases}$$

Solve the system and verify that the sum of the three resulting fractions is the original fraction.

Sports   In Exercises 83 and 84, use the following models, which give the number of males M (in thousands) and the number of females F (in thousands) participating in high school athletic programs from 1995 through 2001.

$$M = \frac{463.76t + 2911.4}{0.09t + 1.0} \quad \text{and} \quad F = \frac{3183.41t - 4827.2}{t}$$

In these models, t represents the year, with $t = 5$ corresponding to 1995. (Source: 2001 High School Participation Survey)

83. Find a rational model that represents the total number of participants in high school athletic programs.

84. Use the model you found in Exercise 83 to complete the table showing the total number of participants in high school athletic programs.

Year	1995	1996	1997	1998
Participants				

Year	1999	2000	2001
Participants			

Explaining Concepts

85. ✪ Answer parts (a)–(c) of Motivating the Chapter on page 398.

86. *Writing* ✏ In your own words, describe how to add or subtract rational expressions with like denominators.

87. *Writing* ✏ In your own words, describe how to add or subtract rational expressions with unlike denominators.

88. *Writing* ✏ Is it possible for the least common denominator of two fractions to be the same as one of the fraction's denominators? If so, give an example.

89. *Error Analysis* Describe the error.

$$\frac{x - 1}{x + 4} - \frac{4x - 11}{x + 4} = \frac{x - 1 - 4x - 11}{x + 4}$$
$$= \frac{-3x - 12}{x + 4} = \frac{-3(x + 4)}{x + 4}$$
$$= -3$$

90. *Error Analysis*   Describe the error.

$$\frac{2}{x} - \frac{3}{x + 1} + \frac{x + 1}{x^2}$$
$$= \frac{2x(x + 1) - 3x^2 + (x + 1)^2}{x^2(x + 1)}$$
$$= \frac{2x^2 + x - 3x^2 + x^2 + 1}{x^2(x + 1)}$$
$$= \frac{x + 1}{x^2(x + 1)} = \frac{1}{x^2}$$

91. *Writing* ✏   Evaluate each expression at the given value of the variable in two different ways: (1) combine and simplify the rational expressions first and then evaluate the simplified expression at the given variable value, and (2) substitute the given value of the variable first and then simplify the resulting expression. Do you get the same result with each method? Discuss which method you prefer and why. List any advantages and/or disadvantages of each method.

(a) $\dfrac{1}{m - 4} - \dfrac{1}{m + 4} + \dfrac{3m}{m^2 - 16}$, $m = 2$

(b) $\dfrac{x - 2}{x^2 - 9} + \dfrac{3x + 2}{x^2 - 5x + 6}$, $x = 4$

(c) $\dfrac{3y^2 + 16y - 8}{y^2 + 2y - 8} - \dfrac{y - 1}{y - 2} + \dfrac{y}{y + 4}$, $y = 3$

Mid-Chapter Quiz

Take this quiz as you would take a quiz in class. After you are done, check your work against the answers in the back of the book.

1. Determine the domain of $f(y) = \dfrac{y + 2}{y(y - 4)}$.

2. Evaluate $f(x) = \dfrac{2x - 1}{x^2 + 1}$ for the indicated values of x and simplify.
   If it is not possible, state the reason.
   (a) $f(3)$ (b) $f(1)$ (c) $f(-1)$ (d) $f(\tfrac{1}{2})$

3. Evaluate $h(x) = (x^2 - 9)/(x^2 - x - 2)$ for the indicated values of x and simplify. If it is not possible, state the reason.
   (a) $h(-3)$ (b) $h(0)$ (c) $h(-1)$ (d) $h(5)$

In Exercises 4–9, simplify the rational expression.

4. $\dfrac{9y^2}{6y}$

5. $\dfrac{8u^3v^2}{36uv^3}$

6. $\dfrac{4x^2 - 1}{x - 2x^2}$

7. $\dfrac{(z + 3)^2}{2z^2 + 5z - 3}$

8. $\dfrac{7ab + 3a^2b^2}{a^2b}$

9. $\dfrac{2mn^2 - n^3}{2m^2 + mn - n^2}$

In Exercises 10–20, perform the indicated operations and simplify.

10. $\dfrac{11t^2}{6} \cdot \dfrac{9}{33t}$

11. $(x^2 + 2x) \cdot \dfrac{5}{x^2 - 4}$

12. $\dfrac{4}{3(x - 1)} \cdot \dfrac{12x}{6(x^2 + 2x - 3)}$

13. $\dfrac{80z^4}{49x^5y^7} \div \dfrac{25z^5}{14x^{12}y^5}$

14. $\dfrac{a - b}{9a + 9b} \div \dfrac{a^2 - b^2}{a^2 + 2a + 1}$

15. $\dfrac{10}{x^2 + 2x} \div \dfrac{15}{x^2 + 3x + 2}$

16. $\dfrac{3x}{x + 5} \cdot \dfrac{x + 4x - 5}{x^2} \div \dfrac{x - 1}{2x}$

17. $\dfrac{5u}{3(u + v)} \cdot \dfrac{2(u^2 - v^2)}{3v} \div \dfrac{25u^2}{18(u - v)}$

18. $\dfrac{5x - 6}{x - 2} + \dfrac{2x - 5}{x - 2}$

19. $\dfrac{x}{x^2 - 9} - \dfrac{4(x - 3)}{x + 3}$

20. $\dfrac{x^2 + 2}{x^2 - x - 2} + \dfrac{1}{x + 1} - \dfrac{x}{x - 2}$

21. You open a floral shop with a setup cost of $25,000. The cost of creating one dozen floral arrangements is $144.

    (a) Write the total cost C as a function of x, the number of floral arrangements (in dozens) created.

    (b) Write the average cost per dozen $\overline{C} = C/x$ as a function of x, the number of floral arrangements (in dozens) created.

7.4 Complex Fractions

David Forbert/SuperStock, Inc.

What You Should Learn

1 Simplify complex fractions using rules for dividing rational expressions.

2 Simplify complex fractions having a sum or difference in the numerator and/or denominator.

Why You Should Learn It

Complex fractions can be used to model real-life situations. For instance, in Exercise 66 on page 438, a complex fraction is used to model the annual percent rate for a home-improvement loan.

1 Simplify complex fractions using rules for dividing rational expressions.

Complex Fractions

Problems involving the division of two rational expressions are sometimes written as **complex fractions.** A complex fraction is a fraction that has a fraction in its numerator or denominator, or both. The rules for dividing rational expressions still apply. For instance, consider the following complex fraction.

$$\dfrac{\left(\dfrac{x+2}{3}\right)}{\left(\dfrac{x-2}{x}\right)} \longrightarrow$$

Numerator fraction

Main fraction line

Denominator fraction

To perform the division implied by this complex fraction, invert the denominator fraction and multiply, as follows.

$$\dfrac{\left(\dfrac{x+2}{3}\right)}{\left(\dfrac{x-2}{x}\right)} = \dfrac{x+2}{3} \cdot \dfrac{x}{x-2}$$

$$= \dfrac{x(x+2)}{3(x-2)}, \quad x \neq 0$$

Note that for complex fractions you make the main fraction line slightly longer than the fraction lines in the numerator and denominator.

Example 1   Simplifying a Complex Fraction

$$\dfrac{\left(\dfrac{5}{14}\right)}{\left(\dfrac{25}{8}\right)} = \dfrac{5}{14} \cdot \dfrac{8}{25}$$ Invert divisor and multiply.

$$= \dfrac{5 \cdot 2 \cdot 2 \cdot 2}{2 \cdot 7 \cdot 5 \cdot 5}$$ Multiply, factor, and divide out common factors.

$$= \dfrac{4}{35}$$ Simplified form

Example 2 Simplifying a Complex Fraction

Simplify the complex fraction.

$$\frac{\left(\dfrac{4y^3}{(5x)^2}\right)}{\left(\dfrac{(2y)^2}{10x^3}\right)}$$

Solution

$$\frac{\left(\dfrac{4y^3}{(5x)^2}\right)}{\left(\dfrac{(2y)^2}{10x^3}\right)} = \frac{4y^3}{25x^2} \cdot \frac{10x^3}{4y^2}$$ Invert divisor and multiply.

$$= \frac{4y^2 \cdot y \cdot 2 \cdot 5x^2 \cdot x}{5 \cdot 5x^2 \cdot 4y^2}$$ Multiply and factor.

$$= \frac{\cancel{4y^2} \cdot y \cdot 2 \cdot \cancel{5x^2} \cdot x}{5 \cdot \cancel{5x^2} \cdot \cancel{4y^2}}$$ Divide out common factors.

$$= \frac{2xy}{5}, \quad x \neq 0, \ y \neq 0$$ Simplified form

Example 3 Simplifying a Complex Fraction

Simplify the complex fraction.

$$\frac{\left(\dfrac{x+1}{x+2}\right)}{\left(\dfrac{x+1}{x+5}\right)}$$

Solution

$$\frac{\left(\dfrac{x+1}{x+2}\right)}{\left(\dfrac{x+1}{x+5}\right)} = \frac{x+1}{x+2} \cdot \frac{x+5}{x+1}$$ Invert divisor and multiply.

$$= \frac{(x+1)(x+5)}{(x+2)(x+1)}$$ Multiply numerators and denominators.

$$= \frac{\cancel{(x+1)}(x+5)}{(x+2)\cancel{(x+1)}}$$ Divide out common factors.

$$= \frac{x+5}{x+2}, \quad x \neq -1, \ x \neq -5$$ Simplified form

In Example 3, the domain of the complex fraction is restricted by the denominators in the original expression and by the denominators in the original expression after the divisor has been inverted. So, the domain of the original expression is all real values of x such that $x \neq -2$, $x \neq -5$, and $x \neq -1$.

Example 4   Simplifying a Complex Fraction

Simplify the complex fraction.

$$\frac{\left(\dfrac{x^2 + 4x + 3}{x - 2}\right)}{2x + 6}$$

Solution

Begin by writing the denominator in fractional form.

$$\frac{\left(\dfrac{x^2 + 4x + 3}{x - 2}\right)}{2x + 6} = \frac{\left(\dfrac{x^2 + 4x + 3}{x - 2}\right)}{\left(\dfrac{2x + 6}{1}\right)} \qquad \text{Rewrite denominator.}$$

$$= \frac{x^2 + 4x + 3}{x - 2} \cdot \frac{1}{2x + 6} \qquad \text{Invert divisor and multiply.}$$

$$= \frac{(x + 1)(x + 3)}{(x - 2)(2)(x + 3)} \qquad \text{Multiply and factor.}$$

$$= \frac{(x + 1)(x + 3)}{(x - 2)(2)(x + 3)} \qquad \text{Divide out common factor.}$$

$$= \frac{x + 1}{2(x - 2)}, \quad x \neq -3 \qquad \text{Simplified form}$$

2 Simplify complex fractions having a sum or difference in the numerator and/or denominator.

Study Tip

Another way of simplifying the complex fraction in Example 5 is to multiply the numerator and denominator by the least common denominator for all fractions in the numerator and denominator. For this fraction, when you multiply the numerator and denominator by $3x$, you obtain the same result.

$$\frac{\left(\dfrac{x}{3} + \dfrac{2}{3}\right)}{\left(1 - \dfrac{2}{x}\right)} = \frac{\left(\dfrac{x}{3} + \dfrac{2}{3}\right)}{\left(1 - \dfrac{2}{x}\right)} \cdot \frac{3x}{3x}$$

$$= \frac{\dfrac{x}{3}(3x) + \dfrac{2}{3}(3x)}{(1)(3x) - \dfrac{2}{x}(3x)}$$

$$= \frac{x^2 + 2x}{3x - 6}$$

$$= \frac{x(x + 2)}{3(x - 2)}, \quad x \neq 0$$

Complex Fractions with Sums or Differences

Complex fractions can have numerators and/or denominators that are sums or differences of fractions. To simplify a complex fraction, combine its numerator and its denominator into single fractions. Then divide by inverting the denominator and multiplying.

Example 5   Simplifying a Complex Fraction

$$\frac{\left(\dfrac{x}{3} + \dfrac{2}{3}\right)}{\left(1 - \dfrac{2}{x}\right)} = \frac{\left(\dfrac{x}{3} + \dfrac{2}{3}\right)}{\left(\dfrac{x}{x} - \dfrac{2}{x}\right)} \qquad \text{Rewrite with least common denominators.}$$

$$= \frac{\left(\dfrac{x + 2}{3}\right)}{\left(\dfrac{x - 2}{x}\right)} \qquad \text{Add fractions.}$$

$$= \frac{x + 2}{3} \cdot \frac{x}{x - 2} \qquad \text{Invert divisor and multiply.}$$

$$= \frac{x(x + 2)}{3(x - 2)}, \quad x \neq 0 \qquad \text{Simplified form}$$

Example 6   Simplifying a Complex Fraction

$$\frac{\left(\dfrac{2}{x + 2}\right)}{\left(\dfrac{3}{x + 2} + \dfrac{2}{x}\right)} = \frac{\left(\dfrac{2}{x + 2}\right)(x)(x + 2)}{\left(\dfrac{3}{x + 2}\right)(x)(x + 2) + \left(\dfrac{2}{x}\right)(x)(x + 2)}$$ $x(x + 2)$ is the least common denominator.

$$= \frac{2x}{3x + 2(x + 2)}$$ Multiply and simplify.

$$= \frac{2x}{3x + 2x + 4}$$ Distributive Property

$$= \frac{2x}{5x + 4}, \quad x \neq -2, \quad x \neq 0$$ Simplify.

Notice that the numerator and denominator of the complex fraction were multiplied by $(x)(x + 2)$, which is the least common denominator of the fractions in the original complex fraction.

When simplifying a rational expression containing negative exponents, first rewrite the expression with positive exponents and then proceed with simplifying the expression. This is demonstrated in Example 7.

Example 7   Simplifying a Complex Fraction

$$\frac{5 + x^{-2}}{8x^{-1} + x} = \frac{\left(5 + \dfrac{1}{x^2}\right)}{\left(\dfrac{8}{x} + x\right)}$$ Rewrite with positive exponents.

$$= \frac{\left(\dfrac{5x^2}{x^2} + \dfrac{1}{x^2}\right)}{\left(\dfrac{8}{x} + \dfrac{x^2}{x}\right)}$$ Rewrite with least common denominators.

$$= \frac{\left(\dfrac{5x^2 + 1}{x^2}\right)}{\left(\dfrac{x^2 + 8}{x}\right)}$$ Add fractions.

$$= \frac{5x^2 + 1}{x^2} \cdot \frac{x}{x^2 + 8}$$ Invert divisor and multiply.

$$= \frac{\cancel{x}(5x^2 + 1)}{\cancel{x}(x)(x^2 + 8)}$$ Divide out common factor.

$$= \frac{5x^2 + 1}{x(x^2 + 8)}$$ Simplified form

7.4 Exercises

Review   Concepts, Skills, and Problem Solving

Keep mathematically in shape by doing these exercises *before* the problems of this section.

Properties and Definitions

1. *Writing* ✎   In your own words, explain how to use the zero and negative exponent rules.

2. *Writing* ✎   Explain how to determine the exponent when writing 0.00000237 in scientific notation.

Simplifying Expressions

In Exercises 3–6, simplify the expression. (Assume that any variables in the expression are nonzero.)

3. $(7x^2)(2x^{-3})$

4. $(y^0z^3)(z^2)^{-4}$

5. $\dfrac{a^4b^{-2}}{a^{-1}b^5}$

6. $(x + 2)^4 \div (x + 2)^3$

Graphing Equations

In Exercises 7–10, sketch a graph of the equation and label the coordinates of at least three solution points.

7. $y = 3x - 1$

8. $y = 4x + 9$

9. $y = |x + 2|$

10. $y = |1 - x|$

Problem Solving

11. *Number Problem*   The sum of three positive numbers is 33. The second number is three greater than the first, and the third is four times the first. Find the three numbers.

12. *Nut Mixture*   A grocer wishes to mix peanuts costing $3 per pound, almonds costing $4 per pound, and pistachios costing $6 per pound to make a 50-pound mixture priced at $4.10 per pound. Three-quarters of the mixture should be composed of peanuts and almonds. How many pounds of each type of nut should the grocer use in the mixture?

Developing Skills

In Exercises 1–26, simplify the complex fraction. See Examples 1–4.

1. $\dfrac{\left(\dfrac{x^3}{4}\right)}{\left(\dfrac{x}{8}\right)}$

2. $\dfrac{\left(\dfrac{y^4}{12}\right)}{\left(\dfrac{y}{16}\right)}$

3. $\dfrac{\left(\dfrac{x^2}{12}\right)}{\left(\dfrac{5x}{18}\right)}$

4. $\dfrac{\left(\dfrac{3u^2}{6v^3}\right)}{\left(\dfrac{u}{3v}\right)}$

5. $\dfrac{\left(\dfrac{8x^2y}{3z^2}\right)}{\left(\dfrac{4xy}{9z^5}\right)}$

6. $\dfrac{\left(\dfrac{36x^4}{5y^4z^5}\right)}{\left(\dfrac{9xy^2}{20z^5}\right)}$

7. $\dfrac{\left(\dfrac{6x^3}{(5y)^2}\right)}{\left(\dfrac{(3x)^2}{15y^4}\right)}$

8. $\dfrac{\left(\dfrac{(3r)^3}{10t^4}\right)}{\left(\dfrac{9r}{(2t)^2}\right)}$

9. $\dfrac{\left(\dfrac{y}{3 - y}\right)}{\left(\dfrac{y^2}{y - 3}\right)}$

10. $\dfrac{\left(\dfrac{x}{x - 4}\right)}{\left(\dfrac{x}{4 - x}\right)}$

11. $\dfrac{\left(\dfrac{25x^2}{x-5}\right)}{\left(\dfrac{10x}{5+4x-x^2}\right)}$

12. $\dfrac{\left(\dfrac{5x}{x+7}\right)}{\left(\dfrac{10}{x^2+8x+7}\right)}$

13. $\dfrac{\left(\dfrac{2x-10}{x+1}\right)}{\left(\dfrac{x-5}{x+1}\right)}$

14. $\dfrac{\left(\dfrac{a+5}{6a-15}\right)}{\left(\dfrac{a+5}{2a-5}\right)}$

15. $\dfrac{\left(\dfrac{x^2+3x-10}{x+4}\right)}{3x-6}$

16. $\dfrac{\left(\dfrac{x^2-2x-8}{x-1}\right)}{5x-20}$

17. $\dfrac{2x-14}{\left(\dfrac{x^2-9x+14}{x+3}\right)}$

18. $\dfrac{4x+16}{\left(\dfrac{x^2+9x+20}{x-1}\right)}$

19. $\dfrac{\left(\dfrac{6x^2-17x+5}{3x^2+3x}\right)}{\left(\dfrac{3x-1}{3x+1}\right)}$

20. $\dfrac{\left(\dfrac{6x^2-13x-5}{5x^2+5x}\right)}{\left(\dfrac{2x-5}{5x+1}\right)}$

21. $\dfrac{16x^2+8x+1}{3x^2+8x-3} \div \dfrac{4x^2-3x-1}{x^2+6x+9}$

22. $\dfrac{9x^2-24x+16}{x^2+10x+25} \div \dfrac{6x^2-5x-4}{2x^2+3x-35}$

23. $\dfrac{x^2+3x-2x-6}{x^2-4} \div \dfrac{x+3}{x^2+4x+4}$

24. $\dfrac{t^3+t^2-9t-9}{t^2-5t+6} \div \dfrac{t^2+6t+9}{t-2}$

25. $\dfrac{\left(\dfrac{x^2-3x-10}{x^2-4x+4}\right)}{\left(\dfrac{21+4x-x^2}{x^2-5x-14}\right)}$

26. $\dfrac{\left(\dfrac{x^2+5x+6}{4x^2-20x+25}\right)}{\left(\dfrac{x^2-5x-24}{4x^2-25}\right)}$

In Exercises 27–46, simplify the complex fraction. See Examples 5 and 6.

27. $\dfrac{\left(1+\dfrac{3}{y}\right)}{y}$

28. $\dfrac{x}{\left(\dfrac{5}{x}+2\right)}$

29. $\dfrac{\left(\dfrac{x}{2}\right)}{\left(2+\dfrac{3}{x}\right)}$

30. $\dfrac{\left(1-\dfrac{2}{x}\right)}{\left(\dfrac{x}{2}\right)}$

31. $\dfrac{\left(\dfrac{4}{x}+3\right)}{\left(\dfrac{4}{x}-3\right)}$

32. $\dfrac{\left(\dfrac{1}{t}-1\right)}{\left(\dfrac{1}{t}+1\right)}$

33. $\dfrac{\left(3+\dfrac{9}{x-3}\right)}{\left(4+\dfrac{12}{x-3}\right)}$

34. $\dfrac{\left(x+\dfrac{2}{x-3}\right)}{\left(x+\dfrac{6}{x-3}\right)}$

35. $\dfrac{\left(\dfrac{3}{x^2} + \dfrac{1}{x}\right)}{\left(2 - \dfrac{4}{5x}\right)}$

36. $\dfrac{\left(16 - \dfrac{1}{x^2}\right)}{\left(\dfrac{1}{4x^2} - 4\right)}$

In Exercises 47–54, simplify the expression. See Example 7.

47. $\dfrac{2y - y^{-1}}{10 - y^{-2}}$

48. $\dfrac{9x - x^{-1}}{3 + x^{-1}}$

49. $\dfrac{7x^2 + 2x^{-1}}{5x^{-3} + x}$ **50.** $\dfrac{3x^{-2} - x}{4x^{-1} + 6x}$

37. $\dfrac{\left(\dfrac{y}{x} - \dfrac{x}{y}\right)}{\left(\dfrac{x + y}{xy}\right)}$

38. $\dfrac{\left(x - \dfrac{2y^2}{x - y}\right)}{x - 2y}$

51. $\dfrac{x^{-1} + y^{-1}}{x^{-1} - y^{-1}}$

52. $\dfrac{x^{-1} - y^{-1}}{x^{-2} - y^{-2}}$

39. $\dfrac{\left(1 - \dfrac{1}{y}\right)}{\left(\dfrac{1 - 4y}{y - 3}\right)}$

40. $\dfrac{\left(\dfrac{x + 1}{x + 2} - \dfrac{1}{x}\right)}{\left(\dfrac{2}{x + 2}\right)}$

53. $\dfrac{x^{-2} - y^{-2}}{(x + y)^2}$

54. $\dfrac{x - y}{x^{-2} - y^{-2}}$

41. $\dfrac{\left(\dfrac{10}{x + 1}\right)}{\left(\dfrac{1}{2x + 2} + \dfrac{3}{x + 1}\right)}$

42. $\dfrac{\left(\dfrac{2}{x + 5}\right)}{\left(\dfrac{2}{x + 5} + \dfrac{1}{4x + 20}\right)}$

In Exercises 55 and 56, use the function to find and simplify the expression for

$$\dfrac{f(2 + h) - f(2)}{h}.$$

55. $f(x) = \dfrac{1}{x}$ **56.** $f(x) = \dfrac{x}{x - 1}$

43. $\dfrac{\left(\dfrac{1}{x} - \dfrac{1}{x + 1}\right)}{\left(\dfrac{1}{x + 1}\right)}$

44. $\dfrac{\left(\dfrac{5}{y} - \dfrac{6}{2y + 1}\right)}{\left(\dfrac{5}{2y + 1}\right)}$

45. $\dfrac{\left(\dfrac{x}{x - 3} - \dfrac{2}{3}\right)}{\left(\dfrac{10}{3x} + \dfrac{x^2}{x - 3}\right)}$

46. $\dfrac{\left(\dfrac{1}{2x} - \dfrac{6}{x + 5}\right)}{\left(\dfrac{x}{x - 5} + \dfrac{1}{x}\right)}$

Solving Problems

57. *Average of Two Numbers*   Determine the average of the two real numbers $x/5$ and $x/6$.

58. *Average of Two Numbers*   Determine the average of the two real numbers $2x/3$ and $3x/5$.

59. *Average of Two Numbers*   Determine the average of the two real numbers $2x/3$ and $x/4$.

60. *Average of Two Numbers*   Determine the average of the two real numbers $4/a^2$ and $2/a$.

61. *Average of Two Numbers*   Determine the average of the two real numbers $(b + 5)/4$ and $2/b$.

62. *Average of Two Numbers*   Determine the average of the two real numbers $5/2s$ and $(s + 1)/5$.

63. *Number Problem* Find three real numbers that divide the real number line between $x/9$ and $x/6$ into four equal parts (see figure).

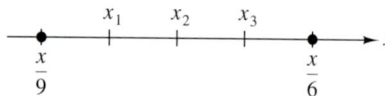

64. *Number Problem* Find two real numbers that divide the real number line between $x/3$ and $5x/4$ into three equal parts (see figure).

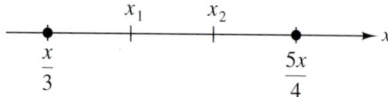

65. *Electrical Resistance* When two resistors of resistance R_1 and R_2 are connected in parallel, the total resistance is modeled by

$$\frac{1}{\left(\dfrac{1}{R_1} + \dfrac{1}{R_2}\right)}.$$

Simplify this complex fraction.

66. *Monthly Payment* The approximate annual percent rate r of a monthly installment loan is

$$r = \frac{\left[\dfrac{24(MN - P)}{N}\right]}{\left(P + \dfrac{MN}{12}\right)}$$

where N is the total number of payments, M is the monthly payment, and P is the amount financed.

(a) Simplify the expression.

(b) Approximate the annual percent rate for a four-year home-improvement loan of $15,000 with monthly payments of $350.

In Exercises 67 and 68, use the following models, which give the number N (in thousands) of cellular telephone subscribers and the annual service revenue R (in millions of dollars) generated by subscribers in the United States from 1994 through 2000.

$$N = \frac{4568.33t + 1042.7}{-0.06t + 1.0} \quad \text{and} \quad R = \frac{1382.16t + 5847.9}{-0.06t + 1.0}$$

In these models, t represents the year, with $t = 4$ corresponding to 1994. (Source: Cellular Telecommunications and Internet Association)

67. (a) ▦ Use a graphing calculator to graph the two models in the same viewing window.

(b) Find a model for the average monthly bill per subscriber. (*Note:* Modify the revenue model from years to months.)

68. (a) Use the model in Exercise 67(b) to complete the table.

Year, t	4	6	8	10
Monthly bill				

(b) The number of subscribers and the revenue were increasing over the last few years, and yet the average monthly bill was decreasing. Explain how this is possible.

Explaining Concepts

69. *Writing* ✐ Define the term *complex fraction*. Give an example and show how to simplify the fraction.

70. What are the numerator and denominator of each complex fraction?

(a) $\dfrac{5}{\left(\dfrac{3}{x^2 + 5x + 6}\right)}$ (b) $\dfrac{\left(\dfrac{5}{3}\right)}{x^2 + 5x + 6}$

71. What are the numerator and denominator of each complex fraction?

(a) $\dfrac{\left(\dfrac{x - 1}{5}\right)}{\left(\dfrac{2}{x^2 + 2x - 35}\right)}$ (b) $\dfrac{\left(\dfrac{1}{2y} + x\right)}{\left(\dfrac{3}{y} + x\right)}$

72. *Writing* ✐ Of the two methods discussed in this section for simplifying complex fractions, select the method you prefer and explain the method in your own words.

7.5 Solving Rational Equations

Thinkstock/Getty Images

What You Should Learn

① Solve rational equations containing constant denominators.

② Solve rational equations containing variable denominators.

Why You Should Learn It

Rational equations can be used to model and solve real-life applications. For instance, in Exercise 85 on page 446, you will use a rational equation to determine the speeds of two runners.

Equations Containing Constant Denominators

In Section 3.2, you studied a strategy for solving equations that contain fractions with *constant* denominators. That procedure is reviewed here because it is the basis for solving more general equations involving fractions. Recall from Section 3.2 that you can "clear an equation of fractions" by multiplying each side of the equation by the least common denominator (LCD) of the fractions in the equation. Note how this is done in the next three examples.

① Solve rational equations containing constant denominators.

Example 1   An Equation Containing Constant Denominators

Solve $\dfrac{3}{5} = \dfrac{x}{2} + 1$.

Solution

The least common denominator of the fractions is 10, so begin by multiplying each side of the equation by 10.

$$\frac{3}{5} = \frac{x}{2} + 1 \qquad \text{Write original equation.}$$

$$10\left(\frac{3}{5}\right) = 10\left(\frac{x}{2} + 1\right) \qquad \text{Multiply each side by LCD of 10.}$$

$$6 = 5x + 10 \qquad \text{Distribute and simplify.}$$

$$-4 = 5x \quad \Longrightarrow \quad -\frac{4}{5} = x \qquad \text{Subtract 10 from each side, then divide each side by 5.}$$

The solution is $x = -\frac{4}{5}$. You can check this in the original equation as follows.

Check

$$\frac{3}{5} \stackrel{?}{=} \frac{-4/5}{2} + 1 \qquad \text{Substitute } -\tfrac{4}{5} \text{ for } x \text{ in the original equation.}$$

$$\frac{3}{5} \stackrel{?}{=} -\frac{4}{5} \cdot \frac{1}{2} + 1 \qquad \text{Invert divisor and multiply.}$$

$$\frac{3}{5} = -\frac{2}{5} + 1 \qquad \text{Solution checks. } \checkmark$$

Example 2   An Equation Containing Constant Denominators

Solve $\dfrac{x-3}{6} = 7 - \dfrac{x}{12}$.

Solution

The least common denominator of the fractions is 12, so begin by multiplying each side of the equation by 12.

$$\frac{x-3}{6} = 7 - \frac{x}{12}$$
Write original equation.

$$12\left(\frac{x-3}{6}\right) = 12\left(7 - \frac{x}{12}\right)$$
Multiply each side by LCD of 12.

$$2x - 6 = 84 - x$$
Distribute and simplify.

$$3x - 6 = 84$$
Add x to each side.

$$3x = 90 \implies x = 30$$
Add 6 to each side, then divide each side by 3.

Check

$$\frac{30-3}{6} \stackrel{?}{=} 7 - \frac{30}{12}$$
Substitute 30 for x in the original equation.

$$\frac{27}{6} = \frac{42}{6} - \frac{15}{6}$$
Solution checks. ✔

Example 3   An Equation That Has Two Solutions

Solve $\dfrac{x^2}{3} + \dfrac{x}{2} = \dfrac{5}{6}$.

Solution

The least common denominator of the fractions is 6, so begin by multiplying each side of the equation by 6.

$$\frac{x^2}{3} + \frac{x}{2} = \frac{5}{6}$$
Write original equation.

$$6\left(\frac{x^2}{3} + \frac{x}{2}\right) = 6\left(\frac{5}{6}\right)$$
Multiply each side by LCD of 6.

$$\frac{6x^2}{3} + \frac{6x}{2} = \frac{30}{6}$$
Distributive Property

$$2x^2 + 3x = 5$$
Simplify.

$$2x^2 + 3x - 5 = 0$$
Subtract 5 from each side.

$$(2x + 5)(x - 1) = 0$$
Factor.

$$2x + 5 = 0 \implies x = -\frac{5}{2}$$
Set 1st factor equal to 0.

$$x - 1 = 0 \implies x = 1$$
Set 2nd factor equal to 0.

The solutions are $x = -\frac{5}{2}$ and $x = 1$. Check these in the original equation.

② Solve rational equations containing variable denominators.

Equations Containing Variable Denominators

Remember that you always *exclude* those values of a variable that make the denominator of a rational expression equal to zero. This is especially critical in solving equations that contain variable denominators. You will see why in the examples that follow.

Example 4   An Equation Containing Variable Denominators

Solve the equation.

$$\frac{7}{x} - \frac{1}{3x} = \frac{8}{3}$$

Solution

The least common denominator of the fractions is $3x$, so begin by multiplying each side of the equation by $3x$.

$$\frac{7}{x} - \frac{1}{3x} = \frac{8}{3}$$ Write original equation.

$$3x\left(\frac{7}{x} - \frac{1}{3x}\right) = 3x\left(\frac{8}{3}\right)$$ Multiply each side by LCD of $3x$.

$$\frac{21x}{x} - \frac{3x}{3x} = \frac{24x}{3}$$ Distributive Property

$$21 - 1 = 8x$$ Simplify.

$$\frac{20}{8} = x$$ Combine like terms and divide each side by 8.

$$\frac{5}{2} = x$$ Simplify.

The solution is $x = \frac{5}{2}$. You can check this in the original equation as follows.

Check

$$\frac{7}{x} - \frac{1}{3x} = \frac{8}{3}$$ Write original equation.

$$\frac{7}{5/2} - \frac{1}{3(5/2)} \stackrel{?}{=} \frac{8}{3}$$ Substitute $\frac{5}{2}$ for x.

$$7\left(\frac{2}{5}\right) - \left(\frac{1}{3}\right)\left(\frac{2}{5}\right) \stackrel{?}{=} \frac{8}{3}$$ Invert divisors and multiply.

$$\frac{14}{5} - \frac{2}{15} \stackrel{?}{=} \frac{8}{3}$$ Simplify.

$$\frac{40}{15} \stackrel{?}{=} \frac{8}{3}$$ Combine like terms.

$$\frac{8}{3} = \frac{8}{3}$$ Solution checks. ✔

Technology: Tip

You can use a graphing calculator to approximate the solution of the equation in Example 4. To do this, graph the left side of the equation and the right side of the equation in the same viewing window.

$$y_1 = \frac{7}{x} - \frac{1}{3x} \text{ and } y_2 = \frac{8}{3}$$

The solution of the equation is the x-coordinate of the point at which the two graphs intersect, as shown below. You can use the *intersect* feature of the graphing calculator to approximate the point of intersection to be $\left(\frac{5}{2}, \frac{8}{3}\right)$. So, the solution is $x = \frac{5}{2}$, which is the same solution obtained in Example 4.

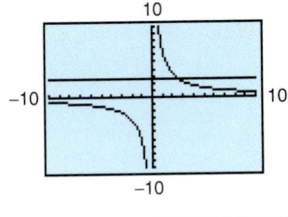

Throughout the text, the importance of checking solutions is emphasized. Up to this point, the main reason for checking has been to make sure that you did not make arithmetic errors in the solution process. In the next example, you will see that there is another reason for checking solutions in the *original* equation. That is, even with no mistakes in the solution process, it can happen that a "trial solution" does not satisfy the original equation. This type of solution is called an **extraneous solution.** An extraneous solution of an equation does not, by definition, satisfy the original equation, and so *must not* be listed as an actual solution.

Example 5　An Equation with No Solution

Solve $\dfrac{5x}{x-2} = 7 + \dfrac{10}{x-2}$.

Solution

The least common denominator of the fractions is $x - 2$, so begin by multiplying each side of the equation by $x - 2$.

$$\frac{5x}{x-2} = 7 + \frac{10}{x-2}$$ Write original equation.

$$(x-2)\left(\frac{5x}{x-2}\right) = (x-2)\left(7 + \frac{10}{x-2}\right)$$ Multiply each side by $x - 2$.

$$5x = 7(x-2) + 10, \quad x \neq 2$$ Distribute and simplify.

$$5x = 7x - 14 + 10$$ Distributive Property

$$5x = 7x - 4$$ Combine like terms.

$$-2x = -4$$ Subtract $7x$ from each side.

$$x = 2$$ Divide each side by -2.

At this point, the solution appears to be $x = 2$. However, by performing a check, you can see that this "trial solution" is extraneous.

Check

$$\frac{5x}{x-2} = 7 + \frac{10}{x-2}$$ Write original equation.

$$\frac{5(2)}{2-2} \overset{?}{=} 7 + \frac{10}{2-2}$$ Substitute 2 for x.

$$\frac{10}{0} \overset{?}{=} 7 + \frac{10}{0}$$ Solution does not check. ✗

Because the check results in *division by zero,* you can conclude that 2 is extraneous. So, the original equation has no solution.

Notice that $x = 2$ is excluded from the domains of the two fractions in the original equation in Example 5. You may find it helpful when solving these types of equations to list the domain restrictions *before* beginning the solution process.

An equation with a single fraction on each side can be cleared of fractions by **cross-multiplying,** which is equivalent to multiplying by the LCD and then dividing out. To do this, multiply the left numerator by the right denominator and multiply the right numerator by the left denominator, as shown below.

$$\frac{a}{b} = \frac{c}{d} \implies ad = bc, b \neq 0, d \neq 0$$

Example 6   Cross-Multiplying

Solve $\dfrac{2x}{x + 4} = \dfrac{3}{x - 1}$.

Solution

The domain is all real values of x such that $x \neq -4$ and $x \neq 1$. You can use cross-multiplication to solve this equation.

$\dfrac{2x}{x + 4} = \dfrac{3}{x - 1}$	Write original equation.
$2x(x - 1) = 3(x + 4), \ x \neq -4, x \neq 1$	Cross-multiply.
$2x^2 - 2x = 3x + 12$	Distributive Property
$2x^2 - 5x - 12 = 0$	Subtract $3x$ and 12 from each side.
$(2x + 3)(x - 4) = 0$	Factor.
$2x + 3 = 0 \implies x = -\frac{3}{2}$	Set 1st factor equal to 0.
$x - 4 = 0 \implies x = 4$	Set 2nd factor equal to 0.

The solutions are $x = -\frac{3}{2}$ and $x = 4$. Check these in the original equation.

Example 7   An Equation That Has Two Solutions

Solve $\dfrac{3x}{x + 1} = \dfrac{12}{x^2 - 1} + 2$.

The domain is all real values of x such that $x \neq 1$ and $x \neq -1$. The least common denominator is $(x + 1)(x - 1) = x^2 - 1$.

$(x^2 - 1)\left(\dfrac{3x}{x + 1}\right) = (x^2 - 1)\left(\dfrac{12}{x^2 - 1} + 2\right)$	Multiply each side of original equation by LCD of $x^2 - 1$.
$(x - 1)(3x) = 12 + 2(x^2 - 1), \quad x \neq \pm 1$	Simplify.
$3x^2 - 3x = 12 + 2x^2 - 2$	Distributive Property
$x^2 - 3x - 10 = 0$	Subtract $2x^2$ and 10 from each side.
$(x + 2)(x - 5) = 0$	Factor.
$x + 2 = 0 \implies x = -2$	Set 1st factor equal to 0.
$x - 5 = 0 \implies x = 5$	Set 2nd factor equal to 0.

The solutions are $x = -2$ and $x = 5$. Check these in the original equation.

7.5 Exercises

Review   Concepts, Skills, and Problem Solving

Keep mathematically in shape by doing these exercises *before* the problems of this section.

Properties and Definitions

In Exercises 1 and 2, determine the quadrants in which the point must be located.

1. $(-2, y)$, y is a real number.

2. $(x, 3)$, x is a real number.

3. Give the positions of points whose y-coordinates are 0.

4. Find the coordinates of the point nine units to the right of the y-axis and six units below the x-axis.

Solving Inequalities

In Exercises 5–10, solve the inequality.

5. $7 - 3x > 4 - x$

6. $2(x + 6) - 20 < 2$

7. $|x - 3| < 2$

8. $|x - 5| > 3$

9. $\left|\frac{1}{4}x - 1\right| \geq 3$

10. $\left|2 - \frac{1}{3}x\right| \leq 10$

Problem Solving

11. *Distance*   A jogger leaves a point on a fitness trail running at a rate of 6 miles per hour. Five minutes later, a second jogger leaves from the same location running at 8 miles per hour. How long will it take the second runner to overtake the first, and how far will each have run at that point?

12. *Investment*   An inheritance of \$24,000 is invested in two bonds that pay 7.5% and 9% simple interest. The annual interest is \$1935. How much is invested in each bond?

Developing Skills

In Exercises 1–4, determine whether each value of x is a solution to the equation.

Equation	Values
1. $\frac{x}{3} - \frac{x}{5} = \frac{4}{3}$	(a) $x = 0$ (b) $x = -1$ (c) $x = \frac{1}{8}$ (d) $x = 10$
2. $x = 4 + \frac{21}{x}$	(a) $x = 0$ (b) $x = -3$ (c) $x = 7$ (d) $x = -1$
3. $\frac{x}{4} + \frac{3}{4x} = 1$	(a) $x = -1$ (b) $x = 1$ (c) $x = 3$ (d) $x = -3$
4. $5 - \frac{1}{x - 3} = 2$	(a) $x = \frac{10}{3}$ (b) $x = -\frac{1}{3}$ (c) $x = 0$ (d) $x = 1$

In Exercises 5–22, solve the equation. See Examples 1–3.

5. $\frac{x}{6} - 1 = \frac{2}{3}$

6. $\frac{y}{8} + 7 = -\frac{1}{2}$

7. $\frac{1}{4} = \frac{z + 1}{8}$

8. $\frac{a}{5} = \frac{a - 3}{2}$

9. $\frac{x}{3} - \frac{3x}{4} = \frac{5x}{12}$

10. $\frac{x}{4} - \frac{x}{6} = \frac{1}{4}$

11. $\frac{z + 2}{3} = 4 - \frac{z}{12}$

12. $\frac{x - 5}{5} + 3 = -\frac{x}{4}$

13. $\frac{2y - 9}{6} = 3y - \frac{3}{4}$

14. $\frac{4x - 2}{7} - \frac{5}{14} = 2x$

15. $\frac{t}{2} = 12 - \frac{3t^2}{2}$

16. $\frac{x^2}{2} - \frac{3x}{5} = -\frac{1}{10}$

17. $\frac{5y - 1}{12} + \frac{y}{3} = -\frac{1}{4}$

18. $\frac{z - 4}{9} - \frac{3z + 1}{18} = \frac{3}{2}$

19. $\dfrac{h + 2}{5} - \dfrac{h - 1}{9} = \dfrac{2}{3}$ **20.** $\dfrac{u - 2}{6} + \dfrac{2u + 5}{15} = 3$

21. $\dfrac{x + 5}{4} - \dfrac{3x - 8}{3} = \dfrac{4 - x}{12}$

22. $\dfrac{2x - 7}{10} - \dfrac{3x + 1}{5} = \dfrac{6 - x}{5}$

In Exercises 23–66, solve the equation. (Check for extraneous solutions.) See Examples 4–7.

23. $\dfrac{9}{25 - y} = -\dfrac{1}{4}$ **24.** $\dfrac{2}{u + 4} = \dfrac{5}{8}$

25. $5 - \dfrac{12}{a} = \dfrac{5}{3}$ **26.** $\dfrac{6}{b} + 22 = 24$

27. $\dfrac{4}{x} - \dfrac{7}{5x} = -\dfrac{1}{2}$ **28.** $\dfrac{5}{3} = \dfrac{6}{7x} + \dfrac{2}{x}$

29. $\dfrac{12}{y + 5} + \dfrac{1}{2} = 2$ **30.** $\dfrac{7}{8} - \dfrac{16}{t - 2} = \dfrac{3}{4}$

31. $\dfrac{5}{x} = \dfrac{25}{3(x + 2)}$ **32.** $\dfrac{10}{x + 4} = \dfrac{15}{4(x + 1)}$

33. $\dfrac{8}{3x + 5} = \dfrac{1}{x + 2}$ **34.** $\dfrac{500}{3x + 5} = \dfrac{50}{x - 3}$

35. $\dfrac{3}{x + 2} - \dfrac{1}{x} = \dfrac{1}{5x}$ **36.** $\dfrac{12}{x + 5} + \dfrac{5}{x} = \dfrac{20}{x}$

37. $\dfrac{1}{2} = \dfrac{18}{x^2}$ **38.** $\dfrac{1}{4} = \dfrac{16}{z^2}$

39. $\dfrac{32}{t} = 2t$ **40.** $\dfrac{20}{u} = \dfrac{u}{5}$

41. $x + 1 = \dfrac{72}{x}$ **42.** $\dfrac{48}{x} = x - 2$

43. $1 = \dfrac{16}{y} - \dfrac{39}{y^2}$ **44.** $x - \dfrac{24}{x} = 5$

45. $\dfrac{2x}{3x - 10} - \dfrac{5}{x} = 0$ **46.** $\dfrac{x + 42}{x} = x$

47. $\dfrac{2x}{5} = \dfrac{x^2 - 5x}{5x}$ **48.** $\dfrac{3x}{4} = \dfrac{x^2 + 3x}{8x}$

49. $\dfrac{y + 1}{y + 10} = \dfrac{y - 2}{y + 4}$ **50.** $\dfrac{x - 3}{x + 1} = \dfrac{x - 6}{x + 5}$

51. $\dfrac{15}{x} + \dfrac{9x - 7}{x + 2} = 9$

52. $\dfrac{3z - 2}{z + 1} = 4 - \dfrac{z + 2}{z - 1}$

53. $\dfrac{2}{6q + 5} - \dfrac{3}{4(6q + 5)} = \dfrac{1}{28}$

54. $\dfrac{10}{x(x - 2)} + \dfrac{4}{x} = \dfrac{5}{x - 2}$

55. $\dfrac{4}{2x + 3} + \dfrac{17}{5x - 3} = 3$

56. $\dfrac{1}{x - 1} + \dfrac{3}{x + 1} = 2$

57. $\dfrac{2}{x - 10} - \dfrac{3}{x - 2} = \dfrac{6}{x^2 - 12x + 20}$

58. $\dfrac{5}{x + 2} + \dfrac{2}{x^2 - 6x - 16} = \dfrac{-4}{x - 8}$

59. $\dfrac{x + 3}{x^2 - 9} + \dfrac{4}{3 - x} - 2 = 0$

60. $1 - \dfrac{6}{4 - x} = \dfrac{x + 2}{x^2 - 16}$

61. $\dfrac{x}{x - 2} + \dfrac{3x}{x - 4} = \dfrac{-2(x - 6)}{x^2 - 6x + 8}$

62. $\dfrac{2(x + 1)}{x^2 - 4x + 3} + \dfrac{6x}{x - 3} = \dfrac{3x}{x - 1}$

63. $\dfrac{5}{x^2 + 4x + 3} + \dfrac{2}{x^2 + x - 6} = \dfrac{3}{x^2 - x - 2}$

64. $\dfrac{2}{x^2 + 2x - 8} - \dfrac{1}{x^2 + 9x + 20} = \dfrac{4}{x^2 + 3x - 10}$

65. $\dfrac{x}{2} = \dfrac{2 - \dfrac{3}{x}}{1 - \dfrac{1}{x}}$ **66.** $\dfrac{2x}{3} = \dfrac{1 + \dfrac{2}{x}}{1 + \dfrac{1}{x}}$

In Exercises 67–70, (a) use the graph to determine any x-intercepts of the graph and (b) set $y = 0$ and solve the resulting rational equation to confirm the result of part (a).

67. $y = \dfrac{x + 2}{x - 2}$ **68.** $y = \dfrac{2x}{x + 4}$

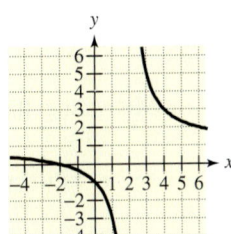

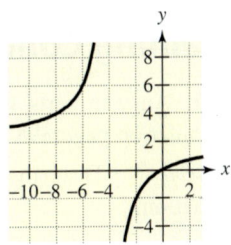

69. $y = x - \dfrac{1}{x}$

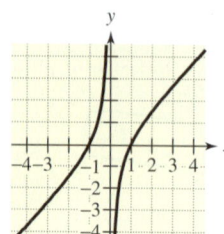

70. $y = x - \dfrac{2}{x} - 1$

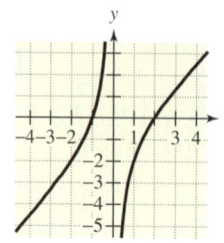

73. $y = \dfrac{1}{x} + \dfrac{4}{x - 5}$

74. $y = 20\left(\dfrac{2}{x} - \dfrac{3}{x - 1}\right)$

75. $y = (x + 1) - \dfrac{6}{x}$

76. $y = \dfrac{x^2 - 4}{x}$

Think About   In Exercises 77–80, if the exercise is an equation, solve it; if it is an expression, simplify it.

77. $\dfrac{16}{x^2 - 16} + \dfrac{x}{2x - 8} = \dfrac{1}{2}$

In Exercises 71–76, (a) use a graphing calculator to graph the equation and determine any *x*-intercepts of the graph and (b) set $y = 0$ and solve the resulting rational equation to confirm the result of part (a).

78. $\dfrac{5}{x + 3} + \dfrac{5}{3} + 3$

79. $\dfrac{16}{x^2 - 16} + \dfrac{x}{2x - 8} + \dfrac{1}{2}$

71. $y = \dfrac{x - 4}{x + 5}$

72. $y = \dfrac{1}{x} - \dfrac{3}{x + 4}$

80. $\dfrac{5}{x + 3} + \dfrac{5}{3} = 3$

Solving Problems

81. *Number Problem*   Find a number such that the sum of the number and its reciprocal is $\frac{65}{8}$.

82. *Number Problem*   Find a number such that the sum of two times the number and three times its reciprocal is $\frac{97}{4}$.

83. *Wind Speed*   A plane has a speed of 300 miles per hour in still air. The plane travels a distance of 680 miles with a tail wind in the same time it takes to travel 520 miles into a head wind. Find the speed of the wind.

84. *Average Speed*   During the first part of a six-hour trip, you travel 240 miles at an average speed of *r* miles per hour. For the next 72 miles, you increase your speed by 10 miles per hour. What are your two average speeds?

85. *Speed*   One person runs 2 miles per hour faster than a second person. The first person runs 5 miles in the same time the second person runs 4 miles. Find the speed of each person.

86. *Speed*   The speed of a commuter plane is 150 miles per hour lower than that of a passenger jet. The commuter plane travels 450 miles in the same time the jet travels 1150 miles. Find the speed of each plane.

Explaining Concepts

87. Answer parts (d) and (e) of Motivating the Chapter on page 398.

88. *Writing*   Describe how to solve a rational equation.

89. *Writing*   Define the term *extraneous solution*. How do you identify an extraneous solution?

90. *Writing*   Explain how you can use a graphing calculator to estimate the solution of a rational equation.

91. *Writing*   When can you use cross-multiplication to solve a rational equation? Explain.

7.6 Applications and Variation

NASA

What You Should Learn

1. Solve application problems involving rational equations.
2. Solve application problems involving direct variation.
3. Solve application problems involving inverse variation.
4. Solve application problems involving joint variation.

Why You Should Learn It

You can use mathematical models in a wide variety of applications including variation. For instance, in Exercise 56 on page 458, you will use direct variation to model the weight of a person on the moon.

Rational Equation Applications

The examples that follow are types of application problems that you have seen earlier in the text. The difference now is that the variable appears in the denominator of a rational expression.

1. Solve application problems involving rational equations.

Example 1   Average Speeds

You and your friend travel to separate colleges in the same amount of time. You drive 380 miles and your friend drives 400 miles. Your friend's average speed is 3 miles per hour faster than your average speed. What is your average speed and what is your friend's average speed?

Solution

Begin by setting your time equal to your friend's time. Then use an alternative version for the formula for distance that gives the time in terms of the distance and the rate.

Verbal Model:

$$\text{Your time} = \text{Your friend's time}$$

$$\frac{\text{Your distance}}{\text{Your rate}} = \frac{\text{Friend's distance}}{\text{Friend's rate}}$$

Labels:

Your distance = 380	(miles)
Your rate = r	(miles per hour)
Friend's distance = 400	(miles)
Friend's rate = $r + 3$	(miles per hour)

Equation:

$$\frac{380}{r} = \frac{400}{r + 3} \qquad \text{Original equation.}$$

$$380(r + 3) = 400(r), \quad r \neq 0, r \neq -3 \qquad \text{Cross-multiply.}$$

$$380r + 1140 = 400r \qquad \text{Distributive Property}$$

$$1140 = 20r \quad \Longrightarrow \quad 57 = r \qquad \text{Simplify.}$$

Your average speed is 57 miles per hour and your friend's average speed is $57 + 3 = 60$ miles per hour. Check this in the original statement of the problem.

Study Tip

When determining the domain of a real-life problem, you must also consider the context of the problem. For instance, in Example 2, the time it takes to fill the tub with water could not be a negative number. The problem implies that the domain must be all real numbers greater than 0.

Example 2   A Work-Rate Problem

With the cold water valve open, it takes 8 minutes to fill a washing machine tub. With both the hot and cold water valves open, it takes 5 minutes to fill the tub. How long will it take to fill the tub with only the hot water valve open?

Solution

Verbal Model:

Rate for cold water	+	Rate for hot water	=	Rate for warm water

Labels:

Rate for cold water $= \dfrac{1}{8}$ (tub per minute)

Rate for hot water $= \dfrac{1}{t}$ (tub per minute)

Rate for warm water $= \dfrac{1}{5}$ (tub per minute)

Equation:

$$\frac{1}{8} + \frac{1}{t} = \frac{1}{5}$$ Original equation

$$5t + 40 = 8t$$ Multiply each side by LCD of $40t$ and simplify.

$$40 = 3t \quad \Longrightarrow \quad \frac{40}{3} = t$$ Simplify.

So, it takes $13\frac{1}{3}$ minutes to fill the tub with hot water. Check this solution.

Example 3   Cost-Benefit Model

A utility company burns coal to generate electricity. The cost C (in dollars) of removing $p\%$ of the pollutants from smokestack emissions is modeled by

$$C = \frac{80{,}000p}{100 - p}, \quad 0 \le p < 100.$$

Determine the percent of air pollutants in the stack emissions that can be removed for $420,000.

Solution

To determine the percent of air pollutants in the stack emissions that can be removed for $420,000, substitute 420,000 for C in the model.

$$420{,}000 = \frac{80{,}000p}{100 - p}$$ Substitute 420,000 for C.

$$420{,}000(100 - p) = 80{,}000p$$ Cross-multiply.

$$42{,}000{,}000 - 420{,}000p = 80{,}000p$$ Distributive Property

$$42{,}000{,}000 = 500{,}000p$$ Add $420{,}000p$ to each side.

$$84 = p$$ Divide each side by 500,000.

So, 84% of air pollutants in the stack emissions can be removed for $420,000. Check this in the original statement of the problem.

② Solve application problems involving direct variation.

Direct Variation

In the mathematical model for **direct variation,** y is a *linear* function of x. Specifically,

$$y = kx.$$

To use this mathematical model in applications involving direct variation, you need to use the given values of x and y to find the value of the constant k.

Direct Variation

The following statements are equivalent.

    1. y **varies directly** as x.

    2. y is **directly proportional** to x.

    3. $y = kx$ for some constant k.

The number k is called the **constant of proportionality.**

Example 4   Direct Variation

The total revenue R (in dollars) obtained from selling x ice show tickets is directly proportional to the number of tickets sold x. When 10,000 tickets are sold, the total revenue is $142,500.

a. Find a mathematical model that relates the total revenue R to the number of tickets sold x.

b. Find the total revenue obtained from selling 12,000 tickets.

Solution

a. Because the total revenue is directly proportional to the number of tickets sold, the linear model is $R = kx$. To find the value of the constant k, use the fact that $R = 142,500$ when $x = 10,000$. Substituting those values into the model produces

$$142,500 = k(10,000) \qquad \text{Substitute for } R \text{ and } x.$$

which implies that

$$k = \frac{142,500}{10,000} = 14.25.$$

So, the equation relating the total revenue to the total number of tickets sold is

$$R = 14.25x. \qquad \text{Direct variation model}$$

The graph of this equation is shown in Figure 7.2.

b. When $x = 12,000$, the total revenue is

$$R = 14.25(12,000) = \$171,000.$$

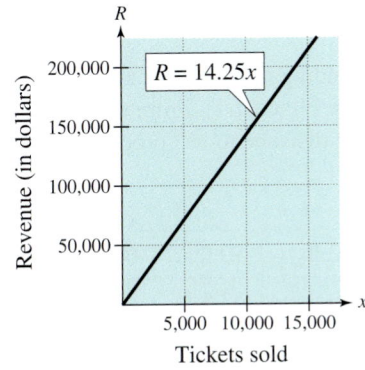

Figure 7.2

Example 5 Direct Variation

Hooke's Law for springs states that the distance a spring is stretched (or compressed) is proportional to the force on the spring. A force of 20 pounds stretches a spring 5 inches.

a. Find a mathematical model that relates the distance the spring is stretched to the force applied to the spring.

b. How far will a force of 30 pounds stretch the spring?

Solution

a. For this problem, let d represent the distance (in inches) that the spring is stretched and let F represent the force (in pounds) that is applied to the spring. Because the distance d is proportional to the force F, the model is

$$d = kF.$$

To find the value of the constant k, use the fact that $d = 5$ when $F = 20$. Substituting these values into the model produces

$$5 = k(20)$$ Substitute 5 for d and 20 for F.

$$\frac{5}{20} = k$$ Divide each side by 20.

$$\frac{1}{4} = k.$$ Simplify.

So, the equation relating distance and force is

$$d = \frac{1}{4}F.$$ Direct variation model

b. When $F = 30$, the distance is

$$d = \frac{1}{4}(30) = 7.5 \text{ inches.}$$ See Figure 7.3.

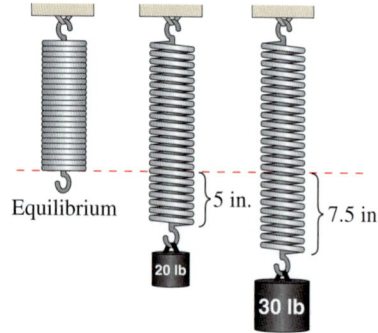

Figure 7.3

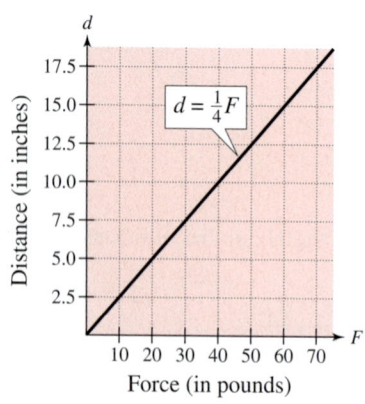

Figure 7.4

In Example 5, you can get a clearer understanding of Hooke's Law by using the model $d = \frac{1}{4}F$ to create a table or a graph (see Figure 7.4). From the table or from the graph, you can see what it means for the distance to be "proportional to the force."

Force, F	10 lb	20 lb	30 lb	40 lb	50 lb	60 lb
Distance, d	2.5 in.	5.0 in.	7.5 in.	10.0 in.	12.5 in.	15.0 in.

In Examples 4 and 5, the direct variations are such that an *increase* in one variable corresponds to an *increase* in the other variable. There are, however, other applications of direct variation in which an increase in one variable corresponds to a *decrease* in the other variable. For instance, in the model $y = -2x$, an increase in x will yield a decrease in y.

Another type of direct variation relates one variable to a power of another.

Direct Variation as *n*th Power

The following statements are equivalent.

   1. *y* **varies directly as the *n*th power** of *x*.

   2. *y* is **directly proportional to the *n*th power** of *x*.

   3. $y = kx^n$ for some constant *k*.

Example 6   Direct Variation as a Power

The distance a ball rolls down an inclined plane is directly proportional to the square of the time it rolls. During the first second, a ball rolls down a plane a distance of 6 feet.

a. Find a mathematical model that relates the distance traveled to the time.

b. How far will the ball roll during the first 2 seconds?

Solution

a. Letting *d* be the distance (in feet) that the ball rolls and letting *t* be the time (in seconds), you obtain the model

$$d = kt^2.$$

Because *d* = 6 when *t* = 1, you obtain

$d = kt^2$	Write original equation.
$6 = k(1)^2$ ⟹ $6 = k.$	Substitute 6 for *d* and 1 for *t*.

So, the equation relating distance to time is

$$d = 6t^2. \qquad \text{Direct variation as 2nd power model}$$

The graph of this equation is shown in Figure 7.5.

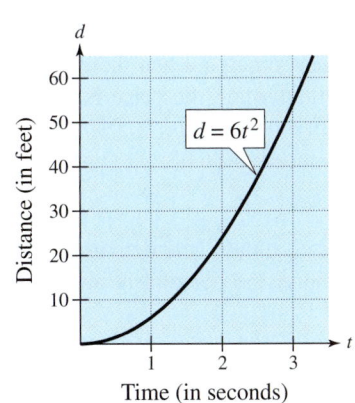

Figure 7.5

b. When *t* = 2, the distance traveled is

$$d = 6(2)^2 = 6(4) = 24 \text{ feet.} \qquad \text{See Figure 7.6.}$$

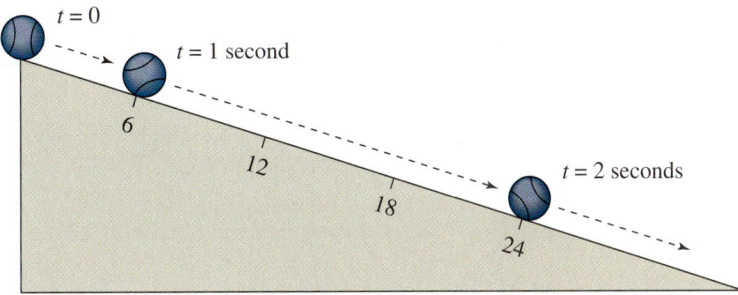

Figure 7.6

③ Solve application problems involving inverse variation.

Inverse Variation

A second type of variation is called **inverse variation.** With this type of variation, one of the variables is said to be inversely proportional to the other variable.

Inverse Variation

1. The following three statements are equivalent.

   a. y **varies inversely** as x.

   b. y is **inversely proportional** to x.

   c. $y = \dfrac{k}{x}$ for some constant k.

2. If $y = \dfrac{k}{x^n}$, then y is inversely proportional to the nth power of x.

Example 7   Inverse Variation

The marketing department of a large company has found that the demand for one of its hand tools varies inversely as the price of the product. (When the price is low, more people are willing to buy the product than when the price is high.) When the price of the tool is $7.50, the monthly demand is 50,000 tools. Approximate the monthly demand if the price is reduced to $6.00.

Solution

Let x represent the number of tools that are sold each month (the demand), and let p represent the price per tool (in dollars). Because the demand is inversely proportional to the price, the model is

$$x = \frac{k}{p}.$$

By substituting $x = 50{,}000$ when $p = 7.50$, you obtain

$$50{,}000 = \frac{k}{7.50} \qquad \text{Substitute 50,000 for } x \text{ and 7.50 for } p.$$

$$375{,}000 = k. \qquad \text{Multiply each side by 7.50.}$$

So, the inverse variation model is $x = \dfrac{375{,}000}{p}$.

The graph of this equation is shown in Figure 7.7. To find the demand that corresponds to a price of $6.00, substitute 6 for p in the equation and obtain

$$x = \frac{375{,}000}{6} = 62{,}500 \text{ tools.}$$

So, if the price is lowered from $7.50 per tool to $6.00 per tool, you can expect the monthly demand to increase from 50,000 tools to 62,500 tools.

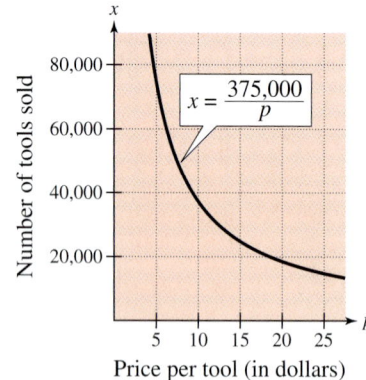

Figure 7.7

Some applications of variation involve problems with *both* direct and inverse variation in the same model. These types of models are said to have **combined variation.**

Example 8   Direct and Inverse Variation

An electronics manufacturer determines that the demand for its portable radio is directly proportional to the amount spent on advertising and inversely proportional to the price of the radio. When $40,000 is spent on advertising and the price per radio is $20, the monthly demand is 10,000 radios.

a. If the amount of advertising were increased to $50,000, how much could the price be increased to maintain a monthly demand of 10,000 radios?

b. If you were in charge of the advertising department, would you recommend this increased expense in advertising?

Solution

a. Let x represent the number of radios that are sold each month (the demand), let a represent the amount spent on advertising (in dollars), and let p represent the price per radio (in dollars). Because the demand is directly proportional to the advertising expense and inversely proportional to the price, the model is

$$x = \frac{ka}{p}.$$

By substituting 10,000 for x when $a = 40,000$ and $p = 20$, you obtain

$$10,000 = \frac{k(40,000)}{20} \qquad \text{Substitute 10,000 for } x, \text{ 40,000 for } a, \text{ and 20 for } p.$$

$$200,000 = 40,000k \qquad \text{Multiply each side by 20.}$$

$$5 = k. \qquad \text{Divide each side by 40,000.}$$

So, the model is

$$x = \frac{5a}{p}. \qquad \text{Direct and inverse variation model}$$

To find the price that corresponds to a demand of 10,000 and an advertising expense of $50,000, substitute 10,000 for x and 50,000 for a into the model and solve for p.

$$10,000 = \frac{5(50,000)}{p} \qquad p = \frac{5(50,000)}{10,000} = \$25$$

So, the price increase would be $25 - $20 = $5.

b. The total revenue for selling 10,000 radios at $20 each is $200,000, and the revenue for selling 10,000 radios at $25 each is $250,000. So, increasing the advertising expense from $40,000 to $50,000 would increase the revenue by $50,000. This implies that you should recommend the increased expense in advertising.

④ Solve application problems involving joint variation.

Joint Variation

The model used in Example 8 involved both direct and inverse variation, and the word "and" was used to couple the two types of variation together. To describe two different *direct* variations in the same statement, the word "jointly" is used. For instance, the model $z = kxy$ can be described by saying that z is *jointly* proportional to x and y.

Joint Variation

1. The following three statements are equivalent.

   a. z **varies jointly** as x and y.

   b. z is **jointly proportional** to x and y.

   c. $z = kxy$ for some constant k.

2. If $z = kx^ny^m$, then z is jointly proportional to the nth power of x and the mth power of y.

Example 9 Joint Variation

The *simple interest* for a savings account is jointly proportional to the time and the principal. After one quarter (3 months), the interest for a principal of $6000 is $120. How much interest would a principal of $7500 earn in 5 months?

Solution

To begin, let I represent the interest earned (in dollars), let P represent the principal (in dollars), and let t represent the time (in years). Because the interest is jointly proportional to the time and the principal, the model is

$$I = ktP.$$

Because $I = 120$ when $P = 6000$ and $t = \frac{1}{4}$, you have

$$120 = k\left(\frac{1}{4}\right)(6000) \qquad \text{Substitute 120 for } I, \tfrac{1}{4} \text{ for } t, \text{ and 6000 for } P.$$

$$120 = 1500\,k \qquad \text{Simplify.}$$

$$0.08 = k. \qquad \text{Divide each side by 1500.}$$

So, the model that relates interest to time and principal is

$$I = 0.08tP. \qquad \text{Joint variation model}$$

To find the interest earned on a principal of $7500 over a five-month period of time, substitute $P = 7500$ and $t = \frac{5}{12}$ into the model and obtain an interest of

$$I = 0.08\left(\frac{5}{12}\right)(7500) = \$250.$$

7.6 Exercises

Review   Concepts, Skills, and Problem Solving

Keep mathematically in shape by doing these exercises *before* the problems of this section.

Properties and Definitions

1. Determine the domain of $f(x) = x^2 - 4x + 9$.

2. Determine the domain of $h(x) = \dfrac{x - 1}{x^2(x^2 + 1)}$.

Functions

In Exercises 3–6, consider the function

$f(x) = 2x^3 - 3x^2 - 18x + 27$
$\quad\quad = (2x - 3)(x + 3)(x - 3)$.

3. ▦ Use a graphing calculator to graph both expressions for the function. Are the graphs the same?

4. Verify the factorization by multiplying the polynomials in the factored form of f.

5. Verify the factorization by performing the long division

$\dfrac{2x^3 - 3x^2 - 18x + 27}{2x - 3}$

and then factoring the quotient.

6. Verify the factorization by performing the long division

$\dfrac{2x^3 - 3x^2 - 18x + 27}{x^2 - 9}$.

In Exercises 7 and 8, use the function to find and simplify the expression for

$\dfrac{f(2 + h) - f(2)}{h}$.

7. $f(x) = x^2 - 3$

8. $f(x) = \dfrac{3}{x + 5}$

Modeling

9. *Cost*   The inventor of a new game believes that the variable cost for producing the game is $5.75 per unit and the fixed costs are $12,000. Write the total cost C as a function of x, the number of games produced.

10. ▲ *Geometry*   The length of a rectangle is one and one-half times its width. Write the perimeter P of the rectangle as a function of the rectangle's width w.

Developing Skills

In Exercises 1–14, write a model for the statement.

1. I varies directly as V.

2. C varies directly as r.

3. V is directly proportional to t.

4. s varies directly as the cube of t.

5. u is directly proportional to the square of v.

6. V varies directly as the cube root of x.

7. p varies inversely as d.

8. S is inversely proportional to the square of v.

9. A varies inversely as the fourth power of t.

10. P is inversely proportional to the square root of $1 + r$.

11. A varies jointly as l and w.

12. V varies jointly as h and the square of r.

13. *Boyle's Law*   If the temperature of a gas is not allowed to change, its absolute pressure P is inversely proportional to its volume V.

14. *Newton's Law of Universal Gravitation*   The gravitational attraction F between two particles of masses m_1 and m_2 is directly proportional to the product of the masses and inversely proportional to the square of the distance r between the particles.

In Exercises 15–20, write a verbal sentence using varia-tion terminology to describe the formula.

15. *Area of a Triangle:* $A = \frac{1}{2}bh$

16. *Area of a Rectangle:* $A = lw$

17. *Volume of a Right Circular Cylinder:* $V = \pi r^2 h$

18. *Volume of a Sphere:* $V = \frac{4}{3}\pi r^3$

19. *Average Speed:* $r = \dfrac{d}{t}$

20. *Height of a Cylinder:* $h = \dfrac{V}{\pi r^2}$

In Exercises 21–32, find the constant of proportionality and write an equation that relates the variables.

21. s varies directly as t, and $s = 20$ when $t = 4$.

22. h is directly proportional to r, and $h = 28$ when $r = 12$.

23. F is directly proportional to the square of x, and $F = 500$ when $x = 40$.

24. M varies directly as the cube of n, and $M = 0.012$ when $n = 0.2$.

25. n varies inversely as m, and $n = 32$ when $m = 1.5$.

26. q is inversely proportional to p, and $q = \frac{3}{2}$ when $p = 50$.

27. g varies inversely as the square root of z, and $g = \frac{4}{5}$ when $z = 25$.

28. u varies inversely as the square of v, and $u = 40$ when $v = \frac{1}{2}$.

29. F varies jointly as x and y, and $F = 500$ when $x = 15$ and $y = 8$.

30. V varies jointly as h and the square of b, and $V = 288$ when $h = 6$ and $b = 12$.

31. d varies directly as the square of x and inversely with r, and $d = 3000$ when $x = 10$ and $r = 4$.

32. z is directly proportional to x and inversely propor-tional to the square root of y, and $z = 720$ when $x = 48$ and $y = 81$.

Solving Problems

33. *Current Speed*   A boat travels at a speed of 20 miles per hour in still water. It travels 48 miles upstream and then returns to the starting point in a total of 5 hours. Find the speed of the current.

34. *Average Speeds*   You and your college roommate travel to your respective hometowns in the same amount of time. You drive 210 miles and your friend drives 190 miles. Your friend's average speed is 6 miles per hour lower than your average speed. What are your average speed and your friend's average speed?

35. *Partnership Costs*   A group plans to start a new business that will require $240,000 for start-up capital. The individuals in the group share the cost equally. If two additional people join the group, the cost per person will decrease by $4000. How many people are presently in the group?

36. *Partnership Costs*   A group of people share equally the cost of a $150,000 endowment. If they could find four more people to join the group, each person's share of the cost would decrease by $6250. How many people are presently in the group?

37. *Population Growth*   A biologist introduces 100 insects into a culture. The population P of the culture is approximated by the model below, where t is the time in hours. Find the time required for the popula-tion to increase to 1000 insects.

$$P = \frac{500(1 + 3t)}{5 + t}$$

38. *Pollution Removal*   The cost C in dollars of remov-ing $p\%$ of the air pollutants in the stack emissions of a utility company is modeled by the equation below. Determine the percent of air pollutants in the stack emissions that can be removed for $680,000.

$$C = \frac{120,000p}{100 - p}.$$

39. *Work Rate*   One landscaper works $1\frac{1}{2}$ times as fast as another landscaper. It takes them 9 hours working together to complete a job. Find the time it takes each landscaper to complete the job working alone.

40. *Flow Rate*   The flow rate for one pipe is $1\frac{1}{4}$ times that of another pipe. A swimming pool can be filled in 5 hours using both pipes. Find the time required to fill the pool using only the pipe with the lower flow rate.

41. *Nail Sizes*   The unit for determining the size of a nail is a *penny*. For example, 8d represents an 8-penny nail. The number N of finishing nails per pound can be modeled by

$$N = -139.1 + \frac{2921}{x}$$

where x is the size of the nail.

(a) What is the domain of the function?

(b) 🖩 Use a graphing calculator to graph the function.

(c) Use the graph to determine the size of the finishing nail if there are 153 nails per pound.

(d) Verify the result of part (c) algebraically.

42. *Learning Curve*   A psychologist observed that a four-year-old child could memorize N lines of a poem, where N depended on the number x of short sessions that the psychologist worked with the child. The number of lines N memorized can be easily modeled by

$$N = \frac{20x}{x+1}, \qquad x \geq 0.$$

(a) 🖩 Use a graphing calculator to graph the function.

(b) Use the graph to determine the number of sessions if the child can memorize 18 lines of the poem.

(c) Verify the result of part (b) algebraically.

43. *Revenue*   The total revenue R is directly proportional to the number of units sold x. When 500 units are sold, the revenue is $3875. Find the revenue when 635 units are sold. Then interpret the constant of proportionality.

44. *Revenue*   The total revenue R is directly proportional to the number of units sold x. When 25 units are sold, the revenue is $300. Find the revenue when 42 units are sold. Then interpret the constant of proportionality.

45. *Hooke's Law*   A force of 50 pounds stretches a spring 5 inches.

(a) How far will a force of 20 pounds stretch the spring?

(b) What force is required to stretch the spring 1.5 inches?

46. *Hooke's Law*   A force of 50 pounds stretches a spring 3 inches.

(a) How far will a force of 20 pounds stretch the spring?

(b) What force is required to stretch the spring 1.5 inches?

47. *Hooke's Law*   A baby weighing $10\frac{1}{2}$ pounds compresses the spring of a baby scale 7 millimeters. Determine the weight of a baby that compresses the spring 12 millimeters.

48. *Hooke's Law*   A force of 50 pounds stretches the spring of a scale 1.5 inches.

(a) Write the force F as a function of the distance x the spring is stretched.

(b) Graph the function in part (a) where $0 \leq x \leq 5$. Identify the graph.

49. *Free-Falling Object*   The velocity v of a free-falling object is proportional to the time that the object has fallen. The constant of proportionality is the acceleration due to gravity. The velocity of a falling object is 96 feet per second after the object has fallen for 3 seconds. Find the acceleration due to gravity.

50. *Free-Falling Object*   Neglecting air resistance, the distance d that an object falls varies directly as the square of the time t it has fallen. An object falls 64 feet in 2 seconds. Determine the distance it will fall in 6 seconds.

51. *Stopping Distance* The stopping distance d of an automobile is directly proportional to the square of its speed s. On a road surface, a car requires 75 feet to stop when its speed is 30 miles per hour. The brakes are applied when the car is traveling at 50 miles per hour under similar road conditions. Estimate the stopping distance.

52. *Frictional Force* The frictional force F between the tires and the road that is required to keep a car on a curved section of a highway is directly proportional to the square of the speed s of the car. If the speed of the car is doubled, the force will change by what factor?

53. *Power Generation* The power P generated by a wind turbine varies directly as the cube of the wind speed w. The turbine generates 750 watts of power in a 25-mile-per-hour wind. Find the power it generates in a 40-mile-per-hour wind.

54. *Demand* A company has found that the daily demand x for its boxes of chocolates is inversely proportional to the price p. When the price is $5, the demand is 800 boxes. Approximate the demand when the price is increased to $6.

55. *Predator-Prey* The number N of prey t months after a natural predator is introduced into a test area is inversely proportional to $t + 1$. If $N = 500$ when $t = 0$, find N when $t = 4$.

56. *Weight of an Astronaut* A person's weight on the moon varies directly with his or her weight on Earth. An astronaut weighs 360 pounds on Earth, including heavy equipment. On the moon the astronaut weighs only 60 pounds with the equipment. If the first woman in space, Valentina Tereshkova, had landed on the moon and weighed 54 pounds with equipment, how much would she have weighed on Earth with her equipment?

57. *Pressure* When a person walks, the pressure P on each sole varies inversely with the area A of the sole. A person is trudging through deep snow, wearing boots that have a sole area of 29 square inches each. The sole pressure is 4 pounds per square inch. If the person was wearing snowshoes, each with an area 11 times that of their boot soles, what would be the pressure on each snowshoe? The constant of variation in this problem is the weight of the person. How much does the person weigh?

58. *Environment* The graph shows the percent p of oil that remained in Chedabucto Bay, Nova Scotia, after an oil spill. The cleaning of the spill was left primarily to natural actions such as wave motion, evaporation, photochemical decomposition, and bacterial decomposition. After about a year, the percent that remained varied inversely as time. Find a model that relates p and t, where t is the number of years since the spill. Then use it to find the percent of oil that remained $6\frac{1}{2}$ years after the spill, and compare the result with the graph.

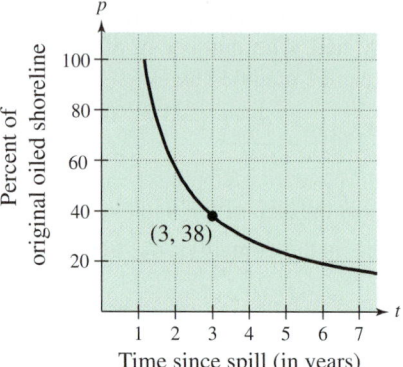

59. *Meteorology* The graph shows the temperature of the water in the north central Pacific Ocean. At depths greater than 900 meters, the water temperature varies inversely with the water depth. Find a model that relates the temperature T to the depth d. Then use it to find the water temperature at a depth of 4385 meters, and compare the result with the graph.

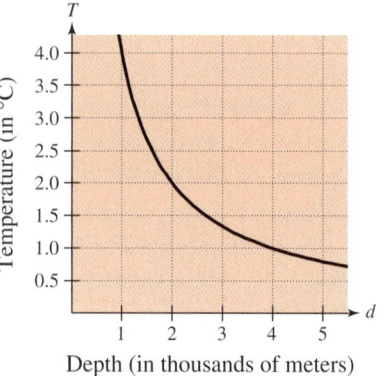

60. *Engineering*   The load P that can be safely supported by a horizontal beam varies jointly as the product of the width W of the beam and the square of the depth D and inversely as the length L. (See figure).

(a) Write a model for the statement.

(b) How does P change when the width and length of the beam are both doubled?

(c) How does P change when the width and depth of the beam are doubled?

(d) How does P change when all three of the dimensions are doubled?

(e) How does P change when the depth of the beam is cut in half?

(f) A beam with width 3 inches, depth 8 inches, and length 10 feet can safely support 2000 pounds. Determine the safe load of a beam made from the same material if its depth is increased to 10 inches.

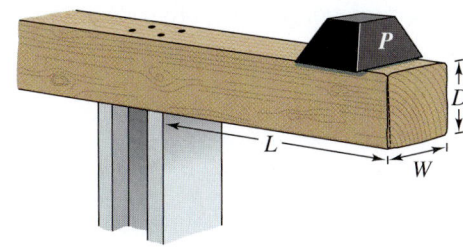

In Exercises 61–64, complete the table and plot the resulting points.

x	2	4	6	8	10
$y = kx^2$					

61. $k = 1$ **62.** $k = 2$

63. $k = \frac{1}{2}$ **64.** $k = \frac{1}{4}$

In Exercises 65–68, complete the table and plot the resulting points.

x	2	4	6	8	10
$y = \dfrac{k}{x^2}$					

65. $k = 2$ **66.** $k = 5$

67. $k = 10$ **68.** $k = 20$

In Exercises 69 and 70, determine whether the variation model is of the form $y = kx$ or $y = k/x$, and find k.

69.

x	10	20	30	40	50
y	$\frac{2}{5}$	$\frac{1}{5}$	$\frac{2}{15}$	$\frac{1}{10}$	$\frac{2}{25}$

70.

x	10	20	30	40	50
y	-3	-6	-9	-12	-15

Explaining Concepts

71. *Writing*   Suppose the constant of proportionality is positive and y varies directly as x. If one of the variables increases, how will the other change? Explain.

72. *Writing*   Suppose the constant of proportionality is positive and y varies inversely as x. If one of the variables increases, how will the other change? Explain.

73. *Writing*   If y varies directly as the square of x and x is doubled, how will y change? Use the rules of exponents to explain your answer.

74. *Writing*   If y varies inversely as the square of x and x is doubled, how will y change? Use the rules of exponents to explain your answer.

75. *Writing*   Describe a real life problem for each type of variation (direct, inverse, and joint).

What Did You Learn?

Key Terms

rational expression, *p. 400*
rational function, *p. 400*
domain (of a rational
 function), *p. 400*
simplified form, *p. 403*
least common multiple, *p. 422*

least common
 denominator, *p. 423*
complex fraction, *p. 431*
extraneous solution, *p. 442*
cross-multiplying, *p. 443*
direct variation, *p. 449*

constant of proportionality,
 p. 449
inverse variation, *p. 452*
combined variation, *p. 453*

Key Concepts

7.1 ◐ Simplifying rational expressions

Let u, v, and w represent real numbers, variables, or algebraic expressions such that $v \neq 0$ and $w \neq 0$. Then the following is valid.

$$\frac{uw}{vw} = \frac{u\cancel{w}}{v\cancel{w}} = \frac{u}{v}$$

7.2 ◐ Multiplying rational expressions

Let u, v, w, and z represent real numbers, variables, or algebraic expressions such that $v \neq 0$ and $z \neq 0$. Then the product of u/v and w/z is

$$\frac{u}{v} \cdot \frac{w}{z} = \frac{uw}{vz}.$$

7.2 ◐ Dividing rational expressions

Let u, v, w, and z represent real numbers, variables, or algebraic expressions such that $v \neq 0$, $w \neq 0$, and $z \neq 0$. Then the quotient of u/v and w/z is

$$\frac{u}{v} \div \frac{w}{z} = \frac{u}{v} \cdot \frac{z}{w} = \frac{uz}{vw}.$$

7.3 ◐ Adding or subtracting with like denominators

If u, v, and w are real numbers, variables, or algebraic expressions, and $w \neq 0$, the following rules are valid.

1. $\dfrac{u}{w} + \dfrac{v}{w} = \dfrac{u + v}{w}$ 2. $\dfrac{u}{w} - \dfrac{v}{w} = \dfrac{u - v}{w}$

7.3 ◐ Adding or subtracting with unlike denominators

Rewrite the rational expressions with like denominators by finding the least common denominator. Then add or subtract as with like denominators.

7.5 ◐ Solving rational equations

1. Determine the domain of each of the fractions in the equation.

2. Obtain an equivalent equation by multiplying each side of the equation by the least common denominator of all the fractions in the equation.

3. Solve the resulting equation.

4. Check your solution(s) in the original equation.

7.6 ◐ Variation models

In the following, k is a constant.

1. *Direct variation:* $y = kx$

2. *Direct variation as nth power:* $y = kx^n$

3. *Inverse variation:* $y = k/x$

4. *Inverse variation as nth power:* $y = k/x^n$

5. *Joint variation:* $z = kxy$

6. *Joint variation as nth and mth powers:* $z = kx^n y^m$

Review Exercises

7.1 Rational Expressions and Functions

1 Find the domain of a rational function.

In Exercises 1–4, find the domain of the rational function.

1. $f(y) = \dfrac{3y}{y - 8}$

2. $g(t) = \dfrac{t + 4}{t + 12}$

3. $g(u) = \dfrac{u}{u^2 - 7u + 6}$

4. $f(x) = \dfrac{x - 12}{x(x^2 - 16)}$

5. ▲ *Geometry* A rectangle with a width of w inches has an area of 36 square inches. The perimeter P of the rectangle is given by

$$P = 2\left(w + \dfrac{36}{w}\right).$$

Describe the domain of the function.

6. *Average Cost* The average cost $\overline{C}$ for a manufacturer to produce x units of a product is given by

$$\overline{C} = \dfrac{15{,}000 + 0.75x}{x}.$$

Describe the domain of the function.

2 Simplify rational expressions.

In Exercises 7-14, simplify the rational expression.

7. $\dfrac{6x^4y^2}{15xy^2}$

8. $\dfrac{2(y^3z)^2}{28(yz^2)^2}$

9. $\dfrac{5b - 15}{30b - 120}$

10. $\dfrac{4a}{10a^2 + 26a}$

11. $\dfrac{9x - 9y}{y - x}$

12. $\dfrac{x + 3}{x^2 - x - 12}$

13. $\dfrac{x^2 - 5x}{2x^2 - 50}$

14. $\dfrac{x^2 + 3x + 9}{x^3 - 27}$

7.2 Multiplying and Dividing Rational Expressions

1 Multiply rational expressions and simplify.

In Exercises 15–22, multiply and simplify.

15. $3x(x^2y)^2$

16. $2b(-3b)^3$

17. $\dfrac{7}{8} \cdot \dfrac{2x}{y} \cdot \dfrac{y^2}{14x^2}$

18. $\dfrac{15(x^2y)^3}{3y^3} \cdot \dfrac{12y}{x}$

19. $\dfrac{60z}{z + 6} \cdot \dfrac{z^2 - 36}{5}$

20. $\dfrac{x^2 - 16}{6} \cdot \dfrac{3}{x^2 - 8x + 16}$

21. $\dfrac{u}{u - 3} \cdot \dfrac{3u - u^2}{4u^2}$

22. $x^2 \cdot \dfrac{x + 1}{x^2 - x} \cdot \dfrac{(5x - 5)^2}{x^2 + 6x + 5}$

② Divide rational expressions and simplify.

In Exercises 23–30, divide and simplify.

23. $\dfrac{24x^4}{15x}$

24. $\dfrac{8u^2v}{6v}$

25. $25y^2 \div \dfrac{xy}{5}$

26. $\dfrac{6}{z^2} \div 4z^2$

27. $\dfrac{x^2 + 3x + 2}{3x^2 + x - 2} \div (x + 2)$

28. $\dfrac{x^2 - 14x + 48}{x^2 - 6x} \div (3x - 24)$

29. $\dfrac{x^2 - 7x}{x + 1} \div \dfrac{x^2 - 14x + 49}{x^2 - 1}$

30. $\dfrac{x^2 - x}{x + 1} \div \dfrac{5x - 5}{x^2 + 6x + 5}$

7.3 Adding and Subtracting Rational Expressions

① Add or subtract rational expressions with like denominators and simplify.

In Exercises 31–38, combine and simplify.

31. $\dfrac{4x}{5} + \dfrac{11x}{5}$

32. $\dfrac{7y}{12} - \dfrac{4y}{12}$

33. $\dfrac{15}{3x} - \dfrac{3}{3x}$

34. $\dfrac{4}{5x} + \dfrac{1}{5x}$

35. $\dfrac{2(3y + 4)}{2y + 1} + \dfrac{3 - y}{2y + 1}$

36. $\dfrac{4x - 2}{3x + 1} - \dfrac{x + 1}{3x + 1}$

37. $\dfrac{4x}{x + 2} + \dfrac{3x - 7}{x + 2} - \dfrac{9}{x + 2}$

38. $\dfrac{3}{2y - 3} - \dfrac{y - 10}{2y - 3} + \dfrac{5y}{2y - 3}$

② Add or subtract rational expressions with unlike denominators and simplify.

In Exercises 39–46, combine and simplify.

39. $\dfrac{1}{x + 5} + \dfrac{3}{x - 12}$

40. $\dfrac{2}{x - 10} + \dfrac{3}{4 - x}$

41. $5x + \dfrac{2}{x - 3} - \dfrac{3}{x + 2}$

42. $4 - \dfrac{4x}{x + 6} + \dfrac{7}{x - 5}$

43. $\dfrac{6}{x - 5} - \dfrac{4x + 7}{x^2 - x - 20}$

44. $\dfrac{5}{x + 2} + \dfrac{25 - x}{x^2 - 3x - 10}$

45. $\dfrac{5}{x + 3} - \dfrac{4x}{(x + 3)^2} - \dfrac{1}{x - 3}$

46. $\dfrac{8}{y} - \dfrac{3}{y + 5} + \dfrac{4}{y - 2}$

In Exercises 47 and 48, use a graphing calculator to graph the two equations in the same viewing window. Use the graphs to verify that the expressions are equivalent. Verify the results algebraically.

47. $y_1 = \dfrac{1}{x} - \dfrac{3}{x + 3}$

 $y_2 = \dfrac{3 - 2x}{x(x + 3)}$

48. $y_1 = \dfrac{5x}{x - 5} + \dfrac{7}{x + 1}$

 $y_2 = \dfrac{5x^2 + 12x - 35}{x^2 - 4x - 5}$

7.4 Complex Fractions

① Simplify complex fractions using rules for dividing rational expressions.

In Exercises 49–52, simplify the complex fraction.

49. $\dfrac{\left(\dfrac{6}{x}\right)}{\left(\dfrac{2}{x^3}\right)}$

50. $\dfrac{xy}{\left(\dfrac{5x^2}{2y}\right)}$

51. $\dfrac{\left(\dfrac{6x^2}{x^2 + 2x - 35}\right)}{\left(\dfrac{x^3}{x^2 - 25}\right)}$

52. $\dfrac{\left[\dfrac{24 - 18x}{(2 - x)^2}\right]}{\left(\dfrac{60 - 45x}{x^2 - 4x + 4}\right)}$

② Simplify complex fractions having a sum or difference in the numerator and/or denominator.

In Exercises 53–58, simplify the complex fraction.

53. $\dfrac{3t}{\left(5 - \dfrac{2}{t}\right)}$

54. $\dfrac{\left(\dfrac{1}{x} - \dfrac{1}{2}\right)}{2x}$

55. $\dfrac{\left(x - 3 + \dfrac{2}{x}\right)}{\left(1 - \dfrac{2}{x}\right)}$

56. $\dfrac{3x - 1}{\left(\dfrac{2}{x^2} + \dfrac{5}{x}\right)}$

57. $\dfrac{\left(\dfrac{1}{a^2 - 16} - \dfrac{1}{a}\right)}{\left(\dfrac{1}{a^2 + 4a} + 4\right)}$

58. $\dfrac{\left(\dfrac{1}{x^2} - \dfrac{1}{y^2}\right)}{\left(\dfrac{1}{x} + \dfrac{1}{y}\right)}$

7.5 Solving Rational Equations

① Solve rational equations containing constant denominators.

In Exercises 59 and 60, solve the equation.

59. $\dfrac{3x}{8} = -15 + \dfrac{x}{4}$

60. $\dfrac{t + 1}{6} = \dfrac{1}{2} - 2t$

② Solve rational equations containing variable denominators.

In Exercises 61–74, solve the equation.

61. $8 - \dfrac{12}{t} = \dfrac{1}{3}$

62. $5 + \dfrac{2}{x} = \dfrac{1}{4}$

63. $\dfrac{2}{y} - \dfrac{1}{3y} = \dfrac{1}{3}$

64. $\dfrac{7}{4x} - \dfrac{6}{8x} = 1$

65. $r = 2 + \dfrac{24}{r}$

66. $\dfrac{2}{x} - \dfrac{x}{6} = \dfrac{2}{3}$

67. $8\left(\dfrac{6}{x} - \dfrac{1}{x + 5}\right) = 15$

68. $\dfrac{3}{y + 1} - \dfrac{8}{y} = 1$

69. $\dfrac{4x}{x - 5} + \dfrac{2}{x} = -\dfrac{4}{x - 5}$

70. $\dfrac{2x}{x - 3} - \dfrac{3}{x} = 0$

71. $\dfrac{12}{x^2 + x - 12} - \dfrac{1}{x - 3} = -1$

72. $\dfrac{3}{x - 1} + \dfrac{6}{x^2 - 3x + 2} = 2$

73. $\dfrac{5}{x^2 - 4} - \dfrac{6}{x - 2} = -5$

74. $\dfrac{3}{x^2 - 9} + \dfrac{4}{x + 3} = 1$

7.6 Applications and Variation

1 Solve application problems involving rational equations.

75. *Average Speed*   You drive 56 miles on a service call for your company. On the return trip, which takes 10 minutes less than the original trip, your average speed is 8 miles per hour greater. What is your average speed on the return trip?

76. *Average Speed*   You drive 220 miles to see a friend. On the return trip, which takes 20 minutes less than the original trip, your average speed is 5 miles per hour faster. What is your average speed on the return trip?

77. *Partnership Costs*   A group of people starting a business agree to share equally in the cost of a $60,000 piece of machinery. If they could find two more people to join the group, each person's share of the cost would decrease by $5000. How many people are presently in the group?

78. *Work Rate*   One painter works $1\frac{1}{2}$ times as fast as another painter. It takes them 4 hours working together to paint a room. Find the time it takes each painter to paint the room working alone.

79. *Population Growth*   The Parks and Wildlife Commission introduces 80,000 fish into a large lake. The population P (in thousands) of the fish is approximated by the model

$$P = \dfrac{20(4 + 3t)}{1 + 0.05t}$$

where t is the time in years. Find the time required for the population to increase to 400,000 fish.

80. *Average Cost*   The average cost $\overline{C}$ for producing x units of a product is given by

$$\overline{C} = 1.5 + \dfrac{4200}{x}.$$

Determine the number of units that must be produced to obtain an average cost of $2.90 per unit.

2 Solve application problems involving direct variation.

81. *Hooke's Law*   A force of 100 pounds stretches a spring 4 inches. Find the force required to stretch the spring 6 inches.

82. *Stopping Distance*   The stopping distance d of an automobile is directly proportional to the square of its speed s. How will the stopping distance be changed by doubling the speed of the car?

3 Solve application problems involving inverse variation.

83. *Travel Time*   The travel time between two cities is inversely proportional to the average speed. A train travels between the cities in 3 hours at an average speed of 65 miles per hour. How long would it take to travel between the cities at an average speed of 80 miles per hour?

84. *Demand*   A company has found that the daily demand x for its cordless telephones is inversely proportional to the price p. When the price is $25, the demand is 1000 telephones. Approximate the demand when the price is increased to $28.

4 Solve application problems involving joint variation.

85. *Simple Interest*   The simple interest for a savings account is jointly proportional to the time and the principal. After three quarters (9 months), the interest for a principal of $12,000 is $675. How much interest would a principal of $8200 earn in 18 months?

86. *Cost*   The cost of constructing a wooden box with a square base varies jointly as the height of the box and the square of the width of the box. A box of height 16 inches and of width 6 inches costs $28.80. How much would a box of height 14 inches and of width 8 inches cost?

Chapter Test

Take this test as you would take a test in class. After you are done, check your work against the answers in the back of the book.

1. Find the domain of $f(x) = \dfrac{2x}{x^2 - 5x + 6}$.

In Exercises 2 and 3, simplify the rational expression.

2. $\dfrac{2 - x}{3x - 6}$

3. $\dfrac{2a^2 - 5a - 12}{5a - 20}$

4. Find the least common multiple of x^2, $3x^3$, and $(x + 4)^2$.

In Exercises 5–16, perform the operation and simplify.

5. $\dfrac{4z^3}{5} \cdot \dfrac{25}{12z^2}$

6. $\dfrac{y^2 + 8y + 16}{2(y - 2)} \cdot \dfrac{8y - 16}{(y + 4)^3}$

7. $(4x^2 - 9) \div \dfrac{2x + 3}{2x^2 - x - 3}$

8. $\dfrac{(2xy^2)^3}{15} \div \dfrac{12x^3}{21}$

9. $2x + \dfrac{1 - 4x^2}{x + 1}$

10. $\dfrac{5x}{x + 2} - \dfrac{2}{x^2 - x - 6}$

11. $\dfrac{3}{x} - \dfrac{5}{x^2} + \dfrac{2x}{x^2 + 2x + 1}$

12. $\dfrac{4}{x + 1} + \dfrac{4x}{x + 1}$

13. $\dfrac{\left(\dfrac{3x}{x + 2}\right)}{\left(\dfrac{12}{x^3 + 2x^2}\right)}$

14. $\dfrac{\left(9x - \dfrac{1}{x}\right)}{\left(\dfrac{1}{x} - 3\right)}$

15. $\dfrac{3x^{-2} + y^{-1}}{(x + y)^{-1}}$

16. $\dfrac{2b - a}{4ba^{-1} - ab^{-1}}$

In Exercises 17–19, solve the equation.

17. $\dfrac{3}{h + 2} = \dfrac{1}{8}$

18. $\dfrac{2}{x + 5} - \dfrac{3}{x + 3} = \dfrac{1}{x}$

19. $\dfrac{1}{x + 1} + \dfrac{1}{x - 1} = \dfrac{2}{x^2 - 1}$

20. Find a mathematical model that relates u and v if v varies directly as the square root of u, and $v = \frac{3}{2}$ when $u = 36$.

21. If the temperature of a gas is not allowed to change, the absolute pressure P of the gas is inversely proportional to its volume V, according to Boyle's Law. A large balloon is filled with 180 cubic meters of helium at atmospheric pressure (1 atm) at sea level. What is the volume of the helium if the balloon rises to an altitude at which the atmospheric pressure is 0.75 atm? (Assume that the temperature does not change.)

Motivating the Chapter

Soccer Club Fundraiser

A collegiate soccer club has a fundraising dinner. Student tickets sell for $8 and nonstudent tickets sell for $15. There are 115 tickets sold and the total revenue is $1445.

See Section 8.1, Exercise 117.

a. Set up a system of linear equations that can be used to determine how many tickets of each type were sold.

b. Solve the system in part (a) by the method of substitution.

The soccer club decides to set goals for the next fundraising dinner. To meet these goals, a "major contributor" category is added. A person donating $100 is considered a major contributor to the soccer club and receives a "free" ticket to the dinner. The club's goals are to have 200 people in attendance, with the number of major contributors being one-fourth the number of students, and to raise $4995.

See Section 8.3, Exercise 55.

c. Set up a system of linear equations to determine how many of each kind of ticket would need to be sold for the second fundraising dinner.

d. Solve the system in part (c) by Gaussian elimination.

e. Would it be possible for the soccer club to meet its goals if only 18 people donated $100? Explain.

See Section 8.5, Exercise 93.

f. Solve the system in part (c) using matrices.

g. Solve the system in part (c) using determinants.

Paul A. Souders/Corbis

8

Systems of Equations and Inequalities

8.1 Solving Systems of Equations by Graphing and Substitution

Brian Hagawara/Foodpix/Getty Images

What You Should Learn

1 Determine if an ordered pair is a solution to a system of equations.

2 Use a coordinate system to solve systems of linear equations graphically.

3 Use the method of substitution to solve systems of equations algebraically.

4 Solve application problems using systems of equations.

Why You Should Learn It

Many businesses use systems of equations to help determine their sales goals. For instance, Example 12 on page 477 shows how to graph a system of equations to determine the break-even point of producing and selling a new energy bar.

1 Determine if an ordered pair is a solution to a system of equations.

Systems of Equations

Many problems in business and science involve **systems of equations.** These systems consist of two or more equations, each containing two or more variables.

$$\begin{cases} ax + by = c & \text{Equation 1} \\ dx + ey = f & \text{Equation 2} \end{cases}$$

A **solution** of such a system is an ordered pair (x, y) of real numbers that satisfies *each* equation in the system. When you find the set of all solutions of the system of equations, you are finding the **solution of the system of equations.**

Example 1   Checking Solutions of a System of Equations

Which of the ordered pairs is a solution of the system: (a) $(3, 3)$ or (b) $(4, 2)$?

$$\begin{cases} x + y = 6 & \text{Equation 1} \\ 2x - 5y = -2 & \text{Equation 2} \end{cases}$$

Solution

a. To determine whether the ordered pair $(3, 3)$ is a solution of the system of equations, you should substitute 3 for x and 3 for y in *each* of the equations. Substituting into Equation 1 produces

$3 + 3 = 6.$ ✔︎   Substitute 3 for x and 3 for y.

Similarly, substituting into Equation 2 produces

$2(3) - 5(3) \neq -2.$ ✘   Substitute 3 for x and 3 for y.

Because the ordered pair $(3, 3)$ fails to check in *both* equations, you can conclude that it *is not* a solution of the system of equations.

b. By substituting 4 for x and 2 for y in the original equations, you can determine that the ordered pair $(4, 2)$ is a solution of both equations.

$4 + 2 = 6$ ✔︎   Substitute 4 for x and 2 for y in Equation 1.

$2(4) - 5(2) = -2$ ✔︎   Substitute 4 for x and 2 for y in Equation 2.

So, $(4, 2)$ *is* a solution of the original system of equations.

2 Use a coordinate system to solve systems of linear equations graphically.

Solving a System of Linear Equations by Graphing

In this chapter you will study three methods of solving a system of linear equations. The first method is *solution by graphing*. With this method, you first sketch the lines representing the equations. Then you try to determine whether the lines intersect at a point, as illustrated in Example 2.

Example 2   Solving a System of Linear Equations

Use the graphical method to solve the system of linear equations.

$$\begin{cases} 2x + 3y = 7 & \text{Equation 1} \\ 2x - 5y = -1 & \text{Equation 2} \end{cases}$$

Solution

Because both equations in the system are linear, you know that they have graphs that are lines. To sketch these lines, first write each equation in slope-intercept form, as follows.

$$\begin{cases} y = -\dfrac{2}{3}x + \dfrac{7}{3} & \text{Slope-intercept form of Equation 1} \\ y = \dfrac{2}{5}x + \dfrac{1}{5} & \text{Slope-intercept form of Equation 2} \end{cases}$$

The lines corresponding to these two equations are shown in Figure 8.1.

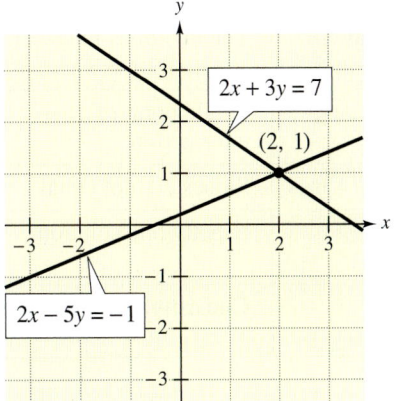

Figure 8.1

It appears that the two lines intersect at a single point, $(2, 1)$. To verify this, substitute the coordinates of the point into each of the two original equations.

Substitute in 1st Equation		*Substitute in 2nd Equation*	
$2x + 3y = 7$	Equation 1	$2x - 5y = -1$	Equation 2
$2(2) + 3(1) \overset{?}{=} 7$	Substitute 2 for x and 1 for y.	$2(2) - 5(1) \overset{?}{=} -1$	Substitute 2 for x and 1 for y.
$7 = 7$	Solution checks. ✓	$-1 = -1$	Solution checks. ✓

Because *both* equations are satisfied, the point $(2, 1)$ is the solution of the system.

> ### Technology: Discovery
>
> Rewrite each system of equations in slope-intercept form and graph the equations using a graphing calculator. What is the relationship between the slopes of the two lines and the number of points of intersection?
>
> **a.** $\begin{cases} 3x + 4y = 12 \\ 2x - 3y = -9 \end{cases}$ **b.** $\begin{cases} -x + 2y = 8 \\ 2x - 4y = 5 \end{cases}$ **c.** $\begin{cases} x + y = 6 \\ 3x + 3y = 18 \end{cases}$

A system of linear equations can have exactly one solution, infinitely many solutions, or no solution. To see why this is true, consider the graphical interpretations of three systems of two linear equations shown below.

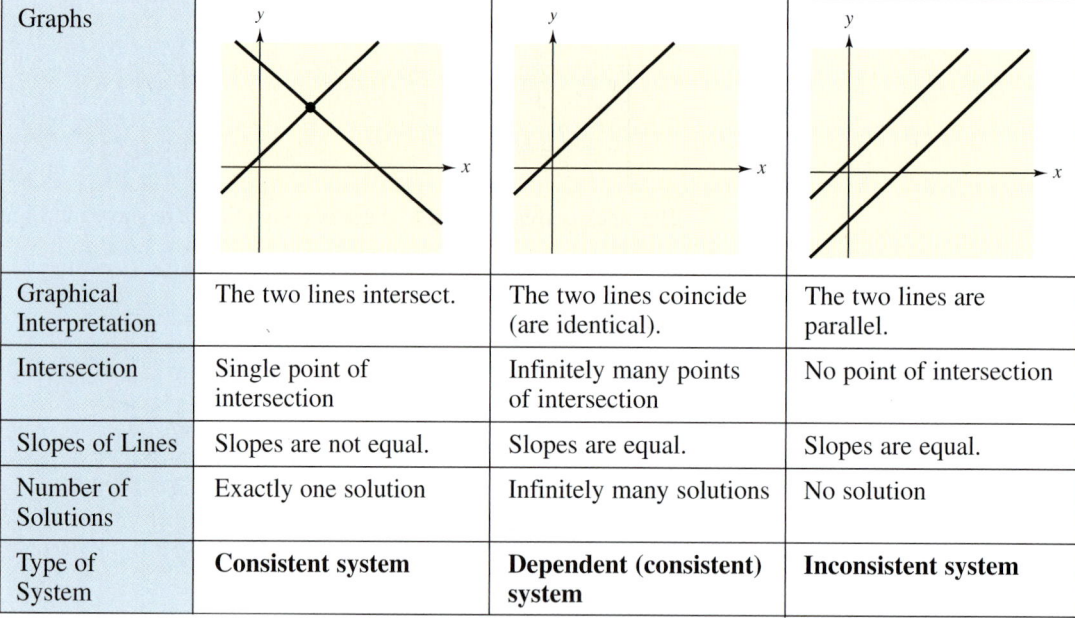

Graphs			
Graphical Interpretation	The two lines intersect.	The two lines coincide (are identical).	The two lines are parallel.
Intersection	Single point of intersection	Infinitely many points of intersection	No point of intersection
Slopes of Lines	Slopes are not equal.	Slopes are equal.	Slopes are equal.
Number of Solutions	Exactly one solution	Infinitely many solutions	No solution
Type of System	**Consistent system**	**Dependent (consistent) system**	**Inconsistent system**

Note that for dependent systems, the slopes of the lines and the y-intercepts are equal. For inconsistent systems, the slopes are equal, but the y-intercepts of the two lines are different. Also, note that the word *consistent* is used to mean that the system of linear equations has at least one solution, whereas the word *inconsistent* is used to mean that the system of linear equations has no solution.

You can see from the graphs above that a comparison of the slopes of two lines gives useful information about the number of solutions of the corresponding system of equations. So, to solve a system of equations graphically, it helps to begin by writing the equations in slope-intercept form,

$$y = mx + b. \qquad \text{\color{red}Slope-intercept form}$$

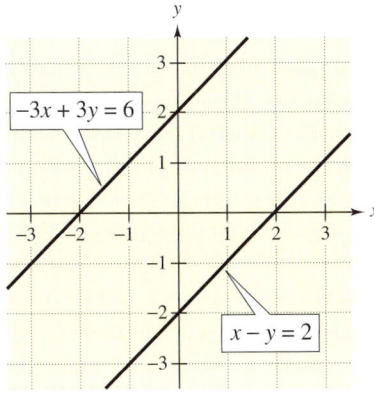

Figure 8.2

Example 3   A System with No Solution

Solve the system of linear equations.

$$\begin{cases} x - y = 2 & \text{Equation 1} \\ -3x + 3y = 6 & \text{Equation 2} \end{cases}$$

Solution

Begin by writing each equation in slope-intercept form.

$$\begin{cases} y = x - 2 & \text{Slope-intercept form of Equation 1} \\ y = x + 2 & \text{Slope-intercept form of Equation 2} \end{cases}$$

From these slope-intercept forms, you can see that the lines representing the two equations are parallel (each has a slope of 1), as shown in Figure 8.2. So, the original system of linear equations has no solution and is an inconsistent system. Try constructing tables of values for the two equations. The tables should help convince you that there is no solution.

Example 4   A System with Infinitely Many Solutions

Solve the system of linear equations.

$$\begin{cases} x - y = 2 & \text{Equation 1} \\ -3x + 3y = -6 & \text{Equation 2} \end{cases}$$

Solution

Begin by writing each equation in slope-intercept form.

$$\begin{cases} y = x - 2 & \text{Slope-intercept form of Equation 1} \\ y = x - 2 & \text{Slope-intercept form of Equation 2} \end{cases}$$

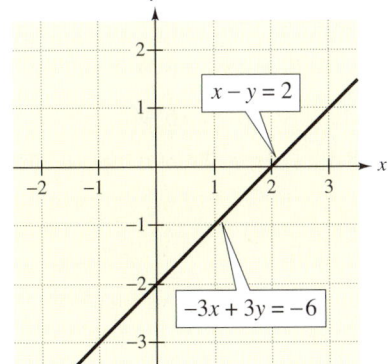

Figure 8.3

From these forms, you can see that the lines representing the two equations are the same (see Figure 8.3). So, the original system of linear equations is dependent and has infinitely many solutions. You can describe the solution set by saying that each point on the line $y = x - 2$ is a solution of the system of linear equations.

Note in Examples 3 and 4 that if the two lines representing a system of linear equations have the same slope, the system must have either no solution or infinitely many solutions. On the other hand, if the two lines have different slopes, they must intersect at a single point and the corresponding system has a single solution.

There are two things you should note as you read through Examples 5 and 6. First, your success in applying the graphical method of solving a system of linear equations depends on sketching accurate graphs. Second, once you have made a graph and estimated the point of intersection, it is critical that you check in the original system to see whether the point you have chosen is the correct solution.

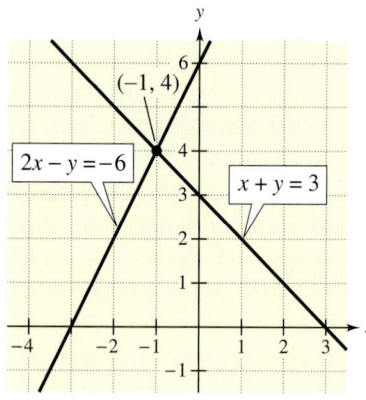

Figure 8.4

Example 5   A System with a Single Solution

Solve the system of linear equations.

$$\begin{cases} x + y = 3 & \text{Equation 1} \\ 2x - y = -6 & \text{Equation 2} \end{cases}$$

Solution

Begin by writing each equation in slope-intercept form.

$$\begin{cases} y = -x + 3 & \text{Slope-intercept form of Equation 1} \\ y = 2x + 6 & \text{Slope-intercept form of Equation 2} \end{cases}$$

Because the lines do not have the same slope, you know that they intersect. To find the point of intersection, sketch both lines on the same rectangular coordinate system, as shown in Figure 8.4. From this sketch, it appears that the solution occurs at the point $(-1, 4)$. To check this solution, substitute the coordinates of the point into each of the two original equations.

Substitute in 1st Equation

$$x + y = 3$$
$$-1 + 4 \overset{?}{=} 3$$
$$3 = 3 \ \checkmark$$

Substitute in 2nd Equation

$$2x - y = -6$$
$$2(-1) - 4 \overset{?}{=} -6$$
$$-2 - 4 \overset{?}{=} -6$$
$$-6 = -6 \ \checkmark$$

Because *both* equations are satisfied, the point $(-1, 4)$ is the solution of the system.

Example 6   A System with a Single Solution

Solve the system of linear equations.

$$\begin{cases} 2x + y = 4 & \text{Equation 1} \\ 4x + 3y = 9 & \text{Equation 2} \end{cases}$$

Solution

Begin by writing each equation in slope-intercept form.

$$\begin{cases} y = -2x + 4 & \text{Slope-intercept form of Equation 1} \\ y = -\tfrac{4}{3}x + 3 & \text{Slope-intercept form of Equation 2} \end{cases}$$

Because the lines do not have the same slope, you know that they intersect. To find the point of intersection, sketch both lines on the same rectangular coordinate system, as shown in Figure 8.5. From this sketch, it appears that the solution occurs at the point $\left(\tfrac{3}{2}, 1\right)$. To check this solution, substitute the coordinates of the point into each of the two original equations.

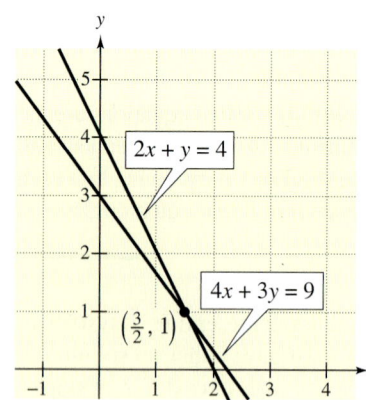

Figure 8.5

Substitute in 1st Equation

$$2x + y = 4$$
$$2\left(\tfrac{3}{2}\right) + 1 \overset{?}{=} 4$$
$$3 + 1 = 4 \ \checkmark$$

Substitute in 2nd Equation

$$4x + 3y = 9$$
$$4\left(\tfrac{3}{2}\right) + 3(1) \overset{?}{=} 4$$
$$6 + 3 = 9 \ \checkmark$$

Because *both* equations are satisfied, the point $\left(\tfrac{3}{2}, 1\right)$ is the solution of the system.

The Method of Substitution

Solving systems of equations by graphing is useful but less accurate than using algebraic methods. In this section, you will study an algebraic method called the **method of substitution.** The goal of the method of substitution is to *reduce a system of two linear equations in two variables to a single equation in one variable.* Examples 7 and 8 illustrate the basic steps of the method.

3 Use the method of substitution to solve systems of equations algebraically.

Example 7   The Method of Substitution

Solve the system of linear equations.

$$\begin{cases} -x + y = 1 & \text{Equation 1} \\ 2x + y = -2 & \text{Equation 2} \end{cases}$$

Solution

Begin by solving for y in Equation 1.

$$-x + y = 1 \qquad \text{Original Equation 1}$$

$$y = x + 1 \qquad \text{Revised Equation 1}$$

Next, substitute this expression for y in Equation 2.

$$2x + y = -2 \qquad \text{Equation 2}$$
$$2x + (x + 1) = -2 \qquad \text{Substitute } x + 1 \text{ for } y.$$
$$3x + 1 = -2 \qquad \text{Combine like terms.}$$
$$3x = -3 \qquad \text{Subtract 1 from each side.}$$
$$x = -1 \qquad \text{Divide each side by 3.}$$

At this point, you know that the x-coordinate of the solution is -1. To find the y-coordinate, *back-substitute* the x-value in the revised Equation 1.

$$y = x + 1 \qquad \text{Revised Equation 1}$$
$$y = -1 + 1 \qquad \text{Substitute } -1 \text{ for } x.$$
$$y = 0 \qquad \text{Simplify.}$$

So, the solution is $(-1, 0)$. Check this solution by substituting $x = -1$ and $y = 0$ in both of the original equations.

Study Tip

The term **back-substitute** implies that you work backwards. After solving for one of the variables, substitute that value back into one of the equations in the original (or revised) system to find the value of the other variable.

When you use substitution, it does not matter which variable you choose to solve for first. You should choose the variable and equation that are easier to work with. For instance, in the system below on the left, it is best to solve for x in Equation 2, whereas for the system on the right, it is best to solve for y in Equation 1.

$$\begin{cases} 3x - 2y = 1 & \text{Equation 1} \\ x + 4y = 3 & \text{Equation 2} \end{cases} \qquad \begin{cases} 2x + y = 5 & \text{Equation 1} \\ 3x - 2y = 11 & \text{Equation 2} \end{cases}$$

Example 8   The Method of Substitution

Solve the system of linear equations.

$$\begin{cases} 5x + 7y = 1 & \text{Equation 1} \\ x + 4y = -5 & \text{Equation 2} \end{cases}$$

Solution

For this system, it is convenient to begin by solving for x in the second equation.

$x + 4y = -5$	Original Equation 2
$x = -4y - 5$	Revised Equation 2

Substituting this expression for x into the first equation produces the following.

$5(-4y - 5) + 7y = 1$	Substitute $-4y - 5$ for x in Equation 1.
$-20y - 25 + 7y = 1$	Distributive Property
$-13y - 25 = 1$	Combine like terms.
$-13y = 26$	Add 25 to each side.
$y = -2$	Divide each side by -13.

Finally, back-substitute this y-value into the revised second equation.

$x = -4(-2) - 5 = 3$	Substitute -2 for y in revised Equation 2.

The solution is $(3, -2)$. Check this by substituting $x = 3$ and $y = -2$ in both of the original equations, as follows.

Substitute in Equation 1	*Substitute in Equation 2*
$5x + 7y = 1$	$x + 4y = -5$
$5(3) + 7(-2) \overset{?}{=} 1$	$(3) + 4(-2) \overset{?}{=} -5$
$15 - 14 = 1 \checkmark$	$3 - 8 = -5 \checkmark$

The steps for using the method of substitution to solve a system of equations involving two variables are summarized as follows.

The Method of Substitution

1. Solve one of the equations for one variable in terms of the other.

2. Substitute the expression obtained in Step 1 in the other equation to obtain an equation in one variable.

3. Solve the equation obtained in Step 2.

4. Back-substitute the solution from Step 3 in the expression obtained in Step 1 to find the value of the other variable.

5. Check the solution to see that it satisfies *both* of the original equations.

If neither variable has a coefficient of 1 in a system of linear equations, you can still use the method of substitution. However, you may have to work with some fractions in the solution steps.

Example 9   The Method of Substitution

Solve the system of linear equations.

$$\begin{cases} 5x + 3y = 18 & \text{Equation 1} \\ 2x - 7y = -1 & \text{Equation 2} \end{cases}$$

Solution

Step 1 Because neither variable has a coefficient of 1, you can choose to solve for either variable. For instance, you can begin by solving for x in Equation 1 to obtain $x = -\frac{3}{5}y + \frac{18}{5}$.

Step 2 Substitute for x in Equation 2 and solve for y.

$$2x - 7y = -1 \qquad\qquad \text{Equation 2}$$
$$2\left(-\tfrac{3}{5}y + \tfrac{18}{5}\right) - 7y = -1 \qquad\qquad \text{Substitute } -\tfrac{3}{5}y + \tfrac{18}{5} \text{ for } x.$$

Step 3
$$-\tfrac{6}{5}y + \tfrac{36}{5} - 7y = -1 \qquad\qquad \text{Distributive Property}$$
$$-6y + 36 - 35y = -5 \qquad\qquad \text{Multiply each side by 5.}$$
$$y = 1 \qquad\qquad \text{Solve for } y.$$

Step 4 Back-substitute for y in the revised first equation.

$$x = -\tfrac{3}{5}y + \tfrac{18}{5} \qquad\qquad \text{Revised Equation 1}$$
$$x = -\tfrac{3}{5}(1) + \tfrac{18}{5} = 3 \qquad\qquad \text{Substitute 1 for } y.$$

Step 5 The solution is $(3, 1)$. Check this in the original system.

Example 10   The Method of Substitution: No-Solution Case

Solve the system of linear equations.

$$\begin{cases} x - 3y = 2 & \text{Equation 1} \\ -2x + 6y = 2 & \text{Equation 2} \end{cases}$$

Solution

Begin by solving for x in Equation 1 to obtain $x = 3y + 2$. Then, substitute this expression for x in Equation 2.

$$-2x + 6y = 2 \qquad\qquad \text{Equation 2}$$
$$-2(3y + 2) + 6y = 2 \qquad\qquad \text{Substitute } 3y + 2 \text{ for } x.$$
$$-6y - 4 + 6y = 2 \qquad\qquad \text{Distributive Property}$$
$$-4 \neq 2 \qquad\qquad \text{Simplify.}$$

Because -4 does not equal 2, you can conclude that the original system is inconsistent and has no solution. The graphs in Figure 8.6 confirm this result.

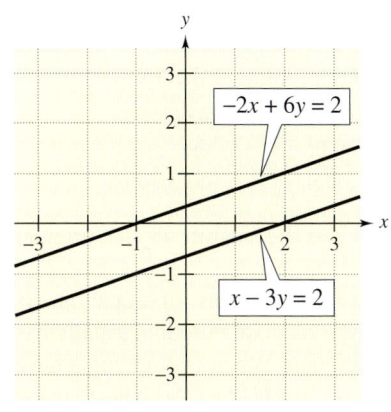

Figure 8.6

<u>Example 11</u>   **The Method of Substitution: Many-Solution Case**

Solve the system of linear equations.

$$\begin{cases} 9x + 3y = 15 & \text{Equation 1} \\ 3x + y = 5 & \text{Equation 2} \end{cases}$$

Solution

Begin by solving for y in Equation 2 to obtain $y = -3x + 5$. Then, substitute this expression for y in Equation 1.

$9x + 3y = 15$	Equation 1
$9x + 3(-3x + 5) = 15$	Substitute $-3x + 5$ for y.
$9x - 9x + 15 = 15$	Distributive Property
$15 = 15$	Simplify.

The equation $15 = 15$ is true for any value of x. This implies that any solution of Equation 2 is also a solution of Equation 1. In other words, the original system of linear equations is *dependent* and has infinitely many solutions. The solutions consist of all ordered pairs (x, y) lying on the line $3x + y = 5$. Some sample solutions are $(-1, 8)$, $(0, 5)$, and $(1, 2)$. Check these as follows:

Solution Point	*Substitute into* $3x + y = 5$
$(-1, 8)$	$3(-1) + 8 = -3 + 8 = 5$ ✓
$(0, 5)$	$3(0) + 5 = 0 + 5 = 5$ ✓
$(1, 2)$	$3(1) + 2 = 3 + 2 = 5$ ✓

By writing both equations in Example 11 in slope-intercept form, you will get identical equations. This means that the lines coincide and the system has infinitely many solutions.

④ Solve application problems using systems of equations.

Applications

To model a real-life situation with a system of equations, you can use the same basic problem-solving strategy that has been used throughout the text.

$$\boxed{\text{Write a verbal model.}} \rightarrow \boxed{\text{Assign labels.}} \rightarrow \boxed{\text{Write an algebraic model.}} \rightarrow \boxed{\text{Solve the algebraic model.}} \rightarrow \boxed{\text{Answer the question.}}$$

After answering the question, remember to check the answer in the original statement of the problem.

A common business application that involves systems of equations is break-even analysis. The total cost C of producing x units of a product usually has two components—the *initial cost* and the *cost per unit*. When enough units have been sold so that the total revenue R equals the total cost C, the sales have reached the **break-even point.** You can find this break-even point by finding the point of intersection of the cost and revenue graphs.

Example 12   Break-Even Analysis

A small business invests $14,000 to produce a new energy bar. Each bar costs $0.80 to produce and is sold for $1.50. How many energy bars must be sold before the business breaks even?

Solution

Verbal Model:

| Total cost | = | Cost per bar | · | Number of bars | + | Initial cost |

| Total revenue | = | Price per bar | · | Number of bars |

Labels:

Total cost = C	(dollars)
Cost per bar = 0.80	(dollars per bar)
Number of bars = x	(bars)
Initial cost = 14,000	(dollars)
Total revenue = R	(dollars)
Price per bar = 1.50	(dollars per bar)

System:
$$\begin{cases} C = 0.80x + 14{,}000 & \text{Equation 1} \\ R = 1.50x & \text{Equation 2} \end{cases}$$

The two equations are in slope-intercept form and because the lines do not have the same slope, you know that they intersect. So, to find the break-even point, graph both equations and determine the point of intersection of the two graphs, as shown in Figure 8.7. From this graph, it appears that the break-even point occurs at the point (20,000, 30,000). To check this solution, substitute the coordinates of the point in each of the two original equations.

Substitute in Equation 1

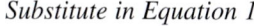

$C = 0.80x + 14{,}000$	Equation 1
$30{,}000 \overset{?}{=} 0.80(20{,}000) + 14{,}000$	Substitute 20,000 for x and 30,000 for C.
$30{,}000 \overset{?}{=} 16{,}000 + 14{,}000$	Multiply.
$30{,}000 = 30{,}000$ ✓	Simplify.

Substitute in Equation 2

$R = 1.50x$	Equation 2
$30{,}000 \overset{?}{=} 1.50(20{,}000)$	Substitute 20,000 for x and 30,000 for R.
$30{,}000 = 30{,}000$ ✓	Simplify.

Because both equations are satisfied, the business must sell 20,000 energy bars before it breaks even.

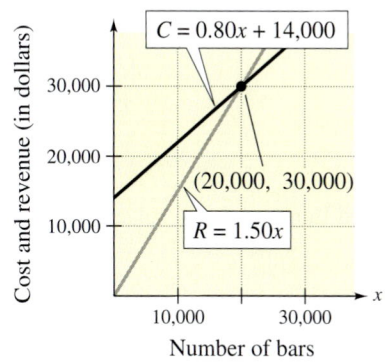

Figure 8.7

Profit P (or loss) for the business can be determined by the equation $P = R - C$. Note in Figure 8.7 that sales less than the break-even point correspond to a loss for the business, whereas sales greater than the break-even point correspond to a profit for the business.

Example 13 Simple Interest

A total of $12,000 is invested in two funds paying 6% and 8% simple interest. The combined annual interest for the two funds is $880. How much of the $12,000 is invested at each rate?

Solution

Verbal Model:

$$\boxed{\text{Amount in 6\% fund}} + \boxed{\text{Amount in 8\% fund}} = \boxed{12{,}000}$$

$$\boxed{6\%} \cdot \boxed{\text{Amount in 6\% fund}} + \boxed{8\%} \cdot \boxed{\text{Amount in 8\% fund}} = \boxed{880}$$

Labels: Amount in 6% fund = x (dollars)
Amount in 8% fund = y (dollars)

System:
$$\begin{cases} x + y = 12{,}000 & \text{Equation 1} \\ 0.06x + 0.08y = 880 & \text{Equation 2} \end{cases}$$

To begin, it is convenient to multiply each side of the second equation by 100. This eliminates the need to work with decimals.

$0.06x + 0.08y = 880$	Equation 2
$6x + 8y = 88{,}000$	Multiply each side by 100.

Then solve for x in Equation 1.

$x = 12{,}000 - y$	Revised Equation 1

Next, substitute this expression for x in the revised Equation 2 and solve for y.

$6x + 8y = 88{,}000$	Revised Equation 2
$6(12{,}000 - y) + 8y = 88{,}000$	Substitute $12{,}000 - y$ for x.
$72{,}000 - 6y + 8y = 88{,}000$	Distributive Property
$72{,}000 + 2y = 88{,}000$	Combine like terms.
$2y = 16{,}000$	Subtract 72,000 from each side.
$y = 8000$	Divide each side by 2.

Back-substitute the value $y = 8000$ in the revised Equation 1 and solve for x.

$x = 12{,}000 - y$	Revised Equation 1
$x = 12{,}000 - 8000$	Substitute 8000 for y.
$x = 4000$	Simplify.

So, $4000 was invested at 6% simple interest and $8000 was invested at 8% simple interest. Check this in the original statement of the problem as follows.

Substitute in Equation 1

$$x + y = 12{,}000$$

$$4000 + 8000 = 12{,}000 \checkmark$$

Substitute in Equation 2

$$0.06x + 0.08y = 880$$

$$0.06(4000) + 0.08(8000) = 880$$

$$240 + 640 = 800 \checkmark$$

8.1 Exercises

Review   Concepts, Skills, and Problem Solving

Keep mathematically in shape by doing these exercises *before* the problems of this section.

Properties and Definitions

1. How many solutions does a linear equation of the form $2x + 8 = 7$ have?

2. What is a helpful usual first step when solving an equation such as $\dfrac{x}{6} + \dfrac{3}{2} = \dfrac{7}{4}$?

Solving Equations

In Exercises 3–8, solve the equation and check your solution.

3. $x - 6 = 5x$

4. $2 - 3x = 14 + x$

5. $y - 3(4y - 2) = 1$

6. $y + 6(3 - 2y) = 4$

7. $\dfrac{x}{2} - \dfrac{x}{5} = 15$

8. $\dfrac{x - 4}{10} = 6$

Models

In Exercises 9 and 10, translate the phrase into an algebraic expression.

9. The time to travel 250 miles at an average speed of r miles per hour

10. The perimeter of a rectangle of length L and width $L/2$

Developing Skills

In Exercises 1–6, determine whether each ordered pair is a solution to the system of equations. See Example 1.

	System		Ordered Pairs	

1. $\begin{cases} x + 3y = 11 \\ -x + 3y = 7 \end{cases}$ (a) $(2, 3)$ (b) $(5, 4)$

2. $\begin{cases} 3x - y = -2 \\ x - 3y = 2 \end{cases}$ (a) $(0, 2)$ (b) $(-1, -1)$

3. $\begin{cases} 2x - 3y = -8 \\ x + y = 1 \end{cases}$ (a) $(5, -3)$ (b) $(-1, 2)$

4. $\begin{cases} 5x - 3y = -12 \\ x - 4y = 1 \end{cases}$ (a) $(-3, -1)$ (b) $(3, 1)$

5. $\begin{cases} 5x - 6y = -2 \\ 7x + y = -31 \end{cases}$ (a) $(-4, -3)$ (b) $(-3, -4)$

6. $\begin{cases} -x - y = 6 \\ -5x - 2y = 3 \end{cases}$ (a) $(7, -13)$ (b) $(3, -9)$

In Exercises 7–14, use the graphs of the equations to determine the solution (if any) of the system of linear equations. Check your solution.

7. $\begin{cases} 2x + y = 4 \\ x - y = 2 \end{cases}$

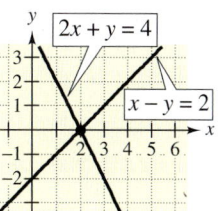

8. $\begin{cases} x + 3y = 2 \\ -x + 2y = 3 \end{cases}$

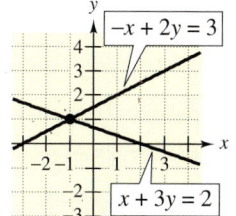

9. $\begin{cases} x - y = 0 \\ 3x - 2y = -1 \end{cases}$

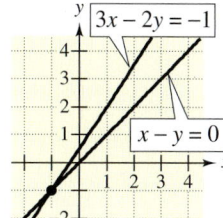

10. $\begin{cases} 2x - y = 2 \\ 4x + 3y = 24 \end{cases}$

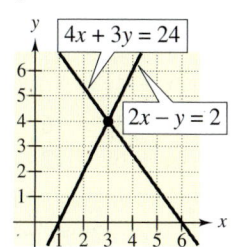

11. $\begin{cases} x - 2y = -4 \\ -0.5x + y = 2 \end{cases}$ **12.** $\begin{cases} 2x - 5y = 10 \\ 6x - 15y = 75 \end{cases}$

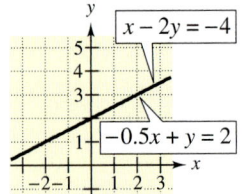   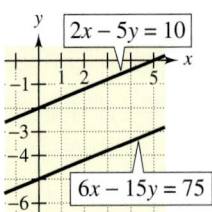

13. $\begin{cases} x + 4y = 8 \\ 3x + 12y = 12 \end{cases}$ **14.** $\begin{cases} 2x - y = -3 \\ -4x + 2y = 6 \end{cases}$

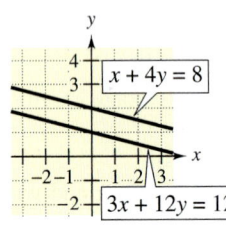   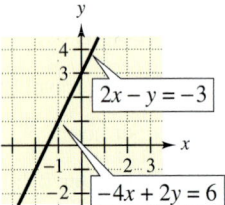

In Exercises 15–40, sketch the graphs of the equations and approximate any solutions of the system of linear equations. See Examples 2–6.

15. $\begin{cases} y = -x + 3 \\ y = x + 1 \end{cases}$ **16.** $\begin{cases} y = 2x - 1 \\ y = x + 1 \end{cases}$

17. $\begin{cases} y = 2x - 4 \\ y = -\frac{1}{2}x + 1 \end{cases}$ **18.** $\begin{cases} y = \frac{1}{2}x + 2 \\ y = -x + 8 \end{cases}$

19. $\begin{cases} x - y = 2 \\ x + y = 2 \end{cases}$ **20.** $\begin{cases} x - y = 0 \\ x + y = 4 \end{cases}$

21. $\begin{cases} -x + 2y = 4 \\ x - 2y = 4 \end{cases}$ **22.** $\begin{cases} 3x - y = 1 \\ -3x + y = 1 \end{cases}$

23. $\begin{cases} 4x - 5y = 0 \\ 6x - 5y = 10 \end{cases}$

24. $\begin{cases} 3x + 2y = -6 \\ 3x - 2y = 6 \end{cases}$

25. $\begin{cases} x - 2y = 4 \\ 2x - 4y = 8 \end{cases}$

26. $\begin{cases} 2x + 3y = 6 \\ 4x + 6y = 12 \end{cases}$

27. $\begin{cases} -2x + y = -1 \\ x - 2y = -4 \end{cases}$

28. $\begin{cases} 2x + y = -4 \\ 4x - 2y = 8 \end{cases}$

29. $\begin{cases} 4x - 3y = 3 \\ 4x - 3y = 0 \end{cases}$

30. $\begin{cases} 2x + 5y = 5 \\ -2x - 5y = -5 \end{cases}$

31. $\begin{cases} x + 2y = 3 \\ x - 3y = 13 \end{cases}$

32. $\begin{cases} -x + 10y = 30 \\ x + 10y = 10 \end{cases}$

33. $\begin{cases} x + 7y = -5 \\ 3x - 2y = 8 \end{cases}$

34. $\begin{cases} x + 2y = 4 \\ 2x - 2y = -1 \end{cases}$

35. $\begin{cases} -3x + 10y = 15 \\ 3x - 10y = 15 \end{cases}$

36. $\begin{cases} 4x - 9y = 12 \\ -4x + 9y = 12 \end{cases}$

37. $\begin{cases} 4x + 5y = 20 \\ \frac{4}{5}x + y = 4 \end{cases}$

38. $\begin{cases} 3x + 7y = 15 \\ x + \frac{7}{3}y = 5 \end{cases}$

39. $\begin{cases} 8x - 6y = -12 \\ x - \frac{3}{4}y = -2 \end{cases}$

40. $\begin{cases} -x + \frac{2}{3}y = 5 \\ 9x - 6y = 6 \end{cases}$

In Exercises 41–44, use a graphing calculator to graph the equations and approximate any solutions of the system of linear equations. Check your solution.

41. $\begin{cases} y = 2x - 1 \\ y = -3x + 9 \end{cases}$

42. $\begin{cases} y = \frac{3}{4}x + 2 \\ y = x + 1 \end{cases}$

43. $\begin{cases} y = x - 1 \\ y = -2x + 8 \end{cases}$

44. $\begin{cases} y = 2x + 3 \\ y = -x - 3 \end{cases}$

In Exercises 45–52, write the equations of the lines in slope-intercept form. What can you conclude about the number of solutions of the system?

45. $\begin{cases} 2x - 3y = -12 \\ -8x + 12y = -12 \end{cases}$

46. $\begin{cases} -5x + 8y = 8 \\ 7x - 4y = 14 \end{cases}$

47. $\begin{cases} -x + 4y = 7 \\ 3x - 12y = -21 \end{cases}$

48. $\begin{cases} 3x + 8y = 28 \\ -4x + 9y = 1 \end{cases}$

49. $\begin{cases} -2x + 3y = 4 \\ 2x + 3y = 8 \end{cases}$

50. $\begin{cases} 2x + 5y = 15 \\ 2x - 5y = 5 \end{cases}$

51. $\begin{cases} -6x + 8y = 9 \\ 3x - 4y = -6 \end{cases}$

52. $\begin{cases} -6x + 8y = 9 \\ 3x - 4y = -4.5 \end{cases}$

In Exercises 53–58, solve the system by the method of substitution. Use the graph to check the solution.

53. $\begin{cases} x + y = 1 \\ 2x - y = 2 \end{cases}$

54. $\begin{cases} 2x + y = 4 \\ -x + y = 1 \end{cases}$

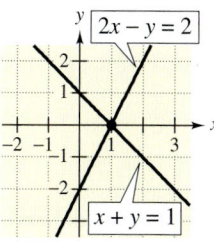

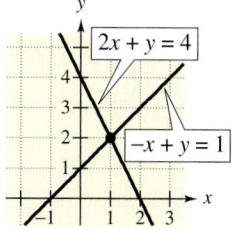

55. $\begin{cases} -x + y = 1 \\ x - y = 1 \end{cases}$

56. $\begin{cases} x + 2y = 6 \\ x + 2y = 2 \end{cases}$

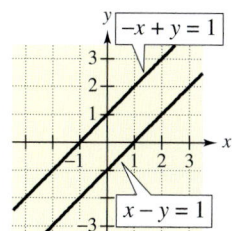

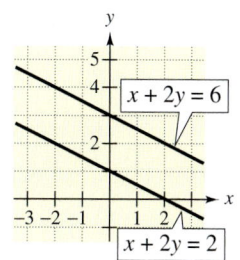

57. $\begin{cases} 2x + y = 3 \\ 4x + 2y = 6 \end{cases}$

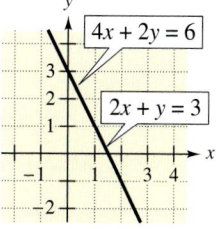

58. $\begin{cases} -4x + 3y = 6 \\ 8x - 6y = -12 \end{cases}$

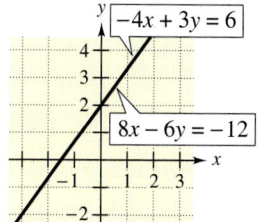

In Exercises 59–96, solve the system by the method of substitution. See Examples 7–11.

59. $\begin{cases} y = 2x - 1 \\ y = -x + 5 \end{cases}$

60. $\begin{cases} y = -2x + 9 \\ y = 3x - 1 \end{cases}$

61. $\begin{cases} x = 4y - 5 \\ x = 3y \end{cases}$

62. $\begin{cases} x = -5y - 2 \\ x = 2y - 23 \end{cases}$

63. $\begin{cases} 2x = 8 \\ x + y = 1 \end{cases}$

64. $\begin{cases} 3x - y = 0 \\ 3y = 6 \end{cases}$

65. $\begin{cases} x - y = 2 \\ x - y = 1 \end{cases}$

66. $\begin{cases} x + y = 8 \\ x + y = -1 \end{cases}$

67. $\begin{cases} x - y = 0 \\ 2x + y = 0 \end{cases}$

68. $\begin{cases} x - y = 0 \\ 5x - 3y = 10 \end{cases}$

69. $\begin{cases} x - 2y = -10 \\ 3x - y = 0 \end{cases}$

70. $\begin{cases} x - 2y = 0 \\ 3x - y = 0 \end{cases}$

71. $\begin{cases} 2x - y = -2 \\ 4x + y = 5 \end{cases}$

72. $\begin{cases} x + 6y = 7 \\ -x + 4y = -2 \end{cases}$

73. $\begin{cases} x + 2y = 1 \\ 5x - 4y = -23 \end{cases}$

74. $\begin{cases} -3x + 6y = 4 \\ 2x + y = 4 \end{cases}$

75. $\begin{cases} 5x + 3y = 11 \\ x - 5y = 5 \end{cases}$

76. $\begin{cases} -3x + y = 4 \\ -9x + 5y = 10 \end{cases}$

77. $\begin{cases} 5x + 2y = 0 \\ x - 3y = 0 \end{cases}$

78. $\begin{cases} 4x + 3y = 0 \\ 2x - y = 0 \end{cases}$

79. $\begin{cases} 2x + 5y = -4 \\ 3x - y = 11 \end{cases}$

80. $\begin{cases} 2x + 5y = 1 \\ -x + 6y = 8 \end{cases}$

81. $\begin{cases} 4x - y = 2 \\ 2x - \frac{1}{2}y = 1 \end{cases}$ **82.** $\begin{cases} 3x - y = 6 \\ 4x - \frac{2}{3}y = -4 \end{cases}$

83. $\begin{cases} \frac{1}{5}x + \frac{1}{2}y = 8 \\ 2x + y = 20 \end{cases}$ **84.** $\begin{cases} \frac{1}{2}x + \frac{3}{4}y = 10 \\ 4x - y = 4 \end{cases}$

85. $\begin{cases} -5x + 4y = 14 \\ 5x - 4y = 4 \end{cases}$ **86.** $\begin{cases} 3x - 2y = 3 \\ -6x + 4y = -6 \end{cases}$

87. $\begin{cases} 2x + y = 8 \\ 5x + 2.5y = 10 \end{cases}$ **88.** $\begin{cases} 0.5x + 0.5y = 4 \\ x + y = -1 \end{cases}$

89. $\begin{cases} -6x + 1.5y = 6 \\ 8x - 2y = -8 \end{cases}$ **90.** $\begin{cases} 0.3x - 0.3y = 0 \\ x - y = 4 \end{cases}$

91. $\begin{cases} \dfrac{x}{3} - \dfrac{y}{4} = 2 \\ \dfrac{x}{2} + \dfrac{y}{6} = 3 \end{cases}$ **92.** $\begin{cases} -\dfrac{x}{5} + \dfrac{y}{2} = -3 \\ \dfrac{x}{4} - \dfrac{y}{4} = 0 \end{cases}$

93. $\begin{cases} \dfrac{x}{4} + \dfrac{y}{2} = 1 \\ \dfrac{x}{2} - \dfrac{y}{3} = 1 \end{cases}$ **94.** $\begin{cases} -\dfrac{x}{6} + \dfrac{y}{12} = 1 \\ \dfrac{x}{2} + \dfrac{y}{8} = 1 \end{cases}$

95. $\begin{cases} 2(x - 5) = y + 2 \\ 3x = 4(y + 2) \end{cases}$

96. $\begin{cases} 3(x - 2) + 5 = 4(y + 3) - 2 \\ 2x + 7 = 2y + 8 \end{cases}$

In Exercises 97–102, solve the system by the method of substitution. Use a graphing calculator to verify the solution graphically.

97. $\begin{cases} y = -2x + 10 \\ y = x + 4 \end{cases}$ **98.** $\begin{cases} y = \frac{5}{4}x + 3 \\ y = \frac{1}{2}x + 6 \end{cases}$

99. $\begin{cases} 3x + 2y = 12 \\ x - y = 3 \end{cases}$ **100.** $\begin{cases} 2x - y = 1 \\ x - y = -2 \end{cases}$

101. $\begin{cases} 5x + 3y = 15 \\ 2x - 3y = 6 \end{cases}$ **102.** $\begin{cases} 4x - 5y = 0 \\ 2x - 5y = -10 \end{cases}$

Solving Problems

103. *Number Problem*   The sum of two numbers x and y is 20 and the difference of the two numbers is 2. Find the two numbers.

104. *Number Problem*   The sum of two numbers x and y is 35 and the difference of the two numbers is 11. Find the two numbers.

105. *Break-Even Analysis*   A small company produces bird feeders that sell for $23 per unit. The cost of producing each unit is $16.75, and the company has fixed costs of $400.

(a) Use a verbal model to show that the cost C of producing x units is $C = 16.75x + 400$ and the revenue R from selling x units is $R = 23x$.

(b) ⊞ Use a graphing calculator to graph the cost and revenue functions in the same viewing window. Approximate the point of intersection of the graphs and interpret the result.

106. ⊞ *Supply and Demand*   The Law of Supply and Demand states that as the price of a product increases, the demand for the product decreases and the supply increases. The demand and supply equations for a tool set are

$$p = 90 - x$$

and

$$p = 2x - 48$$

respectively, where p is the price in dollars and x represents the number of units. Market equilibrium is the point of intersection of the two equations. Use a graphing calculator to graph the equations in the same viewing window and determine the price of the tool set that yields market equilibrium.

Think About It   In Exercises 107 and 108, the graphs of the two equations appear to be parallel. Are the two lines actually parallel? Does the system have a solution? If so, find the solution.

107. $\begin{cases} x - 200y = -200 \\ x - 199y = 198 \end{cases}$

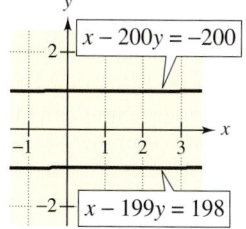

108. $\begin{cases} 25x - 24y = 0 \\ 13x - 12y = 24 \end{cases}$

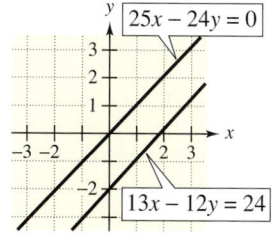

109. *Investment*   A total of $15,000 is invested in two funds paying 5% and 8% simple interest. The combined annual interest for the two funds is $900. Determine how much of the $15,000 is invested at each rate.

110. *Investment*   A total of $10,000 is invested in two funds paying 7% and 10% simple interest. The combined annual interest for the two funds is $775. Determine how much of the $10,000 is invested at each rate.

111. *Dinner Price*   Six people ate dinner for $63.90. The price for adults was $16.95 and the price for children was $7.50. Determine how many adults attended the dinner.

112. *Ticket Sales*   You are selling football tickets. Student tickets cost $2 and general admission tickets cost $3. You sell 1957 tickets and collect $5035. Determine how many of each type of ticket were sold.

113. *Comparing Costs*   Car model ES costs $16,000 and costs an average of $0.26 per mile to maintain. Car model LS costs $18,000 and costs an average of $0.22 per mile to maintain. Determine after how many miles the total costs of the two models will be the same (the two models are driven the same number of miles).

114. *Comparing Costs*   Heating a three-bedroom home using a solar heating system costs $28,500 for installation and $125 per year to operate. Heating the same home using an electric heating system costs $5750 for installation and $1000 per year to operate. Determine after how many years the total costs for solar heating and electric heating will be the same. What will be the cost at that time?

115. ▲ *Geometry*   Find an equation of the line with slope $m = 2$ passing through the intersection of the lines $x - 2y = 3$ and $3x + y = 16$.

116. ▲ *Geometry*   Find an equation of the line with slope $m = -3$ passing through the intersection of the lines $4x + 6y = 26$ and $5x - 2y = -15$.

Explaining Concepts

117. Answer parts (a) and (b) of Motivating the Chapter on page 466.

118. *Writing*   Give geometric descriptions of the three cases for a system of linear equations in two variables.

119. *Writing*   In your own words, explain what is meant by a dependent system of linear equations.

120. *Writing*   In your own words, explain what is meant by an inconsistent system of linear equations.

121. *True or False?* It is possible for a consistent system of linear equations to have exactly two solutions. Justify your answer.

122. *Writing* Explain how you can check the solution of a system of linear equations algebraically and graphically.

123. *Writing* In your own words, explain the basic steps in solving a system of linear equations by the method of substitution.

124. *Writing* When solving a system of linear equations by the method of substitution, how do you recognize that it has no solution?

125. *Writing* When solving a system of linear equations by the method of substitution, how do you recognize that it has infinitely many solutions?

126. *Writing* Describe any advantages of the method of substitution over the graphical method of solving a system of linear equations.

127. *Creating a System* Write a system of linear equations with integer coefficients that has the unique solution $(3, 1)$. (There are many correct answers.)

128. *Creating an Example* Write an example of a system of linear equations that has no solution. (There are many correct answers.)

129. *Creating an Example* Write an example of a system of linear equations that has infinitely many solutions. (There are many correct answers.)

130. Your instructor says, "An equation (not in standard form) such as $2x - 3 = 5x - 9$ can be considered to be a system of equations." Create the system, and find the solution point. How many solution points does the "system" $x^2 - 1 = 2x - 1$ have? Illustrate your results with a graphing calculator.

Think About It In Exercises 131–134, find the value of a or b such that the system of linear equations is inconsistent.

131. $\begin{cases} x + by = 1 \\ x + 2y = 2 \end{cases}$

132. $\begin{cases} ax + 3y = 6 \\ 5x - 5y = 2 \end{cases}$

133. $\begin{cases} -6x + y = 4 \\ 2x + by = 3 \end{cases}$

134. $\begin{cases} 6x - 3y = 4 \\ ax - y = -2 \end{cases}$

8.2 Solving Systems of Equations by Elimination

Staffan Widstrand/Corbis

What You Should Learn

1. Solve systems of linear equations algebraically using the method of elimination.
2. Choose a method for solving systems of equations.

Why You Should Learn It

The method of elimination is one method of solving a system of linear equations. For instance, in Exercise 68 on page 494, this method is convenient for solving a system of linear equations used to find the focal length of a camera.

1. **Solve systems of linear equations algebraically using the method of elimination.**

The Method of Elimination

In this section, you will study another way to solve a system of linear equations algebraically—the **method of elimination.** The key step in this method is to obtain opposite coefficients for one of the variables so that *adding* the two equations eliminates this variable. For instance, by adding the equations

$$\begin{cases} 3x + 5y = 7 & \text{Equation 1} \\ -3x - 2y = -1 & \text{Equation 2} \end{cases}$$
$$3y = 6 \qquad \text{Add equations.}$$

you eliminate the variable x and obtain a single equation in one variable, y.

Example 1   The Method of Elimination

Solve the system of linear equations.

$$\begin{cases} 4x + 3y = 1 & \text{Equation 1} \\ 2x - 3y = 5 & \text{Equation 2} \end{cases}$$

Solution

Begin by noting that the coefficients of y are opposites. So, by adding the two equations, you can eliminate y.

$$\begin{cases} 4x + 3y = 1 & \text{Equation 1} \\ 2x - 3y = 5 & \text{Equation 2} \end{cases}$$
$$6x = 6 \qquad \text{Add equations.}$$

So, $x = 1$. By back-substituting this value into the first equation, you can solve for y, as follows.

$$4(1) + 3y = 1 \qquad \text{Substitute 1 for } x \text{ in Equation 1.}$$
$$3y = -3 \qquad \text{Subtract 4 from each side.}$$
$$y = -1 \qquad \text{Divide each side by 3.}$$

The solution is $(1, -1)$. Check this in both of the original equations.

Study Tip

Try solving the system in Example 1 by substitution. Which method do you think is easier? Many people find that the method of elimination is more efficient.

To obtain opposite coefficients for one of the variables, you often need to multiply one or both of the equations by a suitable constant. This is demonstrated in the following example.

Example 2   The Method of Elimination

Solve the system of linear equations.

$$\begin{cases} 2x - 3y = -7 & \text{Equation 1} \\ 3x + y = -5 & \text{Equation 2} \end{cases}$$

Solution

For this system, you can obtain opposite coefficients of y by multiplying the second equation by 3.

$$\begin{cases} 2x - 3y = -7 \\ 3x + y = -5 \end{cases}$$

$2x - 3y = \quad -7$	Equation 1
$9x + 3y = -15$	Multiply Equation 2 by 3.
$11x \qquad = -22$	Add equations.

So, $x = -2$. By back-substituting this value of x into the second equation, you can solve for y.

$3x + y = -5$	Equation 2
$3(-2) + y = -5$	Substitute -2 for x.
$-6 + y = -5$	Simplify.
$y = \quad 1$	Add 6 to each side.

The solution is $(-2, 1)$. Check this in the original equations, as follows.

Substitute into Equation 1	Substitute into Equation 2
$2x - 3y = -7$	$3x + y = -5$
$2(-2) - 3(1) \overset{?}{=} -7$	$3(-2) + 1 \overset{?}{=} -5$
$-4 - 3 = -7 \checkmark$	$-6 + 1 = -5 \checkmark$

This method is called "elimination" because the first step in the process is to "eliminate" one of the variables. This method is summarized as follows.

The Method of Elimination

1. Obtain opposite coefficients of x (or y) by multiplying all terms of one or both equations by suitable constants.

2. Add the equations to eliminate one variable and solve the resulting equation.

3. Back-substitute the value obtained in Step 2 into either of the original equations and solve for the other variable.

4. Check your solution in *both* of the original equations.

Example 3   The Method of Elimination

Solve the system of linear equations.

$$\begin{cases} 5x + 3y = 6 & \text{Equation 1} \\ 2x - 4y = 5 & \text{Equation 2} \end{cases}$$

Solution

You can obtain opposite coefficients of y by multiplying the first equation by 4 and the second equation by 3.

$$\begin{cases} 5x + 3y = 6 \quad \Longrightarrow \quad 20x + 12y = 24 & \text{Multiply Equation 1 by 4.} \\ 2x - 4y = 5 \quad \Longrightarrow \quad \underline{6x - 12y = 15} & \text{Multiply Equation 2 by 3.} \end{cases}$$
$$26x \qquad = 39 \qquad \text{Add equations.}$$

From this equation, you can see that $x = \frac{3}{2}$. By back-substituting this value of x into the second equation, you can solve for y, as follows.

$$2x - 4y = 5 \qquad\qquad \text{Equation 2}$$

$$2\left(\frac{3}{2}\right) - 4y = 5 \qquad\qquad \text{Substitute } \tfrac{3}{2} \text{ for } x.$$

$$3 - 4y = 5 \qquad\qquad \text{Simplify.}$$

$$-4y = 2 \qquad\qquad \text{Subtract 3 from each side.}$$

$$y = -\frac{1}{2} \qquad\qquad \text{Divide each side by } -4.$$

The solution is $\left(\frac{3}{2}, -\frac{1}{2}\right)$. You can check this as follows.

Substitute into Equation 1   *Substitute into Equation 2*

$$5x + 3y = 6 \qquad\qquad\qquad 2x - 4y = 5$$

$$5\left(\frac{3}{2}\right) + 3\left(-\frac{1}{2}\right) \overset{?}{=} 6 \qquad\qquad 2\left(\frac{3}{2}\right) - 4\left(-\frac{1}{2}\right) \overset{?}{=} 5$$

$$\frac{15}{2} - \frac{3}{2} = 6 \ \checkmark \qquad\qquad\qquad 3 + 2 = 5 \ \checkmark$$

The graph of this system is shown in Figure 8.8. From the graph it appears that the solution $\left(\frac{3}{2}, -\frac{1}{2}\right)$ is reasonable.

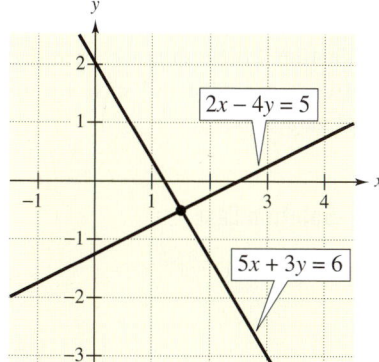

Figure 8.8

In Example 3, the y-variable was eliminated first. You could just as easily have solved the system by eliminating the x-variable first, as follows.

$$\begin{cases} 5x + 3y = 6 \quad \Longrightarrow \quad 10x + 6y = 12 & \text{Multiply Equation 1 by 2.} \\ 2x - 4y = 5 \quad \Longrightarrow \quad \underline{-10x + 20y = -25} & \text{Multiply Equation 2 by } -5. \end{cases}$$
$$26y = -13 \qquad \text{Add equations.}$$

From this equation, $y = -\frac{1}{2}$. By back-substituting this value of y into the second equation, you can solve for x to obtain $x = \frac{3}{2}$.

In the next example, note how the method of elimination can be used to determine that a system of linear equations has no solution. As with substitution, notice that the key is recognizing the occurrence of a *false statement.*

Example 4   The Method of Elimination: No-Solution Case

Solve the system of linear equations.

$$\begin{cases} 2x - 6y = 5 & \text{Equation 1} \\ 3x - 9y = 2 & \text{Equation 2} \end{cases}$$

Solution

To obtain coefficients that differ only in sign, multiply the first equation by 3 and multiply the second equation by -2.

$$\begin{cases} 2x - 6y = 5 \\ 3x - 9y = 2 \end{cases} \quad \Longrightarrow \quad \begin{array}{rl} 6x - 18y = 15 & \text{Multiply Equation 1 by 3.} \\ -6x + 18y = -4 & \text{Multiply Equation 2 by } -2. \\ \hline 0 \neq 11 & \text{Add equations.} \end{array}$$

Because 0 does not equal 11, you can conclude that the system is inconsistent and has no solution. The lines corresponding to the two equations of this system are shown in Figure 8.9. Note that the two lines are parallel and so have no point of intersection.

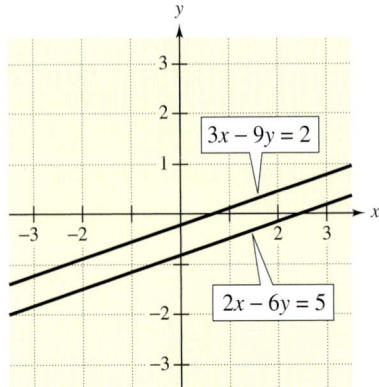

Figure 8.9

Example 5 shows how the method of elimination works with a system that has infinitely many solutions. Notice that you can recognize this case by the occurrence of an equation that is true for all real values of x and y.

Example 5   The Method of Elimination: Many-Solution Case

Solve the system of linear equations.

$$\begin{cases} 2x - 6y = -5 & \text{Equation 1} \\ -4x + 12y = 10 & \text{Equation 2} \end{cases}$$

Solution

To obtain the coefficients of x that differ only in sign, multiply the first equation by 2.

$$\begin{cases} 2x - 6y = -5 \\ -4x + 12y = 10 \end{cases} \quad \Longrightarrow \quad \begin{array}{rl} 4x - 12y = -10 & \text{Multiply Equation 1 by 2.} \\ -4x + 12y = 10 & \text{Equation 2} \\ \hline 0 = 0 & \text{Add equations.} \end{array}$$

Because $0 = 0$ is a true statement, you can conclude that the system is dependent and has infinitely many solutions. The solution set consists of all ordered pairs (x, y) lying on the line $2x - 6y = -5$.

The next example shows how the method of elimination works with a system of linear equations with decimal coefficients.

Example 6   Solving a System with Decimal Coefficients

Solve the system of linear equations.

$$\begin{cases} 0.02x - 0.05y = -0.38 & \text{Equation 1} \\ 0.03x + 0.04y = 1.04 & \text{Equation 2} \end{cases}$$

Solution

Because the coefficients in this system have two decimal places, begin by multiplying each equation by 100. This produces a system in which the coefficients are all integers.

$$\begin{cases} 2x - 5y = -38 & \text{Revised Equation 1} \\ 3x + 4y = 104 & \text{Revised Equation 2} \end{cases}$$

Now, to obtain coefficients of x that differ only in sign, multiply the first equation by 3 and multiply the second equation by -2.

$$\begin{cases} 2x - 5y = -38 \\ 3x + 4y = 104 \end{cases} \implies \begin{array}{rl} 6x - 15y = -114 & \text{Multiply Equation 1 by 3.} \\ -6x - 8y = -208 & \text{Multiply Equation 2 by } -2. \\ \hline -23y = -322 & \text{Add equations.} \end{array}$$

So, the y-coordinate of the solution is

$$y = \frac{-322}{-23} = 14.$$

Back-substituting this value into revised Equation 2 produces the following.

$$3x + 4(14) = 104 \qquad \text{Substitute 14 for } y \text{ in revised Equation 2.}$$
$$3x + 56 = 104 \qquad \text{Simplify.}$$
$$3x = 48 \qquad \text{Subtract 56 from each side.}$$
$$x = 16 \qquad \text{Divide each side by 3.}$$

So, the solution is (16, 14). You can check this solution as follows.

Substitute into Equation 1

$$0.02x - 0.05y = -0.38 \qquad \text{Equation 1}$$
$$0.02(16) - 0.05(14) \overset{?}{=} -0.38 \qquad \text{Substitute 16 for } x \text{ and 14 for } y.$$
$$0.32 - 0.70 = -0.38 \checkmark \qquad \text{Solution checks.}$$

Substitute into Equation 2

$$0.03x + 0.04y = 1.04 \qquad \text{Equation 2}$$
$$0.03(16) + 0.04(14) \overset{?}{=} 1.04 \qquad \text{Substitute 16 for } x \text{ and 14 for } y.$$
$$0.48 + 0.56 = 1.04 \checkmark \qquad \text{Solution checks.}$$

Study Tip

When multiplying an equation by a negative number, be sure to distribute the negative sign to each term of the equation. For instance, in Example 6 the second equation is multiplied by -2.

Example 7   An Application of a System of Linear Equations

A fundraising dinner was held on two consecutive nights. On the first night, 100 adult tickets and 175 children's tickets were sold, for a total of $1225. On the second night, 200 adult tickets and 316 children's tickets were sold, for a total of $2348. The system of linear equations that represents this problem is

$$\begin{cases} 100x + 175y = 1225 & \text{Equation 1} \\ 200x + 316y = 2348 & \text{Equation 2} \end{cases}$$

where x represents the price of the adult tickets and y represents the price of the children's tickets. Solve this system to find the price of each type of ticket.

Solution

To obtain coefficients of x that differ only in sign, multiply Equation 1 by -2.

$$\begin{cases} 100x + 175y = 1225 \\ 200x + 316y = 2348 \end{cases}$$

$$\begin{aligned} -200x - 350y &= -2450 & &\text{Multiply Equation 1 by } -2. \\ 200x + 316y &= 2348 & &\text{Equation 2} \\ \hline -34y &= -102 & &\text{Add equations.} \end{aligned}$$

So, the y-coordinate of the solution is $y = -102/-34 = 3$. Back-substituting this value into Equation 2 produces the following.

$$\begin{aligned} 200x + 316(3) &= 2348 & &\text{Substitute 3 for } y \text{ in Equation 2.} \\ 200x &= 1400 & &\text{Simplify.} \\ x &= 7 & &\text{Divide each side by 200.} \end{aligned}$$

The solution is $(7, 3)$. So the price of the adult tickets was $7 and the price of the children's tickets was $3. Check this solution in both of the original equations.

---

2 Choose a method for solving systems of equations.

Choosing Methods

To decide which of the three methods (graphing, substitution, or elimination) to use to solve a system of two linear equations, use the following guidelines.

> ### Guidelines for Solving a System of Linear Equations
>
> To decide whether to use graphing, substitution, or elimination, consider the following.
>
> 1. The graphing method is useful for approximating the solution and for giving an overall picture of how one variable changes with respect to the other.
>
> 2. To find exact solutions, use either substitution or elimination.
>
> 3. For systems of equations in which one variable has a coefficient of 1, substitution may be more efficient than elimination.
>
> 4. Elimination is usually more efficient. This is especially true when the coefficients of one of the variables are opposites.

8.2 Exercises

Review   Concepts, Skills, and Problem Solving

Keep mathematically in shape by doing these exercises *before* the problems of this section.

Properties and Definitions

In Exercises 1–4, identify the property of real numbers illustrated by the statement.

1. $2ab \cdot \dfrac{1}{2ab} = 1$

2. $8t + 0 = 8t$

3. $2yx = 2xy$

4. $3(2x) = (3 \cdot 2)x$

Algebraic Operations

In Exercises 5–10, plot the points on a rectangular coordinate system. Find the slope of the line passing through the points. If not possible, state why.

5. $(-6, 4), (-3, -4)$

6. $(4, 6), (8, -2)$

7. $\left(\frac{7}{2}, \frac{9}{2}\right), \left(\frac{4}{3}, -3\right)$

8. $\left(-\frac{3}{4}, -\frac{7}{4}\right), \left(-1, \frac{5}{2}\right)$

9. $(-3, 6), (-3, 2)$

10. $(6, 2), (10, 2)$

Problem Solving

11. *Quality Control*   A quality control engineer for a buyer found three defective units in a sample of 100. At this rate, what is the expected number of defective units in a shipment of 5000 units?

12. *Consumer Awareness*   The cost of a long-distance telephone call is $0.70 for the first minute and $0.42 for each additional minute. The total cost of the call cannot exceed $8. Find the interval of time that is available for the call. Round the maximum value to one decimal place.

Developing Skills

In Exercises 1–4, solve the system by the method of elimination. Use the graph to check your solution.

1. $\begin{cases} 2x + y = 4 \\ x - y = 2 \end{cases}$

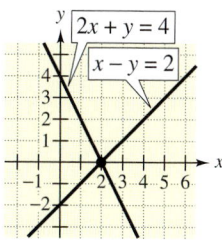

2. $\begin{cases} x + 3y = 2 \\ -x + 2y = 3 \end{cases}$

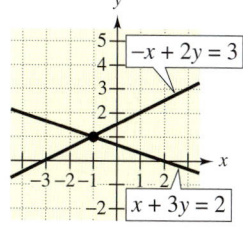

3. $\begin{cases} x - y = 0 \\ 3x - 2y = -1 \end{cases}$

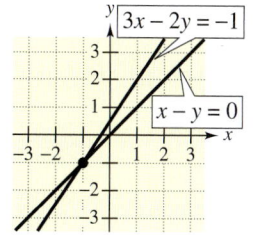

4. $\begin{cases} 2x - y = 2 \\ 4x + 3y = 24 \end{cases}$

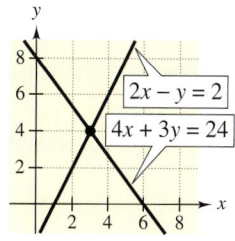

In Exercises 5–32, solve the system by the method of elimination. See Examples 1–6.

5. $\begin{cases} x - y = 4 \\ x + y = 12 \end{cases}$

6. $\begin{cases} x + y = 7 \\ x - y = 3 \end{cases}$

7. $\begin{cases} -x + 2y = 12 \\ x + 6y = 20 \end{cases}$

8. $\begin{cases} x + 2y = 14 \\ x - 2y = 10 \end{cases}$

9. $\begin{cases} 3x - 5y = 1 \\ 2x + 5y = 9 \end{cases}$

10. $\begin{cases} -2x + 3y = -4 \\ 2x - 4y = 6 \end{cases}$

11. $\begin{cases} 3a + 3b = 7 \\ 3a + 5b = 3 \end{cases}$

12. $\begin{cases} 4a + 5b = 9 \\ 2a + 5b = 7 \end{cases}$

13. $\begin{cases} -x + 2y = 12 \\ 3x - 6y = 10 \end{cases}$

14. $\begin{cases} -6x + 3y = 18 \\ 2x - y = 11 \end{cases}$

15. $\begin{cases} 3x - 4y = 11 \\ 2x + 3y = -4 \end{cases}$

16. $\begin{cases} 2x + 3y = 16 \\ 5x - 10y = 30 \end{cases}$

17. $\begin{cases} 3x + 2y = -1 \\ -2x + 7y = 9 \end{cases}$

18. $\begin{cases} 5x + 3y = 27 \\ 7x - 2y = 13 \end{cases}$

19. $\begin{cases} 3x - 4y = 1 \\ 4x + 3y = 1 \end{cases}$

20. $\begin{cases} 2x - 5y = -1 \\ 2x - y = 1 \end{cases}$

21. $\begin{cases} 3x + 2y = 10 \\ 2x + 5y = 3 \end{cases}$

22. $\begin{cases} 4x + 5y = 7 \\ 6x - 2y = -18 \end{cases}$

23. $\begin{cases} 5u + 6v = 14 \\ 3u + 5v = 7 \end{cases}$

24. $\begin{cases} 5x + 3y = 18 \\ 2x - 7y = -1 \end{cases}$

25. $\begin{cases} 6r + 5s = 3 \\ \frac{3}{2}r - \frac{5}{4}s = \frac{3}{4} \end{cases}$

26. $\begin{cases} \frac{2}{3}x + \frac{1}{6}y = \frac{2}{3} \\ 4x + \phantom{\frac{1}{6}}y = 4 \end{cases}$

27. $\begin{cases} \frac{1}{2}s - t = \frac{3}{2} \\ 4s + 2t = 27 \end{cases}$

28. $\begin{cases} 3u + 4v = 14 \\ \frac{1}{6}u - v = -2 \end{cases}$

29. $\begin{cases} 0.4a + 0.7b = 3 \\ 0.2a + 0.6b = 5 \end{cases}$

30. $\begin{cases} 0.2u - 0.1v = 1 \\ -0.8u + 0.4v = 3 \end{cases}$

31. $\begin{cases} 0.02x - 0.05y = -0.19 \\ 0.03x + 0.04y = 0.52 \end{cases}$

32. $\begin{cases} 0.05x - 0.03y = 0.21 \\ 0.01x + 0.01y = 0.09 \end{cases}$

In Exercises 33–38, solve the system by the method of elimination. Use a graphing calculator to verify your solution.

33. $\begin{cases} x + 2y = 3 \\ -x - y = -1 \end{cases}$

34. $\begin{cases} -2x + 2y = 7 \\ 2x + y = 8 \end{cases}$

35. $\begin{cases} 7x + 8y = 6 \\ 3x - 4y = 10 \end{cases}$

36. $\begin{cases} 10x - 11y = 7 \\ 2x - y = 5 \end{cases}$

37. $\begin{cases} 5x + 2y = 7 \\ 3x - 6y = -3 \end{cases}$

38. $\begin{cases} -4x + 5y = 8 \\ 2x + 3y = 18 \end{cases}$

In Exercises 39–52, use the most convenient method (graphing, substitution, or elimination) to solve the system of linear equations. State which method you used.

39. $\begin{cases} x - y = 2 \\ y = 3 \end{cases}$

40. $\begin{cases} y = 7 \\ x - 3y = 0 \end{cases}$

41. $\begin{cases} 6x + 21y = 132 \\ 6x - 4y = 32 \end{cases}$

42. $\begin{cases} -2x + y = 12 \\ 2x + 3y = 20 \end{cases}$

43. $\begin{cases} y = 2x - 1 \\ y = x + 1 \end{cases}$

44. $\begin{cases} 2x - y = 4 \\ y = x \end{cases}$

45. $\begin{cases} -4x + 3y = 11 \\ 3x - 10y = 15 \end{cases}$

46. $\begin{cases} -3x + 5y = -11 \\ 5x - 9y = 19 \end{cases}$

47. $\begin{cases} x + y = 0 \\ 8x + 3y = 15 \end{cases}$

48. $\begin{cases} x - 2y = 0 \\ 0.2x + 0.8y = 2.4 \end{cases}$

49. $\begin{cases} -\dfrac{x}{4} + y = 1 \\ \dfrac{x}{4} + \dfrac{y}{2} = 1 \end{cases}$

50. $\begin{cases} \dfrac{x}{3} - \dfrac{y}{5} = 1 \\ \dfrac{x}{12} + \dfrac{y}{40} = 1 \end{cases}$

51. $\begin{cases} 3(x + 5) - 7 = 2(3 - 2y) \\ 2x + 1 = 4(y + 2) \end{cases}$

52. $\begin{cases} \frac{1}{2}(x - 4) + 9 = y - 10 \\ -5(x + 3) = 8 - 2(y - 3) \end{cases}$

Solving Problems

53. *Number Problem*   The sum of two numbers x and y is 40 and the difference of the two numbers is 10. Find the two numbers.

54. *Number Problem*   The sum of two numbers x and y is 50 and the difference of the two numbers is 20. Find the two numbers.

55. *Number Problem*   The sum of two numbers x and y is 82 and the difference of the numbers is 14. Find the two numbers.

56. *Number Problem*   The sum of two numbers x and y is 154 and the difference of the numbers is 38. Find the two numbers.

57. *Sports*   A basketball player scored 20 points in a game by shooting two-point and three-point baskets. He made a total of 9 baskets. How many of each type did he make?

58. *Sports*   A basketball team scored 84 points in a game by shooting two-point and three-point baskets. The team made a total of 36 baskets. How many of each type did the team make?

59. *Ticket Sales*   Ticket sales for a play were $3799 on the first night and $4905 on the second night. On the first night, 213 student tickets were sold and 632 general admission tickets were sold. On the second night, 275 student tickets were sold and 816 general admission tickets were sold. Determine the price of each type of ticket.

60. *Ticket Sales*   Ticket sales for an annual variety show were $540 the first night and $850 the second night. On the first night, 150 student tickets were sold and 80 general admission tickets were sold. On the second night, 200 student tickets were sold and 150 general admission tickets were sold. Determine the price of each type of ticket.

61. *Investment*   You invest a total of $10,000 in two investments earning 7.5% and 10% simple interest. (There is more risk in the 10% fund.) Your goal is to have a total annual interest income of $850. Determine the smallest amount that you can invest at 10% in order to meet your objective.

62. *Investment*   You invest a total of $12,000 in two investments earning 8% and 11.5% simple interest. (There is more risk in the 11.5% fund.) Your goal is to have a total annual interest income of $1065. Determine the smallest amount that you can invest at 11.5% in order to meet your objective.

63. *Music*   A music intructor charges $25 for a private flute lesson and $18 per student for a group flute lesson. In one day, the instructor earns $265 from 12 students. How many students of each type did the instructor teach?

64. *Dance*   A tap dance instructor charges $20 for a private lesson and $12 per student for a group lesson. In one day, the instructor earns $216 from 14 students. How many students of each type did the instructor teach?

65. *Jewelry*   A bracelet that is supposed to be 18-karat gold weighs 277.92 grams. The volume of the bracelet is 18.52 cubic centimeters. The bracelet is made of gold and copper. Gold weighs 19.3 grams per cubic centimeter and copper weighs 9 grams per cubic centimeter. Determine whether or not the bracelet is 18-karat gold.

18K = 3/4 gold by weight

66. ▲ *Geometry*   Find an equation of the line of slope $m = 3$ passing through the intersection of the lines

$$3x + 4y = 7$$

and

$$5x - 4y = 1.$$

67. ▲ *Geometry*   Find an equation of the line of slope $m = -2$ passing through the intersection of the lines

$$2x + 5y = 11$$

and

$$4x - y = 11.$$

68. *Focal Length*   When parallel rays of light pass through a convex lens, they are bent inward and meet at a *focus* (see figure). The distance from the center of the lens to the focus is called the *focal length*. The equations of the lines representing the two bent rays in the camera are

$$\begin{cases} x + 3y = 1 \\ -x + 3y = -1 \end{cases}$$

where x and y are measured in inches. Which equation is the upper ray? What is the focal length?

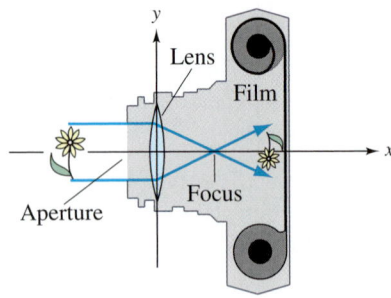

Figure for 68

Explaining Concepts

69. *Writing*   In your own words, describe the basic steps for solving a system of linear equations by the method of elimination.

70. *Writing*   When solving a system by the method of elimination, how do you recognize that it has no solution?

71. *Writing*   When solving a system by the method of elimination, how do you recognize that it has infinitely many solutions?

72. *Creating a System*   Write an example of a system of linear equations that is better solved by the method of elimination than by the method of substitution. There are many correct answers.

73. *Creating a System*   Write an example of a system of linear equations that is better solved by the method of substitution than by the method of elimination. There are many correct answers.

74. *Creating an Example*   Write an example of "clearing" a system of decimals. There are many correct answers.

75. *Writing*   Both $(-2, 3)$ and $(8, 1)$ are solutions to a system of linear equations. How many solutions does the system have? Explain.

76. Consider the system of linear equations.

$$\begin{cases} x + y = 8 \\ 2x + 2y = k \end{cases}$$

(a) Find the value(s) of k for which the system has an infinite number of solutions.

(b) Find one value of k for which the system has no solution. There are many correct answers.

(c) Can the system have a single solution for some value of k? Why or why not?

8.3 Linear Systems in Three Variables

What You Should Learn

1 Solve systems of linear equations using row-echelon form with back-substitution.

2 Solve systems of linear equations using the method of Gaussian elimination.

3 Solve application problems using elimination with back-substitution.

Frank Whitney/Getty Images

Why You Should Learn It

Systems of linear equations in three variables can be used to model and solve real-life problems. For instance, in Exercise 47 on page 505, a system of linear equations can be used to determine a chemical mixture for a pesticide.

1 Solve systems of linear equations using row-echelon form with back-substitution.

Row-Echelon Form

The method of elimination can be applied to a system of linear equations in more than two variables. In fact, this method easily adapts to computer use for solving systems of linear equations with dozens of variables.

When the method of elimination is used to solve a system of linear equations, the goal is to rewrite the system in a form to which back-substitution can be applied. For instance, consider the following two systems of linear equations.

$$\begin{cases} x - 2y + 2z = 9 \\ -x + 3y \quad\quad = -4 \\ 2x - 5y + z = 10 \end{cases} \quad\quad \begin{cases} x - 2y + 2z = 9 \\ y + 2z = 5 \\ z = 3 \end{cases}$$

Which of these two systems do you think is easier to solve? After comparing the two systems, it should be clear that it is easier to solve the system on the right because the value of z is already shown and back-substitution will readily yield the values of x and y. The system on the right is said to be in **row-echelon form,** which means that it has a "stair-step" pattern with leading coefficients of 1.

Example 1   Using Back-Substitution

In the following system of linear equations, you know the value of z from Equation 3.

$$\begin{cases} x - 2y + 2z = 9 & \text{Equation 1} \\ y + 2z = 5 & \text{Equation 2} \\ z = 3 & \text{Equation 3} \end{cases}$$

To solve for y, substitute $z = 3$ in Equation 2 to obtain

$$y + 2(3) = 5 \implies y = -1. \quad\quad \text{Substitute 3 for } z.$$

Finally, substitute $y = -1$ and $z = 3$ in Equation 1 to obtain

$$x - 2(-1) + 2(3) = 9 \implies x = 1. \quad\quad \text{Substitute} -1 \text{ for } y \text{ and 3 for } z.$$

The solution is $x = 1$, $y = -1$, and $z = 3$, which can also be written as the **ordered triple** $(1, -1, 3)$. Check this in the original system of equations.

Study Tip

When checking a solution, remember that the solution must satisfy each equation in the original system.

② Solve systems of linear equations using the method of Gaussian elimination.

The Method of Gaussian Elimination

Two systems of equations are **equivalent systems** if they have the same solution set. To solve a system that is not in row-echelon form, first convert it to an *equivalent* system that is in row-echelon form. To see how this is done, let's take another look at the method of elimination, as applied to a system of two linear equations.

Example 2   The Method of Elimination

Solve the system of linear equations.

$$\begin{cases} 3x - 2y = -1 & \text{Equation 1} \\ x - y = 0 & \text{Equation 2} \end{cases}$$

Solution

$$\begin{cases} x - y = 0 \\ 3x - 2y = -1 \end{cases} \qquad \text{Interchange the two equations in the system.}$$

$$\begin{aligned} -3x + 3y &= 0 & \text{Multiply new Equation 1 by } -3 \text{ and add it to new} \\ \underline{3x - 2y} &= \underline{-1} & \text{Equation 2.} \\ y &= -1 \end{aligned}$$

$$\begin{cases} x - y = 0 \\ \phantom{x - {}} y = -1 \end{cases} \qquad \text{New system in row-echelon form}$$

Using back-substitution, you can determine that the solution is $(-1, -1)$. Check the solution in each equation in the original system, as follows.

$$\begin{array}{cc} \textit{Equation 1} & \textit{Equation 2} \\ 3x - 2y \overset{?}{=} -1 & x - y \overset{?}{=} 0 \\ 3(-1) - 2(-1) = -1 \checkmark & (-1) - (-1) = 0 \checkmark \end{array}$$

Rewriting a system of linear equations in row-echelon form usually involves a chain of equivalent systems, each of which is obtained by using one of the three basic row operations. This process is called **Gaussian elimination.**

Operations That Produce Equivalent Systems

Each of the following **row operations** on a system of linear equations produces an *equivalent* system of linear equations.

1. Interchange two equations.

2. Multiply one of the equations by a nonzero constant.

3. Add a multiple of one of the equations to another equation to replace the latter equation.

Example 3   Using Gaussian Elimination to Solve a System

Solve the system of linear equations.

$$\begin{cases} x - 2y + 2z = 9 & \text{Equation 1} \\ -x + 3y = -4 & \text{Equation 2} \\ 2x - 5y + z = 10 & \text{Equation 3} \end{cases}$$

Solution

Because the leading coefficient of the first equation is 1, you can begin by saving the x in the upper left position and eliminating the other x terms from the first column, as follows.

$$\begin{cases} x - 2y + 2z = 9 \\ y + 2z = 5 \\ 2x - 5y + z = 10 \end{cases}$$

> Adding the first equation to the second equation produces a new second equation.

$$\begin{cases} x - 2y + 2z = 9 \\ y + 2z = 5 \\ -y - 3z = -8 \end{cases}$$

> Adding -2 times the first equation to the third equation produces a new third equation.

Now that all but the first x have been eliminated from the first column, go to work on the second column. (You need to eliminate y from the third equation.)

$$\begin{cases} x - 2y + 2z = 9 \\ y + 2z = 5 \\ -z = -3 \end{cases}$$

> Adding the second equation to the third equation produces a new third equation.

Finally, you need a coefficient of 1 for z in the third equation.

$$\begin{cases} x - 2y + 2z = 9 \\ y + 2z = 5 \\ z = 3 \end{cases}$$

> Multiplying the third equation by -1 produces a new third equation.

This is the same system that was solved in Example 1, and, as in that example, you can conclude by back-substitution that the solution is

$$x = 1, \quad y = -1, \quad \text{and} \quad z = 3.$$

The solution can be written as the ordered triple

$$(1, -1, 3).$$

You can check the solution by substituting 1 for x, -1 for y, and 3 for z in each equation of the original system, as follows.

Check

$$\begin{aligned} \text{Equation 1:} \quad & x - 2y + 2z \overset{?}{=} 9 \\ & 1 - 2(-1) + 2(3) = 9 \checkmark \\ \text{Equation 2:} \quad & -x + 3y \overset{?}{=} -4 \\ & -(1) + 3(-1) = -4 \checkmark \\ \text{Equation 3:} \quad & 2x - 5y + z \overset{?}{=} 10 \\ & 2(1) - 5(-1) + 3 = 10 \checkmark \end{aligned}$$

Example 4 Using Gaussian Elimination to Solve a System

Solve the system of linear equations.

$$\begin{cases} 4x + y - 3z = 11 & \text{Equation 1} \\ 2x - 3y + 2z = 9 & \text{Equation 2} \\ x + y + z = -3 & \text{Equation 3} \end{cases}$$

Solution

$$\begin{cases} x + y + z = -3 \\ 2x - 3y + 2z = 9 \\ 4x + y - 3z = 11 \end{cases}$$

Interchange the first and third equations.

$$\begin{cases} x + y + z = -3 \\ -5y = 15 \\ 4x + y - 3z = 11 \end{cases}$$

Adding -2 times the first equation to the second equation produces a new second equation.

$$\begin{cases} x + y + z = -3 \\ -5y = 15 \\ -3y - 7z = 23 \end{cases}$$

Adding -4 times the first equation to the third equation produces a new third equation.

$$\begin{cases} x + y + z = -3 \\ y = -3 \\ -3y - 7z = 23 \end{cases}$$

Multiplying the second equation by $-\frac{1}{5}$ produces a new second equation.

$$\begin{cases} x + y + z = -3 \\ y = -3 \\ -7z = 14 \end{cases}$$

Adding 3 times the second equation to the third equation produces a new third equation.

$$\begin{cases} x + y + z = -3 \\ y = -3 \\ z = -2 \end{cases}$$

Multiplying the third equation by $-\frac{1}{7}$ produces a new third equation.

Now you can see that $z = -2$ and $y = -3$. Moreover, by back-substituting these values in Equation 1, you can determine that $x = 2$. So, the solution is

$$x = 2, \quad y = -3, \quad \text{and} \quad z = -2$$

which can be written as the ordered triple $(2, -3, -2)$. You can check this solution as follows.

Check

$$\begin{aligned} \text{Equation 1:} \quad 4x + y - 3z &\overset{?}{=} 11 \\ 4(2) + (-3) - 3(-2) &\overset{?}{=} 11 \\ 11 &= 11 \checkmark \end{aligned}$$

$$\begin{aligned} \text{Equation 2:} \quad 2x - 3y + 2z &\overset{?}{=} 9 \\ 2(2) - 3(-3) + 2(-2) &\overset{?}{=} 9 \\ 9 &= 9 \checkmark \end{aligned}$$

$$\begin{aligned} \text{Equation 3:} \quad x + y + z &\overset{?}{=} -3 \\ (2) + (-3) + (-2) &\overset{?}{=} -3 \\ -3 &= -3 \checkmark \end{aligned}$$

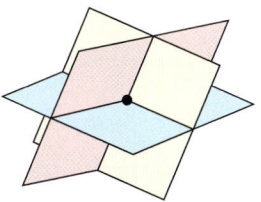

Solution: one point

Figure 8.10

Solution: one line

Figure 8.11

Solution: one plane

Figure 8.12

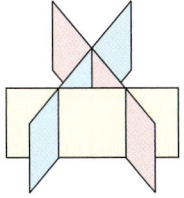

Solution: none

Figure 8.13

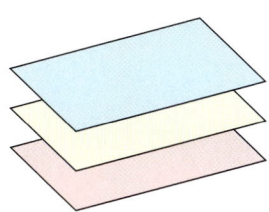

Solution: none

Figure 8.14

The next example involves an inconsistent system—one that has no solution. The key to recognizing an inconsistent system is that at some stage in the elimination process, you obtain a false statement such as $0 = 6$. Watch for such statements as you do the exercises for this section.

Example 5   An Inconsistent System

Solve the system of linear equations.

$$\begin{cases} x - 3y + z = 1 & \text{Equation 1} \\ 2x - y - 2z = 2 & \text{Equation 2} \\ x + 2y - 3z = -1 & \text{Equation 3} \end{cases}$$

Solution

$$\begin{cases} x - 3y + z = 1 \\ 5y - 4z = 0 \\ x + 2y - 3z = -1 \end{cases}$$

> Adding -2 times the first equation to the second equation produces a new second equation.

$$\begin{cases} x - 3y + z = 1 \\ 5y - 4z = 0 \\ 5y - 4z = -2 \end{cases}$$

> Adding -1 times the first equation to the third equation produces a new third equation.

$$\begin{cases} x - 3y + z = 1 \\ 5y - 4z = 0 \\ 0 = -2 \end{cases}$$

> Adding -1 times the second equation to the third equation produces a new third equation.

Because the third "equation" is a false statement, you can conclude that this system is inconsistent and so has no solution. Moreover, because this system is equivalent to the original system, you can conclude that the original system also has no solution.

As with a system of linear equations in two variables, the solution(s) of a system of linear equations in more than two variables must fall into one of three categories.

The Number of Solutions of a Linear System

For a system of linear equations, exactly one of the following is true.

1. There is exactly one solution.

2. There are infinitely many solutions.

3. There is no solution.

The graph of a system of three linear equations in three variables consists of *three planes*. When these planes intersect in a single point, the system has exactly one solution. (See Figure 8.10.) When the three planes intersect in a line or a plane, the system has infinitely many solutions. (See Figures 8.11 and 8.12.) When the three planes have no point in common, the system has no solution. (See Figures 8.13 and 8.14.)

Example 6   A System with Infinitely Many Solutions

Solve the system of linear equations.

$$\begin{cases} x + y - 3z = -1 & \text{Equation 1} \\ \quad\quad y - z = 0 & \text{Equation 2} \\ -x + 2y \quad\quad = 1 & \text{Equation 3} \end{cases}$$

Solution

Begin by rewriting the system in row-echelon form.

$$\begin{cases} x + y - 3z = -1 \\ \quad\quad y - z = 0 \\ \quad\quad 3y - 3z = 0 \end{cases}$$

> Adding the first equation to the third equation produces a new third equation.

$$\begin{cases} x + y - 3z = -1 \\ \quad\quad y - z = 0 \\ \quad\quad\quad 0 = 0 \end{cases}$$

> Adding -3 times the second equation to the third equation produces a new third equation.

This means that Equation 3 depends on Equations 1 and 2 in the sense that it gives us no additional information about the variables. So, the original system is equivalent to the system

$$\begin{cases} x + y - 3z = -1 \\ \quad\quad y - z = 0. \end{cases}$$

In this last equation, solve for y in terms of z to obtain $y = z$. Back-substituting for y in the previous equation produces $x = 2z - 1$. Finally, letting $z = a$, where a is any real number, you can see that solutions to the original system are all of the form

$$x = 2a - 1, \; y = a, \text{ and } z = a.$$

So, every ordered triple of the form

$$(2a - 1, a, a), \quad\quad a \text{ is a real number}$$

is a solution of the system.

In Example 6, there are other ways to write the same infinite set of solutions. For instance, letting $x = b$, the solutions could have been written as

$$\left(b, \frac{1}{2}(b + 1), \frac{1}{2}(b + 1) \right), \quad\quad b \text{ is a real number.}$$

To convince yourself that this description produces the same set of solutions, consider the comparison shown below.

Substitution	Solution	
$a = 0$	$(2(0) - 1, 0, 0) = (-1, 0, 0)$	
$b = -1$	$\left(-1, \frac{1}{2}(-1 + 1), \frac{1}{2}(-1 + 1)\right) = (-1, 0, 0)$	Same solution
$a = 1$	$(2(1) - 1, 1, 1) = (1, 1, 1)$	
$b = 1$	$\left(1, \frac{1}{2}(1 + 1), \frac{1}{2}(1 + 1)\right) = (1, 1, 1)$	Same solution

Study Tip

When comparing descriptions of an infinite solution set, keep in mind that there is more than one way to describe the set.

3 Solve application problems using elimination with back-substitution.

Applications

Example 7   Vertical Motion

The height at time t of an object that is moving in a (vertical) line with constant acceleration a is given by the **position equation**

$$s = \frac{1}{2}at^2 + v_0 t + s_0.$$

The height s is measured in feet, the acceleration a is measured in feet per second squared, the time t is measured in seconds, v_0 is the initial velocity (at time $t = 0$), and s_0 is the initial height. Find the values of a, v_0, and s_0, if $s = 164$ feet at 1 second, $s = 180$ feet at 2 seconds, and $s = 164$ feet at 3 seconds.

Solution

By substituting the three values of t and s into the position equation, you obtain three linear equations in a, v_0, and s_0.

When $t = 1$, $s = 164$: $\frac{1}{2}a(1)^2 + v_0(1) + s_0 = 164$

When $t = 2$, $s = 180$: $\frac{1}{2}a(2)^2 + v_0(2) + s_0 = 180$

When $t = 3$, $s = 164$: $\frac{1}{2}a(3)^2 + v_0(3) + s_0 = 164$

By multiplying the first and third equations by 2, this system can be rewritten as

$$\begin{cases} a + 2v_0 + 2s_0 = 328 & \text{Equation 1} \\ 2a + 2v_0 + s_0 = 180 & \text{Equation 2} \\ 9a + 6v_0 + 2s_0 = 328 & \text{Equation 3} \end{cases}$$

and you can apply Gaussian elimination to obtain

$$\begin{cases} a + 2v_0 + 2s_0 = 328 & \text{Equation 1} \\ \quad -2v_0 - 3s_0 = -476 & \text{Equation 2} \\ \quad\quad\quad 2s_0 = 232. & \text{Equation 3} \end{cases}$$

From the third equation, $s_0 = 116$, so back-substitution in Equation 2 yields

$$-2v_0 - 3(116) = -476$$

$$-2v_0 = -128$$

$$v_0 = 64.$$

Finally, back-substituting $s_0 = 116$ and $v_0 = 64$ in Equation 1 yields

$$a + 2(64) + 2(116) = 328$$

$$a = -32.$$

So, the position equation for this object is $s = -16t^2 + 64t + 116$.

Example 8   A Geometry Application

The sum of the measures of two angles of a triangle is twice the measure of the third angle. The measure of the first angle is $18°$ more than the measure of the third angle. Find the measures of the three angles.

Solution

Let x, y, and z represent the measures of the first, second, and third angles, respectively. The sum of the measures of the three angles of a triangle is $180°$. From the given information, you can write the system of equations as follows.

$$\begin{cases} x + y + z = 180 & \text{Equation 1} \\ \quad\; x + y = 2z & \text{Equation 2} \\ \quad\quad\;\; x = z + 18 & \text{Equation 3} \end{cases}$$

By rewriting this system in the standard form you obtain

$$\begin{cases} x + y + \;\; z = 180 & \text{Equation 1} \\ x + y - 2z = \quad 0 & \text{Equation 2} \\ x \quad\quad - \;\; z = \;\; 18. & \text{Equation 3} \end{cases}$$

Using Gaussian elimination to solve this system yields $x = 78$, $y = 42$, and $z = 60$. So, the measures of the three angles are $78°$, $42°$, and $60°$, respectively. You can check these solutions as follows.

Check

Equation 1: $78 + 42 + 60 = 180$ ✓

Equation 2: $78 + 42 - 2(60) = 0$ ✓

Equation 3: $78 - 60 = 18$ ✓

Example 9   Grades of Paper

A paper manufacturer sells a 50-pound package that consists of three grades of computer paper. Grade A costs $6.00 per pound, grade B costs $4.50 per pound, and grade C costs $3.50 per pound. Half of the 50-pound package consists of the two cheaper grades. The cost of the 50-pound package is $252.50. How many pounds of each grade of paper are there in the 50-pound package?

Solution

Let A represent grade A paper, B represent grade B paper, and C represent grade C paper. From the given information you can write the system of equations as follows.

$$\begin{cases} A + \quad B + \quad\;\; C = 50 & \text{Equation 1} \\ 6A + 4.50B + 3.50C = 252.50 & \text{Equation 2} \\ \quad\quad\;\; B + \quad\;\; C = 25 & \text{Equation 3} \end{cases}$$

Using Gaussian elimination to solve this system yields $A = 25$, $B = 15$, and $C = 10$. So, there are 25 pounds of grade A paper, 15 pounds of grade B paper, and 10 pounds of grade C paper in the 50-pound package. Check this solution in the original statement of the problem.

8.3 Exercises

Review   Concepts, Skills, and Problem Solving

Keep mathematically in shape by doing these exercises *before* the problems of this section.

Properties and Definitions

1. A linear equation of the form $2x + 8 = 7$ has how many solutions?

2. What is the usual first step in solving an equation such as

$$\frac{t}{6} + \frac{5}{8} = \frac{7}{4}?$$

Solving Equations

In Exercises 3–8, solve the equation.

3. $\dfrac{x}{6} + \dfrac{5}{8} = \dfrac{7}{4}$

4. $0.25x + 1.75 = 4.5$

5. $|2x - 4| = 6$

6. $2|7 - x| = 10$

7. $6x - (x + 1) = 5x - 1$

8. $\frac{1}{4}(5 - 2x) = 9x - 7x$

Models and Graphs

9. The length of each edge of a cube is s inches. Write the volume V of the cube as a function of s.

10. Write the area A of a circle as a function of its circumference C.

11. The speed of a ship is 15 knots. Write the distance d the ship travels as a function of time t. Graph the model.

12. Your weekly pay is \$180 plus \$1.25 per sale. Write your weekly pay P as a function of the number of sales n. Graph the model.

Developing Skills

In Exercises 1 and 2, determine whether each ordered triple is a solution of the system of linear equations.

1. $\begin{cases} x + 3y + 2z = 1 \\ 5x - y + 3z = 16 \\ -3x + 7y + z = -14 \end{cases}$

  (a) $(0, 3, -2)$ (b) $(12, 5, -13)$

  (c) $(1, -2, 3)$ (d) $(-2, 5, -3)$

2. $\begin{cases} 3x - y + 4z = -10 \\ -x + y + 2z = 6 \\ 2x - y + z = -8 \end{cases}$

  (a) $(-2, 4, 0)$ (b) $(0, -3, 10)$

  (c) $(1, -1, 5)$ (d) $(7, 19, -3)$

In Exercises 3–6, use back-substitution to solve the system of linear equations. See Example 1.

3. $\begin{cases} x - 2y + 4z = 4 \\ 3y - z = 2 \\ z = -5 \end{cases}$

4. $\begin{cases} 5x + 4y - z = 0 \\ 10y - 3z = 11 \\ z = 3 \end{cases}$

5. $\begin{cases} x - 2y + 4z = 4 \\ y = 3 \\ y + z = 2 \end{cases}$

6. $\begin{cases} x = 10 \\ 3x + 2y = 2 \\ x + y + 2z = 0 \end{cases}$

In Exercises 7 and 8, determine whether the two systems of linear equations are equivalent. Give reasons for your answer.

7. $\begin{cases} x + 3y - z = 6 \\ 2x - y + 2z = 1 \\ 3x + 2y - z = 2 \end{cases}$ $\begin{cases} x + 3y - z = 6 \\ -7y + 4z = 1 \\ -7y - 4z = -16 \end{cases}$

8. $\begin{cases} x - 2y + 3z = 9 \\ -x + 3y = -4 \\ 2x - 5y + 5z = 17 \end{cases}$ $\begin{cases} x - 2y + 3z = 9 \\ y + 3z = 5 \\ -y - z = -1 \end{cases}$

In Exercises 9 and 10, perform the row operation and write the equivalent system of linear equations. See Example 2.

9. Add Equation 1 to Equation 2.

$$\begin{cases} x - 2y = 8 & \text{Equation 1} \\ -x + 3y = 6 & \text{Equation 2} \end{cases}$$

What did this operation accomplish?

10. Add -2 times Equation 1 to Equation 3.

$$\begin{cases} x - 2y + 3z = 5 & \text{Equation 1} \\ -x + y + 5z = 4 & \text{Equation 2} \\ 2x \quad\;\; - 3z = 0 & \text{Equation 3} \end{cases}$$

What did this operation accomplish?

In Exercises 11–34, solve the system of linear equations. See Examples 3–6.

11. $\begin{cases} x \;\; + z = 4 \\ y \quad\;\; = 2 \\ 4x \;\; + z = 7 \end{cases}$ **12.** $\begin{cases} x \quad\qquad = 3 \\ -x + 3y \quad\;\; = 3 \\ \quad\;\; y + 2z = 4 \end{cases}$

13. $\begin{cases} x + y + z = 6 \\ 2x - y + z = 3 \\ 3x \quad\;\; - z = 0 \end{cases}$ **14.** $\begin{cases} x + y + z = 2 \\ -x + 3y + 2z = 8 \\ 4x + y \quad\;\; = 4 \end{cases}$

15. $\begin{cases} x + y + z = -3 \\ 4x + y - 3z = 11 \\ 2x - 3y + 2z = 9 \end{cases}$ **16.** $\begin{cases} x - y + 2z = -4 \\ 3x + y - 4z = -6 \\ 2x + 3y - 4z = 4 \end{cases}$

17. $\begin{cases} x + 2y + 6z = 5 \\ -x + y - 2z = 3 \\ x - 4y - 2z = 1 \end{cases}$ **18.** $\begin{cases} x + 6y + 2z = 9 \\ 3x - 2y + 3z = -1 \\ 5x - 5y + 2z = 7 \end{cases}$

19. $\begin{cases} 2x \quad\;\; + 2z = 2 \\ 5x + 3y \quad\;\; = 4 \\ \quad\; 3y - 4z = 4 \end{cases}$ **20.** $\begin{cases} x + y + 8z = 3 \\ 2x + y + 11z = 4 \\ x \quad\;\; + 3z = 0 \end{cases}$

21. $\begin{cases} \quad\; 6y + 4z = -12 \\ 3x + 3y \quad\;\; = 9 \\ 2x \quad\;\; - 3z = 10 \end{cases}$

22. $\begin{cases} 2x - 4y + z = 0 \\ 3x \quad\;\; + 2z = -1 \\ -6x + 3y + 2z = -10 \end{cases}$

23. $\begin{cases} 2x + y + 3z = 1 \\ 2x + 6y + 8z = 3 \\ 6x + 8y + 18z = 5 \end{cases}$ **24.** $\begin{cases} 3x - y - 2z = 5 \\ 2x + y + 3z = 6 \\ 6x - y - 4z = 9 \end{cases}$

25. $\begin{cases} \quad\; y + z = 5 \\ 2x \quad\;\; + 4z = 4 \\ 2x - 3y \quad\;\; = -14 \end{cases}$ **26.** $\begin{cases} 5x + 2y \quad\;\; = -8 \\ \quad\qquad z = 5 \\ 3x - y + z = 9 \end{cases}$

27. $\begin{cases} 2x \quad\;\; + z = 1 \\ \quad\; 5y - 3z = 2 \\ 6x + 20y - 9z = 11 \end{cases}$ **28.** $\begin{cases} 2x + y - z = 4 \\ \quad\; y + 3z = 2 \\ 3x + 2y \quad\;\; = 4 \end{cases}$

29. $\begin{cases} 3x + y + z = 2 \\ 4x \quad\;\; + 2z = 1 \\ 5x - y + 3z = 0 \end{cases}$

30. $\begin{cases} 2x \quad\;\; + 3z = 4 \\ 5x + y + z = 2 \\ 11x + 3y - 3z = 0 \end{cases}$

31. $\begin{cases} 0.2x + 1.3y + 0.6z = 0.1 \\ 0.1x \quad\;\; + 0.3z = 0.7 \\ 2x + 10y + 8z = 8 \end{cases}$

32. $\begin{cases} 0.3x - 0.1y + 0.2z = 0.35 \\ 2x + y - 2z = -1 \\ 2x + 4y + 3z = 10.5 \end{cases}$

33. $\begin{cases} x + 4y - 2z = 2 \\ -3x + y + z = -2 \\ 5x + 7y - 5z = 6 \end{cases}$

34. $\begin{cases} x - 2y - z = 3 \\ 2x + y - 3z = 1 \\ x + 8y - 3z = -7 \end{cases}$

In Exercises 35 and 36, find a system of linear equations in three variables with integer coefficients that has the given point as a solution. (There are many correct answers.)

35. $(4, -3, 2)$ **36.** $(5, 7, -10)$

Solving Problems

Vertical Motion   In Exercises 37–40, find the position equation $s = \frac{1}{2}at^2 + v_0 t + s_0$ for an object that has the indicated heights at the specified times. See Example 7.

37. $s = 128$ feet at $t = 1$ second

$s = 80$ feet at $t = 2$ seconds

$s = 0$ feet at $t = 3$ seconds

38. $s = 48$ feet at $t = 1$ second

$s = 64$ feet at $t = 2$ seconds

$s = 48$ feet at $t = 3$ seconds

39. $s = 32$ feet at $t = 1$ second

$s = 32$ feet at $t = 2$ seconds

$s = 0$ feet at $t = 3$ seconds

40. $s = 10$ feet at $t = 0$ seconds

$s = 54$ feet at $t = 1$ second

$s = 46$ feet at $t = 3$ seconds

41. ▲ *Geometry*   The sum of the measures of two angles of a triangle is twice the measure of the third angle. The measure of the second angle is $28°$ less than the measure of the third angle. Find the measures of the three angles.

42. ▲ *Geometry*   The measure of the second angle of a triangle is one-half the measure of the first angle. The measure of the third angle is $70°$ less than 2 times the measure of the second angle. Find the measures of the three angles.

43. *Investment*   An inheritance of $80,000 is divided among three investments yielding a total of $8850 in interest per year. The interest rates for the three investments are 6%, 10%, and 15%. The amount invested at 10% is $750 more than the amount invested at 15%. Find the amount invested at each rate.

44. *Investment*   An inheritance of $16,000 is divided among three investments yielding a total of $940 in interest per year. The interest rates for the three investments are 5%, 6%, and 7%. The amount invested at 6% is $3000 less than the amount invested at 5%. Find the amount invested at each rate.

45. *Investment*   You receive a total of $708 a year in interest from three investments. The interest rates for the three investments are 6%, 8%, and 9%. The 8% investment is half of the 6% investment, and the 9% investment is $1000 less than the 6% investment. What is the amount of each investment?

46. *Investment*   You receive a total of $1520 a year in interest from three investments. The interest rates for the three investments are 5%, 7%, and 8%. The 5% investment is half of the 7% investment, and the 7% investment is $1500 less than the 8% investment. What is the amount of each investment?

47. *Chemical Mixture*   A mixture of 12 gallons of chemical A, 16 gallons of chemical B, and 26 gallons of chemical C is required to kill a destructive crop insect. Commercial spray X contains one, two, and two parts of these chemicals. Spray Y contains only chemical C. Spray Z contains only chemicals A and B in equal amounts. How much of each type of commercial spray is needed to obtain the desired mixture?

48. *Fertilizer Mixture*   A mixture of 5 pounds of fertilizer A, 13 pounds of fertilizer B, and 4 pounds of fertilizer C provides the optimal nutrients for a plant. Commercial brand X contains equal parts of fertilizer B and fertilizer C. Brand Y contains one part of fertilizer A and two parts of fertilizer B. Brand Z contains two parts of fertilizer A, five parts of fertilizer B, and two parts of fertilizer C. How much of each fertilizer brand is needed to obtain the desired mixture?

49. *Floral Arrangements*   A florist sells three types of floral arrangements for $40, $30, and $20 per arrangement. In one year the total revenue for the arrangements was $25,000, which corresponds to the sale of 850 arrangements. The florist sold 4 times as many of the $20 arrangements as the $30 arrangements. How many arrangements of each type were sold?

50. *Coffee* A coffee manufacturer sells a 10-pound package of coffee that consists of three flavors of coffee. Vanilla flavored coffee costs $2 per pound, Hazelnut flavored coffee costs $2.50 per pound, and French Roast flavored coffee costs $3 per pound. The package contains the same amount of Hazelnut coffee as French Roast coffee. The cost of the 10-pound package is $26. How many pounds of each type of coffee are in the package?

51. *Mixture Problem* A chemist needs 12 gallons of a 20% acid solution. It is mixed from three solutions whose concentrations are 10%, 15%, and 25%. How many gallons of each solution will satisfy each condition?

(a) Use 4 gallons of the 25% solution.

(b) Use as little as possible of the 25% solution.

(c) Use as much as possible of the 25% solution.

52. *Mixture Problem* A chemist needs 10 liters of a 25% acid solution. It is mixed from three solutions whose concentrations are 10%, 20%, and 50%. How many liters of each solution will satisfy each condition?

(a) Use 2 liters of the 50% solution.

(b) Use as little as possible of the 50% solution.

(c) Use as much as possible of the 50% solution.

53. *School Orchestra* The table shows the percents of each section of the North High School orchestra that were chosen to participate in the city orchestra, the county orchestra, and the state orchestra. Thirty members of the city orchestra, 17 members of the county orchestra, and 10 members of the state orchestra are from North High. How many members are in each section of North High's orchestra?

Orchestra	String	Wind	Percussion
City orchestra	40%	30%	50%
County orchestra	20%	25%	25%
State orchestra	10%	15%	25%

54. *Sports* The table shows the percents of each unit of the North High School football team that were chosen for academic honors, as city all-stars, and as county all-stars. Of all the players on the football team, 5 were awarded with academic honors, 13 were named city all-stars, and 4 were named county all-stars. How many members of each unit are there on the football team?

	Defense	Offense	Special teams
Academic honors	0%	10%	20%
City all-stars	10%	20%	50%
County all-stars	10%	0%	20%

Explaining Concepts

55. ⊘ Answer parts (c)–(e) of Motivating the Chapter on page 466.

56. Give an example of a system of linear equations that is in row-echelon form.

57. Show how to use back-substitution to solve the system you found in Exercise 56.

58. *Writing* Describe the row operations that are performed on a system of linear equations to produce an equivalent system of equations.

59. Write a system of four linear equations in four unknowns, and solve it by elimination.

Mid-Chapter Quiz

Take this quiz as you would take a quiz in class. After you are done, check your work against the answers in the back of the book.

1. Determine whether each ordered pair is a solution of the system of linear equations: (a) $(1, -2)$ (b) $(10, 4)$

$$\begin{cases} 5x - 12y = 2 \\ 2x + 1.5y = 26 \end{cases}$$

In Exercises 2–4, graph the equations in the system. Use the graphs to determine the number of solutions of the system.

2. $\begin{cases} -6x + 9y = 9 \\ 2x - 3y = 6 \end{cases}$ **3.** $\begin{cases} x - 2y = -4 \\ 3x - 2y = 4 \end{cases}$ **4.** $\begin{cases} 0.5x - 1.5y = 7 \\ -2x + 6y = -28 \end{cases}$

In Exercises 5–7, use the graphical method to solve the system of equations.

5. $\begin{cases} x = 4 \\ 2x - y = 6 \end{cases}$ **6.** $\begin{cases} 2x + 7y = 16 \\ 3x + 2y = 24 \end{cases}$ **7.** $\begin{cases} 4x - y = 9 \\ x - 3y = 16 \end{cases}$

In Exercises 8–10, solve the system of equations by the method of substitution.

8. $\begin{cases} 2x - 3y = 4 \\ y = 2 \end{cases}$ **9.** $\begin{cases} 6x - 2y = 2 \\ 9x - 3y = 1 \end{cases}$ **10.** $\begin{cases} 5x - y = 32 \\ 6x - 9y = 18 \end{cases}$

In Exercises 11–14, use elimination or Gaussian elimination to solve the linear system.

11. $\begin{cases} x + 10y = 18 \\ 5x + 2y = 42 \end{cases}$ **12.** $\begin{cases} 3x + 11y = 38 \\ 7x - 5y = -34 \end{cases}$

13. $\begin{cases} a + b + c = 1 \\ 4a + 2b + c = 2 \\ 9a + 3b + c = 4 \end{cases}$ **14.** $\begin{cases} x + 4z = 17 \\ -3x + 2y - z = -20 \\ x - 5y + 3z = 19 \end{cases}$

In Exercises 15 and 16, write a system of linear equations having the given solution. (There are many correct answers.)

15. $(10, -12)$ **16.** $(1, 3, -7)$

17. A small company produces one-time-use cameras that sell for $5.95 per unit. The cost of producing each camera is $3.45 and the company has fixed costs of $16,000. Use a graphing calculator to graph the cost and revenue functions in the same viewing window. Approximate the point of intersection of the graphs and interpret the results.

18. The measure of the second angle of a triangle is $10°$ less than twice the measure of the first angle. The measure of the third angle is $10°$ greater than the measure of the first angle. Find the measures of the three angles.

8.4 Matrices and Linear Systems

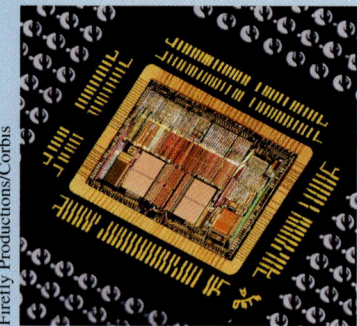

Firefly Productions/Corbis

Firefly Productions/Corbis

What You Should Learn

1. Determine the order of matrices.
2. Form coefficient and augmented matrices and form linear systems from augmented matrices.
3. Perform elementary row operations to solve systems of linear equations.
4. Use matrices and Gaussian elimination with back-substitution to solve systems of linear equations.

Why You Should Learn It

Systems of linear equations that model real-life situations can be solved using matrices. For instance, in Exercise 81 on page 520, the numbers of computer parts a company produces can be found using a matrix.

Matrices

In this section, you will study a streamlined technique for solving systems of linear equations. This technique involves the use of a rectangular array of real numbers called a **matrix.** (The plural of matrix is matrices.) Here is an example of a matrix.

$$
\begin{array}{c}
\\
\text{Row 1} \\
\text{Row 2} \\
\text{Row 3}
\end{array}
\begin{array}{cccc}
\text{Column} & \text{Column} & \text{Column} & \text{Column} \\
1 & 2 & 3 & 4 \\
\begin{bmatrix}
3 & -2 & 4 & 1 \\
0 & 1 & -1 & 2 \\
2 & 0 & -3 & 0
\end{bmatrix}
\end{array}
$$

This matrix has three rows and four columns, which means that its **order** is 3×4, which is read as "3 by 4." Each number in the matrix is an **entry** of the matrix.

Example 1 Order of Matrices

Determine the order of each matrix.

a. $\begin{bmatrix} 1 & -2 & 4 \\ 0 & 1 & -2 \end{bmatrix}$ **b.** $\begin{bmatrix} 0 & 0 \\ 0 & 0 \end{bmatrix}$ **c.** $\begin{bmatrix} 1 & -3 \\ -2 & 0 \\ 4 & -2 \end{bmatrix}$

Solution

a. This matrix has two rows and three columns, so the order is 2×3.

b. This matrix has two rows and two columns, so the order is 2×2.

c. This matrix has three rows and two columns, so the order is 3×2.

Study Tip

The order of a matrix is always given as *row by column.* A matrix with the same number of rows as columns is called a **square matrix.** For instance, the 2×2 matrix in Example 1(b) is square.

1. Determine the order of matrices.

Augmented and Coefficient Matrices

A matrix derived from a system of linear equations (each written in standard form) is the **augmented matrix** of the system. Moreover, the matrix derived from the coefficients of the system (but that does not include the constant terms) is the **coefficient matrix** of the system. Here is an example.

Study Tip

Note the use of 0 for the missing y-variable in the third equation, and also note the fourth column of constant terms in the augmented matrix.

System	Coefficient Matrix	Augmented Matrix
$\begin{cases} x - 4y + 3z = 5 \\ -x + 3y - z = -3 \\ 2x - 4z = 6 \end{cases}$	$\begin{bmatrix} 1 & -4 & 3 \\ -1 & 3 & -1 \\ 2 & 0 & -4 \end{bmatrix}$	$\begin{bmatrix} 1 & -4 & 3 & \vdots & 5 \\ -1 & 3 & -1 & \vdots & -3 \\ 2 & 0 & -4 & \vdots & 6 \end{bmatrix}$

When forming either the coefficient matrix or the augmented matrix of a system, you should begin by vertically aligning the variables in the equations.

Given System	Align Variables	Form Augmented Matrix
$\begin{cases} x + 3y = 9 \\ -y + 4z = -2 \\ x - 5z = 0 \end{cases}$	$\begin{cases} x + 3y = 9 \\ -y + 4z = -2 \\ x - 5z = 0 \end{cases}$	$\begin{bmatrix} 1 & 3 & 0 & \vdots & 9 \\ 0 & -1 & 4 & \vdots & -2 \\ 1 & 0 & -5 & \vdots & 0 \end{bmatrix}$

Example 2 Forming Coefficient and Augmented Matrices

Form the coefficient matrix and the augmented matrix for each system.

a. $\begin{cases} -x + 5y = 2 \\ 7x - 2y = -6 \end{cases}$ **b.** $\begin{cases} 3x + 2y - z = 1 \\ x + 2z = -3 \\ -2x - y = 4 \end{cases}$

Solution

System	Coefficient Matrix	Augmented Matrix
a. $\begin{cases} -x + 5y = 2 \\ 7x - 2y = -6 \end{cases}$	$\begin{bmatrix} -1 & 5 \\ 7 & -2 \end{bmatrix}$	$\begin{bmatrix} -1 & 5 & \vdots & 2 \\ 7 & -2 & \vdots & -6 \end{bmatrix}$
b. $\begin{cases} 3x + 2y - z = 1 \\ x + 2z = -3 \\ -2x - y = 4 \end{cases}$	$\begin{bmatrix} 3 & 2 & -1 \\ 1 & 0 & 2 \\ -2 & -1 & 0 \end{bmatrix}$	$\begin{bmatrix} 3 & 2 & -1 & \vdots & 1 \\ 1 & 0 & 2 & \vdots & -3 \\ -2 & -1 & 0 & \vdots & 4 \end{bmatrix}$

Example 3 Forming Linear Systems from Their Matrices

Write the system of linear equations that is represented by each matrix.

a. $\begin{bmatrix} 3 & -5 & \vdots & 4 \\ -1 & 2 & \vdots & 0 \end{bmatrix}$ **b.** $\begin{bmatrix} 1 & 3 & \vdots & 2 \\ 0 & 1 & \vdots & -3 \end{bmatrix}$ **c.** $\begin{bmatrix} 2 & 0 & -8 & \vdots & 1 \\ -1 & 1 & 1 & \vdots & 2 \\ 5 & -1 & 7 & \vdots & 3 \end{bmatrix}$

Solution

a. $\begin{cases} 3x - 5y = 4 \\ -x + 2y = 0 \end{cases}$ **b.** $\begin{cases} x + 3y = 2 \\ y = -3 \end{cases}$ **c.** $\begin{cases} 2x - 8z = 1 \\ -x + y + z = 2 \\ 5x - y + 7z = 3 \end{cases}$

③ Perform elementary row operations to solve systems of linear equations.

Elementary Row Operations

In Section 8.3, you studied three operations that can be used on a system of linear equations to produce an equivalent system: (1) interchange two equations, (2) multiply an equation by a nonzero constant, and (3) add a multiple of an equation to another equation. In matrix terminology, these three operations correspond to **elementary row operations.**

Elementary Row Operations

Any of the following **elementary row operations** performed on an augmented matrix will produce a matrix that is row-equivalent to the original matrix. Two matrices are **row-equivalent** if one can be obtained from the other by a sequence of elementary row operations.

1. Interchange two rows.

2. Multiply a row by a nonzero constant.

3. Add a multiple of a row to another row.

Example 4   Elementary Row Operations

a. Interchange the first and second rows.

Original Matrix
$$\begin{bmatrix} 0 & 1 & 3 & 4 \\ -1 & 2 & 0 & 3 \\ 2 & -3 & 4 & 1 \end{bmatrix}$$

New Row-Equivalent Matrix
$$\begin{matrix} R_2 \\ R_1 \\ {} \end{matrix} \begin{bmatrix} -1 & 2 & 0 & 3 \\ 0 & 1 & 3 & 4 \\ 2 & -3 & 4 & 1 \end{bmatrix}$$

b. Multiply the first row by $\frac{1}{2}$.

Original Matrix
$$\begin{bmatrix} 2 & -4 & 6 & -2 \\ 1 & 3 & -3 & 0 \\ 5 & -2 & 1 & 2 \end{bmatrix}$$

New Row-Equivalent Matrix
$$\frac{1}{2}R_1 \rightarrow \begin{bmatrix} 1 & -2 & 3 & -1 \\ 1 & 3 & -3 & 0 \\ 5 & -2 & 1 & 2 \end{bmatrix}$$

c. Add -2 times the first row to the third row.

Original Matrix
$$\begin{bmatrix} 1 & 2 & -4 & 3 \\ 0 & 3 & -2 & -1 \\ 2 & 1 & 5 & -2 \end{bmatrix}$$

New Row-Equivalent Matrix
$$\begin{matrix} {} \\ {} \\ -2R_1 + R_3 \rightarrow \end{matrix} \begin{bmatrix} 1 & 2 & -4 & 3 \\ 0 & 3 & -2 & -1 \\ 0 & -3 & 13 & -8 \end{bmatrix}$$

d. Add 6 times the first row to the second row.

Original Matrix
$$\begin{bmatrix} 1 & 2 & 2 & -4 \\ -6 & -11 & 3 & 18 \\ 0 & 0 & 4 & 7 \end{bmatrix}$$

New Row-Equivalent Matrix
$$\begin{matrix} {} \\ 6R_1 + R_2 \rightarrow \\ {} \end{matrix} \begin{bmatrix} 1 & 2 & 2 & -4 \\ 0 & 1 & 15 & -6 \\ 0 & 0 & 4 & 7 \end{bmatrix}$$

In Section 8.3, Gaussian elimination was used with back-substitution to solve systems of linear equations. Example 5 demonstrates the matrix version of Gaussian elimination. The two methods are essentially the same. The basic difference is that with matrices you do not need to keep writing the variables.

Example 5   Solving a System of Linear Equations

Technology: Tip

Most graphing calculators are capable of performing row operations on matrices. Some graphing calculators have a function that will return the reduced row-echelon form of a matrix. Consult the user's guide of your graphing calculator to learn how to perform elementary row operations. Most graphing calculators store the resulting matrix of each step in an answer variable. It is suggested that you store the results of each operation in a matrix variable.

Enter the matrix from Example 5 into your graphing calculator and perform the indicated row operations.

Linear System

$$\begin{cases} x - 2y + 2z = 9 \\ -x + 3y \quad\;\; = -4 \\ 2x - 5y + z = 10 \end{cases}$$

Associated Augmented Matrix

$$\left[\begin{array}{ccc:c} 1 & -2 & 2 & 9 \\ -1 & 3 & 0 & -4 \\ 2 & -5 & 1 & 10 \end{array} \right]$$

Add the first equation to the second equation.

Add the first row to the second row $(R_1 + R_2)$.

$$\begin{cases} x - 2y + 2z = 9 \\ y + 2z = 5 \\ 2x - 5y + z = 10 \end{cases}$$

$R_1 + R_2 \longrightarrow$ $$\left[\begin{array}{ccc:c} 1 & -2 & 2 & 9 \\ 0 & 1 & 2 & 5 \\ 2 & -5 & 1 & 10 \end{array} \right]$$

Add -2 times the first equation to the third equation.

Add -2 times the first row to the third row $(-2R_1 + R_3)$.

$$\begin{cases} x - 2y + 2z = 9 \\ y + 2z = 5 \\ -y - 3z = -8 \end{cases}$$

$-2R_1 + R_3 \longrightarrow$ $$\left[\begin{array}{ccc:c} 1 & -2 & 2 & 9 \\ 0 & 1 & 2 & 5 \\ 0 & -1 & -3 & -8 \end{array} \right]$$

Add the second equation to the third equation.

Add the second row to the third row $(R_2 + R_3)$.

$$\begin{cases} x - 2y + 2z = 9 \\ y + 2z = 5 \\ -z = -3 \end{cases}$$

$R_2 + R_3 \longrightarrow$ $$\left[\begin{array}{ccc:c} 1 & -2 & 2 & 9 \\ 0 & 1 & 2 & 5 \\ 0 & 0 & -1 & -3 \end{array} \right]$$

Multiply the third equation by -1.

Multiply the third row by -1.

$$\begin{cases} x - 2y + 2z = 9 \\ y + 2z = 5 \\ z = 3 \end{cases}$$

$-R_3 \longrightarrow$ $$\left[\begin{array}{ccc:c} 1 & -2 & 2 & 9 \\ 0 & 1 & 2 & 5 \\ 0 & 0 & 1 & 3 \end{array} \right]$$

At this point, you can use back-substitution to find that the solution is $x = 1$, $y = -1$, and $z = 3$. The solution can be written as the ordered triple $(1, -1, 3)$.

Study Tip

The last matrix in Example 5 is in **row-echelon form.** The term *echelon* refers to the stair-step pattern formed by the nonzero elements of the matrix.

Definition of Row-Echelon Form of a Matrix

A matrix in **row-echelon form** has the following properties.

1. All rows consisting entirely of zeros occur at the bottom of the matrix.

2. For each row that does not consist entirely of zeros, the first nonzero entry is 1 (called a **leading 1**).

3. For two successive (nonzero) rows, the leading 1 in the higher row is farther to the left that the leading 1 in the lower row.

④ Use matrices and Gaussian elimination with back-substitution to solve systems of linear equations.

Solving a System of Linear Equations

Gaussian Elimination with Back-Substitution

To use matrices and Gaussian elimination to solve a system of linear equations, use the following steps.

1. Write the augmented matrix of the system of linear equations.

2. Use elementary row operations to rewrite the augmented matrix in row-echelon form.

3. Write the system of linear equations corresponding to the matrix in row-echelon form, and use back-substitution to find the solution.

When you perform Gaussian elimination with back-substitution, you should operate from *left to right by columns*, using elementary row operations to obtain zeros in all entries directly below the leading 1's.

Example 6   Gaussian Elimination with Back-Substitution

Solve the system of linear equations.

$$\begin{cases} 2x - 3y = -2 \\ x + 2y = 13 \end{cases}$$

Solution

$$\begin{bmatrix} 2 & -3 & \vdots & -2 \\ 1 & 2 & \vdots & 13 \end{bmatrix}$$

Augmented matrix for system of linear equations

$$\begin{matrix} R_2 \\ R_1 \end{matrix} \begin{bmatrix} 1 & 2 & \vdots & 13 \\ 2 & -3 & \vdots & -2 \end{bmatrix}$$

First column has leading 1 in upper left corner.

$$-2R_1 + R_2 \rightarrow \begin{bmatrix} 1 & 2 & \vdots & 13 \\ 0 & -7 & \vdots & -28 \end{bmatrix}$$

First column has a zero under its leading 1.

$$-\tfrac{1}{7}R_2 \rightarrow \begin{bmatrix} 1 & 2 & \vdots & 13 \\ 0 & 1 & \vdots & 4 \end{bmatrix}$$

Second column has leading 1 in second row.

The system of linear equations that corresponds to the (row-echelon) matrix is

$$\begin{cases} x + 2y = 13 \\ y = 4. \end{cases}$$

Using back-substitution, you can find that the solution of the system is $x = 5$ and $y = 4$, which can be written as the ordered pair $(5, 4)$. Check this solution in the original system, as follows.

Check

Equation 1: $2(5) - 3(4) = -2$ ✓

Equation 2: $5 + 2(4) = 13$ ✓

Example 7   Gaussian Elimination with Back-Substitution

Solve the system of linear equations.

$$\begin{cases} 3x + 3y & = 9 \\ 2x & - 3z = 10 \\ & 6y + 4z = -12 \end{cases}$$

Solution

$$\begin{bmatrix} 3 & 3 & 0 & \vdots & 9 \\ 2 & 0 & -3 & \vdots & 10 \\ 0 & 6 & 4 & \vdots & -12 \end{bmatrix}$$

Augmented matrix for system of linear equations

$$\frac{1}{3}R_1 \rightarrow \begin{bmatrix} 1 & 1 & 0 & \vdots & 3 \\ 2 & 0 & -3 & \vdots & 10 \\ 0 & 6 & 4 & \vdots & -12 \end{bmatrix}$$

First column has leading 1 in upper left corner.

$$-2R_1 + R_2 \rightarrow \begin{bmatrix} 1 & 1 & 0 & \vdots & 3 \\ 0 & -2 & -3 & \vdots & 4 \\ 0 & 6 & 4 & \vdots & -12 \end{bmatrix}$$

First column has zeros under its leading 1.

$$-\frac{1}{2}R_2 \rightarrow \begin{bmatrix} 1 & 1 & 0 & \vdots & 3 \\ 0 & 1 & \frac{3}{2} & \vdots & -2 \\ 0 & 6 & 4 & \vdots & -12 \end{bmatrix}$$

Second column has leading 1 in second row.

$$-6R_2 + R_3 \rightarrow \begin{bmatrix} 1 & 1 & 0 & \vdots & 3 \\ 0 & 1 & \frac{3}{2} & \vdots & -2 \\ 0 & 0 & -5 & \vdots & 0 \end{bmatrix}$$

Second column has zero under its leading 1.

$$-\frac{1}{5}R_3 \rightarrow \begin{bmatrix} 1 & 1 & 0 & \vdots & 3 \\ 0 & 1 & \frac{3}{2} & \vdots & -2 \\ 0 & 0 & 1 & \vdots & 0 \end{bmatrix}$$

Third column has leading 1 in third row.

The system of linear equations that corresponds to this (row-echelon) matrix is

$$\begin{cases} x + y & = 3 \\ y + \frac{3}{2}z & = -2 \\ z & = 0. \end{cases}$$

Using back-substitution, you can find that the solution is

$$x = 5, \quad y = -2, \quad \text{and} \quad z = 0$$

which can be written as the ordered triple $(5, -2, 0)$. Check this in the original system, as follows.

Check

Equation 1: $3(5) + 3(-2) \quad\quad = 9$ ✓

Equation 2: $2(5) \quad\quad - 3(0) = 10$ ✓

Equation 3: $\quad\quad 6(-2) + 4(0) = -12$ ✓

Example 8   A System with No Solution

Solve the system of linear equations.

$$\begin{cases} 6x - 10y = -4 \\ 9x - 15y = 5 \end{cases}$$

Solution

$$\begin{bmatrix} 6 & -10 & \vdots & -4 \\ 9 & -15 & \vdots & 5 \end{bmatrix}$$
    Augmented matrix for system of linear equations

$$\tfrac{1}{6}R_1 \rightarrow \begin{bmatrix} 1 & -\frac{5}{3} & \vdots & -\frac{2}{3} \\ 9 & -15 & \vdots & 5 \end{bmatrix}$$
    First column has leading 1 in upper left corner.

$$-9R_1 + R_2 \rightarrow \begin{bmatrix} 1 & -\frac{5}{3} & \vdots & -\frac{2}{3} \\ 0 & 0 & \vdots & 11 \end{bmatrix}$$
    First column has a zero under its leading 1.

The "equation" that corresponds to the second row of this matrix is $0 = 11$. Because this is a false statement, the system of equations has no solution.

Example 9   A System with Infinitely Many Solutions

Solve the system of linear equations.

$$\begin{cases} 12x - 6y = -3 \\ -8x + 4y = 2 \end{cases}$$

Solution

$$\begin{bmatrix} 12 & -6 & \vdots & -3 \\ -8 & 4 & \vdots & 2 \end{bmatrix}$$
    Augmented matrix for system of linear equations

$$\tfrac{1}{12}R_1 \rightarrow \begin{bmatrix} 1 & -\frac{1}{2} & \vdots & -\frac{1}{4} \\ -8 & 4 & \vdots & 2 \end{bmatrix}$$
    First column has leading 1 in upper left corner.

$$8R_1 + R_2 \rightarrow \begin{bmatrix} 1 & -\frac{1}{2} & \vdots & -\frac{1}{4} \\ 0 & 0 & \vdots & 0 \end{bmatrix}$$
    First column has a zero under its leading 1.

Because the second row of the matrix is all zeros, you can conclude that the system of equations has an infinite number of solutions, represented by all points (x, y) on the line

$$x - \frac{1}{2}y = -\frac{1}{4}.$$

Because this line can be written as

$$x = \frac{1}{2}y - \frac{1}{4}$$

you can write the solution set as

$$\left(\frac{1}{2}a - \frac{1}{4}, a \right), \quad \text{where } a \text{ is any real number.}$$

Example 10   Investment Portfolio

You have a portfolio totaling $219,000 and want to invest in municipal bonds, blue-chip stocks, and growth or speculative stocks. The municipal bonds pay 6% annually. Over a five-year period, you expect blue-chip stocks to return 10% annually and growth stocks to return 15% annually. You want a combined annual return of 8%, and you also want to have only one-fourth of the portfolio invested in stocks. How much should be allocated to each type of investment?

Solution

Let M represent municipal bonds, B represent blue-chip stocks, and G represent growth stocks. These three equations make up the following system.

$$\begin{cases} M + B + G = 219{,}000 & \text{Equation 1: total investment is \$219,000.} \\ 0.06M + 0.10B + 0.15G = 17{,}520 & \text{Equation 2: combined annual return is 8\%.} \\ B + G = 54{,}750 & \text{Equation 3: } \tfrac{1}{4} \text{ of investment is allocated to stocks.} \end{cases}$$

Form the augmented matrix for this system of equations, and then use elementary row operations to obtain the row-echelon form of the matrix.

$$\begin{bmatrix} 1 & 1 & 1 & \vdots & 219{,}000 \\ 0.06 & 0.10 & 0.15 & \vdots & 17{,}520 \\ 0 & 1 & 1 & \vdots & 54{,}750 \end{bmatrix}$$ Augmented matrix for system of linear equations

$$-0.06R_1 + R_2 \rightarrow \begin{bmatrix} 1 & 1 & 1 & \vdots & 219{,}000 \\ 0 & 0.04 & 0.09 & \vdots & 4{,}380 \\ 0 & 1 & 1 & \vdots & 54{,}750 \end{bmatrix}$$ First column has zeros under its leading 1.

$$25R_2 \rightarrow \begin{bmatrix} 1 & 1 & 1 & \vdots & 219{,}000 \\ 0 & 1 & 2.25 & \vdots & 109{,}500 \\ 0 & 1 & 1 & \vdots & 54{,}750 \end{bmatrix}$$ Second column has leading 1 in second row.

$$-R_2 + R_3 \rightarrow \begin{bmatrix} 1 & 1 & 1 & \vdots & 219{,}000 \\ 0 & 1 & 2.25 & \vdots & 109{,}500 \\ 0 & 0 & -1.25 & \vdots & -54{,}750 \end{bmatrix}$$ Second column has zero under its leading 1.

$$-0.8R_3 \rightarrow \begin{bmatrix} 1 & 1 & 1 & \vdots & 219{,}000 \\ 0 & 1 & 2.25 & \vdots & 109{,}500 \\ 0 & 0 & 1 & \vdots & 43{,}800 \end{bmatrix}$$ Third column has leading 1 in third row and matrix is in row-echelon form.

From the row-echelon form, you can see that $G = 43{,}800$. By back-substituting G into the revised second equation, you can determine the value of B.

$$B + 2.25(43{,}800) = 109{,}500 \quad \Longrightarrow \quad B = 10{,}950$$

By back-substituting B and G into Equation 1, you can solve for M.

$$M + 10{,}950 + 43{,}800 = 219{,}000 \quad \Longrightarrow \quad M = 164{,}250$$

So, you should invest $164,250 in municipal bonds, $10,950 in blue-chip stocks, and $43,800 in growth or speculative stocks. Check this solution by substituting these values into the original system of equations.

8.4 Exercises

Review   Concepts, Skills, and Problem Solving

Keep mathematically in shape by doing these exercises *before* the problems of this section.

Properties and Definitions

In Exercises 1–4, identify the property of real numbers illustrated by the statement.

1. $2ab - 2ab = 0$

2. $8t \cdot 1 = 8t$

3. $b + 3a = 3a + b$

4. $3(2x) = (3 \cdot 2)x$

Algebraic Operations

In Exercises 5–8, plot the points on the rectangular coordinate system. Find the slope of the line passing through the points. If not possible, state why.

5. $(0, -6), (8, 0)$

6. $\left(\frac{5}{2}, \frac{7}{2}\right), \left(\frac{5}{2}, 4\right)$

7. $\left(-\frac{5}{8}, -\frac{3}{4}\right), \left(1, -\frac{9}{2}\right)$

8. $(3, 1.2), (-3, 2.1)$

In Exercises 9 and 10, find the distance between the two points and the midpoint of the line segment joining the two points.

9. $(12, 8), (6, 8)$ **10.** $\left(-3, 2\right), \left(-\frac{3}{2}, -2\right)$

Problem Solving

11. *Membership Drive*   Through a membership drive, the membership of a public television station increased by 10%. The current number of members is 8415. How many members did the station have before the membership drive?

12. *Consumer Awareness*   A sales representative indicates that if a customer waits another month for a new car that currently costs \$23,500, the price will increase by 4%. The customer has a certificate of deposit that comes due in 1 month and will pay a penalty for early withdrawal if the money is withdrawn before the due date. Determine the penalty for early withdrawal that would equal the cost increase of waiting to buy the car.

Developing Skills

In Exercises 1–10, determine the order of the matrix. See Example 1.

1. $\begin{bmatrix} 3 & -2 \\ -4 & 0 \\ 2 & -7 \\ -1 & -3 \end{bmatrix}$

2. $\begin{bmatrix} 4 & 0 & -5 \\ -1 & 8 & 9 \\ 0 & -3 & 4 \end{bmatrix}$

3. $\begin{bmatrix} -2 & 5 \\ 0 & -1 \end{bmatrix}$

4. $\begin{bmatrix} 5 & -8 & 32 \\ 7 & 15 & 28 \end{bmatrix}$

5. $\begin{bmatrix} 4 \\ -2 \\ 0 \\ 1 \end{bmatrix}$

6. $\begin{bmatrix} 1 & -1 & 2 & 3 \end{bmatrix}$

7. $\begin{bmatrix} 5 \end{bmatrix}$

8. $\begin{bmatrix} 3 & 4 \\ 2 & -1 \\ 8 & 10 \\ -6 & -6 \\ 12 & 50 \end{bmatrix}$

9. $\begin{bmatrix} 13 & 12 & -9 & 0 \end{bmatrix}$ **10.** $\begin{bmatrix} 6 \\ -13 \\ 22 \end{bmatrix}$

In Exercises 11–16, form (a) the coefficient matrix and (b) the augmented matrix for the system of linear equations. See Example 2.

11. $\begin{cases} 4x - 5y = -2 \\ -x + 8y = 10 \end{cases}$

12. $\begin{cases} 8x + 3y = 25 \\ 3x - 9y = 12 \end{cases}$

13. $\begin{cases} x + 10y - 3z = 2 \\ 5x - 3y + 4z = 0 \\ 2x + 4y = 6 \end{cases}$

14. $\begin{cases} 9x - 3y + z = 13 \\ 12x - 8z = 5 \\ 3x + 4y - z = 6 \end{cases}$

15. $\begin{cases} 5x + y - 3z = 7 \\ 2y + 4z = 12 \end{cases}$

16. $\begin{cases} 10x + 6y - 8z = -4 \\ -4x - 7y = 9 \end{cases}$

22. $\begin{bmatrix} 0 & 1 & -5 & 8 & \vdots & 10 \\ 2 & 4 & -1 & 0 & \vdots & 15 \\ 1 & 1 & 7 & 9 & \vdots & -8 \end{bmatrix}$

23. $\begin{bmatrix} 13 & 1 & 4 & -2 & \vdots & -4 \\ 5 & 4 & 0 & -1 & \vdots & 0 \\ 1 & 2 & 6 & 8 & \vdots & 5 \\ -10 & 12 & 3 & 1 & \vdots & -2 \end{bmatrix}$

24. $\begin{bmatrix} 7 & 3 & -2 & 4 & \vdots & 2 \\ -1 & 0 & 4 & -1 & \vdots & 6 \\ 8 & 3 & 0 & 0 & \vdots & -4 \\ 0 & 2 & -4 & 3 & \vdots & 12 \end{bmatrix}$

In Exercises 17–24, write the system of linear equations represented by the augmented matrix. (Use variables x, y, z, and w.) See Example 3.

17. $\begin{bmatrix} 4 & 3 & \vdots & 8 \\ 1 & -2 & \vdots & 3 \end{bmatrix}$

18. $\begin{bmatrix} 9 & -4 & \vdots & 0 \\ 6 & 1 & \vdots & -4 \end{bmatrix}$

19. $\begin{bmatrix} 1 & 0 & 2 & \vdots & -10 \\ 0 & 3 & -1 & \vdots & 5 \\ 4 & 2 & 0 & \vdots & 3 \end{bmatrix}$

20. $\begin{bmatrix} 4 & -1 & 3 & \vdots & 5 \\ 2 & 0 & -2 & \vdots & -1 \\ -1 & 6 & 0 & \vdots & 3 \end{bmatrix}$

21. $\begin{bmatrix} 5 & 8 & 2 & 0 & \vdots & -1 \\ -2 & 15 & 5 & 1 & \vdots & 9 \\ 1 & 6 & -7 & 0 & \vdots & -3 \end{bmatrix}$

In Exercises 25–30, fill in the blank(s) by using elementary row operations to form a row-equivalent matrix. See Examples 4 and 5.

25. $\begin{bmatrix} 1 & 4 & 3 \\ 2 & 10 & 5 \end{bmatrix}$

$\begin{bmatrix} 1 & 4 & 3 \\ 0 & & -1 \end{bmatrix}$

26. $\begin{bmatrix} 3 & 6 & 8 \\ 4 & -3 & 6 \end{bmatrix}$

$\begin{bmatrix} 3 & 6 & 8 \\ 1 & -9 & \end{bmatrix}$

27. $\begin{bmatrix} 9 & -18 & 6 \\ 2 & 8 & 15 \end{bmatrix}$

$\begin{bmatrix} 1 & & \\ 2 & 8 & 15 \end{bmatrix}$

28. $\begin{bmatrix} 2 & 3 & -5 & 6 \\ 5 & -7 & 12 & 9 \\ -4 & 6 & 9 & 5 \end{bmatrix}$

$\begin{bmatrix} 2 & 3 & -5 & 6 \\ 5 & -7 & 12 & 9 \\ 0 & 12 & & \end{bmatrix}$

29. $\begin{bmatrix} 1 & 1 & 4 & -1 \\ 3 & 8 & 10 & 3 \\ -2 & 1 & 12 & 6 \end{bmatrix}$

$\begin{bmatrix} 1 & 1 & 4 & -1 \\ 0 & 5 & & \\ 0 & 3 & & \end{bmatrix}$

$\begin{bmatrix} 1 & 1 & 4 & -1 \\ 0 & 1 & -\frac{2}{5} & \frac{6}{5} \\ 0 & 3 & & \end{bmatrix}$

30. $\begin{bmatrix} 2 & 4 & 8 & 3 \\ 1 & -1 & -3 & 2 \\ 2 & 6 & 4 & 9 \end{bmatrix}$

$\begin{bmatrix} 1 & & & \\ 1 & -1 & -3 & 2 \\ 2 & 6 & 4 & 9 \end{bmatrix}$

$\begin{bmatrix} 1 & 2 & 4 & \frac{3}{2} \\ 0 & & -7 & \frac{1}{2} \\ 0 & 2 & & \end{bmatrix}$

In Exercises 31–36, convert the matrix to row-echelon form. (There are many correct answers.)

31. $\begin{bmatrix} 1 & 2 & 3 \\ 2 & -1 & -4 \end{bmatrix}$ **32.** $\begin{bmatrix} 1 & 3 & 6 \\ -4 & -9 & 3 \end{bmatrix}$

33. $\begin{bmatrix} 4 & 6 & 1 \\ -2 & 2 & 5 \end{bmatrix}$ **34.** $\begin{bmatrix} 3 & 2 & 6 \\ 2 & 3 & -3 \end{bmatrix}$

35. $\begin{bmatrix} 1 & 1 & 0 & 5 \\ -2 & -1 & 2 & -10 \\ 3 & 6 & 7 & 14 \end{bmatrix}$ **36.** $\begin{bmatrix} 1 & 2 & -1 & 3 \\ 3 & 7 & -5 & 14 \\ -2 & -1 & -3 & 8 \end{bmatrix}$

▦ In Exercises 37–40, use the matrix capabilities of a graphing calculator to write the matrix in row-echelon form. (There are many correct answers.)

37. $\begin{bmatrix} 1 & -1 & -1 & 1 \\ 4 & -4 & 1 & 8 \\ -6 & 8 & 18 & 0 \end{bmatrix}$

38. $\begin{bmatrix} 1 & -3 & 0 & -7 \\ -3 & 10 & 1 & 23 \\ 4 & -10 & 2 & -24 \end{bmatrix}$

39. $\begin{bmatrix} 1 & 1 & -1 & 3 \\ 2 & 1 & 2 & 5 \\ 3 & 2 & 1 & 8 \end{bmatrix}$

40. $\begin{bmatrix} 1 & -3 & -2 & -8 \\ 1 & 3 & -2 & 17 \\ 1 & 2 & -2 & -5 \end{bmatrix}$

In Exercises 41–46, write the system of linear equations represented by the augmented matrix. Then use back-substitution to find the solution. (Use variables x, y, and z.)

41. $\begin{bmatrix} 1 & -2 & \vdots & 4 \\ 0 & 1 & \vdots & -3 \end{bmatrix}$ **42.** $\begin{bmatrix} 1 & 5 & \vdots & 0 \\ 0 & 1 & \vdots & -1 \end{bmatrix}$

43. $\begin{bmatrix} 1 & 5 & \vdots & 3 \\ 0 & 1 & \vdots & -2 \end{bmatrix}$ **44.** $\begin{bmatrix} 1 & 5 & -3 & \vdots & 0 \\ 0 & 1 & 0 & \vdots & 6 \\ 0 & 0 & 1 & \vdots & -5 \end{bmatrix}$

45. $\begin{bmatrix} 1 & -1 & 2 & \vdots & 4 \\ 0 & 1 & -1 & \vdots & 2 \\ 0 & 0 & 1 & \vdots & -2 \end{bmatrix}$

46. $\begin{bmatrix} 1 & 2 & -2 & \vdots & -1 \\ 0 & 1 & 1 & \vdots & 9 \\ 0 & 0 & 1 & \vdots & -3 \end{bmatrix}$

In Exercises 47–72, use matrices to solve the system of linear equations. See Examples 5–9.

47. $\begin{cases} x + 2y = 7 \\ 3x + y = 8 \end{cases}$ **48.** $\begin{cases} 2x + 6y = 16 \\ 2x + 3y = 7 \end{cases}$

49. $\begin{cases} 6x - 4y = 2 \\ 5x + 2y = 7 \end{cases}$ **50.** $\begin{cases} x - 3y = 5 \\ -2x + 6y = -10 \end{cases}$

51. $\begin{cases} 12x + 10y = -14 \\ 4x - 3y = -11 \end{cases}$ **52.** $\begin{cases} -x - 5y = -10 \\ 2x - 3y = 7 \end{cases}$

53. $\begin{cases} -x + 2y = 1.5 \\ 2x - 4y = 3 \end{cases}$ **54.** $\begin{cases} 2x - y = -0.1 \\ 3x + 2y = 1.6 \end{cases}$

55. $\begin{cases} x - 2y - z = 6 \\ y + 4z = 5 \\ 4x + 2y + 3z = 8 \end{cases}$ **56.** $\begin{cases} x - 3z = -2 \\ 3x + y - 2z = 5 \\ 2x + 2y + z = 4 \end{cases}$

57. $\begin{cases} x + y - 5z = 3 \\ x - 2z = 1 \\ 2x - y - z = 0 \end{cases}$ **58.** $\begin{cases} 2y + z = 3 \\ -4y - 2z = 0 \\ x + y + z = 2 \end{cases}$

59. $\begin{cases} 2x + 4y = 10 \\ 2x + 2y + 3z = 3 \\ -3x + y + 2z = -3 \end{cases}$

60. $\begin{cases} 2x - y + 3z = 24 \\ 2y - z = 14 \\ 7x - 5y = 6 \end{cases}$ **61.** $\begin{cases} x - 3y + 2z = 8 \\ 2y - z = -4 \\ x + z = 3 \end{cases}$

62. $\begin{cases} 2x + 3z = 3 \\ 4x - 3y + 7z = 5 \\ 8x - 9y + 15z = 9 \end{cases}$

63. $\begin{cases} -2x - 2y - 15z = 0 \\ x + 2y + 2z = 18 \\ 3x + 3y + 22z = 2 \end{cases}$

64. $\begin{cases} 2x + 4y + 5z = 5 \\ x + 3y + 3z = 2 \\ 2x + 4y + 4z = 2 \end{cases}$ **65.** $\begin{cases} 2x + 4z = 1 \\ x + y + 3z = 0 \\ x + 3y + 5z = 0 \end{cases}$

66. $\begin{cases} 3x + y - 2z = 2 \\ 6x + 2y - 4z = 1 \\ -3x - y + 2z = 1 \end{cases}$ **67.** $\begin{cases} x + 3y = 2 \\ 2x + 6y = 4 \\ 2x + 5y + 4z = 3 \end{cases}$

68. $\begin{cases} 4x + 3y = 10 \\ 2x - y = 10 \\ -2x + z = -9 \end{cases}$

69. $\begin{cases} 4x - y + z = 4 \\ -6x + 3y - 2z = -5 \\ 2x + 5y - z = 7 \end{cases}$

70. $\begin{cases} 2x + 2y + z = 8 \\ 2x + 3y + z = 7 \\ 6x + 8y + 3z = 22 \end{cases}$

71. $\begin{cases} 2x + y - 2z = 4 \\ 3x - 2y + 4z = 6 \\ -4x + y + 6z = 12 \end{cases}$

72. $\begin{cases} 3x + 3y + z = 4 \\ 2x + 6y + z = 5 \\ -x - 3y + 2z = -5 \end{cases}$

Solving Problems

73. *Investment*   A corporation borrowed $1,500,000 to expand its line of clothing. Some of the money was borrowed at 8%, some at 9%, and the remainder at 12%. The annual interest payment to the lenders was $133,000. The amount borrowed at 8% was 4 times the amount borrowed at 12%. How much was borrowed at each rate?

74. *Investment*   An inheritance of $16,000 was divided among three investments yielding a total of $990 in simple interest per year. The interest rates for the three investments were 5%, 6%, and 7%. The 5% and 6% investments were $3000 and $2000 less than the 7% investment, respectively. Find the amount placed in each investment.

Investment Portfolio   In Exercises 75 and 76, consider an investor with a portfolio totaling $500,000 that is to be allocated among the following types of investments: certificates of deposit, municipal bonds, blue-chip stocks, and growth or speculative stocks. How much should be allocated to each type of investment?

75. The certificates of deposit pay 10% annually, and the municipal bonds pay 8% annually. Over a five-year period, the investor expects the blue-chip stocks to return 12% annually and the growth stocks to return 13% annually. The investor wants a combined annual return of 10% and also wants to have only one-fourth of the portfolio invested in stocks.

76. The certificates of deposit pay 9% annually, and the municipal bonds pay 5% annually. Over a five-year period, the investor expects the blue-chip stocks to return 12% annually and the growth stocks to return 14% annually. The investor wants a combined annual return of 10% and also wants to have only one-fourth of the portfolio invested in stocks.

77. *Nut Mixture*   A grocer wishes to mix three kinds of nuts to obtain 50 pounds of a mixture priced at $4.95 per pound. Peanuts cost $3.50 per pound, almonds cost $4.50 per pound, and pistachios cost $6.00 per pound. Half of the mixture is composed of peanuts and almonds. How many pounds of each variety should the grocer use?

78. *Nut Mixture*   A grocer wishes to mix three kinds of nuts to obtain 50 pounds of a mixture priced at $4.10 per pound. Peanuts cost $3.00 per pound, pecans cost $4.00 per pound, and cashews cost $6.00 per pound. Three-quarters of the mixture is composed of peanuts and pecans. How many pounds of each variety should the grocer use?

79. *Number Problem*   The sum of three positive numbers is 33. The second number is 3 greater than the first, and the third is 4 times the first. Find the three numbers.

80. *Number Problem*   The sum of three positive numbers is 24. The second number is 4 greater than the first, and the third is 3 times the first. Find the three numbers.

81. *Production*   A company produces computer chips, resistors, and transistors. Each computer chip requires 2 units of copper, 2 units of zinc, and 1 unit of glass. Each resistor requires 1 unit of copper, 3 units of zinc, and 2 units of glass. Each transistor requires 3 units of copper, 2 units of zinc, and 2 units of glass. There are 70 units of copper, 80 units of zinc, and 55 units of glass available for use. Find the number of computer chips, resistors, and transistors the company can produce.

82. *Production*   A gourmet baked goods company specializes in chocolate muffins, chocolate cookies, and chocolate brownies. Each muffin requires 2 units of chocolate, 3 units of flour, and 2 units of sugar. Each cookie requires 1 unit of chocolate, 1 unit of flour, and 1 unit of sugar. Each brownie requires 2 units of chocolate, 1 unit of flour, and 1.5 units of sugar. There are 550 units of chocolate, 525 units of flour, and 500 units of sugar available for use. Find the number of chocolate muffins, chocolate cookies, and chocolate brownies the company can produce.

Explaining Concepts

83. *Writing*   Describe the three elementary row operations that can be performed on an augmented matrix.

84. *Writing*   What is the relationship between the three elementary row operations on an augmented matrix and the row operations on a system of linear equations?

85. *Writing*   What is meant by saying that two augmented matrices are *row-equivalent?*

86. Give an example of a matrix in *row-echelon form.* There are many correct answers.

87. *Writing*   Describe the row-echelon form of an augmented matrix that corresponds to a system of linear equations that is inconsistent.

88. *Writing*   Describe the row-echelon form of an augmented matrix that corresponds to a system of linear equations that has an infinite number of solutions.

8.5 Determinants and Linear Systems

Kevin R. Morris/Corbis

What You Should Learn

1. Find determinants of 2×2 matrices and 3×3 matrices.
2. Use determinants and Cramer's Rule to solve systems of linear equations.
3. Use determinants to find areas of triangles, to test for collinear points, and to find equations of lines.

Why You Should Learn It

You can use determinants and matrices to model and solve real-life problems. For instance, in Exercise 71 on page 531, you can use a matrix to estimate the area of a region of land.

The Determinant of a Matrix

Associated with each square matrix is a real number called its **determinant.** The use of determinants arose from special number patterns that occur during the solution of systems of linear equations. For instance, the system

$$\begin{cases} a_1 x + b_1 y = c_1 \\ a_2 x + b_2 y = c_2 \end{cases}$$

has a solution given by

$$x = \frac{b_2 c_1 - b_1 c_2}{a_1 b_2 - a_2 b_1} \quad \text{and} \quad y = \frac{a_1 c_2 - a_2 c_1}{a_1 b_2 - a_2 b_1}$$

① Find determinants of 2×2 matrices and 3×3 matrices.

provided that $a_1 b_2 - a_2 b_1 \neq 0$. Note that the denominator of each fraction is the same. This denominator is called the **determinant** of the coefficient matrix of the system.

Coefficient Matrix   *Determinant*

$$A = \begin{bmatrix} a_1 & b_1 \\ a_2 & b_2 \end{bmatrix} \qquad \det(A) = a_1 b_2 - a_2 b_1$$

The determinant of the matrix A can also be denoted by vertical bars on both sides of the matrix, as indicated in the following definition.

Study Tip

Note that $\det(A)$ and $|A|$ are used interchangeably to represent the determinant of A. Although vertical bars are also used to denote the absolute value of a real number, the context will show which use is intended.

Definition of the Determinant of a 2×2 Matrix

$$\det(A) = |A| = \begin{vmatrix} a_1 & b_1 \\ a_2 & b_2 \end{vmatrix} = a_1 b_2 - a_2 b_1$$

A convenient method for remembering the formula for the determinant of a 2×2 matrix is shown in the diagram below.

$$\det(A) = \begin{vmatrix} a_1 & b_1 \\ a_2 & b_2 \end{vmatrix} = a_1 b_2 - a_2 b_1$$

Note that the determinant is given by the difference of the products of the two diagonals of the matrix.

Example 1 The Determinant of a 2 × 2 Matrix

Find the determinant of each matrix.

a. $A = \begin{bmatrix} 2 & -3 \\ 1 & 4 \end{bmatrix}$ **b.** $B = \begin{bmatrix} -1 & 2 \\ 2 & -4 \end{bmatrix}$ **c.** $C = \begin{bmatrix} 1 & 3 \\ 2 & 5 \end{bmatrix}$

Solution

a. $\det(A) = \begin{vmatrix} 2 & -3 \\ 1 & 4 \end{vmatrix} = 2(4) - 1(-3) = 8 + 3 = 11$

b. $\det(B) = \begin{vmatrix} -1 & 2 \\ 2 & -4 \end{vmatrix} = (-1)(-4) - 2(2) = 4 - 4 = 0$

c. $\det(C) = \begin{vmatrix} 1 & 3 \\ 2 & 5 \end{vmatrix} = 1(5) - 2(3) = 5 - 6 = -1$

Technology: Tip

A graphing calculator with matrix capabilities can be used to evaluate the determinant of a square matrix. Consult the user's guide of your graphing calculator to learn how to evaluate a determinant. Use the graphing calculator to check the result in Example 1(a). Then try to evaluate the determinant of the 3 × 3 matrix at the right using a graphing calculator. Finish the evaluation of the determinant by expanding by minors to check the result.

Notice in Example 1 that the determinant of a matrix can be positive, zero, or negative.

One way to evaluate the determinant of a 3 × 3 matrix, called **expanding by minors,** allows you to write the determinant of a 3 × 3 matrix in terms of three 2 × 2 determinants. The **minor** of an entry in a 3 × 3 matrix is the determinant of the 2 × 2 matrix that remains after deletion of the row and column in which the entry occurs. Here are three examples.

Determinant	Entry	Minor of Entry	Value of Minor
$\begin{vmatrix} 1 & -1 & 3 \\ 0 & 2 & 5 \\ -2 & 4 & -7 \end{vmatrix}$	1	$\begin{vmatrix} 2 & 5 \\ 4 & -7 \end{vmatrix}$	$2(-7) - 4(5) = -34$
$\begin{vmatrix} 1 & -1 & 3 \\ 0 & 2 & 5 \\ -2 & 4 & -7 \end{vmatrix}$	-1	$\begin{vmatrix} 0 & 5 \\ -2 & -7 \end{vmatrix}$	$0(-7) - (-2)(5) = 10$
$\begin{vmatrix} 1 & -1 & 3 \\ 0 & 2 & 5 \\ -2 & 4 & -7 \end{vmatrix}$	3	$\begin{vmatrix} 0 & 2 \\ -2 & 4 \end{vmatrix}$	$0(4) - (-2)(2) = 4$

Expanding by Minors

$$\det(A) = \begin{vmatrix} a_1 & b_1 & c_1 \\ a_2 & b_2 & c_2 \\ a_3 & b_3 & c_3 \end{vmatrix}$$

$$= a_1(\text{minor of } a_1) - b_1(\text{minor of } b_1) + c_1(\text{minor of } c_1)$$

$$= a_1 \begin{vmatrix} b_2 & c_2 \\ b_3 & c_3 \end{vmatrix} - b_1 \begin{vmatrix} a_2 & c_2 \\ a_3 & c_3 \end{vmatrix} + c_1 \begin{vmatrix} a_2 & b_2 \\ a_3 & b_3 \end{vmatrix}$$

This pattern is called **expanding by minors** along the first row. A similar pattern can be used to expand by minors along any row or column.

$$\begin{bmatrix} + & - & + \\ - & + & - \\ + & - & + \end{bmatrix}$$

Figure 8.15   Sign Pattern for a
3×3 Matrix

The *signs* of the terms used in expanding by minors follow the alternating pattern shown in Figure 8.15. For instance, the signs used to expand by minors along the second row are $-$, $+$, $-$, as shown below.

$$\det(A) = \begin{vmatrix} a_1 & b_1 & c_1 \\ a_2 & b_2 & c_2 \\ a_3 & b_3 & c_3 \end{vmatrix}$$

$$= -a_2(\text{minor of } a_2) + b_2(\text{minor of } b_2) - c_2(\text{minor of } c_2)$$

Example 2   Finding the Determinant of a 3×3 Matrix

Find the determinant of $A = \begin{bmatrix} -1 & 1 & 2 \\ 0 & 2 & 3 \\ 3 & 4 & 2 \end{bmatrix}$.

Solution

By expanding by minors along the *first column*, you obtain

$$\det(A) = \begin{vmatrix} -1 & 1 & 2 \\ 0 & 2 & 3 \\ 3 & 4 & 2 \end{vmatrix}$$

$$= (-1)\begin{vmatrix} 2 & 3 \\ 4 & 2 \end{vmatrix} - (0)\begin{vmatrix} 1 & 2 \\ 4 & 2 \end{vmatrix} + (3)\begin{vmatrix} 1 & 2 \\ 2 & 3 \end{vmatrix}$$

$$= (-1)(4 - 12) - (0)(2 - 8) + (3)(3 - 4)$$

$$= 8 - 0 - 3 = 5$$

Example 3   Finding the Determinant of a 3×3 Matrix

Find the determinant of $A = \begin{bmatrix} 1 & 2 & 1 \\ 3 & 0 & 2 \\ 4 & 0 & -1 \end{bmatrix}$.

Solution

By expanding by minors along the *second column*, you obtain

$$\det(A) = \begin{vmatrix} 1 & 2 & 1 \\ 3 & 0 & 2 \\ 4 & 0 & -1 \end{vmatrix}$$

$$= -(2)\begin{vmatrix} 3 & 2 \\ 4 & -1 \end{vmatrix} + (0)\begin{vmatrix} 1 & 1 \\ 4 & -1 \end{vmatrix} - (0)\begin{vmatrix} 1 & 1 \\ 3 & 2 \end{vmatrix}$$

$$= -(2)(-3 - 8) + 0 - 0 = 22$$

Note in the expansions in Examples 2 and 3 that a zero entry will always yield a zero term when expanding by minors. So, when you are evaluating the determinant of a matrix, you should choose to expand along the row or column that has the most zero entries.

② Use determinants and Cramer's Rule to solve systems of linear equations.

Cramer's Rule

So far in this chapter, you have studied four methods for solving a system of linear equations: graphing, substitution, elimination (with equations), and elimination (with matrices). You will now learn one more method, called *Cramer's Rule,* which is named after Gabriel Cramer (1704–1752). This rule uses determinants to write the solution of a system of linear equations.

In Cramer's Rule, the value of a variable is expressed as the quotient of two determinants of the coefficient matrix of the system. The numerator is the determinant of the matrix formed by using the column of constants as replacements for the coefficients of the variable. In the definition below, note the notation for the different determinants.

Study Tip

Cramer's Rule is not as general as the elimination method because Cramer's Rule requires that the coefficient matrix of the system be square *and* that the system have exactly one solution.

Cramer's Rule

1. For the system of linear equations

$$\begin{cases} a_1x + b_1y = c_1 \\ a_2x + b_2y = c_2 \end{cases}$$

the solution is given by

$$x = \frac{D_x}{D} = \frac{\begin{vmatrix} c_1 & b_1 \\ c_2 & b_2 \end{vmatrix}}{\begin{vmatrix} a_1 & b_1 \\ a_2 & b_2 \end{vmatrix}}, \qquad y = \frac{D_y}{D} = \frac{\begin{vmatrix} a_1 & c_1 \\ a_2 & c_2 \end{vmatrix}}{\begin{vmatrix} a_1 & b_1 \\ a_2 & b_2 \end{vmatrix}}$$

provided that $D \neq 0$.

2. For the system of linear equations

$$\begin{cases} a_1x + b_1y + c_1z = d_1 \\ a_2x + b_2y + c_2z = d_2 \\ a_3x + b_3y + c_3z = d_3 \end{cases}$$

the solution is given by

$$x = \frac{D_x}{D} = \frac{\begin{vmatrix} d_1 & b_1 & c_1 \\ d_2 & b_2 & c_2 \\ d_3 & b_3 & c_3 \end{vmatrix}}{\begin{vmatrix} a_1 & b_1 & c_1 \\ a_2 & b_2 & c_2 \\ a_3 & b_3 & c_3 \end{vmatrix}}, \qquad y = \frac{D_y}{D} = \frac{\begin{vmatrix} a_1 & d_1 & c_1 \\ a_2 & d_2 & c_2 \\ a_3 & d_3 & c_3 \end{vmatrix}}{\begin{vmatrix} a_1 & b_1 & c_1 \\ a_2 & b_2 & c_2 \\ a_3 & b_3 & c_3 \end{vmatrix}},$$

$$z = \frac{D_z}{D} = \frac{\begin{vmatrix} a_1 & b_1 & d_1 \\ a_2 & b_2 & d_2 \\ a_3 & b_3 & d_3 \end{vmatrix}}{\begin{vmatrix} a_1 & b_1 & c_1 \\ a_2 & b_2 & c_2 \\ a_3 & b_3 & c_3 \end{vmatrix}}, \quad D \neq 0$$

Example 4   Using Cramer's Rule for a 2 × 2 System

Use Cramer's Rule to solve the system of linear equations.

$$\begin{cases} 4x - 2y = 10 \\ 3x - 5y = 11 \end{cases}$$

Solution

Begin by finding the determinant of the coefficient matrix.

$$D = \begin{vmatrix} 4 & -2 \\ 3 & -5 \end{vmatrix} = -20 - (-6) = -14$$

Then, use the formulas for x and y given by Cramer's Rule.

$$x = \frac{D_x}{D} = \frac{\begin{vmatrix} 10 & -2 \\ 11 & -5 \end{vmatrix}}{-14} = \frac{(-50) - (-22)}{-14} = \frac{-28}{-14} = 2$$

$$y = \frac{D_y}{D} = \frac{\begin{vmatrix} 4 & 10 \\ 3 & 11 \end{vmatrix}}{-14} = \frac{44 - 30}{-14} = \frac{14}{-14} = -1$$

The solution is $(2, -1)$. Check this in the original system of equations.

Example 5   Using Cramer's Rule for a 3 × 3 System

Use Cramer's Rule to solve the system of linear equations.

$$\begin{cases} -x + 2y - 3z = 1 \\ 2x \quad\quad + z = 0 \\ 3x - 4y + 4z = 2 \end{cases}$$

Solution

The determinant of the coefficient matrix is $D = 10$.

$$x = \frac{D_x}{D} = \frac{\begin{vmatrix} 1 & 2 & -3 \\ 0 & 0 & 1 \\ 2 & -4 & 4 \end{vmatrix}}{10} = \frac{8}{10} = \frac{4}{5}$$

$$y = \frac{D_y}{D} = \frac{\begin{vmatrix} -1 & 1 & -3 \\ 2 & 0 & 1 \\ 3 & 2 & 4 \end{vmatrix}}{10} = \frac{-15}{10} = -\frac{3}{2}$$

$$z = \frac{D_z}{D} = \frac{\begin{vmatrix} -1 & 2 & 1 \\ 2 & 0 & 0 \\ 3 & -4 & 2 \end{vmatrix}}{10} = \frac{-16}{10} = -\frac{8}{5}$$

The solution is $\left(\frac{4}{5}, -\frac{3}{2}, -\frac{8}{5}\right)$. Check this in the original system of equations.

Study Tip

When using Cramer's Rule, remember that the method *does not* apply if the determinant of the coefficient matrix is zero.

③ Use determinants to find areas of triangles, to test for collinear points, and to find equations of lines.

Applications of Determinants

In addition to Cramer's Rule, determinants have many other practical applications. For instance, you can use a determinant to find the area of a triangle whose vertices are given by three points on a rectangular coordinate system.

Area of a Triangle

The area of a triangle with vertices (x_1, y_1), (x_2, y_2), and (x_3, y_3) is

$$\text{Area} = \pm\frac{1}{2}\begin{vmatrix} x_1 & y_1 & 1 \\ x_2 & y_2 & 1 \\ x_3 & y_3 & 1 \end{vmatrix}$$

where the symbol $(\pm)$ indicates that the appropriate sign should be chosen to yield a positive area.

Example 6   Finding the Area of a Triangle

Find the area of the triangle whose vertices are $(2, 0)$, $(1, 3)$, and $(3, 2)$, as shown in Figure 8.16.

Solution

Choose $(x_1, y_1) = (2, 0)$, $(x_2, y_2) = (1, 3)$, and $(x_3, y_3) = (3, 2)$. To find the area of the triangle, evaluate the determinant by expanding by minors along the first row.

$$\begin{vmatrix} x_1 & y_1 & 1 \\ x_2 & y_2 & 1 \\ x_3 & y_3 & 1 \end{vmatrix} = \begin{vmatrix} 2 & 0 & 1 \\ 1 & 3 & 1 \\ 3 & 2 & 1 \end{vmatrix}$$

$$= 2\begin{vmatrix} 3 & 1 \\ 2 & 1 \end{vmatrix} - 0\begin{vmatrix} 1 & 1 \\ 3 & 1 \end{vmatrix} + 1\begin{vmatrix} 1 & 3 \\ 3 & 2 \end{vmatrix}$$

$$= 2(1) - 0 + 1(-7)$$

$$= -5$$

Using this value, you can conclude that the area of the triangle is

$$\text{Area} = -\frac{1}{2}\begin{vmatrix} 2 & 0 & 1 \\ 1 & 3 & 1 \\ 3 & 2 & 1 \end{vmatrix}$$

$$= -\frac{1}{2}(-5) = \frac{5}{2}.$$

To see the benefit of the "determinant formula," try finding the area of the triangle in Example 6 using the standard formula:

$$\text{Area} = \frac{1}{2}(\text{Base})(\text{Height}).$$

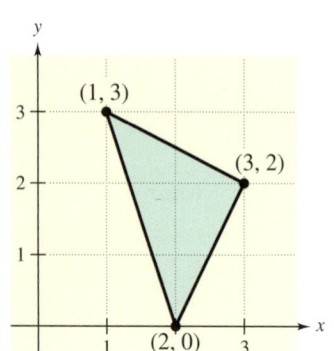

Figure 8.16

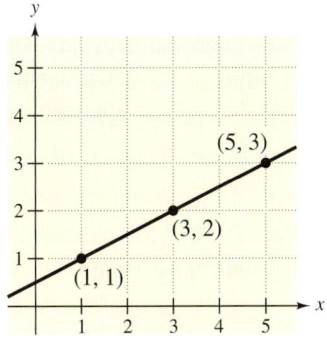

Figure 8.17

Suppose the three points in Example 6 had been on the same line. What would have happened had the area formula been applied to three such points? The answer is that the determinant would have been zero. Consider, for instance, the three collinear points $(1, 1)$, $(3, 2)$, and $(5, 3)$, as shown in Figure 8.17. The area of the "triangle" that has these three points as vertices is

$$\frac{1}{2}\begin{vmatrix} 1 & 1 & 1 \\ 3 & 2 & 1 \\ 5 & 3 & 1 \end{vmatrix} = \frac{1}{2}\left(1\begin{vmatrix} 2 & 1 \\ 3 & 1 \end{vmatrix} - 1\begin{vmatrix} 3 & 1 \\ 5 & 1 \end{vmatrix} + 1\begin{vmatrix} 3 & 2 \\ 5 & 3 \end{vmatrix} \right)$$

$$= \frac{1}{2}[-1 - (-2) + (-1)]$$

$$= 0.$$

This result is generalized as follows.

Test for Collinear Points

Three points (x_1, y_1), (x_2, y_2), and (x_3, y_3) are collinear (lie on the same line) if and only if

$$\begin{vmatrix} x_1 & y_1 & 1 \\ x_2 & y_2 & 1 \\ x_3 & y_3 & 1 \end{vmatrix} = 0$$

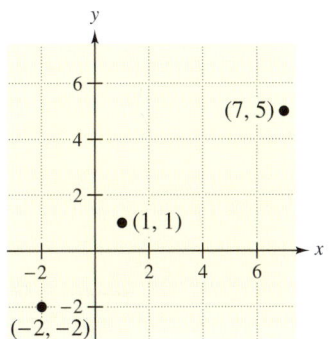

Figure 8.18

Example 7   Testing for Collinear Points

Determine whether the points $(-2, -2)$, $(1, 1)$, and $(7, 5)$ are collinear. (See Figure 8.18.)

Solution

Letting $(x_1, y_1) = (-2, -2)$, $(x_2, y_2) = (1, 1)$, and $(x_3, y_3) = (7, 5)$, you have

$$\begin{vmatrix} x_1 & y_1 & 1 \\ x_2 & y_2 & 1 \\ x_3 & y_3 & 1 \end{vmatrix} = \begin{vmatrix} -2 & -2 & 1 \\ 1 & 1 & 1 \\ 7 & 5 & 1 \end{vmatrix}$$

$$= -2\begin{vmatrix} 1 & 1 \\ 5 & 1 \end{vmatrix} - (-2)\begin{vmatrix} 1 & 1 \\ 7 & 1 \end{vmatrix} + 1\begin{vmatrix} 1 & 1 \\ 7 & 5 \end{vmatrix}$$

$$= -2(-4) - (-2)(-6) + 1(-2)$$

$$= -6.$$

Because the value of this determinant *is not* zero, you can conclude that the three points *do not* lie on the same line and are not collinear.

As a good review, look at how the slope can be used to verify the result in Example 7. Label the points $A(-2, -2)$, $B(1, 1)$, and $C(7, 5)$. Because the slopes from A to B and from A to C are different, the points are not collinear.

You can also use determinants to find the equation of a line through two points. In this case, the first row consists of the variables x and y and the number 1. By expanding by minors along the first row, the resulting 2×2 determinants are the coefficients of the variables x and y and the constant of the linear equation, as shown in Example 8.

Two-Point Form of the Equation of a Line

An equation of the line passing through the distinct points (x_1, y_1) and (x_2, y_2) is given by

$$\begin{vmatrix} x & y & 1 \\ x_1 & y_1 & 1 \\ x_2 & y_2 & 1 \end{vmatrix} = 0.$$

Example 8 Finding an Equation of a Line

Find an equation of the line passing through $(-2, 1)$ and $(3, -2)$.

Solution

Applying the determinant formula for the equation of a line produces

$$\begin{vmatrix} x & y & 1 \\ -2 & 1 & 1 \\ 3 & -2 & 1 \end{vmatrix} = 0.$$

To evaluate this determinant, you can expand by minors along the first row to obtain the following.

$$x \begin{vmatrix} 1 & 1 \\ -2 & 1 \end{vmatrix} - y \begin{vmatrix} -2 & 1 \\ 3 & 1 \end{vmatrix} + 1 \begin{vmatrix} -2 & 1 \\ 3 & -2 \end{vmatrix} = 0$$

$$3x + 5y + 1 = 0$$

So, an equation of the line is $3x + 5y + 1 = 0$.

Note that this method of finding the equation of a line works for all lines, including horizontal and vertical lines, as shown below.

Vertical Line Through	*Horizontal Line Through*
$(2, 0)$ *and* $(2, 2)$:	$(-3, 4)$ *and* $(2, 4)$:

$$\begin{vmatrix} x & y & 1 \\ 2 & 0 & 1 \\ 2 & 2 & 1 \end{vmatrix} = 0 \qquad\qquad \begin{vmatrix} x & y & 1 \\ -3 & 4 & 1 \\ 2 & 4 & 1 \end{vmatrix} = 0$$

$$-2x - 0y + 4 = 0 \qquad\qquad 0x + 5y - 20 = 0$$

$$-2x = -4 \qquad\qquad 5y = 20$$

$$x = 2 \qquad\qquad y = 4$$

8.5 Exercises

Review   Concepts, Skills, and Problem Solving

Keep mathematically in shape by doing these exercises *before* the problems of this section.

Properties and Definitions

1. *Writing* ✐   Explain what is meant by the *domain* of a function.

2. *Writing* ✐   Explain what is meant by the *range* of a function.

3. *Writing* ✐   What distinguishes a *function* from a relation?

4. *Writing* ✐   In your own words, explain what the notation $f(4)$ means when $f(x) = x^2 - x + 2$.

Functions and Graphs

In Exercises 5–8, evaluate the function as indicated.

5. For $f(x) = 3x + 2$, find $f(-2)$.

6. For $f(x) = -x^2 - x + 5$, find $f(-1)$.

7. For $f(x) = |10 - 3x|$, find $f(5)$.

8. For $f(x) = \begin{cases} -2x, & x < 0 \\ x^2 + 4, & x \ge 0 \end{cases}$, find $f(0)$.

In Exercises 9 and 10, sketch the graph of the function, and then determine its domain and range.

9. $f(x) = 2 - \sqrt{x}$

10. $f(x) = x^3 - 3$

Models

In Exercises 11 and 12, translate the phrase into an algebraic expression.

11. The time to travel 320 miles if the average speed is r miles per hour

12. The perimeter of a triangle if the sides are $x + 1$, $\frac{1}{2}x + 5$, and $3x + 1$

Developing Skills

In Exercises 1–12, find the determinant of the matrix. See Example 1.

1. $\begin{bmatrix} 2 & 1 \\ 3 & 4 \end{bmatrix}$

2. $\begin{bmatrix} -3 & 1 \\ 5 & 2 \end{bmatrix}$

3. $\begin{bmatrix} 5 & 2 \\ -6 & 3 \end{bmatrix}$

4. $\begin{bmatrix} 2 & -2 \\ 4 & 3 \end{bmatrix}$

5. $\begin{bmatrix} 5 & -4 \\ -10 & 8 \end{bmatrix}$

6. $\begin{bmatrix} 4 & -3 \\ 0 & 0 \end{bmatrix}$

7. $\begin{bmatrix} 2 & 6 \\ 0 & 3 \end{bmatrix}$

8. $\begin{bmatrix} -2 & 3 \\ 6 & -9 \end{bmatrix}$

9. $\begin{bmatrix} -7 & 6 \\ \frac{1}{2} & 3 \end{bmatrix}$

10. $\begin{bmatrix} \frac{2}{3} & \frac{5}{6} \\ 14 & -2 \end{bmatrix}$

11. $\begin{bmatrix} 0.3 & 0.5 \\ 0.5 & 0.3 \end{bmatrix}$

12. $\begin{bmatrix} -1.2 & 4.5 \\ 0.4 & -0.9 \end{bmatrix}$

In Exercises 13–30, evaluate the determinant of the matrix. Expand by minors along the row or column that appears to make the computation easiest. See Examples 2 and 3.

13. $\begin{bmatrix} 2 & 3 & -1 \\ 6 & 0 & 0 \\ 4 & 1 & 1 \end{bmatrix}$

14. $\begin{bmatrix} 10 & 2 & -4 \\ 8 & 0 & -2 \\ 4 & 0 & 2 \end{bmatrix}$

15. $\begin{bmatrix} 1 & 1 & 2 \\ 3 & 1 & 0 \\ -2 & 0 & 3 \end{bmatrix}$

16. $\begin{bmatrix} 2 & 1 & 3 \\ 1 & 4 & 4 \\ 1 & 0 & 2 \end{bmatrix}$

17. $\begin{bmatrix} 2 & 4 & 6 \\ 0 & 3 & 1 \\ 0 & 0 & -5 \end{bmatrix}$

18. $\begin{bmatrix} 2 & 3 & 1 \\ 0 & 5 & -2 \\ 0 & 0 & -2 \end{bmatrix}$

19. $\begin{bmatrix} -2 & 2 & 3 \\ 1 & -1 & 0 \\ 0 & 1 & 4 \end{bmatrix}$

20. $\begin{bmatrix} 3 & 2 & 2 \\ 2 & 2 & 2 \\ -4 & 4 & 3 \end{bmatrix}$

21. $\begin{bmatrix} 1 & 4 & -2 \\ 3 & 6 & -6 \\ -2 & 1 & 4 \end{bmatrix}$ **22.** $\begin{bmatrix} 2 & -1 & 0 \\ 4 & 2 & 1 \\ 4 & 2 & 1 \end{bmatrix}$

23. $\begin{bmatrix} 1 & 4 & -2 \\ 3 & 2 & 0 \\ -1 & 4 & 3 \end{bmatrix}$ **24.** $\begin{bmatrix} 6 & 8 & -7 \\ 0 & 0 & 0 \\ 4 & -6 & 22 \end{bmatrix}$

25. $\begin{bmatrix} 2 & -5 & 0 \\ 4 & 7 & 0 \\ -7 & 25 & 3 \end{bmatrix}$ **26.** $\begin{bmatrix} 8 & 7 & 6 \\ -4 & 0 & 0 \\ 5 & 1 & 4 \end{bmatrix}$

27. $\begin{bmatrix} 0.1 & 0.2 & 0.3 \\ -0.3 & 0.2 & 0.2 \\ 5 & 4 & 4 \end{bmatrix}$ **28.** $\begin{bmatrix} -0.4 & 0.4 & 0.3 \\ 0.2 & 0.2 & 0.2 \\ 0.3 & 0.2 & 0.2 \end{bmatrix}$

29. $\begin{bmatrix} x & y & 1 \\ 3 & 1 & 1 \\ -2 & 0 & 1 \end{bmatrix}$ **30.** $\begin{bmatrix} x & y & 1 \\ -2 & -2 & 1 \\ 1 & 5 & 1 \end{bmatrix}$

In Exercises 31–36, use a graphing calculator to evaluate the determinant of the matrix.

31. $\begin{bmatrix} 5 & -3 & 2 \\ 7 & 5 & -7 \\ 0 & 6 & -1 \end{bmatrix}$ **32.** $\begin{bmatrix} 3 & -1 & 2 \\ 1 & -1 & 2 \\ -2 & 3 & 10 \end{bmatrix}$

33. $\begin{bmatrix} -\frac{1}{2} & -1 & 6 \\ 8 & -\frac{1}{4} & -4 \\ 1 & 2 & 1 \end{bmatrix}$ **34.** $\begin{bmatrix} \frac{1}{2} & \frac{3}{2} & \frac{1}{2} \\ 4 & 8 & 10 \\ -2 & -6 & 12 \end{bmatrix}$

35. $\begin{bmatrix} 0.2 & 0.8 & -0.3 \\ 0.1 & 0.8 & 0.6 \\ -10 & -5 & 1 \end{bmatrix}$

36. $\begin{bmatrix} 0.4 & 0.3 & 0.3 \\ -0.2 & 0.6 & 0.6 \\ 3 & 1 & 1 \end{bmatrix}$

In Exercises 37–52, use Cramer's Rule to solve the system of linear equations. (If not possible, state the reason.) See Examples 4 and 5.

37. $\begin{cases} x + 2y = 5 \\ -x + y = 1 \end{cases}$ **38.** $\begin{cases} 2x - y = -10 \\ 3x + 2y = -1 \end{cases}$

39. $\begin{cases} 3x + 4y = -2 \\ 5x + 3y = 4 \end{cases}$ **40.** $\begin{cases} 18x + 12y = 13 \\ 30x + 24y = 23 \end{cases}$

41. $\begin{cases} 20x + 8y = 11 \\ 12x - 24y = 21 \end{cases}$ **42.** $\begin{cases} 13x - 6y = 17 \\ 26x - 12y = 8 \end{cases}$

43. $\begin{cases} -0.4x + 0.8y = 1.6 \\ 2x - 4y = 5 \end{cases}$ **44.** $\begin{cases} -0.4x + 0.8y = 1.6 \\ 0.2x + 0.3y = 2.2 \end{cases}$

45. $\begin{cases} 3u + 6v = 5 \\ 6u + 14v = 11 \end{cases}$ **46.** $\begin{cases} 3x_1 + 2x_2 = 1 \\ 2x_1 + 10x_2 = 6 \end{cases}$

47. $\begin{cases} 4x - y + z = -5 \\ 2x + 2y + 3z = 10 \\ 5x - 2y + 6z = 1 \end{cases}$

48. $\begin{cases} 4x - 2y + 3z = -2 \\ 2x + 2y + 5z = 16 \\ 8x - 5y - 2z = 4 \end{cases}$

49. $\begin{cases} 3a + 3b + 4c = 1 \\ 3a + 5b + 9c = 2 \\ 5a + 9b + 17c = 4 \end{cases}$ **50.** $\begin{cases} 2x + 3y + 5z = 4 \\ 3x + 5y + 9z = 7 \\ 5x + 9y + 17z = 13 \end{cases}$

51. $\begin{cases} 5x - 3y + 2z = 2 \\ 2x + 2y - 3z = 3 \\ x - 7y + 8z = -4 \end{cases}$ **52.** $\begin{cases} 3x + 2y + 5z = 4 \\ 4x - 3y - 4z = 1 \\ -8x + 2y + 3z = 0 \end{cases}$

In Exercises 53–56, solve the system of linear equations using a graphing calculator and Cramer's Rule. See Examples 4 and 5.

53. $\begin{cases} -3x + 10y = 22 \\ 9x - 3y = 0 \end{cases}$ **54.** $\begin{cases} 3x + 7y = 3 \\ 7x + 25y = 11 \end{cases}$

55. $\begin{cases} 3x - 2y + 3z = 8 \\ x + 3y + 6z = -3 \\ x + 2y + 9z = -5 \end{cases}$

56. $\begin{cases} 6x + 4y - 8z = -22 \\ -2x + 2y + 3z = 13 \\ -2x + 2y - z = 5 \end{cases}$

In Exercises 57 and 58, solve the equation.

57. $\begin{vmatrix} 5 - x & 4 \\ 1 & 2 - x \end{vmatrix} = 0$

58. $\begin{vmatrix} 4 - x & -2 \\ 1 & 1 - x \end{vmatrix} = 0$

Solving Problems

Area of a Triangle In Exercises 59–66, use a determinant to find the area of the triangle with the given vertices. See Example 6.

59. $(0, 3), (4, 0), (8, 5)$

60. $(2, 0), (0, 5), (6, 3)$

61. $(0, 0), (3, 1), (1, 5)$

62. $(-2, -3), (2, -3), (0, 4)$

63. $(-2, 1), (3, -1), (1, 6)$

64. $(-4, 2), (1, 5), (4, -4)$

65. $\left(0, \frac{1}{2}\right), \left(\frac{5}{2}, 0\right), (4, 3)$

66. $\left(\frac{1}{4}, 0\right), \left(0, \frac{3}{4}\right), (8, -2)$

Area of a Region In Exercises 67–70, find the area of the shaded region of the figure.

67.
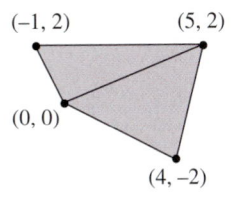

68.

69.
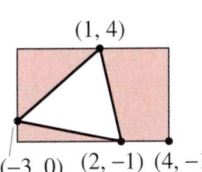

70.

71. *Area of a Region* A large region of forest has been infested with gypsy moths. The region is roughly triangular, as shown in the figure. Approximate the number of square miles in this region.

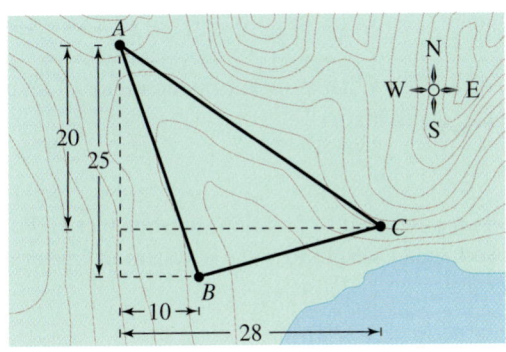

72. *Area of a Region* You have purchased a triangular tract of land, as shown in the figure. How many square feet are there in the tract of land?

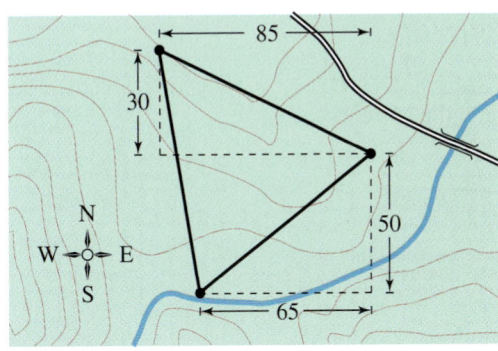

Collinear Points In Exercises 73–78, determine whether the points are collinear. See Example 7.

73. $(-1, 11), (0, 8), (2, 2)$

74. $(-1, -1), (1, 9), (2, 13)$

75. $(-1, -5), (1, -1), (4, 5)$

76. $(-1, 8), (1, 2), (2, 0)$

77. $\left(-2, \frac{1}{3}\right), (2, 1), \left(3, \frac{1}{5}\right)$

78. $\left(0, \frac{1}{2}\right), \left(1, \frac{7}{6}\right), \left(9, \frac{13}{2}\right)$

Equation of a Line In Exercises 79–86, use a determinant to find the equation of the line through the points. See Example 8.

79. $(0, 0), (5, 3)$

80. $(-4, 3), (2, 1)$

81. $(10, 7), (-2, -7)$

82. $(-8, 3), (4, 6)$

83. $\left(-2, \frac{3}{2}\right), (3, -3)$

84. $\left(-\frac{1}{2}, 3\right), \left(\frac{5}{2}, 1\right)$

85. $(2, 3.6), (8, 10)$

86. $(3, 1.6), (5, -2.2)$

87. *Electrical Networks* Laws that deal with electrical currents are known as *Kirchhoff's Laws*. When Kirchhoff's Laws are applied to the electrical network shown in the figure, the currents I_1, I_2, and I_3 are the solution of the system

$$\begin{cases} I_1 - I_2 + I_3 = 0 \\ 3I_1 + 2I_2 = 7 \\ 2I_2 + 4I_3 = 8. \end{cases}$$

Find the currents.

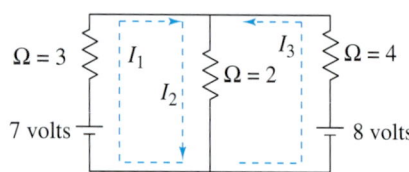

88. *Electrical Networks* When Kirchhoff's Laws are applied to the electrical network shown in the figure, the currents I_1, I_2, and I_3 are the solution of the system

$$\begin{cases} I_1 + I_2 - I_3 = 0 \\ I_1 + 2I_3 = 12 \\ I_1 - 2I_2 = -4. \end{cases}$$

Find the currents.

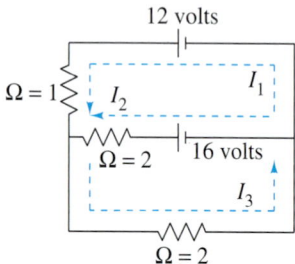

89. *Electrical Networks* When Kirchhoff's Laws are applied to the electrical network shown in the figure, the currents I_1, I_2, and I_3 are the solution of the system

$$\begin{cases} I_1 - I_2 + I_3 = 0 \\ I_2 + 4I_3 = 8 \\ 4I_1 + I_2 = 16. \end{cases}$$

Find the currents.

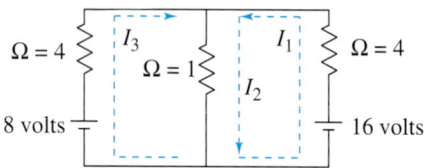

90. *Force* When three forces are applied to a beam, Newton's Laws suggests that the forces F_1, F_2, and F_3 are the solution of the system

$$\begin{cases} 3F_1 + F_2 - F_3 = 2 \\ F_1 - 2F_2 + F_3 = 0 \\ 4F_1 - F_2 + F_3 = 0. \end{cases}$$

Find the forces.

91. (a) Use Cramer's Rule to solve the system of linear equations.

$$\begin{cases} kx + 3ky = 2 \\ (2 + k)x + ky = 5 \end{cases}$$

(b) For what values of k can Cramer's Rule not be used?

92. (a) Use Cramer's Rule to solve the system of linear equations.

$$\begin{cases} kx + (1 - k)y = 1 \\ (1 - k)x + ky = 3 \end{cases}$$

(b) For what value(s) of k will the system be inconsistent?

Explaining Concepts

93. ⌖ Answer parts (f) and (g) of Motivating the Chapter on page 466.

94. *Writing* Explain the difference between a square matrix and its determinant.

95. *Writing* Is it possible to find the determinant of a 2×3 matrix? Explain.

96. *Writing* What is meant by the minor of an entry of a square matrix?

97. *Writing* What conditions must be met in order to use Cramer's Rule to solve a system of linear equations?

8.6 Systems of Linear Inequalities

Bonnie Kamin/PhotoEdit

What You Should Learn

1. Solve systems of linear inequalities in two variables.
2. Use systems of linear inequalities to model and solve real-life problems.

Why You Should Learn It

Systems of linear inequalities can be used to model and solve real-life problems. For instance, in Exercise 65 on page 542, a system of linear inequalities can be used to analyze the compositions of dietary supplements.

1. Solve systems of linear inequalities in two variables.

Systems of Linear Inequalities in Two Variables

You have already graphed linear inequalities in two variables. However, many practical problems in business, science, and engineering involve **systems of linear inequalities.** This type of system arises in problems that have *constraint* statements that contain phrases such as "more than," "less than," "at least," "no more than," "a minimum of," and "a maximum of." A **solution** of a system of linear inequalities in x and y is a point (x, y) that satisfies each inequality in the system.

To sketch the graph of a system of inequalities in two variables, first sketch (on the same coordinate system) the graph of each individual inequality. The **solution set** is the region that is *common* to every graph in the system.

Example 1   Graphing a System of Linear Inequalities

Sketch the graph of the system of linear inequalities: $\begin{cases} 2x - y \le 5 \\ x + 2y \ge 2. \end{cases}$

Solution

Begin by rewriting each inequality in slope-intercept form. Then sketch the line for each corresponding equation of each inequality. See Figures 8.19–8.21.

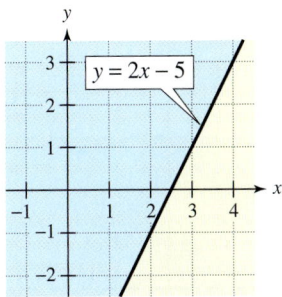

Graph of $2x - y \le 5$ is all points on and above $y = 2x - 5$.
Figure 8.19

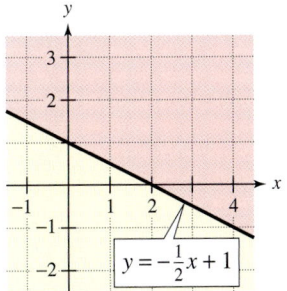

Graph of $x + 2y \ge 2$ is all points on and above $y = -\frac{1}{2}x + 1$.
Figure 8.20

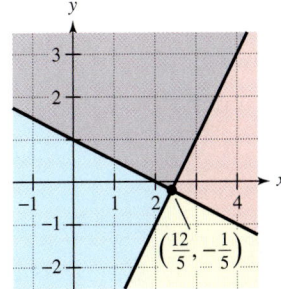

Graph of system is the purple wedge-shaped region.
Figure 8.21

In Figure 8.21, note that the two borderlines of the region

$$y = 2x - 5 \quad \text{and} \quad y = -\frac{1}{2}x + 1$$

intersect at the point $\left(\frac{12}{5}, -\frac{1}{5}\right)$. Such a point is called a **vertex** of the region. The region shown in the figure has only one vertex. Some regions, however, have several vertices. When you are sketching the graph of a system of linear inequalities, it is helpful to find and label any vertices of the region.

Graphing a System of Linear Inequalities

1. Sketch the line that corresponds to each inequality. (Use dashed lines for inequalities with $<$ or $>$ and solid lines for inequalities with $\leq$ or $\geq$.)

2. Lightly shade the half-plane that is the graph of each linear inequality. (Colored pencils may help distinguish different half-planes.)

3. The graph of the system is the intersection of the half-planes. (If you use colored pencils, it is the region that is selected with *every* color.)

Example 2   Graphing a System of Linear Inequalities

Sketch the graph of the system of linear inequalities: $\begin{cases} y < 4 \\ y > 1 \end{cases}$.

Solution

The graph of the first inequality is the half-plane below the horizontal line

$$y = 4. \qquad \text{Upper boundary}$$

The graph of the second inequality is the half-plane above the horizontal line

$$y = 1. \qquad \text{Lower boundary}$$

The graph of the system is the horizontal band that lies *between* the two horizontal lines (where $y < 4$ *and* $y > 1$), as shown in Figure 8.22.

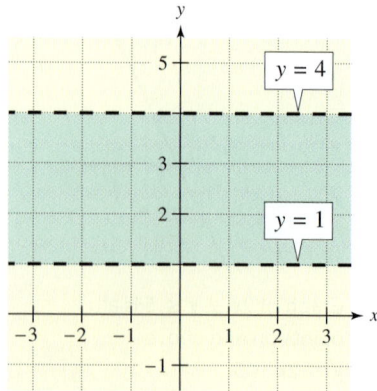

Figure 8.22

Example 3   Graphing a System of Linear Inequalities

Sketch the graph of the system of linear inequalities, and label the vertices.

$$\begin{cases} x - y < & 2 \\ x & > -2 \\ y \leq & 3 \end{cases}$$

Solution

Begin by sketching the half-planes represented by the three linear inequalities. The graph of

$$x - y < 2$$

is the half-plane lying above the line $y = x - 2$, the graph of

$$x > -2$$

is the half-plane lying to the right of the line $x = -2$, and the graph of

$$y \leq 3$$

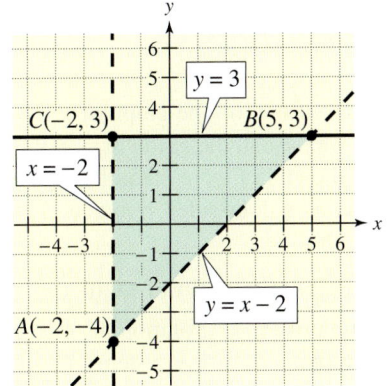

Figure 8.23

is the half-plane lying on or below the line $y = 3$. As shown in Figure 8.23, the region that is common to all three of these half-planes is a triangle. The vertices of the triangle are found as follows.

Vertex A: $(-2, -4)$	Vertex B: $(5, 3)$	Vertex C: $(-2, 3)$
Solution of the system	Solution of the system	Solution of the system
$\begin{cases} x - y = 2 \\ x = -2 \end{cases}$	$\begin{cases} x - y = 2 \\ y = 3 \end{cases}$	$\begin{cases} x = -2 \\ y = 3 \end{cases}$

For the triangular region shown in Figure 8.23, each point of intersection of a pair of boundary lines corresponds to a vertex. With more complicated regions, two border lines can sometimes intersect at a point that is not a vertex of the region, as shown in Figure 8.24. To keep track of which points of intersection are actually vertices of the region, you should sketch the region and refer to your sketch as you find each point of intersection.

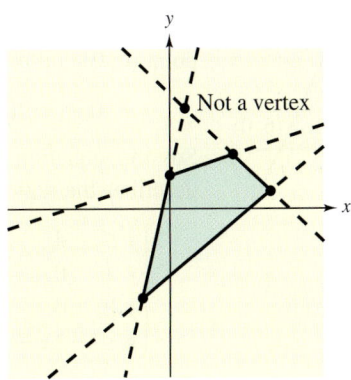

Figure 8.24

Example 4 Graphing a System of Linear Inequalities

Sketch the graph of the system of linear inequalities, and label the vertices.

$$\begin{cases} x + y \le 5 \\ 3x + 2y \le 12 \\ x \ge 0 \\ y \ge 0 \end{cases}$$

Solution

Begin by sketching the half-planes represented by the four linear inequalities. The graph of

$$x + y \le 5$$

is the half-plane lying on and below the line $y = -x + 5$. The graph of

$$3x + 2y \le 12$$

is the half-plane lying on and below the line $y = -\frac{3}{2}x + 6$. The graph of $x \ge 0$ is the half-plane lying on and to the right of the y-axis, and the graph of $y \ge 0$ is the half-plane lying on and above the x-axis. As shown in Figure 8.25, the region that is common to all four of these half-planes is a four-sided polygon. The vertices of the region are found as follows.

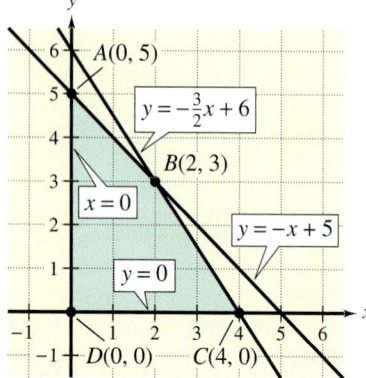

Figure 8.25

Vertex A: $(0, 5)$	*Vertex B:* $(2, 3)$	*Vertex C:* $(4, 0)$	*Vertex D:* $(0, 0)$
Solution of the system	Solution of the system	Solution of the system	Solution of the system
$\begin{cases} x + y = 5 \\ x = 0 \end{cases}$	$\begin{cases} x + y = 5 \\ 3x + 2y = 12 \end{cases}$	$\begin{cases} 3x + 2y = 12 \\ y = 0 \end{cases}$	$\begin{cases} x = 0 \\ y = 0 \end{cases}$

Example 5 Finding the Boundaries of a Region

Find a system of inequalities that defines the region shown in Figure 8.26.

Solution

Three of the boundaries of the region are horizontal or vertical—they are easy to find. To find the diagonal boundary line, use the techniques in Section 4.5 to find the equation of the line passing through the points $(4, 4)$ and $(6, 0)$. You can use the formula for slope to find $m = -2$, and then use the point-slope form with point $(6, 0)$ and $m = -2$ to obtain

$$y - 0 = -2(x - 6).$$

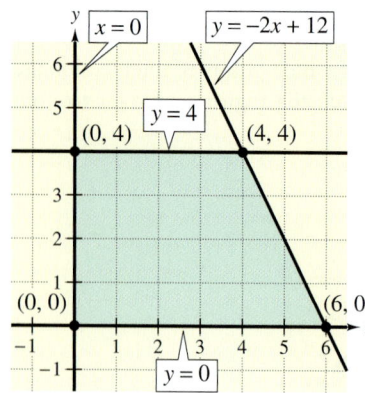

Figure 8.26

So, the equation is $y = -2x + 12$. The system of linear inequalities that describes the region is as follows.

$$\begin{cases} y \le 4 & \text{Region lies on and below line } y = 4. \\ y \ge 0 & \text{Region lies on and above } x\text{-axis.} \\ x \ge 0 & \text{Region lies on and to the right of } y\text{-axis.} \\ y \le -2x + 12 & \text{Region lies on and below line } y = -2x + 12. \end{cases}$$

Technology: Tip

A graphing calculator can be used to graph a system of linear inequalities. The graph of

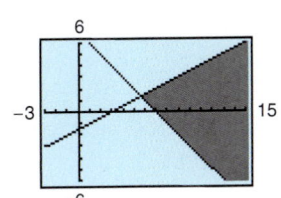

$$\begin{cases} 4y < 2x - 6 \\ x + y \geq 7 \end{cases}$$

is shown at the left. The shaded region, in which all points satisfy both inequalities, is the solution of the system. Try using a graphing calculator to graph

$$\begin{cases} 3x + y < 1 \\ -2x - 2y < 8. \end{cases}$$

---

2 Use systems of linear inequalities to model and solve real-life problems.

Application

Example 6   Nutrition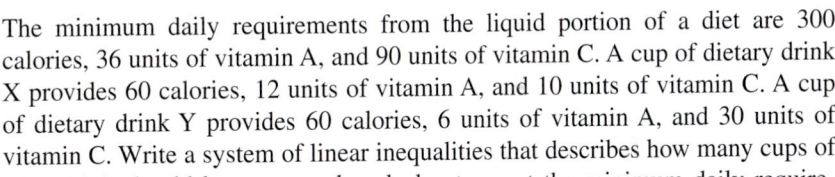

The minimum daily requirements from the liquid portion of a diet are 300 calories, 36 units of vitamin A, and 90 units of vitamin C. A cup of dietary drink X provides 60 calories, 12 units of vitamin A, and 10 units of vitamin C. A cup of dietary drink Y provides 60 calories, 6 units of vitamin A, and 30 units of vitamin C. Write a system of linear inequalities that describes how many cups of each drink should be consumed each day to meet the minimum daily requirements for calories and vitamins.

Solution

Begin by letting x and y represent the following.

x = number of cups of dietary drink X

y = number of cups of dietary drink Y

To meet the minimum daily requirements, the following inequalities must be satisfied.

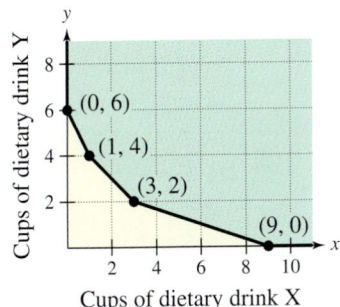

Cups of dietary drink Y

Cups of dietary drink X

Figure 8.27

$$\begin{cases} 60x + 60y \geq 300 & \text{Calories} \\ 12x + 6y \geq 36 & \text{Vitamin A} \\ 10x + 30y \geq 90 & \text{Vitamin C} \\ x \geq 0 \\ y \geq 0 \end{cases}$$

The last two inequalities are included because x and y cannot be negative. The graph of this system of inequalities is shown in Figure 8.27.

8.6 Exercises

Review   Concepts, Skills, and Problem Solving

Keep mathematically in shape by doing these exercises *before* the problems of this section.

Properties and Definitions

1. *Writing* ✎   Given a function $f(x)$, describe how the graph of $h(x) = f(x) + c$ compares with the graph of $f(x)$ for a positive real number c.

2. *Writing* ✎   Given a function $f(x)$, describe how the graph of $h(x) = f(x) - c$ compares with the graph of $f(x)$ for a positive real number c.

3. *Writing* ✎   Given a function $f(x)$, describe how the graph of $h(x) = f(x - c)$ compares with the graph of $f(x)$ for a positive real number c.

4. *Writing* ✎   Given a function $f(x)$, describe how the graph of $h(x) = f(x + c)$ compares with the graph of $f(x)$ for a positive real number c.

5. *Writing* ✎   Given a function $f(x)$, describe how the graph of $h(x) = -f(x)$ compares with the graph of $f(x)$.

6. *Writing* ✎   Given a function $f(x)$, describe how the graph of $h(x) = f(-x)$ compares with the graph of $f(x)$.

Graphing

In Exercises 7 and 8, use the graph of $f(x) = x^2$ to write a function that represents the graph.

7.
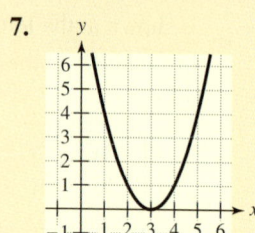

8.
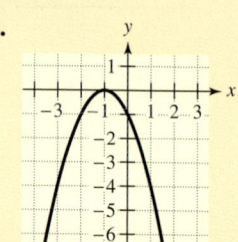

In Exercises 9 and 10, use the graph of f to sketch the graph.

9. $f(x) - 3$

10. $f(x - 1) + 2$

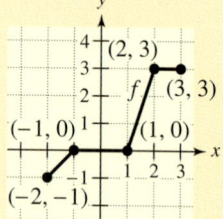

Problem Solving

11. *Retail Price*   A video game system that costs a retailer $169.50 is marked up by 36%. Find the price to the consumer.

12. *Sale Price*   A sweater is listed at $44. Find the sale price if there is a 20%-off sale.

Developing Skills

In Exercises 1–6, match the system of linear inequalities with its graph. [The graphs are labeled (a), (b), (c), (d), (e), and (f).]

(a)
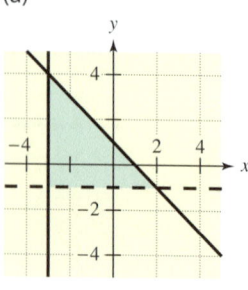

(b)
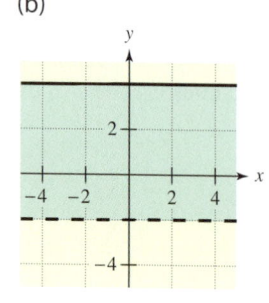

(c)
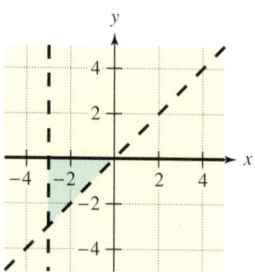

(d)
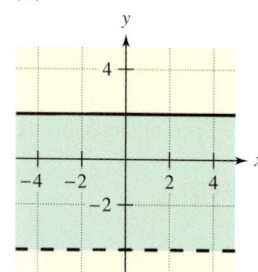

(e)

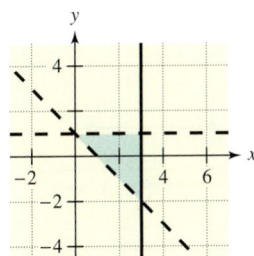

(f)

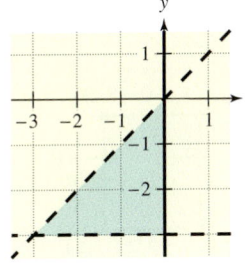

1. $\begin{cases} y > x \\ x > -3 \\ y \le 0 \end{cases}$

2. $\begin{cases} y \le 4 \\ y > -2 \end{cases}$

3. $\begin{cases} y < x \\ y > -3 \\ x \le 0 \end{cases}$

4. $\begin{cases} x \le 3 \\ y < 1 \\ y > -x + 1 \end{cases}$

5. $\begin{cases} y > -1 \\ x \ge -3 \\ y \le -x + 1 \end{cases}$

6. $\begin{cases} y > -4 \\ y \le 2 \end{cases}$

In Exercises 7–44, sketch a graph of the solution of the system of linear inequalities. See Examples 1–4.

7. $\begin{cases} x < 3 \\ x > -2 \end{cases}$

8. $\begin{cases} y > -1 \\ y \le 2 \end{cases}$

9. $\begin{cases} x + y \le 3 \\ x - 1 \le 1 \end{cases}$

10. $\begin{cases} x + y \ge 2 \\ x - y \le 2 \end{cases}$

11. $\begin{cases} 2x - 4y \le 6 \\ x + y \ge 2 \end{cases}$

12. $\begin{cases} 4x + 10y \le 5 \\ x - y \le 4 \end{cases}$

13. $\begin{cases} x + 2y \le 6 \\ x - 2y \le 0 \end{cases}$

14. $\begin{cases} 2x + y \le 0 \\ x - y \le 8 \end{cases}$

15. $\begin{cases} x - 2y > 4 \\ 2x + y > 6 \end{cases}$

16. $\begin{cases} 3x + y < 6 \\ x + 2y > 2 \end{cases}$

17. $\begin{cases} x + y > -1 \\ x + y < 3 \end{cases}$

18. $\begin{cases} x - y > 2 \\ x - y < -4 \end{cases}$

19. $\begin{cases} y \ge \frac{4}{3}x + 1 \\ y \le 5x - 2 \end{cases}$

20. $\begin{cases} y \ge \frac{1}{2}x + \frac{1}{2} \\ y \le 4x - \frac{1}{2} \end{cases}$

21. $\begin{cases} y > x - 2 \\ y > -\frac{1}{3}x + 5 \end{cases}$

22. $\begin{cases} y > x - 4 \\ y > \frac{2}{3}x + \frac{1}{3} \end{cases}$

23. $\begin{cases} y \ge 3x - 3 \\ y \le -x + 1 \end{cases}$

24. $\begin{cases} y \ge 2x - 3 \\ y \le 3x + 1 \end{cases}$

25. $\begin{cases} y > 2x \\ y > -x + 4 \end{cases}$

26. $\begin{cases} y \le -x \\ y \le x + 1 \end{cases}$

27. $\begin{cases} x + 2y \le -4 \\ y \ge x + 5 \end{cases}$

28. $\begin{cases} x + y \le -3 \\ y \ge 3x - 4 \end{cases}$

29. $\begin{cases} x + y \le 4 \\ x \ge 0 \\ y \ge 0 \end{cases}$

30. $\begin{cases} 2x + y \le 6 \\ x \ge 0 \\ y \ge 0 \end{cases}$

31. $\begin{cases} 4x - 2y > 8 \\ x \ge 0 \\ y \le 0 \end{cases}$

32. $\begin{cases} 2x - 6y > 6 \\ x \le 0 \\ y \le 0 \end{cases}$

33. $\begin{cases} y > -5 \\ x \le 2 \\ y \le x + 2 \end{cases}$

34. $\begin{cases} y \ge -1 \\ x \le 2 \\ y \le x + 2 \end{cases}$

35. $\begin{cases} x + y \le 1 \\ -x + y \le 1 \\ y \ge 0 \end{cases}$

36. $\begin{cases} 3x + 2y < 6 \\ x - 3y \ge 1 \\ y \ge 0 \end{cases}$

37. $\begin{cases} x + y \le 5 \\ x - 2y \ge 2 \\ y \ge 3 \end{cases}$

38. $\begin{cases} 2x + y \ge 2 \\ x - 3y \le 2 \\ y \le 1 \end{cases}$

39. $\begin{cases} -3x + 2y < 6 \\ x - 4y > -2 \\ 2x + y < 3 \end{cases}$

40. $\begin{cases} x - 7y > -36 \\ 5x + 2y > 5 \\ 6x + 5y > 6 \end{cases}$

41. $\begin{cases} 2x + y < 2 \\ 6x + 3y > 2 \end{cases}$

42. $\begin{cases} x - 2y < -6 \\ 5x - 3y > -9 \end{cases}$

43. $\begin{cases} x \ge 1 \\ x - 2y \le 3 \\ 3x + 2y \ge 9 \\ x + y \le 6 \end{cases}$

44. $\begin{cases} x + y \le 4 \\ x + y \ge -1 \\ x - y \ge -2 \\ x - y \le 2 \end{cases}$

In Exercises 45–50, use a graphing calculator to graph the solution of the system of linear inequalities.

45. $\begin{cases} 2x - 3y \le 6 \\ \qquad y \le 4 \end{cases}$

46. $\begin{cases} 6x + 3y \ge 12 \\ \qquad y \le 4 \end{cases}$

47. $\begin{cases} 2x - 2y \le 5 \\ \qquad y \le 6 \end{cases}$

48. $\begin{cases} 2x + 3y \ge 12 \\ \qquad y \ge 2 \end{cases}$

49. $\begin{cases} 2x + y \le 2 \\ \qquad y \ge -4 \end{cases}$

50. $\begin{cases} x - 2y \ge -6 \\ \qquad y \le 6 \end{cases}$

53.

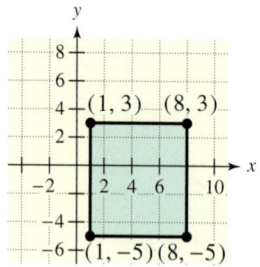

54.

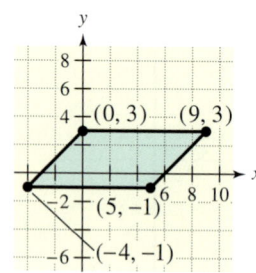

In Exercises 51–56, write a system of linear inequalities that describes the shaded region. See Example 5.

51.

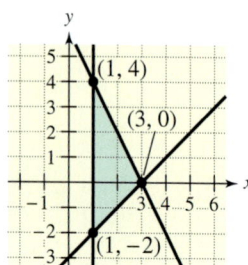

52.

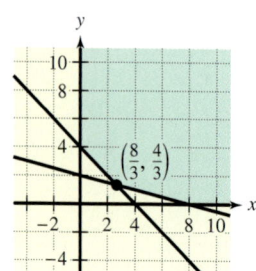

55.

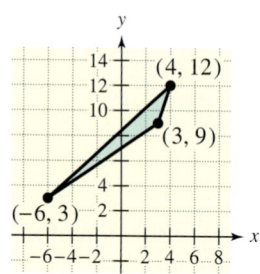

56.

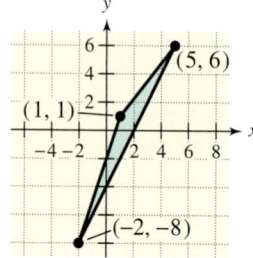

Solving Problems

57. *Production*   A furniture company can sell all the tables and chairs it produces. Each table requires 1 hour in the assembly center and $1\frac{1}{3}$ hours in the finishing center. Each chair requires $1\frac{1}{2}$ hours in the assembly center and $1\frac{1}{2}$ hours in the finishing center. The company's assembly center is available 12 hours per day, and its finishing center is available 15 hours per day. Write a system of linear inequalities describing the different production levels. Graph the system.

58. *Production*   An electronics company can sell all the VCRs and DVD players it produces. Each VCR requires 2 hours on the assembly line and $1\frac{1}{2}$ hours on the testing line. Each DVD player requires $2\frac{1}{2}$ hours on the assembly line and 3 hours on the testing line. The company's assembly line is available 18 hours per day, and its testing line is available 16 hours per day. Write a system of linear inequalities describing the different production levels. Graph the system.

59. *Investment*   A person plans to invest up to $20,000 in two different interest-bearing accounts, account X and account Y. Account X is to contain at least $5000. Moreover, account Y should have at least twice the amount in account X. Write a system of linear inequalities describing the various amounts that can be deposited in each account. Graph the system.

60. *Investment*   A person plans to invest up to $10,000 in two different interest-bearing accounts, account X and account Y. Account Y is to contain at least $3000. Moreover, account X should have at least three times the amount in account Y. Write a system of linear inequalities describing the various amounts that can be deposited in each account. Graph the system.

61. 🔲 *Ticket Sales*   Two types of tickets are to be sold for a concert. General admission tickets cost $15 per ticket and stadium seat tickets cost $25 per ticket. The promoter of the concert must sell at least 15,000 tickets, including at least 8000 general admission tickets and at least 4000 stadium seat tickets. Moreover, the gross receipts must total at least $275,000 in order for the concert to be held. Write a system of linear inequalities describing the different numbers of tickets that can be sold. Use a graphing calculator to graph the system.

62. 🔲 *Ticket Sales*   For a concert event, there are $30 reserved seat tickets and $20 general admission tickets. There are 2000 reserved seats available, and fire regulations limit the number of paid ticket holders to 3000. The promoter must take in at least $75,000 in ticket sales. Write a system of linear inequalities describing the different numbers of tickets that can be sold. Use a graphing calculator to graph the system.

63. 🔺 *Geometry*   The figure shows a cross section of a roped-off swimming area at a beach. Write a system of linear inequalities describing the cross section. (Each unit in the coordinate system represents 1 foot.)

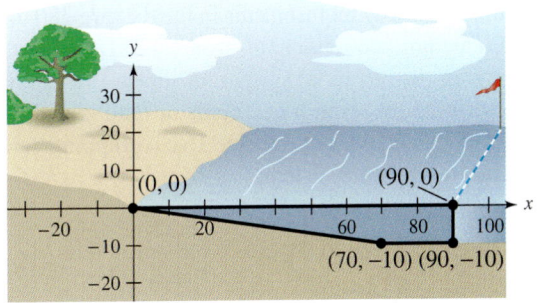

64. 🔺 *Geometry*   The figure shows the chorus platform on a stage. Write a system of linear inequalities describing the part of the audience that can see the full chorus. (Each unit in the coordinate system represents 1 meter.)

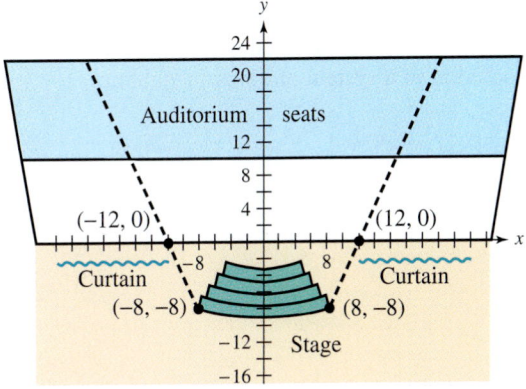

65. ▦ *Nutrition*   A dietitian is asked to design a special diet supplement using two different foods. Each ounce of food X contains 20 units of calcium, 15 units of iron, and 10 units of vitamin B. Each ounce of food Y contains 10 units of calcium, 10 units of iron, and 20 units of vitamin B. The minimum daily requirements in the diet are 280 units of calcium, 160 units of iron, and 180 units of vitamin B. Write a system of linear inequalities describing the different amounts of food X and food Y that can be used in the diet. Use a graphing calculator to graph the system.

66. ▦ *Nutrition*   A veterinarian is asked to design a special canine dietary supplement using two different dog foods. Each ounce of food X contains 12 units of calcium, 8 units of iron, and 6 units of protein. Each ounce of food Y contains 10 units of calcium, 10 units of iron, and 8 units of protein. The minimum daily requirements of the diet are 200 units of calcium, 100 units of iron, and 120 units of protein. Write a system of linear inequalities describing the different amounts of dog food X and dog food Y that can be used. Use a graphing calculator to graph the system.

Explaining Concepts

67. *Writing*   Explain the meaning of the term *half-plane*. Give an example of an inequality whose graph is a half-plane.

68. *Writing*   Explain how you can check any single point (x_1, y_1) to determine whether the point is a solution of a system of linear inequalities.

69. *Writing*   Explain how to determine the vertices of the solution region for a system of linear inequalities.

70. *Writing*   Describe the difference between the solution set of a system of linear equations and the solution set of a system of linear inequalities.

What Did You Learn?

Key Terms

system of equations, *p. 468*
solution of a system of
    equations, *p. 468*
consistent system, *p. 470*
dependent system, *p. 470*
inconsistent system, *p. 470*
back-substitute, *p. 473*
row-echelon form, *p. 495*

equivalent systems, *p. 496*
Gaussian elimination, *p. 496*
row operations, *p. 496*
matrix, *p. 508*
order (of a matrix), *p. 508*
square matrix, *p. 508*
augmented matrix, *p. 509*
coefficient matrix, *p. 509*

row-equivalent matrices, *p. 510*
minor (of an entry), *p. 522*
Cramer's Rule, *p. 524*
system of linear inequalities,
    p. 533
solution of a system of linear
    inequalities, *p. 533*
vertex, *p. 534*

Key Concepts

8.1 ⬤ The method of substitution

1. Solve one of the equations for one variable in terms of the other.

2. Substitute the expression obtained in Step 1 in the other equation to obtain an equation in one variable.

3. Solve the equation obtained in Step 2.

4. Back-substitute the solution from Step 3 in the expression obtained in Step 1 to find the value of the other variable.

5. Check the solution to see that it satisfies both of the original equations.

8.2 ⬤ The method of elimination

1. Obtain opposite coefficients of x (or y) by multiplying all terms of one or both equations by suitable constants.

2. Add the equations to eliminate one variable and solve the resulting equation.

3. Back-substitute the value obtained in Step 2 in either of the original equations and solve for the other variable.

4. Check your solution in *both* of the original equations.

8.3 ⬤ Operations that produce equivalent systems

Each of the following row operations on a system of linear equations produces an equivalent system of linear equations.

1. Interchange two equations.

2. Multiply one of the equations by a nonzero constant.

3. Add a multiple of one of the equations to another equation to replace the latter equation.

8.4 ⬤ Elementary row operations

Two matrices are row-equivalent if one can be obtained from the other by a sequence of elementary row operations.

1. Interchange two rows.

2. Multiply a row by a nonzero constant.

3. Add a multiple of a row to another row.

8.4 ⬤ Gaussian elimination with back-substitution

To use matrices and Gaussian elimination to solve a system of linear equations, use the following steps.

1. Write the augmented matrix of the system of linear equations.

2. Use elementary row operations to rewrite the augmented matrix in row-echelon form.

3. Write the system of linear equations corresponding to the matrix in row-echelon form, and use back-substitution to find the solution.

8.5 ⬤ Determinant of a 2 × 2 matrix

$$\det(A) = |A| = \begin{vmatrix} a_1 & b_1 \\ a_2 & b_2 \end{vmatrix} = a_1 b_2 - a_2 b_1$$

8.6 ⬤ Graphing a system of linear inequalities

1. Sketch the line that corresponds to each inequality. (Use dashed lines for inequalities with < or > and solid lines for inequalities with ≤ or ≥.)

2. Lightly shade the half-plane that is the graph of each linear inequality.

3. The graph of the system is the intersection of the half-planes.

Review Exercises

8.1 Solving Systems of Equations by Graphing and Substitution

1 Determine if an ordered pair is a solution to a system of equations.

In Exercises 1–4, determine whether each ordered pair is a solution to the system of equations.

System	Ordered Pairs

1. $\begin{cases} 3x - 5y = 11 \\ -x + 2y = -4 \end{cases}$ (a) $(2, -1)$ (b) $(3, -2)$

2. $\begin{cases} 10x + 8y = -2 \\ 2x - 5y = 26 \end{cases}$ (a) $(4, -4)$ (b) $(3, -4)$

3. $\begin{cases} 0.2x + 0.4y = 5 \\ x + 3y = 30 \end{cases}$ (a) $(0.5, -0.7)$ (b) $(15, 5)$

4. $\begin{cases} -\frac{1}{2}x - \frac{2}{3}y = \frac{1}{2} \\ x + y = 1 \end{cases}$ (a) $(-5, 6)$ (b) $(7, -3)$

2 Use a coordinate system to solve systems of linear equations graphically.

In Exercises 5–8, use the graphs of the equations to determine the solution (if any) of the system of linear equations. Check your solution.

5. $\begin{cases} 2x + y = 4 \\ 2x - y = 0 \end{cases}$

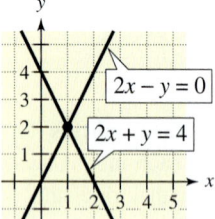

6. $\begin{cases} -x + y = 1 \\ x + y = 5 \end{cases}$

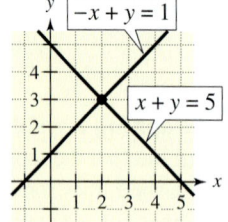

7. $\begin{cases} 3x - 2y = 6 \\ 2x - y = 3 \end{cases}$

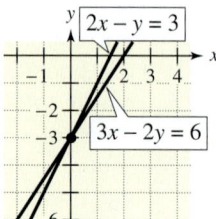

8. $\begin{cases} x + 2y = 6 \\ x + 2y = 2 \end{cases}$

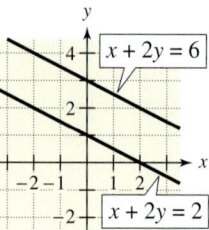

In Exercises 9–14, sketch the graphs of the equations and approximate any solutions of the system of linear equations.

9. $\begin{cases} y = x - 4 \\ y = 2x - 9 \end{cases}$

10. $\begin{cases} y = -\frac{5}{3}x + 6 \\ y = x - 10 \end{cases}$

11. $\begin{cases} x + y = 2 \\ x - y = 0 \end{cases}$

12. $\begin{cases} x - y = 9 \\ -x + y = 1 \end{cases}$

13. $\begin{cases} 2x + 3 = 3y \\ y = \frac{2}{3}x \end{cases}$

14. $\begin{cases} x + y = -1 \\ 3x + 2y = 0 \end{cases}$

3 Use the method of substitution to solve systems of equations algebraically.

In Exercises 15–26, solve the system by the method of substitution.

15. $\begin{cases} y = 2x \\ y = x + 4 \end{cases}$

16. $\begin{cases} x = -2y + 13 \\ x = \frac{y}{2} + 3 \end{cases}$

17. $\begin{cases} x = 3y - 2 \\ x = 6 - y \end{cases}$

18. $\begin{cases} y = -4x + 1 \\ y = x - 4 \end{cases}$

19. $\begin{cases} x - 2y = 6 \\ 3x + 2y = 10 \end{cases}$

20. $\begin{cases} 5x + y = 20 \\ 7x - 5y = -4 \end{cases}$

21. $\begin{cases} 2x - y = 2 \\ 6x + 8y = 39 \end{cases}$

22. $\begin{cases} 3x + 4y = 1 \\ x - 7y = -3 \end{cases}$

23. $\begin{cases} x = y + 3 \\ x = y + 1 \end{cases}$

24. $\begin{cases} y = 3x + 4 \\ 9x = 3y - 12 \end{cases}$

25. $\begin{cases} -6x + y = -3 \\ 12x - 2y = 6 \end{cases}$

26. $\begin{cases} 3x + 4y = 7 \\ 6x + 8y = 10 \end{cases}$

④ Solve application problems using systems of equations.

27. *Investment*   A total of $12,000 is invested in two funds paying 5% and 10% simple interest. The combined annual interest for the two funds is $800. Determine how much of the $12,000 is invested at each rate.

28. *Ticket Sales*   You are selling tickets to your school musical. Adult tickets cost $5 and children's tickets cost $3. You sell 1510 tickets and collect $6138. Determine how many of each type of ticket were sold.

29. *Coin Problem*   A cash register has 15 coins consisting of dimes and quarters. The total value of the coins is $2.85. Find the number of each type of coin.

30. *Video Rental*   You go to the video store to rent five movies for the weekend. Videos rent for $2 and $3. You spend $13. How many $2 videos did you rent?

8.2 Solving Systems of Equations by Elimination

① Solve systems of linear equations algebraically using the method of elimination.

In Exercises 31–38, solve the system by the method of elimination.

31. $\begin{cases} 3x - y = 5 \\ 2x + y = 5 \end{cases}$

32. $\begin{cases} 2x + 4y = 2 \\ -2x - 7y = 4 \end{cases}$

33. $\begin{cases} 5x + 4y = 2 \\ -x + y = -22 \end{cases}$

34. $\begin{cases} 3x - 2y = 9 \\ x + y = 3 \end{cases}$

35. $\begin{cases} 8x - 6y = 4 \\ -4x + 3y = -2 \end{cases}$

36. $\begin{cases} 2x - 5y = 2 \\ 3x - 7y = 1 \end{cases}$

37. $\begin{cases} 0.2x + 0.1y = 0.03 \\ 0.3x - 0.1y = -0.13 \end{cases}$

38. $\begin{cases} 0.2x - 0.1y = 0.07 \\ 0.4x - 0.5y = -0.01 \end{cases}$

② Choose a method for solving systems of equations.

In Exercises 39–56, use the most convenient method to solve the system of linear equations. State which method you used.

39. $\begin{cases} 6x - 5y = 0 \\ y = 6 \end{cases}$

40. $\begin{cases} -x + 2y = 2 \\ x = 4 \end{cases}$

41. $\begin{cases} -x + 4y = 4 \\ x + y = 6 \end{cases}$

42. $\begin{cases} -x + y = 4 \\ x + y = 4 \end{cases}$

43. $\begin{cases} x - y = 0 \\ x - 6y = 5 \end{cases}$

44. $\begin{cases} x + 2y = 2 \\ x - 4y = 20 \end{cases}$

45. $\begin{cases} 5x + 8y = 8 \\ x - 8y = 16 \end{cases}$

46. $\begin{cases} -7x + 9y = 9 \\ 2x + 9y = -18 \end{cases}$

47. $\begin{cases} 6x - 3y = 27 \\ -2x + y = -9 \end{cases}$

48. $\begin{cases} -5x + 2y = -4 \\ x - 6y = 4 \end{cases}$

49. $\begin{cases} \frac{1}{5}x + \frac{3}{2}y = 2 \\ 2x + 13y = 20 \end{cases}$

50. $\begin{cases} -\frac{1}{4}x + \frac{2}{3}y = 1 \\ 3x - 8y = 1 \end{cases}$

51. $\begin{cases} x + y = 0 \\ 2x + y = 0 \end{cases}$

52. $\begin{cases} 2x + 6y = 16 \\ 2x + 3y = 7 \end{cases}$

53. $\begin{cases} \frac{1}{3}x + \frac{4}{7}y = 3 \\ 2x + 3y = 15 \end{cases}$

54. $\begin{cases} \frac{1}{2}x - \frac{1}{3}y = 0 \\ 3x + 2y = 0 \end{cases}$

55. $\begin{cases} 1.2s + 4.2t = -1.7 \\ 3.0s - 1.8t = 1.9 \end{cases}$

56. $\begin{cases} 0.2u + 0.3v = 0.14 \\ 0.4u + 0.5v = 0.20 \end{cases}$

57. *Fuel Costs* You buy 2 gallons of gasoline for your lawn mower and 5 gallons of diesel fuel for your garden tractor. The total bill is $8.59. Diesel fuel costs $0.08 more per gallon than gasoline. Find the price per gallon of each type of fuel.

58. *Seed Mixture* Ten pounds of mixed birdseed sells for $6.97 per pound. The mixture is obtained from two kinds of birdseed, with one variety priced at $5.65 per pound and the other at $8.95 per pound. How many pounds of each variety of birdseed are used in the mixture?

8.3 Linear Systems in Three Variables

① Solve systems of linear equations using row-echelon form with back-substitution.

In Exercises 59–62, use back-substitution to solve the system of linear equations.

59. $\begin{cases} x = 3 \\ x + 2y = 7 \\ -3x - y + 4z = 9 \end{cases}$

60. $\begin{cases} 2x + 3y = 9 \\ 4x - 6z = 12 \\ y = 5 \end{cases}$

61. $\begin{cases} x + 2y = 6 \\ 3y = 9 \\ x + 2z = 12 \end{cases}$

62. $\begin{cases} 5x - 6z = -17 \\ 3x - 4y + 5z = -1 \\ 2z = -6 \end{cases}$

② Solve systems of linear equations using the method of Gaussian elimination.

In Exercises 63–66, solve the system of linear equations.

63. $\begin{cases} -x + y + 2z = 1 \\ 2x + 3y + z = -2 \\ 5x + 4y + 2z = 4 \end{cases}$

64. $\begin{cases} 2x + 3y + z = 10 \\ 2x - 3y - 3z = 22 \\ 4x - 2y + 3z = -2 \end{cases}$

65. $\begin{cases} x - y - z = 1 \\ -2x + y + 3z = -5 \\ 3x + 4y - z = 6 \end{cases}$

66. $\begin{cases} -3x + y + 2z = -13 \\ -x - y + z = 0 \\ 2x + 2y - 3z = -1 \end{cases}$

③ Solve application problems using elimination with back-substitution.

67. *Investment* An inheritance of $20,000 is divided among three investments yielding a total of $1780 in interest per year. The interest rates for the three investments are 7%, 9%, and 11%. The amounts invested at 9% and 11% are $3000 and $1000 less than the amount invested at 7%, respectively. Find the amount invested at each rate.

68. *Vertical Motion* Find the position equation

$$s = \frac{1}{2}at^2 + v_0t + s_0$$

for an object that has the indicated heights at the specified times.

$s = 192$ feet at $t = 1$ second

$s = 128$ feet at $t = 2$ seconds

$s = 80$ feet at $t = 3$ seconds

8.4 Matrices and Linear Systems

① Determine the order of matrices.

In Exercises 69–72, determine the order of the matrix.

69. $[4 \ \ -5]$

70. $\begin{bmatrix} 1 & 5 \\ 3 & -4 \end{bmatrix}$

71. $\begin{bmatrix} 5 & 7 & 9 \\ 11 & -12 & 0 \end{bmatrix}$

72. $\begin{bmatrix} 15 \\ 13 \\ -9 \end{bmatrix}$

② Form coefficient and augmented matrices and form linear systems from augmented matrices.

In Exercises 73 and 74, form the (a) coefficient matrix and (b) augmented matrix for the system of linear equations.

73. $\begin{cases} 3x - 2y = 12 \\ -x + y = -2 \end{cases}$

74. $\begin{cases} x + 2y + z = 4 \\ 3x \quad - z = 2 \\ -x + 5y - 2z = -6 \end{cases}$

In Exercises 75 and 76, write the system of linear equations represented by the matrix. (Use variables x, y, and z.)

75. $\begin{bmatrix} 4 & -1 & 0 & \vdots & 2 \\ 6 & 3 & 2 & \vdots & 1 \\ 0 & 1 & 4 & \vdots & 0 \end{bmatrix}$

76. $\begin{bmatrix} -15 & 2 & \vdots & -7 \\ 3 & 7 & \vdots & 8 \end{bmatrix}$

③ Perform elementary row operations to solve systems of linear equations.

In Exercises 77–80, use matrices and elementary row operations to solve the system.

77. $\begin{cases} 5x + 4y = 2 \\ -x + y = -22 \end{cases}$

78. $\begin{cases} 2x - 5y = 2 \\ 3x - 7y = 1 \end{cases}$

79. $\begin{cases} 0.2x - 0.1y = 0.07 \\ 0.4x - 0.5y = -0.01 \end{cases}$

80. $\begin{cases} 2x + y = 0.3 \\ 3x - y = -1.3 \end{cases}$

④ Use matrices and Gaussian elimination with back-substitution to solve systems of linear equations.

In Exercises 81–86, use matrices to solve the system of linear equations.

81. $\begin{cases} x + 2y + 6z = 4 \\ -3x + 2y - z = -4 \\ 4x + 2z = 16 \end{cases}$

82. $\begin{cases} -x + 3y - z = -4 \\ 2x + 6z = 14 \\ -3x - y + z = 10 \end{cases}$

83. $\begin{cases} 2x_1 + 3x_2 + 3x_3 = 3 \\ 6x_1 + 6x_2 + 12x_3 = 13 \\ 12x_1 + 9x_2 - x_3 = 2 \end{cases}$

84. $\begin{cases} -x_1 + 2x_2 + 3x_3 = 4 \\ 2x_1 - 4x_2 - x_3 = -13 \\ 3x_1 + 2x_2 - 4x_3 = -1 \end{cases}$

85. $\begin{cases} x - 4z = 17 \\ -2x + 4y + 3z = -14 \\ 5x - y + 2z = -3 \end{cases}$

86. $\begin{cases} 2x + 3y - 5z = 3 \\ -x + 2y = 3 \\ 3x + 5y + 2z = 15 \end{cases}$

8.5 Determinants and Linear Systems

① Find determinants of 2×2 matrices and 3×3 matrices.

In Exercises 87–92, find the determinant of the matrix using any appropriate method.

87. $\begin{bmatrix} 7 & 10 \\ 10 & 15 \end{bmatrix}$

88. $\begin{bmatrix} -3.4 & 1.2 \\ -5 & 2.5 \end{bmatrix}$

89. $\begin{bmatrix} 8 & 6 & 3 \\ 6 & 3 & 0 \\ 3 & 0 & 2 \end{bmatrix}$

90. $\begin{bmatrix} 7 & -1 & 10 \\ -3 & 0 & -2 \\ 12 & 1 & 1 \end{bmatrix}$

91. $\begin{bmatrix} 8 & 3 & 2 \\ 1 & -2 & 4 \\ 6 & 0 & 5 \end{bmatrix}$

92. $\begin{bmatrix} 4 & 0 & 10 \\ 0 & 10 & 0 \\ 10 & 0 & 34 \end{bmatrix}$

2 Use determinants and Cramer's Rule to solve systems of linear equations.

In Exercises 93–98, use Cramer's Rule to solve the system of linear equations. (If not possible, state the reason.)

93. $\begin{cases} 7x + 12y = 63 \\ 2x + 3y = 15 \end{cases}$ **94.** $\begin{cases} 12x + 42y = -17 \\ 30x - 18y = 19 \end{cases}$

95. $\begin{cases} 3x - 2y = 16 \\ 12x - 8y = -5 \end{cases}$ **96.** $\begin{cases} 4x + 24y = 20 \\ -3x + 12y = -5 \end{cases}$

97. $\begin{cases} -x + y + 2z = 1 \\ 2x + 3y + z = -2 \\ 5x + 4y + 2z = 4 \end{cases}$

98. $\begin{cases} 2x_1 + x_2 + 2x_3 = 4 \\ 2x_1 + 2x_2 = 5 \\ 2x_1 - x_2 + 6x_3 = 2 \end{cases}$

3 Use determinants to find areas of triangles, to test for collinear points, and to find equations of lines.

Area of a Triangle   **In Exercises 99–102, use a determinant to find the area of the triangle with the given vertices.**

99. $(1, 0), (5, 0), (5, 8)$

100. $(-4, 0), (4, 0), (0, 6)$

101. $(1, 2), (4, -5), (3, 2)$

102. $\left(\frac{3}{2}, 1\right), \left(4, -\frac{1}{2}\right), (4, 2)$

Collinear Points   **In Exercises 103 and 104, determine whether the points are collinear.**

103. $(1, 2), (5, 0), (10, -2)$

104. $(-3, 7), (1, 3), (5, -1)$

Equation of a Line   **In Exercises 105–108, use a determinant to find the equation of the line through the points.**

105. $(-4, 0), (4, 4)$ **106.** $\left(-\frac{5}{2}, 3\right), \left(\frac{7}{2}, 1\right)$

107. $(2, 5), (6, -1)$ **108.** $(-0.8, 0.2), (0.7, 3.2)$

8.6 Systems of Linear Inequalities

1 Solve systems of linear inequalities in two variables.

In Exercises 109–112, sketch a graph of the solution of the system of linear inequalities.

109. $\begin{cases} x + y < 5 \\ x > 2 \\ y \geq 0 \end{cases}$ **110.** $\begin{cases} 2x + y > 2 \\ x < 2 \\ y < 1 \end{cases}$

111. $\begin{cases} x + 2y \leq 160 \\ 3x + y \leq 180 \\ x \geq 0 \\ y \geq 0 \end{cases}$ **112.** $\begin{cases} 2x + 3y \leq 24 \\ 2x + y \leq 16 \\ x \geq 0 \\ y \geq 0 \end{cases}$

2 Use systems of linear inequalities to model and solve real-life problems.

113. *Fruit Distribution*   A Pennsylvania fruit grower has up to 1500 bushels of apples that are to be divided between markets in Harrisburg and Philadelphia. These two markets need at least 400 bushels and 600 bushels, respectively. Write a system of linear inequalities describing the various ways the fruit can be divided between the cities. Graph the system.

114. *Inventory Costs*   A warehouse operator has up to 24,000 square feet of floor space in which to store two products. Each unit of product I requires 20 square feet of floor space and costs $12 per day to store. Each unit of product II requires 30 square feet of floor space and costs $8 per day to store. The total storage cost per day cannot exceed $12,400. Write a system of linear inequalities describing the various ways the two products can be stored. Graph the system.

Chapter Test

Take this test as you would take a test in class. After you are done, check your work against the answers in the back of the book.

1. Determine whether each ordered pair is a solution of the system at the left.
   (a) $(3, -4)$
   (b) $\left(1, \frac{1}{2}\right)$

$$\begin{cases} 2x - 2y = 1 \\ -x + 2y = 0 \end{cases}$$

System for 1

In Exercises 2–10, use the indicated method to solve the system.

2. *Graphical:* $\begin{cases} x - 2y = -1 \\ 2x + 3y = 12 \end{cases}$

3. *Substitution:* $\begin{cases} 5x - y = 6 \\ 4x - 3y = -4 \end{cases}$

4. *Substitution:* $\begin{cases} 2x - 2y = -2 \\ 3x + y = 9 \end{cases}$

5. *Elimination:* $\begin{cases} 3x - 4y = -14 \\ -3x + y = 8 \end{cases}$

6. *Elimination:* $\begin{cases} x + 2y - 4z = 0 \\ 3x + y - 2z = 5 \\ 3x - y + 2z = 7 \end{cases}$

7. *Matrices:* $\begin{cases} x - 3z = -10 \\ -2y + 2z = 0 \\ x - 2y = -7 \end{cases}$

8. *Matrices:* $\begin{cases} x - 3y + z = -3 \\ 3x + 2y - 5z = 18 \\ y + z = -1 \end{cases}$

9. *Cramer's Rule:* $\begin{cases} 2x - 7y = 7 \\ 3x + 7y = 13 \end{cases}$

10. *Any Method:* $\begin{cases} 3x - 2y + z = 12 \\ x - 3y = 2 \\ -3x - 9z = -6 \end{cases}$

11. Evaluate the determinant of the matrix shown at the left.

12. Use a determinant to find the area of the triangle with vertices $(0, 0)$, $(5, 4)$, and $(6, 0)$.

13. Graph the solution of the system of linear inequalities.
$$\begin{cases} x - 2y > -3 \\ 2x + 3y \le 22 \\ y \ge 0 \end{cases}$$

14. A mid-size car costs $24,000 and costs an average of $0.28 per mile to maintain. A minivan costs $26,000 and costs an average of $0.24 per mile to maintain. Determine after how many miles the total costs of the two vehicles will be the same (each model is driven the same number of miles.)

15. An inheritance of $25,000 is divided among three investments yielding a total of $1275 in interest per year. The interest rates for the three investments are 4.5%, 5%, and 8%. The amounts invested at 5% and 8% are $4000 and $10,000 less than the amount invested at 4.5%, respectively. Find the amount invested at each rate.

16. Two types of tickets are sold for a concert. Reserved seat tickets cost $20 per ticket and floor seat tickets cost $30 per ticket. The promoter of the concert must sell at least 16,000 tickets, including at least 9000 reserved seat tickets and at least 4000 floor seat tickets. Moreover, the gross receipts must total at least $400,000 in order for the concert to be held. Write a system of linear inequalities describing the different numbers of tickets that can be sold. Graph the system.

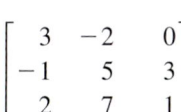

$$\begin{bmatrix} 3 & -2 & 0 \\ -1 & 5 & 3 \\ 2 & 7 & 1 \end{bmatrix}$$

Matrix for 11

Motivating the Chapter

Building a Greenhouse

You are building a greenhouse in the form of a half cylinder. The volume of the greenhouse is to be approximately 35,350 cubic feet.

See Section 9.1, Exercise 153.

a. The formula for the radius r (in feet) of a half cylinder is

$$r = \sqrt{\frac{2V}{\pi l}}$$

where V is the volume (in cubic feet) and l is the length (in feet). Find the radius of the greenhouse shown and round your result to the nearest whole number. Use this value of r in parts (b)–(d).

b. Beams for holding a sprinkler system are to be placed across the building. The formula for the height h at which the beams are to be placed is

$$h = \sqrt{r^2 - \left(\frac{a}{2}\right)^2}$$

where a is the length of the beam. Rewrite h as a function of a.

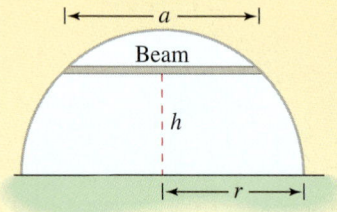

Cross Section of Greenhouse

c. The length of each beam is $a = 25$ feet. Find the height h at which the beams should be placed. Round your answer to two decimal places.

d. The equation from part (b) can be rewritten as

$$a = 2\sqrt{r^2 - h^2}.$$

The height is $h = 8$ feet. What is the length a of each beam? Round your answer to two decimal places.

See Section 9.5, Exercise 103.

e. The cost of building the greenhouse is estimated to be $25,000. The money to pay for the greenhouse was invested in an interest-bearing account 10 years ago at an annual percent rate of 7%. The amount of money earned can be found using the formula

$$r = \left(\frac{A}{P}\right)^{1/n} - 1$$

where r is the annual percent rate (in decimal form), A is the amount in the account after 10 years, P is the initial deposit, and n is the number of years. What initial deposit P would have generated enough money to cover the building cost of $25,000?

John A. Rizzo/Photodisc/Getty Images

9

Radicals and Complex Numbers

9.1 Radicals and Rational Exponents

Jeffrey Blackman/Index Stock

What You Should Learn

1. Determine the *n*th roots of numbers and evaluate radical expressions.
2. Use the rules of exponents to evaluate or simplify expressions with rational exponents.
3. Use a calculator to evaluate radical expressions.
4. Evaluate radical functions and find the domains of radical functions.

Why You Should Learn It

Algebraic equations often involve rational exponents. For instance, in Exercise 147 on page 561, you will use an equation involving a rational exponent to find the depreciation rate for a truck.

① Determine the *n*th roots of numbers and evaluate radical expressions.

Roots and Radicals

A **square root** of a number is defined as one of its two equal factors. For example, 5 is a square root of 25 because 5 is one of the two equal factors of 25. In a similar way, a **cube root** of a number is one of its three equal factors.

Number	Equal Factors	Root	Type
$9 = 3^2$	$3 \cdot 3$	3	Square root
$25 = (-5)^2$	$(-5)(-5)$	-5	Square root
$-27 = (-3)^3$	$(-3)(-3)(-3)$	-3	Cube root
$64 = 4^3$	$4 \cdot 4 \cdot 4$	4	Cube root
$16 = 2^4$	$2 \cdot 2 \cdot 2 \cdot 2$	2	Fourth root

Definition of *n*th Root of a Number

Let a and b be real numbers and let n be an integer such that $n \geq 2$. If

$$a = b^n$$

then b is an **nth root of a**. If $n = 2$, the root is a **square root.** If $n = 3$, the root is a **cube root.**

Some numbers have more than one *n*th root. For example, both 5 and -5 are square roots of 25. To avoid ambiguity about which root you are referring to, the **principal nth root** of a number is defined in terms of a **radical symbol** $\sqrt[n]{}$. So the *principal square root* of 25, written as $\sqrt{25}$, is the positive root, 5.

Principal *n*th Root of a Number

Let a be a real number that has at least one (real number) *n*th root. The **principal nth root of a** is the *n*th root that has the same sign as a, and it is denoted by the **radical symbol**

$$\sqrt[n]{a}. \qquad \text{Principal } n\text{th root}$$

The positive integer n is the **index** of the radical, and the number a is the **radicand**. If $n = 2$, omit the index and write $\sqrt{a}$ rather than $\sqrt[2]{a}$.

Study Tip

In the definition at the right, "the *n*th root that has the same sign as a" means that the principal *n*th root of a is positive if a is positive and negative if a is negative. For example, $\sqrt{4} = 2$ and $\sqrt[3]{-8} = -2$. When a negative root is needed, you must use the negative sign with the square root sign. For example, $-\sqrt{4} = -2$.

Example 1   Finding Roots of Numbers

Find each root.

a. $\sqrt{36}$ b. $-\sqrt{36}$ c. $\sqrt{-4}$ d. $\sqrt[3]{8}$ e. $\sqrt[3]{-8}$

Solution

a. $\sqrt{36} = 6$ because $6 \cdot 6 = 6^2 = 36$.

b. $-\sqrt{36} = -6$ because $6 \cdot 6 = 6^2 = 36$. So, $(-1)(\sqrt{36}) = (-1)(6) = -6$.

c. $\sqrt{-4}$ is not real because there is no real number that when multiplied by itself yields -4.

d. $\sqrt[3]{8} = 2$ because $2 \cdot 2 \cdot 2 = 2^3 = 8$.

e. $\sqrt[3]{-8} = -2$ because $(-2)(-2)(-2) = (-2)^3 = -8$.

Properties of nth Roots

Property	Example
1. If a is a positive real number and n is even, then a has exactly two (real) nth roots, which are denoted by $\sqrt[n]{a}$ and $-\sqrt[n]{a}$.	The two real square roots of 81 are $\sqrt{81} = 9$ and $-\sqrt{81} = -9$.
2. If a is any real number and n is odd, then a has only one (real) nth root, which is denoted by $\sqrt[n]{a}$.	$\sqrt[3]{27} = 3$ $\sqrt[3]{-64} = -4$
3. If a is a negative real number and n is even, then a has no (real) nth root.	$\sqrt{-64}$ is not a real number.

Integers such as 1, 4, 9, 16, 49, and 81 are called **perfect squares** because they have integer square roots. Similarly, integers such as 1, 8, 27, 64, and 125 are called **perfect cubes** because they have integer cube roots.

Example 2   Classifying Perfect nth Powers

> **Study Tip**
>
> The square roots of perfect squares are rational numbers, so $\sqrt{25}$, $\sqrt{49}$, and $\sqrt{100}$ are examples of rational numbers. However, square roots such as $\sqrt{5}$, $\sqrt{19}$, and $\sqrt{34}$ are irrational numbers. Similarly, $\sqrt[3]{27}$ and $\sqrt[4]{16}$ are rational numbers, whereas $\sqrt[3]{6}$ and $\sqrt[4]{21}$ are irrational numbers.

State whether each number is a perfect square, a perfect cube, both, or neither.

a. 81   b. -125 c. 64   d. 32

Solution

a. 81 is a perfect square because $9^2 = 81$. It is not a perfect cube.

b. -125 is a perfect cube because $(-5)^3 = -125$. It is not a perfect square.

c. 64 is a perfect square because $8^2 = 64$, and it is also a perfect cube because $4^3 = 64$.

d. 32 is not a perfect square or a perfect cube. (It is, however, a perfect fifth power, because $2^5 = 32$.)

Raising a number to the nth power and taking the principal nth root of a number can be thought of as *inverse* operations. Here are some examples.

$$\left(\sqrt{4}\right)^2 = (2)^2 = 4 \quad \text{and} \quad \sqrt{4} = \sqrt{2^2} = 2$$

$$\left(\sqrt[3]{27}\right)^3 = (3)^3 = 27 \quad \text{and} \quad \sqrt[3]{27} = \sqrt[3]{3^3} = 3$$

$$\left(\sqrt[4]{16}\right)^4 = (2)^4 = 16 \quad \text{and} \quad \sqrt[4]{16} = \sqrt[4]{2^4} = 2$$

$$\left(\sqrt[5]{-243}\right)^5 = (-3)^5 = -243 \quad \text{and} \quad \sqrt[5]{-243} = \sqrt[5]{(-3)^5} = -3$$

Inverse Properties of nth Powers and nth Roots

Let a be a real number, and let n be an integer such that $n \geq 2$.

Property	Example
1. If a has a principal nth root, then $\left(\sqrt[n]{a}\right)^n = a.$	$\left(\sqrt{5}\right)^2 = 5$
2. If n is odd, then $\sqrt[n]{a^n} = a.$	$\sqrt[3]{5^3} = 5$
If n is even, then $\sqrt[n]{a^n} = \lvert a \rvert.$	$\sqrt{(-5)^2} = \lvert -5 \rvert = 5$

Example 3   Evaluating Radical Expressions

Evaluate each radical expression.

a. $\sqrt[3]{4^3}$ **b.** $\sqrt[3]{(-2)^3}$ **c.** $\left(\sqrt{7}\right)^2$
d. $\sqrt{(-3)^2}$ **e.** $\sqrt{-3^2}$

Solution

a. Because the index of the radical is odd, you can write

$$\sqrt[3]{4^3} = 4.$$

b. Because the index of the radical is odd, you can write

$$\sqrt[3]{(-2)^3} = -2.$$

c. Using the inverse property of powers and roots, you can write

$$\left(\sqrt{7}\right)^2 = 7.$$

d. Because the index of the radical is even, you must include absolute value signs, and write

$$\sqrt{(-3)^2} = \lvert -3 \rvert = 3.$$

e. Because $\sqrt{-3^2} = \sqrt{-9}$ is an even root of a negative number, its value is not a real number.

Study Tip

In parts (d) and (e) of Example 3, notice that the two expressions inside the radical are different. In part (d), the negative sign is part of the base. In part (e), the negative sign is not part of the base.

② Use the rules of exponents to evaluate or simplify expressions with rational exponents.

Rational Exponents

So far in the text you have worked with algebraic expressions involving only integer exponents. Next you will see that algebraic expressions may also contain **rational exponents.**

Definition of Rational Exponents

Let a be a real number, and let n be an integer such that $n \geq 2$. If the principal nth root of a exists, then $a^{1/n}$ is defined as

$$a^{1/n} = \sqrt[n]{a}.$$

If m is a positive integer that has no common factor with n, then

$$a^{m/n} = (a^{1/n})^m = \left(\sqrt[n]{a}\right)^m \quad \text{and} \quad a^{m/n} = (a^m)^{1/n} = \sqrt[n]{a^m}.$$

It does not matter in which order the two operations are performed, provided the nth root exists. Here is an example.

$$8^{2/3} = \left(\sqrt[3]{8}\right)^2 = 2^2 = 4 \qquad \text{Cube root, then second power}$$
$$8^{2/3} = \sqrt[3]{8^2} = \sqrt[3]{64} = 4 \qquad \text{Second power, then cube root}$$

The rules of exponents that were listed in Section 5.1 also apply to rational exponents (provided the roots indicated by the denominators exist). These rules are listed below, with different examples.

Summary of Rules of Exponents

Let r and s be rational numbers, and let a and b be real numbers, variables, or algebraic expressions. (All denominators and bases are nonzero.)

	Product and Quotient Rules	*Example*
1.	$a^r \cdot a^s = a^{r+s}$	$4^{1/2}(4^{1/3}) = 4^{5/6}$
2.	$\dfrac{a^r}{a^s} = a^{r-s}$	$\dfrac{x^2}{x^{1/2}} = x^{2-(1/2)} = x^{3/2}$
	Power Rules	
3.	$(ab)^r = a^r \cdot b^r$	$(2x)^{1/2} = 2^{1/2}(x^{1/2})$
4.	$(a^r)^s = a^{rs}$	$(x^3)^{1/2} = x^{3/2}$
5.	$\left(\dfrac{a}{b}\right)^r = \dfrac{a^r}{b^r}$	$\left(\dfrac{x}{3}\right)^{2/3} = \dfrac{x^{2/3}}{3^{2/3}}$
	Zero and Negative Exponent Rules	
6.	$a^0 = 1$	$(3x)^0 = 1$
7.	$a^{-r} = \dfrac{1}{a^r}$	$4^{-3/2} = \dfrac{1}{4^{3/2}} = \dfrac{1}{(2)^3} = \dfrac{1}{8}$
8.	$\left(\dfrac{a}{b}\right)^{-r} = \left(\dfrac{b}{a}\right)^r$	$\left(\dfrac{x}{4}\right)^{-1/2} = \left(\dfrac{4}{x}\right)^{1/2} = \dfrac{2}{x^{1/2}}$

Study Tip

The numerator of a rational exponent denotes the *power* to which the base is raised, and the denominator denotes the *root* to be taken.

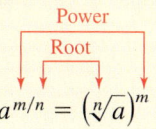

$$a^{m/n} = \left(\sqrt[n]{a}\right)^m$$

Technology: Discovery

Use a calculator to evaluate the expressions below.

$$\dfrac{3.4^{4.6}}{3.4^{3.1}} \quad \text{and} \quad 3.4^{1.5}$$

How are these two expressions related? Use your calculator to verify some of the other rules of exponents.

Example 4   Evaluating Expressions with Rational Exponents

Evaluate each expression.

a. $8^{4/3}$ **b.** $(4^2)^{3/2}$ **c.** $25^{-3/2}$

d. $\left(\dfrac{64}{125}\right)^{2/3}$ **e.** $-16^{1/2}$ **f.** $(-16)^{1/2}$

Solution

a. $8^{4/3} = (8^{1/3})^4 = \left(\sqrt[3]{8}\right)^4 = 2^4 = 16$ Root is 3. Power is 4.

b. $(4^2)^{3/2} = 4^{2 \cdot (3/2)} = 4^{6/2} = 4^3 = 64$ Root is 2. Power is 3.

c. $25^{-3/2} = \dfrac{1}{25^{3/2}} = \dfrac{1}{\left(\sqrt{25}\right)^3} = \dfrac{1}{5^3} = \dfrac{1}{125}$ Root is 2. Power is 3.

d. $\left(\dfrac{64}{125}\right)^{2/3} = \dfrac{64^{2/3}}{125^{2/3}} = \dfrac{\left(\sqrt[3]{64}\right)^2}{\left(\sqrt[3]{125}\right)^2} = \dfrac{4^2}{5^2} = \dfrac{16}{25}$ Root is 3. Power is 2.

e. $-16^{1/2} = -\sqrt{16} = -(4) = -4$ Root is 2. Power is 1.

f. $(-16)^{1/2} = \sqrt{-16}$ is not a real number.   Root is 2. Power is 1.

In parts (e) and (f) of Example 4, be sure that you see the distinction between the expressions $-16^{1/2}$ and $(-16)^{1/2}$.

Example 5   Using Rules of Exponents

Rewrite each expression using rational exponents.

a. $x\sqrt[4]{x^3}$ **b.** $\dfrac{\sqrt[3]{x^2}}{\sqrt{x^3}}$ **c.** $\sqrt[3]{x^2 y}$

Solution

a. $x\sqrt[4]{x^3} = x(x^{3/4}) = x^{1+(3/4)} = x^{7/4}$

b. $\dfrac{\sqrt[3]{x^2}}{\sqrt{x^3}} = \dfrac{x^{2/3}}{x^{3/2}} = x^{(2/3)-(3/2)} = x^{-5/6} = \dfrac{1}{x^{5/6}}$

c. $\sqrt[3]{x^2 y} = (x^2 y)^{1/3} = (x^2)^{1/3} y^{1/3} = x^{2/3} y^{1/3}$

Example 6   Using Rules of Exponents

Use rules of exponents to simplify each expression.

a. $\sqrt{\sqrt[3]{x}}$ **b.** $\dfrac{(2x-1)^{4/3}}{\sqrt[3]{2x-1}}$

Solution

a. $\sqrt{\sqrt[3]{x}} = \sqrt{x^{1/3}} = (x^{1/3})^{1/2} = x^{(1/3)(1/2)} = x^{1/6}$

b. $\dfrac{(2x-1)^{4/3}}{\sqrt[3]{2x-1}} = \dfrac{(2x-1)^{4/3}}{(2x-1)^{1/3}} = (2x-1)^{(4/3)-(1/3)} = (2x-1)^{3/3} = 2x-1$

③ Use a calculator to evaluate radical expressions.

Radicals and Calculators

There are two methods of evaluating radicals on most calculators. For square roots, you can use the *square root key* $\boxed{\checkmark}$ or $\boxed{\sqrt{x}}$. For other roots, you should first convert the radical to exponential form and then use the *exponential key* $\boxed{y^x}$ or $\boxed{\wedge}$.

Technology: Tip

Some calculators have cube root functions $\sqrt[3]{}$ or $\sqrt[3]{x}$ and xth root functions $\sqrt[x]{}$ or $\sqrt[x]{y}$ that can be used to evaluate roots other than square roots. Consult the user's guide of your calculator for specific keystrokes.

Example 7   Evaluating Roots with a Calculator

Evaluate each expression. Round the result to three decimal places.

a. $\sqrt{5}$ **b.** $\sqrt[5]{25}$ **c.** $\sqrt[3]{-4}$ **d.** $(-8)^{3/2}$

Solution

a. 5 $\boxed{\sqrt{x}}$ Scientific

   $\boxed{\checkmark}$ 5 $\boxed{\text{ENTER}}$ Graphing

   The display is 2.236067977. Rounded to three decimal places, $\sqrt{5} \approx 2.236$.

b. First rewrite the expression as $\sqrt[5]{25} = 25^{1/5}$. Then use one of the following keystroke sequences.

   25 $\boxed{y^x}$ $\boxed{(}$ 1 $\boxed{\div}$ 5 $\boxed{)}$ $\boxed{=}$ Scientific

   25 $\boxed{\wedge}$ $\boxed{(}$ 1 $\boxed{\div}$ 5 $\boxed{)}$ $\boxed{\text{ENTER}}$ Graphing

   The display is 1.903653939. Rounded to three decimal places, $\sqrt[5]{25} \approx 1.904$.

c. If your calculator does not have a cube root key, use the fact that

   $$\sqrt[3]{-4} = \sqrt[3]{(-1)(4)} = \sqrt[3]{-1}\sqrt[3]{4} = -\sqrt[3]{4} = -4^{1/3}$$

   and attach the negative sign of the radical as the last keystroke.

   4 $\boxed{+/-}$ $\boxed{y^x}$ $\boxed{(}$ 1 $\boxed{\div}$ 3 $\boxed{)}$ $\boxed{=}$ Scientific

   $\boxed{\sqrt[3]{}}$ $\boxed{(-)}$ 4 $\boxed{)}$ $\boxed{\text{ENTER}}$ Graphing

   The display is -1.587401052. Rounded to three decimal places, $\sqrt[3]{-4} \approx -1.587$.

d. 8 $\boxed{+/-}$ $\boxed{y^x}$ $\boxed{(}$ 3 $\boxed{\div}$ 2 $\boxed{)}$ $\boxed{=}$ Scientific

   $\boxed{(}$ $\boxed{(-)}$ 8 $\boxed{)}$ $\boxed{\wedge}$ $\boxed{(}$ 3 $\boxed{\div}$ 2 $\boxed{)}$ $\boxed{\text{ENTER}}$ Graphing

   The display should indicate an error because an even root of a negative number is not real.

---

④ Evaluate radical functions and find the domains of radical functions.

Radical Functions

A **radical function** is a function that contains a radical such as

$$f(x) = \sqrt{x} \quad \text{or} \quad g(x) = \sqrt[3]{x}.$$

When evaluating a radical function, note that the radical symbol is a symbol of grouping.

Consider the function $f(x) = x^{2/3}$.

a. What is the domain of the function?

b. Use your graphing calculator to graph each equation, in order.

$y_1 = x^{(2 \div 3)}$

$y_2 = (x^2)^{1/3}$ Power, then root

$y_3 = (x^{1/3})^2$ Root, then power

c. Are the graphs all the same? Are their domains all the same?

d. On your graphing calculator, which of the forms properly represent the function $f(x) = x^{m/n}$?

$y_1 = x^{(m \div n)}$

$y_2 = (x^m)^{1/n}$

$y_3 = (x^{1/n})^m$

e. Explain how the domains of $f(x) = x^{2/3}$ and $g(x) = x^{-2/3}$ differ.

Study Tip

In general, the domain of a radical function where the index n is even includes all real values for which the expression under the radical is greater than or equal to zero.

Example 8 Evaluating Radical Functions

Evaluate each radical function when $x = 4$.

a. $f(x) = \sqrt[3]{x - 31}$ **b.** $g(x) = \sqrt{16 - 3x}$

Solution

a. $f(4) = \sqrt[3]{4 - 31} = \sqrt[3]{-27} = -3$

b. $g(4) = \sqrt{16 - 3(4)} = \sqrt{16 - 12} = \sqrt{4} = 2$

The **domain** of the radical function $f(x) = \sqrt[n]{x}$ is the set of all real numbers such that x has a principal nth root.

Domain of a Radical Function

Let n be an integer that is greater than or equal to 2.

1. If n is odd, the domain of $f(x) = \sqrt[n]{x}$ is the set of all real numbers.

2. If n is even, the domain of $f(x) = \sqrt[n]{x}$ is the set of all nonnegative real numbers.

Example 9 Finding the Domains of Radical Functions

Describe the domain of each function.

a. $f(x) = \sqrt[3]{x}$ **b.** $f(x) = \sqrt{x^3}$

Solution

a. The domain of $f(x) = \sqrt[3]{x}$ is the set of all real numbers because for any real number x, the expression $\sqrt[3]{x}$ is a real number.

b. The domain of $f(x) = \sqrt{x^3}$ is the set of all nonnegative real numbers. For instance, 1 is in the domain but -1 is not because $\sqrt{(-1)^3} = \sqrt{-1}$ is not a real number.

Example 10 Finding the Domain of a Radical Function

Find the domain of $f(x) = \sqrt{2x - 1}$.

Solution

The domain of f consists of all x such that $2x - 1 \geq 0$. Using the methods described in Section 3.6, you can solve this inequality as follows.

$2x - 1 \geq 0$ Write original inequality.

$2x \geq 1$ Add 1 to each side.

$x \geq \frac{1}{2}$ Divide each side by 2.

So, the domain is the set of all real numbers x such that $x \geq \frac{1}{2}$.

9.1 Exercises

Developing Skills

In Exercises 1–8, find the root if it exists. See Example 1.

1. $\sqrt{64}$

2. $-\sqrt{100}$

3. $-\sqrt{49}$

4. $\sqrt{-25}$

5. $\sqrt[3]{-27}$

6. $\sqrt[3]{-64}$

7. $\sqrt{-1}$

8. $-\sqrt[3]{1}$

In Exercises 9–14, state whether the number is a perfect square, a perfect cube, or neither. See Example 2.

9. 49

10. -27

11. 1728

12. 964

13. 96

14. 225

In Exercises 15–18, determine whether the square root is a rational or irrational number.

15. $\sqrt{6}$

16. $\sqrt{\frac{9}{16}}$

17. $\sqrt{900}$

18. $\sqrt{72}$

In Exercises 19–48, evaluate the radical expression without using a calculator. If not possible, state the reason. See Example 3.

19. $\sqrt{8^2}$

20. $-\sqrt{10^2}$

21. $\sqrt{(-10)^2}$

22. $\sqrt{(-12)^2}$

23. $\sqrt{-9^2}$

24. $\sqrt{-12^2}$

25. $-\sqrt{\left(\frac{2}{3}\right)^2}$

26. $\sqrt{\left(\frac{3}{4}\right)^2}$

27. $\sqrt{-\left(\frac{3}{10}\right)^2}$

28. $\sqrt{\left(-\frac{3}{5}\right)^2}$

29. $\left(\sqrt{5}\right)^2$

30. $-\left(\sqrt{10}\right)^2$

31. $-\left(\sqrt{23}\right)^2$

32. $\left(-\sqrt{18}\right)^2$

33. $\sqrt[3]{5^3}$

34. $\sqrt[3]{(-7)^3}$

35. $\sqrt[3]{10^3}$

36. $\sqrt[3]{4^3}$

37. $-\sqrt[3]{(-6)^3}$

38. $-\sqrt[3]{9^3}$

39. $\sqrt[3]{\left(-\frac{1}{4}\right)^3}$

40. $-\sqrt[3]{\left(\frac{1}{5}\right)^3}$

41. $\left(\sqrt[3]{11}\right)^3$

42. $\left(\sqrt[3]{-6}\right)^3$

43. $\left(-\sqrt[3]{24}\right)^3$

44. $\left(\sqrt[3]{21}\right)^3$

45. $\sqrt[4]{3^4}$

46. $\sqrt[5]{(-2)^5}$

47. $-\sqrt[4]{-5^4}$

48. $-\sqrt[4]{2^4}$

In Exercises 49–52, fill in the missing description.

Radical Form	Rational Exponent Form
49. $\sqrt{16} = 4$	
50. $\sqrt[3]{27^2} = 9$	
51.	$125^{1/3} = 5$
52.	$256^{3/4} = 64$

In Exercises 53–68, evaluate without using a calculator. See Example 4.

53. $25^{1/2}$

54. $49^{1/2}$

55. $-36^{1/2}$

56. $-121^{1/2}$

57. $32^{-2/5}$

58. $81^{-3/4}$

59. $(-27)^{-2/3}$

60. $(-243)^{-3/5}$

61. $\left(\frac{8}{27}\right)^{2/3}$

62. $\left(\frac{256}{625}\right)^{1/4}$

63. $\left(\frac{121}{9}\right)^{-1/2}$

64. $\left(\frac{27}{1000}\right)^{-4/3}$

65. $(3^3)^{2/3}$

66. $(8^2)^{3/2}$

67. $-(4^4)^{3/4}$

68. $(-2^3)^{5/3}$

In Exercises 69–86, rewrite the expression using rational exponents. See Example 5.

69. $\sqrt{t}$

70. $\sqrt[3]{x}$

71. $x\sqrt[3]{x^6}$

72. $t\sqrt[5]{t^2}$

73. $u^2\sqrt[3]{u}$

74. $y\sqrt[4]{y^2}$

75. $\dfrac{\sqrt{x}}{\sqrt{x^3}}$

76. $\dfrac{\sqrt[3]{x^2}}{\sqrt[3]{x^4}}$

77. $\dfrac{\sqrt[4]{t}}{\sqrt{t^5}}$

78. $\dfrac{\sqrt[3]{x^4}}{\sqrt{x^3}}$

79. $\sqrt[3]{x^2} \cdot \sqrt[3]{x^7}$

80. $\sqrt[5]{z^3} \cdot \sqrt[5]{z^2}$

81. $\sqrt[4]{y^3} \cdot \sqrt[3]{y}$

82. $\sqrt[6]{x^5} \cdot \sqrt[3]{x^4}$

83. $\sqrt[4]{x^3y}$

84. $\sqrt[3]{u^4v^2}$

85. $z^2\sqrt{y^5z^4}$

86. $x^2\sqrt[3]{xy^4}$

In Exercises 87–108, simplify the expression. See Example 6.

87. $3^{1/4} \cdot 3^{3/4}$

88. $2^{2/5} \cdot 2^{3/5}$

89. $(2^{1/2})^{2/3}$

90. $(4^{1/3})^{9/4}$

91. $\dfrac{2^{1/5}}{2^{6/5}}$

92. $\dfrac{5^{-3/4}}{5}$

93. $(c^{3/2})^{1/3}$

94. $(k^{-1/3})^{3/2}$

95. $\dfrac{18y^{4/3}z^{-1/3}}{24y^{-2/3}z}$

96. $\dfrac{a^{3/4} \cdot a^{1/2}}{a^{5/2}}$

97. $(3x^{-1/3}y^{3/4})^2$

98. $(-2u^{3/5}v^{-1/5})^3$

99. $\left(\dfrac{x^{1/4}}{x^{1/6}}\right)^3$

100. $\left(\dfrac{3m^{1/6}n^{1/3}}{4n^{-2/3}}\right)^2$

101. $\sqrt{\sqrt[4]{y}}$

102. $\sqrt[3]{\sqrt{2x}}$

103. $\sqrt[4]{\sqrt{x^3}}$

104. $\sqrt[5]{\sqrt[3]{y^4}}$

105. $\dfrac{(x+y)^{3/4}}{\sqrt[4]{x+y}}$

106. $\dfrac{(a-b)^{1/3}}{\sqrt[3]{a-b}}$

107. $\dfrac{(3u-2v)^{2/3}}{\sqrt{(3u-2v)^3}}$

108. $\dfrac{\sqrt[4]{2x+y}}{(2x+y)^{3/2}}$

In Exercises 109–122, use a calculator to evaluate the expression. Round your answer to four decimal places. If not possible, state the reason. See Example 7.

109. $\sqrt{45}$

110. $\sqrt{-23}$

111. $315^{2/5}$

112. $962^{2/3}$

113. $1698^{-3/4}$

114. $382.5^{-3/2}$

115. $\sqrt[4]{212}$

116. $\sqrt[3]{-411}$

117. $\sqrt[3]{545^2}$

118. $\sqrt[5]{-35^3}$

119. $\dfrac{8 - \sqrt{35}}{2}$

120. $\dfrac{-5 + \sqrt{3215}}{10}$

121. $\dfrac{3 + \sqrt{17}}{9}$

122. $\dfrac{7 - \sqrt{241}}{12}$

In Exercises 123–128, evaluate the function as indicated, if possible, and simplify. See Example 8.

123. $f(x) = \sqrt{2x + 9}$
    (a) $f(0)$ (b) $f(8)$ (c) $f(-6)$ (d) $f(36)$

124. $g(x) = \sqrt{5x - 6}$
    (a) $g(0)$ (b) $g(2)$ (c) $g(30)$ (d) $g\left(\frac{7}{5}\right)$

125. $g(x) = \sqrt[3]{x + 1}$
    (a) $g(7)$ (b) $g(26)$ (c) $g(-9)$ (d) $g(-65)$

126. $f(x) = \sqrt[3]{2x - 1}$
    (a) $f(0)$ (b) $f(-62)$ (c) $f(-13)$ (d) $f(63)$

127. $f(x) = \sqrt[4]{x - 3}$
    (a) $f(19)$ (b) $f(1)$ (c) $f(84)$ (d) $f(4)$

128. $g(x) = \sqrt[4]{x + 1}$
    (a) $g(0)$ (b) $g(15)$ (c) $g(-82)$ (d) $g(80)$

In Exercises 129–138, describe the domain of the function. See Examples 9 and 10.

129. $f(x) = 3\sqrt{x}$

130. $h(x) = \sqrt[4]{x}$

131. $g(x) = \dfrac{2}{\sqrt[4]{x}}$

132. $g(x) = \dfrac{10}{\sqrt[3]{x}}$

133. $f(x) = \sqrt[3]{x^4}$

134. $f(x) = \sqrt{-x}$

135. $h(x) = \sqrt{3x + 7}$

136. $f(x) = \sqrt{8x - 1}$

137. $g(x) = \sqrt{4 - 9x}$

138. $g(x) = \sqrt{10 - 2x}$

In Exercises 139–142, describe the domain of the function algebraically. Use a graphing calculator to graph the function. Did the graphing calculator omit part of the domain? If so, complete the graph by hand.

139. $y = \dfrac{5}{\sqrt[4]{x^3}}$

140. $y = 4\sqrt[3]{x}$

141. $g(x) = 2x^{3/5}$

142. $h(x) = 5x^{2/3}$

In Exercises 143–146, perform the multiplication. Use a graphing calculator to confirm your result.

143. $x^{1/2}(2x - 3)$

144. $x^{4/3}(3x^2 - 4x + 5)$

145. $y^{-1/3}(y^{1/3} + 5y^{4/3})$

146. $(x^{1/2} - 3)(x^{1/2} + 3)$

Solving Problems

Mathematical Modeling In Exercises 147 and 148, use the formula for the *declining balances method*

$$r = 1 - \left(\dfrac{S}{C}\right)^{1/n}$$

to find the depreciation rate r. In the formula, n is the useful life of the item (in years), S is the salvage value (in dollars), and C is the original cost (in dollars).

147. A $75,000 truck depreciates over an eight-year period, as shown in the graph. Find r.

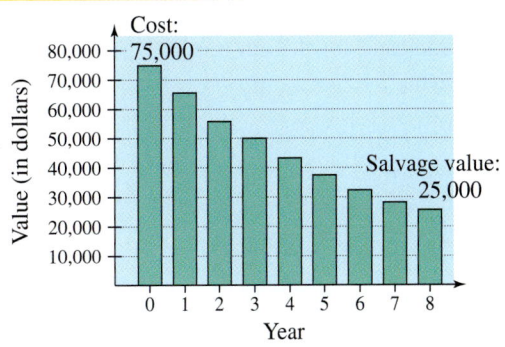

Figure for 147

148. A \$125,000 printing press depreciates over a 10-year period, as shown in the graph. Find r.

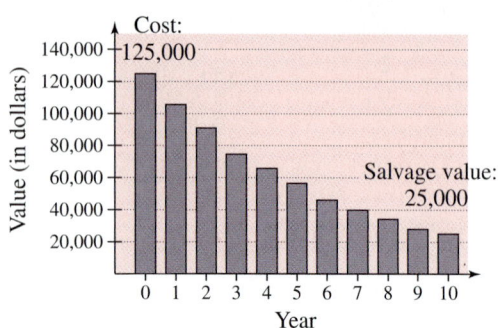

149. ▲ *Geometry*   Find the dimensions of a piece of carpet for a classroom with 529 square feet of floor space, assuming the floor is square.

150. ▲ *Geometry*   Find the dimensions of a square mirror with an area of 1024 square inches.

151. ▲ *Geometry*   The length D of a diagonal of a rectangular solid of length l, width w, and height h is represented by $D = \sqrt{l^2 + w^2 + h^2}$. Approximate to two decimal places the length of D of the solid shown in the figure.

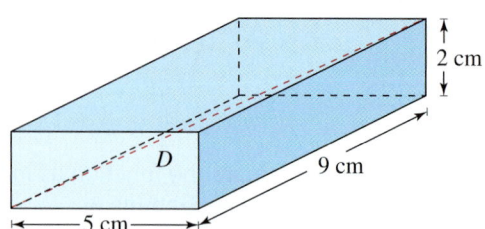

152. *Velocity*   A stream of water moving at a rate of v feet per second can carry particles of size $0.03\sqrt{v}$ inches.

(a) Find the particle size that can be carried by a stream flowing at the rate of $\frac{3}{4}$ foot per second. Round your answer to three decimal places.

(b) Find the particle size that can be carried by a stream flowing at the rate of $\frac{3}{16}$ foot per second. Round your answer to three decimal places.

Explaining Concepts

153. ⚡ Answer parts (a)–(d) of Motivating the Chapter on page 550.

154. *Writing* ✏ In your own words, define the nth root of a number.

155. *Writing* ✏ Define the *radicand* and the *index* of a radical.

156. *Writing* ✏ If n is even, what must be true about the radicand for the nth root to be a real number? Explain.

157. *Writing* ✏   Is it true that $\sqrt{2} = 1.414$? Explain.

158. Given a real number x, state the conditions on n for each of the following.

(a) $\sqrt[n]{x^n} = x$

(b) $\sqrt[n]{x^n} = |x|$

159. *Investigation*   Find all possible "last digits" of perfect squares. (For instance, the last digit of 81 is 1 and the last digit of 64 is 4.) Is it possible that 4,322,788,986 is a perfect square?

9.2 Simplifying Radical Expressions

Fundamental Photographs

What You Should Learn

1 Use the Product and Quotient Rules for Radicals to simplify radical expressions.

2 Use rationalization techniques to simplify radical expressions.

3 Use the Pythagorean Theorem in application problems.

Why You Should Learn It

Algebraic equations often involve radicals. For instance, in Exercise 74 on page 569, you will use a radical equation to find the period of a pendulum.

Simplifying Radicals

In this section, you will study ways to simplify radicals. For instance, the expression $\sqrt{12}$ can be simplified as

$$\sqrt{12} = \sqrt{4 \cdot 3} = \sqrt{4}\sqrt{3} = 2\sqrt{3}.$$

This rewritten form is based on the following rules for multiplying and dividing radicals.

1 Use the Product and Quotient Rules for Radicals to simplify radical expressions.

Product and Quotient Rules for Radicals

Let u and v be real numbers, variables, or algebraic expressions. If the nth roots of u and v are real, the following rules are true.

1. $\sqrt[n]{uv} = \sqrt[n]{u}\sqrt[n]{v}$ Product Rule for Radicals

2. $\sqrt[n]{\dfrac{u}{v}} = \dfrac{\sqrt[n]{u}}{\sqrt[n]{v}}, \quad v \neq 0$ Quotient Rule for Radicals

Study Tip

The Product and Quotient Rules for Radicals can be shown to be true by converting the radicals to exponential form and using the rules of exponents on page 555.

Using Rule 3

$\sqrt[n]{uv} = (uv)^{1/n}$

$= u^{1/n}v^{1/n}$

$= \sqrt[n]{u}\sqrt[n]{v}$

Using Rule 5

$\sqrt[n]{\dfrac{u}{v}} = \left(\dfrac{u}{v}\right)^{1/n}$

$= \dfrac{u^{1/n}}{v^{1/n}} = \dfrac{\sqrt[n]{u}}{\sqrt[n]{v}}$

You can use the Product Rule for Radicals to *simplify* square root expressions by finding the largest perfect square factor and removing it from the radical, as follows.

$$\sqrt{48} = \sqrt{16 \cdot 3} = \sqrt{16}\sqrt{3} = 4\sqrt{3}$$

This process is called **removing perfect square factors from the radical.**

Example 1   Removing Constant Factors from Radicals

Simplify each radical by removing as many factors as possible.

a. $\sqrt{75}$ **b.** $\sqrt{72}$ **c.** $\sqrt{162}$

Solution

a. $\sqrt{75} = \sqrt{25 \cdot 3} = \sqrt{25}\sqrt{3} = 5\sqrt{3}$ 25 is a perfect square factor of 75.

b. $\sqrt{72} = \sqrt{36 \cdot 2} = \sqrt{36}\sqrt{2} = 6\sqrt{2}$ 36 is a perfect square factor of 72.

c. $\sqrt{162} = \sqrt{81 \cdot 2} = \sqrt{81}\sqrt{2} = 9\sqrt{2}$ 81 is a perfect square factor of 162.

When removing *variable* factors from a square root radical, remember that it is not valid to write $\sqrt{x^2} = x$ *unless* you happen to know that x is nonnegative. Without knowing anything about x, the only way you can simplify $\sqrt{x^2}$ is to include absolute value signs when you remove x from the radical.

$$\sqrt{x^2} = |x| \qquad \text{Restricted by absolute value signs}$$

When simplifying the expression $\sqrt{x^3}$, it is not necessary to include absolute value signs because the domain does not include negative numbers.

$$\sqrt{x^3} = \sqrt{x^2(x)} = x\sqrt{x} \qquad \text{Restricted by domain of radical}$$

Example 2   Removing Variable Factors from Radicals

Simplify each radical expression.

a. $\sqrt{25x^2}$ **b.** $\sqrt{12x^3}, \quad x \geq 0$ **c.** $\sqrt{144x^4}$ **d.** $\sqrt{72x^3y^2}$

Solution

a. $\sqrt{25x^2} = \sqrt{5^2x^2} = \sqrt{5^2}\sqrt{x^2}$ 　　　　Product Rule for Radicals

$\qquad\qquad = 5|x|$ 　　　　$\sqrt{x^2} = |x|$

b. $\sqrt{12x^3} = \sqrt{2^2x^2(3x)} = \sqrt{2^2}\sqrt{x^2}\sqrt{3x}$ 　　　　Product Rule for Radicals

$\qquad\qquad = 2x\sqrt{3x}$ 　　　　$\sqrt{2^2}\sqrt{x^2} = 2x, \quad x \geq 0$

c. $\sqrt{144x^4} = \sqrt{12^2(x^2)^2} = \sqrt{12^2}\sqrt{(x^2)^2}$ 　　　　Product Rule for Radicals

$\qquad\qquad = 12x^2$ 　　　　$\sqrt{12^2}\sqrt{(x^2)^2} = 12|x^2| = 12x^2$

d. $\sqrt{72x^3y^2} = \sqrt{6^2x^2y^2} \cdot \sqrt{2x}$ 　　　　Product Rule for Radicals

$\qquad\qquad = \sqrt{6^2}\sqrt{x^2}\sqrt{y^2} \cdot \sqrt{2x}$ 　　　　Product Rule for Radicals

$\qquad\qquad = 6x|y|\sqrt{2x}$ 　　　　$\sqrt{6^2}\sqrt{x^2}\sqrt{y^2} = 6x|y|$

In the same way that perfect squares can be removed from square root radicals, perfect nth powers can be removed from nth root radicals.

Example 3   Removing Factors from Radicals

Simplify each radical expression.

a. $\sqrt[3]{40}$ **b.** $\sqrt[4]{x^5}, \quad x \geq 0$

Solution

a. $\sqrt[3]{40} = \sqrt[3]{8(5)} = \sqrt[3]{2^3} \cdot \sqrt[3]{5}$ 　　　　Product Rule for Radicals

$\qquad\qquad = 2\sqrt[3]{5}$ 　　　　$\sqrt[3]{2^3} = 2$

b. $\sqrt[4]{x^5} = \sqrt[4]{x^4(x)} = \sqrt[4]{x^4}\sqrt[4]{x}$ 　　　　Product Rule for Radicals

$\qquad\qquad = x\sqrt[4]{x}$ 　　　　$\sqrt[4]{x^4} = x, \quad x \geq 0$

Example 4 **Removing Factors from Radicals**

Simplify each radical expression.

a. $\sqrt[5]{486x^7}$ **b.** $\sqrt[3]{128x^3y^5}$

Solution

a. $\sqrt[5]{486x^7} = \sqrt[5]{243x^5(2x^2)}$

$\qquad\qquad = \sqrt[5]{3^5x^5} \cdot \sqrt[5]{2x^2}$ Product Rule for Radicals

$\qquad\qquad = 3x\,\sqrt[5]{2x^2}$ $\sqrt[5]{3^5}\,\sqrt[5]{x^5} = 3x$

b. $\sqrt[3]{128x^3y^5} = \sqrt[3]{64x^3y^3(2y^2)}$

$\qquad\qquad = \sqrt[3]{4^3x^3y^3} \cdot \sqrt[3]{2y^2}$ Product Rule for Radicals

$\qquad\qquad = 4xy\,\sqrt[3]{2y^2}$ $\sqrt[3]{4^3}\,\sqrt[3]{x^3}\,\sqrt[3]{y^3} = 4xy$

Example 5 **Removing Factors from Radicals**

Simplify each radical expression.

a. $\sqrt{\dfrac{81}{25}}$ **b.** $\dfrac{\sqrt{56x^2}}{\sqrt{8}}$

Solution

a. $\sqrt{\dfrac{81}{25}} = \dfrac{\sqrt{81}}{\sqrt{25}} = \dfrac{9}{5}$ Quotient Rule for Radicals

b. $\dfrac{\sqrt{56x^2}}{\sqrt{8}} = \sqrt{\dfrac{56x^2}{8}}$ Quotient Rule for Radicals

$\qquad\qquad = \sqrt{7x^2}$ Simplify.

$\qquad\qquad = \sqrt{7} \cdot \sqrt{x^2}$ Product Rule for Radicals

$\qquad\qquad = \sqrt{7}\,|x|$ $\sqrt{x^2} = |x|$

Example 6 **Removing Factors from Radicals**

Simplify the radical expression.

$$-\sqrt[3]{\dfrac{y^5}{27x^3}}$$

Solution

$$-\sqrt[3]{\dfrac{y^5}{27x^3}} = -\dfrac{\sqrt[3]{y^3y^2}}{\sqrt[3]{27x^3}}$$ Quotient Rule for Radicals

$$= -\dfrac{\sqrt[3]{y^3} \cdot \sqrt[3]{y^2}}{\sqrt[3]{27} \cdot \sqrt[3]{x^3}}$$ Product Rule for Radicals

$$= -\dfrac{y\,\sqrt[3]{y^2}}{3x}$$ Simplify.

2 Use rationalization techniques to simplify radical expressions.

Rationalization Techniques

Removing factors from radicals is only one of two techniques used to simplify radicals. Three conditions must be met in order for a radical expression to be in simplest form. These three conditions are summarized as follows.

> ## Simplifying Radical Expressions
>
> A radical expression is said to be in *simplest form* if all three of the statements below are true.
>
> 1. All possible nth powered factors have been removed from each radical.
>
> 2. No radical contains a fraction.
>
> 3. No denominator of a fraction contains a radical.

To meet the last two conditions, you can use a second technique for simplifying radical expressions called **rationalizing the denominator.** This involves multiplying both the numerator and denominator by a *rationalizing factor* that creates a perfect nth power in the denominator.

Study Tip

When rationalizing a denominator, remember that for square roots you want a perfect square in the denominator, for cube roots you want a perfect cube, and so on. For instance, to find the rationalizing factor needed to create a perfect square in the denominator of Example 7(c) you can write the prime factorization of 18.

$$18 = 2 \cdot 3 \cdot 3$$

$$= 2 \cdot 3^2$$

From its prime factorization, you can see that 3^2 is a square root factor of 18. You need one more factor of 2 to create a perfect square in the denominator:

$$2 \cdot (2 \cdot 3^2) = 2 \cdot 2 \cdot 3^2$$

$$= 2^2 \cdot 3^2$$

$$= 4 \cdot 9 = 36.$$

Example 7   Rationalizing the Denominator

Rationalize the denominator in each expression.

a. $\sqrt{\dfrac{3}{5}}$ **b.** $\dfrac{4}{\sqrt[3]{9}}$ **c.** $\dfrac{8}{3\sqrt{18}}$

Solution

a. $\sqrt{\dfrac{3}{5}} = \dfrac{\sqrt{3}}{\sqrt{5}} = \dfrac{\sqrt{3}}{\sqrt{5}} \cdot \dfrac{\sqrt{5}}{\sqrt{5}} = \dfrac{\sqrt{15}}{\sqrt{5^2}} = \dfrac{\sqrt{15}}{5}$ Multiply by $\sqrt{5}/\sqrt{5}$ to create a perfect square in the denominator.

b. $\dfrac{4}{\sqrt[3]{9}} = \dfrac{4}{\sqrt[3]{9}} \cdot \dfrac{\sqrt[3]{3}}{\sqrt[3]{3}} = \dfrac{4\sqrt[3]{3}}{\sqrt[3]{27}} = \dfrac{4\sqrt[3]{3}}{3}$ Multiply by $\sqrt[3]{3}/\sqrt[3]{3}$ to create a perfect cube in the denominator.

c. $\dfrac{8}{3\sqrt{18}} = \dfrac{8}{3\sqrt{18}} \cdot \dfrac{\sqrt{2}}{\sqrt{2}} = \dfrac{8\sqrt{2}}{3\sqrt{36}} = \dfrac{8\sqrt{2}}{3\sqrt{6^2}}$ Multiply by $\sqrt{2}/\sqrt{2}$ to create a perfect square in the denominator.

$$= \dfrac{8\sqrt{2}}{3(6)} = \dfrac{4\sqrt{2}}{9}$$

Example 8   Rationalizing the Denominator

a. $\sqrt{\dfrac{8x}{12y^5}} = \sqrt{\dfrac{(4)(2)x}{(4)(3)y^5}} = \sqrt{\dfrac{2x}{3y^5}} = \dfrac{\sqrt{2x}}{\sqrt{3y^5}} \cdot \dfrac{\sqrt{3y}}{\sqrt{3y}} = \dfrac{\sqrt{6xy}}{\sqrt{3^2y^6}} = \dfrac{\sqrt{6xy}}{3|y^3|}$

b. $\sqrt[3]{\dfrac{54x^6y^3}{5z^2}} = \dfrac{\sqrt[3]{(3^3)(2)(x^6)(y^3)}}{\sqrt[3]{5z^2}} \cdot \dfrac{\sqrt[3]{25z}}{\sqrt[3]{25z}} = \dfrac{3x^2y\sqrt[3]{50z}}{\sqrt[3]{5^3z^3}} = \dfrac{3x^2y\sqrt[3]{50z}}{5z}$

3 Use the Pythagorean Theorem in application problems.

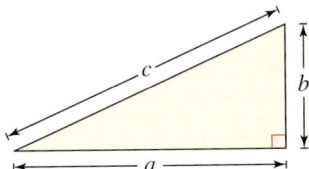

Figure 9.1

Applications of Radicals

Radicals commonly occur in applications involving right triangles. Recall that a right triangle is one that contains a right (or 90°) angle, as shown in Figure 9.1. The relationship among the three sides of a right triangle is described by the **Pythagorean Theorem,** which states that if a and b are the lengths of the legs and c is the length of the hypotenuse, then

$$c = \sqrt{a^2 + b^2} \text{ and } a = \sqrt{c^2 - b^2}. \qquad \text{Pythagorean Theorem: } a^2 + b^2 = c^2$$

Example 9   The Pythagorean Theorem

Find the length of the hypotenuse of the right triangle shown in Figure 9.2.

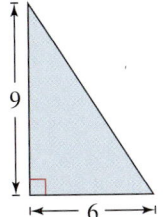

Figure 9.2

Solution

Because you know that $a = 6$ and $b = 9$, you can use the Pythagorean Theorem to find c as follows.

$$
\begin{aligned}
c &= \sqrt{a^2 + b^2} & & \text{Pythagorean Theorem} \\
&= \sqrt{6^2 + 9^2} & & \text{Substitute 6 for } a \text{ and 9 for } b. \\
&= \sqrt{117} & & \text{Simplify.} \\
&= \sqrt{9}\sqrt{13} & & \text{Product Rule for Radicals} \\
&= 3\sqrt{13} & & \text{Simplify.}
\end{aligned}
$$

Example 10   An Application of the Pythagorean Theorem

A softball diamond has the shape of a square with 60-foot sides, as shown in Figure 9.3. The catcher is 5 feet behind home plate. How far does the catcher have to throw to reach second base?

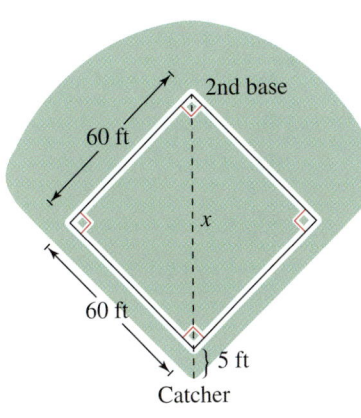

Figure 9.3

Solution

In Figure 9.3, let x be the hypotenuse of a right triangle with 60-foot sides. So, by the Pythagorean Theorem, you have the following.

$$
\begin{aligned}
x &= \sqrt{60^2 + 60^2} & & \text{Pythagorean Theorem} \\
&= \sqrt{7200} & & \text{Simplify.} \\
&= \sqrt{3600}\sqrt{2} & & \text{Product Rule for Radicals} \\
&= 60\sqrt{2} & & \text{Simplify.} \\
&\approx 84.9 \text{ feet} & & \text{Use a calculator.}
\end{aligned}
$$

So, the distance from home plate to second base is approximately 84.9 feet. Because the catcher is 5 feet behind home plate, the catcher must make a throw of

$$x + 5 \approx 84.9 + 5 = 89.9 \text{ feet.}$$

9.2 Exercises

Review   Concepts, Skills, and Problem Solving

Keep mathematically in shape by doing these exercises *before* the problems of this section.

Properties and Definitions

1. *Writing* ✏️   Explain how to determine the half-plane satisfying $x - y > -3$.

2. *Writing* ✏️   Describe the difference between the graphs of $3x + 4y \leq 4$ and $3x + 4y < 4$.

Factoring

In Exercises 3–8, factor the expression completely.

3. $-x^3 + 3x^2 - x + 3$ 4. $4t^2 - 169$

5. $x^2 - 3x + 2$ 6. $2x^2 + 5x - 7$

7. $11x^2 + 6x - 5$ 8. $4x^2 - 28x + 49$

Problem Solving

9. *Ticket Sales*   Twelve hundred tickets were sold for a theater production, and the receipts for the performance totaled $21,120. The tickets for adults and students sold for $20 and $12.50, respectively. How many of each kind of ticket were sold?

10. *Quality Control*   A quality control engineer for a buyer found two defective units in a sample of 75. At that rate, what is the expected number of defective units in a shipment of 10,000 units?

Developing Skills

In Exercises 1–18, simplify the radical. (Do not use a calculator.) See Example 1.

1. $\sqrt{20}$ 2. $\sqrt{27}$

3. $\sqrt{50}$ 4. $\sqrt{125}$

5. $\sqrt{96}$ 6. $\sqrt{84}$

7. $\sqrt{216}$ 8. $\sqrt{147}$

9. $\sqrt{1183}$ 10. $\sqrt{1176}$

11. $\sqrt{0.04}$ 12. $\sqrt{0.25}$

13. $\sqrt{0.0072}$ 14. $\sqrt{0.0027}$

15. $\sqrt{2.42}$ 16. $\sqrt{9.8}$

17. $\sqrt{\frac{13}{25}}$ 18. $\sqrt{\frac{15}{36}}$

In Exercises 19–52, simplify the radical expression. See Examples 2–6.

19. $\sqrt{9x^5}$ 20. $\sqrt{64x^3}$

21. $\sqrt{48y^4}$ 22. $\sqrt{32x}$

23. $\sqrt{117y^6}$ 24. $\sqrt{160x^8}$

25. $\sqrt{120x^2y^3}$ 26. $\sqrt{125u^4v^6}$

27. $\sqrt{192a^5b^7}$ 28. $\sqrt{363x^{10}y^9}$

29. $\sqrt[3]{48}$ 30. $\sqrt[3]{81}$

31. $\sqrt[3]{112}$ 32. $\sqrt[4]{112}$

33. $\sqrt[3]{40x^5}$ 34. $\sqrt[3]{54z^7}$

35. $\sqrt[4]{324y^6}$ 36. $\sqrt[5]{160x^8}$

37. $\sqrt[3]{x^4y^3}$ 38. $\sqrt[3]{a^5b^6}$

39. $\sqrt[4]{3x^4y^2}$ 40. $\sqrt[4]{128u^4v^7}$

41. $\sqrt[5]{32x^5y^6}$ 42. $\sqrt[3]{16x^4y^5}$

43. $\sqrt[3]{\frac{35}{64}}$ 44. $\sqrt[4]{\frac{5}{16}}$

45. $\sqrt[5]{\frac{32x^2}{y^5}}$ 46. $\sqrt[3]{\frac{16z^3}{y^6}}$

47. $\sqrt[3]{\frac{54a^4}{b^9}}$ 48. $\sqrt[4]{\frac{3u^2}{16v^8}}$

49. $\sqrt{\frac{32a^4}{b^2}}$ 50. $\sqrt{\frac{18x^2}{z^6}}$

51. $\sqrt[4]{(3x^2)^4}$ 52. $\sqrt[5]{96x^5}$

In Exercises 53–70, rationalize the denominator and simplify further, if possible. See Examples 7 and 8.

53. $\sqrt{\frac{1}{3}}$ 54. $\sqrt{\frac{1}{5}}$

55. $\frac{1}{\sqrt{7}}$ 56. $\frac{12}{\sqrt{3}}$

57. $\sqrt[4]{\dfrac{5}{4}}$

58. $\sqrt[3]{\dfrac{9}{25}}$

59. $\dfrac{6}{\sqrt[3]{32}}$

60. $\dfrac{10}{\sqrt[5]{16}}$

61. $\dfrac{1}{\sqrt{y}}$

62. $\sqrt{\dfrac{5}{c}}$

63. $\sqrt{\dfrac{4}{x}}$

64. $\sqrt{\dfrac{4}{x^3}}$

65. $\dfrac{1}{\sqrt{2x}}$

66. $\dfrac{5}{\sqrt{8x^5}}$

67. $\dfrac{6}{\sqrt{3b^3}}$

68. $\dfrac{1}{\sqrt{xy}}$

69. $\sqrt[3]{\dfrac{2x}{3y}}$

70. $\sqrt[3]{\dfrac{20x^2}{9y^2}}$

△ *Geometry*   In Exercises 71 and 72, find the length of the hypotenuse of the right triangle. See Example 9.

71.

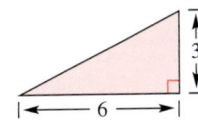

72.

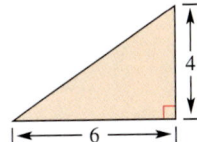

Solving Problems

73. *Frequency*   The frequency f in cycles per second of a vibrating string is given by

$$f = \dfrac{1}{100}\sqrt{\dfrac{400 \times 10^6}{5}}.$$

Use a calculator to approximate this number. (Round the result to two decimal places.)

74. *Period of a Pendulum*   The time t (in seconds) for a pendulum of length L (in feet) to go through one complete cycle (its period) is given by

$$t = 2\pi\sqrt{\dfrac{L}{32}}.$$

Find the period of a pendulum whose length is 4 feet. (Round your answer to two decimal places.)

75. △ *Geometry*   A ladder is to reach a window that is 26 feet high. The ladder is placed 10 feet from the base of the wall (see figure). How long must the ladder be?

Figure for 75

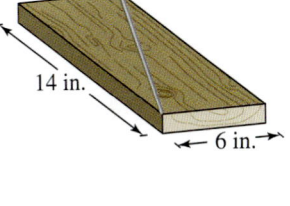

Figure for 76

76. △ *Geometry*   A string is attached to opposite corners of a piece of wood that is 6 inches wide and 14 inches long (see figure). How long must the string be?

77. *Investigation*   Enter any positive real number into your calculator and find its square root. Then repeatedly take the square root of the result.

$$\sqrt{x}, \ \sqrt{\sqrt{x}}, \ \sqrt{\sqrt{\sqrt{x}}}, \ \ldots$$

What real number does the display appear to be approaching?

Explaining Concepts

78. Give an example of multiplying two radicals.

79. *Writing*   Describe the three conditions that characterize a simplified radical expression.

80. *Writing*   Describe how you would simplify $1/\sqrt{3}$.

81. *Writing*   When is $\sqrt{x^2} \neq x$? Explain.

82. Square the real number $5/\sqrt{3}$ and note that the radical is eliminated from the denominator. Is this equivalent to rationalizing the denominator? Why or why not?

9.3 Adding and Subtracting Radical Expressions

Jose Luis Pelaez, Inc./Corbis

What You Should Learn

1. Use the Distributive Property to add and subtract like radicals.

2. Use radical expressions in application problems.

Why You Should Learn It

Radical expressions can be used to model and solve real-life problems. For instance, Example 6 on page 572 shows how to find a radical expression that models the total number of SAT and ACT tests taken.

Adding and Subtracting Radical Expressions

Two or more radical expressions are called **like radicals** if they have the same index and the same radicand. For instance, the expressions $\sqrt{2}$ and $3\sqrt{2}$ are like radicals, whereas the expressions $\sqrt{3}$ and $\sqrt[3]{3}$ are not. Two radical expressions that are like radicals can be added or subtracted by adding or subtracting their coefficients.

1. Use the Distributive Property to add and subtract like radicals.

Example 1   Combining Radical Expressions

Simplify each expression by combining like terms.

a. $\sqrt{7} + 5\sqrt{7} - 2\sqrt{7}$

b. $6\sqrt{x} - \sqrt[3]{4} - 5\sqrt{x} + 2\sqrt[3]{4}$

c. $3\sqrt[3]{x} + 2\sqrt[3]{x} + \sqrt{x} - 8\sqrt{x}$

Solution

a. $\sqrt{7} + 5\sqrt{7} - 2\sqrt{7} = (1 + 5 - 2)\sqrt{7}$ Distributive Property

$\qquad\qquad\qquad\qquad\quad = 4\sqrt{7}$ Simplify.

b. $6\sqrt{x} - \sqrt[3]{4} - 5\sqrt{x} + 2\sqrt[3]{4}$

$\quad = \left(6\sqrt{x} - 5\sqrt{x}\right) + \left(-\sqrt[3]{4} + 2\sqrt[3]{4}\right)$ Group like terms.

$\quad = (6 - 5)\sqrt{x} + (-1 + 2)\sqrt[3]{4}$ Distributive Property

$\quad = \sqrt{x} + \sqrt[3]{4}$ Simplify.

c. $3\sqrt[3]{x} + 2\sqrt[3]{x} + \sqrt{x} - 8\sqrt{x}$

$\quad = (3 + 2)\sqrt[3]{x} + (1 - 8)\sqrt{x}$ Distributive Property

$\quad = 5\sqrt[3]{x} - 7\sqrt{x}$ Simplify.

Study Tip

It is important to realize that the expression $\sqrt{a} + \sqrt{b}$ is not equal to $\sqrt{a + b}$. For instance, you may be tempted to add $\sqrt{6} + \sqrt{3}$ and get $\sqrt{9} = 3$. But remember, you cannot add unlike radicals. So, $\sqrt{6} + \sqrt{3}$ cannot be simplified further.

Before concluding that two radicals cannot be combined, you should first rewrite them in simplest form. This is illustrated in Examples 2 and 3.

Example 2   Simplifying Before Combining Radical Expressions

Simplify each expression by combining like terms.

a. $\sqrt{45x} + 3\sqrt{20x}$

b. $5\sqrt{x^3} - x\sqrt{4x}$

Solution

a. $\sqrt{45x} + 3\sqrt{20x} = 3\sqrt{5x} + 6\sqrt{5x}$ Simplify radicals.

$\qquad\qquad\qquad = 9\sqrt{5x}$ Combine like radicals.

b. $5\sqrt{x^3} - x\sqrt{4x} = 5x\sqrt{x} - 2x\sqrt{x}$ Simplify radicals.

$\qquad\qquad\qquad = 3x\sqrt{x}$ Combine like radicals.

Example 3   Simplifying Before Combining Radical Expressions

Simplify each expression by combining like terms.

a. $\sqrt[3]{54y^5} + 4\sqrt[3]{2y^2}$

b. $\sqrt[3]{6x^4} + \sqrt[3]{48x} - \sqrt[3]{162x^4}$

Solution

a. $\sqrt[3]{54y^5} + 4\sqrt[3]{2y^2} = 3y\sqrt[3]{2y^2} + 4\sqrt[3]{2y^2}$ Simplify radicals.

$\qquad\qquad\qquad = (3y + 4)\sqrt[3]{2y^2}$ Combine like radicals.

b. $\sqrt[3]{6x^4} + \sqrt[3]{48x} - \sqrt[3]{162x^4}$ Write original expression.

$\qquad = x\sqrt[3]{6x} + 2\sqrt[3]{6x} - 3x\sqrt[3]{6x}$ Simplify radicals.

$\qquad = (x + 2 - 3x)\sqrt[3]{6x}$ Distributive Property

$\qquad = (2 - 2x)\sqrt[3]{6x}$ Combine like terms.

In some instances, it may be necessary to rationalize denominators before combining radicals.

Example 4   Rationalizing Denominators Before Simplifying

$$\sqrt{7} - \frac{5}{\sqrt{7}} = \sqrt{7} - \left(\frac{5}{\sqrt{7}} \cdot \frac{\sqrt{7}}{\sqrt{7}}\right)$$ Multiply by $\sqrt{7}/\sqrt{7}$ to create a perfect square in the denominator.

$$= \sqrt{7} - \frac{5\sqrt{7}}{7}$$ Simplify.

$$= \left(1 - \frac{5}{7}\right)\sqrt{7}$$ Distributive Property

$$= \frac{2}{7}\sqrt{7}$$ Simplify.

2 Use radical expressions in application problems.

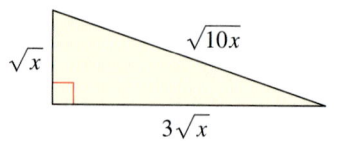

Figure 9.4

Applications

Example 5   Geometry: Perimeter of a Triangle

Write and simplify an expression for the perimeter of the triangle shown in Figure 9.4.

Solution

$$P = a + b + c$$ Formula for perimeter of a triangle

$$= \sqrt{x} + 3\sqrt{x} + \sqrt{10x}$$ Substitute.

$$= (1 + 3)\sqrt{x} + \sqrt{10x}$$ Distributive Property

$$= 4\sqrt{x} + \sqrt{10x}$$ Simplify.

Example 6   SAT and ACT Participants

The number S (in thousands) of SAT tests taken and the number A (in thousands) of ACT tests taken from 1996 to 2001 can be modeled by the equations

$$S = 496 + 239.7\sqrt{t}, \quad 6 \le t \le 11$$ SAT tests

$$A = 494 + 176.5\sqrt{t}, \quad 6 \le t \le 11$$ ACT tests

where t represents the year, with $t = 6$ corresponding to 1996. Find a radical expression that models the total number T (in thousands) of SAT and ACT tests taken from 1996 to 2001. Estimate the total number of SAT and ACT tests taken in 2000.   (Source: College Entrance Examination Board and ACT, Inc.)

Solution

The sum of the two models is as follows.

$$S + A = 496 + 239.7\sqrt{t} + 494 + 176.5\sqrt{t}$$

$$= \left(239.7\sqrt{t} + 176.5\sqrt{t}\right) + (496 + 494)$$

$$= 416.2\sqrt{t} + 990$$

So, the radical expression that models the total number of SAT and ACT tests taken is

$$T = S + A$$

$$= 416.2\sqrt{t} + 990.$$

Using this model, substitute $t = 10$ to estimate the total number of SAT and ACT tests taken in 2000.

$$T = 416.2\sqrt{10} + 990$$

$$\approx 2306$$

So, the total number of SAT and ACT tests taken in 2000 is approximately 2,306,000.

9.3 Exercises

Review   *Concepts, Skills, and Problem Solving*

Keep mathematically in shape by doing these exercises *before* the problems of this section.

Properties and Definitions

1. *Writing* ✎   Explain what is meant by a solution to a system of linear equations.

2. *Writing* ✎   Is it possible for a system of linear equations to have no solution? Explain.

3. *Writing* ✎   Is it possible for a system of linear equations to have infinitely many solutions? Explain.

4. *Writing* ✎   Is it possible for a system of linear equations to have exactly two solutions? Explain.

Solving Systems of Equations

In Exercises 5 and 6, sketch the graphs of the equations and approximate any solutions of the system of linear equations.

5. $\begin{cases} 3x + 2y = -4 \\ y = 3x + 7 \end{cases}$

6. $\begin{cases} 2x + 3y = 12 \\ 4x - y = 10 \end{cases}$

In Exercises 7 and 8, solve the system by the method of substitution.

7. $\begin{cases} x - 3y = -2 \\ 7y - 4x = 6 \end{cases}$

8. $\begin{cases} y = x + 2 \\ y - x = 8 \end{cases}$

In Exercises 9 and 10, solve the system by the method of elimination.

9. $\begin{cases} 1.5x - 3 = -2y \\ 3x + 4y = 6 \end{cases}$

10. $\begin{cases} x + 4y + 3z = 2 \\ 2x + y + z = 10 \\ -x + y + 2z = 8 \end{cases}$

Problem Solving

11. *Cost*   Two DVDs and one videocassette tape cost $72. One DVD and two videocassette tapes cost $57. What is the price of each item?

12. *Money*   A collection of $20, $5, and $1 bills totals $159. There are as many $20 bills as there are $5 and $1 bills combined. There are 14 bills in total. How many of each type of bill are there?

Developing Skills

In Exercises 1–46, combine the radical expressions, if possible. See Examples 1–3.

1. $3\sqrt{2} - \sqrt{2}$

2. $6\sqrt{5} - 2\sqrt{5}$

3. $4\sqrt{32} + 7\sqrt{32}$

4. $3\sqrt{7} + 2\sqrt{7}$

5. $8\sqrt{5} + 9\sqrt[3]{5}$

6. $12\sqrt{8} - 3\sqrt[3]{8}$

7. $9\sqrt[3]{5} - 6\sqrt[3]{5}$

8. $14\sqrt[5]{2} - 6\sqrt[5]{2}$

9. $4\sqrt[3]{y} + 9\sqrt[3]{y}$

10. $13\sqrt{x} + \sqrt{x}$

11. $15\sqrt[4]{s} - \sqrt[4]{s}$

12. $9\sqrt[4]{t} - 3\sqrt[4]{t}$

13. $8\sqrt{2} + 6\sqrt{2} - 5\sqrt{2}$

14. $2\sqrt{6} + 8\sqrt{6} - 3\sqrt{6}$

15. $\sqrt[4]{3} - 5\sqrt[4]{7} - 12\sqrt[4]{3}$

16. $9\sqrt[3]{17} + 7\sqrt[3]{2} - 4\sqrt[3]{17} + \sqrt[3]{2}$

17. $9\sqrt[3]{7} - \sqrt{3} + 4\sqrt[3]{7} + 2\sqrt{3}$

18. $5\sqrt{7} - 8\sqrt[4]{11} + \sqrt{7} + 9\sqrt[4]{11}$

19. $8\sqrt{27} - 3\sqrt{3}$

20. $9\sqrt{50} - 4\sqrt{2}$

21. $3\sqrt{45} + 7\sqrt{20}$

22. $5\sqrt{12} + 16\sqrt{27}$

23. $2\sqrt[3]{54} + 12\sqrt[3]{16}$

24. $4\sqrt[4]{48} - \sqrt[4]{243}$

25. $5\sqrt{9x} - 3\sqrt{x}$

26. $4\sqrt{y} + 2\sqrt{16y}$

27. $3\sqrt{x+1} + 10\sqrt{x+1}$

28. $4\sqrt{a-1} + \sqrt{a-1}$

29. $\sqrt{25y} + \sqrt{64y}$

30. $\sqrt[3]{16t^4} - \sqrt[3]{54t^4}$

31. $10\sqrt[3]{z} - \sqrt[3]{z^4}$

32. $5\sqrt[3]{24u^2} + 2\sqrt[3]{81u^5}$

33. $\sqrt{5a} + 2\sqrt{45a^3}$

34. $4\sqrt{3x^3} - \sqrt{12x}$

35. $\sqrt[3]{6x^4} + \sqrt[3]{48x}$

36. $\sqrt[3]{54x} - \sqrt[3]{2x^4}$

37. $\sqrt{9x-9} + \sqrt{x-1}$

38. $\sqrt{4y+12} + \sqrt{y+3}$

39. $\sqrt{x^3-x^2} + \sqrt{4x-4}$

40. $\sqrt{9x-9} - \sqrt{x^3-x^2}$

41. $2\sqrt[3]{a^4b^2} + 3a\sqrt[3]{ab^2}$

42. $3|y|\sqrt[4]{48x^5} - x\sqrt[4]{3x^5y^4}$

43. $\sqrt{4r^7s^5} + 3r^2\sqrt{r^3s^5} - 2rs\sqrt{r^5s^3}$

44. $x\sqrt[3]{27x^5y^2} - x^2\sqrt[3]{x^2y^2} + z\sqrt[3]{x^8y^2}$

45. $\sqrt[3]{128x^9y^{10}} - 2x^2y\sqrt[3]{16x^3y^7}$

46. $5\sqrt[3]{320x^5y^8} + 2x\sqrt[3]{135x^2y^8}$

In Exercises 47–56, perform the addition or subtraction and simplify your answer. See Example 4.

47. $\sqrt{5} - \dfrac{3}{\sqrt{5}}$

48. $\sqrt{10} + \dfrac{5}{\sqrt{10}}$

49. $\sqrt{20} - \sqrt{\dfrac{1}{5}}$

50. $\sqrt{\dfrac{1}{3}} + \sqrt{48}$

51. $\sqrt{12y} - \dfrac{y}{\sqrt{3y}}$

52. $\dfrac{x}{\sqrt{3x}} + \sqrt{27x}$

53. $\sqrt{2x} - \dfrac{3}{\sqrt{2x}}$

54. $\dfrac{8}{\sqrt{5x}} + \sqrt{5x}$

55. $\sqrt{7y^3} - \sqrt{\dfrac{9}{7y^3}}$

56. $\sqrt{\dfrac{4}{3x^3}} + \sqrt{3x^3}$

In Exercises 57–60, place the correct symbol ($<$, $>$, or $=$) between the numbers.

57. $\sqrt{7} + \sqrt{18}$ ___ $\sqrt{7+18}$

58. $\sqrt{10} - \sqrt{6}$ ___ $\sqrt{10-6}$

59. 5 ___ $\sqrt{3^2+2^2}$

60. 5 ___ $\sqrt{3^2+4^2}$

Solving Problems

⚠ *Geometry*   In Exercises 61–64, write and simplify an expression for the perimeter of the figure.

61.

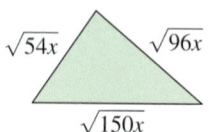

62.

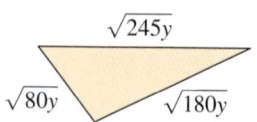

63.

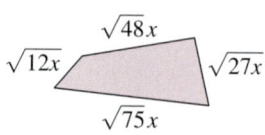

64.

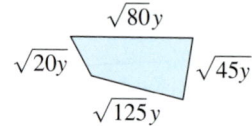

65. ▲ *Geometry* The foundation of a house is 40 feet long and 30 feet wide. The height of the attic is 5 feet (see figure).

(a) Use the Pythagorean Theorem to find the length of the hypotenuse of each of the two right triangles formed by the roof line. (Assume no overhang.)

(b) Use the result of part (a) to determine the total area of the roof.

66. ▲ *Geometry* The four corners are cut from a four-foot-by-eight-foot sheet of plywood, as shown in the figure. Find the perimeter of the remaining piece of plywood.

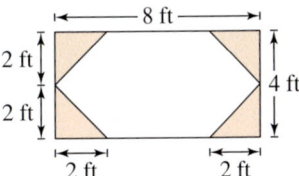

Explaining Concepts

67. *Writing*   Is $\sqrt{2} + \sqrt{18}$ in simplest form? Explain.

68. *Writing*   Explain what it means for two radical expressions to be like radicals.

69. Will the sum of two radicals always be a radical? Give an example to support your answer.

70. Will the difference of two radicals always be a radical? Give an example to support your answer.

71. You are an algebra instructor, and one of your students hands in the following work. Find and correct the errors, and discuss how you can help your student avoid such errors in the future.

(a) $7\sqrt{3} + 4\sqrt{2} = 11\sqrt{5}$

(b) $3\sqrt[3]{k} - 6\sqrt{k} = -3\sqrt{k}$

Mid-Chapter Quiz

Take this quiz as you would take a quiz in class. After you are done, check your work against the answers in the back of the book.

In Exercises 1–4, evaluate the expression.

1. $\sqrt{225}$

2. $\sqrt[4]{\frac{81}{16}}$

3. $64^{1/2}$

4. $(-27)^{2/3}$

In Exercises 5 and 6, evaluate the function as indicated, if possible, and simplify.

5. $f(x) = \sqrt{3x - 5}$

    (a) $f(0)$ (b) $f(2)$ (c) $f(10)$

6. $g(x) = \sqrt{9 - x}$

    (a) $g(0)$ (b) $g(5)$ (c) $g(10)$

In Exercises 7 and 8, describe the domain of the function.

7. $g(x) = \dfrac{12}{\sqrt[3]{x}}$

8. $h(x) = \sqrt{4x - 5}$

In Exercises 9–14, simplify the expression.

9. $\sqrt{27x^2}$

10. $\sqrt[4]{81x^6}$

11. $\sqrt{\dfrac{4u^3}{9}}$

12. $\sqrt[3]{\dfrac{16}{u^6}}$

13. $\sqrt{125x^3y^2z^4}$

14. $2a\sqrt[3]{16a^3b^5}$

In Exercises 15 and 16, rationalize the denominator and simplify further, if possible.

15. $\dfrac{24}{\sqrt{12}}$

16. $\dfrac{10}{\sqrt{5x}}$

In Exercises 17–22, combine the radical expressions, if possible.

17. $2\sqrt{3} - 4\sqrt{7} + \sqrt{3}$

18. $\sqrt{200y} - 3\sqrt{8y}$

19. $5\sqrt{12} + 2\sqrt{3} - \sqrt{75}$

20. $\sqrt{25x + 50} - \sqrt{x + 2}$

21. $6x\sqrt[3]{5x^2} + 2\sqrt[3]{40x^4}$

22. $3\sqrt{x^3y^4z^5} + 2xy^2\sqrt{xz^5} - xz^2\sqrt{xy^4z}$

23. The four corners are cut from an $8\frac{1}{2}$-inch-by-11-inch sheet of paper, as shown in the figure at the left. Find the perimeter of the remaining piece of paper.

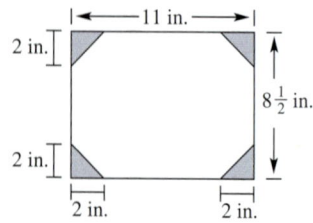

Figure for 23

9.4 Multiplying and Dividing Radical Expressions

Paul A. Souders/Corbis

What You Should Learn

1. Use the Distributive Property or the FOIL Method to multiply radical expressions.
2. Determine the products of conjugates.
3. Simplify quotients involving radicals by rationalizing the denominators.

Why You Should Learn It

Multiplication of radicals is often used in real-life applications. For instance, in Exercise 107 on page 583, you will multiply two radical expressions to find the area of the cross section of a wooden beam.

Multiplying Radical Expressions

You can multiply radical expressions by using the Distributive Property or the FOIL Method. In both procedures, you also make use of the Product Rule for Radicals from Section 9.2, which is given by $\sqrt[n]{uv} = \sqrt[n]{u}\,\sqrt[n]{v}$, where u and v are real numbers whose nth roots are also real numbers.

1. Use the Distributive Property or the FOIL Method to multiply radical expressions.

Example 1 Multiplying Radical Expressions

Find each product and simplify.

a. $\sqrt{6} \cdot \sqrt{3}$ **b.** $\sqrt[3]{5} \cdot \sqrt[3]{16}$

Solution

a. $\sqrt{6} \cdot \sqrt{3} = \sqrt{6 \cdot 3} = \sqrt{18} = \sqrt{9 \cdot 2} = 3\sqrt{2}$

b. $\sqrt[3]{5} \cdot \sqrt[3]{16} = \sqrt[3]{5 \cdot 16} = \sqrt[3]{80} = \sqrt[3]{8 \cdot 10} = 2\sqrt[3]{10}$

Example 2 Multiplying Radical Expressions

Find each product and simplify.

a. $\sqrt{3}\left(2 + \sqrt{5}\right)$ **b.** $\sqrt{2}\left(4 - \sqrt{8}\right)$ **c.** $\sqrt{6}\left(\sqrt{12} - \sqrt{3}\right)$

Solution

a. $\sqrt{3}\left(2 + \sqrt{5}\right) = 2\sqrt{3} + \sqrt{3}\sqrt{5}$ Distributive Property

$\qquad\qquad\quad = 2\sqrt{3} + \sqrt{15}$ Product Rule for Radicals

b. $\sqrt{2}\left(4 - \sqrt{8}\right) = 4\sqrt{2} - \sqrt{2}\sqrt{8}$ Distributive Property

$\qquad\qquad\quad = 4\sqrt{2} - \sqrt{16} = 4\sqrt{2} - 4$ Product Rule for Radicals

c. $\sqrt{6}\left(\sqrt{12} - \sqrt{3}\right) = \sqrt{6}\sqrt{12} - \sqrt{6}\sqrt{3}$ Distributive Property

$\qquad\qquad\quad = \sqrt{72} - \sqrt{18}$ Product Rule for Radicals

$\qquad\qquad\quad = 6\sqrt{2} - 3\sqrt{2} = 3\sqrt{2}$ Find perfect square factors.

In Example 2, the Distributive Property was used to multiply radical expressions. In Example 3, note how the FOIL Method can be used to multiply binomial radical expressions.

Example 3   Using the FOIL Method

$$
\overset{\text{F}\qquad\quad 0\qquad\quad\text{I}\qquad\text{L}}{}
$$

a. $\left(2\sqrt{7} - 4\right)\left(\sqrt{7} + 1\right) = 2\left(\sqrt{7}\right)^2 + 2\sqrt{7} - 4\sqrt{7} - 4$ FOIL Method

$$= 2(7) + (2 - 4)\sqrt{7} - 4$$ Combine like radicals.

$$= 10 - 2\sqrt{7}$$ Combine like terms.

b. $\left(3 - \sqrt{x}\right)\left(1 + \sqrt{x}\right) = 3 + 3\sqrt{x} - \sqrt{x} - \left(\sqrt{x}\right)^2$ FOIL Method

$$= 3 + 2\sqrt{x} - x, \quad x \geq 0$$ Combine like radicals.

---

2 Determine the products of conjugates.

Conjugates

The expressions $3 + \sqrt{6}$ and $3 - \sqrt{6}$ are called **conjugates** of each other. Notice that they differ only in the sign between the terms. The product of two conjugates is the difference of two squares, which is given by the special product formula $(a + b)(a - b) = a^2 - b^2$. Here are some other examples.

Expression	*Conjugate*	*Product*
$1 - \sqrt{3}$	$1 + \sqrt{3}$	$(1)^2 - \left(\sqrt{3}\right)^2 = 1 - 3 = -2$
$\sqrt{5} + \sqrt{2}$	$\sqrt{5} - \sqrt{2}$	$\left(\sqrt{5}\right)^2 - \left(\sqrt{2}\right)^2 = 5 - 2 = 3$
$\sqrt{10} - 3$	$\sqrt{10} + 3$	$\left(\sqrt{10}\right)^2 - (3)^2 = 10 - 9 = 1$
$\sqrt{x} + 2$	$\sqrt{x} - 2$	$\left(\sqrt{x}\right)^2 - (2)^2 = x - 4, x \geq 0$

Example 4   Multiplying Conjugates

Find the conjugate of the expression and multiply the expression by its conjugate.

a. $2 - \sqrt{5}$ **b.** $\sqrt{3} + \sqrt{x}$

Solution

a. The conjugate of $2 - \sqrt{5}$ is $2 + \sqrt{5}$.

$$\left(2 - \sqrt{5}\right)\left(2 + \sqrt{5}\right) = 2^2 - \left(\sqrt{5}\right)^2$$ Special product formula

$$= 4 - 5 = -1$$ Simplify.

b. The conjugate of $\sqrt{3} + \sqrt{x}$ is $\sqrt{3} - \sqrt{x}$.

$$\left(\sqrt{3} + \sqrt{x}\right)\left(\sqrt{3} - \sqrt{x}\right) = \left(\sqrt{3}\right)^2 - \left(\sqrt{x}\right)^2$$ Special product formula

$$= 3 - x, \quad x \geq 0$$ Simplify.

③ Simplify quotients involving radicals by rationalizing the denominators.

Dividing Radical Expressions

To simplify a *quotient* involving radicals, you rationalize the denominator. For single-term denominators, you can use the rationalization process described in Section 9.2. To rationalize a denominator involving two terms, multiply both the numerator and denominator by the *conjugate of the denominator.*

Example 5 Simplifying Quotients Involving Radicals

Simplify each expression.

a. $\dfrac{\sqrt{3}}{1 - \sqrt{5}}$ **b.** $\dfrac{4}{2 - \sqrt{3}}$

Solution

a. $\dfrac{\sqrt{3}}{1 - \sqrt{5}} = \dfrac{\sqrt{3}}{1 - \sqrt{5}} \cdot \dfrac{1 + \sqrt{5}}{1 + \sqrt{5}}$ Multiply numerator and denominator by conjugate of denominator.

$= \dfrac{\sqrt{3}\left(1 + \sqrt{5}\right)}{1^2 - \left(\sqrt{5}\right)^2}$ Special product formula

$= \dfrac{\sqrt{3} + \sqrt{15}}{1 - 5}$ Simplify.

$= -\dfrac{\sqrt{3} + \sqrt{15}}{4}$ Simplify.

b. $\dfrac{4}{2 - \sqrt{3}} = \dfrac{4}{2 - \sqrt{3}} \cdot \dfrac{2 + \sqrt{3}}{2 + \sqrt{3}}$ Multiply numerator and denominator by conjugate of denominator.

$= \dfrac{4\left(2 + \sqrt{3}\right)}{2^2 - \left(\sqrt{3}\right)^2}$ Special product formula

$= \dfrac{8 + 4\sqrt{3}}{4 - 3}$ Simplify.

$= 8 + 4\sqrt{3}$ Simplify.

Example 6 Simplifying a Quotient Involving Radicals

$\dfrac{5\sqrt{2}}{\sqrt{7} + \sqrt{2}} = \dfrac{5\sqrt{2}}{\sqrt{7} + \sqrt{2}} \cdot \dfrac{\sqrt{7} - \sqrt{2}}{\sqrt{7} - \sqrt{2}}$ Multiply numerator and denominator by conjugate of denominator.

$= \dfrac{5\left(\sqrt{14} - \sqrt{4}\right)}{\left(\sqrt{7}\right)^2 - \left(\sqrt{2}\right)^2}$ Special product formula

$= \dfrac{5\left(\sqrt{14} - 2\right)}{7 - 2}$ Simplify.

$= \dfrac{5\left(\sqrt{14} - 2\right)}{5}$ Divide out common factor.

$= \sqrt{14} - 2$ Simplest form

Example 7   Dividing Radical Expressions

Perform each division and simplify.

a. $\dfrac{6}{\sqrt{x} - 2}$

b. $\dfrac{2 - \sqrt{3}}{\sqrt{6} + \sqrt{2}}$

Solution

a. $\dfrac{6}{\sqrt{x} - 2} = \dfrac{6}{\sqrt{x} - 2} \cdot \dfrac{\sqrt{x} + 2}{\sqrt{x} + 2}$ Multiply numerator and denominator by conjugate of denominator.

$= \dfrac{6\left(\sqrt{x} + 2\right)}{\left(\sqrt{x}\right)^2 - 2^2}$ Special product formula

$= \dfrac{6\sqrt{x} + 12}{x - 4}$ Simplify.

b. $\dfrac{2 - \sqrt{3}}{\sqrt{6} + \sqrt{2}} = \dfrac{2 - \sqrt{3}}{\sqrt{6} + \sqrt{2}} \cdot \dfrac{\sqrt{6} - \sqrt{2}}{\sqrt{6} - \sqrt{2}}$ Multiply numerator and denominator by conjugate of denominator.

$= \dfrac{2\sqrt{6} - 2\sqrt{2} - \sqrt{18} + \sqrt{6}}{\left(\sqrt{6}\right)^2 - \left(\sqrt{2}\right)^2}$ FOIL Method and special product formula

$= \dfrac{3\sqrt{6} - 2\sqrt{2} - 3\sqrt{2}}{6 - 2}$ Simplify.

$= \dfrac{3\sqrt{6} - 5\sqrt{2}}{4}$ Simplify.

Example 8   Dividing Radical Expressions

Perform the division and simplify.

$$\dfrac{1}{\sqrt{x} - \sqrt{x + 1}}$$

Solution

$\dfrac{1}{\sqrt{x} - \sqrt{x + 1}} = \dfrac{1}{\sqrt{x} - \sqrt{x + 1}} \cdot \dfrac{\sqrt{x} + \sqrt{x + 1}}{\sqrt{x} + \sqrt{x + 1}}$ Multiply numerator and denominator by conjugate of denominator.

$= \dfrac{\sqrt{x} + \sqrt{x + 1}}{\left(\sqrt{x}\right)^2 - \left(\sqrt{x + 1}\right)^2}$ Special product formula

$= \dfrac{\sqrt{x} + \sqrt{x + 1}}{x - (x + 1)}$ Simplify.

$= \dfrac{\sqrt{x} + \sqrt{x + 1}}{-1}$ Combine like terms.

$= -\sqrt{x} - \sqrt{x + 1}$ Simplify.

9.4 Exercises

Review *Concepts, Skills, and Problem Solving*

Keep mathematically in shape by doing these exercises *before* the problems of this section.

Properties and Definitions

In Exercises 1–4, use $x^2 + bx + c = (x + m)(x + n)$.

1. $mn =$

2. If $c > 0$, then what must be true about the signs of m and n?

3. If $c < 0$, then what must be true about the signs of m and n?

4. If m and n have like signs, then $m + n =$.

Equations of Lines

In Exercises 5–10, find an equation of the line through the two points.

5. $(-1, -2), (3, 6)$

6. $(1, 5), (6, 0)$

7. $(6, 3), (10, 3)$

8. $(4, -2), (4, 5)$

9. $\left(\frac{4}{3}, 8\right), (5, 6)$

10. $(7, 4), (10, 1)$

Models

In Exercises 11 and 12, translate the phrase into an algebraic expression.

11. The time to travel 360 miles if the average speed is r miles per hour

12. The perimeter of a rectangle of length L and width $L/3$

Developing Skills

In Exercises 1–46, multiply and simplify. See Examples 1–3.

1. $\sqrt{2} \cdot \sqrt{8}$

2. $\sqrt{6} \cdot \sqrt{18}$

3. $\sqrt{3} \cdot \sqrt{6}$

4. $\sqrt{5} \cdot \sqrt{10}$

5. $\sqrt[3]{12} \cdot \sqrt[3]{6}$

6. $\sqrt[3]{9} \cdot \sqrt[3]{9}$

7. $\sqrt[4]{8} \cdot \sqrt[4]{6}$

8. $\sqrt[4]{54} \cdot \sqrt[4]{3}$

9. $\sqrt{7}\left(3 - \sqrt{7}\right)$

10. $\sqrt{3}\left(4 + \sqrt{3}\right)$

11. $\sqrt{2}\left(\sqrt{20} + 8\right)$

12. $\sqrt{7}\left(\sqrt{14} + 3\right)$

13. $\sqrt{6}\left(\sqrt{12} - \sqrt{3}\right)$

14. $\sqrt{10}\left(\sqrt{5} + \sqrt{6}\right)$

15. $4\sqrt{2}\left(\sqrt{3} - \sqrt{5}\right)$

16. $3\sqrt{5}\left(\sqrt{5} - \sqrt{2}\right)$

17. $\sqrt{y}\left(\sqrt{y} + 4\right)$

18. $\sqrt{x}\left(5 - \sqrt{x}\right)$

19. $\sqrt{a}\left(4 - \sqrt{a}\right)$

20. $\sqrt{z}\left(\sqrt{z} + 5\right)$

21. $\sqrt[3]{4}\left(\sqrt[3]{2} - 7\right)$

22. $\sqrt[3]{9}\left(\sqrt[3]{3} + 2\right)$

23. $\left(\sqrt{3} + 2\right)\left(\sqrt{3} - 2\right)$

24. $\left(3 - \sqrt{5}\right)\left(3 + \sqrt{5}\right)$

25. $\left(\sqrt{5} + 3\right)\left(\sqrt{3} - 5\right)$

26. $\left(\sqrt{7} + 6\right)\left(\sqrt{2} + 6\right)$

27. $\left(\sqrt{20} + 2\right)^2$

28. $\left(4 - \sqrt{20}\right)^2$

29. $\left(\sqrt[3]{6} - 3\right)\left(\sqrt[3]{4} + 3\right)$

30. $\left(\sqrt[3]{9} + 5\right)\left(\sqrt[3]{5} - 5\right)$

31. $\left(10 + \sqrt{2x}\right)^2$

32. $\left(5 - \sqrt{3v}\right)^2$

33. $\left(9\sqrt{x} + 2\right)\left(5\sqrt{x} - 3\right)$

34. $\left(16\sqrt{u} - 3\right)\left(\sqrt{u} - 1\right)$

35. $\left(3\sqrt{x} - 5\right)\left(3\sqrt{x} + 5\right)$

36. $\left(7 - 3\sqrt{3t}\right)\left(7 + 3\sqrt{3t}\right)$

37. $\left(\sqrt[3]{2x} + 5\right)^2$

38. $\left(\sqrt[3]{3x} - 4\right)^2$

39. $\left(\sqrt[3]{y} + 2\right)\left(\sqrt[3]{y^2} - 5\right)$

40. $\left(\sqrt[3]{2y} + 10\right)\left(\sqrt[3]{4y^2} - 10\right)$

41. $\left(\sqrt[3]{t} + 1\right)\left(\sqrt[3]{t^2} + 4\sqrt[3]{t} - 3\right)$

42. $\left(\sqrt{x} - 2\right)\left(\sqrt{x^3} - 2\sqrt{x^2} + 1\right)$

43. $\sqrt{x^3y^4}\left(2\sqrt{xy^2} - \sqrt{x^3y}\right)$

44. $3\sqrt{xy^3}\left(\sqrt{x^3y} + 2\sqrt{xy^2}\right)$

45. $2\sqrt[3]{x^4y^5}\left(\sqrt[3]{8x^{12}y^4} + \sqrt[3]{16xy^9}\right)$

46. $\sqrt[4]{8x^3y^5}\left(\sqrt[4]{4x^5y^7} - \sqrt[4]{3x^7y^6}\right)$

In Exercises 47–52, complete the statement.

47. $5x\sqrt{3} + 15\sqrt{3} = 5\sqrt{3}($ ▒ $)$

48. $x\sqrt{7} - x^2\sqrt{7} = x\sqrt{7}($ ▒ $)$

49. $4\sqrt{12} - 2x\sqrt{27} = 2\sqrt{3}($ ▒ $)$

50. $5\sqrt{50} + 10y\sqrt{8} = 5\sqrt{2}($ ▒ $)$

51. $6u^2 + \sqrt{18u^3} = 3u($ ▒ $)$

52. $12s^3 - \sqrt{32s^4} = 4s^2($ ▒ $)$

In Exercises 53–66, find the conjugate of the expression. Then multiply the expression by its conjugate and simplify. See Example 4.

53. $2 + \sqrt{5}$

54. $\sqrt{2} - 9$

55. $\sqrt{11} - \sqrt{3}$

56. $\sqrt{10} + \sqrt{7}$

57. $\sqrt{15} + 3$

58. $\sqrt{11} + 3$

59. $\sqrt{x} - 3$

60. $\sqrt{t} + 7$

61. $\sqrt{2u} - \sqrt{3}$

62. $\sqrt{5a} + \sqrt{2}$

63. $2\sqrt{2} + \sqrt{4}$

64. $4\sqrt{3} + \sqrt{2}$

65. $\sqrt{x} + \sqrt{y}$

66. $3\sqrt{u} + \sqrt{3v}$

In Exercises 67–70, evaluate the function as indicated and simplify.

67. $f(x) = x^2 - 6x + 1$

(a) $f\left(2 - \sqrt{3}\right)$ (b) $f\left(3 - 2\sqrt{2}\right)$

68. $g(x) = x^2 + 8x + 11$

(a) $g\left(-4 + \sqrt{5}\right)$ (b) $g\left(-4\sqrt{2}\right)$

69. $f(x) = x^2 - 2x - 1$

(a) $f\left(1 + \sqrt{2}\right)$ (b) $f\left(\sqrt{4}\right)$

70. $g(x) = x^2 - 4x + 1$

(a) $g\left(1 + \sqrt{5}\right)$ (b) $g\left(2 - \sqrt{3}\right)$

In Exercises 71–94, rationalize the denominator of the expression and simplify. See Examples 5–8.

71. $\dfrac{6}{\sqrt{11} - 2}$

72. $\dfrac{8}{\sqrt{7} + 3}$

73. $\dfrac{7}{\sqrt{3} + 5}$

74. $\dfrac{5}{9 - \sqrt{6}}$

75. $\dfrac{3}{2\sqrt{10} - 5}$

76. $\dfrac{4}{3\sqrt{5} - 1}$

77. $\dfrac{2}{\sqrt{6} + \sqrt{2}}$

78. $\dfrac{10}{\sqrt{9} + \sqrt{5}}$

79. $\dfrac{9}{\sqrt{3} - \sqrt{7}}$

80. $\dfrac{12}{\sqrt{5} + \sqrt{8}}$

81. $\left(\sqrt{7} + 2\right) \div \left(\sqrt{7} - 2\right)$

82. $\left(5 - \sqrt{3}\right) \div \left(3 + \sqrt{3}\right)$

83. $\left(\sqrt{x} - 5\right) \div \left(2\sqrt{x} - 1\right)$

84. $\left(2\sqrt{t} + 1\right) \div \left(2\sqrt{t} - 1\right)$

85. $\dfrac{3x}{\sqrt{15} - \sqrt{3}}$

86. $\dfrac{5y}{\sqrt{12} + \sqrt{10}}$

87. $\dfrac{2t^2}{\sqrt{5} - \sqrt{t}}$

88. $\dfrac{5x}{\sqrt{x} - \sqrt{2}}$

89. $\dfrac{8a}{\sqrt{3a} + \sqrt{a}}$

90. $\dfrac{7z}{\sqrt{5z} - \sqrt{z}}$

91. $\dfrac{3(x - 4)}{x^2 - \sqrt{x}}$

92. $\dfrac{6(y + 1)}{y^2 + \sqrt{y}}$

93. $\dfrac{\sqrt{u} + v}{\sqrt{u - v} - \sqrt{u}}$

94. $\dfrac{z}{\sqrt{u + z} - \sqrt{u}}$

In Exercises 95–98, use a graphing calculator to graph the functions in the same viewing window. Use the graphs to verify that the expressions are equivalent. Verify your results algebraically.

95. $y_1 = \dfrac{10}{\sqrt{x} + 1}$

$y_2 = \dfrac{10(\sqrt{x} - 1)}{x - 1}$

96. $y_1 = \dfrac{4x}{\sqrt{x} + 4}$

$y_2 = \dfrac{4x(\sqrt{x} - 4)}{x - 16}$

97. $y_1 = \dfrac{2\sqrt{x}}{2 - \sqrt{x}}$

$y_2 = \dfrac{2(2\sqrt{x} + x)}{4 - x}$

98. $y_1 = \dfrac{\sqrt{2x} + 6}{\sqrt{2x} - 2}$

$y_2 = \dfrac{x + 6 + 4\sqrt{2x}}{x - 2}$

Rationalizing Numerators In the study of calculus, students sometimes rewrite an expression by rationalizing the numerator. In Exercises 99–106, rationalize the numerator. (*Note:* The results will not be in simplest radical form.)

99. $\dfrac{\sqrt{2}}{7}$

100. $\dfrac{\sqrt{10}}{\sqrt{3x}}$

101. $\dfrac{\sqrt{5}}{\sqrt{7x}}$

102. $\dfrac{\sqrt{10}}{5}$

103. $\dfrac{\sqrt{7} + \sqrt{3}}{5}$

104. $\dfrac{\sqrt{2} - \sqrt{5}}{4}$

105. $\dfrac{\sqrt{y} - 5}{\sqrt{3}}$

106. $\dfrac{\sqrt{x} + 6}{\sqrt{2}}$

Solving Problems

107. ▲ *Geometry* The rectangular cross section of a wooden beam cut from a log of diameter 24 inches (see figure) will have maximum strength if its width w and height h are given by

$$w = 8\sqrt{3} \quad \text{and} \quad h = \sqrt{24^2 - \left(8\sqrt{3}\right)^2}.$$

Find the area of the rectangular cross section and write the area in simplest form.

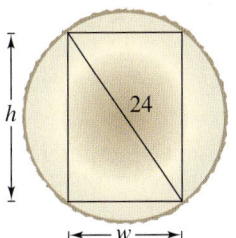

Figure for 107

108. ▲ *Geometry*   The areas of the circles in the figure are 15 square centimeters and 20 square centimeters. Find the ratio of the radius of the small circle to the radius of the large circle.

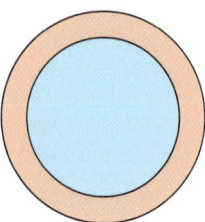

109. *Force*   The force required to slide a steel block weighing 500 pounds across a milling machine is

$$\frac{500k}{\dfrac{1}{\sqrt{k^2+1}}+\dfrac{k^2}{\sqrt{k^2+1}}}$$

where k is the friction constant (see figure). Simplify this expression.

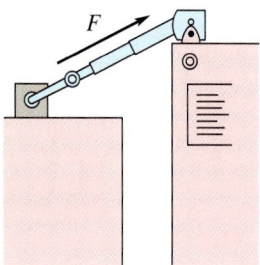

110. The ratio of the width of the Temple of Hephaestus to its height (see figure) is approximately

$$\frac{w}{h}\approx\frac{2}{\sqrt{5}-1}.$$

This number is called the **golden section.** Early Greeks believed that the most aesthetically pleasing rectangles were those whose sides had this ratio.

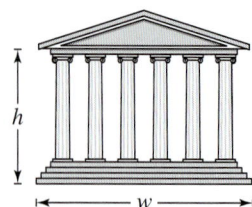

(a) Rationalize the denominator for this expression. Approximate your answer, rounded to two decimal places.

(b) Use the Pythagorean Theorem, a straightedge, and a compass to construct a rectangle whose sides have the golden section as their ratio.

Explaining Concepts

111. Multiply $\sqrt{3}(1-\sqrt{6})$. State an algebraic property to justify each step.

112. *Writing*   Describe the differences and similarities of using the FOIL Method with polynomial expressions and with radical expressions.

113. *Writing*   Multiply $3-\sqrt{2}$ by its conjugate. Explain why the result has no radicals.

114. *Writing*   Is the number $3/(1+\sqrt{5})$ in simplest form? If not, explain the steps for writing it in simplest form.

9.5 Radical Equations and Applications

Jeff Greenberg/The Image Works

Why You Should Learn It

Radical equations can be used to model and solve real-life applications. For instance, in Exercise 100 on page 594, a radical equation is used to model the total monthly cost of daily flights between Chicago and Denver.

What You Should Learn

1. Solve a radical equation by raising each side to the nth power.
2. Solve application problems involving radical equations.

1 Solve a radical equation by raising each side to the nth power.

Solving Radical Equations

Solving equations involving radicals is somewhat like solving equations that contain fractions—first try to eliminate the radicals and obtain a polynomial equation. Then, solve the polynomial equation using the standard procedures. The following property plays a key role.

> **Raising Each Side of an Equation to the nth Power**
>
> Let u and v be real numbers, variables, or algebraic expressions, and let n be a positive integer. If $u = v$, then it follows that
>
> $$u^n = v^n.$$
>
> This is called **raising each side of an equation to the nth power.**

To use this property to solve a radical equation, first try to isolate one of the radicals on one side of the equation. When using this property to solve radical equations, it is critical that you check your solutions in the original equation.

Example 1   Solving an Equation Having One Radical

Solve $\sqrt{x} - 8 = 0$.

Solution

$$\sqrt{x} - 8 = 0 \qquad \text{Write original equation.}$$
$$\sqrt{x} = 8 \qquad \text{Isolate radical.}$$
$$\left(\sqrt{x}\right)^2 = 8^2 \qquad \text{Square each side.}$$
$$x = 64 \qquad \text{Simplify.}$$

Check

$$\sqrt{64} - 8 \overset{?}{=} 0 \qquad \text{Substitute 64 for } x \text{ in original equation.}$$
$$8 - 8 = 0 \qquad \text{Solution checks.} \checkmark$$

So, the equation has one solution: $x = 64$.

Technology: Tip

To use a graphing calculator to check the solution in Example 1, graph

$$y = \sqrt{x} - 8$$

as shown below. Notice that the graph crosses the x-axis at $x = 64$, which confirms the solution that was obtained algebraically.

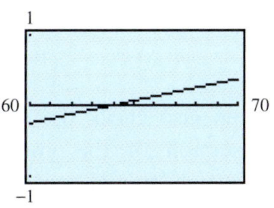

Checking solutions of a radical equation is especially important because raising each side of an equation to the nth power to remove the radical(s) often introduces *extraneous* solutions.

Example 2   Solving an Equation Having One Radical

Solve $\sqrt{3x} + 6 = 0$.

Solution

$$\sqrt{3x} + 6 = 0 \qquad \text{Write original equation.}$$
$$\sqrt{3x} = -6 \qquad \text{Isolate radical.}$$
$$\left(\sqrt{3x}\right)^2 = (-6)^2 \qquad \text{Square each side.}$$
$$3x = 36 \qquad \text{Simplify.}$$
$$x = 12 \qquad \text{Divide each side by 3.}$$

Check

$$\sqrt{3(12)} + 6 \overset{?}{=} 0 \qquad \text{Substitute 12 for } x \text{ in original equation.}$$
$$6 + 6 \neq 0 \qquad \text{Solution does not check.} ✗$$

The solution $x = 12$ is an extraneous solution. So, the original equation has no solution. You can also check this graphically, as shown in Figure 9.5. Notice that the graph does not cross the x-axis and so has no x-intercept.

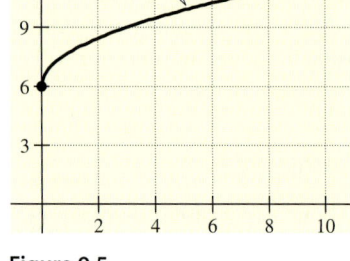

$y = \sqrt{3x} + 6$

Figure 9.5

Example 3   Solving an Equation Having One Radical

Solve $\sqrt[3]{2x + 1} - 2 = 3$.

Solution

$$\sqrt[3]{2x + 1} - 2 = 3 \qquad \text{Write original equation.}$$
$$\sqrt[3]{2x + 1} = 5 \qquad \text{Isolate radical.}$$
$$\left(\sqrt[3]{2x + 1}\right)^3 = 5^3 \qquad \text{Cube each side.}$$
$$2x + 1 = 125 \qquad \text{Simplify.}$$
$$2x = 124 \qquad \text{Subtract 1 from each side.}$$
$$x = 62 \qquad \text{Divide each side by 2.}$$

Check

$$\sqrt[3]{2(62) + 1} - 2 \overset{?}{=} 3 \qquad \text{Substitute 62 for } x \text{ in original equation.}$$
$$\sqrt[3]{125} - 2 \overset{?}{=} 3 \qquad \text{Simplify.}$$
$$5 - 2 = 3 \qquad \text{Solution checks.} ✓$$

So, the equation has one solution: $x = 62$. You can also check the solution graphically by determining the point of intersection of the graphs of $y = \sqrt[3]{2x + 1} - 2$ (left side of equation) and $y = 3$ (right side of equation), as shown in Figure 9.6.

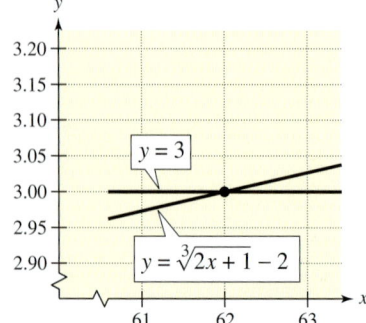

$y = 3$

$y = \sqrt[3]{2x + 1} - 2$

Figure 9.6

In Example 4, you can graphically check the solution of the equation by graphing the left side and right side in the same viewing window. That is, graph the equations

$$y_1 = \sqrt{5x + 3}$$

and

$$y_2 = \sqrt{x + 11}$$

in the same viewing window, as shown below. Using the *intersect* feature of the graphing calculator will enable you to approximate the point(s) at which the graphs intersect. From the figure, you can see that the two graphs intersect at $x = 2$, which is the solution obtained in Example 4.

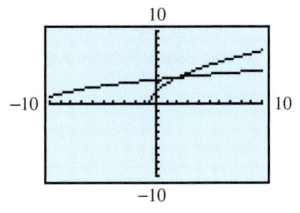

Example 4   Solving an Equation Having Two Radicals

Solve $\sqrt{5x + 3} = \sqrt{x + 11}$.

Solution

$$\sqrt{5x + 3} = \sqrt{x + 11} \qquad \text{Write original equation.}$$

$$\left(\sqrt{5x + 3}\right)^2 = \left(\sqrt{x + 11}\right)^2 \qquad \text{Square each side.}$$

$$5x + 3 = x + 11 \qquad \text{Simplify.}$$

$$4x + 3 = 11 \qquad \text{Subtract } x \text{ from each side.}$$

$$4x = 8 \qquad \text{Subtract 3 from each side.}$$

$$x = 2 \qquad \text{Divide each side by 4.}$$

Check

$$\sqrt{5x + 3} = \sqrt{x + 11} \qquad \text{Write original equation.}$$

$$\sqrt{5(2) + 3} \overset{?}{=} \sqrt{2 + 11} \qquad \text{Substitute 2 for } x.$$

$$\sqrt{13} = \sqrt{13} \qquad \text{Solution checks. } \checkmark$$

So, the equation has one solution: $x = 2$.

Example 5   Solving an Equation Having Two Radicals

Solve $\sqrt[4]{3x} + \sqrt[4]{2x - 5} = 0$.

Solution

$$\sqrt[4]{3x} + \sqrt[4]{2x - 5} = 0 \qquad \text{Write original equation.}$$

$$\sqrt[4]{3x} = -\sqrt[4]{2x - 5} \qquad \text{Isolate radicals.}$$

$$\left(\sqrt[4]{3x}\right)^4 = \left(-\sqrt[4]{2x - 5}\right)^4 \qquad \text{Raise each side to fourth power.}$$

$$3x = 2x - 5 \qquad \text{Simplify.}$$

$$x = -5 \qquad \text{Subtract } 2x \text{ from each side.}$$

Check

$$\sqrt[4]{3x} + \sqrt[4]{2x - 5} = 0 \qquad \text{Write original equation.}$$

$$\sqrt[4]{3(-5)} + \sqrt[4]{2(-5) - 5} \overset{?}{=} 0 \qquad \text{Substitute } -5 \text{ for } x.$$

$$\sqrt[4]{-15} + \sqrt[4]{-15} \neq 0 \qquad \text{Solution does not check. } \times$$

The solution does not check because it yields fourth roots of negative radicands. So, this equation has no solution. Try checking this graphically. If you graph both sides of the equation, you will discover that the graphs do not intersect.

In the next example you will see that squaring each side of the equation results in a quadratic equation. Remember that you must check the solutions in the *original* radical equation.

Example 6 An Equation That Converts to a Quadratic Equation

Solve $\sqrt{x} + 2 = x$.

Solution

$\sqrt{x} + 2 = x$	Write original equation.
$\sqrt{x} = x - 2$	Isolate radical.
$\left(\sqrt{x}\right)^2 = (x - 2)^2$	Square each side.
$x = x^2 - 4x + 4$	Simplify.
$-x^2 + 5x - 4 = 0$	Write in general form.
$(-1)(x - 4)(x - 1) = 0$	Factor.
$x - 4 = 0 \implies x = 4$	Set 1st factor equal to 0.
$x - 1 = 0 \implies x = 1$	Set 2nd factor equal to 0.

Check

First Solution	Second Solution
$\sqrt{4} + 2 \overset{?}{=} 4$	$\sqrt{1} + 2 \overset{?}{=} 1$
$2 + 2 = 4$	$1 + 2 \neq 1$

From the check you can see that $x = 1$ is an extraneous solution. So, the only solution is $x = 4$.

When an equation contains two radicals, it may not be possible to isolate both. In such cases, you may have to raise each side of the equation to a power at *two* different stages in the solution.

Example 7 Repeatedly Squaring Each Side of an Equation

Solve $\sqrt{3t + 1} = 2 - \sqrt{3t}$.

Solution

$\sqrt{3t + 1} = 2 - \sqrt{3t}$	Write original equation.
$\left(\sqrt{3t + 1}\right)^2 = \left(2 - \sqrt{3t}\right)^2$	Square each side (1st time).
$3t + 1 = 4 - 4\sqrt{3t} + 3t$	Simplify.
$-3 = -4\sqrt{3t}$	Isolate radical.
$(-3)^2 = \left(-4\sqrt{3t}\right)^2$	Square each side (2nd time).
$9 = 16(3t)$	Simplify.
$\dfrac{3}{16} = t$	Divide each side by 48 and simplify.

The solution is $t = \frac{3}{16}$. Check this in the original equation.

② Solve application problems involving radical equations.

Applications

Example 8   Electricity

The amount of power consumed by an electrical appliance is given by

$$I = \sqrt{\frac{P}{R}}$$

where I is the current measured in amps, R is the resistance measured in ohms, and P is the power measured in watts. Find the power used by an electric heater for which $I = 10$ amps and $R = 16$ ohms.

Solution

$$10 = \sqrt{\frac{P}{16}}$$ Substitute 10 for I and 16 for R in original equation.

$$10^2 = \left(\sqrt{\frac{P}{16}}\right)^2$$ Square each side.

$$100 = \frac{P}{16} \implies 1600 = P$$ Simplify and multiply each side by 16.

So, the solution is $P = 1600$ watts. Check this in the original equation.

Example 9   An Application of the Pythagorean Theorem

The distance between a house on shore and a playground on shore is 40 meters. The distance between the playground and a house on an island is 50 meters (see Figure 9.7). What is the distance between the two houses?

Solution

From Figure 9.7, you can see that the distances form a right triangle. So, you can use the Pythagorean Theorem to find the distance between the two houses.

$$c = \sqrt{a^2 + b^2}$$ Pythagorean Theorem

$$50 = \sqrt{40^2 + b^2}$$ Substitute 40 for a and 50 for c.

$$50 = \sqrt{1600 + b^2}$$ Simplify.

$$50^2 = (\sqrt{1600 + b^2})^2$$ Square each side.

$$2500 = 1600 + b^2$$ Simplify.

$$0 = b^2 - 900$$ Write in general form.

$$0 = (b + 30)(b - 30)$$ Factor.

$$b + 30 = 0 \implies b = -30$$ Set 1st factor equal to 0.

$$b - 30 = 0 \implies b = 30$$ Set 2nd factor equal to 0.

Choose the positive solution to obtain a distance of 30 meters. Check this solution in the original equation.

Study Tip

An alternative way to solve the problem in Example 8 would be first to solve the equation for P.

$$I = \sqrt{\frac{P}{R}}$$

$$I^2 = \left(\sqrt{\frac{P}{R}}\right)^2$$

$$I^2 = \frac{P}{R}$$

$$I^2 R = P$$

At this stage, you can substitute the known values of I and R to obtain

$$P = (10)^2 16 = 1600.$$

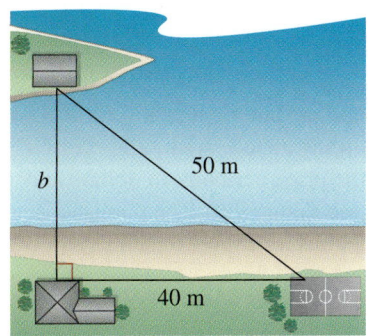

Figure 9.7

Example 10   Velocity of a Falling Object

The velocity of a free-falling object can be determined from the equation $v = \sqrt{2gh}$, where v is the velocity measured in feet per second, $g = 32$ feet per second per second, and h is the distance (in feet) the object has fallen. Find the height from which a rock has been dropped when it strikes the ground with a velocity of 50 feet per second.

Solution

$v = \sqrt{2gh}$	Write original equation.
$50 = \sqrt{2(32)h}$	Substitute 50 for v and 32 for g.
$50^2 = \left(\sqrt{64h}\right)^2$	Square each side.
$2500 = 64h$	Simplify.
$39 \approx h$	Divide each side by 64.

Check

Because the value of h was rounded in the solution, the check will not result in an equality. If the solution is valid, the expressions on each side of the equal sign will be *approximately* equal to each other.

$v = \sqrt{2gh}$	Write original equation.
$50 \overset{?}{\approx} \sqrt{2(32)(39)}$	Substitute 50 for v, 32 for g, and 39 for h.
$50 \overset{?}{\approx} \sqrt{2496}$	Simplify.
$50 \approx 49.96$	Solution checks. ✔

So, the height from which the rock has been dropped is approximately 39 feet.

Example 11   Market Research

The marketing department at a publisher determines that the demand for a book depends on the price of the book in accordance with the formula $p = 40 - \sqrt{0.0001x + 1}$, $x \geq 0$, where p is the price per book in dollars and x is the number of books sold at the given price (see Figure 9.8). The publisher sets the price at $12.95. How many copies can the publisher expect to sell?

Solution

$p = 40 - \sqrt{0.0001x + 1}$	Write original equation.
$12.95 = 40 - \sqrt{0.0001x + 1}$	Substitute 12.95 for p.
$\sqrt{0.0001x + 1} = 27.05$	Isolate radical.
$0.0001x + 1 = 731.7025$	Square each side.
$0.0001x = 730.7025$	Subtract 1 from each side.
$x = 7{,}307{,}025$	Divide each side by 0.0001.

So, by setting the book's price at $12.95, the publisher can expect to sell about 7.3 million copies. Check this in the original equation.

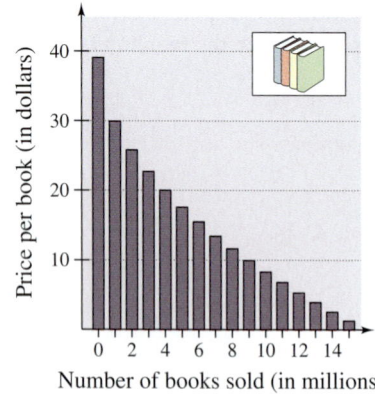

Figure 9.8

9.5 Exercises

Review   Concepts, Skills, and Problem Solving

Keep mathematically in shape by doing these exercises *before* the problems of this section.

Properties and Definitions

1. *Writing*   Explain how to determine the domain of the function

$$f(x) = \frac{4}{(x + 2)(x - 3)}.$$

2. *Writing*   Explain the excluded value $(x \neq -3)$ in the following.

$$\frac{2x^2 + 5x - 3}{x^2 - 9} = \frac{2x - 1}{x - 3}, \quad x \neq 3, \quad x \neq -3$$

Simplifying Expressions

In Exercises 3–6, simplify the expression. (Assume that any variables in the expression are nonzero.)

3. $(-3x^2y^3)^2 \cdot (4xy^2)$

4. $(x^2 - 3xy)^0$

5. $\dfrac{64r^2s^4}{16rs^2}$

6. $\left(\dfrac{3x}{4y^3}\right)^2$

In Exercises 7–10, perform the indicated operation and simplify.

7. $\dfrac{x + 13}{x^3(3 - x)} \cdot \dfrac{x(x - 3)}{5}$

8. $\dfrac{x + 2}{5x + 15} \cdot \dfrac{x - 2}{5(x - 3)}$

9. $\dfrac{2x}{x - 5} - \dfrac{5}{5 - x}$

10. $\dfrac{3}{x - 1} - 5$

Graphs

In Exercises 11 and 12, graph the function and identify any intercepts.

11. $f(x) = 2x - 3$

12. $f(x) = -\frac{3}{4}x + 2$

Developing Skills

In Exercises 1–4, determine whether each value of x is a solution of the equation.

Equation	Values of x

1. $\sqrt{x} - 10 = 0$ (a) $x = -4$ (b) $x = -100$
(c) $x = \sqrt{10}$ (d) $x = 100$

2. $\sqrt{3x} - 6 = 0$ (a) $x = \frac{2}{3}$ (b) $x = 2$
(c) $x = 12$ (d) $x = -\frac{1}{3}\sqrt{6}$

3. $\sqrt[3]{x} - 4 = 4$ (a) $x = -60$ (b) $x = 68$
(c) $x = 20$ (d) $x = 0$

4. $\sqrt[4]{2x} + 2 = 6$ (a) $x = 128$ (b) $x = 2$
(c) $x = -2$ (d) $x = 0$

In Exercises 5–54, solve the equation and check your solution(s). (Some of the equations have no solution.) See Examples 1–7.

5. $\sqrt{x} = 12$

6. $\sqrt{x} = 5$

7. $\sqrt{y} = 7$

8. $\sqrt{t} = 4$

9. $\sqrt[3]{z} = 3$

10. $\sqrt[4]{x} = 2$

11. $\sqrt{y} - 7 = 0$

12. $\sqrt{t} - 13 = 0$

13. $\sqrt{u} + 13 = 0$

14. $\sqrt{y} + 15 = 0$

15. $\sqrt{x} - 8 = 0$

16. $\sqrt{x} - 10 = 0$

17. $\sqrt{10x} = 30$

18. $\sqrt{8x} = 6$

19. $\sqrt{-3x} = 9$

20. $\sqrt{-4y} = 4$

21. $\sqrt{5t} - 2 = 0$

22. $10 - \sqrt{6x} = 0$

23. $\sqrt{3y + 1} = 4$

24. $\sqrt{3 - 2x} = 2$

25. $\sqrt{4 - 5x} = -3$

26. $\sqrt{2t - 7} = -5$

27. $\sqrt{3y + 5} - 3 = 4$ **28.** $\sqrt{a - 3} + 5 = 6$

29. $5\sqrt{x + 2} = 8$ **30.** $2\sqrt{x + 4} = 7$

31. $\sqrt{x + 3} = \sqrt{2x - 1}$ **32.** $\sqrt{3t + 1} = \sqrt{t + 15}$

33. $\sqrt{3y - 5} - 3\sqrt{y} = 0$

34. $\sqrt{2u + 10} - 2\sqrt{u} = 0$

35. $\sqrt[3]{3x - 4} = \sqrt[3]{x + 10}$

36. $2\sqrt[3]{10 - 3x} = \sqrt[3]{2 - x}$

37. $\sqrt[3]{2x + 15} - \sqrt[3]{x} = 0$

38. $\sqrt[4]{2x} + \sqrt[4]{x + 3} = 0$

39. $\sqrt{x^2 - 2} = x + 4$ **40.** $\sqrt{x^2 - 4} = x - 2$

41. $\sqrt{2x} = x - 4$ **42.** $\sqrt{x} = x - 6$

43. $\sqrt{8x + 1} = x + 2$ **44.** $\sqrt{3x + 7} = x + 3$

45. $\sqrt{3x + 4} = \sqrt{4x + 3}$

46. $\sqrt{2x - 7} = \sqrt{3x - 12}$

47. $\sqrt{z + 2} = 1 + \sqrt{z}$

48. $\sqrt{2x + 5} = 7 - \sqrt{2x}$

49. $\sqrt{2t + 3} = 3 - \sqrt{2t}$

50. $\sqrt{x} + \sqrt{x + 2} = 2$

51. $\sqrt{x + 5} - \sqrt{x} = 1$

52. $\sqrt{x + 1} = 2 - \sqrt{x}$

53. $\sqrt{x - 6} + 3 = \sqrt{x + 9}$

54. $\sqrt{x + 3} - \sqrt{x - 1} = 1$

In Exercises 55–62, solve the equation and check your solution(s).

55. $t^{3/2} = 8$ **56.** $v^{2/3} = 25$

57. $3y^{1/3} = 18$ **58.** $2x^{3/4} = 54$

59. $(x + 4)^{2/3} = 4$ **60.** $(u - 2)^{4/3} = 81$

61. $(2x + 5)^{1/3} + 3 = 0$ **62.** $(x - 6)^{3/2} - 27 = 0$

In Exercises 63–72, use a graphing calculator to graph each side of the equation in the same viewing window. Use the graphs to approximate the solution(s). Verify your answer algebraically.

63. $\sqrt{x} = 2(2 - x)$ **64.** $\sqrt{2x + 3} = 4x - 3$

65. $\sqrt{x^2 + 1} = 5 - 2x$ **66.** $\sqrt{8 - 3x} = x$

67. $\sqrt{x + 3} = 5 - \sqrt{x}$ **68.** $\sqrt[3]{5x - 8} = 4 - \sqrt[3]{x}$

69. $4\sqrt[3]{x} = 7 - x$ **70.** $\sqrt[3]{x + 4} = \sqrt{6 - x}$

71. $\sqrt{15 - 4x} = 2x$ **72.** $\dfrac{4}{\sqrt{x}} = 3\sqrt{x} - 4$

In Exercises 73–76, use the given function to find the indicated value of x.

73. For $f(x) = \sqrt{x} - \sqrt{x - 9}$,
    find x such that $f(x) = 1$.

74. For $g(x) = \sqrt{x} + \sqrt{x - 5}$,
    find x such that $g(x) = 5$.

75. For $h(x) = \sqrt{x - 2} - \sqrt{4x + 1}$,
    find x such that $h(x) = -3$.

76. For $f(x) = \sqrt{2x + 7} - \sqrt{x + 15}$,
    find x such that $f(x) = -1$.

Solving Problems

▲ *Geometry* In Exercises 77–80, find the length x of the unknown side of the right triangle. (Round your answer to two decimal places.)

77.

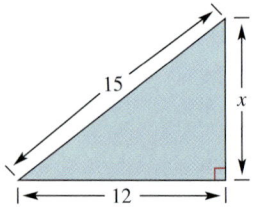

78.

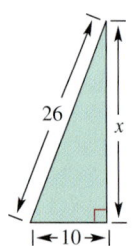

79.

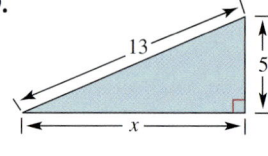

80.

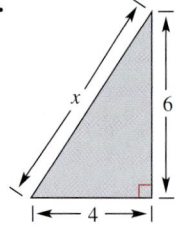

81. ▲ *Geometry* The screen of a computer monitor has a diagonal of 13.75 inches and a width of 8.25 inches. Draw a diagram of the computer monitor and find the length of the screen.

82. ▲ *Geometry* A basketball court is 50 feet wide and 94 feet long. Draw a diagram of the basketball court and find the length of a diagonal of the court.

83. ▲ *Geometry* An extension ladder is placed against the side of a house such that the base of the ladder is 2 meters from the base of the house and the ladder reaches 6 meters up the side of the house. How far is the ladder extended?

84. ▲ *Geometry* A guy wire on a 100-foot radio tower is attached to the top of the tower and to an anchor 50 feet from the base of the tower. Find the length of the guy wire.

85. ▲ *Geometry* A ladder is 17 feet long, and the bottom of the ladder is 8 feet from the side of a house. How far does the ladder reach up the side of the house?

86. ▲ *Geometry* A 10-foot plank is used to brace a basement wall during construction of a home. The plank is nailed to the wall 6 feet above the floor. Find the slope of the plank.

87. ▲ *Geometry* Determine the length and width of a rectangle with a perimeter of 92 inches and a diagonal of 34 inches.

88. ▲ *Geometry* Determine the length and width of a rectangle with a perimeter of 68 inches and a diagonal of 26 inches.

89. ▲ *Geometry* The lateral surface area of a cone (see figure) is given by $S = \pi r \sqrt{r^2 + h^2}$. Solve the equation for h. Then find the height of a cone with a lateral surface area of $364\pi\sqrt{2}$ square centimeters and a radius of 14 centimeters.

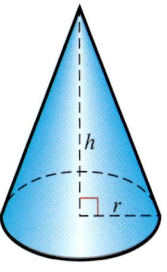

90. ▦ ▲ *Geometry* Write a function that gives the radius r of a circle in terms of the circle's area A. Use a graphing calculator to graph this function.

Height In Exercises 91 and 92, use the formula $t = \sqrt{d/16}$, which gives the time t in seconds for a free-falling object to fall d feet.

91. A construction worker drops a nail from a building and observes it strike a water puddle after approximately 2 seconds. Estimate the height from which the nail was dropped.

92. A construction worker drops a nail from a building and observes it strike a water puddle after approximately 3 seconds. Estimate the height from which the nail was dropped.

Free-Falling Object In Exercises 93–96, use the equation for the velocity of a free-falling object, $v = \sqrt{2gh}$, as described in Example 10.

93. An object is dropped from a height of 50 feet. Estimate the velocity of the object when it strikes the ground.

94. An object is dropped from a height of 200 feet. Estimate the velocity of the object when it strikes the ground.

95. An object strikes the ground with a velocity of 60 feet per second. Estimate the height from which the object was dropped.

96. An object strikes the ground with a velocity of 120 feet per second. Estimate the height from which the object was dropped.

Period of a Pendulum In Exercises 97 and 98, the time t (in seconds) for a pendulum of length L (in feet) to go through one complete cycle (its period) is given by $t = 2\pi\sqrt{L/32}$.

97. How long is the pendulum of a grandfather clock with a period of 1.5 seconds?

98. How long is the pendulum of a mantel clock with a period of 0.75 second?

99. *Demand*   The demand equation for a sweater is

$$p = 50 - \sqrt{0.8(x - 1)}$$

where x is the number of units demanded per day and p is the price per sweater. Find the demand when the price is set at $30.02.

100. *Airline Passengers*   An airline offers daily flights between Chicago and Denver. The total monthly cost C (in millions of dollars) of these flights is

$$C = \sqrt{0.2x + 1}, \quad x \geq 0$$

where x is measured in thousands of passengers (see figure). The total cost of the flights for June is 2.5 million dollars. Approximately how many passengers flew in June?

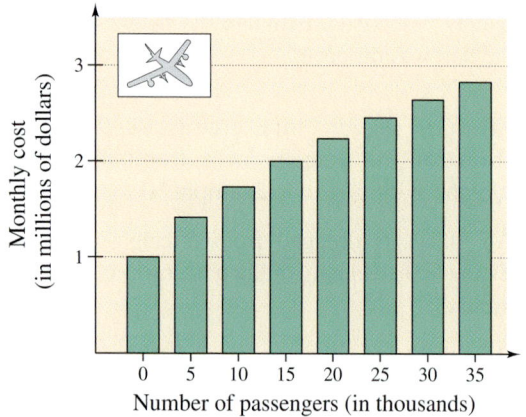

Number of passengers (in thousands)

101. *Consumer Spending*   The total amount m (in dollars) consumers spent per person on movies in theaters in the United States for the years 1995 through 2000 can be modeled by

$$m = 1.63 + 10.463\sqrt{t}, \quad 5 \leq t \leq 10$$

where t represents the year, with $t = 5$ corresponding to 1995. (Source: Veronis, Suhler & Associates Inc.)

(a)   Use a graphing calculator to graph the model.

(b)   In what year did the amount consumers spent per person on movies in theaters reach $34?

102. *Falling Object*   Without using a stopwatch, you can find the length of time an object has been falling by using the equation from physics

$$t = \sqrt{\frac{h}{384}}$$

where t is the time (in seconds) and h is the distance (in inches) the object has fallen. How far does an object fall in 0.25 second? In 0.10 second?

Explaining Concepts

103.   Answer part (e) of Motivating the Chapter on page 550.

104. *Writing*   In your own words, describe the steps that can be used to solve a radical equation.

105. *Writing*   Does raising each side of an equation to the nth power always yield an equivalent equation? Explain.

106. *Writing*   One reason for checking a solution in the original equation is to discover errors that were made when solving the equation. Describe another reason.

107. *Error Analysis*   Describe the error.

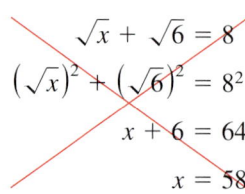

$$\sqrt{x} + \sqrt{6} = 8$$
$$\left(\sqrt{x}\right)^2 + \left(\sqrt{6}\right)^2 = 8^2$$
$$x + 6 = 64$$
$$x = 58$$

108. *Exploration*   The solution of the equation $x + \sqrt{x - a} = b$ is $x = 20$. Discuss how to find a and b. (There are many correct values for a and b.)

9.6 Complex Numbers

What You Should Learn

① Write square roots of negative numbers in *i*-form and perform operations on numbers in *i*-form.

② Determine the equality of two complex numbers.

③ Add, subtract, and multiply complex numbers.

④ Use complex conjugates to write the quotient of two complex numbers in standard form.

Why You Should Learn It

Understanding complex numbers can help you in Section 10.3 to identify quadratic equations that have no real solutions.

The Imaginary Unit *i*

In Section 9.1, you learned that a negative number has no *real* square root. For instance, $\sqrt{-1}$ is not real because there is no real number x such that $x^2 = -1$. So, as long as you are dealing only with real numbers, the equation $x^2 = -1$ has no solution. To overcome this deficiency, mathematicians have expanded the set of numbers by including the **imaginary unit *i*,** defined as

$$i = \sqrt{-1}. \qquad \textcolor{red}{\text{Imaginary unit}}$$

This number has the property that $i^2 = -1$. So, the imaginary unit i is a solution of the equation $x^2 = -1$.

① Write square roots of negative numbers in *i*-form and perform operations on numbers in *i*-form.

The Square Root of a Negative Number

Let c be a positive real number. Then the square root of $-c$ is given by

$$\sqrt{-c} = \sqrt{c(-1)} = \sqrt{c}\sqrt{-1} = \sqrt{c}\,i.$$

When writing $\sqrt{-c}$ in the ***i*-form,** $\sqrt{c}\,i$, note that i is outside the radical.

Example 1   Writing Numbers in *i*-Form

Write each number in *i*-form.

a. $\sqrt{-36}$ **b.** $\sqrt{-\dfrac{16}{25}}$ **c.** $\sqrt{-54}$ **d.** $\dfrac{\sqrt{-48}}{\sqrt{-3}}$

Solution

a. $\sqrt{-36} = \sqrt{36(-1)} = \sqrt{36}\sqrt{-1} = 6i$

b. $\sqrt{-\dfrac{16}{25}} = \sqrt{\dfrac{16}{25}(-1)} = \sqrt{\dfrac{16}{25}}\sqrt{-1} = \dfrac{4}{5}i$

c. $\sqrt{-54} = \sqrt{54(-1)} = \sqrt{54}\sqrt{-1} = 3\sqrt{6}i$

d. $\dfrac{\sqrt{-48}}{\sqrt{-3}} = \dfrac{\sqrt{48}\sqrt{-1}}{\sqrt{3}\sqrt{-1}} = \dfrac{\sqrt{48}i}{\sqrt{3}i} = \sqrt{\dfrac{48}{3}} = \sqrt{16} = 4$

Technology: Discovery

Use a calculator to evaluate each radical. Does one result in an error message? Explain why.

a. $\sqrt{121}$

b. $\sqrt{-121}$

c. $-\sqrt{121}$

To perform operations with square roots of negative numbers, you must *first* write the numbers in *i*-form. Once the numbers have been written in *i*-form, you add, subtract, and multiply as follows.

$$ai + bi = (a + b)i$$ Addition

$$ai - bi = (a - b)i$$ Subtraction

$$(ai)(bi) = ab(i^2) = ab(-1) = -ab$$ Multiplication

Example 2   Operations with Square Roots of Negative Numbers

Perform each operation.

a. $\sqrt{-9} + \sqrt{-49}$ **b.** $\sqrt{-32} - 2\sqrt{-2}$

Solution

a. $\sqrt{-9} + \sqrt{-49} = \sqrt{9}\sqrt{-1} + \sqrt{49}\sqrt{-1}$ Product Rule for Radicals

$$= 3i + 7i$$ Write in *i*-form.

$$= 10i$$ Simplify.

b. $\sqrt{-32} - 2\sqrt{-2} = \sqrt{32}\sqrt{-1} - 2\sqrt{2}\sqrt{-1}$ Product Rule for Radicals

$$= 4\sqrt{2}i - 2\sqrt{2}i$$ Write in *i*-form.

$$= 2\sqrt{2}i$$ Simplify.

Example 3   Multiplying Square Roots of Negative Numbers

Find each product.

a. $\sqrt{-15}\sqrt{-15}$ **b.** $\sqrt{-5}\left(\sqrt{-45} - \sqrt{-4}\right)$

Solution

a. $\sqrt{-15}\sqrt{-15} = \left(\sqrt{15}i\right)\left(\sqrt{15}i\right)$ Write in *i*-form.

$$= \left(\sqrt{15}\right)^2 i^2$$ Multiply.

$$= 15(-1)$$ $i^2 = -1$

$$= -15$$ Simplify.

b. $\sqrt{-5}\left(\sqrt{-45} - \sqrt{-4}\right) = \sqrt{5}i(3\sqrt{5}i - 2i)$ Write in *i*-form.

$$= \left(\sqrt{5}i\right)\left(3\sqrt{5}i\right) - \left(\sqrt{5}i\right)(2i)$$ Distributive Property

$$= 3(5)(-1) - 2\sqrt{5}(-1)$$ Multiply.

$$= -15 + 2\sqrt{5}$$ Simplify.

When multiplying square roots of negative numbers, be sure to write them in *i*-form *before multiplying*. If you do not do this, you can obtain incorrect answers. For instance, in Example 3(a) be sure you see that

$$\sqrt{-15}\sqrt{-15} \neq \sqrt{(-15)(-15)} = \sqrt{225} = 15.$$

② Determine the equality of two complex numbers.

Complex Numbers

A number of the form $a + bi$, where a and b are real numbers, is called a **complex number.** The real number a is called the **real part** of the complex number $a + bi$, and the number bi is called the **imaginary part.**

Definition of Complex Number

If a and b are real numbers, the number $a + bi$ is a **complex number,** and it is said to be written in **standard form.** If $b = 0$, the number $a + bi = a$ is a real number. If $b \neq 0$, the number $a + bi$ is called an **imaginary number.** A number of the form bi, where $b \neq 0$, is called a **pure imaginary number.**

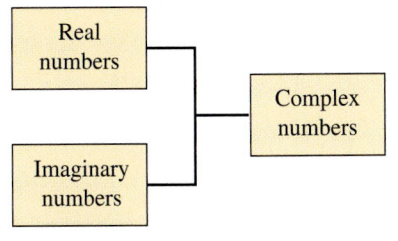

Figure 9.9

A number cannot be both real and imaginary. For instance, the numbers -2, 0, 1, $\frac{1}{2}$, and $\sqrt{2}$ are real numbers (but they are *not* imaginary numbers), and the numbers $-3i$, $2 + 4i$, and $-1 + i$ are imaginary numbers (but they are *not* real numbers). The diagram shown in Figure 9.9 further illustrates the relationship among real, complex, and imaginary numbers.

Two complex numbers $a + bi$ and $c + di$, in standard form, are equal if and only if $a = c$ and $b = d$.

Example 4   Equality of Two Complex Numbers

To determine whether the complex numbers $\sqrt{9} + \sqrt{-48}$ and $3 - 4\sqrt{3}i$ are equal, begin by writing the first number in standard form.

$$\sqrt{9} + \sqrt{-48} = \sqrt{3^2} + \sqrt{4^2(3)(-1)} = 3 + 4\sqrt{3}i$$

From this form, you can see that the two numbers are not equal because they have imaginary parts that differ in sign.

Example 5   Equality of Two Complex Numbers

Find values of x and y that satisfy the equation $3x - \sqrt{-25} = -6 + 3yi$.

Solution

Begin by writing the left side of the equation in standard form.

$$3x - 5i = -6 + 3yi \qquad \text{Each side is in standard form.}$$

For these two numbers to be equal, their real parts must be equal to each other and their imaginary parts must be equal to each other.

Real Parts	*Imaginary Parts*
$3x = -6$	$3yi = -5i$
$x = -2$	$3y = -5$
	$y = -\frac{5}{3}$

So, $x = -2$ and $y = -\frac{5}{3}$.

3 Add, subtract, and multiply complex numbers.

Operations with Complex Numbers

To add or subtract two complex numbers, you add (or subtract) the real and imaginary parts separately. This is similar to combining like terms of a polynomial.

$$(a + bi) + (c + di) = (a + c) + (b + d)i \qquad \text{Addition of complex numbers}$$

$$(a + bi) - (c + di) = (a - c) + (b - d)i \qquad \text{Subtraction of complex numbers}$$

Study Tip

Note in part (b) of Example 6 that the sum of two complex numbers can be a real number.

Example 6   Adding and Subtracting Complex Numbers

a. $(3 - i) + (-2 + 4i) = (3 - 2) + (-1 + 4)i = 1 + 3i$

b. $3i + (5 - 3i) = 5 + (3 - 3)i = 5$

c. $4 - (-1 + 5i) + (7 + 2i) = [4 - (-1) + 7] + (-5 + 2)i = 12 - 3i$

d. $(6 + 3i) + (2 - \sqrt{-8}) - \sqrt{-4} = (6 + 3i) + (2 - 2\sqrt{2}i) - 2i$

$$= (6 + 2) + (3 - 2\sqrt{2} - 2)i$$

$$= 8 + (1 - 2\sqrt{2})i$$

The Commutative, Associative, and Distributive Properties of real numbers are also valid for complex numbers, as is the FOIL Method.

Example 7   Multiplying Complex Numbers

Perform each operation and write the result in standard form.

a. $(7i)(-3i)$ **b.** $(1 - i)(\sqrt{-9})$

c. $(2 - i)(4 + 3i)$ **d.** $(3 + 2i)(3 - 2i)$

Solution

a. $(7i)(-3i) = -21i^2$ Multiply.

$$= -21(-1) = 21 \qquad i^2 = -1$$

b. $(1 - i)(\sqrt{-9}) = (1 - i)(3i)$ Write in i-form.

$$= 3i - 3(i^2) \qquad \text{Distributive Property}$$

$$= 3i - 3(-1) = 3 + 3i \qquad i^2 = -1$$

c. $(2 - i)(4 + 3i) = 8 + 6i - 4i - 3i^2$ FOIL Method

$$= 8 + 6i - 4i - 3(-1) \qquad i^2 = -1$$

$$= 11 + 2i \qquad \text{Combine like terms.}$$

d. $(3 + 2i)(3 - 2i) = 3^2 - (2i)^2$ Special product formula

$$= 9 - 4i^2 \qquad \text{Simplify.}$$

$$= 9 - 4(-1) = 13 \qquad i^2 = -1$$

④ Use complex conjugates to write the quotient of two complex numbers in standard form.

Complex Conjugates

In Example 7(d), note that the product of two complex numbers can be a real number. This occurs with pairs of complex numbers of the form $a + bi$ and $a - bi$, called **complex conjugates.** In general, the product of complex conjugates has the following form.

$$(a + bi)(a - bi) = a^2 - (bi)^2 = a^2 - b^2i^2 = a^2 - b^2(-1) = a^2 + b^2$$

Here are some examples.

Complex Number	Complex Conjugate	Product
$4 - 5i$	$4 + 5i$	$4^2 + 5^2 = 41$
$3 + 2i$	$3 - 2i$	$3^2 + 2^2 = 13$
$-2 = -2 + 0i$	$-2 = -2 - 0i$	$(-2)^2 + 0^2 = 4$
$i = 0 + i$	$-i = 0 - i$	$0^2 + 1^2 = 1$

To write the quotient of $a + bi$ and $c + di$ in standard form, where c and d are not both zero, multiply the numerator and denominator by the *complex conjugate of the denominator*, as shown in Example 8.

Example 8   Writing Quotients of Complex Numbers in Standard Form

a. $\dfrac{2 - i}{4i} = \dfrac{2 - i}{4i} \cdot \dfrac{(-4i)}{(-4i)}$
Multiply numerator and denominator by complex conjugate of denominator.

$= \dfrac{-8i + 4i^2}{-16i^2}$
Multiply fractions.

$= \dfrac{-8i + 4(-1)}{-16(-1)}$
$i^2 = -1$

$= \dfrac{-8i - 4}{16}$
Simplify.

$= -\dfrac{1}{4} - \dfrac{1}{2}i$
Write in standard form.

b. $\dfrac{5}{3 - 2i} = \dfrac{5}{3 - 2i} \cdot \dfrac{3 + 2i}{3 + 2i}$
Multiply numerator and denominator by complex conjugate of denominator.

$= \dfrac{5(3 + 2i)}{(3 - 2i)(3 + 2i)}$
Multiply fractions.

$= \dfrac{5(3 + 2i)}{3^2 + 2^2}$
Product of complex conjugates

$= \dfrac{15 + 10i}{13}$
Simplify.

$= \dfrac{15}{13} + \dfrac{10}{13}i$
Write in standard form.

Example 9 Writing a Quotient of Complex Numbers in Standard Form

$$\frac{8 - i}{8 + i} = \frac{8 - i}{8 + i} \cdot \frac{8 - i}{8 - i}$$

Multiply numerator and denominator by complex conjugate of denominator.

$$= \frac{64 - 16i + i^2}{8^2 + 1^2}$$

Multiply fractions.

$$= \frac{64 - 16i + (-1)}{8^2 + 1^2}$$

$i^2 = -1$

$$= \frac{63 - 16i}{65}$$

Simplify.

$$= \frac{63}{65} - \frac{16}{65}i$$

Write in standard form.

Example 10 Writing a Quotient of Complex Numbers in Standard Form

$$\frac{2 + 3i}{4 - 2i} = \frac{2 + 3i}{4 - 2i} \cdot \frac{4 + 2i}{4 + 2i}$$

Multiply numerator and denominator by complex conjugate of denominator.

$$= \frac{8 + 16i + 6i^2}{4^2 + 2^2}$$

Multiply fractions.

$$= \frac{8 + 16i + 6(-1)}{4^2 + 2^2}$$

$i^2 = -1$

$$= \frac{2 + 16i}{20}$$

Simplify.

$$= \frac{1}{10} + \frac{4}{5}i$$

Write in standard form.

Example 11 Verifying a Complex Solution of an Equation

Show that $x = 2 + i$ is a solution of the equation

$$x^2 - 4x + 5 = 0.$$

Solution

$$x^2 - 4x + 5 = 0$$

Write original equation.

$$(2 + i)^2 - 4(2 + i) + 5 \overset{?}{=} 0$$

Substitute $2 + i$ for x.

$$4 + 4i + i^2 - 8 - 4i + 5 \overset{?}{=} 0$$

Expand.

$$i^2 + 1 \overset{?}{=} 0$$

Combine like terms.

$$(-1) + 1 \overset{?}{=} 0$$

$i^2 = -1$

$$0 = 0$$

Solution checks. ✔

So, $x = 2 + i$ is a solution of the original equation.

9.6 Exercises

Review   *Concepts, Skills, and Problem Solving*

Keep mathematically in shape by doing these exercises *before* the problems of this section.

Properties and Definitions

1. *Writing*🖉 In your own words, describe how to multiply $\dfrac{3t}{5} \cdot \dfrac{8t^2}{15}$.

2. *Writing*🖉 In your own words, describe how to divide $\dfrac{3t}{5} \div \dfrac{8t^2}{15}$.

3. *Writing*🖉 In your own words, describe how to add $\dfrac{3t}{5} + \dfrac{8t^2}{15}$.

4. *Writing*🖉 What is the value of $\dfrac{t-5}{5-t}$? Explain.

Simplifying Expressions

In Exercises 5–10, simplify the expression.

5. $\dfrac{x^2}{2x+3} \div \dfrac{5x}{2x+3}$

6. $\dfrac{x-y}{5x} \div \dfrac{x^2-y^2}{x^2}$

7. $\dfrac{\dfrac{9}{x}}{\left(\dfrac{6}{x}+2\right)}$

8. $\dfrac{\left(1+\dfrac{2}{x}\right)}{\left(x-\dfrac{4}{x}\right)}$

9. $\dfrac{\left(\dfrac{4}{x^2-9}+\dfrac{2}{x-2}\right)}{\left(\dfrac{1}{x+3}+\dfrac{1}{x-3}\right)}$

10. $\dfrac{\left(\dfrac{1}{x+1}+\dfrac{1}{2}\right)}{\left(\dfrac{3}{2x^2+4x+2}\right)}$

Problem Solving

11. *Number Problem* Find two real numbers that divide the real number line between $x/2$ and $4x/3$ into three equal parts.

12. *Capacitance* When two capacitors with capacitances C_1 and C_2 are connected in series, the equivalent capacitance is given by

$$\frac{1}{\left(\dfrac{1}{C_1}+\dfrac{1}{C_2}\right)}.$$

Simplify this complex fraction.

Developing Skills

In Exercises 1–16, write the number in *i*-form. See Example 1.

1. $\sqrt{-4}$

2. $\sqrt{-9}$

3. $-\sqrt{-144}$

4. $\sqrt{-49}$

5. $\sqrt{-\frac{4}{25}}$

6. $-\sqrt{-\frac{36}{121}}$

7. $\sqrt{-0.09}$

8. $\sqrt{-0.0004}$

9. $\sqrt{-8}$

10. $\sqrt{-75}$

11. $\sqrt{-7}$

12. $\sqrt{-15}$

13. $\dfrac{\sqrt{-12}}{\sqrt{-3}}$

14. $\dfrac{\sqrt{-45}}{\sqrt{-5}}$

15. $\sqrt{-\frac{18}{64}}$

16. $\sqrt{-\frac{8}{25}}$

In Exercises 17–38, perform the operation(s) and write the result in standard form. See Examples 2 and 3.

17. $\sqrt{-16}+\sqrt{-36}$

18. $\sqrt{-25}-\sqrt{-9}$

19. $\sqrt{-50}-\sqrt{-8}$

20. $\sqrt{-500}+\sqrt{-45}$

21. $\sqrt{-48}+\sqrt{-12}-\sqrt{-27}$

22. $\sqrt{-32}-\sqrt{-18}+\sqrt{-50}$

23. $\sqrt{-8}\sqrt{-2}$

24. $\sqrt{-25}\sqrt{-6}$

25. $\sqrt{-18}\sqrt{-3}$

26. $\sqrt{-7}\sqrt{-7}$

27. $\sqrt{-0.16}\sqrt{-1.21}$

28. $\sqrt{-0.49}\sqrt{-1.44}$

29. $\sqrt{-3}\left(\sqrt{-3}+\sqrt{-4}\right)$

30. $\sqrt{-12}\left(\sqrt{-3} - \sqrt{-12}\right)$

31. $\sqrt{-5}\left(\sqrt{-16} - \sqrt{-10}\right)$

32. $\sqrt{-24}\left(\sqrt{-9} + \sqrt{-4}\right)$

33. $\sqrt{-2}\left(3 - \sqrt{-8}\right)$ **34.** $\sqrt{-9}\left(1 + \sqrt{-16}\right)$

35. $\left(\sqrt{-16}\right)^2$ **36.** $\left(\sqrt{-2}\right)^2$

37. $\left(\sqrt{-4}\right)^3$ **38.** $\left(\sqrt{-5}\right)^3$

In Exercises 39–46, determine the values of a and b that satisfy the equation. See Examples 4 and 5.

39. $3 - 4i = a + bi$

40. $-8 + 6i = a + bi$

41. $5 - 4i = (a + 3) + (b - 1)i$

42. $-10 + 12i = 2a + (5b - 3)i$

43. $-4 - \sqrt{-8} = a + bi$

44. $\sqrt{-36} - 3 = a + bi$

45. $(a + 5) + (b - 1)i = 7 - 3i$

46. $(2a + 1) + (2b + 3)i = 5 + 12i$

In Exercises 47–60, perform the operation(s) and write the result in standard form. See Example 6.

47. $(4 - 3i) + (6 + 7i)$

48. $(-10 + 2i) + (4 - 7i)$

49. $(-4 - 7i) + (-10 - 33i)$

50. $(15 + 10i) - (2 + 10i)$

51. $13i - (14 - 7i)$

52. $(-21 - 50i) + (21 - 20i)$

53. $(30 - i) - (18 + 6i) + 3i^2$

54. $(4 + 6i) + (15 + 24i) - (1 - i)$

55. $6 - (3 - 4i) + 2i$

56. $22 + (-5 + 8i) + 10i$

57. $\left(\frac{4}{3} + \frac{1}{3}i\right) + \left(\frac{5}{6} + \frac{7}{6}i\right)$

58. $(0.05 + 2.50i) - (6.2 + 11.8i)$

59. $15i - (3 - 25i) + \sqrt{-81}$

60. $(-1 + i) - \sqrt{2} - \sqrt{-2}$

In Exercises 61–88, perform the operation and write the result in standard form. See Example 7.

61. $(3i)(12i)$ **62.** $(-5i)(4i)$

63. $(3i)(-8i)$ **64.** $(-2i)(-10i)$

65. $(-6i)(-i)(6i)$ **66.** $(10i)(12i)(-3i)$

67. $(-3i)^3$ **68.** $(8i)^2$

69. $(-3i)^2$ **70.** $(2i)^4$

71. $-5(13 + 2i)$ **72.** $10(8 - 6i)$

73. $4i(-3 - 5i)$ **74.** $-3i(10 - 15i)$

75. $(9 - 2i)\left(\sqrt{-4}\right)$ **76.** $(11 + 3i)\left(\sqrt{-25}\right)$

77. $(4 + 3i)(-7 + 4i)$ **78.** $(3 + 5i)(2 + 15i)$

79. $(-7 + 7i)(4 - 2i)$ **80.** $(3 + 5i)(2 - 15i)$

81. $\left(-2 + \sqrt{-5}\right)\left(-2 - \sqrt{-5}\right)$

82. $\left(-3 - \sqrt{-12}\right)\left(4 - \sqrt{-12}\right)$

83. $(3 - 4i)^2$ **84.** $(7 + i)^2$

85. $(2 + 5i)^2$ **86.** $(8 - 3i)^2$

87. $(2 + i)^3$ **88.** $(3 - 2i)^3$

In Exercises 89–98, simplify the expression.

89. i^7 **90.** i^{11}

91. i^{24} **92.** i^{35}

93. i^{42} **94.** i^{64}

95. i^9 **96.** i^{71}

97. $(-i)^6$ **98.** $(-i)^4$

In Exercises 99–110, multiply the number by its complex conjugate and simplify.

99. $2 + i$ **100.** $3 + 2i$

101. $-2 - 8i$ **102.** $10 - 3i$

103. $5 - \sqrt{6}i$ **104.** $-4 + \sqrt{2}i$

105. $10i$ **106.** 20

107. $1 + \sqrt{-3}$ **108.** $-3 - \sqrt{-5}$

109. $1.5 + \sqrt{-0.25}$ **110.** $3.2 - \sqrt{-0.04}$

In Exercises 111–124, write the quotient in standard form. See Examples 8–10.

111. $\dfrac{20}{2i}$ **112.** $\dfrac{-5}{-3i}$

113. $\dfrac{2 + i}{-5i}$ **114.** $\dfrac{1 + i}{3i}$

115. $\dfrac{4}{1-i}$

116. $\dfrac{20}{3+i}$

117. $\dfrac{7i+14}{7i}$

118. $\dfrac{6i+3}{3i}$

119. $\dfrac{-12}{2+7i}$

120. $\dfrac{15}{2(1-i)}$

121. $\dfrac{3i}{5+2i}$

122. $\dfrac{4i}{5-3i}$

123. $\dfrac{4+5i}{3-7i}$

124. $\dfrac{5+3i}{7-4i}$

In Exercises 125–130, perform the operation and write the result in standard form.

125. $\dfrac{5}{3+i}+\dfrac{1}{3-i}$

126. $\dfrac{1}{1-2i}+\dfrac{4}{1+2i}$

127. $\dfrac{3i}{1+i}+\dfrac{2}{2+3i}$

128. $\dfrac{i}{4-3i}-\dfrac{5}{2+i}$

129. $\dfrac{1+i}{i}-\dfrac{3}{5-2i}$

130. $\dfrac{3-2i}{i}-\dfrac{1}{7+i}$

In Exercises 131–134, determine whether each number is a solution of the equation. See Example 11.

131. $x^2+2x+5=0$

    (a) $x=-1+2i$ (b) $x=-1-2i$

132. $x^2-4x+13=0$

    (a) $x=2-3i$ (b) $x=2+3i$

133. $x^3+4x^2+9x+36=0$

    (a) $x=-4$ (b) $x=-3i$

134. $x^3-8x^2+25x-26=0$

    (a) $x=2$ (b) $x=3-2i$

135. *Cube Roots* The principal cube root of 125, $\sqrt[3]{125}$, is 5. Evaluate the expression x^3 for each value of x.

    (a) $x=\dfrac{-5+5\sqrt{3}\,i}{2}$

    (b) $x=\dfrac{-5-5\sqrt{3}\,i}{2}$

136. *Cube Roots* The principal cube root of 27, $\sqrt[3]{27}$, is 3. Evaluate the expression x^3 for each value of x.

    (a) $x=\dfrac{-3+3\sqrt{3}\,i}{2}$

    (b) $x=\dfrac{-3-3\sqrt{3}\,i}{2}$

137. *Pattern Recognition* Compare the results of Exercises 135 and 136. Use the results to list possible cube roots of (a) 1, (b) 8, and (c) 64. Verify your results algebraically.

138. *Algebraic Properties* Consider the complex number $1+5i$.

    (a) Find the additive inverse of the number.

    (b) Find the multiplicative inverse of the number.

In Exercises 139–142, perform the operations.

139. $(a+bi)+(a-bi)$

140. $(a+bi)(a-bi)$

141. $(a+bi)-(a-bi)$

142. $(a+bi)^2+(a-bi)^2$

Explaining Concepts

143. *Writing* Define the imaginary unit i.

144. *Writing* Explain why the equation $x^2=-1$ does not have real number solutions.

145. *Writing* Describe the error.

$$\sqrt{-3}\sqrt{-3}=\sqrt{(-3)(-3)}=\sqrt{9}=3$$

146. *True or False?* Some numbers are both real and imaginary. Justify your answer.

147. The polynomial x^2+1 is prime *with respect to the integers*. It is not, however, prime *with respect to the complex numbers*. Show how x^2+1 can be factored using complex numbers.

What Did You Learn?

Key Terms

square root, *p. 552*
cube root, *p. 552*
*n*th root of *a*, *p. 552*
principal *n*th root of *a*, *p. 552*
radical symbol, *p. 552*
index, *p. 552*
radicand, *p. 552*
perfect square, *p. 553*

perfect cube, *p. 553*
rational exponent, *p. 555*
radical function, *p. 557*
rationalizing the
  denominator, *p. 566*
Pythagorean Theorem, *p. 567*
like radicals, *p. 570*
conjugates, *p. 578*

imaginary unit *i*, *p. 595*
i-form, *p. 595*
complex number, *p. 597*
real part, *p. 597*
imaginary part, *p. 597*
imaginary number, *p. 597*
complex conjugates,
  p. 599

Key Concepts

9.1 ◉ Properties of *n*th roots

1. If a is a positive real number and n is even, then a has exactly two (real) *n*th roots, which are denoted by $\sqrt[n]{a}$ and $-\sqrt[n]{a}$.

2. If a is any real number and n is odd, then a has only one (real) *n*th root, which is denoted by $\sqrt[n]{a}$.

3. If a is a negative real number and n is even, then a has no (real) *n*th root.

9.1 ◉ Inverse properties of *n*th powers and *n*th roots

Let a be a real number, and let n be an integer such that $n \geq 2$.

1. If a has a principal *n*th root, then $\left(\sqrt[n]{a}\right)^n = a$.

2. If n is odd, then $\sqrt[n]{a^n} = a$. If n is even, then $\sqrt[n]{a^n} = |a|$.

9.1 ◉ Rules of exponents

Let r and s be rational numbers, and let a and b be real numbers, variables, or algebraic expressions. (All denominators and bases are nonzero.)

1. $a^r \cdot a^s = a^{r+s}$ 2. $\dfrac{a^r}{a^s} = a^{r-s}$

3. $(ab)^r = a^r \cdot b^r$ 4. $(a^r)^s = a^{rs}$

5. $\left(\dfrac{a}{b}\right)^r = \dfrac{a^r}{b^r}$ 6. $a^0 = 1$

7. $a^{-r} = \dfrac{1}{a^r}$ 8. $\left(\dfrac{a}{b}\right)^{-r} = \left(\dfrac{b}{a}\right)^r$

9.1 ◉ Domain of a radical function

Let n be an integer that is greater than or equal to 2.
1. If n is odd, the domain of $f(x) = \sqrt[n]{x}$ is the set of all real numbers.

2. If n is even, the domain of $f(x) = \sqrt[n]{x}$ is the set of all nonnegative real numbers.

9.2 ◉ Product and Quotient Rules for Radicals

Let u and v be real numbers, variables, or algebraic expressions. If the *n*th roots of u and v are real, the following rules are true.

1. $\sqrt[n]{uv} = \sqrt[n]{u}\,\sqrt[n]{v}$ 2. $\sqrt[n]{\dfrac{u}{v}} = \dfrac{\sqrt[n]{u}}{\sqrt[n]{v}}, \quad v \neq 0$

9.2 ◉ Simplifying radical expressions

A radical expression is said to be in simplest form if all three of the statements below are true.
1. All possible *n*th powered factors have been removed from each radical.

2. No radical contains a fraction.

3. No denominator of a fraction contains a radical.

9.5 ◉ Raising each side of an equation to the *n*th power

Let u and v be real numbers, variables, or algebraic expressions, and let n be a positive integer. If $u = v$, then it follows that $u^n = v^n$.

9.6 ◉ The square root of a negative number

Let c be a positive real number. Then the square root of $-c$ is given by

$$\sqrt{-c} = \sqrt{c(-1)} = \sqrt{c}\,\sqrt{-1} = \sqrt{c}\,i.$$

When writing $\sqrt{-c}$ in the *i*-form, $\sqrt{c}\,i$, note that i is outside the radical.

Review Exercises

9.1 Radicals and Rational Exponents

1 Determine the nth roots of numbers and evaluate radical expressions.

In Exercises 1–14, evaluate the radical expression without using a calculator. If not possible, state the reason.

1. $\sqrt{49}$ 2. $\sqrt{25}$

3. $-\sqrt{81}$ 4. $\sqrt{-16}$

5. $\sqrt[3]{-8}$ 6. $\sqrt[3]{-1}$

7. $-\sqrt[3]{64}$ 8. $-\sqrt[3]{125}$

9. $\sqrt{\left(\frac{5}{6}\right)^2}$ 10. $\sqrt{\left(\frac{8}{15}\right)^2}$

11. $\sqrt[3]{-\left(\frac{1}{5}\right)^3}$ 12. $-\sqrt[3]{\left(-\frac{27}{64}\right)^3}$

13. $\sqrt{-2^2}$ 14. $\sqrt{-4^2}$

2 Use the rules of exponents to evaluate or simplify expressions with rational exponents.

In Exercises 15–18, fill in the missing description.

	Radical Form	Rational Exponent Form
15.	$\sqrt{49} = 7$	
16.	$\sqrt[3]{0.125} = 0.5$	
17.		$216^{1/3} = 6$
18.		$16^{1/4} = 2$

In Exercises 19–24, evaluate without using a calculator.

19. $27^{4/3}$ 20. $16^{3/4}$

21. $-(5^2)^{3/2}$ 22. $(-9)^{5/2}$

23. $8^{-4/3}$ 24. $243^{-2/5}$

In Exercises 25–36, rewrite the expression using rational exponents.

25. $x^{3/4} \cdot x^{-1/6}$ 26. $a^{2/3} \cdot a^{3/5}$

27. $z\sqrt[3]{z^2}$ 28. $x^2\sqrt[4]{x^3}$

29. $\dfrac{\sqrt[4]{x^3}}{\sqrt{x^4}}$ 30. $\dfrac{\sqrt{x^3}}{\sqrt[3]{x^2}}$

31. $\sqrt[3]{a^3b^2}$ 32. $\sqrt[5]{x^6y^2}$

33. $\sqrt[4]{\sqrt{x}}$ 34. $\sqrt{\sqrt[3]{x^4}}$

35. $\dfrac{(3x + 2)^{2/3}}{\sqrt[3]{3x + 2}}$ 36. $\dfrac{\sqrt[5]{3x + 6}}{(3x + 6)^{4/5}}$

3 Use a calculator to evaluate radical expressions.

In Exercises 37–40, use a calculator to evaluate the expression. Round the answer to four decimal places.

37. $75^{-3/4}$ 38. $510^{5/3}$

39. $\sqrt{13^2 - 4(2)(7)}$ 40. $\dfrac{-3.7 + \sqrt{15.8}}{2(2.3)}$

4 Evaluate radical functions and find the domains of radical functions.

In Exercises 41–44, evaluate the function as indicated, if possible, and simplify.

41. $f(x) = \sqrt{x - 2}$ 42. $f(x) = \sqrt{6x - 5}$
     (a) $f(11)$ (b) $f(83)$ (a) $f(5)$ (b) $f(-1)$

43. $g(x) = \sqrt[3]{2x - 1}$ 44. $g(x) = \sqrt[4]{x + 5}$
     (a) $g(0)$ (b) $g(14)$ (a) $g(11)$ (b) $g(-10)$

In Exercises 45 and 46, describe the domain of the function.

45. $f(x) = \sqrt{9 - 2x}$ 46. $g(x) = \sqrt[3]{x + 2}$

9.2 Simplifying Radical Expressions

1 Use the Product and Quotient Rules for Radicals to simplify radical expressions.

In Exercises 47–54, simplify the radical expression.

47. $\sqrt{75u^5v^4}$ 48. $\sqrt{24x^3y^4}$

49. $\sqrt{0.25x^4y}$ 50. $\sqrt{0.16s^6t^3}$

51. $\sqrt[4]{64a^2b^5}$ 52. $\sqrt{36x^3y^2}$

53. $\sqrt[3]{48a^3b^4}$ 54. $\sqrt[4]{32u^4v^5}$

2 Use rationalization techniques to simplify radical expressions.

In Exercises 55–60, rationalize the denominator and simplify further, if possible.

55. $\sqrt{\frac{5}{6}}$

56. $\sqrt{\frac{3}{20}}$

57. $\dfrac{3}{\sqrt{12x}}$

58. $\dfrac{4y}{\sqrt{10z}}$

59. $\dfrac{2}{\sqrt[3]{2x}}$

60. $\sqrt[3]{\dfrac{16t}{s^2}}$

3 Use the Pythagorean Theorem in application problems.

△ *Geometry* **In Exercises 61 and 62, find the length of the hypotenuse of the right triangle.**

61.
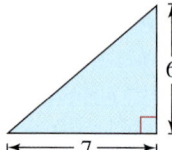

62.
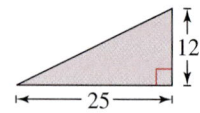

9.3 Adding and Subtracting Radical Expressions

1 Use the Distributive Property to add and subtract like radicals.

In Exercises 63–74, combine the radical expressions, if possible.

63. $2\sqrt{7} - 5\sqrt{7} + 4\sqrt{7}$

64. $3\sqrt{5} - 7\sqrt{5} + 2\sqrt{5}$

65. $3\sqrt{40} - 10\sqrt{90}$

66. $9\sqrt{50} - 5\sqrt{8} + \sqrt{48}$

67. $5\sqrt{x} - \sqrt[3]{x} + 9\sqrt{x} - 8\sqrt[3]{x}$

68. $\sqrt{3x} - \sqrt[4]{6x^2} + 2\sqrt[4]{6x^2} - 4\sqrt{3x}$

69. $10\sqrt[4]{y+3} - 3\sqrt[4]{y+3}$

70. $5\sqrt[3]{x-3} + 4\sqrt[3]{x-3}$

71. $2x\sqrt[3]{24x^2y} - \sqrt[3]{3x^5y}$

72. $4xy^2\sqrt[4]{243x} + 2y^2\sqrt[4]{48x^5}$

73. $x\sqrt{9x^4y^5} - 2x^3\sqrt{8y^5} + 4xy^2\sqrt{4x^4y}$

74. $2x^2y^2\sqrt{75xy} + 5xy\sqrt{3x^3y^3} - xy^2\sqrt{300x^3y}$

2 Use radical expressions in application problems.

75. △ *Geometry*   The four corners are cut from an $8\frac{1}{2}$-inch-by-14-inch sheet of paper (see figure). Find the perimeter of the remaining piece of paper.

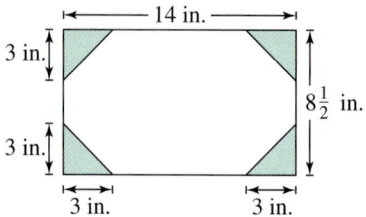

76. △ *Geometry*   Write and simplify an expression for the perimeter of the triangle.

9.4 Multiplying and Dividing Radical Expressions

1 Use the Distributive Property or the FOIL Method to multiply radical expressions.

In Exercises 77–86, multiply and simplify.

77. $\sqrt{15} \cdot \sqrt{20}$

78. $\sqrt{42} \cdot \sqrt{21}$

79. $\sqrt{5}(\sqrt{10} + 3)$

80. $\sqrt{6}(\sqrt{24} - 8)$

81. $\sqrt{10}(\sqrt{2} + \sqrt{5})$

82. $\sqrt{12}(\sqrt{6} - \sqrt{8})$

83. $(\sqrt{5} + 6)^2$

84. $(4 - 3\sqrt{2})^2$

85. $(\sqrt{3} - \sqrt{x})(\sqrt{3} + \sqrt{x})$

86. $(2 + 3\sqrt{5})(2 - 3\sqrt{5})$

2 Determine the products of conjugates.

In Exercises 87–90, find the conjugate of the expression. Then multiply the expression by its conjugate and simplify.

87. $3 - \sqrt{7}$

88. $\sqrt{5} + 10$

89. $\sqrt{x} + 20$

90. $9 - \sqrt{2y}$

3 Simplify quotients involving radicals by rationalizing the denominators.

In Exercises 91–98, rationalize the denominator of the expression and simplify.

91. $\dfrac{3}{1 - \sqrt{2}}$

92. $\dfrac{\sqrt{5}}{\sqrt{10} + 3}$

93. $\dfrac{3\sqrt{8}}{2\sqrt{2} + \sqrt{3}}$

94. $\dfrac{7\sqrt{6}}{\sqrt{3} - 4\sqrt{2}}$

95. $\dfrac{\sqrt{2} - 1}{\sqrt{3} - 4}$

96. $\dfrac{3 + \sqrt{3}}{5 - \sqrt{3}}$

97. $\left(\sqrt{x} + 10\right) \div \left(\sqrt{x} - 10\right)$

98. $\left(3\sqrt{s} + 4\right) \div \left(\sqrt{s} + 2\right)$

9.5 Radical Equations and Applications

1 Solve a radical equation by raising each side to the nth power.

In Exercises 99–114, solve the equation and check your solution(s).

99. $\sqrt{y} = 15$

100. $\sqrt{x} - 3 = 0$

101. $\sqrt{3x} + 9 = 0$

102. $\sqrt{4x} + 6 = 9$

103. $\sqrt{2(a - 7)} = 14$

104. $\sqrt{5(4 - 3x)} = 10$

105. $\sqrt[3]{5x - 7} - 3 = -1$

106. $\sqrt[4]{2x + 3} + 4 = 5$

107. $\sqrt[3]{5x + 2} - \sqrt[3]{7x - 8} = 0$

108. $\sqrt[4]{9x - 2} - \sqrt[4]{8x} = 0$

109. $\sqrt{2(x + 5)} = x + 5$

110. $y - 2 = \sqrt{y + 4}$

111. $\sqrt{v - 6} = 6 - v$

112. $\sqrt{5t} = 1 + \sqrt{5(t - 1)}$

113. $\sqrt{1 + 6x} = 2 - \sqrt{6x}$

114. $\sqrt{2 + 9b} + 1 = 3\sqrt{b}$

2 Solve application problems involving radical equations.

115. ▲ *Geometry* Determine the length and width of a rectangle with a perimeter of 34 inches and a diagonal of 13 inches.

116. ▲ *Geometry* Determine the length and width of a rectangle with a perimeter of 84 inches and a diagonal of 30 inches.

117. ▲ *Geometry* A ladder is 18 feet long, and the bottom of the ladder is 9 feet from the side of a house. How far does the ladder reach up the side of the house?

118. *Period of a Pendulum* The time t (in seconds) for a pendulum of length L (in feet) to go through one complete cycle (its period) is given by

$$t = 2\pi \sqrt{\frac{L}{32}}.$$

How long is the pendulum of a grandfather clock with a period of 1.3 seconds?

119. *Height* The time t (in seconds) for a free-falling object to fall d feet is given by

$$t = \sqrt{\frac{d}{16}}.$$

A child drops a pebble from a bridge and observes it strike the water after approximately 6 seconds. Estimate the height from which the pebble was dropped.

120. *Free-Falling Object*   The velocity of a free-falling object can be determined from the equation

$$v = \sqrt{2gh}$$

where v is the velocity (in feet per second), $g = 32$ feet per second per second, and h is the distance (in feet) the object has fallen. Find the height from which a rock has been dropped when it strikes the ground with a velocity of 25 feet per second.

9.6 Complex Numbers

① Write square roots of negative numbers in i-form and perform operations on numbers in i-form.

In Exercises 121–126, write the number in i-form.

121. $\sqrt{-48}$

122. $\sqrt{-0.16}$

123. $10 - 3\sqrt{-27}$

124. $3 + 2\sqrt{-500}$

125. $\frac{3}{4} - 5\sqrt{-\frac{3}{25}}$

126. $-0.5 + 3\sqrt{-1.21}$

In Exercises 127–134, perform the operation(s) and write the result in standard form.

127. $\sqrt{-81} + \sqrt{-36}$

128. $\sqrt{-49} + \sqrt{-1}$

129. $\sqrt{-121} - \sqrt{-84}$

130. $\sqrt{-169} - \sqrt{-4}$

131. $\sqrt{-5}\sqrt{-5}$

132. $\sqrt{-24}\sqrt{-6}$

133. $\sqrt{-10}\left(\sqrt{-4} - \sqrt{-7}\right)$

134. $\sqrt{-5}\left(\sqrt{-10} + \sqrt{-15}\right)$

② Determine the equality of two complex numbers.

In Exercises 135–138, determine the values of a and b that satisfy the equation.

135. $12 - 5i = (a + 2) + (b - 1)i$

136. $-48 + 9i = (a - 5) + (b + 10)i$

137. $\sqrt{-49} + 4 = a + bi$

138. $-3 - \sqrt{-4} = a + bi$

③ Add, subtract, and multiply complex numbers.

In Exercises 139–146, perform the operation and write the result in standard form.

139. $(-4 + 5i) - (-12 + 8i)$

140. $(-8 + 3i) - (6 + 7i)$

141. $(3 - 8i) + (5 + 12i)$

142. $(-6 + 3i) + (-1 + i)$

143. $(4 - 3i)(4 + 3i)$

144. $(12 - 5i)(2 + 7i)$

145. $(6 - 5i)^2$

146. $(2 - 9i)^2$

④ Use complex conjugates to write the quotient of two complex numbers in standard form.

In Exercises 147–152, write the quotient in standard form.

147. $\dfrac{7}{3i}$

148. $\dfrac{4}{5i}$

149. $\dfrac{4i}{2 - 8i}$

150. $\dfrac{5i}{2 + 9i}$

151. $\dfrac{3 - 5i}{6 + i}$

152. $\dfrac{2 + i}{1 - 9i}$

Chapter Test

Take this test as you would take a test in class. After you are done, check your work against the answers in the back of the book.

In Exercises 1 and 2, evaluate each expression without using a calculator.

1. (a) $16^{3/2}$
   (b) $\sqrt{5}\sqrt{20}$

2. (a) $27^{-2/3}$
   (b) $\sqrt{2}\sqrt{18}$

3. For $f(x) = \sqrt{9 - 5x}$, find $f(-8)$ and $f(0)$.
4. Find the domain of $g(x) = \sqrt{7x - 3}$.

In Exercises 5–7, simplify each expression.

5. (a) $\left(\dfrac{x^{1/2}}{x^{1/3}}\right)^2$
   (b) $5^{1/4} \cdot 5^{7/4}$

6. (a) $\sqrt{\dfrac{32}{9}}$
   (b) $\sqrt[3]{24}$

7. (a) $\sqrt{24x^3}$
   (b) $\sqrt[4]{16x^5y^8}$

In Exercises 8 and 9, rationalize the denominator of the expression and simplify.

8. $\dfrac{2}{\sqrt[3]{9y}}$

9. $\dfrac{10}{\sqrt{6} - \sqrt{2}}$

10. Subtract: $5\sqrt{3x} - 3\sqrt{75x}$
11. Multiply and simplify: $\sqrt{5}\left(\sqrt{15x} + 3\right)$
12. Expand: $\left(4 - \sqrt{2x}\right)^2$
13. Factor: $7\sqrt{27} + 14y\sqrt{12} = 7\sqrt{3}\left(\right)$

In Exercises 14–16, solve the equation.

14. $\sqrt{3y} - 6 = 3$
15. $\sqrt{x^2 - 1} = x - 2$
16. $\sqrt{x} - x + 6 = 0$

In Exercises 17–20, perform the operation(s) and simplify.

17. $(2 + 3i) - \sqrt{-25}$
18. $(2 - 3i)^2$
19. $\sqrt{-16}\left(1 + \sqrt{-4}\right)$
20. $(3 - 2i)(1 + 5i)$

21. Write $\dfrac{5 - 2i}{3 + i}$ in standard form.

22. The velocity v (in feet per second) of an object is given by $v = \sqrt{2gh}$, where $g = 32$ feet per second per second and h is the distance (in feet) the object has fallen. Find the height from which a rock has been dropped when it strikes the ground with a velocity of 80 feet per second.

Cumulative Test: Chapters 7–9

Take this test as you would take a test in class. After you are done, check your work against the answers in the back of the book.

1. Find the domain of $f(x) = \dfrac{3(x-1)}{5x-2}$.

In Exercises 2–5, perform the indicated operation(s) and simplify.

2. $\dfrac{x^2 + 8x + 16}{18x^2} \cdot \dfrac{2x^4 + 4x^3}{x^2 - 16}$

3. $\dfrac{x^3 - 4x}{x^2 + 10x + 24} \div \dfrac{x^2 + 2x}{x^2 + 5x + 4}$

4. $\dfrac{2}{x} - \dfrac{x}{x^3 + 3x^2} + \dfrac{1}{x + 3}$

5. $\dfrac{\left(\dfrac{x}{y} - \dfrac{y}{x}\right)}{\left(\dfrac{x - y}{xy}\right)}$

6. Determine whether each ordered pair is a solution of the system of linear equations.

$$\begin{cases} 2x - y = 2 \\ -x + 3y = 4 \end{cases} \quad \text{(a) } (2, 2) \quad \text{(b) } (0, 0)$$

In Exercises 7–10, match the system of equations with its graph. [The graphs are labeled (a), (b), (c), and (d).]

(a)

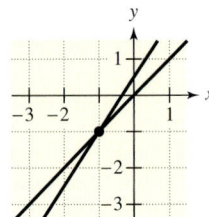

(b)

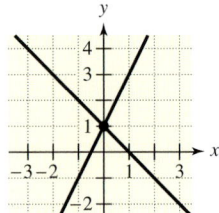

(c)

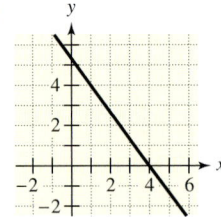

(d)

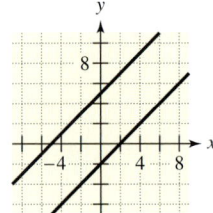

7. $\begin{cases} x + y = 1 \\ 2x - y = -1 \end{cases}$

8. $\begin{cases} 4x + 3y = 16 \\ -8x - 6y = -32 \end{cases}$

9. $\begin{cases} 5x - 5y = 10 \\ -x + y = 5 \end{cases}$

10. $\begin{cases} -x + y = 0 \\ 3x - 2y = -1 \end{cases}$

In Exercises 11–16, use the indicated method to solve the system.

11. *Graphical:* $\begin{cases} x - y = 1 \\ 2x + y = 5 \end{cases}$

12. *Substitution:* $\begin{cases} 4x + 2y = 8 \\ x - 5y = 13 \end{cases}$

13. *Elimination:* $\begin{cases} 4x - 3y = 8 \\ -2x + y = -6 \end{cases}$ **14.** *Elimination:* $\begin{cases} x + y + z = -1 \\ x = 0 \\ 2x + y = 1 \end{cases}$

15. *Matrices:* $\begin{cases} x + y + z = 1 \\ 5x + 4y + 3z = 0 \\ 6x + 3y + 2z = 1 \end{cases}$ **16.** *Cramer's Rule:* $\begin{cases} 2x - y = 4 \\ 3x + y = -5 \end{cases}$

17. Graph the solution of the system of inequalities.

$$\begin{cases} x - 2y < 0 \\ -2x + y > 2 \\ y > 0 \end{cases}$$

In Exercises 18–23, perform the indicated operation and simplify.

18. $\sqrt{-2}\left(\sqrt{-8} + 3\right)$

19. $(3 - 4i)^2$

20. $\left(\dfrac{t^{1/2}}{t^{1/4}}\right)^2$

21. $10\sqrt{20x} + 3\sqrt{125x}$

22. $\left(\sqrt{2x} - 3\right)^2$

23. $\dfrac{6}{\sqrt{10} - 2}$

24. Write the quotient in standard form: $\dfrac{1 - 2i}{4 + i}$.

In Exercises 25–28, solve the equation.

25. $\dfrac{1}{x} + \dfrac{4}{10 - x} = 1$

26. $\dfrac{x - 3}{x} + 1 = \dfrac{x - 4}{x - 6}$

27. $\sqrt{x} - x + 12 = 0$

28. $\sqrt{5 - x} + 10 = 11$

29. The stopping distance d of a car is directly proportional to the square of its speed s. On a certain type of pavement, a car requires 50 feet to stop when its speed is 25 miles per hour. Estimate the stopping distance when the speed of the car is 40 miles per hour. Explain your reasoning.

30. The number N of prey t months after a predator is introduced into an area is inversely proportional to $t + 1$. If $N = 300$ when $t = 0$, find N when $t = 5$.

31. At a local high school city championship basketball game, 1435 tickets were sold. A student admission ticket cost \$1.50 and an adult admission ticket cost \$5.00. The total ticket sales for the basketball game were \$3552.50. How many of each type of ticket were sold?

32. A total of \$50,000 is invested in two funds paying 8% and 8.5% simple interest. The yearly interest is \$4150. How much is invested at each rate?

33. The four corners are cut from a 12-inch-by-12-inch piece of glass, as shown in the figure. Find the perimeter of the remaining piece of glass.

34. The velocity v (in feet per second) of an object is given by $v = \sqrt{2gh}$, where $g = 32$ feet per second per second and h is the distance (in feet) the object has fallen. Find the height from which a rock has dropped if it strikes the ground with a velocity of 65 feet per second.

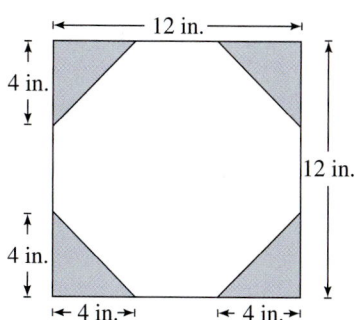

Figure for 33

Motivating the Chapter

Height of a Falling Object

You drop or throw a rock from the Tacoma Narrows Bridge 192 feet above Puget Sound. The height h (in feet) of the rock at any time t (in seconds) is

$$h = -16t^2 + v_0 t + h_0$$

where v_0 is the initial velocity (in feet per second) of the rock and h_0 is the initial height.

See Section 10.1, Exercise 143.

a. You drop the rock ($v_0 = 0$ ft/sec). How long will it take to hit the water? What method did you use to solve the quadratic equation? Explain why you used that method.

b. You throw the rock straight upward with an initial velocity of 32 feet per second. Find the time(s) when h is 192 feet. What method did you use to solve this quadratic equation? Explain why you used this method.

See Section 10.3, Exercise 121.

c. You throw the rock straight upward with an initial velocity of 32 feet per second. Find the time when h is 100 feet. What method did you use to solve this quadratic equation? Explain why you used this method.

d. You move to a lookout point that is 128 feet above the water. You throw the rock straight upward at the same rate as when you were 192 feet above the water. Would you expect it to reach the water in less time? Verify your conclusion algebraically.

See Section 10.6, Exercise 117.

e. You throw a rock straight upward with an initial velocity of 32 feet per second from a height of 192 feet. During what interval of time is the height greater than 144 feet?

Bohemian Nomad Picturemakers/Corbis

10

Quadratic Equations, Functions, and Inequalities

10.1 Solving Quadratic Equations: Factoring and Special Forms

Chris Whitehead/Getty Images

What You Should Learn

1 Solve quadratic equations by factoring.
2 Solve quadratic equations by the Square Root Property.
3 Solve quadratic equations with complex solutions by the Square Root Property.
4 Use substitution to solve equations of quadratic form.

Why You Should Learn It

Quadratic equations can be used to model and solve real-life problems. For instance, in Exercises 141 and 142 on page 622, you will use a quadratic equation to determine national health care expenditures in the United States.

Solving Quadratic Equations by Factoring

In this chapter, you will study methods for solving quadratic equations and equations of quadratic form. To begin, let's review the method of factoring that you studied in Section 6.5.

Remember that the first step in solving a quadratic equation by factoring is to write the equation in general form. Next, factor the left side. Finally, set each factor equal to zero and solve for x. Check each solution in the original equation.

1 Solve quadratic equations by factoring.

Example 1   Solving Quadratic Equations by Factoring

a.

$x^2 + 5x = 24$	Original equation
$x^2 + 5x - 24 = 0$	Write in general form.
$(x + 8)(x - 3) = 0$	Factor.
$x + 8 = 0 \implies x = -8$	Set 1st factor equal to 0.
$x - 3 = 0 \implies x = 3$	Set 2nd factor equal to 0.

b.

$3x^2 = 4 - 11x$	Original equation
$3x^2 + 11x - 4 = 0$	Write in general form.
$(3x - 1)(x + 4) = 0$	Factor.
$3x - 1 = 0 \implies x = \dfrac{1}{3}$	Set 1st factor equal to 0.
$x + 4 = 0 \implies x = -4$	Set 2nd factor equal to 0.

c.

$9x^2 + 12 = 3 + 12x + 5x^2$	Original equation
$4x^2 - 12x + 9 = 0$	Write in general form.
$(2x - 3)(2x - 3) = 0$	Factor.
$2x - 3 = 0 \implies x = \dfrac{3}{2}$	Set factor equal to 0.

Check each solution in its original equation.

Study Tip

In Example 1(c), the quadratic equation produces two identical solutions. This is called a **double** or **repeated solution.**

② Solve quadratic equations by the Square Root Property.

The Square Root Property

Consider the following equation, where $d > 0$ and u is an algebraic expression.

$$u^2 = d \qquad \text{Original equation}$$
$$u^2 - d = 0 \qquad \text{Write in general form.}$$
$$\left(u + \sqrt{d}\right)\left(u - \sqrt{d}\right) = 0 \qquad \text{Factor.}$$
$$u + \sqrt{d} = 0 \implies u = -\sqrt{d} \qquad \text{Set 1st factor equal to 0.}$$
$$u - \sqrt{d} = 0 \implies u = \sqrt{d} \qquad \text{Set 2nd factor equal to 0.}$$

Because the solutions differ only in sign, they can be written together using a "plus or minus sign": $u = \pm\sqrt{d}$. This form of the solution is read as "u is equal to plus or minus the square root of d." When you are solving an equation of the form $u^2 = d$ *without* going through the steps of factoring, you are using the Square Root Property.

Square Root Property

The equation $u^2 = d$, where $d > 0$, has exactly two solutions:

$$u = \sqrt{d} \quad \text{and} \quad u = -\sqrt{d}.$$

These solutions can also be written as $u = \pm\sqrt{d}$. This solution process is also called **extracting square roots.**

Example 2   Square Root Property

a. $3x^2 = 15$ Original equation

$\qquad x^2 = 5$ Divide each side by 3.

$\qquad x = \pm\sqrt{5}$ Square Root Property

The solutions are $x = \sqrt{5}$ and $x = -\sqrt{5}$. Check these in the original equation.

b. $(x - 2)^2 = 10$ Original equation

$\qquad x - 2 = \pm\sqrt{10}$ Square Root Property

$\qquad x = 2 \pm\sqrt{10}$ Add 2 to each side.

The solutions are $x = 2 + \sqrt{10} \approx 5.16$ and $x = 2 - \sqrt{10} \approx -1.16$.

c. $(3x - 6)^2 - 8 = 0$ Original equation

$\qquad (3x - 6)^2 = 8$ Add 8 to each side.

$\qquad 3x - 6 = \pm 2\sqrt{2}$ Square Root Property and rewrite $\sqrt{8}$ as $2\sqrt{2}$.

$\qquad 3x = 6 \pm 2\sqrt{2}$ Add 6 to each side.

$\qquad x = \dfrac{6 \pm 2\sqrt{2}}{3}$ Divide each side by 3.

The solutions are $x = \left(6 + 2\sqrt{2}\right)/3 \approx 2.94$ and $x = \left(6 - 2\sqrt{2}\right)/3 \approx 1.06$.

Technology: Tip

To check graphically the solutions of an equation written in general form, graph the left side of the equation and locate its x-intercepts. For instance, in Example 2(b), write the equation as

$$(x - 2)^2 - 10 = 0$$

and then use a graphing calculator to graph

$$y = (x - 2)^2 - 10$$

as shown below. You can use the *zoom* and *trace* features or the *zero* or *root* feature to approximate the x-intercepts of the graph to be $x \approx 5.16$ and $x \approx -1.16$.

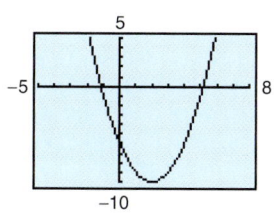

3 Solve quadratic equations with complex solutions by the Square Root Property.

Quadratic Equations with Complex Solutions

Prior to Section 9.6, the only solutions to find were real numbers. But now that you have studied complex numbers, it makes sense to look for other types of solutions. For instance, although the quadratic equation $x^2 + 1 = 0$ has no solutions that are real numbers, it does have two solutions that are complex numbers: i and $-i$. To check this, substitute i and $-i$ for x.

$$(i)^2 + 1 = -1 + 1 = 0 \qquad \text{Solution checks. } ✓$$

$$(-i)^2 + 1 = -1 + 1 = 0 \qquad \text{Solution checks. } ✓$$

One way to find complex solutions of a quadratic equation is to extend the Square Root Property to cover the case in which d is a negative number.

Square Root Property (Complex Square Root)

The equation $u^2 = d$, where $d < 0$, has exactly two solutions:

$$u = \sqrt{|d|}\,i \quad \text{and} \quad u = -\sqrt{|d|}\,i.$$

These solutions can also be written as $u = \pm\sqrt{|d|}\,i$.

Technology: Discovery

Solve each quadratic equation below algebraically. Then use a graphing calculator to check the solutions. Which equations have real solutions and which have complex solutions? Which graphs have x-intercepts and which have no x-intercepts? Compare the type of solution(s) of each quadratic equation with the x-intercept(s) of the graph of the equation.

a. $y = 2x^2 + 3x - 5$

b. $y = 2x^2 + 4x + 2$

c. $y = x^2 + 4$

d. $y = (x + 7)^2 + 2$

Example 3   Square Root Property

a. $x^2 + 8 = 0$ Original equation

$\quad\quad x^2 = -8$ Subtract 8 from each side.

$\quad\quad x = \pm\sqrt{8}\,i = \pm2\sqrt{2}\,i$ Square Root Property

The solutions are $x = 2\sqrt{2}\,i$ and $x = -2\sqrt{2}\,i$. Check these in the original equation.

b. $(x - 4)^2 = -3$ Original equation

$\quad\quad x - 4 = \pm\sqrt{3}\,i$ Square Root Property

$\quad\quad x = 4 \pm \sqrt{3}\,i$ Add 4 to each side.

The solutions are $x = 4 + \sqrt{3}\,i$ and $x = 4 - \sqrt{3}\,i$. Check these in the original equation.

c. $2(3x - 5)^2 + 32 = 0$ Original equation

$\quad\quad 2(3x - 5)^2 = -32$ Subtract 32 from each side.

$\quad\quad (3x - 5)^2 = -16$ Divide each side by 2.

$\quad\quad 3x - 5 = \pm4i$ Square Root Property

$\quad\quad 3x = 5 \pm 4i$ Add 5 to each side.

$$x = \frac{5 \pm 4i}{3} \qquad \text{Divide each side by 3.}$$

The solutions are $x = (5 + 4i)/3$ and $x = (5 - 4i)/3$. Check these in the original equation.

④ Use substitution to solve equations of quadratic form.

Equations of Quadratic Form

Both the factoring method and the Square Root Property can be applied to nonquadratic equations that are of **quadratic form.** An equation is said to be of quadratic form if it has the form

$$au^2 + bu + c = 0$$

where u is an algebraic expression. Here are some examples.

Equation	Written in Quadratic Form
$x^4 + 5x^2 + 4 = 0$	$(x^2)^2 + 5(x^2) + 4 = 0$
$x - 5\sqrt{x} + 6 = 0$	$(\sqrt{x})^2 - 5(\sqrt{x}) + 6 = 0$
$2x^{2/3} + 5x^{1/3} - 3 = 0$	$2(x^{1/3})^2 + 5(x^{1/3}) - 3 = 0$
$18 + 2x^2 + (x^2 + 9)^2 = 8$	$(x^2 + 9)^2 + 2(x^2 + 9) - 8 = 0$

To solve an equation of quadratic form, it helps to make a substitution and rewrite the equation in terms of u, as demonstrated in Examples 4 and 5.

Example 4   Solving an Equation of Quadratic Form

Solve $x^4 - 13x^2 + 36 = 0$.

Solution

Begin by writing the original equation in quadratic form, as follows.

$$x^4 - 13x^2 + 36 = 0 \qquad \text{Write original equation.}$$
$$(x^2)^2 - 13(x^2) + 36 = 0 \qquad \text{Write in quadratic form.}$$

Next, let $u = x^2$ and substitute u into the equation written in quadratic form. Then, factor and solve the equation.

$$u^2 - 13u + 36 = 0 \qquad \text{Substitute } u \text{ for } x^2.$$
$$(u - 4)(u - 9) = 0 \qquad \text{Factor.}$$
$$u - 4 = 0 \quad \Longrightarrow \quad u = 4 \qquad \text{Set 1st factor equal to 0.}$$
$$u - 9 = 0 \quad \Longrightarrow \quad u = 9 \qquad \text{Set 2nd factor equal to 0.}$$

At this point you have found the "u-solutions." To find the "x-solutions," replace u with x^2 and solve for x.

$$u = 4 \quad \Longrightarrow \quad x^2 = 4 \quad \Longrightarrow \quad x = \pm 2$$
$$u = 9 \quad \Longrightarrow \quad x^2 = 9 \quad \Longrightarrow \quad x = \pm 3$$

The solutions are $x = 2$, $x = -2$, $x = 3$, and $x = -3$. Check these in the original equation.

Be sure you see in Example 4 that the u-solutions of 4 and 9 represent only a temporary step. They are not solutions of the original equation and cannot be substituted into the original equation.

Technology: Tip

You may find it helpful to graph the equation with a graphing calculator before you begin. The graph will indicate the number of real solutions an equation has. For instance, the graph shown below is from the equation in Example 4. You can see from the graph that there are four x-intercepts and so there are four real solutions.

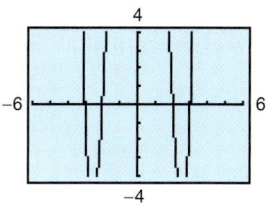

Example 5 Solving an Equation of Quadratic Form

a. $x - 5\sqrt{x} + 6 = 0$ *Original equation*

This equation is of quadratic form with $u = \sqrt{x}$.

$$(\sqrt{x})^2 - 5(\sqrt{x}) + 6 = 0$$ *Write in quadratic form.*

$$u^2 - 5u + 6 = 0$$ *Substitute u for $\sqrt{x}$.*

$$(u - 2)(u - 3) = 0$$ *Factor.*

$$u - 2 = 0 \implies u = 2$$ *Set 1st factor equal to 0.*

$$u - 3 = 0 \implies u = 3$$ *Set 2nd factor equal to 0.*

Now, using the u-solutions of 2 and 3, you obtain the x-solutions as follows.

$$u = 2 \implies \sqrt{x} = 2 \implies x = 4$$

$$u = 3 \implies \sqrt{x} = 3 \implies x = 9$$

b. $x^{2/3} - x^{1/3} - 6 = 0$ *Original equation*

This equation is of quadratic form with $u = x^{1/3}$.

$$(x^{1/3})^2 - (x^{1/3}) - 6 = 0$$ *Write in quadratic form.*

$$u^2 - u - 6 = 0$$ *Substitute u for $x^{1/3}$.*

$$(u + 2)(u - 3) = 0$$ *Factor.*

$$u + 2 = 0 \implies u = -2$$ *Set 1st factor equal to 0.*

$$u - 3 = 0 \implies u = 3$$ *Set 2nd factor equal to 0.*

Now, using the u-solutions of -2 and 3, you obtain the x-solutions as follows.

$$u = -2 \implies x^{1/3} = -2 \implies x = -8$$

$$u = 3 \implies x^{1/3} = 3 \implies x = 27$$

Example 6 Surface Area of a Softball

The surface area of a sphere of radius r is given by $S = 4\pi r^2$. The surface area of a softball is $144/\pi$ square inches. Find the diameter d of the softball.

Solution

$$\frac{144}{\pi} = 4\pi r^2$$ *Substitute $144/\pi$ for S.*

$$\frac{36}{\pi^2} = r^2 \implies \pm\sqrt{\frac{36}{\pi^2}} = r$$ *Divide each side by 4π and use Square Root Property.*

Choosing the positive root, you obtain $r = 6/\pi$, and so the diameter of the softball is

$$d = 2r = 2\left(\frac{6}{\pi}\right) = \frac{12}{\pi} \approx 3.82 \text{ inches.}$$

10.1 Exercises

Review   *Concepts, Skills, and Problem Solving*

Keep mathematically in shape by doing these exercises *before* the problems of this section.

Properties and Definitions

1. *Writing*   Identify the leading coefficient in $5t - 3t^3 + 7t^2$. Explain.

2. *Writing*   State the degree of the product $(y^2 - 2)$ $(y^3 + 7)$. Explain.

3. *Writing*   Sketch a graph for which y is not a function of x. Explain why it is not a function.

4. *Writing*   Sketch a graph for which y is a function of x. Explain why it is a function.

Simplifying Expressions

In Exercises 5–10, simplify the expression.

5. $(x^3 \cdot x^{-2})^{-3}$ 6. $(5x^{-4}y^5)(-3x^2y^{-1})$

7. $\left(\dfrac{2x}{3y}\right)^{-2}$

8. $\left(\dfrac{7u^{-4}}{3v^{-2}}\right)\left(\dfrac{14u}{6v^2}\right)^{-1}$

9. $\dfrac{6u^2v^{-3}}{27uv^3}$

10. $\dfrac{-14r^4s^2}{-98rs^2}$

Problem Solving

11. *Predator-Prey*   The number N of prey t months after a natural predator is introduced into the test area is inversely proportional to the square root of $t + 1$. If $N = 300$ when $t = 0$, find N when $t = 8$.

12. *Travel Time*   The travel time between two cities is inversely proportional to the average speed. A train travels between two cities in 2 hours at an average speed of 58 miles per hour. How long would the trip take at an average speed of 72 miles per hour? What does the constant of proportionality measure in this problem?

Developing Skills

In Exercises 1–20, solve the equation by factoring. See Example 1.

1. $x^2 - 12x + 35 = 0$ 2. $x^2 + 15x + 44 = 0$

3. $x^2 - x - 30 = 0$ 4. $x^2 + 2x - 63 = 0$

5. $x^2 + 4x = 45$

6. $x^2 - 7x = 18$

7. $x^2 - 12x + 36 = 0$

8. $x^2 + 60x + 900 = 0$

9. $9x^2 + 24x + 16 = 0$

10. $8x^2 - 10x + 3 = 0$

11. $4x^2 - 12x = 0$

12. $25y^2 - 75y = 0$

13. $u(u - 9) - 12(u - 9) = 0$

14. $16x(x - 8) - 12(x - 8) = 0$

15. $3x(x - 6) - 5(x - 6) = 0$

16. $3(4 - x) - 2x(4 - x) = 0$

17. $(y - 4)(y - 3) = 6$

18. $(6 + u)(1 - u) = 10$

19. $2x(3x + 2) = 5 - 6x^2$

20. $(2z + 1)(2z - 1) = -4z^2 - 5z + 2$

In Exercises 21–42, solve the equation by using the Square Root Property. See Example 2.

21. $x^2 = 49$ 22. $z^2 = 144$

23. $6x^2 = 54$ 24. $5t^2 = 5$

25. $25x^2 = 16$ 26. $9z^2 = 121$

27. $\dfrac{y^2}{2} = 32$ 28. $\dfrac{x^2}{6} = 24$

29. $4x^2 - 25 = 0$

30. $16y^2 - 121 = 0$

31. $4u^2 - 225 = 0$

32. $16x^2 - 1 = 0$

33. $(x + 4)^2 = 64$

34. $(y - 20)^2 = 25$

35. $(x - 3)^2 = 0.25$

36. $(x + 2)^2 = 0.81$

37. $(x - 2)^2 = 7$

38. $(x + 8)^2 = 28$

39. $(2x + 1)^2 = 50$

40. $(3x - 5)^2 = 48$

41. $(4x - 3)^2 - 98 = 0$

42. $(5x + 11)^2 - 300 = 0$

In Exercises 43–64, solve the equation by using the Square Root Property. See Example 3.

43. $z^2 = -36$

44. $x^2 = -16$

45. $x^2 + 4 = 0$

46. $y^2 + 16 = 0$

47. $9u^2 + 17 = 0$

48. $4v^2 + 9 = 0$

49. $(t - 3)^2 = -25$

50. $(x + 5)^2 = -81$

51. $(3z + 4)^2 + 144 = 0$

52. $(2y - 3)^2 + 25 = 0$

53. $(2x + 3)^2 = -54$

54. $(6y - 5)^2 = -8$

55. $9(x + 6)^2 = -121$

56. $4(x - 4)^2 = -169$

57. $(x - 1)^2 = -27$

58. $(2x + 3)^2 = -54$

59. $(x + 1)^2 + 0.04 = 0$

60. $(x - 3)^2 + 2.25 = 0$

61. $\left(c - \frac{2}{3}\right)^2 + \frac{1}{9} = 0$

62. $\left(u + \frac{5}{8}\right)^2 + \frac{49}{16} = 0$

63. $\left(x + \frac{7}{3}\right)^2 = -\frac{38}{9}$

64. $\left(y - \frac{5}{6}\right)^2 = -\frac{4}{5}$

In Exercises 65–80, find all real and complex solutions of the quadratic equation.

65. $2x^2 - 5x = 0$

66. $3t^2 + 6t = 0$

67. $2x^2 + 5x - 12 = 0$

68. $3x^2 + 8x - 16 = 0$

69. $x^2 - 900 = 0$

70. $y^2 - 225 = 0$

71. $x^2 + 900 = 0$

72. $y^2 + 225 = 0$

73. $\frac{2}{3}x^2 = 6$

74. $\frac{1}{3}x^2 = 4$

75. $(x - 5)^2 - 100 = 0$

76. $(y + 12)^2 - 400 = 0$

77. $(x - 5)^2 + 100 = 0$

78. $(y + 12)^2 + 400 = 0$

79. $(x + 2)^2 + 18 = 0$

80. $(x + 2)^2 - 18 = 0$

In Exercises 81–90, use a graphing calculator to graph the function. Use the graph to approximate any x-intercepts. Set $y = 0$ and solve the resulting equation. Compare the result with the x-intercepts of the graph.

81. $y = x^2 - 9$

82. $y = 5x - x^2$

83. $y = x^2 - 2x - 15$

84. $y = 9 - 4(x - 3)^2$

85. $y = 4 - (x - 3)^2$

86. $y = 4(x + 1)^2 - 9$

87. $y = 2x^2 - x - 6$

88. $y = 4x^2 - x - 14$

89. $y = 3x^2 - 8x - 16$

90. $y = 5x^2 + 9x - 18$

🖩 In Exercises 91–96, use a graphing calculator to graph the function and observe that the graph has no x-intercepts. Set $y = 0$ and solve the resulting equation. Identify the type of solutions of the equation.

91. $y = x^2 + 7$

92. $y = x^2 + 5$

93. $y = (x - 1)^2 + 1$

94. $y = (x + 2)^2 + 3$

95. $y = (x + 3)^2 + 5$

96. $y = (x - 2)^2 + 3$

🖩 In Exercises 97–100, solve for y in terms of x. Let f and g be functions representing, respectively, the positive square root and the negative square root. Use a graphing calculator to graph f and g in the same viewing window.

97. $x^2 + y^2 = 4$

98. $x^2 - y^2 = 4$

99. $x^2 + 4y^2 = 4$

100. $x - y^2 = 0$

In Exercises 101–130, solve the equation of quadratic form. (Find all real *and* complex solutions.) See Examples 4–5.

101. $x^4 - 5x^2 + 4 = 0$

102. $x^4 - 10x^2 + 25 = 0$

103. $x^4 - 5x^2 + 6 = 0$

104. $x^4 - 11x^2 + 30 = 0$

105. $(x^2 - 4)^2 + 2(x^2 - 4) - 3 = 0$

106. $(x^2 - 1)^2 + (x^2 - 1) - 6 = 0$

107. $x - 3\sqrt{x} - 4 = 0$

108. $x - \sqrt{x} - 6 = 0$

109. $x - 7\sqrt{x} + 10 = 0$

110. $x - 11\sqrt{x} + 24 = 0$

111. $x^{2/3} - x^{1/3} - 6 = 0$

112. $x^{2/3} + 3x^{1/3} - 10 = 0$

113. $2x^{2/3} - 7x^{1/3} + 5 = 0$

114. $3x^{2/3} + 8x^{1/3} + 5 = 0$

115. $x^{2/5} - 3x^{1/5} + 2 = 0$

116. $x^{2/5} + 5x^{1/5} + 6 = 0$

117. $2x^{2/5} - 7x^{1/5} + 3 = 0$

118. $2x^{2/5} + 3x^{1/5} + 1 = 0$

119. $x^{1/3} - x^{1/6} - 6 = 0$

120. $x^{1/3} + 2x^{1/6} - 3 = 0$

121. $x^{1/2} - 3x^{1/4} + 2 = 0$

122. $x^{1/2} - 5x^{1/4} + 6 = 0$

123. $\dfrac{1}{x^2} - \dfrac{3}{x} + 2 = 0$

124. $\dfrac{1}{x^2} - \dfrac{1}{x} - 6 = 0$

125. $4x^{-2} - x^{-1} - 5 = 0$

126. $2x^{-2} - x^{-1} - 1 = 0$

127. $(x^2 - 3x)^2 - 2(x^2 - 3x) - 24 = 0$

128. $(x^2 - 6x)^2 - 2(x^2 - 6x) - 35 = 0$

129. $16\left(\dfrac{x - 1}{x - 8}\right)^2 + 8\left(\dfrac{x - 1}{x - 8}\right) + 1 = 0$

130. $9\left(\dfrac{x + 2}{x + 3}\right)^2 - 6\left(\dfrac{x + 2}{x + 3}\right) + 1 = 0$

Solving Problems

131. 🔺 *Geometry*   The surface area S of a spherical float for a parade is 289π square feet. Find the diameter d of the float.

132. 🔺 *Geometry*   The surface area S of a basketball is $900/\pi$ square inches. Find the radius r of the basketball.

Free-Falling Object   In Exercises 133–136, find the time required for an object to reach the ground when it is dropped from a height of s_0 feet. The height h (in feet) is given by

$$h = -16t^2 + s_0$$

where t measures the time (in seconds) after the object is released.

133. $s_0 = 256$ **134.** $s_0 = 48$

135. $s_0 = 128$ **136.** $s_0 = 500$

137. *Free-Falling Object*   The height h (in feet) of an object thrown vertically upward from a tower 144 feet tall is given by $h = 144 + 128t - 16t^2$, where t measures the time in seconds from the time when the object is released. How long does it take for the object to reach the ground?

138. *Revenue*   The revenue R (in dollars) from selling x televisions is given by $R = x\left(120 - \frac{1}{2}x\right)$. Find the number of televisions that must be sold to produce a revenue of $7000.

Compound Interest   The amount A after 2 years when a principal of P dollars is invested at annual interest rate r compounded annually is given by $A = P(1 + r)^2$. In Exercises 139 and 140, find r.

139. $P = \$1500$, $A = \$1685.40$

140. $P = \$5000$, $A = \$5724.50$

National Health Expenditures   In Exercises 141 and 142, the national expenditures for health care in the United States from 1995 through 2001 is given by

$$y = 4.43t^2 + 872, \quad 5 \le t \le 11.$$

In this model, y represents the expenditures (in billions of dollars) and t represents the year, with $t = 5$ corresponding to 1995 (see figure).   (Source: U.S. Centers for Medicare & Medicaid Services)

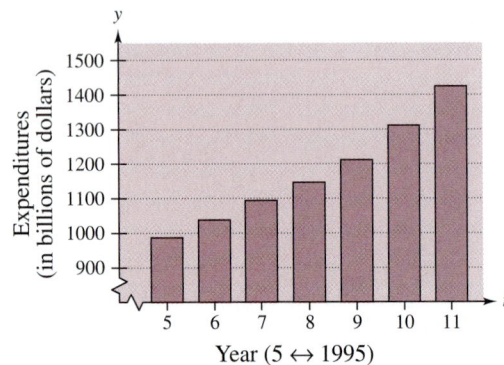

Figure for 141 and 142

141. Algebraically determine the year when expenditures were approximately $1100 billion. Graphically confirm the result.

142. Algebraically determine the year when expenditures were approximately $1200 billion. Graphically confirm the result.

Explaining Concepts

143. ⚡ Answer parts (a) and (b) of Motivating the Chapter on page 612.

144. *Writing* ✏ For a quadratic equation $ax^2 + bx + c = 0$, where a, b, and c are real numbers with $a \neq 0$, explain why b and c can equal 0, but a cannot.

145. *Writing* ✏ Explain the Zero-Factor Property and how it can be used to solve a quadratic equation.

146. Is it possible for a quadratic equation to have only one solution? If so, give an example.

147. *True or False?*   The only solution of the equation $x^2 = 25$ is $x = 5$. Justify your answer.

148. *Writing* ✏ Describe the steps in solving a quadratic equation by using the Square Root Property.

149. *Writing* ✏ Describe the procedure for solving an equation of quadratic form. Give an example.

10.2 Completing the Square

Lon C. Diehl/PhotoEdit, Inc.

What You Should Learn

1. Rewrite quadratic expressions in completed square form.
2. Solve quadratic equations by completing the square.

Why You Should Learn It

You can use techniques such as completing the square to solve quadratic equations that model real-life situations. For instance, Example 7 on page 626 shows how to find the dimensions of a cereal box by completing the square.

1. **Rewrite quadratic expressions in completed square form.**

Constructing Perfect Square Trinomials

Consider the quadratic equation

$$(x - 2)^2 = 10. \qquad \text{Completed square form}$$

You know from Example 2(b) in the preceding section that this equation has two solutions: $x = 2 + \sqrt{10}$ and $x = 2 - \sqrt{10}$. Suppose you were given the equation in its general form

$$x^2 - 4x - 6 = 0. \qquad \text{General form}$$

How could you solve this form of the quadratic equation? You could try factoring, but after attempting to do so you would find that the left side of the equation is not factorable using integer coefficients.

In this section, you will study a technique for rewriting an equation in a completed square form. This technique is called **completing the square.** Note that prior to completing the square, the coefficient of the second-degree term must be 1.

Completing the Square

To **complete the square** for the expression $x^2 + bx$, add $(b/2)^2$, which is the square of half the coefficient of x. Consequently,

$$x^2 + bx + \left(\frac{b}{2}\right)^2 = \left(x + \frac{b}{2}\right)^2.$$

$$\underbrace{\phantom{\left(\frac{b}{2}\right)^2}}_{(\text{half})^2}$$

Example 1   Constructing a Perfect Square Trinomial

What term should be added to $x^2 - 8x$ so that it becomes a perfect square trinomial? To find this term, notice that the coefficient of the x-term is -8. Take half of this coefficient and square the result to get $(-4)^2 = 16$. Add this term to the expression to make it a perfect square trinomial.

$$x^2 - 8x + (-4)^2 = x^2 - 8x + 16 \qquad \text{Add } (-4)^2 = 16 \text{ to the expression.}$$

You can then rewrite the expression as the square of a binomial, $(x - 4)^2$.

2 Solve quadratic equations by completing the square.

Solving Equations by Completing the Square

Completing the square can be used to solve quadratic equations. When using this procedure, remember to *preserve the equality* by adding the same constant to each side of the equation.

Study Tip

In Example 2, completing the square is used for the sake of illustration. This particular equation would be easier to solve by factoring. Try reworking the problem by factoring to see that you obtain the same two solutions.

Example 2 Completing the Square: Leading Coefficient Is 1

Solve $x^2 + 12x = 0$ by completing the square.

Solution

$$x^2 + 12x = 0 \qquad \text{Write original equation.}$$

$$x^2 + 12x + 6^2 = 36 \qquad \text{Add } 6^2 = 36 \text{ to each side.}$$

$$\underbrace{\qquad}_{(\text{half})^2}$$

$$(x + 6)^2 = 36 \qquad \text{Completed square form}$$

$$x + 6 = \pm\sqrt{36} \qquad \text{Square Root Property}$$

$$x = -6 \pm 6 \qquad \text{Subtract 6 from each side.}$$

$$x = -6 + 6 \text{ or } x = -6 - 6 \qquad \text{Separate solutions.}$$

$$x = 0 \qquad\qquad x = -12 \qquad \text{Solutions}$$

The solutions are $x = 0$ and $x = -12$. Check these in the original equation.

Example 3 Completing the Square: Leading Coefficient Is 1

Solve $x^2 - 6x + 7 = 0$ by completing the square.

Solution

$$x^2 - 6x + 7 = 0 \qquad \text{Write original equation.}$$

$$x^2 - 6x = -7 \qquad \text{Subtract 7 from each side.}$$

$$x^2 - 6x + (-3)^2 = -7 + 9 \qquad \text{Add } (-3)^2 = 9 \text{ to each side.}$$

$$\underbrace{\qquad}_{(\text{half})^2}$$

$$(x - 3)^2 = 2 \qquad \text{Completed square form}$$

$$x - 3 = \pm\sqrt{2} \qquad \text{Square Root Property}$$

$$x = 3 \pm \sqrt{2} \qquad \text{Add 3 to each side.}$$

$$x = 3 + \sqrt{2} \text{ or } x = 3 - \sqrt{2} \qquad \text{Solutions}$$

The solutions are $x = 3 + \sqrt{2} \approx 4.41$ and $x = 3 - \sqrt{2} \approx 1.59$. Check these in the original equation.

Technology: Tip

You can use a graphing calculator to check the solution to Example 3. Graph

$$y = x^2 - 6x + 7$$

as shown below. Then use the *zero* or *root* feature of the graphing calculator to approximate the x-intercepts to be $x \approx 4.41$ and $x \approx 1.59$, which are the same solutions obtained in Example 3.

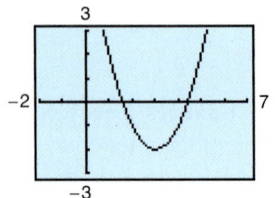

If the leading coefficient of a quadratic equation is not 1, you must divide each side of the equation by this coefficient *before* completing the square. This process is demonstrated in Examples 4 and 5.

Example 4   Completing the Square: Leading Coefficient Is Not 1

$$2x^2 - x - 2 = 0 \qquad \text{Original equation}$$

$$2x^2 - x = 2 \qquad \text{Add 2 to each side.}$$

$$x^2 - \frac{1}{2}x = 1 \qquad \text{Divide each side by 2.}$$

$$x^2 - \frac{1}{2}x + \left(-\frac{1}{4}\right)^2 = 1 + \frac{1}{16} \qquad \text{Add } \left(-\tfrac{1}{4}\right)^2 = \tfrac{1}{16} \text{ to each side.}$$

$$\left(x - \frac{1}{4}\right)^2 = \frac{17}{16} \qquad \text{Completed square form}$$

$$x - \frac{1}{4} = \pm\frac{\sqrt{17}}{4} \qquad \text{Square Root Property}$$

$$x = \frac{1}{4} \pm \frac{\sqrt{17}}{4} \qquad \text{Add } \tfrac{1}{4} \text{ to each side.}$$

The solutions are $x = \frac{1}{4}\left(1 + \sqrt{17}\right)$ and $x = \frac{1}{4}\left(1 - \sqrt{17}\right)$. Check these in the original equation.

Example 5   Completing the Square: Leading Coefficient Is Not 1

$$3x^2 - 6x + 1 = 0 \qquad \text{Original equation}$$

$$3x^2 - 6x = -1 \qquad \text{Subtract 1 from each side.}$$

$$x^2 - 2x = -\frac{1}{3} \qquad \text{Divide each side by 3.}$$

$$x^2 - 2x + (-1)^2 = -\frac{1}{3} + 1 \qquad \text{Add } (-1)^2 = 1 \text{ to each side.}$$

$$(x - 1)^2 = \frac{2}{3} \qquad \text{Completed square form}$$

$$x - 1 = \pm\sqrt{\frac{2}{3}} \qquad \text{Square Root Property}$$

$$x - 1 = \pm\frac{\sqrt{6}}{3} \qquad \text{Rationalize the denominator.}$$

$$x = 1 \pm \frac{\sqrt{6}}{3} \qquad \text{Add 1 to each side.}$$

The solutions are $x = 1 + \sqrt{6}/3$ and $x = 1 - \sqrt{6}/3$. Check these in the original equation.

Example 6 A Quadratic Equation with Complex Solutions

Solve $x^2 - 4x + 8 = 0$ by completing the square.

Solution

$$x^2 - 4x + 8 = 0 \qquad \text{Write original equation.}$$

$$x^2 - 4x = -8 \qquad \text{Subtract 8 from each side.}$$

$$x^2 - 4x + (-2)^2 = -8 + 4 \qquad \text{Add } (-2)^2 = 4 \text{ to each side.}$$

$$(x - 2)^2 = -4 \qquad \text{Completed square form}$$

$$x - 2 = \pm 2i \qquad \text{Square Root Property}$$

$$x = 2 \pm 2i \qquad \text{Add 2 to each side.}$$

The solutions are $x = 2 + 2i$ and $x = 2 - 2i$. Check these in the original equation.

Example 7 Dimensions of a Cereal Box

A cereal box has a volume of 441 cubic inches. Its height is 12 inches and its base has the dimensions x by $x + 7$ (see Figure 10.1). Find the dimensions of the base in inches.

12 in.

$x + 7$

x

Figure 10.1

Solution

$$lwh = V \qquad \text{Formula for volume of a rectangular box}$$

$$(x + 7)(x)(12) = 441 \qquad \text{Substitute 441 for } V, x + 7 \text{ for } l, x \text{ for } w, \text{ and 12 for } h.$$

$$12x^2 + 84x = 441 \qquad \text{Multiply factors.}$$

$$x^2 + 7x = \frac{441}{12} \qquad \text{Divide each side by 12.}$$

$$x^2 + 7x + \left(\frac{7}{2}\right)^2 = \frac{147}{4} + \frac{49}{4} \qquad \text{Add } \left(\frac{7}{2}\right)^2 = \frac{49}{4} \text{ to each side.}$$

$$\left(x + \frac{7}{2}\right)^2 = \frac{196}{4} \qquad \text{Completed square form}$$

$$x + \frac{7}{2} = \pm\sqrt{49} \qquad \text{Square Root Property}$$

$$x = -\frac{7}{2} \pm 7 \qquad \text{Subtract } \tfrac{7}{2} \text{ from each side.}$$

Choosing the positive root, you obtain

$$x = -\frac{7}{2} + 7 = \frac{7}{2} = 3.5 \text{ inches} \qquad \text{Width of base}$$

and

$$x + 7 = 3.5 + 7 = 10.5 \text{ inches.} \qquad \text{Length of base}$$

10.2 Exercises

Review   Concepts, Skills, and Problem Solving

Keep mathematically in shape by doing these exercises *before* the problems of this section.

Properties and Definitions

In Exercises 1–4, complete the rule of exponents and/or simplify.

1. $(ab)^4 = $

2. $(a^r)^s = $

3. $\left(\dfrac{a}{b}\right)^{-r} = $, $a \neq 0, b \neq 0$

4. $a^{-r} = $, $a \neq 0$

Solving Equations

In Exercises 5–8, solve the equation.

5. $\dfrac{4}{x} - \dfrac{2}{3} = 0$

6. $2x - 3[1 + (4 - x)] = 0$

7. $3x^2 - 13x - 10 = 0$

8. $x(x - 3) = 40$

Graphing

In Exercises 9–12, graph the function.

9. $g(x) = \frac{2}{3}x - 5$

10. $h(x) = 5 - \sqrt{x}$

11. $f(x) = \dfrac{4}{x + 2}$

12. $f(x) = 2x + |x - 1|$

Developing Skills

In Exercises 1–16, add a term to the expression so that it becomes a perfect square trinomial. See Example 1.

1. $x^2 + 8x + $

2. $x^2 + 12x + $

3. $y^2 - 20y + $

4. $y^2 - 2y + $

5. $x^2 - 16x + $

6. $x^2 + 18x + $

7. $t^2 + 5t + $

8. $u^2 + 7u + $

9. $x^2 - 9x + $

10. $y^2 - 11y + $

11. $a^2 - \frac{1}{3}a + $

12. $y^2 + \frac{4}{3}y + $

13. $y^2 - \frac{3}{5}y + $

14. $x^2 - \frac{6}{5}x + $

15. $r^2 - 0.4r + $

16. $s^2 + 4.6s + $

In Exercises 17–32, solve the equation first by completing the square and then by factoring. See Examples 2–5.

17. $x^2 - 20x = 0$

18. $x^2 + 32x = 0$

19. $x^2 + 6x = 0$

20. $t^2 - 10t = 0$

21. $y^2 - 5y = 0$

22. $t^2 - 9t = 0$

23. $t^2 - 8t + 7 = 0$

24. $y^2 - 8y + 12 = 0$

25. $x^2 + 7x + 12 = 0$

26. $z^2 + 3z - 10 = 0$

27. $x^2 - 3x - 18 = 0$

28. $t^2 - 5t - 36 = 0$

29. $2x^2 - 14x + 12 = 0$

30. $3x^2 - 3x - 6 = 0$

31. $4x^2 + 4x - 15 = 0$

32. $3x^2 - 13x + 12 = 0$

In Exercises 33–72, solve the equation by completing the square. Give the solutions in exact form and in decimal form rounded to two decimal places. (The solutions may be complex numbers.) See Examples 2–6.

33. $x^2 - 4x - 3 = 0$

34. $x^2 - 6x + 7 = 0$

35. $x^2 + 4x - 3 = 0$

36. $x^2 + 6x + 7 = 0$

37. $x^2 + 6x = 7$

38. $x^2 + 8x = 9$

39. $x^2 - 10x = 22$

40. $x^2 - 4x = -9$

41. $x^2 + 8x + 7 = 0$

42. $x^2 + 10x + 9 = 0$

43. $x^2 - 10x + 21 = 0$ **44.** $x^2 - 10x + 24 = 0$ **61.** $3x^2 + 9x + 5 = 0$ **62.** $5x^2 - 15x + 7 = 0$

45. $y^2 + 5y + 3 = 0$

63. $4y^2 + 4y - 9 = 0$

46. $y^2 + 6y + 7 = 0$ **47.** $x^2 + 10 = 6x$

48. $x^2 + 23 = 10x$ **49.** $z^2 + 4z + 13 = 0$ **64.** $4z^2 - 3z + 2 = 0$

50. $z^2 + 12z + 25 = 0$

65. $5x^2 - 3x + 10 = 0$

51. $-x^2 + x - 1 = 0$ **52.** $1 - x - x^2 = 0$

66. $7x^2 + 4x + 3 = 0$

53. $x^2 - 7x + 12 = 0$
54. $y^2 + 5y + 9 = 0$

67. $x(x - 7) = 2$ **68.** $2x\left(x + \dfrac{4}{3}\right) = 5$

55. $x^2 - \frac{2}{3}x - 3 = 0$ **56.** $x^2 + \frac{4}{5}x - 1 = 0$

69. $0.5t^2 + t + 2 = 0$

57. $v^2 + \frac{3}{4}v - 2 = 0$

70. $0.1x^2 + 0.5x = -0.2$

58. $u^2 - \frac{2}{3}u + 5 = 0$

71. $0.1x^2 + 0.2x + 0.5 = 0$
72. $0.02x^2 + 0.10x - 0.05 = 0$

59. $2x^2 + 8x + 3 = 0$ **60.** $3x^2 - 24x - 5 = 0$

In Exercises 73–78, find the real solutions.

73. $\dfrac{x}{2} - \dfrac{1}{x} = 1$

74. $\dfrac{x}{2} + \dfrac{5}{x} = 4$

75. $\dfrac{x^2}{4} = \dfrac{x+1}{2}$

76. $\dfrac{x^2+2}{24} = \dfrac{x-1}{3}$

77. $\sqrt{2x+1} = x - 3$

78. $\sqrt{3x-2} = x - 2$

🖩 In Exercises 79–86, use a graphing calculator to graph the function. Use the graph to approximate any x-intercepts of the graph. Set $y = 0$ and solve the resulting equation. Compare the result with the x-intercepts of the graph.

79. $y = x^2 + 4x - 1$ **80.** $y = x^2 + 6x - 4$

81. $y = x^2 - 2x - 5$ **82.** $y = 2x^2 - 6x - 5$

83. $y = \frac{1}{3}x^2 + 2x - 6$ **84.** $y = \frac{1}{2}x^2 - 3x + 1$

85. $y = -x^2 - x + 3$ **86.** $y = \sqrt{x} - x + 2$

Solving Problems

87. 🔺 *Geometric Modeling*

(a) Find the area of the two adjoining rectangles and large square in the figure.

(b) Find the area of the small square in the lower right-hand corner of the figure and add it to the area found in part (a).

(c) Find the dimensions and the area of the entire figure after adjoining the small square in the lower right-hand corner of the figure. Note that you have shown geometrically the technique of completing the square.

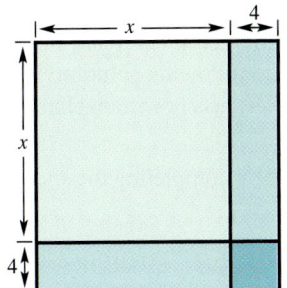

Figure for 87

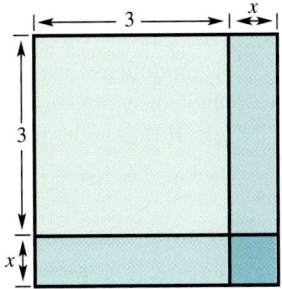

Figure for 88

88. 🔺 *Geometric Modeling*   Repeat Exercise 87 for the model shown above.

89. 🔺 *Geometry*   The area of the triangle in the figure is 12 square centimeters. Find the base and height.

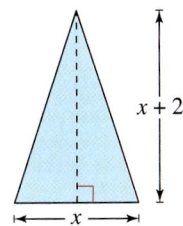

Figure for 89

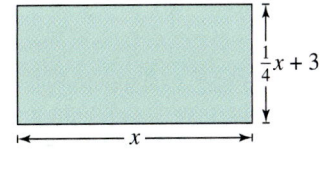

Figure for 90

90. 🔺 *Geometry*   The area of the rectangle in the figure is 160 square feet. Find the rectangle's dimensions.

91. 🔺 *Geometry*   You have 200 meters of fencing to enclose two adjacent rectangular corrals (see figure). The total area of the enclosed region is 1400 square meters. What are the dimensions of each corral? (The corrals are the same size.)

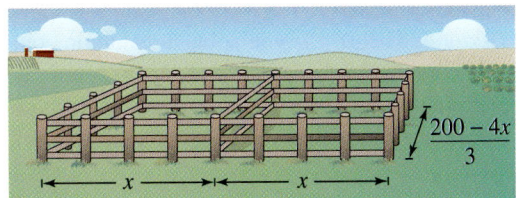

92. ▲ *Geometry* An open box with a rectangular base of x inches by $x + 4$ inches has a height of 6 inches (see figure). The volume of the box is 840 cubic inches. Find the dimensions of the box.

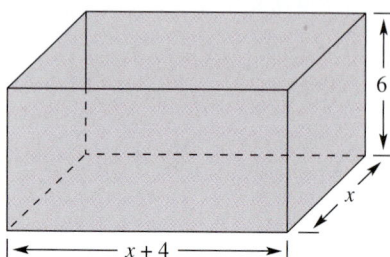

93. *Revenue* The revenue R (in dollars) from selling x pairs of running shoes is given by

$$R = x\left(50 - \frac{1}{2}x\right).$$

Find the number of pairs of running shoes that must be sold to produce a revenue of $1218.

94. *Revenue* The revenue R (in dollars) from selling x golf clubs is given by

$$R = x\left(100 - \frac{1}{10}x\right).$$

Find the number of golf clubs that must be sold to produce a revenue of $11,967.90.

Explaining Concepts

95. *Writing* What is a perfect square trinomial?

96. *Writing* What term must be added to $x^2 + 5x$ to complete the square? Explain how you found the term.

97. *Writing* Explain the use of the Square Root Property when solving a quadratic equation by the method of completing the square.

98. Is it possible for a quadratic equation to have no real number solution? If so, give an example.

99. *Writing* When using the method of completing the square to solve a quadratic equation, what is the first step if the leading coefficient is not 1? Is the resulting equation equivalent to the original equation? Explain.

100. *True or False?* If you solve a quadratic equation by completing the square and obtain solutions that are rational numbers, then you could have solved the equation by factoring. Justify your answer.

101. *Writing* Consider the quadratic equation $(x - 1)^2 = d$.

(a) What value(s) of d will produce a quadratic equation that has exactly one (repeated) solution?

(b) Describe the value(s) of d that will produce two different solutions, both of which are *rational* numbers.

(c) Describe the value(s) of d that will produce two different solutions, both of which are *irrational* numbers.

(d) Describe the value(s) of d that will produce two different solutions, both of which are *complex* numbers.

102. *Writing* You teach an algebra class and one of your students hands in the following solution. Find and correct the error(s). Discuss how to explain the error(s) to your student.

Solve $x^2 + 6x - 13 = 0$ by completing the square.

$$x^2 + 6x = 13$$
$$x^2 + 6x + \left(\frac{6}{2}\right)^2 = 13$$
$$(x + 3)^2 = 13$$
$$x + 3 = \pm\sqrt{13}$$
$$x = -3 \pm \sqrt{13}$$

10.3 The Quadratic Formula

Mug Shots/Corbis

Why You Should Learn It

Knowing the Quadratic Formula can be helpful in solving quadratic equations that model real-life situations. For instance, in Exercise 117 on page 639, you will solve a quadratic equation that models the number of people employed in the construction industry.

What You Should Learn

1. Derive the Quadratic Formula by completing the square for a general quadratic equation.
2. Use the Quadratic Formula to solve quadratic equations.
3. Determine the types of solutions of quadratic equations using the discriminant.
4. Write quadratic equations from solutions of the equations.

The Quadratic Formula

A fourth technique for solving a quadratic equation involves the **Quadratic Formula.** This formula is derived by completing the square for a general quadratic equation.

1. Derive the Quadratic Formula by completing the square for a general quadratic equation.

$$ax^2 + bx + c = 0 \qquad \text{General form, } a \neq 0$$

$$ax^2 + bx = -c \qquad \text{Subtract } c \text{ from each side.}$$

$$x^2 + \frac{b}{a}x = -\frac{c}{a} \qquad \text{Divide each side by } a.$$

$$x^2 + \frac{b}{a}x + \left(\frac{b}{2a}\right)^2 = -\frac{c}{a} + \left(\frac{b}{2a}\right)^2 \qquad \text{Add } \left(\frac{b}{2a}\right)^2 \text{ to each side.}$$

$$\left(x + \frac{b}{2a}\right)^2 = \frac{b^2 - 4ac}{4a^2} \qquad \text{Simplify.}$$

$$x + \frac{b}{2a} = \pm\sqrt{\frac{b^2 - 4ac}{4a^2}} \qquad \text{Square Root Property}$$

$$x = -\frac{b}{2a} \pm \frac{\sqrt{b^2 - 4ac}}{2|a|} \qquad \text{Subtract } \frac{b}{2a} \text{ from each side.}$$

$$x = \frac{-b \pm \sqrt{b^2 - 4ac}}{2a} \qquad \text{Simplify.}$$

Study Tip

The Quadratic Formula is one of the most important formulas in algebra, and you should memorize it. It helps to try to memorize a verbal statement of the rule. For instance, you might try to remember the following verbal statement of the Quadratic Formula: "The opposite of b, plus or minus the square root of b squared minus $4ac$, all divided by $2a$."

The Quadratic Formula

The solutions of $ax^2 + bx + c = 0$, $a \neq 0$, are given by the **Quadratic Formula**

$$x = \frac{-b \pm \sqrt{b^2 - 4ac}}{2a}.$$

The expression inside the radical, $b^2 - 4ac$, is called the **discriminant.**

1. If $b^2 - 4ac > 0$, the equation has *two* real solutions.

2. If $b^2 - 4ac = 0$, the equation has *one* (repeated) real solution.

3. If $b^2 - 4ac < 0$, the equation has *no* real solutions.

② Use the Quadratic Formula to solve quadratic equations.

Solving Equations by the Quadratic Formula

When using the Quadratic Formula, remember that *before* the formula can be applied, you must first write the quadratic equation in general form in order to determine the values of a, b, and c.

Example 1 The Quadratic Formula: Two Distinct Solutions

$x^2 + 6x = 16$	Original equation
$x^2 + 6x - 16 = 0$	Write in general form.
$x = \dfrac{-b \pm \sqrt{b^2 - 4ac}}{2a}$	Quadratic Formula
$x = \dfrac{-6 \pm \sqrt{6^2 - 4(1)(-16)}}{2(1)}$	Substitute 1 for a, 6 for b, and -16 for c.
$x = \dfrac{-6 \pm \sqrt{100}}{2}$	Simplify.
$x = \dfrac{-6 \pm 10}{2}$	Simplify.
$x = 2$ or $x = -8$	Solutions

The solutions are $x = 2$ and $x = -8$. Check these in the original equation.

> **Study Tip**
>
> In Example 1, the solutions are rational numbers, which means that the equation could have been solved by factoring. Try solving the equation by factoring.

> **Study Tip**
>
> If the leading coefficient of a quadratic equation is negative, you should begin by multiplying each side of the equation by -1, as shown in Example 2. This will produce a positive leading coefficient, which is less cumbersome to work with.

Example 2 The Quadratic Formula: Two Distinct Solutions

$-x^2 - 4x + 8 = 0$	Leading coefficient is negative.
$x^2 + 4x - 8 = 0$	Multiply each side by -1.
$x = \dfrac{-b \pm \sqrt{b^2 - 4ac}}{2a}$	Quadratic Formula
$x = \dfrac{-4 \pm \sqrt{4^2 - 4(1)(-8)}}{2(1)}$	Substitute 1 for a, 4 for b, and -8 for c.
$x = \dfrac{-4 \pm \sqrt{48}}{2}$	Simplify.
$x = \dfrac{-4 \pm 4\sqrt{3}}{2}$	Simplify.
$x = \dfrac{2(-2 \pm 2\sqrt{3})}{2}$	Factor numerator.
$x = \dfrac{\cancel{2}(-2 \pm 2\sqrt{3})}{\cancel{2}}$	Divide out common factor.
$x = -2 \pm 2\sqrt{3}$	Solutions

The solutions are $x = -2 + 2\sqrt{3}$ and $x = -2 - 2\sqrt{3}$. Check these in the original equation.

Example 3   The Quadratic Formula: One Repeated Solution

$$18x^2 - 24x + 8 = 0$$ Original equation

$$9x^2 - 12x + 4 = 0$$ Divide each side by 2.

$$x = \frac{-b \pm \sqrt{b^2 - 4ac}}{2a}$$ Quadratic Formula

$$x = \frac{-(-12) \pm \sqrt{(-12)^2 - 4(9)(4)}}{2(9)}$$ Substitute 9 for a, -12 for b, and 4 for c.

$$x = \frac{12 \pm \sqrt{144 - 144}}{18}$$ Simplify.

$$x = \frac{12 \pm \sqrt{0}}{18}$$ Simplify.

$$x = \frac{2}{3}$$ Solution

The only solution is $x = \frac{2}{3}$. Check this in the original equation.

Note in the next example how the Quadratic Formula can be used to solve a quadratic equation that has complex solutions.

Example 4   The Quadratic Formula: Complex Solutions

$$2x^2 - 4x + 5 = 0$$ Original equation

$$x = \frac{-b \pm \sqrt{b^2 - 4ac}}{2a}$$ Quadratic Formula

$$x = \frac{-(-4) \pm \sqrt{(-4)^2 - 4(2)(5)}}{2(2)}$$ Substitute 2 for a, -4 for b, and 5 for c.

$$x = \frac{4 \pm \sqrt{-24}}{4}$$ Simplify.

$$x = \frac{4 \pm 2\sqrt{6}\,i}{4}$$ Write in i-form.

$$x = \frac{2(2 \pm \sqrt{6}\,i)}{2 \cdot 2}$$ Factor numerator and denominator.

$$x = \frac{2(2 \pm \sqrt{6}\,i)}{2 \cdot 2}$$ Divide out common factor.

$$x = \frac{2 \pm \sqrt{6}\,i}{2}$$ Solutions

The solutions are $x = \frac{1}{2}(2 + \sqrt{6}\,i)$ and $x = \frac{1}{2}(2 - \sqrt{6}\,i)$. Check these in the original equation.

Study Tip

Example 3 could have been solved as follows, without dividing each side by 2 in the first step.

$$x = \frac{-(-24) \pm \sqrt{(-24)^2 - 4(18)(8)}}{2(18)}$$

$$x = \frac{24 \pm \sqrt{576 - 576}}{36}$$

$$x = \frac{24 \pm 0}{36}$$

$$x = \frac{2}{3}$$

While the result is the same, dividing each side by 2 simplifies the equation before the Quadratic Formula is applied and so allows you to work with smaller numbers.

③ Determine the types of solutions of quadratic equations using the discriminant.

The Discriminant

The radicand in the Quadratic Formula, $b^2 - 4ac$, is called the discriminant because it allows you to "discriminate" among different types of solutions.

Study Tip

By reexamining Examples 1 through 4, you can see that the equations with rational or repeated solutions could have been solved by *factoring*. In general, quadratic equations (with integer coefficients) for which the discriminant is either zero or a perfect square are factorable using integer coefficients. Consequently, a quick test of the discriminant will help you decide which solution method to use to solve a quadratic equation.

Using the Discriminant

Let a, b, and c be rational numbers such that $a \neq 0$. The discriminant of the quadratic equation $ax^2 + bx + c = 0$ is given by $b^2 - 4ac$, and can be used to classify the solutions of the equation as follows.

Discriminant	Solution Type
1. Perfect square	Two distinct rational solutions (Example 1)
2. Positive nonperfect square	Two distinct irrational solutions (Example 2)
3. Zero	One repeated rational solution (Example 3)
4. Negative number	Two distinct complex solutions (Example 4)

Example 5 Using the Discriminant

Determine the type of solution(s) for each quadratic equation.

a. $x^2 - x + 2 = 0$

b. $2x^2 - 3x - 2 = 0$

c. $x^2 - 2x + 1 = 0$

d. $x^2 - 2x - 1 = 9$

Solution

Equation	Discriminant	Solution Type
a. $x^2 - x + 2 = 0$	$b^2 - 4ac = (-1)^2 - 4(1)(2)$ $= 1 - 8 = -7$	Two distinct complex solutions
b. $2x^2 - 3x - 2 = 0$	$b^2 - 4ac = (-3)^2 - 4(2)(-2)$ $= 9 + 16 = 25$	Two distinct rational solutions
c. $x^2 - 2x + 1 = 0$	$b^2 - 4ac = (-2)^2 - 4(1)(1)$ $= 4 - 4 = 0$	One repeated rational solution
d. $x^2 - 2x - 1 = 9$	$b^2 - 4ac = (-2)^2 - 4(1)(-10)$ $= 4 + 40 = 44$	Two distinct irrational solutions

Technology: Discovery

Use a graphing calculator to graph each equation.

a. $y = x^2 - x + 2$

b. $y = 2x^2 - 3x - 2$

c. $y = x^2 - 2x + 1$

d. $y = x^2 - 2x - 10$

Describe the solution type of each equation and check your results with those shown in Example 5. Why do you think the discriminant is used to determine solution types?

Summary of Methods for Solving Quadratic Equations

Method	*Example*
1. Factoring	$3x^2 + x = 0$
	$x(3x + 1) = 0$ ➡ $x = 0$ and $x = -\dfrac{1}{3}$
2. Square Root Property	$(x + 2)^2 = 7$
	$x + 2 = \pm\sqrt{7}$ ➡ $x = -2 + \sqrt{7}$ and $x = -2 - \sqrt{7}$
3. Completing the square	$x^2 + 6x = 2$
	$x^2 + 6x + 3^2 = 2 + 9$
	$(x + 3)^2 = 11$ ➡ $x = -3 + \sqrt{11}$ and $x = -3 - \sqrt{11}$
4. Quadratic Formula	$3x^2 - 2x + 2 = 0$ ➡ $x = \dfrac{-(-2)\pm\sqrt{(-2)^2 - 4(3)(2)}}{2(3)} = \dfrac{1}{3} \pm \dfrac{\sqrt{5}}{3}i$

4 Write quadratic equations from solutions of the equations.

Writing Quadratic Equations from Solutions

Using the Zero-Factor Property, you know that the equation $(x + 5)(x - 2) = 0$ has two solutions, $x = -5$ and $x = 2$. You can use the Zero-Factor Property in reverse to find a quadratic equation given its solutions. This process is demonstrated in Example 6.

Reverse of Zero-Factor Property

Let a and b be real numbers, variables, or algebraic expressions. If $a = 0$ or $b = 0$, then a and b are factors such that $ab = 0$.

Technology: Tip

A program for several models of graphing calculators that uses the Quadratic Formula to solve quadratic equations can be found at our website, *math.college.hmco.com/students*. The program will display real solutions to quadratic equations.

Example 6   Writing a Quadratic Equation from Its Solutions

Write a quadratic equation that has the solutions $x = 4$ and $x = -7$. Using the solutions $x = 4$ and $x = -7$, you can write the following.

$$x = 4 \qquad \text{and} \qquad x = -7 \quad \text{\small\color{red}Solutions}$$
$$x - 4 = 0 \qquad\qquad\qquad x + 7 = 0 \quad \text{\small\color{red}Obtain zero on one side of each equation.}$$
$$(x - 4)(x + 7) = 0 \quad \text{\small\color{red}Reverse of Zero-Factor Property}$$
$$x^2 + 3x - 28 = 0 \quad \text{\small\color{red}Foil Method}$$

So, a quadratic equation that has the solutions $x = 4$ and $x = -7$ is

$$x^2 + 3x - 28 = 0.$$

This is not the only quadratic equation with the solutions $x = 4$ and $x = -7$. You can obtain other quadratic equations with these solutions by multiplying $x^2 + 3x - 28 = 0$ by any nonzero real number.

10.3 Exercises

Review *Concepts, Skills, and Problem Solving*

Keep mathematically in shape by doing these exercises *before* the problems of this section.

Properties and Definitions

In Exercises 1 and 2, rewrite the expression using the specified property, where *a* and *b* are nonnegative real numbers.

1. Multiplication Property: $\sqrt{ab} =$ ▭

2. Division Property: $\sqrt{\dfrac{a}{b}} =$ ▭

3. *Writing* Is $\sqrt{72}$ in simplest form? Explain.

4. *Writing* Is $10/\sqrt{5}$ in simplest form? Explain.

Simplifying Expressions

In Exercises 5–10, perform the operation and simplify the expression.

5. $\sqrt{128} + 3\sqrt{50}$

6. $3\sqrt{5}\sqrt{500}$

7. $\left(3 + \sqrt{2}\right)\left(3 - \sqrt{2}\right)$

8. $\left(3 + \sqrt{2}\right)^2$

9. $\dfrac{8}{\sqrt{10}}$

10. $\dfrac{5}{\sqrt{12} - 2}$

Problem Solving

11. ▲ *Geometry* Determine the length and width of a rectangle with a perimeter of 50 inches and a diagonal of $5\sqrt{13}$ inches.

12. *Demand* The demand equation for a product is given by $p = 75 - \sqrt{1.2(x - 10)}$, where x is the number of units demanded per day and p is the price per unit. Find the demand when the price is set at $59.90.

Developing Skills

In Exercises 1–4, write the quadratic equation in general form.

1. $2x^2 = 7 - 2x$

2. $7x^2 + 15x = 5$

3. $x(10 - x) = 5$

4. $x(3x + 8) = 15$

In Exercises 5–16, solve the equation first by using the Quadratic Formula and then by factoring. See Examples 1–4.

5. $x^2 - 11x + 28 = 0$

6. $x^2 - 12x + 27 = 0$

7. $x^2 + 6x + 8 = 0$

8. $x^2 + 9x + 14 = 0$

9. $4x^2 + 4x + 1 = 0$

10. $9x^2 + 12x + 4 = 0$

11. $4x^2 + 12x + 9 = 0$

12. $9x^2 - 30x + 25 = 0$

13. $6x^2 - x - 2 = 0$

14. $10x^2 - 11x + 3 = 0$

15. $x^2 - 5x - 300 = 0$

16. $x^2 + 20x - 300 = 0$

In Exercises 17–46, solve the equation by using the Quadratic Formula. (Find all real *and* complex solutions.) See Examples 1–4.

17. $x^2 - 2x - 4 = 0$

18. $x^2 - 2x - 6 = 0$

19. $t^2 + 4t + 1 = 0$

20. $y^2 + 6y + 4 = 0$

21. $x^2 + 6x - 3 = 0$

22. $x^2 + 8x - 4 = 0$

23. $x^2 - 10x + 23 = 0$

24. $u^2 - 12u + 29 = 0$

25. $2x^2 + 3x + 3 = 0$

26. $2x^2 - x + 1 = 0$

27. $3v^2 - 2v - 1 = 0$ **28.** $4x^2 + 6x + 1 = 0$

29. $2x^2 + 4x - 3 = 0$ **30.** $x^2 - 8x + 19 = 0$

31. $9z^2 + 6z - 4 = 0$ **32.** $8y^2 - 8y - 1 = 0$

33. $-4x^2 - 6x + 3 = 0$

34. $-5x^2 - 15x + 10 = 0$

35. $8x^2 - 6x + 2 = 0$

36. $6x^2 + 3x - 9 = 0$

37. $-4x^2 + 10x + 12 = 0$

38. $-15x^2 - 10x + 25 = 0$

39. $9x^2 = 1 + 9x$ **40.** $7x^2 = 3 - 5x$

41. $3x - 2x^2 = 4 - 5x^2$

42. $x - x^2 = 1 - 6x^2$ **43.** $x^2 - 0.4x - 0.16 = 0$

44. $x^2 + 0.6x - 0.41 = 0$

45. $2.5x^2 + x - 0.9 = 0$

46. $0.09x^2 - 0.12x - 0.26 = 0$

In Exercises 47–56, use the discriminant to determine the type of solutions of the quadratic equation. See Example 5.

47. $x^2 + x + 1 = 0$

48. $x^2 + x - 1 = 0$

49. $2x^2 - 5x - 4 = 0$

50. $10x^2 + 5x + 1 = 0$

51. $5x^2 + 7x + 3 = 0$

52. $3x^2 - 2x - 5 = 0$

53. $4x^2 - 12x + 9 = 0$

54. $2x^2 + 10x + 6 = 0$

55. $3x^2 - x + 2 = 0$

56. $9x^2 - 24x + 16 = 0$

In Exercises 57–74, solve the quadratic equation by using the most convenient method. (Find all real *and* complex solutions.)

57. $z^2 - 169 = 0$ **58.** $t^2 = 144$

59. $5y^2 + 15y = 0$ **60.** $7u^2 + 49u = 0$

61. $25(x - 3)^2 - 36 = 0$

62. $9(x + 4)^2 + 16 = 0$

63. $2y(y - 18) + 3(y - 18) = 0$

64. $4y(y + 7) - 5(y + 7) = 0$

65. $x^2 + 8x + 25 = 0$ **66.** $x^2 - 3x - 4 = 0$

67. $x^2 - 24x + 128 = 0$ **68.** $y^2 + 21y + 108 = 0$

69. $3x^2 - 13x + 169 = 0$

70. $2x^2 - 15x + 225 = 0$

71. $18x^2 + 15x - 50 = 0$

72. $14x^2 + 11x - 40 = 0$

73. $7x(x + 2) + 5 = 3x(x + 1)$

74. $5x(x - 1) - 7 = 4x(x - 2)$

In Exercises 75–84, write a quadratic equation having the given solutions. See Example 6.

75. $5, -2$ **76.** $-2, 3$

77. $1, 7$ **78.** $3, 9$

79. $1 + \sqrt{2}, 1 - \sqrt{2}$

80. $-3 + \sqrt{5}, -3 - \sqrt{5}$

81. $5i, -5i$ **82.** $2i, -2i$

83. 12 **84.** -4

In Exercises 85–92, use a graphing calculator to graph the function. Use the graph to approximate any *x*-intercepts of the graph. Set $y = 0$ and solve the resulting equation. Compare the result with the *x*-intercepts of the graph.

85. $y = 3x^2 - 6x + 1$ **86.** $y = x^2 + x + 1$

87. $y = -(4x^2 - 20x + 25)$
88. $y = x^2 - 4x + 3$ **89.** $y = 5x^2 - 18x + 6$

90. $y = 15x^2 + 3x - 105$
91. $y = -0.04x^2 + 4x - 0.8$
92. $y = 3.7x^2 - 10.2x + 3.2$

In Exercises 93–96, use a graphing calculator to determine the number of real solutions of the quadratic equation. Verify your answer algebraically.

93. $2x^2 - 5x + 5 = 0$ **94.** $2x^2 - x - 1 = 0$

95. $\frac{1}{5}x^2 + \frac{6}{5}x - 8 = 0$ **96.** $\frac{1}{3}x^2 - 5x + 25 = 0$

In Exercises 97–100, determine all real values of *x* for which the function has the indicated value.

97. $f(x) = 2x^2 - 7x + 1, \ f(x) = -3$

98. $f(x) = 2x^2 - 7x + 5, \ f(x) = 0$
99. $g(x) = 2x^2 - 3x + 16, \ g(x) = 14$
100. $h(x) = 6x^2 + x + 10, \ h(x) = -2$

In Exercises 101–104, solve the equation.

101. $\dfrac{2x^2}{5} - \dfrac{x}{2} = 1$ **102.** $\dfrac{x^2 - 9x}{6} = \dfrac{x - 1}{2}$

103. $\sqrt{x + 3} = x - 1$ **104.** $\sqrt{2x - 3} = x - 2$

Think About It In Exercises 105–108, describe the values of *c* such that the equation has (a) two real number solutions, (b) one real number solution, and (c) two complex number solutions.

105. $x^2 - 6x + c = 0$

106. $x^2 - 12x + c = 0$

107. $x^2 + 8x + c = 0$

108. $x^2 + 2x + c = 0$

Solving Problems

109. ▲ *Geometry* A rectangle has a width of *x* inches, a length of $x + 6.3$ inches, and an area of 58.14 square inches. Find its dimensions.

110. ▲ *Geometry* A rectangle has a length of $x + 1.5$ inches, a width of *x* inches, and an area of 18.36 square inches. Find its dimensions.

111. *Free-Falling Object* A stone is thrown vertically upward at a velocity of 40 feet per second from a bridge that is 50 feet above the level of the water (see figure). The height *h* (in feet) of the stone at time *t* (in seconds) after it is thrown is

$h = -16t^2 + 40t + 50.$

(a) Find the time when the stone is again 50 feet above the water.

(b) Find the time when the stone strikes the water.

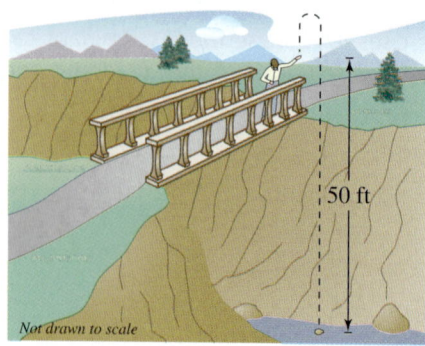

Not drawn to scale

Figure for 111

112. *Free-Falling Object*   A stone is thrown vertically upward at a velocity of 20 feet per second from a bridge that is 40 feet above the level of the water. The height h (in feet) of the stone at time t (in seconds) after it is thrown is

$$h = -16t^2 + 20t + 40.$$

(a) Find the time when the stone is again 40 feet above the water.

(b) Find the time when the stone strikes the water.

113. *Free-Falling Object*   You stand on a bridge and throw a stone upward from 25 feet above a lake with an initial velocity of 20 feet per second. The height h (in feet) of the stone at time t (in seconds) after it is thrown is modeled by

$$h = -16t^2 + 20t + 25.$$

(a) Find the time when the stone is again 25 feet above the lake.

(b) Find the time when the stone strikes the water.

114. *Free-Falling Object*   You stand on a bridge and throw a stone upward from 61 feet above a lake with an initial velocity of 36 feet per second. The height h (in feet) of the stone at time t (in seconds) after it is thrown is modeled by

$$h = -16t^2 + 36t + 61.$$

(a) Find the time when the stone is again 61 feet above the lake.

(b) Find the time when the stone strikes the water.

115. *Free-Falling Object*   From the roof of a building, 100 feet above the ground, you toss a coin upward with an initial velocity of 5 feet per second. The height h (in feet) of the coin at time t (in seconds) after it is tossed is modeled by

$$h = -16t^2 + 5t + 100.$$

(a) Find the time when the coin is again 100 feet above the ground.

(b) Find the time when the coin hits the ground.

116. *Free-Falling Object*   You throw an apple upward from 42 feet above the ground in an apple tree, with an initial velocity of 30 feet per second. The height h (in feet) of the apple at time t (in seconds) after it is thrown is modeled by

$$h = -16t^2 + 30t + 42.$$

(a) Find the time when the apple is again 42 feet above the ground.

(b) Find the time when the apple hits the ground.

117. *Employment*   The number y (in thousands) of people employed in the construction industry in the United States from 1994 through 2001 can be modeled by

$$y = 10.29t^2 + 164.5t + 6624, \ 4 \le t \le 11$$

where t represents the year, with $t = 4$ corresponding to 1994. (Source: U.S. Bureau of Labor Statistics)

(a) ⊞ Use a graphing calculator to graph the model.

(b) ⊞ Use the graph in part (a) to find the year in which there were approximately 9,000,000 employed in the construction industry in the United States. Verify your answer algebraically.

(c) Use the model to estimate the number employed in the construction industry in 2002.

118. ⊞ *Cellular Phone Subscribers*   The number s (in thousands) of cellular phone subscribers in the United States for the years 1994 through 2001 can be modeled by

$$s = 1178.29t^2 - 2816.5t + 17{,}457, \ 4 \le t \le 11$$

where $t = 4$ corresponds to 1994.

(Source: Cellular Telecommunications & Internet Association)

(a) Use a graphing calculator to graph the model.

(b) Use the graph in part (a) to determine the year in which there were 44 million cellular phone subscribers. Verify your answer algebraically.

119. *Exploration*   Determine the two solutions, x_1 and x_2, of each quadratic equation. Use the values of x_1 and x_2 to fill in the boxes.

Equation	x_1, x_2	$x_1 + x_2$	$x_1 x_2$
(a) $x^2 - x - 6 = 0$			
(b) $2x^2 + 5x - 3 = 0$			
(c) $4x^2 - 9 = 0$			
(d) $x^2 - 10x + 34 = 0$			

120. *Think About It*   Consider a general quadratic equation $ax^2 + bx + c = 0$ whose solutions are x_1 and x_2. Use the results of Exercise 119 to determine a relationship among the coefficients a, b, and c, and the sum $(x_1 + x_2)$ and product $(x_1 x_2)$ of the solutions.

Explaining Concepts

121. 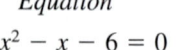   Answer parts (c) and (d) of Motivating the Chapter on page 612.

122. *Writing*   State the Quadratic Formula *in words*.

123. *Writing*   What is the discriminant of $ax^2 + bx + c = 0$? How is the discriminant related to the number and type of solutions of the equation?

124. *Writing*   Explain how completing the square can be used to develop the Quadratic Formula.

125. *Writing*   List the four methods for solving a quadratic equation.

Mid-Chapter Quiz

Take this quiz as you would take a quiz in class. After you are done, check your work against the answers in the back of the book.

In Exercises 1–8, solve the quadratic equation by the specified method.

1. Factoring:

$2x^2 - 72 = 0$

2. Factoring:

$2x^2 + 3x - 20 = 0$

3. Square Root Property:

$3x^2 = 36$

4. Square Root Property:

$(u - 3)^2 - 16 = 0$

5. Completing the square:

$s^2 + 10s + 1 = 0$

6. Completing the square:

$2y^2 + 6y - 5 = 0$

7. Quadratic Formula:

$x^2 + 4x - 6 = 0$

8. Quadratic Formula:

$6v^2 - 3v - 4 = 0$

In Exercises 9–16, solve the equation by using the most convenient method. (Find all real *and* complex solutions.)

9. $x^2 + 5x + 7 = 0$

10. $36 - (t - 4)^2 = 0$

11. $x(x - 10) + 3(x - 10) = 0$

12. $x(x - 3) = 10$

13. $4b^2 - 12b + 9 = 0$

14. $3m^2 + 10m + 5 = 0$

15. $x - 2\sqrt{x} - 24 = 0$

16. $x^4 + 7x^2 + 12 = 0$

In Exercises 17 and 18, solve the equation of quadratic form. (Find all real *and* complex solutions.)

17. $x - 4\sqrt{x} - 1 = 0$

18. $x^4 - 12x^2 + 27 = 0$

In Exercises 19 and 20, use a graphing calculator to graph the function. Use the graph to approximate any *x*-intercepts of the graph. Set $y = 0$ and solve the resulting equation. Compare the results with the *x*-intercepts of the graph.

19. $y = \frac{1}{2}x^2 - 3x - 1$

20. $y = x^2 + 0.45x - 4$

21. The revenue R from selling x alarm clocks is given by

$R = x(20 - 0.2x).$

Find the number of alarm clocks that must be sold to produce a revenue of $500.

22. A rectangle has a length of x meters, a width of $100 - x$ meters, and an area of 2275 square meters. Find its dimensions.

10.4 Graphs of Quadratic Functions

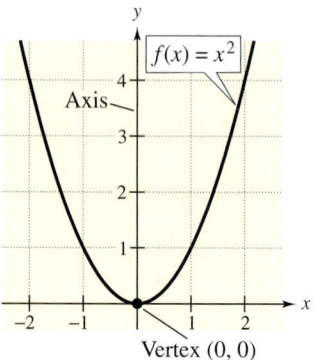

Joaquin Palting/Photodisc/Getty Images

What You Should Learn

1. Determine the vertices of parabolas by completing the square.
2. Sketch parabolas.
3. Write the equation of a parabola given the vertex and a point on the graph.
4. Use parabolas to solve application problems.

Why You Should Learn It

Real-life situations can be modeled by graphs of quadratic functions. For instance, in Exercise 101 on page 650, a quadratic equation is used to model the maximum height of a diver.

Graphs of Quadratic Functions

In this section, you will study graphs of quadratic functions of the form

$$f(x) = ax^2 + bx + c. \qquad \text{Quadratic function}$$

Figure 10.2 shows the graph of a simple quadratic function, $f(x) = x^2$.

1 Determine the vertices of parabolas by completing the square.

> **Graphs of Quadratic Functions**
>
> The graph of $f(x) = ax^2 + bx + c$, $a \neq 0$, is a **parabola.** The completed square form
>
> $$f(x) = a(x - h)^2 + k \qquad \text{Standard form}$$
>
> is the **standard form** of the function. The **vertex** of the parabola occurs at the point (h, k), and the vertical line passing through the vertex is the **axis** of the parabola.

Every parabola is *symmetric* about its axis, which means that if it were folded along its axis, the two parts would match.

If a is positive, the graph of $f(x) = ax^2 + bx + c$ opens upward, and if a is negative, the graph opens downward, as shown in Figure 10.3. Observe in Figure 10.3 that the y-coordinate of the vertex identifies the minimum function value if $a > 0$ and the maximum function value if $a < 0$.

Axis

$f(x) = x^2$

Vertex $(0, 0)$

Figure 10.2

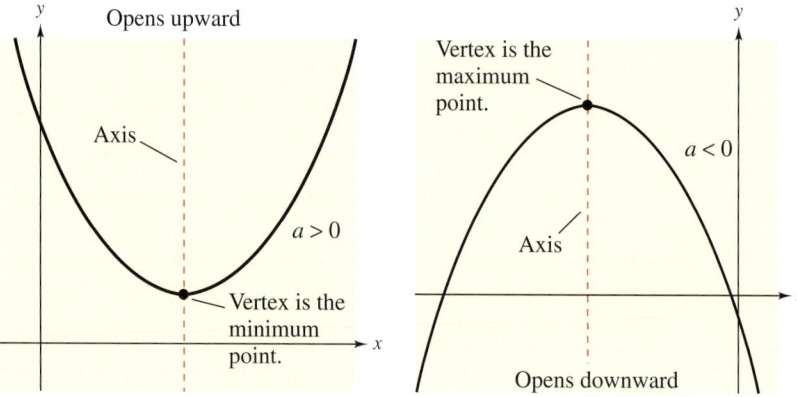

Figure 10.3

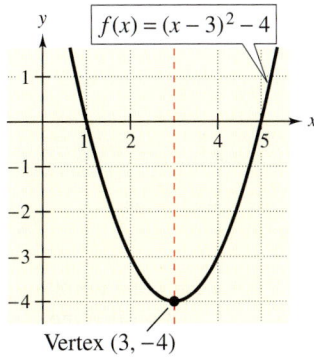

$f(x) = (x - 3)^2 - 4$

Vertex $(3, -4)$

Figure 10.4

Example 1   Finding the Vertex by Completing the Square

Find the vertex of the parabola given by $f(x) = x^2 - 6x + 5$.

Solution

Begin by writing the function in standard form.

$$f(x) = x^2 - 6x + 5 \qquad \text{Original function}$$

$$f(x) = x^2 - 6x + (-3)^2 - (-3)^2 + 5 \qquad \text{Complete the square.}$$

$$f(x) = (x^2 - 6x + 9) - 9 + 5 \qquad \text{Regroup terms.}$$

$$f(x) = (x - 3)^2 - 4 \qquad \text{Standard form}$$

From the standard form, you can see that the vertex of the parabola occurs at the point $(3, -4)$, as shown in Figure 10.4. The minimum value of the function is $f(3) = -4$.

Study Tip

When a number is added to a function and then that same number is subtracted from the function, the value of the function remains unchanged. Notice in Example 1 that $(-3)^2$ is added to the function to complete the square and then $(-3)^2$ is subtracted from the function so that the value of the function remains the same.

In Example 1, the vertex of the graph was found by *completing the square.* Another approach to finding the vertex is to complete the square once for a general function and then use the resulting formula for the vertex.

$$f(x) = ax^2 + bx + c \qquad \text{Quadratic function}$$

$$= a\left(x^2 + \frac{b}{a}x\right) + c \qquad \text{Factor } a \text{ out of first two terms.}$$

$$= a\left[x^2 + \frac{b}{a}x + \left(\frac{b}{2a}\right)^2\right] + c - \frac{b}{4a} \qquad \text{Complete the square.}$$

$$= a\left(x + \frac{b}{2a}\right)^2 + c - \frac{b^2}{4a} \qquad \text{Standard form}$$

From this form you can see that the vertex occurs when $x = -b/(2a)$.

Example 2   Finding the Vertex with a Formula

Find the vertex of the parabola given by $f(x) = 3x^2 - 9x$.

Solution

From the original function, it follows that $a = 3$ and $b = -9$. So, the x-coordinate of the vertex is

$$x = \frac{-b}{2a} = \frac{-(-9)}{2(3)} = \frac{3}{2}.$$

Substitute $\frac{3}{2}$ for x into the original equation to find the y-coordinate.

$$f\left(-\frac{b}{2a}\right) = f\left(\frac{3}{2}\right) = 3\left(\frac{3}{2}\right)^2 - 9\left(\frac{3}{2}\right) = -\frac{27}{4}$$

So, the vertex of the parabola is $\left(\frac{3}{2}, -\frac{27}{4}\right)$, the minimum value of the function is $f\left(\frac{3}{2}\right) = -\frac{27}{4}$, and the parabola opens upward, as shown in Figure 10.5.

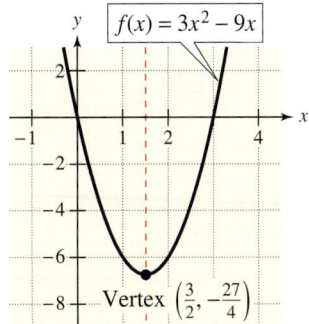

$f(x) = 3x^2 - 9x$

Vertex $\left(\frac{3}{2}, -\frac{27}{4}\right)$

Figure 10.5

2 Sketch parabolas.

Sketching a Parabola

To obtain an accurate sketch of a parabola, the following guidelines are useful.

> ### Sketching a Parabola
>
> 1. Determine the vertex and axis of the parabola by completing the square or by using the formula $x = -b/(2a)$.
>
> 2. Plot the vertex, axis, x- and y-intercepts, and a few additional points on the parabola. (Using the symmetry about the axis can reduce the number of points you need to plot.)
>
> 3. Use the fact that the parabola opens *upward* if $a > 0$ and opens *downward* if $a < 0$ to complete the sketch.

Example 3   Sketching a Parabola

To sketch the parabola given by $y = x^2 + 6x + 8$, begin by writing the equation in standard form.

$$y = x^2 + 6x + 8 \qquad \text{Write original equation.}$$

$$y = (x^2 + 6x + 3^2 - 3^2) + 8 \qquad \text{Complete the square.}$$

$$\text{(half of 6)}^2$$

$$y = (x^2 + 6x + 9) - 9 + 8 \qquad \text{Regroup terms.}$$

$$y = (x + 3)^2 - 1 \qquad \text{Standard form}$$

The vertex occurs at the point $(-3, -1)$ and the axis is the line $x = -3$. After plotting this information, calculate a few additional points on the parabola, as shown in the table. Note that the y-intercept is $(0, 8)$ and the x-intercepts are solutions to the equation

$$x^2 + 6x + 8 = (x + 4)(x + 2) = 0.$$

x	-5	-4	-3	-2	-1
$y = (x + 3)^2 - 1$	3	0	-1	0	3
Solution point	$(-5, 3)$	$(-4, 0)$	$(-3, -1)$	$(-2, 0)$	$(-1, 3)$

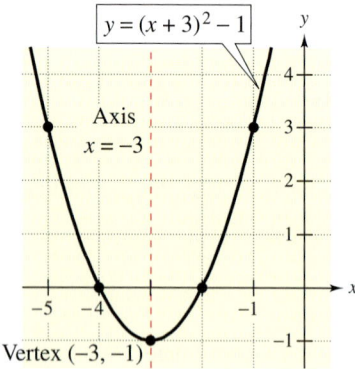

Figure 10.6

The graph of the parabola is shown in Figure 10.6. Note that the parabola opens upward because the leading coefficient (in general form) is positive.

3 Write the equation of a parabola given the vertex and a point on the graph.

Writing the Equation of a Parabola

To write the equation of a parabola with a vertical axis, use the fact that its standard equation has the form $y = a(x - h)^2 + k$, where (h, k) is the vertex.

Example 4   Writing the Equation of a Parabola

Write the equation of the parabola with vertex $(-2, 1)$ and y-intercept $(0, -3)$, as shown in Figure 10.7.

Solution

Because the vertex occurs at $(h, k) = (-2, 1)$, the equation has the form

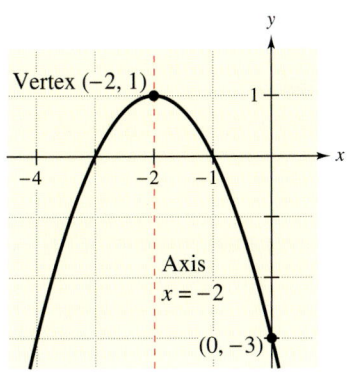

Vertex $(-2, 1)$

Axis
$x = -2$

$(0, -3)$

Figure 10.7

$$y = a(x - h)^2 + k \qquad \text{Standard form}$$
$$y = a[x - (-2)]^2 + 1 \qquad \text{Substitute } -2 \text{ for } h \text{ and } 1 \text{ for } k.$$
$$y = a(x + 2)^2 + 1. \qquad \text{Simplify.}$$

To find the value of a, use the fact that the y-intercept is $(0, -3)$.

$$y = a(x + 2)^2 + 1 \qquad \text{Write standard form.}$$
$$-3 = a(0 + 2)^2 + 1 \qquad \text{Substitute 0 for } x \text{ and } -3 \text{ for } y.$$
$$-1 = a \qquad \text{Simplify.}$$

So, the standard form of the equation of the parabola is

$$y = -(x + 2)^2 + 1.$$

Example 5   Writing the Equation of a Parabola

Write the equation of the parabola with vertex $(3, -4)$ and that passes through the point $(5, -2)$, as shown in Figure 10.8.

Solution

Because the vertex occurs at $(h, k) = (3, -4)$, the equation has the form

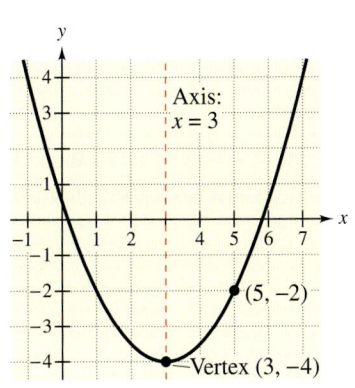

Axis:
$x = 3$

$(5, -2)$

Vertex $(3, -4)$

Figure 10.8

$$y = a(x - h)^2 + k \qquad \text{Standard form}$$
$$y = a(x - 3)^2 + (-4) \qquad \text{Substitute 3 for } h \text{ and } -4 \text{ for } k.$$
$$y = a(x - 3)^2 - 4. \qquad \text{Simplify.}$$

To find the value of a, use the fact that the parabola passes through the point $(5, -2)$.

$$y = a(x - 3)^2 - 4 \qquad \text{Write standard form.}$$
$$-2 = a(5 - 3)^2 - 4 \qquad \text{Substitute 5 for } x \text{ and } -2 \text{ for } y.$$
$$\frac{1}{2} = a \qquad \text{Simplify.}$$

So, the standard form of the equation of the parabola is $y = \frac{1}{2}(x - 3)^2 - 4$.

④ Use parabolas to solve application problems.

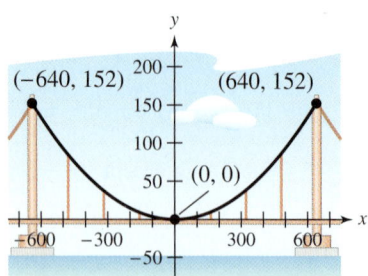

Figure 10.9

Application

Example 6   Golden Gate Bridge

Each cable of the Golden Gate Bridge is suspended (in the shape of a parabola) between two towers that are 1280 meters apart. The top of each tower is 152 meters above the roadway. The cables touch the roadway at the midpoint between the towers (see Figure 10.9).

a. Write an equation that models the cables of the bridge.

b. Find the height of the suspension cables over the roadway at a distance of 320 meters from the center of the bridge.

Solution

a. From Figure 8.9, you can see that the vertex of the parabola occurs at (0, 0). So, the equation has the form

$$y = a(x - h)^2 + k \qquad \text{Standard form}$$

$$y = a(x - 0)^2 + 0 \qquad \text{Substitute 0 for } h \text{ and 0 for } k.$$

$$y = ax^2. \qquad \text{Simplify.}$$

To find the value of a, use the fact that the parabola passes through the point (640, 152).

$$y = ax^2 \qquad \text{Write standard form.}$$

$$152 = a(640)^2 \qquad \text{Substitute 640 for } x \text{ and 152 for } y.$$

$$\frac{19}{51{,}200} = a \qquad \text{Simplify.}$$

So, an equation that models the cables of the bridge is

$$y = \frac{19}{51{,}200}x^2.$$

b. To find the height of the suspension cables over the roadway at a distance of 320 meters from the center of the bridge, evaluate the equation from part (a) when $x = 320$.

$$y = \frac{19}{51{,}200}x^2 \qquad \text{Write original equation.}$$

$$y = \frac{19}{51{,}200}(320)^2 \qquad \text{Substitute 320 for } x.$$

$$y = 38 \qquad \text{Simplify.}$$

So, the height of the suspension cables over the roadway is 38 meters.

10.4 Exercises

Review   *Concepts, Skills, and Problem Solving*

Keep mathematically in shape by doing these exercises *before* the problems of this section.

Properties and Definitions

1. Fill in the blanks: $(x + b)^2 = x^2 + \boxed{}\, x + \boxed{}$.

2. *Writing* ✐ Fill in the blank so that the expression is a perfect square trinomial. Explain how the constant is determined.

$$x^2 + 5x + \boxed{}$$

Simplifying Expressions

In Exercises 3–10, simplify the expression.

3. $(4x + 3y) - 3(5x + y)$

4. $(-15u + 4v) + 5(3u - 9v)$

5. $2x^2 + (2x - 3)^2 + 12x$

6. $y^2 - (y + 2)^2 + 4y$

7. $\sqrt{24x^2y^3}$

8. $\sqrt[3]{9} \cdot \sqrt[3]{15}$

9. $(12a^{-4}b^6)^{1/2}$

10. $(16^{1/3})^{3/4}$

Problem Solving

11. *Alcohol Mixture* How many liters of an 18% alcohol solution must be mixed with a 45% solution to obtain 12 liters of a 36% solution?

12. *Television* During a television show there were 12 commercials. Some of the commercials were 30-seconds long and some were 60-seconds long. The total amount of time for the 30-second commercials was 6 minutes less than the total amount of time for all the commercials during the show. How many 30-second commercials and how many 60-second commercials were there?

Developing Skills

In Exercises 1–6, match the equation with its graph. [The graphs are labeled (a), (b), (c), (d), (e), and (f).]

(a)
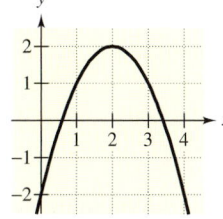

(b)
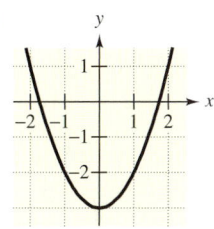

(c)
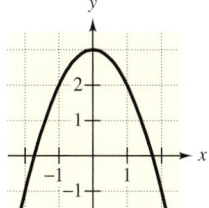

(d)
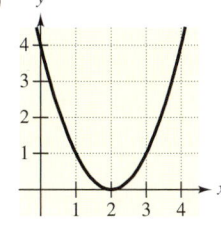

(e)
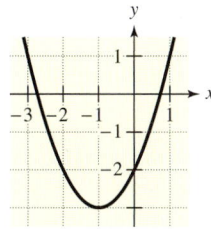

(f)
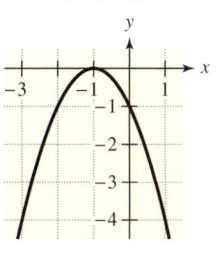

1. $y = (x + 1)^2 - 3$

2. $y = -(x + 1)^2$

3. $y = x^2 - 3$

4. $y = -x^2 + 3$

5. $y = (x - 2)^2$

6. $y = 2 - (x - 2)^2$

In Exercises 7–18, write the equation of the parabola in standard form and find the vertex of its graph. See Example 1.

7. $y = x^2 + 2$

8. $y = x^2 + 2x$

9. $y = x^2 - 4x + 7$

10. $y = x^2 + 6x - 5$

11. $y = x^2 + 6x + 5$

12. $y = x^2 - 4x + 5$

13. $y = -x^2 + 6x - 10$

14. $y = 4 - 8x - x^2$

15. $y = -x^2 + 2x - 7$

16. $y = -x^2 - 10x + 10$

17. $y = 2x^2 + 6x + 2$

18. $y = 3x^2 - 3x - 9$

In Exercises 19–24, find the vertex of the graph of the function by using the formula $x = -b/(2a)$. See Example 2.

19. $f(x) = x^2 - 8x + 15$ **20.** $f(x) = x^2 + 4x + 1$

21. $g(x) = -x^2 - 2x + 1$

22. $h(x) = -x^2 + 14x - 14$

23. $y = 4x^2 + 4x + 4$ **24.** $y = 9x^2 - 12x$

In Exercises 25–34, state whether the graph opens upward or downward and find the vertex.

25. $y = 2(x - 0)^2 + 2$ **26.** $y = -3(x + 5)^2 - 3$

27. $y = 4 - (x - 10)^2$ **28.** $y = 2(x - 12)^2 + 3$

29. $y = x^2 - 6$ **30.** $y = -(x + 1)^2$

31. $y = -(x - 3)^2$ **32.** $y = x^2 - 6x$

33. $y = -x^2 + 6x$ **34.** $y = -x^2 - 5$

In Exercises 35–46, find the x- and y-intercepts of the graph.

35. $y = 25 - x^2$ **36.** $y = x^2 - 49$

37. $y = x^2 - 9x$ **38.** $y = x^2 + 4x$

39. $y = x^2 + 2x - 3$ **40.** $y = -x^2 + 4x - 5$

41. $y = 4x^2 - 12x + 9$ **42.** $y = 10 - x - 2x^2$

43. $y = x^2 - 3x + 3$ **44.** $y = x^2 - 3x - 10$

45. $y = -2x^2 - 6x + 5$

46. $y = -4x^2 + 6x - 9$

In Exercises 47–70, sketch the parabola. Identify the vertex and any x-intercepts. Use a graphing calculator to verify your results. See Example 3.

47. $g(x) = x^2 - 4$

48. $h(x) = x^2 - 9$

49. $f(x) = -x^2 + 4$

50. $f(x) = -x^2 + 9$

51. $f(x) = x^2 - 3x$

52. $g(x) = x^2 - 4x$

53. $y = -x^2 + 3x$

54. $y = -x^2 + 4x$

55. $y = (x - 4)^2$

56. $y = -(x + 4)^2$

57. $y = x^2 - 8x + 15$

58. $y = x^2 + 4x + 2$

59. $y = -(x^2 + 6x + 5)$

60. $y = -x^2 + 2x + 8$

61. $q(x) = -x^2 + 6x - 7$

62. $g(x) = x^2 + 4x + 7$

63. $y = 2(x^2 + 6x + 8)$

64. $y = 3x^2 - 6x + 4$

65. $y = \frac{1}{2}(x^2 - 2x - 3)$

66. $y = -\frac{1}{2}(x^2 - 6x + 7)$

67. $y = \frac{1}{5}(3x^2 - 24x + 38)$

68. $y = \frac{1}{5}(2x^2 - 4x + 7)$

69. $f(x) = 5 - \frac{1}{3}x^2$

70. $f(x) = \frac{1}{3}x^2 - 2$

In Exercises 71–78, identify the transformation of the graph of $f(x) = x^2$ and sketch a graph of h.

71. $h(x) = x^2 + 2$

72. $h(x) = x^2 - 4$

73. $h(x) = (x + 2)^2$

74. $h(x) = (x - 4)^2$

75. $h(x) = -(x + 5)^2$

76. $h(x) = -x^2 - 6$

77. $h(x) = -(x + 1)^2 - 1$

78. $h(x) = -(x - 3)^2 + 2$

85.

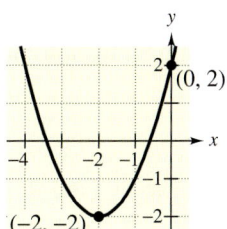

86.

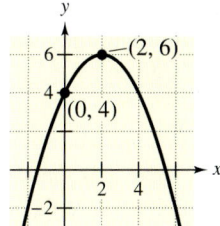

In Exercises 79–82, use a graphing calculator to approximate the vertex of the graph. Verify the result algebraically.

79. $y = \frac{1}{6}(2x^2 - 8x + 11)$

80. $y = -\frac{1}{4}(4x^2 - 20x + 13)$

81. $y = -0.7x^2 - 2.7x + 2.3$

82. $y = 0.75x^2 - 7.50x + 23.00$

In Exercises 83–86, write an equation of the parabola. See Example 4.

83.

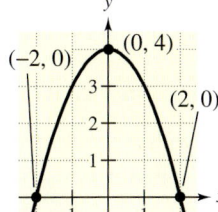

84.

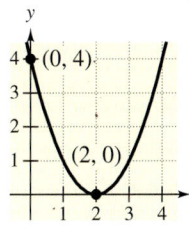

In Exercises 87–94, write an equation of the parabola $y = a(x - h)^2 + k$ that satisfies the conditions. See Example 5.

87. Vertex: $(2, 1)$; $a = 1$

88. Vertex: $(-3, -3)$; $a = 1$

89. Vertex: $(2, -4)$; Point on the graph: $(0, 0)$

90. Vertex: $(-2, -4)$; Point on the graph: $(0, 0)$

91. Vertex: $(3, 2)$; Point on the graph: $(1, 4)$

92. Vertex: $(-1, -1)$; Point on the graph: $(0, 4)$

93. Vertex: $(-1, 5)$; Point on the graph: $(0, 1)$

94. Vertex: $(5, 2)$; Point on the graph: $(10, 3)$

Solving Problems

95. *Path of a Ball*   The height y (in feet) of a ball thrown by a child is given by

$$y = -\frac{1}{12}x^2 + 2x + 4$$

where x is the horizontal distance (in feet) from where the ball is thrown.

(a) How high is the ball when it leaves the child's hand?

(b) How high is the ball when it reaches its maximum height?

(c) How far from the child does the ball strike the ground?

96. *Path of a Ball*   Repeat Exercise 95 if the path of the ball is modeled by

$$y = -\frac{1}{16}x^2 + 2x + 5.$$

97. *Path of an Object*   A child launches a toy rocket from a table. The height y (in feet) of the rocket is given by

$$y = -\frac{1}{5}x^2 + 6x + 3$$

where x is the horizontal distance (in feet) from where the rocket is launched.

(a) Determine the height from which the rocket is launched.

(b) How high is the rocket at its maximum height?

(c) How far away does the rocket land from where it is launched?

98. *Path of an Object* You use a fishing rod to cast a lure into the water. The height y (in feet) of the lure is given by

$$y = -\frac{1}{90}x^2 + \frac{1}{5}x + 9$$

where x is the horizontal distance (in feet) from the point where the lure is released.

(a) Determine the height from which the lure is released.

(b) How high is the lure at its maximum height?

(c) How far away does the lure land from where it is released?

99. *Path of a Ball* The height y (in feet) of a ball that you throw is given by

$$y = -\frac{1}{200}x^2 + x + 6$$

where x is the horizontal distance (in feet) from where you release the ball.

(a) How high is the ball when you release it?

(b) How high is the ball when it reaches its maximum height?

(c) How far away does the ball strike the ground from where you released it?

100. *Path of a Ball* The height y (in feet) of a softball that you hit is given by

$$y = -\frac{1}{70}x^2 + 2x + 2$$

where x is the horizontal distance (in feet) from where you hit the ball.

(a) How high is the ball when you hit it?

(b) How high is the ball at its maximum height?

(c) How far from where you hit the ball does it strike the ground?

101. *Path of a Diver* The path of a diver is given by

$$y = -\frac{4}{9}x^2 + \frac{24}{9}x + 10$$

where y is the height in feet and x is the horizontal distance from the end of the diving board in feet. What is the maximum height of the diver?

102. *Path of a Diver* Repeat Exercise 101 if the path of the diver is modeled by

$$y = -\frac{4}{3}x^2 + \frac{10}{3}x + 10.$$

103. ⊞ *Cost* The cost C of producing x units of a product is given by

$$C = 800 - 10x + \frac{1}{4}x^2, \quad 0 < x < 40.$$

Use a graphing calculator to graph this function and approximate the value of x when C is minimum.

104. ⊞ ▲ *Geometry* The area A of a rectangle is given by the function

$$A = \frac{2}{\pi}(100x - x^2), \quad 0 < x < 100$$

where x is the length of the base of the rectangle in feet. Use a graphing calculator to graph the function and to approximate the value of x when A is maximum.

105. ⊞ *Graphical Estimation* The number N (in thousands) of military reserve personnel in the United States for the years 1992 through 2000 is approximated by the model

$$N = 4.64t^2 - 85.5t + 1263, \quad 2 \le t \le 10$$

where t is the time in years, with $t = 2$ corresponding to 1992. (Source: U.S. Department of Defense)

(a) Use a graphing calculator to graph the model.

(b) Use your graph from part (a) to determine the year when the number of military reserves was greatest. Approximate the number for that year.

106. ⊞ *Graphical Estimation* The profit P (in thousands of dollars) for a landscaping company is given by

$$P = 230 + 20s - \frac{1}{2}s^2$$

where s is the amount (in hundreds of dollars) spent on advertising. Use a graphing calculator to graph the profit function and approximate the amount of advertising that yields a maximum profit. Verify the maximum profit algebraically.

107. *Bridge Design*   A bridge is to be constructed over a gorge with the main supporting arch being a parabola (see figure). The equation of the parabola is $y = 4[100 - (x^2/2500)]$, where x and y are measured in feet.

(a) Find the length of the road across the gorge.

(b) Find the height of the parabolic arch at the center of the span.

(c) Find the lengths of the vertical girders at intervals of 100 feet from the center of the bridge.

108. *Highway Design*   A highway department engineer must design a parabolic arc to create a turn in a freeway around a city. The vertex of the parabola is placed at the origin, and the parabola must connect with roads represented by the equations

$$y = -0.4x - 100, \quad x < -500$$

and

$$y = 0.4x - 100, \quad x > 500$$

(see figure). Find an equation of the parabolic arc.

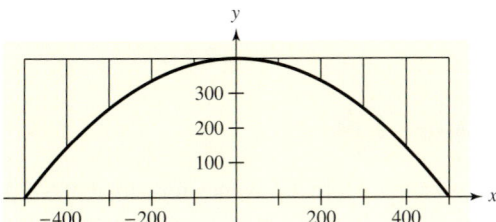

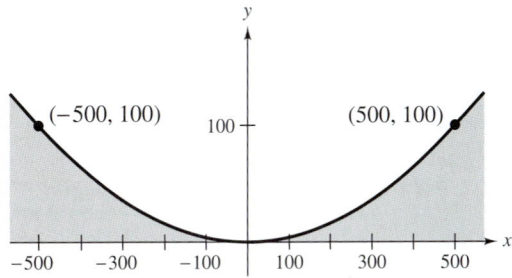

Explaining Concepts

109. *Writing*   In your own words, describe the graph of the quadratic function $f(x) = ax^2 + bx + c$.

110. *Writing*   Explain how to find the vertex of the graph of a quadratic function.

111. *Writing*   Explain how to find any x- or y-intercepts of the graph of a quadratic function.

112. *Writing*   Explain how to determine whether the graph of a quadratic function opens upward or downward.

113. *Writing*   How is the discriminant related to the graph of a quadratic function?

114. *Writing*   Is it possible for the graph of a quadratic function to have two y-intercepts? Explain.

115. *Writing*   Explain how to determine the maximum (or minimum) value of a quadratic function.

10.5 Applications of Quadratic Equations

Lon C. Diehl/PhotoEdit, Inc.

Why You Should Learn It

Quadratic equations are used in a wide variety of real-life problems. For instance, in Exercise 46 on page 661, a quadratic equation is used to model the height of a baseball after you hit the ball.

What You Should Learn

1 Use quadratic equations to solve application problems.

1 Use quadratic equations to solve application problems.

Applications of Quadratic Equations

Example 1   An Investment Problem

A car dealer bought a fleet of cars from a car rental agency for a total of $120,000. By the time the dealer had sold all but four of the cars, at an average profit of $2500 each, the original investment of $120,000 had been regained. How many cars did the dealer sell, and what was the average price per car?

Solution

Although this problem is stated in terms of average price and average profit per car, you can use a model that assumes that each car sold for the same price.

| *Verbal Model:* | Selling price per car | = | Cost per car | + | Profit per car |

Labels:	Number of cars sold $= x$	(cars)
	Number of cars bought $= x + 4$	(cars)
	Selling price per car $= 120{,}000/x$	(dollars per car)
	Cost per car $= 120{,}000/(x + 4)$	(dollars per car)
	Profit per car $= 2500$	(dollars per car)

Equation:

$$\frac{120{,}000}{x} = \frac{120{,}000}{x + 4} + 2500$$

$$120{,}000(x + 4) = 120{,}000x + 2500x(x + 4), \quad x \neq 0, \ x \neq -4$$

$$120{,}000x + 480{,}000 = 120{,}000x + 2500x^2 + 10{,}000x$$

$$0 = 2500x^2 + 10{,}000x - 480{,}000$$

$$0 = x^2 + 4x - 192$$

$$0 = (x - 12)(x + 16)$$

$$x - 12 = 0 \quad \Longrightarrow \quad x = 12$$

$$x + 16 = 0 \quad \Longrightarrow \quad x = -16$$

Choosing the positive value, it follows that the dealer sold 12 cars at an average price of $120{,}000/12 = 10{,}000$ per car. Check this in the original statement.

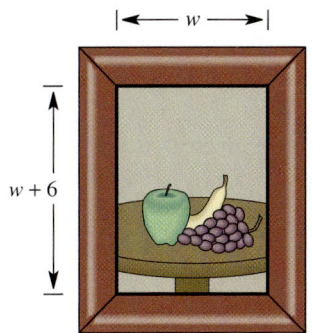

Figure 10.10

Example 2   Geometry: Dimensions of a Picture

A picture is 6 inches taller than it is wide and has an area of 216 square inches. What are the dimensions of the picture?

Solution

Begin by drawing a diagram, as shown in Figure 10.10.

Verbal Model:. | Area of picture $=$ Width $\cdot$ Height |

Labels: Picture width $= w$ (inches)
Picture height $= w + 6$ (inches)
Area $= 216$ (square inches)

Equation: $216 = w(w + 6)$

$0 = w^2 + 6w - 216$

$0 = (w + 18)(w - 12)$

$w + 18 = 0 \implies w = -18$

$w - 12 = 0 \implies w = 12$

Of the two possible solutions, choose the positive value of w and conclude that the picture is $w = 12$ inches wide and $w + 6 = 12 + 6 = 18$ inches tall. Check these dimensions in the original statement of the problem.

Example 3   An Interest Problem

The formula $A = P(1 + r)^2$ represents the amount of money A in an account in which P dollars is deposited for 2 years at an annual interest rate of r (in decimal form). Find the interest rate if a deposit of $6000 increases to $6933.75 over a two-year period.

Solution

$A = P(1 + r)^2$ Write given formula.

$6933.75 = 6000(1 + r)^2$ Substitute 6933.75 for A and 6000 for P.

$1.155625 = (1 + r)^2$ Divide each side by 6000.

$\pm 1.075 = 1 + r$ Square Root Property

$0.075 = r$ Choose positive solution.

The annual interest rate is $r = 0.075 = 7.5\%$.

Check

$A = P(1 + r)^2$ Write given formula.

$6933.75 \stackrel{?}{=} 6000(1 + 0.075)^2$ Substitute 6933.75 for A, 6000 for P, and 0.075 for r.

$6933.75 \stackrel{?}{=} 6000(1.155625)$ Simplify.

$6933.75 = 6933.75$ Solution checks. ✔

Example 4 Reduced Rates

A ski club chartered a bus for a ski trip at a cost of $720. In an attempt to lower the bus fare per skier, the club invited nonmembers to go along. When four nonmembers agreed to go on the trip, the fare per skier decreased by $6. How many club members are going on the trip?

Solution

Verbal Model:

$$\boxed{\text{Cost per skier}} \cdot \boxed{\text{Number of skiers}} = \boxed{\$720}$$

Labels:

Number of ski club members $= x$ (people)

Number of skiers $= x + 4$ (people)

Original cost per skier $= \dfrac{720}{x}$ (dollars per person)

New cost per skier $= \dfrac{720}{x} - 6$ (dollars per person)

Equation:

$$\left(\frac{720}{x} - 6\right)(x + 4) = 720 \qquad \text{Original equation}$$

$$\left(\frac{720 - 6x}{x}\right)(x + 4) = 720 \qquad \text{Rewrite 1st factor.}$$

$$(720 - 6x)(x + 4) = 720x, \ x \neq 0 \qquad \text{Multiply each side by } x.$$

$$720x + 2880 - 6x^2 - 24x = 720x \qquad \text{Multiply factors.}$$

$$-6x^2 - 24x + 2880 = 0 \qquad \text{Subtract } 720x \text{ from each side.}$$

$$x^2 + 4x - 480 = 0 \qquad \text{Divide each side by } -6.$$

$$(x + 24)(x - 20) = 0 \qquad \text{Factor left side of equation.}$$

$$x + 24 = 0 \implies x = -24 \qquad \text{Set 1st factor equal to 0.}$$

$$x - 20 = 0 \implies x = 20 \qquad \text{Set 2nd factor equal to 0.}$$

Choosing the positive value of x, you can conclude that 20 ski club members are going on the trip. Check this solution in the original statement of the problem, as follows.

Check

Original cost for 20 ski club members:

$$\frac{720}{x} = \frac{720}{20} = \$36 \qquad \text{Substitute 20 for } x.$$

New cost with 4 nonmembers:

$$\frac{720}{x + 4} = \frac{720}{24} = \$30$$

Decrease in fare with 4 nonmembers:

$$36 - 30 = \$6 \qquad \text{Solution checks. } \checkmark$$

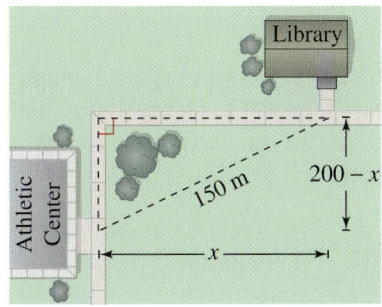

Figure 10.11

Example 5   An Application Involving the Pythagorean Theorem

An L-shaped sidewalk from the athletic center to the library on a college campus is 200 meters long, as shown in Figure 10.11. By cutting diagonally across the grass, students shorten the walking distance to 150 meters. What are the lengths of the two legs of the sidewalk?

Solution

Common
Formula: $a^2 + b^2 = c^2$ Pythagorean Theorem

Labels: Length of one leg $= x$ (meters)
             Length of other leg $= 200 - x$ (meters)
             Length of diagonal $= 150$ (meters)

Equation: $x^2 + (200 - x)^2 = 150^2$

$$x^2 + 40{,}000 - 400x + x^2 = 22{,}500$$

$$2x^2 - 400x + 40{,}000 = 22{,}500$$

$$2x^2 - 400x + 17{,}500 = 0$$

$$x^2 - 200x + 8750 = 0$$

By the Quadratic Formula, you can find the solutions as follows.

$$x = \frac{-(-200) \pm \sqrt{(-200)^2 - 4(1)(8750)}}{2(1)}$$ Substitute 1 for a, -200 for b, and 8750 for c.

$$= \frac{200 \pm \sqrt{5000}}{2}$$

$$= \frac{200 \pm 50\sqrt{2}}{2}$$

$$= \frac{2(100 \pm 25\sqrt{2})}{2}$$

$$= 100 \pm 25\sqrt{2}$$

Both solutions are positive, so it does not matter which you choose. If you let

$$x = 100 + 25\sqrt{2} \approx 135.4 \text{ meters}$$

the length of the other leg is

$$200 - x \approx 200 - 135.4 \approx 64.6 \text{ meters.}$$

In Example 5, notice that you obtain the same dimensions if you choose the other value of x. That is, if the length of one leg is

$$x = 100 - 25\sqrt{2} \approx 64.6 \text{ meters}$$

the length of the other leg is

$$200 - x \approx 200 - 64.6 \approx 135.4 \text{ meters.}$$

Example 6   Work-Rate Problem

An office contains two copy machines. machine B is known to take 12 minutes longer than machine A to copy the company's monthly report. Using both machines together, it takes 8 minutes to reproduce the report. How long would it take each machine alone to reproduce the report?

Solution

Verbal Model:

$$\boxed{\begin{array}{c}\text{Work done by}\\\text{machine A}\end{array}} + \boxed{\begin{array}{c}\text{Work done by}\\\text{machine B}\end{array}} = \boxed{\begin{array}{c}\text{1 complete}\\\text{job}\end{array}}$$

$$\boxed{\begin{array}{c}\text{Rate}\\\text{for A}\end{array}} \cdot \boxed{\begin{array}{c}\text{Time}\\\text{for both}\end{array}} + \boxed{\begin{array}{c}\text{Rate}\\\text{for B}\end{array}} \cdot \boxed{\begin{array}{c}\text{Time}\\\text{for both}\end{array}} = \boxed{1}$$

Labels:

Time for machine A $= t$	(minutes)
Rate for machine A $= \dfrac{1}{t}$	(job per minute)
Time for machine B $= t + 12$	(minutes)
Rate for machine B $= \dfrac{1}{t + 12}$	(job per minute)
Time for both machines $= 8$	(minutes)
Rate for both machines $= \dfrac{1}{8}$	(job per minute)

Equation:

$$\frac{1}{t}(8) + \frac{1}{t + 12}(8) = 1 \qquad \text{Original equation}$$

$$8\left(\frac{1}{t} + \frac{1}{t + 12}\right) = 1 \qquad \text{Distributive Property}$$

$$8\left[\frac{t + 12 + t}{t(t + 12)}\right] = 1 \qquad \text{Rewrite with common denominator.}$$

$$8t(t + 12)\left[\frac{2t + 12}{t(t + 12)}\right] = t(t + 12) \qquad \text{Multiply each side by } t(t + 12).$$

$$8(2t + 12) = t^2 + 12t \qquad \text{Simplify.}$$

$$16t + 96 = t^2 + 12t \qquad \text{Distributive Property}$$

$$0 = t^2 - 4t - 96 \qquad \text{Subtract } 16t + 96 \text{ from each side.}$$

$$0 = (t - 12)(t + 8) \qquad \text{Factor right side of equation.}$$

$$t - 12 = 0 \implies t = 12 \qquad \text{Set 1st factor equal to 0.}$$

$$t + 8 = 0 \implies t = -8 \qquad \text{Set 2nd factor equal to 0.}$$

By choosing the positive value for t, you can conclude that the times for the two machines are

Time for machine A $= t = 12$ minutes

Time for machine B $= t + 12 = 12 + 12 = 24$ minutes.

Check these solutions in the original statement of the problem.

Example 7 The Height of a Model Rocket

A model rocket is projected straight upward from ground level according to the height equation

$$h = -16t^2 + 192t, \, t \geq 0$$

where h is the height in feet and t is the time in seconds.

a. After how many seconds is the height 432 feet?

b. After how many seconds does the rocket hit the ground?

c. What is the maximum height of the rocket?

Solution

a.

$h = -16t^2 + 192t$	Write original equation.
$432 = -16t^2 + 192t$	Substitute 432 for h.
$16t^2 - 192t + 432 = 0$	Write in general form.
$t^2 - 12t + 27 = 0$	Divide each side by 16.
$(t-3)(t-9) = 0$	Factor.
$t - 3 = 0 \implies t = 3$	Set 1st factor equal to 0.
$t - 9 = 0 \implies t = 9$	Set 2nd factor equal to 0.

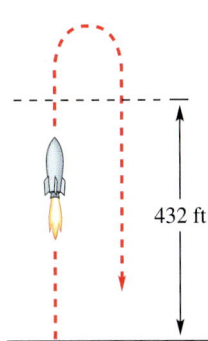

432 ft

Figure 10.12

The rocket attains a height of 432 feet at two different times—once (going up) after 3 seconds, and again (coming down) after 9 seconds. (See Figure 10.12.)

b. To find the time it takes for the rocket to hit the ground, let the height be 0.

$0 = -16t^2 + 192t$	Substitute 0 for h in original equation.
$0 = t^2 - 12t$	Divide each side by -16.
$0 = t(t - 12)$	Factor.
$t = 0 \quad \text{or} \quad t = 12$	Solutions

The rocket hits the ground after 12 seconds. (Note that the time of $t = 0$ seconds corresponds to the time of lift-off.)

c. The maximum value for h in the equation $h = -16t^2 + 192t$ occurs when $t = -\dfrac{b}{2a}$. So, the t-coordinate is

$$t = \frac{-b}{2a} = \frac{-192}{2(-16)} = 6$$

and the h-coordinate is

$$h = -16(6)^2 + 192(6) = 576.$$

So, the maximum height of the rocket is 576 feet.

10.5 Exercises

Keep mathematically in shape by doing these exercises *before* the problems of this section.

Properties and Definitions

1. *Writing* Define the slope of the line through the points (x_1, y_1) and (x_2, y_2).

2. Give the following forms of an equation of a line.

(a) Slope-intercept form

(b) Point-slope form

(c) General form

(d) Horizontal line

Equations of Lines

In Exercises 3–10, find the general form of the equation of the line through the two points.

3. $(0, 0), (4, -2)$

4. $(0, 0), (100, 75)$

5. $(-1, -2), (3, 6)$

6. $(1, 5), (6, 0)$

7. $\left(\frac{3}{2}, 8\right), \left(\frac{11}{2}, \frac{5}{2}\right)$

8. $(0, 2), (7.3, 15.4)$

9. $(0, 8), (5, 8)$

10. $(-3, 2), (-3, 5)$

Problem Solving

11. *Endowment* A group of people agree to share equally in the cost of a \$250,000 endowment to a college. If they could find two more people to join the group, each person's share of the cost would decrease by \$6250. How many people are presently in the group?

12. *Current Speed* A boat travels at a speed of 18 miles per hour in still water. It travels 35 miles upstream and then returns to the starting point in a total of 4 hours. Find the speed of the current.

Solving Problems

1. *Selling Price* A store owner bought a case of eggs for \$21.60. By the time all but 6 dozen of the eggs had been sold at a profit of \$0.30 per dozen, the original investment of \$21.60 had been regained. How many dozen eggs did the owner sell, and what was the selling price per dozen? See Example 1.

2. *Selling Price* A manager of a computer store bought several computers of the same model for \$27,000. When all but three of the computers had been sold at a profit of \$750 per computer, the original investment of \$27,000 had been regained. How many computers were sold, and what was the selling price of each?

3. *Selling Price* A store owner bought a case of video games for \$480. By the time he had sold all but eight of them at a profit of \$10 each, the original investment of \$480 had been regained. How many video games were sold, and what was the selling price of each game?

4. *Selling Price* A math club bought a case of sweatshirts for \$850 to sell as a fundraiser. By the time all but 16 sweatshirts had been sold at a profit of \$8 per sweatshirt, the original investment of \$850 had been regained. How many sweatshirts were sold, and what was the selling price of each sweatshirt?

▲ *Geometry* In Exercises 5–14, complete the table of widths, lengths, perimeters, and areas of rectangles.

	Width	Length	Perimeter	Area
5.	$1.4l$	l	54 in.	
6.	w	$3.5w$	60 m	
7.	w	$2.5w$		250 ft²
8.	w	$1.5w$		216 cm²
9.	$\frac{1}{3}l$	l		192 in.²
10.	$\frac{3}{4}l$	l		2700 in.²
11.	w	$w + 3$	54 km	
12.	$l - 6$	l	108 ft	

	Width	Length	Perimeter	Area
13.	$l - 20$	l		12,000 m²
14.	w	$w + 5$		500 ft²

15. ▲ *Geometry*   A picture frame is 4 inches taller than it is wide and has an area of 192 square inches. What are the dimensions of the picture frame? See Example 2.

16. ▲ *Geometry*   The height of a triangle is 8 inches less than its base. The area of the triangle is 192 square inches. Find the dimensions of the triangle.

17. *Storage Area*   A retail lumberyard plans to store lumber in a rectangular region adjoining the sales office (see figure). The region will be fenced on three sides, and the fourth side will be bounded by the wall of the office building. There is 350 feet of fencing available, and the area of the region is 12,500 square feet. Find the dimensions of the region.

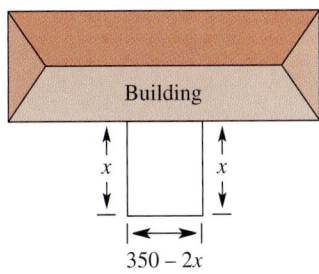

$350 - 2x$

18. ▲ *Geometry*   Your home is on a square lot. To add more space to your yard, you purchase an additional 20 feet along the side of the property (see figure). The area of the lot is now 25,500 square feet. What are the dimensions of the new lot?

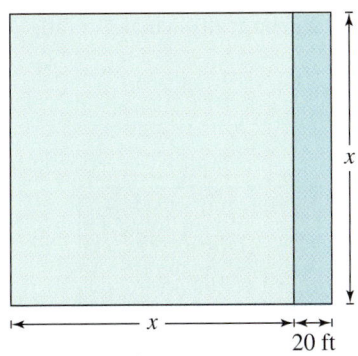

19. *Fenced Area*   A family built a fence around three sides of their property (see figure). In total, they used 550 feet of fencing. By their calculations, the lot is 1 acre (43,560 square feet). Is this correct? Explain your reasoning.

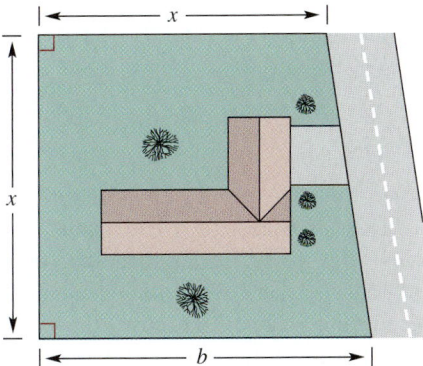

20. *Fenced Area*   You have 100 feet of fencing. Do you have enough to enclose a rectangular region whose area is 630 square feet? Is there enough to enclose a circular area of 630 square feet? Explain.

21. *Open Conduit*   An open-topped rectangular conduit for carrying water in a manufacturing process is made by folding up the edges of a sheet of aluminum 48 inches wide (see figure). A cross section of the conduit must have an area of 288 square inches. Find the width and height of the conduit.

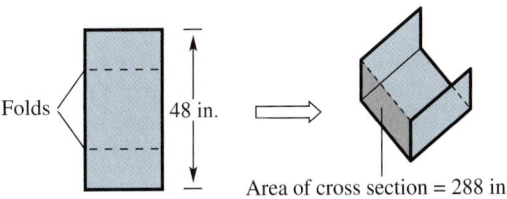

Folds   48 in.   Area of cross section = 288 in.²

22. *Photography*   A photographer has a photograph that is 6 inches by 8 inches. He wishes to reduce the photo by the same amount on each side such that the resulting photo will have an area that is half the area of the original photo. By how much should each side be reduced?

Compound Interest In Exercises 23–28, find the interest rate *r*. Use the formula $A = P(1 + r)^2$, where *A* is the amount after 2 years in an account earning *r* percent (in decimal form) compounded annually, and *P* is the original investment. See Example 3.

23. $P = \$10,000$

$A = \$11,990.25$

24. $P = \$3000$

$A = \$3499.20$

25. $P = \$500$

$A = \$572.45$

26. $P = \$250$

$A = \$280.90$

27. $P = \$6500$

$A = \$7372.46 \approx$

28. $P = \$8000$

$A = \$8421.41 \approx$

29. *Reduced Rates* A service organization pays \$210 for a block of tickets to a ball game. The block contains three more tickets than the organization needs for its members. By inviting three more people to attend (and share in the cost), the organization lowers the price per ticket by \$3.50. How many people are going to the game? See Example 4.

30. *Reduced Rates* A service organization buys a block of tickets to a ball game for \$240. After eight more people decide to go to the game, the price per ticket is decreased by \$1. How many people are going to the game?

31. *Reduced Fares* A science club charters a bus to attend a science fair at a cost of \$480. To lower the bus fare per person, the club invites nonmembers to go along. When two nonmembers join the trip, the fare per person is decreased by \$1. How many people are going on the excursion?

32. *Venture Capital* Eighty thousand dollars is needed to begin a small business. The cost will be divided equally among the investors. Some have made a commitment to invest. If three more investors are found, the amount required from each will decrease by \$6000. How many have made a commitment to invest in the business?

33. *Delivery Route* You are asked to deliver pizza to an insurance office and an apartment complex (see figure), and you keep a log of all the mileages between stops. You forget to look at the odometer at the insurance office, but after getting to the apartment complex you record the total distance traveled from the pizza shop as 18 miles. The return distance from the apartment complex to the pizza shop is 16 miles. The route approximates a right triangle. Estimate the distance from the pizza shop to the insurance office. See Example 5.

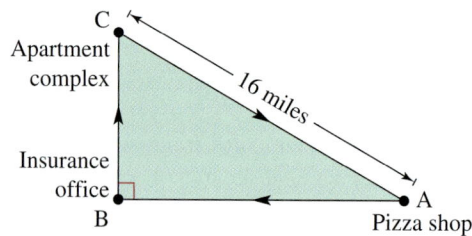

Figure for 33

34. ▲ *Geometry* An L-shaped sidewalk from the library (point A) to the gym (point B) on a high school campus is 100 yards long, as shown in the figure. By cutting diagonally across the grass, students shorten the walking distance to 80 yards. What are the lengths of the two legs of the sidewalk?

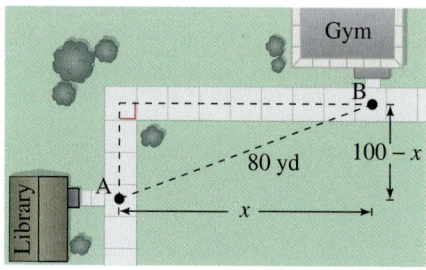

35. ▲ *Geometry* An adjustable rectangular form has minimum dimensions of 3 meters by 4 meters. The length and width can be expanded by equal amounts *x* (see figure).

(a) Write an equation relating the length *d* of the diagonal to *x*.

(b) 🖩 Use a graphing calculator to graph the equation.

(c) 🖩 Use the graph to approximate the value of *x* when $d = 10$ meters.

(d) Find *x* algebraically when $d = 10$.

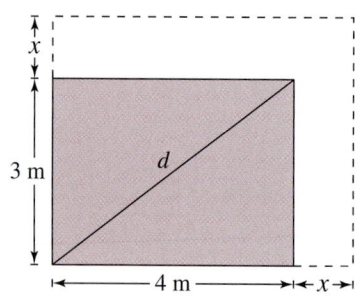

36. *Solving Graphically and Numerically*   A meteorologist is positioned 100 feet from the point where a weather balloon is launched (see figure).

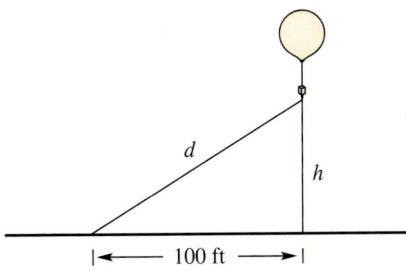

(a) Write an equation relating the distance d between the balloon and the meteorologist to the height h of the balloon.

(b) 🖩 Use a graphing calculator to graph the equation.

(c) 🖩 Use the graph to approximate the value of h when $d = 200$ feet.

(d) Complete the table.

h	0	100	200	300
d				

37. *Work Rate*   Working together, two people can complete a task in 5 hours. Working alone, one person takes 2 hours longer than the other. How long would it take each person to do the task alone? See Example 6.

38. *Work Rate*   Working together, two people can complete a task in 6 hours. Working alone, one person takes 2 hours longer than the other. How long would it take each person to do the task alone?

39. *Work Rate*   An office contains two printers. Machine B is known to take 3 minutes longer than Machine A to produce the company's monthly financial report. Using both machines together, it takes 6 minutes to produce the report. How long would it take each machine to produce the report?

40. *Work Rate*   A builder works with two plumbing companies. Company A is known to take 3 days longer than Company B to do the plumbing in a particular style of house. Using both companies, it takes 4 days. How long would it take to do the plumbing using each company individually?

Free-Falling Object   In Exercises 41–44, find the time necessary for an object to fall to ground level from an initial height of h_0 feet if its height h at any time t (in seconds) is given by $h = h_0 - 16t^2$.

41. $h_0 = 169$ **42.** $h_0 = 729$

43. $h_0 = 1454$ (height of the Sears Tower)

44. $h_0 = 984$ (height of the Eiffel Tower)

45. *Height*   The height h in feet of a baseball hit 3 feet above the ground is given by $h = 3 + 75t - 16t^2$, where t is time in seconds. Find the time when the ball hits the ground. See Example 7.

46. *Height*   You are hitting baseballs. When you toss the ball into the air, your hand is 5 feet above the ground (see figure). You hit the ball when it falls back to a height of 4 feet. You toss the ball with an initial velocity of 25 feet per second. The height h of the ball t seconds after leaving your hand is given by $h = 5 + 25t - 16t^2$. How much time will pass before you hit the ball?

47. *Height*   A model rocket is projected straight upward from ground level according to the height equation $h = -16t^2 + 160t$, where h is the height of the rocket in feet and t is the time in seconds.

(a) After how many seconds is the height 336 feet?

(b) After how many seconds does the rocket hit the ground?

(c) What is the maximum height of the rocket?

48. *Height*   A tennis ball is tossed vertically upward from a height of 5 feet according to the height equation $h = -16t^2 + 21t + 5$, where h is the height of the tennis ball in feet and t is the time in seconds.

(a) After how many seconds is the height 11 feet?

(b) After how many seconds does the tennis ball hit the ground?

(c) What is the maximum height of the ball?

Number Problems In Exercises 49–54, find two positive integers that satisfy the requirement.

49. The product of two consecutive integers is 182.

50. The product of two consecutive integers is 1806.

51. The product of two consecutive even integers is 168.

52. The product of two consecutive even integers is 2808.

53. The product of two consecutive odd integers is 323.

54. The product of two consecutive odd integers is 1443.

55. *Air Speed* An airline runs a commuter flight between two cities that are 720 miles apart. If the average speed of the planes could be increased by 40 miles per hour, the travel time would be decreased by 12 minutes. What air speed is required to obtain this decrease in travel time?

56. *Average Speed* A truck traveled the first 100 miles of a trip at one speed and the last 135 miles at an average speed of 5 miles per hour less. The entire trip took 5 hours. What was the average speed for the first part of the trip?

57. *Speed* A small business uses a minivan to make deliveries. The cost per hour for fuel for the van is $C = v^2/600$, where v is the speed in miles per hour. The driver is paid $5 per hour. Find the speed if the cost for wages and fuel for a 110-mile trip is $20.39.

58. *Distance* Find any points on the line $y = 14$ that are 13 units from the point (1, 2).

59. ▲ *Geometry* The area of an ellipse is given by $A = \pi a b$ (see figure). For a certain ellipse, it is required that $a + b = 20$.

(a) Show that $A = \pi a(20 - a)$.

(b) Complete the table.

a	4	7	10	13	16
A					

(c) Find two values of a such that $A = 300$.

(d) ⊞ Use a graphing calculator to graph the area equation. Then use the graph to verify the results in part (c).

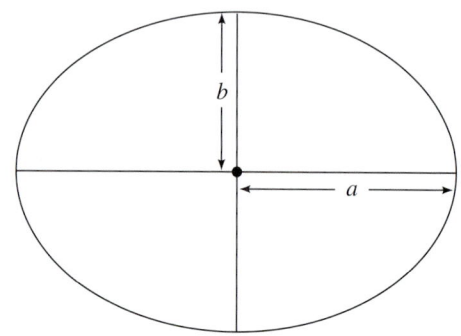

Figure for 59

60. ⊞ *Data Analysis* For the years 1993 through 2000, the sales s (in millions of dollars) of recreational vehicles in the United States can be approximated by $s = 156.45t^2 - 1035.5t + 6875$, $3 \le t \le 10$, where t is time in years, with $t = 3$ corresponding to 1993. (Source: National Sporting Goods Association)

(a) Use a graphing calculator to graph the model.

(b) Use the graph in part (a) to determine the year in which sales were approximately $6.3 billion.

Explaining Concepts

61. *Writing* In your own words, describe strategies for solving word problems.

62. *Writing* List the strategies that can be used to solve a quadratic equation.

63. *Unit Analysis* Describe the units of the product.

$$\frac{9 \text{ dollars}}{\text{hour}} \cdot (20 \text{ hours})$$

64. *Unit Analysis* Describe the units of the product.

$$\frac{20 \text{ feet}}{\text{minute}} \cdot \frac{1 \text{ minute}}{60 \text{ seconds}} \cdot (45 \text{ seconds})$$

10.6 Quadratic and Rational Inequalities

Will Hart/PhotoEdit, Inc.

Why You Should Learn It

Rational inequalities can be used to model and solve real-life problems. For instance, in Exercise 116 on page 672, a rational inequality is used to model the temperature of a metal in a laboratory experiment.

1 Determine test intervals for polynomials.

What You Should Learn

1 Determine test intervals for polynomials.

2 Use test intervals to solve quadratic inequalities.

3 Use test intervals to solve rational inequalities.

4 Use inequalities to solve application problems.

Finding Test Intervals

When working with polynomial inequalities, it is important to realize that the value of a polynomial can change signs only at its **zeros**. That is, a polynomial can change signs only at the x-values for which the value of the polynomial is zero. For instance, the first-degree polynomial $x + 2$ has a zero at $x = -2$, and it changes signs at that zero. You can picture this result on the real number line, as shown in Figure 10.13.

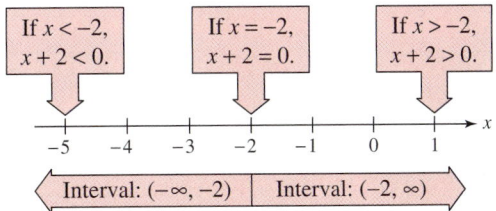

If $x < -2$, $x + 2 < 0$.   If $x = -2$, $x + 2 = 0$.   If $x > -2$, $x + 2 > 0$.

Interval: $(-\infty, -2)$ Interval: $(-2, \infty)$

Figure 10.13

Note in Figure 10.13 that the zero of the polynomial partitions the real number line into two **test intervals**. The value of the polynomial is negative for every x-value in the first test interval $(-\infty, -2)$, and it is positive for every x-value in the second test interval $(-2, \infty)$. You can use the same basic approach to determine the test intervals for any polynomial.

Finding Test Intervals for a Polynomial

1. Find all real zeros of the polynomial, and arrange the zeros in increasing order. The zeros of a polynomial are called its **critical numbers**.

2. Use the critical numbers of the polynomial to determine its test intervals.

3. Choose a representative x-value in each test interval and evaluate the polynomial at that value. If the value of the polynomial is negative, the polynomial will have negative values for *every* x-value in the interval. If the value of the polynomial is positive, the polynomial will have positive values for *every* x-value in the interval.

② Use test intervals to solve quadratic inequalities.

Quadratic Inequalities

The concepts of critical numbers and test intervals can be used to solve nonlinear inequalities, as demonstrated in Examples 1, 2, and 4.

Example 1   Solving a Quadratic Inequality

Solve the inequality $x^2 - 5x < 0$.

Solution

First find the *critical numbers* of $x^2 - 5x < 0$ by finding the solutions of the equation $x^2 - 5x = 0$.

$$x^2 - 5x = 0 \qquad \text{Write corresponding equation.}$$
$$x(x - 5) = 0 \qquad \text{Factor.}$$
$$x = 0, \ x = 5 \qquad \text{Critical numbers}$$

This implies that the test intervals are $(-\infty, 0)$, $(0, 5)$, and $(5, \infty)$. To test an interval, choose a convenient number in the interval and determine if the number satisfies the inequality.

Test interval	Representative x-value	Is inequality satisfied?
$(-\infty, 0)$	$x = -1$	$(-1)^2 - 5(-1) \overset{?}{<} 0$ $6 \not< 0$
$(0, 5)$	$x = 1$	$1^2 - 5(1) \overset{?}{<} 0$ $-4 < 0$
$(5, \infty)$	$x = 6$	$6^2 - 5(6) \overset{?}{<} 0$ $6 \not< 0$

Because the inequality $x^2 - 5x < 0$ is satisfied only by the middle test interval, you can conclude that the solution set of the inequality is the interval $(0, 5)$, as shown in Figure 10.14.

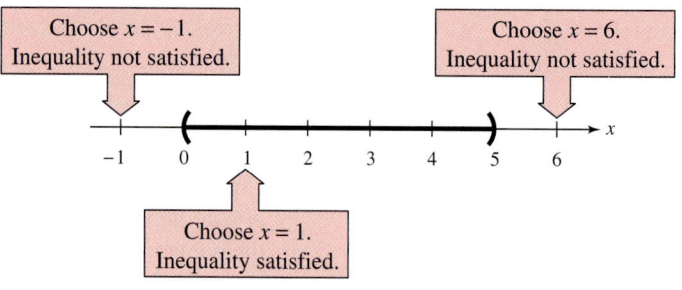

Figure 10.14

Technology: Tip

Most graphing calculators can graph the solution set of a quadratic inequality. Consult the user's guide of your graphing calculator for specific instructions. Notice that the solution set for the quadratic inequality

$$x^2 - 5x < 0$$

shown below appears as a horizontal line above the x-axis.

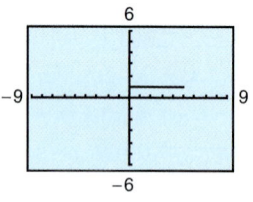

Study Tip

In Example 1, note that you would have used the same basic procedure if the inequality symbol had been $\leq$, $>$, or $\geq$. For instance, in Figure 10.14, you can see that the solution set of the inequality $x^2 - 5x \geq 0$ consists of the union of the half-open intervals $(-\infty, 0]$ and $[5, \infty)$, which is written as $(-\infty, 0] \cup [5, \infty)$.

Just as in solving quadratic *equations*, the first step in solving a quadratic *inequality* is to write the inequality in **general form,** with the polynomial on the left and zero on the right, as demonstrated in Example 2.

Example 2   Solving a Quadratic Inequality

Solve the inequality $2x^2 + 5x \geq 12$.

Solution

Begin by writing the inequality in general form.

$$2x^2 + 5x - 12 \geq 0 \qquad \text{\color{red}Write in general form.}$$

Next, find the critical numbers of $2x^2 + 5x - 12 \geq 0$ by finding the solutions to the equation $2x^2 + 5x - 12 = 0$.

$$2x^2 + 5x - 12 = 0 \qquad \text{\color{red}Write corresponding equation.}$$

$$(x + 4)(2x - 3) = 0 \qquad \text{\color{red}Factor.}$$

$$x = -4, \; x = \frac{3}{2} \qquad \text{\color{red}Critical numbers}$$

This implies that the test intervals are $(-\infty, -4)$, $\left(-4, \frac{3}{2}\right)$, and $\left(\frac{3}{2}, \infty\right)$. To test an interval, choose a convenient number in the interval and determine if the number satisfies the inequality.

Test interval	Representative x-value	Is inequality satisfied?
$(-\infty, -4)$	$x = -5$	$2(-5)^2 + 5(-5) \overset{?}{\geq} 12$ $25 \geq 12$
$\left(-4, \frac{3}{2}\right)$	$x = 0$	$2(0)^2 + 5(0) \overset{?}{\geq} 12$ $0 \ngeq 12$
$\left(\frac{3}{2}, \infty\right)$	$x = 2$	$2(2)^2 + 5(2) \overset{?}{\geq} 12$ $18 \geq 12$

From this table you can see that the inequality $2x^2 + 5x \geq 12$ is satisfied for the intervals $(-\infty, -4]$ and $\left[\frac{3}{2}, \infty\right)$. So, the solution set of the inequality is $(-\infty, -4] \cup \left[\frac{3}{2}, \infty\right)$, as shown in Figure 10.15.

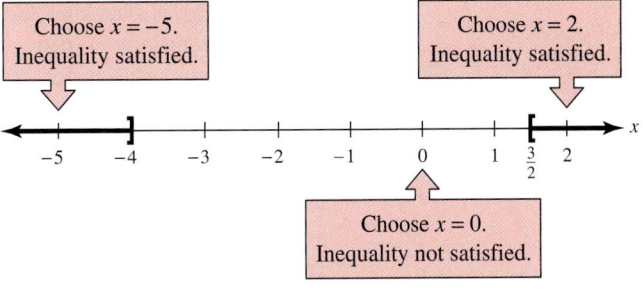

Figure 10.15

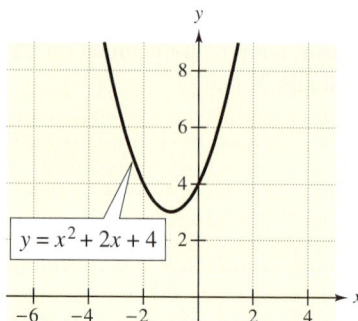

Figure 10.16

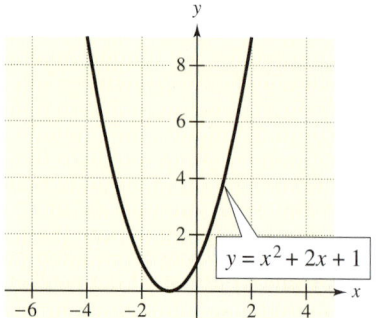

Figure 10.17

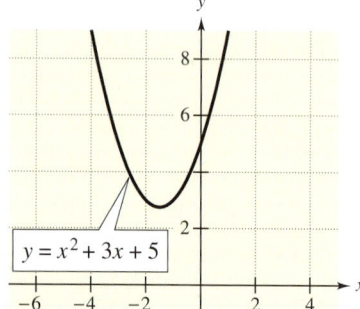

Figure 10.18

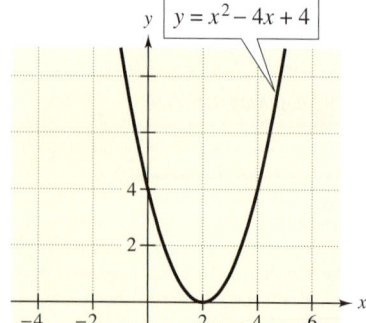

Figure 10.19

The solutions of the quadratic inequalities in Examples 1 and 2 consist, respectively, of a single interval and the union of two intervals. When solving the exercises for this section, you should watch for some unusual solution sets, as illustrated in Example 3.

Example 3   Unusual Solution Sets

Solve each inequality.

a. The solution set of the quadratic inequality

$$x^2 + 2x + 4 > 0$$

consists of the entire set of real numbers, $(-\infty, \infty)$. This is true because the value of the quadratic $x^2 + 2x + 4$ is positive for every real value of x. You can see in Figure 10.16 that the entire parabola lies above the x-axis.

b. The solution set of the quadratic inequality

$$x^2 + 2x + 1 \leq 0$$

consists of the single number $\{-1\}$. This is true because $x^2 + 2x + 1 = (x + 1)^2$ has just one critical number, $x = -1$, and it is the only value that satisfies the inequality. You can see in Figure 10.17 that the parabola meets the x-axis at $x = -1$.

c. The solution set of the quadratic inequality

$$x^2 + 3x + 5 < 0$$

is empty. This is true because the value of the quadratic $x^2 + 3x + 5$ is not less than zero for any value of x. No point on the parabola lies below the x-axis, as shown in Figure 10.18.

d. The solution set of the quadratic inequality

$$x^2 - 4x + 4 > 0$$

consists of all real numbers *except* the number 2. In interval notation, this solution set can be written as $(-\infty, 2) \cup (2, \infty)$. You can see in Figure 10.19 that the parabola lies above the x-axis *except* at $x = 2$, where it meets the x-axis.

Remember that checking the solution set of an inequality is not as straightforward as checking the solutions of an equation, because inequalities tend to have infinitely many solutions. Even so, you should check several x-values in your solution set to confirm that they satisfy the inequality. Also try checking x-values that are not in the solution set to verify that they do not satisfy the inequality.

For instance, the solution set of $x^2 - 5x < 0$ is the interval $(0, 5)$. Try checking some numbers in this interval to verify that they satisfy the inequality. Then check some numbers outside the interval to verify that they do not satisfy the inequality.

3 Use test intervals to solve rational inequalities.

Rational Inequalities

The concepts of critical numbers and test intervals can be extended to inequalities involving rational expressions. To do this, use the fact that the value of a rational expression can change sign only at its *zeros* (the x-values for which its numerator is zero) and its *undefined values* (the x-values for which its denominator is zero). These two types of numbers make up the **critical numbers** of a rational inequality. For instance, the critical numbers of the inequality

$$\frac{x - 2}{(x - 1)(x + 3)} < 0$$

are $x = 2$ (the numerator is zero), and $x = 1$ and $x = -3$ (the denominator is zero). From these three critical numbers you can see that the inequality has *four* test intervals: $(-\infty, -3)$, $(-3, 1)$, $(1, 2)$, and $(2, \infty)$.

Example 4   Solving a Rational Inequality

To solve the inequality $\dfrac{x}{x - 2} > 0$, first find the critical numbers. The numerator is zero when $x = 0$, and the denominator is zero when $x = 2$. So, the two critical numbers are 0 and 2, which implies that the test intervals are $(-\infty, 0)$, $(0, 2)$, and $(2, \infty)$. To test an interval, choose a convenient number in the interval and determine if the number satisfies the inequality, as shown in the table.

Test interval	Representative x-value	Is inequality satisfied?	
$(-\infty, 0)$	$x = -1$	$\dfrac{-1}{-1 - 2} \overset{?}{>} 0$	$\dfrac{1}{3} > 0$
$(0, 2)$	$x = 1$	$\dfrac{1}{1 - 2} \overset{?}{>} 0$	$-1 \not> 0$
$(2, \infty)$	$x = 3$	$\dfrac{3}{3 - 2} \overset{?}{>} 0$	$3 > 0$

You can see that the inequality is satisfied for the intervals $(-\infty, 0)$ and $(2, \infty)$. So, the solution set of the inequality is $(-\infty, 0) \cup (2, \infty)$. See Figure 10.20.

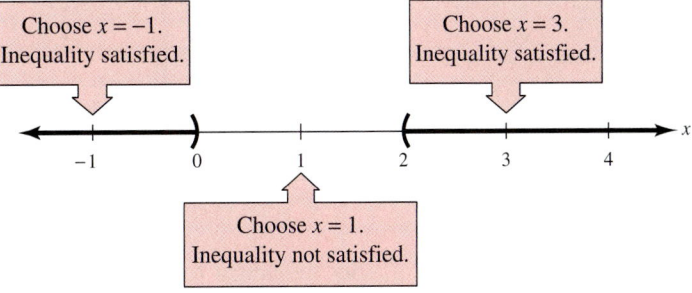

Choose $x = -1$.
Inequality satisfied.

Choose $x = 3$.
Inequality satisfied.

Choose $x = 1$.
Inequality not satisfied.

Figure 10.20

Study Tip

When solving a rational inequality, you should begin by writing the inequality in general form, with the rational expression (as a single fraction) on the left and zero on the right. For instance, the first step in solving

$$\frac{2x}{x + 3} < 4$$

is to write it as

$$\frac{2x}{x + 3} - 4 < 0$$

$$\frac{2x - 4(x + 3)}{x + 3} < 0$$

$$\frac{-2x - 12}{x + 3} < 0.$$

Try solving this inequality. You should find that the solution set is $(-\infty, -6) \cup (-3, \infty)$.

④ **Use inequalities to solve application problems.**

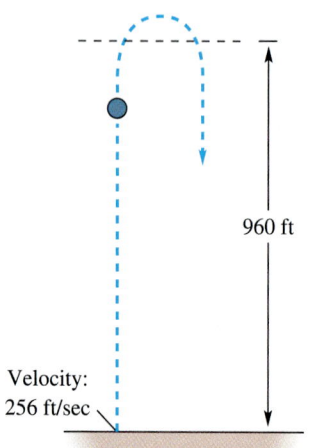

Figure 10.21

Application

Example 5   The Height of a Projectile

A projectile is fired straight upward from ground level with an initial velocity of 256 feet per second, as shown in Figure 10.21, so that its height h at any time t is given by

$$h = -16t^2 + 256t$$

where h is measured in feet and t is measured in seconds. During what interval of time will the height of the projectile exceed 960 feet?

Solution

To solve this problem, begin by writing the inequality in general form.

$-16t^2 + 256t > 960$	Write original inequality.
$-16t^2 + 256t - 960 > 0$	Write in general form.

Next, find the critical numbers for $-16t^2 + 256t - 960 > 0$ by finding the solution to the equation $-16t^2 + 256t - 960 = 0$.

$-16t^2 + 256t - 960 = 0$	Write corresponding equation.
$t^2 - 16t + 60 = 0$	Divide each side by -16.
$(t - 6)(t - 10) = 0$	Factor.
$t = 6, t = 10$	Critical numbers

This implies that the test intervals are

$(-\infty, 6), (6, 10),$ and $(10, \infty)$.	Test intervals

To test an interval, choose a convenient number in the interval and determine if the number satisfies the inequality.

Test interval	Representative x-value	Is inequality satisfied?
$(-\infty, 6)$	$t = 0$	$-16(0)^2 + 256(0) \overset{?}{>} 960$ $0 \not> 960$
$(6, 10)$	$t = 7$	$-16(7)^2 + 256(7) \overset{?}{>} 960$ $1008 > 960$
$(10, \infty)$	$t = 11$	$-16(11)^2 + 256(11) \overset{?}{>} 960$ $880 \not> 960$

So, the height of the projectile will exceed 960 feet for values of t such that $6 < t < 10$.

10.6 Exercises

Review   Concepts, Skills, and Problem Solving

Keep mathematically in shape by doing these exercises *before* the problems of this section.

Properties and Definitions

1. *Writing* ✏ Is 36.83×10^8 written in scientific notation? Explain.

2. *Writing* ✏ The numbers $n_1 \times 10^2$ and $n_2 \times 10^4$ are in scientific notation. The product of these two numbers must lie in what interval? Explain.

Simplifying Expressions

In Exercises 3–8, factor the expression.

3. $6u^2v - 192v^2$

4. $5x^{2/3} - 10x^{1/3}$

5. $x(x - 10) - 4(x - 10)$

6. $x^3 + 3x^2 - 4x - 12$

7. $16x^2 - 121$

8. $4x^3 - 12x^2 + 16x$

Mathematical Modeling

▲ *Geometry*   In Exercises 9–12, write an expression for the area of the figure.

9.
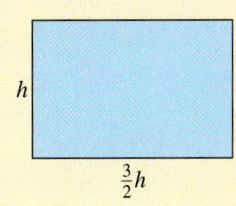

10.
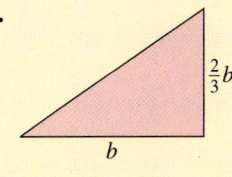

11.
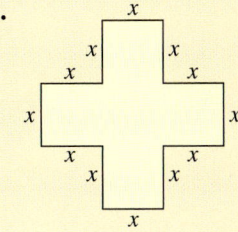

12.
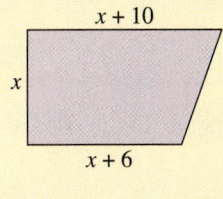

Developing Skills

In Exercises 1–10, find the critical numbers.

1. $x(2x - 5)$

2. $5x(x - 3)$

3. $4x^2 - 81$

4. $9y^2 - 16$

5. $x(x + 3) - 5(x + 3)$

6. $y(y - 4) - 3(y - 4)$

7. $x^2 - 4x + 3$

8. $3x^2 - 2x - 8$

9. $4x^2 - 20x + 25$

10. $4x^2 - 4x - 3$

In Exercises 11–20, determine the intervals for which the polynomial is entirely negative and entirely positive.

11. $x - 4$

12. $3 - x$

13. $3 - \frac{1}{2}x$

14. $\frac{2}{3}x - 8$

15. $2x(x - 4)$

16. $7x(3 - x)$

17. $4 - x^2$

18. $x^2 - 9$

19. $x^2 - 4x - 5$

20. $2x^2 - 4x - 3$

In Exercises 21–60, solve the inequality and graph the solution on the real number line. (Some of the inequalities have no solution.) See Examples 1–3.

21. $3x(x - 2) < 0$

22. $2x(x - 6) > 0$

23. $3x(2 - x) \geq 0$

24. $2x(6 - x) > 0$

25. $x^2 > 4$

26. $z^2 \leq 9$

27. $x^2 - 3x - 10 \geq 0$

28. $x^2 + 8x + 7 < 0$

29. $x^2 + 4x > 0$

30. $x^2 - 5x \geq 0$

31. $x^2 + 3x \leq 10$

32. $t^2 - 4t > 12$

33. $u^2 + 2u - 2 > 1$

34. $t^2 - 15t + 50 < 0$

35. $x^2 + 4x + 5 < 0$

36. $x^2 + 6x + 10 > 0$

37. $x^2 + 2x + 1 \geq 0$

38. $y^2 - 5y + 6 > 0$

39. $x^2 - 4x + 2 > 0$

40. $-x^2 + 8x - 11 \leq 0$

41. $x^2 - 6x + 9 \geq 0$

42. $x^2 + 8x + 16 < 0$

43. $u^2 - 10u + 25 < 0$

44. $y^2 + 16y + 64 \leq 0$

45. $3x^2 + 2x - 8 \leq 0$

46. $2t^2 - 3t - 20 \geq 0$

47. $-6u^2 + 19u - 10 > 0$

48. $4x^2 - 4x - 63 < 0$

49. $-2u^2 + 7u + 4 < 0$

50. $-3x^2 - 4x + 4 \leq 0$

51. $4x^2 + 28x + 49 \leq 0$

52. $9x^2 - 24x + 16 \geq 0$

53. $(x - 5)^2 < 0$

54. $(y + 3)^2 \geq 0$

55. $6 - (x - 5)^2 < 0$

56. $(y + 3)^2 - 6 \geq 0$

57. $16 \leq (u + 5)^2$

58. $25 \geq (x - 3)^2$

59. $x(x - 2)(x + 2) > 0$

60. $x(x - 1)(x + 4) \leq 0$

▦ In Exercises 61–68, use a graphing calculator to solve the inequality. Verify your result algebraically.

61. $x^2 - 6x < 0$

62. $2x^2 + 5x > 0$

63. $0.5x^2 + 1.25x - 3 > 0$

64. $\frac{1}{3}x^2 - 3x < 0$

65. $x^2 + 4x + 4 \geq 9$

66. $x^2 - 6x + 9 < 16$

67. $9 - 0.2(x - 2)^2 < 4$

68. $8x - x^2 > 12$

In Exercises 69–72, find the critical numbers.

69. $\dfrac{5}{x - 3}$

70. $\dfrac{-6}{x + 2}$

71. $\dfrac{2x}{x + 5}$

72. $\dfrac{x - 2}{x - 10}$

In Exercises 73–94, solve the inequality and graph the solution on the real number line. See Example 4.

73. $\dfrac{5}{x - 3} > 0$

74. $\dfrac{3}{4 - x} > 0$

75. $\dfrac{-5}{x - 3} > 0$

76. $\dfrac{-3}{4 - x} > 0$

77. $\dfrac{3}{y - 1} \leq -1$

78. $\dfrac{2}{x - 3} \geq -1$

79. $\dfrac{x + 3}{x - 1} > 0$

80. $\dfrac{x - 5}{x + 2} < 0$

81. $\dfrac{y - 4}{y - 1} \leq 0$

82. $\dfrac{y + 6}{y + 2} \geq 0$

83. $\dfrac{4x - 2}{2x - 4} > 0$

84. $\dfrac{3x + 4}{2x - 1} < 0$

85. $\dfrac{x + 2}{4x + 6} \leq 0$

86. $\dfrac{u - 6}{3u - 5} \le 0$

87. $\dfrac{3(u - 3)}{u + 1} < 0$

88. $\dfrac{2(4 - t)}{4 + t} > 0$

89. $\dfrac{6}{x - 4} > 2$

90. $\dfrac{1}{x + 2} > -3$

91. $\dfrac{4x}{x + 2} < -1$

92. $\dfrac{6x}{x - 4} < 5$

93. $\dfrac{x - 1}{x - 3} \le 2$

94. $\dfrac{x + 4}{x - 5} \ge 10$

In Exercises 95–102, use a graphing calculator to solve the rational inequality. Verify your result algebraically.

95. $\dfrac{1}{x} - x > 0$

96. $\dfrac{1}{x} - 4 < 0$

97. $\dfrac{x + 6}{x + 1} - 2 < 0$

98. $\dfrac{x + 12}{x + 2} - 3 \ge 0$

99. $\dfrac{6x - 3}{x + 5} < 2$

100. $\dfrac{3x - 4}{x - 4} < -5$

101. $x + \dfrac{1}{x} > 3$

102. $4 - \dfrac{1}{x^2} > 1$

Graphical Analysis   In Exercises 103–106, use a graphing calculator to graph the function. Use the graph to approximate the values of x that satisfy the specified inequalities.

Function	Inequalities	
103. $f(x) = \dfrac{3x}{x - 2}$	(a) $f(x) \le 0$	(b) $f(x) \ge 6$
104. $f(x) = \dfrac{2(x - 2)}{x + 1}$	(a) $f(x) \le 0$	(b) $f(x) \ge 8$
105. $f(x) = \dfrac{2x^2}{x^2 + 4}$	(a) $f(x) \ge 1$	(b) $f(x) \le 2$
106. $f(x) = \dfrac{5x}{x^2 + 4}$	(a) $f(x) \ge 1$	(b) $f(x) \ge 0$

Solving Problems

107. *Height*   A projectile is fired straight upward from ground level with an initial velocity of 128 feet per second, so that its height h at any time t is given by $h = -16t^2 + 128t$, where h is measured in feet and t is measured in seconds. During what interval of time will the height of the projectile exceed 240 feet?

108. *Height*   A projectile is fired straight upward from ground level with an initial velocity of 88 feet per second, so that its height h at any time t is given by $h = -16t^2 + 88t$, where h is measured in feet and t is measured in seconds. During what interval of time will the height of the projectile exceed 50 feet?

109. *Compound Interest*   You are investing $1000 in a certificate of deposit for 2 years and you want the interest for that time period to exceed $150. The interest is compounded annually. What interest rate should you have? [*Hint:* Solve the inequality $1000(1 + r)^2 > 1150$.]

110. *Compound Interest*   You are investing $500 in a certificate of deposit for 2 years and you want the interest for that time to exceed $50. The interest is compounded annually. What interest rate should you have? [*Hint:* Solve the inequality $500(1 + r)^2 > 550$.]

111. ▲ *Geometry* You have 64 feet of fencing to enclose a rectangular region. Determine the interval for the length such that the area will exceed 240 square feet.

112. ▲ *Geometry* A rectangular playing field with a perimeter of 100 meters is to have an area of at least 500 square meters. Within what bounds must the length of the field lie?

113. *Cost, Revenue, and Profit* The revenue and cost equations for a computer desk are given by

$$R = x(50 - 0.0002x) \text{ and } C = 12x + 150,000$$

where R and C are measured in dollars and x represents the number of desks sold. How many desks must be sold to obtain a profit of at least $1,650,000?

114. *Cost, Revenue, and Profit* The revenue and cost equations for a digital camera are given by

$$R = x(125 - 0.0005x) \text{ and } C = 3.5x + 185,000$$

where R and C are measured in dollars and x represents the number of cameras sold. How many cameras must be sold to obtain a profit of at least $6,000,000?

115. *Average Cost* The cost C of producing x calendars is $C = 3000 + 0.75x$, $x > 0$.

(a) Write the average cost $\overline{C} = C/x$ as a function of x.

(b) ▦ Use a graphing calculator to graph the average cost function in part (a).

(c) How many calendars must be produced if the average cost per unit is to be less than $2?

116. *Data Analysis* The temperature T (in degrees Fahrenheit) of a metal in a laboratory experiment was recorded every 2 minutes for a period of 16 minutes. The table shows the experimental data, where t is the time in minutes.

t	0	2	4	6	8
T	250	290	338	410	498

t	10	12	14	16
T	560	530	370	160

A model for this data is

$$T = \frac{248.5 - 13.72t}{1.0 - 0.13t + 0.005t^2}.$$

(a) ▦ Use a graphing calculator to plot the data and graph the model in the same viewing window.

(b) Use the graph to approximate the times when the temperature was at least $400°$F.

Explaining Concepts

117. ⚡ Answer part (e) of Motivating the Chapter on page 612.

118. *Writing* ✎ Explain the change in an inequality when each side is multiplied by a negative real number.

119. *Writing* ✎ Give a verbal description of the intervals $(-\infty, 5] \cup (10, \infty)$.

120. *Writing* ✎ Define the term *critical number* and explain how critical numbers are used in solving quadratic and rational inequalities.

121. *Writing* ✎ In your own words, describe the procedure for solving quadratic inequalities.

122. Give an example of a quadratic inequality that has no real solution.

What Did You Learn?

Key Terms

double or repeated solution, *p. 614*

quadratic form, *p. 617*

discriminant, *p. 631*

parabola, *p. 642*

standard form of a quadratic function, *p. 642*

vertex of a parabola, *p. 642*

axis of a parabola, *p. 642*

zeros of a polynomial, *p. 663*

test intervals, *p. 663*

critical numbers, *p. 663*

Key Concepts

10.1 ◯ Square Root Property

The equation $u^2 = d$, where $d > 0$, has exactly two solutions:

$$u = \sqrt{d} \quad \text{and} \quad u = -\sqrt{d}.$$

10.1 ◯ Square Root Property (complex square root)

The equation $u^2 = d$, where $d < 0$, has exactly two solutions:

$$u = \sqrt{|d|}\,i \quad \text{and} \quad u = -\sqrt{|d|}\,i.$$

10.2 ◯ Completing the square

To complete the square for the expression $x^2 + bx$, add $(b/2)^2$, which is the square of half the coefficient of x. Consequently

$$x^2 + bx + \left(\frac{b}{2}\right)^2 = \left(x + \frac{b}{2}\right)^2.$$

10.3 ◯ The Quadratic Formula

The solutions of $ax^2 + bx + c = 0$, $a \neq 0$, are given by the Quadratic Formula

$$x = \frac{-b \pm \sqrt{b^2 - 4ac}}{2a}.$$

The expression inside the radical, $b^2 - 4ac$, is called the discriminant.

1. If $b^2 - 4ac > 0$, the equation has two real solutions.

2. If $b^2 - 4ac = 0$, the equation has one (repeated) real solution.

3. If $b^2 - 4ac < 0$, the equation has no real solutions.

10.3 ◯ Using the discriminant

The discriminant of the quadratic equation

$$ax^2 + bx + c = 0, a \neq 0$$

can be used to classify the solutions of the equation as follows.

Discriminant	Solution Type
1. Perfect square	Two distinct rational solutions
2. Positive nonperfect square	Two distinct irrational solutions
3. Zero	One repeated rational solution
4. Negative number	Two distinct complex solutions

10.4 ◯ Sketching a parabola

1. Determine the vertex and axis of the parabola by completing the square or by using the formula $x = -b/(2a)$.

2. Plot the vertex, axis, x- and y-intercepts, and a few additional points on the parabola. (Using the symmetry about the axis can reduce the number of points you need to plot.)

3. Use the fact that the parabola opens upward if $a > 0$ and opens downward if $a < 0$ to complete the sketch.

10.6 ◯ Finding test intervals for inequalities.

1. For a polynomial expression, find all the real zeros. For a rational expression, find all the real zeros and those x-values for which the function is undefined.

2. Arrange the numbers found in Step 1 in increasing order. These numbers are called critical numbers.

3. Use the critical numbers to determine the test intervals.

4. Choose a representative x-value in each test interval and evaluate the expression at that value. If the value of the expression is negative, the expression will have negative values for every x-value in the interval. If the value of the expression is positive, the expression will have positive values for every x-value in the interval.

Review Exercises

10.1 Solving Quadratic Equations: Factoring and Special Forms

① Solve quadratic equations by factoring.

In Exercises 1–10, solve the equation by factoring.

1. $x^2 + 12x = 0$ **2.** $u^2 - 18u = 0$

3. $4y^2 - 1 = 0$ **4.** $2z^2 - 72 = 0$

5. $4y^2 + 20y + 25 = 0$

6. $x^2 + \frac{8}{3}x + \frac{16}{9} = 0$

7. $2x^2 - 2x - 180 = 0$

8. $15x^2 - 30x - 45 = 0$

9. $6x^2 - 12x = 4x^2 - 3x + 18$

10. $10x - 8 = 3x^2 - 9x + 12$

② Solve quadratic equations by the Square Root Property.

In Exercises 11–16, solve the equation by using the Square Root Property.

11. $4x^2 = 10{,}000$ **12.** $2x^2 = 98$

13. $y^2 - 12 = 0$ **14.** $y^2 - 8 = 0$

15. $(x - 16)^2 = 400$ **16.** $(x + 3)^2 = 900$

③ Solve quadratic equations with complex solutions by the Square Root Property.

In Exercises 17–22, solve the equation by using the Square Root Property.

17. $z^2 = -121$ **18.** $u^2 = -25$

19. $y^2 + 50 = 0$ **20.** $x^2 + 48 = 0$

21. $(y + 4)^2 + 18 = 0$

22. $(x - 2)^2 + 24 = 0$

④ Use substitution to solve equations of quadratic form.

In Exercises 23–30, solve the equation of quadratic form. (Find all real *and* complex solutions.)

23. $x^4 - 4x^2 - 5 = 0$

24. $x^4 - 10x^2 + 9 = 0$

25. $x - 4\sqrt{x} + 3 = 0$

26. $x + 2\sqrt{x} - 3 = 0$

27. $(x^2 - 2x)^2 - 4(x^2 - 2x) - 5 = 0$

28. $\left(\sqrt{x} - 2\right)^2 + 2\left(\sqrt{x} - 2\right) - 3 = 0$

29. $x^{2/3} + 3x^{1/3} - 28 = 0$

30. $x^{2/5} + 4x^{1/5} + 3 = 0$

10.2 Completing the Square

① Rewrite quadratic expressions in completed square form.

In Exercises 31–36, add a term to the expression so that it becomes a perfect square trinomial.

31. $x^2 + 24x +$ **32.** $y^2 - 80y +$

33. $x^2 - 15x +$ **34.** $x^2 + 21x +$

35. $y^2 + \frac{2}{5}y +$ **36.** $x^2 - \frac{3}{4}x +$

② Solve quadratic equations by completing the square.

In Exercises 37–42, solve the equation by completing the square. Give the solutions in exact form and in decimal form rounded to two decimal places. (The solutions may be complex numbers.)

37. $x^2 - 6x - 3 = 0$

38. $x^2 + 12x + 6 = 0$

39. $x^2 - 3x + 3 = 0$

40. $u^2 - 5u + 6 = 0$

41. $y^2 - \frac{2}{3}y + 2 = 0$

42. $t^2 + \frac{1}{2}t - 1 = 0$

10.3 The Quadratic Formula

② Use the Quadratic Formula to solve quadratic equations.

In Exercises 43–48, solve the equation by using the Quadratic Formula. (Find all real *and* complex solutions.)

43. $y^2 + y - 30 = 0$

44. $x^2 - x - 72 = 0$

45. $2y^2 + y - 21 = 0$

46. $2x^2 - 3x - 20 = 0$

47. $5x^2 - 16x + 2 = 0$

48. $3x^2 + 12x + 4 = 0$

③ Determine the types of solutions of quadratic equations using the discriminant.

In Exercises 49–56, use the discriminant to determine the type of solutions of the quadratic equation.

49. $x^2 + 4x + 4 = 0$

50. $y^2 - 26y + 169 = 0$

51. $s^2 - s - 20 = 0$

52. $r^2 - 5r - 45 = 0$

53. $3t^2 + 17t + 10 = 0$

54. $7x^2 + 3x - 18 = 0$

55. $v^2 - 6v + 21 = 0$

56. $9y^2 + 1 = 0$

④ Write quadratic equations from solutions of the equations.

In Exercises 57–62, write a quadratic equation having the given solutions.

57. $3, -7$

58. $-1, 10$

59. $5 + \sqrt{7}, 5 - \sqrt{7}$

60. $2 + \sqrt{2}, 2 - \sqrt{2}$

61. $6 + 2i, 6 - 2i$

62. $3 + 4i, 3 - 4i$

10.4 Graphs of Quadratic Functions

① Determine the vertices of parabolas by completing the square.

In Exercises 63–66, write the equation of the parabola in standard form and find the vertex of its graph.

63. $y = x^2 - 8x + 3$

64. $y = x^2 + 12x - 9$

65. $y = 2x^2 - x + 3$

66. $y = 3x^2 + 2x - 6$

② Sketch parabolas.

In Exercises 67–70, sketch the parabola. Identify the vertex and any x-intercepts. Use a graphing calculator to verify your results.

67. $y = x^2 + 8x$

68. $y = -x^2 + 3x$

69. $f(x) = x^2 - 6x + 5$

70. $f(x) = x^2 + 3x - 10$

③ Write the equation of a parabola given the vertex and a point on the graph.

In Exercises 71–74, write an equation of the parabola $y = a(x - h)^2 + k$ that satisfies the conditions.

71. Vertex: $(2, -5)$; Point on the graph: $(0, 3)$

72. Vertex: $(-4, 0)$; Point on the graph: $(0, -6)$

73. Vertex: $(5, 0)$; Point on the graph: $(1, 1)$

74. Vertex: $(-2, 5)$; Point on the graph: $(0, 1)$

④ Use parabolas to solve application problems.

75. *Path of a Ball*   The height y (in feet) of a ball thrown by a child is given by $y = -\frac{1}{10}x^2 + 3x + 6$, where x is the horizontal distance (in feet) from where the ball is thrown.

(a) 🖩 Use a graphing calculator to graph the path of the ball.

(b) How high is the ball when it leaves the child's hand?

(c) How high is the ball when it reaches its maximum height?

(d) How far from the child does the ball strike the ground?

76. 🖩 *Graphical Estimation*   The number N (in thousands) of bankruptcies filed in the United States for the years 1996 through 2000 is approximated by $N = -66.36t^2 + 1116.2t - 3259$, $6 \le t \le 10$, where t is the time in years, with $t = 6$ corresponding to 1996.   (Source: Administrative Office of the U.S. Courts)

(a) Use a graphing calculator to graph the model.

(b) Use the graph from part (a) to approximate the maximum number of bankruptcies filed during 1996 through 2000. During what year did this maximum occur?

10.5 Applications of Quadratic Equations

1 Use quadratic equations to solve application problems.

77. *Selling Price*   A car dealer bought a fleet of used cars for a total of $80,000. By the time all but four of the cars had been sold, at an average profit of $1000 each, the original investment of $80,000 had been regained. How many cars were sold, and what was the average price per car?

78. *Selling Price*   A manager of a computer store bought several computers of the same model for $27,000. When all but five of the computers had been sold at a profit of $900 per computer, the original investment of $27,000 had been regained. How many computers were sold, and what was the selling price of each computer?

79. ▲ *Geometry*   The length of a rectangle is 12 inches greater than its width. The area of the rectangle is 108 square inches. Find the dimensions of the rectangle.

80. *Compound Interest*   You want to invest $35,000 for 2 years at an annual interest rate of r (in decimal form). Interest on the account is compounded annually. Find the interest rate if a deposit of $35,000 increases to $38,955.88 over a two-year period.

81. *Reduced Rates*   A Little League baseball team obtains a block of tickets to a ball game for $96. After three more people decide to go to the game, the price per ticket is decreased by $1.60. How many people are going to the game?

82. ▲ *Geometry*   A corner lot has an L-shaped sidewalk along its sides. The total length of the sidewalk is 51 feet. By cutting diagonally across the lot, the walking distance is shortened to 39 feet. What are the lengths of the two legs of the sidewalk?

83. *Work-Rate Problem*   Working together, two people can complete a task in 10 hours. Working alone, one person takes 2 hours longer than the other. How long would it take each person to do the task alone?

84. *Height*   An object is projected vertically upward at an initial velocity of 64 feet per second from a height of 192 feet, so that the height h at any time is given by $h = -16t^2 + 64t + 192$, where t is the time in seconds.

(a) After how many seconds is the height 256 feet?

(b) After how many seconds does the object hit the ground?

10.6 Quadratic and Rational Inequalities

1 Determine test intervals for polynomials.

In Exercises 85–88, find the critical numbers.

85. $2x(x + 7)$ **86.** $5x(x - 13)$

87. $x^2 - 6x - 27$

88. $2x^2 + 11x + 5$

2 Use test intervals to solve quadratic inequalities.

In Exercises 89–94, solve the inequality and graph the solution on the real number line.

89. $5x(7 - x) > 0$

90. $-2x(x - 10) \le 0$

91. $16 - (x - 2)^2 \le 0$

92. $(x - 5)^2 - 36 > 0$

93. $2x^2 + 3x - 20 < 0$

94. $3x^2 - 2x - 8 > 0$

3 Use test intervals to solve rational inequalities.

In Exercises 95–98, solve the inequality and graph the solution on the real number line.

95. $\dfrac{x + 3}{2x - 7} \ge 0$

96. $\dfrac{3x + 2}{x - 3} > 0$

97. $\dfrac{x + 4}{x - 1} < 0$ **98.** $\dfrac{2x - 9}{x - 1} \le 0$

4 Use inequalities to solve application problems.

99. *Height*   A projectile is fired straight upward from ground level with an initial velocity of 312 feet per second, so that its height h at any time t is given by $h = -16t^2 + 312t$, where h is measured in feet and t is measured in seconds. During what interval of time will the height of the projectile exceed 1200 feet?

100. *Average Cost*   The cost C of producing x notebooks is $C = 100{,}000 + 0.9x$, $x > 0$. Write the average cost $\overline{C} = C/x$ as a function of x. Then determine how many notebooks must be produced if the average cost per unit is to be less than $2.

Chapter Test

Take this test as you would take a test in class. After you are done, check your work against the answers in the back of the book.

In Exercises 1–6, solve the equation by the specified method.

1. Factoring:

   $x(x - 3) - 10(x - 3) = 0$

2. Factoring:

   $8x^2 - 21x - 9 = 0$

3. Square Root Property:

   $(x - 2)^2 = 0.09$

4. Square Root Property:

   $(x + 4)^2 + 100 = 0$

5. Completing the square:

   $2x^2 - 6x + 3 = 0$

6. Quadratic Formula:

   $2y(y - 2) = 7$

In Exercises 7 and 8, solve the equation of quadratic form.

7. $\dfrac{1}{x^2} - \dfrac{6}{x} + 4 = 0$

8. $x^{2/3} - 9x^{1/3} + 8 = 0$

9. Find the discriminant and explain what it means in terms of the type of solutions of the quadratic equation $5x^2 - 12x + 10 = 0$.

10. Find a quadratic equation having the solutions -4 and 5.

In Exercises 11 and 12, sketch the parabola. Identify the vertex and any x-intercepts. Use a graphing calculator to verify your results.

11. $y = -x^2 + 7$

12. $y = x^2 - 2x - 15$

In Exercises 13–15, solve the inequality and sketch its solution.

13. $16 \leq (x - 2)^2$

14. $2x(x - 3) < 0$

15. $\dfrac{x + 1}{x - 5} \leq 0$

16. The width of a rectangle is 8 feet less than its length. The area of the rectangle is 240 square feet. Find the dimensions of the rectangle.

17. An English club chartered a bus trip to a Shakespearean festival. The cost of the bus was $1250. To lower the bus fare per person, the club invited nonmembers to go along. When 10 nonmembers joined the trip, the fare per person decreased by $6.25. How many club members are going on the trip?

18. An object is dropped from a height of 75 feet. Its height h (in feet) at any time t is given by $h = -16t^2 + 75$, where t is measured in seconds. Find the time required for the object to fall to a height of 35 feet.

19. Two buildings are connected by an L-shaped protected walkway. The total length of the walkway is 140 feet. By cutting diagonally across the grass, the walking distance is shortened to 100 feet. What are the lengths of the two legs of the walkway?

Motivating the Chapter

 Choosing the Best Investment

You receive an inheritance of $5000 and want to invest it.

See Section 11.1, Exercise 99.

a. Complete the table by finding the amount A of the $5000 investment after 3 years with an annual interest rate of $r = 6\%$. Which form of compounding gives you the greatest balance?

Compounding	Amount, A
Annual	
Quarterly	
Monthly	
Daily	
Hourly	
Continuous	

b. You are considering two different investment options. The first investment option has an interest rate of 7% compounded continuously. The second investment option has an interest rate of 8% compounded quarterly. Which investment should you choose? Explain.

See Section 11.5, Exercise 139.

c. What annual percentage rate is needed to obtain a balance of $6200 in 3 years when the interest is compounded monthly?

d. With an interest rate of 6%, compounded continuously, how long will it take for your inheritance to grow to $7500?

e. What is the *effective yield* on your investment when the interest rate is 8% compounded quarterly?

f. With an interest rate of 6%, compounded continuously, how long will it take your inheritance to double? How long will it take your inheritance to quadruple (reach four times the original amount)?

Monika Graff/The Image Works

Exponential and Logarithmic Functions

11.1 Exponential Functions

What You Should Learn

1 Evaluate exponential functions.

2 Graph exponential functions.

3 Evaluate the natural base e and graph natural exponential functions.

4 Use exponential functions to solve application problems

Why You Should Learn It

Exponential functions can be used to model and solve real-life problems. For instance, in Exercise 96 on page 691, you will use an exponential function to model the median price of a home in the United States.

1 Evaluate exponential functions.

Exponential Functions

In this section, you will study a new type of function called an **exponential function.** Whereas polynomial and rational functions have terms with variable bases and constant exponents, exponential functions have terms with *constant bases* and *variable exponents.* Here are some examples.

Polynomial or Rational Function

Constant Exponents

$$f(x) = x^2, \quad f(x) = x^{-3}$$

Variable Bases

Exponential Function

Variable Exponents

$$f(x) = 2^x, \quad f(x) = 3^{-x}$$

Constant Bases

Definition of Exponential Function

The **exponential function** f with base a is denoted by

$$f(x) = a^x$$

where $a > 0$, $a \neq 1$, and x is any real number.

The base $a = 1$ is excluded because $f(x) = 1^x = 1$ is a constant function, *not* an exponential function.

In Chapters 5 and 9, you learned to evaluate a^x for integer and rational values of x. For example, you know that

$$a^3 = a \cdot a \cdot a, \quad a^{-4} = \frac{1}{a^4}, \quad \text{and} \quad a^{5/3} = \left(\sqrt[3]{a}\right)^5.$$

However, to evaluate a^x for any real number x, you need to interpret forms with *irrational* exponents, such as $a^{\sqrt{2}}$ or a^π. For the purposes of this text, it is sufficient to think of a number such as $a^{\sqrt{2}}$, where $\sqrt{2} \approx 1.414214$, as the number that has the successively closer approximations

$$a^{1.4}, a^{1.41}, a^{1.414}, a^{1.4142}, a^{1.41421}, a^{1.414214}, \ldots .$$

The rules of exponents that were discussed in Section 5.1 can be extended to cover exponential functions, as described on the following page.

Study Tip

Rule 4 of the rules of exponential functions indicates that 2^{-x} can be written as

$$2^{-x} = \frac{1}{2^x}.$$

Similarly, $\frac{1}{3^{-x}}$ can be written as

$$\frac{1}{3^{-x}} = 3^x.$$

In other words, you can move a *factor* from the numerator to the denominator (or from the denominator to the numerator) by changing the sign of its exponent.

Rules of Exponential Functions

Let a be a positive real number, and let x and y be real numbers, variables, or algebraic expressions.

1. $a^x \cdot a^y = a^{x+y}$ Product rule

2. $\dfrac{a^x}{a^y} = a^{x-y}$ Quotient rule

3. $(a^x)^y = a^{xy}$ Power rule

4. $a^{-x} = \dfrac{1}{a^x} = \left(\dfrac{1}{a}\right)^x$ Negative exponent rule

To evaluate exponential functions with a calculator, you can use the exponential key $\boxed{y^x}$ (where y is the base and x is the exponent) or $\boxed{\wedge}$. For example, to evaluate $3^{-1.3}$, you can use the following keystrokes.

Keystrokes	Display	
3 $\boxed{y^x}$ 1.3 $\boxed{+/-}$ $\boxed{=}$	0.239741	Scientific
3 $\boxed{\wedge}$ $\boxed{(}$ $\boxed{(-)}$ 1.3 $\boxed{)}$ $\boxed{\text{ENTER}}$	0.239741	Graphing

Example 1   Evaluating Exponential Functions

Evaluate each function. Use a calculator only if it is necessary or more efficient.

Function	Values
a. $f(x) = 2^x$	$x = 3, x = -4, x = \pi$
b. $g(x) = 12^x$	$x = 3, x = -0.1, x = \frac{5}{7}$
c. $h(x) = (1.04)^{2x}$	$x = 0, x = -2, x = \sqrt{2}$

Solution

Evaluation	Comment
a. $f(3) = 2^3 = 8$	Calculator is not necessary.
$f(-4) = 2^{-4} = \dfrac{1}{2^4} = \dfrac{1}{16}$	Calculator is not necessary.
$f(\pi) = 2^\pi \approx 8.825$	Calculator is necessary.
b. $g(3) = 12^3 = 1728$	Calculator is more efficient.
$g(-0.1) = 12^{-0.1} \approx 0.780$	Calculator is necessary.
$g\left(\dfrac{5}{7}\right) = 12^{5/7} \approx 5.900$	Calculator is necessary.
c. $h(0) = (1.04)^{2 \cdot 0} = (1.04)^0 = 1$	Calculator is not necessary.
$h(-2) = (1.04)^{2(-2)} \approx 0.855$	Calculator is more efficient.
$h(\sqrt{2}) = (1.04)^{2\sqrt{2}} \approx 1.117$	Calculator is necessary.

2 Graph exponential functions.

Graphs of Exponential Functions

The basic nature of the graph of an exponential function can be determined by the point-plotting method or by using a graphing calculator.

Example 2   The Graphs of Exponential Functions

In the same coordinate plane, sketch the graph of each function. Determine the domain and range.

a. $f(x) = 2^x$

b. $g(x) = 4^x$

Solution

The table lists some values of each function, and Figure 11.1 shows the graph of each function. From the graphs, you can see that the domain of each function is the set of all real numbers and that the range of each function is the set of all positive real numbers.

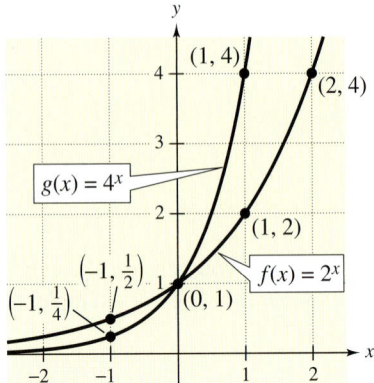

Figure 11.1

x	-2	-1	0	1	2	3
2^x	$\frac{1}{4}$	$\frac{1}{2}$	1	2	4	8
4^x	$\frac{1}{16}$	$\frac{1}{4}$	1	4	16	64

Note in the next example that a graph of the form $f(x) = a^x$ (as shown in Example 2) is a reflection in the y-axis of a graph of the form $g(x) = a^{-x}$.

Example 3   The Graphs of Exponential Functions

In the same coordinate plane, sketch the graph of each function.

a. $f(x) = 2^{-x}$

b. $g(x) = 4^{-x}$

Solution

The table lists some values of each function, and Figure 11.2 shows the graph of each function.

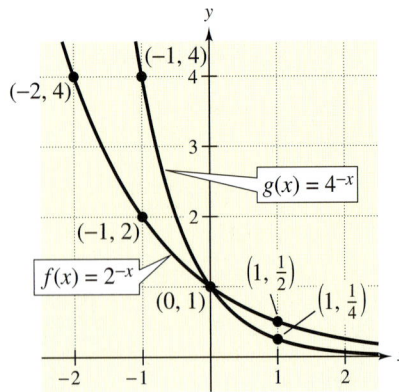

Figure 11.2

x	-3	-2	-1	0	1	2
2^{-x}	8	4	2	1	$\frac{1}{2}$	$\frac{1}{4}$
4^{-x}	64	16	4	1	$\frac{1}{4}$	$\frac{1}{16}$

Study Tip

An **asymptote** of a graph is a line to which the graph becomes arbitrarily close as $|x|$ or $|y|$ increases without bound. In other words, if a graph has an asymptote, it is possible to move far enough out on the graph so that there is almost no difference between the graph and the asymptote.

Examples 2 and 3 suggest that for $a > 1$, the values of the function of $y = a^x$ increase as x increases and the values of the function $y = a^{-x} = (1/a)^x$ decrease as x increases. The graphs shown in Figure 11.3 are typical of the graphs of exponential functions. Note that each graph has a y-intercept at $(0, 1)$ and a **horizontal asymptote** of $y = 0$ (the x-axis).

Graph of $y = a^x$

- Domain: $(-\infty, \infty)$
- Range: $(0, \infty)$
- Intercept: $(0, 1)$
- Increasing
  (moves up to the right)
- Asymptote: x-axis

Graph of $y = a^{-x} = \left(\dfrac{1}{a}\right)^x$

- Domain: $(-\infty, \infty)$
- Range: $(0, \infty)$
- Intercept: $(0, 1)$
- Decreasing
  (moves down to the right)
- Asymptote: x-axis

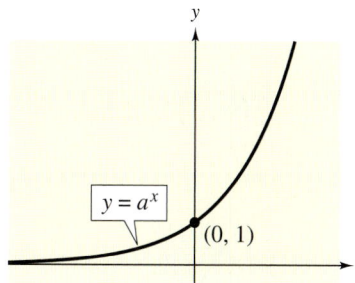

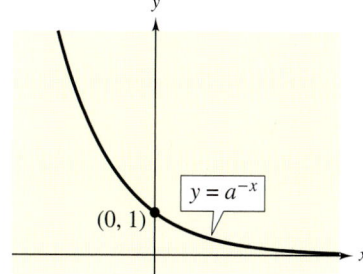

Figure 11.3   Characteristics of the exponential functions $y = a^x$ and $y = a^{-x}(a > 1)$

In the next two examples, notice how the graph of $y = a^x$ can be used to sketch the graphs of functions of the form $f(x) = b \pm a^{x+c}$. Also note that the transformation in Example 4(a) keeps the x-axis as a horizontal asymptote, but the transformation in Example 4(b) yields a new horizontal asymptote of $y = -2$. Also, be sure to note how the y-intercept is affected by each transformation.

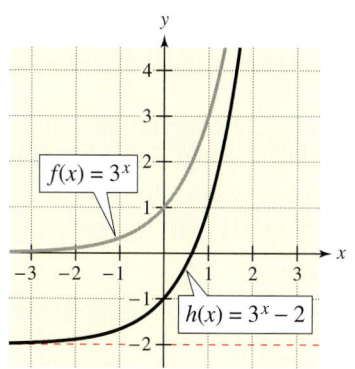

Figure 11.4

Figure 11.5

Example 4   Transformations of Graphs of Exponential Functions

Use transformations to analyze and sketch the graph of each function.

a. $g(x) = 3^{x+1}$ **b.** $h(x) = 3^x - 2$

Solution

Consider the function $f(x) = 3^x$.

a. The function g is related to f by $g(x) = f(x + 1)$. To sketch the graph of g, shift the graph of f one unit to the left, as shown in Figure 11.4. Note that the y-intercept of g is $(0, 3)$.

b. The function h is related to f by $h(x) = f(x) - 2$. To sketch the graph of g, shift the graph of f two units downward, as shown in Figure 11.5. Note that the y-intercept of h is $(0, -1)$ and the horizontal asymptote is $y = -2$.

Example 5   Reflections of Graphs of Exponential Functions

Use transformations to analyze and sketch the graph of each function.

a. $g(x) = -3^x$ **b.** $h(x) = 3^{-x}$

Solution

Consider the function $f(x) = 3^x$.

a. The function g is related to f by $g(x) = -f(x)$. To sketch the graph of g, reflect the graph of f about the x-axis, as shown in Figure 11.6. Note that the y-intercept of g is $(0, -1)$.

b. The function h is related to f by $h(x) = f(-x)$. To sketch the graph of h, reflect the graph of f about the y-axis, as shown in Figure 11.7.

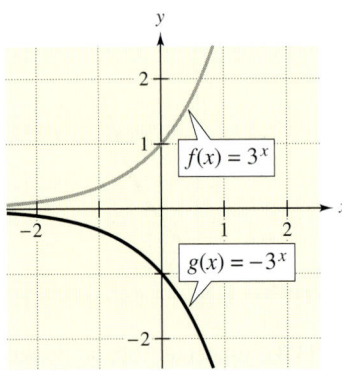

Figure 11.6   Figure 11.7

---

③ Evaluate the natural base e and graph natural exponential functions.

The Natural Exponential Function

So far, integers or rational numbers have been used as bases of exponential functions. In many applications of exponential functions, the convenient choice for a base is the following irrational number, denoted by the letter "e."

$$e \approx 2.71828 \ldots \qquad \text{Natural base}$$

This number is called the **natural base.** The function

$$f(x) = e^x \qquad \text{Natural exponential function}$$

is called the **natural exponential function.** To evaluate the natural exponential function, you need a calculator, preferably one having a natural exponential key $\boxed{e^x}$. Here are some examples of how to use such a calculator to evaluate the natural exponential function.

Value	Keystrokes	Display	
e^2	2 $\boxed{e^x}$	7.3890561	Scientific
e^2	$\boxed{e^x}$ 2 $\boxed{ENTER}$	7.3890561	Graphing
e^{-3}	3 $\boxed{+/-}$ $\boxed{e^x}$	0.0497871	Scientific
e^{-3}	$\boxed{e^x}$ $\boxed{(-)}$ 3 $\boxed{)}$ $\boxed{ENTER}$	0.0497871	Graphing

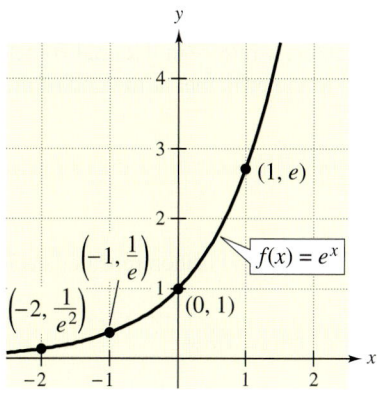

Figure 11.8

When evaluating the natural exponential function, remember that e is the constant number 2.71828 . . . , and x is a variable. After evaluating this function at several values, as shown in the table, you can sketch its graph, as shown in Figure 11.8.

x	-2	-1.5	-1.0	-0.5	0.0	0.5	1.0	1.5
$f(x) = e^x$	0.135	0.223	0.368	0.607	1.000	1.649	2.718	4.482

From the graph, notice the following characteristics of the natural exponential function.

- Domain: $(-\infty, \infty)$
- Range: $(0, \infty)$
- Intercept: $(0, 1)$
- Increasing (moves up to the right)
- Asymptote: x-axis

Notice that these characteristics are consistent with those listed for the exponential function $y = a^x$ on page 683.

④ Use exponential functions to solve application problems.

Applications

A common scientific application of exponential functions is **radioactive decay.**

Example 6   Radioactive Decay

After t years, the remaining mass y (in grams) of 10 grams of a radioactive element whose half-life is 25 years is given by

$$y = 10\left(\frac{1}{2}\right)^{t/25}, \quad t \geq 0.$$

How much of the initial mass remains after 120 years?

Solution

When $t = 120$, the mass is given by

$$y = 10\left(\frac{1}{2}\right)^{120/25} \qquad \text{Substitute 120 for } t.$$

$$= 10\left(\frac{1}{2}\right)^{4.8} \qquad \text{Simplify.}$$

$$\approx 0.359. \qquad \text{Use a calculator.}$$

So, after 120 years, the mass has decayed from an initial amount of 10 grams to only 0.359 gram. Note in Figure 11.9 that the graph of the function shows the 25-year half-life. That is, after 25 years the mass is 5 grams (half of the original), after another 25 years the mass is 2.5 grams, and so on.

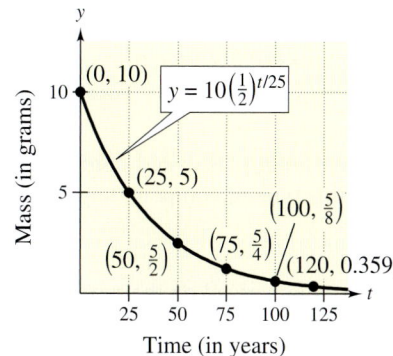

Figure 11.9

One of the most familiar uses of exponential functions involves **compound interest.** A principal P is invested at an annual interest rate r (in decimal form), compounded once a year. If the interest is added to the principal at the end of the year, the balance is

$$A = P + Pr$$

$$= P(1 + r).$$

This pattern of multiplying the previous principal by $(1 + r)$ is then repeated each successive year, as shown below.

Time in Years	Balance at Given Time
0	$A = P$
1	$A = P(1 + r)$
2	$A = P(1 + r)(1 + r) = P(1 + r)^2$
3	$A = P(1 + r)^2(1 + r) = P(1 + r)^3$
$\vdots$	$\vdots$
t	$A = P(1 + r)^t$

To account for more frequent compounding of interest (such as quarterly or monthly compounding), let n be the number of compoundings per year and let t be the number of years. Then the rate per compounding is r/n and the account balance after t years is

$$A = P\left(1 + \frac{r}{n}\right)^{nt}.$$

Example 7   Finding the Balance for Compound Interest

A sum of $10,000 is invested at an annual interest rate of 7.5%, compounded monthly. Find the balance in the account after 10 years.

Solution

Using the formula for compound interest, with $P = 10,000$, $r = 0.075$, $n = 12$ (for monthly compounding), and $t = 10$, you obtain the following balance.

$$A = 10{,}000\left(1 + \frac{0.075}{12}\right)^{12(10)}$$

$$\approx \$21{,}120.65$$

A second method that banks use to compute interest is called **continuous compounding.** The formula for the balance for this type of compounding is

$$A = Pe^{rt}.$$

The formulas for both types of compounding are summarized on the next page.

Formulas for Compound Interest

After t years, the balance A in an account with principal P and annual interest rate r (in decimal form) is given by the following formulas.

1. For n compoundings per year: $A = P\left(1 + \dfrac{r}{n}\right)^{nt}$

2. For continuous compounding: $A = Pe^{rt}$

Technology: Discovery

Use a graphing calculator to evaluate

$$A = 15{,}000\left(1 + \frac{0.08}{n}\right)^{n(6)}$$

for $n = 1000$, $10{,}000$, and $100{,}000$. Compare these values with those found in parts (a) and (b) of Example 8.

As n gets larger and larger, do you think that the value of A will ever exceed the value found in Example 8(c)? Explain.

Example 8 Comparing Three Types of Compounding

A total of $15,000 is invested at an annual interest rate of 8%. Find the balance after 6 years for each type of compounding.

a. Quarterly

b. Monthly

c. Continuous

Solution

a. Letting $P = 15{,}000$, $r = 0.08$, $n = 4$, and $t = 6$, the balance after 6 years at quarterly compounding is

$$A = 15{,}000\left(1 + \frac{0.08}{4}\right)^{4(6)}$$

$$\approx \$24{,}126.56.$$

b. Letting $P = 15{,}000$, $r = 0.08$, $n = 12$, and $t = 6$, the balance after 6 years at monthly compounding is

$$A = 15{,}000\left(1 + \frac{0.08}{12}\right)^{12(6)}$$

$$\approx \$24{,}202.53.$$

c. Letting $P = 15{,}000$, $r = 0.08$, and $t = 6$, the balance after 6 years at continuous compounding is

$$A = 15{,}000e^{0.08(6)}$$

$$\approx \$24{,}241.12.$$

Note that the balance is greater with continuous compounding than with quarterly or monthly compounding.

———

Example 8 illustrates the following general rule. For a given principal, interest rate, and time, the more often the interest is compounded per year, the greater the balance will be. Moreover, the balance obtained by continuous compounding is larger than the balance obtained by compounding n times per year.

11.1 Exercises

Review Concepts, Skills, and Problem Solving

Keep mathematically in shape by doing these exercises *before* the problems of this section.

Properties and Definitions

1. *Writing✏* Explain how to determine the half-plane satisfying $x + y < 5$.

2. *Writing✏* Describe the difference between the graphs of $3x - 5y \leq 15$ and $3x - 5y < 15$.

Graphing Inequalities

In Exercises 3–10, graph the inequality.

3. $y > x - 2$

4. $y \leq 5 - \frac{3}{2}x$

5. $y < \frac{2}{3}x - 1$

6. $x > 6 - y$

7. $y \leq -2$

8. $x > 7$

9. $2x + 3y \geq 6$

10. $5x - 2y < 5$

Problem Solving

11. *Work Rate* Working together, two people can complete a task in 10 hours. Working alone, one person takes 3 hours longer than the other. How long would it take each person to do the task alone?

12. ▲ *Geometry* A family is setting up the boundaries for a backyard volleyball court. The court is to be 60 feet long and 30 feet wide. To be assured that the court is rectangular, someone suggests that they measure the diagonals of the court. What should be the length of each diagonal?

Developing Skills

In Exercises 1–8, simplify the expression.

1. $2^x \cdot 2^{x-1}$

2. $10e^{2x} \cdot e^{-x}$

3. $\dfrac{e^{x+2}}{e^x}$

4. $\dfrac{3^{2x+3}}{3^{x+1}}$

5. $(2e^x)^3$

6. $-4e^{-2x}$

7. $\sqrt[3]{-8e^{3x}}$

8. $\sqrt{4e^{6x}}$

In Exercises 9–16, evaluate the expression. (Round your answer to three decimal places.)

9. $4^{\sqrt{3}}$

10. $6^{-\pi}$

11. $e^{1/3}$

12. $e^{-1/3}$

13. $4(3e^4)^{1/2}$

14. $(9e^2)^{3/2}$

15. $\dfrac{4e^3}{12e^2}$

16. $\dfrac{6e^5}{10e^7}$

In Exercises 17–30, evaluate the function as indicated. Use a calculator only if it is necessary or more efficient. (Round your answer to three decimal places.) See Example 1.

17. $f(x) = 3^x$
  (a) $x = -2$
  (b) $x = 0$
  (c) $x = 1$

18. $F(x) = 3^{-x}$
  (a) $x = -2$
  (b) $x = 0$
  (c) $x = 1$

19. $g(x) = 3.8^x$
  (a) $x = -1$
  (b) $x = 3$
  (c) $x = \sqrt{5}$

20. $G(x) = 1.1^{-x}$
  (a) $x = -1$
  (b) $x = 1$
  (c) $x = \sqrt{3}$

21. $f(t) = 500\left(\frac{1}{2}\right)^t$
  (a) $t = 0$
  (b) $t = 1$
  (c) $t = \pi$

22. $g(s) = 1200\left(\frac{2}{3}\right)^s$
  (a) $s = 0$
  (b) $s = 2$
  (c) $s = \sqrt{2}$

23. $f(x) = 1000(1.05)^{2x}$
  (a) $x = 0$
  (b) $x = 5$
  (c) $x = 10$

24. $g(t) = 10{,}000(1.03)^{4t}$
  (a) $t = 1$
  (b) $t = 3$
  (c) $t = 5.5$

25. $h(x) = \dfrac{5000}{(1.06)^{8x}}$

   (a) $x = 5$

   (b) $x = 10$

   (c) $x = 20$

27. $g(x) = 10e^{-0.5x}$

   (a) $x = -4$

   (b) $x = 4$

   (c) $x = 8$

29. $g(x) = \dfrac{1000}{2 + e^{-0.12x}}$

   (a) $x = 0$

   (b) $x = 10$

   (c) $x = 50$

26. $P(t) = \dfrac{10{,}000}{(1.01)^{12t}}$

   (a) $t = 2$

   (b) $t = 10$

   (c) $t = 20$

28. $A(t) = 200e^{0.1t}$

   (a) $t = 10$

   (b) $t = 20$

   (c) $t = 40$

30. $f(z) = \dfrac{100}{1 + e^{-0.05z}}$

   (a) $z = 0$

   (b) $z = 10$

   (c) $z = 20$

In Exercises 31–36, match the function with its graph.
[The graphs are labeled (a), (b), (c), (d), (e), and (f).]

(a)

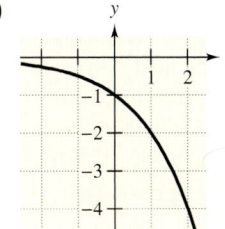

(b)

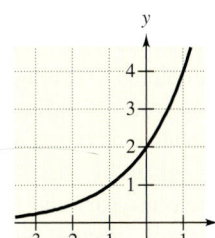

(c)

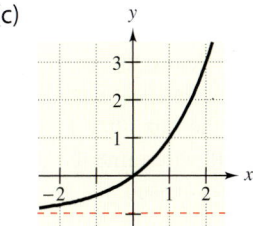

(d)

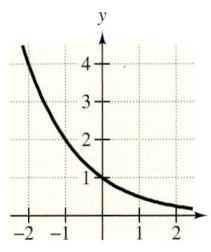

(e)

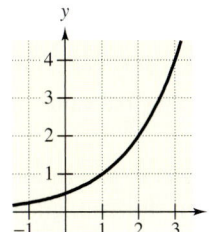

(f)

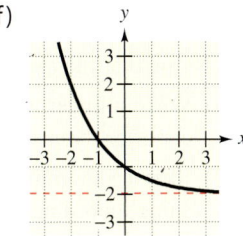

31. $f(x) = -2^x$

33. $f(x) = 2^x - 1$

35. $f(x) = 2^{x+1}$

32. $f(x) = 2^{-x}$

34. $f(x) = 2^{x-1}$

36. $f(x) = \left(\frac{1}{2}\right)^x - 2$

In Exercises 37–56, sketch the graph of the function.
Identify the horizontal asymptote. See Examples 2 and 3.

37. $f(x) = 3^x$

38. $f(x) = 3^{-x} = \left(\frac{1}{3}\right)^x$

39. $h(x) = \frac{1}{2}(3^x)$

40. $h(x) = \frac{1}{2}(3^{-x})$

41. $g(x) = 3^x - 2$

42. $g(x) = 3^x + 1$

43. $f(x) = 5^x + 2$

44. $f(x) = 5^x - 4$

45. $g(x) = 5^{x-1}$

46. $g(x) = 5^{x+3}$

47. $f(t) = 2^{-t^2}$

48. $f(t) = 2^{t^2}$

49. $f(x) = -2^{0.5x}$

50. $h(t) = -2^{-0.5t}$

51. $h(x) = 2^{0.5x}$

52. $g(x) = 2^{-0.5x}$

53. $f(x) = -\left(\frac{1}{3}\right)^x$

54. $f(x) = \left(\frac{3}{4}\right)^x + 1$

55. $g(t) = 200\left(\frac{1}{2}\right)^t$

56. $h(x) = 27\left(\frac{2}{3}\right)^x$

 In Exercises 57–68, use a graphing calculator to
graph the function.

57. $y = 7^{x/2}$

59. $y = 7^{-x/2} + 5$

61. $y = 500(1.06)^t$

63. $y = 3e^{0.2x}$

65. $P(t) = 100e^{-0.1t}$

67. $y = 6e^{-x^2/3}$

58. $y = 7^{-x/2}$

60. $y = 7^{(x-3)/2}$

62. $y = 100(1.06)^{-t}$

64. $y = 50e^{-0.05x}$

66. $A(t) = 1000e^{0.08t}$

68. $g(x) = 7e^{(x+1)/2}$

In Exercises 69–74, identify the transformation of the
graph of $f(x) = 4^x$ and sketch the graph of h. See
Examples 4 and 5.

69. $h(x) = 4^x - 1$

71. $h(x) = 4^{x+2}$

73. $h(x) = -4^x$

70. $h(x) = 4^x + 2$

72. $h(x) = 4^{x-4}$

74. $h(x) = -4^x + 2$

Solving Problems

75. *Radioactive Decay* After t years, the remaining mass y (in grams) of 16 grams of a radioactive element whose half-life is 30 years is given by

$$y = 16\left(\frac{1}{2}\right)^{t/30}, \quad t \geq 0.$$

How much of the initial mass remains after 80 years?

76. *Radioactive Decay* After t years, the remaining mass y (in grams) of 23 grams of a radioactive element whose half-life is 45 years is given by

$$y = 23\left(\frac{1}{2}\right)^{t/45}, \quad t \geq 0.$$

How much of the initial mass remains after 150 years?

Compound Interest In Exercises 77–80, complete the table to determine the balance A for P dollars invested at rate r for t years, compounded n times per year.

n	1	4	12	365	Continuous compounding
A					

Principal	Rate	Time
77. $P = \$100$	$r = 7\%$	$t = 15$ years
78. $P = \$400$	$r = 7\%$	$t = 20$ years
79. $P = \$2000$	$r = 9.5\%$	$t = 10$ years
80. $P = \$1500$	$r = 6.5\%$	$t = 20$ years

Compound Interest In Exercises 81–84, complete the table to determine the principal P that will yield a balance of A dollars when invested at rate r for t years, compounded n times per year.

n	1	4	12	365	Continuous compounding
P					

Balance	Rate	Time
81. $A = \$5000$	$r = 7\%$	$t = 10$ years
82. $A = \$100,000$	$r = 9\%$	$t = 20$ years
83. $A = \$1,000,000$	$r = 10.5\%$	$t = 40$ years
84. $A = \$2500$	$r = 7.5\%$	$t = 2$ years

85. *Demand* The daily demand x and the price p for a collectible are related by $p = 25 - 0.4e^{0.02x}$. Find the prices for demands of (a) $x = 100$ units and (b) $x = 125$ units.

86. *Population Growth* The population P (in millions) of the United States from 1970 to 2000 can be approximated by the exponential function $P(t) = 203.0(1.0107)^t$, where t is the time in years, with $t = 0$ corresponding to 1970. Use the model to estimate the population in the years (a) 2010 and (b) 2020.   (Source: U.S. Census Bureau)

87. *Property Value* The value of a piece of property doubles every 15 years. You buy the property for $64,000. Its value t years after the date of purchase should be $V(t) = 64,000(2)^{t/15}$. Use the model to approximate the value of the property (a) 5 years and (b) 20 years after it is purchased.

88. *Inflation Rate* The annual rate of inflation is predicted to average 5% over the next 10 years. With this rate of inflation, the approximate cost C of goods or services during any year in that decade will be given by $C(t) = P(1.05)^t$, $0 \leq t \leq 10$, where t is time in years and P is the present cost. The price of an oil change for your car is presently $24.95. Estimate the price 10 years from now.

89. *Depreciation* After t years, the value of a car that originally cost $16,000 depreciates so that each year it is worth $\frac{3}{4}$ of its value for the previous year. Find a model for $V(t)$, the value of the car after t years. Sketch a graph of the model and determine the value of the car 2 years after it was purchased.

90. *Depreciation* Straight-line depreciation is used to determine the value of the car in Exercise 89. Assume that the car depreciates $3000 per year.

(a) Write a linear equation for $V(t)$, the value of the car after t years.

(b) Sketch the graph of the model in part (a) on the same coordinate axes used for the graph in Exercise 89.

(c) If you were selling the car after 2 years, which depreciation model would you prefer?

(d) If you were selling the car after 4 years, which model would you prefer?

91. *Graphical Interpretation*   Investments of $500 in two different accounts with interest rates of 6% and 8% are compounded continuously.

(a) For each account, write an exponential function that represents the balance after t years.

(b) ⊞ Use a graphing calculator to graph each of the models in the same viewing window.

(c) ⊞ Use a graphing calculator to graph the function $A_2 - A_1$ in the same viewing window with the graphs in part (b).

(d) ⊞ Use the graphs to discuss the rates of increase of the balances in the two accounts.

92. *Savings Plan*   You decide to start saving pennies according to the following pattern. You save 1 penny the first day, 2 pennies the second day, 4 the third day, 8 the fourth day, and so on. Each day you save twice the number of pennies you saved on the previous day. Write an exponential function that models this problem. How many pennies do you save on the thirtieth day? (In the next chapter you will learn how to find the total number saved.)

93. *Parachute Drop*   A parachutist jumps from a plane and opens the parachute at a height of 2000 feet (see figure). The distance h between the parachutist and the ground is $h = 1950 + 50e^{-0.4433t} - 22t$, where h is in feet and t is the time in seconds. (The time $t = 0$ corresponds to the time when the parachute is opened.)

(a) ⊞ Use a graphing calculator to graph the function.

(b) Find the distances between the parachutist and the ground when $t = 0, 25, 50,$ and 75.

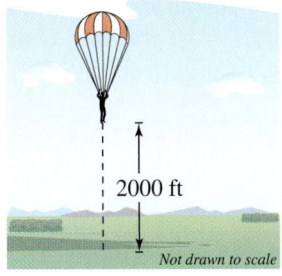

2000 ft

Not drawn to scale

94. *Parachute Drop*   A parachutist jumps from a plane and opens the parachute at a height of 3000 feet. The distance h (in feet) between the parachutist and the ground is $h = 2940 + 60e^{-0.4021t} - 24t$, where t is the time in seconds. (The time $t = 0$ corresponds to the time when the parachute is opened.)

(a) ⊞ Use a graphing calculator to graph the function.

(b) Find the distances between the parachutist and the ground when $t = 0, 50,$ and 100.

95. ⊞ *Data Analysis*   A meteorologist measures the atmospheric pressure P (in kilograms per square meter) at altitude h (in kilometers). The data are shown in the table.

h	0	5	10	15	20
P	10,332	5583	2376	1240	517

(a) Use a graphing calculator to plot the data points.

(b) A model for the data is given by $P = 10,958e^{-0.15h}$. Use a graphing calculator to graph the model in the same viewing window as in part (a). How well does the model fit the data?

(c) Use a graphing calculator to create a table comparing the model with the data points.

(d) Estimate the atmospheric pressure at an altitude of 8 kilometers.

(e) Use the graph to estimate the altitude at which the atmospheric pressure is 2000 kilograms per square meter.

96. *Data Analysis*   For the years 1994 through 2001, the median prices of a one-family home in the United States are shown in the table.   (Source: U.S. Census Bureau and U.S. Department of Housing and Urban Development)

Year	1994	1995	1996	1997
Price	$130,000	$133,900	$140,000	$146,000

Year	1998	1999	2000	2001
Price	$152,500	$161,000	$169,000	$175,200

A model for this data is given by $y = 107{,}773e^{0.0442t}$, where t is time in years, with $t = 4$ representing 1994.

(a) Use the model to complete the table and compare the results with the actual data.

Year	1994	1995	1996	1997
Price				

Year	1998	1999	2000	2001
Price				

(b) 🖩 Use a graphing calculator to graph the model.

(c) If the model were used to predict home prices in the years ahead, would the predictions be increasing at a higher rate or a lower rate with increasing t? Do you think the model would be reliable for predicting the future prices of homes? Explain.

97. *Calculator Experiment*

(a) Use a calculator to complete the table.

x	1	10	100	1000	10,000
$\left(1 + \dfrac{1}{x}\right)^{x}$					

(b) Use the table to sketch the graph of the function
$$f(x) = \left(1 + \frac{1}{x}\right)^{x}.$$

Does this graph appear to be approaching a horizontal asymptote?

(c) From parts (a) and (b), what conclusions can you make about the value of
$$\left(1 + \frac{1}{x}\right)^{x}$$
as x gets larger and larger?

98. Identify the graphs of $y_1 = e^{0.2x}$, $y_2 = e^{0.5x}$, and $y_3 = e^{x}$ in the figure. Describe the effect on the graph of $y = e^{kx}$ when $k > 0$ is changed.

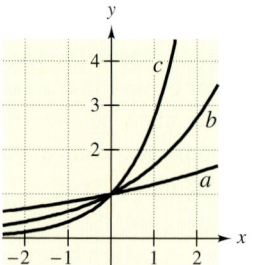

Explaining Concepts

99. ⚡ Answer parts (a) and (b) of Motivating the Chapter on page 678.

100. *Writing* ✏ Describe the differences between exponential functions and polynomial functions.

101. *Writing* ✏ Explain why 1^{x} is not an exponential function.

102. *Writing* ✏ Compare the graphs of $f(x) = 3^{x}$ and $g(x) = \left(\frac{1}{3}\right)^{x}$.

103. *Writing* ✏ Is $e = \dfrac{271{,}801}{99{,}990}$? Explain.

104. *Writing* ✏ Without using a calculator, explain why $2^{\sqrt{2}}$ is greater than 2 but less than 4?

105. 🖩 Use a graphing calculator to investigate the function $f(x) = k^{x}$ for $0 < k < 1$, $k = 1$, and $k > 1$. Discuss the effect that k has on the shape of the graph.

11.2 Composite and Inverse Functions

Superstock, Inc.

What You Should Learn

1. Form compositions of two functions and find the domains of composite functions.
2. Use the Horizontal Line Test to determine whether functions have inverse functions.
3. Find inverse functions algebraically.
4. Graphically verify that two functions are inverse functions of each other.

Why You Should Learn It

Inverse functions can be used to model and solve real-life problems. For instance, in Exercise 108 on page 706, you will use an inverse function to determine the number of units produced for a certain hourly wage.

Composite Functions

Two functions can be combined to form another function called the **composition** of the two functions. For instance, if $f(x) = 2x^2$ and $g(x) = x - 1$, the composition of f with g is denoted by $f \circ g$ and is given by

$$f(g(x)) = f(x - 1) = 2(x - 1)^2.$$

1 Form compositions of two functions and find the domains of composite functions.

Definition of Composition of Two Functions

The **composition** of the functions f and g is given by

$$(f \circ g)(x) = f(g(x)).$$

The domain of the **composite function** $(f \circ g)$ is the set of all x in the domain of g such that $g(x)$ is in the domain of f. See Figure 11.10.

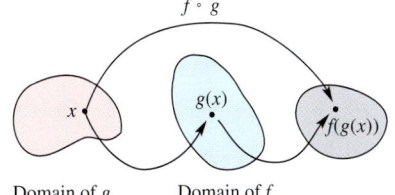

Domain of g Domain of f

Figure 11.10

Study Tip

A composite function can be viewed as a function within a function, where the composition

$$(f \circ g)(x) = f(g(x))$$

has f as the "outer" function and g as the "inner" function. This is reversed in the composition

$$(g \circ f)(x) = g(f(x)).$$

Example 1 Forming the Composition of Two Functions

Given $f(x) = 2x + 4$ and $g(x) = 3x - 1$, find the composition of f with g. Then evaluate the composite function when $x = 1$ and when $x = -3$.

Solution

$$
\begin{aligned}
(f \circ g)(x) &= f(g(x)) &&\text{Definition of } f \circ g \\
&= f(3x - 1) &&g(x) = 3x - 1 \text{ is the inner function.} \\
&= 2(3x - 1) + 4 &&\text{Input } 3x - 1 \text{ into the outer function } f. \\
&= 6x - 2 + 4 &&\text{Distributive Property} \\
&= 6x + 2 &&\text{Simplify.}
\end{aligned}
$$

When $x = 1$, the value of this composite function is

$$(f \circ g)(1) = 6(1) + 2 = 8.$$

When $x = -3$, the value of this composite function is

$$(f \circ g)(-3) = 6(-3) + 2 = -16.$$

The composition of f with g is generally *not* the same as the composition of g with f. This is illustrated in Example 2.

Example 2   Comparing the Compositions of Functions

Given $f(x) = 2x - 3$ and $g(x) = x^2 + 1$, find each composition.

a. $(f \circ g)(x)$ **b.** $(g \circ f)(x)$

Solution

a. $(f \circ g)(x) = f(g(x))$ Definition of $f \circ g$

$\qquad\qquad = f(x^2 + 1)$ $g(x) = x^2 + 1$ is the inner function.

$\qquad\qquad = 2(x^2 + 1) - 3$ Input $x^2 + 1$ into the outer function f.

$\qquad\qquad = 2x^2 + 2 - 3$ Distributive Property

$\qquad\qquad = 2x^2 - 1$ Simplify.

b. $(g \circ f)(x) = g(f(x))$ Definition of $g \circ f$

$\qquad\qquad = g(2x - 3)$ $f(x) = 2x - 3$ is the inner function.

$\qquad\qquad = (2x - 3)^2 + 1$ Input $2x - 3$ into the outer function g.

$\qquad\qquad = 4x^2 - 12x + 9 + 1$ Expand.

$\qquad\qquad = 4x^2 - 12x + 10$ Simplify.

Note that $(f \circ g)(x) \neq (g \circ f)(x)$.

To determine the domain of a composite function, first write the composite function in simplest form. Then use the fact that its domain *either is equal to or is a restriction of the domain of the "inner" function.* This is demonstrated in Example 3.

Example 3   Finding the Domain of a Composite Function

Find the domain of the composition of f with g when $f(x) = x^2$ and $g(x) = \sqrt{x}$.

Solution

$\qquad (f \circ g)(x) = f(g(x))$ Definition of $f \circ g$

$\qquad\qquad\qquad = f(\sqrt{x})$ $g(x) = \sqrt{x}$ is the inner function.

$\qquad\qquad\qquad = (\sqrt{x})^2$ Input $\sqrt{x}$ into the outer function f.

$\qquad\qquad\qquad = x, \; x \geq 0$ Domain of $f \circ g$ is all $x \geq 0$.

The domain of the inner function $g(x) = \sqrt{x}$ is the set of all nonnegative real numbers. The simplified form of $f \circ g$ has no restriction on this set of numbers. So, the restriction $x \geq 0$ must be added to the composition of this function. The domain of $f \circ g$ is the set of all nonnegative real numbers.

② Use the Horizontal Line Test to determine whether functions have inverse functions.

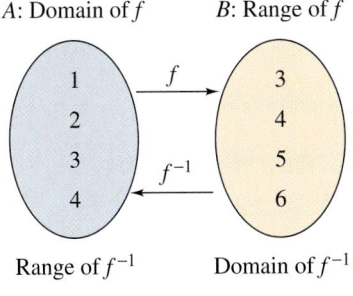

A: Domain of f B: Range of f

Range of f^{-1} Domain of f^{-1}

Figure 11.11 f is one-to-one and has inverse function f^{-1}.

One-to-One and Inverse Functions

In Section 4.3, you learned that a function can be represented by a set of ordered pairs. For instance, the function $f(x) = x + 2$ from the set $A = \{1, 2, 3, 4\}$ to the set $B = \{3, 4, 5, 6\}$ can be written as follows.

$$f(x) = x + 2: \quad \{(1, 3), (2, 4), (3, 5), (4, 6)\}$$

By interchanging the first and second coordinates of each of these ordered pairs, you can form another function that is called the **inverse function** of f, denoted by f^{-1}. It is a function from the set B to the set A, and can be written as follows.

$$f^{-1}(x) = x - 2: \quad \{(3, 1), (4, 2), (5, 3), (6, 4)\}$$

Interchanging the ordered pairs for a function f will only produce another function when f is one-to-one. A function f is **one-to-one** if each value of the dependent variable corresponds to exactly one value of the independent variable. Figure 11.11 shows that the domain of f is the range of f^{-1} and the range of f is the domain of f^{-1}.

> ### Horizontal Line Test for Inverse Functions
>
> A function f has an inverse function f^{-1} if and only if f is one-to-one. Graphically, a function f has an inverse function f^{-1} if and only if no *horizontal* line intersects the graph of f at more than one point.

Example 4   Applying the Horizontal Line Test

Use the Horizontal Line Test to determine if the function is one-to-one and so has an inverse function.

a. The graph of the function $f(x) = x^3 - 1$ is shown in Figure 11.12. Because no horizontal line intersects the graph of f at more than one point, you can conclude that f *is* a one-to-one function and *does* have an inverse function.

b. The graph of the function $f(x) = x^2 - 1$ is shown in Figure 11.13. Because it is possible to find a horizontal line that intersects the graph of f at more than one point, you can conclude that f *is not* a one-to-one function and *does not* have an inverse function.

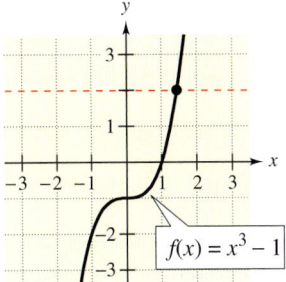

Figure 11.12

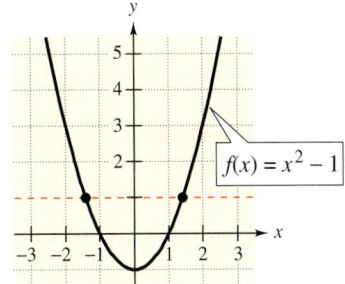

Figure 11.13

The formal definition of an inverse function is given as follows.

Definition of Inverse Function

Let f and g be two functions such that

$$f(g(x)) = x \quad \text{for every } x \text{ in the domain of } g$$

and

$$g(f(x)) = x \quad \text{for every } x \text{ in the domain of } f.$$

The function g is called the **inverse function** of the function f, and is denoted by f^{-1} (read "f-inverse"). So, $f(f^{-1}(x)) = x$ and $f^{-1}(f(x)) = x$. The domain of f must be equal to the range of f^{-1}, and vice versa.

Do not be confused by the use of -1 to denote the inverse function f^{-1}. Whenever f^{-1} is written, it *always* refers to the inverse function f and *not* to the reciprocal of $f(x)$.

If the function g is the inverse function of the function f, it must also be true that the function f is the inverse function of the function g. For this reason, you can refer to the functions f and g as being *inverse functions of each other.*

Example 5   Verifying Inverse Functions

Show that $f(x) = 2x - 4$ and $g(x) = \dfrac{x + 4}{2}$ are inverse functions of each other.

Solution

Begin by noting that the domain and range of both functions are the entire set of real numbers. To show that f and g are inverse functions of each other, you need to show that $f(g(x)) = x$ and $g(f(x)) = x$, as follows.

$$f(g(x)) = f\left(\frac{x + 4}{2}\right) \qquad g(x) = (x + 4)/2 \text{ is the inner function.}$$

$$= 2\left(\frac{x + 4}{2}\right) - 4 \qquad \text{Input } (x + 4)/2 \text{ into the outer function } f.$$

$$= x + 4 - 4 = x \qquad \text{Simplify.}$$

$$g(f(x)) = g(2x - 4) \qquad f(x) = 2x - 4 \text{ is the inner function.}$$

$$= \frac{(2x - 4) + 4}{2} \qquad \text{Input } 2x - 4 \text{ into the outer function } g.$$

$$= \frac{2x}{2} = x \qquad \text{Simplify.}$$

Note that the two functions f and g "undo" each other in the following verbal sense. The function f first *multiplies* the input x by 2 and then *subtracts* 4, whereas the function g first *adds* 4 and then *divides* the result by 2.

Example 6   Verifying Inverse Functions

Show that the functions

$$f(x) = x^3 + 1 \text{ and } g(x) = \sqrt[3]{x - 1}$$

are inverse functions of each other.

Solution

Begin by noting that the domain and range of both functions are the entire set of real numbers. To show that f and g are inverse functions of each other, you need to show that $f(g(x)) = x$ and $g(f(x)) = x$, as follows.

$$f(g(x)) = f\left(\sqrt[3]{x - 1}\right) \qquad g(x) = \sqrt[3]{x - 1} \text{ is the inner function.}$$

$$= \left(\sqrt[3]{x - 1}\right)^3 + 1 \qquad \text{Input } \sqrt[3]{x - 1} \text{ into the outer function } f.$$

$$= (x - 1) + 1 = x \qquad \text{Simplify.}$$

$$g(f(x)) = g(x^3 + 1) \qquad f(x) = x^3 + 1 \text{ is the inner function.}$$

$$= \sqrt[3]{(x^3 + 1) - 1} \qquad \text{Input } x^3 + 1 \text{ into the outer function } g.$$

$$= \sqrt[3]{x^3} = x \qquad \text{Simplify.}$$

Note that the two functions f and g "undo" each other in the following verbal sense. The function f first *cubes* the input x and then *adds* 1, whereas the function g first *subtracts* 1 and then *takes the cube root* of the result.

---

3 Find inverse functions algebraically.

Finding an Inverse Function Algebraically

You can find the inverse function of a simple function by inspection. For instance, the inverse function of $f(x) = 10x$ is $f^{-1}(x) = x/10$. For more complicated functions, however, it is best to use the following steps for finding an inverse function. The key step in these guidelines is switching the roles of x and y. This step corresponds to the fact that inverse functions have ordered pairs with the coordinates reversed.

Study Tip

You can graph a function and use the Horizontal Line Test to see if the function is one-to-one before trying to find its inverse function.

Finding an Inverse Function Algebraically

1. In the equation for $f(x)$, replace $f(x)$ with y.

2. Interchange the roles of x and y.

3. If the new equation does not represent y as a function of x, the function f does not have an inverse function. If the new equation does represent y as a function of x, solve the new equation for y.

4. Replace y with $f^{-1}(x)$.

5. Verify that f and f^{-1} are inverse functions of each other by showing that $f(f^{-1}(x)) = x = f^{-1}(f(x))$.

Example 7   Finding an Inverse Function

Determine whether each function has an inverse function. If it does, find its inverse function.

a. $f(x) = 2x + 3$ **b.** $f(x) = x^3 + 3$

Solution

a.

$f(x) = 2x + 3$	Write original function.
$y = 2x + 3$	Replace $f(x)$ with y.
$x = 2y + 3$	Interchange x and y.
$y = \dfrac{x - 3}{2}$	Solve for y.
$f^{-1}(x) = \dfrac{x - 3}{2}$	Replace y with $f^{-1}(x)$.

You can verify that $f(f^{-1}(x)) = x = f^{-1}(f(x))$, as follows.

$$f(f^{-1}(x)) = f\left(\frac{x - 3}{2}\right) = 2\left(\frac{x - 3}{2}\right) + 3 = (x - 3) + 3 = x$$

$$f^{-1}(f(x)) = f^{-1}(2x + 3) = \frac{(2x + 3) - 3}{2} = \frac{2x}{2} = x$$

b.

$f(x) = x^3 + 3$	Write original function.
$y = x^3 + 3$	Replace $f(x)$ with y.
$x = y^3 + 3$	Interchange x and y.
$\sqrt[3]{x - 3} = y$	Solve for y.
$f^{-1}(x) = \sqrt[3]{x - 3}$	Replace y with $f^{-1}(x)$.

You can verify that $f(f^{-1}(x)) = x = f^{-1}(f(x))$, as follows.

$$f(f^{-1}(x)) = f(\sqrt[3]{x - 3}) = (\sqrt[3]{x - 3})^3 + 3 = (x - 3) + 3 = x$$

$$f^{-1}(f(x)) = f^{-1}(x^3 + 3) = \sqrt[3]{(x^3 + 3) - 3} = \sqrt[3]{x^3} = x$$

Example 8   A Function That Has No Inverse Function

$f(x) = x^2$	Original equation
$y = x^2$	Replace $f(x)$ with y.
$x = y^2$	Interchange x and y.

Recall from Section 4.3 that the equation $x = y^2$ does not represent y as a function of x because you can find two different y-values that correspond to the same x-value. Because the equation does not represent y as a function of x, you can conclude that the original function f does not have an inverse function.

Technology: Discovery

Use a graphing calculator to graph $f(x) = x^3 + 1$, $f^{-1}(x) = \sqrt[3]{x - 1}$, and $y = x$ in the same viewing window.

a. Relative to the line $y = x$, how do the graphs of f and f^{-1} compare?

b. For the graph of f, complete the table.

x	-1	0	1
f			

For the graph of f^{-1}, complete the table.

x	0	1	2
f^{-1}			

What can you conclude about the coordinates of the points on the graph of f compared with those on the graph of f^{-1}?

④ Graphically verify that two functions are inverse functions of each other.

Graphs of Inverse Functions

The graphs of f and f^{-1} are related to each other in the following way. If the point (a, b) lies on the graph of f, the point (b, a) must lie on the graph of f^{-1}, and vice versa. This means that the graph of f^{-1} is a reflection of the graph of f in the line $y = x$, as shown in Figure 11.14. This "reflective property" of the graphs of f and f^{-1} is illustrated in Examples 9 and 10.

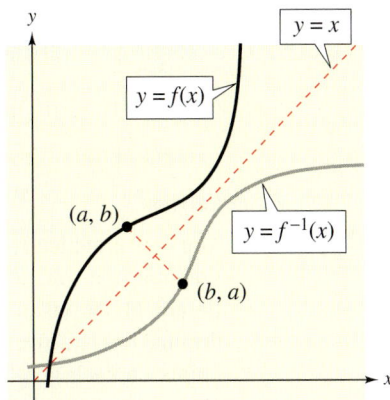

Figure 11.14   The graph of f^{-1} is a reflection of the graph of f in the line $y = x$.

Example 9   The Graphs of f and f^{-1}

Sketch the graphs of the inverse functions $f(x) = 2x - 3$ and $f^{-1}(x) = \frac{1}{2}(x + 3)$ on the same rectangular coordinate system, and show that the graphs are reflections of each other in the line $y = x$.

Solution

The graphs of f and f^{-1} are shown in Figure 11.15. Visually, it appears that the graphs are reflections of each other. You can further verify this reflective property by testing a few points on each graph. Note in the following list that if the point (a, b) is on the graph of f, the point (b, a) is on the graph of f^{-1}.

$f(x) = 2x - 3$	$f^{-1}(x) = \frac{1}{2}(x + 3)$
$(-1, -5)$	$(-5, -1)$
$(0, -3)$	$(-3, 0)$
$(1, -1)$	$(-1, 1)$
$(2, 1)$	$(1, 2)$
$(3, 3)$	$(3, 3)$

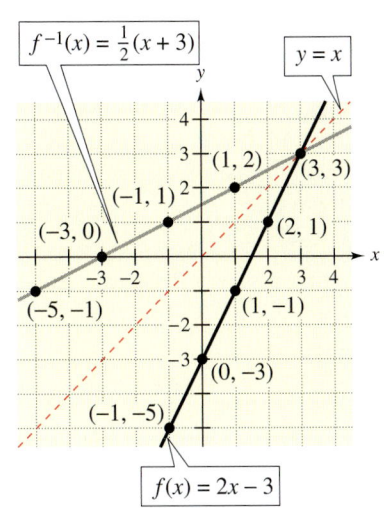

Figure 11.15

You can sketch the graph of an inverse function without knowing the equation of the inverse function. Simply find the coordinates of points that lie on the original function. By interchanging the x- and y-coordinates, you have points that lie on the graph of the inverse function. Plot these points and sketch the graph of the inverse function.

In Example 8, you saw that the function

$$f(x) = x^2$$

has no inverse function. A more complete way of saying this is "*assuming that the domain of f is the entire real line,* the function $f(x) = x^2$ has no inverse function." If, however, you *restrict* the domain of f to the nonnegative real numbers, then f does have an inverse function, as demonstrated in Example 10.

Example 10   Verifying Inverse Functions Graphically

Graphically verify that f and g are inverse functions of each other.

$$f(x) = x^2, \quad x \geq 0 \quad \text{and} \quad g(x) = \sqrt{x}$$

Solution

You can graphically verify that f and g are inverse functions of each other by graphing the functions on the same rectangular coordinate system, as shown in Figure 11.16. Visually, it appears that the graphs are reflections of each other in the line $y = x$. You can further verify this reflective property by testing a few points on each graph. Note in the following list that if the point (a, b) is on the graph of f, the point (b, a) is on the graph of g.

$f(x) = x^2, \quad x \geq 0$	$g(x) = f^{-1}(x) = \sqrt{x}$
$(0, 0)$	$(0, 0)$
$(1, 1)$	$(1, 1)$
$(2, 4)$	$(4, 2)$
$(3, 9)$	$(9, 3)$

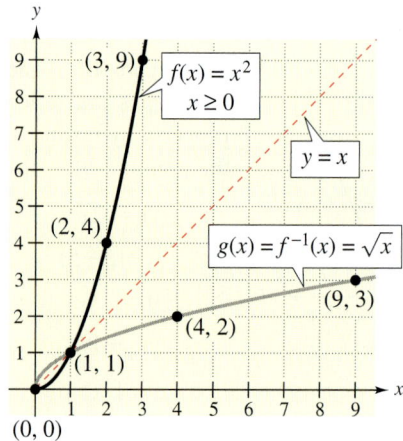

Figure 11.16

So, f and g are inverse functions of each other.

11.2 Exercises

Review   Concepts, Skills, and Problem Solving

Keep mathematically in shape by doing these exercises *before* the problems of this section.

Properties and Definitions

1. *Writing*✎   Decide whether $x - y^2 = 0$ represents y as a function of x. Explain.

2. *Writing*✎   Decide whether $|x| - 2y = 4$ represents y as a function of x. Explain.

3. *Writing*✎   Explain why the domains of f and g are not the same.

$$f(x) = \sqrt{4 - x^2} \qquad g(x) = \frac{6}{\sqrt{4 - x^2}}$$

4. Determine the range of $h(x) = 8 - \sqrt{x}$ over the domain $\{0, 4, 9, 16\}$.

Simplifying Expressions

In Exercises 5–10, perform the indicated operations and simplify.

5. $-(5x^2 - 1) + (3x^2 - 5)$

6. $(-2x)(-5x)(3x + 4)$

7. $(u - 4v)(u + 4v)$

8. $(3a - 2b)^2$

9. $(t - 2)^3$

10. $\dfrac{6x^3 - 3x^2}{12x}$

Problem Solving

11. *Free-Falling Object*   The velocity of a free-falling object is given by $v = \sqrt{2gh}$, where v is the velocity measured in feet per second, $g = 32$ feet per second per second, and h is the distance (in feet) the object has fallen. Find the distance an object has fallen if its velocity is 80 feet per second.

12. *Consumer Awareness*   The cost of a long-distance telephone call is $0.95 for the first minute and $0.35 for each additional minute. The total cost of a call is $5.15. Find the length of the call.

Developing Skills

In Exercises 1–10, find the compositions. See Examples 1 and 2.

1. $f(x) = 2x + 3, \quad g(x) = x - 6$
   (a) $(f \circ g)(x)$
   (b) $(g \circ f)(x)$
   (c) $(f \circ g)(4)$
   (d) $(g \circ f)(7)$

2. $f(x) = x - 5, \quad g(x) = 6 + 2x$
   (a) $(f \circ g)(x)$
   (b) $(g \circ f)(x)$
   (c) $(f \circ g)(3)$
   (d) $(g \circ f)(3)$

3. $f(x) = x^2 + 3, \quad g(x) = 2x + 1$
   (a) $(f \circ g)(x)$
   (b) $(g \circ f)(x)$
   (c) $(f \circ g)(2)$
   (d) $(g \circ f)(-3)$

4. $f(x) = 2x + 1, \quad g(x) = x^2 - 5$
   (a) $(f \circ g)(x)$
   (b) $(g \circ f)(x)$
   (c) $(f \circ g)(-1)$
   (d) $(g \circ f)(3)$

5. $f(x) = |x - 3|, \quad g(x) = 3x$
   (a) $(f \circ g)(x)$
   (b) $(g \circ f)(x)$
   (c) $(f \circ g)(1)$
   (d) $(g \circ f)(2)$

6. $f(x) = |x|, \quad g(x) = 2x + 5$
   (a) $(f \circ g)(x)$
   (b) $(g \circ f)(x)$
   (c) $(f \circ g)(-2)$
   (d) $(g \circ f)(-4)$

7. $f(x) = \sqrt{x - 4}, \quad g(x) = x + 5$
   (a) $(f \circ g)(x)$
   (b) $(g \circ f)(x)$
   (c) $(f \circ g)(3)$
   (d) $(g \circ f)(8)$

8. $f(x) = \sqrt{x + 6}, \quad g(x) = 2x - 3$
   (a) $(f \circ g)(x)$
   (b) $(g \circ f)(x)$
   (c) $(f \circ g)(3)$
   (d) $(g \circ f)(-2)$

9. $f(x) = \dfrac{1}{x - 3}, \quad g(x) = \dfrac{2}{x^2}$
   (a) $(f \circ g)(x)$
   (b) $(g \circ f)(x)$
   (c) $(f \circ g)(-1)$
   (d) $(g \circ f)(2)$

10. $f(x) = \dfrac{4}{x^2 - 4}, \quad g(x) = \dfrac{1}{x}$

    (a) $(f \circ g)(x)$ (b) $(g \circ f)(x)$

    (c) $(f \circ g)(-2)$ (d) $(g \circ f)(1)$

In Exercises 11–14, use the functions f and g to find the indicated values.

$f = \{(-2, 3), (-1, 1), (0, 0), (1, -1), (2, -3)\},$

$g = \{(-3, 1), (-1, -2), (0, 2), (2, 2), (3, 1)\}$

11. (a) $f(1)$ **12.** (a) $g(0)$

      (b) $g(-1)$ (b) $f(2)$

      (c) $(g \circ f)(1)$ (c) $(f \circ g)(0)$

13. (a) $(f \circ g)(-3)$ **14.** (a) $(f \circ g)(2)$

      (b) $(g \circ f)(-2)$ (b) $(g \circ f)(2)$

In Exercises 15–18, use the functions f and g to find the indicated values.

$f = \{(0, 1), (1, 2), (2, 5), (3, 10), (4, 17)\},$

$g = \{(5, 4), (10, 1), (2, 3), (17, 0), (1, 2)\}$

15. (a) $f(3)$ **16.** (a) $g(2)$

      (b) $g(10)$ (b) $f(0)$

      (c) $(g \circ f)(3)$ (c) $(f \circ g)(10)$

17. (a) $(g \circ f)(4)$ **18.** (a) $(f \circ g)(1)$

      (b) $(f \circ g)(2)$ (b) $(g \circ f)(0)$

In Exercises 19–26, find the compositions (a) $f \circ g$ and (b) $g \circ f$. Then find the domain of each composition. See Example 3.

19. $f(x) = 3x + 4$ **20.** $f(x) = x + 5$

      $g(x) = x - 7$ $g(x) = 4x - 1$

21. $f(x) = \sqrt{x}$ **22.** $f(x) = \sqrt{x - 5}$

      $g(x) = x - 2$ $g(x) = x + 3$

23. $f(x) = x^2 + 3$

      $g(x) = \sqrt{x - 1}$

24. $f(x) = \sqrt{3x + 1}$

      $g(x) = x^2 - 8$

25. $f(x) = \dfrac{x}{x + 5}$ **26.** $f(x) = \dfrac{x}{x - 4}$

      $g(x) = \sqrt{x - 1}$ $g(x) = \sqrt{x}$

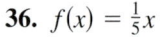 In Exercises 27–34, use a graphing calculator to graph the function and determine whether the function is one-to-one.

27. $f(x) = x^3 - 1$ **28.** $f(x) = (2 - x)^3$

29. $f(t) = \sqrt[3]{5 - t}$ **30.** $h(t) = 4 - \sqrt[3]{t}$

31. $g(x) = x^4 - 6$ **32.** $f(x) = (x + 2)^5$

33. $h(t) = \dfrac{5}{t}$ **34.** $g(t) = \dfrac{5}{t^2}$

In Exercises 35–40, use the Horizontal Line Test to determine if the function is one-to-one and so has an inverse function. See Example 4.

35. $f(x) = x^2 - 2$ **36.** $f(x) = \frac{1}{5}x$

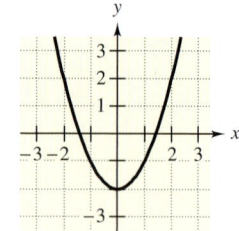   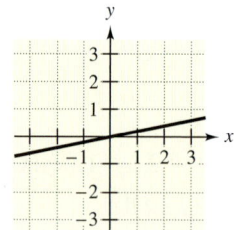

37. $f(x) = x^2, \quad x \geq 0$

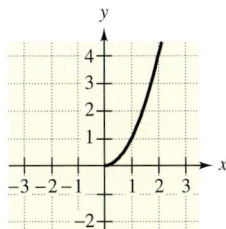

38. $f(x) = \sqrt{-x}$

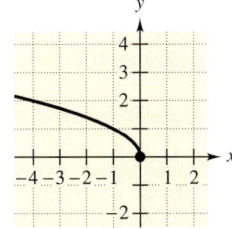

48. $f(x) = -\frac{1}{4}x + 3, \ g(x) = -4(x - 3)$

49. $f(x) = \sqrt[3]{x + 1}, \ g(x) = x^3 - 1$

50. $f(x) = x^7, \ g(x) = \sqrt[7]{x}$

39. $g(x) = \sqrt{25 - x^2}$

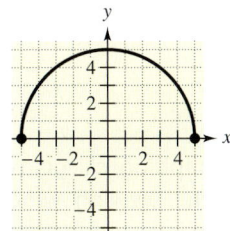

40. $g(x) = |x - 4|$

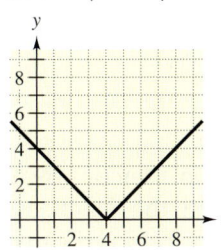

51. $f(x) = \frac{1}{x}, \ g(x) = \frac{1}{x}$

52. $f(x) = \frac{1}{x + 1}, \ g(x) = \frac{1 - x}{x}$

In Exercises 41–52, verify algebraically that the functions *f* and *g* are inverse functions of each other. See Examples 5 and 6.

41. $f(x) = -6x, \ g(x) = -\frac{1}{6}x$

42. $f(x) = \frac{2}{3}x, \ g(x) = \frac{3}{2}x$

43. $f(x) = x + 15, \ g(x) = x - 15$

44. $f(x) = 3 - x, \ g(x) = 3 - x$

45. $f(x) = 1 - 2x, \ g(x) = \frac{1}{2}(1 - x)$

46. $f(x) = 2x - 1, \ g(x) = \frac{1}{2}(x + 1)$

47. $f(x) = 2 - 3x, \ g(x) = \frac{1}{3}(2 - x)$

In Exercises 53–64, find the inverse function of *f*. Verify that $f(f^{-1}(x))$ and $f^{-1}(f(x))$ are equal to the identity function. See Example 7.

53. $f(x) = 5x$

54. $f(x) = -3x$

55. $f(x) = -\frac{2}{5}x$

56. $f(x) = \frac{1}{3}x$

57. $f(x) = x + 10$

58. $f(x) = x - 5$

59. $f(x) = 3 - x$

60. $f(x) = 8 - x$

61. $f(x) = x^7$

62. $f(x) = x^5$

63. $f(x) = \sqrt[3]{x}$

64. $f(x) = x^{1/5}$

In Exercises 65–78, find the inverse function. See Example 7.

65. $f(x) = 8x$

66. $f(x) = \frac{1}{10}x$

67. $g(x) = x + 25$

68. $f(x) = 7 - x$

69. $g(x) = 3 - 4x$

70. $g(t) = 6t + 1$

71. $g(t) = \frac{1}{4}t + 2$

72. $h(s) = 5 - \frac{3}{2}s$

73. $h(x) = \sqrt{x}$

74. $h(x) = \sqrt{x + 5}$

75. $f(t) = t^3 - 1$

76. $h(t) = t^5 + 8$

77. $f(x) = \sqrt{x + 3}, \quad x \geq -3$

78. $f(x) = \sqrt{x^2 - 4}, \quad x \geq 2$

In Exercises 79–82, match the graph with the graph of its inverse function. [The graphs of the inverse functions are labeled (a), (b), (c), and (d).]

(a)
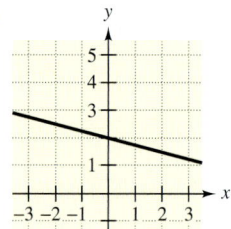

(b)
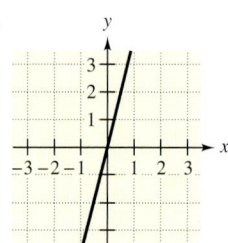

(c)
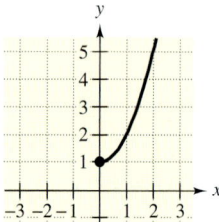

(d)
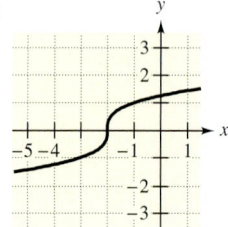

79.
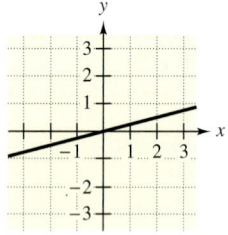

80.
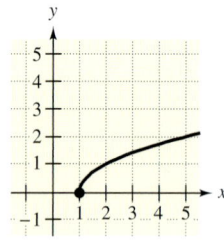

81.
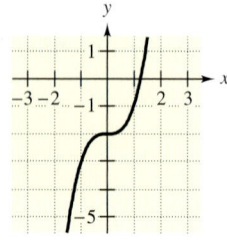

82.
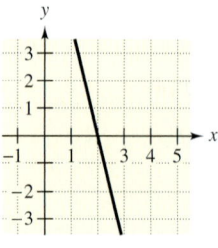

In Exercises 83–88, sketch the graphs of f and f^{-1} on the same rectangular coordinate system. Show that the graphs are reflections of each other in the line $y = x$. See Example 9.

83. $f(x) = x + 4, \ f^{-1}(x) = x - 4$

84. $f(x) = x - 7, \ f^{-1}(x) = x + 7$

85. $f(x) = 3x - 1, \ f^{-1}(x) = \frac{1}{3}(x + 1)$

86. $f(x) = 5 - 4x, \ f^{-1}(x) = -\frac{1}{4}(x - 5)$

87. $f(x) = x^2 - 1, \ x \geq 0,$
$\quad f^{-1}(x) = \sqrt{x + 1}$

88. $f(x) = (x + 2)^2, \ x \geq -2,$
$\quad f^{-1}(x) = \sqrt{x} - 2$

In Exercises 89–96, use a graphing calculator to graph the functions in the same viewing window. Graphically verify that f and g are inverse functions of each other.

89. $f(x) = \frac{1}{3}x$
$\quad g(x) = 3x$

90. $f(x) = \frac{1}{5}x - 1$
$\quad g(x) = 5x + 5$

91. $f(x) = \sqrt{x + 1}$
$\quad g(x) = x^2 - 1, \ x \geq 0$

92. $f(x) = \sqrt{4 - x}$
$\quad g(x) = 4 - x^2, \ x \geq 0$

93. $f(x) = \frac{1}{8}x^3$
$\quad g(x) = 2\sqrt[3]{x}$

94. $f(x) = \sqrt[3]{x + 2}$
$\quad g(x) = x^3 - 2$

95. $f(x) = |3 - x|, \ x \geq 3$
$\quad g(x) = 3 + x, \ x \geq 0$

96. $f(x) = |x - 2|, \ x \geq 2$
$\quad g(x) = x + 2, \ x \geq 0$

In Exercises 97–100, delete part of the graph of the function so that the remaining part is one-to-one. Find the inverse function of the remaining part and find the domain of the inverse function. (*Note:* There is more than one correct answer.) See Example 10.

97. $f(x) = (x - 2)^2$

98. $f(x) = 9 - x^2$

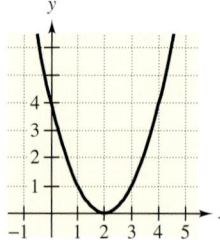

99. $f(x) = |x| + 1$ **100.** $f(x) = |x - 2|$

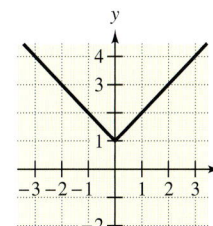   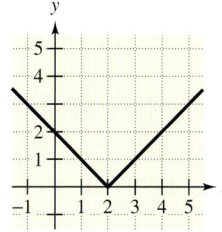

In Exercises 101 and 102, consider the function $f(x) = 3 - 2x$.

101. Find $f^{-1}(x)$.   **102.** Find $(f^{-1})^{-1}(x)$.

Solving Problems

103. ▲ *Geometry* You are standing on a bridge over a calm pond and drop a pebble, causing ripples of concentric circles in the water. The radius (in feet) of the outer ripple is given by $r(t) = 0.6t$, where t is time in seconds after the pebble hits the water. The area of the circle is given by the function $A(r) = \pi r^2$. Find an equation for the composition $A(r(t))$. What are the input and output of this composite function?

104. *Sales Bonus* You are a sales representative for a clothing manufacturer. You are paid an annual salary plus a bonus of 2% of your sales over $200,000. Consider the two functions $f(x) = x - 200,000$ and $g(x) = 0.02x$. If x is greater than $200,000, find each composition and determine which represents your bonus. Explain.

(a) $f(g(x))$

(b) $g(f(x))$

105. *Daily Production Cost* The daily cost of producing x units in a manufacturing process is $C(x) = 8.5x + 300$. The number of units produced in t hours during a day is given by $x(t) = 12t$, $0 \le t \le 8$. Find, simplify, and interpret $(C \circ x)(t)$.

106. *Rebate and Discount* The suggested retail price of a new car is p dollars. The dealership advertised a factory rebate of $2000 and a 5% discount.

(a) Write a function R in terms of p, giving the cost of the car after receiving the factory rebate.

(b) Write a function S in terms of p, giving the cost of the car after receiving the dealership discount.

(c) Form the composite functions $(R \circ S)(p)$ and $(S \circ R)(p)$ and interpret each.

(d) Find $(R \circ S)(26,000)$ and $(S \circ R)(26,000)$. Which yields the smaller cost for the car? Explain.

107. *Rebate and Discount* The suggested retail price of a plasma television is p dollars. The electronics store is offering a manufacturer's rebate of $500 and a 10% discount.

(a) Write a function R in terms of p, giving the cost of the television after receiving the manufacturer's rebate.

(b) Write a function S in terms of p, giving the cost of the television after receiving the 10% discount.

(c) Form the composite functions $(R \circ S)(p)$ and $(S \circ R)(p)$ and interpret each.

(d) Find $(R \circ S)(6000)$ and $(S \circ R)(6000)$. Which yields the smaller cost for the plasma television? Explain.

108. *Hourly Wage*   Your wage is $9.00 per hour plus $0.65 for each unit produced per hour. So, your hourly wage y in terms of the number of units produced x is $y = 9 + 0.65x$.

(a) Find the inverse function.

(b) What does each variable represent in the inverse function?

(c) Determine the number of units produced when your hourly wage averages $14.20.

109. *Cost*   You need 100 pounds of two fruits: oranges that cost $0.75 per pound and apples that cost $0.95 per pound.

(a) Verify that your total cost is $y = 0.75x + 0.95(100 - x)$, where x is the number of pounds of oranges.

(b) Find the inverse function. What does each variable represent in the inverse function?

(c) Use the context of the problem to determine the domain of the inverse function.

(d) Determine the number of pounds of oranges purchased if the total cost is $84.

110. *Exploration*   Consider the functions $f(x) = 4x$ and $g(x) = x + 6$.

(a) Find $(f \circ g)(x)$.

(b) Find $(f \circ g)^{-1}(x)$.

(c) Find $f^{-1}(x)$ and $g^{-1}(x)$.

(d) Find $(g^{-1} \circ f^{-1})(x)$ and compare the result with that of part (b).

(e) Repeat parts (a) through (d) for $f(x) = x^3 + 1$ and $g(x) = 2x$.

(f) Make a conjecture about $(f \circ g)^{-1}(x)$ and $(g^{-1} \circ f^{-1})(x)$.

Explaining Concepts

True or False?   In Exercises 111–114, decide whether the statement is true or false. If true, explain your reasoning. If false, give an example.

111. If the inverse function of f exists, the y-intercept of f is an x-intercept of f^{-1}. Explain.

112. There exists no function f such that $f = f^{-1}$.

113. If the inverse function of f exists, the domains of f and f^{-1} are the same.

114. If the inverse function of f exists and its graph passes through the point $(2, 2)$, the graph of f^{-1} also passes through the point $(2, 2)$.

115. *Writing*   Describe how to find the inverse of a function given by a set of ordered pairs. Give an example.

116. *Writing*   Describe how to find the inverse of a function given by an equation in x and y. Give an example.

117. Give an example of a function that does not have an inverse function.

118. *Writing*   Explain the Horizontal Line Test. What is the relationship between this test and a function being one-to-one?

119. *Writing*   Describe the relationship between the graph of a function and its inverse function.

11.3 Logarithmic Functions

A.T. Willett/Alamy

What You Should Learn

1. Evaluate logarithmic functions.
2. Graph logarithmic functions.
3. Graph and evaluate natural logarithmic functions.
4. Use the change-of-base formula to evaluate logarithms.

Why You Should Learn It

Logarithmic functions can be used to model and solve real-life problems. For instance, in Exercise 128 on page 717, you will use a logarithmic function to determine the speed of the wind near the center of a tornado.

1 Evaluate logarithmic functions.

Logarithmic Functions

In Section 11.2, you were introduced to the concept of an inverse function. Moreover, you saw that if a function has the property that no horizontal line intersects the graph of the function more than once, the function must have an inverse function. By looking back at the graphs of the exponential functions introduced in Section 11.1, you will see that every function of the form

$$f(x) = a^x$$

passes the Horizontal Line Test, and so must have an inverse function. To describe the inverse function of $f(x) = a^x$, follow the steps used in Section 11.2.

$y = a^x$ Replace $f(x)$ by y.

$x = a^y$ Interchange x and y.

At this point, there is no way to solve for y. A verbal description of y in the equation $x = a^y$ is "y equals the exponent needed on base a to get x." This inverse of $f(x) = a^x$ is denoted by the **logarithmic function with base a**

$$f^{-1}(x) = \log_a x.$$

Definition of Logarithmic Function

Let a and x be positive real numbers such that $a \neq 1$. The **logarithm of x with base a** is denoted by $\log_a x$ and is defined as follows.

$$y = \log_a x \quad \text{if and only if} \quad x = a^y$$

The function $f(x) = \log_a x$ is the **logarithmic function with base a.**

From the definition it is clear that

Logarithmic Equation   *Exponential Equation*

$\quad y = \log_a x$ is equivalent to $x = a^y.$

So, to find the value of $\log_a x$, *think*

"$\log_a x$ = the exponent needed on base a to get x."

For instance,

$$y = \log_2 8 \qquad \text{Think: "The exponent needed on 2 to get 8."}$$

$$y = 3.$$

That is,

$$3 = \log_2 8. \qquad \text{This is equivalent to } 2^3 = 8.$$

By now it should be clear that *a logarithm is an exponent*.

Example 1   Evaluating Logarithms

Evaluate each logarithm.

a. $\log_2 16$ **b.** $\log_3 9$ **c.** $\log_4 2$

Solution

In each case you should answer the question, "To what power must the base be raised to obtain the given number?"

a. The power to which 2 must be raised to obtain 16 is 4. That is,

$$2^4 = 16 \quad \Longrightarrow \quad \log_2 16 = 4.$$

b. The power to which 3 must be raised to obtain 9 is 2. That is,

$$3^2 = 9 \quad \Longrightarrow \quad \log_3 9 = 2.$$

c. The power to which 4 must be raised to obtain 2 is $\frac{1}{2}$. That is,

$$4^{1/2} = 2 \quad \Longrightarrow \quad \log_4 2 = \frac{1}{2}.$$

Study Tip

Study the results in Example 2 carefully. Each of the logarithms illustrates an important special property of logarithms that you should know.

Example 2   Evaluating Logarithms

Evaluate each logarithm.

a. $\log_5 1$ **b.** $\log_{10} \dfrac{1}{10}$ **c.** $\log_3(-1)$ **d.** $\log_4 0$

Solution

a. The power to which 5 must be raised to obtain 1 is 0. That is,

$$5^0 = 1 \quad \Longrightarrow \quad \log_5 1 = 0.$$

b. The power to which 10 must be raised to obtain $\frac{1}{10}$ is -1. That is,

$$10^{-1} = \frac{1}{10} \quad \Longrightarrow \quad \log_{10} \frac{1}{10} = -1.$$

c. There is no power to which 3 can be raised to obtain -1. The reason for this is that for any value of x, 3^x is a positive number. So, $\log_3(-1)$ is undefined.

d. There is no power to which 4 can be raised to obtain 0. So, $\log_4 0$ is undefined.

The following properties of logarithms follow directly from the definition of the logarithmic function with base a.

Properties of Logarithms

Let a and x be positive real numbers such that $a \neq 1$. Then the following properties are true.

1. $\log_a 1 = 0$ because $a^0 = 1$.

2. $\log_a a = 1$ because $a^1 = a$.

3. $\log_a a^x = x$ because $a^x = a^x$.

The logarithmic function with base 10 is called the **common logarithmic function.** On most calculators, this function can be evaluated with the common logarithmic key $\boxed{\text{LOG}}$, as illustrated in the next example.

Example 3   Evaluating Common Logarithms

Evaluate each logarithm. Use a calculator only if necessary.

a. $\log_{10} 100$ **b.** $\log_{10} 0.01$

c. $\log_{10} 5$ **d.** $\log_{10} 2.5$

Solution

a. The power to which 10 must be raised to obtain 100 is 2. That is,

$$10^2 = 100 \quad \Longrightarrow \quad \log_{10} 100 = 2.$$

b. The power to which 10 must be raised to obtain 0.01 or $\frac{1}{100}$ is -2. That is,

$$10^{-2} = \tfrac{1}{100} \quad \Longrightarrow \quad \log_{10} 0.01 = -2.$$

c. There is no simple power to which 10 can be raised to obtain 5, so you should use a calculator to evaluate $\log_{10} 5$.

Keystrokes	Display	
5 $\boxed{\text{LOG}}$	0.69897	Scientific
$\boxed{\text{LOG}}$ 5 $\boxed{\text{ENTER}}$	0.69897	Graphing

So, rounded to three decimal places, $\log_{10} 5 \approx 0.699$.

d. There is no simple power to which 10 can be raised to obtain 2.5, so you should use a calculator to evaluate $\log_{10} 2.5$.

Keystrokes	Display	
2.5 $\boxed{\text{LOG}}$	0.39794	Scientific
$\boxed{\text{LOG}}$ 2.5 $\boxed{\text{ENTER}}$	0.39794	Graphing

So, rounded to three decimal places, $\log_{10} 2.5 \approx 0.398$.

Study Tip

Be sure you see that the value of a logarithm can be zero or negative, as in Example 3(b), *but* you cannot take the logarithm of zero or a negative number. This means that the logarithms $\log_{10}(-10)$ and $\log_5 0$ are not valid.

② Graph logarithmic functions.

Graphs of Logarithmic Functions

To sketch the graph of

$$y = \log_a x$$

you can use the fact that the graphs of inverse functions are reflections of each other in the line $y = x$.

Example 4   Graphs of Exponential and Logarithmic Functions

On the same rectangular coordinate system, sketch the graph of each function.

a. $f(x) = 2^x$ **b.** $g(x) = \log_2 x$

Solution

a. Begin by making a table of values for $f(x) = 2^x$.

x	-2	-1	0	1	2	3
$f(x) = 2^x$	$\frac{1}{4}$	$\frac{1}{2}$	1	2	4	8

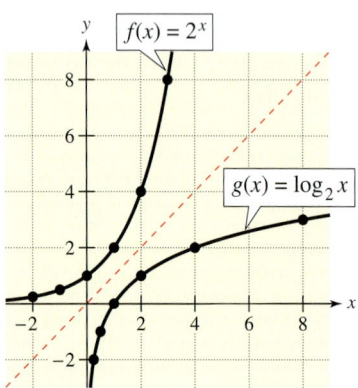

Figure 11.17   Inverse Functions

By plotting these points and connecting them with a smooth curve, you obtain the graph shown in Figure 11.17.

b. Because $g(x) = \log_2 x$ is the inverse function of $f(x) = 2^x$, the graph of g is obtained by reflecting the graph of f in the line $y = x$, as shown in Figure 11.17.

Study Tip

In Example 4, the inverse property of logarithmic functions is used to sketch the graph of $g(x) = \log_2 x$. You could also use a standard point-plotting approach or a graphing calculator.

Notice from the graph of $g(x) = \log_2 x$, shown in Figure 11.17, that the domain of the function is the set of positive numbers and the range is the set of all real numbers. The basic characteristics of the graph of a logarithmic function are summarized in Figure 11.18. In this figure, note that the graph has one x-intercept at $(1, 0)$. Also note that $x = 0$ (y-axis) is a vertical asymptote of the graph.

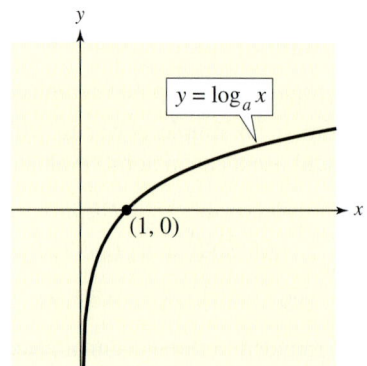

Graph of $y = \log_a x$, $a > 1$
- Domain: $(0, \infty)$
- Range: $(-\infty, \infty)$
- Intercept: $(1, 0)$
- Increasing (moves up to the right)
- Asymptote: y-axis

Figure 11.18   Characteristics of logarithmic function $y = \log_a x \, (a > 1)$

In the following example, the graph of $\log_a x$ is used to sketch the graphs of functions of the form $y = b \pm \log_a(x + c)$. Notice how each transformation affects the vertical asymptote.

Example 5   Sketching the Graphs of Logarithmic Functions

The graph of each function is similar to the graph of $f(x) = \log_{10} x$, as shown in Figure 11.19. From the graph you can determine the domain of the function.

a. Because $g(x) = \log_{10}(x - 1) = f(x - 1)$, the graph of g can be obtained by shifting the graph of f one unit to the right. The vertical asymptote of the graph of g is $x = 1$. The domain of g is $(1, \infty)$.

b. Because $h(x) = 2 + \log_{10} x = 2 + f(x)$, the graph of h can be obtained by shifting the graph of f two units upward. The vertical asymptote of the graph of h is $x = 0$. The domain of h is $(0, \infty)$.

c. Because $k(x) = -\log_{10} x = -f(x)$, the graph of k can be obtained by reflecting the graph of f in the x-axis. The vertical asymptote of the graph of k is $x = 0$. The domain of k is $(0, \infty)$.

d. Because $j(x) = \log_{10}(-x) = f(-x)$, the graph of j can be obtained by reflecting the graph of f in the y-axis. The vertical asymptote of the graph of j is $x = 0$. The domain of j is $(-\infty, 0)$.

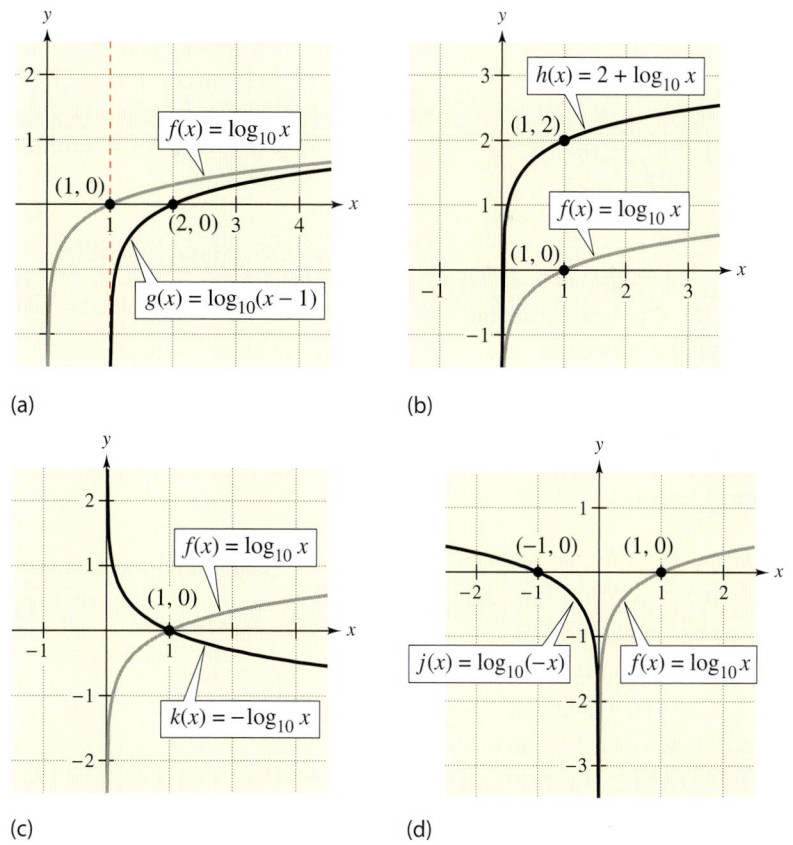

(a)

(b)

(c)

(d)

Figure 11.19

③ Graph and evaluate natural logarithmic functions.

The Natural Logarithmic Function

As with exponential functions, the most widely used base for logarithmic functions is the number e. The logarithmic function with base e is the **natural logarithmic function** and is denoted by the special symbol ln x, which is read as "el en of x."

The Natural Logarithmic Function

The function defined by

$$f(x) = \log_e x = \ln x$$

where $x > 0$, is called the **natural logarithmic function.**

The definition above implies that the natural logarithmic function and the natural exponential function are inverse functions of each other. So, every logarithmic equation can be written in an equivalent exponential form and every exponential equation can be written in logarithmic form.

Because the functions $f(x) = e^x$ and $g(x) = \ln x$ are inverse functions of each other, their graphs are reflections of each other in the line $y = x$. This reflective property is illustrated in Figure 11.20. The figure also contains a summary of several characteristics of the graph of the natural logarithmic function.

Notice that the domain of the natural logarithmic function, as with every other logarithmic function, is the set of *positive real numbers*—be sure you see that ln x is not defined for zero or for negative numbers.

The three properties of logarithms listed earlier in this section are also valid for natural logarithms.

Graph of $g(x) = \ln x$
- Domain: $(0, \infty)$
- Range: $(-\infty, \infty)$
- Intercept: $(1, 0)$
- Increasing (moves up to the right)
- Asymptote: y-axis

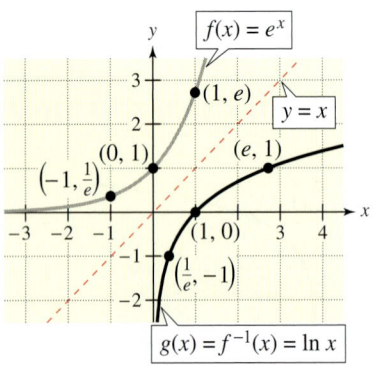

Figure 11.20 Characteristics of the natural logarithmic function $g(x) = \ln x$

Properties of Natural Logarithms

Let x be a positive real number. Then the following properties are true.

1. $\ln 1 = 0$ because $e^0 = 1$.

2. $\ln e = 1$ because $e^1 = e$.

3. $\ln e^x = x$ because $e^x = e^x$.

Technology: Tip

On most calculators, the natural logarithm key is denoted by $\boxed{\text{LN}}$. For instance, on a scientific calculator, you can evaluate ln 2 as 2 $\boxed{\text{LN}}$ and on a graphing calculator, you can evaluate it as $\boxed{\text{LN}}$ 2 $\boxed{\text{ENTER}}$. In either case, you should obtain a display of 0.6931472.

Example 6   Evaluating Natural Logarithmic Functions

Evaluate each expression.

a. $\ln e^2$ **b.** $\ln \dfrac{1}{e}$

Solution

Using the property that $\ln e^x = x$, you obtain the following.

a. $\ln e^2 = 2$ **b.** $\ln \dfrac{1}{e} = \ln e^{-1} = -1$

④ Use the change-of-base formula to evaluate logarithms.

Change of Base

Although 10 and e are the most frequently used bases, you occasionally need to evaluate logarithms with other bases. In such cases the following **change-of-base formula** is useful.

Change-of-Base Formula

Let a, b, and x be positive real numbers such that $a \neq 1$ and $b \neq 1$. Then $\log_a x$ is given as follows.

$$\log_a x = \frac{\log_b x}{\log_b a} \qquad \text{or} \qquad \log_a x = \frac{\ln x}{\ln a}$$

The usefulness of this change-of-base formula is that you can use a calculator that has only the common logarithm key [LOG] and the natural logarithm key [LN] to evaluate logarithms to any base.

Example 7   Changing Bases to Evaluate Logarithms

a. Use *common* logarithms to evaluate $\log_3 5$.

b. Use *natural* logarithms to evaluate $\log_6 2$.

Solution

Using the change-of-base formula, you can convert to common and natural logarithms by writing

$$\log_3 5 = \frac{\log_{10} 5}{\log_{10} 3} \qquad \text{and} \qquad \log_6 2 = \frac{\ln 2}{\ln 6}.$$

Now, use the following keystrokes.

a.

Keystrokes	Display	
5 [LOG] [÷] 3 [LOG] [=]	1.4649735	Scientific
[LOG] 5 [)] [÷] [LOG] 3 [)] [ENTER]	1.4649735	Graphing

So, $\log_3 5 \approx 1.465$.

b.

Keystrokes	Display	
2 [LN] [÷] 6 [LN] [=]	0.3868528	Scientific
[LN] 2 [)] [÷] [LN] 6 [)] [ENTER]	0.3868528	Graphing

So, $\log_6 2 \approx 0.387$.

At this point, you have been introduced to all the basic types of functions that are covered in this course: polynomial functions, radical functions, rational functions, exponential functions, and logarithmic functions. The only other common types of functions are *trigonometric functions*, which you will study if you go on to take a course in trigonometry or precalculus.

Technology: Tip

You can use a graphing calculator to graph logarithmic functions that do not have a base of 10 by using the change-of-base formula. Use the change-of-base formula to rewrite $g(x) = \log_2 x$ in Example 4 on page 710 (with $b = 10$) and graph the function. You should obtain a graph similar to the one below.

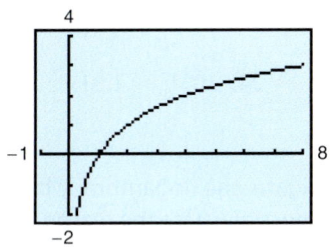

Study Tip

In Example 7(a), $\log_3 5$ could have been evaluated using natural logarithms in the change-of-base formula.

$$\log_3 5 = \frac{\ln 5}{\ln 3} \approx 1.465$$

Notice that you get the same answer whether you use natural logarithms or common logarithms in the change-of-base formula.

11.3 Exercises

Review   Concepts, Skills, and Problem Solving

Keep mathematically in shape by doing these exercises *before* the problems of this section.

Properties and Definitions

1. *Writing* In your own words, explain how to solve a quadratic equation by completing the square.

2. Write the quadratic equation $2x^2 + 4 = 4x - 1$ in general form.

3. *Writing* Explain how to determine the type of solution of a quadratic equation using the discriminant.

4. Write the equation of the parabola $y = x^2 + 4x - 6$ in standard form.

Factoring

In Exercises 5–8, factor the expression completely.

5. $2x^3 - 6x$

6. $16 - (y + 2)^2$

7. $t^2 + 10t + 25$

8. $5 - u + 5u^2 - u^3$

Graphing

In Exercises 9–12, graph the equation.

9. $y = 3 - \frac{1}{2}x$

10. $3x - 4y = 6$

11. $y = x^2 - 6x + 5$

12. $y = -(x - 2)^2 + 1$

Developing Skills

In Exercises 1–12, write the logarithmic equation in exponential form.

1. $\log_7 49 = 2$

2. $\log_{11} 121 = 2$

3. $\log_2 \frac{1}{32} = -5$

4. $\log_3 \frac{1}{27} = -3$

5. $\log_3 \frac{1}{243} = -5$

6. $\log_{10} 10{,}000 = 4$

7. $\log_{36} 6 = \frac{1}{2}$

8. $\log_{32} 4 = \frac{2}{5}$

9. $\log_8 4 = \frac{2}{3}$

10. $\log_{16} 8 = \frac{3}{4}$

11. $\log_2 2.462 \approx 1.3$

12. $\log_3 1.179 \approx 0.15$

In Exercises 13–24, write the exponential equation in logarithmic form.

13. $6^2 = 36$

14. $3^5 = 243$

15. $4^{-2} = \frac{1}{16}$

16. $6^{-4} = \frac{1}{1296}$

17. $8^{2/3} = 4$

18. $81^{3/4} = 27$

19. $25^{-1/2} = \frac{1}{5}$

20. $6^{-3} = \frac{1}{216}$

21. $4^0 = 1$

22. $6^1 = 6$

23. $5^{1.4} \approx 9.518$

24. $10^{0.12} \approx 1.318$

In Exercises 25–46, evaluate the logarithm without using a calculator. (If not possible, state the reason.) See Examples 1 and 2.

25. $\log_2 8$

26. $\log_3 27$

27. $\log_{10} 1000$

28. $\log_{10} 0.00001$

29. $\log_2 \frac{1}{4}$

30. $\log_3 \frac{1}{9}$

31. $\log_4 \frac{1}{64}$

32. $\log_5 \frac{1}{125}$

33. $\log_{10} \frac{1}{10{,}000}$

34. $\log_{10} \frac{1}{100}$

35. $\log_2(-3)$

36. $\log_4(-4)$

37. $\log_4 1$

38. $\log_3 1$

39. $\log_5(-6)$

40. $\log_2 0$

41. $\log_9 3$

42. $\log_{25} 125$

43. $\log_{16} 8$

44. $\log_{144} 12$

45. $\log_7 7^4$

46. $\log_5 5^3$

In Exercises 47–52, use a calculator to evaluate the common logarithm. (Round your answer to four decimal places.) See Example 3.

47. $\log_{10} 42$

48. $\log_{10} 6281$

49. $\log_{10} 0.023$

50. $\log_{10} 0.149$

51. $\log_{10}\left(\sqrt{2} + 4\right)$

52. $\log_{10} \dfrac{\sqrt{3}}{2}$

In Exercises 53–56, match the function with its graph. [The graphs are labeled (a), (b), (c), and (d).]

(a)

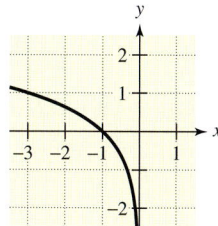

(b)

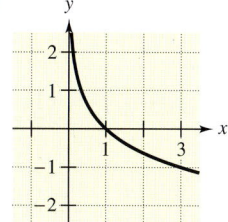

(c)

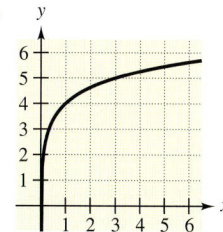

(d)

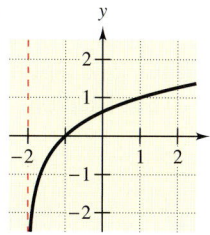

53. $f(x) = 4 + \log_3 x$

54. $f(x) = -\log_3 x$

55. $f(x) = \log_3(-x)$

56. $f(x) = \log_3(x + 2)$

In Exercises 57–60, sketch the graphs of f and g on the same set of coordinate axes. What can you conclude about the relationship between f and g? See Example 4.

57. $f(x) = \log_3 x$
     $g(x) = 3^x$

58. $f(x) = \log_4 x$
     $g(x) = 4^x$

59. $f(x) = \log_6 x$
     $g(x) = 6^x$

60. $f(x) = \log_{1/2} x$
     $g(x) = \left(\tfrac{1}{2}\right)^x$

In Exercises 61–66, identify the transformation of the graph of $f(x) = \log_2 x$ and sketch the graph of h. See Example 5.

61. $h(x) = 3 + \log_2 x$

62. $h(x) = -4 + \log_2 x$

63. $h(x) = \log_2(x - 2)$

64. $h(x) = \log_2(x + 4)$

65. $h(x) = \log_2(-x)$

66. $h(x) = -\log_2(x)$

In Exercises 67–76, sketch the graph of the function. Identify the vertical asymptote.

67. $f(x) = \log_5 x$

68. $g(x) = \log_8 x$

69. $g(t) = -\log_2 t$

70. $h(s) = -2 \log_3 s$

71. $f(x) = 3 + \log_2 x$

72. $f(x) = -2 + \log_3 x$

73. $g(x) = \log_2(x - 3)$

74. $h(x) = \log_3(x + 1)$

75. $f(x) = \log_{10}(10x)$

76. $g(x) = \log_4(4x)$

In Exercises 77–82, find the domain and vertical asymptote of the function. Sketch its graph.

77. $f(x) = \log_4 x$

78. $g(x) = \log_6 x$

79. $h(x) = \log_4(x - 3)$

80. $f(x) = -\log_6(x + 2)$

81. $y = -\log_3 x + 2$

82. $y = \log_5(x - 1) + 4$

 In Exercises 83–88, use a graphing calculator to graph the function. Determine the domain and the vertical asymptote.

83. $y = 5 \log_{10} x$

84. $y = 5 \log_{10}(x - 3)$

85. $y = -3 + 5 \log_{10} x$

86. $y = 5 \log_{10}(3x)$

87. $y = \log_{10}\left(\dfrac{x}{5}\right)$

88. $y = \log_{10}(-x)$

In Exercises 89–94, use a calculator to evaluate the natural logarithm. (Round your answer to four decimal places.) See Example 6.

89. ln 38

90. ln 14.2

91. ln 0.15

92. ln 0.002

93. $\ln\left(\dfrac{1 + \sqrt{5}}{3}\right)$

94. $\ln\left(1 + \dfrac{0.10}{12}\right)$

In Exercises 95–98, match the function with its graph. [The graphs are labeled (a), (b), (c), and (d).]

(a)
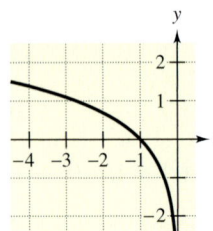

(b)
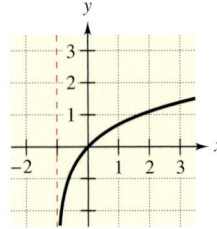

(c)
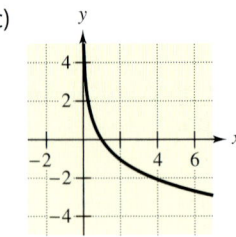

(d)
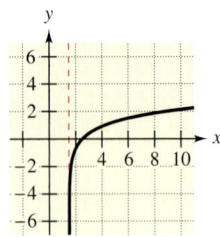

95. $f(x) = \ln(x + 1)$

96. $f(x) = \ln(-x)$

97. $f(x) = \ln\left(x - \frac{3}{2}\right)$

98. $f(x) = -\frac{3}{2}\ln x$

In Exercises 99–106, sketch the graph of the function. Identify the vertical asymptote.

99. $f(x) = -\ln x$

100. $f(x) = -2 \ln x$

101. $f(x) = 3 \ln x$

102. $h(t) = 4 \ln t$

103. $f(x) = 1 + \ln\ x$

104. $h(x) = 2 + \ln x$

105. $g(t) = 2 \ln(t - 4)$

106. $g(x) = -3 \ln(x + 3)$

 In Exercises 107–110, use a graphing calculator to graph the function. Determine the domain and the vertical asymptote.

107. $g(x) = -\ln(x + 1)$

108. $h(x) = \ln(x + 5)$

109. $f(t) = 7 + 3 \ln t$

110. $g(t) = \ln(5 - t)$

In Exercises 111–124, use a calculator to evaluate the logarithm by means of the change-of-base formula. Use (a) the common logarithm key and (b) the natural logarithm key. (Round your answer to four decimal places.) See Example 7.

111. $\log_9 36$

112. $\log_7 411$

113. $\log_4 6$

114. $\log_6 9$

115. $\log_2 0.72$

116. $\log_{12} 0.6$

117. $\log_{15} 1250$

118. $\log_{20} 125$

119. $\log_{1/2} 4$

120. $\log_{1/3} 18$

121. $\log_4 \sqrt{42}$

122. $\log_3 \sqrt{26}$

123. $\log_2(1 + e)$

124. $\log_4(2 + e^3)$

Solving Problems

125. *American Elk*   The antler spread a (in inches) and shoulder height h (in inches) of an adult male American elk are related by the model

$$h = 116 \log_{10}(a + 40) - 176.$$

Approximate the shoulder height of a male American elk with an antler spread of 55 inches.

126. *Sound Intensity*   The relationship between the number of decibels B and the intensity of a sound I in watts per centimeter squared is given by

$$B = 10 \log_{10}\left(\frac{I}{10^{-16}}\right).$$

Determine the number of decibels of a sound with an intensity of 10^{-4} watts per centimeter squared.

127. *Compound Interest* The time t in years for an investment to double in value when compounded continuously at interest rate r is given by

$$t = \frac{\ln 2}{r}.$$

Complete the table, which shows the "doubling times" for several annual percent rates.

r	0.07	0.08	0.09	0.10	0.11	0.12
t						

128. *Meteorology* Most tornadoes last less than 1 hour and travel about 20 miles. The speed of the wind S (in miles per hour) near the center of the tornado and the distance d (in miles) the tornado travels are related by the model $S = 93 \log_{10} d + 65$. On March 18, 1925, a large tornado struck portions of Missouri, Illinois, and Indiana, covering a distance of 220 miles. Approximate the speed of the wind near the center of this tornado.

129. *Tractrix* A person walking along a dock (the y-axis) drags a boat by a 10-foot rope (see figure). The boat travels along a path known as a *tractrix*. The equation of the path is

$$y = 10 \ln\left(\frac{10 + \sqrt{100 - x^2}}{x}\right) - \sqrt{100 - x^2}.$$

(a) ⊞ Use a graphing calculator to graph the function. What is the domain of the function?

(b) ⊞ Identify any asymptotes.

(c) Determine the position of the person when the x-coordinate of the position of the boat is $x = 2$.

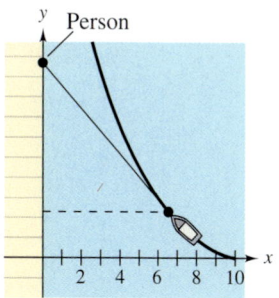

Figure for 129

130. ⊞ *Home Mortgage* The model

$$t = 10.042 \ln\left(\frac{x}{x - 1250}\right), \quad x > 1250$$

approximates the length t (in years) of a home mortgage of $150,000 at 10% interest in terms of the monthly payment x.

(a) Use a graphing calculator to graph the model. Describe the change in the length of the mortgage as the monthly payment increases.

(b) Use the graph in part (a) to approximate the length of the mortgage when the monthly payment is $1316.35.

(c) Use the result of part (b) to find the total amount paid over the term of the mortgage. What amount of the total is interest costs?

Explaining Concepts

131. Write "logarithm of x with base 5" symbolically.

132. *Writing* Explain the relationship between the functions $f(x) = 2^x$ and $g(x) = \log_2 x$.

133. *Writing* Explain why $\log_a a = 1$.

134. *Writing* Explain why $\log_a a^x = x$.

135. *Writing* What are common logarithms and natural logarithms?

136. *Writing* Describe how to use a calculator to find the logarithm of a number if the base is not 10 or e.

Think About It In Exercises 137–142, answer the question for the function $f(x) = \log_{10} x$. (Do not use a calculator.)

137. What is the domain of f?

138. Find the inverse function of f.

139. Describe the values of $f(x)$ for $1000 \le x \le 10{,}000$.

140. Describe the values of x, given that $f(x)$ is negative.

141. By what amount will x increase, given that $f(x)$ is increased by 1 unit?

142. Find the ratio of a to b when $f(a) = 3 + f(b)$.

Mid-Chapter Quiz

Take this quiz as you would take a quiz in class. After you are done, check your work against the answers in the back of the book.

1. Given $f(x) = \left(\frac{4}{3}\right)^x$, find (a) $f(2)$, (b) $f(0)$, (c) $f(-1)$, and (d) $f(1.5)$.

2. Find the domain and range of $g(x) = 2^{-0.5x}$.

In Exercises 3–6, sketch the graph of the function. Identify the horizontal asymptote. Use a graphing calculator for Exercises 5 and 6.

3. $y = \frac{1}{2}(4^x)$

4. $y = 5(2^{-x})$

5. 🖩 $f(t) = 12e^{-0.4t}$

6. 🖩 $g(x) = 100(1.08)^x$

7. Given $f(x) = 2x - 3$ and $g(x) = x^3$, find the indicated composition.

(a) $(f \circ g)(x)$ (b) $(g \circ f)(x)$ (c) $(f \circ g)(-2)$ (d) $(g \circ f)(4)$

8. Verify algebraically and graphically that $f(x) = 3 - 5x$ and $g(x) = \frac{1}{5}(3 - x)$ are inverse functions of each other.

In Exercises 9 and 10, find the inverse function.

9. $h(x) = 10x + 3$

10. $g(t) = \frac{1}{2}t^3 + 2$

11. Write the logarithmic equation $\log_9 \frac{1}{81} = -2$ in exponential form.

12. Write the exponential equation $3^4 = 81$ in logarithmic form.

13. Evaluate $\log_5 125$ without a calculator.

🖩 In Exercises 14 and 15, use a graphing calculator to graph the function. Identify the vertical asymptote.

14. $f(t) = -2 \ln(t + 3)$

15. $h(x) = 5 + \frac{1}{2} \ln x$

16. Use the graph of f shown at the left to determine h and k if $f(x) = \log_5(x - h) + k$.

17. Use a calculator and the change-of-base formula to evaluate $\log_3 782$.

18. You deposit $750 in an account at an annual interest rate of $7\frac{1}{2}\%$. Complete the table showing the balance A in the account after 20 years for several types of compounding.

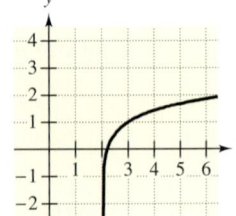

Figure for 16

n	1	4	12	365	Continuous compounding
A					

19. After t years, the remaining mass y (in grams) of 14 grams of a radioactive element whose half-life is 40 years is given by $y = 14\left(\frac{1}{2}\right)^{t/40}, t \geq 0$. How much of the initial mass remains after 125 years?

11.4 Properties of Logarithms

Charles Gupton/Corbis

What You Should Learn

1. Use the properties of logarithms to evaluate logarithms.
2. Use the properties of logarithms to rewrite, expand, or condense logarithmic expressions.
3. Use the properties of logarithms to solve application problems.

Why You Should Learn It

Logarithmic equations are often used to model scientific observations. For instance, in Example 8 on page 723, a logarithmic equation is used to model human memory.

1. Use the properties of logarithms to evaluate logarithms.

Properties of Logarithms

You know from the preceding section that the logarithmic function with base a is the *inverse function* of the exponential function with base a. So, it makes sense that each property of exponents should have a corresponding property of logarithms. For instance, the exponential property

$$a^0 = 1 \qquad \text{Exponential property}$$

has the corresponding logarithmic property

$$\log_a 1 = 0. \qquad \text{Corresponding logarithmic property}$$

In this section you will study the logarithmic properties that correspond to the following three exponential properties:

	Base a	*Natural Base*
1.	$a^m a^n = a^{m+n}$	$e^m e^n = e^{m+n}$
2.	$\dfrac{a^m}{a^n} = a^{m-n}$	$\dfrac{e^m}{e^n} = e^{m-n}$
3.	$(a^m)^n = a^{mn}$	$(e^m)^n = e^{mn}$

Properties of Logarithms

Let a be a positive real number such that $a \neq 1$, and let n be a real number. If u and v are real numbers, variables, or algebraic expressions such that $u > 0$ and $v > 0$, the following properties are true.

		Logarithm with Base a	*Natural Logarithm*
1.	Product Property:	$\log_a(uv) = \log_a u + \log_a v$	$\ln(uv) = \ln u + \ln v$
2.	Quotient Property:	$\log_a \dfrac{u}{v} = \log_a u - \log_a v$	$\ln \dfrac{u}{v} = \ln u - \ln v$
3.	Power Property:	$\log_a u^n = n \log_a u$	$\ln u^n = n \ln u$

There is no general property of logarithms that can be used to simplify $\log_a(u + v)$. Specifically,

$$\log_a(u + v) \text{ does not equal } \log_a u + \log_a v.$$

Example 1   Using Properties of Logarithms

Use $\ln 2 \approx 0.693$, $\ln 3 \approx 1.099$, and $\ln 5 \approx 1.609$ to approximate each expression.

a. $\ln \dfrac{2}{3}$ **b.** $\ln 10$ **c.** $\ln 30$

Solution

a. $\ln \dfrac{2}{3} = \ln 2 - \ln 3$ Quotient Property

$\approx 0.693 - 1.099 = -0.406$ Substitute for $\ln 2$ and $\ln 3$.

b. $\ln 10 = \ln(2 \cdot 5)$ Factor.

$= \ln 2 + \ln 5$ Product Property

$\approx 0.693 + 1.609$ Substitute for $\ln 2$ and $\ln 5$.

$= 2.302$ Simplify.

c. $\ln 30 = \ln(2 \cdot 3 \cdot 5)$ Factor.

$= \ln 2 + \ln 3 + \ln 5$ Product Property

$\approx 0.693 + 1.099 + 1.609$ Substitute for $\ln 2$, $\ln 3$, and $\ln 5$.

$= 3.401$ Simplify.

When using the properties of logarithms, it helps to state the properties *verbally*. For instance, the verbal form of the Product Property

$$\ln(uv) = \ln u + \ln v$$

is: *The log of a product is the sum of the logs of the factors.*
Similarly, the verbal form of the Quotient Property

$$\ln \frac{u}{v} = \ln u - \ln v$$

is: *The log of a quotient is the difference of the logs of the numerator and denominator.*

Example 2   Using Properties of Logarithms

Use the properties of logarithms to verify that $-\ln 2 = \ln \frac{1}{2}$.

Solution

Using the Power Property, you can write the following.

$-\ln 2 = (-1)\ln 2$ Rewrite coefficient as -1.

$= \ln 2^{-1}$ Power Property

$= \ln \dfrac{1}{2}$ Rewrite 2^{-1} as $\frac{1}{2}$.

② Use the properties of logarithms to rewrite, expand, or condense logarithmic expressions.

Rewriting Logarithmic Expressions

In Examples 1 and 2, the properties of logarithms were used to rewrite logarithmic expressions involving the log of a *constant*. A more common use of these properties is to rewrite the log of a *variable expression*.

Example 3   Rewriting Logarithmic Expressions

Use the properties of logarithms to rewrite each expression.

a. $\log_{10} 7x^3 = \log_{10} 7 + \log_{10} x^3$ Product Property

              $= \log_{10} 7 + 3 \log_{10} x$ Power Property

b. $\ln \dfrac{8x^3}{y} = \ln 8x^3 - \ln y$ Quotient Property

             $= \ln 8 + \ln x^3 - \ln y$ Product Property

             $= \ln 8 + 3 \ln x - \ln y$ Power Property

When you rewrite a logarithmic expression as in Example 3, you are **expanding** the expression. The reverse procedure is demonstrated in Example 4, and is called **condensing** a logarithmic expression.

Example 4   Condensing Logarithmic Expressions

Use the properties of logarithms to condense each expression.

a. $\ln x - \ln 3$ **b.** $2 \log_3 x + \log_3 5$

Solution

a. $\ln x - \ln 3 = \ln \dfrac{x}{3}$ Quotient Property

b. $2 \log_3 x + \log_3 5 = \log_3 x^2 + \log_3 5$ Power Property

                        $= \log_3 5x^2$ Product Property

Technology: Tip

When you are rewriting a logarithmic expression, remember that you can use a graphing calculator to check your result graphically. For instance, in Example 4(a), try graphing the functions.

$$y_1 = \ln x - \ln 3 \quad \text{and} \quad y_2 = \ln \frac{x}{3}$$

in the same viewing window. You should obtain the same graph for each function.

Example 5   Expanding Logarithmic Expressions

Use the properties of logarithms to expand each expression.

a. $\log_6 3xy^2,\ x > 0,\ y > 0$ **b.** $\ln \dfrac{\sqrt{3x - 5}}{7}$

Solution

a. $\log_6 3xy^2 = \log_6 3 + \log_6 x + \log_6 y^2$ Product Property

$\qquad\qquad\quad = \log_6 3 + \log_6 x + 2 \log_6 y$ Power Property

b. $\ln \dfrac{\sqrt{3x - 5}}{7} = \ln\left[\dfrac{(3x - 5)^{1/2}}{7} \right]$ Rewrite using rational exponent.

$\qquad\qquad\quad = \ln(3x - 5)^{1/2} - \ln 7$ Quotient Property

$\qquad\qquad\quad = \dfrac{1}{2} \ln(3x - 5) - \ln 7$ Power Property

Sometimes expanding or condensing logarithmic expressions involves several steps. In the next example, be sure that you can justify each step in the solution. Notice how different the expanded expression is from the original.

Example 6   Expanding a Logarithmic Expression

Use the properties of logarithms to expand $\ln \sqrt{x^2 - 1},\ \ x > 1$.

Solution

$\ln \sqrt{x^2 - 1} = \ln(x^2 - 1)^{1/2}$ Rewrite using rational exponent.

$\qquad\quad = \tfrac{1}{2} \ln(x^2 - 1)$ Power Property

$\qquad\quad = \tfrac{1}{2} \ln[(x - 1)(x + 1)]$ Factor.

$\qquad\quad = \tfrac{1}{2}[\ln(x - 1) + \ln(x + 1)]$ Product Property

$\qquad\quad = \tfrac{1}{2} \ln(x - 1) + \tfrac{1}{2} \ln(x + 1)$ Distributive Property

Example 7   Condensing Logarithmic Expressions

Use the properties of logarithms to condense each expression.

a. $\ln 2 - 2 \ln x = \ln 2 - \ln x^2,\ \ x > 0$ Power Property

$\qquad\qquad\quad = \ln \dfrac{2}{x^2},\ \ x > 0$ Quotient Property

b. $3(\ln 4 + \ln x) = 3(\ln 4x)$ Product Property

$\qquad\qquad\quad = \ln (4x)^3$ Power Property

$\qquad\qquad\quad = \ln 64x^3,\ \ x \geq 0$ Simplify.

When you expand or condense a logarithmic expression, it is possible to change the domain of the expression. For instance, the domain of the function

$$f(x) = 2 \ln x \qquad \text{Domain is the set of positive real numbers.}$$

is the set of positive real numbers, whereas the domain of

$$g(x) = \ln x^2 \qquad \text{Domain is the set of nonzero real numbers.}$$

is the set of nonzero real numbers. So, when you expand or condense a logarithmic expression, you should check to see whether the rewriting has changed the domain of the expression. In such cases, you should restrict the domain appropriately. For instance, you can write

$$f(x) = 2 \ln x$$
$$= \ln x^2, \, x > 0.$$

3 Use the properties of logarithms to solve application problems.

Application

Example 8   Human Memory Model

In an experiment, students attended several lectures on a subject. Every month for a year after that, the students were tested to see how much of the material they remembered. The average scores for the group are given by the **human memory model**

$$f(t) = 80 - \ln(t + 1)^9, \quad 0 \leq t \leq 12$$

where t is the time in months. Find the average scores for the group after 2 months and 8 months.

Solution

To make the calculations easier, rewrite the model using the Power Property, as follows.

$$f(t) = 80 - 9 \ln(t + 1), \quad 0 \leq t \leq 12$$

After 2 months, the average score was

$$f(2) = 80 - 9 \ln(2 + 1) \qquad \text{Substitute 2 for } t.$$
$$\approx 80 - 9.9 \qquad \text{Simplify.}$$
$$= 70.1 \qquad \text{Average score after 2 months}$$

and after 8 months, the average score was

$$f(8) = 80 - 9 \ln(8 + 1) \qquad \text{Substitute 8 for } t.$$
$$\approx 80 - 19.8 \qquad \text{Simplify.}$$
$$= 60.2. \qquad \text{Average score after 8 months}$$

The graph of the function is shown in Figure 11.21.

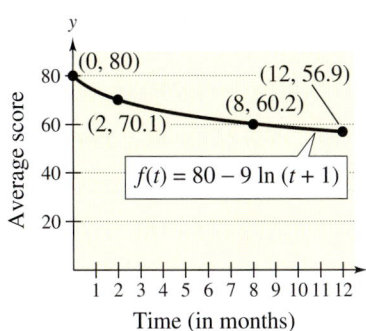

Human Memory Model
Figure 11.21

11.4 Exercises

Review　Concepts, Skills, and Problem Solving

Keep mathematically in shape by doing these exercises *before* the problems of this section.

Properties and Definitions

In Exercises 1 and 2, use the rule for radicals to fill in the blank.

1. Product Rule: $\sqrt[n]{u}\,\sqrt[n]{v} = $ [blank]

2. Quotient Rule: $\dfrac{\sqrt[n]{u}}{\sqrt[n]{v}} = $ [blank]

3. *Writing* ✎　Explain why the radicals $\sqrt{2x}$ and $\sqrt[3]{2x}$ cannot be added.

4. *Writing* ✎　Is $1/\sqrt{2x}$ in simplest form? Explain.

Simplifying Expressions

In Exercises 5–10, perform the indicated operations and simplify. (Assume all variables are positive.)

5. $25\sqrt{3x} - 3\sqrt{12x}$

6. $\left(\sqrt{x} + 3\right)\left(\sqrt{x} - 3\right)$

7. $\sqrt{u}\left(\sqrt{20} - \sqrt{5}\right)$

8. $\left(2\sqrt{t} + 3\right)^2$

9. $\dfrac{50x}{\sqrt{2}}$

10. $\dfrac{12}{\sqrt{t+2} + \sqrt{t}}$

Problem Solving

11. *Demand*　The demand equation for a product is given by $p = 30 - \sqrt{0.5(x - 1)}$, where x is the number of units demanded per day and p is the price per unit. Find the demand when the price is set at $26.76.

12. *List Price*　The sale price of a computer is $1955. The discount is 15% of the list price. Find the list price.

Developing Skills

In Exercises 1–24, use properties of logarithms to evaluate the expression without a calculator. (If not possible, state the reason.)

1. $\log_{12} 12^3$

2. $\log_5 125$

3. $\log_4\left(\frac{1}{16}\right)^2$

4. $\log_7\left(\frac{1}{49}\right)^3$

5. $\log_5 \sqrt[3]{5}$

6. $\ln \sqrt{e}$

7. $\ln 14^0$

8. $\ln\left(\dfrac{7.14}{7.14}\right)$

9. $\ln e^{-6}$

10. $\ln e^7$

11. $\log_4 8 + \log_4 2$

12. $\log_6 2 + \log_6 3$

13. $\log_8 4 + \log_8 16$

14. $\log_{10} 5 + \log_{10} 20$

15. $\log_4 8 - \log_4 2$

16. $\log_5 50 - \log_5 2$

17. $\log_6 72 - \log_6 2$

18. $\log_3 324 - \log_3 4$

19. $\log_2 5 - \log_2 40$

20. $\log_3\left(\frac{2}{3}\right) + \log_3\left(\frac{1}{2}\right)$

21. $\ln e^8 + \ln e^4$

22. $\ln e^5 - \ln e^2$

23. $\ln \dfrac{e^3}{e^2}$

24. $\ln(e^2 \cdot e^4)$

In Exercises 25–36, use $\log_4 2 = 0.5000$, $\log_4 3 \approx 0.7925$, and the properties of logarithms to approximate the expression. Do not use a calculator. See Example 1.

25. $\log_4 4$

26. $\log_4 8$

27. $\log_4 6$

28. $\log_4 24$

29. $\log_4 \frac{3}{2}$

30. $\log_4 \frac{9}{2}$

31. $\log_4 \sqrt{2}$

32. $\log_4 \sqrt[3]{9}$

33. $\log_4(3 \cdot 2^4)$

34. $\log_4 \sqrt{3 \cdot 2^5}$

35. $\log_4 3^0$

36. $\log_4 4^3$

In Exercises 37–42, use $\ln 3 \approx 1.0986$, $\ln 12 \approx 2.4849$, and the properties of logarithms to approximate the expression. Use a calculator to verify your result.

37. $\ln 9$

38. $\ln \frac{1}{4}$

39. $\ln 36$

40. $\ln 144$

41. $\ln \sqrt{36}$

42. $\ln 5^0$

In Exercises 43–76, use the properties of logarithms to expand the expression. See Examples 3, 5, and 6.

43. $\log_3 11x$

44. $\log_2 3x$

45. $\log_7 x^2$

46. $\log_3 x^3$

47. $\log_5 x^{-2}$

48. $\log_2 s^{-4}$

49. $\log_4 \sqrt{3x}$

50. $\log_3 \sqrt[3]{5y}$

51. $\ln 3y$

52. $\ln 5x$

53. $\log_2 \frac{z}{17}$

54. $\log_{10} \frac{7}{y}$

55. $\log_9 \frac{\sqrt{x}}{12}$

56. $\ln \frac{\sqrt{x}}{x-9}$

57. $\ln x^2(y-2)$

58. $\ln y(y-1)^2$

59. $\log_4[x^6(x-7)^2]$

60. $\log_8[(x-y)^4 z^6]$

61. $\log_3 \sqrt[3]{x+1}$

62. $\log_5 \sqrt{xy}$

63. $\ln \sqrt{x(x+2)}$

64. $\ln \sqrt[3]{x(x+5)}$

65. $\ln\left(\frac{x+1}{x-1}\right)^2$

66. $\log_2\left(\frac{x^2}{x-3}\right)^3$

67. $\ln \sqrt[3]{\frac{x^2}{x+1}}$

68. $\ln \sqrt{\frac{3x}{x-5}}$

69. $\ln \frac{xy^2}{z^3}$

70. $\log_5 \frac{x^2 y^5}{z^7}$

71. $\log_3 \sqrt{\frac{x^7}{y^5 z^8}}$

72. $\ln \sqrt[3]{\frac{x^4}{y^3 z^2}}$

73. $\log_6[a\sqrt{b}(c-d)^3]$

74. $\ln[(xy)^2(x+3)^4]$

75. $\ln\left[(x+y)\frac{\sqrt[5]{w+2}}{3t}\right]$

76. $\ln\left[(u-v)\frac{\sqrt[3]{u-4}}{3v}\right]$

In Exercises 77–108, use the properties of logarithms to condense the expression. See Examples 4 and 7.

77. $\log_{12} x - \log_{12} 3$

78. $\log_6 12 - \log_6 y$

79. $\log_2 3 + \log_2 x$

80. $\log_5 2x + \log_5 3y$

81. $\log_{10} 4 - \log_{10} x$

82. $\ln 10x - \ln z$

83. $4 \ln b$

84. $10 \log_4 z$

85. $-2 \log_5 2x$

86. $-5 \ln(x + 3)$

87. $7 \log_2 x + 3 \log_2 z$

88. $2 \log_{10} x + \frac{1}{2} \log_{10} y$

89. $\log_3 2 + \frac{1}{2} \log_3 y$

90. $\ln 6 - 3 \ln z$

91. $2 \ln x + 3 \ln y - \ln z$

92. $4 \ln 3 - 2 \ln x - \ln y$

93. $5 \ln 2 - \ln x + 3 \ln y$

94. $4 \ln 2 + 2 \ln x - \frac{1}{2} \ln y$

95. $4(\ln x + \ln y)$

96. $\frac{1}{2}(\ln 8 + \ln 2x)$

97. $2[\ln x - \ln(x + 1)]$

98. $5\left[\ln x - \frac{1}{2} \ln(x + 4)\right]$

99. $\log_4(x + 8) - 3 \log_4 x$

100. $5 \log_3 x + \log_3(x - 6)$

101. $\frac{1}{2} \log_5(x + 2) - \log_5(x - 3)$

102. $\frac{1}{4} \log_6(x + 1) - 5 \log_6(x - 4)$

103. $5 \log_6(c + d) - \frac{1}{2} \log_6(m - n)$

104. $2 \log_5(x + y) + 3 \log_5 w$

105. $\frac{1}{5}(3 \log_2 x - 4 \log_2 y)$

106. $\frac{1}{3}[\ln(x - 6) - 4 \ln y - 2 \ln z]$

107. $\frac{1}{5} \log_6(x - 3) - 2 \log_6 x - 3 \log_6(x + 1)$

108. $3\left[\frac{1}{2} \log_9(a + 6) - 2 \log_9(a - 1)\right]$

In Exercises 109–114, simplify the expression.

109. $\ln 3e^2$

110. $\log_3(3^2 \cdot 4)$

111. $\log_5 \sqrt{50}$

112. $\log_2 \sqrt{22}$

113. $\log_4 \dfrac{4}{x^2}$

114. $\ln \dfrac{6}{e^5}$

⊞ In Exercises 115–118, use a graphing calculator to graph the two equations in the same viewing window. Use the graphs to verify that the expressions are equivalent. Assume $x > 0$.

115. $y_1 = \ln\left(\dfrac{10}{x^2 + 1}\right)^2$

$y_2 = 2[\ln 10 - \ln(x^2 + 1)]$

116. $y_1 = \ln \sqrt{x(x + 1)}$

$y_2 = \frac{1}{2}[\ln x + \ln(x + 1)]$

117. $y_1 = \ln[x^2(x + 2)]$

$y_2 = 2 \ln x + \ln(x + 2)$

118. $y_1 = \ln\left(\dfrac{\sqrt{x}}{x - 3}\right)$

$y_2 = \frac{1}{2} \ln x - \ln(x - 3)$

Solving Problems

119. *Sound Intensity* The relationship between the number of decibels B and the intensity of a sound I in watts per centimeter squared is given by

$$B = 10 \log_{10}\left(\dfrac{I}{10^{-16}}\right).$$

Use properties of logarithms to write the formula in simpler form, and determine the number of decibels of a sound with an intensity of 10^{-10} watts per centimeter squared.

120. *Human Memory Model*   Students participating in an experiment attended several lectures on a subject. Every month for a year after that, the students were tested to see how much of the material they remembered. The average scores for the group are given by the human memory model

$$f(t) = 80 - \log_{10}(t + 1)^{12}, \quad 0 \le t \le 12$$

where t is the time in months.

(a) Find the average scores for the group after 2 months and 8 months.

(b) 🖩 Use a graphing calculator to graph the function.

Molecular Transport   In Exercises 121 and 122, use the following information. The energy E (in kilocalories per gram molecule) required to transport a substance from the outside to the inside of a living cell is given by

$$E = 1.4(\log_{10} C_2 - \log_{10} C_1)$$

where C_1 and C_2 are the concentrations of the substance outside and inside the cell, respectively.

121. Condense the expression.

122. The concentration of a substance inside a cell is twice the concentration outside the cell. How much energy is required to transport the substance from outside to inside the cell?

Explaining Concepts

True or False?   In Exercises 123–130, use properties of logarithms to determine whether the equation is true or false. If it is false, state why or give an example to show that it is false.

123. $\ln e^{2-x} = 2 - x$

124. $\log_2 8x = 3 + \log_2 x$

125. $\log_8 4 + \log_8 16 = 2$

126. $\log_3(u + v) = \log_3 u + \log_3 v$

127. $\log_3(u + v) = \log_3 u \cdot \log_3 v$

128. $\dfrac{\log_6 10}{\log_6 3} = \log_6 10 - \log_6 3$

129. If $f(x) = \log_a x$, then $f(ax) = 1 + f(x)$.

130. If $f(x) = \log_a x$, then $f(a^n) = n$.

True or False?   In Exercises 131–136, determine whether the statement is true or false given that $f(x) = \ln x$. If it is false, state why or give an example to show that the statement is false.

131. $f(0) = 0$

132. $f(2x) = \ln 2 + \ln x$

133. $f(x - 3) = \ln x - \ln 3, \quad x > 3$

134. $\sqrt{f(x)} = \frac{1}{2} \ln x$

135. If $f(u) = 2f(v)$, then $v = u^2$.

136. If $f(x) > 0$, then $x > 1$.

137. *Error Analysis*   Describe the error.

$$\log_b\left(\frac{1}{x}\right) = \log_b\left(\frac{x}{xx}\right)$$
$$= \log_b x - \log_b x + \log_b x$$
$$= \log_b x$$

138. *Think About It*   Explain how you can show that

$$\frac{\ln x}{\ln y} \ne \ln \frac{x}{y}.$$

139. *Think About It*   Without a calculator, approximate the natural logarithms of as many integers as possible between 1 and 20 using $\ln 2 \approx 0.6931$, $\ln 3 \approx 1.0986$, $\ln 5 \approx 1.6094$, and $\ln 7 \approx 1.9459$. Explain the method you used. Then verify your results with a calculator and explain any differences in the results.

11.5 Solving Exponential and Logarithmic Equations

Susumu Sato/Corbis

What You Should Learn

1️⃣ Solve basic exponential and logarithmic equations.

2️⃣ Use inverse properties to solve exponential equations.

3️⃣ Use inverse properties to solve logarithmic equations.

4️⃣ Use exponential or logarithmic equations to solve application problems.

Why You Should Learn It

Exponential and logarithmic equations occur in many scientific applications. For instance, in Exercise 137 on page 737, you will use a logarithmic equation to determine how long it will take for ice cubes to form.

Exponential and Logarithmic Equations

In this section, you will study procedures for *solving equations* that involve exponential or logarithmic expressions. As a simple example, consider the exponential equation $2^x = 16$. By rewriting this equation in the form $2^x = 2^4$, you can see that the solution is $x = 4$. To solve this equation, you can use one of the following properties, which result from the fact that exponential and logarithmic functions are one-to-one functions.

1️⃣ Solve basic exponential and logarithmic equations.

One-to-One Properties of Exponential and Logarithmic Equations

Let a be a positive real number such that $a \neq 1$, and let x and y be real numbers. Then the following properties are true.

1. $a^x = a^y$ if and only if $x = y$.

2. $\log_a x = \log_a y$ if and only if $x = y$ $(x > 0, y > 0)$.

Example 1   Solving Exponential and Logarithmic Equations

Solve each equation.

a. $4^{x+2} = 64$ Original equation

     $4^{x+2} = 4^3$ Rewrite with like bases.

   $x + 2 = 3$ One-to-one property

       $x = 1$ Subtract 2 from each side.

The solution is $x = 1$. Check this in the original equation.

b. $\ln(2x - 3) = \ln 11$ Original equation

     $2x - 3 = 11$ One-to-one property

        $2x = 14$ Add 3 to each side.

          $x = 7$ Divide each side by 2.

The solution is $x = 7$. Check this in the original equation.

2 Use inverse properties to solve exponential equations.

Solving Exponential Equations

In Example 1(a), you were able to use a one-to-one property to solve the original equation because each side of the equation was written in exponential form with the same base. However, if only one side of the equation is written in exponential form or if both sides cannot be written with the same base, it is more difficult to solve the equation. For example, to solve the equation $2^x = 7$, you must find the power to which 2 can be raised to obtain 7. To do this, *rewrite the exponential equation in logarithmic form* by taking the logarithm of each side and use one of the following inverse properties of exponents and logarithms.

Solving Exponential Equations

To solve an exponential equation, first isolate the exponential expression, then **take the logarithm of each side of the equation** (or write the equation in logarithmic form) and solve for the variable.

Inverse Properties of Exponents and Logarithms

Base a	Natural Base e
1. $\log_a(a^x) = x$	$\ln(e^x) = x$
2. $a^{(\log_a x)} = x$	$e^{(\ln x)} = x$

Technology: Discovery

Use a graphing calculator to graph each side of each equation. What does this tell you about the inverse properties of exponents and logarithms?

1. (a) $\log_{10}(10^x) = x$

    (b) $10^{(\log_{10} x)} = x$

2. (a) $\ln(e^x) = x$

    (b) $e^{(\ln x)} = x$

Study Tip

Remember that to evaluate a logarithm such as $\log_2 7$ you need to use the change-of-base formula.

$$\log_2 7 = \frac{\ln 7}{\ln 2} \approx 2.807$$

Similarly,

$$\log_4 9 + 3 = \frac{\ln 9}{\ln 4} + 3$$

$$\approx 1.585 + 3$$

$$\approx 4.585$$

Example 2   Solving Exponential Equations

Solve each exponential equation.

a. $2^x = 7$ **b.** $4^{x-3} = 9$ **c.** $2e^x = 10$

Solution

a. To isolate the x, take the $\log_2$ of each side of the equation or write the equation in logarithmic form, as follows.

$2^x = 7$	Write original equation.
$x = \log_2 7$	Inverse property

The solution is $x = \log_2 7 \approx 2.807$. Check this in the original equation.

b. $4^{x-3} = 9$	Write original equation.
$x - 3 = \log_4 9$	Inverse property
$x = \log_4 9 + 3$	Add 3 to each side.

The solution is $x = \log_4 9 + 3 \approx 4.585$. Check this in the original equation.

c. $2e^x = 10$	Write original equation.
$e^x = 5$	Divide each side by 2.
$x = \ln 5$	Inverse property.

The solution is $x = \ln 5 \approx 1.609$. Check this in the original equation.

Technology: Tip

Remember that you can use a graphing calculator to solve equations graphically or check solutions that are obtained algebraically. For instance, to check the solutions in Examples 2(a) and 2(c), graph each side of the equations, as shown below.

Graph $y_1 = 2^x$ and $y_2 = 7$. Then use the *intersect* feature of the graphing calculator to approximate the intersection of the two graphs to be $x \approx 2.807$.

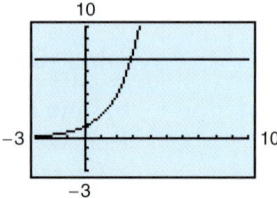

Graph $y_1 = 2e^x$ and $y_2 = 10$. Then use the *intersect* feature of the graphing calculator to approximate the intersection of the two graphs to be $x \approx 1.609$.

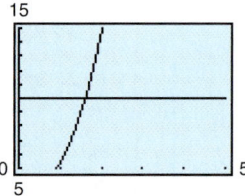

Example 3 Solving an Exponential Equation

Solve $5 + e^{x+1} = 20$.

Solution

$5 + e^{x+1} = 20$	Write original equation.
$e^{x+1} = 15$	Subtract 5 from each side.
$\ln e^{x+1} = \ln 15$	Take the logarithm of each side.
$x + 1 = \ln 15$	Inverse property
$x = -1 + \ln 15$	Subtract 1 from each side.

The solution is $x = -1 + \ln 15 \approx 1.708$. You can check this as follows.

Check

$5 + e^{x+1} = 20$	Write original equation.
$5 + e^{-1 + \ln 15 + 1} \stackrel{?}{=} 20$	Substitute $-1 + \ln 15$ for x.
$5 + e^{\ln 15} \stackrel{?}{=} 20$	Simplify.
$5 + 15 = 20$	Solution checks. ✓

③ Use inverse properties to solve logarithmic equations.

Solving Logarithmic Equations

You know how to solve an exponential equation by *taking the logarithm of each side*. To solve a logarithmic equation, you need to **exponentiate** each side. For instance, to solve a logarithmic equation such as $\ln x = 2$ you can exponentiate each side of the equation as follows.

$$\ln x = 2 \qquad \text{Write original equation.}$$

$$e^{\ln x} = e^2 \qquad \text{Exponentiate each side.}$$

$$x = e^2 \qquad \text{Inverse property}$$

Notice that you obtain the same result by writing the equation in exponential form. This procedure is demonstrated in the next three examples. The following guideline can be used for solving logarithmic equations.

Solving Logarithmic Equations

To solve a logarithmic equation, first isolate the logarithmic expression, then **exponentiate each side of the equation** (or write the equation in exponential form) and solve for the variable.

Example 4   Solving Logarithmic Equations

a. $2 \log_4 x = 5$ — Original equation

$$\log_4 x = \frac{5}{2} \qquad \text{Divide each side by 2.}$$

$$4^{\log_4 x} = 4^{5/2} \qquad \text{Exponentiate each side.}$$

$$x = 4^{5/2} \qquad \text{Inverse property}$$

$$x = 32 \qquad \text{Simplify.}$$

The solution is $x = 32$. Check this in the original equation, as follows.

$$2 \log_4 x = 5 \qquad \text{Original equation}$$

$$2 \log_4 (32) \overset{?}{=} 5 \qquad \text{Substitute 32 for } x.$$

$$2(2.5) \overset{?}{=} 5 \qquad \text{Use a calculator.}$$

$$5 = 5 \qquad \text{Solution checks. ✓}$$

b. $\dfrac{1}{4} \log_2 x = \dfrac{1}{2}$ — Original equation

$$\log_2 x = 2 \qquad \text{Multiply each side by 4.}$$

$$2^{\log_2 x} = 2^2 \qquad \text{Exponentiate each side.}$$

$$x = 4 \qquad \text{Inverse property}$$

Example 5   Solving a Logarithmic Equation

Solve $3 \log_{10} x = 6$.

Solution

$$3 \log_{10} x = 6 \qquad \text{Write original equation.}$$

$$\log_{10} x = 2 \qquad \text{Divide each side by 3.}$$

$$x = 10^2 \qquad \text{Exponential form}$$

$$x = 100 \qquad \text{Simplify.}$$

The solution is $x = 100$. Check this in the original equation.

Example 6   Solving a Logarithmic Equation

Solve $20 \ln 0.2x = 30$.

Solution

$$20 \ln 0.2x = 30 \qquad \text{Write original equation.}$$

$$\ln 0.2x = 1.5 \qquad \text{Divide each side by 20.}$$

$$e^{\ln 0.2x} = e^{1.5} \qquad \text{Exponentiate each side.}$$

$$0.2x = e^{1.5} \qquad \text{Inverse property}$$

$$x = 5e^{1.5} \qquad \text{Divide each side by 0.2.}$$

The solution is $x = 5e^{1.5} \approx 22.408$. Check this in the original equation.

Study Tip

When checking approximate solutions to exponential and logarithmic equations, be aware of the fact that because the solution is approximate, the check will not be exact.

The next two examples use logarithmic properties as part of the solutions.

Example 7   Solving a Logarithmic Equation

Solve $\log_3 2x - \log_3(x - 3) = 1$.

Solution

$$\log_3 2x - \log_3(x - 3) = 1 \qquad \text{Write original equation.}$$

$$\log_3 \frac{2x}{x - 3} = 1 \qquad \text{Condense the left side.}$$

$$\frac{2x}{x - 3} = 3^1 \qquad \text{Exponential form}$$

$$2x = 3x - 9 \qquad \text{Multiply each side by } x - 3.$$

$$-x = -9 \qquad \text{Subtract } 3x \text{ from each side.}$$

$$x = 9 \qquad \text{Divide each side by } -1.$$

The solution is $x = 9$. Check this in the original equation.

Example 8   Checking For Extraneous Solutions

$$\log_6 x + \log_6(x - 5) = 2 \qquad \text{Original equation}$$

$$\log_6 [x(x - 5)] = 2 \qquad \text{Condense the left side.}$$

$$x(x - 5) = 6^2 \qquad \text{Exponential form}$$

$$x^2 - 5x - 36 = 0 \qquad \text{Write in general form.}$$

$$(x - 9)(x + 4) = 0 \qquad \text{Factor.}$$

$$x - 9 = 0 \quad \Longrightarrow \quad x = 9 \qquad \text{Set 1st factor equal to 0.}$$

$$x + 4 = 0 \quad \Longrightarrow \quad x = -4 \qquad \text{Set 2nd factor equal to 0.}$$

From this, it appears that the solutions are $x = 9$ and $x = -4$. To be sure, you need to check each solution in the original equation, as follows.

First Solution	*Second Solution*

$$\log_6 (9) + \log_6 (9 - 5) \overset{?}{=} 2 \qquad\qquad \log_6 (-4) + \log_6 (-4 - 5) \overset{?}{=} 2$$

$$\log_6 (9 \cdot 4) \overset{?}{=} 2 \qquad\qquad\qquad \log_6 (-4) + \log_6 (-9) \neq 2 \;\times$$

$$\log_6 36 = 2 \;\checkmark$$

Of the two possible solutions, only $x = 9$ checks. So, $x = -4$ is extraneous.

---

4 Use exponential or logarithmic equations to solve application problems.

Application

Example 9   Compound Interest

A deposit of \$5000 is placed in a savings account for 2 years. The interest on the account is compounded continuously. At the end of 2 years, the balance in the account is \$5867.55. What is the annual interest rate for this account?

Solution

Using the formula for continuously compounded interest, $A = Pe^{rt}$, you have the following solution.

Formula: $A = Pe^{rt}$

Labels: Principal $= P = 5000$ (dollars)
Amount $= A = 5867.55$ (dollars)
Time $= t = 2$ (years)
Annual interest rate $= r$ (percent in decimal form)

Equation: $5867.55 = 5000e^{2r} \qquad$ Substitute for A, P, and t.

$$1.17351 = e^{2r} \qquad \text{Divide each side by 5000 and simplify.}$$

$$\ln 1.17351 = \ln(e^{2r}) \qquad \text{Take logarithm of each side.}$$

$$0.16 \approx 2r \quad \Longrightarrow \quad 0.08 \approx r \qquad \text{Inverse property}$$

The annual interest rate is approximately 8%. Check this solution.

11.5 Exercises

Review Concepts, Skills, and Problem Solving

Keep mathematically in shape by doing these exercises *before* the problems of this section.

Properties and Definitions

1. *Writing* Is it possible for the system
$$\begin{cases} 7x - 2y = 8 \\ x + y = 4 \end{cases}$$
to have exactly two solutions? Explain.

2. *Writing* Explain why the following system has no solution.
$$\begin{cases} 8x - 4y = 5 \\ -2x + y = 1 \end{cases}$$

Solving Equations

In Exercises 3–8, solve the equation.

3. $\frac{2}{3}x + \frac{2}{3} = 4x - 6$

4. $x^2 - 10x + 17 = 0$

5. $\frac{5}{2x} - \frac{4}{x} = 3$

6. $\frac{1}{x} + \frac{2}{x-5} = 0$

7. $|x - 4| = 3$

8. $\sqrt{x + 2} = 7$

Models and Graphing

9. *Distance* A train is traveling at 73 miles per hour. Write the distance d the train travels as a function of the time t. Graph the function.

10. ▲ *Geometry* The diameter of a right circular cylinder is 10 centimeters. Write the volume V of the cylinder as a function of its height h if the formula for its volume is $V = \pi r^2 h$. Graph the function.

11. ▲ *Geometry* The height of a right circular cylinder is 10 centimeters. Write the volume V of the cylinder as a function of its radius r if the formula for its volume is $V = \pi r^2 h$. Graph the function.

12. *Force* A force of 100 pounds stretches a spring 4 inches. Write the force F as a function of the distance x that the spring is stretched. Graph the function.

Developing Skills

In Exercises 1–6, determine whether the value of x is a solution of the equation.

1. $3^{2x-5} = 27$
   (a) $x = 1$
   (b) $x = 4$

2. $4^{x+3} = 16$
   (a) $x = -1$
   (b) $x = 0$

3. $e^{x+5} = 45$
   (a) $x = -5 + \ln 45$
   (b) $x = -5 + e^{45}$

4. $2^{3x-1} = 324$
   (a) $x \approx 3.1133$
   (b) $x \approx 2.4327$

5. $\log_9(6x) = \frac{3}{2}$
   (a) $x = 27$
   (b) $x = \frac{9}{2}$

6. $\ln(x + 3) = 2.5$
   (a) $x = -3 + e^{2.5}$
   (b) $x \approx 9.1825$

In Exercises 7–34, solve the equation. (Do not use a calculator.) See Example 1.

7. $7^x = 7^3$

8. $4^x = 4^6$

9. $e^{1-x} = e^4$

10. $9^{x+3} = 9^{10}$

11. $5^{x+6} = 25^5$

12. $2^{x-4} = 8^2$

13. $6^{2x} = 36$

14. $5^{3x} = 25$

15. $3^{2-x} = 81$

16. $4^{2x-1} = 64$

17. $5^x = \frac{1}{125}$

18. $3^x = \frac{1}{243}$

19. $2^{x+2} = \frac{1}{16}$

20. $3^{2-x} = 9$

21. $4^{x+3} = 32^x$

22. $9^{x-2} = 243^{x+1}$

23. $\ln 5x = \ln 22$

24. $\ln 3x = \ln 24$

25. $\log_6 3x = \log_6 18$

26. $\log_5 2x = \log_5 36$

27. $\ln(2x - 3) = \ln 15$

28. $\ln(2x - 3) = \ln 17$

29. $\log_2(x + 3) = \log_2 7$

30. $\log_4(x - 4) = \log_4 12$

31. $\log_5(2x - 3) = \log_5(4x - 5)$

32. $\log_3(4 - 3x) = \log_3(2x + 9)$

33. $\log_3(2 - x) = 2$ **34.** $\log_2(3x - 1) = 5$

In Exercises 35–38, simplify the expression.

35. $\ln e^{2x-1}$ **36.** $\log_3 3^{x^2}$

37. $10^{\log_{10} 2x}$

38. $e^{\ln(x+1)}$

In Exercises 39–82, solve the exponential equation. (Round your answer to two decimal places.) See Examples 2 and 3.

39. $3^x = 91$ **40.** $4^x = 40$

41. $5^x = 8.2$ **42.** $2^x = 3.6$

43. $6^{2x} = 205$ **44.** $4^{3x} = 168$

45. $7^{3y} = 126$ **46.** $5^{5y} = 305$

47. $3^{x+4} = 6$ **48.** $5^{3-x} = 15$

49. $10^{x+6} = 250$ **50.** $12^{x-1} = 324$

51. $3e^x = 42$ **52.** $6e^{-x} = 3$

53. $\frac{1}{4}e^x = 5$ **54.** $\frac{2}{3}e^x = 1$

55. $\frac{1}{2}e^{3x} = 20$ **56.** $4e^{-3x} = 6$

57. $250(1.04)^x = 1000$

58. $32(1.5)^x = 640$

59. $300e^{x/2} = 9000$

60. $6000e^{-2t} = 1200$

61. $1000^{0.12x} = 25{,}000$

62. $10{,}000e^{-0.1t} = 4000$

63. $\frac{1}{5}(4^{x+2}) = 300$

64. $3(2^{t+4}) = 350$

65. $6 + 2^{x-1} = 1$

66. $5^{x+6} - 4 = 12$

67. $7 + e^{2-x} = 28$

68. $9 + e^{5-x} = 32$

69. $8 - 12e^{-x} = 7$

70. $4 - 2e^x = -23$

71. $4 + e^{2x} = 10$ **72.** $10 + e^{4x} = 18$

73. $32 + e^{7x} = 46$

74. $50 - e^{x/2} = 35$

75. $23 - 5e^{x+1} = 3$

76. $2e^x + 5 = 115$

77. $4(1 + e^{x/3}) = 84$

78. $50(3 - e^{2x}) = 125$

79. $\dfrac{8000}{(1.03)^t} = 6000$

80. $\dfrac{5000}{(1.05)^x} = 250$

81. $\dfrac{300}{2 - e^{-0.15t}} = 200$

82. $\dfrac{500}{1 + e^{-0.1t}} = 400$

In Exercises 83–118, solve the logarithmic equation. (Round your answer to two decimal places.) See Examples 4–8.

83. $\log_{10} x = -1$ **84.** $\log_{10} x = 3$

85. $\log_3 x = 4.7$

86. $\log_5 x = 9.2$

87. $4 \log_3 x = 28$

88. $6 \log_2 x = 18$ **89.** $16 \ln x = 30$

90. $12 \ln x = 20$ **91.** $\log_{10} 4x = 2$

92. $\log_3 6x = 4$ **93.** $\ln 2x = 3$

94. $\ln(0.5t) = \frac{1}{4}$ **95.** $\ln x^2 = 6$

96. $\ln \sqrt{x} = 6.5$

97. $2 \log_4(x + 5) = 3$

98. $5 \log_{10}(x + 2) = 15$

99. $2 \log_8(x + 3) = 3$

100. $\frac{2}{3} \ln(x + 1) = -1$

101. $1 - 2 \ln x = -4$

102. $5 - 4 \log_2 x = 2$

103. $-1 + 3 \log_{10} \dfrac{x}{2} = 8$

104. $-5 + 2 \ln 3x = 5$

105. $\log_4 x + \log_4 5 = 2$

106. $\log_5 x - \log_5 4 = 2$

107. $\log_6(x + 8) + \log_6 3 = 2$

108. $\log_7(x - 1) - \log_7 4 = 1$

109. $\log_5(x + 3) - \log_5 x = 1$

110. $\log_3(x - 2) + \log_3 5 = 3$

111. $\log_{10} x + \log_{10}(x - 3) = 1$

112. $\log_{10} x + \log_{10}(x + 1) = 0$

113. $\log_2(x - 1) + \log_2(x + 3) = 3$

114. $\log_6(x - 5) + \log_6 x = 2$

115. $\log_4 3x + \log_4(x - 2) = \frac{1}{2}$

116. $\log_{10}(25x) - \log_{10}(x - 1) = 2$

117. $\log_2 x + \log_2(x + 2) - \log_2 3 = 4$

118. $\log_3 2x + \log_3(x - 1) - \log_3 4 = 1$

In Exercises 119–122, use a graphing calculator to approximate the x-intercept of the graph.

119. $y = 10^{x/2} - 5$

120. $y = 2e^x - 21$

121. $y = 6 \ln(0.4x) - 13$

122. $y = 5 \log_{10}(x + 1) - 3$

In Exercises 123–126, use a graphing calculator to solve the equation. (Round your answer to two decimal places.)

123. $e^x = 2$

124. $\ln x = 2$

125. $2 \ln(x + 3) = 3$

126. $1000e^{-x/2} = 200$

Solving Problems

127. *Compound Interest* A deposit of $10,000 is placed in a savings account for 2 years. The interest for the account is compounded continuously. At the end of 2 years, the balance in the account is $11,972.17. What is the annual interest rate for this account?

128. *Compound Interest* A deposit of $2500 is placed in a savings account for 2 years. The interest for the account is compounded continuously. At the end of 2 years, the balance in the account is $2847.07. What is the annual interest rate for this account?

129. *Doubling Time* Solve the exponential equation $5000 = 2500e^{0.09t}$ for t to determine the number of years for an investment of $2500 to double in value when compounded continuously at the rate of 9%.

130. *Doubling Rate* Solve the exponential equation $10,000 = 5000e^{10r}$ for r to determine the interest rate required for an investment of $5000 to double in value when compounded continuously for 10 years.

131. *Sound Intensity* The relationship between the number of decibels B and the intensity of a sound I in watts per centimeter squared is given by

$$B = 10 \log_{10}\left(\frac{I}{10^{-16}}\right).$$

Determine the intensity of a sound I if it registers 75 decibels on a decibel meter.

132. *Sound Intensity* The relationship between the number of decibels B and the intensity of a sound I in watts per centimeter squared is given by

$$B = 10 \log_{10}\left(\frac{I}{10^{-16}}\right).$$

Determine the intensity of a sound I if it registers 90 decibels on a decibel meter.

133. *Muon Decay* A muon is an elementary particle that is similar to an electron, but much heavier. Muons are unstable—they quickly decay to form electrons and other particles. In an experiment conducted in 1943, the number of muon decays m (of an original 5000 muons) was related to the time T by the model $T = 15.7 - 2.48 \ln m$, where T is in microseconds. How many decays were recorded when $T = 2.5$?

134. *Friction* In order to restrain an untrained horse, a person partially wraps a rope around a cylindrical post in a corral (see figure). The horse is pulling on the rope with a force of 200 pounds. The force F required by the person is $F = 200e^{-0.5\pi\theta/180}$, where F is in pounds and θ is the angle of wrap in degrees. Find the smallest value of θ if F cannot exceed 80 pounds.

135. *Human Memory Model*   The average score A for a group of students who took a test t months after the completion of a course is given by the human memory model $A = 80 - \log_{10}(t + 1)^{12}$. How long after completing the course will the average score fall to $A = 72$?

(a) Answer the question algebraically by letting $A = 72$ and solving the resulting equation.

(b) ▦ Answer the question graphically by using a graphing calculator to graph the equations $y_1 = 80 - \log_{10}(t + 1)^{12}$ and $y_2 = 72$, and finding the point(s) of intersection.

(c) Which strategy works better for this problem? Explain.

136. ▦ *Car Sales*   The number N (in billions of dollars) of car sales at new car dealerships for the years 1994 through 2001 is modeled by the equation $N = 322.2e^{0.0689t}$, for $4 \le t \le 11$, where t is time in years, with $t = 4$ corresponding to 1994. (Source: National Automobile Dealers Association)

(a) Use a graphing calculator to graph the equation over the specified domain.

(b) Use the graph in part (a) to estimate the value of t when $N = 580$.

137. *Newton's Law of Cooling*   You place a tray of water at 60°F in a freezer that is set at 0°F. The water cools according to Newton's Law of Cooling

$$kt = \ln \frac{T - S}{T_0 - S}$$

where T is the temperature of the water (in °F), t is the number of hours the tray is in the freezer, S is the temperature of the surrounding air, and T_0 is the original temperature of the water.

(a) The water freezes in 4 hours. What is the constant k? (*Hint:* Water freezes at 32°F.)

(b) You lower the temperature in the freezer to -10°F. At this temperature, how long will it take for the ice cubes to form?

(c) The initial temperature of the water is 50°F. The freezer temperature is 0°F. How long will it take for the ice cubes to form?

138. *Oceanography*   Oceanographers use the density d (in grams per cubic centimeter) of seawater to obtain information about the circulation of water masses and the rates at which waters of different densities mix. For water with a salinity of 30%, the water temperature T (in °C) is related to the density by

$$T = 7.9 \ln(1.0245 - d) + 61.84.$$

Find the densities of the subantarctic water and the antarctic bottom water shown in the figure.

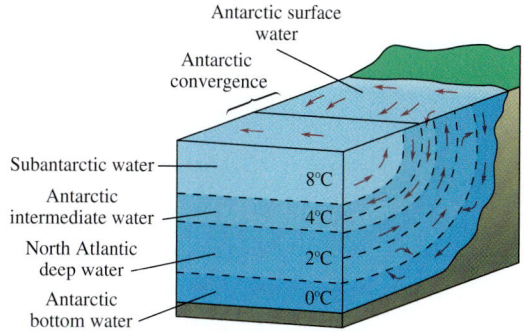

Figure for 138   This cross section shows complex currents at various depths in the South Atlantic Ocean off Antarctica.

Explaining Concepts

139. ◆ Answer parts (c)–(f) of Motivating the Chapter on page 678.

140. *Writing*   State the three basic properties of logarithms.

141. Which equation requires logarithms for its solution: $2^{x-1} = 32$ or $2^{x-1} = 30$?

142. *Writing*   Explain how to solve $10^{2x-1} = 5316$.

143. *Writing*   In your own words, state the guidelines for solving exponential and logarithmic equations.

11.6 Applications

Frank Siteman/PhotoEdit, Inc.

What You Should Learn

1 Use exponential equations to solve compound interest problems.

2 Use exponential equations to solve growth and decay problems.

3 Use logarithmic equations to solve intensity problems.

Why You Should Learn It

Exponential growth and decay models can be used in many real-life situations. For instance, in Exercise 55 on page 746, you will use an exponential growth model to represent the spread of a computer virus.

1 Use exponential equations to solve compound interest problems.

Compound Interest

In Section 11.1, you were introduced to two formulas for compound interest. Recall that in these formulas, A is the balance, P is the principal, r is the annual interest rate (in decimal form), and t is the time in years.

n Compoundings per Year

$$A = P\left(1 + \frac{r}{n}\right)^{nt}$$

Continuous Compounding

$$A = Pe^{rt}$$

Example 1   Finding the Annual Interest Rate

An investment of $50,000 is made in an account that compounds interest quarterly. After 4 years, the balance in the account is $71,381.07. What is the annual interest rate for this account?

Solution

Formula: $A = P\left(1 + \dfrac{r}{n}\right)^{nt}$

Labels:

Principal $= P = 50{,}000$	(dollars)
Amount $= A = 71{,}381.07$	(dollars)
Time $= t = 4$	(years)
Number of compoundings per year $= n = 4$	
Annual interest rate $= r$	(percent in decimal form)

Study Tip

Solving a power equation often requires "getting rid of" the exponent on the variable expression. This can be accomplished by raising each side of the equation to the *reciprocal* power. For instance, in Example 1 the variable expression had power 16, so each side was raised to the reciprocal power $\frac{1}{16}$.

Equation:

$$71{,}381.07 = 50{,}000\left(1 + \frac{r}{4}\right)^{(4)(4)}$$ Substitute for A, P, n, and t.

$$1.42762 \approx \left(1 + \frac{r}{4}\right)^{16}$$ Divide each side by 50,000.

$$(1.42762)^{1/16} \approx 1 + \frac{r}{4}$$ Raise each side to $\frac{1}{16}$ power.

$$1.0225 \approx 1 + \frac{r}{4}$$ Simplify.

$$0.09 \approx r$$ Subtract 1 from each side and then multiply each side by 4.

The annual interest rate is approximately 9%. Check this in the original problem.

Example 2   Doubling Time for Continuous Compounding

An investment is made in a trust fund at an annual interest rate of 8.75%, compounded continuously. How long will it take for the investment to double?

Solution

$A = Pe^{rt}$ — Formula for continuous compounding

$2P = Pe^{0.0875t}$ — Substitute known values.

$2 = e^{0.0875t}$ — Divide each side by P.

$\ln 2 = 0.0875t$ — Inverse property

$\dfrac{\ln 2}{0.0875} = t$ — Divide each side by 0.0875.

$7.92 \approx t$ — Use a calculator.

It will take approximately 7.92 years for the investment to double.

Check

$A = Pe^{rt}$ — Formula for continuous compounding

$2P \overset{?}{=} Pe^{0.0875(7.92)}$ — Substitute $2P$ for A, 0.0875 for r, and 7.92 for t.

$2P \overset{?}{=} Pe^{0.693}$ — Simplify.

$2P \approx 1.9997P$ — Solution checks. ✓

Example 3   Finding the Type of Compounding

You deposit $1000 in an account. At the end of 1 year, your balance is $1077.63. The bank tells you that the annual interest rate for the account is 7.5%. How was the interest compounded?

Solution

If the interest had been compounded continuously at 7.5%, the balance would have been

$$A = 1000e^{(0.075)(1)} = \$1077.88.$$

Because the actual balance is slightly less than this, you should use the formula for interest that is compounded n times per year.

$$A = 1000\left(1 + \frac{0.075}{n}\right)^n = 1077.63$$

At this point, it is not clear what you should do to solve the equation for n. However, by completing a table like the one shown below, you can see that $n = 12$. So, the interest was compounded monthly.

n	1	4	12	365
$1000\left(1 + \dfrac{0.075}{n}\right)^n$	1075	1077.14	1077.63	1077.88

In Example 3, notice that an investment of $1000 compounded monthly produced a balance of $1077.63 at the end of 1 year. Because $77.63 of this amount is interest, the **effective yield** for the investment is

$$\text{Effective yield} = \frac{\text{Year's interest}}{\text{Amount invested}} = \frac{77.63}{1000} = 0.07763 = 7.763\%.$$

In other words, the effective yield for an investment collecting compound interest is the *simple interest rate* that would yield the same balance at the end of 1 year.

Example 4   Finding the Effective Yield

An investment is made in an account that pays 6.75% interest, compounded continuously. What is the effective yield for this investment?

Solution

Notice that you do not have to know the principal or the time that the money will be left in the account. Instead, you can choose an arbitrary principal, such as $1000. Then, because effective yield is based on the balance at the end of 1 year, you can use the following formula.

$$A = Pe^{rt}$$

$$= 1000e^{0.0675(1)}$$

$$= 1069.83$$

Now, because the account would earn $69.83 in interest after 1 year for a principal of $1000, you can conclude that the effective yield is

$$\text{Effective yield} = \frac{69.83}{1000} = 0.06983 = 6.983\%.$$

---

2 Use exponential equations to solve growth and decay problems.

Growth and Decay

The balance in an account earning *continuously* compounded interest is one example of a quantity that increases over time according to the **exponential growth model** $y = Ce^{kt}$.

Exponential Growth and Decay

The mathematical model for exponential growth or decay is given by

$$y = Ce^{kt}.$$

For this model, t is the time, C is the original amount of the quantity, and y is the amount after time t. The number k is a constant that is determined by the rate of growth. If $k > 0$, the model represents **exponential growth,** and if $k < 0$, it represents **exponential decay.**

One common application of exponential growth is in modeling the growth of a population. Example 5 illustrates the use of the growth model

$$y = Ce^{kt}, \quad k > 0.$$

Example 5   Population Growth

The population of Texas was 17 million in 1990 and 21 million in 2000. What would you predict the population of Texas to be in 2010?   (Source: U.S. Census Bureau)

Solution

If you assumed a *linear growth model*, you would simply predict the population in the year 2010 to be 25 million because the population would increase by 4 million every 10 years. However, social scientists and demographers have discovered that *exponential growth models* are better than linear growth models for representing population growth. So, you can use the exponential growth model

$$y = Ce^{kt}.$$

In this model, let $t = 0$ represent 1990. The given information about the population can be described by the following table.

t (year)	0	10	20
Ce^{kt} (million)	$Ce^{k(0)} = 17$	$Ce^{k(10)} = 21$	$Ce^{k(20)} = ?$

To find the population when $t = 20$, you must first find the values of C and k. From the table, you can use the fact that $Ce^{k(0)} = Ce^0 = 17$ to conclude that $C = 17$. Then, using this value of C, you can solve for k as follows.

$Ce^{k(10)} = 21$	From table
$17e^{10k} = 21$	Substitute value of C.
$e^{10k} = \dfrac{21}{17}$	Divide each side by 17.
$10k = \ln \dfrac{21}{17}$	Inverse property
$k = \dfrac{1}{10} \ln \dfrac{21}{17}$	Divide each side by 10.
$k \approx 0.0211$	Simplify.

Finally, you can use this value of k in the model from the table for 2010 (for $t = 20$) to predict the population in the year 2010 to be

$$17e^{0.0211(20)} \approx 17(1.53) = 26.01 \text{ million.}$$

Figure 11.22 graphically compares the exponential growth model with a linear growth model.

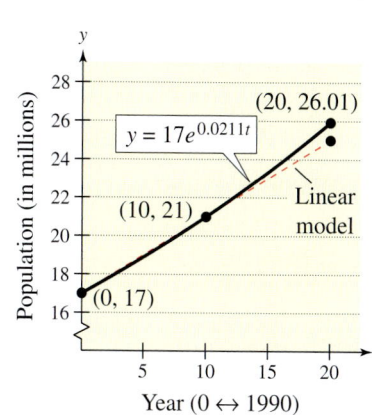

Population Models
Figure 11.22

Example 6   Radioactive Decay

Radioactive iodine is a by-product of some types of nuclear reactors. Its **half-life** is 60 days. That is, after 60 days, a given amount of radioactive iodine will have decayed to half the original amount. A nuclear accident occurs and releases 20 grams of radioactive iodine. How long will it take for the radioactive iodine to decay to a level of 1 gram?

Solution

To solve this problem, use the model for exponential decay.

$$y = Ce^{kt}$$

Next, use the information given in the problem to set up the following table.

t (days)	0	60	?
Ce^{kt} (grams)	$Ce^{k(0)} = 20$	$Ce^{k(60)} = 10$	$Ce^{k(t)} = 1$

Because $Ce^{k(0)} = Ce^0 = 20$, you can conclude that $C = 20$. Then, using this value of C, you can solve for k as follows.

$Ce^{k(60)} = 10$	From table
$20e^{60k} = 10$	Substitute value of C.
$e^{60k} = \dfrac{1}{2}$	Divide each side by 20.
$60k = \ln\dfrac{1}{2}$	Inverse property
$k = \dfrac{1}{60}\ln\dfrac{1}{2}$	Divide each side by 60.
$k \approx -0.01155$	Simplify.

Finally, you can use this value of k in the model from the table to find the time when the amount is 1 gram, as follows.

$Ce^{kt} = 1$	From table
$20e^{-0.01155t} = 1$	Substitute values of C and k.
$e^{-0.01155t} = \dfrac{1}{20}$	Divide each side by 20.
$-0.01155t = \ln\dfrac{1}{20}$	Inverse property
$t = \dfrac{1}{-0.01155}\ln\dfrac{1}{20}$	Divide each side by -0.01155.
$t \approx 259.4$ days	Simplify.

So, 20 grams of radioactive iodine will have decayed to 1 gram after about 259.4 days. This solution is shown graphically in Figure 11.23.

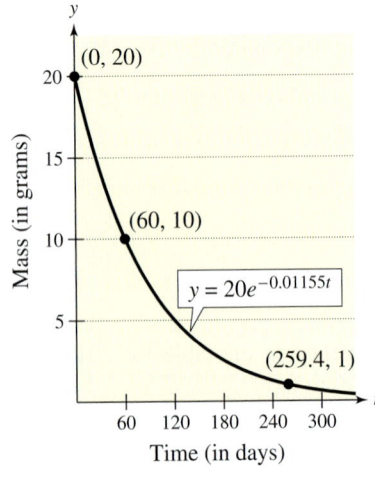

Radioactive Decay

Figure 11.23

Example 7   Website Growth

You created an algebra tutoring website in 2000. You have been keeping track of the number of hits (the number of visits to the site) for each year. In 2000, your website had 4080 hits, and in 2002, your website had 6120 hits. Use an exponential growth model to determine how many hits you can expect in 2008.

Solution

In the exponential growth model $y = Ce^{kt}$, let $t = 0$ represent 2000. Next, use the information given in the problem to set up the table shown at the left. Because $Ce^{k(0)} = Ce^0 = 4080$, you can conclude that $C = 4080$. Then, using this value of C, you can solve for k, as follows

t (year)	Ce^{kt}
0	$Ce^{k(0)} = 4080$
2	$Ce^{k(2)} = 6120$
8	$Ce^{k(8)} = ?$

$$Ce^{k(2)} = 6120 \qquad \text{From table}$$

$$4080e^{2k} = 6120 \qquad \text{Substitute value of } C.$$

$$e^{2k} = \tfrac{3}{2} \qquad \text{Divide each side by 4080.}$$

$$2k = \ln \tfrac{3}{2} \qquad \text{Inverse property}$$

$$k = \tfrac{1}{2} \ln \tfrac{3}{2} \approx 0.2027 \qquad \text{Divide each side by 2 and simplify.}$$

Finally, you can use this value of k in the model from the table to predict the number of hits in 2008 to be $4080e^{0.2027(8)} \approx 4080(5.06) \approx 20{,}645$ hits.

---

③ Use logarithmic equations to solve intensity problems.

Intensity Models

On the **Richter scale,** the magnitude R of an earthquake can be measured by the **intensity model** $R = \log_{10} I$, where I is the intensity of the shock wave.

Example 8   Earthquake Intensity

In 2001, Java, Indonesia experienced an earthquake that measured 5.0 on the Richter scale. In 2002, central Alaska experienced an earthquake that measured 7.9 on the Richter scale. Compare the intensities of these two earthquakes.

Solution

The intensity of the 2001 earthquake is given as follows.

$$5.0 = \log_{10} I \quad \Longrightarrow \quad 10^{5.0} = I \qquad \text{Inverse property}$$

The intensity of the 2002 earthquake can be found in a similar way.

$$7.9 = \log_{10} I \quad \Longrightarrow \quad 10^{7.9} = I \qquad \text{Inverse property}$$

The ratio of these two intensities is

$$\frac{I \text{ for } 2002}{I \text{ for } 2001} = \frac{10^{7.9}}{10^{5.0}} = 10^{7.9-5.0} = 10^{2.9} \approx 794.$$

So, the 2002 earthquake had an intensity that was about 794 times greater than the intensity of the 2001 earthquake.

11.6 Exercises

Review Concepts, Skills, and Problem Solving

Keep mathematically in shape by doing these exercises *before* the problems of this section.

Properties and Definitions

In Exercises 1–4, identify the type of variation given in the model.

1. $y = kx^2$

2. $y = \dfrac{k}{x}$

3. $z = kxy$

4. $z = \dfrac{kx}{y}$

Solving Systems

In Exercises 5–10, solve the system of equations.

5. $\begin{cases} x - y = 0 \\ x + 2y = 9 \end{cases}$

6. $\begin{cases} 2x + 5y = 15 \\ 3x + 6y = 20 \end{cases}$

7. $\begin{cases} y = x^2 \\ -3x + 2y = 2 \end{cases}$

8. $\begin{cases} x - y^3 = 0 \\ x - 2y^2 = 0 \end{cases}$

9. $\begin{cases} x - y = -1 \\ x + 2y - 2z = 3 \\ 3x - y + 2z = 1 \end{cases}$

10. $\begin{cases} 2x + y - 2z = 1 \\ x - z = 1 \\ 3x + 3y + z = 12 \end{cases}$

Graphs

In Exercises 11 and 12, use the function $y = -x^2 + 4x$.

11. (a) Does the graph open upward or downward? Explain.

(b) Find the x-intercepts algebraically.

(c) Find the coordinates of the vertex of the parabola.

12. 🖩 Use a graphing calculator to graph the function and verify the results of Exercise 11.

Solving Problems

Compound Interest In Exercises 1–6, find the annual interest rate. See Example 1.

Principal	Balance	Time	Compounding
1. $500	$1004.83	10 years	Monthly
2. $3000	$21,628.70	20 years	Quarterly
3. $1000	$36,581.00	40 years	Daily
4. $200	$314.85	5 years	Yearly
5. $750	$8267.38	30 years	Continuous
6. $2000	$4234.00	10 years	Continuous

Doubling Time In Exercises 7–12, find the time for the investment to double. Use a graphing calculator to verify the result graphically. See Example 2.

Principal	Rate	Compounding
7. $2500	7.5%	Monthly
8. $900	$5\frac{3}{4}\%$	Quarterly
9. $18,000	8%	Continuous
10. $250	6.5%	Yearly

Principal	Rate	Compounding
11. $1500	$7\frac{1}{4}\%$	Monthly
12. $600	9.75%	Continuous

Compound Interest In Exercises 13–16, determine the type of compounding. Solve the problem by trying the more common types of compounding. See Example 3.

Principal	Balance	Time	Rate
13. $750	$1587.75	10 years	7.5%
14. $10,000	$73,890.56	20 years	10%
15. $100	$141.48	5 years	7%
16. $4000	$4788.76	2 years	9%

Effective Yield In Exercises 17–24, find the effective yield. See Example 4.

Rate	Compounding
17. 8%	Continuous
18. 9.5%	Daily

Rate	Compounding
19. 7%	Monthly
20. 8%	Yearly
21. 6%	Quarterly
22. 9%	Quarterly
23. 8%	Monthly
24. $5\frac{1}{4}\%$	Daily

25. *Doubling Time* Is it necessary to know the principal P to find the doubling time in Exercises 7–12? Explain.

26. *Effective Yield*

(a) Is it necessary to know the principal P to find the effective yield in Exercises 17–24? Explain.

(b) When the interest is compounded more frequently, what inference can you make about the difference between the effective yield and the stated annual percentage rate?

Compound Interest In Exercises 27–34, find the principal that must be deposited in an account to obtain the given balance.

	Balance	Rate	Time	Compounding
27.	$10,000	9%	20 years	Continuous
28.	$5000	8%	5 years	Continuous
29.	$750	6%	3 years	Daily
30.	$3000	7%	10 years	Monthly
31.	$25,000	7%	30 years	Monthly
32.	$8000	6%	2 years	Monthly
33.	$1000	5%	1 year	Daily
34.	$100,000	9%	40 years	Daily

Monthly Deposits In Exercises 35–38, you make monthly deposits of P dollars in a savings account at an annual interest rate r, compounded continuously. Find the balance A after t years given that

$$A = \frac{P(e^{rt} - 1)}{e^{r/12} - 1}.$$

	Principal	Rate	Time
35.	$P = 30$	$r = 8\%$	$t = 10$ years
36.	$P = 100$	$r = 9\%$	$t = 30$ years
37.	$P = 50$	$r = 10\%$	$t = 40$ years
38.	$P = 20$	$r = 7\%$	$t = 20$ years

Monthly Deposits In Exercises 39 and 40, you make monthly deposits of $30 in a savings account at an annual interest rate of 8%, compounded continuously (see figure).

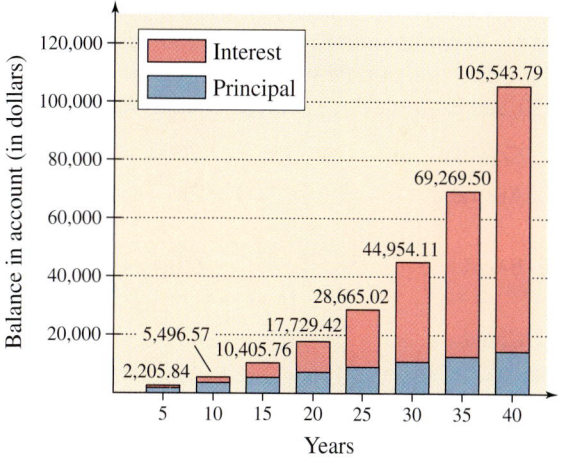

39. Find the total amount that has been deposited in the account in 20 years and the total interest earned.

40. Find the total amount that has been deposited in the account in 40 years and the total interest earned.

Exponential Growth and Decay In Exercises 41–44, find the constant k such that the graph of $y = Ce^{kt}$ passes through the points.

41.

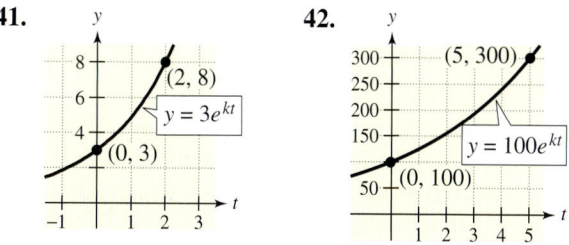

42.

43.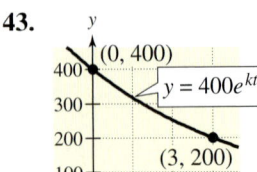

44.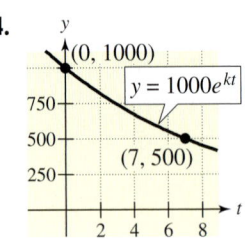

Population of a Country   In Exercises 45–52, the population (in millions) of a country for 2000 and the predicted population (in millions) for the year 2025 are given. Find the constants C and k to obtain the exponential growth model $y = Ce^{kt}$ for the population. (Let $t = 0$ correspond to the year 2000.) Use the model to predict the population of the region in the year 2030. See Example 5.   (Source: United Nations)

Country	2000	2025
45. Australia	19.1	23.5
46. Brazil	170.4	219.0
47. China	1275.1	1470.8
48. Japan	127.1	123.8
49. Ireland	3.8	4.7
50. Uruguay	3.3	3.9
51. United States of America	283.2	346.8
52. United Kingdom	59.4	61.2

53. *Rate of Growth*

(a) Compare the values of k in Exercises 47 and 49. Which is larger? Explain.

(b) What variable in the continuous compound interest formula is equivalent to k in the model for population growth? Use your answer to give an interpretation of k.

54. *World Population*   The figure shows the population P (in billions) of the world as projected by the U.S. Census Bureau. The bureau's projection can be modeled by

$$P = \frac{10.8}{1 + 1.03e^{-0.0283t}}$$

where $t = 0$ represents 1990. Use the model to estimate the population in 2020.

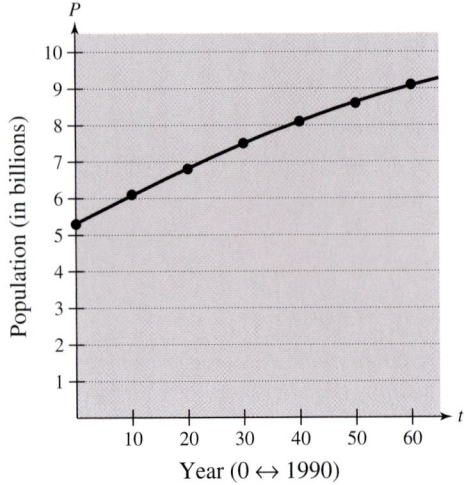

55. *Computer Virus*   A computer virus tends to spread at an exponential rate. In 2000, the number of computers infected by the "Love Bug" virus spread from 100 to about 1,000,000 in 2 hours.

(a) Find the constants C and k to obtain the exponential growth model $y = Ce^{kt}$ for the "Love Bug."

(b) Use your model from part (a) to estimate how long it took the "Love Bug" virus to spread to 80,000 computers.

56. *Painting Value*   In 1990, $82.5 million was paid for Vincent VanGogh's *Portrait of Dr. Gachet*. The same painting was sold for $58 million in 1987. Assume that the value of the painting increases at an exponential rate.

(a) Find the constants C and k to obtain the exponential growth model $y = Ce^{kt}$ for the value of Van Gogh's *Portrait of Dr. Gachet*.

(b) Use your model from part (a) to estimate the value of the painting in 2007.

(c) When will the value of the painting be $1 billion?

57. *Cellular Phones*   The number of users of cellular phones in the United States in 1995 was 33,786,000. In 2000, the number of users was 109,478,000. If it is assumed that the number of users of cellular phones grows at an exponential rate, how many users will there be in 2010? (Source: Cellular Telecommunications & Internet Association)

58. *DVDs*   The number of DVDs shipped by DVD manufacturers in the United States in 1998 was 500,000. In 2000, the number had increased to 3,300,000. If it is assumed that the number of DVDs shipped by manufacturers grows at an exponential rate, how many DVDs will be shipped in 2010? (Source: Recording Industry Association of America)

Radioactive Decay   In Exercises 59–64, complete the table for the radioactive isotopes. See Example 6.

Isotope	Half-Life (Years)	Initial Quantity	Amount After 1000 Years
59. ^{226}Ra	1620	6 g	▨ g
60. ^{226}Ra	1620	▨ g	0.25 g
61. ^{14}C	5730	▨ g	4.0 g
62. ^{14}C	5730	10 g	▨ g
63. ^{230}Pu	24,360	4.2 g	▨ g
64. ^{230}Pu	24,360	▨ g	1.5 g

65. *Radioactive Decay*   Radioactive radium (^{226}Ra) has a half-life of 1620 years. If you start with 5 grams of the isotope, how much will remain after 1000 years?

66. *Carbon 14 Dating*   Carbon 14 dating assumes that the carbon dioxide on Earth today has the same radioactive content as it did centuries ago. If this is true, the amount of ^{14}C absorbed by a tree that grew several centuries ago should be the same as the amount of ^{14}C absorbed by a tree growing today. A piece of ancient charcoal contains only 15% as much of the radioactive carbon as a piece of modern charcoal. How long ago did the tree burn to make the ancient charcoal if the half-life of ^{14}C is 5730 years? (Round your answer to the nearest 100 years.)

67. *Radioactive Decay*   The isotope ^{230}Pu has a half-life of 24,360 years. If you start with 10 grams of the isotope, how much will remain after 10,000 years?

68. *Radioactive Decay*   Carbon 14 (^{14}C) has a half-life of 5730 years. If you start with 5 grams of this isotope, how much will remain after 1000 years?

69. *Depreciation*   A sport utility vehicle that cost $34,000 new has a depreciated value of $26,000 after 1 year. Find the value of the sport utility vehicle when it is 3 years old by using the exponential model $y = Ce^{kt}$.

70. ▦ *Depreciation*   After x years, the value y of a recreational vehicle that cost $8000 new is given by

$$y = 8000(0.8)^x.$$

(a) Use a graphing calculator to graph the model.

(b) Graphically approximate the value of the recreational vehicle after 1 year.

(c) Graphically approximate the time when the recreational vehicle's value will be $4000.

Earthquake Intensity   In Exercises 71–74, compare the intensities of the two earthquakes. See Example 8.

Location	Date	Magnitude
71. Central Alaska	10/23/2002	6.7
Southern Italy	10/31/2002	5.9
72. Southern California	10/16/1999	7.2
Morocco	2/29/1960	5.8
73. Mexico City, Mexico	9/19/1985	8.1
New York	4/20/2002	5.0
74. Peru	6/23/2001	8.4
Armenia, USSR	12/7/1988	6.8

Acidity   In Exercises 75–78, use the acidity model pH $= -\log_{10}[H^+]$, where acidity (pH) is a measure of the hydrogen ion concentration $[H^+]$ (measured in moles of hydrogen per liter) of a solution.

75. Find the pH of a solution that has a hydrogen ion concentration of 9.2×10^{-8}.

76. Compute the hydrogen ion concentration if the pH of a solution is 4.7.

77. A blueberry has a pH of 2.5 and an antacid tablet has a pH of 9.5. The hydrogen ion concentration of the fruit is how many times the concentration of the tablet?

78. If the pH of a solution is decreased by 1 unit, the hydrogen ion concentration is increased by what factor?

79. *Population Growth*   The population p of a species of wild rabbit t years after it is introduced into a new habitat is given by

$$p(t) = \frac{5000}{1 + 4e^{-t/6}}.$$

(a) ⊞ Use a graphing calculator to graph the population function.

(b) Determine the size of the population of rabbits that was introduced into the habitat.

(c) Determine the size of the population of rabbits after 9 years.

(d) After how many years will the size of the population of rabbits be 2000?

80. ⊞ *Sales Growth*   Annual sales y of a personal digital assistant x years after it is introduced are approximated by

$$y = \frac{2000}{1 + 4e^{-x/2}}.$$

(a) Use a graphing calculator to graph the model.

(b) Use the graph in part (a) to approximate annual sales of this personal digital assistant model when $x = 4$.

(c) Use the graph in part (a) to approximate the time when annual sales of this personal digital assistant model are $y = 1100$ units.

(d) Use the graph in part (a) to estimate the maximum level that annual sales of this model will approach.

81. *Advertising Effect*   The sales S (in thousands of units) of a brand of jeans after spending x hundred dollars in advertising are given by

$$S = 10(1 - e^{kx}).$$

(a) Write S as a function of x if 2500 jeans are sold when $500 is spent on advertising.

(b) How many jeans will be sold if advertising expenditures are raised to $700?

82. *Advertising Effect*   The sales S of a video game after spending x thousand dollars in advertising are given by

$$S = 4500(1 - e^{kx}).$$

(a) Write S as a function of x if 2030 copies of the video game are sold when $10,000 is spent on advertising.

(b) How many copies of the video game will be sold if advertising expenditures are raised to $25,000?

Explaining Concepts

83. If the equation $y = Ce^{kt}$ models exponential growth, what must be true about k?

84. If the equation $y = Ce^{kt}$ models exponential decay, what must be true about k?

85. *Writing*   The formulas for periodic and continuous compounding have the four variables A, P, r, and t in common. Explain what each variable measures.

86. *Writing*   What is meant by the effective yield of an investment? Explain how it is computed.

87. *Writing*   In your own words, explain what is meant by the half-life of a radioactive isotope.

88. If the reading on the Richter scale is increased by 1, the intensity of the earthquake is increased by what factor?

What Did You Learn?

Key Terms

exponential function, *p. 680*
natural base, *p. 684*
natural exponential function,
  p. 684
composition, *p. 693*

inverse function, *p. 695*
one-to-one, *p. 695*
logarithmic function with base *a*,
  p. 707
common logarithmic function,
  p. 709

natural logarithmic function,
  p. 712
exponentiate, *p. 731*
exponential growth, *p. 740*
exponential decay, *p. 740*

Key Concepts

11.1 ⬤ Rules of exponential functions

1. $a^x \cdot a^y = a^{x+y}$

2. $\dfrac{a^x}{a^y} = a^{x-y}$

3. $(a^x)^y = a^{xy}$

4. $a^{-x} = \dfrac{1}{a^x} = \left(\dfrac{1}{a}\right)^x$

11.2 ⬤ Composition of two functions

The composition of two functions f and g is given by $(f \circ g)(x) = f(g(x))$. The domain of the composite function $(f \circ g)$ is the set of all x in the domain of g such that $g(x)$ is in the domain of f.

11.2 ⬤ Horizontal Line Test for inverse functions

A function f has an inverse function f^{-1} if and only if f is one-to-one. Graphically, a function f has an inverse function f^{-1} if and only if no horizontal line intersects the graph of f at more than one point.

11.2 ⬤ Finding an inverse function algebraically

1. In the equation for $f(x)$, replace $f(x)$ with y.

2. Interchange the roles of x and y.

3. If the new equation does not represent y as a function of x, the function f does not have an inverse function. If the new equation does represent y as a function of x, solve the new equation for y.

4. Replace y with $f^{-1}(x)$.

5. Verify that f and f^{-1} are inverse functions of each other by showing that $f(f^{-1}(x)) = x = f^{-1}(f(x))$.

11.3 ⬤ Properties of logarithms and natural logarithms

Let a and x be positive real numbers such that $a \neq 1$. Then the following properties are true.

1. $\log_a 1 = 0$ because $a^0 = 1$.
   $\ln 1 = 0$ because $e^0 = 1$.

2. $\log_a a = 1$ because $a^1 = a$.
   $\ln e = 1$ because $e^1 = e$.

3. $\log_a a^x = x$ because $a^x = a^x$.
   $\ln e^x = x$ because $e^x = e^x$.

11.3 ⬤ Change-of-base formula

Let a, b, and x be positive real numbers such that $a \neq 1$ and $b \neq 1$. Then $\log_a x = \dfrac{\log_b x}{\log_b a}$ or $\log_a x = \dfrac{\ln x}{\ln a}$.

11.4 ⬤ Properties of logarithms

Let a be a positive real number such that $a \neq 1$, and let n be a real number. If u and v are real numbers, variables, or algebraic expressions such that $u > 0$ and $v > 0$, the following properties are true.

Logarithm with base a	*Natural logarithm*
1. $\log_a(uv) = \log_a u + \log_a v$	$\ln(uv) = \ln u + \ln v$
2. $\log_a \dfrac{u}{v} = \log_a u - \log_a v$	$\ln \dfrac{u}{v} = \ln u - \ln v$
3. $\log_a u^n = n \log_a u$	$\ln u^n = n \ln u$

11.5 ⬤ One-to-one properties of exponential and logarithmic equations

Let a be a positive real number such that $a \neq 1$, and let x and y be real numbers. Then the following properties are true.

1. $a^x = a^y$ if and only if $x = y$.

2. $\log_a x = \log_a y$ if and only if $x = y (x > 0, y > 0)$.

11.5 ⬤ Inverse properties of exponents and logarithms

Base a	*Natural base e*
1. $\log_a(a^x) = x$	$\ln(e^x) = x$
2. $a^{(\log_a x)} = x$	$e^{(\ln x)} = x$

Review Exercises

11.1 Exponential Functions

1 Evaluate exponential functions.

In Exercises 1–4, evaluate the exponential function as indicated. (Round your answer to three decimal places.)

1. $f(x) = 2^x$
   (a) $x = -3$
   (b) $x = 1$
   (c) $x = 2$

2. $g(x) = 2^{-x}$
   (a) $x = -2$
   (b) $x = 0$
   (c) $x = 2$

3. $g(t) = 5^{-t/3}$
   (a) $t = -3$
   (b) $t = \pi$
   (c) $t = 6$

4. $h(s) = 1 - 3^{0.2s}$
   (a) $s = 0$
   (b) $s = 2$
   (c) $s = \sqrt{10}$

2 Graph exponential functions.

In Exercises 5–14, sketch the graph of the function. Identify the horizontal asymptote.

5. $f(x) = 3^x$
6. $f(x) = 3^{-x}$
7. $f(x) = 3^x - 1$
8. $f(x) = 3^x + 2$
9. $f(x) = 3^{(x+1)}$
10. $f(x) = 3^{(x-1)}$
11. $f(x) = 3^{x/2}$
12. $f(x) = 3^{-x/2}$
13. $f(x) = 3^{x/2} - 2$
14. $f(x) = 3^{x/2} + 3$

In Exercises 15–18, use a graphing calculator to graph the function.

15. $f(x) = 2^{-x^2}$
16. $g(x) = 2^{|x|}$
17. $y = 10(1.09)^t$
18. $y = 250(1.08)^t$

3 Evaluate the natural base e and graph natural exponential functions.

In Exercises 19 and 20, evaluate the exponential function as indicated. (Round your answer to three decimal places.)

19. $f(x) = 3e^{-2x}$
    (a) $x = 3$
    (b) $x = 0$
    (c) $x = -19$

20. $g(x) = e^{x/5} + 11$
    (a) $x = 12$
    (b) $x = -8$
    (c) $x = 18.4$

In Exercises 21–24, use a graphing calculator to graph the function.

21. $y = 5e^{-x/4}$
22. $y = 6 - e^{x/2}$
23. $f(x) = e^{x+2}$
24. $h(t) = \dfrac{8}{1 + e^{-t/5}}$

4 Use exponential functions to solve application problems.

Compound Interest In Exercises 25 and 26, complete the table to determine the balance A for P dollars invested at interest rate r for t years, compounded n times per year.

n	1	4	12	365	Continuous compounding
A					

Principal	*Rate*	*Time*
25. $P = \$5000$	$r = 10\%$	$t = 40$ years
26. $P = \$10,000$	$r = 9.5\%$	$t = 30$ years

27. *Radioactive Decay*   After t years, the remaining mass y (in grams) of 21 grams of a radioactive element whose half-life is 25 years is given by $y = 21(\frac{1}{2})^{t/25}$, $t \geq 0$. How much of the initial mass remains after 58 years?

28. *Depreciation*   After t years, the value of a truck that originally cost $29,000 depreciates so that each year it is worth $\frac{2}{3}$ of its value for the previous year. Find a model for $V(t)$, the value of the truck after t years. Sketch a graph of the model and determine the value of the truck 4 years after it was purchased.

11.2 Composite and Inverse Functions

① Form compositions of two functions and find the domains of composite functions.

In Exercises 29–32, find the compositions.

29. $f(x) = x + 2$, $g(x) = x^2$

   (a) $(f \circ g)(2)$ (b) $(g \circ f)(-1)$

30. $f(x) = \sqrt[3]{x}$, $g(x) = x + 2$

   (a) $(f \circ g)(6)$ (b) $(g \circ f)(64)$

31. $f(x) = \sqrt{x + 1}$, $g(x) = x^2 - 1$

   (a) $(f \circ g)(5)$ (b) $(g \circ f)(-1)$

32. $f(x) = \dfrac{1}{x - 5}$, $g(x) = \dfrac{5x + 1}{x}$

   (a) $(f \circ g)(1)$ (b) $(g \circ f)\left(\dfrac{1}{5}\right)$

In Exercises 33 and 34, find the compositions (a) $f \circ g$ and (b) $g \circ f$. Then find the domain of each composition.

33. $f(x) = \sqrt{x - 4}$, $g(x) = 2x$

34. $f(x) = \dfrac{2}{x - 4}$, $g(x) = x^2$

② Use the Horizontal Line Test to determine whether functions have inverse functions.

In Exercises 35–38, use the Horizontal Line Test to determine if the function is one-to-one and so has an inverse function.

35. $f(x) = x^2 - 25$

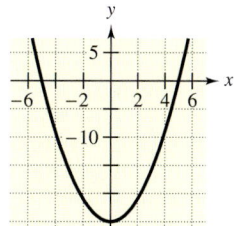

36. $f(x) = \frac{1}{4}x^3$

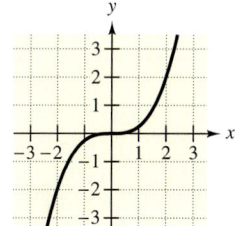

37. $h(x) = 4\sqrt[3]{x}$

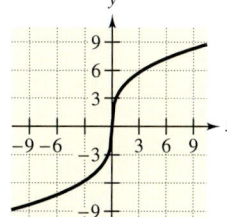

38. $g(x) = \sqrt{9 - x^2}$

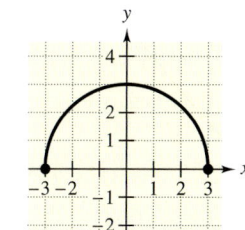

③ Find inverse functions algebraically.

In Exercises 39–44, find the inverse function.

39. $f(x) = 3x + 4$

40. $f(x) = 2x - 3$

41. $h(x) = \sqrt{x}$

42. $g(x) = x^2 + 2$, $x \geq 0$

43. $f(t) = t^3 + 4$

44. $h(t) = \sqrt[3]{t - 1}$

④ Graphically verify that two functions are inverse functions of each other.

In Exercises 45 and 46, use a graphing calculator to graph the functions in the same viewing window. Graphically verify that f and g are inverse functions of each other.

45. $f(x) = 3x + 4$

   $g(x) = \frac{1}{3}(x - 4)$

46. $f(x) = \frac{1}{3}\sqrt[3]{x}$

   $g(x) = 27x^3$

In Exercises 47–50, use the graph of f to sketch the graph of f^{-1}.

47.

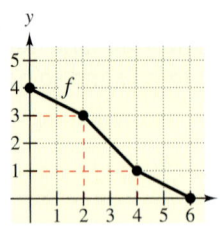

48.

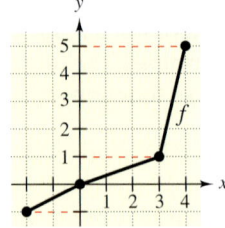

49.

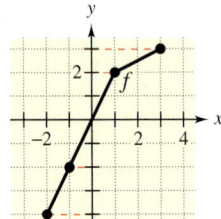

50.

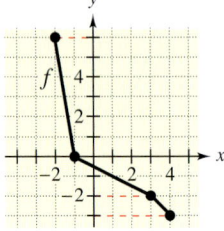

11.3 Logarithmic Functions

1 Evaluate logarithmic functions.

In Exercises 51–58, evaluate the logarithm.

51. $\log_{10} 1000$

52. $\log_9 3$

53. $\log_3 \frac{1}{9}$

54. $\log_4 \frac{1}{16}$

55. $\log_2 64$

56. $\log_{10} 0.01$

57. $\log_3 1$

58. $\log_2 \sqrt{4}$

2 Graph logarithmic functions.

In Exercises 59–64, sketch the graph of the function. Identify the vertical asymptote.

59. $f(x) = \log_3 x$

60. $f(x) = -\log_3 x$

61. $f(x) = -2 + \log_3 x$

62. $f(x) = 2 + \log_3 x$

63. $y = \log_2(x - 4)$

64. $y = \log_4(x + 1)$

3 Graph and evaluate natural logarithmic functions.

In Exercises 65 and 66, use your calculator to evaluate the natural logarithm.

65. $\ln e^7$

66. $\ln \frac{1}{e}$

In Exercises 67–70, sketch the graph of the function. Identify the vertical asymptote.

67. $y = \ln(x - 3)$

68. $y = -\ln(x + 2)$

69. $y = 5 - \ln x$

70. $y = 3 + \ln x$

4 Use the change-of-base formula to evaluate logarithms.

In Exercises 71–74, use a calculator to evaluate the logarithm by means of the change-of-base formula. Round your answer to four decimal places.

71. $\log_4 9$

72. $\log_{1/2} 5$

73. $\log_{12} 200$

74. $\log_3 0.28$

11.4 Properties of Logarithms

1 Use the properties of logarithms to evaluate logarithms.

In Exercises 75–80, use $\log_5 2 \approx 0.4307$ and $\log_5 3 \approx 0.6826$ to approximate the expression. Do not use a calculator.

75. $\log_5 18$

76. $\log_5 \sqrt{6}$

77. $\log_5 \frac{1}{2}$

78. $\log_5 \frac{2}{3}$

79. $\log_5(12)^{2/3}$

80. $\log_5(5^2 \cdot 6)$

2 Use the properties of logarithms to rewrite, expand, or condense logarithmic expressions.

In Exercises 81–88, use the properties of logarithms to expand the expression.

81. $\log_4 6x^4$

82. $\log_{10} 2x^{-3}$

83. $\log_5 \sqrt{x + 2}$

84. $\ln \sqrt[3]{\dfrac{x}{5}}$

85. $\ln \dfrac{x + 2}{x - 2}$

86. $\ln x(x - 3)^2$

87. $\ln\left[\sqrt{2x}(x + 3)^5 \right]$

88. $\log_3 \dfrac{a^2 \sqrt{b}}{cd^5}$

In Exercises 89–98, use the properties of logarithms to condense the expression.

89. $-\frac{2}{3} \ln 3y$

90. $5 \log_2 y$

91. $\log_8 16x + \log_8 2x^2$

92. $\log_4 6x - \log_4 10$

93. $-2(\ln 2x - \ln 3)$

94. $4(1 + \ln x + \ln x)$

95. $4\left[\log_2 k - \log_2(k - t)\right]$

96. $\frac{1}{3}(\log_8 a + 2 \log_8 b)$

97. $3 \ln x + 4 \ln y + \ln z$

98. $\ln(x + 4) - 3 \ln x - \ln y$

True or False? **In Exercises 99–104, use properties of logarithms to determine whether the equation is true or false. If it is false, state why or give an example to show that it is false.**

99. $\log_2 4x = 2 \log_2 x$

100. $\dfrac{\ln 5x}{\ln 10x} = \ln \dfrac{1}{2}$

101. $\log_{10} 10^{2x} = 2x$

102. $e^{\ln t} = t$

103. $\log_4 \dfrac{16}{x} = 2 - \log_4 x$

104. $6 \ln x + 6 \ln y = \ln(xy)^6$

3 Use the properties of logarithms to solve application problems.

105. *Light Intensity*   The intensity of light y as it passes through a medium is given by

$$y = \ln\left(\dfrac{I_0}{I}\right)^{0.83}.$$

Use properties of logarithms to write the formula in simpler form, and determine the intensity of light passing through this medium when $I_0 = 4.2$ and $I = 3.3$.

106. *Human Memory Model*   A psychologist finds that the percent p of retention in a group of subjects can be modeled by

$$p = \dfrac{\log_{10}(10^{68})}{\log_{10}(t + 1)^{20}}$$

where t is the time in months from the subjects' initial testing. Use properties of logarithms to write the formula in simpler form, and determine the percent of retention after 5 months.

11.5 Solving Exponential and Logarithmic Equations

1 Solve basic exponential and logarithmic equations.

In Exercises 107–112, solve the equation.

107. $2^x = 64$

108. $5^x = 25$

109. $4^{x-3} = \frac{1}{16}$

110. $3^{x-2} = 81$

111. $\log_7(x + 6) = \log_7 12$

112. $\ln(5 - x) = \ln(8)$

2 Use inverse properties to solve exponential equations.

In Exercises 113–118, solve the exponential equation. (Round your answer to two decimal places.)

113. $3^x = 500$

114. $8^x = 1000$

115. $2e^{0.5x} = 45$

116. $100e^{-0.6x} = 20$

117. $12(1 - 4^x) = 18$

118. $25(1 - e^t) = 12$

3 Use inverse properties to solve logarithmic equations.

In Exercises 119–128, solve the logarithmic equation. (Round your answer to two decimal places.)

119. $\ln x = 7.25$

120. $\ln x = -0.5$

121. $\log_{10} 2x = 1.5$

122. $\log_2 2x = -0.65$

123. $\log_3(2x + 1) = 2$

124. $\log_5(x - 10) = 2$

125. $\frac{1}{3} \log_2 x + 5 = 7$

126. $4 \log_5(x + 1) = 4.8$

127. $\log_2 x + \log_2 3 = 3$

128. $2 \log_4 x - \log_4(x - 1) = 1$

4 Use exponential or logarithmic equations to solve application problems.

129. *Compound Interest*   A deposit of $5000 is placed in a savings account for 2 years. The interest for the account is compounded continuously. At the end of 2 years, the balance in the account is $5751.37. What is the annual interest rate for this account?

130. *Sound Intensity*   The relationship between the number of decibels B and the intensity of a sound I in watts per centimeter squared is given by

$$B = 10 \log_{10}\left(\frac{I}{10^{-16}}\right).$$

Determine the intensity of a sound I if it registers 125 decibels on a decibel meter.

11.6 Applications

1 Use exponential equations to solve compound interest problems.

Annual Interest rate   **In Exercises 131–136, find the annual interest rate.**

	Principal	Balance	Time	Compounding
131.	$250	$410.90	10 years	Quarterly
132.	$1000	$1348.85	5 years	Monthly
133.	$5000	$15,399.30	15 years	Daily
134.	$10,000	$35,236.45	20 years	Yearly

	Principal	Balance	Time	Compounding
135.	$1500	$24,666.97	40 years	Continuous
136.	$7500	$15,877.50	15 years	Continuous

Effective Yield   **In Exercises 137–142, find the effective yield.**

	Rate	Compounding
137.	5.5%	Daily
138.	6%	Monthly
139.	7.5%	Quarterly
140.	8%	Yearly
141.	7.5%	Continuously
142.	4%	Continuously

2 Use exponential equations to solve growth and decay problems.

Radioactive Decay   **In Exercises 143–148, complete the table for the radioactive isotopes.**

	Isotope	Half-Life (Years)	Initial Quantity	Amount After 1000 Years
143.	^{226}Ra	1620	3.5 g	▨ g
144.	^{226}Ra	1620	▨ g	0.5 g
145.	^{14}C	5730	▨ g	2.6 g
146.	^{14}C	5730	10 g	▨ g
147.	^{230}Pu	24,360	5 g	▨ g
148.	^{230}Pu	24,360	▨ g	2.5 g

3 Use logarithmic equations to solve intensity problems.

In Exercises 149 and 150, compare the intensities of the two earthquakes.

	Location	Date	Magnitude
149.	San Francisco, California	4/18/1906	8.3
	Napa, California	9/3/2000	4.9
150.	El Salvador	2/13/2001	6.5
	Colombia	1/25/1999	5.7

Chapter Test

Take this test as you would take a test in class. After you are done, check your work against the answers in the back of the book.

1. Evaluate $f(t) = 54\left(\frac{2}{3}\right)^t$ when $t = -1, 0, \frac{1}{2}$, and 2.

2. Sketch a graph of the function $f(x) = 2^{x/3}$ and identify the horizontal asymptote.

3. Find the compositions (a) $f \circ g$ and (b) $g \circ f$. Then find the domain of each composition.

$$f(x) = 2x^2 + x \qquad g(x) = 5 - 3x$$

4. Find the inverse function of $f(x) = 5x + 6$.

5. Verify algebraically that the functions f and g are inverse functions of each other.

$$f(x) = -\tfrac{1}{2}x + 3, \qquad g(x) = -2x + 6$$

6. Evaluate $\log_8 2$ without a calculator.

7. Describe the relationship between the graphs of $f(x) = \log_5 x$ and $g(x) = 5^x$.

8. Use the properties of logarithms to expand $\log_4\left(5x^2/\sqrt{y}\right)$.

9. Use the properties of logarithms to condense $\ln x - 4 \ln y$.

In Exercises 10–17, solve the equation. Round your answer to two decimal places, if necessary.

10. $\log_2 x = 5$

11. $9^{2x} = 182$

12. $400e^{0.08t} = 1200$

13. $3 \ln(2x - 3) = 10$

14. $8(2 - 3^x) = -56$

15. $\log_2 x + \log_2 4 = 5$

16. $\ln x - \ln 2 = 4$

17. $30(e^x + 9) = 300$

18. Determine the balance after 20 years if $2000 is invested at 7% compounded (a) quarterly and (b) continuously.

19. Determine the principal that will yield $100,000 when invested at 9% compounded quarterly for 25 years.

20. A principal of $500 yields a balance of $1006.88 in 10 years when the interest is compounded continuously. What is the annual interest rate?

21. A car that cost $18,000 new has a depreciated value of $14,000 after 1 year. Find the value of the car when it is 3 years old by using the exponential model $y = Ce^{kt}$.

In Exercises 22–24, the population p of a species of fox t years after it is introduced into a new habitat is given by

$$p(t) = \frac{2400}{1 + 3e^{-t/4}}.$$

22. Determine the population size that was introduced into the habitat.

23. Determine the population after 4 years.

24. After how many years will the population be 1200?

755

Motivating the Chapter

⟳ Postal Delivery Route

You are a mail carrier for a post office that receives mail for everyone living within a five-mile radius. Your route covers the portions of Anderson Road and Murphy Road that pass through this region.

See Section 12.1, Exercise 101.

a. Assume that the post office is located at the point $(0, 0)$. Write an equation for the circle that bounds the region where the mail is delivered.

b. Sketch the graph of the circular region serviced by the post office.

See Section 12.3, Exercise 45.

c. Assume that Anderson Road follows one branch of a hyperbolic path given by $x^2 - y^2 - 4x - 23 = 0$. Find the center and vertices of this hyperbola.

d. On the same set of coordinate axes as the circular region, sketch the graph of the hyperbola that represents Anderson Road.

See Section 12.4, Exercise 85.

e. You begin your delivery on Anderson Road at the point $(-4, -3)$. Where on Anderson Road will you end your delivery? Explain.

f. You finish delivery on Anderson Road at the point where it intersects both the circular boundary of the post office and Murphy Road. At the intersection, you begin delivering on Murphy Road, which is a straight road that cuts through the center of the circular boundary and continues past the post office. Find the equation that represents Murphy Road.

g. Where on Murphy Road will you end your delivery? Explain.

Bill Aron/PhotoEdit, Inc.

Conics

12.1 Circles and Parabolas

What You Should Learn

1 Recognize the four basic conics: circles, parabolas, ellipses, and hyperbolas.

2 Graph and write equations of circles centered at the origin.

3 Graph and write equations of circles centered at (h, k).

4 Graph and write equations of parabolas.

Why You Should Learn It

Circles can be used to model and solve scientific problems. For instance, in Exercise 93 on page 768, you will write an equation of the circular orbit of a satellite.

1 Recognize the four basic conics: circles, parabolas, ellipses, and hyperbolas.

The Conics

In Section 10.4, you saw that the graph of a second-degree equation of the form $y = ax^2 + bx + c$ is a parabola. A parabola is one of four types of **conics** or **conic sections.** The other three types are circles, ellipses, and hyperbolas. All four types have equations of second degree. Each figure can be obtained by intersecting a plane with a double-napped cone, as shown in Figure 12.1.

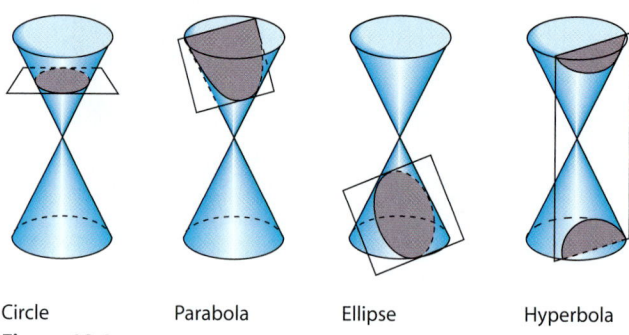

Circle   Parabola   Ellipse   Hyperbola
Figure 12.1

Conics occur in many practical applications. Reflective surfaces in satellite dishes, flashlights, and telescopes often have a parabolic shape. The orbits of planets are elliptical, and the orbits of comets are usually elliptical or hyperbolic. Ellipses and parabolas are also used in building archways and bridges.

2 Graph and write equations of circles centered at the origin.

Circles Centered at the Origin

Definition of a Circle

A **circle** in the rectangular coordinate system consists of all points (x, y) that are a given positive distance r from a fixed point, called the **center** of the circle. The distance r is called the **radius** of the circle.

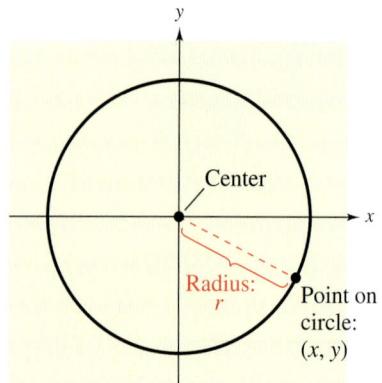

Figure 12.2

If the center of the circle is the origin, as shown in Figure 12.2, the relationship between the coordinates of any point (x, y) on the circle and the radius r is

$$r = \sqrt{(x - 0)^2 + (y - 0)^2} = \sqrt{x^2 + y^2}. \quad \text{Distance Formula}$$

By squaring each side of this equation, you obtain the equation below, which is called the **standard form of the equation of a circle centered at the origin.**

> ### Standard Equation of a Circle (Center at Origin)
>
> The **standard form of the equation of a circle centered at the origin** is
>
> $$x^2 + y^2 = r^2.$$ Circle with center at $(0, 0)$
>
> The positive number r is called the **radius** of the circle.

Example 1   Writing an Equation of a Circle

Write an equation of the circle that is centered at the origin and has a radius of 2 (See Figure 12.3).

Solution

Using the standard form of the equation of a circle (with center at the origin) and $r = 2$, you obtain

$x^2 + y^2 = r^2$ Standard form with center at $(0, 0)$

$x^2 + y^2 = 2^2$ Substitute 2 for r.

$x^2 + y^2 = 4.$ Equation of circle

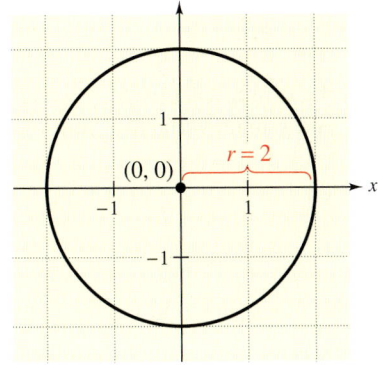

Figure 12.3

To sketch the circle for a given equation, first write the equation in standard form. Then, from the standard form, you can identify the center and radius and sketch the circle.

Example 2   Sketching a Circle

Identify the center and radius of the circle given by the equation $4x^2 + 4y^2 - 25 = 0$. Then sketch the circle.

Solution

Begin by writing the equation in standard form.

$4x^2 + 4y^2 - 25 = 0$ Write original equation.

$4x^2 + 4y^2 = 25$ Add 25 to each side.

$x^2 + y^2 = \dfrac{25}{4}$ Divide each side by 4.

$x^2 + y^2 = \left(\dfrac{5}{2}\right)^2$ Standard form

Now, from this standard form, you can see that the graph of the equation is a circle that is centered at the origin and has a radius of $\frac{5}{2}$. The graph of the equation of the circle is shown in Figure 12.4.

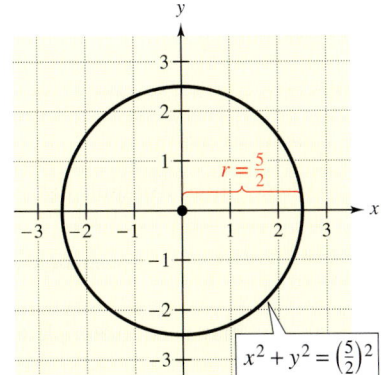

Figure 12.4

③ Graph and write equations of circles centered at (h, k).

Circles Centered at (h, k)

Consider a circle whose radius is r and whose center is the point (h, k), as shown in Figure 12.5. Let (x, y) be any point on the circle. To find an equation for this circle, you can use a variation of the Distance Formula and write

$$\text{Radius} = r = \sqrt{(x - h)^2 + (y - k)^2}. \qquad \text{Distance Formula}$$

By squaring each side of this equation, you obtain the equation shown below, which is called the **standard form of the equation of a circle centered at (h, k).**

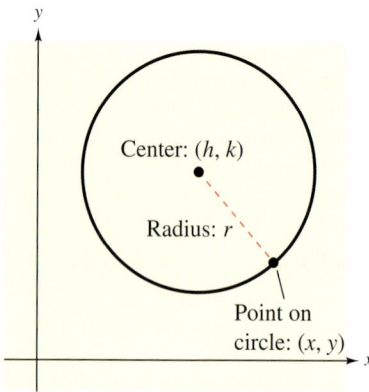

Figure 12.5

Standard Equation of a Circle [Center at (h, k)]

The **standard form of the equation of a circle centered at (h, k)** is

$$(x - h)^2 + (y - k)^2 = r^2.$$

When $h = 0$ and $k = 0$, the circle is centered at the origin. Otherwise, you can shift the center of the circle h units horizontally and k units vertically from the origin.

Example 3   Writing an Equation of a Circle

The point $(2, 5)$ lies on a circle whose center is $(5, 1)$, as shown in Figure 12.6. Write the standard form of the equation of this circle.

Solution

The radius r of the circle is the distance between $(2, 5)$ and $(5, 1)$.

$$r = \sqrt{(2 - 5)^2 + (5 - 1)^2} \qquad \text{Distance Formula}$$

$$= \sqrt{(-3)^2 + 4^2} \qquad \text{Simplify.}$$

$$= \sqrt{9 + 16} \qquad \text{Simplify.}$$

$$= \sqrt{25} \qquad \text{Simplify.}$$

$$= 5 \qquad \text{Radius}$$

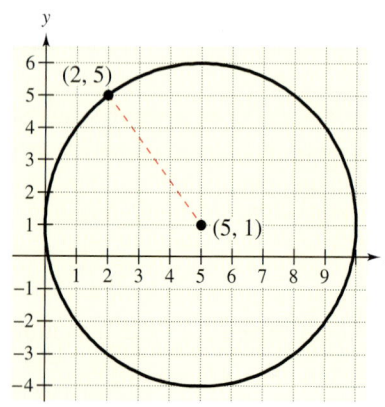

Figure 12.6

Using $(h, k) = (5, 1)$ and $r = 5$, the equation of the circle is

$$(x - h)^2 + (y - k)^2 = r^2 \qquad \text{Standard form}$$

$$(x - 5)^2 + (y - 1)^2 = 5^2 \qquad \text{Substitute for } h, k, \text{ and } r.$$

$$(x - 5)^2 + (y - 1)^2 = 25. \qquad \text{Equation of circle}$$

From the graph, you can see that the center of the circle is shifted five units to the right and one unit upward from the origin.

To write the equation of a circle in standard form, you may need to complete the square, as demonstrated in Example 4.

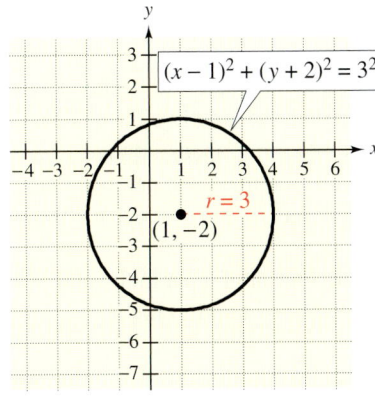

Figure 12.7

Example 4   Writing an Equation in Standard Form

Write the equation

$$x^2 + y^2 - 2x + 4y - 4 = 0$$

in standard form, and sketch the circle represented by the equation.

Solution

$x^2 + y^2 - 2x + 4y - 4 = 0$	Write original equation.
$(x^2 - 2x + \quad) + (y^2 + 4y + \quad) = 4$	Group terms.
$[x^2 - 2x + (-1)^2] + (y^2 + 4y + 2^2) = 4 + 1 + 4$	Complete the square.
	(half)² (half)²
$(x - 1)^2 + (y + 2)^2 = 3^2$	Standard form

From this standard form, you can see that the circle has a radius of 3 and that the center of the circle is $(1, -2)$. The graph of the equation of the circle is shown in Figure 12.7. From the graph you can see that the center of the circle is shifted one unit to the right and two units downward from the origin.

Example 5   An Application: Mechanical Drawing

You are in a mechanical drawing class and are asked to help program a computer to model the metal piece shown in Figure 12.8. Part of your assignment is to find an equation for the semicircular upper portion of the hole in the metal piece. What is the equation?

Solution

From the drawing, you can see that the center of the circle is $(h, k) = (5, 2)$ and that the radius of the circle is $r = 1.5$. This implies that the equation of the entire circle is

$(x - h)^2 + (y - k)^2 = r^2$	Standard form
$(x - 5)^2 + (y - 2)^2 = 1.5^2$	Substitute for h, k, and r.
$(x - 5)^2 + (y - 2)^2 = 2.25.$	Equation of circle

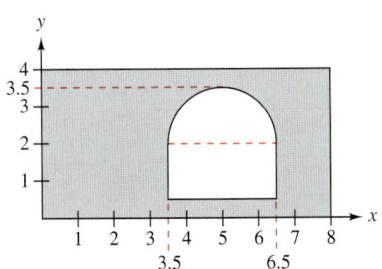

Figure 12.8

To find the equation of the upper portion of the circle, solve this standard equation for y.

$$(x - 5)^2 + (y - 2)^2 = 2.25$$
$$(y - 2)^2 = 2.25 - (x - 5)^2$$
$$y - 2 = \pm\sqrt{2.25 - (x - 5)^2}$$
$$y = 2 \pm \sqrt{2.25 - (x - 5)^2}$$

Finally, take the positive square root to obtain the equation of the upper portion of the circle.

$$y = 2 + \sqrt{2.25 - (x - 5)^2}$$

Study Tip

In Example 5, if you had wanted the equation of the lower portion of the circle, you would have taken the negative square root

$$y = 2 - \sqrt{2.25 - (x - 5)^2}.$$

④ Graph and write equations of parabolas.

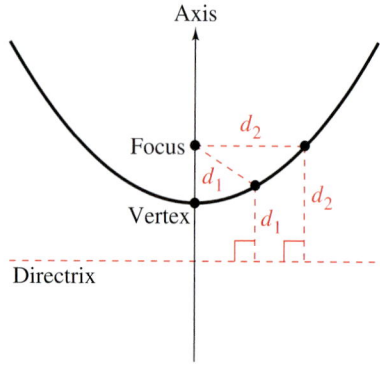

Axis

Focus

d_2

d_1

d_1

d_2

Vertex

Directrix

Figure 12.9

Equations of Parabolas

The second basic type of conic is a **parabola.** In Section 10.4, you studied some of the properties of parabolas. There you saw that the graph of a quadratic function of the form $y = ax^2 + bx + c$ is a parabola that opens upward if a is positive and downward if a is negative. You also learned that each parabola has a vertex and that the vertex of the graph of $y = ax^2 + bx + c$ occurs when $x = -b/(2a)$.

In this section, you will study the technical definition of a parabola, and you will study the equations of parabolas that open to the right and to the left.

Definition of a Parabola

A **parabola** is the set of all points (x, y) that are equidistant from a fixed line (**directrix**) and a fixed point (**focus**) not on the line.

The midpoint between the focus and the directrix is called the **vertex,** and the line passing through the focus and the vertex is called the **axis** of the parabola. Note in Figure 12.9 that a parabola is symmetric with respect to its axis. Using the definition of a parabola, you can derive the **standard form of the equation of a parabola** whose directrix is parallel to the x-axis or to the y-axis.

Standard Equation of a Parabola

The **standard form of the equation of a parabola** with vertex at the origin $(0, 0)$ is

$x^2 = 4py, \quad p \neq 0$ Vertical axis

$y^2 = 4px, \quad p \neq 0.$ Horizontal axis

The focus lies on the axis p units (*directed distance*) from the vertex. If the vertex is at (h, k), then the standard form of the equation is

$(x - h)^2 = 4p(y - k), \quad p \neq 0$ Vertical axis; directrix: $y = k - p$

$(y - k)^2 = 4p(x - h), \quad p \neq 0.$ Horizontal axis; directrix: $x = h - p$

(See Figure 12.10.)

Study Tip

If the focus of a parabola is above or to the right of the vertex, p is positive. If the focus is below or to the left of the vertex, p is negative.

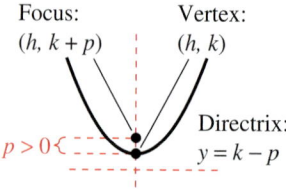

Focus: $(h, k + p)$ Vertex: (h, k)

$p > 0$ Directrix: $y = k - p$

Parabola with vertical axis

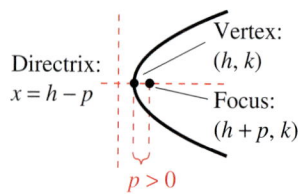

Directrix: $x = h - p$ Vertex: (h, k)

Focus: $(h + p, k)$

$p > 0$

Parabola with horizontal axis

Figure 12.10

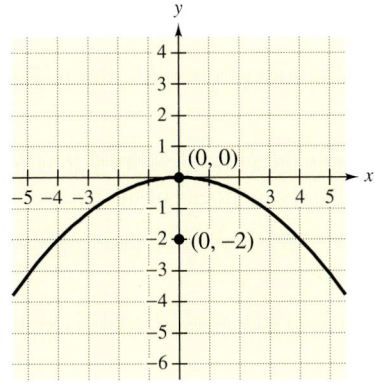

Figure 12.11

Example 6   Writing the Standard Equation of a Parabola

Write the standard form of the equation of the parabola with vertex $(0, 0)$ and focus $(0, -2)$, as shown in Figure 12.11.

Solution

Because the vertex is at the origin and the axis of the parabola is vertical, consider the equation

$$x^2 = 4py$$

where p is the directed distance from the vertex to the focus. Because the focus is two units *below* the vertex, you have $p = -2$. So, the equation of the parabola is

$x^2 = 4py$	Standard form
$x^2 = 4(-2)y$	Substitute for p.
$x^2 = -8y.$	Equation of parabola

Example 7   Writing the Standard Equation of a Parabola

Write the standard form of the equation of the parabola with vertex $(3, -2)$ and focus $(4, -2)$, as shown in Figure 12.12.

Solution

Because the vertex is at $(h, k) = (3, -2)$ and the axis of the parabola is horizontal, consider the equation

$$(y - k)^2 = 4p(x - h)$$

where $h = 3$, $k = -2$, and $p = 1$. So, the equation of the parabola is

$(y - k)^2 = 4p(x - h)$	Standard form
$[y - (-2)]^2 = 4(1)(x - 3)$	Substitute for h, k, and p.
$(y + 2)^2 = 4(x - 3).$	

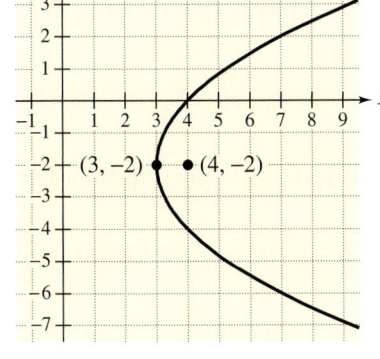

Figure 12.12

Technology: Tip

You cannot represent a circle or a parabola as a single function of x. You can, however, represent it by two functions of x. For instance, try using a graphing calculator to graph the equations below in the same viewing window. Use a viewing window in which $-1 \le x \le 10$ and $-8 \le y \le 4$. Do you obtain a parabola? Does the graphing calculator connect the two portions of the parabola?

$y_1 = -2 + 2\sqrt{x - 3}$	Upper portion of parabola
$y_2 = -2 - 2\sqrt{x - 3}$	Lower portion of parabola

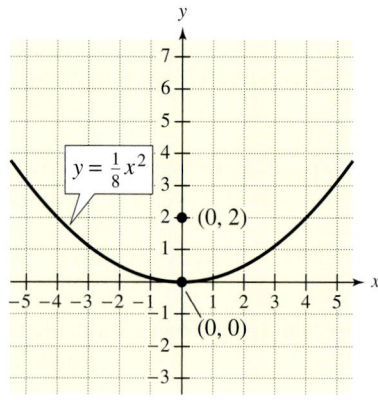

Figure 12.13

Example 8   Analyzing a Parabola

Sketch the graph of the parabola $y = \frac{1}{8}x^2$ and identify its vertex and focus.

Solution

Because the equation can be written in the standard form $x^2 = 4py$, it is a parabola whose vertex is at the origin. You can identify the focus of the parabola by writing its equation in standard form.

$y = \frac{1}{8}x^2$	Write original equation.
$\frac{1}{8}x^2 = y$	Interchange sides of the equation.
$x^2 = 8y$	Multiply each side by 8.
$x^2 = 4(2)y$	Rewrite 8 in the form $4p$.

From this standard form, you can see that $p = 2$. Because the parabola opens upward, as shown in Figure 12.13, you can conclude that the focus lies $p = 2$ units above the vertex. So, the focus is $(0, 2)$.

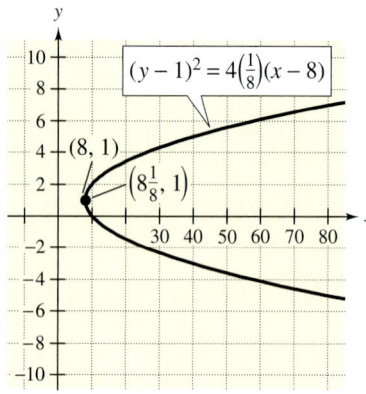

Figure 12.14

Example 9   Analyzing a Parabola

Sketch the parabola $x = 2y^2 - 4y + 10$ and identify its vertex and focus.

Solution

This equation can be written in the standard form $(y - k)^2 = 4p(x - h)$. To do this, you can complete the square, as follows.

$x = 2y^2 - 4y + 10$	Write original equation.
$2y^2 - 4y + 10 = x$	Interchange sides of the equation.
$y^2 - 2y + 5 = \frac{1}{2}x$	Divide each side by 2.
$y^2 - 2y = \frac{1}{2}x - 5$	Subtract 5 from each side.
$y^2 - 2y + 1 = \frac{1}{2}x - 5 + 1$	Complete the square on left side.
$(y - 1)^2 = \frac{1}{2}x - 4$	Simplify.
$(y - 1)^2 = \frac{1}{2}(x - 8)$	Factor.
$(y - 1)^2 = 4\left(\frac{1}{8}\right)(x - 8)$	Rewrite $\frac{1}{2}$ in the form $4p$.

From this standard form, you can see that the vertex is $(h, k) = (8, 1)$ and $p = \frac{1}{8}$. Because the parabola opens to the right, as shown in Figure 12.14, the focus lies $p = \frac{1}{8}$ unit to the right of the vertex. So, the focus is $\left(8\frac{1}{8}, 1\right)$.

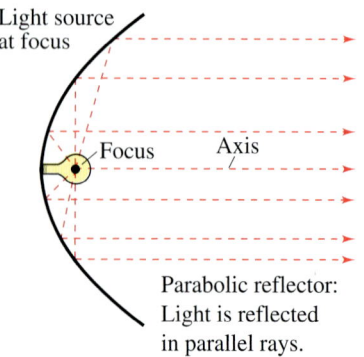

Figure 12.15

Parabolas occur in a wide variety of applications. For instance, a parabolic reflector can be formed by revolving a parabola around its axis. The resulting surface has the property that all incoming rays parallel to the axis are reflected through the focus of the parabola. This is the principle behind the construction of the parabolic mirrors used in reflecting telescopes. Conversely, the light rays emanating from the focus of a parabolic reflector used in a flashlight are all parallel to one another, as shown in Figure 12.15.

12.1 Exercises

Review   Concepts, Skills, and Problem Solving

Keep mathematically in shape by doing these exercises *before* the problems of this section.

Expressions and Equations

In Exercises 1–4, expand and simplify the expression.

1. $(x + 6)^2 - 5$

2. $(x - 7)^2 - 2$

3. $12 - (x - 8)^2$

4. $16 - (x + 1)^2$

In Exercises 5–8, complete the square for the quadratic expression.

5. $x^2 - 4x + 1$

6. $x^2 + 12x - 3$

7. $-x^2 + 6x + 5$

8. $-2x^2 + 10x - 14$

In Exercises 9 and 10, find an equation of the line passing through the point with the specified slope.

9. $(-2, 5)$, $m = \frac{5}{8}$

10. $(1, 7)$, $m = -\frac{2}{5}$

Problem Solving

11. *Reduced Rates*   A service organization paid $288 for a block of tickets to a ball game. The block contained three more tickets than the organization needed for its members. By inviting three more people to attend and share the cost, the organization lowered the price per ticket by $8. How many people are going to the game?

12. *Investment*   To begin a small business, $135,000 is needed. The cost will be divided equally among the investors. Some people have made a commitment to invest. If three more investors could be found, the amount required from each would decrease by $1500. How many people have made a commitment to invest in the business?

Developing Skills

In Exercises 1–6, match the equation with its graph. [The graphs are labeled (a), (b), (c), (d), (e), and (f).]

(a)
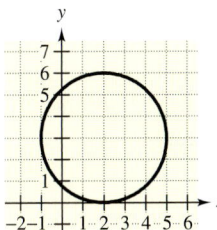

(b)
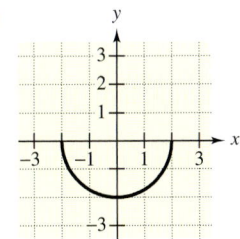

(e)
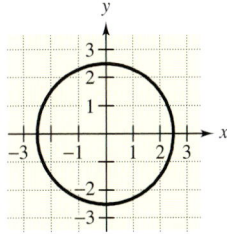

(f)
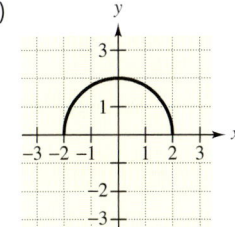

(c)
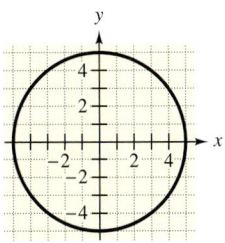

(d)
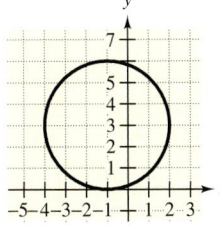

1. $x^2 + y^2 = 25$

2. $4x^2 + 4y^2 = 25$

3. $(x - 2)^2 + (y - 3)^2 = 9$

4. $(x + 1)^2 + (y - 3)^2 = 9$

5. $y = -\sqrt{4 - x^2}$

6. $y = \sqrt{4 - x^2}$

In Exercises 7–14, write the standard form of the equation of the circle with center at $(0, 0)$ that satisfies the criterion. See Example 1.

7. Radius: 5

8. Radius: 7

9. Radius: $\frac{2}{3}$

10. Radius: $\frac{5}{2}$

11. Passes through the point $(0, 8)$

12. Passes through the point $(-2, 0)$

13. Passes through the point $(5, 2)$

14. Passes through the point $(-1, -4)$

In Exercises 15–22, write the standard form of the equation of the circle with center at (h, k) that satisfies the criteria. See Example 3.

15. Center: $(4, 3)$
Radius: 10

16. Center: $(-2, 5)$
Radius: 6

17. Center: $(5, -3)$
Radius: 9

18. Center: $(-5, -2)$
Radius: $\frac{5}{2}$

19. Center: $(-2, 1)$
Passes through the point $(0, 1)$

20. Center: $(8, 2)$
Passes through the point $(8, 0)$

21. Center: $(3, 2)$
Passes through the point $(4, 6)$

22. Center: $(-3, -5)$
Passes through the point $(0, 0)$

In Exercises 23–42, identify the center and radius of the circle and sketch the circle. See Examples 2 and 4.

23. $x^2 + y^2 = 16$

24. $x^2 + y^2 = 1$

25. $x^2 + y^2 = 36$

26. $x^2 + y^2 = 10$

27. $4x^2 + 4y^2 = 1$

28. $9x^2 + 9y^2 = 64$

29. $25x^2 + 25y^2 - 144 = 0$

30. $\dfrac{x^2}{4} + \dfrac{y^2}{4} - 1 = 0$

31. $(x + 1)^2 + (y - 5)^2 = 64$

32. $(x + 10)^2 + (y + 1)^2 = 100$

33. $(x - 2)^2 + (y - 3)^2 = 4$

34. $(x + 4)^2 + (y - 3)^2 = 25$

35. $\left(x + \frac{5}{2}\right)^2 + (y + 3)^2 = 9$

36. $(x - 5)^2 + \left(y + \frac{3}{4}\right)^2 = 1$

37. $x^2 + y^2 - 4x - 2y + 1 = 0$

38. $x^2 + y^2 + 6x - 4y - 3 = 0$

39. $x^2 + y^2 + 2x + 6y + 6 = 0$

40. $x^2 + y^2 - 2x + 6y - 15 = 0$

41. $x^2 + y^2 + 8x + 4y - 5 = 0$

42. $x^2 + y^2 - 14x + 8y + 56 = 0$

In Exercises 43–46, use a graphing calculator to graph the circle. (*Note:* Solve for *y*. Use the square setting so the circles appear correct.)

43. $x^2 + y^2 = 30$

44. $4x^2 + 4y^2 = 45$

45. $(x - 2)^2 + y^2 = 10$

46. $(x + 3)^2 + y^2 = 15$

In Exercises 47–52, match the equation with its graph. [The graphs are labeled (a), (b), (c), (d), (e), and (f).]

(a)
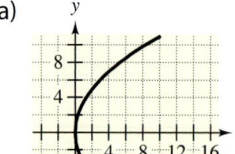

(b)
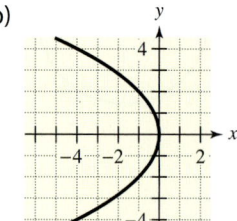

(c)

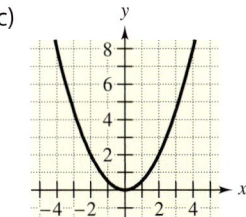

(d)

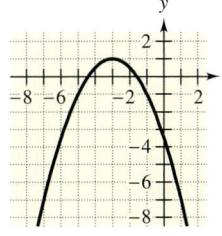

(e)

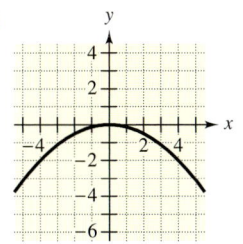

(f)

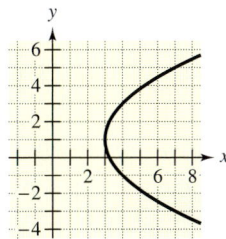

47. $y^2 = -4x$

48. $x^2 = 2y$

49. $x^2 = -8y$

50. $y^2 = 12x$

51. $(y - 1)^2 = 4(x - 3)$

52. $(x + 3)^2 = -2(y - 1)$

In Exercises 53–64, write the standard form of the equation of the parabola with its vertex at the origin. See Example 6.

53.

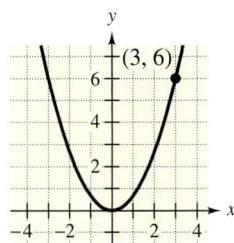

54.

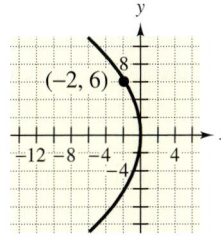

55. Focus: $\left(0, -\frac{3}{2}\right)$

56. Focus: $(2, 0)$

57. Focus: $(-2, 0)$

58. Focus: $(0, -2)$

59. Focus: $(0, 1)$

60. Focus: $(-3, 0)$

61. Focus: $(4, 0)$

62. Focus: $(0, 2)$

63. Horizontal axis and passes through the point $(4, 6)$

64. Vertical axis and passes through the point $(-2, -2)$

In Exercises 65–74, write the standard form of the equation of the parabola with its vertex at (h, k). See Example 7.

65.

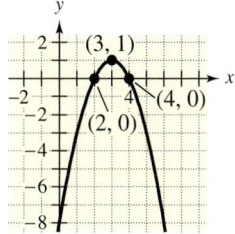

66.

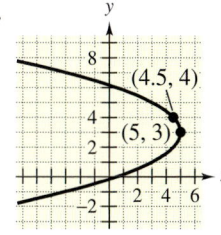

67.

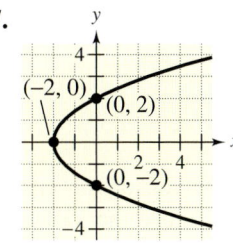

68.

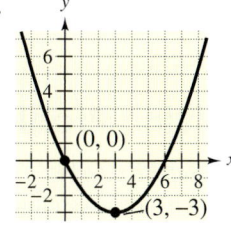

69. Vertex: $(3, 2)$;   Focus: $(1, 2)$

70. Vertex: $(-1, 2)$;   Focus: $(-1, 0)$

71. Vertex: $(0, 4)$;   Focus: $(0, 6)$

72. Vertex: $(-2, 1)$;   Focus: $(-5, 1)$

73. Vertex: $(0, 2)$;
    Horizontal axis and passes through $(1, 3)$

74. Vertex: $(0, 2)$;
    Vertical axis and passes through $(6, 0)$

In Exercises 75–88, identify the vertex and focus of the parabola and sketch the parabola. See Examples 8 and 9.

75. $y = \frac{1}{2}x^2$

76. $y = 2x^2$

77. $y^2 = -6x$

78. $y^2 = 3x$

79. $x^2 + 8y = 0$

80. $x + y^2 = 0$

81. $(x - 1)^2 + 8(y + 2) = 0$

82. $(x + 3) + (y - 2)^2 = 0$

83. $\left(y + \frac{1}{2}\right)^2 = 2(x - 5)$

84. $\left(x + \frac{1}{2}\right)^2 = 4(y - 3)$

85. $y = \frac{1}{4}(x^2 - 2x + 5)$

86. $4x - y^2 - 2y - 33 = 0$

87. $y^2 + 6y + 8x + 25 = 0$

88. $y^2 - 4y - 4x = 0$

In Exercises 89–92, use a graphing calculator to graph the parabola. Identify the vertex and focus.

89. $y = -\frac{1}{6}(x^2 + 4x - 2)$

90. $x^2 - 2x + 8y + 9 = 0$

91. $y^2 + x + y = 0$

92. $y^2 - 4x - 4 = 0$

Solving Problems

93. *Satellite Orbit* Write an equation of the circular orbit of a satellite 500 miles above the surface of Earth. Place the origin of the rectangular coordinate system at the center of Earth and assume the radius of Earth is 4000 miles.

94. *Architecture* The top portion of a stained glass window is in the form of a pointed Gothic arch (see figure). Each side of the arch is an arc of a circle of radius 12 feet and center at the base of the opposite arch. Write an equation of one of the circles and use it to determine the height of the point of the arch above the horizontal base of the window.

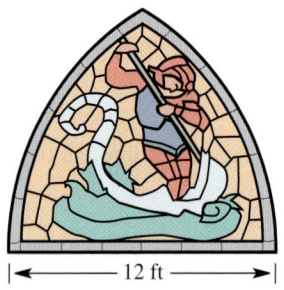

|←——— 12 ft ———→|

95. *Architecture* A semicircular arch for a tunnel under a river has a diameter of 100 feet (see figure). Write an equation of the semicircle. Determine the height of the arch 5 feet from the edge of the tunnel.

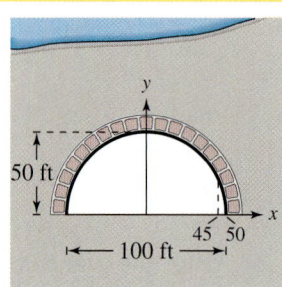

Figure for 95

96. *Graphical Estimation* A rectangle centered at the origin with sides parallel to the coordinate axes is placed in a circle of radius 25 inches centered at the origin (see figure). The length of the rectangle is $2x$ inches.

(a) Show that the width and area of the rectangle are given by $2\sqrt{625 - x^2}$ and $4x\sqrt{625 - x^2}$, respectively.

(b) Use a graphing calculator to graph the area function. Approximate the value of x for which the area is maximum.

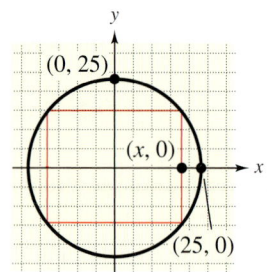

97. *Suspension Bridge*   Each cable of a suspension bridge is suspended (in the shape of a parabola) between two towers that are 120 meters apart, and the top of each tower is 20 meters above the road way. The cables touch the roadway at the midpoint between the two towers (see figure).

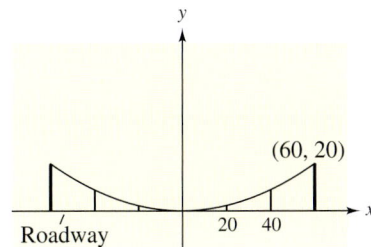

(a) Write an equation for the parabolic shape of each cable.

(b) Complete the table by finding the height of the suspension cables y over the roadway at a distance of x meters from the center of the bridge.

x	0	20	40	60
y				

98. *Beam Deflection*   A simply supported beam is 16 meters long and has a load at the center (see figure). The deflection of the beam at its center is 3 centimeters. Assume that the shape of the deflected beam is parabolic.

(a) Write an equation of the parabola. (Assume that the origin is at the center of the deflected beam.)

(b) How far from the center of the beam is the deflection equal to 1 centimeter?

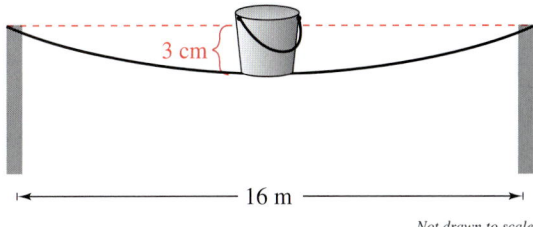

Figure for 98

99. ▦ *Revenue*   The revenue R generated by the sale of x computer desks is given by $R = 375x - \frac{3}{2}x^2$.

(a) Use a graphing calculator to graph the function.

(b) Use the graph to approximate the number of sales that will maximize revenue.

100. ▦ *Path of a Softball*   The path of a softball is given by $y = -0.08x^2 + x + 4$. The coordinates x and y are measured in feet, with $x = 0$ corresponding to the position from which the ball was thrown.

(a) Use a graphing calculator to graph the path of the softball.

(b) Move the cursor along the path to approximate the highest point and the range of the path.

Explaining Concepts

101. ⚡ Answer parts (a) and (b) of Motivating the Chapter on page 756.

102. *Writing*   Name the four types of conics.

103. *Writing*   Define a circle and write the standard form of the equation of a circle centered at the origin.

104. *Writing*   Explain how to use the method of completing the square to write an equation of a circle in standard form.

105. *Writing*   Explain the significance of a parabola's directrix and focus.

106. *Writing*   Is y a function of x in the equation $y^2 = 6x$? Explain.

107. *Writing*   Is it possible for a parabola to intersect its directrix? Explain.

108. *Writing*   If the vertex and focus of a parabola are on a horizontal line, is the directrix of the parabola vertical? Explain.

12.2 Ellipses

David Young-Wolff/PhotoEdit, Inc.

What You Should Learn

1 Graph and write equations of ellipses centered at the origin.

2 Graph and wrire equations of ellipses centered at (h, k).

Why You Should Learn It

Equations of ellipses can be used to model and solve real-life problems. For instance, in Exercise 58 on page 779, you will use an equation of an ellipse to model a chainwheel.

1 Graph and write equations of ellipses centered at the origin.

Ellipses Centered at the Origin

The third type of conic is called an *ellipse* and is defined as follows.

Definition of an Ellipse

An **ellipse** in the rectangular coordinate system consists of all points (x, y) such that the sum of the distances between (x, y) and two distinct fixed points is a constant, as shown in Figure 12.16. Each of the two fixed points is called a **focus** of the ellipse. (The plural of focus is *foci*.)

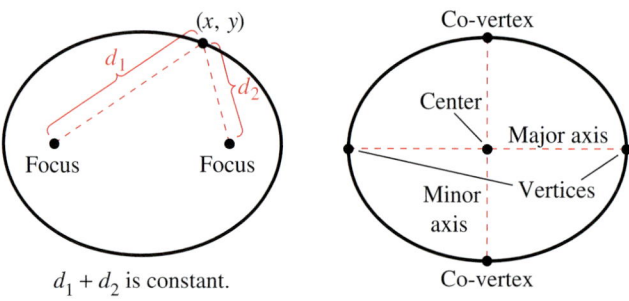

$d_1 + d_2$ is constant.

Figure 12.16   **Figure 12.17**

The line through the foci intersects the ellipse at two points, called the **vertices**, as shown in Figure 12.17. The line segment joining the vertices is called the **major axis,** and its midpoint is called the **center** of the ellipse. The line segment perpendicular to the major axis at the center is called the **minor axis** of the ellipse, and the points at which the minor axis intersects the ellipse are called **co-vertices.**

To trace an ellipse, place two thumbtacks at the foci, as shown in Figure 12.18. If the ends of a fixed length of string are fastened to the thumbtacks and the string is drawn taut with a pencil, the path traced by the pencil will be an ellipse.

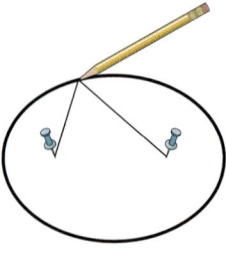

Figure 12.18

The standard form of the equation of an ellipse takes one of two forms, depending on whether the major axis is horizontal or vertical.

Standard Equation of an Ellipse (Center at Origin)

The standard form of the equation of an ellipse centered at the origin with major and minor axes of lengths $2a$ and $2b$ is

$$\frac{x^2}{a^2} + \frac{y^2}{b^2} = 1 \qquad \text{or} \qquad \frac{x^2}{b^2} + \frac{y^2}{a^2} = 1, \qquad 0 < b < a.$$

The vertices lie on the major axis, a units from the center, and the co-vertices lie on the minor axis, b units from the center, as shown in Figure 12.19.

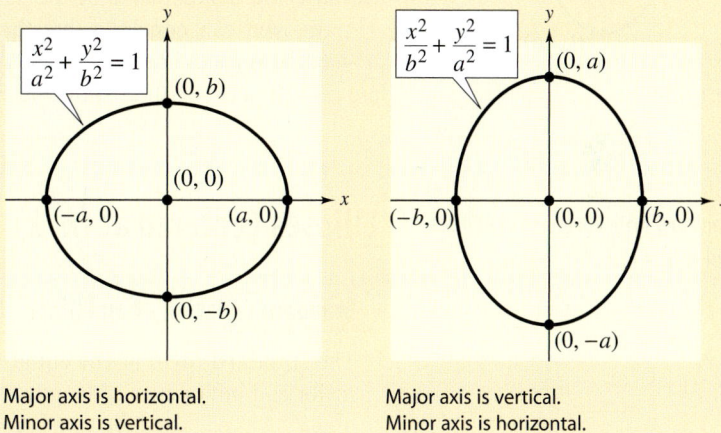

Major axis is horizontal.
Minor axis is vertical.

Major axis is vertical.
Minor axis is horizontal.

Figure 12.19

Example 1   Writing the Standard Equation of an Ellipse

Write an equation of the ellipse that is centered at the origin, with vertices $(-3, 0)$ and $(3, 0)$ and co-vertices $(0, -2)$ and $(0, 2)$.

Solution

Begin by plotting the vertices and co-vertices, as shown in Figure 12.20. The center of the ellipse is $(0, 0)$. So, the equation of the ellipse has the form

$$\frac{x^2}{a^2} + \frac{y^2}{b^2} = 1. \qquad \text{Major axis is horizontal.}$$

For this ellipse, the major axis is horizontal. So, a is the distance between the center and either vertex, which implies that $a = 3$. Similarly, b is the distance between the center and either co-vertex, which implies that $b = 2$. So, the standard form of the equation of the ellipse is

$$\frac{x^2}{3^2} + \frac{y^2}{2^2} = 1. \qquad \text{Standard form}$$

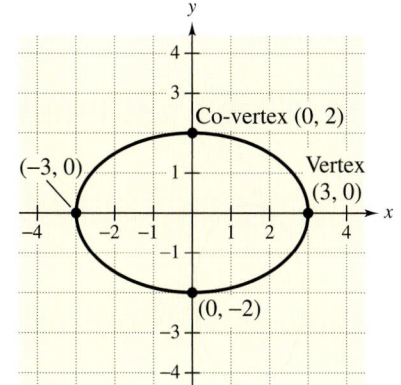

Figure 12.20

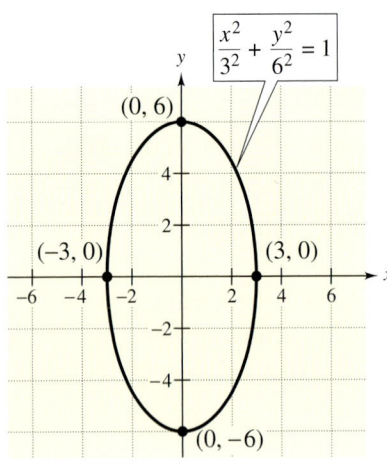

Figure 12.21

Example 2 Sketching an Ellipse

Sketch the ellipse given by $4x^2 + y^2 = 36$. Identify the vertices and co-vertices.

Solution

To sketch an ellipse, it helps first to write its equation in standard form.

$$4x^2 + y^2 = 36 \qquad \text{Write original equation.}$$

$$\frac{x^2}{9} + \frac{y^2}{36} = 1 \qquad \text{Divide each side by 36 and simplify.}$$

$$\frac{x^2}{3^2} + \frac{y^2}{6^2} = 1 \qquad \text{Standard form}$$

Because the denominator of the y^2-term is larger than the denominator of the x^2-term, you can conclude that the major axis is vertical. Moreover, because $a = 6$, the vertices are $(0, -6)$ and $(0, 6)$. Finally, because $b = 3$, the co-vertices are $(-3, 0)$ and $(3, 0)$, as shown in Figure 12.21.

---

2 Graph and write equations of ellipses centered at (h, k).

Ellipses Centered at (h, k)

Standard Equation of an Ellipse [Center at (h, k)]

The **standard form of the equation of an ellipse centered at** (h, k) with major and minor axes of lengths $2a$ and $2b$, where $0 < b < a$, is

$$\frac{(x - h)^2}{a^2} + \frac{(y - k)^2}{b^2} = 1 \qquad \text{Major axis is horizontal.}$$

or

$$\frac{(x - h)^2}{b^2} + \frac{(y - k)^2}{a^2} = 1. \qquad \text{Major axis is vertical.}$$

The foci lie on the major axis, c units from the center, with $c^2 = a^2 - b^2$.

Figure 12.22 shows the horizontal and vertical orientations for an ellipse.

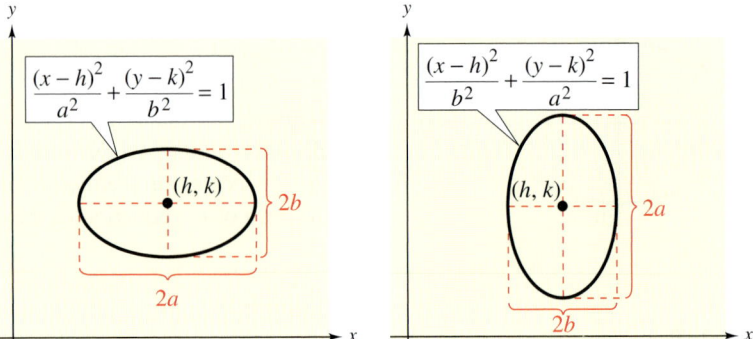

Figure 12.22

When $h = 0$ and $k = 0$, the ellipse is centered at the origin. Otherwise, you can shift the center of the ellipse h units horizontally and k units vertically from the origin.

Example 3   Writing the Standard Equation of an Ellipse

Write the standard form of the equation of the ellipse with vertices $(-2, 2)$ and $(4, 2)$ and co-vertices $(1, 3)$ and $(1, 1)$, as shown in Figure 12.23.

Solution

Because the vertices are $(-2, 2)$ and $(4, 2)$, the center of the ellipse is $(h, k) = (1, 2)$. The distance from the center to either vertex is $a = 3$, and the distance to either co-vertex is $b = 1$. Because the major axis is horizontal, the standard form of the equation is

$$\frac{(x - h)^2}{a^2} + \frac{(y - k)^2}{b^2} = 1. \qquad \text{Major axis is horizontal.}$$

Substitute the values of h, k, a, and b to obtain

$$\frac{(x - 1)^2}{3^2} + \frac{(y - 2)^2}{1^2} = 1. \qquad \text{Standard form}$$

From the graph, you can see that the center of the ellipse is shifted one unit to the right and two units upward from the origin.

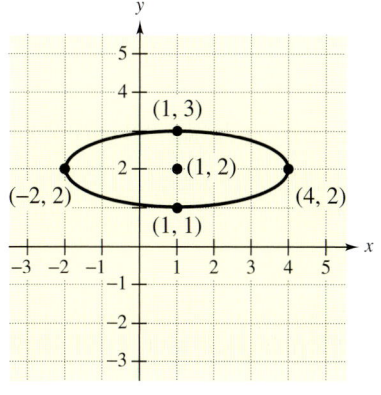

Figure 12.23

Technology: Tip

You can use a graphing calculator to graph an ellipse by graphing the upper and lower portions in the same viewing window. For instance, to graph the ellipse $x^2 + 4y^2 = 4$, first solve for y to obtain

$$y_1 = \frac{1}{2}\sqrt{4 - x^2}$$

and

$$y_2 = -\frac{1}{2}\sqrt{4 - x^2}.$$

Use a viewing window in which $-3 \le x \le 3$ and $-2 \le y \le 2$. You should obtain the graph shown below.

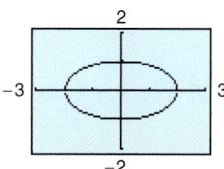

Use this information to graph the ellipse in Example 3 on your graphing calculator.

To write an equation of an ellipse in standard form, you must group the *x*-terms and the *y*-terms and then complete each square, as shown in Example 4.

Example 4 Sketching an Ellipse

Sketch the ellipse given by $4x^2 + y^2 - 8x + 6y + 9 = 0$.

Solution

Begin by writing the equation in standard form. In the fourth step, note that 9 and 4 are added to *each* side of the equation.

$4x^2 + y^2 - 8x + 6y + 9 = 0$	Write original equation.
$(4x^2 - 8x + \quad) + (y^2 + 6y + \quad) = -9$	Group terms.
$4(x^2 - 2x + \quad) + (y^2 + 6y + \quad) = -9$	Factor 4 out of *x*-terms.
$4(x^2 - 2x + 1) + (y^2 + 6y + 9) = -9 + 4(1) + 9$	Complete each square.
$4(x - 1)^2 + (y + 3)^2 = 4$	Simplify.
$\dfrac{(x - 1)^2}{1} + \dfrac{(y + 3)^2}{4} = 1$	Divide each side by 4.
$\dfrac{(x - 1)^2}{1^2} + \dfrac{(y + 3)^2}{2^2} = 1$	Standard form

Now you can see that the center of the ellipse is at $(h, k) = (1, -3)$. Because the denominator of the y^2-term is larger than the denominator of the x^2-term, you can conclude that the major axis is vertical. Because the denominator of the x^2-term is $b^2 = 1^2$, you can locate the endpoints of the minor axis one unit to the right and left of the center, and because the denominator of the y^2-term is $a^2 = 2^2$, you can locate the endpoints of the major axis two units upward and downward from the center, as shown in Figure 12.24. To complete the graph, sketch an oval shape that is determined by the vertices and co-vertices, as shown in Figure 12.25.

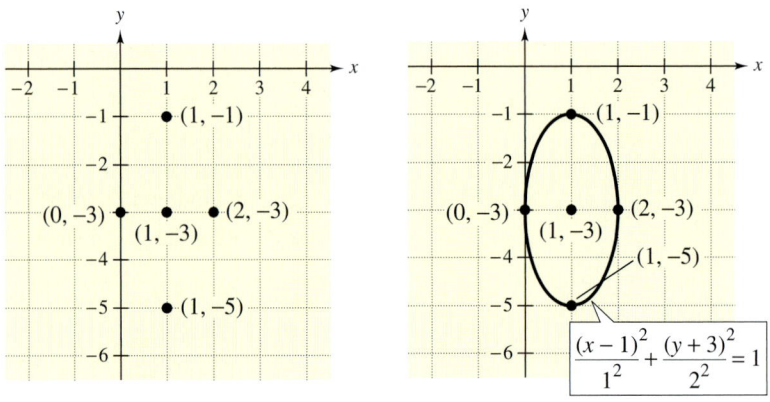

Figure 12.24   **Figure 12.25**

From Figure 12.25, you can see that the center of the ellipse is shifted one unit to the right and three units downward from the origin.

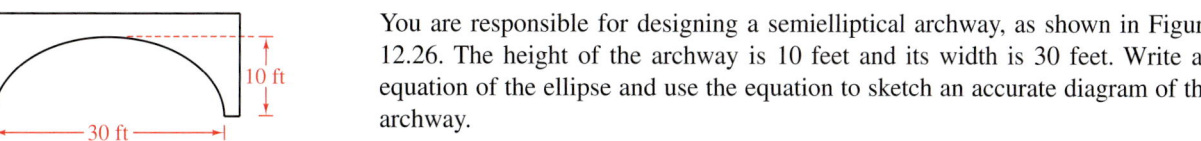

Figure 12.26

Example 5   An Application: Semielliptical Archway

You are responsible for designing a semielliptical archway, as shown in Figure 12.26. The height of the archway is 10 feet and its width is 30 feet. Write an equation of the ellipse and use the equation to sketch an accurate diagram of the archway.

Solution

To make the equation simple, place the origin at the center of the ellipse. This means that the standard form of the equation is

$$\frac{x^2}{a^2} + \frac{y^2}{b^2} = 1.$$ Major axis is horizontal.

Because the major axis is horizontal, it follows that $a = 15$ and $b = 10$, which implies that the equation is

$$\frac{x^2}{15^2} + \frac{y^2}{10^2} = 1.$$ Standard form

To make an accurate sketch of the ellipse, solve this equation for y as follows.

$$\frac{x^2}{225} + \frac{y^2}{100} = 1$$ Simplify denominators.

$$\frac{y^2}{100} = 1 - \frac{x^2}{225}$$ Subtract $\frac{x^2}{225}$ from each side.

$$y^2 = 100\left(1 - \frac{x^2}{225}\right)$$ Multiply each side by 100.

$$y = 10\sqrt{1 - \frac{x^2}{225}}$$ Take the positive square root of each side.

Next, calculate several y-values for the archway, as shown in the table. Then use the values in the table to sketch the archway, as shown in Figure 12.27.

x	± 15	± 12.5	± 10	± 7.5	± 5	± 2.5	0
y	0	5.53	7.45	8.66	9.43	9.86	10

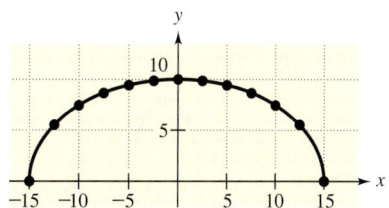

Figure 12.27

12.2 Exercises

Review   *Concepts, Skills, and Problem Solving*

Keep mathematically in shape by doing these exercises *before* the problems of this section.

Simplifying Expressions

In Exercises 1–4, simplify the expression.

1. $\dfrac{15y^{-3}}{10y^2}$

2. $\left(\dfrac{3x^2}{2y}\right)^{-2}$

3. $\dfrac{3x^2y^3}{18x^{-1}y^2}$

4. $(x^2 + 1)^0$

Solving Equations

In Exercises 5 and 6, solve the quadratic equation by completing the square.

5. $x^2 + 6x - 4 = 0$

6. $2x^2 - 16x + 5 = 0$

In Exercises 7–10, sketch the graph of the equation.

7. $y = \frac{2}{5}x + 3$

8. $y = -2x - 1$

9. $y = x^2 - 12x + 36$

10. $y = 25 - x^2$

Problem Solving

11. *Test Scores*   A student has test scores of 90, 74, 82, and 90. The next examination is the final examination, which counts as two tests. What score does the student need on the final examination to produce an average score of 85?

12. *Simple Interest*   An investment of $2500 is made at an annual simple interest rate of 5.5%. How much additional money must be invested at an annual simple interest rate of 8% so that the total interest earned is 7% of the total investment?

Developing Skills

In Exercises 1–6, match the equation with its graph. [The graphs are labeled (a), (b), (c), (d), (e), and (f).]

(a)
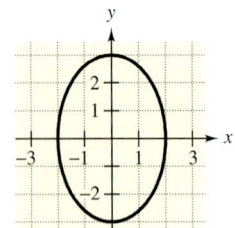

(b)
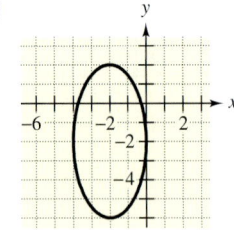

(c)
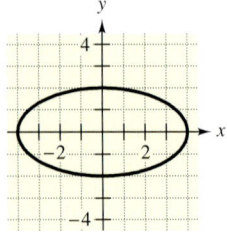

(d)
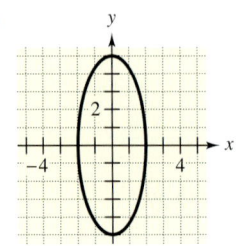

(e)
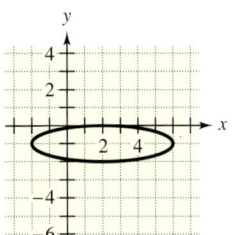

(f)
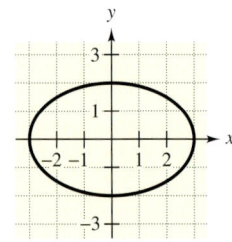

1. $\dfrac{x^2}{4} + \dfrac{y^2}{9} = 1$

2. $\dfrac{x^2}{9} + \dfrac{y^2}{4} = 1$

3. $\dfrac{x^2}{4} + \dfrac{y^2}{25} = 1$

4. $\dfrac{y^2}{4} + \dfrac{x^2}{16} = 1$

5. $\dfrac{(x - 2)^2}{16} + \dfrac{(y + 1)^2}{1} = 1$

6. $\dfrac{(x + 2)^2}{4} + \dfrac{(y + 2)^2}{16} = 1$

In Exercises 7–18, write the standard form of the equation of the ellipse centered at the origin. See Example 1.

Vertices	Co-vertices
7. $(-4, 0), (4, 0)$	$(0, -3), (0, 3)$
8. $(-4, 0), (4, 0)$	$(0, -1), (0, 1)$
9. $(-2, 0), (2, 0)$	$(0, -1), (0, 1)$
10. $(-10, 0), (10, 0)$	$(0, -4), (0, 4)$
11. $(0, -4), (0, 4)$	$(-3, 0), (3, 0)$
12. $(0, -5), (0, 5)$	$(-1, 0), (1, 0)$
13. $(0, -2), (0, 2)$	$(-1, 0), (1, 0)$
14. $(0, -8), (0, 8)$	$(-4, 0), (4, 0)$

15. Major axis (vertical) 10 units, minor axis 6 units

16. Major axis (horizontal) 24 units, minor axis 10 units

17. Major axis (horizontal) 20 units, minor axis 12 units

18. Major axis (horizontal) 50 units, minor axis 30 units

In Exercises 19–32, sketch the ellipse. Identify the vertices and co-vertices. See Example 2.

19. $\dfrac{x^2}{16} + \dfrac{y^2}{4} = 1$

20. $\dfrac{x^2}{25} + \dfrac{y^2}{9} = 1$

21. $\dfrac{x^2}{4} + \dfrac{y^2}{16} = 1$

22. $\dfrac{x^2}{9} + \dfrac{y^2}{25} = 1$

23. $\dfrac{x^2}{25/9} + \dfrac{y^2}{16/9} = 1$

24. $\dfrac{x^2}{1} + \dfrac{y^2}{1/4} = 1$

25. $\dfrac{9x^2}{4} + \dfrac{25y^2}{16} = 1$

26. $\dfrac{36x^2}{49} + \dfrac{16y^2}{9} = 1$

27. $16x^2 + 25y^2 - 9 = 0$

28. $64x^2 + 36y^2 - 49 = 0$

29. $4x^2 + y^2 - 4 = 0$

30. $4x^2 + 9y^2 - 36 = 0$

31. $10x^2 + 16y^2 - 160 = 0$

32. $16x^2 + 4y^2 - 64 = 0$

In Exercises 33–36, use a graphing calculator to graph the ellipse. Identify the vertices. (Note: Solve for y.)

33. $x^2 + 2y^2 = 4$
34. $9x^2 + y^2 = 64$
35. $3x^2 + y^2 - 12 = 0$
36. $5x^2 + 2y^2 - 10 = 0$

In Exercises 37–40, write the standard form of the equation of the ellipse. See Example 3.

37.

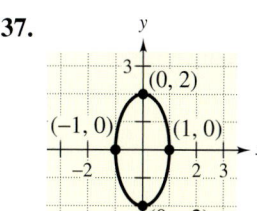

38.

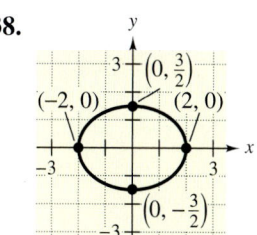

39.

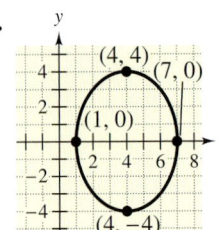

40.

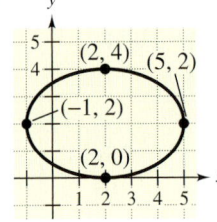

44. $\dfrac{(x + 2)^2}{1/4} + \dfrac{(y + 4)^2}{1} = 1$

45. $9x^2 + 4y^2 + 36x - 24y + 36 = 0$

46. $9x^2 + 4y^2 - 36x + 8y + 31 = 0$

47. $4(x - 2)^2 + 9(y + 2)^2 = 36$

48. $(x + 3)^2 + 9(y + 1)^2 = 81$

In Exercises 41–54, find the center and vertices of the ellipse and sketch the ellipse. See Example 4.

49. $12(x + 4)^2 + 3(y - 1)^2 = 48$

50. $16(x - 2)^2 + 4(y + 3)^2 = 16$

41. $\dfrac{(x + 5)^2}{16} + y^2 = 1$

51. $25x^2 + 9y^2 - 200x + 54y + 256 = 0$

52. $25x^2 + 16y^2 - 150x - 128y + 81 = 0$

42. $\dfrac{(x - 2)^2}{4} + \dfrac{(y - 3)^2}{9} = 1$

53. $x^2 + 4y^2 - 4x - 8y - 92 = 0$

43. $\dfrac{(x - 1)^2}{9} + \dfrac{(y - 5)^2}{25} = 1$

54. $x^2 + 4y^2 + 6x + 16y - 11 = 0$

Solving Problems

55. *Architecture*   A semielliptical arch for a tunnel under a river has a width of 100 feet and a height of 40 feet (see figure). Determine the height of the arch 5 feet from the edge of the tunnel.

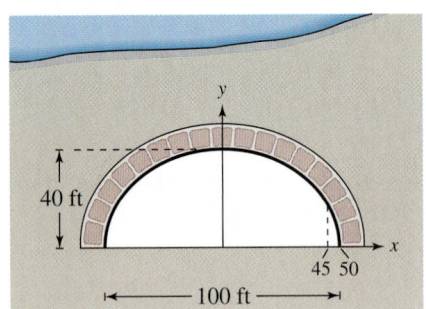

56. *Wading Pool*   You are building a wading pool that is in the shape of an ellipse. Your plans give an equation for the elliptical shape of the pool measured in feet as

$$\dfrac{x^2}{324} + \dfrac{y^2}{196} = 1.$$

Find the longest distance and shortest distance across the pool.

57. *Sports*   In Australia, football by *Australian Rules* (or rugby) is played on elliptical fields. The field can be a maximum of 170 yards wide and a maximum of 200 yards long. Let the center of a field of maximum size be represented by the point $(0, 85)$. Write an equation of the ellipse that represents this field. (Source: Oxford Companion to World Sports and Games)

58. *Bicycle Chainwheel*   The pedals of a bicycle drive a chainwheel, which drives a smaller sprocket wheel on the rear axle (see figure). Many chainwheels are circular. Some, however, are slightly elliptical, which tends to make pedaling easier. Write an equation of an elliptical chainwheel that is 8 inches in diameter at its widest point and $7\frac{1}{2}$ inches in diameter at its narrowest point.

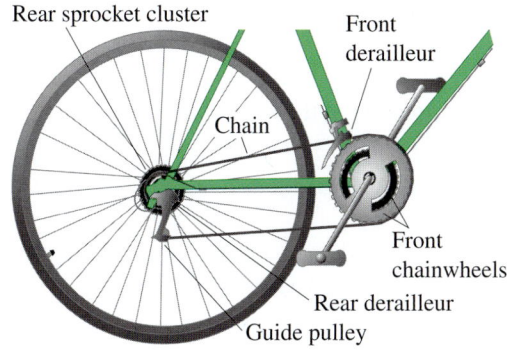

Rear sprocket cluster

Front derailleur

Chain

Front chainwheels

Rear derailleur

Guide pulley

59. *Area*   The area A of the ellipse

$$\frac{x^2}{a^2} + \frac{y^2}{b^2} = 1$$

is given by $A = \pi a b$. Write the equation of an ellipse with an area of 301.59 square units and $a + b = 20$.

60. Sketch a graph of the ellipse that consists of all points (x, y) such that the sum of the distances between (x, y) and two fixed points is 15 units and for which the foci are located at the centers of the two sets of concentric circles in the figure.

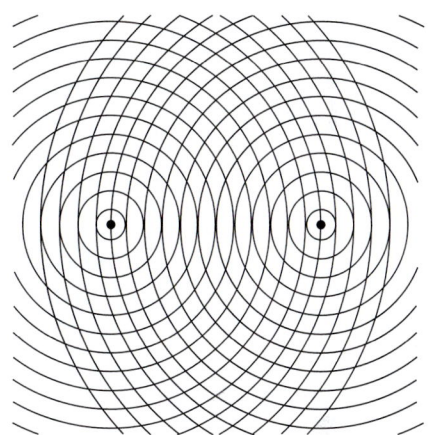

Explaining Concepts

61. *Writing*   Describe the relationship between circles and ellipses. How are they similar? How do they differ?

62. *Writing*   Define an ellipse and write the standard form of the equation of an ellipse centered at the origin.

63. *Writing*   Explain the significance of the foci in an ellipse.

64. *Writing*   Explain how to write an equation of an ellipse if you know the coordinates of the vertices and co-vertices.

65. *Writing*   From its equation, how can you determine the lengths of the axes of an ellipse?

Mid-Chapter Quiz

Take this quiz as you would take a quiz in class. After you are done, check your work against the answers in the back of the book.

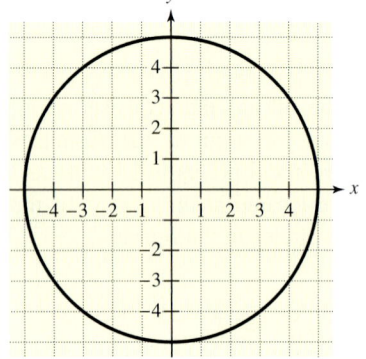

Figure for 1

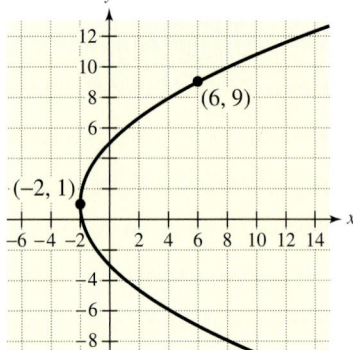

Figure for 2

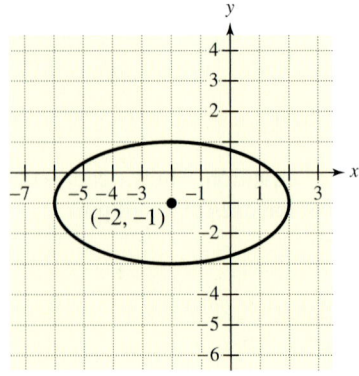

Figure for 3

1. Write the standard form of the equation of the circle shown in the figure.

2. Write the standard form of the equation of the parabola shown in the figure.

3. Write the standard form of the equation of the ellipse shown in the figure.

4. Write the standard form of the equation of the circle with center $(3, -5)$ and passing through the point $(0, -1)$.

5. Write the standard form of the equation of the parabola with vertex $(2, 3)$ and focus $(2, 1)$.

6. Write the standard form of the equation of the ellipse with vertices $(0, -10)$ and $(0, 10)$ and co-vertices $(-3, 0)$ and $(3, 0)$.

In Exercises 7 and 8, write the equation of the circle in standard form, then find the center and the radius of the circle.

7. $x^2 + y^2 - 10x + 16 = 0$

8. $x^2 + y^2 + 2x - 4y + 4 = 0$

In Exercises 9 and 10, write the equation of the parabola in standard form, then find the vertex and the focus of the parabola.

9. $x = y^2 - 6y - 7$

10. $x^2 - 8x + y + 12 = 0$

In Exercises 11 and 12, write the equation of the ellipse in standard form, then find the center and the vertices of the ellipse.

11. $20x^2 + 9y^2 - 180 = 0$ 12. $4x^2 + 9y^2 - 48x + 36y + 144 = 0$

In Exercises 13–18, sketch the graph of the equation.

13. $(x + 5)^2 + (y - 1)^2 = 9$ 14. $\dfrac{x^2}{9} + \dfrac{y^2}{16} = 1$

15. $x = -y^2 - 4y$ 16. $x^2 + (y + 4)^2 = 1$

17. $y = x^2 - 2x + 1$ 18. $4(x + 3)^2 + (y - 2)^2 = 16$

12.3 Hyperbolas

Jonathan Nourok/PhotoEdit, Inc.

What You Should Learn

① Graph and write equations of hyperbolas centered at the origin.

② Graph and write equations of hyperbolas centered at (h, k).

Why You Should Learn It

Equations of hyperbolas are often used in navigation. For instance, in Exercise 43 on page 788, a hyperbola is used to model long-distance radio navigation for a ship.

① Graph and write equations of hyperbolas centered at the origin.

Hyperbolas Centered at the Origin

The fourth basic type of conic is called a **hyperbola** and is defined as follows.

Definition of a Hyperbola

A **hyperbola** on the rectangular coordinate system consists of all points (x, y) such that the *difference* of the distances between (x, y) and two fixed points is a positive constant, as shown in Figure 12.28. The two fixed points are called the **foci** of the hyperbola. The line on which the foci lie is called the **transverse axis** of the hyperbola.

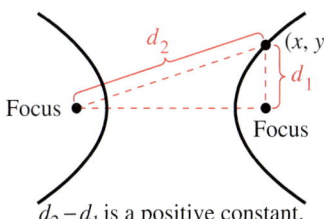

$d_2 - d_1$ is a positive constant.

Figure 12.28

Standard Equation of a Hyperbola (Center at Origin)

The **standard form of the equation of a hyperbola centered at the origin** is

$$\frac{x^2}{a^2} - \frac{y^2}{b^2} = 1 \qquad \text{or} \qquad \frac{y^2}{a^2} - \frac{x^2}{b^2} = 1$$

Transverse axis   Transverse axis
is horizontal.   is vertical.

where a and b are positive real numbers. The **vertices** of the hyperbola lie on the transverse axis, a units from the center, as shown in Figure 12.29.

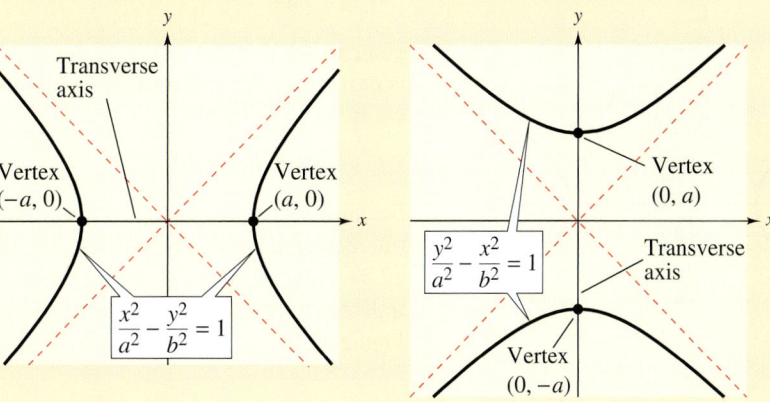

Figure 12.29

A hyperbola has two disconnected parts, each of which is called a **branch** of the hyperbola. The two branches approach a pair of intersecting lines called the **asymptotes** of the hyperbola. The two asymptotes intersect at the center of the hyperbola. To sketch a hyperbola, form a **central rectangle** that is centered at the origin and has side lengths of $2a$ and $2b$. Note in Figure 12.30 that the asymptotes pass through the corners of the central rectangle and that the vertices of the hyperbola lie at the centers of opposite sides of the central rectangle.

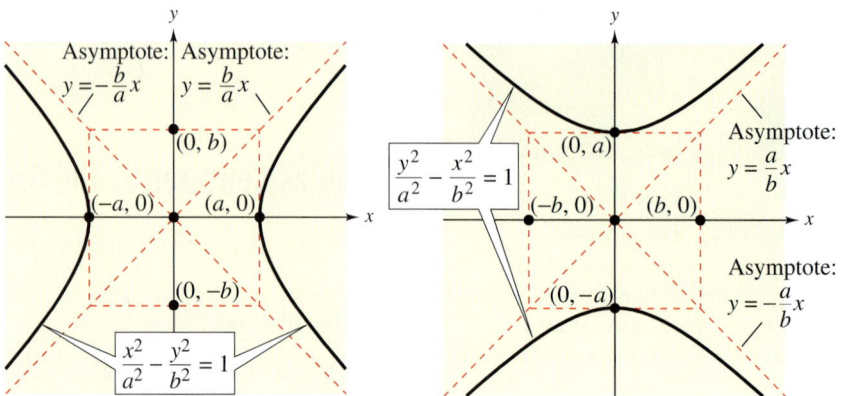

Tranverse axis is horizontal.   Tranverse axis is vertical.

Figure 12.30

Example 1   Sketching a Hyperbola

Identify the vertices of the hyperbola given by the equation, and sketch the hyperbola.

$$\frac{x^2}{36} - \frac{y^2}{16} = 1$$

Solution

From the standard form of the equation

$$\frac{x^2}{6^2} - \frac{y^2}{4^2} = 1$$

you can see that the center of the hyperbola is the origin and the transverse axis is horizontal. So, the vertices lie six units to the left and right of the center at the points

$$(-6, 0) \text{ and } (6, 0).$$

Because $a = 6$ and $b = 4$, you can sketch the hyperbola by first drawing a central rectangle with a width of $2a = 12$ and a height of $2b = 8$, as shown in Figure 12.31. Next, draw the asymptotes of the hyperbola through the corners of the central rectangle and plot the vertices. Finally, draw the hyperbola, as shown in Figure 12.32.

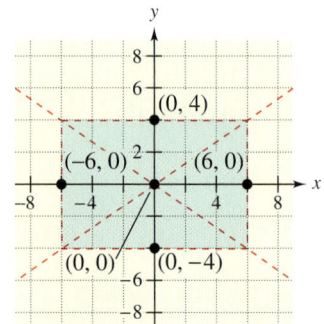

Figure 12.31

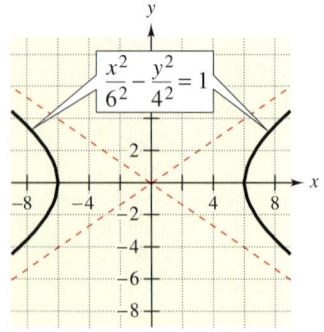

Figure 12.32

Writing the equation of a hyperbola is a little more difficult than writing equations of the other three types of conics. However, if you know the vertices and the asymptotes, you can find the values of a and b, which enable you to write the equation. Notice in Example 2 that the key to this procedure is knowing that the central rectangle has a width of $2b$ and a height of $2a$.

Example 2   Writing the Equation of a Hyperbola

Write the standard form of the equation of the hyperbola with a vertical transverse axis and vertices $(0, 3)$ and $(0, -3)$. The equations of the asymptotes of the hyperbola are $y = \frac{3}{5}x$ and $y = -\frac{3}{5}x$.

Solution

To begin, sketch the lines that represent the asymptotes, as shown in Figure 12.33. Note that these two lines intersect at the origin, which implies that the center of the hyperbola is $(0, 0)$. Next, plot the two vertices at the points $(0, 3)$ and $(0, -3)$. Because you know where the vertices are located, you can sketch the central rectangle of the hyperbola, as shown in Figure 10.33. Note that the corners of the central rectangle occur at the points

$$(-5, 3), (5, 3), (-5, -3), \text{ and } (5, -3).$$

Because the width of the central rectangle is $2b = 10$, it follows that $b = 5$. Similarly, because the height of the central rectangle is $2a = 6$, it follows that $a = 3$. Now that you know the values of a and b, you can use the standard form of the equation of the hyperbola to write the equation.

$$\frac{y^2}{a^2} - \frac{x^2}{b^2} = 1 \qquad \text{Transverse axis is vertical.}$$

$$\frac{y^2}{3^2} - \frac{x^2}{5^2} = 1 \qquad \text{Substitute 3 for } a \text{ and 5 for } b.$$

$$\frac{y^2}{9} - \frac{x^2}{25} = 1 \qquad \text{Simplify.}$$

The graph is shown in Figure 12.34.

Study Tip

For a hyperbola, note that a and b are not determined by size as with an ellipse, where a is always greater than b. In the standard form of the equation of a hyperbola, a^2 is always the denominator of the positive term.

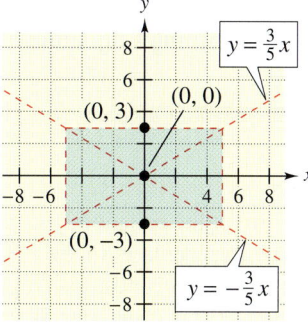

Figure 12.33

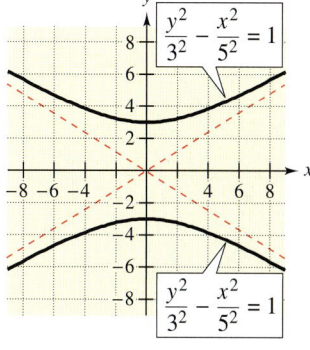

Figure 12.34

2 Graph and write equations of hyperbolas centered at (h, k).

Hyperbolas Centered at (h, k)

Standard Equation of a Hyperbola [Center at (h, k)]

The **standard form of the equation of a hyperbola centered at (h, k) is**

$$\frac{(x - h)^2}{a^2} - \frac{(y - k)^2}{b^2} = 1 \qquad \text{Transverse axis is horizontal.}$$

or

$$\frac{(y - k)^2}{a^2} - \frac{(x - h)^2}{b^2} = 1 \qquad \text{Transverse axis is vertical.}$$

where a and b are positive real numbers. The vertices lie on the transverse axis, a units from the center, as shown in Figure 12.35.

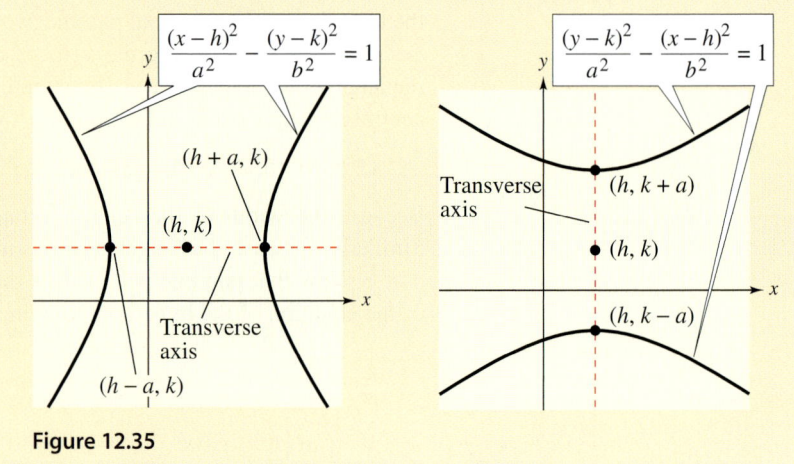

Figure 12.35

When $h = 0$ and $k = 0$, the hyperbola is centered at the origin. Otherwise, you can shift the center of the hyperbola h units horizontally and k units vertically from the origin.

Example 3   Sketching a Hyperbola

Sketch the hyperbola given by $\dfrac{(y - 1)^2}{9} - \dfrac{(x + 2)^2}{4} = 1$.

Solution

From the form of the equation, you can see that the transverse axis is vertical. The center of the hyperbola is $(h, k) = (-2, 1)$. Because $a = 3$ and $b = 2$, you can begin by sketching a central rectangle that is six units high and four units wide, centered at $(-2, 1)$. Then, sketch the asymptotes by drawing lines through the corners of the central rectangle. Sketch the hyperbola, as shown in Figure 12.36. From the graph, you can see that the center of the hyperbola is shifted two units to the left and one unit upward from the origin.

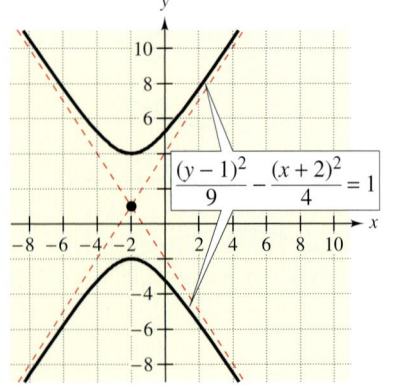

Figure 12.36

Example 4   Sketching a Hyperbola

Sketch the hyperbola given by $x^2 - 4y^2 + 8x + 16y - 4 = 0$.

Solution

Complete the square to write the equation in standard form.

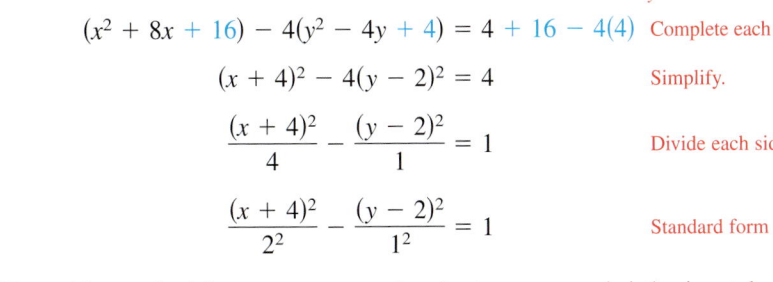

$$x^2 - 4y^2 + 8x + 16y - 4 = 0 \qquad \text{Write original equation.}$$

$$(x^2 + 8x +) - (4y^2 - 16y +) = 4 \qquad \text{Group terms.}$$

$$(x^2 + 8x +) - 4(y^2 - 4y +) = 4 \qquad \text{Factor 4 out of } y\text{-terms.}$$

$$(x^2 + 8x + 16) - 4(y^2 - 4y + 4) = 4 + 16 - 4(4) \qquad \text{Complete each square.}$$

$$(x + 4)^2 - 4(y - 2)^2 = 4 \qquad \text{Simplify.}$$

$$\frac{(x + 4)^2}{4} - \frac{(y - 2)^2}{1} = 1 \qquad \text{Divide each side by 4.}$$

$$\frac{(x + 4)^2}{2^2} - \frac{(y - 2)^2}{1^2} = 1 \qquad \text{Standard form}$$

From this standard form, you can see that the transverse axis is horizontal and the center of the hyperbola is $(h, k) = (-4, 2)$. Because $a = 2$ and $b = 1$, you can begin by sketching a central rectangle that is four units wide and two units high, centered at $(-4, 2)$. Then, sketch the asymptotes by drawing lines through the corners of the central rectangle. Sketch the hyperbola, as shown in Figure 12.37. From the graph, you can see that the center of the hyperbola is shifted four units to the left and two units upward from the origin.

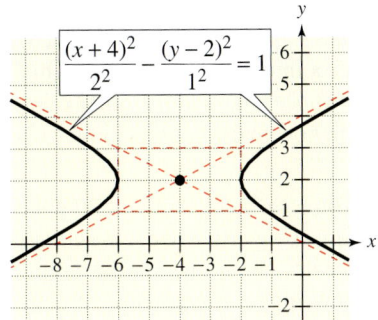

Figure 12.37

Technology: Tip

You can use a graphing calculator to graph a hyperbola. For instance, to graph the hyperbola $4y^2 - 9x^2 = 36$, first solve for y to obtain

$$y_1 = 3\sqrt{\frac{x^2}{4} + 1}$$

and

$$y_2 = -3\sqrt{\frac{x^2}{4} + 1}.$$

Use a viewing window in which $-6 \le x \le 6$ and $-8 \le y \le 8$. You should obtain the graph shown below.

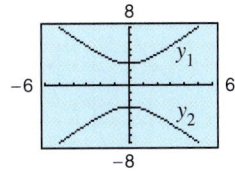

12.3 Exercises

Review Concepts, Skills, and Problem Solving

Keep mathematically in shape by doing these exercises *before* the problems of this section.

Distance Formula

In Exercises 1 and 2, find the distance between the points.

1. $(5, 2), (-1, 4)$
2. $(-4, -3), (6, 10)$

Graphs

In Exercises 3–6, graph the lines on the same set of coordinate axes.

3. $y = \pm 4x$
4. $y = 6 \pm \frac{1}{3}x$
5. $y = 5 \pm \frac{1}{2}(x - 2)$
6. $y = \pm \frac{1}{3}(x - 6)$

Solving Equations

In Exercises 7–10, find the unknown in the equation $c^2 = a^2 - b^2$. (Assume that a, b, and c are positive.)

7. $a = 25$, $b = 7$
8. $a = \sqrt{41}$, $c = 4$
9. $b = 5$, $c = 12$
10. $a = 6$, $b = 3$

Problem Solving

11. *Average Speed* From a point on a straight road, two people ride bicycles in opposite directions. One person rides at 10 miles per hour and the other rides at 12 miles per hour. In how many hours will they be 55 miles apart?

12. *Mixture Problem* You have a collection of 30 gold coins. Some of the coins are worth $10 each, and the rest are worth $20 each. The value of the entire collection is $540. How many of each type of coin do you have?

Developing Skills

In Exercises 1–6, match the equation with its graph. [The graphs are labeled (a), (b), (c), (d), (e), and (f).]

(a)
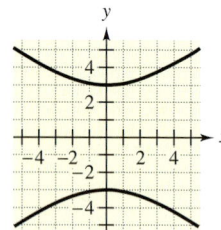

(b)
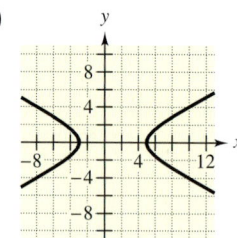

(c)
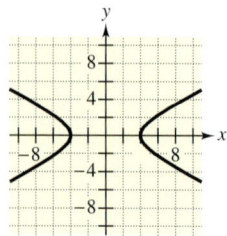

(d)
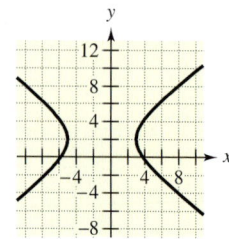

(e)
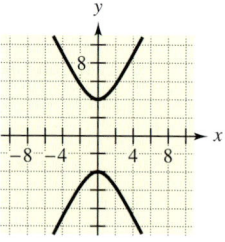

(f)
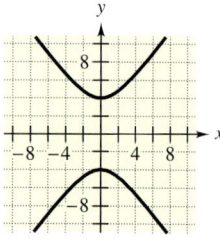

1. $\dfrac{x^2}{16} - \dfrac{y^2}{4} = 1$

2. $\dfrac{y^2}{16} - \dfrac{x^2}{4} = 1$

3. $\dfrac{y^2}{9} - \dfrac{x^2}{16} = 1$

4. $\dfrac{y^2}{16} - \dfrac{x^2}{9} = 1$

5. $\dfrac{(x - 1)^2}{16} - \dfrac{y^2}{4} = 1$

6. $\dfrac{(x + 1)^2}{16} - \dfrac{(y - 2)^2}{9} = 1$

In Exercises 7–18, sketch the hyperbola. Identify the vertices and asymptotes. See Example 1.

7. $x^2 - y^2 = 9$ **8.** $x^2 - y^2 = 1$

9. $y^2 - x^2 = 9$ **10.** $y^2 - x^2 = 1$

11. $\dfrac{x^2}{9} - \dfrac{y^2}{25} = 1$ **12.** $\dfrac{x^2}{4} - \dfrac{y^2}{9} = 1$

13. $\dfrac{y^2}{9} - \dfrac{x^2}{25} = 1$ **14.** $\dfrac{y^2}{4} - \dfrac{x^2}{9} = 1$

15. $\dfrac{x^2}{1} - \dfrac{y^2}{9/4} = 1$ **16.** $\dfrac{y^2}{1/4} - \dfrac{x^2}{25/4} = 1$

17. $4y^2 - x^2 + 16 = 0$ **18.** $4y^2 - 9x^2 - 36 = 0$

In Exercises 19–26, write the standard form of the equation of the hyperbola centered at the origin. See Example 2.

Vertices	Asymptotes	
19. $(-4, 0), (4, 0)$	$y = 2x$	$y = -2x$
20. $(-2, 0), (2, 0)$	$y = \frac{1}{3}x$	$y = -\frac{1}{3}x$
21. $(0, -4), (0, 4)$	$y = \frac{1}{2}x$	$y = -\frac{1}{2}x$
22. $(0, -2), (0, 2)$	$y = 3x$	$y = -3x$
23. $(-9, 0), (9, 0)$	$y = \frac{2}{3}x$	$y = -\frac{2}{3}x$
24. $(-1, 0), (1, 0)$	$y = \frac{1}{2}x$	$y = -\frac{1}{2}x$
25. $(0, -1), (0, 1)$	$y = 2x$	$y = -2x$
26. $(0, -5), (0, 5)$	$y = x$	$y = -x$

In Exercises 27–30, use a graphing calculator to graph the equation. (*Note:* Solve for *y*.)

27. $\dfrac{x^2}{16} - \dfrac{y^2}{4} = 1$

28. $\dfrac{y^2}{16} - \dfrac{x^2}{4} = 1$

29. $5x^2 - 2y^2 + 10 = 0$

30. $x^2 - 2y^2 - 4 = 0$

In Exercises 31–38, find the center and vertices of the hyperbola and sketch the hyperbola. See Examples 3 and 4.

31. $(y + 4)^2 - (x - 3)^2 = 25$

32. $(y + 6)^2 - (x - 2)^2 = 1$

33. $\dfrac{(x - 1)^2}{4} - \dfrac{(y + 2)^2}{1} = 1$

34. $\dfrac{(x - 2)^2}{4} - \dfrac{(y - 3)^2}{9} = 1$

35. $9x^2 - y^2 - 36x - 6y + 18 = 0$

36. $x^2 - 9y^2 + 36y - 72 = 0$

37. $4x^2 - y^2 + 24x + 4y + 28 = 0$

38. $25x^2 - 4y^2 + 100x + 8y + 196 = 0$

In Exercises 39–42, write the standard form of the equation of the hyperbola.

39.

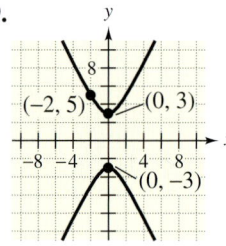

40.

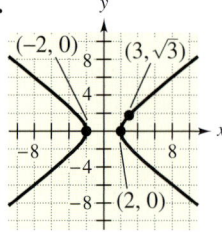

41.

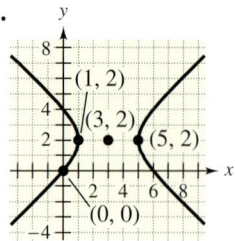

42.

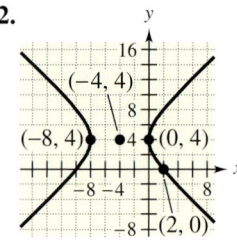

Solving Problems

43. *Navigation*   Long-distance radio navigation for aircraft and ships uses synchronized pulses transmitted by widely separated transmitting stations. These pulses travel at the speed of light (186,000 miles per second). The difference in the times of arrival of these pulses at an aircraft or ship is constant on a hyperbola having the transmitting stations as foci. Assume that two stations 300 miles apart are positioned on a rectangular coordinate system at points with coordinates $(-150, 0)$ and $(150, 0)$ and that a ship is traveling on a path with coordinates $(x, 75)$, as shown in the figure. Find the x-coordinate of the position of the ship if the time difference between the pulses from the transmitting stations is 1000 microseconds (0.001 second).

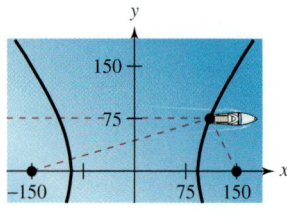

Figure for 43

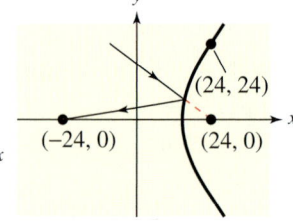

Figure for 44

44. *Optics*   A hyperbolic mirror (used in some telescopes) has the property that a light ray directed at the focus will be reflected to the other focus. The focus of a hyperbolic mirror (see figure) has coordinates $(24, 0)$. Find the vertex of the mirror if its mount at the top edge of the mirror has coordinates $(24, 24)$.

Explaining Concepts

45. ⬢ Answer parts (c) and (d) of Motivating the Chapter on page 756.

46. *Writing*   Define a hyperbola and write the standard form of the equation of a hyperbola centered at the origin.

47. *Writing*   Explain the significance of the foci in a hyperbola.

48. *Writing*   Explain how the central rectangle of a hyperbola can be used to sketch its asymptotes.

49. *Think About It*   Describe the part of the hyperbola

$$\frac{(x-3)^2}{4} - \frac{(y-1)^2}{9} = 1$$

given by each equation.
(a) $x = 3 - \frac{2}{3}\sqrt{9 + (y-1)^2}$
(b) $y = 1 + \frac{3}{2}\sqrt{(x-3)^2 - 4}$

50. Cut cone-shaped pieces of styrofoam to demonstrate how to obtain each type of conic section: circle, parabola, ellipse, and hyperbola. Discuss how you could write directions for someone else to form each conic section. Compile a list of real-life situations and/or everyday objects in which conic sections may be seen.

12.4 Solving Nonlinear Systems of Equations

What You Should Learn

1. Solve nonlinear systems of equations graphically.
2. Solve nonlinear systems of equations by substitution.
3. Solve nonlinear systems of equations by elimination.
4. Use nonlinear systems of equations to model and solve real-life problems.

Why You Should Learn It

Nonlinear systems of equations can be used to analyze real-life data. For instance, in Exercise 84 on page 799, nonlinear models are used to represent the populations of two states in the United States.

1 Solve nonlinear systems of equations graphically.

Solving Nonlinear Systems of Equations by Graphing

In Chapter 8, you studied several methods for solving systems of linear equations. For instance, the following linear system has one solution, $(2, -1)$, which means that $(2, -1)$ is a point of intersection of the two lines represented by the system.

$$\begin{cases} 2x - 3y = 7 \\ x + 4y = -2 \end{cases}$$

In Chapter 8, you also learned that a linear system can have no solution, exactly one solution, or infinitely many solutions. A **nonlinear system of equations** is a system that contains at least one nonlinear equation. Nonlinear systems of equations can have no solution, one solution, or two or more solutions. For instance, the hyperbola and line in Figure 12.38(a) have no point of intersection, the circle and line in Figure 12.38(b) have one point of intersection, and the parabola and line in Figure 12.38(c) have two points of intersection.

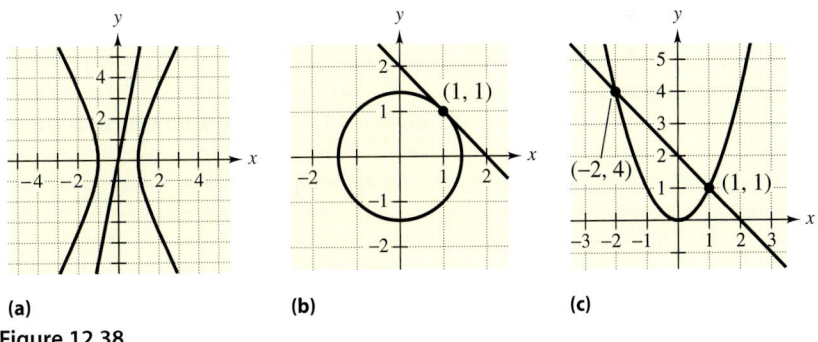

(a)   (b)   (c)
Figure 12.38

You can solve a nonlinear system of equations graphically, as follows.

Solving a Nonlinear System Graphically

1. Sketch the graph of each equation in the system.

2. Locate the point(s) of intersection of the graphs (if any) and graphically approximate the coordinates of the points.

3. Check the coordinates by substituting them into each equation in the original system. If the coordinates do not check, you may have to use an algebraic approach, as discussed later in this section.

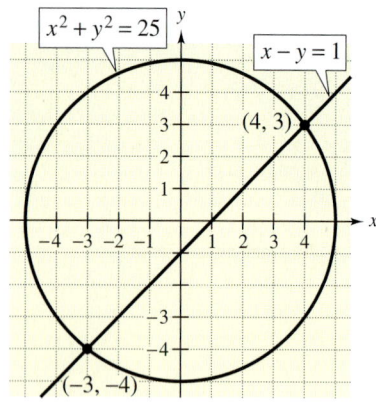

Figure 12.39

Technology: Tip

Try using a graphing calculator to solve the system described in Example 1. When you do this, remember that the circle needs to be entered as two separate equations.

$y_1 = \sqrt{25 - x^2}$ Top half of circle

$y_2 = -\sqrt{25 - x^2}$ Bottom half of circle

$y_3 = x - 1$ Line

Example 1   Solving a Nonlinear System Graphically

Find all solutions of the nonlinear system of equations.

$$\begin{cases} x^2 + y^2 = 25 & \text{Equation 1} \\ x - y = 1 & \text{Equation 2} \end{cases}$$

Solution

Begin by sketching the graph of each equation. The first equation graphs as a circle centered at the origin and having a radius of 5. The second equation, which can be written as $y = x - 1$, graphs as a line with a slope of 1 and a y-intercept of $(0, -1)$. From the graphs shown in Figure 12.39, you can see that the system appears to have two solutions: $(-3, -4)$ and $(4, 3)$. You can check these solutions by substituting for x and y in the original system, as follows.

Check

To check $(-3, -4)$, substitute -3 for x and -4 for y in each equation.

$(-3)^2 + (-4)^2 \overset{?}{=} 25$ Substitute -3 for x and -4 for y in Equation 1.

$9 + 16 = 25$ Solution checks in Equation 1. ✓

$(-3) - (-4) \overset{?}{=} 1$ Substitute -3 for x and -4 for y in Equation 2.

$-3 + 4 = 1$ Solution checks in Equation 2. ✓

To check $(4, 3)$, substitute 4 for x and 3 for y in each equation.

$4^2 + 3^2 \overset{?}{=} 25$ Substitute 4 for x and 3 for y in Equation 1.

$16 + 9 = 25$ Solution checks in Equation 1. ✓

$4 - 3 \overset{?}{=} 1$ Substitute 4 for x and 3 for y in Equation 2.

$1 = 1$ Solution checks in Equation 2. ✓

Example 2   Solving a Nonlinear System Graphically

Find all solutions of the nonlinear system of equations.

$$\begin{cases} x = (y - 3)^2 & \text{Equation 1} \\ x + y = 5 & \text{Equation 2} \end{cases}$$

Solution

Begin by sketching the graph of each equation. Solve the first equation for y.

$x = (y - 3)^2$ Write original equation.

$\pm\sqrt{x} = y - 3$ Take the square root of each side.

$3 \pm \sqrt{x} = y$ Add 3 to each side.

The graph of $y = 3 \pm \sqrt{x}$ is a parabola with its vertex at $(0, 3)$. The second equation, which can be written as $y = -x + 5$, graphs as a line with a slope of -1 and a y-intercept of $(0, 5)$. The system appears to have two solutions: $(4, 1)$ and $(1, 4)$, as shown in Figure 12.40. Check these solutions in the original system.

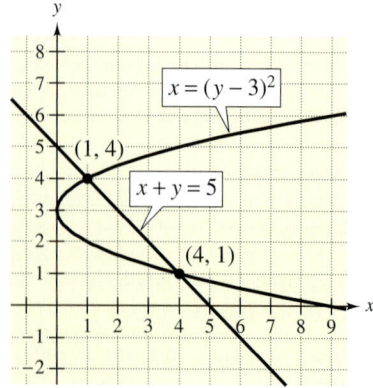

Figure 12.40

Solving Nonlinear Systems of Equations by Substitution

2 Solve nonlinear systems of equations by substitution.

The graphical approach to solving any type of system (linear or nonlinear) in two variables is very useful for helping you see the number of solutions and their approximate coordinates. For systems with solutions having "messy" coordinates, however, a graphical approach is usually not accurate enough to produce exact solutions. In such cases, you should use an algebraic approach. (With an algebraic approach, you should still sketch the graph of each equation in the system.)

As with systems of *linear* equations, there are two basic algebraic approaches: substitution and elimination. Substitution usually works well for systems in which one of the equations is linear, as shown in Example 3.

Example 3   Using Substitution to Solve a Nonlinear System

Solve the nonlinear system of equations.

$$\begin{cases} 4x^2 + y^2 = 4 & \text{Equation 1} \\ -2x + y = 2 & \text{Equation 2} \end{cases}$$

Solution

Begin by solving for y in Equation 2 to obtain $y = 2x + 2$. Next, substitute this expression for y into Equation 1.

$4x^2 + y^2 = 4$	Write Equation 1.
$4x^2 + (2x + 2)^2 = 4$	Substitute $2x + 2$ for y.
$4x^2 + 4x^2 + 8x + 4 = 4$	Expand.
$8x^2 + 8x = 0$	Simplify.
$8x(x + 1) = 0$	Factor.
$8x = 0$ $x = 0$	Set 1st factor equal to 0.
$x + 1 = 0$ $x = -1$	Set 2nd factor equal to 0.

Finally, back-substitute these values of x into the revised Equation 2 to solve for y.

For $x = 0$: $y = 2(0) + 2 = 2$

For $x = -1$: $y = 2(-1) + 2 = 0$

So, the system of equations has two solutions: $(0, 2)$ and $(-1, 0)$. Figure 12.41 shows the graph of the system. You can check the solutions as follows.

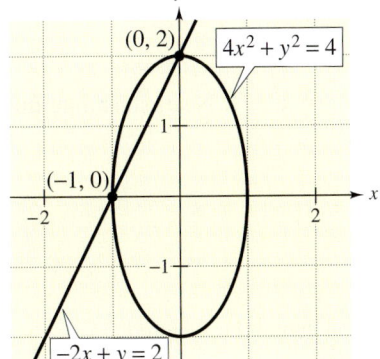

Figure 12.41

Check First Solution

$4(0)^2 + 2^2 \overset{?}{=} 4$

$0 + 4 = 4$ ✓

$-2(0) + 2 \overset{?}{=} 2$

$2 = 2$ ✓

Check Second Solution

$4(-1)^2 + 0^2 \overset{?}{=} 4$

$4 + 0 = 4$ ✓

$-2(-1) + 0 \overset{?}{=} 2$

$2 = 2$ ✓

The steps for using the method of substitution to solve a system of two equations involving two variables are summarized as follows.

Method of Substitution

To solve a system of two equations in two variables, use the steps below.

1. Solve one of the equations for one variable in terms of the other variable.

2. Substitute the expression found in Step 1 into the other equation to obtain an equation in one variable.

3. Solve the equation obtained in Step 2.

4. Back-substitute the solution from Step 3 into the expression obtained in Step 1 to find the value of the other variable.

5. Check the solution to see that it satisfies *both* of the original equations.

Example 4 shows how the method of substitution and graphing can be used to determine that a nonlinear system of equations has no solution.

Example 4   Solving a Nonlinear System: No-Solution Case

Solve the nonlinear system of equations.

$$\begin{cases} x^2 - y = 0 & \text{Equation 1} \\ x - y = 1 & \text{Equation 2} \end{cases}$$

Solution

Begin by solving for y in Equation 2 to obtain $y = x - 1$. Next, substitute this expression for y into Equation 1.

$$x^2 - y = 0 \qquad \text{Write Equation 1.}$$

$$x^2 - (x - 1) = 0 \qquad \text{Substitute } x - 1 \text{ for } y.$$

$$x^2 - x + 1 = 0 \qquad \text{Distributive Property}$$

Use the Quadratic Formula, because this equation cannot be factored.

$$x = \frac{-(-1) \pm \sqrt{(-1)^2 - 4(1)(1)}}{2(1)} \qquad \text{Use Quadratic Formula.}$$

$$= \frac{1 \pm \sqrt{1 - 4}}{2} = \frac{1 \pm \sqrt{-3}}{2} \qquad \text{Simplify.}$$

Now, because the Quadratic Formula yields a negative number inside the radical, you can conclude that the equation $x^2 - x + 1 = 0$ has no (real) solution. So, the system has no (real) solution. Figure 12.42 shows the graph of this system. From the graph, you can see that the parabola and the line have no point of intersection, and so the system has no solution.

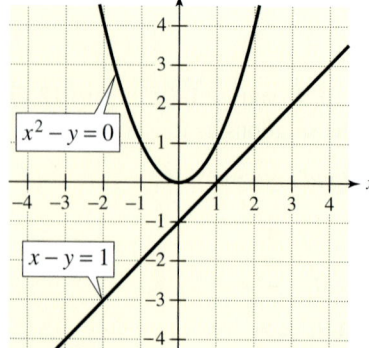

Figure 12.42

③ Solve nonlinear systems of equations by elimination.

Solving Nonlinear Systems of Equations by Elimination

In Section 8.2, you learned how to use the method of elimination to solve a linear system. This method can also be used with special types of nonlinear systems, as demonstrated in Example 5.

Example 5   Using Elimination to Solve a Nonlinear System

Solve the nonlinear system of equations.

$$\begin{cases} 4x^2 + y^2 = 64 & \text{Equation 1} \\ x^2 + y^2 = 52 & \text{Equation 2} \end{cases}$$

Solution

Because both equations have y^2 as a term (and no other terms containing y), you can eliminate y by subtracting Equation 2 from Equation 1.

$$\begin{array}{rl} 4x^2 + y^2 = & 64 \\ -x^2 - y^2 = & -52 \\ \hline 3x^2 \quad\;\; = & 12 \end{array}$$ Subtract Equation 2 from Equation 1.

After eliminating y, solve the remaining equation for x.

$$3x^2 = 12$$ Write resulting equation.

$$x^2 = 4$$ Divide each side by 3.

$$x = \pm 2$$ Take square root of each side.

To find the corresponding values of y, substitute these values of x into either of the original equations. By substituting $x = 2$, you obtain

$$x^2 + y^2 = 52$$ Write Equation 2.

$$(2)^2 + y^2 = 52$$ Substitute 2 for x.

$$y^2 = 48$$ Subtract 4 from each side.

$$y = \pm 4\sqrt{3}.$$ Take square root of each side and simplify.

By substituting $x = -2$, you obtain the same values of y, as follows.

$$x^2 + y^2 = 52$$ Write Equation 2.

$$(-2)^2 + y^2 = 52$$ Substitute -2 for x.

$$y^2 = 48$$ Subtract 4 from each side.

$$y = \pm 4\sqrt{3}$$ Take square root of each side and simplify.

This implies that the system has four solutions:

$$\left(2, 4\sqrt{3}\right),\quad \left(2, -4\sqrt{3}\right),\quad \left(-2, 4\sqrt{3}\right),\quad \left(-2, -4\sqrt{3}\right).$$

Check these in the original system. Figure 12.43 shows the graph of the system. Notice that the graph of Equation 1 is an ellipse and the graph of Equation 2 is a circle.

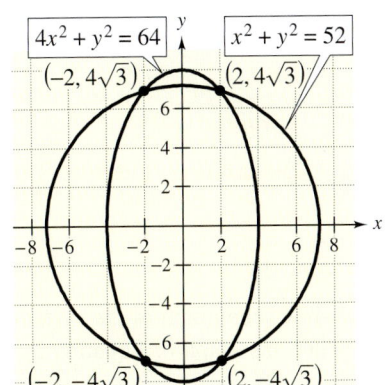

Figure 12.43

Example 6 Using Elimination to Solve a Nonlinear System

Solve the nonlinear system of equations.

$$\begin{cases} x^2 - 2y = 4 & \text{Equation 1} \\ x^2 - y^2 = 1 & \text{Equation 2} \end{cases}$$

Solution

Because both equations have x^2 as a term (and no other terms containing x), you can eliminate x by subtracting Equation 2 from Equation 1.

$$\begin{aligned} x^2 - 2y &= 4 \\ \underline{-x^2 + y^2 = -1} & \qquad \text{Subtract Equation 2 from Equation 1.} \\ y^2 - 2y &= 3 \end{aligned}$$

After eliminating x, solve the remaining equation for y.

$y^2 - 2y = 3$	Write resulting equation.
$y^2 - 2y - 3 = 0$	Write in general form.
$(y + 1)(y - 3) = 0$	Factor.
$y + 1 = 0 \implies y = -1$	Set 1st factor equal to 0.
$y - 3 = 0 \implies y = 3$	Set 2nd factor equal to 0.

When $y = -1$, you obtain

$x^2 - y^2 = 1$	Write Equation 2.
$x^2 - (-1)^2 = 1$	Substitute -1 for y.
$x^2 - 1 = 1$	Simplify.
$x^2 = 2$	Add 1 to each side.
$x = \pm\sqrt{2}.$	Take square root of each side.

When $y = 3$, you obtain

$x^2 - y^2 = 1$	Write Equation 2.
$x^2 - (3)^2 = 1$	Substitute 3 for y.
$x^2 - 9 = 1$	Simplify.
$x^2 = 10$	Add 9 to each side.
$x = \pm\sqrt{10}.$	Take square root of each side.

This implies that the system has four solutions:

$$\left(\sqrt{2}, -1\right), \quad \left(-\sqrt{2}, -1\right), \quad \left(\sqrt{10}, 3\right), \quad \left(-\sqrt{10}, 3\right).$$

Check these in the original system. Figure 12.44 shows the graph of the system. Notice that the graph of Equation 1 is a parabola and the graph of Equation 2 is a hyperbola.

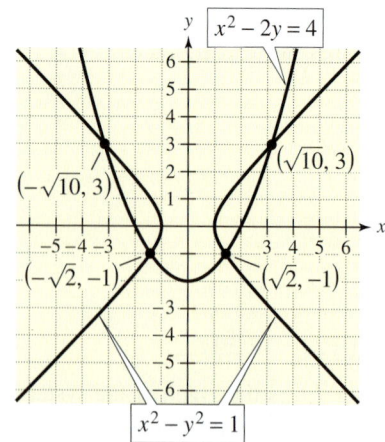

Figure 12.44

In Example 6, the method of elimination yields the four exact solutions $\left(\sqrt{2}, -1\right)$, $\left(-\sqrt{2}, -1\right)$, $\left(\sqrt{10}, 3\right)$, and $\left(-\sqrt{10}, 3\right)$. You can use a calculator to approximate these solutions to be $(1.41, -1)$, $(-1.41, -1)$, $(3.16, 3)$, and $(-3.16, 3)$. If you use the decimal approximations to check your solutions in the original system, be aware that they may not check.

4 Use nonlinear systems of equations to model and solve real-life problems.

Application

There are many examples of the use of nonlinear systems of equations in business and science. For instance, in Example 7 a nonlinear system of equations is used to compare the revenues of two companies.

Example 7   Comparing the Revenues of Two Companies

From 1990 through 2005, the revenues R (in millions of dollars) of a restaurant chain and a sportswear manufacturer can be modeled by

$$\begin{cases} R = 0.1t + 2.5 \\ R = 0.02t^2 - 0.2t + 3.2 \end{cases}$$

respectively, where t represents the year, with $t = 0$ corresponding to 1990. Sketch the graphs of these two models. During which two years did the companies have approximately equal revenues?

Solution

The graphs of the two models are shown in Figure 12.45. From the graph, you can see that the restaurant chain's revenue followed a linear pattern. It had a revenue of $2.5 million in 1990 and had an increase of $0.1 million each year. The sportswear manufacturer's revenue followed a quadratic pattern. From 1990 to 1995, the company's revenue was decreasing. Then, from 1995 through 2005, the revenue was increasing. From the graph, you can see that the two companies had approximately equal revenues in 1993 (the restaurant chain had $2.8 million and the sportswear manufacturer had $2.78 million) and again in 2002 (the restaurant chain had $3.7 million and the sportswear manufacturer had $3.68 million).

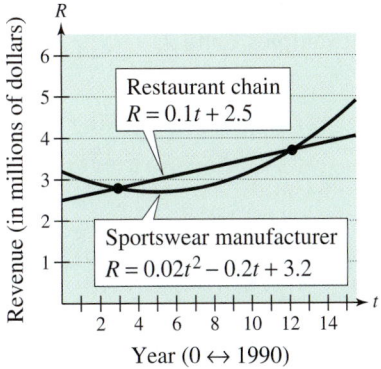

Figure 12.45

12.4 Exercises

Review Concepts, Skills, and Problem Solving

Keep mathematically in shape by doing these exercises *before* the problems of this section.

Properties and Definitions

In Exercises 1–3, identify the row operation performed on the matrix to produce the equivalent matrix.

Original Matrix   *New Row-Equivalent Matrix*

1. $\begin{bmatrix} 2 & 4 & 5 \\ 1 & -3 & 8 \end{bmatrix}$ $\begin{bmatrix} 2 & 4 & 5 \\ -2 & 6 & -16 \end{bmatrix}$

2. $\begin{bmatrix} 7 & -2 & 0 \\ 1 & 0 & 5 \end{bmatrix}$ $\begin{bmatrix} 1 & 0 & 5 \\ 7 & -2 & 0 \end{bmatrix}$

3. $\begin{bmatrix} 1 & 6 & 2 \\ 3 & 1 & -1 \end{bmatrix}$ $\begin{bmatrix} 1 & 6 & 2 \\ 0 & -17 & -7 \end{bmatrix}$

4. *Writing* In your own words, explain the process of Gaussian elimination when using matrices to represent a system of linear equations.

Solving Systems of Equations

In Exercises 5–10, solve the system of linear equations.

5. $\begin{cases} 3x + 5y = 9 \\ 2x - 3y = -13 \end{cases}$ **6.** $\begin{cases} 2x + 6y = -6 \\ 3x + 5y = 7 \end{cases}$

7. $\begin{cases} 5x + 6y = -12 \\ 3x + 9y = 15 \end{cases}$ **8.** $\begin{cases} 3x - 7y = 5 \\ 7x - 3y = 8 \end{cases}$

9. $\begin{cases} x - 2y + z = 7 \\ 2x + y - z = 0 \\ 3x + 2y - 2z = -2 \end{cases}$ **10.** $\begin{cases} x - 3y + 7z = 13 \\ x + y + z = 1 \\ x - 2y + 3z = 4 \end{cases}$

Problem Solving

11. *Simple Interest* An investment of $4600 is made at an annual simple interest rate of 6.8%. How much additional money must be invested at an annual simple interest rate of 9% so that the total interest earned is 8% of the total investment?

12. *Nut Mixture* Cashews sell for $6.75 per pound and Brazil nuts sell for $5.00 per pound. How much of each type of nut should be used to make a 50-pound mixture that sells for $5.70 per pound?

Developing Skills

In Exercises 1–12, graph the equations to determine whether the system has any solutions. Find any solutions that exist. See Examples 1 and 2.

1. $\begin{cases} x + y = 2 \\ x^2 - y = 0 \end{cases}$ **2.** $\begin{cases} 2x + y = 10 \\ x^2 + y^2 = 25 \end{cases}$

3. $\begin{cases} x^2 + y = 9 \\ x - y = -3 \end{cases}$ **4.** $\begin{cases} x - y^2 = 0 \\ x - y = 2 \end{cases}$

5. $\begin{cases} y = \sqrt{x - 2} \\ x - 2y = 1 \end{cases}$ **6.** $\begin{cases} x - 2y = 4 \\ x^2 - y = 0 \end{cases}$

7. $\begin{cases} x^2 + y^2 = 100 \\ x + y = 2 \end{cases}$ **8.** $\begin{cases} x^2 + y^2 = 169 \\ x + y = 7 \end{cases}$

9. $\begin{cases} x^2 + y^2 = 25 \\ 2x - y = -5 \end{cases}$ **10.** $\begin{cases} x^2 - y^2 = 16 \\ 3x - y = 12 \end{cases}$

11. $\begin{cases} 9x^2 - 4y^2 = 36 \\ 5x - 2y = 0 \end{cases}$ **12.** $\begin{cases} 9x^2 + 4y^2 = 36 \\ 3x - 2y + 6 = 0 \end{cases}$

⊞ In Exercises 13–26, use a graphing calculator to graph the equations and find any solutions of the system.

13. $\begin{cases} y = 2x^2 \\ y = -2x + 12 \end{cases}$
14. $\begin{cases} y = 5x^2 \\ y = -15x - 10 \end{cases}$

15. $\begin{cases} y = x \\ y = x^3 \end{cases}$
16. $\begin{cases} y = x^2 \\ y = x + 2 \end{cases}$

17. $\begin{cases} y = x^2 \\ y = -x^2 + 4x \end{cases}$
18. $\begin{cases} y = 8 - x^2 \\ y = 6 - x \end{cases}$

19. $\begin{cases} x^2 - y = 2 \\ 3x + y = 2 \end{cases}$
20. $\begin{cases} x^2 + 2y = 6 \\ x - y = -4 \end{cases}$

21. $\begin{cases} y = x^2 + 2 \\ y = -x^2 + 4 \end{cases}$
22. $\begin{cases} \sqrt{x} + 1 = y \\ 2x + y = 4 \end{cases}$

23. $\begin{cases} x^2 - y^2 = 12 \\ x - 2y = 0 \end{cases}$
24. $\begin{cases} x^2 + y = 4 \\ x + y = 6 \end{cases}$

25. $\begin{cases} y = x^3 \\ y = x^3 - 3x^2 + 3x \end{cases}$
26. $\begin{cases} y = -2(x^2 - 1) \\ y = 2(x^4 - 2x^2 + 1) \end{cases}$

In Exercises 27–54, solve the system by the method of substitution. See Examples 3 and 4.

27. $\begin{cases} y = 2x^2 \\ y = 6x - 4 \end{cases}$

28. $\begin{cases} y = 5x^2 \\ y = -5x + 10 \end{cases}$

29. $\begin{cases} x^2 + y = 5 \\ 2x + y = 5 \end{cases}$

30. $\begin{cases} x - y^2 = 0 \\ x - y = 2 \end{cases}$

31. $\begin{cases} x^2 + y = 1 \\ x + y = -4 \end{cases}$

32. $\begin{cases} x^2 + y^2 = 36 \\ x = 8 \end{cases}$

33. $\begin{cases} x^2 + y^2 = 25 \\ y = 5 \end{cases}$

34. $\begin{cases} x^2 + y^2 = 1 \\ x + y = 7 \end{cases}$

35. $\begin{cases} x^2 + y^2 = 64 \\ -3x + y = 8 \end{cases}$

36. $\begin{cases} x^2 + y^2 = 81 \\ x + 3y = 27 \end{cases}$

37. $\begin{cases} 4x + y^2 = 2 \\ 2x - y = -11 \end{cases}$

38. $\begin{cases} x^2 + y^2 = 10 \\ 2x - y = 5 \end{cases}$

39. $\begin{cases} x^2 + y^2 = 9 \\ x + 2y = 3 \end{cases}$

40. $\begin{cases} x^2 + y^2 = 4 \\ x - 2y = 4 \end{cases}$

41. $\begin{cases} 2x^2 - y^2 = -8 \\ x - y = 6 \end{cases}$

42. $\begin{cases} y^2 = -x + 4 \\ x^2 + y^2 = 6 \end{cases}$

43. $\begin{cases} y = x^2 - 5 \\ 3x + 2y = 10 \end{cases}$

44. $\begin{cases} x + y = 4 \\ x^2 - y^2 = 4 \end{cases}$

45. $\begin{cases} y = \sqrt{4 - x} \\ x + 3y = 6 \end{cases}$

46. $\begin{cases} y = \sqrt[3]{x} \\ y = x \end{cases}$

47. $\begin{cases} x^2 - 4y^2 = 16 \\ x^2 + y^2 = 1 \end{cases}$

48. $\begin{cases} 2x^2 + y^2 = 16 \\ x^2 - y^2 = -4 \end{cases}$

49. $\begin{cases} y = x^2 - 3 \\ x^2 + y^2 = 9 \end{cases}$

50. $\begin{cases} x^2 + y^2 = 25 \\ x - 3y = -5 \end{cases}$

51. $\begin{cases} 16x^2 + 9y^2 = 144 \\ 4x + 3y = 12 \end{cases}$

52. $\begin{cases} y = 2x^2 \\ y = x^4 - 2x^2 \end{cases}$

53. $\begin{cases} x^2 - y^2 = 9 \\ x^2 + y^2 = 1 \end{cases}$

54. $\begin{cases} x^2 - y^2 = 4 \\ x - y = 2 \end{cases}$

In Exercises 55–76, solve the system by the method of elimination. See Examples 5 and 6.

55. $\begin{cases} x^2 + 2y = 1 \\ x^2 + y^2 = 4 \end{cases}$

56. $\begin{cases} x + y^2 = 5 \\ 2x^2 + y^2 = 6 \end{cases}$

57. $\begin{cases} -x + y^2 = 10 \\ x^2 - y^2 = -8 \end{cases}$

58. $\begin{cases} x^2 + y = 9 \\ x^2 - y^2 = 7 \end{cases}$

59. $\begin{cases} x^2 + y^2 = 7 \\ x^2 - y^2 = 1 \end{cases}$

60. $\begin{cases} x^2 + y^2 = 25 \\ y^2 - x^2 = 7 \end{cases}$

61. $\begin{cases} x^2 - y^2 = 4 \\ x^2 + y^2 = 4 \end{cases}$

62. $\begin{cases} x^2 + y^2 = 25 \\ x^2 - 2y^2 = 7 \end{cases}$

63. $\begin{cases} x^2 + y^2 = 13 \\ 2x^2 + 3y^2 = 30 \end{cases}$

64. $\begin{cases} 3x^2 - y^2 = 4 \\ x^2 + 4y^2 = 10 \end{cases}$

65. $\begin{cases} 4x^2 + 9y^2 = 36 \\ 2x^2 - 9y^2 = 18 \end{cases}$

66. $\begin{cases} 5x^2 - 2y^2 = -13 \\ 3x^2 + 4y^2 = 39 \end{cases}$

67. $\begin{cases} 2x^2 + 3y^2 = 21 \\ x^2 + 2y^2 = 12 \end{cases}$

68. $\begin{cases} 2x^2 + y^2 = 11 \\ x^2 + 3y^2 = 28 \end{cases}$

69. $\begin{cases} -x^2 - 2y^2 = 6 \\ 5x^2 + 15y^2 = 20 \end{cases}$

70. $\begin{cases} x^2 - 2y^2 = 7 \\ x^2 + y^2 = 34 \end{cases}$

71. $\begin{cases} x^2 + y^2 = 9 \\ 16x^2 - 4y^2 = 64 \end{cases}$

72. $\begin{cases} 3x^2 + 4y^2 = 35 \\ 2x^2 + 5y^2 = 42 \end{cases}$

73. $\begin{cases} \dfrac{x^2}{4} + y^2 = 1 \\ x^2 + \dfrac{y^2}{4} = 1 \end{cases}$

74. $\begin{cases} x^2 - y^2 = 1 \\ \dfrac{x^2}{2} + y^2 = 1 \end{cases}$

75. $\begin{cases} y^2 - x^2 = 10 \\ x^2 + y^2 = 16 \end{cases}$

76. $\begin{cases} x^2 + y^2 = 25 \\ x^2 + 2y^2 = 36 \end{cases}$

Solving Problems

77. *Hyperbolic Mirror* In a hyperbolic mirror, light rays directed to one focus are reflected to the other focus. The mirror in the figure has the equation

$$\frac{x^2}{9} - \frac{y^2}{16} = 1.$$

At which point on the mirror will light from the point (0, 10) reflect to the focus?

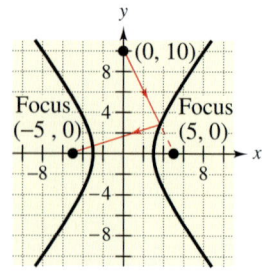

Figure for 77

78. *Sports* You are playing miniature golf and your golf ball is at $(-15, 25)$ (see figure). A wall at the end of the enclosed area is part of a hyperbola whose equation is

$$\frac{x^2}{19} - \frac{y^2}{81} = 1.$$

Using the reflective property of hyperbolas given in Exercise 77, at which point on the wall must your ball hit for it to go into the hole? (The ball bounces off the wall only once.)

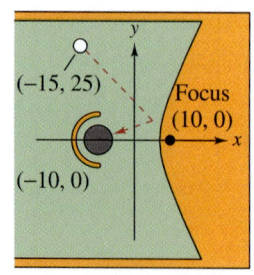

Figure for 78

79. *Geometry* A high-definition rectangular television screen has a picture area of 762 square inches and a diagonal measurement of 42 inches. Find the dimensions of the television.

80. ▲ *Geometry*   A rectangular ice rink has an area of 3000 square feet. The diagonal across the rink is 85 feet. Find the dimensions of the rink.

81. ▲ *Geometry*   A rectangular piece of wood has a diagonal that measures 17 inches. The perimeter of each triangle formed by the diagonal is 40 inches. Find the dimensions of the piece of wood.

82. ▲ *Geometry*   A sail for a sailboat is shaped like a right triangle that has a perimeter of 36 meters and a hypotenuse of 15 meters. Find the dimensions of the sail.

83. *Busing Boundary*   To be eligible to ride the school bus to East High School, a student must live at least 1 mile from the school (see figure). Describe the portion of Clarke Street for which the residents are *not* eligible to ride the school bus. Use a coordinate system in which the school is at $(0, 0)$ and each unit represents 1 mile.

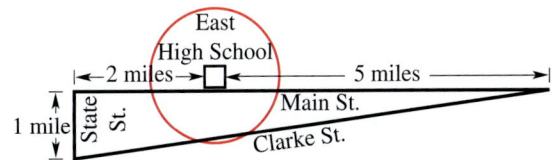

Figure for 83

84. ⊞ *Data Analysis*   From 1991 through 2001, the population of North Carolina grew at a lower rate than the population of Georgia. Two models that represent the populations of the two states are

$$P = 7.66t^2 + 78.0t + 6573 \quad \text{Georgia}$$
$$P = 7.52t^2 + 49.6t + 6715 \quad \text{North Carolina}$$

where P is the population in thousands and t is the year, with $t = 1$ corresponding to 1991. Use a graphing calculator to determine the year in which the population of Georgia overtook the population of North Carolina.   (Source: U.S. Census Bureau)

Explaining Concepts

85. ⊘   Answer parts (e)–(g) of Motivating the Chapter on page 756.

86. *Writing*✏   Explain how to solve a nonlinear system of equations using the method of substitution.

87. *Writing*✏   Explain how to solve a nonlinear system of equations using the method of elimination.

88. A circle and a parabola can have 0, 1, 2, 3, or 4 points of intersection. Sketch the circle given by $x^2 + y^2 = 4$. Discuss how this circle could intersect a parabola with an equation of the form $y = x^2 + C$. Then find the values of C for each of the five cases described below.

(a) No points of intersection

(b) One point of intersection

(c) Two points of intersection

(d) Three points of intersection

(e) Four points of intersection

Use a graphing calculator to confirm your results.

What Did You Learn?

Key Terms

conics (conic sections), *p. 758*
circle, *p. 758*
center (of a circle), *p. 758*
radius, *p. 758*
parabola, *p. 762*
directrix (of a parabola), *p. 762*
focus (of a parabola), *p. 762*
vertex (of a parabola), *p. 762*
axis (of a parabola), *p. 762*

ellipse, *p. 770*
focus (of an ellipse), *p. 770*
vertices (of an ellipse), *p. 770*
major axis (of an ellipse), *p. 770*
center (of an ellipse), *p. 770*
minor axis (of an ellipse), *p. 770*
co-vertices (of an ellipse), *p. 770*
hyperbola, *p. 781*
foci (of a hyperbola), *p. 781*

transverse axis (of a hyperbola),
  p. 781
vertices (of a hyperbola), *p. 781*
branch (of a hyperbola), *p. 782*
asymptotes, *p. 782*
central rectangle, *p. 782*
nonlinear system of equations,
  p. 789

Key Concepts

12.1 ⬤ **Standard forms of the equations of circles**

1. Center at origin and radius r: $x^2 + y^2 = r^2$
2. Center at (h, k) and radius r:
$$(x - h)^2 + (y - k)^2 = r^2$$

12.1 ⬤ **Standard forms of the equations of parabolas**

1. Vertex at the origin:

  $x^2 = 4py, \ p \neq 0$ Vertical axis

  $y^2 = 4px, \ p \neq 0$ Horizontal axis

2. Vertex at (h, k):

  $(x - h)^2 = 4p(y - k), \ p \neq 0$ Vertical axis

  $(y - k)^2 = 4p(x - h), \ p \neq 0$ Horizontal axis

12.2 ⬤ **Standard forms of the equations of ellipses**

1. Center at the origin $(0 < b < a)$:
$$\frac{x^2}{a^2} + \frac{y^2}{b^2} = 1 \ \text{ or } \ \frac{x^2}{b^2} + \frac{y^2}{a^2} = 1$$

2. Center at (h, k) $(0 < b < a)$:
$$\frac{(x - h)^2}{a^2} + \frac{(y - k)^2}{b^2} = 1 \ \text{ or}$$
$$\frac{(x - h)^2}{b^2} + \frac{(y - k)^2}{a^2} = 1$$

12.3 ⬤ **Standard forms of the equations of hyperbolas**

1. Center at the origin $(a > 0, b > 0)$:
$$\frac{x^2}{a^2} - \frac{y^2}{b^2} = 1 \ \text{ or } \ \frac{y^2}{a^2} - \frac{x^2}{b^2} = 1$$

2. Center at (h, k) $(a > 0, b > 0)$:
$$\frac{(x - h)^2}{a^2} - \frac{(y - k)^2}{b^2} = 1 \ \text{ or}$$
$$\frac{(y - k)^2}{a^2} - \frac{(x - h)^2}{b^2} = 1$$

12.4 ⬤ **Solving a nonlinear system graphically**

1. Sketch the graph of each equation in the system.
2. Locate the point(s) of intersection of the graphs (if any) and graphically approximate the coordinates of the points.
3. Check the coordinate values by substituting them into each equation in the original system. If the coordinate values do not check, you may have to use an algebraic approach.

12.4 ⬤ **Method of substitution**

To solve a system of two equations in two variables, use the steps below.

1. Solve one of the equations for one variable in terms of the other variable.
2. Substitute the expression found in Step 1 into the other equation to obtain an equation in one variable.
3. Solve the equation obtained in Step 2.
4. Back-substitute the solution from Step 3 into the expression obtained in Step 1 to find the value of the other variable.
5. Check the solution to see that it satisfies *both* of the original equations.

Review Exercises

12.1 Circles and Parabolas

1 Recognize the four basic conics: circles, parabolas, ellipses, and hyperbolas.

In Exercises 1–8, identify the conic.

1.
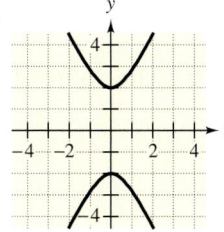

2.
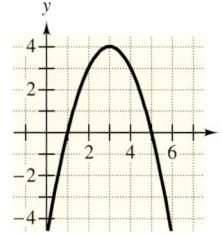

3.
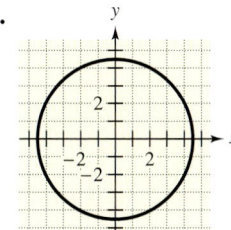

4.
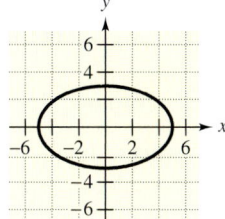

5.
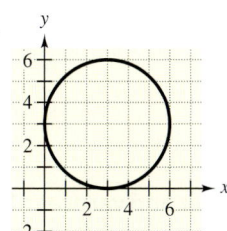

6.
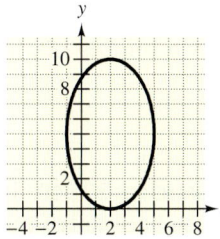

7.
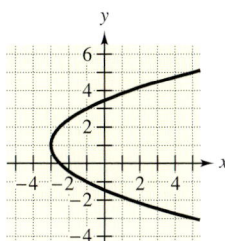

8.
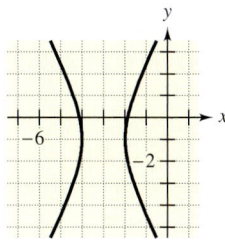

2 Graph and write equations of circles centered at the origin.

In Exercises 9 and 10, write the standard form of the equation of the circle with center at $(0, 0)$ that satisfies the criterion.

9. Radius: 12

10. Passes through the point $(-1, 3)$

In Exercises 11 and 12, identify the center and radius of the circle and sketch the circle.

11. $x^2 + y^2 = 64$

12. $4x^2 + 4y^2 - 9 = 0$

3 Graph and write equations of circles centered at (h, k).

In Exercises 13 and 14, write the standard form of the equation of the circle with center at (h, k) that satisfies the criteria.

13. Center: $(3, 5)$;   Radius: 5

14. Center: $(-2, 3)$;   Passes through the point $(1, 1)$

In Exercises 15 and 16, identify the center and radius of the circle and sketch the circle.

15. $x^2 + y^2 + 6x + 8y + 21 = 0$

16. $x^2 + y^2 + 14x - 10y + 73 = 0$

4 Graph and write equations of parabolas.

In Exercises 17–22, write the standard form of the equation of the parabola. Then sketch the parabola.

17. Vertex: $(0, 0)$;   Focus: $(-2, 0)$

18. Vertex: $(0, 0)$;   Focus: $(0, 4)$

19. Vertex: $(-6, 4)$;   Focus: $(-6, -1)$

20. Vertex: $(0, 5)$;   Focus: $(2, 5)$

21. Vertex: $(-1, 3)$;

Vertical axis and passes through $(-2, 5)$

22. Vertex: $(5, 0)$;

Horizontal axis and passes through $(3, -1)$

In Exercises 23 and 24, identify the vertex and focus of the parabola and sketch the parabola.

23. $y = x^2 - 4x + 2$

24. $x = y^2 + 10y - 4$

12.2 Ellipses

① Graph and write equations of ellipses centered at the origin.

In Exercises 25–28, write the standard form of the equation of the ellipse centered at the origin.

25. Vertices: $(0, -5), (0, 5)$;

Co-vertices: $(-2, 0), (2, 0)$

26. Vertices: $(-10, 0), (10, 0)$;

Co-vertices: $(0, -6), (0, 6)$

27. Major axis (vertical) 6 units, minor axis 4 units

28. Major axis (horizontal) 12 units, minor axis 2 units

In Exercises 29–32, sketch the ellipse. Identify the vertices and co-vertices.

29. $\dfrac{x^2}{64} + \dfrac{y^2}{16} = 1$

30. $\dfrac{x^2}{9} + y^2 = 1$

31. $16x^2 + 4y^2 - 16 = 0$

32. $100x^2 + 4y^2 - 4 = 0$

② Graph and write equations of ellipses centered at (h, k).

In Exercises 33–36, write the standard form of the equation of the ellipse.

33. Vertices: $(-2, 4), (8, 4)$;

Co-vertices: $(3, 0), (3, 8)$

34. Vertices: $(0, 3), (10, 3)$;

Co-vertices: $(5, 0), (5, 6)$

35. Vertices: $(0, 0), (0, 8)$;

Co-vertices: $(-3, 4), (3, 4)$

36. Vertices: $(5, -3), (5, 13)$;

Co-vertices: $(3, 5), (7, 5)$

In Exercises 37–40, find the center and vertices of the ellipse and sketch the ellipse.

37. $9x^2 + 4y^2 - 18x + 16y - 299 = 0$

38. $x^2 + 25y^2 - 4x - 21 = 0$

39. $16x^2 + y^2 + 6y - 7 = 0$

40. $x^2 + 4y^2 + 10x - 24y + 57 = 0$

12.3 Hyperbolas

① Graph and write equations of hyperbolas centered at the origin.

In Exercises 41–44, sketch the hyperbola. Identify the vertices and asymptotes.

41. $x^2 - y^2 = 25$

42. $y^2 - x^2 = 4$

43. $\dfrac{y^2}{25} - \dfrac{x^2}{4} = 1$

44. $\dfrac{x^2}{16} - \dfrac{y^2}{25} = 1$

In Exercises 45–48, write the standard form of the equation of the hyperbola centered at the origin.

Vertices	Asymptotes	
45. $(-2, 0), (2, 0)$	$y = \frac{3}{2}x$	$y = -\frac{3}{2}x$
46. $(0, -6), (0, 6)$	$y = 3x$	$y = -3x$
47. $(0, -5), (0, 5)$	$y = \frac{5}{2}x$	$y = -\frac{5}{2}x$
48. $(-3, 0), (3, 0)$	$y = \frac{4}{3}x$	$y = -\frac{4}{3}x$

② Graph and write equations of hyperbolas centered at (h, k).

In Exercises 49–52, find the center and vertices of the hyperbola and sketch the hyperbola.

49. $\dfrac{(y + 1)^2}{4} - \dfrac{(x - 3)^2}{9} = -1$

50. $\dfrac{(x + 4)^2}{25} - \dfrac{(y - 7)^2}{64} = 1$

51. $8y^2 - 2x^2 + 48y + 16x + 8 = 0$

52. $25x^2 - 4y^2 - 200x - 40y = 0$

In Exercises 53 and 54, write the standard form of the equation of the hyperbola.

53.

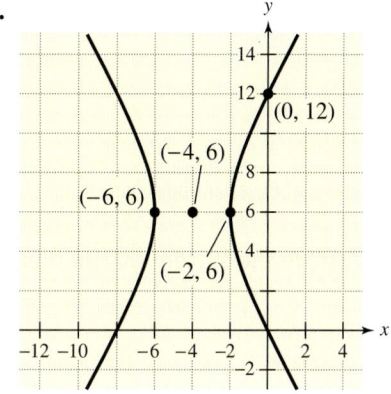

54.

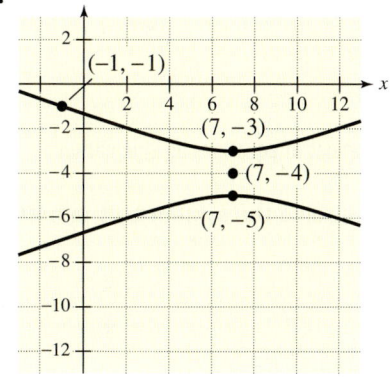

12.4 Solving Nonlinear Systems of Equations

① Solve nonlinear systems of equations graphically.

🖩 **In Exercises 55–58, use a graphing calculator to graph the equations and find any solutions of the system.**

55. $\begin{cases} y = x^2 \\ y = 3x \end{cases}$

56. $\begin{cases} y = 2 + x^2 \\ y = 8 - x \end{cases}$

57. $\begin{cases} x^2 + y^2 = 16 \\ -x + y = 4 \end{cases}$

58. $\begin{cases} 2x^2 - y^2 = -8 \\ y = x + 6 \end{cases}$

② Solve nonlinear systems of equations by substitution.

In Exercises 59–62, solve the system by the method of substitution.

59. $\begin{cases} y = 5x^2 \\ y = -15x - 10 \end{cases}$

60. $\begin{cases} y^2 = 16x \\ 4x - y = -24 \end{cases}$

61. $\begin{cases} x^2 + y^2 = 1 \\ x + y = -1 \end{cases}$

62. $\begin{cases} x^2 + y^2 = 100 \\ x + y = 0 \end{cases}$

③ Solve nonlinear systems of equations by elimination.

In Exercises 63–66, solve the system by the method of elimination.

63. $\begin{cases} \dfrac{x^2}{16} + \dfrac{y^2}{4} = 1 \\ y = x + 2 \end{cases}$

64. $\begin{cases} \dfrac{x^2}{100} + \dfrac{y^2}{25} = 1 \\ y = -x - 5 \end{cases}$

65. $\begin{cases} \dfrac{x^2}{25} + \dfrac{y^2}{9} = 1 \\ \dfrac{x^2}{25} - \dfrac{y^2}{9} = 1 \end{cases}$

66. $\begin{cases} x^2 + y^2 = 16 \\ -x^2 + \dfrac{y^2}{16} = 1 \end{cases}$

④ Use nonlinear systems of equations to model and solve real-life problems.

67. ▲ *Geometry* A rectangle has an area of 20 square inches and a perimeter of 18 inches. Find the dimensions of the rectangle.

68. ▲ *Geometry* A rectangle has an area of 300 square feet and a diagonal of 25 feet. Find the dimensions of the rectangle.

69. ▲ *Geometry* A computer manufacturer needs a circuit board with a perimeter of 28 centimeters and a diagonal of length 10 centimeters. What should the dimensions of the circuit board be?

70. ▲ *Geometry* A home interior decorator wants to find a ceramic tile with a perimeter of 6 inches and a diagonal of length $\sqrt{5}$ inches. What should the dimensions of the tile be?

71. ▲ *Geometry* A piece of wire 100 inches long is to be cut into two pieces. Each of the two pieces is to then be bent into a square. The area of one square is to be 144 square inches greater than the area of the other square. How should the wire be cut?

72. ▲ *Geometry* You have 250 feet of fencing to enclose two corrals of equal size (see figure). The combined area of the corrals is 2400 square feet. Find the dimensions of each corral.

Chapter Test

Take this test as you would take a test in class. After you are done, check your
work against the answers in the back of the book.

1. Write the standard form of the equation of the circle shown in the figure.

**In Exercises 2 and 3, write the equation of the circle in standard form.
Then sketch the circle.**

2. $x^2 + y^2 - 2x - 6y + 1 = 0$

3. $x^2 + y^2 + 4x - 6y + 4 = 0$

4. Identify the vertex and the focus of the parabola $x = -3y^2 + 12y - 8$. Then
sketch the parabola.

5. Write the standard form of the equation of the parabola with vertex $(7, -2)$
and focus $(7, 0)$.

6. Write the standard form of the equation of the ellipse shown in the figure.

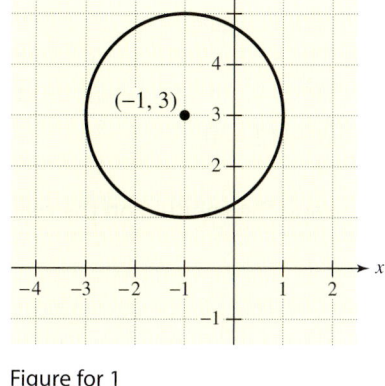

Figure for 1

**In Exercises 7 and 8, find the center and vertices of the ellipse. Then sketch
the ellipse.**

7. $16x^2 + 4y^2 = 64$

8. $9x^2 + 4y^2 - 36x + 32y + 64 = 0$

**In Exercises 9 and 10, write the standard form of the equation of the
hyperbola.**

9. Vertices: $(-3, 0), (3, 0)$; Asymptotes: $y = \pm\frac{2}{3}x$

10. Vertices: $(0, -2), (0, 2)$; Asymptotes: $y = \pm 2x$

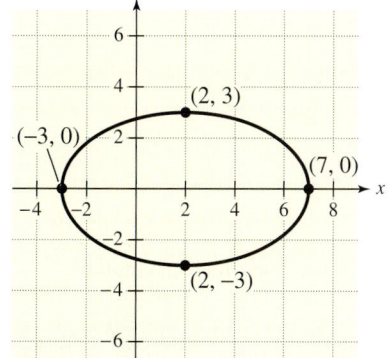

Figure for 6

**In Exercises 11 and 12, find the center and vertices of the hyperbola.
Then sketch the hyperbola.**

11. $9x^2 - 4y^2 + 24y - 72 = 0$

12. $16y^2 - 25x^2 + 64y + 200x - 736 = 0$

In Exercises 13–15, solve the nonlinear system of equations.

13. $\begin{cases} x^2/16 + y^2/9 = 1 \\ 3x + 4y = 12 \end{cases}$ 14. $\begin{cases} x^2 + y^2 = 16 \\ x^2/16 - y^2/9 = 1 \end{cases}$ 15. $\begin{cases} x^2 + y^2 = 10 \\ x^2 = y^2 + 2 \end{cases}$

16. Write the equation of the circular orbit of a satellite 1000 miles above the
surface of Earth. Place the origin of the rectangular coordinate system at the
center of Earth and assume the radius of Earth to be 4000 miles.

17. A rectangle has a perimeter of 56 inches and a diagonal of length 20 inches.
Find the dimensions of the rectangle.

Take this test as you would take a test in class. After you are done, check your work against the answers in the back of the book.

In Exercises 1–4, solve the equation by the specified method.

1. Factoring:

$$4x^2 - 9x - 9 = 0$$

2. Square Root Property:

$$(x - 5)^2 - 64 = 0$$

3. Completing the square:

$$x^2 - 10x - 25 = 0$$

4. Quadratic Formula:

$$3x^2 + 6x + 2 = 0$$

5. Solve the equation of quadratic form: $x - \sqrt{x} - 12 = 0$.

In Exercises 6 and 7, solve the inequality and graph the solution on the real number line.

6. $3x^2 + 5x \leq 3$

7. $\dfrac{3x + 4}{2x - 1} < 0$

8. Find a quadratic equation having the solutions -2 and 6.

9. Find the compositions (a) $f \circ g$ and (b) $g \circ f$. Then find the domain of each composition. $f(x) = 2x^2 - 3, \quad g(x) = 5x - 1$

10. Find the inverse function of $f(x) = \dfrac{2x + 3}{8}$.

11. Evaluate $f(x) = 7 + 2^{-x}$ when $x = 1, 0.5$, and 3.

12. Sketch the graph of $f(x) = 4^{x-1}$ and identify the horizontal asymptote.

13. Describe the relationship between the graphs of $f(x) = e^x$ and $g(x) = \ln x$.

14. Sketch the graph of $\log_3(x - 1)$ and identify the vertical asymptote.

15. Evaluate $\log_4 \frac{1}{16}$ without using a calculator.

16. Use the properties of logarithms to condense $3(\log_2 x + \log_2 y) - \log_2 z$.

17. Use the properties of logarithms to expand $\ln \dfrac{5x}{(x + 1)^2}$.

In Exercises 18–21, solve the equation.

18. $\log_x\left(\frac{1}{9}\right) = -2$

19. $4 \ln x = 10$

20. $500(1.08)^t = 2000$

21. $3(1 + e^{2x}) = 20$

22. If the inflation rate averages 3.5% over the next 5 years, the approximate cost C of goods and services t years from now is given by

$$C(t) = P(1.035)^t, \ 0 \le t \le 5$$

where P is the present cost. The price of an oil change is presently \$24.95. Estimate the price 5 years from now.

23. Determine the effective yield of an 8% interest rate compounded continuously.

24. Determine the length of time for an investment of \$1000 to quadruple in value if the investment earns 9% compounded continuously.

25. Write the equation of the circle in standard form and sketch the circle:

$$x^2 + y^2 - 6x + 14y - 6 = 0$$

26. Identify the vertex and focus of the parabola and sketch the parabola:

$$y = 2x^2 - 20x + 5.$$

27. Write the standard form of the equation of the ellipse shown in the figure.

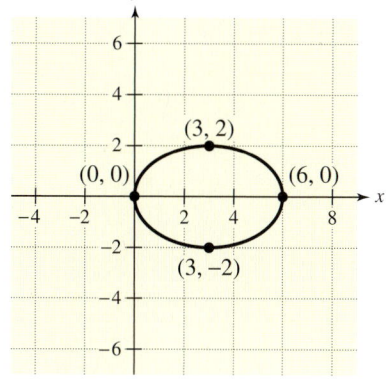

Figure for 27

28. Find the center and vertices of the ellipse and sketch the ellipse: $4x^2 + y^2 = 4.$

29. Write the standard form of the equation of the hyperbola with vertices $(-1, 0)$ and $(1, 0)$ and asymptotes $y = \pm 2x.$

30. Find the center and vertices of the hyperbola and sketch the hyperbola:

$$x^2 - 9y^2 + 18y = 153.$$

In Exercises 31 and 32, solve the nonlinear system of equations.

31. $\begin{cases} y = x^2 - x - 1 \\ 3x - y = 4 \end{cases}$

32. $\begin{cases} 2x^2 + 3y^2 = 6 \\ 5x^2 + 4y^2 = 15 \end{cases}$

33. A rectangle has an area of 32 square feet and a perimeter of 24 feet. Find the dimensions of the rectangle.

34. ▦ The path of a ball is given by $y = -0.1x^2 + 3x + 6.$ The coordinates x and y are measured in feet, with $x = 0$ corresponding to the position from which the ball was thrown.

(a) Use a graphing calculator to graph the path of the ball.

(b) Move the cursor along the path to approximate the highest point and the range of the path.

Motivating the Chapter

⚡ Ancestors and Descendants

See Section 13.3, Exercise 123.

a. Your ancestors consist of your two parents (first generation), your four grandparents (second generation), your eight great-grandparents (third generation), and so on. Write a geometric sequence that describes the number of ancestors for each generation.

b. If your ancestry could be traced back 66 generations (approximately 2000 years), how many different ancestors would you have?

c. A common ancestor is one to whom you are related in more than one way. (See figure.) From the results of part (b), do you think that you have had no common ancestors in the last 2000 years?

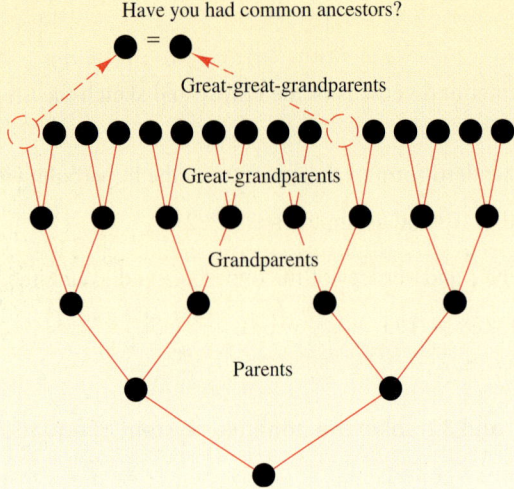

Have you had common ancestors?

Great-great-grandparents

Great-grandparents

Grandparents

Parents

13

Stewart Cohen/Index Stock

Sequences, Series, and the Binomial Theorem

13.1 Sequences and Series

Macduff Everton/Corbis

What You Should Learn

1 Use sequence notation to write the terms of sequences.
2 Write the terms of sequences involving factorials.
3 Find the apparent nth term of a sequence.
4 Sum the terms of sequences to obtain series and use sigma notation to represent partial sums.

Why You Should Learn It

Sequences and series are useful in modeling sets of values in order to identify patterns. For instance, in Exercise 110 on page 819, you will use a sequence to model the depreciation of a sport utility vehicle.

1 **Use sequence notation to write the terms of sequences.**

Year	Contract A	Contract B
1	$20,000	$20,000
2	$22,200	$22,000
3	$24,400	$24,200
4	$26,600	$26,620
5	$28,800	$29,282
Total	$122,000	$122,102

Sequences

You are given the following choice of contract offers for the next 5 years of employment.

Contract A $20,000 the first year and a $2200 raise each year

Contract B $20,000 the first year and a 10% raise each year

Which contract offers the largest salary over the five-year period? The salaries for each contract are shown in the table at the left. Notice that after 5 years contract B represents a better contract offer than contract A. The salaries for each contract option represent a sequence.

A mathematical **sequence** is simply an ordered list of numbers. Each number in the list is a **term** of the sequence. A sequence can have a finite number of terms or an infinite number of terms. For instance, the sequence of positive odd integers that are less than 15 is a *finite* sequence

$$1, 3, 5, 7, 9, 11, 13 \qquad \text{Finite sequence}$$

whereas the sequence of positive odd integers is an *infinite* sequence.

$$1, 3, 5, 7, 9, 11, 13, \ldots \qquad \text{Infinite sequence}$$

Note that the three dots indicate that the sequence continues and has an infinite number of terms.

Because each term of a sequence is matched with its location, a sequence can be defined as a function whose domain is a subset of positive integers.

Sequences

An **infinite sequence** $a_1, a_2, a_3, \ldots, a_n, \ldots$ is a function whose domain is the set of positive integers.
A **finite sequence** $a_1, a_2, a_3, \ldots, a_n$ is a function whose domain is the finite set $\{1, 2, 3, \ldots, n\}$.

In some cases it is convenient to begin subscripting a sequence with 0 instead of 1. Then the domain of the infinite sequence is the set of nonnegative integers and the domain of the finite sequence is the set $\{0, 1, 2, \ldots, n\}$. The terms of the sequence are denoted by $a_0, a_1, a_2, a_3, a_4, \ldots, a_n, \ldots$.

$a_{(\ )} = 2(\ ) + 1$

$a_{(1)} = 2(1) + 1 = 3$

$a_{(2)} = 2(2) + 1 = 5$

$\vdots$

$a_{(51)} = 2(51) + 1 = 103$

The subscripts of a sequence are used in place of function notation. For instance, if parentheses replaced the n in $a_n = 2n + 1$, the notation would be similar to function notation, as shown at the left.

Example 1   Writing the Terms of a Sequence

Write the first six terms of the sequence whose nth term is

$$a_n = n^2 - 1. \qquad \text{Begin sequence with } n = 1.$$

Solution

$$a_1 = (1)^2 - 1 = 0 \qquad a_2 = (2)^2 - 1 = 3 \qquad a_3 = (3)^2 - 1 = 8$$

$$a_4 = (4)^2 - 1 = 15 \qquad a_5 = (5)^2 - 1 = 24 \qquad a_6 = (6)^2 - 1 = 35$$

The entire sequence can be written as follows.

$$0, 3, 8, 15, 24, 35, \ldots, n^2 - 1, \ldots$$

Example 2   Writing the Terms of a Sequence

Write the first six terms of the sequence whose nth term is

$$a_n = 3(2^n). \qquad \text{Begin sequence with } n = 0.$$

Solution

$$a_0 = 3(2^0) = 3 \cdot 1 = 3 \qquad a_1 = 3(2^1) = 3 \cdot 2 = 6$$

$$a_2 = 3(2^2) = 3 \cdot 4 = 12 \qquad a_3 = 3(2^3) = 3 \cdot 8 = 24$$

$$a_4 = 3(2^4) = 3 \cdot 16 = 48 \qquad a_5 = 3(2^5) = 3 \cdot 32 = 96$$

The entire sequence can be written as follows.

$$3, 6, 12, 24, 48, 96, \ldots, 3(2^n), \ldots$$

Example 3   A Sequence Whose Terms Alternate in Sign

Write the first six terms of the sequence whose nth term is

$$a_n = \frac{(-1)^n}{2n - 1}. \qquad \text{Begin sequence with } n = 1.$$

Solution

$$a_1 = \frac{(-1)^1}{2(1) - 1} = -\frac{1}{1} \qquad a_2 = \frac{(-1)^2}{2(2) - 1} = \frac{1}{3} \qquad a_3 = \frac{(-1)^3}{2(3) - 1} = -\frac{1}{5}$$

$$a_4 = \frac{(-1)^4}{2(4) - 1} = \frac{1}{7} \qquad a_5 = \frac{(-1)^5}{2(5) - 1} = -\frac{1}{9} \qquad a_6 = \frac{(-1)^6}{2(6) - 1} = \frac{1}{11}$$

The entire sequence can be written as follows.

$$-1, \frac{1}{3}, -\frac{1}{5}, \frac{1}{7}, -\frac{1}{9}, \frac{1}{11}, \ldots, \frac{(-1)^n}{2n - 1}, \ldots$$

Technology: Tip

Most graphing calculators have a "sequence graphing mode" that allows you to plot the terms of a sequence as points on a rectangular coordinate system. For instance, the graph of the first six terms of the sequence given by

$$a_n = n^2 - 1$$

is shown below.

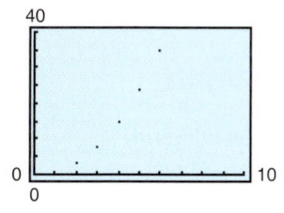

2 Write the terms of sequences involving factorials.

Factorial Notation

Some very important sequences in mathematics involve terms that are defined with special types of products called **factorials.**

> ### Definition of Factorial
>
> If n is a positive integer, n **factorial** is defined as
>
> $$n! = 1 \cdot 2 \cdot 3 \cdot 4 \cdot \cdots \cdot (n-1) \cdot n.$$
>
> As a special case, zero factorial is defined as $0! = 1$.

The first several factorial values are as follows.

$$0! = 1 \qquad\qquad 1! = 1$$

$$2! = 1 \cdot 2 = 2 \qquad\qquad 3! = 1 \cdot 2 \cdot 3 = 6$$

$$4! = 1 \cdot 2 \cdot 3 \cdot 4 = 24 \qquad\qquad 5! = 1 \cdot 2 \cdot 3 \cdot 4 \cdot 5 = 120$$

Many calculators have a factorial key, denoted by $\boxed{n!}$. If your calculator has such a key, try using it to evaluate $n!$ for several values of n. You will see that the value of n does not have to be very large before the value of $n!$ becomes huge. For instance

$$10! = 3{,}628{,}800.$$

Example 4 A Sequence Involving Factorials

Write the first six terms of the sequence with the given nth term.

a. $a_n = \dfrac{1}{n!}$ Begin sequence with $n = 0$.

b. $a_n = \dfrac{2^n}{n!}$ Begin sequence with $n = 0$.

Solution

a. $a_0 = \dfrac{1}{0!} = \dfrac{1}{1} = 1$ $a_1 = \dfrac{1}{1!} = \dfrac{1}{1} = 1$

$a_2 = \dfrac{1}{2!} = \dfrac{1}{1 \cdot 2} = \dfrac{1}{2}$ $a_3 = \dfrac{1}{3!} = \dfrac{1}{1 \cdot 2 \cdot 3} = \dfrac{1}{6}$

$a_4 = \dfrac{1}{4!} = \dfrac{1}{1 \cdot 2 \cdot 3 \cdot 4} = \dfrac{1}{24}$ $a_5 = \dfrac{1}{5!} = \dfrac{1}{1 \cdot 2 \cdot 3 \cdot 4 \cdot 5} = \dfrac{1}{120}$

b. $a_0 = \dfrac{2^0}{0!} = \dfrac{1}{1} = 1$ $a_1 = \dfrac{2^1}{1!} = \dfrac{2}{1} = 2$

$a_2 = \dfrac{2^2}{2!} = \dfrac{2 \cdot 2}{1 \cdot 2} = \dfrac{4}{2} = 2$ $a_3 = \dfrac{2^3}{3!} = \dfrac{8}{1 \cdot 2 \cdot 3} = \dfrac{8}{6} = \dfrac{4}{3}$

$a_4 = \dfrac{2^4}{4!} = \dfrac{2 \cdot 2 \cdot 2 \cdot 2}{1 \cdot 2 \cdot 3 \cdot 4} = \dfrac{2}{3}$ $a_5 = \dfrac{2^5}{5!} = \dfrac{2 \cdot 2 \cdot 2 \cdot 2 \cdot 2}{1 \cdot 2 \cdot 3 \cdot 4 \cdot 5} = \dfrac{4}{15}$

③ Find the apparent nth term of a sequence.

Finding the nth Term of a Sequence

Sometimes you will have the first several terms of a sequence and need to find a formula (the nth term) that will generate those terms. Pattern recognition is crucial in finding a form for the nth term.

Study Tip

Simply listing the first few terms is not sufficient to define a unique sequence—the nth term must be given. Consider the sequence

$$\frac{1}{2}, \frac{1}{4}, \frac{1}{8}, \frac{1}{15}, \cdots$$

The first three terms are identical to the first three terms of the sequence in Example 5(a). However, the nth term of this sequence is defined as

$$a_n = \frac{6}{(n+1)(n^2-n+6)}.$$

Example 5   Finding the nth Term of a Sequence

Write an expression for the nth term of each sequence.

a. $\dfrac{1}{2}, \dfrac{1}{4}, \dfrac{1}{8}, \dfrac{1}{16}, \dfrac{1}{32}, \ldots$ **b.** $1, -4, 9, -16, 25, \ldots$

Solution

a.

n:	1	2	3	4	5	$\ldots$	n
Terms:	$\dfrac{1}{2}$	$\dfrac{1}{4}$	$\dfrac{1}{8}$	$\dfrac{1}{16}$	$\dfrac{1}{32}$	$\ldots$	a_n

Pattern: The numerator is 1 and the denominators are increasing powers of 2.

$$a_n = \frac{1}{2^n}$$

b.

n:	1	2	3	4	5	$\ldots$	n
Terms:	1	-4	9	-16	25	$\ldots$	a_n

Pattern: The terms have alternating signs, with those in the even positions being negative. The absolute value of each term is the square of n.

$$a_n = (-1)^{n+1}n^2$$

④ Sum the terms of sequences to obtain series and use sigma notation to represent partial sums.

Series

In the opening illustration of this section, the terms of the finite sequence were *added*. If you add all the terms of an infinite sequence, you obtain a **series.**

Definition of a Series

For an infinite sequence $a_1, a_2, a_3, \ldots, a_n, \ldots$

1. the sum of the first n terms

$$S_n = a_1 + a_2 + a_3 + \cdots + a_n$$

is called a **partial sum,** and

2. the sum of all the terms

$$a_1 + a_2 + a_3 + \cdots + a_n + \cdots$$

is called an **infinite series,** or simply a **series.**

Example 6   Finding Partial Sums

Find the indicated partial sums for each sequence.

a. Find S_1, S_2, and S_5 for $a_n = 3n - 1$.

b. Find S_2, S_3, and S_4 for $a_n = \dfrac{(-1)^n}{n + 1}$.

Solution

a. The first five terms of the sequence $a_n = 3n - 1$ are

$$a_1 = 2, a_2 = 5, a_3 = 8, a_4 = 11, \text{ and } a_5 = 14.$$

So, the partial sums are

$$S_1 = 2, S_2 = 2 + 5 = 7, \text{ and } S_5 = 2 + 5 + 8 + 11 + 14 = 40.$$

b. The first four terms of the sequence $a_n = \dfrac{(-1)^n}{n + 1}$ are

$$a_1 = -\frac{1}{2}, a_2 = \frac{1}{3}, a_3 = -\frac{1}{4}, \text{ and } a_4 = \frac{1}{5}.$$

So, the partial sums are

$$S_2 = -\frac{1}{2} + \frac{1}{3} = -\frac{1}{6},$$

$$S_3 = -\frac{1}{2} + \frac{1}{3} - \frac{1}{4} = -\frac{5}{12},$$

and

$$S_4 = -\frac{1}{2} + \frac{1}{3} - \frac{1}{4} + \frac{1}{5} = -\frac{13}{60}.$$

A convenient shorthand notation for denoting a partial sum is called **sigma notation.** This name comes from the use of the uppercase Greek letter sigma, written as Σ.

Definition of Sigma Notation

The sum of the first n terms of the sequence whose nth term is a_n is

$$\sum_{i=1}^{n} a_i = a_1 + a_2 + a_3 + a_4 + \cdots + a_n$$

where i is the **index of summation,** n is the **upper limit of summation,** and 1 is the **lower limit of summation.**

Summation notation is an instruction to add the terms of a sequence. From the definition above, the upper limit of summation tells you where to end the sum. Summation notation helps you generate the appropriate terms of the sequence prior to finding the actual sum.

Example 7   **Sigma Notation for Sums**

Find the sum $\displaystyle\sum_{i=1}^{6} 2i$.

Solution

$$\sum_{i=1}^{6} 2i = 2(1) + 2(2) + 2(3) + 2(4) + 2(5) + 2(6)$$

$$= 2 + 4 + 6 + 8 + 10 + 12$$

$$= 42$$

Study Tip

In Example 7, the index of summation is i and the summation begins with $i = 1$. Any letter can be used as the index of summation, and the summation can begin with any integer. For instance, in Example 8, the index of summation is k and the summation begins with $k = 0$.

Example 8   **Sigma Notation for Sums**

Find the sum $\displaystyle\sum_{k=0}^{8} \frac{1}{k!}$.

Solution

$$\sum_{k=0}^{8} \frac{1}{k!} = \frac{1}{0!} + \frac{1}{1!} + \frac{1}{2!} + \frac{1}{3!} + \frac{1}{4!} + \frac{1}{5!} + \frac{1}{6!} + \frac{1}{7!} + \frac{1}{8!}$$

$$= 1 + 1 + \frac{1}{2} + \frac{1}{6} + \frac{1}{24} + \frac{1}{120} + \frac{1}{720} + \frac{1}{5040} + \frac{1}{40{,}320}$$

$$\approx 2.71828$$

Note that this sum is approximately $e = 2.71828.\ldots$

Example 9   **Writing a Sum in Sigma Notation**

Write each sum in sigma notation.

a. $\dfrac{2}{2} + \dfrac{2}{3} + \dfrac{2}{4} + \dfrac{2}{5} + \dfrac{2}{6}$ **b.** $1 - \dfrac{1}{3} + \dfrac{1}{9} - \dfrac{1}{27} + \dfrac{1}{81}$

Solution

a. To write this sum in sigma notation, you must find a pattern for the terms. After examining the terms, you can see that they have numerators of 2 and denominators that range over the integers from 2 to 6. So, one possible sigma notation is

$$\sum_{i=1}^{5} \frac{2}{i+1} = \frac{2}{2} + \frac{2}{3} + \frac{2}{4} + \frac{2}{5} + \frac{2}{6}.$$

b. To write this sum in sigma notation, you must find a pattern for the terms. After examining the terms, you can see that the numerators alternate in sign and the denominators are integer powers of 3, starting with 3^0 and ending with 3^4. So, one possible sigma notation is

$$\sum_{i=0}^{4} \frac{(-1)^i}{3^i} = \frac{1}{3^0} + \frac{-1}{3^1} + \frac{1}{3^2} + \frac{-1}{3^3} + \frac{1}{3^4}.$$

13.1 Exercises

Review   *Concepts, Skills, and Problem Solving*

Keep mathematically in shape by doing these exercises *before* the problems of this section.

Properties and Definitions

1. Demonstrate the Multiplication Property of Equality for the equation $-7x = 35$.

2. Demonstrate the Addition Property of Equality for the equation $7x + 63 = 35$.

3. How do you determine whether $t = -3$ is a solution of the equation $t^2 + 4t + 3 = 0$?

4. What is the usual first step in solving an equation such as the one below?

$$\frac{3}{x} - \frac{1}{x+1} = 10$$

Simplifying Expressions

In Exercises 5–10, simplify the expression.

5. $(x + 10)^{-2}$

6. $\dfrac{18(x-3)^5}{(x-3)^2}$

7. $(a^2)^{-4}$

8. $(8x^3)^{1/3}$

9. $\sqrt{128x^3}$

10. $\dfrac{5}{\sqrt{x}-2}$

Graphs and Models

 ▲ *Geometry* In Exercises 11 and 12, (a) write a function that represents the area of the region, (b) use a graphing calculator to graph the function, and (c) approximate the value of x if the area of the region is 200 square units.

11.

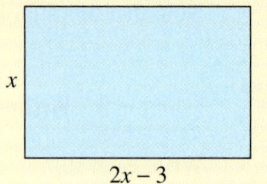

x

$2x - 3$

12.

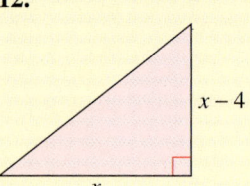

$x - 4$

x

Developing Skills

In Exercises 1–22, write the first five terms of the sequence. (Assume that n begins with 1.) See Examples 1–4. ∧

1. $a_n = 2n$

2. $a_n = 3n$

3. $a_n = (-1)^n 2n$

4. $a_n = (-1)^{n+1} 3n$

5. $a_n = \left(\frac{1}{2}\right)^n$

6. $a_n = \left(\frac{1}{3}\right)^n$

7. $a_n = \left(-\frac{1}{2}\right)^{n+1}$

8. $a_n = \left(\frac{2}{3}\right)^{n-1}$

9. $a_n = 5n - 2$

10. $a_n = 2n + 3$

11. $a_n = \dfrac{4}{n+3}$

12. $a_n = \dfrac{5}{4+2n}$

13. $a_n = \dfrac{3n}{5n-1}$

14. $a_n = \dfrac{2n}{6n-3}$

15. $a_n = \dfrac{(-1)^n}{n^2}$

16. $a_n = \dfrac{1}{\sqrt{n}}$

17. $a_n = 5 - \dfrac{1}{2^n}$

18. $a_n = 7 + \dfrac{1}{3^n}$

19. $a_n = \dfrac{(n + 1)!}{n!}$

20. $a_n = \dfrac{n!}{(n - 1)!}$

21. $a_n = \dfrac{2 + (-2)^n}{n!}$

22. $a_n = \dfrac{1 + (-1)^n}{n^2}$

In Exercises 23–26, find the indicated term of the sequence.

23. $a_n = (-1)^n(5n - 3)$

$a_{15} = $ ▭

24. $a_n = (-1)^{n-1}(2n + 4)$

$a_{14} = $ ▭

25. $a_n = \dfrac{n^2 - 2}{(n - 1)!}$

$a_8 = $ ▭

26. $a_n = \dfrac{n^2}{n!}$

$a_{12} = $ ▭

In Exercises 27–38, simplify the expression.

27. $\dfrac{5!}{4!}$

28. $\dfrac{18!}{17!}$

29. $\dfrac{10!}{12!}$

30. $\dfrac{5!}{8!}$

31. $\dfrac{25!}{20!5!}$

32. $\dfrac{20!}{15! \cdot 5!}$

33. $\dfrac{n!}{(n + 1)!}$

34. $\dfrac{(n + 2)!}{n!}$

35. $\dfrac{(n + 1)!}{(n - 1)!}$

36. $\dfrac{(3n)!}{(3n + 2)!}$

37. $\dfrac{(2n)!}{(2n - 1)!}$

38. $\dfrac{(2n + 2)!}{(2n)!}$

In Exercises 39–42, match the sequence with the graph of its first 10 terms. [The graphs are labeled (a), (b), (c), and (d).]

(a)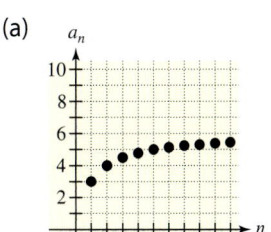

(b)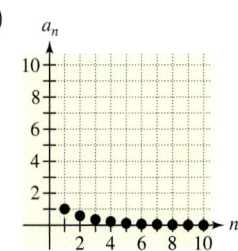

(c)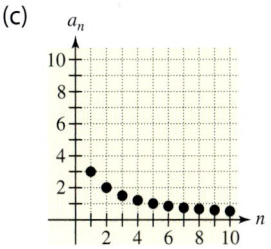

(d)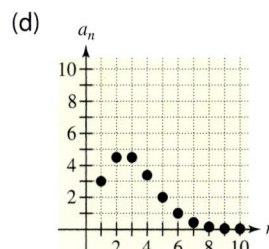

39. $a_n = \dfrac{6}{n + 1}$

40. $a_n = \dfrac{6n}{n + 1}$

41. $a_n = (0.6)^{n-1}$

42. $a_n = \dfrac{3^n}{n!}$

⊞ In Exercises 43–48, use a graphing calculator to graph the first 10 terms of the sequence.

43. $a_n = \dfrac{4n^2}{n^2 - 2}$

44. $a_n = \dfrac{2n^2}{n^2 + 1}$

45. $a_n = 3 - \dfrac{4}{n}$

46. $a_n = \dfrac{n + 2}{n}$

47. $a_n = (-0.8)^{n-1}$

48. $a_n = 10\left(\dfrac{3}{4}\right)^{n-1}$

In Exercises 49–66, write an expression for the *n*th term of the sequence. (Assume that *n* begins with 1.) See Example 5.

49. 1, 3, 5, 7, 9, . . .

50. 2, 4, 6, 8, 10, . . .

51. 2, 6, 10, 14, 18, . . .

52. 5, 8, 11, 14, 17, . . .

53. 0, 3, 8, 15, 24, . . .

54. 1, 8, 27, 64, 125, . . .

55. 2, −4, 6, −8, 10, . . .

56. 1, −1, 1, −1, 1, . . .

57. $\frac{2}{3}, \frac{3}{4}, \frac{4}{5}, \frac{5}{6}, \frac{6}{7}, \ldots$

58. $\frac{2}{1}, \frac{3}{3}, \frac{4}{5}, \frac{5}{7}, \frac{6}{9}, \ldots$

59. $\frac{1}{2}, \frac{-1}{4}, \frac{1}{8}, \frac{-1}{16}, \ldots$

60. $1, \frac{1}{4}, \frac{1}{9}, \frac{1}{16}, \frac{1}{25}, \ldots$

61. $1, \frac{1}{2}, \frac{1}{4}, \frac{1}{8}, \ldots$

62. $\frac{1}{3}, \frac{2}{9}, \frac{4}{27}, \frac{8}{81}, \ldots$

63. $1 + \frac{1}{1}, 1 + \frac{1}{2}, 1 + \frac{1}{3}, 1 + \frac{1}{4}, 1 + \frac{1}{5}, \ldots$

64. $1 + \frac{1}{2}, 1 + \frac{3}{4}, 1 + \frac{7}{8}, 1 + \frac{15}{16}, 1 + \frac{31}{32}, \ldots$

65. $1, \frac{1}{2}, \frac{1}{6}, \frac{1}{24}, \frac{1}{120}, \ldots$

66. $1, 2, \frac{2^2}{2}, \frac{2^3}{6}, \frac{2^4}{24}, \frac{2^5}{120}, \ldots$

In Exercises 67–82, find the partial sum. See Examples 6–8.

67. $\displaystyle\sum_{k=1}^{6} 3k$

68. $\displaystyle\sum_{k=1}^{4} 5k$

69. $\displaystyle\sum_{i=0}^{6} (2i + 5)$

70. $\displaystyle\sum_{i=0}^{4} (2i + 3)$

71. $\displaystyle\sum_{j=3}^{7} (6j - 10)$

72. $\displaystyle\sum_{i=2}^{7} (4i - 1)$

73. $\displaystyle\sum_{j=1}^{5} \frac{(-1)^{j+1}}{j^2}$

74. $\displaystyle\sum_{j=0}^{3} \frac{1}{j^2 + 1}$

75. $\displaystyle\sum_{m=1}^{8} \frac{m}{m + 1}$

76. $\displaystyle\sum_{k=1}^{4} \frac{k - 2}{k + 3}$

77. $\displaystyle\sum_{k=1}^{6} (-8)$

78. $\displaystyle\sum_{n=3}^{12} 10$

79. $\displaystyle\sum_{i=1}^{8} \left(\frac{1}{i} - \frac{1}{i + 1} \right)$

80. $\displaystyle\sum_{k=1}^{5} \left(\frac{2}{k} - \frac{2}{k + 2} \right)$

81. $\displaystyle\sum_{n=0}^{5} \left(-\frac{1}{3} \right)^n$

82. $\displaystyle\sum_{n=0}^{6} \left(\frac{3}{2} \right)^n$

In Exercises 83–90, use a graphing calculator to find the partial sum.

83. $\displaystyle\sum_{n=1}^{6} 3n^2$

84. $\displaystyle\sum_{n=0}^{5} 2n^2$

85. $\displaystyle\sum_{j=2}^{6} (j! - j)$

86. $\displaystyle\sum_{i=0}^{4} (i! + 4)$

87. $\displaystyle\sum_{j=0}^{4} \frac{6}{j!}$

88. $\displaystyle\sum_{k=1}^{6} \left(\frac{1}{2k} - \frac{1}{2k - 1} \right)$

89. $\displaystyle\sum_{k=1}^{6} \ln k$

90. $\displaystyle\sum_{k=2}^{4} \frac{k}{\ln k}$

In Exercises 91–108, write the sum using sigma notation. (Begin with $k = 0$ or $k = 1$.) See Example 9.

91. $1 + 2 + 3 + 4 + 5$

92. $8 + 9 + 10 + 11 + 12 + 13 + 14$

93. $2 + 4 + 6 + 8 + 10$

94. $24 + 30 + 36 + 42$

95. $\dfrac{1}{2(1)} + \dfrac{1}{2(2)} + \dfrac{1}{2(3)} + \dfrac{1}{2(4)} + \cdots + \dfrac{1}{2(10)}$

96. $\dfrac{3}{1 + 1} + \dfrac{3}{1 + 2} + \dfrac{3}{1 + 3} + \cdots + \dfrac{3}{1 + 50}$

97. $\dfrac{1}{1^2} + \dfrac{1}{2^2} + \dfrac{1}{3^2} + \dfrac{1}{4^2} + \cdots + \dfrac{1}{20^2}$

98. $\dfrac{1}{2^0} + \dfrac{1}{2^1} + \dfrac{1}{2^2} + \dfrac{1}{2^3} + \cdots + \dfrac{1}{2^{12}}$

99. $\dfrac{1}{3^0} - \dfrac{1}{3^1} + \dfrac{1}{3^2} - \dfrac{1}{3^3} + \cdots - \dfrac{1}{3^9}$

100. $\left(-\dfrac{2}{3}\right)^0 + \left(-\dfrac{2}{3}\right)^1 + \left(-\dfrac{2}{3}\right)^2 + \cdots + \left(-\dfrac{2}{3}\right)^{20}$

101. $\dfrac{4}{1+3} + \dfrac{4}{2+3} + \dfrac{4}{3+3} + \cdots + \dfrac{4}{20+3}$

102. $\dfrac{1}{2^3} - \dfrac{1}{4^3} + \dfrac{1}{6^3} - \dfrac{1}{8^3} + \cdots + \dfrac{1}{14^3}$

103. $\dfrac{1}{2} + \dfrac{2}{3} + \dfrac{3}{4} + \dfrac{4}{5} + \dfrac{5}{6} + \cdots + \dfrac{11}{12}$

104. $\dfrac{2}{4} + \dfrac{4}{7} + \dfrac{6}{10} + \dfrac{8}{13} + \dfrac{10}{16} + \cdots + \dfrac{20}{31}$

105. $\dfrac{2}{4} + \dfrac{4}{5} + \dfrac{6}{6} + \dfrac{8}{7} + \cdots + \dfrac{40}{23}$

106. $\left(2 + \dfrac{1}{1}\right) + \left(2 + \dfrac{1}{2}\right) + \left(2 + \dfrac{1}{3}\right) + \cdots + \left(2 + \dfrac{1}{25}\right)$

107. $1 + 1 + 2 + 6 + 24 + 120 + 720$

108. $1 + 1 + \dfrac{1}{2} + \dfrac{1}{6} + \dfrac{1}{24} + \dfrac{1}{120} + \dfrac{1}{720}$

Solving Problems

109. *Compound Interest*   A deposit of $500 is made in an account that earns 7% interest compounded yearly. The balance in the account after N years is given by

$$A_N = 500(1 + 0.07)^N, \quad N = 1, 2, 3, \ldots.$$

(a) Compute the first eight terms of the sequence.

(b) Find the balance in this account after 40 years by computing A_{40}.

(c) ▦ Use a graphing calculator to graph the first 40 terms of the sequence.

(d) The terms are increasing. Is the rate of growth of the terms increasing? Explain.

110. *Depreciation*   At the end of each year, the value of a sport utility vehicle with an initial cost of $32,000 is three-fourths what it was at the beginning of the year. After n years, its value is given by

$$a_n = 32,000\left(\dfrac{3}{4}\right)^n, \quad n = 1, 2, 3, \ldots.$$

(a) Find the value of the sport utility vehicle 3 years after it was purchased by computing a_3.

(b) Find the value of the sport utility vehicle 6 years after it was purchased by computing a_6. Is this value half of what it was after 3 years? Explain.

111. *Sports*   The number of degrees a_n in each angle of a regular n-sided polygon is

$$a_n = \dfrac{180(n - 2)}{n}, \quad n \geq 3.$$

The surface of a soccer ball is made of regular hexagons and pentagons. When a soccer ball is taken apart and flattened, as shown in the figure, the sides don't meet each other. Use the terms a_5 and a_6 to explain why there are gaps between adjacent hexagons.

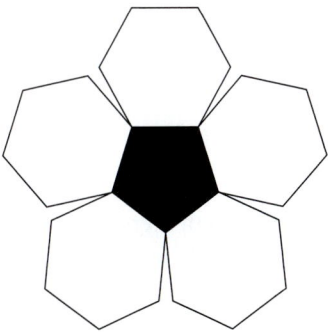

112. *Stars*   The number of degrees d_n in the angle at each point of each of the six n-pointed stars in the figure (on the next page) is given by

$$d_n = \dfrac{180(n - 4)}{n}, \quad n \geq 5.$$

Write the first six terms of this sequence.

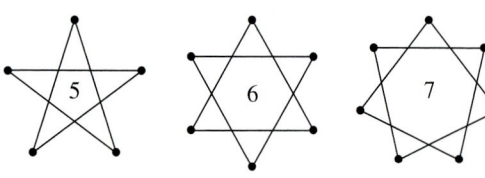

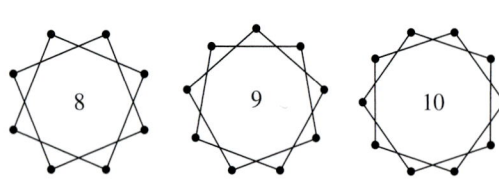

Figure for 112

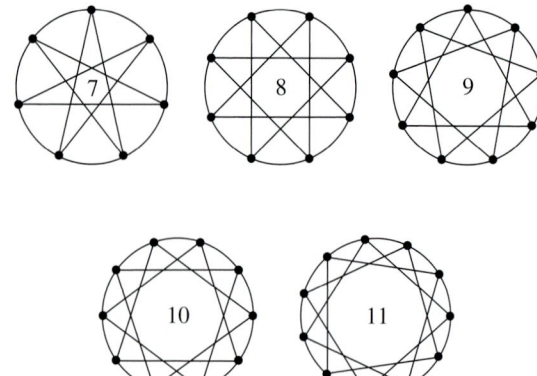

Figure for 113

113. *Stars*   The stars in Exercise 112 were formed by placing n equally spaced points on a circle and connecting each point with the second point from it on the circle. The stars in the figure for this exercise were formed in a similar way except that each point was connected with the third point from it. For these stars, the number of degrees d_n in the angle at each point is given by

$$d_n = \frac{180(n - 6)}{n}, \quad n \geq 7.$$

Write the first five terms of this sequence.

114. 🖩 *Number of Stores*   The number a_n of Home Depot stores for the years 1991 through 2001 is modeled by

$$a_n = 9.73n^2 - 3.0n + 180, \quad n = 1, 2, \ldots, 11$$

where n is the year, with $n = 1$ corresponding to 1991. Find the terms of this finite sequence and use a graphing calculator to construct a bar graph that represents the sequence. (Source: The Home Depot)

Explaining Concepts

115. Give an example of an infinite sequence.

116. *Writing*✎   State the definition of n factorial.

117. *Writing*✎   The nth term of a sequence is $a_n = (-1)^n n$. Which terms of the sequence are negative? Explain.

118. You learned in this section that a sequence is an ordered list of numbers. Study the following sequence and see if you can guess what its next term should be.

Z, O, T, T, F, F, S, S, E, N, T, E, T, . . .

In Exercises 119–121, decide whether the statement is true or false. Justify your answer.

119. $\displaystyle\sum_{i=1}^{4} (i^2 + 2i) = \sum_{i=1}^{4} i^2 + \sum_{i=1}^{4} 2i$

120. $\displaystyle\sum_{k=1}^{4} 3k = 3\sum_{k=1}^{4} k$

121. $\displaystyle\sum_{j=1}^{4} 2^j = \sum_{j=3}^{6} 2^{j-2}$

13.2 Arithmetic Sequences

Lynn Goldsmith/Corbis

Why You Should Learn It

An arithmetic sequence can reduce the amount of time it takes to find the sum of a sequence of numbers with a common difference. For instance, in Exercise 109 on page 828, you will use an arithmetic sequence to determine how much to charge for tickets for a concert at an outdoor arena.

1 Recognize, write, and find the *n*th terms of arithmetic sequences.

What You Should Learn

1 Recognize, write, and find the *n*th terms of arithmetic sequences.

2 Find the *n*th partial sum of an arithmetic sequence.

3 Use arithmetic sequences to solve application problems.

Arithmetic Sequences

A sequence whose consecutive terms have a common difference is called an **arithmetic sequence.**

Definition of an Arithmetic Sequence

A sequence is called **arithmetic** if the differences between consecutive terms are the same. So, the sequence

$$a_1, a_2, a_3, a_4, \ldots, a_n, \ldots$$

is arithmetic if there is a number d such that

$$a_2 - a_1 = d, \quad a_3 - a_2 = d, \quad a_4 - a_3 = d$$

and so on. The number d is the **common difference** of the sequence.

Example 1   Examples of Arithmetic Sequences

a. The sequence whose *n*th term is $3n + 2$ is arithmetic. For this sequence, the common difference between consecutive terms is 3.

$$5, 8, 11, 14, \ldots, 3n + 2, \ldots \qquad \text{Begin with } n = 1.$$

$$8 - 5 = 3$$

b. The sequence whose *n*th term is $7 - 5n$ is arithmetic. For this sequence, the common difference between consecutive terms is -5.

$$2, -3, -8, -13, \ldots, 7 - 5n, \ldots \qquad \text{Begin with } n = 1.$$

$$-3 - 2 = -5$$

c. The sequence whose *n*th term is $\frac{1}{4}(n + 3)$ is arithmetic. For this sequence, the common difference between consecutive terms is $\frac{1}{4}$.

$$1, \frac{5}{4}, \frac{3}{2}, \frac{7}{4}, \ldots, \frac{1}{4}(n + 3), \ldots \qquad \text{Begin with } n = 1.$$

$$\tfrac{5}{4} - 1 = \tfrac{1}{4}$$

The nth Term of an Arithmetic Sequence

The nth term of an arithmetic sequence has the form

$$a_n = a_1 + (n - 1)d$$

where d is the common difference between the terms of the sequence, and a_1 is the first term.

Example 2   Finding the nth Term of an Arithmetic Sequence

Find a formula for the nth term of the arithmetic sequence whose common difference is 2 and whose first term is 5.

Solution

You know that the formula for the nth term is of the form $a_n = a_1 + (n - 1)d$. Moreover, because the common difference is $d = 2$, and the first term is $a_1 = 5$, the formula must have the form

$$a_n = 5 + 2(n - 1).$$

So, the formula for the nth term is

$$a_n = 2n + 3.$$

The sequence therefore has the following form.

$$5, 7, 9, 11, 13, \ldots, 2n + 3, \ldots$$

If you know the nth term and the common difference of an arithmetic sequence, you can find the $(n + 1)$th term by using the **recursion formula**

$$a_{n+1} = a_n + d.$$

Example 3   Using a Recursion Formula

The 12th term of an arithmetic sequence is 52 and the common difference is 3.

a. What is the 13th term of the sequence?   **b.** What is the first term?

Solution

a. You know that $a_{12} = 52$ and $d = 3$. So, using the recursion formula $a_{13} = a_{12} + d$, you can determine that the 13th term of the sequence is

$$a_{13} = 52 + 3 = 55.$$

b. Using $n = 12, d = 3$, and $a_{12} = 52$ in the formula $a_n = a_1 + (n - 1)d$ yields

$$52 = a_1 + (12 - 1)(3)$$

$$19 = a_1.$$

② Find the *n*th partial sum of an arithmetic sequence.

The Partial Sum of an Arithmetic Sequence

The sum of the first *n* terms of an arithmetic sequence is called the **nth partial sum** of the sequence. For instance, the fifth partial sum of the arithmetic sequence whose *n*th term is $3n + 4$ is

$$\sum_{i=1}^{5} (3i + 4) = 7 + 10 + 13 + 16 + 19 = 65.$$

To find a formula for the *n*th partial sum S_n of an arithmetic sequence, write out S_n forwards and backwards and then add the two forms, as follows.

$$S_n = a_1 + (a_1 + d) + (a_1 + 2d) + \cdots + [a_1 + (n - 1)d] \qquad \text{Forwards}$$
$$S_n = a_n + (a_n - d) + (a_n - 2d) + \cdots + [a_n - (n - 1)d] \qquad \text{Backwards}$$
$$2S_n = (a_1 + a_n) + (a_1 + a_n) + (a_1 + a_n) + \cdots + [a_1 + a_n] \qquad \begin{array}{l}\text{Sum of two} \\ \text{equations}\end{array}$$
$$= n(a_1 + a_n) \qquad \begin{array}{l} n \text{ groups of} \\ (a_1 + a_n)\end{array}$$

Dividing each side by 2 yields the following formula.

Study Tip

You can use the formula for the *n*th partial sum of an arithmetic sequence to find the sum of consecutive numbers. For instance, the sum of the integers from 1 to 100 is

$$\sum_{i=1}^{100} i = \frac{100}{2}(1 + 100)$$
$$= 50(101)$$
$$= 5050.$$

The *n*th Partial Sum of an Arithmetic Sequence

The *n*th partial sum of the arithmetic sequence whose *n*th term is a_n is

$$\sum_{i=1}^{n} a_i = a_1 + a_2 + a_3 + a_4 + \cdots + a_n$$

$$= \frac{n}{2}(a_1 + a_n).$$

Or equivalently, you can find the sum of the first *n* terms of an arithmetic sequence, by finding the average of the first and *n*th terms, and multiply by *n*.

Example 4   Finding the *n*th Partial Sum

Find the sum of the first 20 terms of the arithmetic sequence whose *n*th term is $4n + 1$.

Solution

The first term of this sequence is $a_1 = 4(1) + 1 = 5$ and the 20th term is $a_{20} = 4(20) + 1 = 81$. So, the sum of the first 20 terms is given by

$$\sum_{i=1}^{n} a_i = \frac{n}{2}(a_1 + a_n) \qquad \textit{n}\text{th partial sum formula}$$

$$\sum_{i=1}^{20} (4i + 1) = \frac{20}{2}(a_1 + a_{20}) \qquad \text{Substitute 20 for } n.$$

$$= 10(5 + 81) \qquad \text{Substitute 5 for } a_1 \text{ and 81 for } a_{20}.$$

$$= 10(86) \qquad \text{Simplify.}$$

$$= 860. \qquad \textit{n}\text{th partial sum}$$

Example 5   Finding the *n*th Partial Sum

Find the sum of the even integers from 2 to 100.

Solution

Because the integers

$$2, 4, 6, 8, \ldots, 100$$

form an arithmetic sequence, you can find the sum as follows.

$$\sum_{i=1}^{n} a_i = \frac{n}{2}(a_1 + a_n)$$ *n*th partial sum formula

$$\sum_{i=1}^{50} 2i = \frac{50}{2}(a_1 + a_{50})$$ Substitute 50 for *n*.

$$= 25(2 + 100)$$ Substitute 2 for a_1 and 100 for a_{50}.

$$= 25(102)$$ Simplify.

$$= 2550$$ *n*th partial sum

---

3 Use arithmetic sequences to solve application problems.

Application

Example 6   Total Sales

Your business sells $100,000 worth of handmade furniture during its first year. You have a goal of increasing annual sales by $25,000 each year for 9 years. If you meet this goal, how much will you sell during your first 10 years of business?

Solution

The annual sales during the first 10 years form the following arithmetic sequence.

$100,000, \quad $125,000, \quad $150,000, \quad $175,000, \quad $200,000,$
$225,000, \quad $250,000, \quad $275,000, \quad $300,000, \quad $325,000$

Using the formula for the *n*th partial sum of an arithmetic sequence, you find the total sales during the first 10 years as follows.

$$\text{Total sales} = \frac{n}{2}(a_1 + a_n)$$ *n*th partial sum formula

$$= \frac{10}{2}(100,000 + 325,000)$$ Substitute for *n*, a_1, and a_n.

$$= 5(425,000)$$ Simplify.

$$= \$2,125,000$$ Simplify.

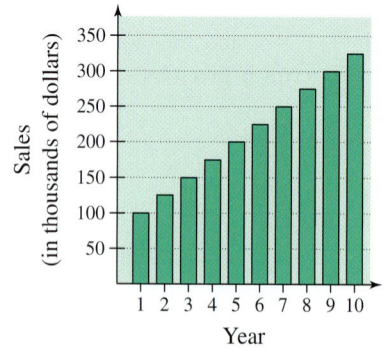

Figure 13.1

From the bar graph shown in Figure 13.1, notice that the annual sales for your company follows a *linear growth* pattern. In other words, saying that a quantity increases arithmetically is the same as saying that it increases linearly.

13.2 Exercises

Review   Concepts, Skills, and Problem Solving

Keep mathematically in shape by doing these exercises *before* the problems of this section.

Properties and Definitions

1. *Writing* ✎   In your own words, state the definition of an algebraic expression.

2. *Writing* ✎   In your own words, state the definition of the terms of an algebraic expression.

3. Give an example of a trinomial of degree 3.

4. Give an example of a monomial of degree 4.

Domain

In Exercises 5–10, find the domain of the function.

5. $f(x) = x^3 - 2x$

6. $g(x) = \sqrt[3]{x}$

7. $h(x) = \sqrt{16 - x^2}$

8. $A(x) = \dfrac{3}{36 - x^2}$

9. $g(t) = \ln(t - 2)$

10. $f(s) = 630e^{-0.2s}$

Problem Solving

11. *Compound Interest*   Determine the balance in an account when $10,000 is invested at $7\frac{1}{2}\%$ compounded daily for 15 years.

12. *Compound Interest*   Determine the amount after 5 years if $4000 is invested in an account earning 6% compounded monthly.

Developing Skills

In Exercises 1–10, find the common difference of the arithmetic sequence. See Example 1.

1. $2, 5, 8, 11, \ldots$

2. $-8, 0, 8, 16, \ldots$

3. $100, 94, 88, 82, \ldots$

4. $3200, 2800, 2400, 2000, \ldots$

5. $10, -2, -14, -26, -38, \ldots$

6. $4, \frac{9}{2}, 5, \frac{11}{2}, 6, \ldots$

7. $1, \frac{5}{3}, \frac{7}{3}, 3, \ldots$

8. $\frac{1}{2}, \frac{5}{4}, 2, \frac{11}{4}, \ldots$

9. $\frac{7}{2}, \frac{9}{4}, 1, -\frac{1}{4}, -\frac{3}{2}, \ldots$

10. $\frac{5}{2}, \frac{11}{6}, \frac{7}{6}, \frac{1}{2}, -\frac{1}{6}, \ldots$

In Exercises 11–26, determine whether the sequence is arithmetic. If so, find the common difference.

11. $2, 4, 6, 8, \ldots$

12. $1, 2, 4, 8, 16, \ldots$

13. $10, 8, 6, 4, 2, \ldots$

14. $2, 6, 10, 14, \ldots$

15. $32, 16, 0, -16, \ldots$

16. $32, 16, 8, 4, \ldots$

17. $3.2, 4, 4.8, 5.6, \ldots$

18. $8, 4, 2, 1, 0.5, 0.25, \ldots$

19. $2, \frac{7}{2}, 5, \frac{13}{2}, \ldots$

20. $3, \frac{5}{2}, 2, \frac{3}{2}, 1, \ldots$

21. $\frac{1}{3}, \frac{2}{3}, \frac{4}{3}, \frac{8}{3}, \frac{16}{3}, \ldots$

22. $\frac{9}{4}, 2, \frac{7}{4}, \frac{3}{2}, \frac{5}{4}, \ldots$

23. $1, \sqrt{2}, \sqrt{3}, 2, \sqrt{5}, \ldots$

24. $1, 4, 9, 16, 25, \ldots$

25. $\ln 4, \ln 8, \ln 12, \ln 16, \ldots$

26. $e, e^2, e^3, e^4, \ldots$

In Exercises 27–36, write the first five terms of the arithmetic sequence. (Assume that n begins with 1.)

27. $a_n = 3n + 4$

28. $a_n = 5n - 4$

29. $a_n = -2n + 8$

30. $a_n = -10n + 100$

31. $a_n = \frac{5}{2}n - 1$

32. $a_n = \frac{2}{3}n + 2$

33. $a_n = \frac{3}{5}n + 1$

34. $a_n = \frac{3}{4}n - 2$

35. $a_n = -\frac{1}{4}(n - 1) + 4$

36. $a_n = 4(n + 2) + 24$

In Exercises 37–54, find a formula for the nth term of the arithmetic sequence. See Example 2.

37. $a_1 = 4, \quad d = 3$

38. $a_1 = 7, \quad d = 2$

39. $a_1 = \frac{1}{2}, \quad d = \frac{3}{2}$

40. $a_1 = \frac{5}{3}, \quad d = \frac{1}{3}$

41. $a_1 = 100, \quad d = -5$

42. $a_1 = -6, \quad d = -1$

43. $a_1 = 3, \quad d = \frac{3}{2}$

44. $a_6 = 5, \quad d = \frac{3}{2}$

45. $a_1 = 5, \quad a_5 = 15$

46. $a_2 = 93, \quad a_6 = 65$

47. $a_3 = 16, \quad a_4 = 20$

48. $a_5 = 30, \quad a_4 = 25$

49. $a_1 = 50, \quad a_3 = 30$

50. $a_{10} = 32, \quad a_{12} = 48$

51. $a_2 = 10, \quad a_6 = 8$

52. $a_7 = 8, \quad a_{13} = 6$

53. $a_1 = 0.35, \quad a_2 = 0.30$

54. $a_1 = 0.08, \quad a_2 = 0.082$

In Exercises 55–62, write the first five terms of the arithmetic sequence defined recursively. See Example 3.

55. $a_1 = 14$

$a_{k+1} = a_k + 6$

56. $a_1 = 3$

$a_{k+1} = a_k - 2$

57. $a_1 = 23$

$a_{k+1} = a_k - 5$

58. $a_1 = 12$

$a_{k+1} = a_k + 6$

59. $a_1 = -16$

$a_{k+1} = a_k + 5$

60. $a_1 = -22$

$a_{k+1} = a_k - 4$

61. $a_1 = 3.4$

$a_{k+1} = a_k - 1.1$

62. $a_1 = 10.9$

$a_{k+1} = a_k + 0.7$

In Exercises 63–72, find the partial sum. See Example 4.

63. $\displaystyle\sum_{k=1}^{20} k$

64. $\displaystyle\sum_{k=1}^{30} 4k$

65. $\displaystyle\sum_{k=1}^{50} (k + 3)$

66. $\displaystyle\sum_{n=1}^{30} (n + 2)$

67. $\displaystyle\sum_{k=1}^{10} (5k - 2)$

68. $\displaystyle\sum_{k=1}^{100} (4k - 1)$

69. $\displaystyle\sum_{n=1}^{500} \frac{n}{2}$

70. $\displaystyle\sum_{n=1}^{600} \frac{2n}{3}$

71. $\displaystyle\sum_{n=1}^{30} \left(\tfrac{1}{3}n - 4\right)$

72. $\displaystyle\sum_{n=1}^{75} (0.3n + 5)$

In Exercises 73–84, find the nth partial sum of the arithmetic sequence. See Example 5.

73. 5, 12, 19, 26, 33, . . . , $n = 12$

74. 2, 12, 22, 32, 42, . . . , $n = 20$

75. 2, 8, 14, 20, . . . , $n = 25$

76. 500, 480, 460, 440, . . . , $n = 20$

77. 200, 175, 150, 125, 100, . . . , $n = 8$

78. 800, 785, 770, 755, 740, . . . , $n = 25$

79. $-50, -38, -26, -14, -2, \ldots,$ $n = 50$

80. $-16, -8, 0, 8, 16, \ldots,$ $n = 30$

81. 1, 4.5, 8, 11.5, 15, . . . , $n = 12$

82. 2.2, 2.8, 3.4, 4.0, 4.6, . . . , $n = 12$

83. $a_1 = 0.5,\ a_4 = 1.7, \ldots,$ $n = 10$

84. $a_1 = 15,\ a_{100} = 307, \ldots,$ $n = 100$

In Exercises 85–90, match the arithmetic sequence with its graph. [The graphs are labeled (a), (b), (c), (d), (e), and (f).]

(a)

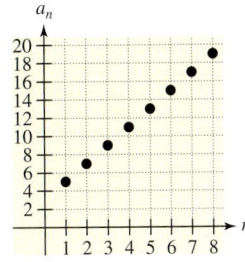

(b)

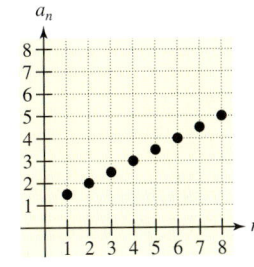

(c)

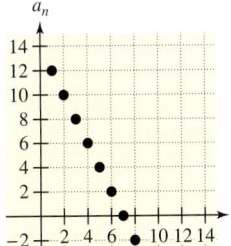

(d)

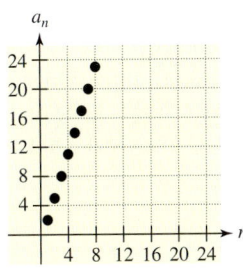

(e)

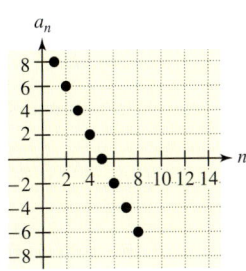

(f)

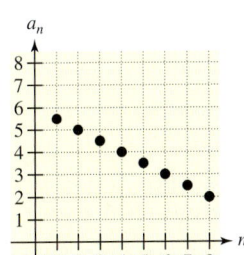

85. $a_n = \tfrac{1}{2}n + 1$

86. $a_n = -\tfrac{1}{2}n + 6$

87. $a_n = -2n + 10$

88. $a_n = 2n + 3$

89. $a_1 = 12$

   $a_{n+1} = a_n - 2$

90. $a_1 = 2$

   $a_{n+1} = a_n + 3$

In Exercises 91–96, use a graphing calculator to graph the first 10 terms of the sequence.

91. $a_n = -2n + 21$

92. $a_n = -25n + 500$

93. $a_n = \tfrac{3}{5}n + \tfrac{3}{2}$

94. $a_n = \tfrac{3}{2}n + 1$

95. $a_n = 2.5n - 8$

96. $a_n = 6.2n + 3$

In Exercises 97–102, use a graphing calculator to find the partial sum.

97. $\displaystyle\sum_{j=1}^{25} (750 - 30j)$

98. $\displaystyle\sum_{n=1}^{40} (1000 - 25n)$

99. $\displaystyle\sum_{i=1}^{60} \left(300 - \tfrac{8}{3}i\right)$

100. $\displaystyle\sum_{n=1}^{20} \left(500 - \tfrac{1}{10}n\right)$

101. $\displaystyle\sum_{n=1}^{50} (2.15n + 5.4)$

102. $\displaystyle\sum_{n=1}^{60} (200 - 3.4n)$

Solving Problems

103. *Number Problem*   Find the sum of the first 75 positive integers.

104. *Number Problem*   Find the sum of the integers from 35 to 100.

105. *Number Problem*   Find the sum of the first 50 positive odd integers.

106. *Number Problem*   Find the sum of the first 100 positive even integers.

107. *Salary*   In your new job as an actuary you are told that your starting salary will be $36,000 with an increase of $2000 at the end of each of the first 5 years. How much will you be paid through the end of your first six years of employment with the company?

108. *Wages*   You earn 25 cents on the first day of the month, 50 cents on the second day, 75 cents on the third day, and so on. Determine the total amount that you will earn during a 30-day month.

109. *Ticket Prices*   There are 20 rows of seats on the main floor of a an outdoor arena: 20 seats in the first row, 21 seats in the second row, 22 seats in the third row, and so on (see figure). How much should you charge per ticket in order to obtain $15,000 for the sale of all the seats on the main floor?

22 seats
21 seats
20 seats

110. *Pile of Logs*   Logs are stacked in a pile as shown in the figure. The top row has 15 logs and the bottom row has 21 logs. How many logs are in the pile?

— 15

— 21

111. *Baling Hay*   In the first two trips baling hay around a large field (see figure), a farmer obtains 93 bales and 89 bales, respectively. The farmer estimates that the same pattern will continue. Estimate the total number of bales made if there are another six trips around the field.

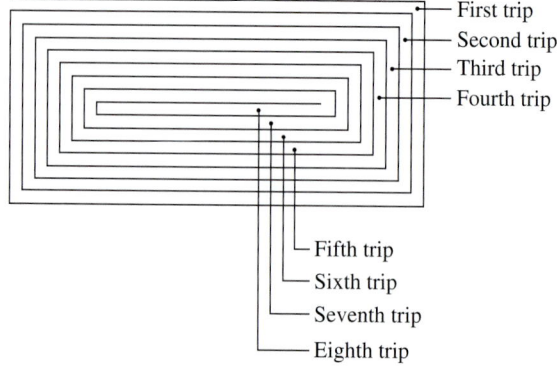

First trip
Second trip
Third trip
Fourth trip
Fifth trip
Sixth trip
Seventh trip
Eighth trip

112. *Baling Hay*   In the first two trips baling hay around a field (see figure), a farmer obtains 64 bales and 60 bales, respectively. The farmer estimates that the same pattern will continue. Estimate the total number of bales made if there are another four trips around the field.

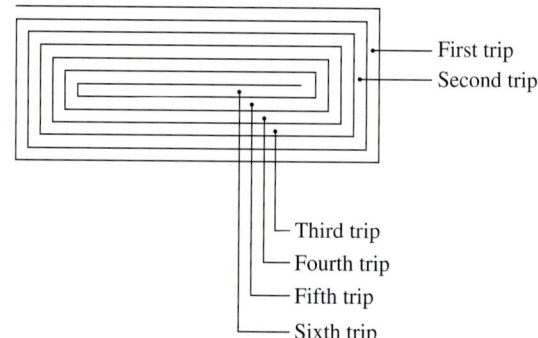

First trip
Second trip
Third trip
Fourth trip
Fifth trip
Sixth trip

113. *Clock Chimes*   A clock chimes once at 1:00, twice at 2:00, three times at 3:00, and so on. The clock also chimes once at 15-minute intervals that are not on the hour. How many times does the clock chime in a 12-hour period?

114. *Clock Chimes*   A clock chimes once at 1:00, twice at 2:00, three times at 3:00, and so on. The clock also chimes once on the half-hour. How many times does the clock chime in a 12-hour period?

115. *Free-Falling Object*   A free-falling object will fall 16 feet during the first second, 48 more feet during the second second, 80 more feet during the third second, and so on. What is the total distance the object will fall in 8 seconds if this pattern continues?

116. *Free-Falling Object*   A free-falling object will fall 4.9 meters during the first second, 14.7 more meters during the second second, 24.5 more meters during the third second, and so on. What is the total distance the object will fall in 5 seconds if this pattern continues?

Explaining Concepts

117. *Pattern Recognition*

(a) Complete the table.

Figure	Number of Sides	Sum of Interior Angles
Triangle	3	180°
Quadrilateral	4	
Pentagon	5	
Hexagon	6	

(b) Use the pattern in part (a) to determine the sum of the interior angles of a figure with n sides.

(c) Determine whether the sequence formed by the entries in the third column of the table in part (a) is an arithmetic sequence. If so, find the common difference.

118. *Writing*   In your own words, explain what makes a sequence arithmetic.

119. The second and third terms of an arithmetic sequence are 12 and 15, respectively. What is the first term?

120. *Writing*   Explain how the first two terms of an arithmetic sequence can be used to find the nth term.

121. *Writing*   Explain what is meant by a recursion formula.

122. *Writing*   Explain what is meant by the nth partial sum of a sequence.

123. *Writing*   Explain how to find the sum of the integers from 100 to 200.

124. *Pattern Recognition*

(a) Compute the sums of positive odd integers.

$1 + 3 =$

$1 + 3 + 5 =$

$1 + 3 + 5 + 7 =$

$1 + 3 + 5 + 7 + 9 =$

$1 + 3 + 5 + 7 + 9 + 11 =$

(b) Use the sums in part (a) to make a conjecture about the sums of positive odd integers. Check your conjecture for the sum

$1 + 3 + 5 + 7 + 9 + 11 + 13 =$.

(c) Verify your conjecture in part (b) analytically.

125. *Writing*   Each term of an arithmetic sequence is multiplied by a constant C. Is the resulting sequence arithmetic? If so, how does the common difference compare with the common difference of the original sequence?

Mid-Chapter Quiz

Take this quiz as you would take a quiz in class. After you are done, check your work against the answers in the back of the book.

In Exercises 1–4, write the first five terms of the sequence. (Assume that n begins with 1.)

1. $a_n = (-2)^{n+1}$

2. $a_n = n(n + 2)$

3. $a_n = 32\left(\dfrac{1}{4}\right)^{n-1}$

4. $a_n = \dfrac{(-3)^n n}{n + 4}$

In Exercises 5–10, find the sum.

5. $\displaystyle\sum_{k=1}^{4} 10k$

6. $\displaystyle\sum_{i=1}^{10} 4$

7. $\displaystyle\sum_{j=1}^{5} \dfrac{60}{j + 1}$

8. $\displaystyle\sum_{n=1}^{8} 8\left(-\dfrac{1}{2}\right)$

9. $\displaystyle\sum_{n=1}^{5} (3n - 1)$

10. $\displaystyle\sum_{k=1}^{4} (k^2 - 1)$

In Exercises 11–14, write the sum using sigma notation. (Begin with $k = 1$.)

11. $\dfrac{2}{3(1)} + \dfrac{2}{3(2)} + \dfrac{2}{3(3)} + \cdots + \dfrac{2}{3(20)}$

12. $\dfrac{1}{1^3} - \dfrac{1}{2^3} + \dfrac{1}{3^3} - \cdots + \dfrac{1}{25^3}$

13. $0 + \dfrac{1}{2} + \dfrac{2}{3} + \dfrac{3}{4} + \cdots + \dfrac{19}{20}$

14. $\dfrac{1}{2} + \dfrac{4}{2} + \dfrac{9}{2} + \cdots + \dfrac{100}{2}$

In Exercises 15 and 16, find the common difference of the arithmetic sequence.

15. $1, \frac{3}{2}, 2, \frac{5}{2}, 3, \ldots$

16. $100, 94, 88, 82, 76, \ldots$

In Exercises 17 and 18, find a formula for the nth term of the arithmetic sequence.

17. $a_1 = 20, \quad a_4 = 11$

18. $a_1 = 32, \quad d = -4$

19. Find the sum of the first 50 positive even numbers.

20. You save $.50 on one day, $1.00 the next day, $1.50 the next day, and so on. How much will you have accumulated at the end of one year (365 days)?

13.3 Geometric Sequences and Series

Paul A. Souders/Corbis

What You Should Learn

1. Recognize, write, and find the nth terms of geometric sequences.
2. Find the nth partial sum of a geometric sequence.
3. Find the sum of an infinite geometric series.
4. Use geometric sequences to solve application problems.

Why You Should Learn It

A geometric sequence can reduce the amount of time it takes to find the sum of a sequence of numbers with a common ratio. For instance, in Exercise 121 on page 840, you will use a geometric sequence to find the total distance traveled by a bungee jumper.

1. Recognize, write, and find the nth terms of geometric sequences.

Geometric Sequences

In Section 13.2, you studied sequences whose consecutive terms have a common *difference*. In this section, you will study sequences whose consecutive terms have a common *ratio*.

Definition of a Geometric Sequence

A sequence is called **geometric** if the ratios of consecutive terms are the same. So, the sequence $a_1, a_2, a_3, a_4, \ldots, a_n, \ldots$ is geometric if there is a number r, $r \neq 0$, such that

$$\frac{a_2}{a_1} = r, \quad \frac{a_3}{a_2} = r, \quad \frac{a_4}{a_3} = r$$

and so on. The number r is the **common ratio** of the sequence.

Example 1   Examples of Geometric Sequences

a. The sequence whose nth term is 2^n is geometric. For this sequence, the common ratio between consecutive terms is 2.

$$2, 4, 8, 16, \ldots, 2^n, \ldots \qquad \text{Begin with } n = 1.$$

$$\tfrac{4}{2} = 2$$

b. The sequence whose nth term is $4(3^n)$ is geometric. For this sequence, the common ratio between consecutive terms is 3.

$$12, 36, 108, 324, \ldots, 4(3^n), \ldots \qquad \text{Begin with } n = 1.$$

$$\tfrac{36}{12} = 3$$

c. The sequence whose nth term is $\left(-\tfrac{1}{3}\right)^n$ is geometric. For this sequence, the common ratio between consecutive terms is $-\tfrac{1}{3}$.

$$-\frac{1}{3}, \frac{1}{9}, -\frac{1}{27}, \frac{1}{81}, \ldots, \left(-\frac{1}{3}\right)^n, \ldots \qquad \text{Begin with } n = 1.$$

$$\tfrac{1/9}{-1/3} = -\tfrac{1}{3}$$

Study Tip

If you know the nth term of a geometric sequence, the $(n + 1)$th term can be found by multiplying by r. That is, $a_{n+1} = ra_n$.

The nth Term of a Geometric Sequence

The nth term of a geometric sequence has the form

$$a_n = a_1 r^{n-1}$$

where r is the common ratio of consecutive terms of the sequence. So, every geometric sequence can be written in the following form.

$$a_1, a_1 r, a_1 r^2, a_1 r^3, a_1 r^4, \ldots, a_1 r^{n-1}, \ldots$$

Example 2   Finding the nth Term of a Geometric Sequence

a. Find a formula for the nth term of the geometric sequence whose common ratio is 3 and whose first term is 1.

b. What is the eighth term of the sequence found in part (a)?

Solution

a. The formula for the nth term is of the form $a_n = a_1 r^{n-1}$. Moreover, because the common ratio is $r = 3$ and the first term is $a_1 = 1$, the formula must have the form

$$a_n = a_1 r^{n-1} \qquad \text{Formula for geometric sequence}$$
$$= (1)(3)^{n-1} \qquad \text{Substitute 1 for } a_1 \text{ and 3 for } r.$$
$$= 3^{n-1}. \qquad \text{Simplify.}$$

The sequence therefore has the following form.

$$1, 3, 9, 27, 81, \ldots, 3^{n-1}, \ldots$$

b. The eighth term of the sequence is $a_8 = 3^{8-1} = 3^7 = 2187$.

Example 3   Finding the nth Term of a Geometric Sequence

Find a formula for the nth term of the geometric sequence whose first two terms are 4 and 2.

Solution

Because the common ratio is

$$r = \frac{a_2}{a_1} = \frac{2}{4} = \frac{1}{2}$$

the formula for the nth term must be

$$a_n = a_1 r^{n-1} \qquad \text{Formula for geometric sequence}$$
$$= 4\left(\frac{1}{2}\right)^{n-1}. \qquad \text{Substitute 4 for } a_1 \text{ and } \tfrac{1}{2} \text{ for } r.$$

The sequence therefore has the form $4, 2, 1, \dfrac{1}{2}, \dfrac{1}{4}, \ldots, 4\left(\dfrac{1}{2}\right)^{n-1}, \ldots$.

2 Find the *n*th partial sum of a geometric sequence.

The Partial Sum of a Geometric Sequence

The *n*th Partial Sum of a Geometric Sequence

The *n*th partial sum of the geometric sequence whose *n*th term is $a_n = a_1 r^{n-1}$ is given by

$$\sum_{i=1}^{n} a_1 r^{i-1} = a_1 + a_1 r + a_1 r^2 + a_1 r^3 + \cdots + a_1 r^{n-1} = a_1 \left(\frac{r^n - 1}{r - 1} \right).$$

Example 4   Finding the *n*th Partial Sum

Find the sum $1 + 2 + 4 + 8 + 16 + 32 + 64 + 128$.

Solution
This is a geometric sequence whose common ratio is $r = 2$. Because the first term of the sequence is $a_1 = 1$, it follows that the sum is

$$\sum_{i=1}^{8} 2^{i-1} = (1) \left(\frac{2^8 - 1}{2 - 1} \right) = \frac{256 - 1}{2 - 1} = 255.$$ Substitute 1 for a_1 and 2 for r.

Example 5   Finding the *n*th Partial Sum

Find the sum of the first five terms of the geometric sequence whose *n*th term is $a_n = \left(\frac{2}{3} \right)^n$.

Solution

$$\sum_{i=1}^{5} \left(\frac{2}{3} \right)^i = \frac{2}{3} \left[\frac{(2/3)^5 - 1}{(2/3) - 1} \right]$$ Substitute $\frac{2}{3}$ for a_1 and $\frac{2}{3}$ for r.

$$= \frac{2}{3} \left[\frac{(32/243) - 1}{-1/3} \right]$$ Simplify.

$$= \frac{2}{3} \left(-\frac{211}{243} \right) (-3)$$ Simplify.

$$= \frac{422}{243} \approx 1.737$$ Use a calculator to simplify.

3 Find the sum of an infinite geometric series.

Geometric Series

Suppose that in Example 5, you were to find the sum of all the terms of the infinite geometric sequence

$$\frac{2}{3}, \frac{4}{9}, \frac{8}{27}, \frac{16}{81}, \ldots, \left(\frac{2}{3} \right)^n, \ldots$$

A summation of all the terms of an infinite geometric sequence is called an **infinite geometric series,** or simply a **geometric series.**

In your mind, would this sum be infinitely large or would it be a finite number? Consider the formula for the nth partial sum of a geometric sequence.

$$S_n = a_1\left(\frac{r^n - 1}{r - 1}\right) = a_1\left(\frac{1 - r^n}{1 - r}\right)$$

Suppose that $|r| < 1$ and you let n become larger and larger. It follows that r^n gets closer and closer to 0, so that the term r^n drops out of the formula above. You then get the sum

$$S = a_1\left(\frac{1}{1 - r}\right) = \frac{a_1}{1 - r}.$$

Notice that this sum is not dependent on the nth term of the sequence. In the case of Example 5, $r = \left(\frac{2}{3}\right) < 1$, and so the sum of the infinite geometric sequence is

$$S = \sum_{i=1}^{\infty} \left(\frac{2}{3}\right)^i = \frac{a_1}{1 - r} = \frac{2/3}{1 - (2/3)} = \frac{2/3}{1/3} = 2.$$

Sum of an Infinite Geometric Series

If $a_1, a_1r, a_1r^2, \ldots, a_1r^n, \ldots$ is an infinite geometric sequence, then for $|r| < 1$, the sum of the terms of the corresponding infinite geometric series is

$$S = \sum_{i=0}^{\infty} a_1r^i = \frac{a_1}{1 - r}.$$

Example 6   Finding the Sum of an Infinite Geometric Series

Find each sum.

a. $\displaystyle\sum_{i=1}^{\infty} 5\left(\frac{3}{4}\right)^{i-1}$ **b.** $\displaystyle\sum_{n=0}^{\infty} 4\left(\frac{3}{10}\right)^n$ **c.** $\displaystyle\sum_{i=0}^{\infty} \left(-\frac{3}{5}\right)^i$

Solution

a. The series is geometric, with $a_1 = 5\left(\frac{3}{4}\right)^{1-1} = 5$ and $r = \frac{3}{4}$. So,

$$\sum_{i=1}^{\infty} 5\left(\frac{3}{4}\right)^{i-1} = \frac{5}{1 - (3/4)}$$

$$= \frac{5}{1/4} = 20.$$

b. The series is geometric, with $a_1 = 4\left(\frac{3}{10}\right)^0 = 4$ and $r = \frac{3}{10}$. So,

$$\sum_{n=0}^{\infty} 4\left(\frac{3}{10}\right)^n = \frac{4}{1 - (3/10)} = \frac{4}{7/10} = \frac{40}{7}.$$

c. The series is geometric, with $a_1 = \left(-\frac{3}{5}\right)^0 = 1$ and $r = -\frac{3}{5}$. So,

$$\sum_{i=0}^{\infty} \left(-\frac{3}{5}\right)^i = \frac{1}{1 - (-3/5)} = \frac{1}{1 + (3/5)} = \frac{5}{8}.$$

4 Use geometric sequences to solve application problems.

Applications

Example 7   A Lifetime Salary

You have accepted a job as a meteorologist that pays a salary of $28,000 the first year. During the next 39 years, suppose you receive a 6% raise each year. What will your total salary be over the 40-year period?

Solution

Using a geometric sequence, your salary during the first year will be $a_1 = 28,000$. Then, with a 6% raise each year, your salary for the next 2 years will be as follows.

$$a_2 = 28,000 + 28,000(0.06) = 28,000(1.06)^1$$

$$a_3 = 28,000(1.06) + 28,000(1.06)(0.06) = 28,000(1.06)^2$$

From this pattern, you can see that the common ratio of the geometric sequence is $r = 1.06$. Using the formula for the nth partial sum of a geometric sequence, you will find that the total salary over the 40-year period is given by

$$\text{Total salary} = a_1\left(\frac{r^n - 1}{r - 1}\right)$$

$$= 28,000\left[\frac{(1.06)^{40} - 1}{1.06 - 1}\right]$$

$$= 28,000\left[\frac{(1.06)^{40} - 1}{0.06}\right] \approx \$4,333,335.$$

Example 8   Increasing Annuity

You deposit $100 in an account each month for 2 years. The account pays an annual interest rate of 9%, compounded monthly. What is your balance at the end of 2 years? (This type of savings plan is called an **increasing annuity.**)

Solution

The first deposit would earn interest for the full 24 months, the second deposit would earn interest for 23 months, the third deposit would earn interest for 22 months, and so on. Using the formula for compound interest, you can see that the total of the 24 deposits would be

$$\text{Total} = a_1 + a_2 + \cdots + a_{24}$$

$$= 100\left(1 + \frac{0.09}{12}\right)^1 + 100\left(1 + \frac{0.09}{12}\right)^2 + \cdots + 100\left(1 + \frac{0.09}{12}\right)^{24}$$

$$= 100(1.0075)^1 + 100(1.0075)^2 + \cdots + 100(1.0075)^{24}$$

$$= 100(1.0075)\left(\frac{1.0075^{24} - 1}{1.0075 - 1}\right) \qquad a_1\left(\frac{r^n - 1}{r - 1}\right)$$

$$= \$2638.49.$$

13.3 Exercises

Review   Concepts, Skills, and Problem Solving

Keep mathematically in shape by doing these exercises *before* the problems of this section.

Properties and Definitions

1. *Writing* ✏ Relative to the x- and y-axes, explain the meaning of each coordinate of the point $(-6, 4)$.

2. A point lies five units from the x-axis and ten units from the y-axis. Give the ordered pair for such a point in each quadrant.

3. *Writing* ✏ In your own words, define the graph of the function $y = f(x)$.

4. *Writing* ✏ Describe the procedure for finding the x- and y-intercepts of the graph of $f(x) = 2\sqrt{x} + 4$.

Solving Inequalities

In Exercises 5–10, solve the inequality.

5. $3x - 5 > 0$

6. $\frac{3}{2}y + 11 < 20$

7. $100 < 2x + 30 < 150$

8. $-5 < -\frac{x}{6} < 2$

9. $2x^2 - 7x + 5 > 0$

10. $2x - \frac{5}{x} > 3$

Problem Solving

11. ▲ *Geometry* A television set is advertised as having a 19-inch screen. Determine the dimensions of the square screen if its diagonal is 19 inches.

12. ▲ *Geometry* A construction worker is building the forms for the rectangular foundation of a home that is 25 feet wide and 40 feet long. To make sure that the corners are square, the worker measures the diagonal of the foundation. What should that measurement be?

Developing Skills

In Exercises 1–12, find the common ratio of the geometric sequence. See Example 1.

1. $7, 14, 28, 56, \ldots$

2. $2, 6, 18, 54, \ldots$

3. $5, -5, 5, -5, \ldots$

4. $-5, -0.5, -0.05, -0.005, \ldots$

5. $\frac{1}{2}, -\frac{1}{4}, \frac{1}{8}, -\frac{1}{16}, \ldots$

6. $\frac{2}{3}, -\frac{4}{3}, \frac{8}{3}, -\frac{16}{3}, \ldots$

7. $75, 15, 3, \frac{3}{5}, \ldots$

8. $12, -4, \frac{4}{3}, -\frac{4}{9}, \ldots$

9. $1, \pi, \pi^2, \pi^3, \ldots$

10. $e, e^2, e^3, e^4, \ldots$

11. $500(1.06), 500(1.06)^2, 500(1.06)^3, 500(1.06)^4, \ldots$

12. $1.1, (1.1)^2, (1.1)^3, (1.1)^4, \ldots$

In Exercises 13–24, determine whether the sequence is geometric. If so, find the common ratio.

13. $64, 32, 16, 8, \ldots$

14. $64, 32, 0, -32, \ldots$

15. $10, 15, 20, 25, \ldots$

16. $10, 20, 40, 80, \ldots$

17. $5, 10, 20, 40, \ldots$

18. $54, -18, 6, -2, \ldots$

19. $1, 8, 27, 64, 125, \ldots$

20. $12, 7, 2, -3, -8, \ldots$

21. $1, -\frac{2}{3}, \frac{4}{9}, -\frac{8}{27}, \ldots$

22. $\frac{1}{3}, -\frac{2}{3}, \frac{4}{3}, -\frac{8}{3}, \ldots$

23. $10(1 + 0.02), 10(1 + 0.02)^2, 10(1 + 0.02)^3, \ldots$

24. $1, 0.2, 0.04, 0.008, \ldots$

In Exercises 25–38, write the first five terms of the geometric sequence. If necessary, round your answers to two decimal places.

25. $a_1 = 4, \quad r = 2$

26. $a_1 = 3, \quad r = 4$

27. $a_1 = 6, \quad r = \frac{1}{2}$

28. $a_1 = 90, \quad r = \frac{1}{3}$

29. $a_1 = 5, \quad r = -2$

30. $a_1 = -12, \quad r = -1$

31. $a_1 = 1, \quad r = -\frac{1}{2}$

32. $a_1 = 3, \quad r = -\frac{3}{2}$

33. $a_1 = 1000, \quad r = 1.01$

34. $a_1 = 200, \quad r = 1.07$

35. $a_1 = 4000, \quad r = \frac{1}{1.01}$

36. $a_1 = 1000, \quad r = \frac{1}{1.05}$

37. $a_1 = 10, \quad r = \frac{3}{5}$

38. $a_1 = 36, \quad r = \frac{2}{3}$

In Exercises 39–52, find the specified term of the geometric sequence.

39. $a_1 = 6, \quad r = \frac{1}{2}, \quad a_{10} =$

40. $a_1 = 8, \quad r = \frac{3}{4}, \quad a_8 =$

41. $a_1 = 3, \quad r = \sqrt{2}, \quad a_{10} =$

42. $a_1 = 5, \quad r = \sqrt{3}, \quad a_9 =$

43. $a_1 = 200, \quad r = 1.2, \quad a_{12} =$

44. $a_1 = 500, \quad r = 1.06, \quad a_{40} =$

45. $a_1 = 120, \quad r = -\frac{1}{3}, \quad a_{10} =$

46. $a_1 = 240, \quad r = -\frac{1}{4}, \quad a_{13} =$

47. $a_1 = 4, \quad a_2 = 3, \quad a_5 =$

48. $a_1 = 1, \quad a_2 = 9, \quad a_7 =$

49. $a_1 = 1, \quad a_3 = \frac{9}{4}, \quad a_6 =$

50. $a_3 = 6, \quad a_5 = \frac{8}{3}, \quad a_6 =$

51. $a_2 = 12, \quad a_3 = 16, \quad a_4 =$

52. $a_4 = 100, \quad a_5 = -25, \quad a_7 =$

In Exercises 53–66, find a formula for the nth term of the geometric sequence. (Assume that n begins with 1.) See Examples 2 and 3.

53. $a_1 = 2, \quad r = 3$

54. $a_1 = 5, \quad r = 4$

55. $a_1 = 1, \quad r = 2$

56. $a_1 = 25, \quad r = 4$

57. $a_1 = 1, \quad r = -\frac{1}{5}$

58. $a_1 = 12, \quad r = -\frac{4}{3}$

59. $a_1 = 4, \quad r = -\frac{1}{2}$

60. $a_1 = 9, \quad r = \frac{2}{3}$

61. $a_1 = 8, \quad a_2 = 2$

62. $a_1 = 18, \quad a_2 = 8$

63. $a_1 = 14, \quad a_2 = \frac{21}{2}$

64. $a_1 = 36, \quad a_2 = \frac{27}{2}$

65. $4, -6, 9, -\frac{27}{2}, \ldots$

66. $1, \frac{3}{2}, \frac{9}{4}, \frac{27}{8}, \ldots$

In Exercises 67–70, match the geometric sequence with its graph. [The graphs are labeled (a), (b), (c), and (d).]

(a)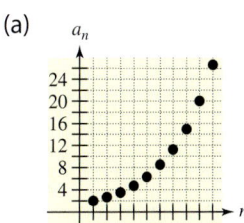

(b)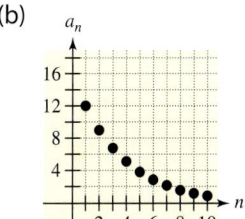

(c)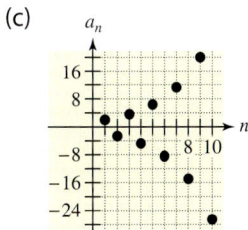

(d)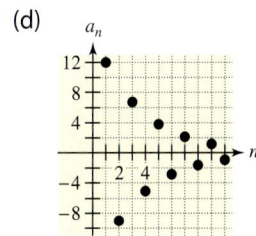

67. $a_n = 12\left(\frac{3}{4}\right)^{n-1}$

68. $a_n = 12\left(-\frac{3}{4}\right)^{n-1}$

69. $a_n = 2\left(\frac{4}{3}\right)^{n-1}$

70. $a_n = 2\left(-\frac{4}{3}\right)^{n-1}$

In Exercises 71–80, find the partial sum. See Examples 4 and 5.

71. $\displaystyle\sum_{i=1}^{10} 2^{i-1}$

72. $\displaystyle\sum_{i=1}^{6} 3^{i-1}$

73. $\displaystyle\sum_{i=1}^{12} 3\left(\frac{3}{2}\right)^{i-1}$

74. $\displaystyle\sum_{i=1}^{20} 12\left(\frac{2}{3}\right)^{i-1}$

75. $\displaystyle\sum_{i=1}^{15} 3\left(-\frac{1}{3}\right)^{i-1}$

76. $\displaystyle\sum_{i=1}^{8} 8\left(-\frac{1}{4}\right)^{i-1}$

77. $\displaystyle\sum_{i=1}^{12} 4(-2)^{i-1}$

78. $\displaystyle\sum_{i=1}^{20} 16\left(\frac{1}{2}\right)^{i-1}$

79. $\displaystyle\sum_{i=1}^{8} 6(0.1)^{i-1}$

80. $\displaystyle\sum_{i=1}^{24} 1000(1.06)^{i-1}$

In Exercises 81–92, find the *n*th partial sum of the geometric sequence.

81. $1, -3, 9, -27, 81, \ldots, n = 10$

82. $3, -6, 12, -24, 48, \ldots, n = 12$

83. $8, 4, 2, 1, \frac{1}{2}, \ldots, n = 15$

84. $9, 6, 4, \frac{8}{3}, \frac{16}{9}, \ldots, n = 10$

85. $4, 12, 36, 108, \ldots, n = 8$

86. $\frac{1}{36}, -\frac{1}{12}, \frac{1}{4}, -\frac{3}{4}, \ldots, n = 20$

87. $60, -15, \frac{15}{4}, -\frac{15}{16}, \ldots, n = 12$

88. $40, -10, \frac{5}{2}, -\frac{5}{8}, \frac{5}{32}, \ldots, n = 10$

89. $30, 30(1.06), 30(1.06)^2, 30(1.06)^3, \ldots, n = 20$

90. $100, 100(1.08), 100(1.08)^2, 100(1.08)^3, \ldots,$
$n = 40$

91. $500, 500(1.04), 500(1.04)^2, 500(1.04)^3, \ldots,$
$n = 18$

92. $1, \sqrt{2}, 2, 2\sqrt{2}, 4, \ldots, n = 12$

In Exercises 93–100, find the sum. See Example 6.

93. $\displaystyle\sum_{n=0}^{\infty} \left(\frac{1}{2}\right)^{n}$

94. $\displaystyle\sum_{n=0}^{\infty} 2\left(\frac{2}{3}\right)^{n}$

95. $\displaystyle\sum_{n=0}^{\infty} \left(-\frac{1}{2}\right)^{n}$

96. $\displaystyle\sum_{n=0}^{\infty} \left(\frac{1}{10}\right)^{n}$

97. $\displaystyle\sum_{n=0}^{\infty} 2\left(-\frac{2}{3}\right)^{n}$

98. $\displaystyle\sum_{n=0}^{\infty} 4\left(\frac{1}{4}\right)^{n}$

99. $8 + 6 + \frac{9}{2} + \frac{27}{8} + \cdots$

100. $3 - 1 + \frac{1}{3} - \frac{1}{9} + \cdots$

In Exercises 101–104, use a graphing calculator to graph the first 10 terms of the sequence.

101. $a_n = 20(-0.6)^{n-1}$

102. $a_n = 4(1.4)^{n-1}$

103. $a_n = 15(0.6)^{n-1}$

104. $a_n = 8(-0.6)^{n-1}$

Solving Problems

105. *Depreciation*　A company buys a machine for $250,000. During the next 5 years, the machine depreciates at the rate of 25% per year. (That is, at the end of each year, the depreciated value is 75% of what it was at the beginning of the year.)

　(a) Find a formula for the *n*th term of the geometric sequence that gives the value of the machine *n* full years after it was purchased.

　(b) Find the depreciated value of the machine at the end of 5 full years.

　(c) During which year did the machine depreciate the most?

106. *Population Increase*　A city of 500,000 people is growing at the rate of 1% per year. (That is, at the end of each year, the population is 1.01 times the population at the beginning of the year.)

　(a) Find a formula for the *n*th term of the geometric sequence that gives the population *n* years from now.

　(b) Estimate the population 20 years from now.

107. *Salary*　You accept a job as an archaeologist that pays a salary of $30,000 the first year. During the next 39 years, you receive a 5% raise each year. What would your total salary be over the 40-year period?

108. *Salary*   You accept a job as a biologist that pays a salary of $30,000 the first year. During the next 39 years, you receive a 5.5% raise each year.

(a) What would your total salary be over the 40-year period?

(b) How much more income did the extra 0.5% provide than the result in Exercise 107?

Increasing Annuity   In Exercises 109–114, find the balance A in an increasing annuity in which a principal of P dollars is invested each month for t years, compounded monthly at rate r.

109. $P = \$100$ $t = 10$ years $r = 9\%$

110. $P = \$50$ $t = 5$ years $r = 7\%$

111. $P = \$30$ $t = 40$ years $r = 8\%$

112. $P = \$200$ $t = 30$ years $r = 10\%$

113. $P = \$75$ $t = 30$ years $r = 6\%$

114. $P = \$100$ $t = 25$ years $r = 8\%$

115. *Wages*   You start work at a company that pays $.01 for the first day, $.02 for the second day, $.04 for the third day, and so on. The daily wage keeps doubling. What would your total income be for working (a) 29 days and (b) 30 days?

116. *Wages*   You start work at a company that pays $.01 for the first day, $.03 for the second day, $.09 for the third day, and so on. The daily wage keeps tripling. What would your total income be for working (a) 25 days and (b) 26 days?

117. *Power Supply*   The electrical power for an implanted medical device decreases by 0.1% each day.

(a) Find a formula for the nth term of the geometric sequence that gives the percent of the initial power n days after the device is implanted.

(b) What percent of the initial power is still available 1 year after the device is implanted?

(c) ▦ The power supply needs to be changed when half the power is depleted. Use a graphing calculator to graph the first 750 terms of the sequence and estimate when the power source should be changed.

118. *Cooling*   The temperature of water in an ice cube tray is 70°F when it is placed in a freezer. Its temperature n hours after being placed in the freezer is 20% less than 1 hour earlier.

(a) Find a formula for the nth term of the geometric sequence that gives the temperature of the water n hours after being placed in the freezer.

(b) Find the temperature of the water 6 hours after it is placed in the freezer.

(c) ▦ Use a graphing calculator to estimate the time when the water freezes. Explain your reasoning.

119. ▲ *Geometry*   A square has 12-inch sides. A new square is formed by connecting the midpoints of the sides of the square. Then two of the triangles are shaded (see figure). This process is repeated five more times. What is the total area of the shaded region?

120. ▲ *Geometry*   A square has 12-inch sides. The square is divided into nine smaller squares and the center square is shaded (see figure). Each of the eight unshaded squares is then divided into nine smaller squares and each center square is shaded. This process is repeated four more times. What is the total area of the shaded region?

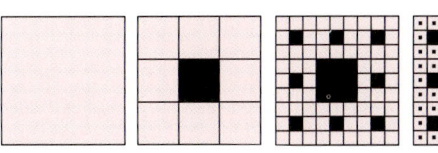

121. *Bungee Jumping*   A bungee jumper jumps from a bridge and stretches a cord 100 feet. Successive bounces stretch the cord 75% of each previous length (see figure). Find the total distance traveled by the bungee jumper during 10 bounces.

$$100 + 2(100)(0.75) + \cdots + 2(100)(0.75)^{10}$$

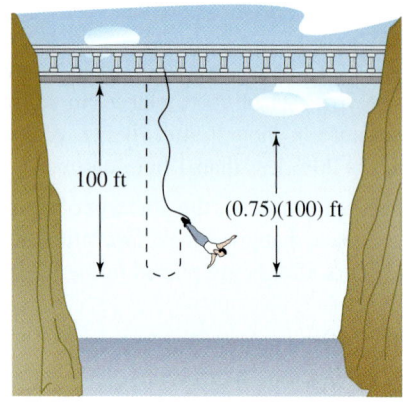

100 ft

(0.75)(100) ft

122. *Distance*   A ball is dropped from a height of 16 feet. Each time it drops h feet, it rebounds $0.81h$ feet.

(a) Find the total distance traveled by the ball.

(b) The ball takes the following time for each fall.

$$s_1 = -16t^2 + 16, \qquad\qquad s_1 = 0 \text{ if } t = 1$$
$$s_2 = -16t^2 + 16(0.81), \qquad s_2 = 0 \text{ if } t = 0.9$$
$$s_3 = -16t^2 + 16(0.81)^2, \qquad s_3 = 0 \text{ if } t = (0.9)^2$$
$$s_4 = -16t^2 + 16(0.81)^3, \qquad s_4 = 0 \text{ if } t = (0.9)^3$$
$$\vdots \qquad\qquad\qquad\qquad\qquad \vdots$$
$$s_n = -16t^2 + 16(0.81)^{n-1}, \; s_n = 0 \text{ if } t = (0.9)^{n-1}$$

Beginning with s_2, the ball takes the same amount of time to bounce up as it does to fall, and so the total time elapsed before it comes to rest is

$$t = 1 + 2\sum_{n=1}^{\infty} (0.9)^n.$$

Find this total.

Explaining Concepts

123. Answer parts (a)–(c) of Motivating the Chapter on page 808.

124. *Writing*   In your own words, explain what makes a sequence geometric.

125. What is the general formula for the nth term of a geometric sequence?

126. The second and third terms of a geometric sequence are 6 and 3, respectively. What is the first term?

127. Give an example of a geometric sequence whose terms alternate in sign.

128. *Writing*   Explain why the terms of a geometric sequence decrease when $a_1 > 0$ and $0 < r < 1$.

129. *Writing*   In your own words, describe an increasing annuity.

130. *Writing*   Explain what is meant by the nth partial sum of a sequence.

13.4 The Binomial Theorem

What You Should Learn

1. Use the Binomial Theorem to calculate binomial coefficients.
2. Use Pascal's Triangle to calculate binomial coefficients.
3. Expand binomial expressions.

Why You Should Learn It

You can use the Binomial Theorem to expand quantities used in probability. See Exercises 71–74 on page 848.

1 Use the Binomial Theorem to calculate binomial coefficients.

Binomial Coefficients

Recall that a **binomial** is a polynomial that has two terms. In this section, you will study a formula that provides a quick method of raising a binomial to a power. To begin, let's look at the expansion of $(x + y)^n$ for several values of n.

$$(x + y)^0 = 1$$

$$(x + y)^1 = x + y$$

$$(x + y)^2 = x^2 + 2xy + y^2$$

$$(x + y)^3 = x^3 + 3x^2y + 3xy^2 + y^3$$

$$(x + y)^4 = x^4 + 4x^3y + 6x^2y^2 + 4xy^3 + y^4$$

$$(x + y)^5 = x^5 + 5x^4y + 10x^3y^2 + 10x^2y^3 + 5xy^4 + y^5$$

There are several observations you can make about these expansions.

1. In each expansion, there are $n + 1$ terms.

2. In each expansion, x and y have symmetrical roles. The powers of x decrease by 1 in successive terms, whereas the powers of y increase by 1.

3. The sum of the powers of each term is n. For instance, in the expansion of $(x + y)^5$, the sum of the powers of each term is 5.

$$4 + 1 = 5 \quad 3 + 2 = 5$$

$$(x + y)^5 = x^5 + 5x^4y^1 + 10x^3y^2 + 10x^2y^3 + 5xy^4 + y^5$$

4. The coefficients increase and then decrease in a symmetrical pattern.

The coefficients of a binomial expansion are called **binomial coefficients**. To find them, you can use the **Binomial Theorem**.

Study Tip

Other notations that are commonly used for $_nC_r$ are

$$\binom{n}{r} \text{ and } C(n, r).$$

The Binomial Theorem

In the expansion of $(x + y)^n$

$$(x + y)^n = x^n + nx^{n-1}y + \cdots + {_nC_r}x^{n-r}y^r + \cdots + nxy^{n-1} + y^n$$

the coefficient of $x^{n-r}y^r$ is given by

$$_nC_r = \frac{n!}{(n - r)!r!}.$$

Example 1   Finding Binomial Coefficients

Find each binomial coefficient.

a. $_8C_2$ **b.** $_{10}C_3$ **c.** $_7C_0$ **d.** $_8C_8$ **e.** $_9C_6$

Solution

a. $_8C_2 = \dfrac{8!}{6! \cdot 2!} = \dfrac{(8 \cdot 7) \cdot 6!}{6! \cdot 2!} = \dfrac{8 \cdot 7}{2 \cdot 1} = 28$

b. $_{10}C_3 = \dfrac{10!}{7! \cdot 3!} = \dfrac{(10 \cdot 9 \cdot 8) \cdot 7!}{7! \cdot 3!} = \dfrac{10 \cdot 9 \cdot 8}{3 \cdot 2 \cdot 1} = 120$

c. $_7C_0 = \dfrac{7!}{7! \cdot 0!} = 1$

d. $_8C_8 = \dfrac{8!}{0! \cdot 8!} = 1$

e. $_9C_6 = \dfrac{9!}{3! \cdot 6!} = \dfrac{(9 \cdot 8 \cdot 7) \cdot 6!}{3! \cdot 6!} = \dfrac{9 \cdot 8 \cdot 7}{3 \cdot 2 \cdot 1} = 84$

> **Technology: Tip**
>
> The formula for the binomial coefficient is the same as the formula for combinations in the study of probability. Most graphing calculators have the capability to evaluate a binomial coefficient. Consult the user's guide for your graphing calculator.

When $r \neq 0$ and $r \neq n,$ as in parts (a) and (b) of Example 1, there is a simple pattern for evaluating binomial coefficients. Note how this is used in parts (a) and (b) of Example 2.

Example 2   Finding Binomial Coefficients

Find each binomial coefficient.

a. $_7C_3$ **b.** $_7C_4$ **c.** $_{12}C_1$ **d.** $_{12}C_{11}$

Solution

a. $_7C_3 = \dfrac{7 \cdot 6 \cdot 5}{3 \cdot 2 \cdot 1} = 35$

b. $_7C_4 = \dfrac{7 \cdot 6 \cdot 5 \cdot 4}{4 \cdot 3 \cdot 2 \cdot 1} = 35$ $_7C_4 = {_7C_3}$

c. $_{12}C_1 = \dfrac{12!}{11! \cdot 1!} = \dfrac{(12) \cdot 11!}{11! \cdot 1!} = \dfrac{12}{1} = 12$

d. $_{12}C_{11} = \dfrac{12!}{1! \cdot 11!} = \dfrac{(12) \cdot 11!}{1! \cdot 11!} = \dfrac{12}{1} = 12$ $_{12}C_{11} = {_{12}C_1}$

In Example 2, it is not a coincidence that the answers to parts (a) and (b) are the same and that the answers to parts (c) and (d) are the same. In general, it is true that

$$_nC_r = {_nC_{n-r}}.$$

This shows the symmetric property of binomial coefficients.

2 Use Pascal's Triangle to calculate binomial coefficients.

Pascal's Triangle

There is a convenient way to remember a pattern for binomial coefficients. By arranging the coefficients in a triangular pattern, you obtain the following array, which is called **Pascal's Triangle.** This triangle is named after the famous French mathematician Blaise Pascal (1623–1662).

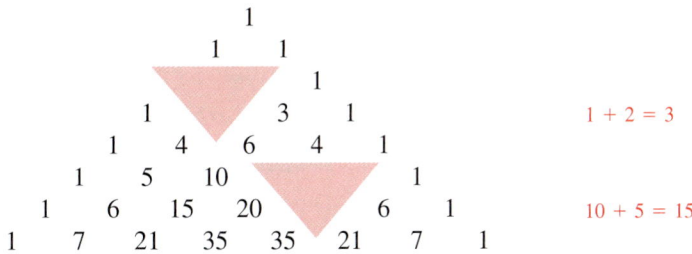

$$1 + 2 = 3$$

$$10 + 5 = 15$$

Study Tip

The top row in Pascal's Triangle is called the *zeroth row* because it corresponds to the binomial expansion

$$(x + y)^0 = 1.$$

Similarly, the next row is called the *first row* because it corresponds to the binomial expansion

$$(x + y)^1 = 1(x) + 1(y).$$

In general, the *nth row* in Pascal's Triangle gives the coefficients of $(x + y)^n$.

The first and last numbers in each row of Pascal's Triangle are 1. As shown above, every other number in each row is formed by adding the two numbers immediately above the number. Pascal noticed that numbers in this triangle are precisely the same numbers that are the coefficients of binomial expansions.

$$(x + y)^0 = 1 \qquad \text{0th row}$$
$$(x + y)^1 = 1x + 1y \qquad \text{1st row}$$
$$(x + y)^2 = 1x^2 + 2xy + 1y^2 \qquad \text{2nd row}$$
$$(x + y)^3 = 1x^3 + 3x^2y + 3xy^2 + 1y^3 \qquad \text{3rd row}$$
$$(x + y)^4 = 1x^4 + 4x^3y + 6x^2y^2 + 4xy^3 + 1y^4 \qquad \vdots$$
$$(x + y)^5 = 1x^5 + 5x^4y + 10x^3y^2 + 10x^2y^3 + 5xy^4 + 1y^5$$
$$(x + y)^6 = 1x^6 + 6x^5y + 15x^4y^2 + 20x^3y^3 + 15x^2y^4 + 6xy^5 + 1y^6$$
$$(x + y)^7 = 1x^7 + 7x^6y + 21x^5y^2 + 35x^4y^3 + 35x^3y^4 + 21x^2y^5 + 7xy^6 + 1y^7$$

You can use the seventh row of Pascal's Triangle to find the binomial coefficients of the eighth row.

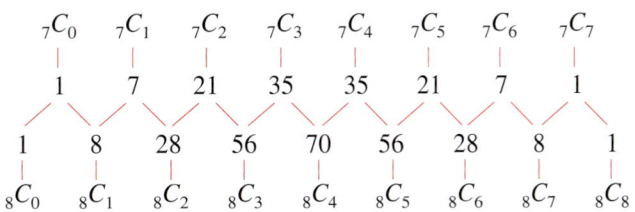

Example 3   Using Pascal's Triangle

Use the fifth row of Pascal's Triangle to evaluate $_5C_2$.

Solution

1	5	10	10	5	1
$_5C_0$	$_5C_1$	$_5C_2$	$_5C_3$	$_5C_4$	$_5C_5$

So, $_5C_2 = 10$.

3 Expand binomial expressions.

Binomial Expansions

As mentioned at the beginning of this section, when you write out the coefficients for a binomial that is raised to a power, you are **expanding a binomial.** The formulas for binomial coefficients give you an easy way to expand binomials, as demonstrated in the next four examples.

Example 4 Expanding a Binomial

Write the expansion of the expression $(x + 1)^5$.

Solution

The binomial coefficients from the fifth row of Pascal's Triangle are

$$1, 5, 10, 10, 5, 1.$$

So, the expansion is as follows.

$$(x + 1)^5 = (1)x^5 + (5)x^4(1) + (10)x^3(1^2) + (10)x^2(1^3) + (5)x(1^4) + (1)(1^5)$$

$$= x^5 + 5x^4 + 10x^3 + 10x^2 + 5x + 1$$

To expand binomials representing *differences*, rather than sums, you alternate signs. Here are two examples.

$$(x - 1)^3 = x^3 - 3x^2 + 3x - 1$$

$$(x - 1)^4 = x^4 - 4x^3 + 6x^2 - 4x + 1$$

Example 5 Expanding a Binomial

Write the expansion of each expression.

a. $(x - 3)^4$ **b.** $(2x - 1)^3$

Solution

a. The binomial coefficients from the fourth row of Pascal's Triangle are

$$1, 4, 6, 4, 1.$$

So, the expansion is as follows.

$$(x - 3)^4 = (1)x^4 - (4)x^3(3) + (6)x^2(3^2) - (4)x(3^3) + (1)(3^4)$$

$$= x^4 - 12x^3 + 54x^2 - 108x + 81$$

b. The binomial coefficients from the third row of Pascal's Triangle are

$$1, 3, 3, 1.$$

So, the expansion is as follows.

$$(2x - 1)^3 = (1)(2x)^3 - (3)(2x)^2(1) + (3)(2x)(1^2) - (1)(1^3)$$

$$= 8x^3 - 12x^2 + 6x - 1$$

Example 6   Expanding a Binomial

Write the expansion of the expression.

$(x - 2y)^4$

Solution

Use the fourth row of Pascal's Triangle, as follows.

$$(x - 2y)^4 = (1)x^4 - (4)x^3(2y) + (6)x^2(2y)^2 - (4)x(2y)^3 + (1)(2y)^4$$

$$= x^4 - 8x^3y + 24x^2y^2 - 32xy^3 + 16y^4$$

Example 7   Expanding a Binomial

Write the expansion of the expression.

$(x^2 + 4)^3$

Solution

Use the third row of Pascal's Triangle, as follows.

$$(x^2 + 4)^3 = (1)(x^2)^3 + (3)(x^2)^2(4) + (3)x^2(4^2) + (1)(4^3)$$

$$= x^6 + 12x^4 + 48x^2 + 64$$

Sometimes you will need to find a specific term in a binomial expansion. Instead of writing out the entire expansion, you can use the fact that from the Binomial Theorem, the $(r + 1)$th term is

$$_nC_r x^{n-r}y^r.$$

Example 8   Finding a Term in the Binomial Expansion

a. Find the sixth term of $(a + 2b)^8$.

b. Find the coefficient of the term a^6b^5 in the expansion of $(3a - 2b)^{11}$.

Solution

a. In this case, $6 = r + 1$ means that $r = 5$. Because $n = 8$, $x = a$, and $y = 2b$, the sixth term in the binomial expansion is

$$_8C_5 a^{8-5}(2b)^5 = 56 \cdot a^3 \cdot (2b)^5$$

$$= 56(2^5)a^3b^5$$

$$= 1792\,a^3b^5.$$

b. In this case, $n = 11$, $r = 5$, $x = 3a$, and $y = -2b$. Substitute these values to obtain

$$_nC_r x^{n-r}y^r = {}_{11}C_5(3a)^6(-2b)^5$$

$$= 462(729a^6)(-32b^5)$$

$$= -10{,}777{,}536a^6b^5.$$

So, the coefficient is $-10{,}777{,}536$.

13.4 Exercises

Review   Concepts, Skills, and Problem Solving

Keep mathematically in shape by doing these exercises *before* the problems of this section.

Properties and Definitions

1. Is it possible to find the determinant of the following matrix? Explain.

$$\begin{bmatrix} 3 & 2 & 6 \\ 1 & -4 & 7 \end{bmatrix}$$

2. State the three elementary row operations that can be used to transform a matrix into a second, row-equivalent matrix.

3. Is the matrix in row-echelon form? Explain.

$$\begin{bmatrix} 1 & 2 & 6 \\ 0 & 1 & 7 \end{bmatrix}$$

4. Form the (a) coefficient matrix and (b) augmented matrix for the system of linear equations.

$$\begin{cases} x + 3y = -1 \\ 4x - y = 2 \end{cases}$$

Determinants

In Exercises 5–8, find the determinant of the matrix.

5. $\begin{bmatrix} 10 & 25 \\ 6 & -5 \end{bmatrix}$

6. $\begin{bmatrix} 3 & 7 \\ -2 & 6 \end{bmatrix}$

7. $\begin{bmatrix} 3 & -2 & 1 \\ 0 & 5 & 3 \\ 6 & 1 & 1 \end{bmatrix}$

8. $\begin{bmatrix} 4 & 3 & 5 \\ 3 & 2 & -2 \\ 5 & -2 & 0 \end{bmatrix}$

Problem Solving

9. Use determinants to find the equation of the line through $(2, -1)$ and $(4, 7)$.

10. Use a determinant to find the area of the triangle with vertices $(-5, 8)$, $(10, 0)$, and $(3, -4)$.

Developing Skills

In Exercises 1–12, evaluate the binomial coefficient $_nC_r$. See Examples 1 and 2.

1. $_6C_4$
2. $_7C_3$
3. $_{10}C_5$
4. $_{12}C_9$
5. $_{20}C_{20}$
6. $_{15}C_0$
7. $_{13}C_0$
8. $_{200}C_1$
9. $_{50}C_1$
10. $_{12}C_{12}$
11. $_{25}C_4$
12. $_{18}C_5$

In Exercises 13–22, use a graphing calculator to evaluate $_nC_r$.

13. $_{30}C_6$
14. $_{25}C_{10}$
15. $_{12}C_7$
16. $_{40}C_5$
17. $_{52}C_5$
18. $_{100}C_6$
19. $_{200}C_{195}$
20. $_{500}C_4$
21. $_{800}C_{797}$
22. $_{1000}C_2$

In Exercises 23–28, use Pascal's Triangle to evaluate $_nC_r$. See Example 3.

23. $_6C_2$

24. $_9C_3$

25. $_7C_3$

26. $_9C_5$

27. $_8C_4$

28. $_{10}C_6$

In Exercises 29–38, use Pascal's Triangle to expand the expression. See Examples 4–7.

29. $(a + 2)^3$

30. $(x + 3)^5$

31. $(m - n)^5$

32. $(r - s)^7$

33. $(2x - 1)^5$

34. $(4 - 3y)^3$

35. $(2y + z)^6$

36. $(3c + d)^6$

37. $(x^2 + 2)^4$

38. $(5 + y^2)^5$

In Exercises 39–50, use the Binomial Theorem to expand the expression.

39. $(x + 3)^6$

40. $(x - 5)^4$

41. $(x + y)^4$

42. $(u + v)^6$

43. $(u - 2v)^3$

44. $(2x + y)^5$

45. $(3a + 2b)^4$

46. $(4u - 3v)^3$

47. $\left(x + \dfrac{2}{y}\right)^4$

48. $\left(3s + \dfrac{1}{t}\right)^5$

49. $(2x^2 - y)^5$

50. $(x - 4y^3)^4$

In Exercises 51–58, find the specified term in the expansion of the binomial. See Example 8.

51. $(x + y)^{10}$, 4th term

52. $(x - y)^6$, 7th term

53. $(a + 4b)^9$, 6th term

54. $(a + 5b)^{12}$, 8th term

55. $(3a - b)^{12}$, 10th term

56. $(8x - y)^4$, 3rd term

57. $(3x + 2y)^{15}$, 7th term

58. $(4a - 3b)^9$, 8th term

In Exercises 59–66, find the coefficient of the term in the expansion of the binomial. See Example 8.

	Expression	Term
59.	$(x + 1)^{10}$	x^7
60.	$(x + 3)^{12}$	x^9
61.	$(x - y)^{15}$	x^4y^{11}
62.	$(x - 3y)^{14}$	x^3y^{11}
63.	$(2x + y)^{12}$	x^3y^9
64.	$(x + y)^{10}$	x^7y^3
65.	$(x^2 - 3)^4$	x^4
66.	$(3 - y^3)^5$	y^9

In Exercises 67–70, use the Binomial Theorem to approximate the quantity accurate to three decimal places. For example:

$$(1.02)^{10} = (1 + 0.02)^{10} \approx 1 + 10(0.02) + 45(0.02)^2.$$

67. $(1.02)^8$

68. $(2.005)^{10}$

69. $(2.99)^{12}$

70. $(1.98)^9$

Solving Problems

Probability   In Exercises 71–74, use the Binomial Theorem to expand the expression. In the study of probability, it is sometimes necessary to use the expansion $(p + q)^n$, where $p + q = 1$.

71. $\left(\frac{1}{2} + \frac{1}{2}\right)^5$

72. $\left(\frac{2}{3} + \frac{1}{3}\right)^4$

73. $\left(\frac{1}{4} + \frac{3}{4}\right)^4$

74. $\left(\frac{2}{5} + \frac{3}{5}\right)^3$

75. *Pascal's Triangle*   Describe the pattern.

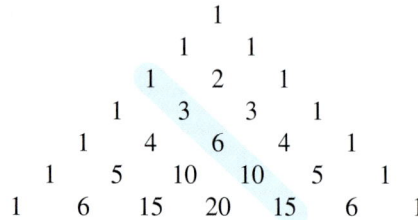

76. *Pascal's Triangle*   Use each encircled group of numbers to form a 2×2 matrix. Find the determinant of each matrix. Describe the pattern.

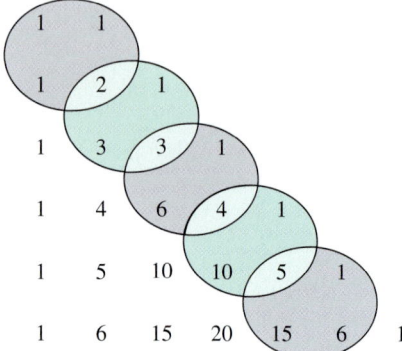

Explaining Concepts

77. How many terms are in the expansion of $(x + y)^n$?

78. How do the expansions of $(x + y)^n$ and $(x - y)^n$ differ?

79. Which of the following is equal to $_{11}C_5$? Explain.

    (a) $\dfrac{11 \cdot 10 \cdot 9 \cdot 8 \cdot 7}{5 \cdot 4 \cdot 3 \cdot 2 \cdot 1}$ (b) $\dfrac{11 \cdot 10 \cdot 9 \cdot 8 \cdot 7}{6 \cdot 5 \cdot 4 \cdot 3 \cdot 2 \cdot 1}$

80. *Writing*   What is the relationship between $_nC_r$ and $_nC_{n-r}$? Explain.

81. *Writing*   In your own words, explain how to form the rows in Pascal's Triangle.

What Did You Learn?

Key Terms

sequence, *p. 810*

term (of a sequence), *p. 810*

infinite sequence, *p. 810*

finite sequence, *p. 810*

factorials, *p. 812*

series, *p. 813*

partial sum, *p. 813*

infinite series, *p. 813*

sigma notation, *p. 814*

index of summation, *p. 814*

upper limit of summation, *p. 814*

lower limit of summation, *p. 814*

arithmetic sequence, *p. 821*

common difference, *p. 821*

recursion formula, *p. 822*

*n*th partial sum, *pp. 823, 833*

geometric sequence, *p. 831*

common ratio, *p. 831*

infinite geometric series, *p. 833*

increasing annuity, *p. 835*

binomial coefficients, *p. 841*

Pascal's Triangle, *p. 843*

expanding a binomial, *p. 844*

Key Concepts

13.1 ● Definition of factorial

If n is a positive integer, n factorial is defined as

$$n! = 1 \cdot 2 \cdot 3 \cdot 4 \cdot \cdots \cdot (n - 1) \cdot n.$$

As a special case, zero factorial is defined as $0! = 1$.

13.1 ● Definition of series

For an infinite sequence, $a_1, a_2, a_3, \ldots, a_n, \ldots$

1. the sum of the first n terms

$$S_n = a_1 + a_2 + a_3 + \cdots + a_n$$

is called a partial sum, and

2. the sum of all terms

$$a_1 + a_2 + a_3 + \cdots + a_n + \cdots$$

is called an infinite series, or simply a series.

13.1 ● Definition of sigma notation

The sum of the first n terms of the sequence whose nth term is a_n is

$$\sum_{i=1}^{n} a_i = a_1 + a_2 + a_3 + a_4 + \cdots + a_n$$

where i is the index of summation, n is the upper limit of summation, and 1 is the lower limit of summation.

13.2 ● The *n*th term of an arithmetic sequence

The nth term of an arithmetic sequence has the form $a_n = a_1 + (n - 1)d$, where d is the common difference of the sequence, and a_1 is the first term.

13.2 ● The *n*th partial sum of an arithmetic sequence

The nth partial sum of the arithmetic sequence whose nth term is a_n is

$$\sum_{i=1}^{n} a_i = a_1 + a_2 + a_3 + a_4 + \cdots + a_n$$

$$= \frac{n}{2}(a_1 + a_n).$$

13.3 ● The *n*th term of a geometric sequence

The nth term of a geometric sequence has the form $a_n = a_1 r^{n-1}$, where r is the common ratio of consecutive terms of the sequence. So, every geometric sequence can be written in the following form.

$$a_1, a_1 r, a_1 r^2, a_1 r^3, a_1 r^4, \ldots, a_1 r^{n-1}, \ldots$$

13.3 ● The *n*th partial sum of a geometric sequence

The nth partial sum of the geometric sequence whose nth term is $a_n = a_1 r^{n-1}$ is given by

$$\sum_{i=1}^{n} a_1 r^{i-1} = a_1 + a_1 r + a_1 r^2 + a_1 r^3 + \cdots$$

$$+ a_1 r^{n-1} = a_1 \left(\frac{r^n - 1}{r - 1} \right).$$

13.3 ● Sum of an infinite geometric series

If $a_1, a_1 r, a_1 r^2, \ldots, a_1 r^n, \ldots$ is an infinite geometric sequence, then for $|r| < 1$, the sum of the terms of the corresponding infinite geometric series is

$$S = \sum_{i=0}^{\infty} a_1 r^i = \frac{a_1}{1 - r}.$$

13.4 ● The Binomial Theorem

In the expansion of $(x + y)^n$

$$(x + y)^n = x^n + nx^{n-1}y + \cdots + {}_nC_r x^{n-r}y^r + \cdots + nxy^{n-1} + y^n$$

the coefficient of $x^{n-r}y^r$ is given by

$${}_nC_r = \frac{n!}{(n - r)!r!}.$$

Review Exercises

13.1 Sequences and Series

① Use sequence notation to write the terms of sequences.

In Exercises 1–4, write the first five terms of the sequence. (Assume that n begins with 1.)

1. $a_n = 3n + 5$ **2.** $a_n = \frac{1}{2}n - 4$

3. $a_n = \frac{1}{2^n} + \frac{1}{2}$ **4.** $a_n = 3^n + n$

② Write the terms of sequences involving factorials.

In Exercises 5–8, write the first five terms of the sequence. (Assume that n begins with 1.)

5. $a_n = (n + 1)!$ **6.** $a_n = n! - 2$

7. $a_n = \frac{n!}{2n}$ **8.** $a_n = \frac{(n + 1)!}{(2n)!}$

③ Find the apparent nth term of a sequence.

In Exercises 9–18, write an expression for the nth term of the sequence. (Assume that n begins with 1.)

9. $1, 3, 5, 7, 9, \ldots$ **10.** $3, -6, 9, -12, 15, \ldots$

11. $\frac{1}{4}, \frac{2}{9}, \frac{3}{16}, \frac{4}{25}, \frac{5}{36}, \ldots$ **12.** $\frac{0}{2}, \frac{1}{3}, \frac{2}{4}, \frac{3}{5}, \frac{4}{6}, \ldots$

13. $4, \frac{9}{2}, \frac{14}{3}, \frac{19}{4}, \frac{24}{5}, \ldots$ **14.** $12, 4, \frac{4}{3}, \frac{4}{9}, \frac{4}{27}, \ldots$

15. $3, 1, -1, -3, -5, \ldots$ **16.** $3, 7, 11, 15, 19, \ldots$

17. $\frac{3}{2}, \frac{12}{5}, \frac{27}{10}, \frac{48}{17}, \frac{75}{26}, \ldots$ **18.** $-1, \frac{1}{2}, -\frac{1}{4}, \frac{1}{8}, -\frac{1}{16}, \ldots$

④ Sum the terms of sequences to obtain series and use sigma notation to represent partial sums.

In Exercises 19–22, find the partial sum.

19. $\sum_{k=1}^{4} 7$ **20.** $\sum_{k=1}^{4} \frac{(-1)^k}{k}$

21. $\sum_{n=1}^{4} \left(\frac{1}{n} - \frac{1}{n + 1} \right)$ **22.** $\sum_{n=1}^{4} \left(\frac{1}{n} - \frac{1}{n + 2} \right)$

In Exercises 23–26, write the sum using sigma notation. (Begin with $k = 0$ or $k = 1$.)

23. $[5(1) - 3] + [5(2) - 3] + [5(3) - 3] + [5(4) - 3]$

24. $[9 - 10(1)] + [9 - 10(2)] + [9 - 10(3)] + [9 - 10(4)]$

25. $\frac{1}{3(1)} + \frac{1}{3(2)} + \frac{1}{3(3)} + \frac{1}{3(4)} + \frac{1}{3(5)} + \frac{1}{3(6)}$

26. $\left(-\frac{1}{3}\right)^0 + \left(-\frac{1}{3}\right)^1 + \left(-\frac{1}{3}\right)^2 + \left(-\frac{1}{3}\right)^3 + \left(-\frac{1}{3}\right)^4$

13.2 Arithmetic Sequences

① Recognize, write, and find the nth terms of arithmetic sequences.

In Exercises 27 and 28, find the common difference of the arithmetic sequence.

27. $30, 27.5, 25, 22.5, 20, \ldots$

28. $9, 12, 15, 18, 21, \ldots$

In Exercises 29–36, write the first five terms of the arithmetic sequence. (Assume that n begins with 1.)

29. $a_n = 132 - 5n$ **30.** $a_n = 2n + 3$

31. $a_n = \frac{3}{4}n + \frac{1}{2}$ **32.** $a_n = -\frac{3}{5}n + 1$

33. $a_1 = 5$ **34.** $a_1 = 12$
$\quad a_{k+1} = a_k + 3$ $\quad a_{k+1} = a_k + 1.5$

35. $a_1 = 80$ **36.** $a_1 = 25$
$\quad a_{k+1} = a_k - \frac{5}{2}$ $\quad a_{k+1} = a_k - 6$

In Exercises 37–40, find a formula for the nth term of the arithmetic sequence.

37. $a_1 = 10, \quad d = 4$

38. $a_1 = 32, \quad d = -2$

39. $a_1 = 1000, \quad a_2 = 950$

40. $a_1 = 12, \quad a_2 = 20$

2 Find the nth partial sum of an arithmetic sequence.

In Exercises 41–44, find the partial sum.

41. $\displaystyle\sum_{k=1}^{12} (7k - 5)$ **42.** $\displaystyle\sum_{k=1}^{10} (100 - 10k)$

43. $\displaystyle\sum_{j=1}^{100} \frac{j}{4}$ **44.** $\displaystyle\sum_{j=1}^{50} \frac{3j}{2}$

In Exercises 45 and 46, use a graphing calculator to find the partial sum.

45. $\displaystyle\sum_{i=1}^{60} (1.25i + 4)$ **46.** $\displaystyle\sum_{i=1}^{100} (5000 - 3.5i)$

3 Use arithmetic sequences to solve application problems.

47. *Number Problem*   Find the sum of the first 50 positive integers that are multiples of 4.

48. *Number Problem*   Find the sum of the integers from 225 to 300.

49. *Auditorium Seating*   Each row in a small auditorium has three more seats than the preceding row. The front row seats 22 people and there are 12 rows of seats. Find the seating capacity of the auditorium.

50. *Pile of Logs*   A pile of logs has 20 logs on the bottom layer and one log on the top layer. Each layer has one log less than the layer below it. How many logs are in the pile?

13.3 Geometric Sequences and Series

1 Recognize, write, and find the nth terms of geometric sequences.

In Exercises 51 and 52, find the common ratio of the geometric sequence.

51. $8, 12, 18, 27, \frac{81}{2}, \ldots$

52. $27, -18, 12, -8, \frac{16}{3}, \ldots$

In Exercises 53–58, write the first five terms of the geometric sequence.

53. $a_1 = 10, \quad r = 3$ **54.** $a_1 = 2, \quad r = -5$

55. $a_1 = 100, \quad r = -\frac{1}{2}$

56. $a_1 = 12, \quad r = \frac{1}{6}$

57. $a_1 = 4, \quad r = \frac{3}{2}$ **58.** $a_1 = 32, \quad r = -\frac{3}{4}$

In Exercises 59–64, find a formula for the nth term of the geometric sequence. (Assume that n begins with 1.)

59. $a_1 = 1, \quad r = -\frac{2}{3}$

60. $a_1 = 100, \quad r = 1.07$

61. $a_1 = 24, \quad a_2 = 48$

62. $a_1 = 16, \quad a_2 = -4$

63. $a_1 = 12, \quad a_4 = -\frac{3}{2}$

64. $a_2 = 1, \quad a_3 = \frac{1}{3}$

2 Find the nth partial sum of a geometric sequence.

In Exercises 65–72, find the partial sum.

65. $\displaystyle\sum_{n=1}^{12} 2^n$ **66.** $\displaystyle\sum_{n=1}^{12} (-2)^n$

67. $\displaystyle\sum_{k=1}^{8} 5\left(-\frac{3}{4}\right)^k$ **68.** $\displaystyle\sum_{k=1}^{10} 4\left(\frac{3}{2}\right)^k$

69. $\displaystyle\sum_{i=1}^{8} (1.25)^{i-1}$ **70.** $\displaystyle\sum_{i=1}^{8} (-1.25)^{i-1}$

71. $\displaystyle\sum_{n=1}^{120} 500(1.01)^n$ **72.** $\displaystyle\sum_{n=1}^{40} 1000(1.1)^n$

In Exercises 73 and 74, use a graphing calculator to find the partial sum.

73. $\displaystyle\sum_{k=1}^{50} 50(1.2)^{k-1}$ **74.** $\displaystyle\sum_{j=1}^{60} 25(0.9)^{j-1}$

3 Find the sum of an infinite geometric series.

In Exercises 75–78, find the sum.

75. $\displaystyle\sum_{i=1}^{\infty} \left(\frac{7}{8}\right)^{i-1}$ **76.** $\displaystyle\sum_{i=1}^{\infty} \left(\frac{1}{3}\right)^{i-1}$

77. $\displaystyle\sum_{k=1}^{\infty} 4\left(\frac{2}{3}\right)^{k-1}$ **78.** $\displaystyle\sum_{k=1}^{\infty} 1.3\left(\frac{1}{10}\right)^{k-1}$

④ **Use geometric sequences to solve application problems.**

79. *Depreciation*   A company pays \$120,000 for a machine. During the next 5 years, the machine depreciates at the rate of 30% per year. (That is, at the end of each year, the depreciated value is 70% of what it was at the beginning of the year.)

    (a) Find a formula for the nth term of the geometric sequence that gives the value of the machine n full years after it was purchased.

    (b) Find the depreciated value of the machine at the end of 5 full years.

80. *Population Increase*   A city of 85,000 people is growing at the rate of 1.2% per year. (That is, at the end of each year, the population is 1.012 times what it was at the beginning of the year.)

    (a) Find a formula for the nth term of the geometric sequence that gives the population n years from now.

    (b) Estimate the population 50 years from now.

81. *Salary*   You accept a job as an architect that pays a salary of \$32,000 the first year. During the next 39 years, you receive a 5.5% raise each year. What would your total salary be over the 40-year period?

82. *Increasing Annuity*   You deposit \$200 in an account each month for 10 years. The account pays an annual interest rate of 8%, compounded monthly. What is your balance at the end of 10 years?

13.4 The Binomial Theorem

① **Use the Binomial Theorem to calculate binomial coefficients.**

In Exercises 83–86, evaluate the binomial coefficient $_nC_r$.

83. $_8C_3$ **84.** $_{12}C_2$

85. $_{12}C_0$ **86.** $_{100}C_1$

▦ **In Exercises 87–90, use a graphing calculator to evaluate** $_nC_r$.

87. $_{40}C_4$ **88.** $_{15}C_9$

89. $_{25}C_6$ **90.** $_{32}C_2$

② **Use Pascal's Triangle to calculate binomial coefficients.**

In Exercises 91–94, use Pascal's Triangle to evaluate $_nC_r$.

91. $_5C_3$ **92.** $_9C_9$

93. $_8C_4$ **94.** $_6C_5$

③ **Expand binomial expressions.**

In Exercises 95–98, use Pascal's Triangle to expand the expression.

95. $(x - 5)^4$

96. $(x + y)^7$

97. $(2x + 1)^3$

98. $(x - 3y)^4$

In Exercises 99–104, use the Binomial Theorem to expand the expression.

99. $(x + 1)^{10}$

100. $(y - 2)^6$

101. $(3x - 2y)^4$

102. $(2u + 5v)^4$

103. $(u^2 + v^3)^9$

104. $(x^4 - y^5)^8$

In Exercises 105 and 106, find the specified term in the expansion of the binomial.

105. $(x + 4)^9$,   4th term

106. $(2x - 3y)^5$,   4th term

In Exercises 107 and 108, find the coefficient of the term in the expansion of the binomial.

	Expression	*Term*
107.	$(x - 3)^{10}$	x^5
108.	$(x + 2y)^7$	x^4y^3

Chapter Test

Take this test as you would take a test in class. After you are done, check your work against the answers in the back of the book.

1. Write the first five terms of the sequence $a_n = \left(-\frac{2}{3}\right)^{n-1}$. (Assume that n begins with 1.)

2. Write the first five terms of the sequence $a_n = 3n^2 - n$. (Assume that n begins with 1.)

In Exercises 3–5, find the partial sum.

3. $\displaystyle\sum_{n=1}^{12} 5$

4. $\displaystyle\sum_{j=0}^{4} (3j + 1)$

5. $\displaystyle\sum_{n=1}^{5} (3 - 4n)$

6. Use sigma notation to write $\dfrac{2}{3(1) + 1} + \dfrac{2}{3(2) + 1} + \cdots + \dfrac{2}{3(12) + 1}$.

7. Use sigma notation to write

$$\left(\frac{1}{2}\right)^0 + \left(\frac{1}{2}\right)^2 + \left(\frac{1}{2}\right)^4 + \left(\frac{1}{2}\right)^6 + \left(\frac{1}{2}\right)^8 + \left(\frac{1}{2}\right)^{10}.$$

8. Write the first five terms of the arithmetic sequence whose first term is $a_1 = 12$ and whose common difference is $d = 4$.

9. Find a formula for the nth term of the arithmetic sequence whose first term is $a_1 = 5000$ and whose common difference is $d = -100$.

10. Find the sum of the first 50 positive integers that are multiples of 3.

11. Find the common ratio of the geometric sequence: $2, -3, \frac{9}{2}, -\frac{27}{4}, \ldots$.

12. Find a formula for the nth term of the geometric sequence whose first term is $a_1 = 4$ and whose common ratio is $r = \frac{1}{2}$.

In Exercises 13 and 14, find the partial sum.

13. $\displaystyle\sum_{n=1}^{8} 2(2^n)$

14. $\displaystyle\sum_{n=1}^{10} 3\left(\frac{1}{2}\right)^n$

In Exercises 15 and 16, find the sum of the infinite geometric series.

15. $\displaystyle\sum_{i=1}^{\infty} \left(\frac{1}{2}\right)^i$

16. $\displaystyle\sum_{i=1}^{\infty} 4\left(\frac{2}{3}\right)^{i-1}$

17. Evaluate: $_{20}C_3$

18. Use Pascal's Triangle to expand $(x - 2)^5$.

19. Find the coefficient of the term x^3y^5 in the expansion of $(x + y)^8$.

20. A free-falling object will fall 4.9 meters during the first second, 14.7 more meters during the second second, 24.5 more meters during the third second, and so on. What is the total distance the object will fall in 10 seconds if this pattern continues?

21. Fifty dollars is deposited each month in an increasing annuity that pays 8%, compounded monthly. What is the balance after 25 years?

Review of Elementary Algebra Topics

A.1 The Real Number System

Sets and Real Numbers   •   Operations with Real Numbers   •   Properties of Real Numbers

Sets and Real Numbers

Real numbers are used in everyday life to describe quantities such as age, miles per gallon, container size, and population. Real numbers are represented by symbols such as

$$-5, 9, 0, \tfrac{4}{3}, 0.666\ldots, 28.21, \sqrt{2}, \pi, \text{ and } \sqrt[3]{-32}.$$

Here are some important subsets of the set of real numbers.

$$\{1, 2, 3, 4, \ldots\} \qquad \text{Set of natural numbers}$$

$$\{0, 1, 2, 3, 4, \ldots\} \qquad \text{Set of whole numbers}$$

$$\{\ldots, -3, -2, -1, 0, 1, 2, 3, \ldots\} \qquad \text{Set of integers}$$

A real number is rational if it can be written as the ratio p/q of two integers, where $q \neq 0$. For instance, the numbers

$$\tfrac{1}{3} = 0.3333\ldots = 0.\overline{3}, \tfrac{1}{8} = 0.125, \text{ and } \tfrac{125}{111} = 1.126126\ldots = 1.\overline{126}$$

are rational. The decimal representation of a rational number either repeats or terminates.

$$\tfrac{173}{55} = 3.1\overline{45} \qquad \text{A rational number that repeats}$$

$$\tfrac{1}{2} = 0.5 \qquad \text{A rational number that terminates}$$

A real number that cannot be written as the ratio of two integers is called irrational. Irrational numbers have infinite nonrepeating decimal representations. For instance, the numbers

$$\sqrt{2} \approx 1.4142136 \quad \text{and} \quad \pi \approx 3.1415927$$

are irrational. (The symbol $\approx$ means "is approximately equal to.")

Real numbers are represented graphically by a real number line. The point 0 on the real number line is the origin. Numbers to the right of 0 are positive, and numbers to the left are negative, as shown in Figure A.1. The term "nonnegative" describes a number that is either positive or zero.

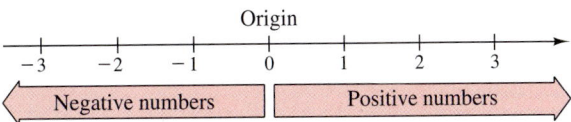

Figure A.1 The Real Number Line

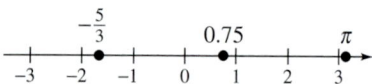

Every real number corresponds to exactly one point on the real number line.

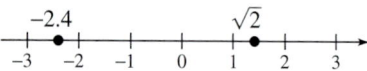

Every point on the real number line corresponds to exactly one real number.

Figure A.2

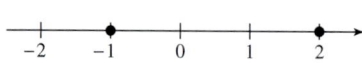

Figure A.3

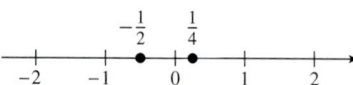

Figure A.4

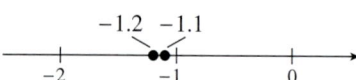

Figure A.5

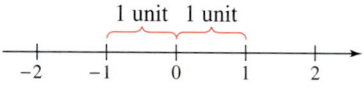

−1 is the opposite of 1.

Figure A.6

As illustrated in Figure A.2, there is a *one-to-one correspondence* between real numbers and points on the real number line.

The real number line provides you with a way of comparing any two real numbers. For any two (different) numbers on the real number line, one of the numbers must be to the left of the other number. A "less than" comparison is denoted by the inequality symbol <, a "greater than" comparison is denoted by >, a "less than or equal to" comparison is denoted by ≤, and a "greater than or equal to" comparison is denoted by ≥. When you are asked to order two numbers, you are simply being asked to say which of the two numbers is greater.

Example 1　Ordering Real Numbers

Place the correct inequality symbol (< or >) between each pair of numbers.

a. 2 ___ -1　**b.** $-\frac{1}{2}$ ___ $\frac{1}{4}$　**c.** -1.1 ___ -1.2

Solution

a. $2 > -1$, because 2 lies to the *right* of -1.　　See Figure A.3.

b. $-\frac{1}{2} < \frac{1}{4}$, because $-\frac{1}{2} = -\frac{2}{4}$ lies to the *left* of $\frac{1}{4}$.　　See Figure A.4.

c. $-1.1 > -1.2$, because -1.1 lies to the *right* of -1.2.　　See Figure A.5.

Two real numbers are opposites of each other if they lie the same distance from, but on opposite sides of, zero. For instance, -1 is the opposite of 1, as shown in Figure A.6.

The absolute value of a real number is its distance from zero on the real number line. A pair of vertical bars, $|\ |$, is used to denote absolute value. The absolute value of a real number is either positive or zero (never negative).

Example 2　Evaluating Absolute Values

a. $|5| = 5$, because the distance between 5 and 0 is 5.

b. $|0| = 0$, because the distance between 0 and itself is 0.

c. $\left|-\frac{2}{3}\right| = \frac{2}{3}$, because the distance between $-\frac{2}{3}$ and 0 is $\frac{2}{3}$.

Operations with Real Numbers

There are four basic arithmetic operations with real numbers: addition, subtraction, multiplication, and division.

The result of adding two real numbers is called the sum of the two numbers. Subtraction of one real number from another can be described as adding the opposite of the second number to the first number. For instance,

$$7 - 5 = 7 + (-5) = 2 \text{ and } 10 - (-13) = 10 + 13 = 23.$$

The result of subtracting one real number from another is called the difference of the two numbers.

Example 3   Adding and Subtracting Real Numbers

a. $-25 + 12 = -13$

b. $5 + (-10) = -5$

c. $-13.8 - 7.02 = -13.8 + (-7.02) = -20.82$

d. To add two fractions with unlike denominators, you must first rewrite one (or both) of the fractions so that they have a common denominator. To do this, find the least common multiple (LCM) of the denominators.

$$\frac{1}{3} + \frac{2}{9} = \frac{1(3)}{3(3)} + \frac{2}{9} \qquad \textcolor{red}{\text{LCM of 3 and 9 is 9.}}$$

$$= \frac{3}{9} + \frac{2}{9} = \frac{5}{9} \qquad \textcolor{red}{\text{Rewrite with like denominators and add numerators.}}$$

The result of multiplying two real numbers is called their *product*, and each of the numbers is called a *factor* of the product. The product of zero and any other number is zero. Multiplication is denoted in a variety of ways. For instance,

$$3 \times 2, \; 3 \cdot 2, \; 3(2), \text{ and } (3)(2)$$

all denote the product of "3 times 2," which you know is 6.

Example 4   Multiplying Real Numbers

a. $(6)(-4) = -24$ **b.** $(-1.2)(-0.4) = 0.48$

c. To find the product of more than two numbers, find the product of their absolute values. If there is an *even* number of negative factors, the product is positive. If there is an *odd* number of negative factors, the product is negative. For instance, in the product $6(2)(-5)(-8)$, there are two negative factors, so the product must be positive, and you can write $6(2)(-5)(-8) = 480$.

d. To multiply two fractions, multiply their numerators and their denominators. For instance, the product of $\frac{2}{3}$ and $\frac{4}{5}$ is

$$\left(\frac{2}{3}\right)\left(\frac{4}{5}\right) = \frac{(2)(4)}{(3)(5)} = \frac{8}{15}.$$

The *reciprocal* of a nonzero real number a is defined as the number by which a must be multiplied to obtain 1. The reciprocal of the fraction a/b is b/a.

To divide one real number by a second (nonzero) real number, multiply the first number by the reciprocal of the second number. The result of dividing two real numbers is called the *quotient* of the two numbers. Division is denoted in a variety of ways. For instance,

$$12 \div 4, \; 12/4, \; \frac{12}{4}, \text{ and } 4\overline{)12}$$

all denote the quotient of "12 divided by 4," which you know is 3.

Example 5   Dividing Real Numbers

a. $-30 \div 5 = -30\left(\dfrac{1}{5}\right) = -\dfrac{30}{5} = -6$ **b.** $-\dfrac{9}{14} \div -\dfrac{1}{3} = -\dfrac{9}{14}\left(-\dfrac{3}{1}\right) = \dfrac{27}{14}$

c. $\dfrac{5}{16} \div 2\dfrac{3}{4} = \dfrac{5}{16} \div \dfrac{11}{4} = \dfrac{5}{16}\left(\dfrac{4}{11}\right) = \dfrac{5(4)}{4(4)(11)} = \dfrac{5}{44}$

Let n be a positive integer and let a be a real number. Then the product of n factors of a is given by

$$a^n = \underbrace{a \cdot a \cdot a \cdot \cdots \cdot a}_{n \text{ factors}}.$$

In the exponential form a^n, a is called the base and n is called the exponent.

Example 6   Evaluating Exponential Expressions

a. $(-2)^5 = (-2)(-2)(-2)(-2)(-2) = -32$

b. $\left(\dfrac{1}{5}\right)^3 = \left(\dfrac{1}{5}\right)\left(\dfrac{1}{5}\right)\left(\dfrac{1}{5}\right) = \dfrac{1}{125}$ **c.** $(-7)^2 = (-7)(-7) = 49$

One way to help avoid confusion when communicating algebraic ideas is to establish an order of operations. This order is summarized below.

Order of Operations

1. Perform operations inside *symbols of grouping*—( ) or [ ]—or *absolute value symbols*, starting with the innermost symbol.

2. Evaluate all *exponential* expressions.

3. Perform all *multiplications* and *divisions* from left to right.

4. Perform all *additions* and *subtractions* from left to right.

Example 7   Order of Operations

a. $20 - 2 \cdot 3^2 = 20 - 2 \cdot 9 = 20 - 18 = 2$

b. $-4 + 2(-2 + 5)^2 = -4 + 2(3)^2 = -4 + 2(9) = -4 + 18 = 14$

c. $\dfrac{2 \cdot 5^2 - 10}{3^2 - 4} = (2 \cdot 5^2 - 10) \div (3^2 - 4)$ Rewrite using parentheses.

$= (50 - 10) \div (9 - 4)$ Evaluate exponential expressions and multiply within symbols of grouping.

$= 40 \div 5 = 8$ Simplify.

Properties of Real Numbers

Below is a review of the properties of real numbers. In this list, a verbal description of each property is given, as well as an example.

Properties of Real Numbers: Let a, b, and c be real numbers.

Property	Example
1. *Commutative Property of Addition:* Two real numbers can be added in either order. $a + b = b + a$	$3 + 5 = 5 + 3$
2. *Commutative Property of Multiplication:* Two real numbers can be multiplied in either order. $ab = ba$	$4 \cdot (-7) = -7 \cdot 4$
3. *Associative Property of Addition:* When three real numbers are added, it makes no difference which two are added first. $(a + b) + c = a + (b + c)$	$(2 + 6) + 5 = 2 + (6 + 5)$
4. *Associative Property of Multiplication:* When three real numbers are multiplied, it makes no difference which two are multiplied first. $(ab)c = a(bc)$	$(3 \cdot 5) \cdot 2 = 3 \cdot (5 \cdot 2)$
5. *Distributive Property:* Multiplication distributes over addition. $a(b + c) = ab + ac$ $(a + b)c = ac + bc$	$3(8 + 5) = 3 \cdot 8 + 3 \cdot 5$ $(3 + 8)5 = 3 \cdot 5 + 8 \cdot 5$
6. *Additive Identity Property:* The sum of zero and a real number equals the number itself. $a + 0 = 0 + a = a$	$3 + 0 = 0 + 3 = 3$
7. *Multiplicative Identity Property:* The product of 1 and a real number equals the number itself. $a \cdot 1 = 1 \cdot a = a$	$4 \cdot 1 = 1 \cdot 4 = 4$
8. *Additive Inverse Property:* The sum of a real number and its opposite is zero. $a + (-a) = 0$	$3 + (-3) = 0$
9. *Multiplicative Inverse Property:* The product of a nonzero real number and its reciprocal is 1. $a \cdot \dfrac{1}{a} = 1, \ a \neq 0$	$8 \cdot \dfrac{1}{8} = 1$

A.2 Fundamentals of Algebra

Algebraic Expressions • Constructing Verbal Models • Equations

Algebraic Expressions

One characteristic of algebra is the use of letters to represent numbers. The letters are variables, and combinations of letters and numbers are algebraic expressions. The terms of an algebraic expression are those parts separated by addition. For example, in the expression $-x^2 + 5x + 8$, $-x^2$ and $5x$ are the variable terms and 8 is the constant term. The coefficient of the variable term $-x^2$ is -1 and the coefficient of $5x$ is 5.

To evaluate an algebraic expression, substitute numerical values for each of the variables in the expression.

Example 1 Evaluating Algebraic Expressions

a. Evaluate the expression $-3x + 5$ when $x = 3$.

$$-3(3) + 5 = -9 + 5 = -4$$

b. Evaluate the expression $3x^2 + 2xy - y^2$ when $x = 3$ and $y = -1$.

$$3(3)^2 + 2(3)(-1) - (-1)^2 = 3(9) + (-6) - 1 = 20$$

The properties of real numbers listed on page A5 can be used to rewrite and simplify algebraic expressions. To simplify an algebraic expression generally means to remove symbols of grouping such as parentheses or brackets and combine like terms. In an algebraic expression, two terms are said to be like terms if they are both constant terms or if they have the same variable factor(s). To combine like terms in an algebraic expression, add their respective coefficients and attach the common variable factor.

Example 2 Combining Like Terms

a. $2x + 3y - 6x - y = (2x - 6x) + (3y - y)$ Group like terms.

$$= (2 - 6)x + (3 - 1)y$$ Distributive Property

$$= -4x + 2y$$ Simplest form

b. $4x^2 + 5x - x^2 - 8x = (4x^2 - x^2) + (5x - 8x)$ Group like terms.

$$= (4 - 1)x^2 + (5 - 8)x$$ Distributive Property

$$= 3x^2 - 3x$$ Simplest form

Example 3   Removing Symbols of Grouping

a. $-2(a + 5) + 4(a - 8) = -2a - 10 + 4a - 32$ Distributive Property

$= (-2a + 4a) + (-10 - 32)$ Group like terms.

$= 2a - 42$ Combine like terms.

b. $3x^2 - [9x + 3x(2x - 1)] = 3x^2 - [9x + 6x^2 - 3x]$ Distributive Property

$= 3x^2 - [6x^2 + 6x]$ Combine like terms.

$= 3x^2 - 6x^2 - 6x$ Distributive Property

$= -3x^2 - 6x$ Combine like terms.

Constructing Verbal Models

When you translate a verbal sentence or phrase into an algebraic expression, watch for key words and phrases that indicate the four different operations of arithmetic.

Example 4   Translating Verbal Phrases

a. *Verbal Description:* Seven more than 3 times x

Algebraic Expression: $3x + 7$

b. *Verbal Description:* Four times the sum of y and 9

Algebraic Expression: $4(y + 9)$

c. *Verbal Description:* Five decreased by the product of 2 and a number

Label: The number $= x$ *Algebraic Expression:* $5 - 2x$

d. *Verbal Description:* One more than the product of 8 and a number, all divided by 6

Label: The number $= x$ *Algebraic Expression:* $\dfrac{8x + 1}{6}$

Example 5   Constructing Verbal Models

A cash register contains x quarters. Write an expression for this amount of money in dollars.

Solution

Verbal Model:	Value of coin	·	Number of coins

Labels: Value of coin $= 0.25$ (dollars per quarter)
Number of coins $= x$ (quarters)

Expression: $0.25x$ (dollars)

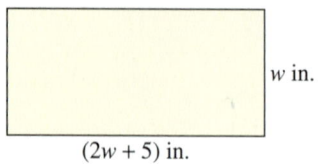

w in.

$(2w + 5)$ in.

Figure A.7

Example 6　Constructing Verbal Models

The width of a rectangle is w inches. The length of the rectangle is 5 inches more than twice its width. Write an expression for the perimeter of the rectangle.

Solution

Draw a rectangle, as shown in Figure A.7. Next, use a verbal model to solve the problem. Use the formula (perimeter) = 2(length) + 2(width).

Verbal Model:	$2 \cdot$ Length $+ 2 \cdot$ Width	
Labels:	Length $= 2w + 5$	(inches)
	Width $= w$	(inches)
Expression:	$2(2w + 5) + 2w = 4w + 10 + 2w = 6w + 10$	(inches)

Equations

An equation is a statement that equates two algebraic expressions. Solving an equation involving x means finding all values of x for which the equation is true. Such values are solutions and are said to satisfy the equation. Example 7 shows how to check whether a given value is a solution of an equation.

Example 7　Checking a Solution of an Equation

Determine whether $x = -3$ is a solution of $-3x - 5 = 4x + 16$.

$$-3(-3) - 5 \overset{?}{=} 4(-3) + 16 \qquad \text{Substitute } -3 \text{ for } x \text{ in original equation.}$$

$$9 - 5 \overset{?}{=} -12 + 16 \qquad \text{Simplify.}$$

$$4 = 4 \qquad \text{Solution checks.} ✓$$

Example 8　Using a Verbal Model to Construct an Equation

You are given a speeding ticket for $80 for speeding on a road where the speed limit is 45 miles per hour. You are fined $10 for each mile per hour over the speed limit. How fast were you driving? Write an algebraic equation that models the situation.

Solution

Verbal Model:	Fine $\cdot$ Speed over limit $=$ Amount of ticket	
Labels:	Fine $= 10$	(dollars per mile per hour)
	Your speed $= x$	(miles per hour)
	Speed over limit $= x - 45$	(miles per hour)
	Amount of ticket $= 80$	(dollars)
Algebraic Model:	$10(x - 45) = 80$	

Study Tip

For more review on the fundamentals of algebra, refer to Chapter 2.

A.3 Equations, Inequalities, and Problem Solving

Equations   •   Inequalities   •   Problem Solving

Equations

A linear equation in one variable x is an equation that can be written in the standard form

$$ax + b = 0$$

where a and b are real numbers with $a \neq 0$. To solve a linear equation, you want to isolate x on one side of the equation by a sequence of equivalent equations, each having the same solution(s) as the original equation. The operations that yield equivalent equations are as follows.

Operations That Yield Equivalent Equations

1. Remove symbols of grouping, combine like terms, or simplify fractions on one or both sides of the equation.

2. Add (or subtract) the same quantity to (from) each side of the equation.

3. Multiply (or divide) each side of the equation by the same nonzero quantity.

4. Interchange the two sides of the equation.

Example 1   Solving a Linear Equation in Standard Form

Solve $3x - 6 = 0$. Then check the solution.

Solution

$3x - 6 = 0$	Write original equation.
$3x - 6 + 6 = 0 + 6$	Add 6 to each side.
$3x = 6$	Combine like terms.
$\dfrac{3x}{3} = \dfrac{6}{3}$	Divide each side by 3.
$x = 2$	Simplify.

Check

$3x - 6 = 0$	Write original equation.
$3(2) - 6 \stackrel{?}{=} 0$	Substitute 2 for x.
$6 - 6 \stackrel{?}{=} 0$	Simplify.
$0 = 0$	Solution checks. ✓

So, the solution is $x = 2$.

Example 2   Solving a Linear Equation in Nonstandard Form

Solve $5x + 4 = 3x - 8$.

Solution

$$5x + 4 = 3x - 8 \qquad \text{Write original equation.}$$
$$5x - 3x + 4 = 3x - 3x - 8 \qquad \text{Subtract } 3x \text{ from each side.}$$
$$2x + 4 = -8 \qquad \text{Combine like terms.}$$
$$2x + 4 - 4 = -8 - 4 \qquad \text{Subtract 4 from each side.}$$
$$2x = -12 \qquad \text{Combine like terms.}$$
$$\frac{2x}{2} = \frac{-12}{2} \qquad \text{Divide each side by 2.}$$
$$x = -6 \qquad \text{Simplify.}$$

The solution is $x = -6$. Check this in the original equation.

Linear equations often contain parentheses or other symbols of grouping. In most cases, it helps to remove symbols of grouping as a first step to solving an equation. This is illustrated in Example 3.

Example 3   Solving a Linear Equation Involving Parentheses

Solve $2(x + 4) = 5(x - 8)$.

Solution

$$2(x + 4) = 5(x - 8) \qquad \text{Write original equation.}$$
$$2x + 8 = 5x - 40 \qquad \text{Distributive Property}$$
$$2x - 5x + 8 = 5x - 5x - 40 \qquad \text{Subtract } 5x \text{ from each side.}$$
$$-3x + 8 = -40 \qquad \text{Combine like terms.}$$
$$-3x + 8 - 8 = -40 - 8 \qquad \text{Subtract 8 from each side.}$$
$$-3x = -48 \qquad \text{Combine like terms.}$$
$$\frac{-3x}{-3} = \frac{-48}{-3} \qquad \text{Divide each side by } -3.$$
$$x = 16 \qquad \text{Simplify.}$$

The solution is $x = 16$. Check this in the original equation.

Study Tip

Recall that when finding the least common multiple of a set of numbers, you should first consider all multiples of each number. Then, you should choose the smallest of the common multiples of the numbers.

To solve an equation involving fractional expressions, find the least common multiple (LCM) of the denominators and multiply each side by the LCM.

Example 4   Solving a Linear Equation Involving Fractions

Solve $\dfrac{x}{3} + \dfrac{3x}{4} = 2$.

Solution

$$12\left(\frac{x}{3} + \frac{3x}{4}\right) = 12(2)$$ Multiply each side of original equation by LCM 12.

$$12 \cdot \frac{x}{3} + 12 \cdot \frac{3x}{4} = 24$$ Distributive Property

$$4x + 9x = 24$$ Clear fractions.

$$13x = 24$$ Combine like terms.

$$x = \frac{24}{13}$$ Divide each side by 13.

The solution is $x = \frac{24}{13}$. Check this in the original equation.

To solve an equation involving an absolute value, remember that the expression inside the absolute value signs can be positive or negative. This results in two separate equations, each of which must be solved.

Example 5   Solving an Equation Involving Absolute Value

Solve $|4x - 3| = 13$.

Solution

$$|4x - 3| = 13$$ Write original equation.

$$4x - 3 = -13 \quad \text{or} \quad 4x - 3 = 13$$ Equivalent equations

$$4x = -10 \qquad\qquad 4x = 16$$ Add 3 to each side.

$$x = -\frac{5}{2} \qquad\qquad x = 4$$ Divide each side by 4.

The solutions are $x = -\frac{5}{2}$ and $x = 4$. Check these in the original equation.

Inequalities

The simplest type of inequality is a linear inequality in one variable. For instance, $2x + 3 > 4$ is a linear inequality in x. The procedures for solving linear inequalities in one variable are much like those for solving linear equations, as described on page A9. The exception is that when each side of an inequality is multiplied or divided by a negative number, the direction of the inequality symbol *must be reversed*.

Example 6   Solving a Linear Inequality

Solve and graph the inequality $-5x - 7 > 3x + 9$.

Solution

$$-5x - 7 > 3x + 9 \qquad \text{Write original inequality.}$$

$$-8x - 7 > 9 \qquad \text{Subtract } 3x \text{ from each side.}$$

$$-8x > 16 \qquad \text{Add 7 to each side.}$$

$$x < -2 \qquad \text{Divide each side by } -8 \text{ and reverse the direction of the inequality symbol.}$$

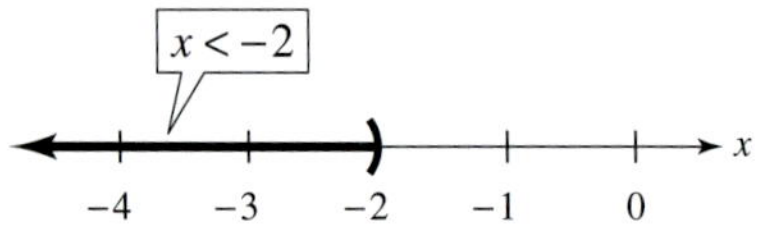

The solution set in interval notation is $(-\infty, -2)$ and in set notation is $\{x \mid x < -2\}$. The graph of the solution set is shown in Figure A.8.

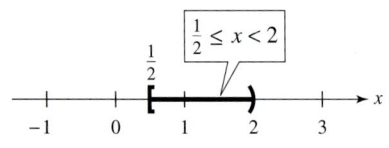

$x < -2$

Figure A.8

Two inequalities joined by the word *and* or the word *or* constitute a compound inequality. Sometimes it is possible to write a compound inequality as a double inequality. For instance, you can write $-3 < 6x - 1$ *and* $6x - 1 < 3$ more simply as $-3 < 6x - 1 < 3$. A compound inequality formed by the word *and* is called conjunctive and may be rewritten as a double inequality. A compound inequality joined by the word *or* is called disjunctive and cannot be rewritten as a double inequality.

Example 7   Solving a Conjunctive Inequality

Solve and graph the inequality $2x + 3 \geq 4$ and $3x - 8 < -2$.

Solution

$$2x + 3 \geq 4 \quad \text{and} \quad 3x - 8 < -2$$

$$2x \geq 1 \qquad\qquad 3x < 6$$

$$x \geq \tfrac{1}{2} \qquad\qquad x < 2$$

The solution set in interval notation is $\left[\tfrac{1}{2}, 2\right)$ and in set notation is $\left\{x \mid \tfrac{1}{2} \leq x < 2\right\}$. The graph of the solution set is shown in Figure A.9.

$\tfrac{1}{2} \leq x < 2$

Figure A.9

Study Tip

Recall that the word *or* is represented by the symbol $\cup$, which is read as *union*.

Example 8   Solving a Disjunctive Inequality

Solve and graph the inequality $x - 8 > -3$ or $-6x + 1 \geq -5$.

Solution

$$x - 8 > -3 \quad \text{or} \quad -6x + 1 \geq -5$$

$$x > 5 \qquad\qquad -6x \geq -6$$

$$\qquad\qquad\qquad x \leq 1$$

The solution set in interval notation is $(-\infty, 1] \cup (5, \infty)$ and in set notation is $\{x \mid x > 5 \text{ or } x \leq 1\}$. The graph of the solution set is shown in Figure A.10.

$x \leq 1$ $x > 5$

Figure A.10

To solve an absolute value inequality, use the following rules.

Solving an Absolute Value Inequality

Let x be a variable or an algebraic expression and let a be a real number such that $a > 0$.

1. The solutions of $|x| < a$ are all values of x that lie between $-a$ and a.

$$|x| < a \text{ if and only if } -a < x < a$$

2. The solutions of $|x| > a$ are all values of x that are less than $-a$ or greater than a.

$$|x| > a \text{ if and only if } x < -a \text{ or } x > a$$

These rules are also valid if $<$ is replaced by $\le$ and $>$ is replaced by $\ge$.

Example 9   Solving Absolute Value Inequalities

Solve and graph each inequality.

a. $|4x + 3| > 9$ **b.** $|2x - 7| \le 1$

Solution

a. $|4x + 3| > 9$ Write original inequality.

$\qquad 4x + 3 < -9 \quad \text{or} \quad 4x + 3 > 9$ Equivalent inequalities

$\qquad\qquad 4x < -12 \qquad\qquad 4x > 6$ Subtract 3 from each side.

$\qquad\qquad x < -3 \qquad\qquad x > \frac{3}{2}$ Divide each side by 4.

The solution set consists of all real numbers that are less than -3 or greater than $\frac{3}{2}$. The solution set in interval notation is $(-\infty, -3) \cup \left(\frac{3}{2}, \infty\right)$ and in set notation is $\left\{x \mid x < -3 \text{ or } x > \frac{3}{2}\right\}$. The graph is shown in Figure A.11.

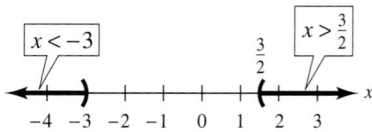

Figure A.11

b. $|2x - 7| \le 1$ Write original inequality.

$\qquad -1 \le 2x - 7 \le 1$ Equivalent double inequality

$\qquad\qquad 6 \le 2x \le 8$ Add 7 to all three parts.

$\qquad\qquad 3 \le x \le 4$ Divide all three parts by 2.

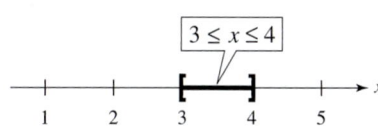

Figure A.12

The solution set consists of all real numbers that are greater than or equal to 3 and less than or equal to 4. The solution set in interval notation is $[3, 4]$ and in set notation is $\{x \mid 3 \le x \le 4\}$. The graph is shown in Figure A.12.

Problem Solving

Algebra is used to solve word problems that relate to real-life situations. The following guidelines summarize the problem-solving strategy that you should use when solving word problems.

Guidelines for Solving Word Problems

1. Write a *verbal model* that describes the problem.

2. Assign *labels* to fixed quantities and variable quantities.

3. Rewrite the verbal model as an *algebraic equation* using the assigned labels.

4. *Solve* the resulting algebraic equation.

5. *Check* to see that your solution satisfies the original problem as stated.

Example 10   Finding the Percent of Monthly Expenses

Your family has an annual income of $57,000 and the following monthly expenses: mortgage ($1100), car payment ($375), food ($295), utilities ($240), and credit cards ($220). The total expenses for one year represent what percent of your family's annual income?

Solution

The total amount of your family's monthly expenses is

$$1100 + 375 + 295 + 240 + 220 = \$2230.$$

The total monthly expenses for one year are

$$2230 \cdot 12 = \$26{,}760.$$

Verbal Model: Expenses $=$ Percent $\cdot$ Income

Labels: Expenses $= 26{,}760$ (dollars)
            Percent $= p$ (in decimal form)
            Income $= 57{,}000$ (dollars)

Equation: $26{,}760 = p \cdot 57{,}000$ Original equation

$$\frac{26{,}760}{57{,}000} = p$$ Divide each side by 57,000.

$$0.469 \approx p$$ Use a calculator.

Your family's total expenses for one year are approximately 0.469 or 46.9% of your family's annual income.

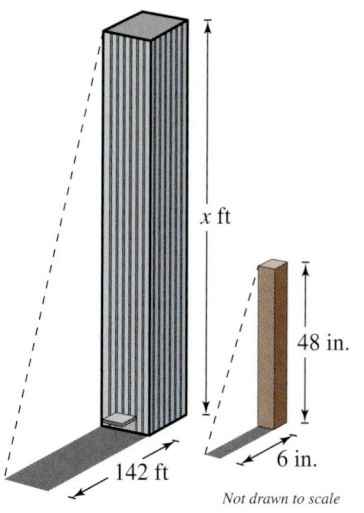

Figure A.13

Example 11   Geometry: Similar Triangles

To determine the height of the Aon Center Building (in Chicago), you measure the shadow cast by the building and find it to be 142 feet long, as shown in Figure A.13. Then you measure the shadow cast by a four-foot post and find it to be 6 inches long. Estimate the building's height.

Solution

To solve this problem, you use a result from geometry that states that the ratios of corresponding sides of similar triangles are equal.

Verbal Model:
$$\frac{\text{Height of building}}{\text{Length of building's shadow}} = \frac{\text{Height of post}}{\text{Length of post's shadow}}$$

Labels: Height of building $= x$ (feet)
           Length of building's shadow $= 142$ (feet)
           Height of post $= 4$ feet $= 48$ inches   (inches)
           Length of post's shadow $= 6$ (inches)

Proportion: $\dfrac{x}{142} = \dfrac{48}{6}$ Original proportion

$x \cdot 6 = 142 \cdot 48$ Cross-multiply

$x = 1136$ Divide each side by 6.

So, you can estimate the Aon Center Building to be 1136 feet high.

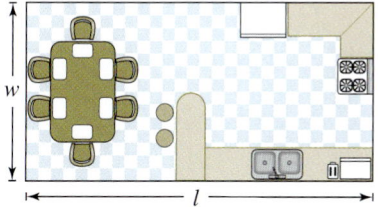

Figure A.14

Example 12   Geometry: Dimensions of a Room

A rectangular kitchen is twice as long as it is wide, and its perimeter is 84 feet. Find the dimensions of the kitchen.

Solution

For this problem, it helps to sketch a diagram, as shown in Figure A.14.

Verbal Model: $2 \cdot$ Length $+ 2 \cdot$ Width $=$ Perimeter

Labels: Length $= l = 2w$ (feet)
           Width $= w$ (feet)
           Perimeter $= 84$ (feet)

Equation: $2(2w) + 2w = 84$ Original equation

$6w = 84$ Combine like terms.

$w = 14$ Divide each side by 6.

Because the length is twice the width, you have

$l = 2w$ Length is twice width.

$= 2(14) = 28.$ Substitute and simplify.

So, the dimensions of the room are 14 feet by 28 feet.

A.4 Graphs and Functions

The Rectangular Coordinate System • Graphs of Equations • Functions
• Slope and Linear Equations • Graphs of Linear Inequalities

The Rectangular Coordinate System

You can represent ordered pairs of real numbers by points in a plane. This plane is called a rectangular coordinate system. A rectangular coordinate system is formed by two real lines, the x-axis (horizontal line) and the y-axis (vertical line), intersecting at right angles. The point of intersection of the axes is called the origin, and the axes divide the plane into four regions called quadrants.

Each point in the plane corresponds to an ordered pair (x, y) of real numbers x and y, called the coordinates of the point. The x-coordinate tells how far to the left or right the point is from the vertical axis, and the y-coordinate tells how far up or down the point is from the horizontal axis, as shown in Figure A.15.

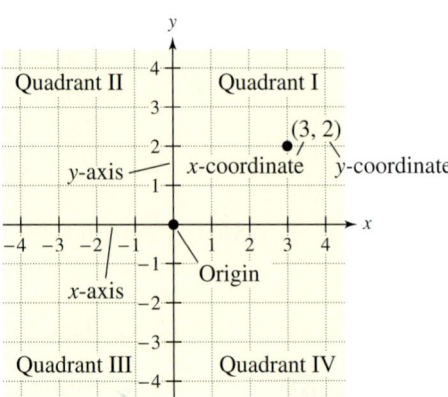

Figure A.15

Example 1   Finding Coordinates of Points

Determine the coordinates of each of the points shown in Figure A.16, and then determine the quadrant in which each point is located.

Solution

Point A lies two units to the *right* of the vertical axis and one unit *below* the horizontal axis. So, the point A must be given by $(2, -1)$. The coordinates of the other four points can be determined in a similar way. The results are as follows.

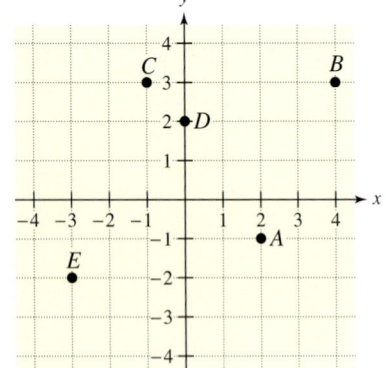

Figure A.16

Point	Coordinates	Quadrant
A	$(2, -1)$	IV
B	$(4, 3)$	I
C	$(-1, 3)$	II
D	$(0, 2)$	None
E	$(-3, -2)$	III

Graphs of Equations

The solutions of an equation involving two variables can be represented by points on a rectangular coordinate system. The graph of an equation is the set of all points that are solutions of the equation.

The simplest way to sketch the graph of an equation is the point-plotting method. With this method, you construct a table of values that consists of several solution points of the equation, plot these points, and then connect the points with a smooth curve or line.

Example 2   Sketching the Graph of an Equation

Sketch the graph of $y = x^2 - 2$.

Solution

Begin by choosing several x-values and then calculating the corresponding y-values. For example, if you choose $x = -2$, the corresponding y-value is

$$y = x^2 - 2 \qquad \text{Original equation}$$

$$y = (-2)^2 - 2 \qquad \text{Substitute } -2 \text{ for } x.$$

$$y = 4 - 2 = 2. \qquad \text{Simplify.}$$

Then, create a table using these values, as shown below.

x	-2	-1	0	1	2	3
$y = x^2 - 2$	2	-1	-2	-1	2	7
Solution point	$(-2, 2)$	$(-1, -1)$	$(0, -2)$	$(1, -1)$	$(2, 2)$	$(3, 7)$

Next, plot the solution points, as shown in Figure A.17. Finally, connect the points with a smooth curve, as shown in Figure A.18.

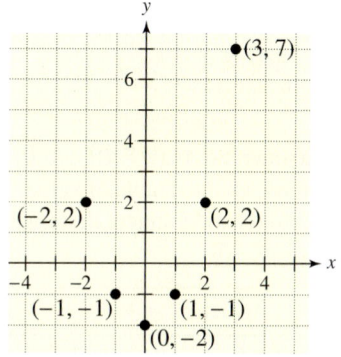

Figure A.17

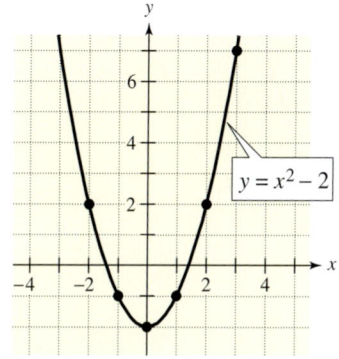

Figure A.18

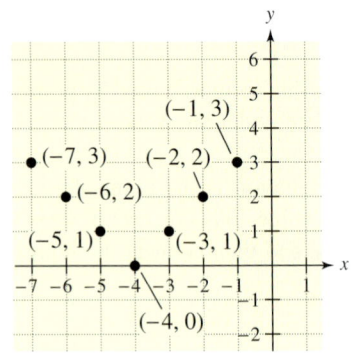

Figure A.19

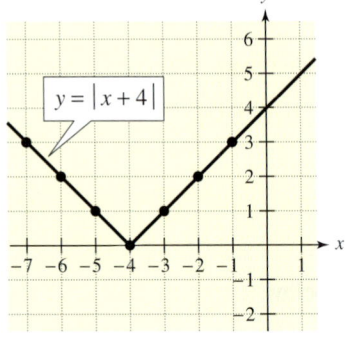

Figure A.20

Example 3   Sketching the Graph of an Equation

Sketch the graph of $y = |x + 4|$.

Solution

Begin by creating a table of values, as shown below. Plot the solution points as shown in Figure A.19. It appears that the points lie in a "V-shaped" pattern, with the point $(-4, 0)$ lying at the bottom of the "V." Following this pattern, connect the points to form the graph shown in Figure A.20.

x	-7	-6	-5	-4	-3	-2	-1
$y = |x + 4|$	3	2	1	0	1	2	3
Solution point	$(-7, 3)$	$(-6, 2)$	$(-5, 1)$	$(-4, 0)$	$(-3, 1)$	$(-2, 2)$	$(-1, 3)$

Intercepts of a graph are the points at which the graph intersects the x- or y-axis. To find x-intercepts, let $y = 0$ and solve the equation for x. To find y-intercepts, let $x = 0$ and solve the equation for y.

Example 4   Finding the Intercepts of a Graph

Find the intercepts and sketch the graph of $y = 3x + 4$.

Solution

To find any x-intercepts, let $y = 0$ and solve the resulting equation for x.

$$y = 3x + 4 \qquad \text{Write original equation.}$$

$$0 = 3x + 4 \qquad \text{Let } y = 0.$$

$$-\frac{4}{3} = x \qquad \text{Solve equation for } x.$$

To find any y-intercepts, let $x = 0$ and solve the resulting equation for y.

$$y = 3x + 4 \qquad \text{Write original equation.}$$

$$y = 3(0) + 4 \qquad \text{Let } x = 0.$$

$$y = 4 \qquad \text{Solve equation for } y.$$

So, the x-intercept is $\left(-\frac{4}{3}, 0\right)$ and the y-intercept is $(0, 4)$. To sketch the graph of the equation, create a table of values (including intercepts), as shown below. Then plot the points and connect them with a line, as shown in Figure A.21.

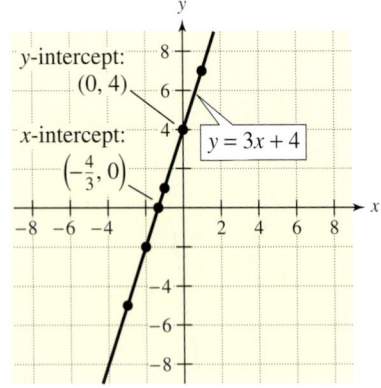

Figure A.21

x	-3	-2	$-\frac{4}{3}$	-1	0	1
$y = 3x + 4$	-5	-2	0	1	4	7
Solution point	$(-3, -5)$	$(-2, -2)$	$\left(-\frac{4}{3}, 0\right)$	$(-1, 1)$	$(0, 4)$	$(1, 7)$

Functions

A relation is any set of ordered pairs, which can be thought of as (input, output).
A function is a relation in which no two ordered pairs have the same first component and different second components.

Example 5   Testing Whether Relations Are Functions

Decide whether the relation represents a function.

a.

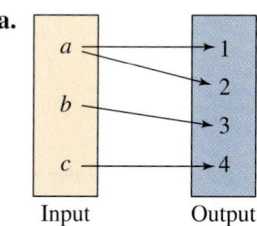

Input   Output

b. Input: 2, 5, 7
   Output: 1, 2, 3
   {(2, 1), (5, 2), (7, 3)}

Solution

a. This diagram *does not* represent a function. The first component *a* is paired
with two different second components, 1 and 2.

b. This set of ordered pairs *does* represent a function. No first component has two
different second components.

The graph of an equation represents *y* as a function of *x* if and only if no
vertical line intersects the graph more than once. This is called the Vertical Line Test.

Example 6   Using the Vertical Line Test for Functions

Use the Vertical Line Test to determine whether *y* is a function of *x*.

a.

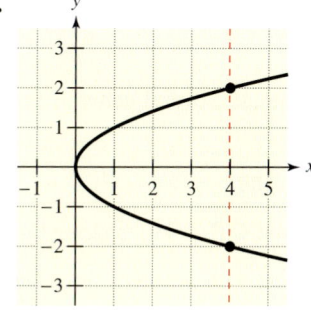

b.

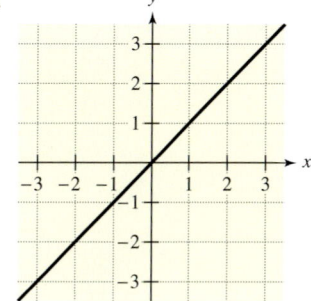

Solution

a. From the graph, you can see that a vertical line intersects more than one point
on the graph. So, the relation *does not* represent *y* as a function of *x*.

b. From the graph, you can see that no vertical line intersects more than one point
on the graph. So, the relation *does* represent *y* as a function of *x*.

Slope and Linear Equations

The graph in Figure A.21 on page A18 is an example of a graph of a linear equation. The equation is written in slope-intercept form, $y = mx + b$, where m is the slope and $(0, b)$ is the y-intercept. Linear equations can be written in other forms, as shown below.

Forms of Linear Equations

1. General form: $ax + by + c = 0$

2. Slope-intercept form: $y = mx + b$

3. Point-slope form: $y - y_1 = m(x - x_1)$

The slope of a nonvertical line is the number of units the line rises or falls vertically for each unit of horizontal change from left to right. To find the slope m of the line through (x_1, y_1) and (x_2, y_2), use the following formula.

$$m = \frac{y_2 - y_1}{x_2 - x_1} = \frac{\text{Change in } y}{\text{Change in } x}$$

Example 7   Finding the Slope of a Line Through Two Points

Find the slope of the line passing through $(3, 1)$ and $(-6, 0)$.

Solution

Let $(x_1, y_1) = (3, 1)$ and $(x_2, y_2) = (-6, 0)$. The slope of the line through these points is

$$m = \frac{y_2 - y_1}{x_2 - x_1} = \frac{0 - 1}{-6 - 3} = \frac{-1}{-9} = \frac{1}{9}.$$

The graph of the line is shown in Figure A.22.

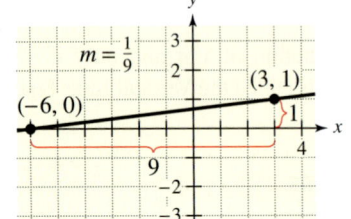

Figure A.22

You can make several generalizations about the slopes of lines.

Slope of a Line

1. A line with positive slope ($m > 0$) rises from left to right.

2. A line with negative slope ($m < 0$) falls from left to right.

3. A line with zero slope ($m = 0$) is horizontal.

4. A line with undefined slope is vertical.

5. Parallel lines have equal slopes: $m_1 = m_2$

6. Perpendicular lines have negative reciprocal slopes: $m_1 = -\dfrac{1}{m_2}$

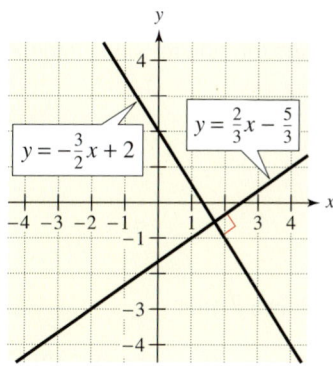

Figure A.23

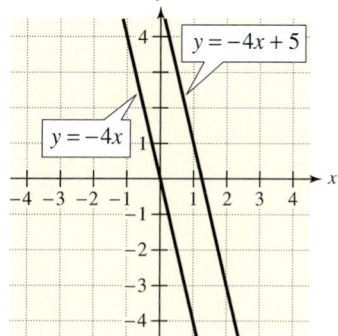

Figure A.24

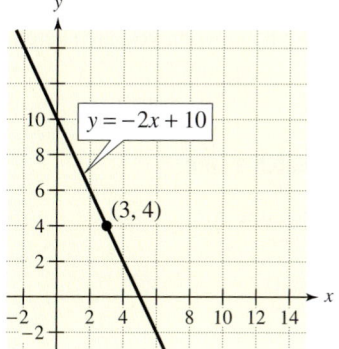

Figure A.25

Example 8 Parallel or Perpendicular?

Determine whether the pairs of lines are parallel, perpendicular, or neither.

a. $y = \frac{2}{3}x - \frac{5}{3}$

$y = -\frac{3}{2}x + 2$

b. $4x + y = 5$

$-8x - 2y = 0$

Solution

a. The first line has a slope of $m_1 = \frac{2}{3}$ and the second line has a slope of $m_2 = -\frac{3}{2}$. Because these slopes are negative reciprocals of each other, the two lines must be perpendicular, as shown in Figure A.23.

b. To begin, write each equation in slope-intercept form.

$$4x + y = 5 \qquad \text{Write first equation.}$$

$$y = -4x + 5 \qquad \text{Slope-intercept form}$$

So, the first line has a slope of $m_1 = -4$.

$$-8x - 2y = 0 \qquad \text{Write second equation.}$$

$$-2y = 8x \qquad \text{Add 8x to each side.}$$

$$y = -4x. \qquad \text{Slope-intercept form}$$

So, the second line has a slope of $m_2 = -4$. Because both lines have the same slope, they must be parallel, as shown in Figure A.24.

You can use the point-slope form of the equation of a line to write the equation of a line when you are given its slope and a point on the line.

Example 9 Writing an Equation of a Line

Write an equation of the line that passes through the point $(3, 4)$ and has slope $m = -2$.

Solution

Use the point-slope form with $(x_1, y_1) = (3, 4)$ and $m = -2$.

$$y - y_1 = m(x - x_1) \qquad \text{Point-slope form}$$

$$y - 4 = -2(x - 3) \qquad \text{Substitute 4 for } y_1, \text{ 3 for } x_1, \text{ and } -2 \text{ for } m.$$

$$y - 4 = -2x + 6 \qquad \text{Simplify.}$$

$$y = -2x + 10 \qquad \text{Equation of line}$$

So, an equation of the line in slope-intercept form is $y = -2x + 10$. The graph of this line is shown in Figure A.25.

The point-slope form can also be used to write the equation of a line passing through any two points. To use this form, substitute the formula for slope into the point-slope form, as follows.

$$y - y_1 = m(x - x_1) \qquad \text{Point-slope form}$$

$$y - y_1 = \frac{y_2 - y_1}{x_2 - x_1}(x - x_1) \qquad \text{Substitute formula for slope.}$$

Example 10 An Equation of a Line Passing Through Two Points

Write an equation of the line that passes through the points $(5, -1)$ and $(2, 0)$.

Solution

Let $(x_1, y_1) = (5, -1)$ and $(x_2, y_2) = (2, 0)$. The slope of a line passing through these points is

$$m = \frac{y_2 - y_1}{x_2 - x_1} = \frac{0 - (-1)}{2 - 5} = \frac{1}{-3} = -\frac{1}{3}.$$

Now, use the point-slope form to find an equation of the line.

$$y - y_1 = m(x - x_1) \qquad \text{Point-slope form}$$

$$y - (-1) = -\tfrac{1}{3}(x - 5) \qquad \text{Substitute } -1 \text{ for } y_1, \ 5 \text{ for } x_1, \text{ and } -\tfrac{1}{3} \text{ for } m.$$

$$y + 1 = -\tfrac{1}{3}x + \tfrac{5}{3} \qquad \text{Simplify.}$$

$$y = -\tfrac{1}{3}x + \tfrac{2}{3} \qquad \text{Equation of line}$$

The graph of this line is shown in Figure A.26.

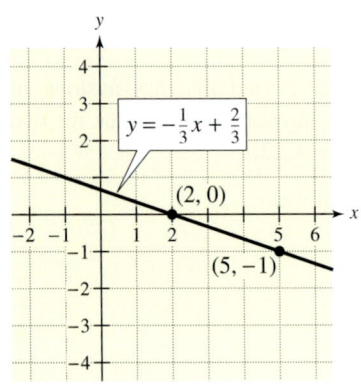

Figure A.26

The slope and y-intercept of a line can be used as an aid when you are sketching a line.

Example 11 Using the Slope and y-Intercept to Sketch a Line

Use the slope and y-intercept to sketch the graph of $-x + 2y = -4$.

Solution

First, write the equation in slope-intercept form.

$$-x + 2y = -4 \qquad \text{Write original equation.}$$

$$2y = x - 4 \qquad \text{Add } x \text{ to each side.}$$

$$y = \tfrac{1}{2}x - 2 \qquad \text{Slope-intercept form}$$

So, the slope of the line is $m = \tfrac{1}{2}$ and the y-intercept is $(0, b) = (0, -2)$. Now, plot the y-intercept and locate a second point by using the slope. Because the slope is $m = \tfrac{1}{2}$, move two units to the right and one unit upward from the y-intercept. Then draw a line through the two points, as shown in Figure A.27.

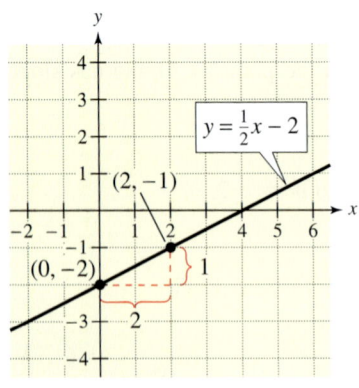

Figure A.27

You know that a horizontal line has a slope of $m = 0$. So, the equation of a horizontal line is $y = b$. A vertical line has an undefined slope, so it has an equation of the form $x = a$.

Example 12   Equations of Horizontal and Vertical Lines

a. Write an equation of the horizontal line passing through $(-1, -1)$.

b. Write an equation of the vertical line passing through $(2, 3)$.

Solution

a. The line is horizontal and passes through the point $(-1, -1)$, so every point on the line has a y-coordinate of -1. The equation of the line is $y = -1$.

b. The line is vertical and passes through the point $(2, 3)$, so every point on the line has an x-coordinate of 2. The equation of the line is $x = 2$.

The graphs of these two lines are shown in Figure A.28.

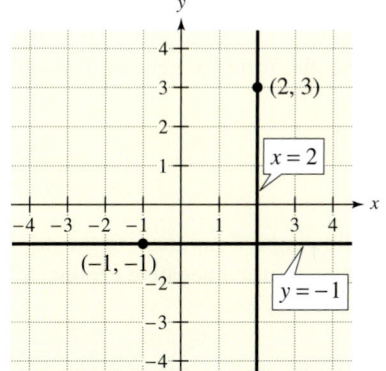

Figure A.28

Graphs of Linear Inequalities

The statements $3x - 2y < 6$ and $2x + 3y \geq 1$ are linear inequalities in two variables. An ordered pair (x_1, y_1) is a solution of a linear inequality in x and y if the inequality is true when x_1 and y_1 are substituted for x and y, respectively. The graph of a linear inequality is the collection of all solution points of the inequality. To sketch the graph of a linear inequality, begin by sketching the graph of the corresponding linear equation (use a dashed line for $<$ and $>$ and a solid line for $\leq$ and $\geq$). The graph of the equation separates the plane into two regions, called half-planes. In each half plane, either *all* points in the half-plane are solutions of the inequality or *no* point in the half-plane is a solution of the inequality. To determine whether the points in an entire half-plane satisfy the inequality, simply test one point in the region. If the point satisfies the inequality, then shade the entire half-plane to denote that every point in the region satisfies the inequality.

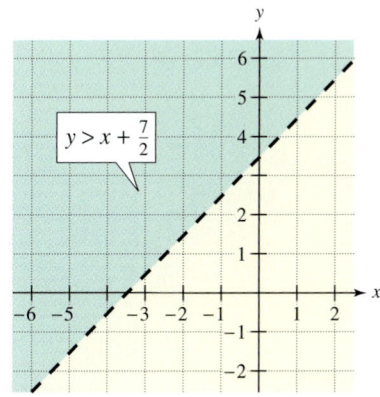

Figure A.29

Example 13   Sketching the Graph of a Linear Inequality

Use the slope-intercept form of a linear equation to graph $-2x + 2y > 7$.

Solution

To begin, rewrite the inequality in slope-intercept form.

$$2y > 2x + 7 \qquad \text{\textcolor{red}{Add $2x$ to each side.}}$$

$$y > x + \frac{7}{2} \qquad \text{\textcolor{red}{Write in slope-intercept form.}}$$

From this form, you can conclude that the solution is the half-plane lying above the line $y = x + \frac{7}{2}$. The graph is shown in Figure A.29.

Study Tip

For more review on graphs and functions, refer to Chapter 4.

A.5 Exponents and Polynomials

Exponents • Polynomials • Operations with Polynomials

Exponents

Repeated multiplication can be written in exponential form. In general, if a is a real number and n is a positive integer, then

$$a^n = \underbrace{a \cdot a \cdot a \cdot \cdots \cdot a}_{n \text{ factors}}$$

where n is the exponent and a is the base. The following is a summary of the rules of exponents. In Rule 6 below, be sure you see how to use a negative exponent.

Summary of Rules of Exponents

Let m and n be integers, and let a and b be real numbers, variables, or algebraic expressions, such that $a \neq 0$ and $b \neq 0$.

Rule	*Example*
1. *Product Rule:* $a^m a^n = a^{m+n}$	$y^2 \cdot y^4 = y^{2+4} = y^6$
2. *Quotient Rule:* $\dfrac{a^m}{a^n} = a^{m-n}$	$\dfrac{x^7}{x^4} = x^{7-4} = x^3$
3. *Product-to-Power Rule:* $(ab)^m = a^m b^m$	$(5x)^4 = 5^4 x^4$
4. *Quotient-to-Power Rule:* $\left(\dfrac{a}{b}\right)^m = \dfrac{a^m}{b^m}$	$\left(\dfrac{2}{x}\right)^3 = \dfrac{2^3}{x^3}$
5. *Power-to-Power Rule:* $(a^m)^n = a^{mn}$	$(y^3)^{-4} = y^{3(-4)} = y^{-12}$
6. *Negative Exponent Rule:* $a^{-n} = \dfrac{1}{a^n}$	$y^{-4} = \dfrac{1}{y^4}$
7. *Zero Exponent Rule:* $a^0 = 1$	$(x^2 + 1)^0 = 1$

Example 1   Using Rules of Exponents

Use the rules of exponents to simplify each expression.

a. $(a^2 b^4)(3ab^{-2})$ **b.** $(2xy^2)^3$ **c.** $3a(-4a^2)^0$ **d.** $\left(\dfrac{4x}{y^3}\right)^3$

Solution

a. $(a^2 b^4)(3ab^{-2}) = (3)(a^2)(a)(b^4)(b^{-2})$ Regroup factors.

$\qquad\qquad\qquad = (3)(a^{2+1})(b^{4-2})$ Apply rules of exponents.

$\qquad\qquad\qquad = 3a^3 b^2$ Simplify.

b. $(2xy^2)^3 = (2)^3(x)^3(y^2)^3$ Apply rules of exponents.

$$= 8x^3y^{2\cdot3}$$ Apply rules of exponents.

$$= 8x^3y^6$$ Simplify.

c. $3a(-4a^2)^0 = (3a)(-4)^0(a^2)^0$ Apply rules of exponents.

$$= 3a(1)(a^{2\cdot0})$$ Apply rules of exponents.

$$= 3a, \, a \neq 0$$ Simplify.

d. $\left(\dfrac{4x}{y^3}\right)^3 = \dfrac{4^3x^3}{(y^3)^3}$ Apply rules of exponents.

$$= \dfrac{64x^3}{y^{3\cdot3}}$$ Apply rules of exponents.

$$= \dfrac{64x^3}{y^9}$$ Simplify.

Example 2   Rewriting with Positive Exponents

Use rules of exponents to simplify each expression using only positive exponents. (Assume that no variable is equal to zero.)

a. x^{-1} **b.** $\dfrac{1}{3x^{-2}}$

c. $\dfrac{25a^3b^{-4}}{5a^{-2}b}$ **d.** $\left(\dfrac{2x^{-1}}{xy^0}\right)^{-2}$

Solution

a. $x^{-1} = \dfrac{1}{x}$ Apply rules of exponents.

b. $\dfrac{1}{3x^{-2}} = \dfrac{x^2}{3}$ Apply rules of exponents.

c. $\dfrac{25a^3b^{-4}}{5a^{-2}b} = 5a^{3-(-2)}b^{-4-1}$ Apply rules of exponents.

$$= 5a^5b^{-5}$$ Simplify.

$$= \dfrac{5a^5}{b^5}$$ Apply rules of exponents.

d. $\left(\dfrac{2x^{-1}}{xy^0}\right)^{-2} = \dfrac{(2)^{-2}x^2}{x^{-2}y^0}$ Apply rules of exponents.

$$= \dfrac{x^2 \cdot x^2}{2^2}$$ Simplify.

$$= \dfrac{x^4}{4}$$ Apply rules of exponents.

It is convenient to write very large or very small numbers in scientific notation. This notation has the form $c \times 10^n$, where $1 \leq c < 10$ and n is an integer. A positive exponent indicates that the number is large (10 or more) and a negative exponent indicates that the number is small (less than 1).

Example 3   Scientific Notation

Write each number in scientific notation.

a. 0.0000782   **b.** 836,100,000

Solution

a. $0.0000782 = 7.82 \times 10^{-5}$ **b.** $836{,}100{,}000 = 8.361 \times 10^8$

Example 4   Decimal Notation

Write each number in decimal notation.

a. 9.36×10^{-6} **b.** 1.345×10^2

Solution

a. $9.36 \times 10^{-6} = 0.00000936$ **b.** $1.345 \times 10^2 = 134.5$

Polynomials

The most common type of algebraic expression is the polynomial. Some examples are

$$-x + 1, \quad 2x^2 - 5x + 4, \quad \text{and} \quad 3x^3.$$

A polynomial in x is an expression of the form

$$a_n x^n + a_{n-1} x^{n-1} + \cdots + a_2 x^2 + a_1 x + a_0$$

where $a_n, a_{n-1}, \ldots, a_2, a_1, a_0$ are real numbers, n is a nonnegative integer, and $a_n \neq 0$. The polynomial is of degree n, a_n is called the leading coefficient, and a_0 is called the constant term.

Polynomials with one, two, and three terms are called monomials, binomials, and trinomials, respectively. In standard form, a polynomial is written with descending powers of x.

Example 5   Determining Degrees and Leading Coefficients

	Polynomial	Standard Form	Degree	Leading Coefficient
a.	$3x - x^2 + 4$	$-x^2 + 3x + 4$	2	-1
b.	-5	-5	0	-5
c.	$8 - 4x^3 + 7x + 2x^5$	$2x^5 - 4x^3 + 7x + 8$	5	2

Operations with Polynomials

You can add and subtract polynomials in much the same way that you add and subtract real numbers. Simply add or subtract the like terms (terms having the same variables to the same powers) by adding their coefficients. For instance, $-3xy^2$ and $5xy^2$ are like terms and their sum is

$$-3xy^2 + 5xy^2 = (-3 + 5)xy^2 = 2xy^2.$$

To subtract one polynomial from another, add the opposite by changing the sign of each term of the polynomial that is being subtracted and then adding the resulting like terms. You can add and subtract polynomials using either a horizontal or vertical format.

Example 6   Adding and Subtracting Polynomials

a. $3x^2 - 2x + 4$
$\underline{-x^2 + 7x - 9}$
$2x^2 + 5x - 5$

b. $(5x^3 - 7x^2 - 3) + (x^3 + 2x^2 - x + 8)$ Original polynomials

    $= (5x^3 + x^3) + (-7x^2 + 2x^2) - x + (-3 + 8)$ Group like terms.

    $= 6x^3 - 5x^2 - x + 5$ Combine like terms.

c. $(4x^3 + 3x - 6)$ ⟹ $4x^3 + 3x - 6$
$\underline{-(3x^3 + x + 10)}$ ⟹ $\underline{-3x^3 - x - 10}$
                                    $x^3 + 2x - 16$

d. $(7x^4 - x^2 - x + 2) - (3x^4 - 4x^2 + 3x)$ Original polynomials

    $= 7x^4 - x^2 - x + 2 - 3x^4 + 4x^2 - 3x$ Distributive Property

    $= (7x^4 - 3x^4) + (-x^2 + 4x^2) + (-x - 3x) + 2$ Group like terms.

    $= 4x^4 + 3x^2 - 4x + 2$ Combine like terms.

The simplest type of polynomial multiplication involves a monomial multiplier. The product is obtained by direct application of the Distributive Property.

Example 7   Finding a Product with a Monomial Multiplier

Find the product of $4x^2$ and $-2x^3 + 3x + 1$.

Solution

$$4x^2(-2x^3 + 3x + 1) = (4x^2)(-2x^3) + (4x^2)(3x) + (4x^2)(1)$$
$$= -8x^5 + 12x^3 + 4x^2$$

To multiply two binomials, use the FOIL Method illustrated below.

$$(ax + b)(cx + d) = ax(cx) + ax(d) + b(cx) + b(d)$$
$$\qquad\qquad\qquad\qquad \text{F} \qquad \text{O} \qquad \text{I} \qquad \text{L}$$

Example 8   Using the FOIL Method

Use the FOIL Method to find the product of $x - 3$ and $x - 9$.

Solution

$$\begin{array}{cccc} \text{F} & \text{O} & \text{I} & \text{L} \end{array}$$
$$(x - 3)(x - 9) = x^2 - 9x - 3x + 27$$
$$ = x^2 - 12x + 27 \qquad \text{Combine like terms.}$$

Example 9   Using the FOIL Method

Use the FOIL Method to find the product of $2x - 4$ and $x + 5$.

Solution

$$\begin{array}{cccc} \text{F} & \text{O} & \text{I} & \text{L} \end{array}$$
$$(2x - 4)(x + 5) = 2x^2 + 10x - 4x - 20$$
$$ = 2x^2 + 6x - 20 \qquad \text{Combine like terms.}$$

To multiply two polynomials that have three or more terms, you can use the same basic principle that you use when multiplying monomials and binomials. That is, each term of one polynomial must be multiplied by each term of the other polynomial. This can be done using either a vertical or a horizontal format.

Example 10   Multiplying Polynomials (Vertical Format)

Multiply $x^2 - 2x + 2$ by $x^2 + 3x + 4$ using a vertical format.

Solution

$$
\begin{array}{r}
x^2 - 2x + 2 \\
\times \quad x^2 + 3x + 4 \\
\hline
4x^2 - 8x + 8 \\
3x^3 - 6x^2 + 6x \\
x^4 - 2x^3 + 2x^2 \\
\hline
x^4 + x^3 + 0x^2 - 2x + 8
\end{array}
$$

$4(x^2 - 2x + 2)$

$3x(x^2 - 2x + 2)$

$x^2(x^2 - 2x + 2)$

Combine like terms.

So, $(x^2 - 2x + 2)(x^2 + 3x + 4) = x^4 + x^3 - 2x + 8$.

Example 11   Multiplying Polynomials (Horizontal Format)

$(4x^2 - 3x - 1)(2x - 5)$

$\quad = 4x^2(2x - 5) - 3x(2x - 5) - 1(2x - 5)$ Distributive Property

$\quad = 8x^3 - 20x^2 - 6x^2 + 15x - 2x + 5$ Distributive Property

$\quad = 8x^3 - 26x^2 + 13x + 5$ Combine like terms.

Some binomial products have special forms that occur frequently in algebra. These special products are listed below.

1. Sum and Difference of Two Terms: $(a + b)(a - b) = a^2 - b^2$

2. Square of a Binomial: $(a + b)^2 = a^2 + 2ab + b^2$

$$(a - b)^2 = a^2 - 2ab + b^2$$

Example 12   Finding Special Products

a. $(3x - 2)(3x + 2) = (3x)^2 - 2^2$ Special product

$\quad\quad\quad\quad\quad\quad = 9x^2 - 4$ Simplify.

b. $(2x - 7)^2 = (2x)^2 - 2(2x)(7) + 7^2$ Special product

$\quad\quad\quad\quad = 4x^2 - 28x + 49$ Simplify.

c. $(4a + 5b)^2 = (4a)^2 + 2(4a)(5b) + (5b)^2$ Special product

$\quad\quad\quad\quad = 16a^2 + 40ab + 25b^2$ Simplify.

To divide a polynomial by a monomial, separate the original division problem into multiple division problems, each involving the division of a monomial by a monomial.

Example 13   Dividing a Polynomial by a Monomial

Perform the division and simplify.

$$\frac{7x^3 - 12x^2 + 4x + 1}{4x}$$

Solution

$$\frac{7x^3 - 12x^2 + 4x + 1}{4x} = \frac{7x^3}{4x} - \frac{12x^2}{4x} + \frac{4x}{4x} + \frac{1}{4x}$$ Divide each term separately.

$$= \frac{7x^2}{4} - 3x + 1 + \frac{1}{4x}$$ Use rules of exponents.

Polynomial division is similar to long division of integers. To use polynomial long division, write the dividend and divisor in descending powers of the variable, insert placeholders with zero coefficients for missing powers of the variable, and divide as you would with integers. Continue this process until the degree of the remainder is less than that of the divisor.

Example 14   Long Division Algorithm for Polynomials

Use the long division algorithm to divide $x^2 + 2x + 4$ by $x - 1$.

Solution

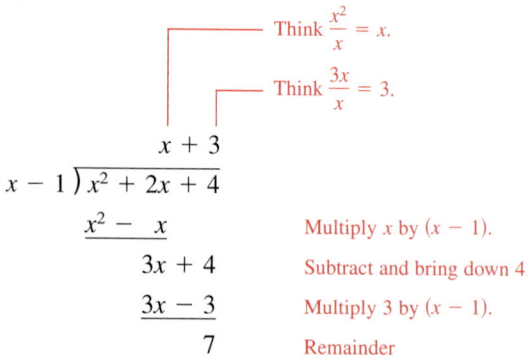

Think $\dfrac{x^2}{x} = x$.

Think $\dfrac{3x}{x} = 3$.

$$
\begin{array}{r}
x + 3 \\
x - 1 \,\overline{)\, x^2 + 2x + 4} \\
\underline{x^2 - x} \\
3x + 4 \\
\underline{3x - 3} \\
7
\end{array}
$$

Multiply x by $(x - 1)$.

Subtract and bring down 4.

Multiply 3 by $(x - 1)$.

Remainder

Considering the remainder as a fractional part of the divisor, the result is

Dividend   Quotient   Remainder

$$
\frac{x^2 + 2x + 4}{x - 1} = x + 3 + \frac{7}{x - 1}.
$$

Divisor   Divisor

Example 15   Accounting for Missing Powers of x

Divide $x^3 - 2$ by $x - 1$.

Solution

Note how the missing x^2- and x-terms are accounted for.

$$
\begin{array}{r}
x^2 + x + 1 \\
x - 1 \,\overline{)\, x^3 + 0x^2 + 0x - 2} \\
\underline{x^3 - x^2} \\
x^2 + 0x \\
\underline{x^2 - x} \\
x - 2 \\
\underline{x - 1} \\
-1
\end{array}
$$

Insert $0x^2$ and $0x$.

Multiply x^2 by $(x - 1)$.

Subtract and bring down $0x$.

Multiply x by $(x - 1)$.

Subtract and bring down -2.

Multiply 1 by $(x - 1)$.

Remainder

So, you have $\dfrac{x^3 - 2}{x - 1} = x^2 + x + 1 - \dfrac{1}{x - 1}.$

Synthetic division is a nice shortcut for dividing by polynomials of the form $x - k$.

Synthetic Division of a Third-Degree Polynomial

Use synthetic division to divide $ax^3 + bx^2 + cx + d$ by $x - k$, as follows.

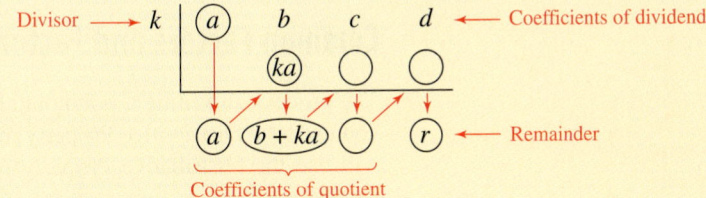

Vertical Pattern: Add terms.
Diagonal Pattern: Multiply by k.

Example 16   Using Synthetic Division

Use synthetic division to divide $x^3 - 6x^2 + 4$ by $x - 3$.

Solution

You should set up the array as follows. Note that a zero is included for the missing x-term in the dividend.

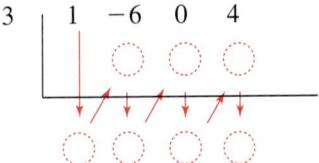

Then, use the synthetic division pattern by adding terms in columns and multiplying the results by 3.

Divisor: $x - 3$ Dividend: $x^3 - 6x^2 + 4$

$$
\begin{array}{c|cccc}
3 & 1 & -6 & 0 & 4 \\
  &   & 3 & -9 & -27 \\
\hline
  & 1 & -3 & -9 & -23
\end{array}
\quad \leftarrow \text{Remainder: } -23
$$

Quotient: $x^2 - 3x - 9$

So, you have

$$\frac{x^3 - 6x^2 + 4}{x - 3} = x^2 - 3x - 9 - \frac{23}{x - 3}.$$

For more review on exponents and polynomials, refer to Chapter 5.

A.6 Factoring and Solving Equations

Common Factors and Factoring by Grouping • Factoring Trinomials •
Factoring Special Polynomial Forms • Solving Polynomial Equations by
Factoring

Common Factors and Factoring by Grouping

The process of writing a polynomial as a product is called factoring. Previously, you used the Distributive Property to *multiply* and *remove* parentheses. Now, you will use the Distributive Property in the reverse direction to *factor* and *create* parentheses.

Removing the common monomial factor is the first step in completely factoring a polynomial. When you use the Distributive Property to remove this factor from each term of the polynomial, you are factoring out the greatest common monomial factor.

Example 1   Common Monomial Factors

Factor out the greatest common monomial factor from each polynomial.

a. $3x + 9$ **b.** $6x^3 - 4x$ **c.** $-4y^2 + 12y - 16$

Solution

a. $3x + 9 = 3(x) + 3(3) = 3(x + 3)$ Greatest common monomial factor is 3.

b. $6x^3 - 4x = 2x(3x^2) - 2x(2) = 2x(3x^2 - 2)$ Greatest common monomial factor is $2x$.

c.

$-4y^2 + 12y - 16 = -4(y^2) + (-4)(-3y) + (-4)(4)$ Greatest common monomial factor is -4.

$= -4(y^2 - 3y + 4)$ Factor -4 out of each term.

Some expressions have common factors that are not simple monomials. For instance, the expression $2x(x - 2) + 3(x - 2)$ has the common binomial factor $(x - 2)$. Factoring out this common factor produces

$$2x(x - 2) + 3(x - 2) = (x - 2)(2x + 3).$$

This type of factoring is called factoring by grouping.

Example 2   Common Binomial Factor

Factor $7a(3a + 4b) + 2(3a + 4b)$.

Solution

Each of the terms of this expression has a binomial factor of $(3a + 4b)$.

$$7a(3a + 4b) + 2(3a + 4b) = (3a + 4b)(7a + 2)$$

In Example 2, the expression was already grouped, and so it was easy to determine the common binomial factor. In practice, you will have to do the grouping *and* the factoring.

Example 3   Factoring by Grouping

Factor $x^3 - 5x^2 + x - 5$.

Solution

$$x^3 - 5x^2 + x - 5 = (x^3 - 5x^2) + (x - 5) \qquad \text{Group terms.}$$

$$= x^2(x - 5) + 1(x - 5) \qquad \text{Factoring out common monomial factor in each group.}$$

$$= (x - 5)(x^2 + 1) \qquad \text{Factored form}$$

Factoring Trinomials

To factor a trinomial $x^2 + bx + c$ into a product of two binomials, you must find two numbers m and n whose product is c and whose sum is b. If c is positive, then m and n have like signs that match the sign of b. If c is negative, then m and n have unlike signs. If $|b|$ is small relative to $|c|$, first try those factors of c that are close to each other in absolute value.

Example 4   Factoring Trinomials

Factor each trinomial.

a. $x^2 - 7x + 12$ **b.** $x^2 - 2x - 8$

Solution

a. You need to find two numbers whose product is 12 and whose sum is -7.

The product of -3 and -4 is 12.

$$x^2 - 7x + 12 = (x - 3)(x - 4)$$

The sum of -3 and -4 is -7.

b. You need to find two numbers whose product is -8 and whose sum is -2.

The product of -4 and 2 is -8.

$$x^2 - 2x - 8 = (x - 4)(x + 2)$$

The sum of -4 and 2 is -2.

Applications of algebra sometimes involve trinomials that have a common monomial factor. To factor such trinomials completely, first factor out the common monomial factor. Then try to factor the resulting trinomial by the methods given in this section.

Example 5 Factoring Completely

Factor the trinomial $5x^3 + 20x^2 + 15x$ completely.

Solution

$$5x^3 + 20x^2 + 15x = 5x(x^2 + 4x + 3)$$

Factor out common monomial factor $5x$.

$$= 5x(x + 1)(x + 3)$$

Factor trinomial.

To factor a trinomial whose leading coefficient is not 1, use the following pattern.

Factors of a

$$ax^2 + bx + c = (\quad x + \quad)(\quad x + \quad)$$

Factors of c

Use the following guidelines to help shorten the list of possible factorizations of a trinomial.

Guidelines for Factoring $ax^2 + bx + c \, (a > 0)$

1. If the trinomial has a common monomial factor, you should factor out the common factor before trying to find the binomial factors.

2. Because the resulting trinomial has no common monomial factors, you do not have to test any binomial factors that have a common monomial factor.

3. Switch the signs of the factors of c when the middle term (O + I) is correct except in sign.

Example 6 Factor a Trinomial of the Form $ax^2 + bx + c$

Factor the trinomial $6x^2 + 17x + 5$.

Solution

First, observe that $6x^2 + 17x + 5$ has no common monomial factor. For this trinomial, $a = 6$, which factors as (1)(6) or (2)(3), and $c = 5$, which factors as (1)(5).

$$(x + 1)(6x + 5) = 6x^2 + 11x + 5$$

$$(x + 5)(6x + 1) = 6x^2 + 31x + 5$$

$$(2x + 1)(3x + 5) = 6x^2 + 13x + 5$$

$$(2x + 5)(3x + 1) = 6x^2 + 17x + 5$$ Correct factorization

So, the correct factorization is $6x^2 + 17x + 5 = (2x + 5)(3x + 1)$.

Example 7   Factoring a Trinomial of the Form $ax^2 + bx + c$

Factor the trinomial.

$$3x^2 - 16x - 35$$

Solution

First, observe that $3x^2 - 16x - 35$ has no common monomial factor. For this trinomial,

$$a = 3 \text{ and } c = -35.$$

The possible factorizations of this trinomial are as follows.

$$(3x - 1)(x + 35) = 3x^2 + 104x - 35$$

$$(3x - 35)(x + 1) = 3x^2 - 32x - 35$$

$$(3x - 5)(x + 7) = 3x^2 + 16x - 35 \qquad \text{Middle term has opposite sign.}$$

$$(3x + 5)(x - 7) = 3x^2 - 16x - 35 \quad \Longleftarrow \quad \text{Correct factorization}$$

So, the correct factorization is

$$3x^2 - 16x - 35 = (3x + 5)(x - 7).$$

Example 8   Factoring Completely

Factor the trinomial completely.

$$6x^2y + 16xy + 10y$$

Solution

Begin by factoring out the common monomial factor $2y$.

$$6x^2y + 16xy + 10y = 2y(3x^2 + 8x + 5)$$

Now, for the new trinomial $3x^2 + 8x + 5$,

$$a = 3 \text{ and } c = 5.$$

The possible factorizations of this trinomial are as follows.

$$(3x + 1)(x + 5) = 3x^2 + 16x + 5$$

$$(3x + 5)(x + 1) = 3x^2 + 8x + 5 \quad \Longleftarrow \quad \text{Correct factorization}$$

So, the correct factorization is

$$6x^2y + 16xy + 10y = 2y(3x^2 + 8x + 5)$$
$$= 2y(3x + 5)(x + 1).$$

Factoring a trinomial can involve quite a bit of trial and error. Some of this trial and error can be lessened by using factoring by grouping. The key to this method of factoring is knowing how to rewrite the middle term. In general, to factor a trinomial $ax^2 + bx + c$ by grouping, choose factors of the product ac that add up to b and use these factors to rewrite the middle term. This technique is illustrated in Example 9.

Example 9   Factoring a Trinomial by Grouping

Use factoring by grouping to factor the trinomial $6y^2 + 5y - 4$.

Solution

In the trinomial $6y^2 + 5y - 4$, $a = 6$ and $c = -4$, which implies that the product of ac is -24. Now, because -24 factors as $(8)(-3)$, and $8 - 3 = 5 = b$, you can rewrite the middle term as $5y = 8y - 3y$. This produces the following result.

$$6y^2 + 5y - 4 = 6y^2 + 8y - 3y - 4 \qquad \text{Rewrite middle term.}$$

$$= (6y^2 + 8y) - (3y + 4) \qquad \text{Group terms.}$$

$$= 2y(3y + 4) - (3y + 4) \qquad \text{Factor out common monomial factor in first group.}$$

$$= (3y + 4)(2y - 1) \qquad \text{Distributive Property}$$

So, the trinomial factors as $6y^2 + 5y - 4 = (3y + 4)(2y - 1)$.

Factoring Special Polynomial Forms

Some polynomials have special forms. You should learn to recognize these forms so that you can factor such polynomials easily.

Factoring Special Polynomial Forms

Let a and b be real numbers, variables, or algebraic expressions.

1. Difference of Two Squares: $a^2 - b^2 = (a + b)(a - b)$

2. Perfect Square Trinomial: $a^2 + 2ab + b^2 = (a + b)^2$
$$a^2 - 2ab + b^2 = (a - b)^2$$

3. Sum or Difference of Two Cubes: $a^3 + b^3 = (a + b)(a^2 - ab + b^2)$
$$a^3 - b^3 = (a - b)(a^2 + ab + b^2)$$

Example 10   Factoring the Difference of Two Squares

Factor each polynomial.

a. $x^2 - 144$ **b.** $4a^2 - 9b^2$

Solution

a. $x^2 - 144 = x^2 - 12^2 \qquad \text{Write as difference of two squares.}$

$$= (x + 12)(x - 12) \qquad \text{Factored form}$$

b. $4a^2 - 9b^2 = (2a)^2 - (3b)^2 \qquad \text{Write as difference of two squares.}$

$$= (2a + 3b)(2a - 3b) \qquad \text{Factored form}$$

To recognize perfect square terms, look for coefficients that are squares of integers and for variables raised to even powers.

Example 11   Factoring Perfect Square Trinomials

Factor each trinomial.

a. $x^2 - 10x + 25$

b. $4y^2 + 4y + 1$

Solution

a. $x^2 - 10x + 25 = x^2 - 2(5x) + 5^2$ Recognize the pattern.

$\qquad\qquad\qquad\quad = (x - 5)^2$ Write in factored form.

b. $4y^2 + 4y + 1 = (2y)^2 + 2(2y)(1) + 1^2$ Recognize the pattern.

$\qquad\qquad\qquad\quad = (2y + 1)^2$ Write in factored form.

Example 12   Factoring Sum or Difference of Two Cubes

Factor each polynomial.

a. $x^3 + 1$

b. $27x^3 - 64y^3$

Solution

a. $x^3 + 1 = x^3 + 1^3$ Write as sum of two cubes.

$\qquad\quad = (x + 1)[x^2 - (x)(1) + 1^2]$ Factored form.

$\qquad\quad = (x + 1)(x^2 - x + 1)$ Simplify.

b. $27x^3 - 64y^3 = (3x)^3 - (4y)^3$ Write as difference of two cubes.

$\qquad\qquad\qquad = (3x - 4y)[(3x)^2 + (3x)(4y) + (4y)^2]$ Factored form

$\qquad\qquad\qquad = (3x - 4y)(9x^2 + 12xy + 16y^2)$ Simplify.

Solving Polynomial Equations by Factoring

A quadratic equation is an equation that can be written in the general form

$$ax^2 + bx + c = 0, \quad a \neq 0.$$

You can combine your factoring skills with the Zero-Factor Property to solve quadratic equations.

Zero-Factor Property

Let a and b be real numbers, variables, or algebraic expressions. If a and b are factors such that

$$ab = 0$$

then $a = 0$ or $b = 0$. This property also applies to three or more factors.

In order for the Zero-Factor Property to be used, a quadratic equation must be written in general form.

Example 13   Using Factoring to Solve a Quadratic Equation

Solve the equation.

$$x^2 - x - 12 = 0$$

Solution

First, check to see that the right side of the equation is zero. Next, factor the left side of the equation. Finally, apply the Zero-Factor Property to find the solutions.

$x^2 - x - 12 = 0$	Write original equation.
$(x + 3)(x - 4) = 0$	Factor left side of equation.
$x + 3 = 0 \quad \Longrightarrow \quad x = -3$	Set 1st factor equal to 0 and solve for x.
$x - 4 = 0 \quad \Longrightarrow \quad x = 4$	Set 2nd factor equal to 0 and solve for x.

So, the equation has two solutions: $x = -3$ and $x = 4$.

Remember to check your solutions in the original equation, as follows.

Check First Solution

$x^2 - x - 12 = 0$	Write original equation.
$(-3)^2 - (-3) - 12 \stackrel{?}{=} 0$	Substitute -3 for x.
$9 + 3 - 12 \stackrel{?}{=} 0$	Simplify.
$0 = 0$	Solution checks. ✓

Check Second Solution

$x^2 - x - 12 = 0$	Write original equation.
$4^2 - 4 - 12 \stackrel{?}{=} 0$	Substitute 4 for x.
$16 - 4 - 12 \stackrel{?}{=} 0$	Simplify.
$0 = 0$	Solution checks. ✓

Example 14   Using Factoring to Solve a Quadratic Equation

Solve $2x^2 - 3 = 7x + 1$.

Solution

$$2x^2 - 3 = 7x + 1 \qquad \text{Write original equation.}$$

$$2x^2 - 7x - 4 = 0 \qquad \text{Write in general form.}$$

$$(2x + 1)(x - 4) = 0 \qquad \text{Factor.}$$

$$2x + 1 = 0 \implies x = -\tfrac{1}{2} \qquad \text{Set 1st factor equal to 0 and solve for } x.$$

$$x - 4 = 0 \implies x = 4 \qquad \text{Set 2nd factor equal to 0 and solve for } x.$$

So, the equation has two solutions: $x = -\tfrac{1}{2}$ and $x = 4$. Check these in the original equation, as follows.

Check First Solution

$$2\left(-\tfrac{1}{2}\right)^2 - 3 \overset{?}{=} 7\left(-\tfrac{1}{2}\right) + 1 \qquad \text{Substitute } -\tfrac{1}{2} \text{ for } x \text{ in original equation.}$$

$$\tfrac{1}{2} - 3 \overset{?}{=} -\tfrac{7}{2} + 1 \qquad \text{Simplify.}$$

$$-\tfrac{5}{2} = -\tfrac{5}{2} \qquad \text{Solution checks. ✔}$$

Check Second Solution

$$2(4)^2 - 3 \overset{?}{=} 7(4) + 1 \qquad \text{Substitute 4 for } x \text{ in original equation.}$$

$$32 - 3 \overset{?}{=} 28 + 1 \qquad \text{Simplify.}$$

$$29 = 29 \qquad \text{Solution checks. ✔}$$

The Zero-Factor Property can be used to solve polynomial equations of degree three or higher. To do this, use the same strategy you used with quadratic equations.

Study Tip

The solution $x = -6$ from Example 15 is called a repeated solution.

Example 15   Solving a Polynomial Equation with Three Factors

Solve $x^3 + 12x^2 + 36x = 0$.

Solution

$$x^3 + 12x^2 + 36x = 0 \qquad \text{Write original equation.}$$

$$x(x^2 + 12x + 36) = 0 \qquad \text{Factor out common monomial factor.}$$

$$x(x + 6)(x + 6) = 0 \qquad \text{Factor perfect square trinomial.}$$

$$x = 0 \qquad \text{Set 1st factor equal to 0.}$$

$$x + 6 = 0 \implies x = -6 \qquad \text{Set 2nd factor equal to 0 and solve for } x.$$

$$x + 6 = 0 \implies x = -6 \qquad \text{Set 3rd factor equal to 0 and solve for } x.$$

Study Tip

For more review on factoring and solving equations, refer to Chapter 6.

Note that even though the left side of the equation has three factors, two of the factors are the same. So, you conclude that the solutions of the equation are $x = 0$ and $x = -6$. Check these in the original equation.

Appendix B

Introduction to Graphing Calculators

Introduction • Using a Graphing Calculator • Using Special Features of a Graphing Calculator

Introduction

In Section 4.2, you studied the point-plotting method for sketching the graph of an equation. One of the disadvantages of the point-plotting method is that to get a good idea about the shape of a graph you need to plot *many* points. By plotting only a few points, you can badly misrepresent the graph.

For instance, consider the equation $y = x^3$. To graph this equation, suppose you calculated only the following three points.

x	-1	0	1
$y = x^3$	-1	0	1
Solution point	$(-1, -1)$	$(0, 0)$	$(1, 1)$

By plotting these three points, as shown in Figure B.1, you might assume that the graph of the equation is a line. This, however, is not correct. By plotting several more points, as shown in Figure B.2, you can see that the actual graph is not straight at all.

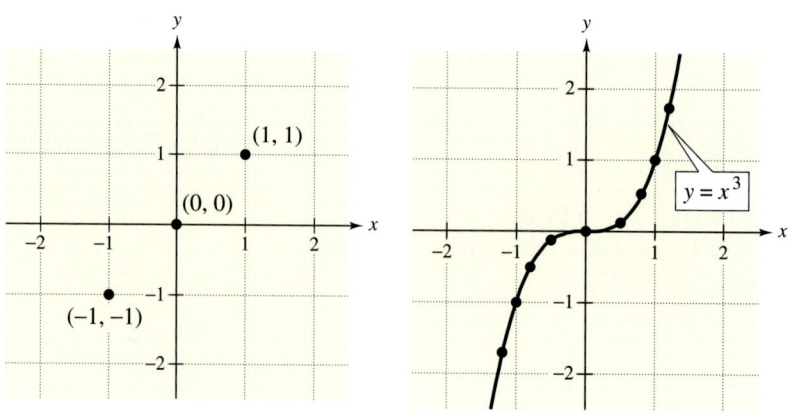

Figure B.1   Figure B.2

So, the point-plotting method leaves you with a dilemma. On the one hand, the method can be very inaccurate if only a few points are plotted. But, on the other hand, it is very time-consuming to plot a dozen (or more) points. Technology can help you solve this dilemma. Plotting several points (or even hundreds of points) on a rectangular coordinate system is something that a computer or graphing calculator can do easily.

Using a Graphing Calculator

There are many different graphing utilities: some are graphing packages for computers and some are hand-held graphing calculators. In this appendix, the steps used to graph an equation with a *TI-83* or *TI-83 Plus* graphing calculator are described. Keystroke sequences are often given for illustration; however, these may not agree precisely with the steps required by *your* calculator.*

Graphing an Equation with a *TI-83* or *TI-83 Plus* Graphing Calculator

Before performing the following steps, set your calculator so that all of the standard defaults are active. For instance, all of the options at the left of the [MODE] screen should be highlighted.

1. Set the viewing window for the graph. (See Example 3.) To set the standard viewing window, press [ZOOM] 6.

2. Rewrite the equation so that y is isolated on the left side of the equation.

3. Press the [Y=] key. Then enter the right side of the equation on the first line of the display. (The first line is labeled $Y_1 = .$)

4. Press the [GRAPH] key.

Example 1   Graphing a Linear Equation

Use a graphing calculator to graph $2y + x = 4$.

Solution

To begin, solve the equation for y in terms of x.

$$2y + x = 4 \qquad \text{Write original equation.}$$

$$2y = -x + 4 \qquad \text{Subtract } x \text{ from each side.}$$

$$y = -\frac{1}{2}x + 2 \qquad \text{Divide each side by 2.}$$

Press the [Y=] key, and enter the following keystrokes.

[(-)] [X,T,θ,n] [÷] 2 [+] 2

The top row of the display should now be as follows.

$$Y_1 = \text{-X}/2 + 2$$

Press the [GRAPH] key, and the screen should look like that shown in Figure B.3.

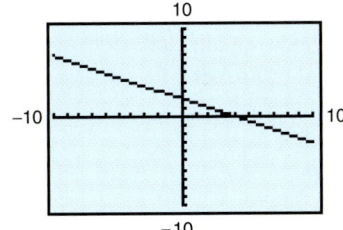

Figure B.3

---

*The graphing calculator keystrokes given in this section correspond to the *TI-83* and *TI-83 Plus* graphing calculators by Texas Instruments. For other graphing calculators, the keystrokes may differ. Consult your user's guide.

In Figure B.3, notice that the calculator screen does not label the tick marks on the *x*-axis or the *y*-axis. To see what the tick marks represent, you can press WINDOW. If you set your calculator to the standard graphing defaults before working Example 1, the screen should show the following values.

Xmin = -10	The minimum *x*-value is − 10.
Xmax = 10	The maximum *x*-value is 10.
Xscl = 1	The *x*-scale is 1 unit per tick mark.
Ymin = -10	The minimum *y*-value is − 10.
Ymax = 10	The maximum *y*-value is 10.
Yscl = 1	The *y*-scale is 1 unit per tick mark.
Xres = 1	Sets the pixel resolution

These settings are summarized visually in Figure B.4.

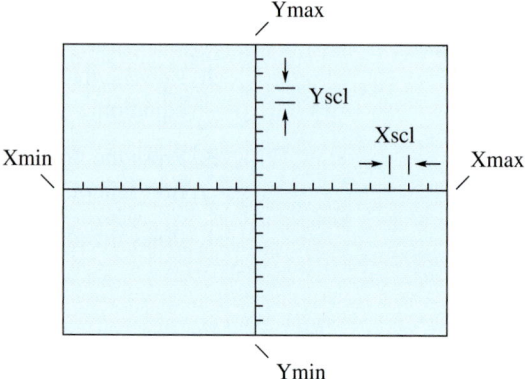

Figure B.4

Example 2 Graphing an Equation Involving Absolute Value

Use a graphing calculator to graph

$$y = |x - 3|.$$

Solution

This equation is already written so that *y* is isolated on the left side of the equation. Press the Y= key, and enter the following keystrokes.

   MATH ► 1 X,T,θ,n − 3)

The top row of the display should now be as follows.

   $Y_1 = \text{abs}(X - 3)$

Press the GRAPH key, and the screen should look like that shown in Figure B.5.

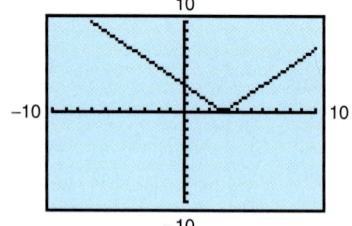

Figure B.5

Using Special Features of a Graphing Calculator

To use your graphing calculator to its best advantage, you must learn to set the viewing window, as illustrated in the next example.

Example 3   Setting the Viewing Window

Use a graphing calculator to graph

$$y = x^2 + 12.$$

Solution

Press $\boxed{Y=}$ and enter $x^2 + 12$ on the first line.

$\boxed{X,T,\theta,n}$ $\boxed{x^2}$ $\boxed{+}$ 12

Press the $\boxed{GRAPH}$ key. If your calculator is set to the standard viewing window, nothing will appear on the screen. The reason for this is that the lowest point on the graph of $y = x^2 + 12$ occurs at the point $(0, 12)$. Using the standard viewing window, you obtain a screen whose largest y-value is 10. In other words, none of the graph is visible on a screen whose y-values vary between -10 and 10, as shown in Figure B.6. To change these settings, press $\boxed{WINDOW}$ and enter the following values.

Xmin = -10	The minimum x-value is -10.
Xmax = 10	The maximum x-value is 10.
Xscl = 1	The x-scale is 1 unit per tick mark.
Ymin = -10	The minimum y-value is -10.
Ymax = 30	The maximum y-value is 30.
Yscl = 5	The y-scale is 5 units per tick mark.
Xres = 1	Sets the pixel resolution

Press $\boxed{GRAPH}$ and you will obtain the graph shown in Figure B.7. On this graph, note that each tick mark on the y-axis represents five units because you changed the y-scale to 5. Also note that the highest point on the y-axis is now 30 because you changed the maximum value of y to 30.

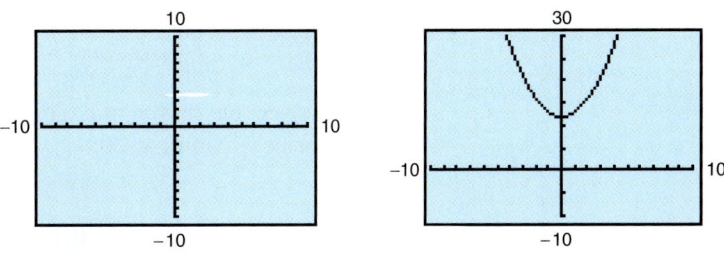

Figure B.6   Figure B.7

If you changed the y-maximum and y-scale on your calculator as indicated in Example 3, you should return to the standard setting before working Example 4. To do this, press $\boxed{ZOOM}$ 6.

Example 4   Using a Square Setting

Use a graphing calculator to graph $y = x$. The graph of this equation is a line that makes a 45° angle with the x-axis and with the y-axis. From the graph on your calculator, does the angle appear to be 45°?

Solution

Press $\boxed{Y=}$ and enter x on the first line.

$$Y_1 = X$$

Press the $\boxed{GRAPH}$ key and you will obtain the graph shown in Figure B.8. Notice that the angle the line makes with the x-axis doesn't appear to be 45°. The reason for this is that the screen is wider than it is tall. This makes the tick marks on the x-axis farther apart than the tick marks on the y-axis. To obtain the same distance between tick marks on both axes, you can change the graphing settings from "standard" to "square." To do this, press the following keys.

$\boxed{ZOOM}$ 5   Square setting

The screen should look like that shown in Figure B.9. Note in this figure that the square setting has changed the viewing window so that the x-values vary from -15 to 15.

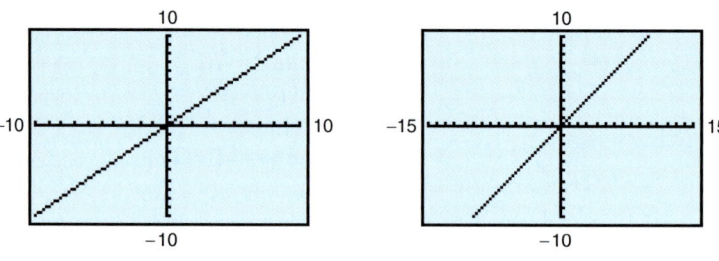

Figure B.8   Figure B.9

There are many possible square settings on a graphing calculator. To create a square setting, you need the following ratio to be $\frac{2}{3}$.

$$\frac{Ymax - Ymin}{Xmax - Xmin}$$

For instance, the setting in Example 4 is square because $(Ymax - Ymin) = 20$ and $(Xmax - Xmin) = 30$.

Example 5   Graphing More than One Equation in the Same Viewing Window

Use a graphing calculator to graph each equation in the same viewing window.

$$y = -x + 4, \quad y = -x, \quad \text{and} \quad y = -x - 4$$

Solution

To begin, press $\boxed{Y=}$ and enter all three equations on the first three lines. The display should now be as follows.

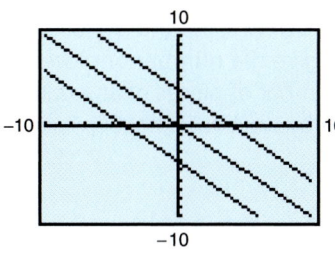

Figure B.10

$Y_1 = -X + 4$ $\boxed{(-)}$ $\boxed{X,T,\theta,n}$ $\boxed{+}$ 4

$Y_2 = -X$ $\boxed{(-)}$ $\boxed{X,T,\theta,n}$

$Y_3 = -X - 4$ $\boxed{(-)}$ $\boxed{X,T,\theta,n}$ $\boxed{-}$ 4

Press the $\boxed{\text{GRAPH}}$ key and you will obtain the graph shown in Figure B.10. Note that the graph of each equation is a line, and that the lines are parallel to each other.

Another special feature of a graphing calculator is the *trace* feature. This feature is used to find solution points of an equation. For example, you can approximate the x- and y-intercepts of $y = 3x + 6$ by first graphing the equation, then pressing the $\boxed{\text{TRACE}}$ key, and finally pressing the $\boxed{\blacktriangleleft}\boxed{\blacktriangleright}$ keys. To get a better approximation of a solution point, you can use the following keystrokes repeatedly.

$\boxed{\text{ZOOM}}$ 2 $\boxed{\text{ENTER}}$

Check to see that you get an x-intercept of $(-2, 0)$ and a y-intercept of $(0, 6)$. Use the *trace* feature to find the x- and y-intercepts of $y = \frac{1}{2}x - 4$.

Appendix B   Exercises

▦ In Exercises 1–12, use a graphing calculator to graph the equation. (Use a standard setting.) See Examples 1 and 2.

1. $y = -3x$

2. $y = x - 4$

3. $y = \frac{3}{4}x - 6$

4. $y = -3x + 2$

5. $y = \frac{1}{2}x^2$

6. $y = -\frac{2}{3}x^2$

7. $y = x^2 - 4x + 2$

8. $y = -0.5x^2 - 2x + 2$

9. $y = |x - 3|$

10. $y = |x + 4|$

11. $y = |x^2 - 4|$

12. $y = |x - 2| - 5$

▦ In Exercises 13–16, use a graphing calculator to graph the equation using the given window settings. See Example 3.

13. $y = 27x + 100$

Xmin = 0
Xmax = 5
Xscl = .5
Ymin = 75
Ymax = 250
Yscl = 25
Xres = 1

14. $y = 50,000 - 6000x$

Xmin = 0
Xmax = 7
Xscl = .5
Ymin = 0
Ymax = 50000
Yscl = 5000
Xres = 1

15. $y = 0.001x^2 + 0.5x$

Xmin = -500
Xmax = 200
Xscl = 50
Ymin = -100
Ymax = 100
Yscl = 20
Xres = 1

16. $y = 100 - 0.5|x|$

Xmin = -300
Xmax = 300
Xscl = 60
Ymin = -100
Ymax = 100
Yscl = 20
Xres = 1

▦ In Exercises 17–20, find a viewing window that shows the important characteristics of the graph.

17. $y = 15 + |x - 12|$

18. $y = 15 + (x - 12)^2$

19. $y = -15 + |x + 12|$

20. $y = -15 + (x + 12)^2$

▦ In Exercises 21–24, use a graphing calculator to graph both equations in the same viewing window. Are the graphs identical? If so, what basic rule of algebra is being illustrated? See Example 5.

21. $y_1 = 2x + (x + 1)$
   $y_2 = (2x + x) + 1$

22. $y_1 = \frac{1}{2}(3 - 2x)$
   $y_2 = \frac{3}{2} - x$

23. $y_1 = 2\left(\frac{1}{2}\right)$

$\quad\ y_2 = 1$

24. $y_1 = x(0.5x)$

$\quad\ y_2 = (0.5x)x$

In Exercises 25–32, use the *trace* feature of a graphing calculator to approximate the x- and y-intercepts of the graph.

25. $y = 9 - x^2$

26. $y = 3x^2 - 2x - 5$

27. $y = 6 - |x + 2|$

28. $y = |x - 2|^2 - 3$

29. $y = 2x - 5$

30. $y = 4 - |x|$

31. $y = x^2 + 1.5x - 1$

32. $y = x^3 - 4x$

▲ *Geometry*   In Exercises 33–36, use a graphing calculator to graph the equations in the same viewing window. Using a "square setting," determine the geometrical shape bounded by the graphs.

33. $y = -4, \quad y = -|x|$

34. $y = |x|, \quad y = 5$

35. $y = |x| - 8, \quad y = -|x| + 8$

36. $y = -\frac{1}{2}x + 7, \quad y = \frac{8}{3}(x + 5), \quad y = \frac{2}{7}(3x - 4)$

Modeling Data   In Exercises 37 and 38, use the following models, which give the number of pieces of first-class mail and the number of pieces of Standard A (third-class) mail handled by the U.S. Postal Service.

First Class

$y = 0.008x^2 + 1.42x + 88.7 \quad 0 \le x \le 10$

Standard A (Third Class)

$y = 0.246x^2 + 0.36x + 62.5 \quad 0 \le x \le 10$

In these models, y is the number of pieces handled (in billions) and x is the year, with $x = 0$ corresponding to 1990.   (Source: U.S. Postal Service)

37. Use the following window setting to graph both models in the same viewing window of a graphing calculator.

$$
\begin{array}{l}
\text{Xmin} = 0 \\
\text{Xmax} = 10 \\
\text{Xscl} = 1 \\
\text{Ymin} = 0 \\
\text{Ymax} = 120 \\
\text{Yscl} = 10 \\
\text{Xres} = 1
\end{array}
$$

38. (a) Were the numbers of pieces of first-class mail and Standard A mail increasing or decreasing over time?

   (b) Is the distance between the graphs increasing or decreasing over time? What does this mean to the U.S. Postal Service?

Answers to Reviews, Odd-Numbered Exercises, Quizzes, and Tests

Chapter 1

Section 1.1   *(page 9)*

1. (a) $20, \frac{9}{3}$ (b) $-3, 20, \frac{9}{3}$ (c) $-3, 20, -\frac{3}{2}, \frac{9}{3}, 4.5$

3. (a) $\frac{8}{4}$ (b) $\frac{8}{4}$ (c) $-\frac{5}{2}, 6.5, -4.5, \frac{8}{4}, \frac{3}{4}$

5.

7.
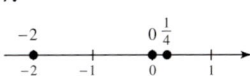

9. $>$
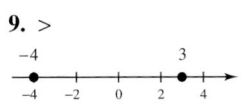

11. $>$
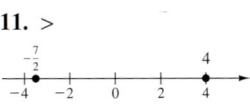

13. $>$

15. $<$

17. $<$

19. 2   **21.** 4

23. -5; Distance: 5

25. 3.8; Distance: 3.8

27. $\frac{5}{2}$; Distance: $\frac{5}{2}$

29. $\frac{5}{2}, \frac{5}{2}$ **31.** 3, 3   **33.** 7   **35.** 11   **37.** 3.4

39. $\frac{7}{2}$ **41.** -4.09 **43.** -23.6 **45.** 0   **47.** $=$

49. $>$ **51.** $<$ **53.** $<$ **55.** $<$ **57.** $>$

59.
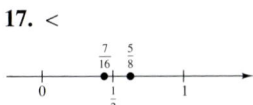

61.
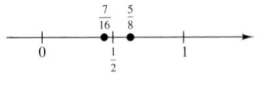

63. $-4, 20$ **65.** 15.3, 27.3   **67.** $-5.5, 1.5$

69. Sample answers: $-3, -100, -\frac{4}{1}$

71. Sample answers: $\sqrt{2}, \pi, -3\sqrt{3}$

73. Sample answers: $-7, 1, 341$

75. Sample answers: $\frac{1}{2}, 10, 20\frac{1}{5}$

77. Sample answers: $\frac{1}{7}, 0.25, 10\frac{1}{2}$

79. (a) $A = \{7°, 12°\}$

(b) $B = \{11.5°, 12°, 7.6°, 6.8°, 7.4°, 7.3°, 6.2°, 2.6°, 2.4°,$
$2.2°, 7°, -\frac{2}{9}°, 1\frac{7}{10}°\}$

(c) $C = \{7.9°, 2.1°, 3.6°, 1.4°, \frac{5}{9}°, 2.2°, 1.8°, 2.6°\}$

(d) $-\frac{2}{9}°, 1\frac{7}{10}°, 2.2°, 2.4°, 2.6°, 6.2°, 6.8°, 7°, 7.3°,$
$7.4°, 7.6°, 11.5°, 12°$

(e) $7.9°, 3.6°, 2.6°, 2.2°, 2.1°, 1.8°, 1.4°, \frac{5}{9}°, -1.3°,$
$-2\frac{1}{2}°, -2.6°, -2.7°, -3.9°, -4.3°$

(f) December 25

81. Two. They are -3 and 3.

83. $-25; |-25| > |10|$

85. The smaller number is located to the left of the larger number on the real number line.

87. True. $-5 > -13$ because -5 lies to the right of -13.

89. False. $6 > -17$ because 6 lies to the right of -17.

91. False. $|0| = 0$

93. True. For example, $\left|\frac{2}{3}\right| = \frac{2}{3}$.

95. True. Definition of opposite.

Section 1.2   *(page 17)*

1. 9

3. 7

5. -2

7. -11

9. 16   **11.** 0   **13.** 0   **15.** 27   **17.** -27

19. 6   **21.** 175   **23.** -5 **25.** 38   **27.** -10

29. 21   **31.** 20   **33.** 3   **35.** -135 **37.** -520

39. -352 **41.** -882 **43.** 3   **45.** 26   **47.** 5

49. 25   **51.** 36   **53.** 50   **55.** -30 **57.** -24

59. -109 **61.** -9 **63.** -11 **65.** -21 **67.** 0

69. -6 **71.** -103 **73.** -610 **75.** -80

77. -21 **79.** 500   **81.** -15 **83.** -36

85. 16   **87.** -4 **89.** -2 **91.** 12°F

93. 462 meters   **95.** \$1,012,000   **97.** \$2356.42

99. 23°   **101.** (a) 2.7 million   (b) 0.8 million

103. (a) $3 + 2 = 5$ (b) Adding two integers with like signs

105. To add two negative integers, add their absolute values and attach the negative sign.

107. $5 - 8$ **109.** $9 + (-12)$

Section 1.3   *(page 29)*

1. $2 + 2 + 2 = 6$

3. $(-3) + (-3) + (-3) + (-3) + (-3) = -15$

5. 21   **7.** 0   **9.** -32 **11.** -930

13. -35 **15.** 72   **17.** 3000   **19.** 90

21. 21   **23.** -30 **25.** 12   **27.** 90   **29.** 480

31. 338   **33.** -336 **35.** -4725 **37.** 260

39. 9009   **41.** 3   **43.** -6 **45.** -7 **47.** 7

49. Division by zero is undefined.   **51.** 0   **53.** 27

55. -6 **57.** -7 **59.** 1   **61.** 32   **63.** -32

65. -82 **67.** 110   **69.** 331   **71.** 86   **73.** 1045

75. 14,400   **77.** $-532,000$ **79.** Composite

81. Prime   **83.** Prime   **85.** Composite

87. Composite   **89.** $2 \cdot 2 \cdot 3$ **91.** $3 \cdot 11 \cdot 17$

93. $2 \cdot 3 \cdot 5 \cdot 7$ **95.** $3 \cdot 5 \cdot 13 \cdot 13$

97. $2 \cdot 2 \cdot 2 \cdot 2 \cdot 2 \cdot 2 \cdot 3$ **99.** 3   **101.** 0

103. $-24°$ **105.** \$6000   **107.** 57,600 square feet

109. 65 miles per hour

111. (a) 82

(b)
```
   73  77     82     87   91
  +●+ +●+ + +●+ + +●+ +●+ →
  72  76  80  84  88  92
```

(c) $5, -9, -5,$ and 9; Sum is 0; Explanations will vary.

113. 594 cubic inches

115. 2; it is divisible only by 1 and itself. Any other even number is divisible by 1, itself, and 2.

117. $\sqrt{1997} < 45$ **119.** Positive

121. The product (or quotient) of two nonzero real numbers of like signs is positive. The product (or quotient) of two nonzero real numbers of unlike signs is negative.

123. $(2m)n = 2(mn)$. The product of two odd integers is odd.

125. n is a multiple of 4. $\frac{12}{2} = 6$

127. Perfect (< 25): 6
Abundant (< 25): 12, 18, 20, 24
First perfect greater than 25 is 28.

Mid-Chapter Quiz   *(page 33)*

1. $<$

```
      3   3
      16  8
 +----●--●--+-------+--→
 0         1        1
           2
```

2. $>$

```
  -4      -2.5
 ●--------●---+----+----+--→
 -4   -3    -2   -1    0
```

3. $<$

```
  -7                  3
 +●-+--+--+--+--+--●-+--→
 -8  -6  -4  -2  0  2  4
```

4. $>$

```
              6  2π
 +----+-------●--●----+--→
 5         6        7
```

5. -0.75 **6.** $\frac{17}{19}$ **7.** $=$ **8.** $>$

9. $-22 - (-13) = -9$ **10.** $|-54 + 26| = 28$

11. 99   **12.** -75 **13.** -27 **14.** -25

15. -50 **16.** -62 **17.** 8   **18.** -5

19. -60 **20.** 91   **21.** 15   **22.** -4

23. $2 \cdot 2 \cdot 2 \cdot 2 \cdot 3 \cdot 3$ **24.** \$450,450

25. 128 cubic feet   **26.** 15 feet

27. \$367   **28.** $65°$

Section 1.4   *(page 44)*

1. 2   **3.** 5   **5.** 45   **7.** 6   **9.** 60   **11.** 1

13. $\frac{1}{4}$ **15.** $\frac{2}{3}$ **17.** $\frac{5}{16}$ **19.** $\frac{2}{25}$ **21.** $\frac{3}{5}$

23. $\frac{6}{10} = \frac{3}{5}$ **25.** 6   **27.** 10   **29.** $\frac{3}{5}$ **31.** $\frac{14}{11}$

33. $\frac{3}{8}$ **35.** -1 **37.** $-\frac{1}{2}$ **39.** $\frac{2}{5}$ **41.** $\frac{7}{5}$

43. $\frac{5}{6}$ **45.** $-\frac{1}{12}$ **47.** $\frac{9}{16}$ **49.** $-\frac{7}{24}$ **51.** $\frac{4}{3}$

53. $-\frac{41}{24}$ **55.** $\frac{7}{20}$ **57.** $-\frac{1}{12}$ **59.** $\frac{55}{6}$ **61.** $-\frac{17}{16}$

63. $-\frac{53}{12}$ **65.** $-\frac{121}{12}$ **67.** $\frac{17}{48}$ **69.** $\frac{64}{9}$ **71.** $-\frac{35}{12}$

73. $\frac{3}{10}$ **75.** $\frac{13}{60}$ **77.** $\frac{3}{8}$ **79.** $-\frac{10}{21}$ **81.** $-\frac{3}{8}$

83. $\frac{1}{3}$ **85.** $\frac{5}{24}$ **87.** $-\frac{3}{16}$ **89.** 1   **91.** $\frac{12}{5}$

93. $\frac{27}{40}$ **95.** 1   **97.** $\frac{121}{12}$ **99.** $-\frac{51}{2}$

101. $\frac{1}{7}$; $7 \cdot \frac{1}{7} = 1$ **103.** $\frac{7}{4}$; $\frac{4}{7} \cdot \frac{7}{4} = 1$ **105.** $\frac{1}{2}$

107. $-\frac{8}{27}$ **109.** $\frac{3}{7}$ **111.** $\frac{25}{24}$ **113.** -90

115. 0   **117.** Division by zero is undefined.   **119.** $\frac{5}{2}$

121. $\frac{10}{7}$ **123.** 0.75   **125.** 0.5625   **127.** $0.\overline{6}$

129. $0.58\overline{3}$ **131.** $0.\overline{45}$ **133.** 106.65   **135.** 2.27

137. -1.90 **139.** -57.02 **141.** 4.30

143. 39.08   **145.** -0.51 **147.** ≈ 1

149. 255.80 points   **151.** $\frac{1013}{40} = 25.325$ tons

153. $\frac{17}{48}$ **155.** \$9.78   **157.** 48   **159.** \$677.49

161. (a) Answers will vary.   (b) \$30,601

163. \$1.302

165. (g) December 19   (h) December 29 and 30

(i) December 29 and 30   (j) $\approx 0.3°$

(k) g, h, i

167. No. Rewrite both fractions with like denominators. Then add their numerators and write the sum over the common denominator.

169. No. $\frac{2}{3} = 0.\overline{6}$ (nonterminating)

171. 4; Divide $\frac{2}{3}$ by $\frac{1}{6}$.

173. False. The reciprocal of 5 is $\frac{1}{5}$.

175. True. The product can always be written as a ratio of two integers.

177. True. If you move v units to the left of u on the number line, the result will be to the right of zero.

179. The product is greater than 20, because the factors are greater than factors that yield a product of 20.

181. N. Since P and R are between 0 and 1, their product PR is less than the smaller of P and R but positive.

Section 1.5 *(page 55)*

1. 2^5 **3.** $(-5)^4$ **5.** $\left(-\frac{1}{4}\right)^3$ **7.** $(1.6)^5$

9. $(-3)(-3)(-3)(-3)(-3)(-3)$ **11.** $\left(\frac{3}{8}\right)\left(\frac{3}{8}\right)\left(\frac{3}{8}\right)\left(\frac{3}{8}\right)\left(\frac{3}{8}\right)$

13. $\left(-\frac{1}{2}\right)\left(-\frac{1}{2}\right)\left(-\frac{1}{2}\right)\left(-\frac{1}{2}\right)\left(-\frac{1}{2}\right)$ **15.** $(9.8)(9.8)(9.8)$

17. 9   **19.** 64   **21.** -125 **23.** -16 **25.** $\frac{1}{64}$

27. -1.728 **29.** 8   **31.** 12   **33.** -9

35. 27   **37.** 17   **39.** $-\frac{11}{2}$ **41.** 36   **43.** 9

45. 34   **47.** 17   **49.** -64 **51.** $\frac{7}{3}$ **53.** 21

55. $\frac{7}{80}$ **57.** $-\frac{1}{8}$ **59.** -1 **61.** $\frac{5}{6}$ **63.** 4

65. Division by zero is undefined.   **67.** 13   **69.** 0

71. 366.12   **73.** 10.69

75. Commutative Property of Multiplication

77. Commutative Property of Addition

79. Additive Identity Property

81. Additive Inverse Property

83. Associative Property of Addition

85. Associative Property of Multiplication

87. Multiplicative Inverse Property

89. Distributive Property   **91.** Distributive Property

93. $5 + y$ **95.** $-3(10)$ **97.** $6x + 12$

99. $100 + 25y$ **101.** $(3x + 2y) + 5$

103. $(12 \cdot 3)4$ **105.** (a) -50 (b) $\frac{1}{50}$

107. (a) 1 (b) -1 **109.** (a) $-2x$ (b) $\frac{1}{2x}$

111. (a) $-ab$ (b) $\frac{1}{ab}$ **113.** (a) 48 (b) 48

115. (a) 22 (b) 22

117. $7x + 9 + 2x$

$= 7x + 2x + 9$	Commutative Property of Addition
$= (7x + 2x) + 9$	Associative Property of Addition
$= (7 + 2)x + 9$	Distributive Property
$= 9x + 9$	Addition of Real Numbers
$= 9(x + 1)$	Distributive Property

119. $3 + 10(x + 1)$

$= 3 + 10x + 10$	Distributive Property
$= 3 + 10 + 10x$	Commutative Property of Addition
$= (3 + 10) + 10x$	Associative Property of Addition
$= 13 + 10x$	Addition of Real Numbers

121. 36 square units

123. (a) $x(1 + 0.06) = 1.06x$ (b) \$27.51

125. $30(30 - 8) = 30(30) - 30(8) = 660$ square units

127. $a - 2 + b + 11 + 2c + 3 = a + b + 2c + 12$

129. $a(b - c) = ab - ac$; Explanations will vary.

131. No

133. (a) Base (b) Exponent

135. No. $2 \cdot 5^2 = 2 \cdot 25 = 50, \ 10^2 = 100$

137. Associative Property of Addition:
$(a + b) + c = a + (b + c)$,
$(x + 3) + 4 = x + (3 + 4)$

Associative Property of Multiplication:
$(ab)c = a(bc), \ (3 \cdot 4)x = 3(4x)$

139. $24^2 = (4 \cdot 6)^2 = 4^2 \cdot 6^2$

141. $-3^2 = -(3)(3) = -9$

143. $-9 + \dfrac{9 + 20}{3(5)} - (-3) = -9 + \dfrac{29}{15} + 3$

$$= -6 + \frac{29}{15}$$

$$= \frac{-90 + 29}{15}$$

$$= -\frac{61}{15}$$

145. $5(x + 3) = 5x + 15$

147. Division by zero is undefined.

149.

Expression	Value
$(6 + 2) \cdot (5 + 3)$	$= 64$
$(6 + 2) \cdot 5 + 3$	$= 43$
$6 + 2 \cdot 5 + 3$	$= 19$
$6 + 2 \cdot (5 + 3)$	$= 22$

151. (a) $2 \cdot 2 + 2 \cdot 3 = 4 + 6 = 10$

(b) $2 \cdot 5 = 10$

(c) $2 \cdot 2 + 2 \cdot 3 = 2(2 + 3) = 2 \cdot 5 = 10$

Review Exercises *(page 61)*

1. (a) none (b) $-1, \sqrt{4}$

(c) $-1, 4.5, \frac{2}{5}, -\frac{1}{7}, \sqrt{4}$ (d) $\sqrt{5}$

3.

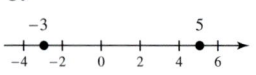

5.

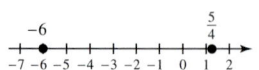

7.

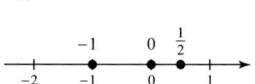

9. $<$

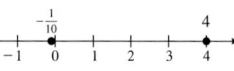

11. $>$;

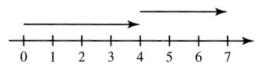

13. 0.6 **15.** $-152, 152$ **17.** $\frac{7}{3}, \frac{7}{3}$ **19.** 8.5

21. -8.5 **23.** $=$ **25.** $>$

27. 7 **29.** -5

31. 11 **33.** -95 **35.** -89 **37.** 5

39. -29 **41.** \$82,400

43. The sum can be positive or negative. The sign is determined by the integer with the greater absolute value.

45. 21 **47.** -7 **49.** 33 **51.** -22 **53.** -5

55. 1162 **57.** 45 **59.** -72 **61.** -48

63. 45 **65.** -54 **67.** -40 **69.** 9

71. -12 **73.** -15 **75.** 13

77. Division by zero is undefined. **79.** 32,000 miles

81. Prime **83.** Composite **85.** $2 \cdot 3 \cdot 3 \cdot 3 \cdot 7$

87. $2 \cdot 2 \cdot 13 \cdot 31$ **89.** -36 **91.** 7 **93.** 18

95. 21 **97.** 10 **99.** 15 **101.** $\frac{2}{5}$

103. $\frac{3}{4}$ **105.** $\frac{1}{9}$ **107.** $-\frac{103}{96}$ **109.** $\frac{5}{4}$ **111.** $\frac{17}{8}$

113. $2\frac{3}{4}$ inches **115.** $-\frac{1}{12}$ **117.** 1 **119.** $-\frac{1}{36}$

121. $\frac{2}{3}$ **123.** $\frac{6}{7}$ **125.** Division by zero is undefined.

127. $\frac{27}{32}$ inches per hour **129.** 5.65 **131.** -1.38

133. -0.75 **135.** 21 **137.** \$3.52 **139.** 6^5

141. $(-7)(-7)(-7)(-7)$ **143.** 16 **145.** $-\frac{27}{64}$

147. 6 **149.** 21 **151.** 52 **153.** 160

155. 81 **157.** $\frac{37}{8}$ **159.** 140 **161.** -3

163. 7 **165.** 0 **167.** 796.11 **169.** 1841.74

171. (a) \$6750 (b) \$9250

173. Additive Inverse Property

175. Commutative Property of Multiplication

177. Multiplicative Identity Property

179. Distributive Property

181. $0 + (z + 1) = z + 1$ **183.** $1 + 2y$

185. $x(y - z) = xy - xz$; Explanations will vary.

Chapter Test *(page 65)*

1. (a) 4 (b) $4, -6, 0$ (c) $4, -6, \frac{1}{2}, 0, \frac{7}{9}$

2. $>$ **3.** -4 **4.** 10 **5.** 10 **6.** 47

7. -160 **8.** 8 **9.** -3 **10.** 1 **11.** $\frac{17}{24}$

12. $\frac{2}{15}$ **13.** $\frac{7}{12}$ **14.** -27 **15.** -0.64

16. 33 **17.** 235 **18.** -2

19. Distributive Property

20. Multiplicative Inverse Property

21. Associative Property of Addition

22. Commutative Property of Multiplication

23. $\frac{2}{9}$ **24.** $2 \cdot 2 \cdot 3 \cdot 3 \cdot 3 \cdot 3$

25. 7.25 minutes per mile **26.** \$12.32

Chapter 2

Section 2.1 *(page 74)*

Review *(page 74)*

1. Commutative Property of Multiplication

2. Additive Inverse Property

3. Distributive Property

4. Associative Property of Addition

5. 3 **6.** 8 **7.** $\frac{9}{2}$ **8.** $\frac{2}{7}$ **9.** $-\frac{7}{11}$

10. $\frac{10}{3}$ **11.** \$6000 **12.** \$15 feet

1. $60t$ **3.** $2.19m$ **5.** Variable: x; Constant: 3

7. Variables: x, z; Constants: none

9. Variable: x; Constant: 2^3 **11.** $4x, 3$ **13.** $6x, -1$

15. $\frac{5}{3}, -3y^3$ **17.** $a^2, 4ab, b^2$ **19.** $3(x + 5), 10$

21. $15, \dfrac{5}{x}$ **23.** $\dfrac{3}{x + 2}, -3x, 4$ **25.** 14 **27.** $-\frac{1}{3}$

29. $\frac{2}{5}$ **31.** 2π **33.** -3.06 **35.** $y \cdot y \cdot y \cdot y \cdot y$

37. $2 \cdot 2 \cdot x \cdot x \cdot x \cdot x$ **39.** $4 \cdot y \cdot y \cdot z \cdot z \cdot z$

41. $a^2 \cdot a^2 \cdot a^2 = a \cdot a \cdot a \cdot a \cdot a \cdot a$

43. $4 \cdot x \cdot x \cdot x \cdot x \cdot x \cdot x \cdot x$

45. $9 \cdot a \cdot a \cdot a \cdot b \cdot b \cdot b$

47. $(x + y)(x + y)$ **49.** $\left(\dfrac{a}{3s}\right)\left(\dfrac{a}{3s}\right)\left(\dfrac{a}{3s}\right)\left(\dfrac{a}{3s}\right)$

51. $2 \cdot 2 \cdot (a - b)(a - b)(a - b)(a - b)(a - b)$

53. $2u^4$ **55.** $(2u)^4$ **57.** $a^3 b^2$ **59.** $3^3(x - y)^2$

61. $\left(\dfrac{x + y}{4}\right)^3$ **63.** (a) 0   (b) -9

65. (a) 3   (b) 13   **67.** (a) 6   (b) 4

69. (a) 3   (b) -20 **71.** (a) 33   (b) 112

73. (a) 0   (b) Division by zero is undefined.

75. (a) $-\dfrac{1}{5}$ (b) $\dfrac{3}{10}$

77. (a) $\dfrac{15}{2}$ (b) 10   **79.** (a) 72   (b) 320

81. (a)

x	-1	0	1	2	3	4
$3x - 2$	-5	-2	1	4	7	10

(b) 3   (c) $\dfrac{2}{3}$

83. $(n - 5)^2$, 9 square units   **85.** $a(a + b)$, 45 square units

87. (a) $\dfrac{3(4)}{2} = 6 = 1 + 2 + 3$

(b) $\dfrac{6(7)}{2} = 21 = 1 + 2 + 3 + 4 + 5 + 6$

(c) $\dfrac{10(11)}{2} = 55 = 1 + 2 + 3 + 4 + 5 + 6 + 7$

$+ 8 + 9 + 10$

89. (a) 4, 5, 5.5, 5.75, 5.875, 5.938, 5.969; Approaches 6.

(b) 9, 7.5, 6.75, 6.375, 6.188, 6.094, 6.047; Approaches 6.

91. (a) $(15 \cdot 12)c = 180c$; Plastic chairs: $351; Wood chairs: $531

(b) Canopy 1: $215; Canopy 2: $265; Canopy 3: $415; Canopy 4: $565; Canopy 5: $715

93. No. The term includes the minus sign and is $-3x$.

95. No. When $y = 3$, the expression is undefined.

97. $y - 2(x - y) = -4 - 2[2 - (-4)]$

$= -4 - 2(2 + 4)$

$= -4 - 2(6)$

$= -4 - 12$

$= -16$

Discussions will vary.

Section 2.2   *(page 85)*

Review   *(page 85)*

1. To find the prime factorization of a number is to write the number as a product of prime factors.

2. Distributive Property   **3.** 12   **4.** 120

5. -11 **6.** -5760 **7.** 35   **8.** -350

9. $\dfrac{1}{80}$ **10.** $\dfrac{45}{16}$ **11.** 2,362,000

12. 52 miles per hour

1. Commutative Property of Addition

3. Associative Property of Multiplication

5. Additive Identity Property

7. Multiplicative Identity Property

9. Associative Property of Addition

11. Commutative Property of Multiplication

13. Distributive Property

15. Additive Inverse Property

17. Multiplicative Inverse Property

19. Distributive Property

21. Additive Inverse Property, Additive Identity Property

23. $(-5r)s = -5(rs)$ Associative Property of Multiplication

25. $v(2) = 2v$ Commutative Property of Multiplication

27. $5(t - 2) = 5(5t) + 5(-2)$ Distributive Property

29. $(2z - 3) + [-(2z - 3)] = 0$ Additive Inverse Property

31. $5x \cdot \dfrac{1}{5x} = 1$ Multiplicative Inverse Property

33. $12 + (8 - x) = (12 + 8) - x$

Associative Property of Addition

35. $32 + 16z$ **37.** $-24 + 40m$ **39.** $90 - 60x$

41. $-16 - 40t$ **43.** $-10x + 5y$ **45.** $3x + 6$

47. $-24 + 6t$ **49.** $4x + 4xy + 4y^2$ **51.** $3x^2 + 3x$

53. $8y^2 - 4y$ **55.** $-5z + 2z^2$ **57.** $-12y^2 + 16y$

59. $-u + v$ **61.** $3x^2 - 4xy$

63. ab; ac; $a(b + c) = ab + ac$

65. $2a$; $2(b - a)$; $2a + 2(b - a) = 2b$

67. $6x^2, -3xy, y^2$; $6, -3, 1$

69. $-ab, 5ac, -7bc$; $-1, 5, -7$ **71.** $16t^3, 3t^3$; $4t, -5t$

73. $4rs^2, 12rs^2$ **75.** $4x^2y, 10x^2y$; $x^3, 3x^3$ **77.** $-2y$

79. $-2x + 5$ **81.** $11x + 4$ **83.** $3r + 7$

85. $x^2 - xy + 4$ **87.** $17z + 11$

89. $z^3 + 3z^2 + 3z + 1$ **91.** $-x^2y + 4xy + 12xy^2$

93. $2\left(\dfrac{1}{x}\right) + 8$ **95.** $11\left(\dfrac{1}{t}\right) - 2t$

97. False. $3(x - 4) = 3x - 12$ **99.** True. $6x - 4x = 2x$

101. 416   **103.** 432   **105.** -236 **107.** 39.9

109. $12x$ **111.** $-4x$ **113.** $6x^2$ **115.** $-10z^3$

117. $9a$ **119.** $-3x^3$ **121.** $-24x^4y^4$

123. $2x$ **125.** $13s - 2$ **127.** $-2m + 15$

129. $44 + 2x$ **131.** $8x + 26$ **133.** $2x - 17$

135. $10x - 7x^2$ **137.** $3x^2 + 5x$ **139.** $4t^2 - 11t$

141. $26t - 2t^2$ **143.** $\dfrac{x}{3}$ **145.** $\dfrac{7z}{5}$ **147.** $-\dfrac{11x}{12}$

149. x **151.** $5x + 9$

153. (a) $8x + 14$ (b) $3x^2 + 21x$

155. (a) Answers will vary.   (b) $\frac{57}{2}$

157. 9375 square feet

159. $(6x)^4 = (6x)(6x)(6x)(6x);\; 6x^4 = 6x \cdot x \cdot x \cdot x$

161. Two terms are like terms if they are both constant or if they have the same variable factor(s). Like terms: $3x^2, -5x^2$; unlike terms: $3x^2, 5x$

163. The corresponding exponents of x and y are not raised to the same power.

165. $\dfrac{x}{3} + \dfrac{4x}{3} = \dfrac{5x}{3}$

167. It does not change if the parentheses are removed because multiplication is a higher-order operation than subtraction. It does change if the brackets are removed because the division would be performed before the subtraction.

Mid-Chapter Quiz *(page 90)*

1. (a) 0   (b) 10   (c) 0

2. (a) 2   (b) 0   (c) Division by zero is undefined.

3. $4x^2, -2x;\; 4, -2$ **4.** $5x, 3y, -12z;\; 5, 3, -12$

5. (a) $(3y)^4$ (b) $2^3(x - 3)^2$ **6.** $20y^2$ **7.** x

8. $9y^5$ **9.** $\dfrac{10z^3}{21y}$

10. Associative Property of Multiplication

11. Distributive Property

12. Multiplicative Inverse Property

13. Commutative Property of Addition   **14.** $6x^2 - 2x$

15. $-8y + 12$ **16.** $y^2 + 4xy + y$ **17.** $3\left(\dfrac{1}{u}\right) + 3u$

18. $8a - 7b$ **19.** $-8x - 66$

20. $8 + (x + 6) + (3x + 1) = 4x + 15$ **21.** 45,700

Section 2.3 *(page 101)*

Review *(page 101)*

1. Negative   **2.** 15, 3

3. False. $-4^2 = -1 \cdot 4 \cdot 4 = -16$

4. True. $(-4)^2 = (-4)(-4) = 16$ **5.** 78

6. 120   **7.** $\frac{3}{4}$ **8.** $\frac{14}{3}$ **9.** $\frac{23}{9}$ **10.** $-\frac{111}{10}$

11. 5 weeks, \$16.50   **12.** 40 meters

1. d   **2.** a   **3.** e   **4.** f   **5.** b   **6.** c

7. $x + 5$ **9.** $x - 25$ **11.** $x - 6$ **13.** $2x$

15. $\dfrac{x}{3}$ **17.** $\dfrac{x}{50}$ **19.** $\dfrac{3}{10}x$ **21.** $3x + 5$

23. $5x + 8$ **25.** $10(x + 4)$

27. $|x + 4|$ **29.** $x^2 + 1$

31. A number decreased by 10

33. The product of 3 and a number, increased by 2

35. One-half a number decreased by 6

37. Three times the difference of 2 and a number

39. The sum of a number and 1, divided by 2

41. One-half decreased by a number divided by 5

43. The square of a number, increased by 5

45. $(x + 3)x = x^2 + 3x$ **47.** $(25 + x) + x = 25 + 2x$

49. $(x - 9)(3) = 3x - 27$ **51.** $\dfrac{8(x + 24)}{2} = 4x + 96$

53. $0.10d$ **55.** $0.06L$ **57.** $\dfrac{100}{r}$ **59.** $15m + 2n$

61. $t = 10.2$ years   **63.** $t = 11.9$ years

65.

n	0	1	2	3	4	5
$2n - 1$	-1	1	3	5	7	9
Differences		2	2	2	2	2

67. $a = 5, b = 4$ **69.** $3x(6x - 1) = 18x^2 - 3x$

71. $\left(\frac{1}{2}\right)(14x + 3)(2x) = 14x^2 + 3x$

73. $\frac{1}{2}(2x^2)[(9x + 4) + (8 - x)] = 8x^3 + 12x^2$

75.

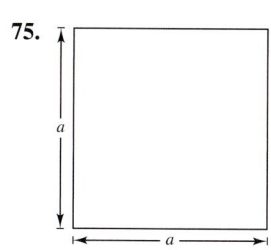 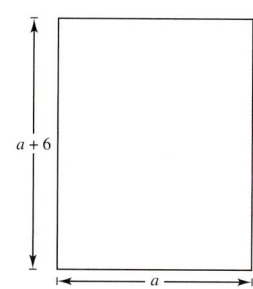

Perimeter of the square: $4a$ centimeters; Area of the square: a^2 square centimeters; Perimeter of the rectangle: $4a + 12$ centimeters; Area of the rectangle: $a(a + 6)$ square centimeters

77. $3w^2$ **79.** $5w$

81. The start time is missing.

83. The amount of the paycheck and the number of hours worked on the paycheck are missing.

85. (c) $3x$; $2x$

(d) 14 in. $= \frac{7}{6}$ ft; $7x + 14$

(e) $2x$; $3x + 7$

(f) 12 in. $= 1$ ft; $19x + 22$

(g) 6 feet; 28 feet; 60 feet; 30 by 60 feet; $1096.00

(h) Rent plastic chairs.

87. Division

89. (a) No. Addition is commutative.

(b) Yes. Subtraction is not commutative.

(c) No. Multiplication is commutative.

(d) Yes. Division is not commutative.

Section 2.4   *(page 111)*

Review   *(page 111)*

1. Negative

2. Positive. The product of an even number of negative factors is positive.

3. $10 + 6$ **4.** Multiplicative Inverse Property

5. t^7 **6.** $-3y^5$ **7.** $15x$ **8.** $4 - 2t$

9. $8b$ **10.** $-70x$ **11.** Perimeter: $6x$; Area: $\dfrac{9x^2}{4}$

12. Perimeter: $9x - 2$; Area: $5x^2 - 4x$

---

1. (a) Solution   (b) Not a solution

3. (a) Not a solution   (b) Solution

5. (a) Not a solution   (b) Solution

7. (a) Solution   (b) Not a solution

9. (a) Solution   (b) Not a solution

11. (a) Solution   (b) Solution

13. (a) Not a solution   (b) Not a solution

15. (a) Not a solution   (b) Not a solution

17. (a) Not a solution   (b) Solution

19. (a) Solution   (b) Not a solution

21. (a) Solution   (b) Not a solution

23. (a) Not a solution   (b) Solution

25. (a) Solution   (b) Not a solution

27.

$5x + 12 = 22$	Original equation
$5x + 12 - 12 = 22 - 12$	Subtract 12 from each side.
$5x = 10$	Combine like terms.
$\dfrac{5x}{5} = \dfrac{10}{5}$	Divide each side by 5.
$x = 2$	Solution

29.

$\frac{2}{3}x = 12$	Original equation
$\frac{3}{2}\left(\frac{2}{3}x\right) = \frac{3}{2}(12)$	Multiply each side by $\frac{3}{2}$.
$x = 18$	Solution

31.

$2(x - 1) = x + 3$	Original equation
$2x - 2 = x + 3$	Distributive Property
$2x - x - 2 = x - x + 3$	Subtract x from each side.
$x - 2 = 3$	Combine like terms.
$x - 2 + 2 = 3 + 2$	Add 2 to each side.
$x = 5$	Solution

33.

$x = -2(x + 3)$	Original equation
$x = -2x - 6$	Distributive Property
$x + 2x = -2x + 2x - 6$	Add $2x$ to each side.
$3x = 0 - 6$	Additive Inverse Property
$3x = -6$	Combine like terms.
$\dfrac{3x}{3} = \dfrac{-6}{3}$	Divide each side by 3.
$x = -2$	Solution

35. 13   **37.** 10

39. Twice a number increased by 5 is 21.

41. Ten times the difference of a number and 3 is 8 times the number.

43. The sum of a number and 1 divided by 3 is 8.

45. $x + 12 = 45$ **47.** $4(x + 6) = 100$

49. $2x - 14 = \dfrac{x}{3}$ **51.** $x + 6 = 94$

53. $1044 + x = 1926$ **55.** $0.35x = 148.05$

57. $24h = 72$ **59.** $3r + 25 = 160$ **61.** $135 = 2.5x$

63. $p + 45 = 375$ **65.** $750,000 - 3D = 75,000$

67. $1.75n = 2000$ **69.** 15 dollars   **71.** 150 pounds

73. 6000 feet   **75.** 240 centimeters

77. Substitute the real number into the equation. If the equation is true, the real number is a solution. Given the equation $2x - 3 = 5$, $x = 4$ is a solution and $x = -2$ is not a solution.

79. Simplifying an expression means removing all symbols of grouping and combining like terms. Solving an equation means finding all values of the variable for which the equation is true.

Simplify: $3(x - 2) - 4(x + 1) = 3x - 6 - 4x - 4$

$$= -x - 10$$

Solve: $3(x - 2) = 6$

$$3x - 6 = 6$$

$$3x = 12 \rightarrow x = 4$$

81. (a) Simplify each side by removing symbols of grouping, combining like terms, and reducing fractions on one or both sides.

(b) Add (or subtract) the same quantity to (from) each side of the equation.

(c) Multiply (or divide) each side of the equation by the same nonzero real number.

(d) Interchange the two sides of the equation.

Review Exercises *(page 117)*

1. $-x, 15$ **3.** $12y, y^2; 12, 1$

5. $5x^2, -3xy, 10y^2; 5, -3, 10$ **7.** $\dfrac{2y}{3}, -\dfrac{4x}{y}; \dfrac{2}{3}, -4$

9. $(5z)^3$ **11.** $6^2(b - c)^2$

13. (a) 5 (b) 5   **15.** (a) 4 (b) -2

17. (a) 0 (b) -7

19. Multiplicative Inverse Property

21. Commutative Property of Multiplication

23. Associative Property of Addition

25. $4x + 12y$ **27.** $-10u + 15v$ **29.** $8x^2 + 5xy$

31. $a - 3b$ **33.** $-2a$ **35.** $11p - 3q$ **37.** $\frac{15}{4}s - 5t$

39. $x^2 + 2xy + 4$ **41.** $3x - 3y + 3xy$

43. $3\left(1 + \dfrac{r}{n}\right)^2$ **45.** $48t$ **47.** $45x^2$ **49.** $-12x^3$

51. $8x$ **53.** $5u - 10$ **55.** $5s - r$ **57.** $10z - 1$

59. $2z - 2$ **61.** $8x - 32$ **63.** $-2x + 4y$

65. $P\left(\frac{9}{10}\right)^5$

67. $(2n - 1) + (2n + 1) + (2n + 3) = 6n + 3$ **69.** $58x^2$

71. *Verbal model:*

$$\begin{array}{c} \text{Base pay} \\ \text{per hour} \end{array} + \begin{array}{c} \text{Additional} \\ \text{pay per unit} \end{array} \cdot \begin{array}{c} \text{Number of units} \\ \text{produced per hour} \end{array}$$

Algebraic expression: $8.25 + 0.60x$

73. $\frac{2}{3}x + 5$ **75.** $2x - 10$ **77.** $50 + 7x$

79. $\dfrac{x + 10}{8}$ **81.** $x^2 + 64$ **83.** A number plus 3

85. A number decreased by 2, divided by 3

87. $0.05x$ **89.** $625n$

91. (a)

n	0	1	2	3	4	5
$n^2 + 3n + 2$	2	6	12	20	30	42
Differences		4	6	8	10	12
Differences			2	2	2	2

(b) Third row: entries increase by 2; Fourth row: constant 2

93. (a) Not a solution   (b) Solution

95. (a) Not a solution   (b) Solution

97. (a) Solution   (b) Not a solution

99. (a) Not a solution   (b) Solution

101. (a) Solution   (b) Solution

103.

$3(x - 2) = x + 2$	Original equation
$3x - 6 = x + 2$	Distributive Property
$3x - x - 6 = x - x + 2$	Subtract x from each side.
$2x - 6 = 2$	Combine like terms.
$2x - 6 + 6 = 2 + 6$	Add 6 to each side.
$2x = 8$	Combine like terms.
$\dfrac{2x}{2} = \dfrac{8}{2}$	Divide each side by 2.
$x = 4$	Solution.

105. $x + \dfrac{1}{x} = \dfrac{37}{6}$ **107.** $6x - \dfrac{1}{2}(6x) = \dfrac{1}{2}(6x) = 24$

Chapter Test *(page 121)*

1. $2x^2, 2; -7xy, -7; 3y^3, 3$ **2.** $x^3(x + y)^2$

3. Associative Property of Multiplication

4. Commutative Property of Addition

5. Additive Identity Property

6. Multiplicative Inverse Property

7. $3x + 24$ **8.** $20r - 5s$ **9.** $-3y + 2y^2$

10. $-36 + 18x - 9x^2$ **11.** $-a - 7b$ **12.** $8u - 8v$

13. $4z - 4$ **14.** $18 - 2t$ **15.** (a) 25 (b) -31

16. Division by zero is undefined. **17.** $\frac{1}{5}n + 2$

18. (a) Perimeter: $2w + 2(2w - 4)$; Area: $w(2x - 4)$

(b) Perimeter: $6w - 8$; Area: $2w^2 - 4w$

(c) Perimeter: unit of length; Area: square units

(d) Perimeter: 64 feet; Area: 240 square feet

19. $15m + 10n$ **20.** (a) Not a solution (b) Solution

Chapter 3

Section 3.1 *(page 132)*

Review *(page 132)*

1. Multiplicative Identity Property

2. Associative Property of Addition

3. $5x + 17$ **4.** $-2b^2 + 7ab + 4a$

5. $-3(x - 5)^5$ **6.** $-40r^3s^4$ **7.** $\dfrac{2m^3}{5n^4}$

8. $\dfrac{(x + 3)^2}{2(x + 8)}$ **9.** $-9x + 11y$ **10.** $8v - 4$

11. $\frac{1}{12}$ mile **12.** $30\frac{23}{30}$ tons

---

1. -6 **3.** 13 **5.** 4 **7.** -9

9.

$5x + 15 = 0$	Original equation
$5x + 15 - 15 = 0 - 15$	Subtract 15 from each side.
$5x = -15$	Combine like terms.
$\dfrac{5x}{5} = \dfrac{-15}{5}$	Divide each side by 5.
$x = -3$	Simplify.

11.

$-2x + 5 = 13$	Original equation
$-2x + 5 - 5 = 13 - 5$	Subtract 5 from each side.
$-2x = 8$	Combine like terms.
$\dfrac{-2x}{-2} = \dfrac{8}{-2}$	Divide each side by -2.
$x = -4$	Simplify.

13. 2 **15.** -7 **17.** 6 **19.** $-\frac{7}{3}$

21. 3 **23.** 2 **25.** 4 **27.** $\frac{2}{3}$ **29.** 2 **31.** $\frac{1}{3}$

33. -2 **35.** No solution **37.** 1 **39.** 4

41. -2 **43.** Identity **45.** $\frac{2}{5}$ **47.** 0 **49.** $\frac{2}{3}$

51. 30 **53.** $\frac{5}{3}$ **55.** $\frac{5}{6}$ **57.** Identity **59.** 2

61. 80 inches $\times$ 40 inches **63.** 75 centimeters

65. Yes. Subtract the cost of parts from the total to find the cost of labor. Then divide by 32 to find the number of hours spent on labor. $2\frac{1}{4}$ hours

67. 150 seats **69.** 7 hours 20 minutes **71.** 4

73. 35, 37 **75.** 51, 53, 55

77. The red box weighs 6 ounces. If you removed three blue boxes from each side, the scale would still balance. The Addition (or Subtraction) Property of Equality

79. Subtract 5 from each side of the equation. Addition Property of Equality

81. True. Subtracting 0 from each side does not change any values. The equation remains the same.

83. (a)

t	1	1.5	2
Width	300	240	200
Length	300	360	400
Area	90,000	86,400	80,000

t	3	4	5
Width	150	120	100
Length	450	480	500
Area	67,500	57,600	50,000

(b) The area decreases.

Section 3.2 *(page 142)*

Review *(page 142)*

1. (a) Add the numerators and write the sum over the like denominator. The result is $\frac{8}{5}$.

(b) Find equivalent fractions with a common denominator. Add the numerators and write the sum over the like denominator. The result is $\frac{38}{15}$.

2. Answers will vary. Examples are given.
$$3x^2 + 2\sqrt{x}; \quad \frac{4}{x^2 + 1}$$

3. $4x^6$ **4.** $8y^5$ **5.** $5z^5$ **6.** $a^2 + a - 2$

7. $x - 4$ **8.** $-x^2 + 1$ **9.** $-y^4 + 2y^2$

10. $10t - 4t^2$ **11.** 7.5 gallons

12. (a) \$24,300 (b) \$4301

---

1. 4 **3.** -5 **5.** $\frac{22}{5}$ **7.** 2 **9.** -10

11. 2 **13.** 3 **15.** No solution **17.** -4

19. No solution **21.** $\frac{8}{5}$ **23.** 1 **25.** No solution

27. $\frac{2}{9}$ **29.** 1 **31.** 3 **33.** $-\frac{3}{2}$ **35.** $\frac{5}{2}$ **37.** $-\frac{2}{5}$

39. $\frac{35}{2}$ **41.** $-\frac{10}{3}$ **43.** No solution **45.** $\frac{1}{6}$ **47.** 50

49. $\frac{32}{5}$ **51.** 10 **53.** 0 **55.** $\frac{16}{3}$ **57.** $\frac{4}{3}$ **59.** $\frac{4}{11}$

61. 6 **63.** 5.00 **65.** 7.71 **67.** 123.00

69. 3.51 **71.** 8.99 **73.** 4.8 hours **75.** 97

77. 25 quarts **79.** $1\frac{1}{3}$ quarts **81.** 2038

83. Use the Distributive Property to remove symbols of grouping. Remove the innermost symbols first and combine like terms. Symbols of grouping preceded by a minus sign can be removed by changing the sign of each term within the symbols.

$$2x - [3 + (x - 1)] = 2x - [3 + x - 1]$$
$$= 2x - [2 + x]$$
$$= 2x - 2 - x = x - 2$$

85. $-2(x - 5) = -2x + 10$

87. The least common multiple of the denominators is the simplest expression that is a multiple of all the denominators. The least common multiple of the denominators contains each prime factor of the denominators repeated the maximum number of times it occurs in any one of the factorizations of the denominators.

89. Because the expression is not an equation, there are not two sides to multiply by the least common multiple of the denominators.

Section 3.3 *(page 153)*

Review *(page 153)*

1. Plot the numbers on a number line. -28 is less than 63 because -28 is to the left of 63.

2. 0 **3.** 0 **4.** -38 **5.** -530 **6.** 29

7. $8x - 20$ **8.** $-xz^2 + 2y^2z$

9. (a) 7 (b) 16

10. (a) Division by zero is undefined. (b) 2

11. $14.67 **12.** $5r$

Percent	Parts out of 100	Decimal	Fraction
1. 40%	40	0.40	$\frac{2}{5}$
3. 7.5%	7.5	0.075	$\frac{3}{40}$
5. 63%	63	0.63	$\frac{63}{100}$

Percent	Parts out of 100	Decimal	Fraction
7. 15.5%	15.5	0.155	$\frac{31}{200}$
9. 60%	60	0.60	$\frac{3}{5}$
11. 150%	150	1.50	$\frac{3}{2}$

13. 62% **15.** 20% **17.** 7.5% **19.** 238%

21. 0.125 **23.** 1.25 **25.** 0.085 **27.** 0.0075

29. 80% **31.** 125% **33.** $83\frac{1}{3}\%$ **35.** 105%

37. $37\frac{1}{2}\%$ **39.** $41\frac{2}{3}\%$ **41.** 45 **43.** 544

45. 0.42 **47.** 176 **49.** 2100 **51.** 2200

53. 132 **55.** 360 **57.** 72% **59.** 12.5%

61. 2.75% **63.** 500%

Cost	Selling Price	Markup	Markup Rate
65. $26.97	$49.95	$22.98	85.2%
67. $40.98	$74.38	$33.40	81.5%
69. $69.29	$125.98	$56.69	81.8%
71. $13,250.00	$15,900.00	$2650.00	20%
73. $107.97	$199.96	$91.99	85.2%

List Price	Sale Price	Discount	Discount Rate
75. $39.95	$29.95	$10.00	25%
77. $23.69	$18.95	$4.74	20%
79. $189.99	$159.99	$30.00	15.8%
81. $119.96	$59.98	$59.98	50%
83. $995.00	$695.00	$300.00	30.2%

85. $544 **87.** $3435 **89.** 74.7% **91.** 7.2%

93. $312.50 **95.** $24,409 **97.** 10,210 eligible votes

99. 0.107%

101. < 15 135.45 million; 15–44 278.47 million 45–64 222.94 million; > 64 204.44 million

103. (a) 2,074,000 (b) 98,000 (c) 114,000

105. (a) $3(19.50) + x = 99$, $40.50

(b) $24 + x = 80$, $56.00, Second package

(c) $40.50 = p(99)$, 40.9%

(d) $56 = p(80)$, 70%

(e) $x = 99 + 0.05(99)$, $103.95

(f) $19.50 + 60(3.2x) = 92.46$, $0.38

107. A rate is a fixed ratio.

109. No. $\frac{1}{2}\% = 0.5\% = 0.005$

Section 3.4 *(page 164)*

Review *(page 164)*

1. Divide both the numerator and denominator by 3.

2. Multiply $\dfrac{3}{5}$ by $\dfrac{2}{x}$.   **3.** $3(xy)$

4. Additive Identity Property   **5.** 13   **6.** -122

7. 9,300,000   **8.** -4 **9.** $\frac{77}{5}$ or 15.4

10. 8   **11.** $2(n - 10)$ **12.** $\frac{1}{4}b(b + 6)$

---

1. $\frac{4}{1}$ **3.** $\frac{1}{2}$ **5.** $\frac{2}{3}$ **7.** $\frac{9}{1}$ **9.** $\frac{2}{1}$ **11.** $\frac{2}{3}$

13. $\frac{1}{4}$ **15.** $\frac{7}{15}$ **17.** $\frac{2}{1}$ **19.** $\frac{3}{8}$ **21.** $\frac{3}{50}$

23. $\frac{3}{4}$ **25.** $\frac{3}{10}$ **27.** \$0.049   **29.** \$0.073

31. 32-ounce jar

33. 16-ounce package   **35.** 2-liter bottle

37. 12   **39.** 50   **41.** 30   **43.** 16   **45.** $\frac{3}{16}$

47. $\frac{1}{2}$ **49.** 27   **51.** $\frac{14}{5}$ **53.** $\frac{2}{3}$ **55.** $\frac{46}{9}$

57. $\frac{20}{1}$ **59.** $\frac{3}{2}$ **61.** $\frac{100}{49}$ **63.** 16 gallons

65. 250 blocks   **67.** \$1142   **69.** 22,691

71. $46\frac{2}{3}$ minutes   **73.** 20 pints   **75.** 384 miles

77. $\frac{5}{2}$ **79.** $6\frac{2}{3}$ feet   **81.** 80%   **83.** \$7346

85. \$0.68   **87.** (g) \$57.00

89. No. It is necessary to know one of the following: the number of men in the class or the number of women in the class.

91. A proportion is a statement that equates two ratios.

Mid-Chapter Quiz *(page 168)*

1. 6   **2.** 8   **3.** $\frac{19}{2}$ **4.** 0   **5.** $-\frac{1}{3}$ **6.** $\frac{40}{13}$

7. 36   **8.** -2 **9.** 5   **10.** -2

11. 2.06 ; Substitute 2.06 for x. After simplifying, the equation should be an identity.

12. 51.23; Substitute 51.23 for x. After simplifying, the equation should be an identity.

13. 15.5   **14.** 42   **15.** 200%   **16.** 455

17. 10 hours

18. 6 square meters, 12 square meters, 24 square meters

19. 93   **20.** 17%   **21.** 3 hours   **22.** $\frac{225}{64}$

23. 26.25 gallons

Section 3.5 *(page 177)*

Review *(page 177)*

1. $2n$ is an even integer and $2n + 1$ is an odd integer.

2. $\begin{aligned} 2x - 3 &= 10 \\ 2x - 3 + 3 &= 10 + 3 \\ 2x &= 13 \end{aligned}$

3. $-28y^3$ **4.** $-3x^6$ **5.** $\dfrac{20u^3}{3}$ **6.** $2y$

7. $13x - 5x^2$ **8.** $-5t + 32$ **9.** $10v - 40$

10. $60 - 10x$ **11.** 6%   **12.** 20% off

---

1. $\dfrac{2A}{b}$ **3.** $\dfrac{A - P}{Pt}$ **5.** $\dfrac{V}{wh}$ **7.** $\dfrac{S}{1 + R}$

9. $\dfrac{Fr^2}{\alpha m_1}$ **11.** $\dfrac{2A - ah}{h}$ **13.** $\dfrac{2(h - v_0 t)}{t^2}$

15. 100π cubic meters   **17.** $\dfrac{150}{11}$ watts per volt

19. 48 meters   **21.** 16 hours   **23.** $114.\overline{1}$ m / sec

25. 784 square feet   **27.** 8 meters

29. 30 inches

31. Radius: 3.98 inches; Area: 49.74 square inches

33. 24 square inches   **35.** 96 cubic inches

37. \$540   **39.** \$15,975   **41.** 11%

43. 10.51 years   **45.** 0.17 hour

47. 1154 miles per hour   **49.** 28 miles

51. $\frac{1}{3}$ hour

53. (a) Answers will vary.

   (b) 48 miles per hour; Answers will vary.

55. Solution 1: 25 gallons; Solution 2: 75 gallons

57. Solution 1: 5 quarts; Solution 2: 5 quarts

59. 46 stamps at 24¢, 54 stamps at 37¢

61. 8 nickels, 12 dimes

63. 30 pounds at \$2.49 per pound, 70 pounds at \$3.89 per pound

65. $\frac{8}{7}$ gallons   **67.** \$2000 at 7%, \$4000 at 9%

69. (a)

Corn, x	Soybeans, $100 - x$	Price per ton of the mixture
0	100	$200
20	80	$185
40	60	$170
60	40	$155
80	20	$140
100	0	$125

(b) Decreases   (c) Decreases

(d) Average of the two prices

71. $1\frac{1}{5}$ hours   **73.** Answers will vary.   **75.** 15 years

77. Candidate A: 250 votes, Candidate B: 250 votes, Candidate C: 500 votes

79. Perimeter: linear units—inches, feet, meters; Area: square units—square inches, square meters; Volume: cubic units—cubic inches, cubic centimeters

81. The circumference would double; the area would quadruple. Circumference: $C = 2\pi r$, Area: $A = \pi r^2$
If r is doubled, you have $C = 2\pi(2r) = 2(2\pi r)$ and $A = \pi(2r)^2 = 4\pi r^2$.

83. $\frac{1}{5}$

Section 3.6   *(page 191)*

Review   *(page 191)*

1. Commutative Property of Multiplication

2. Additive Inverse Property

3. Distributive Property

4. Additive Identity Property

5. 7   **6.** 0   **7.** 0   **8.** 1   **9.** 4   **10.** 9

11. 19.8 square meters   **12.** 104 square feet

---

1. (a) Yes   (b) No   (c) Yes   (d) No

3. (a) No   (b) Yes   (c) Yes   (d) No

5. a   **6.** e   **7.** d   **8.** b   **9.** f   **10.** c

11.

13.

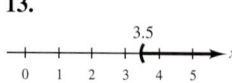

15.

17.

19.

21.

23.

25. $-15 + x < -24$

27. $x \geq 4$

29. $x \leq 2$

31. $x < 4$

33. $x \leq -4$

35. $x > 8$

37. $x \geq 7$

39. $x > 7.55$

41. $x > -\frac{2}{3}$

43. $x > \frac{9}{2}$

45. $x > \frac{20}{11}$

47. $x > \frac{8}{3}$

49. $x \leq -8$

51. $x > -15$

53. $\frac{5}{2} < x < 7$

55. $-3 \leq x < -1$

57. $-5 < x < 5$

59. $-\frac{3}{2} < x < \frac{9}{2}$

61. $1 < x < 10$

63. $-1 < x \le 4$

65. $-5 \le x < 1$

67. $x < \frac{52}{11}$ or $x > \frac{67}{12}$

69. $x < -\frac{8}{3}$ or $x \ge \frac{5}{2}$

71. $y \le -10$

73. $-5 < x \le 0$

75. $x < -3$ or $x \ge 2$, $\{x \mid x < -3\} \cup \{x \mid x \ge 2\}$

77. $-5 \le x < 4$, $\{x \mid x \ge -5\} \cap \{x \mid x < 4\}$

79. $x \le -2.5$ or $x \ge -0.5$, $\{x \mid x \le -2.5\} \cup \{x \mid x \ge -0.5\}$

81. $\{x \mid x \ge -7\} \cap \{x \mid x < 0\}$

83. $\{x \mid x < -5\} \cup \{x \mid x > 3\}$

85. $\left\{x \mid x > -\frac{9}{2}\right\} \cap \left\{x \mid x \le -\frac{3}{2}\right\}$

87. $x \ge 0$ **89.** $z \ge 8$ **91.** $10 \le n \le 16$

93. x is at least $\frac{5}{2}$.   **95.** y is at least 3 and less than 5.

97. z is more than 0 and no more than π.   **99.** $2600

101. The average temperature in Miami is greater than the average temperature in New York.

103. 26,000 miles   **105.** $x \ge 31$

107. The call must be less than or equal to 6.38 minutes. If a portion of a minute is billed as a full minute, the call must be less than or equal to 6 minutes.

109. $2 \le x \le 16$ **111.** $3 \le n \le \frac{15}{2}$

113. $12.50 < 8 + 0.75n$; $n > 6$

115. 1994, 1995

117. (h) $0.35x + 19.50 \le 75.00$; $x \le 158.57$ minutes

119. Yes. By definition, dividing by a number is the same as multiplying by its reciprocal.

121. $3x - 2 \le 4$, $-(3x - 2) \ge -4$

Section 3.7   (page 202)

Review   (page 202)

1. $2n$ is an even integer and $2n - 1$ is an odd integer.

2. No. $-2x^4 \ne 16x^4 = (-2x)^4$

3. $\frac{35}{14} = \frac{7 \cdot 5}{7 \cdot 2} = \frac{5}{2}$

4. $\frac{4}{5} \div \frac{z}{3} = \frac{4}{5} \cdot \frac{3}{z} = \frac{12}{5z}$

5. <   **6.** >   **7.** >   **8.** >   **9.** >

10. <   **11.** More than $500   **12.** Less than $500

1. Not a solution   **3.** Solution

5. $x - 10 = 17$; $x - 10 = -17$

7. $4x + 1 = \frac{1}{2}$; $4x + 1 = -\frac{1}{2}$ **9.** $4, -4$

11. No solution   **13.** 0   **15.** $3, -3$ **17.** $4, -6$

19. $11, -14$ **21.** $\frac{16}{3}, 16$ **23.** No solution   **25.** $\frac{4}{3}$

27. $\frac{15}{2}, -\frac{39}{2}$ **29.** $18.75, -6.25$ **31.** $\frac{17}{5}, -\frac{11}{5}$

33. $-\frac{5}{3}, -\frac{13}{3}$ **35.** 3   **37.** $\frac{11}{5}, -1$ **39.** $\frac{15}{4}, -\frac{1}{4}$

41. $2, 3$ **43.** $7, -3$ **45.** $\frac{3}{2}, -\frac{1}{4}$ **47.** $11, 13$

49. $28, -\frac{12}{5}$ **51.** $\frac{1}{2}$ **53.** $|x - 5| = 3$

55. (a) Solution   (b) Not a solution
       (c) Not a solution   (d) Solution

57. (a) Not a solution   (b) Solution
       (c) Solution   (d) Not a solution

59. $-3 < y + 5 < 3$ **61.** $7 - 2h \ge 9$ or $7 - 2h \le -9$

63.    **65.**

67. $-4 < y < 4$ **69.** $x \le -6$ or $x \ge 6$

71. $-7 < x < 7$ **73.** $-9 \le y \le 9$

75. $-2 \le y \le 6$ **77.** $x < -16$ or $x > 4$

79. $-3 \le x \le 4$ **81.** $t \le -\frac{15}{2}$ or $t \ge \frac{5}{2}$

83. $-\infty < x < \infty$ **85.** No solution

87. $-82 \le x \le 78$ **89.** $-104 < y < 136$

91. $z < -50$ or $z > 110$ **93.** $x < -\frac{11}{3}$ or $x > 1$

95. $-5 < x < 35$ **97.** $\frac{28}{3} \le x \le \frac{32}{3}$

99. $-\infty < x < \infty$ **101.** $-4 \le x \le 40$

103. $-\infty < x < \infty$

105. $-2 < x < \frac{2}{3}$

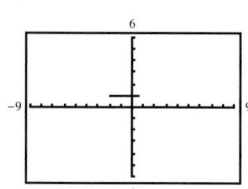

107. $x < -6$ or $x > 3$

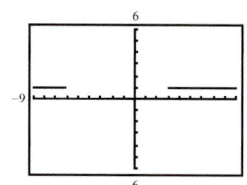

109. $3 \le x \le 7$

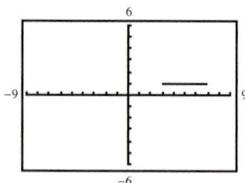

111. d **112.** c **113.** b **114.** a **115.** $|x| \le 2$

117. $|x - 19| < 3$ **119.** $|x| < 3$ **121.** $|x - 5| > 6$

123.

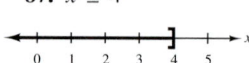

Maximum: 82 degrees Fahrenheit; Minimum: 62 degrees Fahrenheit

125. $|x - 98.6| \le 1$

127. The absolute value of a real number measures the distance of the real number from zero.

129. The solutions of $|x| = a$ are $x = a$ and $x = -a$. $|x - 3| = 5$ means $x - 3 = 5$ or $x - 3 = -5$. Thus, $x = 8$ or $x = -2$.

131. $|2x - 6| \le 6$

133. Because $|3x - 4|$ is always nonnegative, the inequality is always true for *all* values of x. The student's solution eliminates the values $-\frac{1}{3} < x < 3$.

Review Exercises *(page 207)*

1. 5 **3.** -4 **5.** 3 **7.** 3 **9.** 5 **11.** 4

13. 3 **15.** $\frac{4}{3}$ **17.** 20 **19.** 12 units

21. 80×50 meters **23.** 20 **25.** 6 **27.** 1

29. 7 **31.** $\frac{19}{3}$ **33.** 20 **35.** 3 **37.** 23.26

39. 224.31

Percent	Parts out of 100	Decimal	Fraction
41. 35%	35	0.35	$\frac{7}{20}$

43. 20 **45.** 400 **47.** 60% **49.** $77.76

51. 12.1% **53.** $\frac{1}{8}$ **55.** $\frac{4}{3}$

57. 24-ounce container **59.** $\frac{7}{2}$ **61.** $-\frac{10}{3}$ **63.** 9

65. $133 **67.** $w = \dfrac{P - 2l}{2}$

Distance, d	Rate, r	Time, t
69. 520 mi	65 mi/hr	8 hr
71. 3000 mi	60 mph	50 hr

73. 1108.3 miles **75.** 30×26 feet **77.** $475

79. 13 dimes, 17 quarters **81.** $\frac{30}{11} \approx 2.7$ hours

83.  **85.**

87. $x \le 4$ **89.** $x > -6$

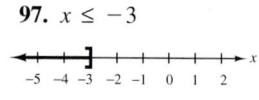

91. $x > 3$ **93.** $x \le 6$

95. $y > -\frac{70}{3}$ **97.** $x \le -3$

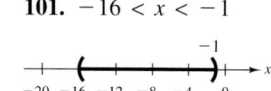

99. $-7 \le x < -2$ **101.** $-16 < x < -1$

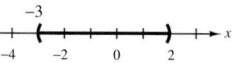

103. $-3 < x < 2$

105. At least $8333.33 **107.** ± 6 **109.** $4, -\frac{4}{3}$

111. $0, -\frac{8}{5}$ **113.** $-10, -\frac{10}{3}$ **115.** $\frac{1}{2}, 3$ **117.** $-\frac{19}{3}, 1$

119. $x < 1$ or $x > 7$ **121.** $-6 \le x \le 24$

123. $x < -3$ or $x > 3$ **125.** $-4 < x < 11$

127. $m \le 0$ or $m \ge 1$ **129.** $b < -9$ or $b > 5$

131. $x \le 2$ or $x \ge 3$

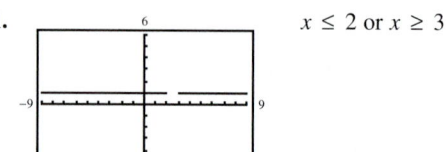

133. $|x - 3| < 2$ **135.** $|x| > 2$

137.

Maximum: 116.6 degrees Fahrenheit

Minimum: 40 degrees Fahrenheit

Chapter Test *(page 211)*

1. -13 **2.** $\frac{21}{4}$ **3.** 7   **4.** 7   **5.** -10 **6.** 10

7. $5, -11$ **8.** $\frac{2}{3}, -\frac{4}{3}$

9. $x \geq -6$

10. $x > 2$

11. $-7 < x \leq 1$

12. $-1 \leq x < \frac{5}{4}$

13. $1 \leq x \leq 5$

14. $x < -\frac{9}{5}$ or $x > 3$

15. 11.03   **16.** $2\frac{1}{2}$ hours   **17.** $37\frac{1}{2}\%$, 0.375

18. 1200   **19.** 36%

20. $\frac{5}{9}$; 2 yards = 6 feet = 72 inches   **21.** $\frac{12}{7}$ **22.** 5

23. 48 mph   **24.** $\frac{36}{7} \approx 5.1$ hours   **25.** $\dfrac{S - C}{C}$

26. $6250   **27.** $t \geq 8$ **28.** 25,000 miles

Cumulative Test: Chapters 1–3 *(page 212)*

1. $<$ **2.** 1200   **3.** $-\frac{11}{24}$ **4.** $-\frac{25}{12}$ **5.** 8

6. 14   **7.** 28   **8.** -30 **9.** 20   **10.** $3^3(x + y)^2$

11. $-2x^2 + 6x$ **12.** Associative Property of Addition

13. $15x^7$ **14.** a^4b^3 **15.** $7x^2 - 6x - 2$

16. $5x^2 - x$ **17.** (a) Not a solution   (b) Solution

18. 6   **19.** $\frac{52}{3}$ **20.** 5   **21.** $\frac{4}{3}, -2$

22. $x \geq 9$

23. $-5 \leq x < 1$

24. $x \geq -1$ or $x \leq -\frac{6}{5}$

25. $x \leq -\frac{5}{4}$ or $x \geq 2$

26. $\dfrac{15{,}000 \text{ miles}}{1 \text{ year}} \cdot \dfrac{1 \text{ gallon}}{28.3 \text{ miles}} \cdot \dfrac{\$1.179}{1 \text{ gallon}} \approx \$624.91$ per year

27. Length: 18 meters; Width: 12 meters   **28.** $495.37

29. $\frac{3}{4}$ **30.** 246, 248   **31.** $920   **32.** $3.34

33. $57,000   **34.** $9\frac{11}{30}$ hours   **35.** 51.7 miles per hour

Chapter 4

Section 4.1 *(page 223)*

Review *(page 223)*

1. $3x = 7$ is a linear equation since it has the form $ax + b = c$. $x^2 + 3x = 2$ is not of that form, and therefore is not linear.

2. Substitute 3 for x in the equation to verify that it satisfies the equation.

3. -10 **4.** 4   **5.** 14   **6.** 4   **7.** 6   **8.** $\frac{1}{9}$

9. 144   **10.** 200   **11.** $19,250

12. 8 hours 45 minutes

1.

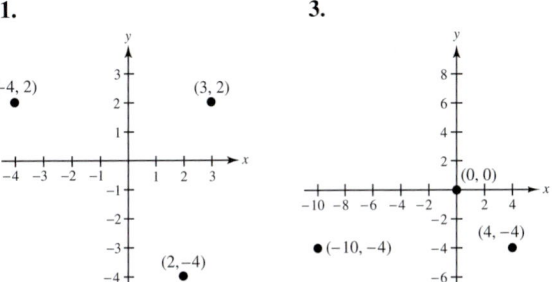

3.

5.

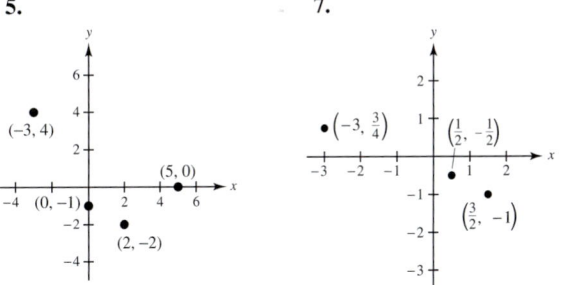

7.

9.

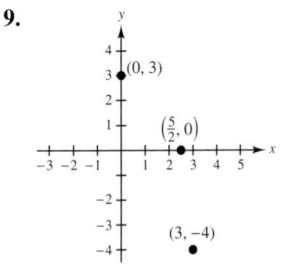

11. A: $(5, 2)$, B: $(-3, 4)$, C: $(2, -5)$, D: $(-2, -2)$

13. A: $(-1, 3)$, B: $(5, 0)$, C: $(2, 1)$, D: $(-1, -2)$

15. Quadrant II   **17.** Quadrant III   **19.** Quadrant III

21. Quadrant II or III   **23.** Quadrant III or IV

25. Quadrant II or IV

27.

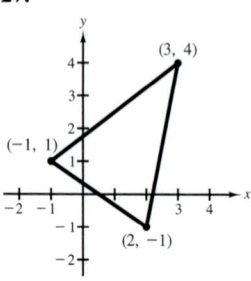

29.

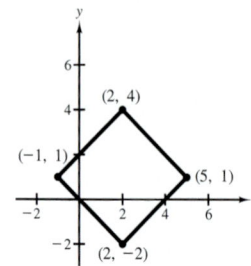

31.

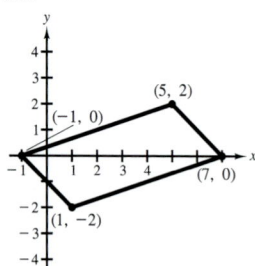

33.

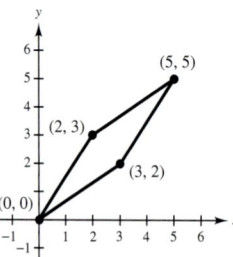

35.

x	-2	0	2	4	6
$y = 3x - 4$	-10	-4	2	8	14

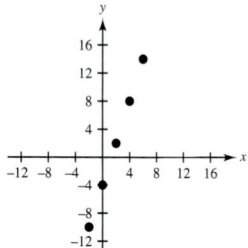

37.

x	-4	-2	4	6	8
$y = -\frac{3}{2}x + 5$	11	8	-1	-4	-7

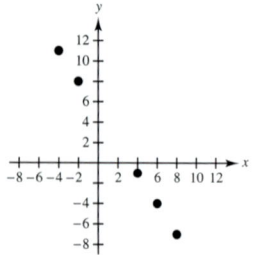

39.

x	-2	-1	0	1	2
$y = 2x - 1$	-5	-3	-1	1	3

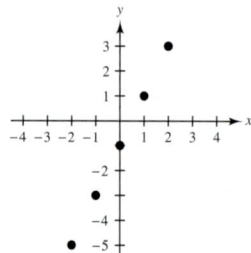

41. $y = -7x + 8$ **43.** $y = 10x - 2$

45. $y = 2x - 1$ **47.** $y = -\frac{1}{4}x + 2$ **49.** $y = \frac{4}{5}x - \frac{3}{5}$

51. (a) Solution   (b) Not a solution

     (c) Not a solution   (d) Solution

53. (a) Solution   (b) Solution

     (c) Not a solution   (d) Solution

55. (a) Not a solution   (b) Solution

     (c) Solution   (d) Not a solution

57. (a) Solution   (b) Not a solution

     (c) Not a solution   (d) Not a solution

59.

x	20	40	60	80	100
$y = 0.066x$	1.32	2.64	3.96	5.28	6.60

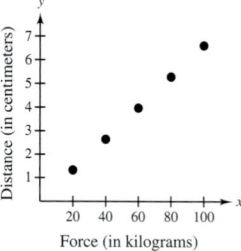

61. $y = 25x + 5000$

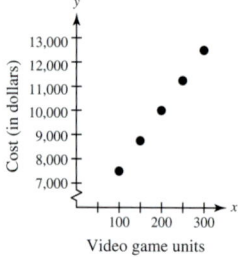

63. (a)

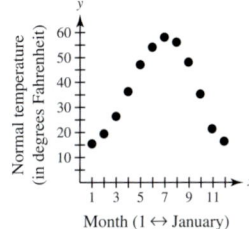

No, because there are only 12 months, but the temperature ranges from 15°F to 58°F.

(b) June, July, August

65. (a)

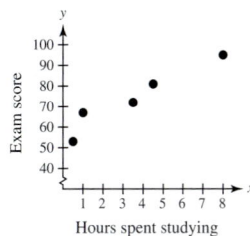

(b) Scores increase with increased study time.

67. 1,380,000   **69.** 150,000; 10%   **71.** $22,500

73. 5%   **75.** 5%

77. (a) (24,000, 980), (7000, 640), (0, 500), (36,000, 1220)

(b)

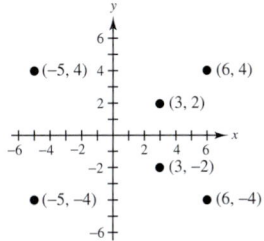

79. First quadrant: $(+, +)$, Second quadrant: $(-, +)$

81. (a) and (b)

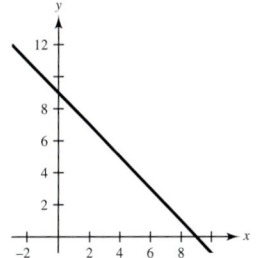

(c) Reflection in the x-axis

83. Order is significant because each number in the pair has a particular interpretation. The first measures horizontal distance and the second measures vertical distance.

85. No. The scales are determined by the magnitudes of the quantities being measured by x and y. If y is measuring revenue for a product and x is measuring time in years, the scale on the y-axis may be in units of $100,000 and the scale on the x-axis may be in units of 1 year.

Section 4.2   *(page 234)*

Review   *(page 234)*

1. $x - 2 + c > 5 + c$ **2.** $(x - 2)c > 5c$

3. 1   **4.** Commutative Property of Addition

5. $-9x + 11y$ **6.** $8z - 4$ **7.** $-y^4 + 2y^2$

8. $10t - 4t^2$ **9.** $3x + 30$ **10.** 0

11. (a) 65 miles   (b) 1 day, since $2(30) > 52.75$

12. 25×15 inches

1. g   **2.** b   **3.** a   **4.** e   **5.** h

6. c   **7.** d   **8.** f

9.

x	-2	-1	0	1	2
y	11	10	9	8	7

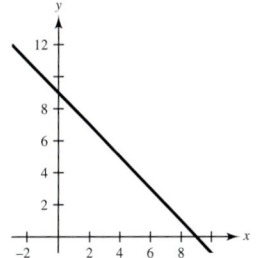

11.

x	-2	0	2	4	6
y	3	2	1	0	-1

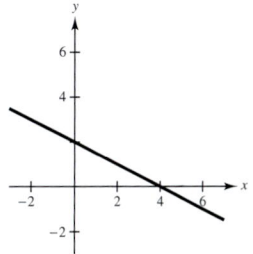

13.

x	-1	0	1	2	3
y	4	1	0	1	4

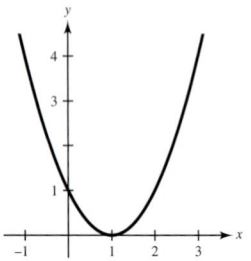

15.

x	-3	-2	-1	0	1
y	2	1	0	1	2

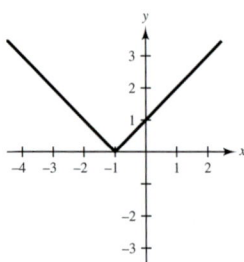

17. $(-2, 0), (0, 4)$ **19.** $(6, 0), (0, 2)$

21. $(-3, 0), (3, 0), (0, -3)$ **23.** $(-4, 0), (4, 0), (0, 16)$

25. $\left(\frac{7}{2}, 0\right), (0, 7)$ **27.** $(2, 0), (0, -1)$

29. $(1, 0), (0, -1)$ **31.** $(2, 0), (0, 4)$

33. $\left(\frac{9}{2}, 0\right), \left(0, \frac{3}{2}\right)$ **35.** $(4, 0), (0, -6)$

37.

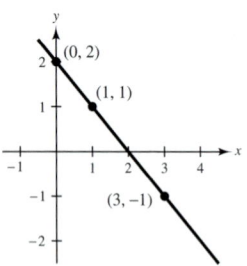

39.

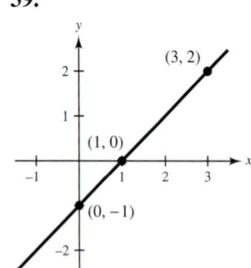

41.

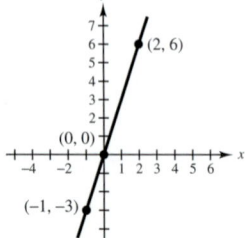

43.

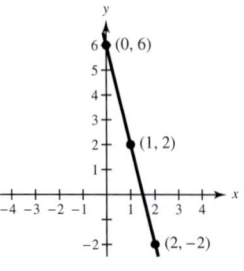

45.

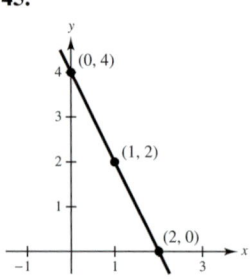

47.

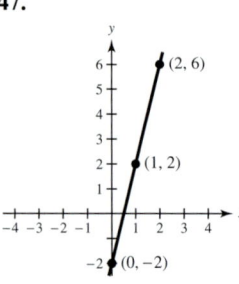

49.

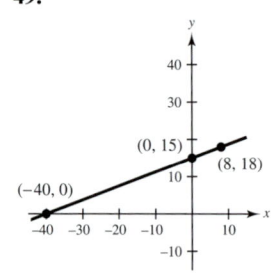

51.

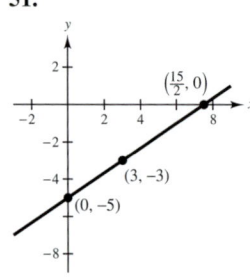

53.

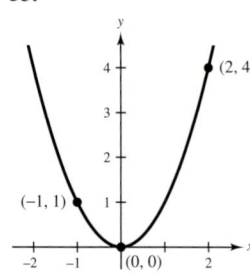

55.

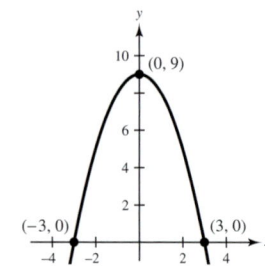

57.

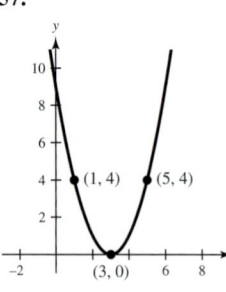

59.

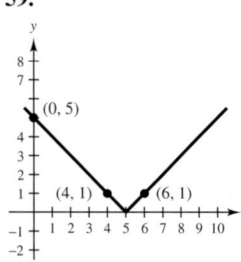

61.

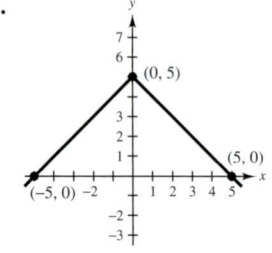

63.

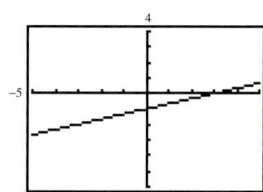

65.

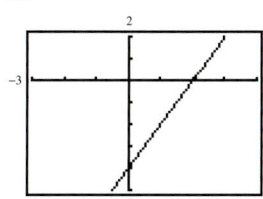

Yes; Commutative
Property of Addition

Yes; Distributive
Property

67.

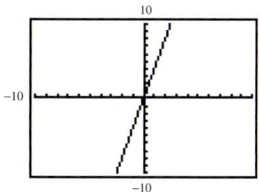

69.

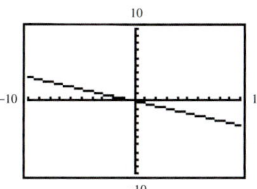

71.

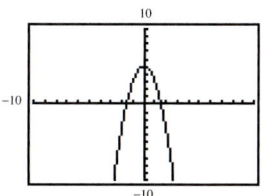

73.

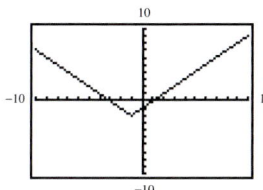

75.

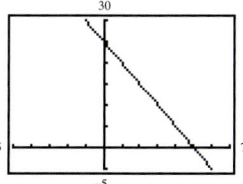

77.

| Xmin = -15 |
| Xmax = 15 |
| Xscl = 1 |
| Ymin = -10 |
| Ymax = 10 |
| Yscl = 1 |

79.

| Xmin = -5 |
| Xmax = 20 |
| Xscl = 5 |
| Ymin = -5 |
| Ymax = 20 |
| Yscl = 5 |

81. $y = 35t$

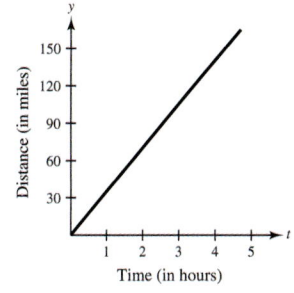

83. (a)

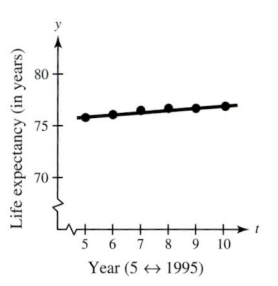

(b) 79.0 years

85. $1000

87. The set of all solutions of an equation plotted on a rectangular coordinate system is called its graph.

89. Substitute the coordinates for the respective variables in the equation and determine if the equation is true.

91.

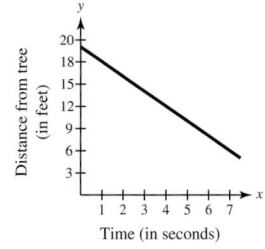

Section 4.3 *(page 243)*

Review *(page 243)*

1. $a < c$ Transitive Property

2. $\dfrac{7x}{7} = \dfrac{21}{7}$; $x = 3$ **3.** $11s - 5t$ **4.** $-x^2 + 1$

5. $x - 4$ **6.** $3x^2y - xy^2 - 5xy$ **7.** $x = -3$

8. $x = \dfrac{3}{5}$ **9.** $x = 28$ **10.** $x = 16$

11. 9.2%   **12.** 21, 23

1. Domain: $\{-4, 1, 2, 4\}$; Range: $\{-3, 2, 3, 5\}$

3. Domain: $\left\{-9, \frac{1}{2}, 2\right\}$; Range: $\{-10, 0, 16\}$

5. Domain: $\{-1, 1, 5, 8\}$; Range: $\{-7, -2, 3, 4\}$

7. Function   **9.** Not a function   **11.** Function

13. Not a function   **15.** Not a function   **17.** Function

19. Function   **21.** Not a function   **23.** Function

25. Function   **27.** Function   **29.** Function

31. Function   **33.** Not a function   **35.** Not a function

37. (a) 1 (b) $\frac{5}{2}$ (c) -2 (d) $-\frac{1}{3}$

39. (a) -1 (b) 5 (c) -7 (d) -2

41. (a) 5 (b) -3 (c) -15 (d) $-\frac{13}{3}$

43. (a) 49   (b) -4 (c) 1   (d) $-\frac{13}{8}$

45. (a) 8   (b) 8   (c) 0   (d) 2

47. (a) 1   (b) 15   (c) 0   (d) 0

49. (a) 4   (b) 0   (c) 12   (d) $\frac{1}{2}$

51. (a) -1 (b) 0   (c) 26   (d) $-\frac{7}{8}$

53. $D = \{0, 1, 2, 3, 4\}$ **55.** $D = \{-2, -1, 0, 1, 2\}$

57. $D = \{-5, -4, -3, -2, -1\}$

59. The set of all real numbers r such that $r > 0$.

61. (a) $f(10) = 15$, $f(15) = 12.5$

     (b) Demand decreases.

63. (a) 100 miles   (b) 200 miles   (c) 500 miles

65. High school enrollment is a function of the year.

67. $f(1996) \approx 14,100,000$

69. $P = 4s$; P is a function of s.

71. (a) L is a function of t.   (b) $9.5 \le L \le 16.5$

73. (c) Yes. Independent variable x represents "Weekly sales." Dependent variable y represents "Weekly earnings."

     (d) Domain: $x \ge 0$; Range: $y \ge 500$

75. No.

77. Check to see that no vertical line intersects the graph at two (or more) points. If this is true, then the equation represents y as a function of x.

79. Yes. For example, $f(x) = 10$ has a domain of $(-\infty, \infty)$, an infinite number of elements, whereas the range has only one element, 10.

Mid-Chapter Quiz   *(page 248)*

1.

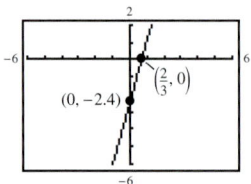

2. Quadrants I and II

3. (a) Solution   (b) Solution   (c) Not a solution

4. 1995: 340 million

     1996: 410 million

     1997: 530 million

     1998: 670 million

     1999: 810 million

     2000: 1040 million

5. $(12, 0), (0, -4)$ **6.** $\left(\frac{2}{7}, 0\right), (0, 2)$

7.   **8.**

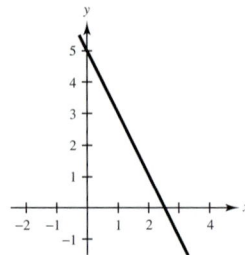 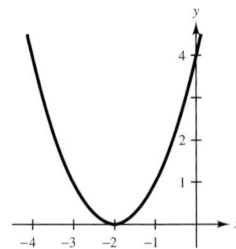

9.

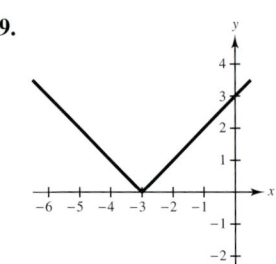

10. Domain: $\{1, 2, 3\}$; Range: $\{0, 4, 6, 10, 14\}$

11. Domain: $\{-3, -2, -1, 0\}$; Range: $\{6\}$

12. Not a function   **13.** (a) 2   (b) -7

14. (a) 3   (b) -60 **15.** $D = \{10, 15, 20, 25\}$

16. Substitute the coordinates for the respective variables in the equation and determine if the equation is true.

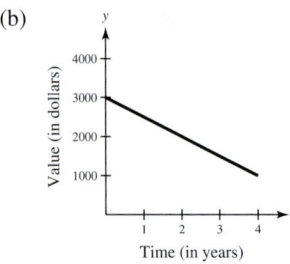

17. (a) $y = 3000 - 500t$

     (b)

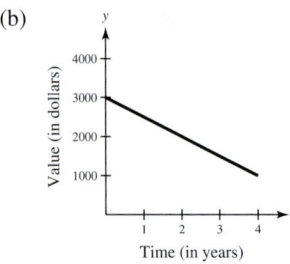

     (c) $(0, 3000)$; The value of the computer system when it is first introduced into the market

Section 4.4   (page 258)

Review   (page 258)

1. equivalent equations   **2.** 2   **3.** x^5 **4.** z^4

5. $-y^3$ **6.** $5x^7$ **7.** $50x^5$ **8.** $18y^2z^4$

9. $x + 2$ **10.** $x^2 - 4x - 2$

11. 2 feet, 2 feet, 6 feet   **12.** 1.5 hours

1. 1   **3.** 0   **5.** $-\frac{1}{3}$ **7.** Undefined

9. $\frac{5}{4}$ **11.** (a) L_2 (b) L_3 (c) L_4 (d) L_1

13.

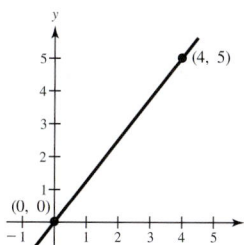

$m = \frac{5}{4}$; The line rises.

15.

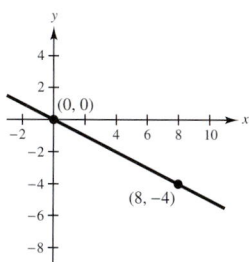

$m = -\frac{1}{2}$; The line falls.

17.

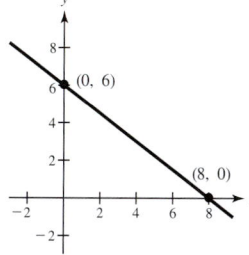

$m = -\frac{3}{4}$; The line falls.

19.

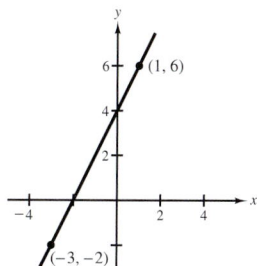

$m = 2$; The line rises.

21.

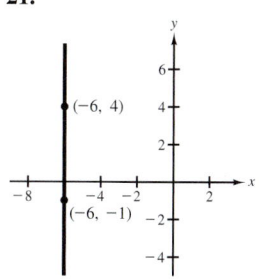

m is undefined.
The line is vertical.

23.

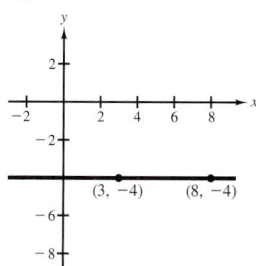

$m = 0$
The line is horizontal.

25.

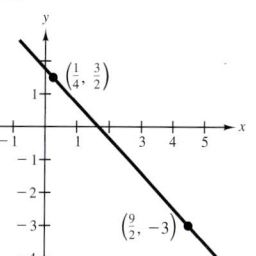

$m = -\frac{18}{17}$; The line falls.

27.

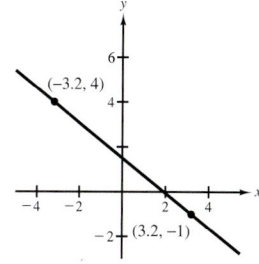

$m = -\frac{25}{32}$; The line falls.

29.

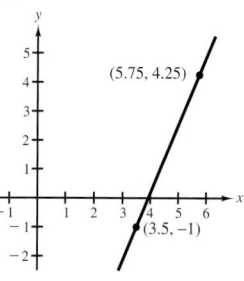

$m = \frac{7}{3}$; The line rises.

31.

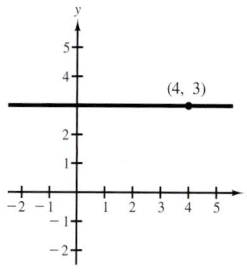

$m = 0$; The line is horizontal.

33. $m = -2$

x	-2	0	2	4
y	2	-2	-6	-10
Solution point	$(-2, 2)$	$(0, -2)$	$(2, -6)$	$(4, -10)$

35. $y = 22$ **37.** $y = -\frac{43}{2}$

39.

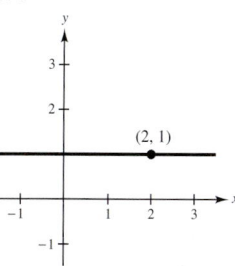

$(0, 1), (1, 1)$

41.

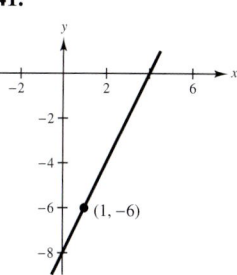

$(2, -4), (3, -2)$

43.

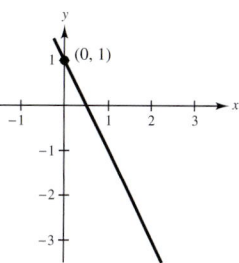

$(1, -1), (2, -3)$

45.

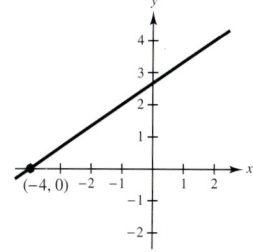

$(-1, 2), (2, 4)$

47.

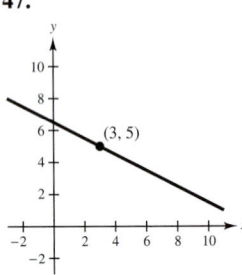

(5, 4), (7, 3)

49.

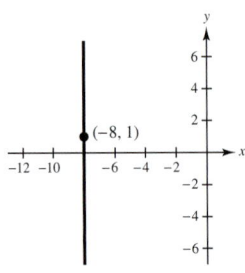

(−8, 0), (−8, −1)

67. $y = 2x - 3$

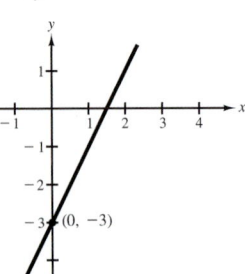

69. $y = \frac{1}{3}x + 2$

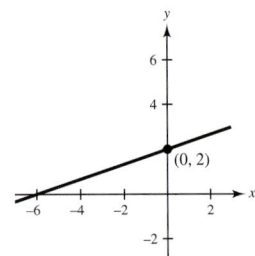

51.

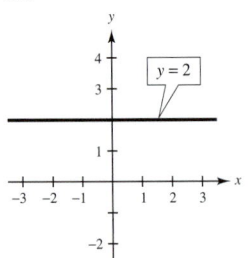

53.

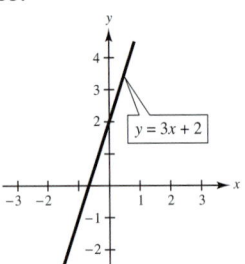

71. $y = -\frac{1}{2}x + 1$

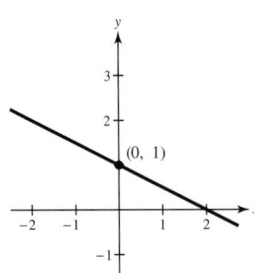

73. $y = \frac{3}{4}x + \frac{1}{2}$

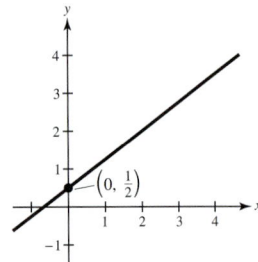

55.

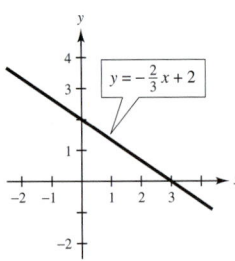

57.

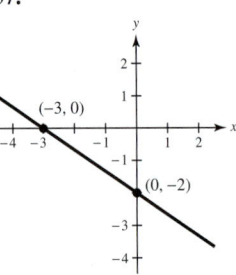

75. $y = -5$

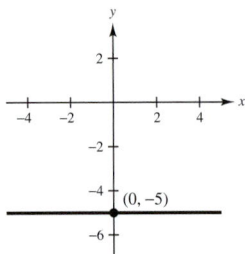

59.

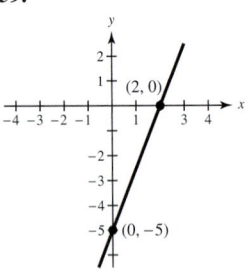

61.

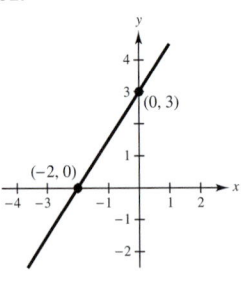

77. Perpendicular   **79.** Parallel

81.   **83.**

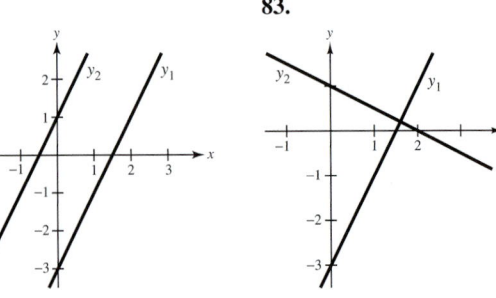

Parallel   Perpendicular

85. $\frac{2}{5}$

87. (a)

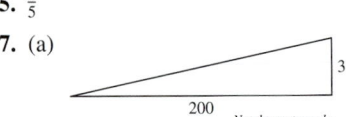

Not drawn to scale

(b) $\frac{3}{200}$ (c) Yes; $\left|\frac{3}{100}\right| > \left|\frac{3}{200}\right|$

89. $\frac{1}{5}$

63. $y = x$

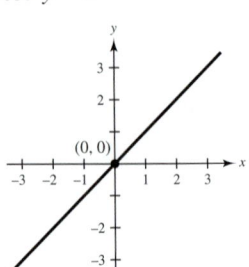

65. $y = -\frac{1}{2}x$

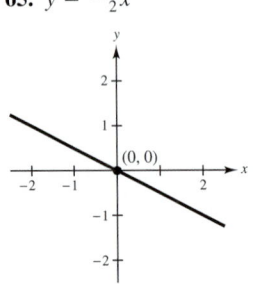

91. (a) 11.3, 13.1, 19.6, 27.4

   (b) 17.85 is the average annual increase in net sales from 1996 to 2000.

93. Sales increase by 76 units.

95. Sales increase by 18 units.

97. Sales decrease by 14 units.

99. Yes. The slope is the ratio of the change in y to the change in x.

101. False. Both the x- and y-intercepts of the line $y = -x + 5$ are positive, but the slope is negative.

103. No. The slopes of nonvertical perpendicular lines have opposite signs. The slopes are the negative reciprocals of each other.

105. The slope

107. Yes. You are free to label either one of the points as (x_1, y_1) and the other as (x_2, y_2). However, once this is done, you must form the numerator and denominator using the same order of subtraction.

Section 4.5   *(page 270)*

Review   *(page 270)*

1. 60; The greatest common factor is the product of the common prime factors.

2. 900; The least common multiple is the product of the highest powers of the prime factors of the numbers.

3. $12 - 8x$ **4.** x^3y^3 **5.** $x + 10$ **6.** 1

7. $y = -3x + 4$ **8.** $y = x + 4$ **9.** $y = \frac{4}{5}x + \frac{2}{5}$

10. $y = -\frac{3}{4}x + \frac{5}{4}$

1. $2x + y = 0$

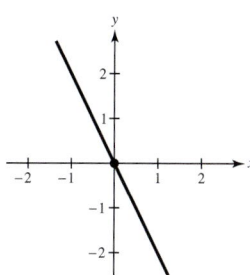

3. $x - 2y = 6$

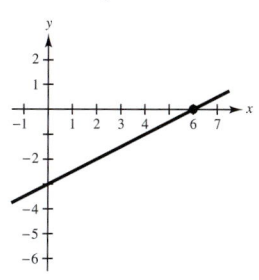

5. $2x - y = -5$

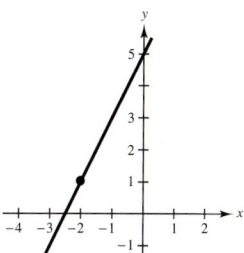

7. $x + 4y = -12$

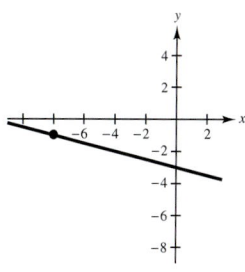

9. $y = -3$

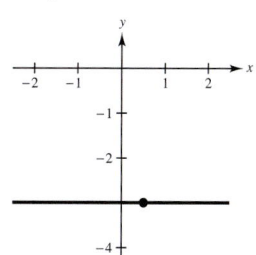

11. $4x - 6y = -9$

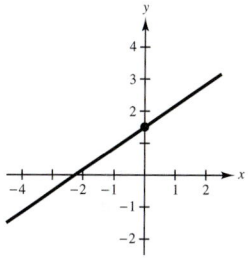

13. $4x + 5y = 28$

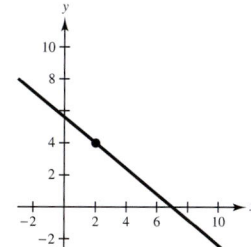

15. $y = 3x - 4$ **17.** $y = -2x$ **19.** $y = -\frac{1}{3}x + 3$

21. $y = 4$ **23.** $y = -\frac{3}{4}x + 7$ **25.** $y = \frac{2}{3}x + \frac{7}{3}$

27. $\frac{3}{8}$ **29.** 5   **31.** $\frac{2}{3}$ **33.** 0   **35.** Undefined

37. $\frac{3}{2}$ **39.** $y = \frac{1}{2}x + 2$ **41.** $y = -3x - 1$

43. $y - 2 = -\frac{1}{2}(x + 1)$

45. $y + 1 = \frac{1}{3}(x + 2)$ or $y - 1 = \frac{1}{3}(x - 4)$

47. $y = -x$

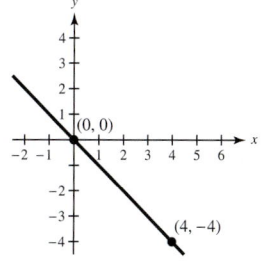

49. $y = -2x$

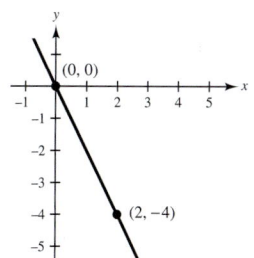

51. $y = -x + 1$

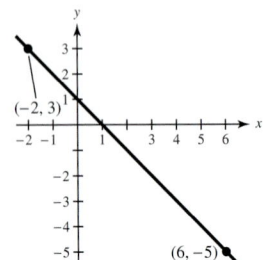

53. $y = \frac{1}{3}x + 4$

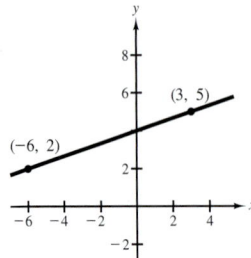

55. $y = -\frac{3}{2}x + \frac{13}{2}$

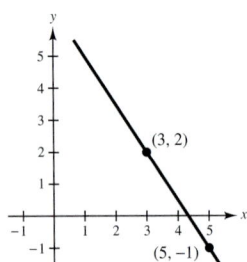

57. $y = 4x - 11$

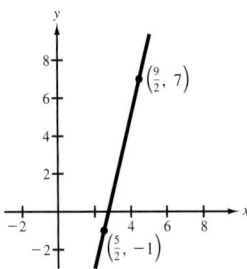

59. $x + y - 3 = 0$ **61.** $3x + 5y - 10 = 0$

63. $2x - y - 6 = 0$ **65.** $3x + 2y - 13 = 0$

67. $3x + 5y - 31 = 0$ **69.** $8x + 6y - 19 = 0$

71. $6x + 5y - 9 = 0$

73. (a) $x - y - 1 = 0$ (b) $x + y - 3 = 0$

75. (a) $3x + 4y + 20 = 0$ (b) $4x - 3y + 60 = 0$

77. (a) $2x + y - 5 = 0$ (b) $x - 2y + 5 = 0$

79. (a) $y = 0$ (b) $x + 1 = 0$

81. (a) $2x - 3y - 11 = 0$ (b) $3x + 2y - 10 = 0$

83. $x = -2$ **85.** $y = \frac{2}{3}$ **87.** $x = 4$

89. $y = -8$

91.

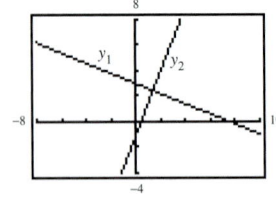

Perpendicular

93.

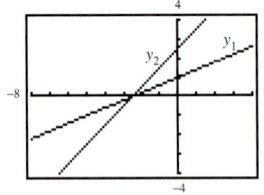

Neither

95.

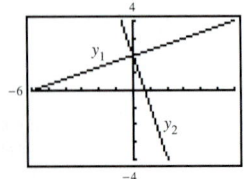

y_1 and y_2 are perpendicular.

97. $W = 2000 + 0.02S$ **99.** $C = 225 + 0.35x$

101. $d = 50t$

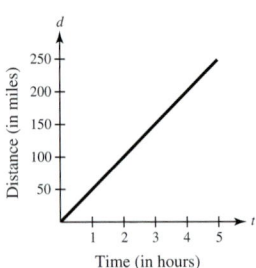

103. (a) $V = -2300t + 25{,}000$ (b) $\$18{,}100$

105. (a) $(50, 580), (47, 625)$

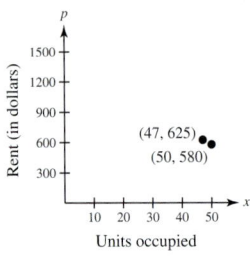

(b) $p = -15x + 1330$; As the rent increases, the demand decreases.

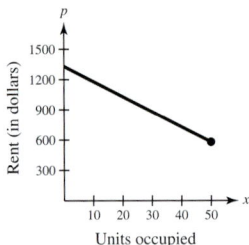

(c) 45 units   (d) 49 units

107. (a) (f): $m = -10$; Loan decreases by $\$10$ per week.

(b) (e): $m = 1.50$; Pay increases by $\$1.50$ per unit.

(c) (g): $m = 0.32$; Amount increases by $\$0.32$ per mile.

(d) (h): $m = -100$; Annual depreciation is $\$100$.

109. (e) The function is linear if the slopes are the same between the points (x, y), where x is the weekly sales and y is weekly earnings.

(f) $m = 0.02$; 2%; Commission rate

(g) $y = 500 + 0.02x$

(h)

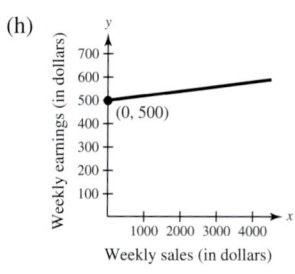

Weekly earnings (in dollars) vs Weekly sales (in dollars)

(0, 500)

(0, 500) The y-intercept is the weekly earnings when no ads are sold. (−25,000, 0) The x-intercept does not have meaning.

111. No. The slope is undefined.

113. The coordinates of a point on the line

115. $-\dfrac{5}{7}, -\dfrac{a}{b}$ **117.** Answers will vary.

Section 4.6 *(page 280)*

Review *(page 280)*

1. <   **2.** <   **3.** >   **4.** <

5. $x > -3$

−5 −4 −3 −2 −1   0

6. $x \le 2$

−1   0   1   2   3   4

7. $t \le 8$

0   2   4   6   8   10

8. $y < 8$

0   2   4   6   8   10

9. $x > 11.5$

9   10   11   12   13   14

10. $x \le 2$

0   1   2   3   4

11. $12,100.00   **12.** $\frac{12}{7}$ hours

1. (a) Not a solution   (b) Solution

   (c) Not a solution   (d) Solution

3. (a) Solution   (b) Solution

   (c) Solution   (d) Not a solution

5. (a) Solution   (b) Not a solution

   (c) Solution   (d) Not a solution

7. (a) Solution   (b) Solution

   (c) Solution   (d) Solution

9. Dashed   **11.** Solid   **13.** b   **14.** c   **15.** d

16. a   **17.** c   **18.** a   **19.** b   **20.** d

21.

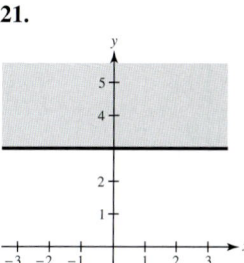

23.

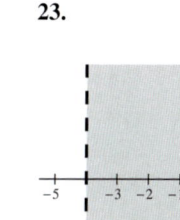

25.

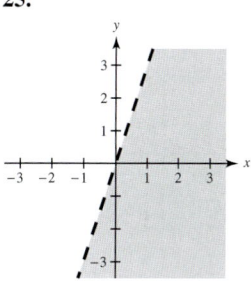

27.

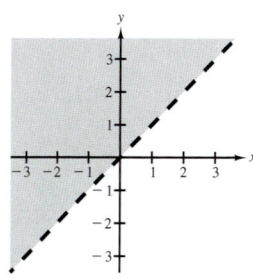

29.

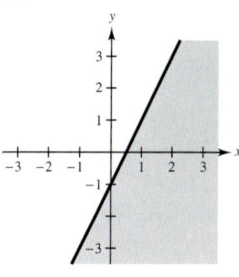

31.

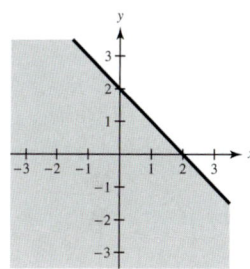

33.

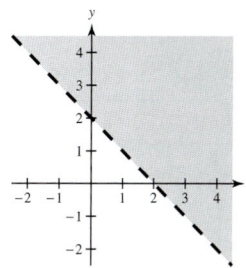

35.

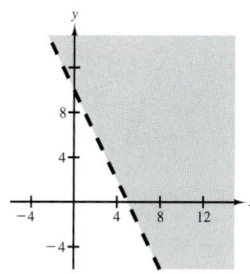

37.

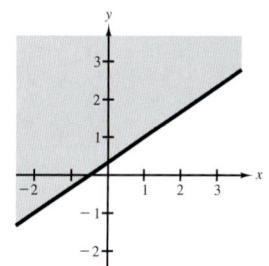

39.

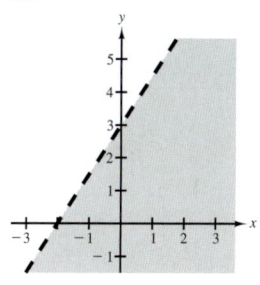

41.

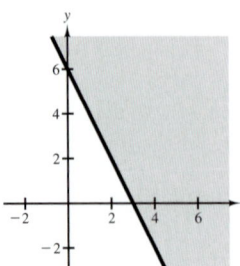

43.

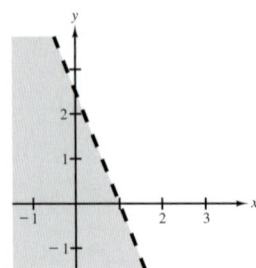

45.

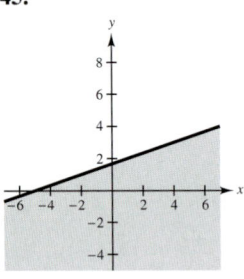

47.

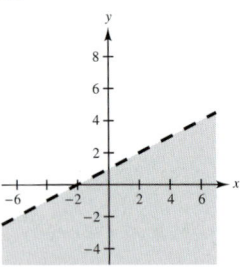

49.

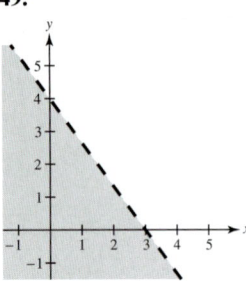

51.

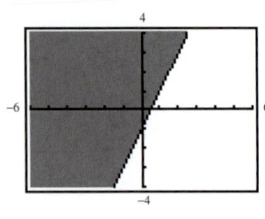

53.

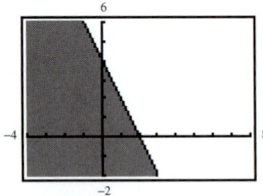

55.

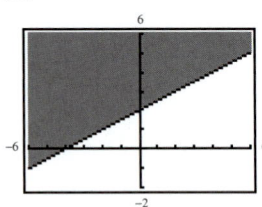

57.

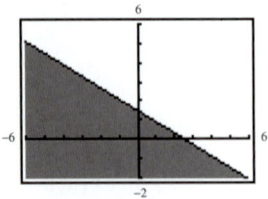

59. $y \geq 2$ **61.** $2x + y \leq 2$ **63.** $2x - y > 0$

65. $9x + 6y \geq 150$; (x, y): $(20, 0)$, $(10, 15)$, $(5, 30)$

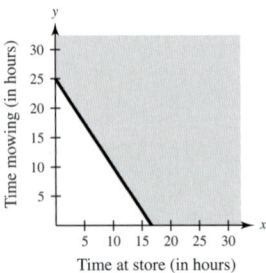

67. $T + \frac{3}{2}C \leq 12$; (T, C): $(5, 4)$, $(2, 6)$, $(0, 8)$

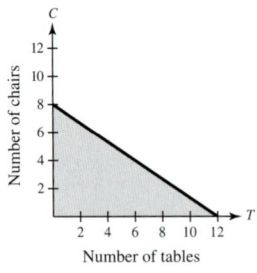

69. $2w + t \geq 60$; (w, t): $(30, 0)$, $(20, 25)$, $(0, 60)$

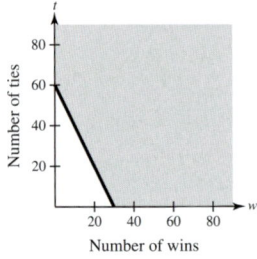

71. (i) At least $17,000

73. The inequality is true when x_1 and y_1 are substituted for x and y, respectively.

75. Test a point in one of the half-planes.

77. $y > 0$ **79.** $x + y < 0$

Review Exercises *(page 285)*

1.

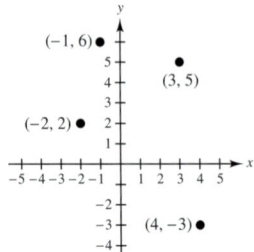

3.

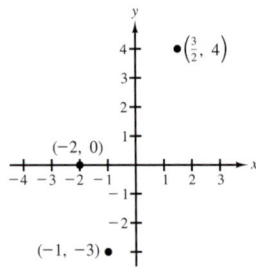

5. A: $(3, -2)$; B: $(0, 5)$; C: $(-1, 3)$; D: $(-5, -2)$

7. Quadrant II   **9.** x-axis   **11.** Quadrant II

13. Quadrant II or III

15.

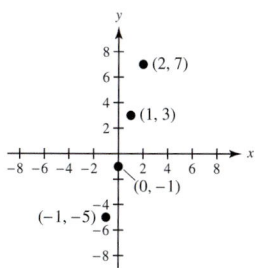

x	-1	0	1	2
$y = 4x - 1$	-5	-1	3	7

17. $y = -\frac{3}{4}x + 3$ **19.** $y = \frac{1}{2}x - 4$

21. (a) Solution   (b) Not a solution

(c) Not a solution   (d) Not a solution

23. (a) Solution   (b) Solution

(c) Not a solution   (d) Solution

25. (a)

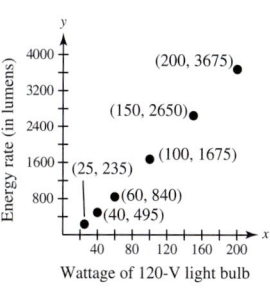

(b) Approximately linear

27.

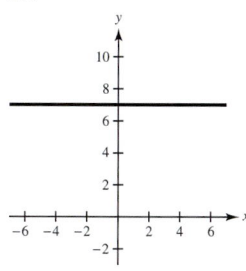

29.

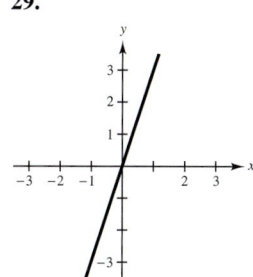

31.

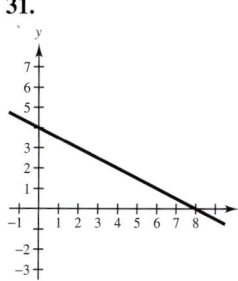

33.

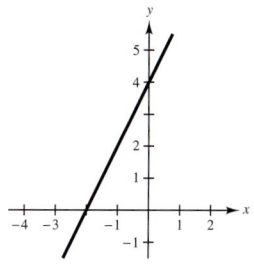

35.

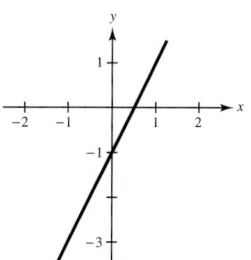

37.

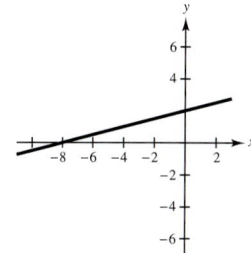

39.

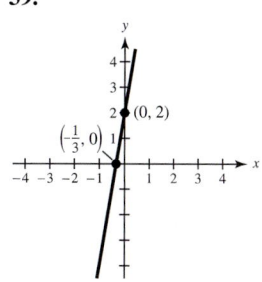

41.

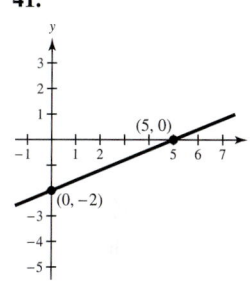

43.

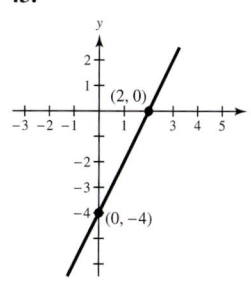

45.

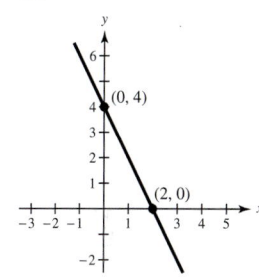

47. $C = 3x + 125$

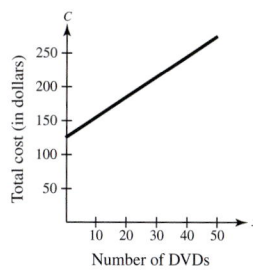

49. Domain: $\{-2, 3, 5, 8\}$; Range: $\{1, 3, 7, 8\}$

51. Domain: $\{-4, -2, 2, 7\}$; Range: $\{-3, -2, 0, 3\}$

53. Function   **55.** Not a function   **57.** Function

59. Not a function   **61.** Function

63. (a) -25 (b) 175 (c) 250 (d) $-\frac{100}{3}$

65. (a) 64 (b) 63 (c) 48 (d) 0

67. (a) 3 (b) 13 (c) 5 (d) 0

69. (a) 38 (b) 30 (c) 20 **71.** $D = \{1, 2, 3, 4, 5\}$

73. $D = \{-2, -1, 0, 1, 2\}$ **75.** $\frac{1}{2}$

77.

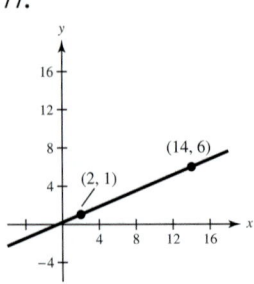

$m = \frac{5}{12}$; The line rises.

79.

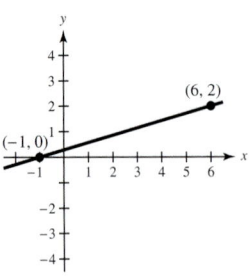

$m = \frac{2}{7}$; The line rises.

m is undefined; The line is vertical.

81.

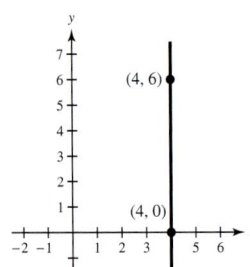

83.

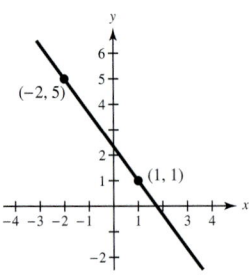

$m = -\frac{4}{3}$; The line falls.

85.

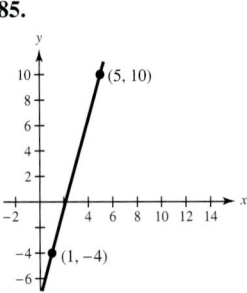

$m = \frac{7}{2}$; The line rises.

$m = -3$; The line falls.

87.

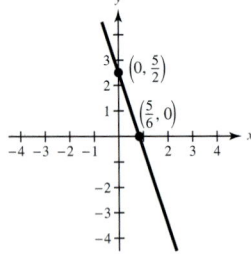

89. $\frac{2}{3}$

91. $y = 2x + 1$

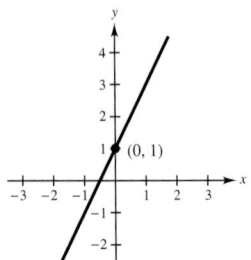

93. $y = -3x + 2$

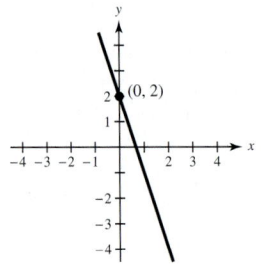

95. $y = -\frac{1}{2}x + 2$

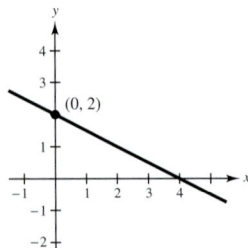

97. $y = \frac{2}{5}x + 1$

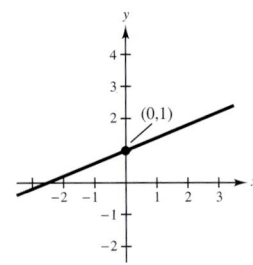

99. Parallel   **101.** Neither   **103.** $y = 2x - 9$

105. $y = -4x + 6$ **107.** $y = \frac{4}{5}x + 2$

109. $y = -\frac{8}{3}x + \frac{1}{3}$ **111.** $x = 3$

113. $x + 2y + 4 = 0$ **115.** $y - 8 = 0$

117. $x - y + 3 = 0$ **119.** $25x - 20y + 6 = 0$

121. (a) $2x + 3y + 3 = 0$ (b) $3x - 2y + 24 = 0$

123. (a) $8x + 6y - 27 = 0$ (b) $24x - 32y + 119 = 0$

125. $y = 5$ **127.** $x = 5$ **129.** $W = 2500 + 0.07S$

131. (a) Not a solution   (b) Not a solution

(c) Solution   (d) Solution

133.

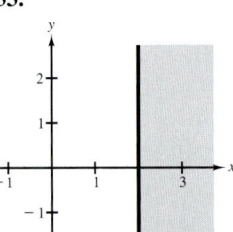

135.

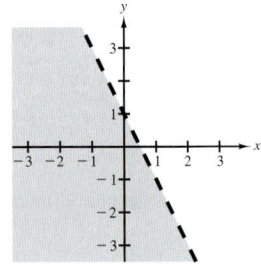

137.

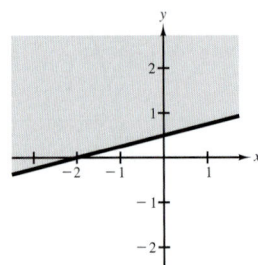

139. $y < 2$ **141.** $y \leq x + 1$

143. $2x + 3y \leq 120$; (x, y): $(10, 15)$, $(20, 20)$, $(30, 20)$

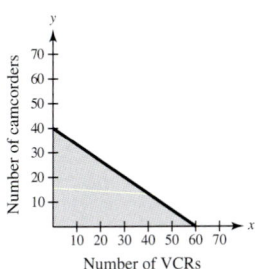

Chapter Test *(page 291)*

1.

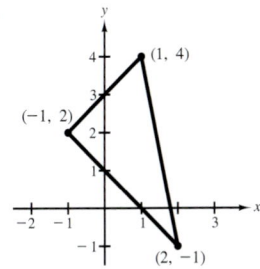

2. (a) Not a solution   (b) Solution

(c) Solution   (d) Not a solution

3. 0   **4.** $(-4, 0)$, $(0, 3)$

5.

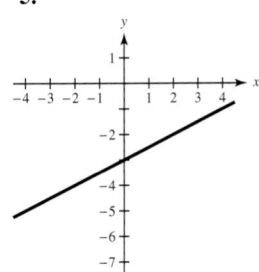

6.

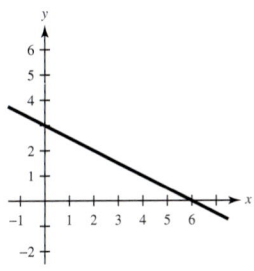

7.

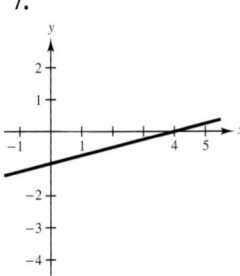

8.

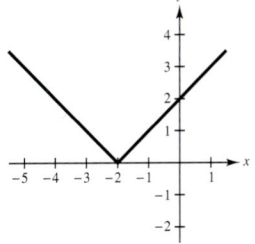

9.

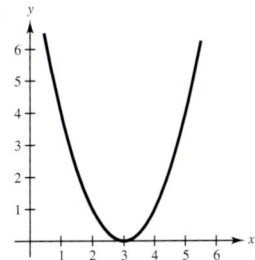

10. No, some input values, 0 and 1, have two different output values.

11. Yes, because it passes the Vertical Line Test.

12. (a) 0   (b) 0   (c) -16 (d) $-\frac{3}{8}$ **13.** $\frac{3}{14}$

14. $(-2, 2)$, $(-1, 0)$

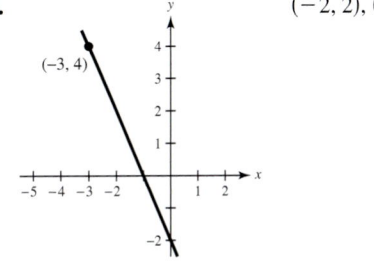

15. $-\frac{5}{3}$

16. $3x + 8y - 48 = 0$

17. $x = 3$

18. (a) Solution   (b) Solution

(c) Solution   (d) Solution

19.   **20.**

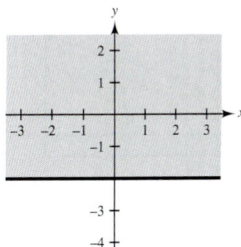   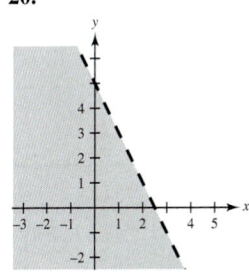

21.   **22.**

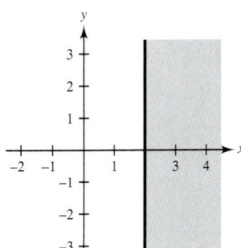   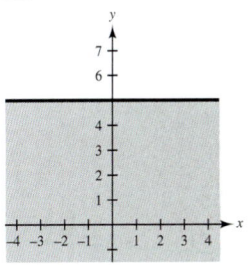

23. Sales are increasing at a rate of 230 units per year.

Chapter 5

Section 5.1 *(page 300)*

Review *(page 300)*

1. The graph of an equation is the set of solution points of the equation on a rectangular coordinate system.

2. Create a table of solution points of the equation, plot those points on a rectangular coordinate system, and connect the points with a smooth curve or line.

3. $(2, 0), (6, 2)$

4. To find the x-intercept, let $y = 0$ and solve the equation for x. To find the y-intercept, let $x = 0$ and solve the equation for y.

5. (a) -15 (b) $-\frac{15}{2}$ **6.** (a) 20 (b) 2

7. (a) 0 (b) $-t^2 + 4t + 5$

8. (a) $\frac{2}{3}$ (b) $\frac{2-z}{6-z}$

9.

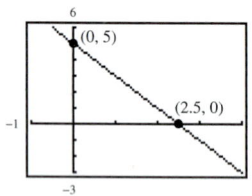

10.

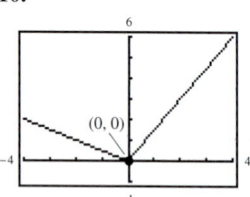

11.

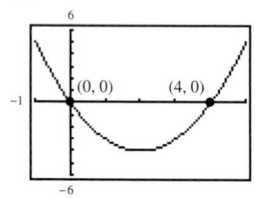

12.

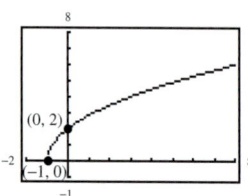

1. (a) $-3x^8$ (b) $9x^7$ **3.** (a) $-125z^6$ (b) $25z^8$

5. (a) $2u^3v^3$ (b) $-4u^9v$ **7.** (a) $-15u^8$ (b) $64u^5$

9. (a) $-m^{19}n^7$ (b) $-m^7n^3$ **11.** (a) $3m^4n^3$ (b) $3m^2n^3$

13. (a) $\frac{9x^2}{16y^2}$ (b) $\frac{125u^3}{27v^3}$ **15.** (a) $\frac{8x^4y}{9}$ (b) $-\frac{2x^2y^4}{3}$

17. (a) $\frac{25u^8v^2}{4}$ (b) $\frac{u^8v^2}{4}$

19. (a) $x^{2n-1}y^{2n-1}$ (b) $x^{2n-2}y^{n-12}$ **21.** $\frac{1}{25}$

23. $-\frac{1}{1000}$ **25.** 1 **27.** 64 **29.** -32 **31.** $\frac{3}{2}$

33. 1 **35.** 1 **37.** 729 **39.** $100,000$ **41.** $\frac{1}{16}$

43. $\frac{1}{64}$ **45.** $\frac{3}{16}$ **47.** $\frac{64}{121}$ **49.** $\frac{16}{15}$ **51.** y^2 **53.** z^2

55. $\frac{7}{x^4}$ **57.** $\frac{1}{64x^3}$ **59.** x^6 **61.** $\frac{4}{3}a$ **63.** t^2

65. $\frac{1}{4x^4}$ **67.** $-\frac{12}{xy^3}$ **69.** $\frac{y^4}{9x^4}$ **71.** $\frac{10}{x}$ **73.** $\frac{x^5}{2y^4}$

75. $\frac{81v^8}{u^6}$ **77.** $\frac{b^5}{a^5}$ **79.** $\frac{1}{2x^8y^3}$ **81.** $6u$ **83.** x^8y^{12}

85. $\frac{2b^{11}}{25a^{12}}$ **87.** $\frac{v^2}{uv^2+1}$ **89.** $\frac{ab}{b-a}$ **91.** 3.6×10^6

93. 4.762×10^7 **95.** 3.1×10^{-4} **97.** 3.81×10^{-8}

99. 5.73×10^7 **101.** 9.4608×10^{12} **103.** 8.99×10^{-2}

105. $60,000,000$ **107.** 0.0000001359

109. $38,757,000,000$ **111.** $15,000,000$

113. 0.00000000048 **115.** 6.8×10^5 **117.** 2.5×10^9

119. 6×10^6 **121.** 9×10^{15} **123.** 1.6×10^{12}

125. 3.46×10^{10} **127.** 4.70×10^{11} **129.** 1.67×10^{14}

131. 2.74×10^{20} **133.** 9.3×10^7 miles

135. 1.59×10^{-5} year ≈ 8.4 minutes   **137.** 3.33×10^5

139. \$20,469   **141.** $3x$ is the base and 4 is the exponent.

143. Change the sign of the exponent of the factor.

145. When the numbers are very large or very small

Section 5.2 *(page 310)*

Review *(page 310)*

1. An algebraic expression is a collection of letters (variables) and real numbers (constants) combined by using addition, subtraction, multiplication, or division.

2. The terms of an algebraic expression are those parts separated by addition.

3. $10x - 10$ **4.** $12 - 8z$ **5.** $-2 + 3x$

6. $-50x + 75$ **7.** $5x - 2y$ **8.** $\frac{1}{6}x + 8$

9. $7x - 16$ **10.** $-12x - 6$

11.   **12.**

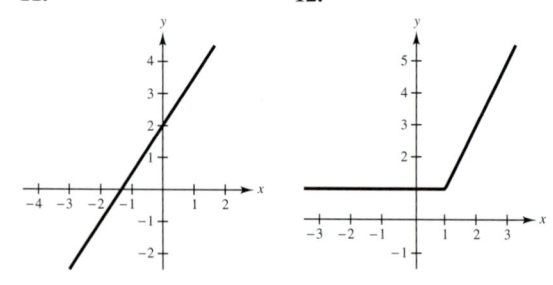

1. Polynomial

3. Not a polynomial because the exponent in the first term is not an integer.

5. Not a polynomial because the exponent is negative.

7. Polynomial

9. Standard form: $12x + 9$
Degree: 1
Leading coefficient: 12

11. Standard form: $-5x^2 + 7x + 10$
Degree: 2
Leading coefficient: -5

13. Standard form: $2x^5 - x^2 + 8x - 1$
Degree: 5
Leading coefficient: 2

15. Standard form: 10
Degree: 0
Leading coefficient: 10

17. Standard form: $-16t^2 + v_0 t$
Degree: 2
Leading coefficient: -16

19. Binomial   **21.** Monomial   **23.** Trinomial

25. $5x^3 - 10$ **27.** $3y^2$ **29.** $x^6 - 4x^3 - 2$

31. $14x + 6$ **33.** $4z^2 - z - 2$ **35.** $2b^3 - b^2$

37. $4b^2 - 3$ **39.** $\frac{3}{2}y^2 + \frac{5}{4}$ **41.** $1.6t^3 - 3.4t^2 - 7.3$

43. $5x + 13$ **45.** $-x - 28$

47. $2x^3 + 2x^2 + 8$ **49.** $3x^4 - 2x^3 - 3x^2 - 5x$

51. $3x^2 + 2$ **53.** $4x^2 + 2x + 2$ **55.** $5y^3 + 12$

57. $9x - 11$ **59.** $x^2 - 2x + 2$ **61.** $-3x^3 + 1$

63. $-u^2 + 5$ **65.** $-3x^5 - 3x^4 + 2x^3 - 6x + 6$

67. $x - 1$ **69.** $-x^2 - 2x + 3$

71. $-2x^4 - 5x^3 - 4x^2 + 6x - 10$ **73.** $-2x^3$

75. $4t^3 - 3t^2 + 15$ **77.** $5x^3 - 6x^2$

79. $3x^3 + 4x + 10$ **81.** $-2x - 20$

83. $3x^3 - 2x + 2$ **85.** $2x^4 + 9x + 2$

87. $8x^3 + 29x^2 + 11$ **89.** $12z + 8$

91. $4t^2 + 20$ **93.** $6v^2 + 90v + 30$ **95.** $10z + 4$

97. $2x^2 - 2x$ **99.** $21x^2 - 8x$

101. (a) $T = -0.29t^2 - 7.1t + 1585, \quad 5 \le t \le 10$

(b)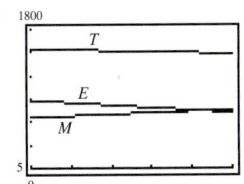

(c) Increasing, decreasing, decreasing

103. (a) Length: $(2x^2)$ inches; Width: $(3x + 5)$ inches

(b) $(4x^2 + 6x + 10)$ inches

(c) Girth: $(8x + 10)$ inches; Length and girth: $(2x^2 + 8x + 10)$ inches; Yes. Substituting 5 for x in $2x^2$, you find that the length is 50 inches. Substituting 5 for x in $2x^2 + 8x + 10$, you find that the sum of the length and girth is 100 inches.

105. (a) Sometimes true. $x^3 - 2x^2 + x + 1$ is a polynomial that is not a trinomial.

(b) True

107. Add (or subtract) their respective coefficients and attach the common variable factor.

109. No. $(x^2 - 2) + (5 - x^2) = 3$

111. To subtract one polynomial from another, add the opposite. You can do this by changing the sign of each of the terms of the polynomial that is being subtracted and then adding the resulting like terms. Examples will vary.

Mid-Chapter Quiz   *(page 314)*

1. $9a^4 b^2$ **2.** $72x^8 y^5$ **3.** $-\dfrac{4}{3x^2 y}$ **4.** $\dfrac{t}{12}$

5. $\dfrac{5}{x^2 y^3}$ **6.** $\dfrac{3yz}{5x^2}$ **7.** $\dfrac{a^6}{9b^4}$ **8.** 1   **9.** 9.46×10^9

10. 0.00000005021

11. Because the exponent of the third term is negative.

12. Degree: 4   **13.** $3x^5 - 3x + 1$
Leading coefficient: -3

14. $y^2 + 6y + 3$ **15.** $-v^3 + v^2 + 6v - 5$

16. $3s - 11$ **17.** $3x^2 + 5x - 4$

18. $5x^4 + 3x^3 - 2x + 2$ **19.** $2x^3 + 6x^2 - 3x - 17$

20. $10x + 36$

Section 5.3   *(page 323)*

Review   *(page 323)*

1. The point represented by $(3, -2)$ is located three units to the right of the y-axis and two units below the x-axis.

2. $(3, 4), (-3, 4), (-3, -4), (3, -4)$ **3.** $\frac{9}{4}x - \frac{5}{2}$

4. $2x - 2$ **5.** $7x - 8$ **6.** $-2y + 14$

7. $-4z + 12$ **8.** $-5u - 5$ **9.** \$29,090.91

10. 1 hour; 5 miles

11.   **12.**

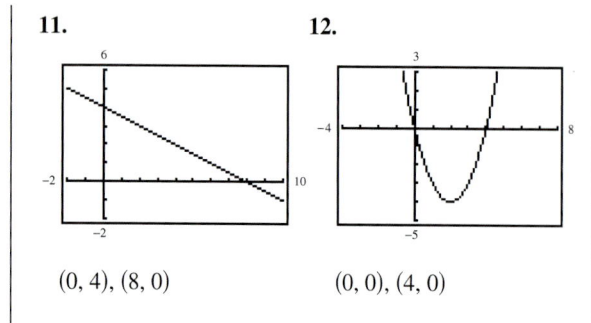

$(0, 4), (8, 0)$ $(0, 0), (4, 0)$

1. $-2x^2$ **3.** $4t^3$ **5.** $\frac{5}{2}x^2$ **7.** $6b^3$ **9.** $3y - y^2$

11. $-x^3 + 4x$ **13.** $-6x^3 - 15x$

15. $-12x - 12x^3 + 24x^4$ **17.** $3x^3 - 6x^2 + 3x$

19. $2x^3 - 4x^2 + 16x$ **21.** $4t^4 - 12t^3$

23. $4x^4 - 3x^3 + x^2$ **25.** $-12x^5 + 18x^4 - 6x^3$

27. $30x^3 + 12x^2$ **29.** $12x^5 - 6x^4$ **31.** $x^2 + 7x + 12$

33. $6x^2 - 7x - 5$ **35.** $2x^2 - 5xy + 2y^2$

37. $2x^2 + 6x + 4$ **39.** $-8x^2 + 18x + 18$

41. $3x^2 - 5xy + 2y^2$ **43.** $3x^3 + 6x^2 - 4x - 8$

45. $2x^4 + 16x^2 + 24$ **47.** $15s + 4$

49. $-x^6 + 8x^3 + 32x^2 - 2x - 8$ **51.** $x^2 + 12x + 20$

53. $2x^2 - x - 10$ **55.** $x^3 + 3x^2 + x - 1$

57. $x^4 - 5x^3 - 2x^2 + 11x - 5$ **59.** $x^3 - 8$

61. $x^4 - 6x^3 + 5x^2 - 18x + 6$

63. $3x^4 - 12x^3 - 5x^2 - 4x - 2$ **65.** $x^2 + x - 6$

67. $8x^5 + 12x^4 - 12x^3 - 18x^2 + 18x + 27$

69. $x^4 - x^2 + 4x - 4$

71. $x^5 + 5x^4 - 3x^3 + 8x^2 + 11x - 12$

73. $x^3 - 6x^2 + 12x - 8$ **75.** $x^4 - 4x^3 + 6x^2 - 4x + 1$

77. $x^3 - 12x - 16$ **79.** $4u^3 + 4u^2 - 5u - 3$

81. $x^2 - 9$ **83.** $x^2 - 400$ **85.** $4u^2 - 9$

87. $16t^2 - 36$ **89.** $4x^2 - 9y^2$ **91.** $16u^2 - 9v^2$

93. $4x^4 - 25$ **95.** $x^2 + 12x + 36$ **97.** $t^2 - 6t + 9$

99. $9x^2 + 12x + 4$ **101.** $64 - 48z + 9z^2$

103. $4x^2 - 20xy + 25y^2$ **105.** $36t^2 + 60st + 25s^2$

107. $x^2 + y^2 + 2xy + 2x + 2y + 1$

109. $u^2 + v^2 - 2uv + 6u - 6v + 9$ **111.** $8x$

113. Yes   **115.** $x^3 + 6x^2 + 12x + 8$

117. $(x^2 + 10x)$ square feet

119. $x^2 + 5x + 4 = (x + 1)(x + 4)$

121. $4x^2 + 6x + 2 = (2x + 1)(2x + 2)$

123. $(x + 2)^2 = x^2 + 4x + 4$; Square of a binomial

125. $(x + 4)(x + 5) = x^2 + 9x + 20$

127. $4x = (x + 4)(x + 3) - x^2 - 3x - 12$

129. (a) $T = 0.00512t^3 - 1.2496t^2 - 14.665t + 6077.43$,
     $0 \le t \le 10$

     (b)

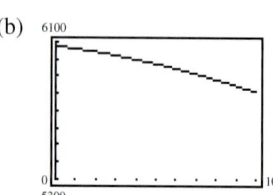

     (c) Approximately 5883 million gallons

131. $500r^2 + 1000r + 500$

133. (d) $(16x^3 + 26x^2 + 10x)$ square inches

     (e) $(3x + 5)^2 = (9x^2 + 30x + 25)$ square inches

     (f) $(9x^3 + 30x^2 + 25x)$ cubic inches

135. Product Rule: $3^2 \cdot 3^4 = 3^{2+4}$

     Product-to-Power Rule: $(5 \cdot 2)^8 = 5^8 \cdot 2^8$

     Power-to-Power Rule: $(2^3)^2 = 2^{3 \cdot 2}$

137. First, Outer, Inner, Last

139. mn. Each term of the first factor must be multiplied by each term of the second factor.

141. (a) $x^2 - 1$ (b) $x^3 - 1$ (c) $x^4 - 1$ (d) $x^5 - 1$

Section 5.4   *(page 334)*

Review   *(page 334)*

1. $\dfrac{120y}{90} = \dfrac{30 \cdot 4y}{30 \cdot 3} = \dfrac{4y}{3}$

2. $(2n + 1)(2n + 3) = 4n^2 + 8n + 3$

3. $(2n + 1) + (2n + 3) = 4n + 4$

4. $2n(2n + 2) = 4n^2 + 4n$

5. $\frac{3}{4}$ **6.** $\frac{5}{2}$ **7.** $\pm\frac{5}{2}$ **8.** $0, 8$

9. $-7, 6$ **10.** 5

11. $y = 1500 + 0.12x$ **12.** $N = 3500 + 60t$

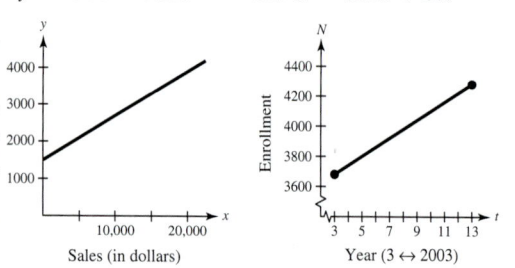

1. $7x^2 - 2x$, $x \neq 0$ **3.** $-4x + 2$, $x \neq 0$

5. $m^3 + 2m - \dfrac{7}{m}$ **7.** $-10z^2 - 6$, $z \neq 0$

9. $4z^2 + \frac{3}{2}z - 1$, $z \neq 0$ **11.** $x^3 - \frac{3}{2}x^2 + 3x - 2$, $x \neq 0$

13. $\frac{5}{2}x - 4 + \frac{7}{2}y$, $x \neq 0$, $y \neq 0$ **15.** $x - 5$, $x \neq 3$

17. $x + 10$, $x \neq -5$ **19.** $x - 3 + \dfrac{2}{x - 2}$

21. $x + 7$, $x \neq 3$ **23.** $5x - 8 + \dfrac{19}{x + 2}$

25. $4x + 3 - \dfrac{11}{3x + 2}$ **27.** $6t - 5$, $t \neq \frac{5}{2}$

29. $y + 3$, $y \neq -\frac{1}{2}$ **31.** $x^2 + 4$, $x \neq 2$

33. $3x^2 - 3x + 1 + \dfrac{2}{3x + 2}$ **35.** $2 + \dfrac{5}{x + 2}$

37. $x - 4 + \dfrac{32}{x + 4}$ **39.** $\frac{6}{5}z + \frac{41}{25} + \dfrac{41}{25(5z - 1)}$

41. $4x - 1$, $x \neq -\frac{1}{4}$ **43.** $x^2 - 5x + 25$, $x \neq -5$

45. $x + 2$ **47.** $4x^2 + 12x + 25 + \dfrac{52x - 55}{x^2 - 3x + 2}$

49. $x^5 + x^4 + x^3 + x^2 + x + 1$, $x \neq 1$

51. $x^3 - x + \dfrac{x}{x^2 + 1}$ **53.** $2x$, $x \neq 0$

55. $7uv$, $u \neq 0$, $v \neq 0$ **57.** $x + 3$, $x \neq 2$

59. $x^2 - x + 4 - \dfrac{17}{x + 4}$

61. $x^3 - 2x^2 - 4x - 7 - \dfrac{4}{x - 2}$

63. $5x^2 + 14x + 56 + \dfrac{232}{x - 4}$

65. $10x^3 + 10x^2 + 60x + 360 + \dfrac{1360}{x - 6}$

67. $0.1x + 0.82 + \dfrac{1.164}{x - 0.2}$ **69.** $(x - 3)(x^2 + 3x - 4)$

71. $(x - 1)(6x^2 - 7x + 2)$

73. $(x + 3)(x^3 + 4x^2 - 9x - 36)$

75. $\left(x - \frac{4}{5}\right)(15x + 10)$ **77.** -8

79.
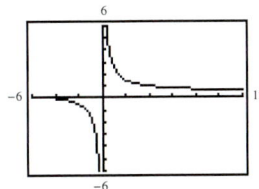

81. $x^{2n} + x^n + 4$, $x^n \neq -2$ **83.** $x^3 - 5x^2 - 5x - 10$

85. $f(k)$ equals the remainder when dividing by $(x - k)$.

k	$f(k)$	Divisor $(x - k)$	Remainder
-2	-15	$x + 2$	-15
-1	0	$x + 1$	0
0	1	x	1
$\frac{1}{2}$	0	$x - \frac{1}{2}$	0
1	0	$x - 1$	0
2	9	$x - 2$	9

87. $x^2 - 3$ **89.** $2x + 8$

91. x is not a factor of the numerator.

93. The remainder is 0 and the divisor is a factor of the dividend.

95. True. If $\dfrac{n(x)}{d(x)} = q(x)$, then $n(x) = d(x) \cdot q(x)$.

97.
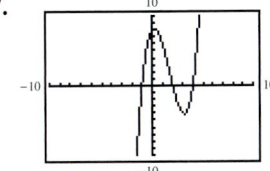

The polynomials in parts (a), (b), and (c) are all equivalent. The x-intercepts are $(-1, 0)$, $(2, 0)$, and $(4, 0)$.

Review Exercises   *(page 339)*

1. x^5 **3.** u^6 **5.** $-8z^3$ **7.** $4u^7v^3$ **9.** $2z^3$

11. $8u^2v^2$ **13.** $144x^4$ **15.** $\frac{1}{72}$ **17.** $\frac{125}{8}$ **19.** $12y$

21. $\dfrac{2}{x^3}$ **23.** 1 **25.** $\dfrac{b^9}{2a^8}$ **27.** $\dfrac{4x^6}{y^5}$ **29.** $\dfrac{405u^5}{v}$

31. 5.38×10^{-5} **33.** $483,300,000$ **35.** 3.6×10^7

37. 500

39. Standard form: $-5x^3 + 10x - 4$;
Degree: 3; Leading coefficient: -5

41. Standard form: $5x^4 + 4x^3 - 7x^2 - 2x$;
Degree: 4; Leading coefficient: 5

43. Standard form: $7x^4 - 1$; Degree: 4; Leading coefficient: 7

45. Standard form: -2; Degree: 0; Leading coefficient: -2

47. $x^4 + x^2 + 2$ **49.** $-2x + 3$ **51.** $3x - 1$

53. $\frac{9}{2}x + 1$ **55.** $3x^3 - 2x + 3$ **57.** $-x^2 + 5x - 24$

59. $7u^2 + 8u + 5$ **61.** $2x^4 - 7x^2 + 3$

63. $(4x - 6)$ units   **65.** $-2t - 4$ **67.** $-\frac{1}{4}x + \frac{16}{3}$

69. $2x^2 - 3x - 6$ **71.** $4x^2 - 7$ **73.** $-2z^2$

75. $7y^2 - y + 6$ **77.** $3x^2 + 4x - 14$

79. (a) $-\frac{1}{2}x^2 + 14x - 15$

    (b)

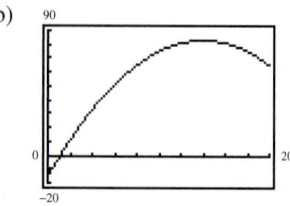

    (c) \$83; When x is less than or greater than 14, the profit is less than \$83.

81. $2x^2 + 8x$ **83.** $-12x^3 + 6x^2$ **85.** $x^2 + 2x - 24$

87. $2x^2 + 2x - 12$ **89.** $12x^2 + 7x - 12$

91. $2x^3 + 13x^2 + 19x + 6$ **93.** $2t^3 - 7t^2 + 9t - 3$

95. $6x^3 - x^2 - 5x + 2$

97. $y^4 - 2y^3 - 6y^2 + 22y - 15$

99. $8x^3 + 12x^2 + 6x + 1$

101. $(6x^2 + 38x + 60)$ square inches   **103.** $x^2 + 6x + 9$

105. $16x^2 - 56x + 49$ **107.** $\frac{1}{4}x^2 - 4x + 16$

109. $u^2 - 36$ **111.** $4x^2 - 4xy + y^2$ **113.** $4x^2 - 16y^2$

115. $2x^2 - \frac{1}{2}$, $x \neq 0$ **117.** $3xy - y + 1$, $x \neq 0$, $y \neq 0$

119. $2x^2 + \frac{4}{3}x - \frac{8}{9} + \frac{10}{9(3x - 1)}$

121. $x^2 - 2$, $x \neq \pm 1$

123. $x^2 - x - 3 - \frac{3x^2 - 2x - 3}{x^3 - 2x^2 + x - 1}$

125. $x^2 + 5x - 7$, $x \neq -2$

127. $x^3 + 3x^2 + 6x + 18 + \frac{29}{x - 3}$

129. $(x - 2)(x^2 + 4x + 3)$

Chapter Test *(page 343)*

1. Degree: 3; Leading coefficient: -5.2

2. The variable appears in the denominator.

3. (a) 3.2×10^{-5} (b) 60,400,000

4. (a) $\dfrac{20y^5}{x^5}$ (b) $\dfrac{1}{4x^4y^2z^6}$

5. (a) $-24u^9v^5$ (b) $\dfrac{27x^6}{2y^4}$

6. (a) $-\dfrac{48}{x}$ (b) $\dfrac{25x^2}{16y^4}$

7. (a) $6a^2 - 3a$ (b) $-2y^2 - 2y$

8. (a) $8x^2 - 4x + 10$ (b) $11t + 7$

9. (a) $-3x^2 + 12x$ (b) $2x^2 + 7xy - 15y^2$

10. (a) $3x^2 - 6x + 3$ (b) $6s^3 - 17s^2 + 26s - 21$

11. (a) $3x + 5 - \dfrac{9}{x}$ (b) $-3y^2 + y - \dfrac{4}{y}$

12. (a) $t^2 + 3 - \dfrac{6t - 6}{t^2 - 2}$

    (b) $2x^3 + 6x^2 + 3x + 9 + \dfrac{20}{x - 3}$

13. $2x(x + 15) - x(x + 4) = x^2 + 26x$

14. $x - 3$ **15.** $P = x^2 - 47x - 150$

Chapter 6

Section 6.1 *(page 351)*

Review *(page 351)*

1. 6; The greatest common factor is the product of the common prime factors.

2. 15; The greatest common factor is the product of the common prime factors.

3. $-3x + 17$ **4.** $-2x - 14$ **5.** $\dfrac{9y^2}{4x^6}$ **6.** $\dfrac{a^{10}}{16b^8}$

7.

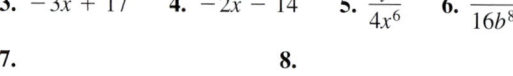

8.

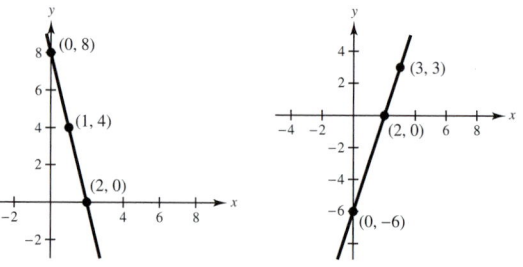

9.   **10.**

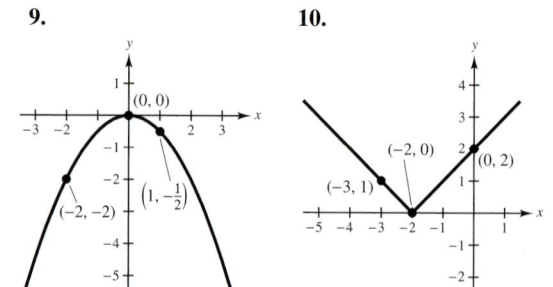

11. 3%   **12.** 3 hours 45 minutes

1. z^2 **3.** $2x$ **5.** u^2v **7.** $3yz^2$ **9.** 1

11. $14a^2b^2$ **13.** $x + 3$ **15.** $7x + 5$

17. $3(x + 1)$ **19.** $6(z - 1)$ **21.** $8(t - 2)$

23. $-5(5x + 2)$ **25.** $6(4y^2 - 3)$ **27.** $x(x + 1)$

29. $u(25u - 14)$ **31.** $2x^3(x + 3)$

33. No common factor **35.** $2x(6x - 1)$

37. $-5r(2r^2 + 7)$ **39.** $8a^3b^3(2 + 3a)$

41. $10ab(1 + a)$ **43.** $4(3x^2 + 4x - 2)$

45. $25(4 + 3z - 2z^2)$ **47.** $3x^2(3x^2 + 2x + 6)$

49. $5u(2u + 1)$ **51.** $(x - 3)(x + 5)$

53. $(s + 10)(t - 8)$ **55.** $(b + 2)(a^2 - b)$

57. $z^2(z + 5)(z + 1)$ **59.** $(a + b)(2a - b)$

61. $-5(2x - 1)$ **63.** $-3(x - 1000)$

65. $-(x^2 - 2x - 4)$ **67.** $-2(x^2 - 6x - 2)$

69. $(x + 10)(x + 1)$ **71.** $(a - 4)(a + 1)$

73. $(x + 3)(x + 4)$ **75.** $(x + 2)(x + 5)$

77. $(x + 3)(x - 5)$ **79.** $(2x - 7)(2x + 7)$

81. $(2x + 1)(3x - 1)$ **83.** $(x + 4)(8x + 1)$

85. $(3x - 2)(x + 1)$ **87.** $(x - 2)(2x - 3)$

89. $(y - 4)(ky + 2)$ **91.** $(t - 3)(t^2 + 2)$

93. $(x + 2)(x^2 + 1)$ **95.** $(2z + 1)(3z^2 - 1)$

97. $(x - 1)(x^2 - 3)$ **99.** $(4 - x)(x^2 - 2)$

101. $(x + 3)$ **103.** $(10y - 1)$

105. $(14x + 5y)$

107. **109.**

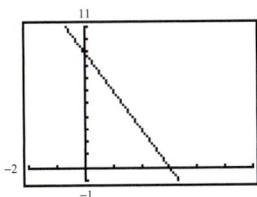

$y_1 = y_2$

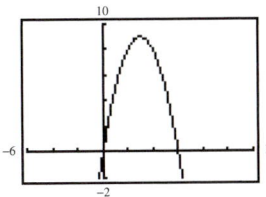

$y_1 = y_2$

111. $x + 1$ **113.** $6x^2$ **115.** $2\pi r(r + h)$

117. $kx(Q - x)$ **119.** $x^2 + x - 6 = (x - 2)(x + 3)$

121. Multiply the factors.

123. Noun: Any one of the expressions that, when multiplied together, yield the product. Verb: To find the expressions that, when multiplied together, yield the given product.

125. $x^3 - 3x^2 - 5x + 15 = (x^3 - 3x^2) + (-5x + 15)$
$$= x^2(x - 3) - 5(x - 3)$$
$$= (x - 3)(x^2 - 5)$$

Section 6.2 *(page 359)*

Review *(page 359)*

1. Intercepts are the points at which the graph intersects the x- or y-axis. To find the x-intercept(s), let y be zero and solve the equation for x. To find the y-intercept(s), let x be zero and solve the equation for y.

2. 4 **3.** $y^2 + 2y$ **4.** $-a^3 + a^2$

5. $x^2 - 7x + 10$ **6.** $v^2 + 3v - 28$

7. $4x^2 - 25$ **8.** $x^3 - 4x^2 + 10$ **9.** $\$3,975,000$

10. $\$717$ **11.** $x \geq 46$

12. 140 miles $\leq x \leq 227.5$ miles

1. $(x + 1)$ **3.** $(a - 2)$ **5.** $(y - 5)$ **7.** $(z - 2)$

9. $(x + 1)(x + 11)$; $(x - 1)(x - 11)$

11. $(x + 14)(x + 1)$; $(x - 14)(x - 1)$; $(x + 7)(x + 2)$; $(x - 7)(x - 2)$

13. $(x + 12)(x + 1)$; $(x - 12)(x - 1)$; $(x + 6)(x + 2)$; $(x - 6)(x - 2)$; $(x + 4)(x + 3)$; $(x - 4)(x - 3)$

15. $(x + 2)(x + 4)$ **17.** $(x - 5)(x - 8)$

19. $(z - 3)(z - 4)$ **21.** Prime **23.** $(x + 2)(x - 3)$

25. $(x - 3)(x + 5)$ **27.** Prime **29.** $(u + 2)(u - 24)$

31. $(x + 15)(x + 4)$ **33.** $(x - 8)(x - 9)$

35. $(x + 12)(x - 20)$ **37.** $(x + 2y)(x - y)$

39. $(x + 5y)(x + 3y)$ **41.** $(x - 9z)(x + 2z)$

43. $(a + 5b)(a - 3b)$ **45.** $3(x + 5)(x + 2)$

47. $4(y - 3)(y + 1)$ **49.** Prime **51.** $9(x^2 + 2x - 2)$

53. $x(x - 10)(x - 3)$ **55.** $x^2(x - 2)(x - 3)$

57. $-3x(y - 3)(y + 6)$ **59.** $x(x + 2y)(x + 3y)$

61. $2xy(x + 3y)(x - y)$ **63.** $x^2y^2(x + 2y)(x + y)$

65. $\pm9, \pm11, \pm19$ **67.** $\pm4, \pm20$

69. $\pm12, \pm13, \pm15, \pm20, \pm37$

71. $2, -10$ **73.** $3, 4$ **75.** $8, -10$

77. **79.**

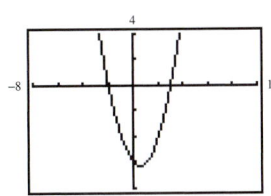

$y_1 = y_2$

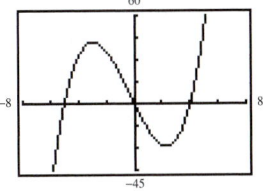

$y_1 = y_2$

81. (a) $4x(x - 2)(x - 3)$;   This   is   equivalent   to $x(4 - 2x)(6 - 2x)$, where x, $4 - 2x$, and $6 - 2x$ are the dimensions of the box. The model was found by expanding this expression.

    (b) 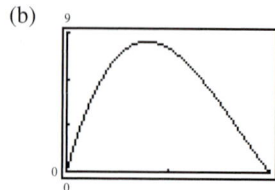   0.785 foot

83. 200 square units

85. (a) and (d); (a) Not completely factored; (d) Completely factored

87. When attempting to factor $x^2 + bx + c$, find factors of c whose sum is b. $x^2 + 7x + 10 = (x + 2)(x + 5)$

89. No. The factorization into prime factors is unique.

Section 6.3   *(page 367)*

Review   *(page 367)*

1. Prime   **2.** The sum of the digits is divisible by 3.

3. $2^2 \cdot 5^3$ **4.** $3^2 \cdot 5 \cdot 7$ **5.** $2^3 \cdot 3^2 \cdot 11$

6. $5^2 \cdot 7 \cdot 13$ **7.** $2x^2 + 9x - 35$

8. $9x^2 - 12x + 4$

9.

10.

11. (a)   (b) 6.6 inches

1. $(x + 4)$ **3.** $(t + 3)$ **5.** $(x + 2)$ **7.** $(5x + 3)$

9. $(5a - 3)$ **11.** $(4z - 1)$ **13.** $(2x - 7)$

15. $(3a - 4)$ **17.** $(3t - 2)$

19. $(5x + 3)(x + 1)$; $(5x - 3)(x - 1)$; $(5x + 1)(x + 3)$; $(5x - 1)(x - 3)$

21. $(5x + 12)(x + 1)$; $(5x - 12)(x - 1)$; $(5x + 6)(x + 2)$; $(5x - 6)(x - 2)$; $(5x + 4)(x + 3)$; $(5x - 4)(x - 3)$; $(5x + 1)(x + 12)$; $(5x - 1)(x - 12)$; $(5x + 2)(x + 6)$; $(5x - 2)(x - 6)$; $(5x + 3)(x + 4)$; $(5x - 3)(x - 4)$

23. $(2x + 3)(x + 1)$ **25.** $(4y + 1)(y + 1)$

27. $(2y - 1)(y - 1)$ **29.** $(2x - 3)(x + 1)$

31. Prime   **33.** Prime   **35.** Prime

37. $(x + 4)(4x - 3)$ **39.** $(3x - 2)(3x - 4)$

41. $(3u - 2)(6u + 1)$ **43.** $(5a - 2)(3a + 4)$

45. $(5t + 6)(2t - 3)$ **47.** $(5m - 3)(3m + 5)$

49. $(8z - 5)(2z - 3)$ **51.** $-(2x - 3)(x + 1)$

53. $-(3x - 2)(x + 2)$ **55.** $-(6x + 5)(x - 2)$

57. $-(10x - 1)(6x + 1)$ **59.** $-(5x - 4)(3x + 4)$

61. $3x(2x - 1)$ **63.** $3y(5y + 6)$ **65.** $(u - 3)(u + 9)$

67. $2(v + 7)(v - 3)$ **69.** $-3(x^2 + x + 20)$

71. $3(z - 1)(3z - 5)$ **73.** $2(2x^2 + 2x + 1)$

75. $-x^2(5x + 4)(3x - 2)$ **77.** $x(3x^2 + 4x + 2)$

79. $6x(x - 4)(x + 8)$ **81.** $9u^2(2u^2 + 2u - 3)$

83. $\pm 11, \pm 13, \pm 17, \pm 31$ **85.** $\pm 1, \pm 4, \pm 11$

87. $\pm 22, \pm 23, \pm 26, \pm 29, \pm 34, \pm 43, \pm 62, \pm 121$

89. $-1, -7$ **91.** $-8, 3$ **93.** $-6, -1$

95. $(3x + 1)(x + 2)$ **97.** $(2x + 3)(x - 1)$

99. $(3x + 4)(2x - 1)$ **101.** $(5x - 2)(3x - 1)$

103. $(3a + 5)(a + 2)$ **105.** $(8x - 3)(2x + 1)$

107. $(3x - 2)(4x - 3)$ **109.** $(u - 2)(6u + 7)$

111. $l = 2x + 3$ **113.** $2x + 10$

115. (a) $y_1 = y_2$

    (b)

    (c) $\left(-\frac{5}{2}, 0\right), (0, 0), (1, 0)$

117. (a) $3x^2 + 16x - 12$ (b) $3x - 2$ and $x + 6$

119. The product of the last terms of the binomials is 15, not -15.

121. Four. $(ax + 1)(x + c)$, $(ax + c)(x + 1)$, $(ax - 1)(x - c)$, $(ax - c)(x - 1)$

123. $2x^3 + 2x^2 + 2x$

125. Factoring by grouping:

$$6x^2 - 13x + 6 = 6x^2 - (4x + 9x) + 6$$
$$= (6x^2 - 4x) - (9x - 6)$$
$$= 2x(3x - 2) - 3(3x - 2)$$
$$= (3x - 2)(2x - 3)$$

$$2x^2 + 5x - 12 = 2x^2 + (8x - 3x) - 12$$
$$= (2x^2 + 8x) - (3x + 12)$$
$$= 2x(x + 4) - 3(x + 4)$$
$$= (x + 4)(2x - 3)$$

$$3x^2 + 11x - 4 = 3x^2 + (12x - x) - 4$$
$$= (3x^2 + 12x) - (x + 4)$$
$$= 3x(x + 4) - (x + 4)$$
$$= (x + 4)(3x - 1)$$

Preferences, advantages, and disadvantages will vary.

Mid-Chapter Quiz   *(page 371)*

1. $(2x - 3)$ **2.** $(x - y)$ **3.** $(y - 6)$

4. $(y + 3)$ **5.** $10(x^2 + 7)$ **6.** $2a^2b(a - 2b)$

7. $(x + 2)(x - 3)$ **8.** $(t - 3)(t^2 + 1)$

9. $(y + 6)(y + 5)$ **10.** $(u + 6)(u - 5)$

11. $x(x - 6)(x + 5)$ **12.** $2y(x + 8)(x - 4)$

13. $(2y - 9)(y + 3)$ **14.** $(3 + z)(2 - 5z)$

15. $(3x - 2)(2x + 1)$ **16.** $2s^2(5s^2 - 7s + 1)$

17. $\pm 7, \pm 8, \pm 13$; These integers are the sums of the factors of 12.

18. 16, 21; The factors of c have a sum of -10.

19. m and n are factors of 6.

$(3x + 1)(x + 6)$ $(3x - 1)(x - 6)$
$(3x + 6)(x + 1)$ $(3x - 6)(x - 1)$
$(3x + 2)(x + 3)$ $(3x - 2)(x - 3)$
$(3x + 3)(x + 2)$ $(3x - 3)(x - 2)$

20. $10(2x + 8)$

21.

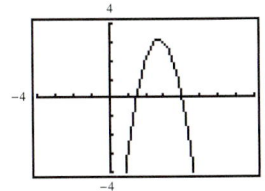 $y_1 = y_2$

Section 6.4   *(page 378)*

1. $(x + 6)(x - 6)$ **3.** $(u + 8)(u - 8)$

5. $(7 + x)(7 - x)$ **7.** $\left(u + \frac{1}{2}\right)\left(u - \frac{1}{2}\right)$

9. $\left(v + \frac{2}{3}\right)\left(v - \frac{2}{3}\right)$ **11.** $(4y + 3)(4y - 3)$

13. $(10 + 7x)(10 - 7x)$ **15.** $(x + 1)(x - 3)$

17. $-z(10 + z)$ **19.** $(x + y)(x - y)$

21. $(3y + 5z)(3y - 5z)$ **23.** $2(x + 6)(x - 6)$

25. $x(2 + 5x)(2 - 5x)$ **27.** $2(2y + 5z)(2y - 5z)$

29. $(y^2 + 9)(y + 3)(y - 3)$

31. $(1 + x^2)(1 + x)(1 - x)$

33. $3(x + 2)(x - 2)(x^2 + 4)$

35. $(9x^2 + 4y^2)(3x + 2y)(3x - 2y)$ **37.** $(x - 2)^2$

39. $(z + 3)^2$ **41.** $(2t + 1)^2$ **43.** $(5y - 1)^2$

45. $\left(b + \frac{1}{2}\right)^2$ **47.** $\left(2x - \frac{1}{4}\right)^2$ **49.** $(x - 3y)^2$

51. $(2y + 5z)^2$ **53.** $(3a - 2b)^2$ **55.** ± 2 **57.** $\pm \frac{8}{5}$

59. ± 36 **61.** 9   **63.** 4   **65.** $(x - 2)(x^2 + 2x + 4)$

67. $(y + 4)(y^2 - 4y + 16)$ **69.** $(1 + 2t)(1 - 2t + 4t^2)$

71. $(3u - 2)(9u^2 + 6u + 4)$ **73.** $(x - y)(x^2 + xy + y^2)$

75. $(3x + 4y)(9x^2 - 12xy + 16y^2)$

77. $6(x - 6)$ **79.** $u(u + 3)$ **81.** $5y(y - 5)$

83. $5(y + 5)(y - 5)$ **85.** $y^2(y + 5)(y - 5)$

87. $(x - 2y)^2$ **89.** $(x - 1)^2$

91. $(9x + 1)(x + 1)$ **93.** $2x(x - 2y)(x + y)$

95. $(3t + 4)(3t - 4)$ **97.** $-z(z + 12)$

99. $(t + 10)(t - 12)$ **101.** $u(u^2 + 2u + 3)$

103. Prime   **105.** $2(t - 2)(t^2 + 2t + 4)$

107. $2(a - 2b)(a^2 + 2ab + 4b^2)$

109. $(x^2 + 9)(x + 3)(x - 3)$

111. $(x^2 + y^2)(x + y)(x - y)$

113. $(x + 1)(x - 1)(x - 4)$

115. $x(x + 3)(x + 4)(x - 4)$

117. $(2 + y)(2 - y)(y^2 + 2y + 4)(y^2 - 2y + 4)$

119.

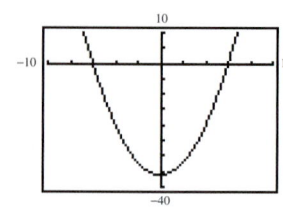

$y_1 = y_2$

121.

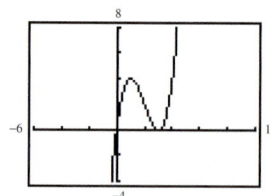

$y_1 = y_2$

123. 441   **125.** 3599

127. $\pi(R - r)(R + r)$

129. $(x + 3)^2 - 1^2 = (x + 4)(x + 2)$

131. Box 1: $(a - b)a^2$; Box 2: $(a - b)ab$; Box 3: $(a - b)b^2$
The sum of the volumes of boxes 1, 2, and 3 equals the volume of the large cube minus the volume of the small cube, which is the difference of two cubes.

133. $a^2 - b^2 = (a + b)(a - b)$

135. No. $(x + 2)(x - 2)$

137. False. $a^3 + b^3 = (a + b)(a^2 - ab + b^2)$

Section 6.5   *(page 388)*

Review   *(page 388)*

1. Additive Inverse Property

2. Multiplicative Identity Property

3. Distributive Property

4. Associative Property of Multiplication

5. -4 **6.** $353.\overline{33}$ **7.** No solution

8. -19 **9.** 40   **10.** 24

11. (a) $P = -\frac{1}{4}x^2 + 8x - 12$

(b)
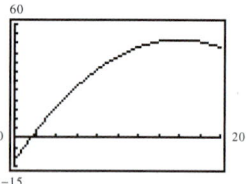

(c) $52

12. $832

1. 0, 8   **3.** $-10, 3$ **5.** $-4, 2$ **7.** $-\frac{5}{2}, -\frac{1}{3}$

9. $-\frac{25}{2}, 0, \frac{3}{2}$ **11.** $-4, -\frac{1}{2}, 3$ **13.** 0, 5   **15.** $-\frac{5}{3}, 0$

17. 0, 16   **19.** 0, 3   **21.** ±5 **23.** ±4 **25.** $-2, 5$

27. 4, 6   **29.** $-5, \frac{5}{4}$ **31.** $-\frac{1}{2}, 7$ **33.** $-1, \frac{2}{3}$

35. 4, 9   **37.** 4   **39.** -8 **41.** $\frac{3}{2}$ **43.** $-2, 10$

45. ±3 **47.** $-4, 9$ **49.** $-12, 6$ **51.** $-\frac{7}{2}, 5$

53. $-7, 0$ **55.** $-6, 5$ **57.** $-2, 6$ **59.** $-5, 1$

61. $-2, 8$ **63.** $-13, 5$ **65.** 0, 7, 12   **67.** $-\frac{1}{3}, 0, \frac{1}{2}$

69. ±2 **71.** $\pm3, -2$ **73.** ±3 **75.** $\pm1, 0, 3$

77. $\pm2, -\frac{3}{2}, 0$

79. $(-3, 0), (3, 0)$; the x-intercepts are solutions of the polynomial equation.

81. $(0, 0), (3, 0)$; the x-intercepts are solutions of the polynomial equation.

83. $(0, 0), (6, 0)$ **85.** $(2, 0), (6, 0)$

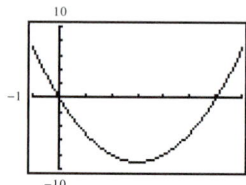 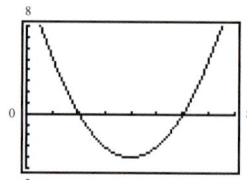

87. $(-4, 0), \left(\frac{3}{2}, 0\right)$ **89.** $\left(-\frac{3}{2}, 0\right), (0, 0), (4, 0)$

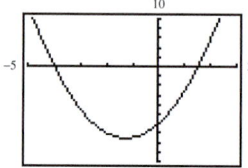 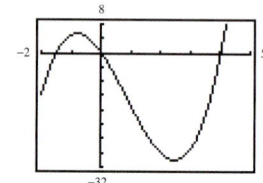

91. $-\frac{b}{a}, 0$ **93.** $x^2 - 2x - 15 = 0$ **95.** 15

97. 15 feet × 22 feet

99. Base: 8 inches; Height: 12 inches

101. (a) Length $= 5 - 2x$; Width $= 4 - 2x$; Height $= x$

Volume $=$ (Length)(Width)(Height)

$V = (5 - 2x)(4 - 2x)(x)$

(b) $0, 2, \frac{5}{2}$; $0 < x < 2$

(c)

x	0.25	0.50	0.75	1.00	1.25	1.50	1.75
V	3.94	6	6.56	6	4.69	3	1.31

(d) 1.50

(e) 0.74

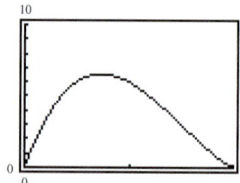

103. 9.75 seconds

105. 4 seconds   **107.** 3 seconds   **109.** 10 units, 20 units

111. (a) $-6, -\frac{1}{2}$ (b) $-6, -\frac{1}{2}$ (c) Answers will vary.

113. (c) 4 feet × 8 feet × 8 feet

(d) Yes. When the area of the base is 32 square feet, $x = 2$ and the dimensions of the bin are 4 feet × 8 feet × 8 feet.

(e) The volume of the bin is twice the volume of the truck bin. So, it takes two truckloads to fill the bin.

115. False. This is not an application of the Zero-Factor Property, because there are an unlimited number of factors whose product is 1.

117. Maximum number: n. The third-degree equation $(x + 1)^3 = 0$ has only one real solution: $x = -1$.

Review Exercises   *(page 393)*

1. t^2 **3.** $3x^2$ **5.** $7x^2y^3$ **7.** $4xy$ **9.** $3(x - 2)$

11. $t(3 - t)$ **13.** $5x^2(1 + 2x)$ **15.** $4a(2 - 3a^2)$

17. $5x(x^2 + x - 1)$ **19.** $x(3x + 4)$

21. $(x + 1)(x - 3)$ **23.** $(u - 2)(2u + 5)$

25. $(y + 3)(y^2 + 2)$ **27.** $(x^2 + 1)(x + 2)$

29. $(x + 3)(x - 4)$ **31.** $(x - 7)(x + 4)$

33. $(u - 4)(u + 9)$ **35.** $(x - 6)(x + 4)$

37. $(y + 7)(y + 3)$ **39.** $\pm 6, \pm 10$ **41.** ± 12

43. $(x - y)(x + 10y)$ **45.** $(y + 3x)(y - 9x)$

47. $(x + 2y)(x - 4y)$ **49.** $4(x - 2)(x - 4)$

51. $x(x + 3)(x + 6)$ **53.** $4x(x + 2)(x + 7)$

55. $(1 - x)(5 + 3x)$ **57.** $(10 + x)(5 - x)$

59. $(3x + 2)(2x + 1)$ **61.** $(4y + 1)(y - 1)$

63. $(3x - 2)(x + 3)$ **65.** $(3x - 1)(x + 2)$

67. $(2x - 1)(x - 1)$ **69.** $\pm 2, \pm 5, \pm 10, \pm 23$

71. $2, -6$ **73.** $3u(2u + 5)(u - 2)$

75. $4y(2y - 3)(y - 1)$ **77.** $2x(3x - 2)(x + 3)$

79. $3x + 1$ **81.** $(2x - 7)(x - 3)$

83. $(4y - 3)(y + 1)$ **85.** $(3x - 2)(2x + 5)$

87. $(a + 10)(a - 10)$ **89.** $(5 + 2y)(5 - 2y)$

91. $3(2x + 3)(2x - 3)$ **93.** $(u + 3)(u - 1)$

95. $(x + 1)(x - 1)$ **97.** $st(s + t)(s - t)$

99. $(x^2 + y^2)(x + y)(x - y)$ **101.** $(x - 4)^2$

103. $(3s + 2)^2$ **105.** $(y + 2z)^2$

107. $(a + 1)(a^2 - a + 1)$ **109.** $(3 - 2t)(9 + 6t + 4t^2)$

111. $(2x + y)(4x^2 - 2xy + y^2)$

113. $0, 2$ **115.** $-\frac{1}{2}, 3$ **117.** $-10, -\frac{9}{5}, \frac{1}{4}$

119. $-\frac{4}{3}, 2$ **121.** $0, 3$ **123.** $-4, 9$ **125.** ± 10

127. $-4, 0, 3$ **129.** $0, 2, 9$ **131.** $\pm 1, 6$

133. $\pm 3, 0, 5$ **135.** $13, 15$ **137.** 45 inches × 20 inches

Chapter Test   *(page 396)*

1. $7x^2(1 - 2x)$ **2.** $(z + 7)(z - 3)$

3. $(t - 5)(t + 1)$ **4.** $(3x - 4)(2x - 1)$

5. $3y(y - 1)(y + 25)$ **6.** $(2 + 5v)(2 - 5v)$

7. $(2x - 5)^2$ **8.** $(-z - 5)(z + 13)$

9. $(x + 2)(x + 3)(x - 3)$ **10.** $(4 + z^2)(2 + z)(2 - z)$

11. $2x - 3$ **12.** ± 6 **13.** 36

14. $3x^2 - 3x - 6 = 3(x + 1)(x - 2)$

15. $-4, \frac{3}{2}$ **16.** $-3, \frac{2}{3}$ **17.** $-\frac{3}{2}, 2$ **18.** $-1, 4$

19. $-2, 0, 6$ **20.** $\pm\sqrt{3}, -7, 0$ **21.** $x + 4$

22. 7 inches × 12 inches   **23.** 8.875 seconds; 5 seconds

24. 24, 26   **25.** 300 feet × 100 feet

Cumulative Test: Chapters 4–6   *(page 397)*

1. Because $x = -2$, the point must lie in Quadrant II or Quadrant III.

2. (a) Not a solution   (b) Solution

(c) Solution   (d) Not a solution

3.   **4.**

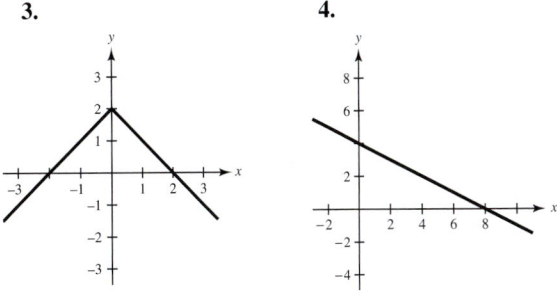

$(-2, 0), (2, 0), (0, 2)$ $(8, 0), (0, 4)$

5. Not a function

6. $(-2, 2)$; There are infinitely many points on a line.

7. $y = \frac{5}{6}x - \frac{3}{2}$

8.   **9.**

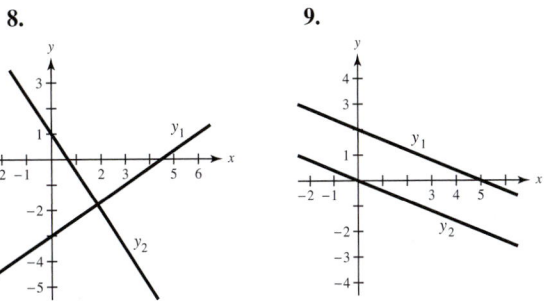

Perpendicular   Parallel

10. $-5x^2 + 5$ **11.** $-42z^4$ **12.** $3x^2 - 7x - 20$

13. $25x^2 - 9$ **14.** $25x^2 + 60x + 36$ **15.** $x + 12$

16. $x + 1 + \dfrac{2}{x - 4}$ **17.** $\dfrac{x}{54y^4}$ **18.** $2u(u - 3)$

19. $(x + 2)(x - 6)$ **20.** $x(x + 4)^2$

21. $(x + 2)^2(x - 2)$ **22.** $0, 12$ **23.** $0, -3, 7$

24. $C = 125 + 0.35x$; $149.50   **25.** $39{,}142$ miles

Chapter 7

Section 7.1   *(page 407)*

Review   *(page 407)*

1. $m = \dfrac{y_2 - y_1}{x_2 - x_1}$

2. (a) $m > 0$ (b) $m < 0$

  (c) $m = 0$ (d) m is undefined.

3. 10 **4.** 12 **5.** $-8x - 10$ **6.** $-2x^2 + 14x$

7. $\dfrac{25}{x^4}$ **8.** $\dfrac{4u^3}{3}$

9. 30% solution: $13\frac{1}{3}$ gallons; 60% solution: $6\frac{2}{3}$ gallons

10. $500

---

1. $(-\infty, \infty)$ **3.** $(-\infty, 5) \cup (5, \infty)$

5. $(-\infty, 4) \cup (4, \infty)$ **7.** $(-\infty, -10) \cup (-10, \infty)$

9. $(-\infty, \infty)$ **11.** $(-\infty, -3) \cup (-3, 0) \cup (0, \infty)$

13. $(-\infty, -4) \cup (-4, 4) \cup (4, \infty)$

15. $(-\infty, 0) \cup (0, 3) \cup (3, \infty)$

17. $(-\infty, 2) \cup (2, 3) \cup (3, \infty)$

19. $(-\infty, -1) \cup \left(-1, \frac{5}{3}\right) \cup \left(\frac{5}{3}, \infty\right)$

21. (a) 1 (b) -8

  (c) Undefined (division by 0)   (d) 0

23. (a) 0 (b) 0

  (c) Undefined (division by 0)

  (d) Undefined (division by 0)

25. (a) $\frac{25}{22}$ (b) 0

  (c) Undefined (division by 0)

  (d) Undefined (division by 0)

27. $(0, \infty)$ **29.** $\{1, 2, 3, 4, \ldots\}$ **31.** $[0, 100]$

33. $x + 3$ **35.** $(3)(x + 16)^2$ **37.** $(x)(x - 2)$

39. $x + 2$ **41.** $\dfrac{x}{5}$ **43.** $6y$, $y \neq 0$ **45.** $\dfrac{6x}{5y^3}$, $x \neq 0$

47. $\dfrac{x - 3}{4x}$ **49.** x, $x \neq 8$, $x \neq 0$ **51.** $\dfrac{1}{2}$, $x \neq \dfrac{3}{2}$

53. $-\dfrac{1}{3}$, $x \neq 5$ **55.** $\dfrac{1}{a + 3}$ **57.** $\dfrac{x}{x - 7}$

59. $\dfrac{y(y + 2)}{y + 6}$, $y \neq 2$ **61.** $\dfrac{x(x + 2)}{x - 3}$, $x \neq 2$

63. $-\dfrac{3x + 5}{x + 3}$, $x \neq 4$ **65.** $\dfrac{x + 8}{x - 3}$, $x \neq -\dfrac{3}{2}$

67. $\dfrac{3x - 1}{5x - 4}$, $x \neq -\dfrac{4}{5}$ **69.** $\dfrac{3y^2}{y^2 + 1}$, $x \neq 0$

71. $\dfrac{y - 8x}{15}$, $y \neq -8x$ **73.** $\dfrac{5 + 3xy}{y^2}$, $x \neq 0$

75. $\dfrac{u - 2v}{u - v}$, $u \neq -2v$

77. $\dfrac{3(m - 2n)}{m + 2n}$

79.

x	-2	-1	0	1	2	3	4
$\dfrac{x^2 - x - 2}{x - 2}$	-1	0	1	2	Undef.	4	5
$x + 1$	-1	0	1	2	3	4	5

$$\dfrac{x^2 - x - 2}{x - 2} = \dfrac{(x - 2)(x + 1)}{x - 2} = x + 1, \; x \neq 2$$

81. $\dfrac{x}{x + 3}$, $x > 0$ **83.** $\dfrac{1}{4}$, $x > 0$

85. (a) $C = 2500 + 9.25x$

  (b) $\overline{C} = \dfrac{2500 + 9.25x}{x}$

  (c) $\{1, 2, 3, 4, \ldots\}$

  (d) $34.25

87. (a) Van: $45(t + 3)$; Car: $60t$ (b) $d = |15(9 - t)|$

  (c) $\dfrac{4t}{3(t + 3)}$

89. π **91.** $\dfrac{1531.1t + 9358}{1.33t + 54.6}$

93. Let u and v be polynomials. The algebraic expression u/v is a rational expression.

95. The rational expression is in simplified form if the numerator and denominator have no factors in common (other than ± 1).

97. You can divide out only common factors.

99. (a) The student forgot to divide each term of the numerator by the denominator.

Correct solution:

$$\frac{3x^2 + 5x - 4}{x} = \frac{3x^2}{x} + \frac{5x}{x} - \frac{4}{x} = 3x + 5 - \frac{4}{x}$$

(b) The student incorrectly divided out; the denominator may not be split up.

Correct solution:

$$\frac{x^2 + 7x}{x + 7} = \frac{x(x + 7)}{x + 7} = x$$

Section 7.2 *(page 417)*

Review *(page 417)*

1. $u^2 - v^2 = (u + v)(u - v)$

    $9t^2 - 4 = (3t + 2)(3t - 2)$

2. $u^2 - 2uv + v^2 = (u - v)^2$

    $4x^2 - 12x + 9 = (2x - 3)^2$

3. $u^3 + v^3 = (u + v)(u^2 - uv + v^2)$

    $8x^3 + 64 = (2x + 4)(4x^2 - 8x + 16)$

4. $(3x - 2)(x + 5)$. Multiply the binomial factors to see whether you obtain the original expression.

5. $5x(1 - 4x)$ **6.** $(2 + x)(14 - x)$

7. $(3x - 5)(5x + 3)$ **8.** $(4t + 1)^2$

9. $(y - 4)(y^2 + 4y + 16)$

10. $(2x + 1)(4x^2 - 2x + 1)$

11.

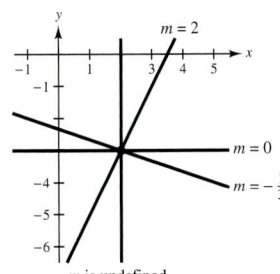

m is undefined.

12.

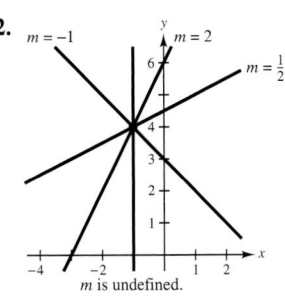

m is undefined.

1. x^2 **3.** $(x + 2)^2$ **5.** $u + 1$ **7.** $(-1)(2 + x)$

9. $\dfrac{9}{2}$ **11.** $\dfrac{s^3}{6}$, $s \neq 0$ **13.** $24u^2$, $u \neq 0$

15. 24, $x \neq -\dfrac{3}{4}$ **17.** $\dfrac{2uv(u + v)}{3(3u + v)}$, $u \neq 0$

19. -1, $r \neq 12$ **21.** $-\dfrac{x + 8}{x^2}$, $x \neq \dfrac{3}{2}$

23. $4(r + 2)$, $r \neq 3$, $r \neq 2$ **25.** $2t + 5$, $t \neq 3$, $t \neq -2$

27. $\dfrac{xy(x + 2y)}{x - 2y}$ **29.** $\dfrac{(x - y)^2}{x + y}$, $x \neq -3y$

31. $\dfrac{(x - 1)(2x + 1)}{(3x - 2)(x + 2)}$, $x \neq \pm 5$, $x \neq -1$

33. $\dfrac{x^2(x^2 - 9)(2x + 5)(3x - 1)}{2(2x + 1)(2x + 3)(3 - 2x)}$, $x \neq 0$, $x \neq \dfrac{1}{2}$

35. $\dfrac{(x + 3)^2}{x}$, $x \neq 3$, $x \neq 4$ **37.** $\dfrac{4x}{3}$, $x \neq 0$ **39.** $\dfrac{6}{x}$

41. $\dfrac{3y^2}{2ux^2}$, $v \neq 0$ **43.** $\dfrac{3}{2(a + b)}$

45. $x^4y(x + 2y)$, $x \neq 0$, $y \neq 0$, $x \neq -2y$

47. $-\dfrac{y - 5}{4}$, $y \neq \pm 3$ **49.** $\dfrac{x - 4}{x - 5}$, $x \neq -6$, $x \neq -5$, $x \neq 3$

51. $\dfrac{x + 4}{3}$, $x \neq -2$, $x \neq 0$ **53.** $\dfrac{1}{4}$, $x \neq -1$, $x \neq 0$, $y \neq 0$

55. $\dfrac{(x + 1)(2x - 5)}{x}$, $x \neq -1$, $x \neq -5$, $x \neq -\dfrac{2}{3}$

57. $\dfrac{x^4}{(x^n + 1)^2}$, $x^n \neq -3$, $x^n \neq 3$, $x \neq 0$

59.

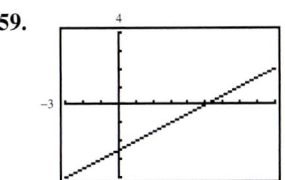

61. $\dfrac{2w^2 + 3w}{6}$ **63.** $\dfrac{x}{4(2x + 1)}$ **65.** $\dfrac{\pi x}{4(2x + 1)}$

67. (a) $\dfrac{1}{20}$ minute (b) $\dfrac{x}{20}$ minutes (c) $\dfrac{7}{4}$ minutes

69. Invert the divisor and multiply.

71. Invert the divisor, not the dividend.

Section 7.3   *(page 426)*

Review   *(page 426)*

1. (a) $y = \frac{3}{5}x + \frac{4}{5}$ (b) $y - 2 = \frac{3}{5}(x - 2)$

2. If the line rises from left to right, $m > 0$.
If the line falls from left to right, $m < 0$.

3. $42x^2 - 60x$ **4.** $6 + y - 2y^2$ **5.** $121 - x^2$

6. $16 - 25z^2$ **7.** $x^2 + 2x + 1$ **8.** $2t$

9. $x^3 - 8$ **10.** $2t^3 - 5t^2 - 12t$

11. $P = 12x + 6$; $A = 5x^2 + 9x$

12. $P = 12x$; $A = 6x^2$

---

1. $-\dfrac{x}{4}$ **3.** $-\dfrac{3}{a}$ **5.** $-\dfrac{2}{9}$ **7.** $\dfrac{2z^2 - 2}{3}$ **9.** $\dfrac{x + 6}{3x}$

11. $-\dfrac{4}{3}$ **13.** 1, $y \neq 6$ **15.** $\dfrac{1}{x - 3}$, $x \neq 0$ **17.** $20x^3$

19. $36y^3$ **21.** $15x^2(x + 5)$ **23.** $126z^2(z + 1)^4$

25. $56t(t + 2)(t - 2)$ **27.** $6x(x + 2)(x - 2)$ **29.** x^2

31. $(u + 1)$ **33.** $-(x + 2)$ **35.** $\dfrac{2n^2(n + 8)}{6n^2(n - 4)}, \dfrac{10(n - 4)}{6n^2(n - 4)}$

37. $\dfrac{2(x + 3)}{x^2(x + 3)(x - 3)}, \dfrac{5x(x - 3)}{x^2(x + 3)(x - 3)}$

39. $\dfrac{3v^2}{6v^2(v + 1)}, \dfrac{8(v + 1)}{6v^2(v + 1)}$

41. $\dfrac{(x - 8)(x - 5)}{(x + 5)(x - 5)^2}, \dfrac{9x(x + 5)}{(x + 5)(x - 5)^2}$ **43.** $\dfrac{25 - 12x}{20x}$

45. $\dfrac{7(a + 2)}{a^2}$ **47.** 0, $x \neq 4$ **49.** $\dfrac{3(x + 2)}{x - 8}$

51. $\dfrac{5(5x + 22)}{x + 4}$ **53.** 1, $x \neq \dfrac{2}{3}$ **55.** $\dfrac{9x - 14}{2x(x - 2)}$

57. $\dfrac{x^2 - 7x - 15}{(x + 3)(x - 2)}$ **59.** $-\dfrac{2}{(x + 3)}$, $x \neq 3$

61. $\dfrac{5(x + 1)}{(x + 5)(x - 5)}$ **63.** $\dfrac{4}{x^2(x^2 + 1)}$

65. $\dfrac{x^2 + x + 9}{(x - 2)(x - 3)(x + 3)}$ **67.** $\dfrac{4x}{(x - 4)^2}$

69. $\dfrac{y - x}{xy}$, $x \neq -y$ **71.** $\dfrac{2(4x^2 + 5x - 3)}{x^2(x + 3)}$

73. $-\dfrac{u^2 - uv - 5u + 2v}{(u - v)^2}$ **75.** $\dfrac{x}{x - 1}$, $x \neq -6$

77. 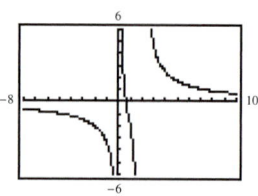   **79.** $\dfrac{5t}{12}$

81. $A = -4$, $B = 2$, $C = 2$

83. $\dfrac{750.27t^2 + 5660.36t - 4827.2}{t(0.09t + 1.0)}$ (in thousands)

85. (a) Upstream: $\dfrac{10}{5 - x}$; Downstream: $\dfrac{10}{5 + x}$

(b) $f(x) = \dfrac{10}{5 - x} + \dfrac{10}{5 + x}$ (c) $f(x) = \dfrac{100}{(5 + x)(5 - x)}$

87. Rewrite each fraction in terms of the lowest common denominator, combine the numerators, and place the result over the lowest common denominator.

89. When the numerators are subtracted, the result should be
$(x - 1) - (4x - 11) = x - 1 - 4x + 11$.

91. (a) $-\frac{7}{6}$ (b) $\frac{51}{7}$ (c) 8

Results are the same. Answers will vary.

Mid-Chapter Quiz   *(page 430)*

1. $(-\infty, 0) \cup (0, 4) \cup (4, \infty)$

2. (a) $\frac{1}{2}$ (b) $\frac{1}{2}$ (c) $-\frac{3}{2}$ (d) 0

3. (a) 0 (b) $\frac{9}{2}$ (c) Undefined (d) $\frac{8}{9}$

4. $\dfrac{3}{2}y$, $y \neq 0$ **5.** $\dfrac{2u^2}{9v}$, $u \neq 0$ **6.** $-\dfrac{2x + 1}{x}$, $x \neq \dfrac{1}{2}$

7. $\dfrac{z + 3}{2z - 1}$, $z \neq -3$ **8.** $\dfrac{7 + 3ab}{a}$, $b \neq 0$

9. $\dfrac{n^2}{m + n}$, $2m - n \neq 0$ **10.** $\dfrac{t}{2}$, $t \neq 0$

11. $\dfrac{5x}{x - 2}$, $x \neq -2$ **12.** $\dfrac{8x}{3(x - 1)(x + 3)(x - 1)}$

13. $\dfrac{32x^7}{35y^2z}$, $x \neq 0$ **14.** $\dfrac{(a + 1)^2}{9(a + b)^2}$, $a \neq b$

15. $\dfrac{2(x + 1)}{3x}$, $x \neq -2$, $x \neq -1$ **16.** $\dfrac{30}{x + 5}$, $x \neq 0$, $x \neq 1$

17. $\dfrac{4(u - v)^2}{5uv}$, $u \neq \pm v$ **18.** $\dfrac{7x - 11}{x - 2}$

19. $-\dfrac{4x^2 - 25x + 36}{(x - 3)(x + 3)}$ **20.** 0, $x \neq 2$, $x \neq -1$

21. (a) $C = 25{,}000 + 144x$ (b) $\overline{C} = \dfrac{25{,}000 + 144x}{x}$

Section 7.4   *(page 435)*

Review   *(page 435)*

1. Any expression with a zero exponent equals 1. Any expression with a negative exponent equals 1 divided by the expression raised to the positive exponent.

2. The exponent is -6 since the decimal needs to be moved six positions to the right.

3. $\dfrac{14}{x}$ 4. $\dfrac{1}{z^5}$ 5. $\dfrac{a^5}{b^7}$ 6. $x + 2$

7.

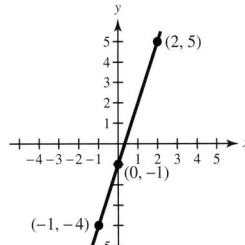

8.

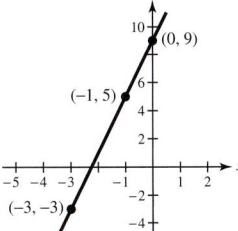

9.

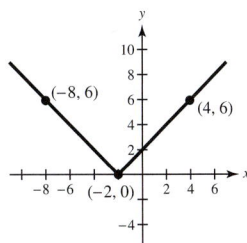

10.

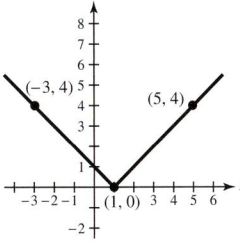

11. 5, 8, 20

12. Peanuts: 20 pounds; Almonds: 17.5 pounds; Pistachios: 12.5 pounds

1. $2x^2$, $x \neq 0$ 3. $\dfrac{3x}{10}$, $x \neq 0$

5. $6xz^3$, $x \neq 0$, $y \neq 0$, $z \neq 0$ 7. $\dfrac{2xy^2}{5}$, $x \neq 0$, $y \neq 0$

9. $-\dfrac{1}{y}$, $y \neq 3$ 11. $-\dfrac{5x(x + 1)}{2}$, $x \neq 0$, $x \neq 5$, $x \neq -1$

13. 2, $x \neq -1$, $x \neq 5$ 15. $\dfrac{x + 5}{3(x + 4)}$, $x \neq 2$

17. $\dfrac{2(x + 3)}{x - 2}$, $x \neq 7$, $x \neq -3$

19. $\dfrac{(2x - 5)(3x + 1)}{3x(x + 1)}$, $x \neq \pm\dfrac{1}{3}$

21. $\dfrac{(x + 3)(4x + 1)}{(3x - 1)(x - 1)}$, $x \neq -3$, $x \neq -\dfrac{1}{4}$

23. $x + 2$, $x \neq \pm 2$, $x \neq -3$

25. $-\dfrac{(x + 2)^2(x - 5)}{(x - 2)^2(x + 3)}$, $x \neq -2$, $x \neq 7$

27. $\dfrac{y + 3}{y^2}$ 29. $\dfrac{x^2}{2(2x + 3)}$, $x \neq 0$ 31. $\dfrac{4 + 3x}{4 - 3x}$, $x \neq 0$

33. $\dfrac{3}{4}$, $x \neq 0$, $x \neq 3$ 35. $\dfrac{5(x + 3)}{2x(5x - 2)}$

37. $y - x$, $x \neq 0$, $y \neq 0$, $x \neq -y$ 39. $-\dfrac{(y - 1)(y - 3)}{y(4y - 1)}$

41. $\dfrac{20}{7}$, $x \neq -1$ 43. $\dfrac{1}{x}$, $x \neq -1$

45. $\dfrac{x(x + 6)}{3x^3 + 10x - 30}$, $x \neq 0$, $x \neq 3$

47. $\dfrac{y(2y^2 - 1)}{10y^2 - 1}$, $y \neq 0$ 49. $\dfrac{x^2(7x^3 + 2)}{x^4 + 5}$, $x \neq 0$

51. $\dfrac{y + x}{y - x}$, $x \neq 0$, $y \neq 0$ 53. $\dfrac{y - x}{(y + x)(x^2y^2)}$

55. $-\dfrac{1}{2(h + 2)}$ 57. $11x/60$ 59. $11x/24$

61. $(b^2 + 5b + 8)/(8b)$ 63. $x/8$, $(5x)/36$, $(11x)/72$

65. $(R_1R_2)/(R_1 + R_2)$

67. (a)

(b) $[250(1382.16t + 5847.9)]/[3(4568.33t + 1042.7)]$

69. A complex fraction is a fraction with a fraction in its numerator or denominator, or both.

$$\dfrac{\left(\dfrac{x + 1}{2}\right)}{\left(\dfrac{x + 1}{3}\right)}$$

Simplify by inverting the denominator and multiplying:

$$\left(\dfrac{x + 1}{2}\right) \cdot \left(\dfrac{3}{x + 1}\right) = \dfrac{3}{2}, \quad x \neq -1.$$

71. (a) Numerator: $\left(\dfrac{x - 1}{5}\right)$; Denominator: $\left(\dfrac{2}{x^2 + 2x - 35}\right)$

(b) Numerator: $\left(\dfrac{1}{2y} + x\right)$; Denominator: $\left(\dfrac{3}{y} + x\right)$

Section 7.5   *(page 444)*

> ### Review   *(page 444)*
>
> **1.** Quadrant II or III   **2.** Quadrant I or II
>
> **3.** x-axis   **4.** $(9, -6)$ **5.** $x < \frac{3}{2}$ **6.** $x < 5$
>
> **7.** $1 < x < 5$ **8.** $x < 2$ or $x > 8$
>
> **9.** $x \le -8$ or $x \ge 16$ **10.** $-24 \le x \le 36$
>
> **11.** 15 minutes, 2 miles
>
> **12.** 7.5%: \$15,000; 9%: \$9000

1. (a) Not a solution   (b) Not a solution

 (c) Not a solution   (d) Solution

3. (a) Not a solution   (b) Solution

 (c) Solution   (d) Not a solution

5. 10   **7.** 1   **9.** 0   **11.** 8   **13.** $-\frac{9}{32}$

15. $-3, \frac{8}{3}$ **17.** $-\frac{2}{9}$ **19.** $\frac{7}{4}$ **21.** $\frac{43}{8}$ **23.** 61

25. $\frac{18}{5}$ **27.** $-\frac{26}{5}$ **29.** 3   **31.** 3   **33.** $-\frac{11}{5}$

35. $\frac{4}{3}$ **37.** ± 6 **39.** ± 4 **41.** $-9, 8$ **43.** 3, 13

45. No solution   **47.** -5 **49.** 8   **51.** 3   **53.** 5

55. $-\frac{11}{10}, 2$ **57.** 20   **59.** $\frac{3}{2}$ **61.** $3, -1$ **63.** $\frac{17}{4}$

65. 2, 3   **67.** (a) and (b) $(-2, 0)$

69. (a) and (b) $(-1, 0), (1, 0)$

71. (a)   (b) $(4, 0)$

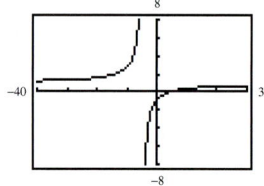

73. (a)   (b) $(1, 0)$

75. (a)   (b) $(-3, 0), (2, 0)$

77. -12 **79.** $\dfrac{x^2 + 2x + 8}{(x + 4)(x - 4)}$ **81.** $8, \dfrac{1}{8}$

83. 40 miles per hour

85. 8 miles per hour; 10 miles per hour

87. (d) 3 miles per hour, obtained by solving

$$\frac{10}{5 - x} + \frac{10}{5 + x} = 6.25$$

 (e) Yes. When $x = 4$, the time for the entire trip is

$$f(4) = \frac{10}{5 - 4} + \frac{10}{5 + 4} \approx 11.1 \text{ hours.}$$

 Because the result is less than 12 hours, you will be able to make the trip.

89. An extraneous solution is an extra solution found by multiplying both sides of the original equation by an expression containing the variable. It is identified by checking all solutions in the original equation.

91. When the equation involves only two fractions, one on each side of the equation, the equation can be solved by cross-multiplication.

Section 7.6   *(page 455)*

> ### Review   *(page 455)*
>
> **1.** $(-\infty, \infty)$ **2.** $(-\infty, 0) \cup (0, \infty)$
>
> **3.**   Yes, the graphs are the same.
>
> **4.** Answers will vary.   **5.** Answers will vary.
>
> **6.** Answers will vary.   **7.** $h + 4, \ h \ne 0$
>
> **8.** $-\dfrac{3}{7(h + 7)}, \ h \ne 0$ **9.** $C = 12,000 + 5.75x$
>
> **10.** $P = 5w$

1. $I = kV$ **3.** $V = kt$ **5.** $u = kv^2$ **7.** $p = k/d$

9. $A = k/t^4$ **11.** $A = klw$ **13.** $P = k/V$

15. Area varies jointly as the base and the height.

17. Volume varies jointly as the square of the radius and the height.

19. Average speed varies directly as the distance and inversely as the time.

21. $s = 5t$ **23.** $F = \frac{5}{16}x^2$ **25.** $n = 48/m$

27. $g = 4/\sqrt{z}$ **29.** $F = \frac{25}{6}xy$ **31.** $d = (120x^2)/r$

33. 4 miles per hour   **35.** 10 people   **37.** 9 hours

39. 15 hours; $22\frac{1}{2}$ hours

41. (a) $\{1, 2, 3, 4, \ldots\}$

(b)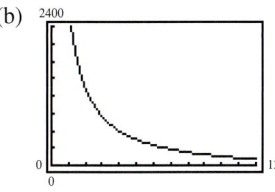

(c) 10d

(d) $153 = -139.1 + 2921/x$

   $292.1 = 2921/x$

   $x = 10$

43. \$4921.25; Price per unit

45. (a) 2 inches   (b) 15 pounds   **47.** 18 pounds

49. 32 feet per second per second   **51.** $208\frac{1}{3}$ feet

53. 3072 watts   **55.** 100

57. 0.36 pounds per square inch; 116 pounds

59. $T = \dfrac{4000}{d}$, 0.91°C

61.

x	2	4	6	8	10
$y = kx^2$	4	16	36	64	100

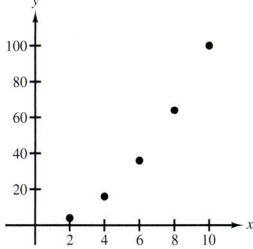

63.

x	2	4	6	8	10
$y = kx^2$	2	8	18	32	50

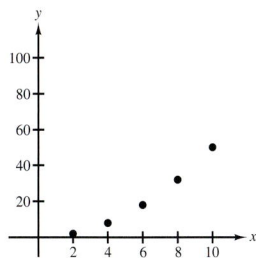

65.

x	2	4	6	8	10
$y = \dfrac{k}{x^2}$	$\frac{1}{2}$	$\frac{1}{8}$	$\frac{1}{18}$	$\frac{1}{32}$	$\frac{1}{50}$

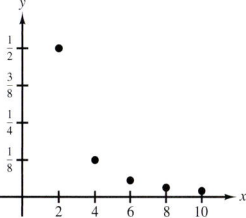

67.

x	2	4	6	8	10
$y = \dfrac{k}{x^2}$	$\frac{5}{2}$	$\frac{5}{8}$	$\frac{5}{18}$	$\frac{5}{32}$	$\frac{1}{10}$

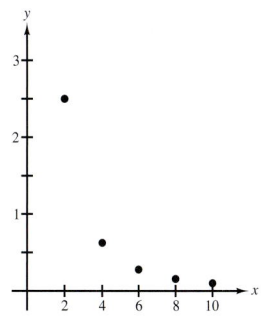

69. $y = k/x$ with $k = 4$

71. Increase. Because $y = kx$ and $k > 0$, the variables increase or decrease together.

73. The variable y will quadruple. If $y = kx^2$ and x is replaced with $2x$, you have $y = k(2x)^2 = 4kx^2$.

75. Answers will vary.

Review Exercises   *(page 461)*

1. $(-\infty, 8) \cup (8, \infty)$ **3.** $(-\infty, 1) \cup (1, 6) \cup (6, \infty)$

5. $(0, \infty)$ **7.** $\dfrac{2x^3}{5}$, $x \neq 0$, $y \neq 0$ **9.** $\dfrac{b - 3}{6(b - 4)}$

11. -9, $x \neq y$ **13.** $\dfrac{x}{2(x + 5)}$, $x \neq 5$ **15.** $3x^5y^2$

17. $\dfrac{y}{8x}$, $y \neq 0$ **19.** $12z(z - 6)$, $z \neq -6$

21. $-\frac{1}{4}$, $u \neq 0$, $u \neq 3$ **23.** $\frac{8}{5}x^3$, $x \neq 0$

25. $\dfrac{125y}{x}$, $y \neq 0$ **27.** $\dfrac{1}{3x - 2}$, $x \neq -2$, $x \neq -1$

29. $\dfrac{x(x - 1)}{x - 7}$, $x \neq -1$, $x \neq 1$ **31.** $3x$

33. $\dfrac{4}{x}$ **35.** $\dfrac{5y + 11}{2y + 1}$ **37.** $\dfrac{7x - 16}{x + 2}$

39. $\dfrac{4x + 3}{(x + 5)(x - 12)}$ **41.** $\dfrac{5x^3 - 5x^2 - 31x + 13}{(x - 3)(x + 2)}$

43. $\dfrac{2x + 17}{(x - 5)(x + 4)}$ **45.** $\dfrac{6(x - 9)}{(x + 3)^2(x - 3)}$

47.

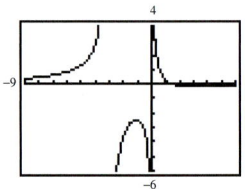

49. $3x^2,\ x \neq 0$ **51.** $\dfrac{6(x + 5)}{x(x + 7)},\ x \neq \pm 5$

53. $\dfrac{3t^2}{5t - 2},\ t \neq 0,\ t \neq \dfrac{2}{5}$ **55.** $x - 1,\ x \neq 0,\ x \neq 2$

57. $\dfrac{-a^2 + a + 16}{(4a^2 + 16a + 1)(a - 4)},\ a \neq 0,\ a \neq -4$

59. -120 **61.** $\dfrac{36}{23}$

63. 5 **65.** $-4, 6$ **67.** $-\dfrac{16}{3}, 3$ **69.** $-\dfrac{5}{2}, 1$

71. $-2, 2$ **73.** $-\dfrac{9}{5}, 3$ **75.** 56 miles per hour

77. 4 people   **79.** 8 years   **81.** 150 pounds

83. 2.44 hours   **85.** \$922.50

Chapter Test   *(page 465)*

1. $(-\infty, 2) \cup (2, 3) \cup (3, \infty)$ **2.** $-\dfrac{1}{3},\ x \neq 2$

3. $\dfrac{2a + 3}{5},\ a \neq 4$ **4.** $3x^3(x + 4)^2$ **5.** $\dfrac{5z}{3},\ z \neq 0$

6. $\dfrac{4}{y + 4},\ y \neq 2$ **7.** $(2x - 3)^2(x + 1),\ x \neq -\dfrac{3}{2}, x \neq -1$

8. $\dfrac{14y^6}{15},\ x \neq 0$ **9.** $\dfrac{-2x^2 + 2x + 1}{x + 1}$

10. $\dfrac{5x^2 - 15x - 2}{(x - 3)(x + 2)}$ **11.** $\dfrac{5x^3 + x^2 - 7x - 5}{x^2(x + 1)^2}$

12. $4,\ x \neq -1$ **13.** $\dfrac{x^3}{4},\ x \neq 0,\ x \neq -2$

14. $-(3x + 1),\ x \neq 0,\ x \neq \dfrac{1}{3}$

15. $\dfrac{(3y + x^2)(x + y)}{x^2 y},\ x \neq -y$

16. $\dfrac{ab}{2b + a},\ a \neq 0, b \neq 0, a \neq 2b$ **17.** 22

18. $-1, -\dfrac{15}{2}$ **19.** No solution   **20.** $V = \dfrac{1}{4}\sqrt{u}$

21. 240 cubic meters

Chapter 8

Section 8.1   *(page 479)*

Review   *(page 479)*

1. One   **2.** Multiply each side of the equation by the lowest common denominator.

3. $-\dfrac{3}{2}$ **4.** -3 **5.** $\dfrac{5}{11}$ **6.** $\dfrac{14}{11}$

7. 50   **8.** 64   **9.** $\dfrac{250}{r}$ **10.** $3L$

1. (a) Solution   (b) Not a solution

3. (a) Not a solution   (b) Solution

5. (a) Solution   (b) Not a solution

7. $(2, 0)$ **9.** $(-1, -1)$ **11.** Infinitely many solutions

13. No solution

15.

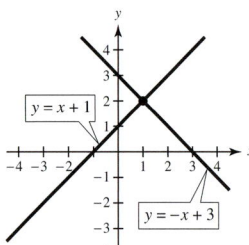

$(1, 2)$

17.

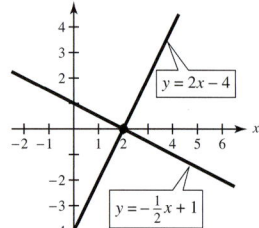

$(2, 0)$

19.

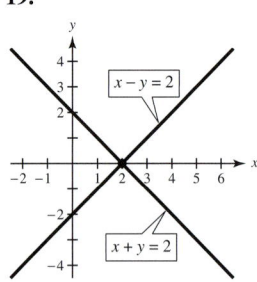

$(2, 0)$

21.

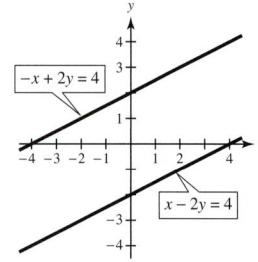

No solution

23.

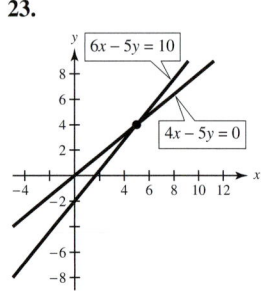

$(5, 4)$

25.

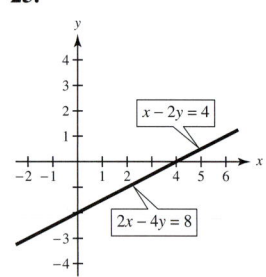

Infinitely many solutions

27.

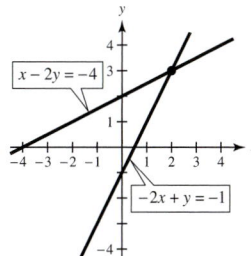

(2, 3)

29.

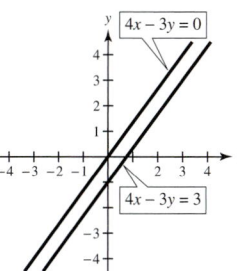

No solution

31.

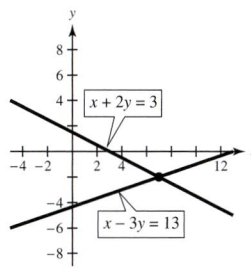

(7, −2)

33.

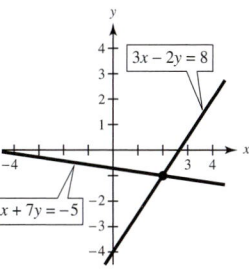

(2, −1)

35.

No solution

37.

Infinitely many solutions

39.

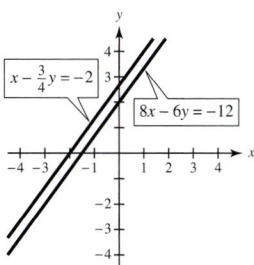

No solution

41.

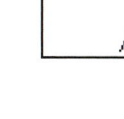

(2, 3)

43.

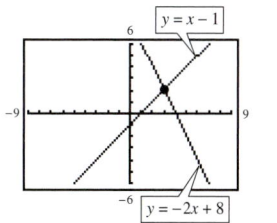

(3, 2)

45. $y = \frac{2}{3}x + 4$, $y = \frac{2}{3}x - 1$, No solution

47. $y = \frac{1}{4}x + \frac{7}{4}$, $y = \frac{1}{4}x + \frac{7}{4}$, Infinitely many solutions

49. $y = \frac{2}{3}x + \frac{4}{3}$, $y = -\frac{2}{3}x + \frac{8}{3}$, One solution

51. $y = \frac{3}{4}x + \frac{9}{8}$, $y = \frac{3}{4}x + \frac{3}{2}$, No solution

53. $(1, 0)$ **55.** No solution

57. Infinitely many solutions   **59.** $(2, 3)$ **61.** $(15, 5)$

63. $(4, -3)$ **65.** No solution   **67.** $(0, 0)$ **69.** $(2, 6)$

71. $\left(\frac{1}{2}, 3\right)$ **73.** $(-3, 2)$ **75.** $\left(\frac{5}{2}, -\frac{1}{2}\right)$ **77.** $(0, 0)$

79. $(3, -2)$ **81.** Infinitely many solutions   **83.** $\left(\frac{5}{2}, 15\right)$

85. No solution   **87.** No solution

89. Infinitely many solutions   **91.** $(6, 0)$ **93.** $\left(\frac{5}{2}, \frac{3}{4}\right)$

95. $(8, 4)$ **97.** $(2, 6)$ **99.** $\left(\frac{18}{5}, \frac{3}{5}\right)$ **101.** $(3, 0)$

103. 9, 11

105. (a) $\text{Cost} = \boxed{\begin{array}{c}\text{Cost per}\\\text{unit}\end{array}} \cdot \boxed{\begin{array}{c}\text{Number}\\\text{of units}\end{array}} + \boxed{\begin{array}{c}\text{Fixed}\\\text{costs}\end{array}}$

$\text{Revenue} = \boxed{\begin{array}{c}\text{Price per}\\\text{unit}\end{array}} \cdot \boxed{\begin{array}{c}\text{Number}\\\text{of units}\end{array}}$

(b) $x = 64$ units, $C = R = \$1472$; This means that the company must sell 64 feeders to cover their cost. Sales over 64 feeders will generate profit.

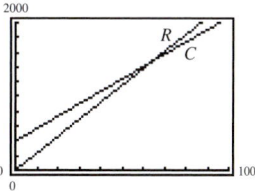

107. Because the slopes of the two lines are not equal, the lines intersect and the system has one solution: $(79,400, 398)$.

109. 5%: $10,000; 8%: $5000   **111.** 2 adults

113. 50,000 miles   **115.** $2x - y - 9 = 0$

117. (a) $\begin{cases} x + y = 115 \\ 8x + 15y = 1445 \end{cases}$

(b) 40 student tickets, 75 nonstudent tickets

119. A system that has an infinite number of solutions

121. False. It may have one solution or infinitely many solutions.

123. (a) Solve one of the equations for one variable in terms of the other.

    (b) Substitute the expression found in Step (a) in the other equation to obtain an equation in one variable.

    (c) Solve the equation obtained in Step (b).

    (d) Back-substitute the solution from Step (c) in the expression obtained in Step (a) to find the value of the other variable.

    (e) Check the solution in the original system.

125. Solve one of the equations for one variable in terms of the other variable. Substitute that expression in the other equation. If an identity statement results, the system has infinitely many solutions.

127. $\begin{cases} x + 2y = 5 \\ -x + 3y = 0 \end{cases}$ **129.** $\begin{cases} x + y = 3 \\ 2x + 2y = 6 \end{cases}$ **131.** $b = 2$

133. $b = -\frac{1}{3}$

Section 8.2   *(page 491)*

Review *(page 491)*

1. Multiplicative Inverse Property

2. Additive Identity Property

3. Commutative Property of Multiplication

4. Associative Property of Multiplication

5.

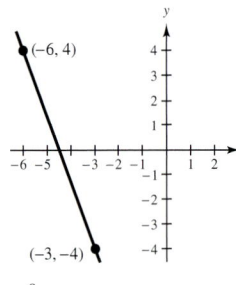

$-\frac{8}{3}$

6.

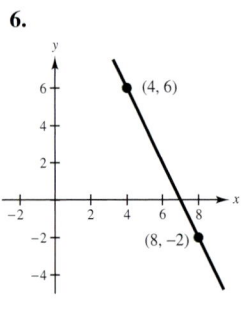

-2

7.

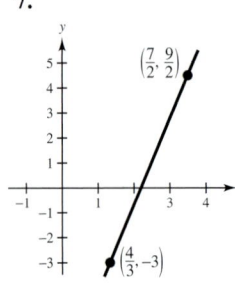

$\frac{45}{13}$

8.

-17

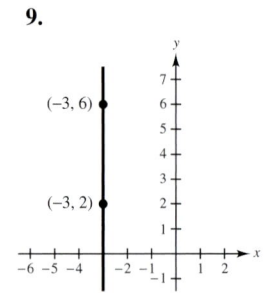

9. Undefined   **10.** 0

11. 150 defective units   **12.** $0 < t \le 18.4$

---

1. $(2, 0)$ **3.** $(-1, -1)$ **5.** $(8, 4)$ **7.** $(-4, 4)$

9. $(2, 1)$ **11.** $\left(\frac{13}{3}, -2\right)$ **13.** No solution

15. $(1, -2)$ **17.** $(-1, 1)$ **19.** $\left(\frac{7}{25}, -\frac{1}{25}\right)$

21. $(4, -1)$ **23.** $(4, -1)$ **25.** $\left(\frac{1}{2}, 0\right)$ **27.** $\left(6, \frac{3}{2}\right)$

29. $(-17, 14)$ **31.** $(8, 7)$ **33.** $(-1, 2)$

35. $(2, -1)$ **37.** $(1, 1)$ **39.** $(5, 3)$ **41.** $(8, 4)$

43. $(2, 3)$ **45.** $(-5, -3)$ **47.** $(3, -3)$ **49.** $\left(\frac{4}{3}, \frac{4}{3}\right)$

51. $\left(1, -\frac{5}{4}\right)$ **53.** 15, 25   **55.** 34, 48

57. Two-point baskets: 7; Three-point baskets: 2

59. Student ticket: \$3; General admission ticket: \$5

61. \$4000

63. Private-lesson students: 7; Group lesson students: 5

65. Yes, it is.   **67.** $2x + y - 7 = 0$

69. Obtain opposite coefficients for x (or y), add the equations and solve the resulting equation, back-substitute the value you just obtained in either of the original equations and solve for the other variable, and check your solution in each of the original equations.

71. When you add the equations to eliminate one variable, both variables are eliminated, yielding an identity. For example, adding the equations in the system $x - y = 3$ and $-x + y = -3$ yields $0 = 0$.

73. $\begin{cases} x - 4y = 3 \\ 7x + 9y = 11 \end{cases}$

75. Infinitely many solutions. Because two solutions are given, the system is dependent.

Section 8.3   *(page 503)*

Review   *(page 503)*

1. One solution

2. Multiply each side of the equation by the lowest common denominator, 24.

3. $\frac{27}{4}$ **4.** 11   **5.** $-1, 5$ **6.** 2, 12

7. $-\infty < x < \infty$ **8.** $\frac{1}{2}$ **9.** $V = s^3$

10. $A = \dfrac{C^2}{4\pi}$

11. $d = 15t$ **12.** $P = 180 + 1.25n$

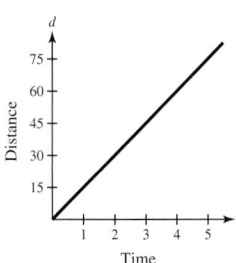

1. (a) Not a solution   (b) Solution
   (c) Solution   (d) Not a solution

3. $(22, -1, -5)$ **5.** $(14, 3, -1)$

7. No. When the first equation was multiplied by -2 and added to the second equation, the constant term should have been -11. Also, when the first equation was multiplied by -3 and added to the third equation, the coefficient of z should have been 2.

9. $\begin{cases} x - 2y = 8 \\ \quad\quad y = 14 \end{cases}$ Eliminated the x-term in Equation 2

11. $(1, 2, 3)$ **13.** $(1, 2, 3)$ **15.** $(2, -3, -2)$

17. No solution   **19.** $(-4, 8, 5)$ **21.** $(5, -2, 0)$

23. $\left(\frac{3}{10}, \frac{2}{5}, 0\right)$ **25.** $(-4, 2, 3)$ **27.** $\left(-\frac{1}{2}a + \frac{1}{2}, \frac{3}{5}a + \frac{2}{5}, a\right)$

29. $\left(-\frac{1}{2}a + \frac{1}{4}, \frac{1}{2}a + \frac{5}{4}, a\right)$ **31.** $(1, -1, 2)$

33. $\left(\frac{6}{13}a + \frac{10}{13}, \frac{5}{13}a + \frac{4}{13}, a\right)$ **35.** $\begin{cases} x + 2y - z = -4 \\ \quad\quad y + 2z = 1 \\ 3x + y + 3z = 15 \end{cases}$

37. $s = -16t^2 + 144$ **39.** $s = -16t^2 + 48t$

41. $88°, 32°, 60°$

43. \$17,404 at 6%, \$31,673 at 10%, \$30,923 at 15%

45. \$4200 at 6%, \$2100 at 8%, \$3200 at 9%

47. 20 gallons of spray X, 18 gallons of spray Y, 16 gallons of spray Z

49. \$20 arrangements: 400; \$30 arrangements: 100; \$40 arrangements: 350

51. (a) Not possible
   (b) 10% solution: 0 gallons; 15% solution: 6 gallons; 25% solution: 6 gallons
   (c) 10% solution: 4 gallons; 15% solution: 0 gallons; 25% solution: 8 gallons

53. Strings: 50; Winds: 20; Percussion: 8

55. (c) $\begin{cases} x + y + z = 200 \\ 8x + 15y + 100z = 4995 \\ x \quad\quad - 4z = 0 \end{cases}$

   (d) Students: 140; Nonstudents: 25; Major contributors: 35

   (e) No, it is not possible. To verify this, let $z = 18$ and solve the system of linear equations. The resulting values do not fulfill all the club's goals.

57. Substitute $y = 3$ in the first equation to obtain $x + 2(3) = 2$ or $x = 2 - 6 = -4$.

59. Answers will vary.

Mid-Chapter Quiz   *(page 507)*

1. (a) Not a solution   (b) Solution

2.

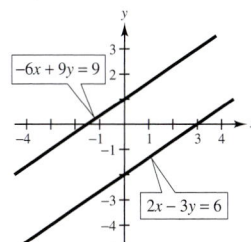

No solution

3.

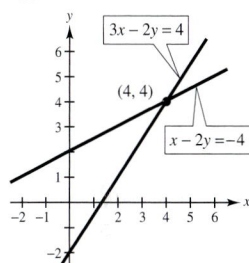

One solution

4.

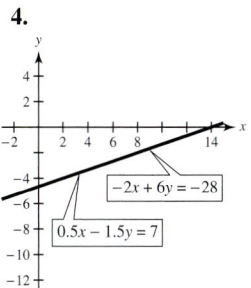

Infinitely many solutions

5.

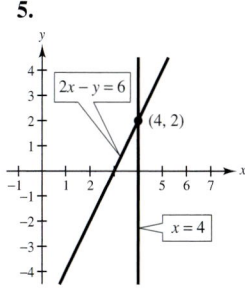

$(4, 2)$

6.

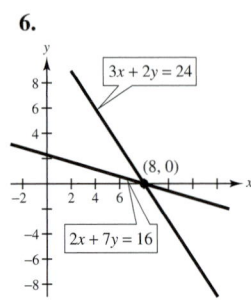

7.

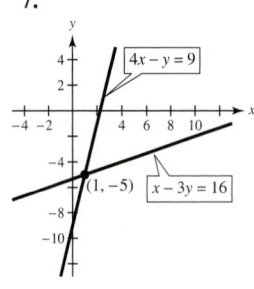

$(8, 0)$

$(1, -5)$

8. $(5, 2)$ **9.** No solution   **10.** $\left(\frac{90}{13}, \frac{34}{13}\right)$

11. $(8, 1)$ **12.** $(-2, 4)$ **13.** $\left(\frac{1}{2}, -\frac{1}{2}, 1\right)$

14. $(5, -1, 3)$ **15.** $\begin{cases} x + y = -2 \\ 2x - y = 32 \end{cases}$

16. $\begin{cases} x + y - z = 11 \\ x + 2y - z = 14 \\ -2x + y + z = -6 \end{cases}$

17.

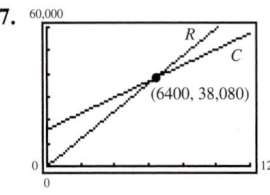

The point of intersection of the two functions, $(6400, 38{,}080)$, represents the break-even point.

18. $45°; 80°; 55°$

Section 8.4   *(page 516)*

Review   *(page 516)*

1. Additive Inverse Property

2. Multiplicative Identity Property

3. Commutative Property of Addition

4. Associative Property of Multiplication

5.

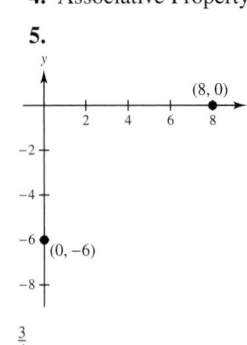

6.

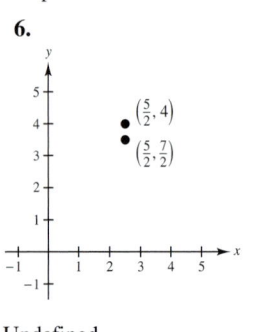

$\frac{3}{4}$ Undefined

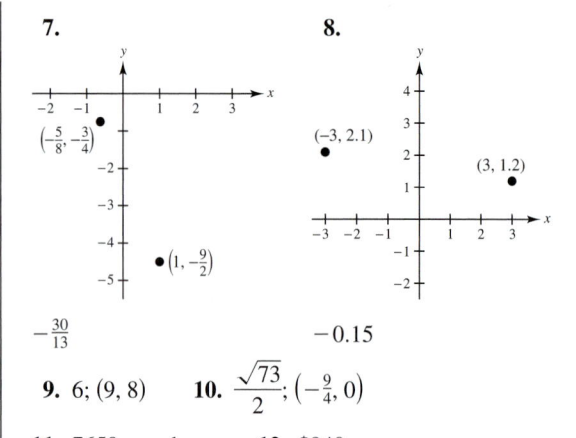

7.   **8.**

$-\frac{30}{13}$ -0.15

9. $6; (9, 8)$ **10.** $\dfrac{\sqrt{73}}{2}; \left(-\frac{9}{4}, 0\right)$

11. 7650 members   **12.** $\$940$

1. 4×2 **3.** 2×2 **5.** 4×1 **7.** 1×1

9. 1×4 **11.** (a) $\begin{bmatrix} 4 & -5 \\ -1 & 8 \end{bmatrix}$ (b) $\begin{bmatrix} 4 & -5 & \vdots & -2 \\ -1 & 8 & \vdots & 10 \end{bmatrix}$

13. (a) $\begin{bmatrix} 1 & 10 & -3 \\ 5 & -3 & 4 \\ 2 & 4 & 0 \end{bmatrix}$ (b) $\begin{bmatrix} 1 & 10 & -3 & \vdots & 2 \\ 5 & -3 & 4 & \vdots & 0 \\ 2 & 4 & 0 & \vdots & 6 \end{bmatrix}$

15. (a) $\begin{bmatrix} 5 & 1 & -3 \\ 0 & 2 & 4 \end{bmatrix}$ (b) $\begin{bmatrix} 5 & 1 & -3 & \vdots & 7 \\ 0 & 2 & 4 & \vdots & 12 \end{bmatrix}$

17. $\begin{cases} 4x + 3y = 8 \\ x - 2y = 3 \end{cases}$ **19.** $\begin{cases} x \quad\quad + 2z = -10 \\ \quad 3y - z = 5 \\ 4x + 2y \quad\quad = 3 \end{cases}$

21. $\begin{cases} 5x + 8y + 2z \quad\quad = -1 \\ -2x + 15y + 5z + w = 9 \\ x + 6y - 7z \quad\quad = -3 \end{cases}$

23. $\begin{cases} 13x + y + 4z - 2w = -4 \\ 5x + 4y \quad\quad - w = 0 \\ x + 2y + 6z + 8w = 5 \\ -10x + 12y + 3z + w = -2 \end{cases}$

25. 2 **27.** $-2, \frac{2}{3}$ **29.** $-2, 6, 20, 4, 20, 4$

31. $\begin{bmatrix} 1 & 2 & 3 \\ 0 & 1 & 2 \end{bmatrix}$ **33.** $\begin{bmatrix} 1 & 0 & -\frac{7}{5} \\ 0 & 1 & \frac{11}{10} \end{bmatrix}$

35. $\begin{bmatrix} 1 & 1 & 0 & 5 \\ 0 & 1 & 2 & 0 \\ 0 & 0 & 1 & -1 \end{bmatrix}$ **37.** $\begin{bmatrix} 1 & -1 & -1 & 1 \\ 0 & 1 & 6 & 3 \\ 0 & 0 & 1 & \frac{4}{5} \end{bmatrix}$

39. $\begin{bmatrix} 1 & 1 & -1 & 3 \\ 0 & 1 & -4 & 1 \\ 0 & 0 & 0 & 0 \end{bmatrix}$

41. $\begin{cases} x - 2y = 4 \\ y = -3 \end{cases}$ **43.** $\begin{cases} x + 5y = 3 \\ y = -2 \end{cases}$

$(-2, -3)$ $(13, -2)$

45. $\begin{cases} x - y + 2z = 4 \\ y - z = 2 \\ z = -2 \end{cases}$ **47.** $\left(\frac{9}{5}, \frac{13}{5}\right)$ **49.** $(1, 1)$

$(8, 0, -2)$

51. $(-2, 1)$ **53.** No solution   **55.** $(2, -3, 2)$

57. $(2a + 1, 3a + 2, a)$ **59.** $(1, 2, -1)$ **61.** $(1, -1, 2)$

63. $(34, -4, -4)$ **65.** No solution

67. $(-12a - 1, 4a + 1, a)$ **69.** $\left(\frac{1}{2}, 2, 4\right)$ **71.** $\left(2, 5, \frac{5}{2}\right)$

73. 8%: $800,000, 9%: $500,000, 12%: $200,000

75. Certificates of deposit: $250,000 - \frac{1}{2}s$

Municipal bonds: $125,000 + \frac{1}{2}s$
Blue-chip stocks: $125,000 - s$
Growth stocks: s
If $s = \$100,000$, then
Certificates of deposit: $200,000
Municipal bonds: $175,000
Blue-chip stocks: $25,000
Growth stocks: $100,000

77. Peanuts: 15 pounds; Almonds: 10 pounds;

Pistachios: 25 pounds

79. 5, 8, 20

81. 15 computer chips, 10 resistors, 10 transistors

83. (a) Interchange two rows.
(b) Multiply a row by a nonzero constant.
(c) Add a multiple of a row to another row.

85. The one matrix can be obtained from the other by using the elementary row operations.

87. There will be a row in the matrix with all zero entries except in the last column.

Section 8.5 *(page 529)*

Review *(page 529)*

1. The set of inputs of the function.

2. The set of outputs of the function.

3. Relations may have ordered pairs with the same first component and different second components, while functions cannot.

4. The value of the function when $x = 4$.

5. -4 **6.** 5   **7.** 5   **8.** 4

9.

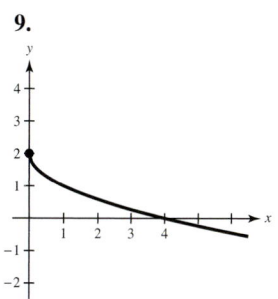

Domain: $0 \leq x < \infty$; Range: $-\infty < y \leq 2$

10.

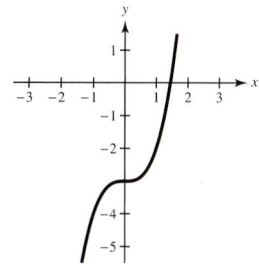

Domain: $-\infty < x < \infty$; Range: $-\infty < y < \infty$

11. $\dfrac{320}{r}$ **12.** $\dfrac{9}{2}x + 7$

1. 5   **3.** 27   **5.** 0   **7.** 6   **9.** -24

11. -0.16 **13.** -24 **15.** -2 **17.** -30 **19.** 3

21. 0   **23.** -58 **25.** 102   **27.** -0.22

29. $x - 5y + 2$ **31.** 248   **33.** 105.625   **35.** -6.37

37. $(1, 2)$ **39.** $(2, -2)$ **41.** $\left(\frac{3}{4}, -\frac{1}{2}\right)$

43. Not possible, $D = 0$ **45.** $\left(\frac{2}{3}, \frac{1}{2}\right)$ **47.** $(-1, 3, 2)$

49. $(1, -2, 1)$ **51.** Not possible, $D = 0$ **53.** $\left(\frac{22}{27}, \frac{22}{9}\right)$

55. $\left(\frac{51}{16}, -\frac{7}{16}, -\frac{13}{16}\right)$ **57.** 1, 6   **59.** 16   **61.** 7

63. $\frac{31}{2}$ **65.** $\frac{33}{8}$ **67.** 16   **69.** $\frac{53}{2}$

71. 250 square miles   **73.** Collinear   **75.** Collinear

77. Not collinear   **79.** $3x - 5y = 0$

81. $7x - 6y - 28 = 0$ **83.** $9x + 10y + 3 = 0$

85. $16x - 15y + 22 = 0$ **87.** $I_1 = 1, I_2 = 2, I_3 = 1$

89. $I_1 = 3, I_2 = 4, I_3 = 1$

91. (a) $\left(\dfrac{13}{2k + 6}, \dfrac{4 - 3k}{2k^2 + 6k}\right)$ (b) $0, -3$

93. (f) and (g) 140 students, 25 nonstudents, 35 major contributors

95. No. The matrix must be square.

97. The coefficient matrix of the system must be square and its determinant must be a nonzero real number.

Section 8.6 *(page 538)*

Review *(page 538)*

1. The graph of $h(x)$ is a vertical shift of $f(x)$ upward c units.

2. The graph of $h(x)$ is a vertical shift of $f(x)$ downward c units.

3. The graph of $h(x)$ is a horizontal shift of $f(x)$ to the right c units.

4. The graph of $h(x)$ is a horizontal shift of $f(x)$ to the left c units.

5. The graph of $h(x)$ is a reflection of $f(x)$ in the x-axis.

6. The graph of $h(x)$ is a reflection of $f(x)$ in the y-axis.

7. $h(x) = (x - 3)^2$ **8.** $h(x) = -(x + 1)^2$

9.

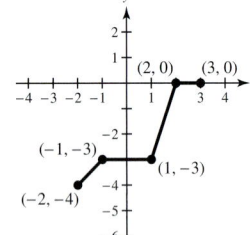

10.

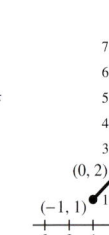

11. \$230.52   **12.** \$35.20

1. c   **2.** b   **3.** f   **4.** e   **5.** a   **6.** d

7.

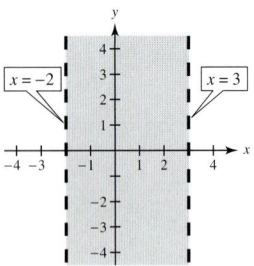

9.

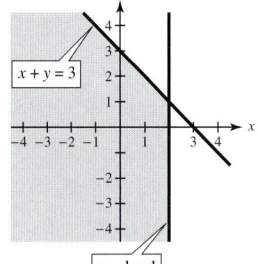

11.

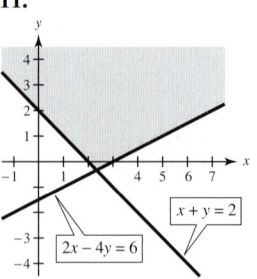

13.

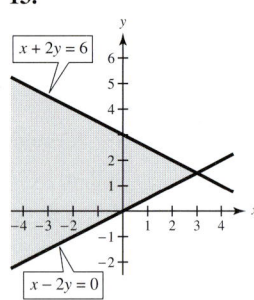

15.

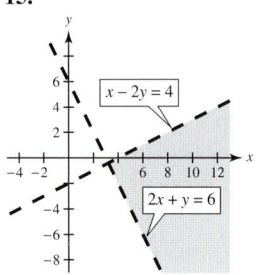

17.

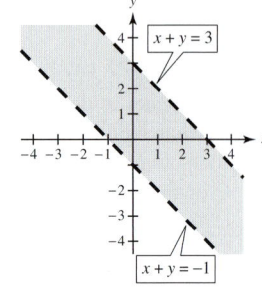

19.

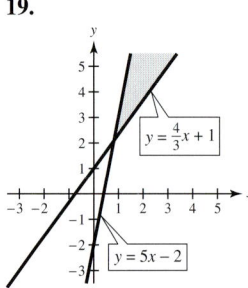

21.

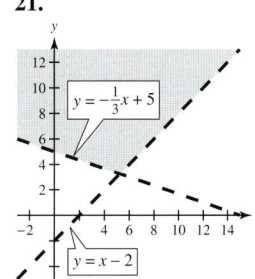

23.

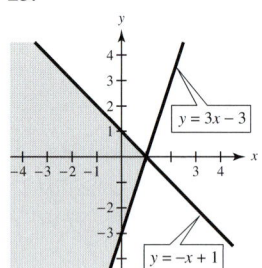

25.

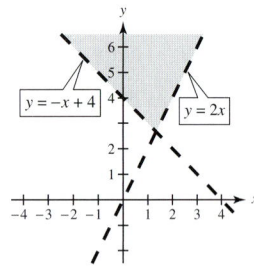

27.

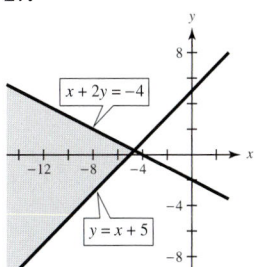

29.

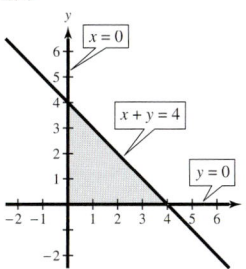

31.

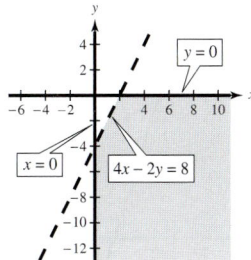

33.

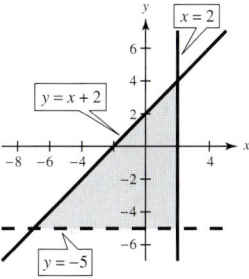

35.

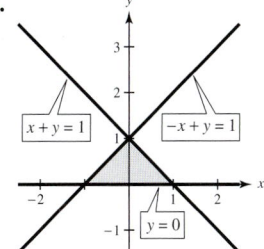

37. No solution

39.

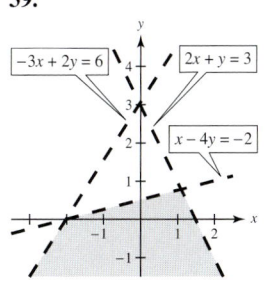

41.

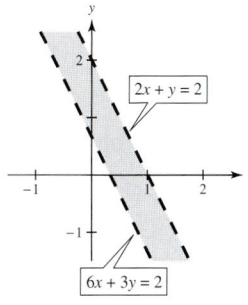

43.

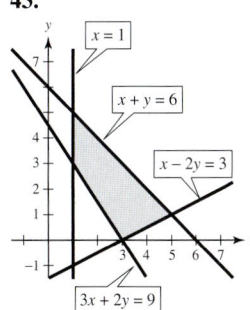

45.

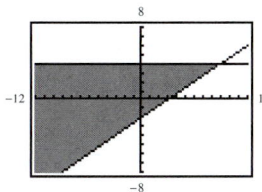

47.

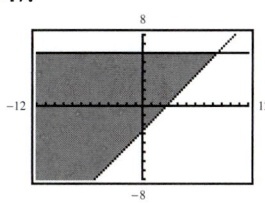

49.

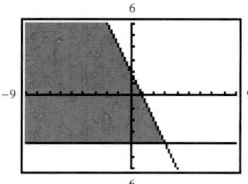

51. $\begin{cases} x \geq & 1 \\ y \geq & x - 3 \\ y \leq & -2x + 6 \end{cases}$

53. $\begin{cases} x \geq & 1 \\ x \leq & 8 \\ y \geq & -5 \\ y \leq & 3 \end{cases}$

55. $\begin{cases} y \leq \frac{9}{10}x + \frac{42}{5} \\ y \geq 3x \\ y \geq \frac{2}{3}x + 7 \end{cases}$

57. $\begin{cases} x + \frac{3}{2}y \leq 12 \\ \frac{4}{3}x + \frac{3}{2}y \leq 15 \\ x \qquad \geq 0 \\ y \qquad \geq 0 \end{cases}$

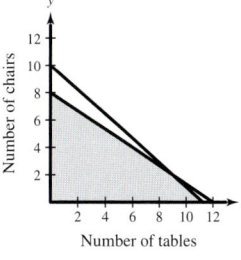

59. $\begin{cases} x + y \leq 20,000 \\ x \qquad \geq 5000 \\ y \geq 2x \end{cases}$

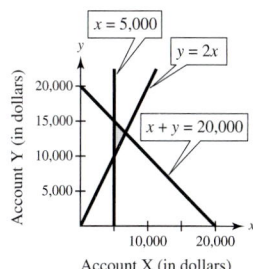

61. $\begin{cases} x + y \geq 15,000 \\ 15x + 25y \geq 275,000 \\ x \qquad \geq 8000 \\ y \geq 4000 \end{cases}$

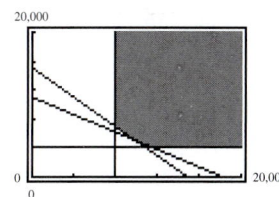

63. $\begin{cases} x \leq 90 \\ y \leq 0 \\ y \geq -10 \\ y \geq -\frac{1}{7}x \end{cases}$

65. $\begin{cases} 20x + 10y \geq 280 \\ 15x + 10y \geq 160 \\ 10x + 20y \geq 180 \\ x \qquad \geq 0 \\ y \geq 0 \end{cases}$

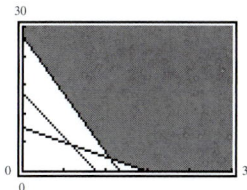

67. The graph of a linear equation splits the *xy*-plane into two parts, each of which is a half-plane. $y < 5$ is a half-plane.

69. Find all intersections between the lines corresponding to the inequalities.

Review Exercises *(page 544)*

1. (a) Solution   (b) Not a solution

3. (a) Not a solution   (b) Solution   **5.** $(1, 2)$

7. $(0, -3)$

9.

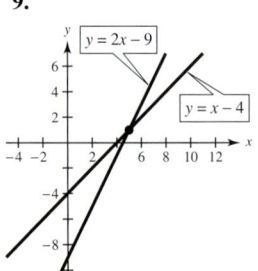

$(5, 1)$

11.

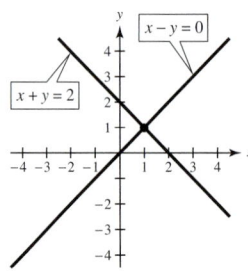

$(1, 1)$

13.

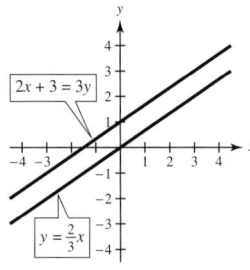

No solution

15. $(4, 8)$ **17.** $(4, 2)$ **19.** $(4, -1)$ **21.** $\left(\frac{5}{2}, 3\right)$

23. No solution   **25.** Infinitely many solutions

27. 5%: $8000; 10%: $4000   **29.** 6 dimes, 9 quarters

31. $(2, 1)$ **33.** $(10, -12)$

35. Infinitely many solutions   **37.** $(-0.2, 0.7)$

39. $(5, 6)$ **41.** $(4, 2)$ **43.** $(-1, -1)$ **45.** $\left(4, -\frac{3}{2}\right)$

47. Infinitely many solutions   **49.** $(10, 0)$ **51.** $(0, 0)$

53. $(-3, 7)$ **55.** $\left(\frac{1}{3}, -\frac{1}{2}\right)$

57. Gasoline: $1.17 per gallon; Diesel fuel: $1.25 per gallon

59. $(3, 2, 5)$ **61.** $(0, 3, 6)$ **63.** $(2, -3, 3)$

65. $(0, 1, -2)$

67. $8000 at 7%, $5000 at 9%, $7000 at 11%   **69.** 1×2

71. 2×3

73. (a) $\begin{bmatrix} 3 & -2 \\ -1 & 1 \end{bmatrix}$ (b) $\begin{bmatrix} 3 & -2 & \vdots & 12 \\ -1 & 1 & \vdots & -2 \end{bmatrix}$

75. $\begin{cases} 4x - y = 2 \\ 6x + 3y + 2z = 1 \\ y + 4z = 0 \end{cases}$ **77.** $(10, -12)$

79. $(0.6, 0.5)$ **81.** $\left(\frac{24}{5}, \frac{22}{5}, -\frac{8}{5}\right)$ **83.** $\left(\frac{1}{2}, -\frac{1}{3}, 1\right)$

85. $(1, 0, -4)$ **87.** 5   **89.** -51 **91.** 1

93. $(-3, 7)$ **95.** Not possible, $D = 0$ **97.** $(2, -3, 3)$

99. 16   **101.** 7   **103.** Not collinear

105. $x - 2y + 4 = 0$ **107.** $3x + 2y - 16 = 0$

109.

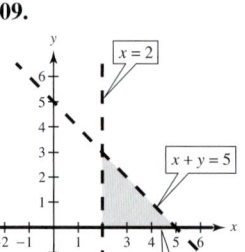

111.

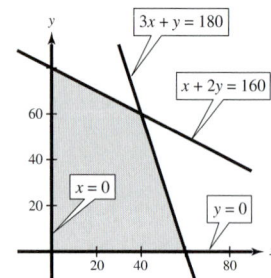

113. $\begin{cases} x + y \le 1500 \\ x \ge 400 \\ y \ge 600 \end{cases}$

where *x* represents the number of bushels for Harrisburg and *y* for Philadelphia.

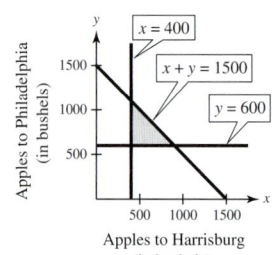

Chapter Test *(page 549)*

1. (a) Not a solution   (b) Solution

2. $(3, 2)$ **3.** $(2, 4)$ **4.** $(2, 3)$ **5.** $(-2, 2)$

6. $(2, 2a - 1, a)$ **7.** $(-1, 3, 3)$ **8.** $(2, 1, -2)$

9. $\left(4, \frac{1}{7}\right)$ **10.** $(5, 1, -1)$ **11.** -62 **12.** 12

13.

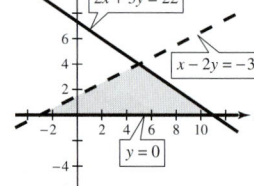

14. 50,000 miles

15. \$13,000 at 4.5%, \$9000 at 5%, \$3000 at 8%

16.
$$\begin{cases} 20x + 30y \geq 400,000 \\ x + y \geq 16,000 \\ y \geq 4000 \\ x \geq 9000 \end{cases}$$

where x is the number of reserved tickets and y is the number of floor tickets.

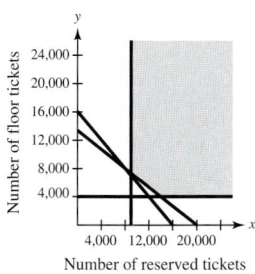

Number of reserved tickets

Chapter 9

Section 9.1 *(page 559)*

Review *(page 559)*

1. a^{m+n} **2.** $a^m b^m$ **3.** a^{mn} **4.** a^{m-n}

5. $y = 4 - 3x$ **6.** $y = \frac{2}{3}(1 - x)$

7. $y = \frac{1}{7}(x - 7)$ **8.** $y = \frac{3}{2}(8x - 1)$ **9.** $-\frac{5}{3}, 0$

10. $0, 10$ **11.** $-4, -2$ **12.** $-6, 7$

---

1. 8   **3.** -7 **5.** -3 **7.** Not a real number

9. Perfect square   **11.** Perfect cube   **13.** Neither

15. Irrational   **17.** Rational   **19.** 8   **21.** 10

23. Not a real number   **25.** $-\frac{2}{3}$ **27.** Not a real number

29. 5   **31.** -23 **33.** 5   **35.** 10   **37.** 6

39. $-\frac{1}{4}$ **41.** 11   **43.** -24 **45.** 3

47. Not a real number

Radical Form	*Rational Exponent Form*
49. $\sqrt{16} = 4$	$16^{1/2} = 4$
51. $\sqrt[3]{125} = 5$	$125^{1/3} = 5$

53. 5   **55.** -6 **57.** $\frac{1}{4}$ **59.** $\frac{1}{9}$ **61.** $\frac{4}{9}$ **63.** $\frac{3}{11}$

65. 9   **67.** -64 **69.** $t^{1/2}$ **71.** x^3 **73.** $u^{7/3}$

75. $x^{-1} = \dfrac{1}{x}$ **77.** $t^{-9/4} = \dfrac{1}{t^{9/4}}$ **79.** x^3 **81.** $y^{13/12}$

83. $x^{3/4}y^{1/4}$ **85.** $y^{5/2}z^4$ **87.** 3   **89.** $\sqrt[3]{2}$ **91.** $\frac{1}{2}$

93. $\sqrt{c}$ **95.** $\dfrac{3y^2}{4z^{4/3}}$ **97.** $\dfrac{9y^{3/2}}{x^{2/3}}$ **99.** $x^{1/4}$

101. $\sqrt[8]{y}$ **103.** $x^{3/8}$ **105.** $\sqrt{x + y}$

107. $\dfrac{1}{(3u - 2v)^{5/6}}$ **109.** 6.7082   **111.** 9.9845

113. 0.0038   **115.** 3.8158   **117.** 66.7213

119. 1.0420   **121.** 0.7915

123. (a) 3   (b) 5   (c) Not a real number   (d) 9

125. (a) 2   (b) 3   (c) -2 (d) -4

127. (a) 2   (b) Not a real number   (c) 3   (d) 1

129. $[0, \infty)$ **131.** $(0, \infty)$ **133.** $(-\infty, \infty)$

135. $\left[-\frac{7}{3}, \infty\right)$ **137.** $\left(-\infty, \frac{4}{9}\right]$

139. Domain: $(0, \infty)$ **141.** Domain: $(-\infty, \infty)$

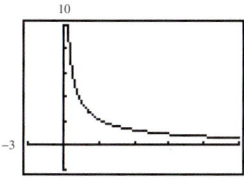   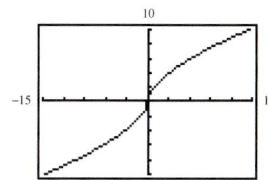

143. $2x^{3/2} - 3x^{1/2}$ **145.** $1 + 5y$ **147.** 0.128

149. 23 feet × 23 feet   **151.** 10.49 centimeters

153. (a) 15 feet   (b) $h = \sqrt{15^2 - \left(\dfrac{a}{2}\right)^2}$

(c) 8.29 feet   (d) 25.38 feet

155. Given $\sqrt[n]{a}$, a is the radicand and n is the index.

157. No. $\sqrt{2}$ is an irrational number. Its decimal representation is a nonterminating, nonrepeating decimal.

159. 0, 1, 4, 5, 6, 9; Yes

Section 9.2 *(page 568)*

Review *(page 568)*

1. Replace the inequality sign with an equal sign and sketch the graph of the resulting equation. (Use a dashed line for < or >, and a solid line for ≤ or ≥.) Test one point in each of the half-planes formed by the graph. If the point satisfies the inequality, shade the entire half-plane to denote that every point in the region satisfies the inequality.

2. The first includes the points on the line $3x + 4y = 4$, whereas the second does not.

3. $-(x - 3)(x^2 + 1)$ **4.** $(2t + 13)(2t - 13)$

5. $(x - 1)(x - 2)$ **6.** $(x - 1)(2x + 7)$

7. $(x + 1)(11x - 5)$ **8.** $(2x - 7)^2$

9. 816 adults; 384 students   **10.** 267 units

---

1. $2\sqrt{5}$ **3.** $5\sqrt{2}$ **5.** $4\sqrt{6}$ **7.** $6\sqrt{6}$ **9.** $13\sqrt{7}$

11. 0.2 **13.** $0.06\sqrt{2}$ **15.** $1.1\sqrt{2}$ **17.** $\dfrac{\sqrt{13}}{5}$

19. $3x^2\sqrt{x}$ **21.** $4y^2\sqrt{3}$ **23.** $3\sqrt{13}|y^3|$

25. $2|x|y\sqrt{30y}$ **27.** $8a^2b^3\sqrt{3ab}$ **29.** $2\sqrt[3]{6}$

31. $2\sqrt[3]{14}$ **33.** $2x\sqrt[3]{5x^2}$ **35.** $3|y|\sqrt{2y}$ **37.** $xy\sqrt[3]{x}$

39. $|x|\sqrt[4]{3y^2}$ **41.** $2xy\sqrt[5]{y}$ **43.** $\dfrac{\sqrt[3]{35}}{4}$ **45.** $\dfrac{2\sqrt[5]{x^2}}{y}$

47. $\dfrac{3a\sqrt[3]{2a}}{b^3}$ **49.** $\dfrac{4a^2\sqrt{2}}{|b|}$ **51.** $3x^2$ **53.** $\dfrac{\sqrt{3}}{3}$

55. $\dfrac{\sqrt{7}}{7}$ **57.** $\dfrac{\sqrt[4]{20}}{2}$ **59.** $\dfrac{3\sqrt[3]{2}}{2}$ **61.** $\dfrac{\sqrt{y}}{y}$

63. $\dfrac{2\sqrt{x}}{x}$ **65.** $\dfrac{\sqrt{2x}}{2x}$ **67.** $\dfrac{2\sqrt{3b}}{b^2}$ **69.** $\dfrac{\sqrt[3]{18xy^2}}{3y}$

71. $3\sqrt{5}$ **73.** 89.44 cycles per second

75. $\sqrt{776} \approx 27.86$ feet   **77.** 1

79. (a) All possible nth powered factors have been removed from each radical.

    (b) No radical contains a fraction.

    (c) No denominator of a fraction contains a radical.

81. When $x < 0$. $\sqrt{x^2}$ will always result in a positive value regardless of the sign of x. So, $\sqrt{x^2} \neq x$ when $x < 0$.

    Example: $\sqrt{(-8)^2} = \sqrt{64} = 8$

Section 9.3   *(page 573)*

Review   *(page 573)*

1. An ordered pair (x, y) of real numbers that satisfies each equation in the system.

2. Yes, if the system is inconsistent.

3. Yes, if the system is dependent.

4. No, it must have one solution, no solution, or infinitely many solutions.

5.

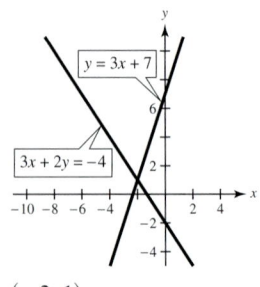

$(-2, 1)$

6.

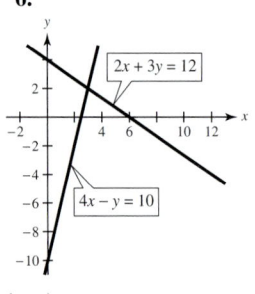

$(3, 2)$

---

7. $\left(-\frac{4}{5}, \frac{2}{5}\right)$ **8.** No solution

9. Infinitely many solutions   **10.** $(4, -8, 10)$

11. DVD: \$29; Videocassette tape: \$14

12. \$20 bills: 7; \$5 bills: 3; \$1 bills: 4

---

1. $2\sqrt{2}$ **3.** $44\sqrt{2}$ **5.** Cannot combine   **7.** $3\sqrt[3]{5}$

9. $13\sqrt[3]{y}$ **11.** $14\sqrt[4]{s}$ **13.** $9\sqrt{2}$

15. $-11\sqrt[4]{3} - 5\sqrt[4]{7}$ **17.** $13\sqrt[3]{7} + \sqrt{3}$ **19.** $21\sqrt{3}$

21. $23\sqrt{5}$ **23.** $30\sqrt[3]{2}$ **25.** $12\sqrt{x}$ **27.** $13\sqrt{x+1}$

29. $13\sqrt{y}$ **31.** $(10 - z)\sqrt[3]{z}$ **33.** $(6a + 1)\sqrt{5a}$

35. $(x + 2)\sqrt[3]{6x}$ **37.** $4\sqrt{x-1}$ **39.** $(x + 2)\sqrt{x-1}$

41. $5a\sqrt[3]{ab^2}$ **43.** $3r^3s^2\sqrt{rs}$ **45.** 0 **47.** $\dfrac{2\sqrt{5}}{5}$

49. $\dfrac{9\sqrt{5}}{5}$ **51.** $\dfrac{5\sqrt{3y}}{3}$ **53.** $\dfrac{\sqrt{2x}(2x - 3)}{2x}$

55. $\dfrac{\sqrt{7y}(7y^3 - 3)}{7y^2}$ **57.** $>$ **59.** $>$ **61.** $12\sqrt{6x}$

63. $9x\sqrt{3} + 5\sqrt{3x}$

65. (a) $5\sqrt{10}$ feet   (b) $400\sqrt{10} \approx 1264.9$ square feet

67. No; $\sqrt{2} + \sqrt{18} = \sqrt{2} + 3\sqrt{2} = 4\sqrt{2}$

69. No; $\sqrt{5} + \left(-\sqrt{5}\right) = 0$

71. (a) The student combined terms with unlike radicands; can be simplified no further.

    (b) The student combined terms with unlike indices; can be simplified no further.

Mid-Chapter Quiz   *(page 576)*

1. 15   **2.** $\frac{3}{2}$ **3.** 8   **4.** 9

5. (a) Not a real number   (b) 1   (c) 5

6. (a) 3   (b) 2   (c) Not a real number

7. $(-\infty, 0) \cup (0, \infty)$ **8.** $\left[\frac{5}{4}, \infty\right)$ **9.** $3|x|\sqrt{3}$

10. $3|x|\sqrt{x}$ **11.** $\dfrac{2u\sqrt{u}}{3}$ **12.** $\dfrac{2\sqrt[3]{2}}{u^2}$ **13.** $5x|y|z^2\sqrt{5x}$

14. $4a^2b\sqrt[3]{2b^2}$ **15.** $4\sqrt{3}$ **16.** $\dfrac{2\sqrt{5x}}{x}$

17. $3\sqrt{3} - 4\sqrt{7}$ **18.** $4\sqrt{2y}$ **19.** $7\sqrt{3}$

20. $4\sqrt{x+2}$ **21.** $6x\sqrt[3]{5x^2} + 4x\sqrt[3]{5x}$ **22.** $4xy^2z^2\sqrt{xz}$

23. $23 + 8\sqrt{2}$ inches

Section 9.4 *(page 581)*

Review *(page 581)*

1. c 2. The signs are the same.

3. The signs are different.   4. b 5. $2x - y = 0$

6. $x + y - 6 = 0$ 7. $y - 3 = 0$

8. $x - 4 = 0$ 9. $6x + 11y - 96 = 0$

10. $x + y - 11 = 0$ 11. $\dfrac{360}{r}$ 12. $2L + 2\left(\dfrac{L}{3}\right)$

---

1. 4 3. $3\sqrt{2}$ 5. $2\sqrt[3]{9}$ 7. $2\sqrt[4]{3}$ 9. $3\sqrt{7} - 7$

11. $2\sqrt{10} + 8\sqrt{2}$ 13. $3\sqrt{2}$ 15. $4\sqrt{6} - 4\sqrt{10}$

17. $y + 4\sqrt{y}$ 19. $4\sqrt{a} - a$ 21. $2 - 7\sqrt[3]{4}$

23. -1 25. $\sqrt{15} - 5\sqrt{5} + 3\sqrt{3} - 15$

27. $8\sqrt{5} + 24$ 29. $2\sqrt[3]{3} + 3\sqrt[3]{6} - 3\sqrt[3]{4} - 9$

31. $2x + 20\sqrt{2x} + 100$ 33. $45x - 17\sqrt{x} - 6$

35. $9x - 25$ 37. $\sqrt[3]{4x^2} + 10\sqrt[3]{2x} + 25$

39. $y - 5\sqrt[3]{y} + 2\sqrt[3]{y^2} - 10$ 41. $t + 5\sqrt[3]{t^2} + \sqrt[3]{t} - 3$

43. $x^2y^2\left(2y - x\sqrt{y}\right)$ 45. $4xy^3\left(x^4\sqrt[3]{x} + y\sqrt[3]{2x^2y^2}\right)$

47. $x + 3$ 49. $4 - 3x$ 51. $2u + \sqrt{2u}$

53. $2 - \sqrt{5}, -1$ 55. $\sqrt{11} + \sqrt{3}, 8$

57. $\sqrt{15} - 3, 6$ 59. $\sqrt{x} + 3, x - 9$

61. $\sqrt{2u} + \sqrt{3}, 2u - 3$ 63. $2\sqrt{2} - \sqrt{4}, 4$

65. $\sqrt{x} - \sqrt{y}, x - y$ 67. (a) $2\sqrt{3} - 4$ (b) 0

69. (a) 0 (b) -1 71. $\dfrac{6\left(\sqrt{11} + 2\right)}{7}$

73. $\dfrac{7\left(5 - \sqrt{3}\right)}{22}$ 75. $\dfrac{5 + 2\sqrt{10}}{5}$ 77. $\dfrac{\sqrt{6} - \sqrt{2}}{2}$

79. $-\dfrac{9\left(\sqrt{3} + \sqrt{7}\right)}{4}$ 81. $\dfrac{4\sqrt{7} + 11}{3}$

83. $\dfrac{2x - 9\sqrt{x} - 5}{4x - 1}$ 85. $\dfrac{\left(\sqrt{15} + \sqrt{3}\right)x}{4}$

87. $\dfrac{2t^2\left(\sqrt{5} + \sqrt{t}\right)}{5 - t}$ 89. $4\left(\sqrt{3a} - \sqrt{a}\right), a \neq 0$

91. $\dfrac{3(x - 4)\left(x^2 + \sqrt{x}\right)}{x(x - 1)(x^2 + x + 1)}$

93. $-\dfrac{\sqrt{u + v}\left(\sqrt{u - v} + \sqrt{u}\right)}{v}$

95.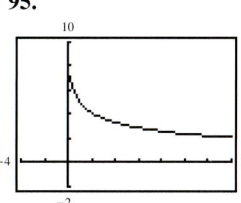

97.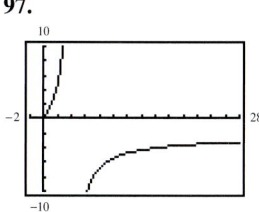

99. $\dfrac{2}{7\sqrt{2}}$ 101. $\dfrac{5}{\sqrt{35x}}$ 103. $\dfrac{4}{5\left(\sqrt{7} - \sqrt{3}\right)}$

105. $\dfrac{y - 25}{\sqrt{3}\left(\sqrt{y} + 5\right)}$ 107. $192\sqrt{2}$ square inches

109. $\dfrac{500k\sqrt{k^2 + 1}}{k^2 + 1}$

111. $\sqrt{3}\left(1 - \sqrt{6}\right)$

$\quad = \sqrt{3} - \sqrt{3} \cdot \sqrt{6}$ Distributive Property

$\quad = \sqrt{3} - \sqrt{9 \cdot 2}$

$\quad = \sqrt{3} - 3\sqrt{2}$ Simplify radicals.

113. $\left(3 - \sqrt{2}\right)\left(3 + \sqrt{2}\right) = 9 - 2 = 7$

Multiplying the number by its conjugate yields the difference of two squares. Squaring a square root eliminates the radical.

Section 9.5 *(page 591)*

Review *(page 591)*

1. The function is undefined when the denominator is zero. The domain is all real numbers x such that $x \neq -2$ and $x \neq 3$.

2. $\dfrac{2x^2 + 5x - 3}{x^2 - 9}$ is undefined if $x = -3$.

3. $36x^5y^8$ 4. 1 5. $4rs^2$ 6. $\dfrac{9x^2}{16y^6}$

7. $-\dfrac{x + 13}{5x^2}, x \neq 3$ 8. $\dfrac{x^2 - 4}{25(x^2 - 9)}$

9. $\dfrac{2x + 5}{x - 5}$ 10. $-\dfrac{5x - 8}{x - 1}$

11.   12.

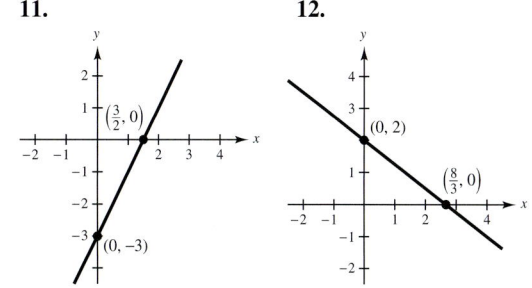

1. (a) Not a solution   (b) Not a solution

(c) Not a solution   (d) Solution

3. (a) Not a solution   (b) Solution

(c) Not a solution   (d) Not a solution

5. 144   **7.** 49   **9.** 27   **11.** 49   **13.** No solution

15. 64   **17.** 90   **19.** -27 **21.** $\frac{4}{5}$ **23.** 5

25. No solution   **27.** $\frac{44}{3}$ **29.** $\frac{14}{25}$ **31.** 4

33. No solution   **35.** 7   **37.** -15 **39.** $-\frac{9}{4}$

41. 8   **43.** 1, 3   **45.** 1   **47.** $\frac{1}{4}$ **49.** $\frac{1}{2}$ **51.** 4

53. 7   **55.** 4   **57.** 216   **59.** 4, -12 **61.** -16

63.

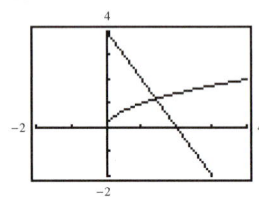

1.407

65.

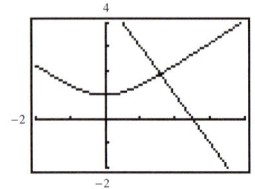

1.569

67.

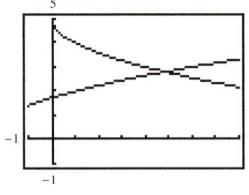

4.840

69.

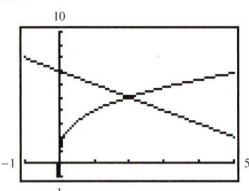

1.978

71.

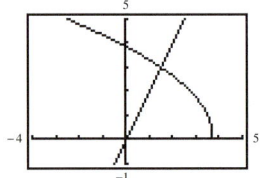

; 1.500

73. 25   **75.** 2, 6   **77.** 9.00   **79.** 12.00

81.

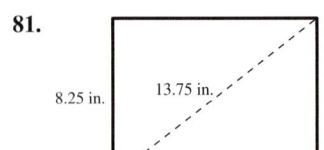

; 11 inches

83. $2\sqrt{10} \approx 6.32$ meters   **85.** 15 feet

87. 30 inches $\times$ 16 inches

89. $h = \dfrac{\sqrt{S^2 - \pi^2 r^4}}{\pi r}$; 34 centimeters

91. 64 feet   **93.** 56.57 feet per second

95. 56.25 feet   **97.** 1.82 feet   **99.** 500 units

101. (a)

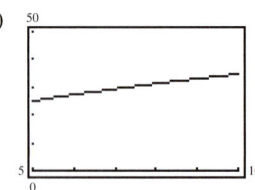

(b) 1999

103. (e) \$12,708.73

105. No. It is not an operation that necessarily yields an equivalent equation. There may be extraneous solutions.

107. $\left(\sqrt{x} + \sqrt{6}\right)^2 \neq \left(\sqrt{x}\right)^2 + \left(\sqrt{6}\right)^2$

Section 9.6   *(page 601)*

Review   *(page 601)*

1. Use the rule $\dfrac{u}{v} \cdot \dfrac{w}{z} = \dfrac{uw}{vz}$. That is, you multiply the numerators, multiply the denominators, and write the new fraction in simplified form.

2. Use the rule $\dfrac{u}{v} \div \dfrac{w}{z} = \dfrac{u}{v} \cdot \dfrac{z}{w}$. That is, you invert the divisor and multiply.

3. Rewrite the fractions so they have common denominators and then use the rule

$$\frac{u}{w} + \frac{v}{w} = \frac{u + v}{w}.$$

4. $\dfrac{t - 5}{5 - t} = \dfrac{-1(5 - t)}{5 - t} = -1$

5. $\dfrac{x}{5}$, $x \neq 0$, $x \neq -\dfrac{3}{2}$ **6.** $\dfrac{x}{5(x + y)}$, $x \neq 0$, $x \neq y$

7. $\dfrac{9}{2(x + 3)}$, $x \neq 0$ **8.** $\dfrac{1}{x - 2}$, $x \neq 0$, $x \neq -2$

9. $\dfrac{x^2 + 2x - 13}{x(x - 2)}$, $x \neq \pm 3$

10. $\dfrac{(x + 1)(x + 3)}{3}$, $x \neq -1$ **11.** $\dfrac{7x}{9}, \dfrac{19x}{18}$

12. $\dfrac{C_1 C_2}{C_1 + C_2}$

1. $2i$ **3.** $-12i$ **5.** $\frac{2}{5}i$ **7.** $0.3i$ **9.** $2\sqrt{2}i$

11. $\sqrt{7}i$ **13.** 2   **15.** $\dfrac{3\sqrt{2}}{8}i$ **17.** $10i$ **19.** $3\sqrt{2}i$

21. $3\sqrt{3}i$ **23.** -4 **25.** $-3\sqrt{6}$ **27.** -0.44

29. $-2\sqrt{3} - 3$ **31.** $5\sqrt{2} - 4\sqrt{5}$ **33.** $4 + 3\sqrt{2}i$

35. -16 **37.** $-8i$ **39.** $a = 3, b = -4$

41. $a = 2, b = -3$ **43.** $a = -4, b = -2\sqrt{2}$

45. $a = 2, b = -2$ **47.** $10 + 4i$ **49.** $-14 - 40i$

51. $-14 + 20i$ **53.** $9 - 7i$ **55.** $3 + 6i$

57. $\frac{13}{6} + \frac{3}{2}i$ **59.** $-3 + 49i$ **61.** -36 **63.** 24

65. $-36i$ **67.** $27i$ **69.** -9 **71.** $-65 - 10i$

73. $20 - 12i$ **75.** $4 + 18i$ **77.** $-40 - 5i$

79. $-14 + 42i$ **81.** 9 **83.** $-7 - 24i$

85. $-21 + 20i$ **87.** $2 + 11i$ **89.** $-i$ **91.** 1

93. -1 **95.** i **97.** -1 **99.** 5 **101.** 68

103. 31 **105.** 100 **107.** 4 **109.** 2.5

111. $-10i$ **113.** $-\frac{1}{5} + \frac{2}{5}i$ **115.** $2 + 2i$

117. $1 - 2i$ **119.** $-\frac{24}{53} + \frac{84}{53}i$ **121.** $\frac{6}{29} + \frac{15}{29}i$

123. $-\frac{23}{58} + \frac{43}{58}i$ **125.** $\frac{9}{5} - \frac{2}{5}i$ **127.** $\frac{47}{26} + \frac{27}{26}i$

129. $\frac{14}{29} - \frac{35}{29}i$ **131–133.** (a) Solution and (b) Solution

135. (a) $\left(\dfrac{-5 + 5\sqrt{3}i}{2}\right)^3 = 125$

(b) $\left(\dfrac{-5 - 5\sqrt{3}i}{2}\right)^3 = 125$

137. (a) $1, \dfrac{-1 + \sqrt{3}i}{2}, \dfrac{-1 - \sqrt{3}i}{2}$

(b) $2, \dfrac{-2 + 2\sqrt{3}i}{2} = -1 + \sqrt{3}i,$

$\dfrac{-2 - 2\sqrt{3}i}{2} = -1 - \sqrt{3}i$

(c) $4, \dfrac{-4 + 4\sqrt{3}i}{2} = -2 + 2\sqrt{3}i,$

$\dfrac{-4 - 4\sqrt{3}i}{2} = -2 - 2\sqrt{3}i$

139. $2a$ **141.** $2bi$ **143.** $i = \sqrt{-1}$

145. $\sqrt{-3}\sqrt{-3} = (\sqrt{3}i)(\sqrt{3}i) = 3i^2 = -3$

147. $x^2 + 1 = (x + i)(x - i)$

Review Exercises   *(page 605)*

1. 7 **3.** -9 **5.** -2 **7.** -4 **9.** $\frac{5}{6}$ **11.** $-\frac{1}{5}$

13. Not a real number

Radical Form	Rational Exponent Form
15. $\sqrt{49} = 7$	$49^{1/2} = 7$
17. $\sqrt[3]{216} = 6$	$216^{1/3} = 6$

19. 81 **21.** -125 **23.** $\frac{1}{16}$ **25.** $x^{7/12}$ **27.** $z^{5/3}$

29. $\dfrac{1}{x^{5/4}}$ **31.** $ab^{2/3}$ **33.** $x^{1/8}$ **35.** $(3x + 2)^{1/3}$

37. 0.0392 **39.** 10.6301 **41.** (a) 3 (b) 9

43. (a) -1 (b) 3 **45.** $\left(-\infty, \frac{9}{2}\right]$ **47.** $5u^2v^2\sqrt{3u}$

49. $0.5x^2\sqrt{y}$ **51.** $2b\sqrt[4]{4a^2b}$ **53.** $2ab\sqrt[3]{6b}$

55. $\dfrac{\sqrt{30}}{6}$ **57.** $\dfrac{\sqrt{3x}}{2x}$ **59.** $\dfrac{\sqrt[3]{4x^2}}{x}$ **61.** $\sqrt{85}$

63. $\sqrt{7}$ **65.** $-24\sqrt{10}$ **67.** $14\sqrt{x} - 9\sqrt[3]{x}$

69. $7\sqrt[4]{y} + 3$ **71.** $3x\sqrt[3]{3x^2y}$ **73.** $x^3y^2\left(11\sqrt{y} - 4\sqrt{2y}\right)$

75. $21 + 12\sqrt{2}$ inches   **77.** $10\sqrt{3}$ **79.** $5\sqrt{2} + 3\sqrt{5}$

81. $5\sqrt{2} + 2\sqrt{5}$ **83.** $12\sqrt{5} + 41$ **85.** $3 - x$

87. $3 + \sqrt{7}; 2$ **89.** $\sqrt{x} - 20; x - 400$

91. $-3(1 + \sqrt{2})$ **93.** $\dfrac{6(4 - \sqrt{6})}{5}$

95. $-\dfrac{(\sqrt{2} - 1)(\sqrt{3} + 4)}{13}$ **97.** $\dfrac{(\sqrt{x} + 10)^2}{x - 100}$

99. 225 **101.** No real solution   **103.** 105 **105.** 3

107. 5 **109.** $-5, -3$ **111.** 6 **113.** $\frac{3}{32}$

115. 12 inches $\times 5$ inches   **117.** $9\sqrt{3} \approx 15.59$ feet

119. 576 feet   **121.** $4\sqrt{3}i$ **123.** $10 - 9\sqrt{3}i$

125. $\frac{3}{4} - \sqrt{3}i$ **127.** $15i$ **129.** $(11 - 2\sqrt{21})i$

131. -5 **133.** $\sqrt{70} - 2\sqrt{10}$ **135.** $a = 10, b = -4$

137. $a = 4, b = 7$ **139.** $8 - 3i$ **141.** $8 + 4i$

143. 25 **145.** $11 - 60i$ **147.** $-\frac{7}{3}i$

149. $-\frac{8}{17} + \frac{2}{17}i$ **151.** $\frac{13}{37} - \frac{33}{37}i$

Chapter Test   *(page 609)*

1. (a) 64 (b) 10 **2.** (a) $\frac{1}{9}$ (b) 6

3. $f(-8) = 7, f(0) = 3$ **4.** $\left[\frac{3}{7}, \infty\right)$

5. (a) $x^{1/3}$ (b) 25 **6.** (a) $\frac{4}{3}\sqrt{2}$ (b) $2\sqrt[3]{3}$

7. (a) $2x\sqrt{6x}$ (b) $2xy^2\sqrt[4]{x}$ **8.** $\dfrac{\sqrt[3]{3y^2}}{3y}$

9. $\dfrac{5(\sqrt{6} + \sqrt{2})}{2}$ **10.** $-10\sqrt{3x}$ **11.** $5\sqrt{3x} + 3\sqrt{5}$

12. $16 - 8\sqrt{2x} + 2x$ **13.** $3 + 4y$ **14.** 27

15. No solution   **16.** 9 **17.** $2 - 2i$ **18.** $-5 - 12i$

19. $-8 + 4i$ **20.** $13 + 13i$ **21.** $\frac{13}{10} - \frac{11}{10}i$

22. 100 feet

Cumulative Test: Chapters 7–9   *(page 610)*

1. Domain $= \left(-\infty, \frac{2}{5}\right) \cup \left(\frac{2}{5}, \infty\right)$

2. $\dfrac{x(x + 2)(x + 4)}{9(x - 4)}, x \neq -4, 0$

3. $\dfrac{(x + 1)(x - 2)}{(x + 6)}, x \neq -4, -2, -1, 0$ **4.** $\dfrac{3x + 5}{x(x + 3)}$

5. $x + y, x \neq 0, y \neq 0, x \neq y$

6. (a) Solution (b) Not a solution   **7.** b   **8.** c

9. d   **10.** a   **11.** $(2, 1)$ **12.** $(3, -2)$ **13.** $(5, 4)$

14. $(0, 1, -2)$ **15.** $(1, -5, 5)$ **16.** $\left(-\frac{1}{5}, -\frac{22}{5}\right)$

17.

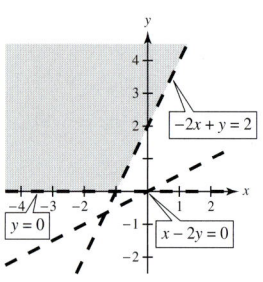

18. $-4 + 3\sqrt{2}\,i$ **19.** $-7 - 24i$ **20.** $t^{1/2}$

21. $35\sqrt{5x}$ **22.** $2x - 6\sqrt{2x} + 9$ **23.** $\sqrt{10} + 2$

24. $\frac{2}{17} - \frac{9}{17}i$ **25.** $2, 5$ **26.** $2, 9$ **27.** 16 **28.** 4

29. 128 feet; $d = 0.08s^2$ **30.** 50

31. Student tickets: 1035; Adult tickets: 400

32. \$20,000 at 8% and \$30,000 at 8.5%

33. $16\left(1 + \sqrt{2}\right) \approx 38.6$ inches   **34.** ≈ 66.02 feet

Chapter 10

Section 10.1   *(page 619)*

Review   *(page 619)*

1. -3. Coefficient of the term of highest degree

2. 5. $(y^2 - 2)(y^3 + 7) = y^5 - 2y^3 + 7y^2 - 14$

3.

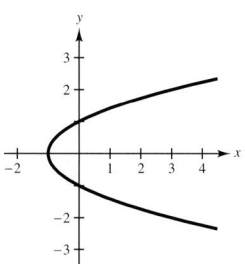

For some values of x there correspond two values of y.

4.

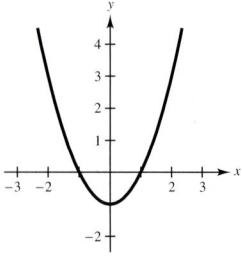

For each value of x there corresponds exactly one value of y.

---

5. $\dfrac{1}{x^3}$ **6.** $-\dfrac{15y^4}{x^2}$ **7.** $\dfrac{9y^2}{4x^2}$ **8.** $\dfrac{v^4}{u^5}$ **9.** $\dfrac{2u}{9v^6}$

10. $\dfrac{r^3}{7}$ **11.** 100 **12.** $\dfrac{29}{18} \approx 1.6$ hours; Distance

---

1. $5, 7$ **3.** $-5, 6$ **5.** $-9, 5$ **7.** 6 **9.** $-\frac{4}{3}$

11. $0, 3$ **13.** $9, 12$ **15.** $\frac{5}{3}, 6$ **17.** $1, 6$ **19.** $-\frac{5}{6}, \frac{1}{2}$

21. ± 7 **23.** ± 3 **25.** $\pm \frac{4}{5}$ **27.** ± 8 **29.** $\pm \frac{5}{2}$

31. $\pm \frac{15}{2}$ **33.** $-12, 4$ **35.** $2.5, 3.5$ **37.** $2 \pm \sqrt{7}$

39. $\dfrac{-1 \pm 5\sqrt{2}}{2}$ **41.** $\dfrac{3 \pm 7\sqrt{2}}{4}$ **43.** $\pm 6i$ **45.** $\pm 2i$

47. $\pm \dfrac{\sqrt{17}}{3}i$ **49.** $3 \pm 5i$ **51.** $-\dfrac{4}{3} \pm 4i$

53. $-\dfrac{3}{2} \pm \dfrac{3\sqrt{6}}{2}i$ **55.** $-6 \pm \dfrac{11}{3}i$ **57.** $1 \pm 3\sqrt{3}\,i$

59. $-1 \pm 0.2i$ **61.** $\dfrac{2}{3} \pm \dfrac{1}{3}i$ **63.** $-\dfrac{7}{3} \pm \dfrac{\sqrt{38}}{3}i$

65. $0, \frac{5}{2}$ **67.** $-4, \frac{3}{2}$ **69.** ± 30 **71.** $\pm 30i$ **73.** ± 3

75. $-5, 15$ **77.** $5 \pm 10i$ **79.** $-2 \pm 3\sqrt{2}\,i$

81.

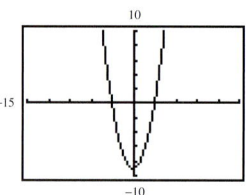

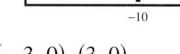

$(-3, 0), (3, 0)$

The result is the same.

83.

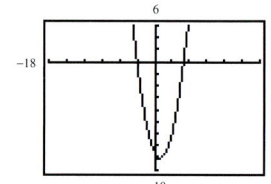

$(-3, 0), (5, 0)$

The result is the same.

85.

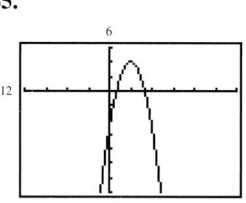

$(1, 0), (5, 0)$

The result is the same.

87.

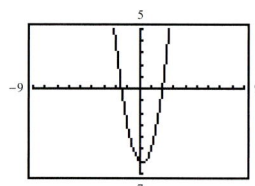

$(2, 0), \left(-\frac{3}{2}, 0\right)$

The result is the same.

89.

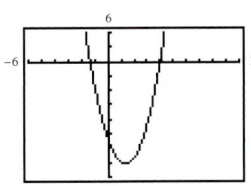

$\left(-\frac{4}{3}, 0\right), (4, 0)$

The result is the same.

91.

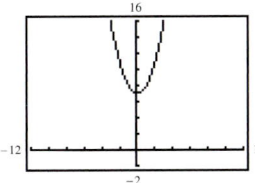

$\pm\sqrt{7}i$, complex solution

93.

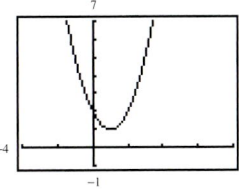

$1 \pm i$, complex solution

95.

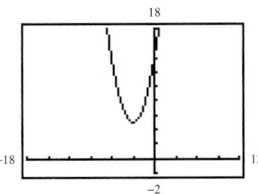

$-3 \pm \sqrt{5}i$, complex solution

97. $f(x) = \sqrt{4 - x^2}$

$g(x) = -\sqrt{4 - x^2}$

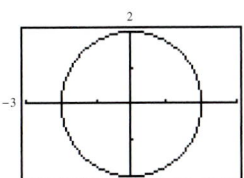

99. $f(x) = \frac{1}{2}\sqrt{4 - x^2}$

$g(x) = -\frac{1}{2}\sqrt{4 - x^2}$

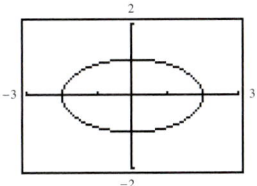

101. $\pm 1, \pm 2$ **103.** $\pm\sqrt{2}, \pm\sqrt{3}$ **105.** $\pm 1, \pm\sqrt{5}$

107. 16   **109.** 4, 25   **111.** $-8, 27$ **113.** $1, \frac{125}{8}$

115. 1, 32   **117.** $\frac{1}{32}$, 243   **119.** 729   **121.** 1, 16

123. $\frac{1}{2}, 1$ **125.** $-1, \frac{4}{5}$ **127.** $\frac{3}{2} \pm \frac{\sqrt{7}i}{2}, \frac{3}{2} \pm \frac{\sqrt{33}}{2}$

129. $\frac{12}{5}$ **131.** 17 feet   **133.** 4 seconds

135. $2\sqrt{2} \approx 2.83$ seconds   **137.** 9 seconds   **139.** 6%

141. 1997

143. (a) $2\sqrt{3} \approx 3.46$ seconds. Square root property, because the quadratic equation did not have a linear term.

(b) 0 seconds, 2 seconds. Factoring, because the quadratic equation did not have a constant term.

145. Factoring and the Zero-Factor Property allow you to solve a quadratic equation by converting it into two linear equations that you already know how to solve.

147. False. The solutions are $x = 5$ and $x = -5$.

149. To solve an equation of quadratic form, determine an algebraic expression u such that substitution yields the quadratic equation $au^2 + bu + c = 0$. Solve this quadratic equation for u and then, through back-substitution, find the solution of the original equation.

Section 10.2 *(page 627)*

Review *(page 627)*

1. a^4b^4 **2.** a^{rs} **3.** b^r/a^r **4.** $1/a^r$ **5.** 6

6. 3   **7.** $-\frac{2}{3}, 5$ **8.** $-5, 8$

9.

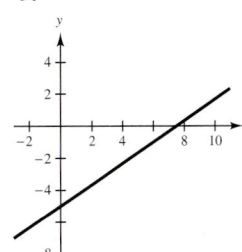

10.

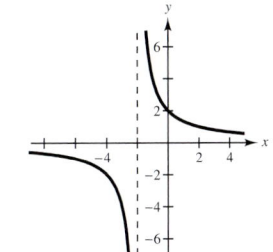

11.

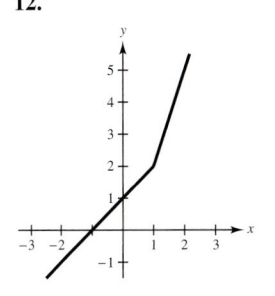

12.

1. 16   **3.** 100   **5.** 64   **7.** $\frac{25}{4}$ **9.** $\frac{81}{4}$ **11.** $\frac{1}{36}$

13. $\frac{9}{100}$ **15.** 0.04   **17.** 0, 20   **19.** $-6, 0$

21. 0, 5   **23.** 1, 7   **25.** $-4, -3$ **27.** $-3, 6$

29. 1, 6   **31.** $-\frac{5}{2}, \frac{3}{2}$

33. $2 + \sqrt{7} \approx 4.65$ **35.** $-2 + \sqrt{7} \approx 0.65$

$2 - \sqrt{7} \approx -0.65$ $-2 - \sqrt{7} \approx -4.65$

37. $-7, 1$

39. $5 + \sqrt{47} \approx 11.86$ **41.** $-7, -1$ **43.** 3, 7

$5 - \sqrt{47} \approx -1.86$

45. $\dfrac{-5 + \sqrt{13}}{2} \approx -0.70$ **47.** $3 \pm i$ **49.** $-2 \pm 3i$

$\dfrac{-5 - \sqrt{13}}{2} \approx -4.30$

51. $\dfrac{1}{2} + \dfrac{\sqrt{3}}{2}i \approx 0.5 + 0.87i$ **53.** 3, 4

$\dfrac{1}{2} - \dfrac{\sqrt{3}}{2}i \approx 0.5 - 0.87i$

55. $\dfrac{1 + 2\sqrt{7}}{3} \approx 2.10$

$\dfrac{1 - 2\sqrt{7}}{3} \approx -1.43$

57. $\dfrac{-3 + \sqrt{137}}{8} \approx 1.09$

$\dfrac{-3 - \sqrt{137}}{8} \approx -1.84$

59. $\dfrac{-4 + \sqrt{10}}{2} \approx -0.42$

$\dfrac{-4 - \sqrt{10}}{2} \approx -3.58$

61. $\dfrac{-9 + \sqrt{21}}{6} \approx -0.74$

$\dfrac{-9 - \sqrt{21}}{6} \approx -2.26$

63. $\dfrac{-1 + \sqrt{10}}{2} \approx 1.08$

$\dfrac{-1 - \sqrt{10}}{2} \approx -2.08$

65. $\dfrac{3}{10} + \dfrac{\sqrt{191}}{10}i \approx 0.30 + 1.38i$

$\dfrac{3}{10} - \dfrac{\sqrt{191}}{10}i \approx 0.30 - 1.38i$

67. $\dfrac{7 + \sqrt{57}}{2} \approx 7.27$

$\dfrac{7 - \sqrt{57}}{2} \approx -0.27$

69. $-1 + \sqrt{3}i \approx -1 + 1.73i$
$-1 - \sqrt{3}i \approx -1 - 1.73i$

71. $-1 \pm 2i$

73. $1 \pm \sqrt{3}$ **75.** $1 \pm \sqrt{3}$ **77.** $4 + 2\sqrt{2}$

79.

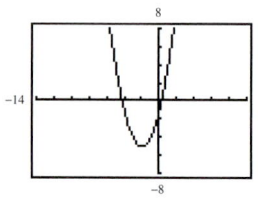

$\left(-2 \pm \sqrt{5}, 0\right)$

The result is the same.

81.

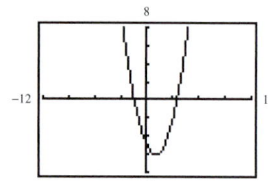

$\left(1 \pm \sqrt{6}, 0\right)$

The result is the same.

83.

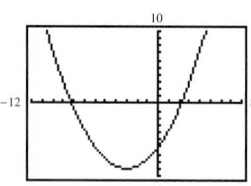

$\left(-3 \pm 3\sqrt{3}, 0\right)$

The result is the same.

85.

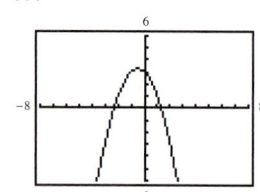

$\left(\dfrac{-1 \pm \sqrt{13}}{2}, 0\right)$

The result is the same.

87. (a) $x^2 + 8x$ (b) $x^2 + 8x + 16$ (c) $(x + 4)^2$

89. 4 centimeters, 6 centimeters

91. 15 meters $\times$ $46\frac{2}{3}$ meters or 20 meters $\times$ 35 meters

93. 42 pairs, 58 pairs

95. A perfect square trinomial is one that can be written in the form $(x + k)^2$.

97. Use the method of completing the square to write the quadratic equation in the form $u^2 = d$. Then use the Square Root Property to simplify.

99. Divide each side of the equation by the leading coefficient. Dividing each side of an equation by a nonzero constant yields an equivalent equation.

101. (a) $d = 0$ (b) d is positive and a perfect square.

(c) d is positive and is not a perfect square.   (d) $d < 0$

Section 10.3   *(page 636)*

> ### Review   *(page 636)*
>
> **1.** $\sqrt{a}\sqrt{b}$ **2.** $\sqrt{a}/\sqrt{b}$
>
> **3.** No. $\sqrt{72} = \sqrt{36 \cdot 2} = 6\sqrt{2}$
>
> **4.** No. $\dfrac{10}{\sqrt{5}} = \dfrac{10\sqrt{5}}{(\sqrt{5})^2} = 2\sqrt{5}$ **5.** $23\sqrt{2}$
>
> **6.** 150   **7.** 7   **8.** $11 + 6\sqrt{2}$ **9.** $\dfrac{4\sqrt{10}}{5}$
>
> **10.** $\dfrac{5(1 + \sqrt{3})}{4}$ **11.** 10 inches $\times$ 15 inches
>
> **12.** 200 units

1. $2x^2 + 2x - 7 = 0$ **3.** $-x^2 + 10x - 5 = 0$

5. 4, 7   **7.** $-2, -4$ **9.** $-\dfrac{1}{2}$ **11.** $-\dfrac{3}{2}$ **13.** $-\dfrac{1}{2}, \dfrac{2}{3}$

15. $-15, 20$ **17.** $1 \pm \sqrt{5}$ **19.** $-2 \pm \sqrt{3}$

21. $-3 \pm 2\sqrt{3}$ **23.** $5 \pm \sqrt{2}$ **25.** $-\dfrac{3}{4} \pm \dfrac{\sqrt{15}}{4}i$

27. $-\dfrac{1}{3}, 1$ **29.** $\dfrac{-2 \pm \sqrt{10}}{2}$ **31.** $\dfrac{-1 \pm \sqrt{5}}{3}$

33. $\dfrac{-3 \pm \sqrt{21}}{4}$ **35.** $\dfrac{3}{8} \pm \dfrac{\sqrt{7}}{8}i$ **37.** $\dfrac{5 \pm \sqrt{73}}{4}$

39. $\dfrac{3 \pm \sqrt{13}}{6}$ **41.** $\dfrac{-3 \pm \sqrt{57}}{6}$ **43.** $\dfrac{1 \pm \sqrt{5}}{5}$

45. $\dfrac{-1 \pm \sqrt{10}}{5}$

47. Two distinct complex solutions

49. Two distinct irrational solutions

51. Two distinct complex solutions

53. One (repeated) rational solution

55. Two distinct complex solutions

57. ± 13 **59.** $-3, 0$ **61.** $\frac{9}{5}, \frac{21}{5}$ **63.** $-\frac{3}{2}, 18$

65. $-4 \pm 3i$ **67.** 8, 16   **69.** $\frac{13}{6} \pm \frac{13\sqrt{11}}{6}i$

71. $\dfrac{-5 \pm 5\sqrt{17}}{12}$ **73.** $\dfrac{-11 \pm \sqrt{41}}{8}$

75. $x^2 - 3x - 10 = 0$ **77.** $x^2 - 8x + 7 = 0$

79. $x^2 - 2x - 1 = 0$ **81.** $x^2 + 25 = 0$

83. $x^2 - 24x + 144 = 0$

85.

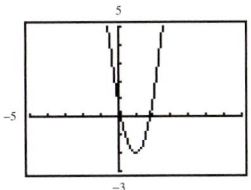

$(0.18, 0), (1.82, 0)$

The result is the same.

87.

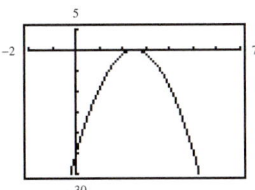

$(2.50, 0)$

The result is the same.

89.

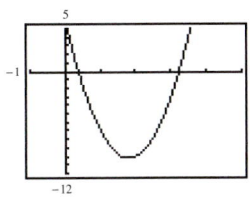

$(3.23, 0), (0.37, 0)$

The result is the same.

91.

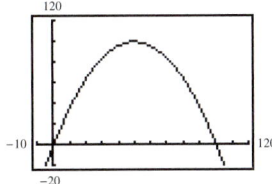

$(99.80, 0), (0.20, 0)$

The result is the same.

93.

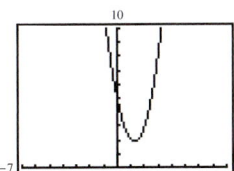

No real solutions

95.

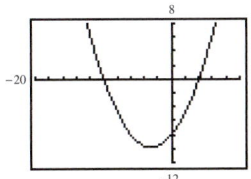

Two real solutions

97. $\dfrac{7 \pm \sqrt{17}}{4}$ **99.** No real values   **101.** $\dfrac{5 \pm \sqrt{185}}{8}$

103. $\dfrac{3 + \sqrt{17}}{2}$ **105.** (a) $c < 9$ (b) $c = 9$ (c) $c > 9$

107. (a) $c < 16$ (b) $c = 16$ (c) $c > 16$

109. 5.1 inches $\times$ 11.4 inches

111. (a) 2.5 seconds   (b) $\dfrac{5 + 5\sqrt{3}}{4} \approx 3.415$ seconds

113. (a) $\frac{5}{4}$ or 1.25 seconds   (b) $\dfrac{5 + 5\sqrt{5}}{8} \approx 2.023$ seconds

115. (a) $\frac{5}{16}$ or 0.3125 second

(b) $\dfrac{5 + 5\sqrt{257}}{32} \approx 2.661$ seconds

117. (a)

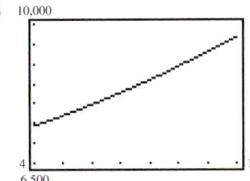

(b) 1999

(c) 10,080,000

119.

	x_1, x_2	$x_1 + x_2$	$x_1 x_2$
(a)	$-2, 3$	1	-6
(b)	$-3, \frac{1}{2}$	$-\frac{5}{2}$	$-\frac{3}{2}$
(c)	$-\frac{3}{2}, \frac{3}{2}$	0	$-\frac{9}{4}$
(d)	$5 + 3i, 5 - 3i$	10	34

121. (c) $\dfrac{2 + 3\sqrt{3}}{2} \approx 3.6$ seconds; Quadratic Formula, because the numbers were large and the equation would not factor.

(d) Yes

123. $b^2 - 4ac$. If the discriminant is positive, the quadratic equation has two real solutions; if it is zero, the equation has one (repeated) real solution; and if it is negative, the equation has no real solutions.

125. The four methods are factoring, the Square Root Property, completing the square, and the Quadratic Formula.

Mid-Chapter Quiz *(page 641)*

1. ± 6 **2.** $-4, \frac{5}{2}$ **3.** $\pm 2\sqrt{3}$ **4.** $-1, 7$

5. $-5 \pm 2\sqrt{6}$ **6.** $\dfrac{-3 \pm \sqrt{19}}{2}$ **7.** $-2 \pm \sqrt{10}$

8. $\dfrac{3 \pm \sqrt{105}}{12}$ **9.** $-\dfrac{5}{2} \pm \dfrac{\sqrt{3}}{2}i$ **10.** $-2, 10$

11. $-3, 10$ **12.** $-2, 5$ **13.** $\frac{3}{2}$ **14.** $\dfrac{-5 \pm \sqrt{10}}{3}$

15. 36   **16.** $\pm 2i, \pm \sqrt{3}i$ **17.** $9 + 4\sqrt{5}$

18. $\pm \sqrt{3}, \pm 3$

19.

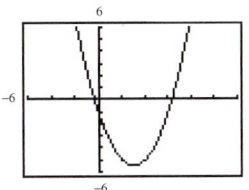

$(-0.32, 0), (6.32, 0)$

The result is the same.

20.

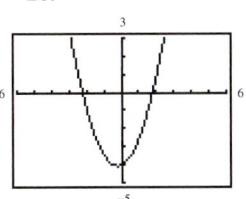

$(-2.24, 0), (1.79, 0)$

The result is the same.

21. 50 alarm clocks   **22.** 35 meters $\times$ 65 meters

Section 10.4 *(page 647)*

Review *(page 647)*

1. $2b, b^2$

2. $\frac{25}{4}$; The constant is found by squaring half the coefficient of x.

3. $-11x$ **4.** $-41v$ **5.** $6x^2 + 9$ **6.** -4

7. $2|x|y\sqrt{6y}$ **8.** $3\sqrt[3]{5}$ **9.** $\dfrac{2\sqrt{3}\,b^3}{a^2}$ **10.** 2

11. 4 liters

12. 30-second commercials: 6; 60-second commercials: 6

---

1. e   **2.** f   **3.** b   **4.** c   **5.** d   **6.** a

7. $y = (x - 0)^2 + 2,\ (0, 2)$ **9.** $y = (x - 2)^2 + 3,\ (2, 3)$

11. $y = (x + 3)^2 - 4,\ (-3, -4)$

13. $y = -(x - 3)^2 - 1,\ (3, -1)$

15. $y = -(x - 1)^2 - 6,\ (1, -6)$

17. $y = 2\left(x + \frac{3}{2}\right)^2 - \frac{5}{2},\ \left(-\frac{3}{2}, -\frac{5}{2}\right)$

19. $(4, -1)$ **21.** $(-1, 2)$ **23.** $\left(-\frac{1}{2}, 3\right)$

25. Upward, $(0, 2)$ **27.** Downward, $(10, 4)$

29. Upward, $(0, -6)$ **31.** Downward, $(3, 0)$

33. Downward, $(3, 9)$ **35.** $(\pm 5, 0), (0, 25)$

37. $(0, 0), (9, 0)$ **39.** $(-3, 0), (1, 0), (0, -3)$

41. $\left(\frac{3}{2}, 0\right), (0, 9)$ **43.** $(0, 3)$

45. $\left(\dfrac{-3 \pm \sqrt{19}}{2}, 0\right), (0, 5)$

47.

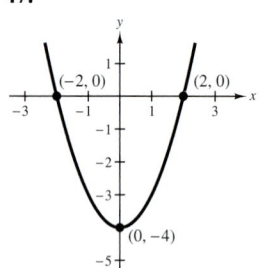

49.

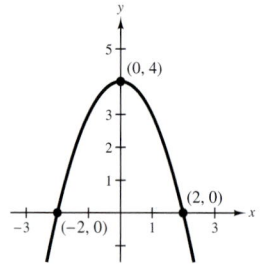

51.

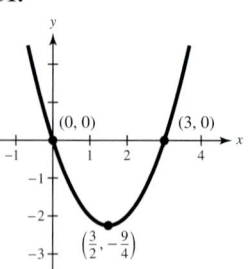

53.

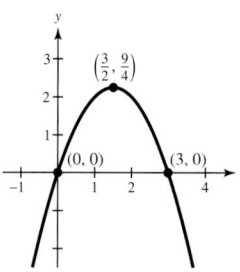

55.

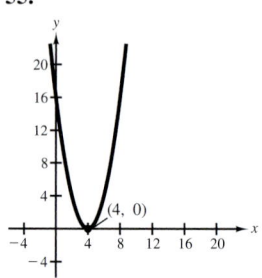

57.

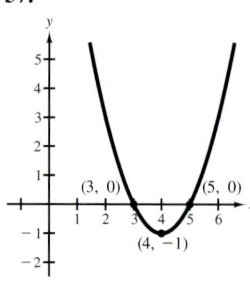

59.

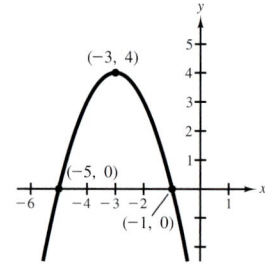

61.

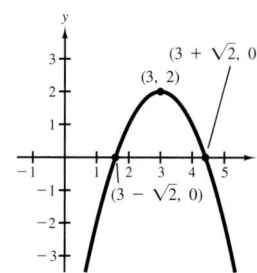

63.

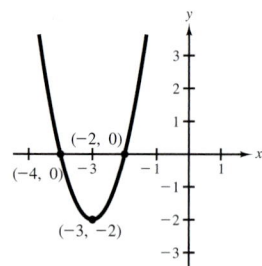

65.

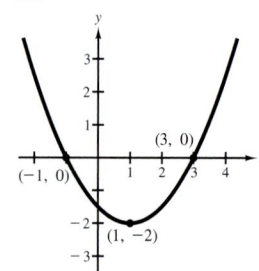

67.

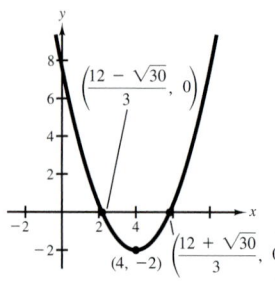

69.

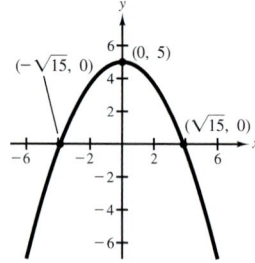

71. Vertical shift

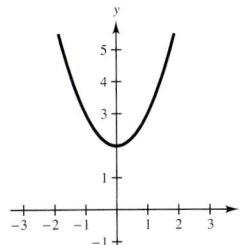

73. Horizontal shift

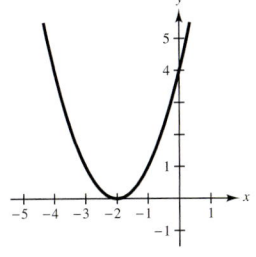

75. Horizontal shift and reflection in the x-axis

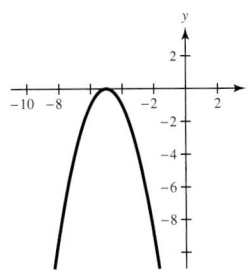

77. Horizontal and vertical shifts, reflection in the x-axis

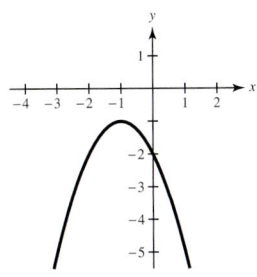

79.

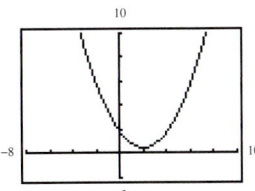

Vertex: $(2, 0.5)$

81.

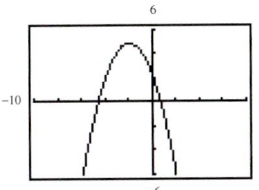

Vertex: $(-1.9, 4.9)$

83. $y = -x^2 + 4$ **85.** $y = x^2 + 4x + 2$

87. $y = x^2 - 4x + 5$ **89.** $y = x^2 - 4x$

91. $y = \frac{1}{2}x^2 - 3x + \frac{13}{2}$ **93.** $y = -4x^2 - 8x + 1$

95. (a) 4 feet (b) 16 feet (c) $12 + 8\sqrt{3} \approx 25.9$ feet

97. (a) 3 feet (b) 48 feet (c) $15 + 4\sqrt{15} \approx 30.5$ feet

99. (a) 6 feet (b) 56 feet (c) $100 + 40\sqrt{7} \approx 205.8$ feet

101. 14 feet

103.

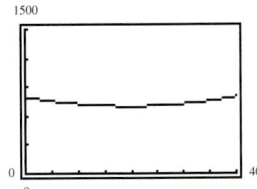

$x = 20$ when C is minimum

105. (a)

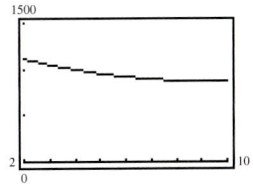

(b) 1992; 1,110,000 military reserves

107. (a) 1000 feet (b) 400 feet

(c)

x	± 100	± 200	± 300	± 400	± 500
y	16	64	144	256	400

109. Parabola

111. To find any x-intercepts, set $y = 0$ and solve the resulting equation for x. To find any y-intercepts, set $x = 0$ and solve the resulting equation for y.

113. If the discriminant is positive, the parabola has two x-intercepts; if it is zero, the parabola has one x-intercept; and if it is negative, the parabola has no x-intercepts.

115. Find the y-coordinate of the vertex of the graph of the function.

Section 10.5 *(page 658)*

> **Review** *(page 658)*
>
> **1.** $m = \dfrac{y_2 - y_1}{x_2 - x_1}$
>
> **2.** (a) $y = mx + b$ (b) $y - y_1 = m(x - x_1)$
>
>   (c) $ax + by + c = 0$ (d) $y - b = 0$
>
> **3.** $x + 2y = 0$ **4.** $3x - 4y = 0$
>
> **5.** $2x - y = 0$ **6.** $x + y - 6 = 0$
>
> **7.** $22x + 16y - 161 = 0$
>
> **8.** $134x - 73y + 146 = 0$
>
> **9.** $y - 8 = 0$ **10.** $x + 3 = 0$ **11.** 8 people
>
> **12.** 3 miles per hour

1. 18 dozen, $1.20 per dozen   **3.** 16 video games, $30

	Width	Length	Perimeter	Area
5.	$1.4l$	l	54 in.	$177\frac{3}{16}$ in.^{2}
7.	w	$2.5w$	70 ft	250 ft^2
9.	$\frac{1}{3}l$	l	64 in.	192 in.^{2}
11.	w	$w + 3$	54 km	180 km^2
13.	$l - 20$	l	440 m	12,000 m^2

15. 12 inches $\times$ 16 inches

17. 50 feet × 250 feet or 100 feet × 125 feet

19. No.

Area $= \frac{1}{2}(b_1 + b_2)h = \frac{1}{2}x[x + (550 - 2x)] = 43{,}560$

This equation has no real solution.

21. Height: 12 inches; Width: 24 inches   **23.** 9.5%

25. 7%   **27.** ≈6.5%   **29.** 15 people   **31.** 32 people

33. 15.86 miles or 2.14 miles

35. (a) $d = \sqrt{(3 + x)^2 + (4 + x)^2}$

(b)
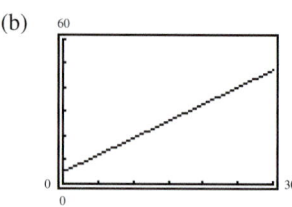

(c) $x \approx 3.55$ when $d = 10$

(d) $\dfrac{-7 + \sqrt{199}}{2} \approx 3.55$ meters

37. 9.1 hours, 11.1 hours   **39.** 10.7 minutes, 13.7 minutes

41. $3\frac{1}{4}$ seconds   **43.** 9.5 seconds   **45.** 4.7 seconds

47. (a) 3 seconds, 7 seconds   (b) 10 seconds   (c) 400 feet

49. 13, 14   **51.** 12, 14   **53.** 17, 19

55. 400 miles per hour

57. 46 miles per hour or 65 miles per hour

59. (a) $b = 20 - a$; $A = \pi ab$; $A = \pi a(20 - a)$

(b)

a	4	7	10	13	16
A	201.1	285.9	314.2	285.9	201.1

(c) 7.9, 12.1

(d)
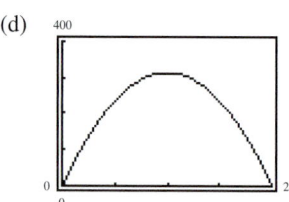

61. (a) Write a verbal model that describes what you need to know.

(b) Assign labels to each part of the verbal model—numbers to the known quantities and letters to the variable quantities.

(c) Use the labels to write an algebraic model based on the verbal model.

(d) Solve the resulting algebraic equation and check your solution.

63. Dollars

Section 10.6   *(page 669)*

Review   *(page 669)*

1. No. 3.683×10^9 **2.** $[10^6, 10^8)$

3. $6v(u^2 - 32v)$ **4.** $5x^{1/3}(x^{1/3} - 2)$

5. $(x - 10)(x - 4)$ **6.** $(x + 2)(x - 2)(x + 3)$

7. $(4x + 11)(4x - 11)$ **8.** $4x(x^2 - 3x + 4)$

9. $\frac{3}{2}h^2$ **10.** $\frac{1}{3}b^2$ **11.** $5x^2$ **12.** $x^2 + 8x$

1. $0, \frac{5}{2}$ **3.** $\pm\frac{9}{2}$ **5.** $-3, 5$ **7.** 1, 3   **9.** $\frac{5}{2}$

11. Negative: $(-\infty, 4)$; Positive: $(4, \infty)$

13. Negative: $(6, \infty)$; Positive: $(-\infty, 6)$

15. Negative: $(0, 4)$; Positive: $(-\infty, 0) \cup (4, \infty)$

17. Negative: $(-\infty, -2) \cup (2, \infty)$; Positive: $(-2, 2)$

19. Negative: $(-1, 5)$; Positive: $(-\infty, -1) \cup (5, \infty)$

21. $(0, 2)$ **23.** $[0, 2]$

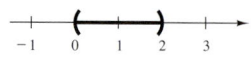   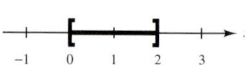

25. $(-\infty, -2) \cup (2, \infty)$ **27.** $(-\infty, -2] \cup [5, \infty)$

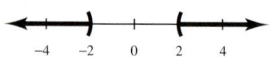   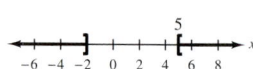

29. $(-\infty, -4) \cup (0, \infty)$ **31.** $[-5, 2]$

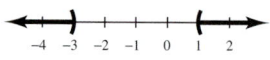

33. $(-\infty, -3) \cup (1, \infty)$ **35.** No solution

37. $(-\infty, \infty)$

39. $\left(-\infty, 2 - \sqrt{2}\right) \cup \left(2 + \sqrt{2}, \infty\right)$

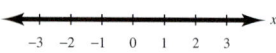

41. $(-\infty, \infty)$ **43.** No solution

45. $\left[-2, \frac{4}{3}\right]$

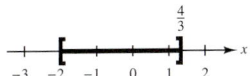

47. $\left(\frac{2}{3}, \frac{5}{2}\right)$

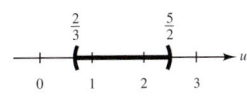

49. $\left(-\infty, -\frac{1}{2}\right) \cup (4, \infty)$

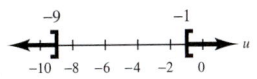

51. $-\frac{7}{2}$

53. No solution

55. $\left(-\infty, 5 - \sqrt{6}\right) \cup \left(5 + \sqrt{6}, \infty\right)$

57. $(-\infty, -9] \cup [-1, \infty)$

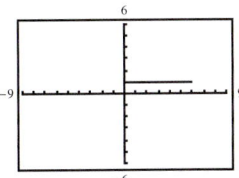

59. $(-2, 0) \cup (2, \infty)$

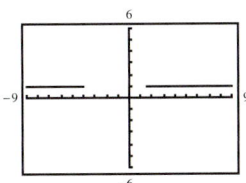

61.

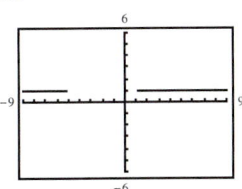

$(0, 6)$

63.

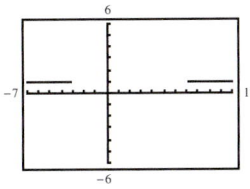

$(-\infty, -4) \cup \left(\frac{3}{2}, \infty\right)$

65.

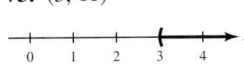

$(-\infty, -5] \cup [1, \infty)$

67.

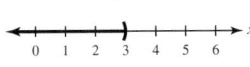

$(-\infty, -3) \cup (7, \infty)$

69. 3   **71.** $0, -5$

73. $(3, \infty)$

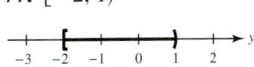

75. $(-\infty, 3)$

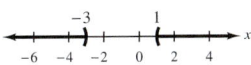

77. $[-2, 1)$

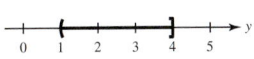

79. $(-\infty, -3) \cup (1, \infty)$

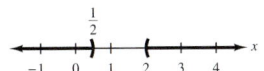

81. $(1, 4]$

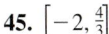

83. $\left(-\infty, \frac{1}{2}\right) \cup (2, \infty)$

85. $\left[-2, -\frac{3}{2}\right)$

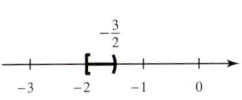

87. $(-1, 3)$

89. $(4, 7)$

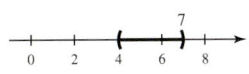

91. $\left(-2, -\frac{2}{5}\right)$

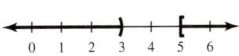

93. $(-\infty, 3) \cup [5, \infty)$

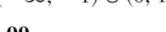

95.

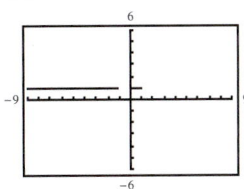

$(-\infty, -1) \cup (0, 1)$

97.

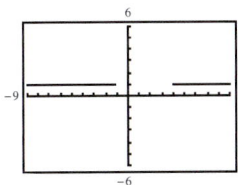

$(-\infty, -1) \cup (4, \infty)$

99.

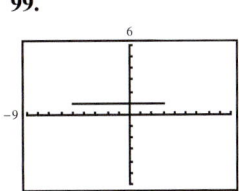

$\left(-5, \frac{13}{4}\right)$

101.

$(0, 0.382) \cup (2.618, \infty)$

103.

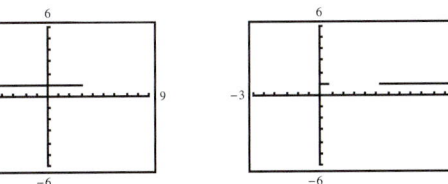

(a) $[0, 2)$ (b) $(2, 4]$

105.

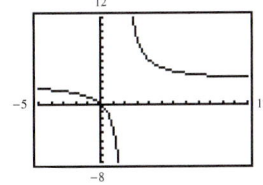

(a) $(-\infty, -2] \cup [2, \infty)$

(b) $(-\infty, \infty)$

107. $(3, 5)$ **109.** $r > 7.24\%$ **111.** $(12, 20)$

113. $90{,}000 \le x \le 100{,}000$

115. (a) $\overline{C} = \dfrac{3000}{x} + 0.75,\ x > 0$

(b)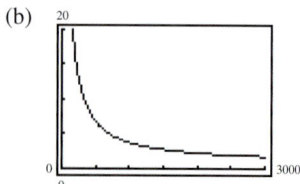

(c) $x > 2400$ calendars

117. (e) $[0, 3)$

119. All real numbers less than or equal to 5, and all real numbers greater than 10.

121. (a) Find the critical numbers of the quadratic polynomial.

(b) Use the critical numbers to determine the test intervals.

(c) Choose a representative x-value from each test interval and evaluate the quadratic polynomial.

Review Exercises   *(page 674)*

1. $-12, 0$ **3.** $\pm\frac{1}{2}$ **5.** $-\frac{5}{2}$ **7.** $-9, 10$

9. $-\frac{3}{2}, 6$ **11.** ± 50 **13.** $\pm 2\sqrt{3}$ **15.** $-4, 36$

17. $\pm 11i$ **19.** $\pm 5\sqrt{2}\,i$ **21.** $-4 \pm 3\sqrt{2}\,i$

23. $\pm\sqrt{5}, \pm i$ **25.** $1, 9$ **27.** $1,\ 1 \pm \sqrt{6}$

29. $-343, 64$ **31.** 144 **33.** $\frac{225}{4}$ **35.** $\frac{1}{25}$

37. $3 + 2\sqrt{3} \approx 6.46;\ 3 - 2\sqrt{3} \approx -0.46$

39. $\dfrac{3}{2} + \dfrac{\sqrt{3}}{2}i \approx 1.5 + 0.87i;\ \dfrac{3}{2} - \dfrac{\sqrt{3}}{2}i \approx 1.5 - 0.87i$

41. $\dfrac{1}{3} + \dfrac{\sqrt{17}}{3}i \approx 0.33 + 1.37i;\ \dfrac{1}{3} - \dfrac{\sqrt{17}}{3}i \approx 0.33 - 1.37i$

43. $-6, 5$ **45.** $-\frac{7}{2}, 3$ **47.** $\dfrac{8 \pm 3\sqrt{6}}{5}$

49. One repeated rational solution

51. Two distinct rational solutions

53. Two distinct rational solutions

55. Two distinct complex solutions

57. $x^2 + 4x - 21 = 0$

59. $x^2 - 10x + 18 = 0$ **61.** $x^2 - 12x + 40 = 0$

63. $y = (x - 4)^2 - 13$; Vertex: $(4, -13)$

65. $y = 2\left(x - \frac{1}{4}\right)^2 + \frac{23}{8}$; Vertex: $\left(\frac{1}{4}, \frac{23}{8}\right)$

67.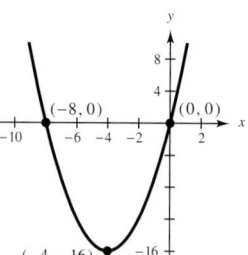

69.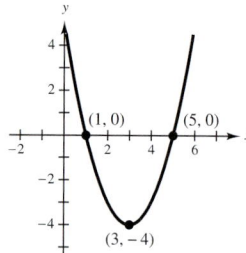

71. $y = 2(x - 2)^2 - 5$ **73.** $y = \frac{1}{16}(x - 5)^2$

75. (a)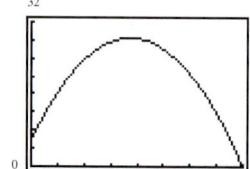

(b) 6 feet

(c) 28.5 feet

(d) 31.9 feet

77. 16 cars; \$5000   **79.** 6 inches $\times$ 18 inches

81. 15 people

83. $9 + \sqrt{101} \approx 19$ hours, $11 + \sqrt{101} \approx 21$ hours

85. $-7, 0$ **87.** $-3, 9$

89. $(0, 7)$ **91.** $(-\infty, -2] \cup [6, \infty)$

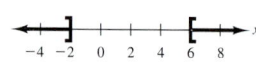

93. $\left(-4, \frac{5}{2}\right)$ **95.** $(-\infty, -3] \cup \left(\frac{7}{2}, \infty\right)$

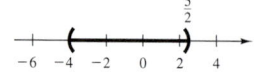

97. $(-4, 1)$ **99.** $(5.3, 14.2)$

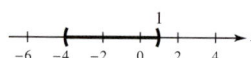

Chapter Test   *(page 677)*

1. $3, 10$ **2.** $-\frac{3}{8}, 3$ **3.** $1.7, 2.3$ **4.** $-4 \pm 10i$

5. $\dfrac{3 \pm \sqrt{3}}{2}$ **6.** $\dfrac{2 \pm 3\sqrt{2}}{2}$ **7.** $\dfrac{3 \pm \sqrt{5}}{4}$ **8.** $1, 512$

9. -56; A negative discriminant tells us the equation has two imaginary solutions.

10. $x^2 - x - 20 = 0$

11.

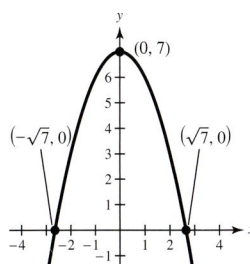

12.

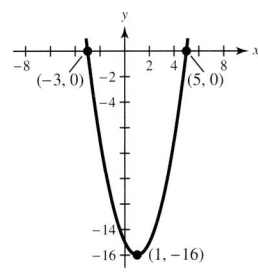

13. $(-\infty, -2] \cup [6, \infty)$

14. $(0, 3)$

15. $[-1, 5)$

17. 40   **18.** $\dfrac{\sqrt{10}}{2} \approx 1.58$ seconds   **19.** 60 feet, 80 feet

Chapter 11

Section 11.1   *(page 688)*

Review   *(page 688)*

1. Test one point in each of the half-planes formed by the graph of $x + y = 5$. If the point satisfies the inequality, shade the entire half-plane to denote that every point in the region satisfies the inequality.

2. The first contains the boundary and the second does not.

3.

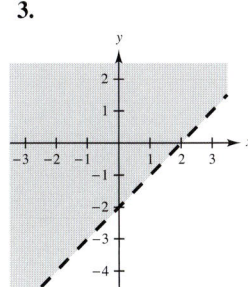

4.

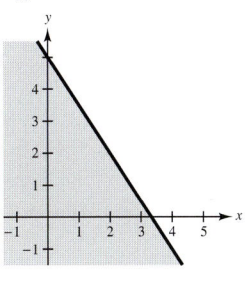

5.

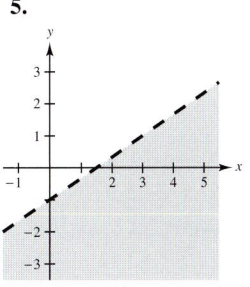

6.

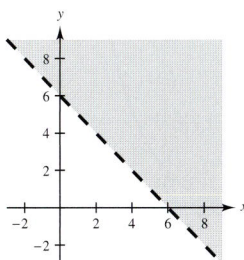

7.

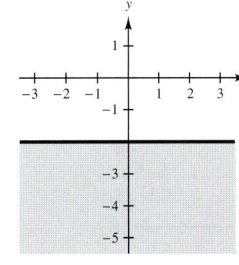

8.

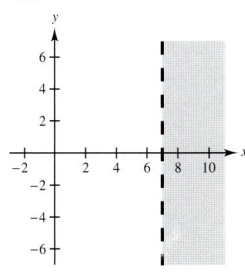

9.

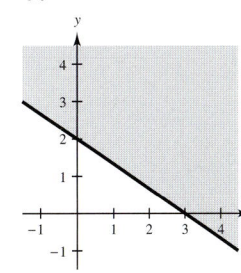

10.

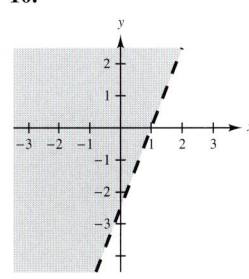

11. 18.6 hours, 21.6 hours

12. $30\sqrt{5} \approx 67.1$ feet

1. 2^{2x-1} **3.** e^2 **5.** $8e^{3x}$ **7.** $-2e^x$ **9.** 11.036

11. 1.396   **13.** 51.193   **15.** 0.906

17. (a) $\frac{1}{9}$ (b) 1   (c) 3

19. (a) 0.263   (b) 54.872   (c) 19.790

21. (a) 500   (b) 250   (c) 56.657

23. (a) 1000   (b) 1628.895   (c) 2653.298

25. (a) 486.111   (b) 47.261   (c) 0.447

27. (a) 73.891   (b) 1.353   (c) 0.183

29. (a) 333.333   (b) 434.557   (c) 499.381

31. a   **32.** d   **33.** c   **34.** e   **35.** b   **36.** f

37.

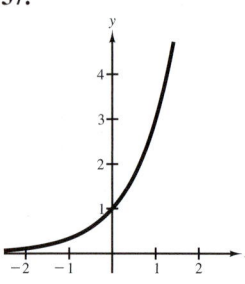

Horizontal asymptote: $y = 0$

39.

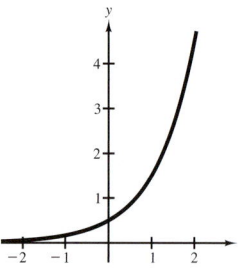

Horizontal asymptote: $y = 0$

53.

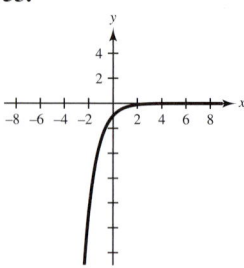

Horizontal asymptote: $y = 0$

55.

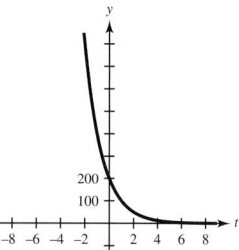

Horizontal asymptote: $y = 0$

41.

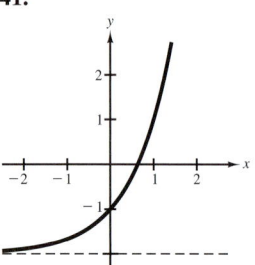

Horizontal asymptote: $y = -2$

43.

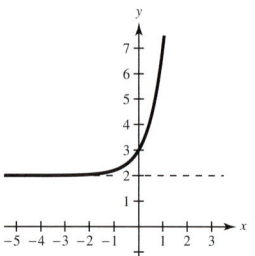

Horizontal asymptote: $y = 2$

57.

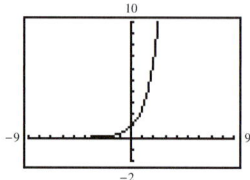

59.

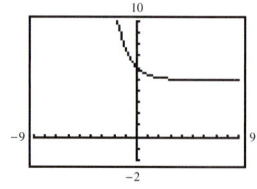

45.

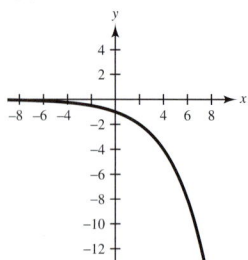

Horizontal asymptote: $y = 0$

47.

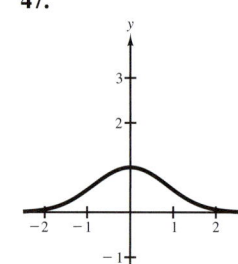

Horizontal asymptote: $y = 0$

61.

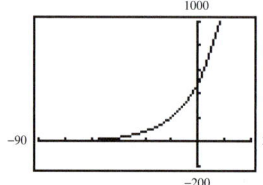

63.

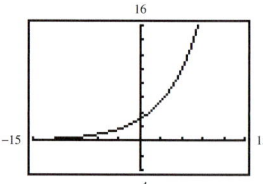

65.

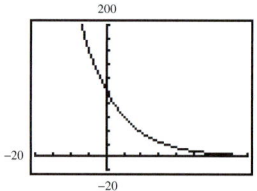

67.

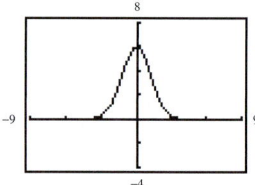

49.

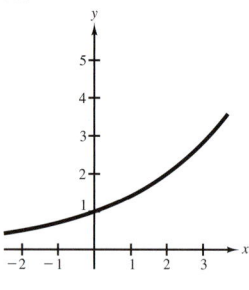

Horizontal asymptote: $y = 0$

51.

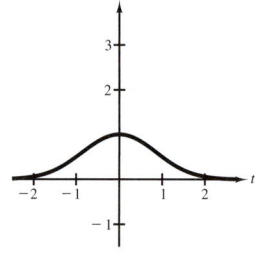

Horizontal asymptote: $y = 0$

69. Vertical shift

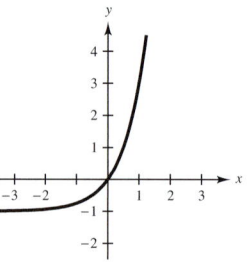

71. Horizontal shift

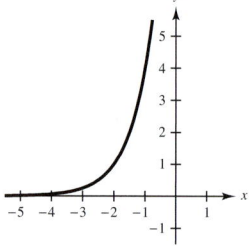

73. Reflection in the x-axis   **75.** 2.520 grams

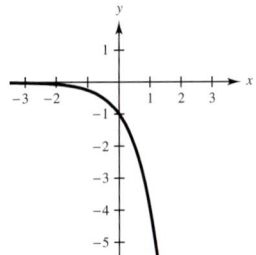

77.

n	1	4	12	365	Continuous
A	\$275.90	\$283.18	\$284.89	\$285.74	\$285.77

79.

n	1	4	12
A	\$4956.46	\$5114.30	\$5152.11

n	365	Continuous
A	\$5170.78	\$5171.42

81.

n	1	4	12
P	\$2541.75	\$2498.00	\$2487.98

n	365	Continuous
P	\$2483.09	\$2482.93

83.

n	1	4	12
P	\$18,429.30	\$15,830.43	\$15,272.04

n	365	Continuous
P	\$15,004.64	\$14,995.58

85. (a) \$22.04   (b) \$20.13
87. (a) \$80,634.95   (b) \$161,269.89
89. $V(t) = 16,000\left(\frac{3}{4}\right)^{t}$ \$9000

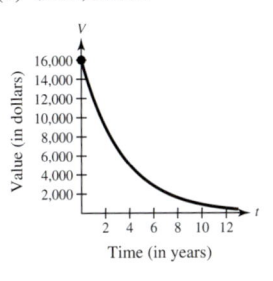

91. (a) $A_1 = 500e^{0.06t}, A_2 = 500e^{0.08t}$

(b)   (c)

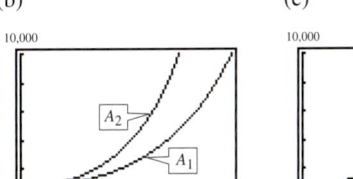   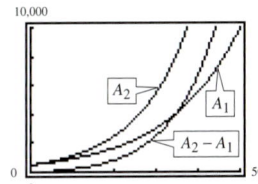

(d) The difference between the functions increases at an increasing rate.

93. (a)

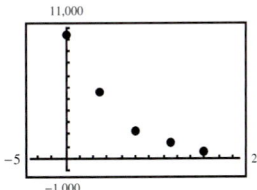

(b)

t	0	25	50	75
h	2000 ft	1400 ft	850 ft	300 ft

95. (a)

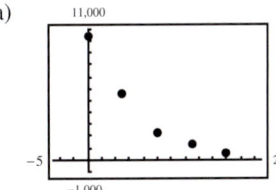

(b)

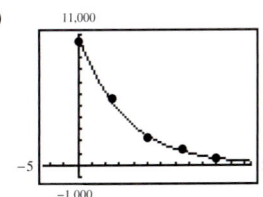

The model is a good fit for the data.

(c)

h	0	5	10	15	20
P	10,332	5583	2376	1240	517
Approx.	10,958	5176	2445	1155	546

(d) 3300 kilograms per square meter

(e) 11.3 kilometers

97. (a)

x	1	10	100	1000	10,000
$\left(1 + \dfrac{1}{x}\right)^x$	2	2.5937	2.7048	2.7169	2.7181

(b)

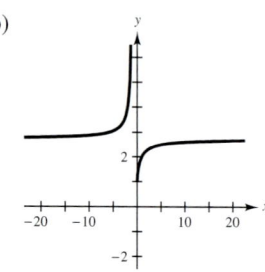

The graph appears to be approaching a horizontal asymptote.

(c) The value approaches e.

99. (a) Continuous

Compounding	Amount, A
Annual	$5955.08
Quarterly	$5978.09
Monthly	$5983.40
Daily	$5986.00
Hourly	$5986.08
Continuous	$5986.09

(b) Compounding quarterly, because the balance is greater at 8% than with continuous compounding at 7%. 7% continuous: $6168.39; 8% quarterly: $6341.21

101. By definition, the base of an exponential function must be positive and not equal to 1. If the base is 1, the function simplifies to the constant function $y = 1$.

103. No; e is an irrational number.

105. When $0 < k < 1$, the graph falls from left to right.

When $k = 1$, the graph is the straight line $y = 1$.

When $k > 1$, the graph rises from left to right.

Section 11.2   *(page 701)*

Review   *(page 701)*

1. y is not a function of x because for some values of x there correspond two values of y. For example, $(4, 2)$ and $(4, -2)$ are solution points.

2. y is a function of x because for each value of x there corresponds exactly one value of y.

3. The domain of f is $-2 \le x \le 2$ and the domain of g is $-2 < x < 2$. g is undefined at $x = \pm 2$.

4. $\{4, 5, 6, 8\}$ **5.** $-2x^2 - 4$ **6.** $30x^3 + 40x^2$

7. $u^2 - 16v^2$ **8.** $9a^2 - 12ab + 4b^2$

9. $t^3 - 6t^2 + 12t - 8$ **10.** $\frac{1}{2}x^2 - \frac{1}{4}x$

11. 100 feet   **12.** 13 minutes

1. (a) $2x - 9$ (b) $2x - 3$ (c) -1 (d) 11

3. (a) $4x^2 + 4x + 4$ (b) $2x^2 + 7$ (c) 28 (d) 25

5. (a) $|3x - 3|$ (b) $3|x - 3|$ (c) 0 (d) 3

7. (a) $\sqrt{x + 1}$ (b) $\sqrt{x - 4} + 5$ (c) 2 (d) 7

9. (a) $\dfrac{x^2}{2 - 3x^2}$ (b) $2(x - 3)^2$ (c) -1 (d) 2

11. (a) -1 (b) -2 (c) -2

13. (a) -1 (b) 1

15. (a) 10 (b) 1 (c) 1

17. (a) 0 (b) 10

19. (a) $(f \circ g)(x) = 3x - 17$
Domain: $(-\infty, \infty)$
(b) $(g \circ f)(x) = 3x - 3$
Domain: $(-\infty, \infty)$

21. (a) $(f \circ g)(x) = \sqrt{x - 2}$
Domain: $[2, \infty)$
(b) $(g \circ f)(x) = \sqrt{x} - 2$
Domain: $[0, \infty)$

23. (a) $(f \circ g)(x) = x + 2$
Domain: $[1, \infty)$
(b) $(g \circ f)(x) = \sqrt{x^2 + 2}$
Domain: $(-\infty, \infty)$

25. (a) $(f \circ g)(x) = \dfrac{\sqrt{x - 1}}{\sqrt{x - 1} + 5}$
Domain: $[1, \infty)$
(b) $(g \circ f)(x) = \sqrt{-\dfrac{5}{x + 5}}$
Domain: $(-\infty, -5)$

27.

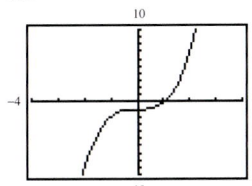

Yes

29.

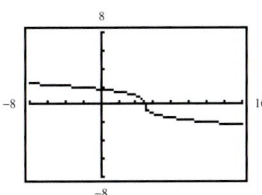

Yes

31.

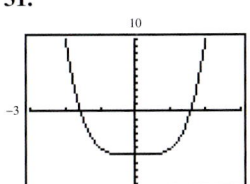

No

33.

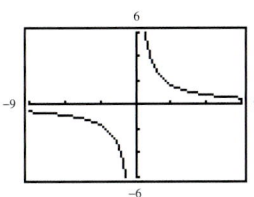

Yes

35. No   **37.** Yes   **39.** No

41. $f(g(x)) = f\left(-\frac{1}{6}x\right) = -6\left(-\frac{1}{6}x\right) = x$
  $g(f(x)) = g(-6x) = -\frac{1}{6}(-6x) = x$

43. $f(g(x)) = f(x - 15) = (x - 15) + 15 = x$
  $g(f(x)) = g(x + 15) = (x + 15) - 15 = x$

45. $f(g(x)) = f\left[\frac{1}{2}(1 - x)\right] = 1 - 2\left[\frac{1}{2}(1 - x)\right]$
  $= 1 - (1 - x) = x$
  $g(f(x)) = g(1 - 2x) = \frac{1}{2}[1 - (1 - 2x)] = \frac{1}{2}(2x) = x$

47. $f(g(x)) = f\left[\frac{1}{3}(2 - x)\right] = 2 - 3\left[\frac{1}{3}(2 - x)\right]$
  $= 2 - (2 - x) = x$
  $g(f(x)) = g(2 - 3x) = \frac{1}{3}[2 - (2 - 3x)] = \frac{1}{3}(3x) = x$

49. $f(g(x)) = f(x^3 - 1) = \sqrt[3]{(x^3 - 1) + 1} = \sqrt[3]{x^3} = x$
  $g(f(x)) = g\left(\sqrt[3]{x + 1}\right) = \left(\sqrt[3]{x + 1}\right)^3 - 1$
  $= x + 1 - 1 = x$

51. $f(g(x)) = f\left(\frac{1}{x}\right) = \frac{1}{(1/x)} = x$
  $g(f(x)) = g\left(\frac{1}{x}\right) = \frac{1}{(1/x)} = x$

53. $f^{-1}(x) = \frac{1}{5}x$ **55.** $f^{-1}(x) = -\frac{5}{2}x$

57. $f^{-1}(x) = x - 10$ **59.** $f^{-1}(x) = 3 - x$

61. $f^{-1}(x) = \sqrt[7]{x}$ **63.** $f^{-1}(x) = x^3$ **65.** $f^{-1}(x) = \frac{x}{8}$

67. $g^{-1}(x) = x - 25$ **69.** $g^{-1}(x) = \frac{3 - x}{4}$

71. $g^{-1}(t) = 4t - 8$ **73.** $h^{-1}(x) = x^2, \ x \geq 0$

75. $f^{-1}(t) = \sqrt[3]{t + 1}$ **77.** $f^{-1}(x) = x^2 - 3, \ x \geq 0$

79. b   **80.** c   **81.** d   **82.** a

83.

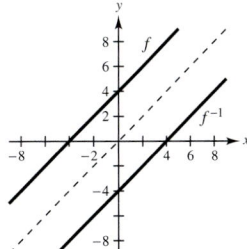

85.

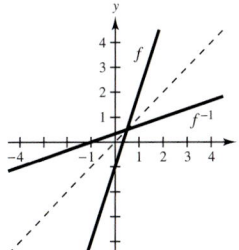

87.

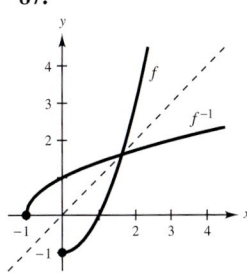

89.

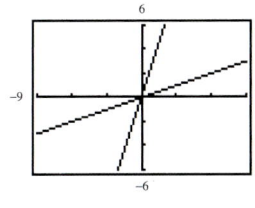

91.

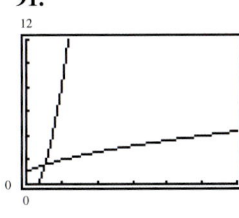

93.

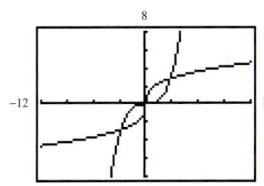

95.

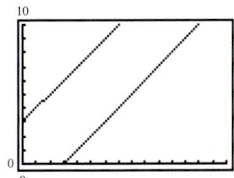

97. $x \geq 2; \ f^{-1}(x) = \sqrt{x} + 2;$
  Domain of f^{-1}: $x \geq 0$

99. $x \geq 0; \ f^{-1}(x) = x - 1$; Domain of f^{-1}: $x \geq 1$

101. $f^{-1}(x) = \frac{1}{2}(3 - x)$

103. $A(r(t)) = 0.36\pi t^2$; Input: time; Output: area

105. $(C \circ x)(t) = 102t + 300$
  Production cost after t hours of operation.

107. (a) $R = p - 500$

(b) $S = 0.9p$

(c) $(R \circ S)(p) = 0.9p - 500$; 10% discount followed by the $500 rebate.

$(S \circ R)(p) = 0.9(p - 500)$; 10% discount after the price is reduced by the rebate.

(d) $(R \circ S)(6000) = 4900$; $(S \circ R)(6000) = 4950$

$R \circ S$ yields a lower cost because the dealer discount is calculated on a larger base.

109. (a) Total cost $=$ Cost of oranges $+$ Cost of apples
$$y = 0.75x + 0.95(100 - x)$$

(b) $y = 5(95 - x)$

x: total cost

y: number of pounds of oranges at $0.75 per pound

(c) $75 \le x \le 95$ (d) 55 pounds

111. True. The x-coordinate of a point on the graph of f becomes the y-coordinate of a point on the graph of f^{-1}.

113. False. $f(x) = \sqrt{x - 1}$; Domain: $[1, \infty)$;

$f^{-1}(x) = x^2 + 1$, $x \ge 0$; Domain: $[0, \infty)$

115. Interchange the coordinates of each ordered pair. The inverse of the function defined by $\{(3, 6), (5, -2)\}$ is $\{(6, 3), (-2, 5)\}$.

117. $f(x) = x^4$

119. They are reflections in the line $y = x$.

Section 11.3 *(page 714)*

Review *(page 714)*

1. To solve a quadratic equation $x^2 + bx$ by completing the square, first add $(b/2)^2$ to each side of the equation, which is the square of half the coefficient of x. So,

$$x^2 + bx + \left(\frac{b}{2}\right)^2 = \left(x + \frac{b}{2}\right)^2.$$

2. $2x^2 - 4x + 5 = 0$

3. To determine the type of solution of a quadratic equation using the discriminant, first determine whether the discriminant is positive, negative, or zero. If the discriminant > 0, the equation has two real solutions. If the discriminant $= 0$, the equation has one (repeated) real solution. If the discriminant < 0, the equation has no real solution(s).

4. $y = (x + 2)^2 - 10$ **5.** $2x(x^2 - 3)$

6. $(2 - y)(6 + y)$ **7.** $(t + 5)^2$

8. $(5 - u)(1 + u^2)$

9.

10.

11.

12.

1. $7^2 = 49$ **3.** $2^{-5} = \frac{1}{32}$ **5.** $3^{-5} = \frac{1}{243}$

7. $36^{1/2} = 6$ **9.** $8^{2/3} = 4$ **11.** $2^{1.3} \approx 2.462$

13. $\log_6 36 = 2$ **15.** $\log_4 \frac{1}{16} = -2$ **17.** $\log_8 4 = \frac{2}{3}$

19. $\log_{25} \frac{1}{5} = -\frac{1}{2}$ **21.** $\log_4 1 = 0$

23. $\log_5 9.518 \approx 1.4$ **25.** 3 **27.** 3 **29.** -2

31. -3 **33.** -4

35. There is no power to which 2 can be raised to obtain -3.

37. 0

39. There is no power to which 5 can be raised to obtain -6.

41. $\frac{1}{2}$ **43.** $\frac{3}{4}$ **45.** 4 **47.** 1.6232

49. -1.6383 **51.** 0.7335

53. c **54.** b **55.** a **56.** d

57.

Inverse functions

59.

Inverse functions

61. The graph is shifted 3 units upward.

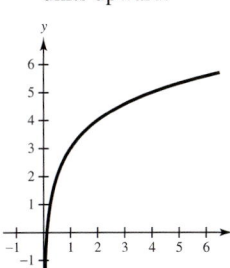

63. The graph is shifted 2 units to the right.

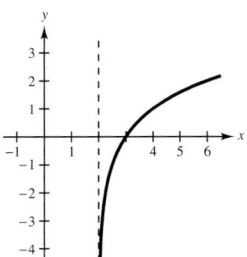

75.

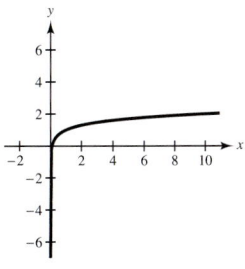

Vertical asymptote: $x = 0$

65. The graph is reflected in the y-axis.

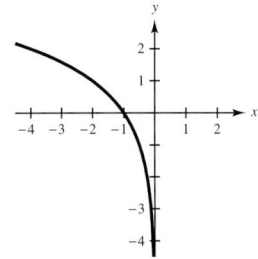

77. Domain: $(0, \infty)$

Vertical asymptote: $x = 0$

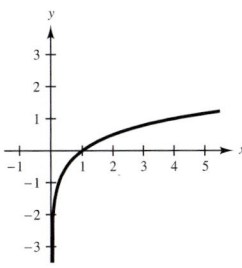

67.

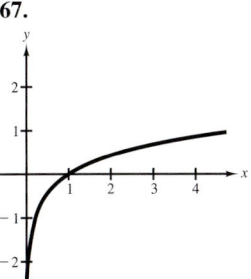

69.

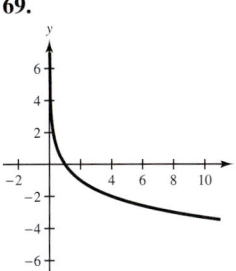

Vertical asymptote: $x = 0$

Vertical asymptote: $t = 0$

79. Domain: $(3, \infty)$

Vertical asymptote: $x = 3$

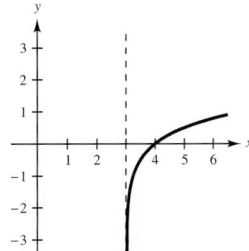

71.

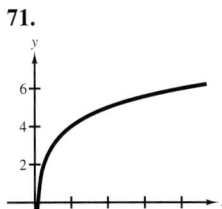

73.

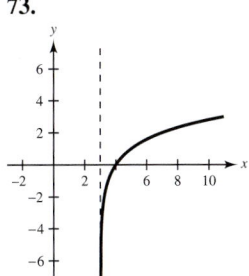

Vertical asymptote: $x = 0$

Vertical asymptote: $x = 3$

81. Domain: $(0, \infty)$

Vertical asymptote: $x = 0$

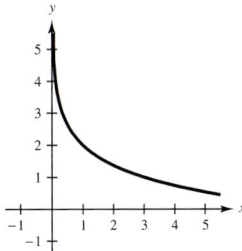

83. Domain: $(0, \infty)$

Vertical asymptote: $x = 0$

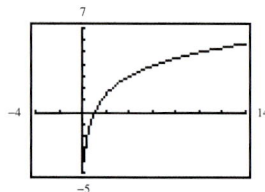

85. Domain: $(0, \infty)$

Vertical asymptote: $x = 0$

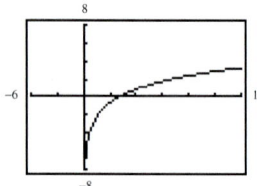

87. Domain: $(0, \infty)$

Vertical asymptote: $x = 0$

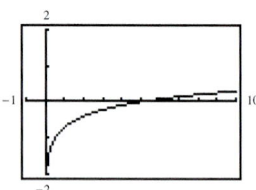

89. 3.6376   **91.** -1.8971 **93.** 0.0757   **95.** b

96. a   **97.** d   **98.** c

99.

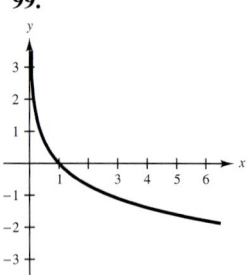

Vertical asymptote: $x = 0$

101.

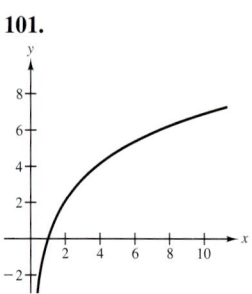

Vertical asymptote: $x = 0$

103.

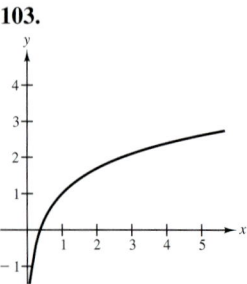

Vertical asymptote: $x = 0$

105.

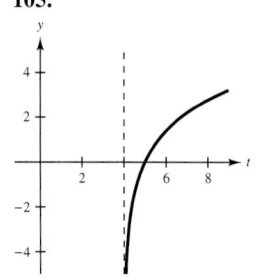

Vertical asymptote: $t = 4$

107. Domain: $(-1, \infty)$

Vertical asymptote: $x = -1$

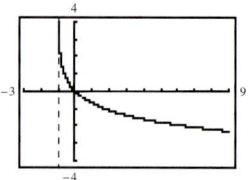

109. Domain: $(0, \infty)$

Vertical asymptote: $t = 0$

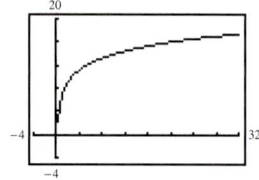

111. 1.6309   **113.** 1.2925   **115.** -0.4739

117. 2.6332   **119.** -2 **121.** 1.3481   **123.** 1.8946

125. 53.4 inches

127.

r	0.07	0.08	0.09	0.10	0.11	0.12
t	9.9	8.7	7.7	6.9	6.3	5.8

129. (a)

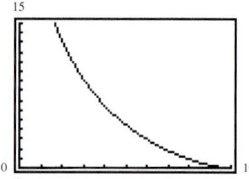

Domain: $(0, 10]$

(b) $x = 0$ (c) $(2, y) \approx (2, 13.1)$

131. $\log_5 x$ **133.** $a^1 = a$

135. Common logarithms are base 10 and natural logarithms are base e.

137. $(0, \infty)$ **139.** $3 \le f(x) \le 4$ **141.** A factor of 10

Mid-Chapter Quiz   *(page 718)*

1. (a) $\dfrac{16}{9}$ (b) 1   (c) $\dfrac{3}{4}$ (d) $\dfrac{8\sqrt{3}}{9}$

2. Domain: $(-\infty, \infty)$; Range: $(0, \infty)$

3.

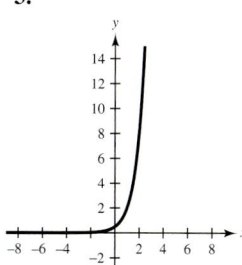

4.

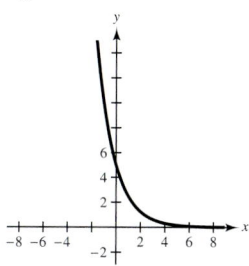

Horizontal asymptote: $y = 0$ Horizontal asymptote: $y = 0$

5.

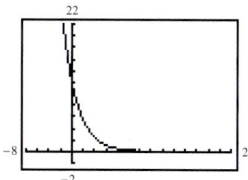

6.

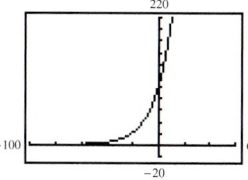

Horizontal asymptote: $y = 0$ Horizontal asymptote: $y = 0$

7. (a) $2x^3 - 3$ (b) $(2x - 3)^3$ (c) -19 (d) 125

8. $f(g(x)) = 3 - 5\left[\frac{1}{5}(3 - x)\right]$

$= 3 - 3 + x = x$

$g(f(x)) = \frac{1}{5}[3 - (3 - 5x)]$

$= \frac{1}{5}(5x) = x$

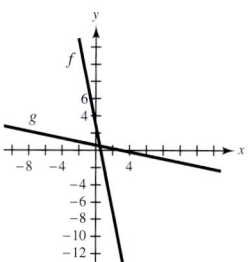

9. $h^{-1}(x) = \frac{1}{10}(x - 3)$ **10.** $g^{-1}(t) = \sqrt[3]{2(t - 2)}$

11. $9^{-2} = \frac{1}{81}$ **12.** $\log_3 81 = 4$ **13.** 3

14.

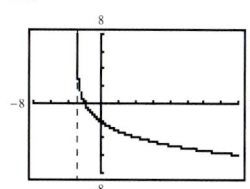

15.

Vertical asymptote: $t = -3$ Vertical asymptote: $x = 0$

16. $h = 2, k = 1$ **17.** 6.0639

18.

n	1	4	12	365	Continuous compounding
A	\$3185.89	\$3314.90	\$3345.61	\$3360.75	\$3361.27

19. 1.60 grams

Section 11.4 *(page 724)*

Review *(page 724)*

1. $\sqrt[n]{uv}$ **2.** $\sqrt[n]{\dfrac{u}{v}}$ **3.** Different indices

4. No; $\dfrac{1}{\sqrt{2x}} = \dfrac{\sqrt{2x}}{2x}$ **5.** $19\sqrt{3x}$ **6.** $x - 9$

7. $\sqrt{5u}$ **8.** $4t + 12\sqrt{t} + 9$ **9.** $25\sqrt{2}x$

10. $6\left(\sqrt{t + 2} - \sqrt{t}\right)$ **11.** 22 units   **12.** \$2300

1. 3 **3.** -4 **5.** $\frac{1}{3}$ **7.** 0 **9.** -6 **11.** 2

13. 2 **15.** 1 **17.** 2 **19.** -3 **21.** 12 **23.** 1

25. 1 **27.** 1.2925 **29.** 0.2925 **31.** 0.2500

33. 2.7925 **35.** 0 **37.** 2.1972 **39.** 3.5835

41. 1.7918 **43.** $\log_3 11 + \log_3 x$ **45.** $2\log_7 x$

47. $-2\log_5 x$ **49.** $\frac{1}{2}(\log_4 3 + \log_4 x)$ **51.** $\ln 3 + \ln y$

53. $\log_2 z - \log_2 17$ **55.** $\frac{1}{2}\log_9 x - \log_9 12$

57. $2\ln x + \ln(y - 2)$ **59.** $6\log_4 x + 2\log_4(x - 7)$

61. $\frac{1}{3}\log_3(x + 1)$ **63.** $\frac{1}{2}[\ln x + \ln(x + 2)]$

65. $2[\ln(x + 1) - \ln(x - 1)]$ **67.** $\frac{1}{3}[2\ln x - \ln(x + 1)]$

69. $\ln x + 2\ln y - 3\ln z$

71. $\frac{1}{2}(7\log_3 x - 5\log_3 y - 8\log_3 z)$

73. $\log_6 a + \frac{1}{2}\log_6 b + 3\log_6(c - d)$

75. $\ln(x + y) + \frac{1}{5}\ln(w + 2) - (\ln 3 + \ln t)$

77. $\log_{12}\dfrac{x}{3}$ **79.** $\log_2 3x$ **81.** $\log_{10}\dfrac{4}{x}$

83. $\ln b^4,\ b > 0$ **85.** $\log_5(2x)^{-2},\ x > 0$

87. $\log_2 x^7 z^3$ **89.** $\log_3 2\sqrt{y}$

91. $\ln\dfrac{x^2 y^3}{z},\ x > 0, y > 0, z > 0$

93. $\ln\dfrac{2^5 y^3}{x},\ x > 0, y > 0$ **95.** $\ln(xy)^4,\ x > 0, y > 0$

97. $\ln\left(\dfrac{x}{x + 1}\right)^2,\ x > 0$ **99.** $\log_4\dfrac{x + 8}{x^3},\ x > 0$

101. $\log_5\dfrac{\sqrt{x + 2}}{x - 3}$ **103.** $\log_6\dfrac{(c + d)^5}{\sqrt{m - n}}$

105. $\log_2\sqrt[5]{\dfrac{x^3}{y^4}},\ y > 0$ **107.** $\log_6\dfrac{\sqrt[5]{x - 3}}{x^2(x + 1)^3},\ x > 3$

109. $2 + \ln 3$ **111.** $1 + \frac{1}{2}\log_5 2$ **113.** $1 - 2\log_4 x$

115.   **117.**

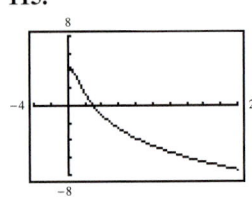   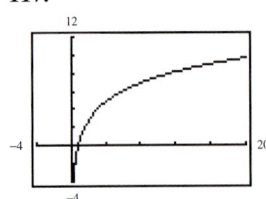

119. $B = 10(\log_{10} I + 16)$; 60 decibels

121. $E = \log_{10}\left(\dfrac{C_2}{C_1}\right)^{1.4}$ **123.** True   **125.** True

127. False; $\log_3(u + v)$ does not simplify.   **129.** True

131. False; 0 is not in the domain of f.

133. False; $f(x - 3) = \ln(x - 3)$.

135. False; If $f(u) = 2f(v)$, then $\ln u = 2 \ln v = \ln v^2 \Rightarrow u = v^2$.

137. $\log_b\left(\dfrac{x}{xx}\right) = \log_b x - (\log_b x + \log_b x)$

139. $\ln 1 = 0$ $\ln 9 \approx 2.1972$

   $\ln 2 \approx 0.6931$ $\ln 10 \approx 2.3025$

   $\ln 3 \approx 1.0986$ $\ln 12 \approx 2.4848$

   $\ln 4 \approx 1.3862$ $\ln 14 \approx 2.6390$

   $\ln 5 \approx 1.6094$ $\ln 15 \approx 2.7080$

   $\ln 6 \approx 1.7917$ $\ln 16 \approx 2.7724$

   $\ln 7 \approx 1.9459$ $\ln 18 \approx 2.8903$

   $\ln 8 \approx 2.0793$ $\ln 20 \approx 2.9956$

   Explanations will vary. Any differences are due to round-off errors.

Section 11.5   *(page 734)*

Review   *(page 734)*

1. No. A system of linear equations has no solutions, one solution, or an infinite number of solutions.

2. The equations represent parallel lines and therefore have no point of intersection.

3. 2   **4.** $5 \pm 2\sqrt{2}$ **5.** $-\frac{1}{2}$ **6.** $\frac{5}{3}$

7. 1, 7   **8.** 47

9. $d = 73t$

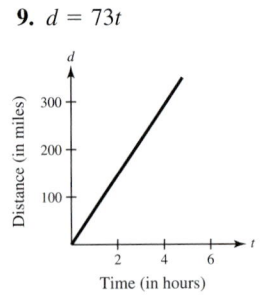

10. $V = 25\pi h$

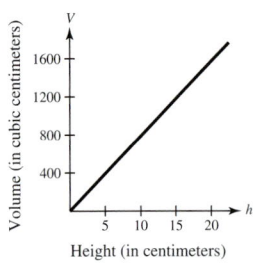

11. $V = 10\pi r^2$

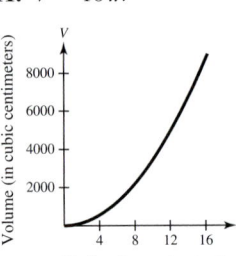

12. $F = 25x$

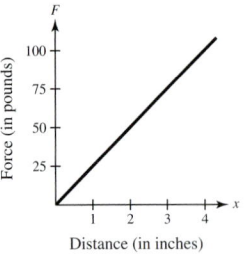

1. (a) Not a solution   (b) Solution

3. (a) Solution   (b) Not a solution

5. (a) Not a solution   (b) Solution

7. 3   **9.** -3 **11.** 4   **13.** 1   **15.** -2

17. -3 **19.** -6 **21.** 2   **23.** $\frac{22}{5}$ **25.** 6

27. 9   **29.** 4   **31.** No solution   **33.** -7

35. $2x - 1$ **37.** $2x, \ x > 0$ **39.** 4.11   **41.** 1.31

43. 1.49   **45.** 0.83   **47.** -2.37 **49.** -3.60

51. 2.64   **53.** 3.00   **55.** 1.23   **57.** 35.35

59. 0.80   **61.** 12.22   **63.** 3.28   **65.** No solution

67. -1.04 **69.** 2.48   **71.** 0.90   **73.** 0.38

75. 0.39   **77.** 8.99   **79.** 9.73   **81.** 4.62

83. 0.10   **85.** 174.77   **87.** 2187.00   **89.** 6.52

91. 25.00   **93.** 10.04   **95.** ± 20.09 **97.** 3.00

99. 19.63   **101.** 12.18   **103.** 2000.00   **105.** 3.20

107. 4.00   **109.** 0.75   **111.** 5.00   **113.** 2.46

115. 2.29   **117.** 6.00

119. $(1.40, 0)$

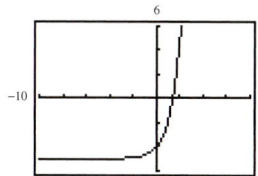

121. $(21.82, 0)$

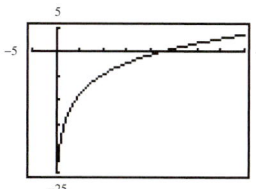

123. 0.69

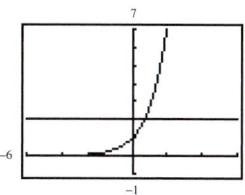

125. 1.48

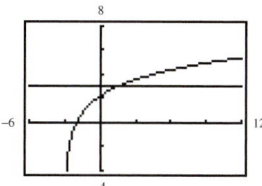

127. 9% **129.** 7.70 years

131. $10^{-8.5}$ watts per square centimeter   **133.** 205

135. (a) 3.64 months

(b)

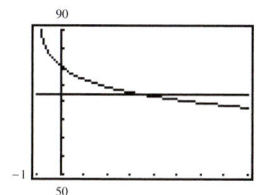

(c) Answers will vary.

137. (a) $k = \frac{1}{4} \ln \frac{8}{15} \approx -0.1572$ (b) ≈ 3.25 hours

(c) ≈ 2.84 hours

139. (c) 7.2% (d) $\dfrac{\ln 1.5}{0.06} \approx 6.7578$ or $6\frac{3}{4}$ years   (e) 8.24%

(f) Double: 11.6 years; Quadruple: 23.1 years

141. $2^{x-1} = 30$

143. To solve an exponential equation, first isolate the exponential expression, then take the logarithms of both sides of the equation, and solve for the variable.

To solve a logarithmic equation, first isolate the logarithmic expression, then exponentiate both sides of the equation, and solve for the variable.

Section 11.6   *(page 744)*

Review   *(page 744)*

1. Direct variation as nth power

2. Inverse variation   3. Joint variation

4. Combined variation   5. $(3, 3)$ 6. $\left(\frac{10}{3}, \frac{5}{3}\right)$

7. $(2, 4), \left(-\frac{1}{2}, \frac{1}{4}\right)$ 8. $(0, 0), (8, 2)$ 9. $\left(\frac{3}{5}, \frac{8}{5}, \frac{2}{5}\right)$

10. $(4, -1, 3)$

11. (a) Downward, because the equation is of quadratic type $y = ax^2 + bx + c$, and $a < 0$.

(b) $(0, 0), (4, 0)$ (c) $(2, 4)$

12.

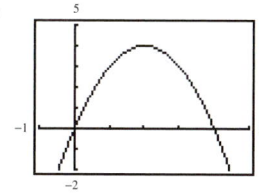

1. 7% 3. 9% 5. 8% 7. 9.27 years

9. 8.66 years   11. 9.59 years   13. Continuous

15. Quarterly   17. 8.33% 19. 7.23% 21. 6.14%

23. 8.30%

25. No. Each time the amount is divided by the principal, the result is always 2.

27. $\$1652.99$ 29. $\$626.46$ 31. $\$3080.15$

33. $\$951.23$ 35. $\$5496.57$ 37. $\$320,250.81$

39. Total deposits: $\$7200.00$; Total interest: $\$10,529.42$

41. $k = \frac{1}{2} \ln \frac{8}{3} \approx 0.4904$ 43. $k = \frac{1}{3} \ln \frac{1}{2} \approx -0.2310$

45. $y = 19.1e^{0.0083t}$; 24.5 million

47. $y = 1275.1e^{0.0057t}$; 1512.9 million

49. $y = 3.8e^{0.0085t}$; 4.9 million

51. $y = 283.2e^{0.0081t}$; 361.1 million

53. (a) k is larger in Exercise 49, because the population of Ireland is increasing faster than the population of China.

(b) k corresponds to r; k gives the annual percent rate of growth.

55. (a) $y = 100e^{4.6052t}$ $(t = 0 \leftrightarrow 2000)$ (b) 1.45 hours

57. $1,149,000,000$ users

Isotope	Half-Life (Years)	Initial Quantity	Amount After 1000 Years
59. ^{226}Ra	1620	6 g	3.91 g
61. ^{14}C	5730	4.51 g	4.0 g
63. ^{230}Pu	$24{,}360$	4.2 g	4.08 g

65. 3.3 grams   67. 7.5 grams   69. $\$15,203$

71. The earthquake in Alaska was about 6.3 times greater.

73. The earthquake in Mexico was about 1259 times greater.

75. 7.04 77. 10^7 times

79. (a)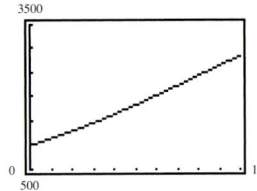

(b) 1000 rabbits   (c) 2642 rabbits   (d) 5.88 years

81. (a) $S = 10(1 - e^{-0.0575x})$ (b) 3314 jeans

83. $k > 0$

85. A is the balance, P is the principal, r is the annual interest rate, and t is the time in years.

87. The time required for the radioactive material to decay to half of its original amount

Review Exercises   *(page 750)*

1. (a) $\frac{1}{8}$ (b) 2   (c) 4   **3.** (a) 5   (b) 0.185   (c) $\frac{1}{25}$

5.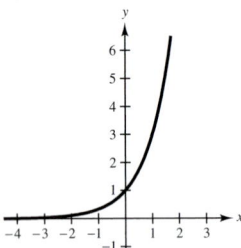

Horizontal asymptote:
$y = 0$

7.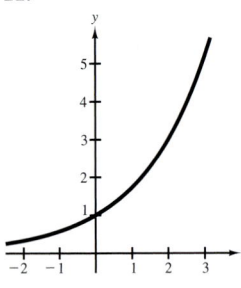

Horizontal asymptote:
$y = -1$

9.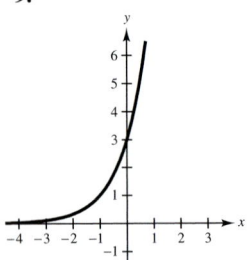

Horizontal asymptote:
$y = 0$

11.

Horizontal asymptote:
$y = 0$

13.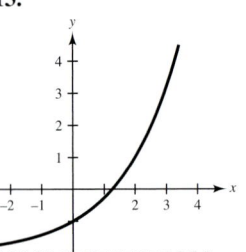

Horizontal asymptote: $y = -2$

15.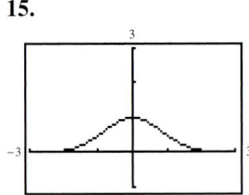

17.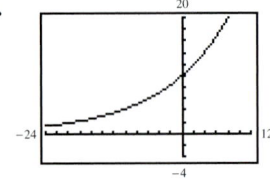

19. (a) 0.007   (b) 3   (c) 9.56×10^{16}

21. 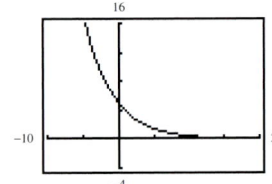   **23.**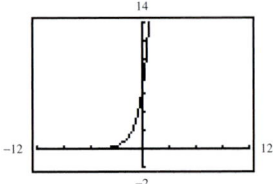

25.

n	1	4	12
A	\$226,296.28	\$259,889.34	\$268,503.32

n	365	Continuous
A	\$272,841.23	\$272,990.75

27. ≈ 4.21 grams   **29.** (a) 6   (b) 1

31. (a) 5   (b) -1

33. (a) $(f \circ g)(x) = \sqrt{2x - 4}$ (b) $(g \circ f)(x) = 2\sqrt{x - 4}$
Domain: $[2, \infty)$ Domain: $[4, \infty)$

35. No   **37.** Yes   **39.** $f^{-1}(x) = \frac{1}{3}(x - 4)$

41. $h^{-1}(x) = x^2,\ x \geq 0$ **43.** $f^{-1}(t) = \sqrt[3]{t - 4}$

45.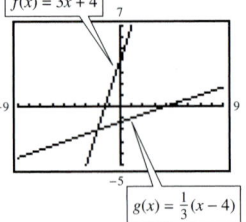

47.

x	0	1	3	4
f^{-1}	6	4	2	0

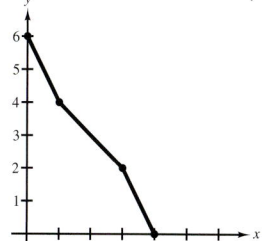

49.

x	-4	-2	2	3
f^{-1}	-2	-1	1	3

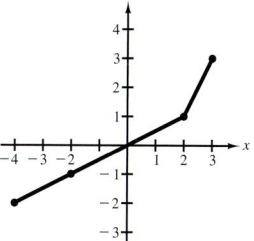

79. 1.0293　**81.** $\log_4 6 + 4\log_4 x$　**83.** $\frac{1}{2}\log_5(x + 2)$

85. $\ln(x + 2) - \ln(x - 2)$

87. $\frac{1}{2}(\ln 2 + \ln x) + 5\ln(x + 3)$　**89.** $\ln\left(\dfrac{1}{3y}\right)^{2/3}$

91. $\log_8 32x^3$　**93.** $\ln\dfrac{9}{4x^2}$, $x > 0$

95. $\log_2\left(\dfrac{k}{k - t}\right)^4$, $k > t$

97. $\ln(x^3 y^4 z)$, $x > 0$, $y > 0$, $z > 0$

99. False; $\log_2 4x = 2 + \log_2 x$.　**101.** True

103. True　**105.** $y = 0.83(\ln I_0 - \ln I)$; 0.20　**107.** 6

109. 1　**111.** 6　**113.** 5.66　**115.** 6.23

117. No solution　**119.** 1408.10　**121.** 15.81

123. 4　**125.** 64　**127.** 2.67　**129.** 7%　**131.** 5%

133. 7.5%　**135.** 7%　**137.** 5.65%　**139.** 7.71%

141. 7.79%

Isotope	Half-Life (Years)	Initial Quantity	Amount After 1000 Years
143. ^{226}Ra	1620	3.5 g	2.282 g
145. ^{14}C	5730	2.934g	2.6 g
147. ^{230}Pu	24,360	5 g	4.860 g

149. The earthquake in San Francisco was about 2512 times greater.

51. 3　**53.** -2　**55.** 6　**57.** 0

59.

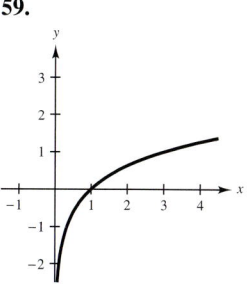

Vertical asymptote: $x = 0$

61.

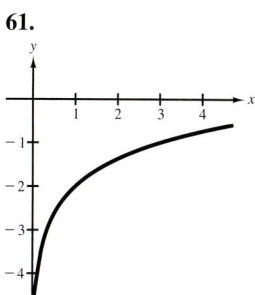

Vertical asymptote: $x = 0$

63.

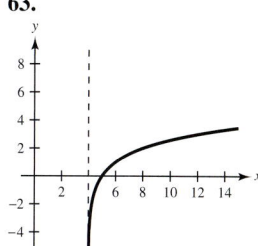

Vertical asymptote: $x = 4$

65. 7

67.

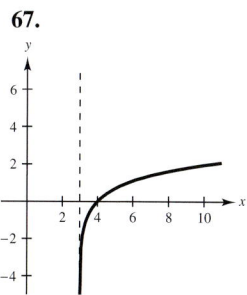

Vertical asymptote: $x = 3$

69.

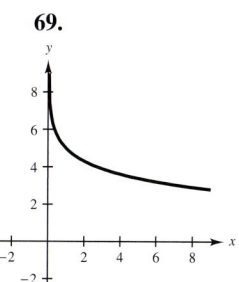

Vertical asymptote: $x = 0$

71. 1.5850　**73.** 2.1322　**75.** 1.7959　**77.** -0.4307

Chapter Test　*(page 755)*

1. $f(-1) = 81$;

$f(0) = 54$;

$f\left(\frac{1}{2}\right) = 18\sqrt{6} \approx 44.09$;

$f(2) = 24$

2.

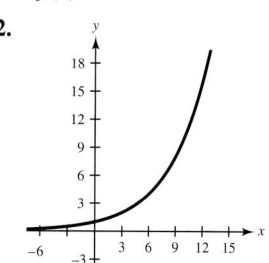

Horizontal asymptote: $y = 0$

3. (a) $(f \circ g)(x) = 18x^2 - 63x + 55$;

Domain: $(-\infty, \infty)$

(b) $(g \circ f)(x) = -6x^2 - 3x + 5$;

Domain: $(-\infty, \infty)$

4. $f^{-1}(x) = \frac{1}{5}(x - 6)$

5. $(f \circ g)(x) = -\frac{1}{2}(-2x + 6) + 3$

$= (x - 3) + 3$

$= x$

$(g \circ f)(x) = -2\left(-\frac{1}{2}x + 3\right) + 6$

$= (x - 6) + 6$

$= x$

6. $\frac{1}{3}$

7. $g = f^{-1}$

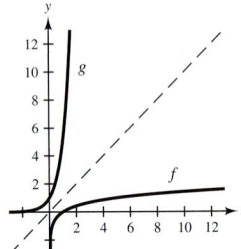

8. $\log_4 5 + 2 \log_4 x - \frac{1}{2} \log_4 y$ **9.** $\ln \frac{x}{y^4}, \; y > 0$

10. 32   **11.** 1.18   **12.** 13.73   **13.** 15.52

14. 2   **15.** 8   **16.** 109.20   **17.** 0

18. (a) \$8012.78   (b) \$8110.40   **19.** \$10,806.08

20. 7%   **21.** \$8469.14   **22.** 600   **23.** 1141

24. 4.4 years

Chapter 12

Section 12.1   *(page 765)*

Review   *(page 765)*

1. $x^2 + 12x + 31$ **2.** $x^2 - 14x + 47$

3. $-x^2 + 16x - 52$ **4.** $-x^2 - 2x + 15$

5. $(x - 2)^2 - 3$ **6.** $(x + 6)^2 - 39$

7. $-(x - 3)^2 + 14$ **8.** $-2\left(x - \frac{5}{2}\right)^2 - \frac{3}{2}$

9. $y = \frac{5}{8}x + \frac{25}{4}$ **10.** $y = -\frac{2}{5}x + \frac{37}{5}$

11. 12 people   **12.** 15 people

---

1. c   **2.** e   **3.** a   **4.** d   **5.** b   **6.** f

7. $x^2 + y^2 = 25$ **9.** $x^2 + y^2 = \frac{4}{9}$ **11.** $x^2 + y^2 = 64$

13. $x^2 + y^2 = 29$ **15.** $(x - 4)^2 + (y - 3)^2 = 100$

17. $(x - 5)^2 + (y + 3)^2 = 81$

19. $(x + 2)^2 + (y - 1)^2 = 4$

21. $(x - 3)^2 + (y - 2)^2 = 17$

23. Center: $(0, 0)$; $r = 4$ **25.** Center: $(0, 0)$; $r = 6$

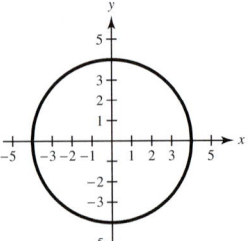   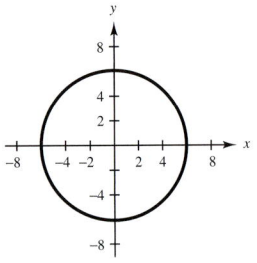

27. Center: $(0, 0)$; $r = \frac{1}{2}$ **29.** Center: $(0, 0)$; $r = \frac{12}{5}$

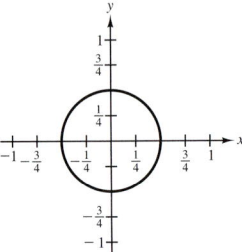   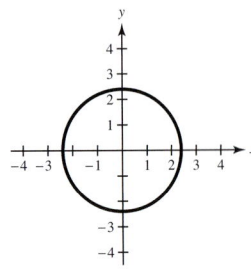

31. Center: $(-1, 5)$; $r = 8$ **33.** Center: $(2, 3)$; $r = 2$

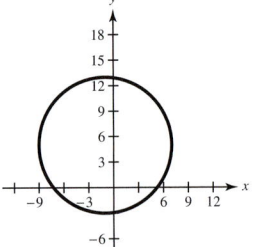   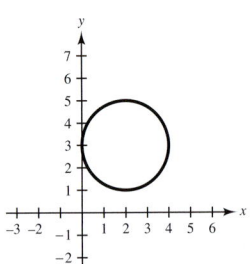

35. Center: $\left(-\frac{5}{2}, -3\right)$; $r = 3$ **37.** Center: $(2, 1)$; $r = 2$

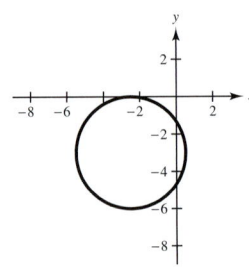   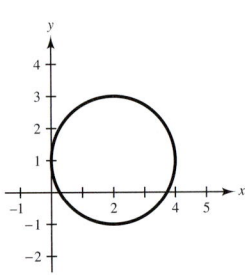

39. Center: $(-1, -3)$;
r = 2

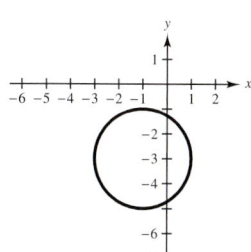

41. Center: $(-4, -2)$;
r = 5

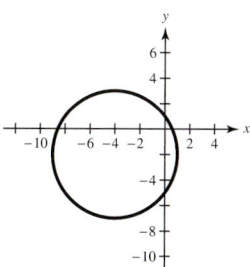

83. Vertex: $\left(5, -\frac{1}{2}\right)$
Focus: $\left(\frac{11}{2}, -\frac{1}{2}\right)$

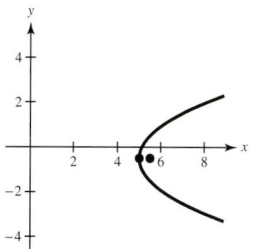

85. Vertex: $(1, 1)$
Focus: $(1, 2)$

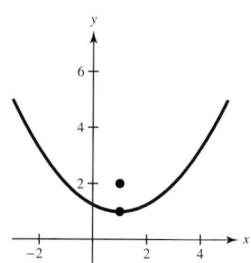

43.

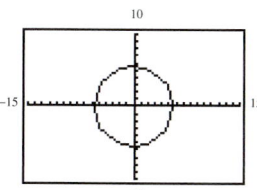

45.

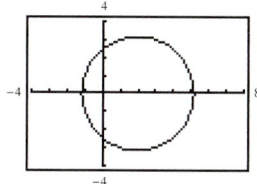

87. Vertex: $(-2, -3)$
Focus: $(-4, -3)$

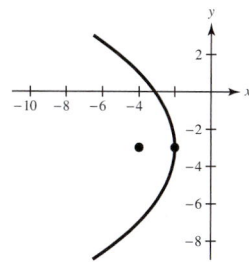

89. Vertex: $(-2, 1)$
Focus: $\left(-2, -\frac{1}{2}\right)$

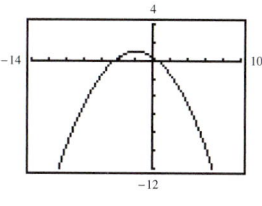

47. b   **48.** c   **49.** e   **50.** a   **51.** f   **52.** d

53. $x^2 = \frac{3}{2}y$ **55.** $x^2 = -6y$ **57.** $y^2 = -8x$

59. $x^2 = 4y$ **61.** $y^2 = 16x$ **63.** $y^2 = 9x$

65. $(x-3)^2 = -(y-1)$ **67.** $y^2 = 2(x+2)$

69. $(y-2)^2 = -8(x-3)$ **71.** $x^2 = 8(y-4)$

73. $(y-2)^2 = x$

91. Vertex: $\left(\frac{1}{4}, -\frac{1}{2}\right)$
Focus: $\left(0, -\frac{1}{2}\right)$

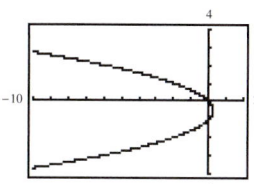

93. $x^2 + y^2 = 4500^2$

75. Vertex: $(0, 0)$
Focus: $\left(0, \frac{1}{2}\right)$

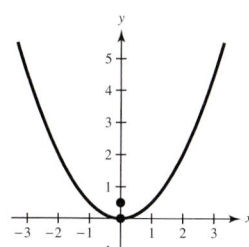

77. Vertex: $(0, 0)$
Focus: $\left(-\frac{3}{2}, 0\right)$

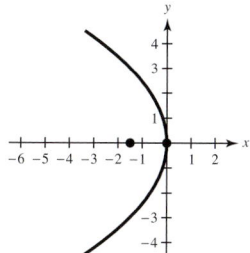

95. $y = \sqrt{2500 - x^2}$; $5\sqrt{19} \approx 21.8$ feet

97. (a) $y = \dfrac{x^2}{180}$ (b)

x	0	20	40	60
y	0	$2\frac{2}{9}$	$8\frac{8}{9}$	20

79. Vertex: $(0, 0)$
Focus: $(0, -2)$

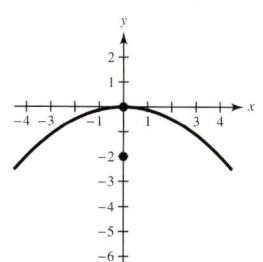

81. Vertex: $(1, -2)$
Focus: $(1, -4)$

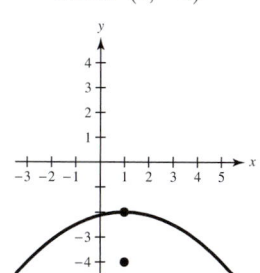

99. (a)

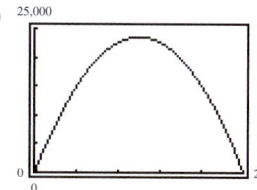

(b) $x = 125$

101. (a) $x^2 + y^2 = 25$ (b)

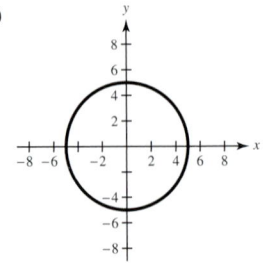

103. A circle is the set of all points (x, y) that are a given positive distance r from a fixed point (h, k) called the center. $x^2 + y^2 = r^2$

105. All points on the parabola are equidistant from the directrix and the focus.

107. No. If the graph intersected the directrix, there would exist points nearer the directrix than the focus.

Section 12.2   *(page 776)*

Review   *(page 776)*

1. $\dfrac{3}{2y^5}$ **2.** $\dfrac{4y^2}{9x^4}$ **3.** $\dfrac{x^3 y}{6}$ **4.** 1

5. $-3 \pm \sqrt{13}$ **6.** $4 \pm \dfrac{3\sqrt{6}}{2}$

7.

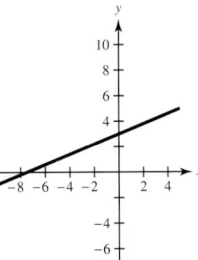

8.

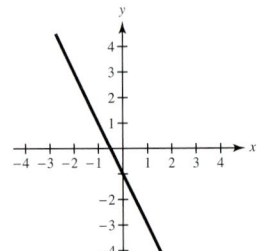

9.

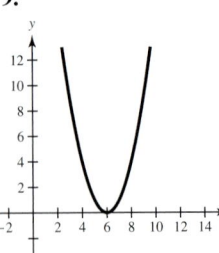

10.

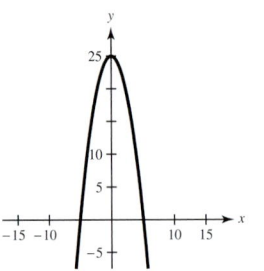

11. 87   **12.** $3750

7. $\dfrac{x^2}{16} + \dfrac{y^2}{9} = 1$ **9.** $\dfrac{x^2}{4} + \dfrac{y^2}{1} = 1$ **11.** $\dfrac{x^2}{9} + \dfrac{y^2}{16} = 1$

13. $\dfrac{x^2}{1} + \dfrac{y^2}{4} = 1$ **15.** $\dfrac{x^2}{9} + \dfrac{y^2}{25} = 1$

17. $\dfrac{x^2}{100} + \dfrac{y^2}{36} = 1$

19.

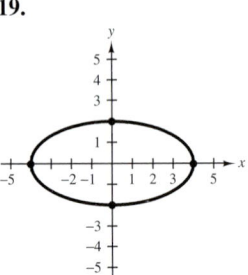

21.

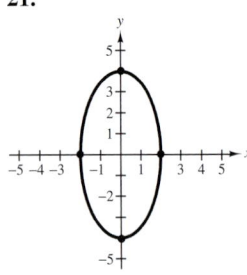

Vertices: $(\pm 4, 0)$
Co-vertices: $(0, \pm 2)$

Vertices: $(0, \pm 4)$
Co-vertices: $(\pm 2, 0)$

23.

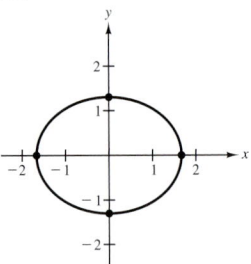

25.

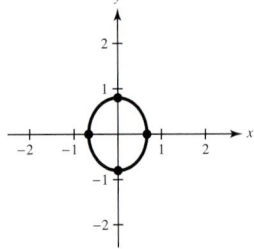

Vertices: $\left(\pm \frac{5}{3}, 0\right)$
Co-vertices: $\left(0, \pm \frac{4}{3}\right)$

Vertices: $\left(0, \pm \frac{4}{5}\right)$
Co-vertices: $\left(\pm \frac{2}{3}, 0\right)$

27.

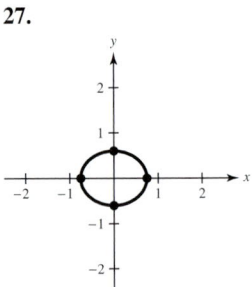

29.

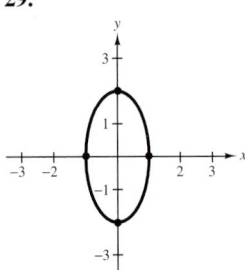

Vertices: $\left(\pm \frac{3}{4}, 0\right)$;
Co-vertices: $\left(0, \pm \frac{3}{5}\right)$

Vertices: $(0, \pm 2)$;
Co-vertices: $(\pm 1, 0)$

1. a   **2.** f   **3.** d   **4.** c   **5.** e   **6.** b

31.

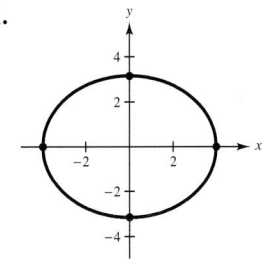

Vertices: $(\pm 4, 0)$

Co-vertices: $\left(0, \pm\sqrt{10}\right)$

33.

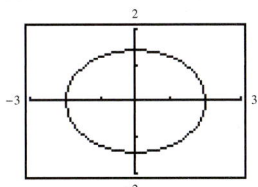

Vertices: $(\pm 2, 0)$

35.

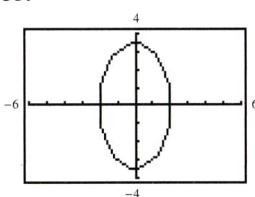

Vertices: $\left(0, \pm 2\sqrt{3}\right)$

37. $\dfrac{x^2}{1} + \dfrac{y^2}{4} = 1$ **39.** $\dfrac{(x-4)^2}{9} + \dfrac{y^2}{16} = 1$

41. Center: $(-5, 0)$

Vertices: $(-9, 0), (-1, 0)$

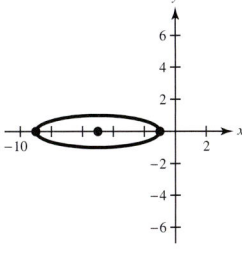

43. Center: $(1, 5)$

Vertices: $(1, 0), (1, 10)$

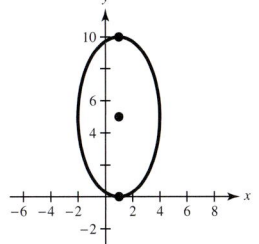

45. Center: $(-2, 3)$

Vertices: $(-2, 6), (-2, 0)$

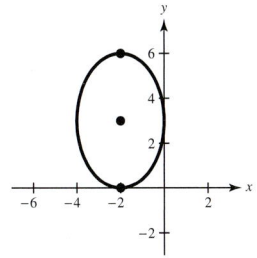

47. Center: $(2, -2)$

Vertices: $(-1, -2), (5, -2)$

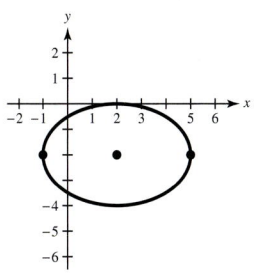

49. Center: $(-4, 1)$

Vertices: $(-4, -3), (-4, 5)$

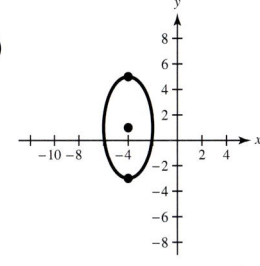

51. Center: $(4, -3)$

Vertices: $(4, -8), (4, 2)$

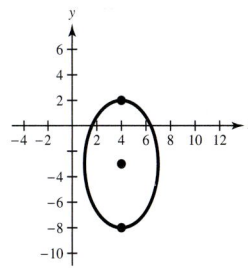

53. Center: $(2, 1)$

Vertices: $(-8, 1), (12, 1)$

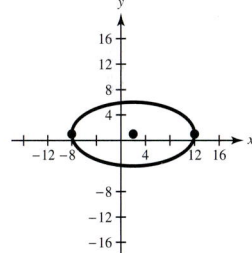

55. $\sqrt{304} \approx 17.4$ feet

57. $\dfrac{x^2}{7225} + \dfrac{(y-85)^2}{10{,}000} = 1$ or $\dfrac{x^2}{10{,}000} + \dfrac{(y-85)^2}{7225} = 1$

59. $\dfrac{x^2}{144} + \dfrac{y^2}{64} = 1$

61. A circle is an ellipse in which the major axis and the minor axis have the same length. Both circles and ellipses have foci; however, in a circle the foci are both at the same point, whereas in an ellipse they are not.

63. The sum of the distances between each point on the ellipse and the two foci is a constant.

65. Major axis: $2a$; Minor axis: $2b$

Mid-Chapter Quiz   *(page 780)*

1. $x^2 + y^2 = 25$ **2.** $(y - 1)^2 = 8(x + 2)$

3. $\dfrac{(x + 2)^2}{16} + \dfrac{(y + 1)^2}{4} = 1$

4. $(x - 3)^2 + (y + 5)^2 = 25$

5. $(x - 2)^2 = -8(y - 3)$ **6.** $\dfrac{x^2}{9} + \dfrac{y^2}{100} = 1$

7. $(x - 5)^2 + y^2 = 9$; Center: $(5, 0)$; $r = 3$

8. $(x + 1)^2 + (y - 2)^2 = 1$; Center: $(-1, 2)$; $r = 1$

9. $(y - 3)^2 = x + 16$; Vertex: $(-16, 3)$; Focus: $\left(-\frac{63}{4}, 3\right)$

10. $(x - 4)^2 = -(y - 4)$; Vertex: $(4, 4)$; Focus: $\left(4, \frac{15}{4}\right)$

11. $\dfrac{x^2}{9} + \dfrac{y^2}{20} = 1$

Center: $(0, 0)$

Vertices: $\left(0, -2\sqrt{5}\right), \left(0, 2\sqrt{5}\right)$

12. $\dfrac{(x - 6)^2}{9} + \dfrac{(y + 2)^2}{4} = 1$

Center: $(6, -2)$

Vertices: $(3, -2), (9, -2)$

13.

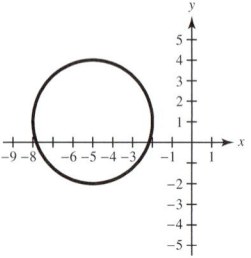

14.

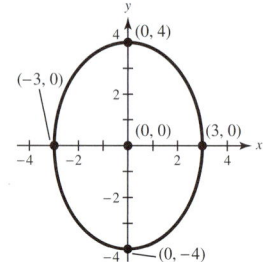

15.

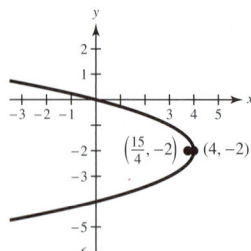

16.

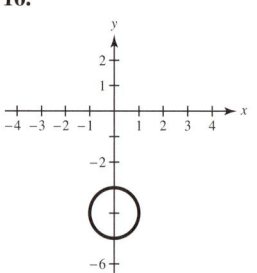

17.

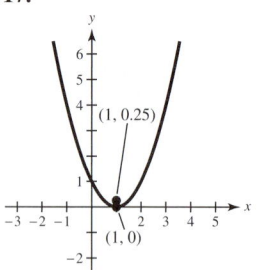

18.

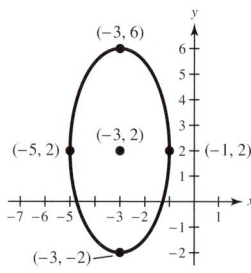

Section 12.3   *(page 786)*

Review   *(page 786)*

1. $2\sqrt{10}$ **2.** $\sqrt{269}$

3.

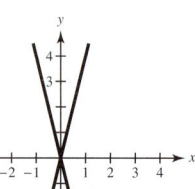

4.

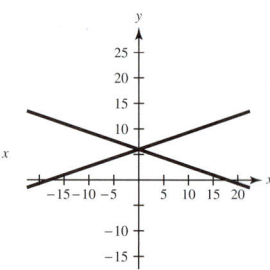

5.

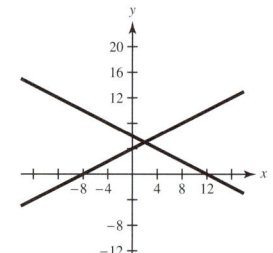

6.

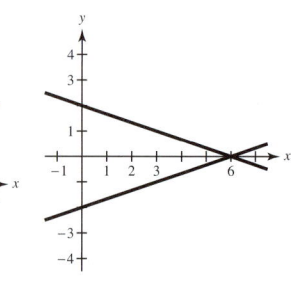

7. 24   **8.** 5   **9.** 13   **10.** $3\sqrt{3}$

11. $2\frac{1}{2}$ hours   **12.** \$10 coins: 6; \$20 coins: 24

1. c   **2.** e   **3.** a   **4.** f   **5.** b   **6.** d

7.

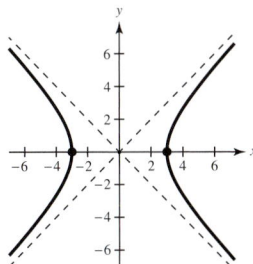

Vertices: $(\pm 3, 0)$

Asymptotes: $y = \pm x$

9.

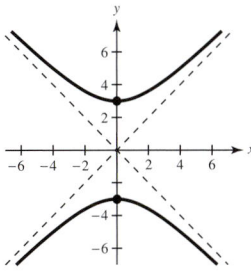

Vertices: $(0, \pm 3)$

Asymptotes: $y = \pm x$

11.

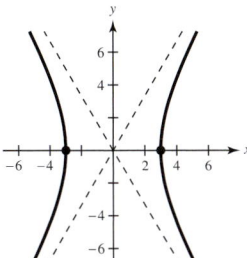

Vertices: $(\pm 3, 0)$

Asymptotes: $y = \pm\frac{5}{3}x$

13.

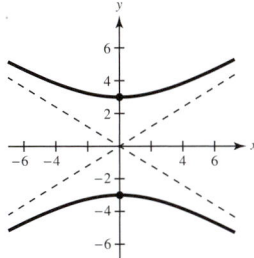

Vertices: $(0, \pm 3)$

Asymptotes: $y = \pm\frac{3}{5}x$

15.

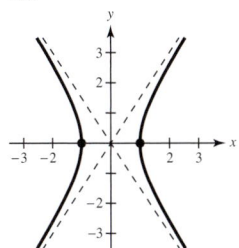

Vertices: $(\pm 1, 0)$

Asymptotes: $y = \pm\frac{3}{2}x$

17.

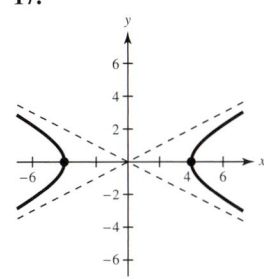

Vertices: $(\pm 4, 0)$

Asymptotes: $y = \pm\frac{1}{2}x$

19. $\dfrac{x^2}{16} - \dfrac{y^2}{64} = 1$ **21.** $\dfrac{y^2}{16} - \dfrac{x^2}{64} = 1$

23. $\dfrac{x^2}{81} - \dfrac{y^2}{36} = 1$ **25.** $\dfrac{y^2}{1} - \dfrac{x^2}{1/4} = 1$

27.

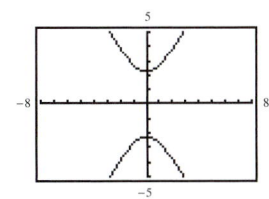

29.

31. Center: $(3, -4)$

Vertices: $(3, 1), (3, -9)$

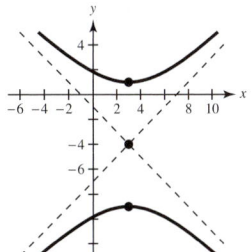

33. Center: $(1, -2)$

Vertices: $(-1, -2), (3, -2)$

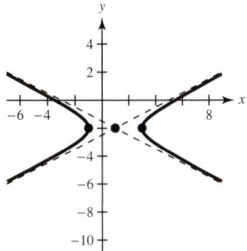

35. Center: $(2, -3)$

Vertices: $(3, -3), (1, -3)$

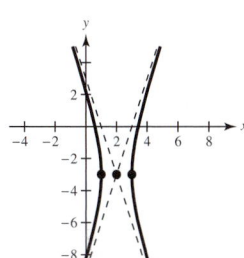

37. Center: $(-3, 2)$

Vertices: $(-4, 2), (-2, 2)$

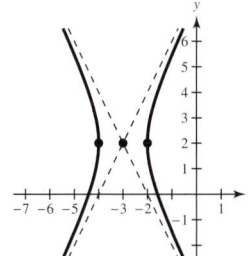

39. $\dfrac{y^2}{9} - \dfrac{x^2}{9/4} = 1$ **41.** $\dfrac{(x-3)^2}{4} - \dfrac{(y-2)^2}{16/5} = 1$

43. $x \approx 110.28$

45. (c) Center: $(2, 0)$

Vertices: $\left(2 + 3\sqrt{3}, 0\right), \left(2 - 3\sqrt{3}, 0\right)$

(d)

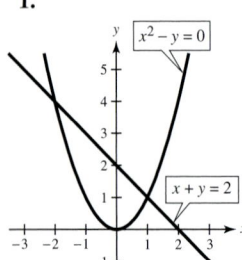

$x^2 + y^2 = 25$

$x^2 - y^2 - 4x - 23 = 0$

47. The difference of the distances between each point on the hyperbola and the two foci is a positive constant.

49. (a) Left half   (b) Top half

Section 12.4   *(page 796)*

Review   *(page 796)*

1. The second row in the new matrix was formed by multiplying the second row of the original matrix by -2.

2. Rows one and two were swapped.

3. The second row in the new matrix was formed by subtracting 3 times the first row from the second row of the original matrix.

4. Gaussian elimination is the process of using elementary row operations to rewrite a matrix representing the system of linear equations in row-echelon form.

5. $x = -2, y = 3$ **6.** $x = 9, y = -4$

7. $x = -\frac{22}{3}, y = \frac{37}{9}$ **8.** $x = \frac{41}{40}, y = -\frac{11}{40}$

9. $x = 2, y = -1, z = 3$

10. $x = -2, y = \frac{3}{5}, z = \frac{12}{5}$ **11.** $\$5520$

12. Cashews: 20 pounds; Brazil nuts: 30 pounds

1.

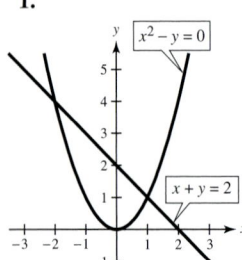

$x^2 - y = 0$

$x + y = 2$

$(-2, 4), (1, 1)$

3.

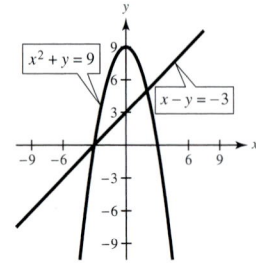

$x^2 + y = 9$

$x - y = -3$

$(2, 5), (-3, 0)$

5.

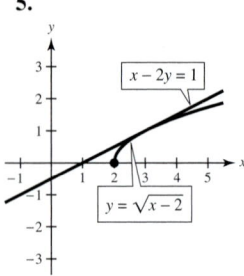

$x - 2y = 1$

$y = \sqrt{x - 2}$

$(3, 1)$

7.

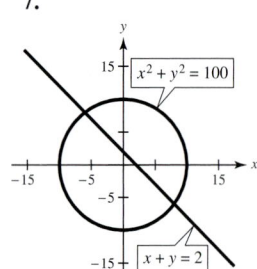

$x^2 + y^2 = 100$

$x + y = 2$

$(-6, 8), (8, -6)$

9.

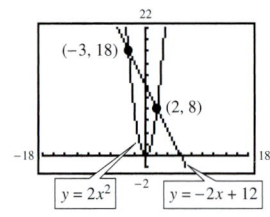

$2x - y = -5$

$x^2 + y^2 = 25$

$(0, 5), (-4, -3)$

11.

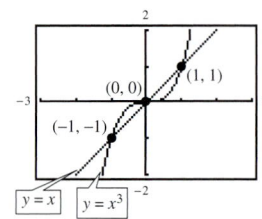

$5x - 2y = 0$

$9x^2 - 4y^2 = 36$

No real solution

13.

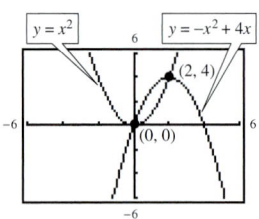

$(-3, 18)$

$(2, 8)$

$y = 2x^2$ $y = -2x + 12$

$(-3, 18), (2, 8)$

15.

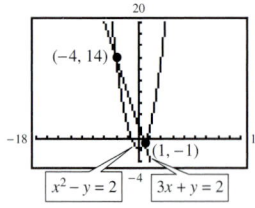

$(0, 0)$ $(1, 1)$

$(-1, -1)$

$y = x$ $y = x^3$

$(0, 0), (1, 1), (-1, -1)$

17.

$y = x^2$ $y = -x^2 + 4x$

$(2, 4)$

$(0, 0)$

$(0, 0), (2, 4)$

19.

$(-4, 14)$

$(1, -1)$

$x^2 - y = 2$ $3x + y = 2$

$(-4, 14), (1, -1)$

21.

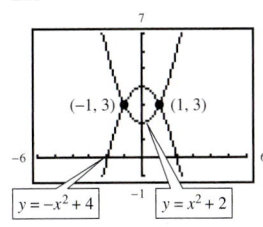

$(-1, 3), (1, 3)$

23.

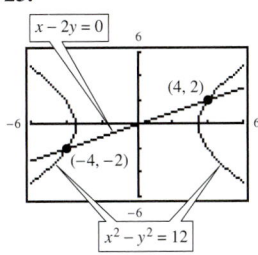

$(4, 2), (-4, -2)$

$(0, 0), (1, 1)$

25.

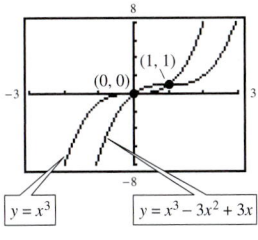

27. $(1, 2), (2, 8)$ **29.** $(0, 5), (2, 1)$ **31.** No real solution

33. $(0, 5)$ **35.** $(0, 8), \left(-\frac{24}{5}, -\frac{32}{5}\right)$

37. $\left(-\frac{17}{2}, -6\right), \left(-\frac{7}{2}, 4\right)$ **39.** $\left(-\frac{9}{5}, \frac{12}{5}\right), (3, 0)$

41. $(-14, -20), (2, -4)$ **43.** $(-4, 11), \left(\frac{5}{2}, \frac{5}{4}\right)$

45. $(0, 2), (3, 1)$ **47.** No real solution

49. $\left(\pm\sqrt{5}, 2\right), (0, -3)$ **51.** $(0, 4), (3, 0)$

53. No real solution   **55.** $\left(\pm\sqrt{3}, -1\right)$

57. $\left(2, \pm 2\sqrt{3}\right), (-1, \pm 3)$ **59.** $\left(\pm 2, \pm\sqrt{3}\right)$

61. $(\pm 2, 0)$ **63.** $(\pm 3, \pm 2)$ **65.** $(\pm 3, 0)$

67. $\left(\pm\sqrt{6}, \pm\sqrt{3}\right)$ **69.** No real solution

71. $\left(\pm\sqrt{5}, \pm 2\right)$ **73.** $\left(\pm\frac{2\sqrt{5}}{5}, \pm\frac{2\sqrt{5}}{5}\right)$

75. $\left(\pm\sqrt{3}, \pm\sqrt{13}\right)$ **77.** $(3.633, 2.733)$

79. ≈ 21 inches $\times$ 36 inches   **81.** 15 inches $\times$ 8 inches

83. Between points $\left(-\frac{3}{5}, -\frac{4}{5}\right)$ and $\left(\frac{4}{5}, -\frac{3}{5}\right)$

85. (e) $(-4, 3)$; This is a point of intersection between the hyperbola and the circle.

(f) $y = -\frac{3}{4}x$

(g) $(4, -3)$; This is the other point at which the line representing Murphy Road intersects the circle.

87. Multiply Equation 2 by a factor that makes the coefficients of one variable equal. Subtract Equation 2 from Equation 1. Write the resulting equation, and solve. Substitute the solution into either equation. Solve for the value of the other variable.

Review Exercises *(page 801)*

1. Hyperbola   **3.** Circle   **5.** Circle   **7.** Parabola

9. $x^2 + y^2 = 144$

11.

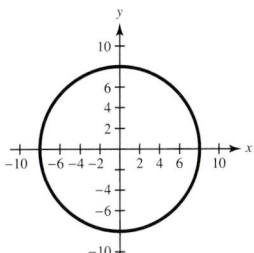

Center: $(0, 0)$; $r = 8$

13. $(x - 3)^2 + (y - 5)^2 = 25$

15.

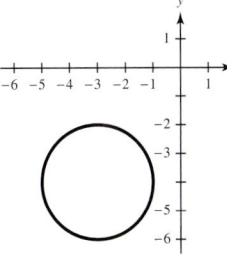

Center: $(-3, -4)$; $r = 2$

17. $y^2 = -8x$

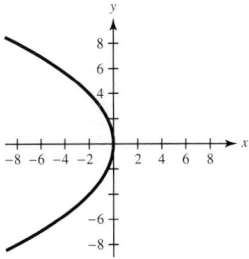

19. $(x + 6)^2 = -20(y - 4)$

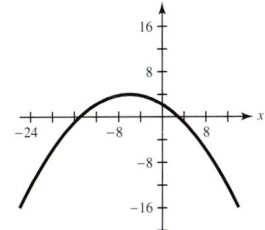

21. $(x + 1)^2 = \frac{1}{2}(y - 3)$

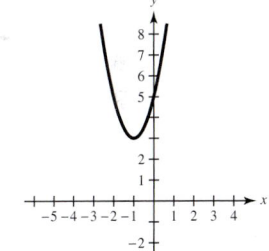

23.

Vertex: $(2, -2)$; Focus: $\left(2, -\frac{7}{4}\right)$

25. $\dfrac{x^2}{4} + \dfrac{y^2}{25} = 1$ **27.** $\dfrac{x^2}{4} + \dfrac{y^2}{9} = 1$

29.

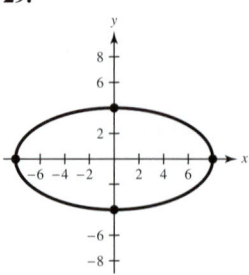

Vertices: $(\pm 8, 0)$

Co-vertices: $(0, \pm 4)$

31.

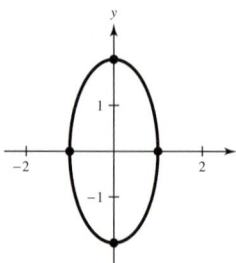

Vertices: $(0, \pm 2)$

Co-vertices: $(\pm 1, 0)$

33. $\dfrac{(x - 3)^2}{25} + \dfrac{(y - 4)^2}{16} = 1$ **35.** $\dfrac{x^2}{9} + \dfrac{(y - 4)^2}{16} = 1$

37.

Center: $(1, -2)$

Vertices: $(1, -11), (1, 7)$

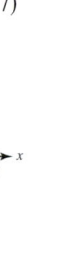

39.

Center: $(0, -3)$

Vertices: $(0, -7), (0, 1)$

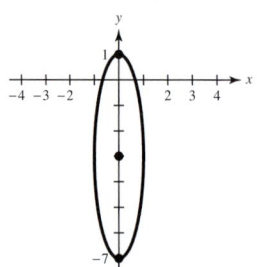

41.

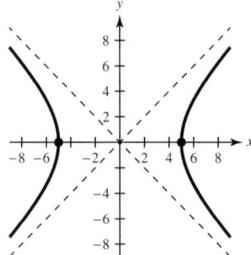

Vertices: $(\pm 5, 0)$

Asymptotes: $y = \pm x$

43.

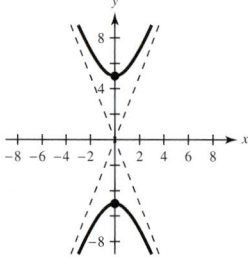

Vertices: $(0, \pm 5)$

Asymptotes: $y = \pm \frac{5}{2}x$

45. $\dfrac{x^2}{4} - \dfrac{y^2}{9} = 1$

47. $\dfrac{y^2}{25} - \dfrac{x^2}{4} = 1$

49. Center: $(3, -1)$

Vertices: $(0, -1), (6, -1)$

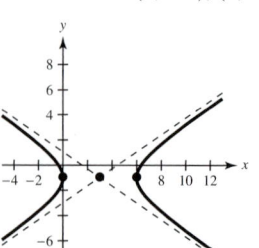

51. Center: $(4, -3)$

Vertices: $(4, -1), (4, -5)$

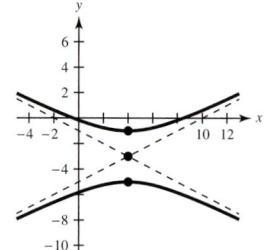

53. $\dfrac{(x + 4)^2}{4} - \dfrac{(y - 6)^2}{12} = 1$

55.

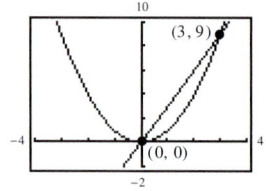

$(0, 0), (3, 9)$

57.

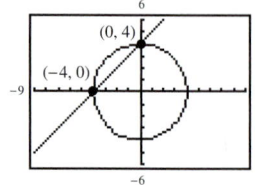

$(-4, 0), (0, 4)$

59. $(-1, 5), (-2, 20)$ **61.** $(-1, 0), (0, -1)$

63. $(0, 2), \left(-\frac{16}{5}, -\frac{6}{5}\right)$ **65.** $(\pm 5, 0)$

67. 4 inches $\times$ 5 inches

69. 6 centimeters $\times$ 8 centimeters

71. Piece 1: 38.48 inches; Piece 2: 61.52 inches

Chapter Test *(page 805)*

1. $(x + 1)^2 + (y - 3)^2 = 4$

2. $(x - 1)^2 + (y - 3)^2 = 9$ **3.** $(x + 2)^2 + (y - 3)^2 = 9$

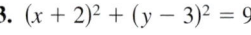

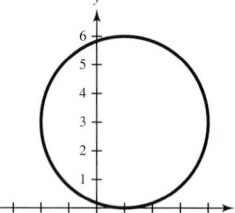

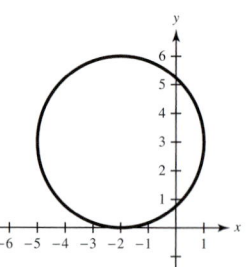

4. Vertex: $(4, 2)$; Focus: $\left(\frac{47}{12}, 2\right)$

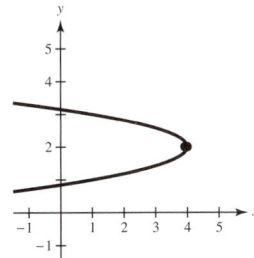

5. $(x - 7)^2 = 8(y + 2)$ **6.** $\dfrac{(x - 2)^2}{25} + \dfrac{y^2}{9} = 1$

7. Center: $(0, 0)$;
    Vertices: $(0, \pm 4)$

8. Center: $(2, -4)$;
    Vertices: $(2, -7), (2, -1)$

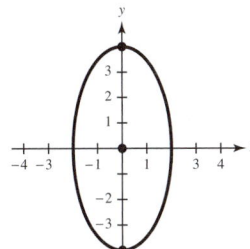
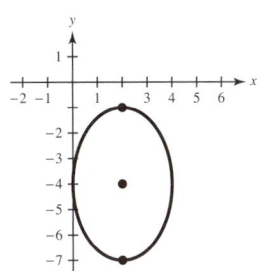

9. $\dfrac{x^2}{9} - \dfrac{y^2}{4} = 1$ **10.** $\dfrac{y^2}{4} - x^2 = 1$

11. Center: $(0, 3)$
    Vertices: $(\pm 2, 3)$

12. Center: $(4, -2)$
    Vertices: $(4, -7), (4, 3)$

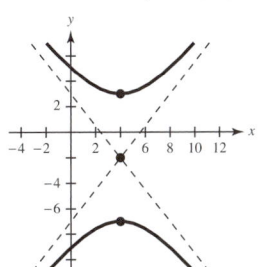

13. $(0, 3), (4, 0)$ **14.** $(\pm 4, 0)$

15. $\left(\sqrt{6}, 2\right), \left(\sqrt{6}, -2\right), \left(-\sqrt{6}, 2\right), \left(-\sqrt{6}, -2\right)$

16. $x^2 + y^2 = 25{,}000{,}000$ **17.** 16 inches $\times$ 12 inches

Cumulative Test: Chapters 10–12 *(page 806)*

1. $x = -\frac{3}{4}, 3$ **2.** $x = -3, 13$

3. $x = 5 \pm 5\sqrt{2}$ **4.** $x = -1 \pm \dfrac{\sqrt{3}}{3}$ **5.** $x = 16$

6. $\dfrac{-5 - \sqrt{61}}{6} \le x \le \dfrac{-5 + \sqrt{61}}{6}$

7. $-\dfrac{4}{3} < x < \dfrac{1}{2}$

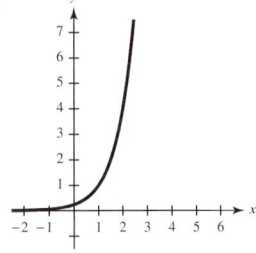

8. $x^2 - 4x - 12 = 0$

9. (a) $(f \circ g)(x) = 50x^2 - 20x - 1$; Domain: $(-\infty, \infty)$

    (b) $(g \circ f)(x) = 10x^2 - 16$; Domain: $(-\infty, \infty)$

10. $f^{-1}(x) = 4x - \dfrac{3}{2}$

11. $f(1) = \dfrac{15}{2}, f(0.5) = \dfrac{14 + \sqrt{2}}{2}, f(3) = \dfrac{57}{8}$

12.

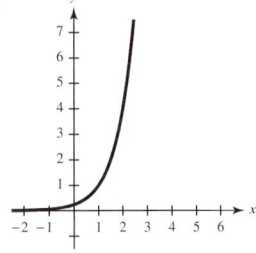

Horizontal asymptote: $y = 0$

13. The graphs are reflections of each other in the line $y = x$.

14.

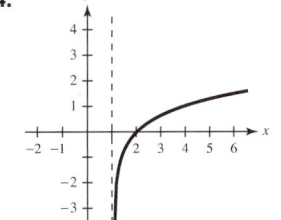

Vertical asymptote: $x = 1$

15. -2 **16.** $\log_2 \dfrac{x^3 y^3}{z}$ **17.** $\ln 5 + \ln x - 2 \ln(x + 1)$

18. $x = 3$ **19.** $x = e^{5/2} \approx 12.182$ **20.** $t \approx 18.013$

21. $x \approx 0.867$ **22.** \$29.63   **23.** $\approx 8.33\%$

24. ≈ 15.403 years

25. $(x - 3)^2 + (y + 7)^2 = 64$ **26.** Vertex: $(5, -45)$

Focus: $\left(5, -\frac{359}{8}\right)$

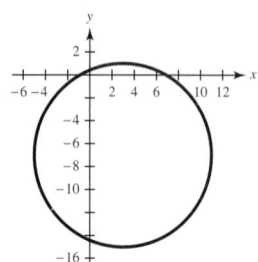   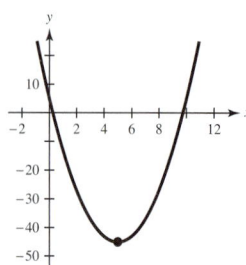

27. $\dfrac{(x - 3)^2}{9} + \dfrac{y^2}{4} = 1$

28. Center: $(0, 0)$ **29.** $x^2 - \dfrac{y^2}{4} = 1$

Vertices: $(0, \pm 2)$

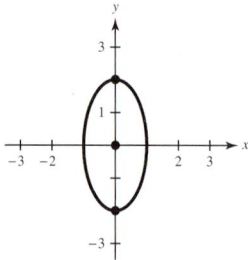

30. Center: $(0, 1)$ **31.** $(1, -1), (3, 5)$

Vertices: $(\pm 12, 1)$

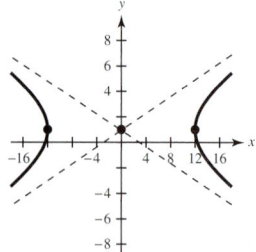

32. $\left(\pm \sqrt{3}, 0\right)$ **33.** 8 feet × 4 feet

34. (a)

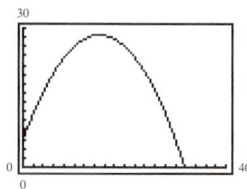

(b) Highest point: 28.5 feet; Range: $[0, 28.5]$

Chapter 13

Section 13.1   *(page 816)*

Review *(page 816)*

1. $-7x = 35$

$$\frac{-7x}{-7} = \frac{35}{-7}$$

$$x = -5$$

2. $7x + 63 = 35$

$7x + 63 - 63 = 35 - 63$

$$7x = -28$$

3. It is a solution if the equation is true when -3 is substituted for t.

4. Multiply each side of the equation by the lowest common denominator; in this example, it is $x(x + 1)$.

5. $\dfrac{1}{(x + 10)^2}$ **6.** $18(x - 3)^3, x \neq 3$ **7.** $\dfrac{1}{a^8}$

8. $2x$ **9.** $8x\sqrt{2x}$ **10.** $\dfrac{5\left(\sqrt{x} + 2\right)}{x - 4}$

11. (a) $A = x(2x - 3)$

(b)

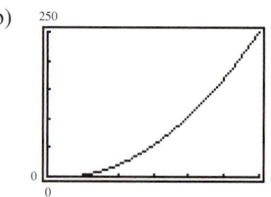

(c) $\dfrac{3 + \sqrt{1609}}{4} \approx 10.8$

12. (a) $A = \frac{1}{2}x(x - 4)$

(b)

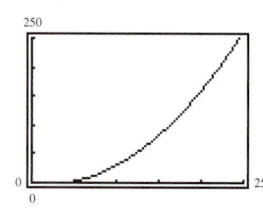

(c) $2\left(1 + \sqrt{101}\right) \approx 22.1$

1. $2, 4, 6, 8, 10$ **3.** $-2, 4, -6, 8, -10$

5. $\frac{1}{2}, \frac{1}{4}, \frac{1}{8}, \frac{1}{16}, \frac{1}{32}$ **7.** $\frac{1}{4}, -\frac{1}{8}, \frac{1}{16}, -\frac{1}{32}, \frac{1}{64}$

9. $3, 8, 13, 18, 23$ **11.** $1, \frac{4}{5}, \frac{2}{3}, \frac{4}{7}, \frac{1}{2}$ **13.** $\frac{3}{4}, \frac{2}{3}, \frac{9}{14}, \frac{12}{19}, \frac{5}{8}$

15. $-1, \frac{1}{4}, -\frac{1}{9}, \frac{1}{16}, -\frac{1}{25}$ **17.** $\frac{9}{2}, \frac{19}{4}, \frac{39}{8}, \frac{79}{16}, \frac{159}{32}$

19. 2, 3, 4, 5, 6   **21.** $0, 3, -1, \frac{3}{4}, -\frac{1}{4}$ **23.** -72

25. $\frac{31}{2520}$ **27.** 5   **29.** $\frac{1}{132}$ **31.** 53,130   **33.** $\frac{1}{n+1}$

35. $n(n+1)$ **37.** $2n$ **39.** c   **40.** a   **41.** b

42. d

43.   **45.**

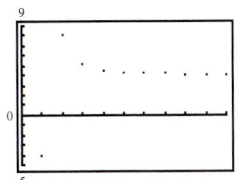   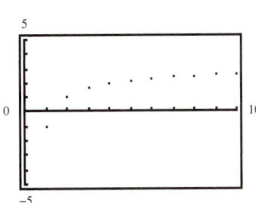

47.   **49.** $a_n = 2n - 1$

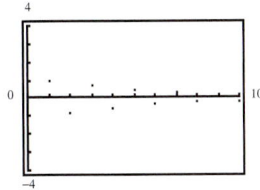

51. $a_n = 4n - 2$ **53.** $a_n = n^2 - 1$

55. $a_n = (-1)^{n+1}2n$ **57.** $a_n = \dfrac{n+1}{n+2}$

59. $a_n = \dfrac{(-1)^{n+1}}{2^n}$ **61.** $a_n = \dfrac{1}{2^{n-1}}$ **63.** $a_n = 1 + \dfrac{1}{n}$

65. $a_n = \dfrac{1}{n!}$ **67.** 63   **69.** 77   **71.** 100   **73.** $\frac{3019}{3600}$

75. $\frac{15,551}{2520}$ **77.** -48 **79.** $\frac{8}{9}$ **81.** $\frac{182}{243}$ **83.** 273

85. 852   **87.** 16.25   **89.** 6.5793   **91.** $\displaystyle\sum_{k=1}^{5} k$

93. $\displaystyle\sum_{k=1}^{5} 2k$ **95.** $\displaystyle\sum_{k=1}^{10} \dfrac{1}{2k}$ **97.** $\displaystyle\sum_{k=1}^{20} \dfrac{1}{k^2}$ **99.** $\displaystyle\sum_{k=0}^{9} \dfrac{1}{(-3)^k}$

101. $\displaystyle\sum_{k=1}^{20} \dfrac{4}{k+3}$ **103.** $\displaystyle\sum_{k=1}^{11} \dfrac{k}{k+1}$ **105.** $\displaystyle\sum_{k=1}^{20} \dfrac{2k}{k+3}$

107. $\displaystyle\sum_{k=0}^{6} k!$

109. (a) $535, $572.45, $612.52, $655.40, $701.28, $750.37,
$802.89, $859.09

(b) $7487.23

(c)

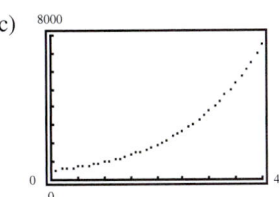

(d) Yes. Investment earning compound interest increases
at an increasing rate.

111. $a_5 = 108°, a_6 = 120°$; At the point where any two hexagons and a pentagon meet, the sum of the three angles is
$a_5 + 2a_6 = 348° < 360°$. Therefore, there is a gap of 12°.

113. 25.7°, 45°, 60°, 72°, 81.8°

115. $a_n = 3n$: 3, 6, 9, 12, . . .

117. Terms in which n is odd, because $(-1)^n = -1$ when n is
odd and $(-1)^n = 1$ when n is even.

119. True.

$$\sum_{i=1}^{4}(i^2 + 2i) = (1 + 2) + (4 + 4) + (9 + 6) + (16 + 8)$$
$$= (1 + 4 + 9 + 16) + (2 + 4 + 6 + 8)$$
$$= \sum_{i=1}^{4} i^2 + \sum_{i=1}^{4} 2i$$

121. True.

$$\sum_{j=1}^{4} 2^j = 2^1 + 2^2 + 2^3 + 2^4$$
$$= 2^{3-2} + 2^{4-2} + 2^{5-2} + 2^{6-2}$$
$$= \sum_{j=3}^{6} 2^{j-2}$$

Section 13.2   *(page 825)*

Review   *(page 825)*

1. A collection of letters (called variables) and real
numbers (called constants) combined with the operations of addition, subtraction, multiplication, and
division is called an algebraic expression.

2. The terms of an algebraic expression are those parts
separated by addition or subtraction.

3. $2x^3 - 3x^2 + 2$ **4.** $7x^4$ **5.** $(-\infty, \infty)$

6. $(-\infty, \infty)$ **7.** $[-4, 4]$

8. $(-\infty, -6) \cup (-6, 6) \cup (6, \infty)$ **9.** $(2, \infty)$

10. $(-\infty, \infty)$ **11.** $30,798.61   **12.** $5395.40

1. 3   **3.** -6 **5.** -12 **7.** $\frac{2}{3}$ **9.** $-\frac{5}{4}$

11. Arithmetic, 2   **13.** Arithmetic, -2

15. Arithmetic, -16 **17.** Arithmetic, 0.8

19. Arithmetic, $\frac{3}{2}$ **21.** Not arithmetic

23. Not arithmetic   **25.** Not arithmetic

27. 7, 10, 13, 16, 19   **29.** 6, 4, 2, 0, -2

31. $\frac{3}{2}, 4, \frac{13}{2}, 9, \frac{23}{2}$ **33.** $\frac{8}{5}, \frac{11}{5}, \frac{14}{5}, \frac{17}{5}, 4$ **35.** $4, \frac{15}{4}, \frac{7}{2}, \frac{13}{4}, 3$

37. $a_n = 3n + 1$ **39.** $a_n = \frac{3}{2}n - 1$

41. $a_n = -5n + 105$ **43.** $a_n = \frac{3}{2}n + \frac{3}{2}$

45. $a_n = \frac{5}{2}n + \frac{5}{2}$ **47.** $a_n = 4n + 4$

49. $a_n = -10n + 60$ **51.** $a_n = -\frac{1}{2}n + 11$

53. $a_n = -0.05n + 0.40$ **55.** 14, 20, 26, 32, 38

57. 23, 18, 13, 8, 3   **59.** $-16, -11, -6, -1, 4$

61. 3.4, 2.3, 1.2, 0.1, -1 **63.** 210   **65.** 1425

67. 255   **69.** 62,625   **71.** 35   **73.** 522

75. 1850   **77.** 900   **79.** 12,200   **81.** 243   **83.** 23

85. b   **86.** f   **87.** e   **88.** a   **89.** c   **90.** d

91.

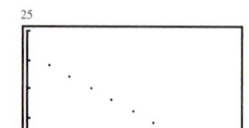

93.

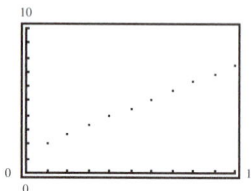

95.

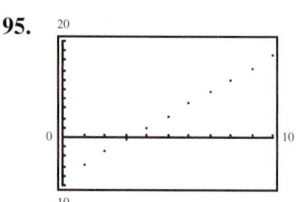

97. 9000

99. 13,120   **101.** 3011.25   **103.** 2850   **105.** 2500

107. $246,000   **109.** $25.43   **111.** 632 bales

113. 114   **115.** 1024 feet

117. (a)

Figure	Number of Sides	Sum of Interior Angles
Triangle	3	180°
Quadrilateral	4	360°
Pentagon	5	540°
Hexagon	6	720°

(b) $180(n - 2)°$ (c) It is arithmetic; $d = 180°$

119. 9

121. A recursion formula gives the relationship between the terms a_{n+1} and a_n.

123. Use the formula for the nth partial sum of an arithmetic sequence to find the sum of the integers from 100 to 200. So,

$$\sum_{i=1}^{101} i + 99 = \frac{101}{2}(100 + 200).$$

125. Yes. C times the common difference of the original sequence.

Mid-Chapter Quiz   *(page 830)*

1. $4, -8, 16, -32, 64$ **2.** 3, 8, 15, 24, 35

3. $32, 8, 2, \frac{1}{2}, \frac{1}{8}$ **4.** $-\frac{3}{5}, 3, -\frac{81}{7}, \frac{81}{2}, -135$

5. 100   **6.** 40   **7.** 87   **8.** -32 **9.** 40

10. 26   **11.** $\sum_{k=1}^{20} \frac{2}{3k}$ **12.** $\sum_{k=1}^{25} \frac{(-1)^{k-1}}{k^3}$

13. $\sum_{k=1}^{20} \frac{k-1}{k}$ **14.** $\sum_{k=1}^{10} \frac{k^2}{2}$ **15.** $\frac{1}{2}$ **16.** -6

17. $-3n + 23$ **18.** $-4n + 36$ **19.** 2550

20. $33,397.50

Section 13.3   *(page 836)*

Review   *(page 836)*

1. The point is 6 units to the left of the y-axis and 4 units above the x-axis.

2. $(10, 5), (-10, 5), (-10, -5), (10, -5)$

3. The graph of f is the set of ordered pairs $(x, f(x))$, where x is in the domain of f.

4. To find the x-intercept(s), set $y = 0$ and solve the equation for x. To find the y-intercept(s), set $x = 0$ and solve the equation for y.

5. $x > \frac{5}{3}$ **6.** $y < 6$ **7.** $35 < x < 60$

8. $-12 < x < 30$ **9.** $x < 1$ or $x > \frac{5}{2}$

10. $-1 < x < 0$ or $x > \frac{5}{2}$

11. $\frac{19\sqrt{2}}{2} \approx 13.4$ inches   **12.** $5\sqrt{89} \approx 47.2$ feet

1. 2   **3.** -1 **5.** $-\frac{1}{2}$ **7.** $\frac{1}{5}$ **9.** π **11.** 1.06

13. Geometric, $\frac{1}{2}$ **15.** Not geometric   **17.** Geometric, 2

19. Not geometric   **21.** Geometric, $-\frac{2}{3}$

23. Geometric, 1.02   **25.** 4, 8, 16, 32, 64

27. $6, 3, \frac{3}{2}, \frac{3}{4}, \frac{3}{8}$ **29.** $5, -10, 20, -40, 80$

31. $1, -\frac{1}{2}, \frac{1}{4}, -\frac{1}{8}, \frac{1}{16}$

33. 1000, 1010, 1020.1, 1030.30, 1040.60

35. 4000, 3960.40, 3921.18, 3882.36, 3843.92

37. $10, 6, \frac{18}{5}, \frac{54}{25}, \frac{162}{125}$ **39.** $\frac{3}{256}$ **41.** $48\sqrt{2}$

43. 1486.02   **45.** -0.00610 **47.** $\frac{81}{64}$ **49.** $\pm\frac{243}{32}$

51. $\frac{64}{3}$ **53.** $a_n = 2(3)^{n-1}$ **55.** $a_n = 2^{n-1}$

57. $a_n = \left(-\frac{1}{5}\right)^{n-1}$ **59.** $a_n = 4\left(-\frac{1}{2}\right)^{n-1}$

61. $a_n = 8\left(\frac{1}{4}\right)^{n-1}$ **63.** $a_n = 14\left(\frac{3}{4}\right)^{n-1}$

65. $a_n = 4\left(-\frac{3}{2}\right)^{n-1}$ **67.** b   **68.** d   **69.** a   **70.** c

71. 1023   **73.** 772.48   **75.** 2.25   **77.** -5460

79. 6.67   **81.** $-14{,}762$ **83.** 16   **85.** 13,120

87. 48   **89.** 1103.57   **91.** 12,822.71   **93.** 2

95. $\frac{2}{3}$ **97.** $\frac{6}{5}$ **99.** 32

101.

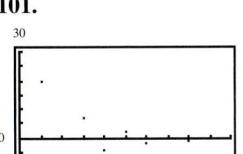

103.

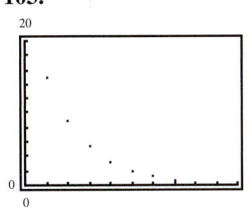

105. (a) There are many correct answers.

$$a_n = 187{,}500(0.75)^{n-1} \text{ or } a_n = 250{,}000(0.75)^n$$

(b) $59,326.17   (c) The first year

107. $3,623,993   **109.** $19,496.56   **111.** $105,428.44

113. $75,715.32

115. (a) $5,368,709.11   (b) $10,737,418.23

117. (a) $P = (0.999)^n$ (b) 69.4%

(c)   693 days

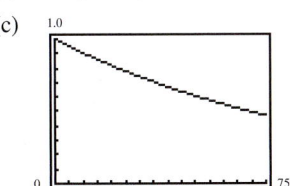

119. 70.875 square inches   **121.** 666.21 feet

123. (a) $a_n = 2^n$

(b) $2 + 2^2 + 2^3 + 2^4 + \cdots + 2^{66} \approx 1.48 \times 10^{20}$

ancestors

(c) It is likely that you have had common ancestors in the last 2000 years.

125. $a_n = a_1 r^{n-1}$ **127.** $a_n = \left(-\frac{2}{3}\right)^{n-1}$

129. An increasing annuity is an investment plan in which equal deposits are made in an account at equal time intervals.

Section 13.4   *(page 846)*

Review   *(page 846)*

1. No. The matrix must be square.

2. Interchange two rows.

Multiply a row by a nonzero constant. Add a multiple of one row to another row.

3. Yes, because the matrix takes on a "stair-step" pattern with leading coefficients of 1.

4. (a) $\begin{bmatrix} 1 & 3 \\ 4 & -1 \end{bmatrix}$ (b) $\begin{bmatrix} 1 & 3 & \vdots & -1 \\ 4 & -1 & \vdots & 2 \end{bmatrix}$

5. -200 **6.** 32   **7.** -60 **8.** -126

9. $y = 4x - 9$ **10.** 58

1. 15   **3.** 252   **5.** 1   **7.** 1   **9.** 50

11. 12,650   **13.** 593,775   **15.** 792   **17.** 2,598,960

19. 2,535,650,040   **21.** 85,013,600   **23.** 15   **25.** 35

27. 70   **29.** $a^3 + 6a^2 + 12a + 8$

31. $m^5 - 5m^4n + 10m^3n^2 - 10m^2n^3 + 5mn^4 - n^5$

33. $32x^5 - 80x^4 + 80x^3 - 40x^2 + 10x - 1$

35. $64y^6 + 192y^5z + 240y^4z^2 + 160y^3z^3 + 60y^2z^4$
$\qquad + 12yz^5 + z^6$

37. $x^8 + 8x^6 + 24x^4 + 32x^2 + 16$

39. $x^6 + 18x^5 + 135x^4 + 540x^3 + 1215x^2 + 1458x + 729$

41. $x^4 + 4x^3y + 6x^2y^2 + 4xy^3 + y^4$

43. $u^3 - 6u^2v + 12uv^2 - 8v^3$

45. $81a^4 + 216a^3b + 216a^2b^2 + 96ab^3 + 16b^4$

47. $x^4 + \dfrac{8x^3}{y} + \dfrac{24x^2}{y^2} + \dfrac{32x}{y^3} + \dfrac{16}{y^4}$

49. $32x^{10} - 80x^8y + 80x^6y^2 - 40x^4y^3 + 10x^2y^4 - y^5$

51. $120x^7y^3$ **53.** $129{,}024a^4b^5$ **55.** $-5940a^3b^9$

57. $6{,}304{,}858{,}560x^9y^6$ **59.** 120   **61.** -1365

63. 1760   **65.** 54   **67.** 1.172   **69.** 510,568.785

71. $\frac{1}{32} + \frac{5}{32} + \frac{10}{32} + \frac{10}{32} + \frac{5}{32} + \frac{1}{32}$

73. $\frac{1}{256} + \frac{12}{256} + \frac{54}{256} + \frac{108}{256} + \frac{81}{256}$

75. The difference between consecutive entries increases by 1.   2, 3, 4, 5

77. $n + 1$ **79.** (a) $_{11}C_5 = \dfrac{11!}{6!5!} = \dfrac{11 \cdot 10 \cdot 9 \cdot 8 \cdot 7}{5 \cdot 4 \cdot 3 \cdot 2 \cdot 1}$

81. The first and last numbers in each row are 1. Every other number in the row is formed by adding the two numbers immediately above the number.

Review Exercises   *(page 850)*

1. 8, 11, 14, 17, 20   **3.** $1, \frac{3}{4}, \frac{5}{8}, \frac{9}{16}, \frac{17}{32}$

5. 2, 6, 24, 120, 720   **7.** $\frac{1}{2}, \frac{1}{2}, 1, 3, 12$ **9.** $a_n = 2n - 1$

11. $a_n = \dfrac{n}{(n+1)^2}$ **13.** $a_n = \dfrac{5n-1}{n}$

15. $a_n = -2n + 5$ **17.** $a_n = \dfrac{3n^2}{n^2 + 1}$ **19.** 28

21. $\frac{4}{5}$ **23.** $\sum_{k=1}^{4} (5k - 3)$ **25.** $\sum_{k=1}^{6} \frac{1}{3k}$ **27.** -2.5

29. 127, 122, 117, 112, 107   **31.** $\frac{5}{4}, 2, \frac{11}{4}, \frac{7}{2}, \frac{17}{4}$

33. 5, 8, 11, 14, 17   **35.** 80, $\frac{155}{2}$, 75, $\frac{145}{2}$, 70

37. $4n + 6$ **39.** $-50n + 1050$ **41.** 486   **43.** $\frac{2525}{2}$

45. 2527.5   **47.** 5100   **49.** 462 seats   **51.** $\frac{3}{2}$

53. 10, 30, 90, 270, 810   **55.** 100, -50, 25, -12.5, 6.25

57. 4, 6, 9, $\frac{27}{2}, \frac{81}{4}$ **59.** $a_n = \left(-\frac{2}{3}\right)^{n-1}$

61. $a_n = 24(2)^{n-1}$ **63.** $a_n = 12\left(-\frac{1}{2}\right)^{n-1}$ **65.** 8190

67. -1.928 **69.** 19.842   **71.** 116,169.54

73. 2.275×10^6 **75.** 8   **77.** 12

79. (a) There are many correct answers. $a_n = 120,000(0.70)^n$

 (b) \$20,168.40

81. \$4,371,379.65   **83.** 56   **85.** 1   **87.** 91,390

89. 177,100   **91.** 10   **93.** 70

95. $x^4 - 20x^3 + 150x^2 - 500x + 625$

97. $8x^3 + 12x^2 + 6x + 1$

99. $x^{10} + 10x^9 + 45x^8 + 120x^7 + 210x^6 + 252x^5$
 $+ 210x^4 + 120x^3 + 45x^2 + 10x + 1$

101. $81x^4 - 216x^3y + 216x^2y^2 - 96xy^3 + 16y^4$

103. $u^{18} + 9u^{16}v^3 + 36u^{14}v^6 + 84u^{12}v^9 + 126u^{10}v^{12}$
 $+ 126u^8v^{15} + 84u^6v^{18} + 36u^4v^{21} + 9u^2v^{24} + v^{27}$

105. $5376x^6$ **107.** $-61,236$

Chapter Test *(page 853)*

1. 1, $-\frac{2}{3}, \frac{4}{9}, -\frac{8}{27}, \frac{16}{81}$ **2.** 2, 10, 24, 44, 70   **3.** 60

4. 35   **5.** -45 **6.** $\sum_{k=1}^{12} \frac{2}{3k+1}$ **7.** $\sum_{k=1}^{6} \left(\frac{1}{2}\right)^{2n-2}$

8. 12, 16, 20, 24, 28   **9.** $a_n = -100n + 5100$

10. 3825   **11.** $-\frac{3}{2}$ **12.** $a_n = 4\left(\frac{1}{2}\right)^{n-1}$ **13.** 1020

14. $\frac{3069}{1024}$ **15.** 1   **16.** 12   **17.** 1140

18. $x^5 - 10x^4 + 40x^3 - 80x^2 + 80x - 32$

19. 56   **20.** 490 meters   **21.** \$47,868.33

Appendix
Appendix B *(page A45)*

1.
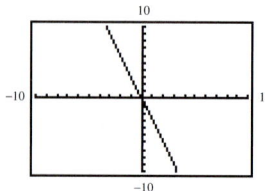

3.
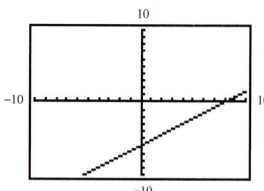

5.
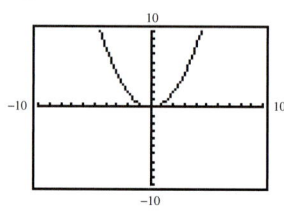

7.
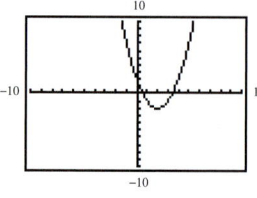

9.
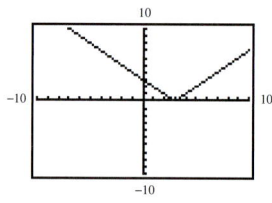

11.
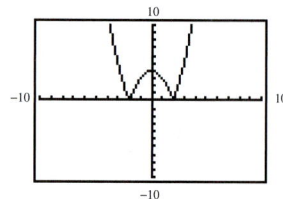

13.
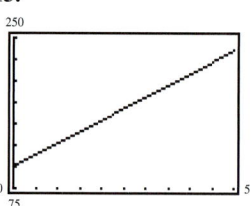

15.
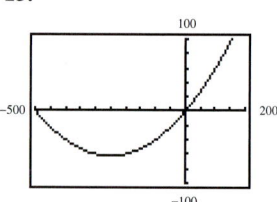

17. Sample answer:

Xmin = 4
Xmax = 20
Xscl = 1
Ymin = 14
Ymax = 22
Yscl = 1

19. Sample answer:

Xmin = -20
Xmax = -4
Xscl = 1
Ymin = -16
Ymax = -8
Yscl = 1

21. Yes, Associative Property of Addition

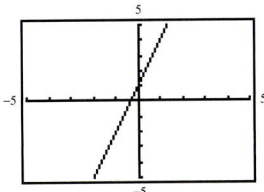

23. Yes, Multiplicative Inverse Property

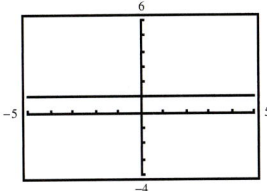

25. $(-3, 0), (3, 0), (0, 9)$ **27.** $(-8, 0), (4, 0), (0, 4)$

29. $\left(\frac{5}{2}, 0\right), (0, -5)$ **31.** $(-2, 0), \left(\frac{1}{2}, 0\right), (0, -1)$

33. Triangle   **35.** Square

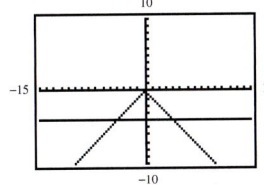

37.

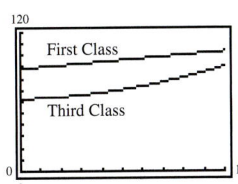

Index of Applications

Index